THE GREEK ALPHABET

A	α	alpha	N	ν	nu
B	β	beta	Ξ	ξ	xi
Γ	γ	gamma	O	o	omicron
Δ	δ	delta	Π	π	pi
E	ϵ	epsilon	P	ρ	rho
Z	ζ	zeta	Σ	σ	sigma
H	η	eta	T	τ	tau
Θ	θ	theta	Υ	υ	upsilon
I	ι	iota	Φ	ϕ	phi
K	κ	kappa	X	χ	chi
Λ	λ	lambda	Ψ	ψ	psi
M	μ	mu	Ω	ω	omega

SI PREFIXES

symbol	prefix	
a	atto	
f	femto	
p	pico	
n	nano	10^{-9}
μ	micro	10^{-6}
m	milli	10^{-3}
c	centi	10^{-2}
d	deci	10^{-1}
da	deka	10
h	hecto	10^{2}
k	kilo	10^{3}
M	mega	10^{6}
G	giga	10^{9}
T	tera	10^{12}
P	peta	10^{15}
E	exa	10^{18}

EQUIVALENT UNITS OF DERIVED AND COMMON SI UNITS

symbol	equivalent units					
A	C/s	W/V	V/Ω	J/s$\cdot$V	N/T$\cdot$m	Wb/H
C	A$\cdot$s	J/V	N$\cdot$m/V	V$\cdot$F		
F	C/V	C^2/J	C^2/N$\cdot$m	A$^2\cdot$s^4/kg$\cdot$m^2		
F/m	C/V$\cdot$m	C^2/J$\cdot$m	C^2/N$\cdot$m^2	A$^2\cdot$s^4/kg$\cdot$m^3		
H	Wb/A	V$\cdot$s/A	T$\cdot$m^2/A	kg$\cdot$m^2/A$^2\cdot$s^2	$\Omega\cdot$s	
Hz	1/s					
J	N$\cdot$m	V$\cdot$C	W$\cdot$s	C^2/F	kg$\cdot$m^2/s^2	
m^2/s^2	V$\cdot$C/kg					
N	J/m	V$\cdot$C/m	kg$\cdot$m/s^2			
N/A^2	Wb/N$\cdot$m^2					
Pa	N/m^2	kg/m$\cdot$s^2				
Ω	V/A	kg$\cdot$m^2/A$^2\cdot$s^3				
S	A/V	A$^2\cdot$s^3/kg$\cdot$m^2				
T	Wb/m^2	N/A$\cdot$m	N$\cdot$s/C$\cdot$m	kg/A$\cdot$s^2		
V	W/A	C/F	J/C	kg$\cdot$m^2/A$\cdot$s^3		
V/m	N/C					
W	J/s	V$\cdot$A	kg$\cdot$m^2/s^3			
Wb	V$\cdot$s	H$\cdot$A	T$\cdot$m^2	kg$\cdot$m^2/A$\cdot$s^2		

Electrical Engineering Reference Manual

for the Electrical and Computer PE Exam

Sixth Edition

John A. Camara, PE

Professional Publications, Inc. • Belmont, CA

How to Locate Errata and Other Updates for This Book

At Professional Publications, we do our best to bring you error-free books. But when errors do occur, we want to make sure that you know about them so they cause as little confusion as possible.

A current list of known errata and other updates for this book is available on the PPI website at **www.ppi2pass.com**. From the website home page, click on "Errata." We update the errata page as often as necessary, so check in regularly. You will also find instructions for submitting suspected errata. We are grateful to every reader who takes the time to help us improve the quality of our books by pointing out an error.

ELECTRICAL ENGINEERING REFERENCE MANUAL
Sixth Edition

Current printing of this edition: 4

Printing History

edition number	printing number	update
6	2	Minor corrections.
6	3	Minor corrections.
6	4	Minor corrections.

Printed in the United States of America

Professional Publications, Inc.
1250 Fifth Avenue, Belmont, CA 94002
(650) 593-9119
www.ppi2pass.com

Library of Congress Cataloging-in-Publication Data
Camara, John A., 1956-
    Electrical engineering reference manual for the electrical and computer PE exam / John
A. Camara.--6th ed.
        p.   cm.
    Fifth ed. published under title: Electrical engineering reference manual for the PE exam /
Raymond B. Yarbrough.
    Includes index.
    ISBN 1-888577-56-8
    1. Electric engineering--Examinations, questions, etc. 2. Electric engineering--United
States--Examinations--Study guides. I. Yarbrough, Raymond B. Electrical engineering
reference manual for the PE exam. II. Title.

TK169.Y37   2001
621.3'076--dc21

                                                        2001048442

Topics

Mathematics

Measurement & Instrumentation

Professional

Theory

Electronics

Support Material

Fields

Computers

Circuit Theory

Communications

Power—Generation

Biomedical Systems

Power—Transmission

Control Systems

Power—Machinery

Electrical Materials

Power—Lightning

Codes and Standards

Where do I find practice problems to test what I've learned in this Reference Manual?

The *Electrical Engineering Reference Manual* provides a knowledge base that will prepare you for the Electrical and Computer PE exam. But there's no better way to exercise your skills than to practice solving problems. To test your knowledge, you need *Practice Problems for the Electrical and Computer Engineering PE Exam: A Companion to the Electrical Engineering Reference Manual*. This essential study aid will challenge you with 445 practice problems, each with a complete, step-by-step solution.

Practice Problems for the Electrical and Computer Engineering PE Exam may be purchased from Professional Publications (800-426-1178 or **www.ppi2pass.com**) or from your favorite bookstore.

Table of Contents

Appendices
Table of Contents

Preface to the Sixth Edition

The sixth edition of the *Electrical Engineering Reference Manual* is completely new, yet it owes much to its predecessors and to the outstanding guidance of Professional Publications, Inc.

This text is written as a broad review of electrical engineering design, analysis, and operational fundamentals. In choosing the subjects and preparing the contents, I was directed by the need for a comprehensive study guide for the newly developed Principles and Practice of Engineering (PE) examination in electrical and computer engineering, as well as an all-inclusive reference book for both the practicing engineer and the undergraduate electrical engineering student.

This book strives to meet the needs of three groups. For PE candidates, it is an efficient resource for exploring the exam topics systematically and exhaustively. For the practicing electrical or electronics engineer, it functions as a comprehensive reference, discussing all aspects of these fields in a realistic manner and incorporating the most common formulas and data. Finally, for the engineering student, it presents a thorough review of the fundamentals of electrical engineering.

The *Reference Manual* covers all the topics on the electrical and computer engineering PE exam, providing sufficient background information to preclude the necessity of referring to other texts. Additional subjects not specifically covered on the exam but essential for full understanding of exam topics are explained as well. The mathematical, theoretical, and practical applications of each topic are explored, so the reader may focus on any or all of these areas.

A number of features enhance the usefulness of the text itself. The introduction gives details on how to use this book efficiently. Lists of codes, handbooks, and references recommended for additional study are provided. The means for accessing online updates (and errata) is given. Appendices are provided to exhibit mathematical, basic theoretical, and practical data. A glossary of common electrical terms is included. A comprehensive index is provided to aid in the search for specific information.

The sources used in assembling the content of this work include: (1) information on the electrical and computer PE exam made public by the National Council of Examiners for Engineering and Surveying (NCEES); (2) PE exam review course material published by the National Society of Professional Engineers; (3) electrical engineering curricula at leading colleges and universities; (4) current literature in the field of electrical engineering; (5) information on the numerous electrical engineering websites on the Internet; and (6) survey comments from those who have recently taken the PE exam and/or have purchased previous editions of this text.

Changes from the previous edition are numerous. All are meant to enhance the enduring usefulness of the book. Here are some of the major differences between the fifth and sixth editions.

- Mathematics chapters are added for those who need an extra review before moving on to the electrical engineering information.

- Theory and Fields topics are added, both to improve readers' understanding of principles and to prepare them for potential changes to the PE exam structure.

- Information from the fifth edition is completely rewritten to reflect the mindset of a practicing engineer, and it is separated into generally smaller, more logical divisions.

- The Power topic is expanded to include Generation Systems; Transmission and Distribution Systems; and Lightning Protection and Grounding.

- The Electronics topic is updated to reflect the latest advances.

- The Computers topic is added. The Digital Systems information is expanded to include the basics of interfaces, protocols, and standards.

- The Communications information is gathered into a single topic and expanded to reflect recent progress in the field.

- A chapter on Biomedical Engineering is incorporated.

- Sections on Electrical Materials, Law, Ethics, and Electrical Engineering Frontiers are added for completeness.

- The National Electrical Code is addressed as a separate topic, and there is complete coverage of the 1999 NEC.

- SI units are used, except where common practice dictates customary U.S. units (as in the National Electrical Code topic).

- Nomenclature used is consistent from chapter to chapter and with current SI system usage. Varying symbology is mentioned to minimize confusion when using other texts.

- Topics and individual chapters are arranged in a logical, progressive manner appropriate to the new organization of the PE exam.

The change in the electrical and computer engineering PE exam to a breadth-and-depth format is mirrored in this book. The *Reference Manual* is meant as a compendium of the *breadth* of electrical engineering, providing the *depth* required in individual sections to enable engineers to gain a solid understanding of theory and practical applications.

The scope of this book is beyond that of the PE exam, as it is intended to be. When the exam is complete, you will still need a resource for electrical engineering information, either in your work or simply to satisfy the curiosity that makes us thinking humans. In both the exam and your career, I hope this book serves you well.

Should you find an error in this book, know that it is mine, and that I regret it. Beyond that, I hope two things happen. First, please let me know about it, either by using the "Errata" section on the PPI website at **www.ppi2pass.com** or by filling out the errata card found in this book. Second, I hope you learn something from the error—I know I will! I would appreciate constructive comments and suggestions for improvement, additional questions, and recommendations for expansion so that new editions or similar texts will more nearly meet the needs of future examinees.

John A. Camara, PE

Acknowledgments
for the Sixth Edition

It is with enduring gratitude that I thank Michael Lindeburg for giving me the opportunity to realize my dream of authorship of a significant engineering text. It is my hope that this book will match the comprehensive and cohesive quality of the best of Professional Publications' family of engineering books, which I have admired for two decades.

The team at Professional Publications contributed substantially to this book, making it better than it ever could have been without them. Though I don't yet know them all by name, I thank them one and all. In particular, I owe this book to the dedication and professional competence of Aline Magee. Without her encouragement, guidance, and gentle chastisement, this book would never have come to fruition. My personal thanks to the very professional Jessica Whitney-Holden, who ably guided me through production, and to her daughter, Emily, who reminded me of the important things in life.

The National Society of Professional Engineers provided questions from previous professional engineering examination review courses for my study and guidance. Michael Lindeburg provided Chapters 1–14 on mathematics, as well as the chapters on economic analysis, law, ethics, and PE registration. This saved considerable time. Additionally, there was little, if anything, I could do to improve them, so I'm grateful for their use. The late Raymond Yarbrough's fifth edition provided the basis for expansion, and numerous figures, tables, and problems were taken, with minor modifications, directly from his book. Steven B. Edington reviewed my initial Communication topic outline and greatly improved its scope and timeliness. John Goularte, Jr., taught me first-hand the applications of biomedical engineering, and the real-world practicality of the questions in this topic is attributable to him. Becky Owens provided guidance on computer fundamentals and showed me practical applications, as well as typed the majority of the index. More importantly, she was patient and supportive through the entire process, even when I wasn't.

Gregg Wagener, PE, reviewed each of the chapters, providing valuable insight, correcting numerous errors and, in general, teaching me a great deal along the way.

A very special thanks to my son, Jac Camara, whose annotated periodic table became the basis for the Electrical Materials topic. Additionally, our discussions about quantum mechanics and a variety of other topics helped stir the intellectual curiosity I thought was fading. His knowledge will one day surpass mine, and I will be proud when that day comes to pass (though he probably believes that day has already come and gone).

Thanks also to my daughter, Cassiopeia, who has been a constant source of inspiration, challenge, and pride. Her knowledge of the healing arts does exceed mine. I am sure that veterinary or medical schools will see the treasure that awaits in her.

The knowledge expressed in this book represents years of training, instruction, and self-study in electrical, electronic, mechanical, nuclear, marine, space, and a variety of other types of engineering—all of which I find fascinating. I would not understand any of it were it not for the teachers, instructors, mentors, family, and friends who've spent countless hours with me and from whom I've learned so much. A few of them include: Mrs. Mary Avila, Capt. C.E. Ellis, Mr. Claude Estes, Capt. Karl Hasslinger, Mr. Jerome Herbeck, Mr. Jack Hunnicutt, Mr. Ralph Loya, Harry Lynch, Mr. Harold Mackey, Michael O'Neal, Mark Richwine, Mr. Charles Taylor, Mrs. Abigail Thyarks, and Mr. Jim Triguerio. There are others who have touched my life in special ways: Jim, Marla, Lora, Todd, Tom, and Rick. There are many others, though here they will remain nameless, to whom I am indebted. Thanks.

Finally, thanks to my mom, for without her steadfast love little in this world would be worthwhile.

John A. Camara, PE

Codes, Handbooks, and References

This edition of the *Electrical Engineering Reference Manual* is based on the following codes, standards, and references. The most current versions or editions available were used. However, the PE examination is not always based on the most currently available codes, as adoption by state and local controlling authorities often lags issuance by several years.

The PPI website (**www.ppi2pass.com**) provides the dates of the codes, standards, and regulations on which NCEES has announced the current exams are based. Use this information to decide which editions of these books should be part of your exam preparation.

The minimum recommended library for the electrical exam consists of this book, the National Electrical Code, a standard handbook of electrical engineering, and two textbooks that cover fundamental circuit theory (both electrical and electronic). Numerous textbooks covering basic electrical engineering topics are listed. These texts, or their equivalent (see the topics in brackets), should be used in preparation for the examination. Adequate preparation, not an extensive portable library, is the key to success.

Codes

Code of Federal Regulations, Title 47—Telecommunications, Ch. 73—"Radio Broadcast Rules." 47CFR73. U.S. Government.[1]

National Electrical Code, NFPA 70. National Fire Protection Association.

National Electric Safety Code, NESC.

Standards Organizations of Interest:

ANSI: American National Standards Institute

EIA: Electronic Industries Alliance

FCC: Federal Communications Commission

IEEE: Institute of Electrical and Electronic Engineers

ISA: Instrument Society of America

ISO: International Organization for Standardization[2]

NEMA: National Electrical Manufacturers Association

---

[1]The chapters within this title of the Code of Federal Regulations are known as the *FCC Rules*.
[2]ISO is not the acronym for the organization. Rather, "iso" is from the Greek word for "equal."

Handbooks

American Electrician's Handbook. Terrell Croft and Wilford I. Summers. McGraw-Hill.

CRC Materials Science and Engineering Handbook. James F. Shackelford, William Alexander, and Jun S. Park, eds. CRC Press, Inc.

Electronic Engineer's Handbook. Donald Christiansen, ed. McGraw-Hill.

McGraw-Hill Internetworking Handbook. Ed Taylor. McGraw-Hill.

National Electrical Code Handbook. Mark W. Earley, Joseph V. Sheehan, and John M. Caloggero. National Fire Protection Association.

Standard Handbook for Electrical Engineers. Donald G. Fink and H. Wayne Beaty. McGraw-Hill.

The Communications Handbook. Jerry D. Gibson, ed. CRC Press, Inc.

The Computer Science and Engineering Handbook. Allen B. Tucker, Jr., ed. CRC Press, Inc.

References

CRC Standard Mathematical Tables. William H. Beyer, ed. CRC Press, Inc.

McGraw-Hill Dictionary of Scientific and Technical Terms. Sybil P. Parker, ed. McGraw-Hill.

Schaum's Outline Series (Electronics and Electrical Engineering). McGraw-Hill.

The Internet for Scientists and Engineers, Brian J. Thomas. SPIE Press & IEEE Press.

Texts

An Introduction to Digital and Analog Integrated Circuits and Applications. Sanjit K. Mitra. Harper & Row, Publishers. [Digital Circuit Fundamentals]

Applied Electromagnetics. Martin A. Plonus. McGraw-Hill. [Electromagnetic Theory]

Introduction to Computer Engineering. Taylor L. Booth. John Wiley & Sons. [Computer Design Basics]

Electrical Power Technology. Theodore Wildi. John Wiley & Sons. [Power Theory and Application]

Linear Circuits. M.E. Van Valkenburg and B.K. Kinariwala. Prentice-Hall, Inc. [AC/DC Fundamentals]

Microelectronics. Jacob Millman. McGraw-Hill. [Electronic Fundamentals]

Introduction

Part 1: How to Use This Book

QUICKSTART

If you're the type who desires information distilled into concise, pointed bits and if you're ready to begin your studies, here's the paragraph for you:

> The chapters in this book are independent. The number of pages devoted to each chapter is approximately consistent with the emphasis given on the electrical and computer engineering Professional Engineering (PE) examination. The Mathematics, Theory, and Electrical Materials topics are primarily meant as background information for that quick review sometimes necessary to recall seldom-used information. The Appendices and Glossary are meant as quick references for common electrical information. The Professional topics are to refine your professional engineering knowledge. You should focus your studies on the fundamentals. The fundamentals of electrical engineering are covered in the Circuit Theory topic and those of electronic engineering are covered in the Electronics topic. The other topics complete the coverage of the exam, both current and proposed. The index provides an excellent starting place for information searches. Commence study immediately, reviewing all examples and working all the problems in the companion volume, *Practice Problems for the Electrical and Computer Engineering PE Exam.* Start with your weakest areas and work toward your strongest. Review each week. Continue to study until the exam date. Remember to enjoy the learning process and revel in the knowledge gained. Good luck!

However, if you're the type who desires a more thorough review, with a deeper understanding of the PE exam before you begin your studies, the rest of this introduction is for you.

FOR THE PE CANDIDATE

When you are preparing for the PE examination in electrical and computer engineering, the following suggestions may help.

- Concentrate primarily on the fundamentals of electrical and electronic engineering, which are covered in the Circuit Theory and Electronics topics. Take time to review the principles of any problem you find difficult.

- Use the subject index extensively. Every significant term, law, theorem, and concept has been indexed. If you don't recognize a term, look for it in the index.

- Know which subjects in this book are not covered on the PE exam. Those chapters that are supportive do not cover exam subjects. They provide background for the other chapters, enhance understanding of fundamentals, and are useful following the exam for quick reference. (The exam subjects known at the time of this printing are provided in Table 2. These are subject to change; the most up-to-date information can be found on PPI's website (**www.ppi2pass.com**).)

- Become intimately familiar with this book. This means knowing the order of the chapters, the approximate location of important figures and tables, what appendices are available, and so on.

- Skim through a chapter and familiarize yourself with the subjects before starting the practice problems in the companion volume, *Practice Problems for the Electrical and Computer Engineering PE Exam.*

- *Experience in problem solving is the key to success on the exam.* Review examples and solve as many problems as possible in the *Practice Problems* manual. It's a good idea to begin with problems in your weakest areas and work into your strongest areas.

- Use the answers to the practice problems to check your work. If your answer doesn't match the given solution, study the solution to determine why.

- Some subjects appear in more than one chapter, and in more than one topic. Use the index liberally to learn all there is to know about a particular subject.

- The majority of this book uses SI units, except where customary U.S. (English) units are the standard, such as in the National Electrical Code. Some equations in the U.S. system use the term g/g_c. For calculations at standard gravity, the numerical value of this fraction is 1.00. Therefore, it is necessary to incorporate this quantity only in problems with a nonstandard gravity, or when being meticulous with units.

FOR THE PRACTICING ENGINEER AND ENGINEERING STUDENT

If you are a practicing engineer or an engineering major and you purchased this book as a general reference, the following suggestions and information may be of assistance.

- The index is extensive and meant as the starting point for any specific information desired.

- The table of contents is similarly extensive and meant as the starting point for concept information searches.

- The entire book is separated into interrelated topics for the broad overview of a given area of electrical engineering, if an introduction to a given topic is desired.

- The mathematics chapters can be used as quick reviews when working any engineering problems.

- The Advanced Engineering Mathematics chapter provides the minimum information necessary to understand the importance and use of various advanced mathematical concepts, in simple but accurate terms. This should allow you to more fully comprehend and appreciate professional journal- or engineering-focused presentations using such concepts without resorting to a complete review of undergraduate math texts.

- The Advanced Engineering Mathematics chapter is also meant as the starting point for delving more thoroughly into such concepts by providing defining terms and the names of often-used mathematical formulas. This is done to help you through those situations where you know you need more mathematical information but don't even know where to begin, what questions to ask, or what mathematical concept is involved.

- The Appendices and Glossary are meant as quick references for common electrical information.

This book is meant as a broad sweep of the fundamentals of electrical engineering. Here you should find most topics and terms you deal with daily, as well as many that you rarely encounter, all with enough depth of information that this book may be your first and last information stop. At the very least, you will find the term you are looking for, its relative place in the spectrum of electrical engineering, and enough information to perform a more intelligent search for details.

FOR THE INSTRUCTOR

This book focuses on the topics covered in the electrical and computer engineering PE exam. Topics not on the exam are included to provide the necessary background, and to enhance understanding of the fundamental concepts. Any review should focus on the topics covered and omit topics not covered. The fundamentals are in the Circuit Theory topic and the Electronics topic.

The emphasis to be given to any exam topic can be deduced from the approximate percentage of problems on the exam (see Table 2) and the depth of coverage provided in this book.

Use the formulas, illustrations, and tables of data in the text to minimize chalkboard time. Avoid proofs and emphasize problem solution. The greater the number of problems solved, the higher the chances of success on the exam.

The practice problems in the companion volume *Practice Problems for the Electrical and Computer Engineering PE Exam* span the spectrum from quick-answer to in-depth probing of knowledge. The format is similar to that used on the exam. The quick-answer questions, which are given first in each chapter, are meant to ensure understanding of fundamentals and can be used for this purpose during review sessions. The more in-depth problems can be used as homework or timed questions.

One possible review course format follows. The sessions are weekly, with the course lasting three months.

Table 1 *Typical PE Review Course Format*

session	subject covered	chapter(s)
1	Introduction to the PE Exam, Engineering Economics, Computer Mathematics	13, 66, 70
2	DC Fundamentals	26
3	AC Fundamentals	27
4	Linear Circuit Analysis	29
5	Transient Analysis, Bode Plots	30–32
6	Electronic Fundamentals	43
7	Amplifiers and Wave Shaping	44, 45
8	Digital Logic and Design	49, 50
9	Synchronous Sequential Networks, Digital Systems, Computers	48, 51, 55
10	Three-Phase Systems, Transformers, Rotating Machinery	34, 28, 36, 39, 40
11	Instrumentation, Communication	42, 57–61
12	National Electrical Code, Miscellaneous Topics and Review	65

Part 2: Everything You Ever Wanted to Know About the PE Exam

WHAT IS THE FORMAT OF THE PE EXAM?

The NCEES PE examination in electrical and computer engineering consists of two four-hour sessions separated by a one-hour lunch period. The exam is structured in a "breadth-and-depth" format, and all questions are multiple-choice. In the four-hour morning session, all examinees work the same "breadth" exam, which consists of 40 multiple-choice questions. In the afternoon, examinees choose to work one of three "depth" exam modules: (1) Computer Engineering, (2) Electronics, Control, and Communication Engineering, and (3) Power Engineering. Each depth exam consists of 40 questions that test knowledge in the areas specified. Examinees must work all 40 questions in the depth exam of their choice.

For each session, you will be given an exam booklet that contains problems only for electrical engineers.

WHAT SUBJECTS ARE ON THE PE EXAM?

Table 2 provides a listing of the exam subjects as known at the time of this printing. The subjects may change from time to time; for current information check PPI's website (**www.ppi2pass.com**).

WHAT IS THE TYPICAL QUESTION FORMAT?

Almost all the questions are standalone—that is, they are completely independent. However, NCEES states that some sets of questions may start with a "situation statement," which may apply to two to five following questions.

NCEES tries hard to make sure that the questions are not interrelated—so making a mistake on one question shouldn't cause you to get a subsequent question wrong. However, considerable time may be required to repeat previous calculations with a new set of given data.

Since the questions are multiple choice, all required data should appear in the situation statement. Generally, you will not be required to come up with numerical data that might affect your success on the question. However, there may be superfluous information given in many of the questions.

Each of the questions will have four answer choices, labeled A, B, C, and D. One of the answer choices is correct (or "most nearly correct," as described in the following section). The remaining answer choices are incorrect and may consist of one or more logical distractors (answers that are arrived at by making common mistakes in logic, computation, or unit conversion).

WHAT DOES "MOST NEARLY" REALLY MEAN?

One of the more disquieting aspects of these questions is that the available answer choices are seldom exact. Answer choices generally have only two or three significant digits. Exam questions ask, "Which answer choice is most nearly the correct value?" or they instruct you to complete the sentence, "The value is approximately ..." A lot of self-confidence is required to move on to the next question when you don't find an exact match for the answer you calculated, or if you have had to split the difference because no available answer choice is close. The NCEES website describes it like this:

> Many of the questions on NCEES exams require calculations to arrive at a numerical answer. Depending on the method of calculation used, it is very possible that examinees working correctly will arrive at a range of answers. The phrase "most nearly" is used to accommodate answers that have been derived correctly but that may be slightly different from the correct answer choice given on the exam. You should use good engineering judgment when selecting your choice of answer. For example, if the question asks you to calculate an electrical current or determine the load on a beam, you should literally select the answer option that is most nearly what you calculated, regardless of whether it is more or less than your calculated value. However, if the question asks you to select a fuse or circuit breaker to protect against a calculated current or to size a beam to carry a load, you should select an answer option that will safely carry the current or load. Typically, this requires selecting a value that is closest to but larger than the current or load.

HOW MUCH MATHEMATICS IS NEEDED FOR THE EXAM?

Generally, only simple algebra, trigonometry, and geometry are needed on the PE exam. You will need to use the trigonometric, logarithm, square root, and similar buttons on your calculator. There is no need to use any other method for these functions.

There are no pure mathematics (algebra, geometry, trigonometry, etc.) problems on the exam. However, you will need to apply your knowledge of these subjects to the exam problems.

Except for simple quadratic equations, you will probably not need to find the roots of high-order equations.

There is little or no use of calculus on the exam. Rarely, you may need to take a simple derivative to find a maximum or minimum of some function. Even more rare is the need to integrate to find an average.

There is minimal need to solve differential equations. Simple linear circuits and components appear that can be solved or explained with differential equations. Often these applications, and others that use such mathematics, can be solved by other means.

Probability and statistics are covered in the most basic fashion as applied to an electrical topic—for example, power generation system outages.

The PE exam is concerned with numerical answers, not with proofs or derivations. You will not be asked to prove or derive formulas.

Table 2 *Subjects on the Electrical and Computer Engineering PE Exam**

MORNING SESSION
(40 multiple-choice problems)

subject	approximate percentage of problems
Basic Electrical Engineering: professionalism and engineering economics, safety and reliability, electric circuits, electric and magnetic field theory and applications, digital logic	45%
Electronics, Electronic Circuits, and Components: components, electrical and electronic materials	20%
Controls and Communications Systems	15%
Power: transmission and distribution, rotating machines and electromagnetic devices	20%

AFTERNOON SESSIONS
(each includes 40 multiple-choice problems)

Computer Engineering Depth Module

subject	approximate percentage of problems
General Computer Systems: interpretation of codes and standards, microprocessor systems	10%
Hardware: digital electronics, design and analysis, systems	45%
Software: system software, development/applications	35%
Networks	10%

Electronics, Controls, and Communications Depth Module

subject	approximate percentage of problems
General Electrical Engineering Knowledge: measurement and instrumentation, interpretation of codes and standards, computer systems	10%
Electronics: electrical circuit theory, electric and magnetic field theory and applications, electronic components and circuits	35%
Controls: control system fundamentals, control system design/implementation, stability	25%
Communications: communications and signal processing, noise and interference, telecommunications	30%

Power Depth Module

subject	approximate percentage of problems
General Power Engineering: measurement, instrumentation, and statistics; special applications; codes and standards	15%
Circuit Analysis: analysis, devices and power electronic circuits, electric and magnetic fields and applications	28%
Rotating Machines and Electromagnetic Devices: rotating machines, electromagnetic devices	27%
Transmission and Distribution: system analysis, power system performance, protection	30%

*This table reflects the subjects as known at the time of this printing. It is possible that the subjects may change. Check **www.ppi2pass.com** for the lastest information.

HOW ABOUT ANY ENGINEERING ECONOMICS PROBLEMS?

For most of the early years of engineering licensing, questions on engineering economics appeared frequently on the exams. This is no longer the case. However, in its outline of exam subjects, NCEES notes: "Some problems may include economic aspects." What this means is that engineering economics can constitute anything from nothing to a full hour's worth of work on the exam.

If engineering economics is incorporated into problems, its "disguise" may be totally transparent. For example, you may be required to compare the economics of operating two electrical generating stations, whose costs can be calculated from electrical properties of three-phase systems, efficiencies, and so on.

While the degree of engineering economics knowledge may have decreased somewhat, the basic economic concepts (e.g., time value of money, present worth, non-annual compounding, comparison of alternatives, etc.) are still valid test subjects.

WHAT ABOUT ETHICS AND PROFESSIONALISM QUESTIONS?

For many years, the NCEES has discussed adding professionalism and ethics to the PE exam. However, these subjects are not part of the test outline, and there has yet to be a question on them on the exam.

IS THE EXAM TRICKY?

The PE exam is not "tricky." The exam does not overtly try to get you to fail. Examinees manage to fail on a regular basis with perfectly straightforward problems. The exam problems are difficult in their own right. NCEES does not need to provide misleading or conflicting statements. However, with the multiple-choice problems, you might find that commonly made mistakes are represented in the available answer choices. Thus, the alternative answers (known as *distracters*) will be logical.

Problems are generally practical, dealing with common and plausible situations that you might experience in an electrical engineering job. Circuit design is restricted to connecting electrical components to create a desired outcome. You will not be asked to design a new IC chip or build an equivalent vacuum tube circuit.

WHAT MAKES THE PROBLEMS DIFFICULT?

Problems can be difficult for any number of reasons, the most common being a lack of knowledge in a given area.

Beyond that fundamental cause, some problems are difficult because the pertinent theory is not obvious. Also, data needed may be hard to find. For example, building requirements are in the National Electrical Code;

without the code, and a knowledge of where given information is located within the code, certain problems will be nearly impossible to solve.

Some problems are difficult because they defy imagination. For example, problems pertaining to the theoretical functioning of certain electronic components are more easily solved if the physical layout of the component can be visualized. Some problems are difficult because the computational time is lengthy. Some problems are difficult because the terminology is obscure. This can occur in almost any subject.

DOES THE PE EXAM USE SI UNITS?

The PE exam in electrical and computer engineering is written primarily in SI units. Nevertheless, customary U.S. units (also known as "English units") are occasionally used, especially when referring to problems associated with applications of the National Electrical Code.

WHAT REFERENCE MATERIAL IS PERMITTED IN THE EXAM?

The PE exam is open-book. Most states do not have any limits on the numbers and types of books you can use. Personal notes in a three-ring binder and other semipermanent covers can usually be used.

The references you bring into the examination room in the morning do not have to be the same as the references you use in the afternoon. However, you cannot share books with other examinees during the exam.

A few states prohibit collections of solved problems, such as *Schaum's Outline Series*, or solutions manuals that are companion volumes to review books. A few states maintain a formal list of banned books.

Strictly speaking, loose paper and scratch pads (including Post-it® notes) are not permitted in the examination room. Certain types of preprinted graphs and logarithmically scaled graph papers (which are almost never needed) should be three-hole punched and brought in a three-ring binder. An exception to this restriction may be made for laminated and oversize charts, graphs, and tables that are commonly needed for particular types of problems.

HOW MANY BOOKS SHOULD YOU BRING?

You actually won't use many books in the exam. The trouble is, you don't know in advance which ones you will need. That's the reason many examinees show up with boxes and boxes of books. Without a doubt, there is information you will need that is not in this book. But there is not so much that you need to bring your company's entire library. The exam is very fast-paced. You will not have time to use books with which you are not thoroughly familiar. The exam doesn't require you to know obscure solution methods or to use difficult-to-find data.

So, it really is unnecessary to bring a large quantity of books with you. The books that are truly useful are identified in the Codes, Handbooks, and References section, and you should be able to decide which of them support the areas in which you intend to work. This book and five to ten other references of your choice should be sufficient for most of the problems you solve.

WHAT THE PREPARED ENGINEER WOULD TAKE TO THE PE EXAM

Certain references provide specifications not covered in this text that may be on the exam, primarily the National Electrical Code. Handbooks are useful as quick references. Textbooks provide detailed information, but the information is usually too dispersed within the text to be useful in an exam situation, though the simple act of having them is sometimes a confidence booster. See the Codes, Handbooks, and References section for a list of recommended supplementary study materials.

The minimum recommended library for the electrical exam would consist of this book, the current National Electrical Code, a standard handbook of electrical engineering, and two textbooks you are most comfortable with that cover fundamental circuit theory (both electrical and electronic). Additional books beyond these will most likely not be needed during the examination and may indeed slow you down. Preparation prior to the exam is a much better tool than a large reference library during the exam.

WHAT YOU WON'T NEED

Generally, people bring too many things to the exam. A general rule is that you shouldn't bring anything to the exam that you didn't use during your preparation for the exam.

Here are some things that you won't need.

- Books on non-exam subjects.

- Books on mathematical analysis, or extensive mathematics tabulations.

- Extensive collections of materials properties.

- Local, county, or state-specific electrical codes.

- Obscure books and materials: Books that are in a different language, doctoral theses, and papers presented at technical societies won't be needed.

- Old textbooks, obsolete, rare, and ancient books: NCEES exam problem writers are aware of which textbooks are currently in use. Material that is available only in out-of-print publications and old editions won't be used.

- Handbooks in other disciplines: You won't need a mechanical engineering or civil engineering handbook.

- *Handbook of Chemistry and Physics*

- Computer science books that cover specific languages.

- Crafts- and trades-oriented books: The exam does not expect you to have detailed knowledge of trades or manufacturing operations.

- Manufacturer's literature and catalogs: No part of the exam requires you to be familiar with products that are proprietary to any manufacturer.

- Government publications.

- Your state's engineering practice laws: The PE exam is a national exam. Nothing unique to your state will appear on it.

WHAT ABOUT CALCULATORS?

It goes without saying that a calculator is needed for the exam. However, it may not be obvious that you should bring a spare calculator with you as well. It is always unfortunate when an examinee is not able to finish because his or her calculator was dropped, was stolen, or stopped working for some unknown reason.

The exam has not been "optimized" for any particular brand or type of calculator. In fact, for most problems, a $15 scientific calculator will produce results as satisfactory as those of a $200 calculator. There are definite advantages to having built-in statistical functions, graphing, unit-conversions, and root-finding capabilities. However, these advantages are not so great as to give anyone an unfair advantage.

In most states, any calculator may be used as long as it is noncommunicating, battery-operated, silent, and nonprinting and does not have any significant word-processing functions (i.e., does not have a QWERTY keyboard). In most states, there are no restrictions on using programmable, preprogrammed, or business/financial calculators. Similarly, nomographs and specialty slide rules are permitted.

It is essential that a calculator used for the electrical PE exam have the following functions.

- trigonometric functions

- inverse trigonometric functions

- pi

- square root and x^2

- both common and natural logarithms

- y^x and e^x

For maximum speed, your calculator should also have or be programmed for the following functions.

- interpolation
- finding standard deviations and variances
- extracting roots of quadratic equations
- calculating determinants of matrices
- linear regression
- calculating factors for economic analysis problems

You cannot share calculators with other examinees.

Laptops are not permitted in the exam. You may not use a walkie-talkie, cellular telephone, interactive pager, or other communications device during the exam.

Be sure to take your calculator with you whenever you leave the examination room for any length of time.

HOW IS THE EXAM GRADED AND SCORED?

The maximum number of points you can earn on the electrical and computer engineering PE exam is 80. The minimum number of points for passing (referred to by NCEES as the "cut score") varies from exam to exam. The cut score is determined through a rational procedure, without the benefit of knowing examinees' performance on the exam. That is, the exam is not graded on a curve. The cut score is selected based on what you are expected to know, not based on allowing a certain percentage of engineers "through."

Each of the questions is worth one point. Grading is straightforward, since a computer grades your score sheet. Either you get the question right or you don't. There is no deduction for incorrect answers, so guessing is encouraged. However, if you mark two or more answers, no credit is given for the question.

You will receive the results of your examination from your state board (not NCEES) by mail. Allow at least four months for notification. Your score may or may not be revealed to you, depending on your state's procedure. Even when the score is reported to you, it may have been scaled or normalized to 100%.

HOW YOU SHOULD GUESS

NCEES produces defensible licensing exams. One of the characteristics is that there is no pattern to the placement of correct responses. Therefore, it is not important whether you randomly guess all "A," "B," "C," or "D."

The proper way to guess is as an engineer. You should use your knowledge of the subject to eliminate illogical answer choices. Illogical answer choices are those that violate good engineering principles, that are outside normal operating ranges, or that require extraordinary assumptions. Of course, this requires you to

have some basic understanding of the subject in the first place. Otherwise, it's back to random guessing. That's the reason that the minimum passing score is higher than 25%.

You won't get any points using the "test-taking skills" that helped you in college—the skills that helped with tests prepared by amateurs. You won't be able to eliminate any [verb] answer choices from "Which [noun] ..." questions. You won't find questions with four answer choices, two of which are of the "more than 50" and "less than 50" variety. You won't find one answer choice among the four that has a different number of significant digits, or whose verb is written in a different tense, or that has some singular/plural discrepancy with the stem. The distractors will always match the stem, and they will be logical.

WHAT IS THE HISTORICAL PASSING RATE?

The historical passing rate for the electrical exam varies from approximately 25% to 50%. The rate for first-time and repeat examinees differs. The rate from one exam to the next differs. Because of the variability, and because the exam is changing to an entirely new format, historical passing rates are essentially irrelevant. The most important considerations determining your probability of success are preparation, problem solution experience, and knowledge of the fundamentals.

HOW IS THE CUT SCORE ESTABLISHED?

The PE exam is not graded on a curve. Rather, the minimum passing score (i.e., the "cut score") is determined independent of examinee performance. The raw cut score may be established before or after the exam is administered.

The NCEES uses a process known as the "modified Angoff" procedure to establish the cut score. This procedure starts with a small group (the "cut score panel") of professional engineers selected by the NCEES. Each individual in the group reviews each question and makes an estimate of its difficulty. Specifically, each individual estimates the number of minimally qualified engineers out of a hundred examinees who should know the correct answer to the question. (This is equivalent to predicting the percentage of minimally qualified engineers who will answer correctly.)

Next, the panel assembles, and the estimates for each question are openly compared and discussed. Eventually, a consensus value is obtained for each question. When the panel has established a consensus value for every question, the values are summed and divided by 100 to establish the cut score.

Various minor adjustments can be made to account for examinee population (as characterized by the average performance on equater questions) and any flawed questions.

CHEATING AND EXAM SUBVERSION

There aren't very many ways to cheat on an open-book test. The proctors are well-trained on the few ways that do exist. It goes without saying that you should not talk to other examinees in the room, nor should you pass notes back and forth. The number of people who are released to use the restroom may be similarly limited to prevent discussions.

NCEES regularly reuses good problems that have appeared on previous exams. Therefore, examination security is a serious issue with NCEES, which goes to great lengths to make sure nobody copies the problems. You may not keep your exam booklet, enter text of problems into your calculator, or copy problems into your own material.

The proctors are concerned about *exam subversion*, which generally means any activity that might invalidate the examination or the examination process. The most common form of exam subversion involves trying to copy exam problems for future use.

Part 3: How To Prepare For and Pass the PE Exam in Electrical and Computer Engineering

WHAT SHOULD YOU STUDY?

The exam covers many diverse subjects. Strictly speaking, you don't have to study every subject on the exam in order to pass. However, the more subjects you study, the better your chances will be of passing the exam. You should decide early in the preparation process which subjects you are going to study. The strategy you select will depend on your background. Here are the four most common strategies.

- A broad approach has been successful for examinees who have recently completed their academic studies. Their strategy has been to review the fundamentals in a broad range of undergraduate electrical and computer engineering subjects. (This means studying all or most of the chapters in this book.) The examination includes enough fundamentals problems to make this strategy worthwhile. This is the best strategy.

- Engineers who have little time to prepare tend to concentrate on the subjects in which they will find the most problems. By studying the list of exam subjects, some have been able to choose those that will give them the highest probability of finding enough problems they can solve. This strategy works as long as the exam "cooperates" and has enough of the types of problems they need. Too

often, though, examinees who "pick and choose" subjects to review can't find enough problems to complete the exam.

- Engineers who have been away from classroom work for a long time tend to concentrate on the subjects in which they have had extensive experience and hope that problems in those subjects will be on the exam. This method is seldom successful.

- Some engineers plan on modeling their solutions from similar problems they have found in textbooks, collections of solved problems, and old exams. These engineers often spend a lot of time indexing the example and sample problem types in all of their books. This is not a legitimate preparation method, and it is almost never successful.

DO YOU NEED A CLASSROOM REVIEW COURSE?

Many first-time PE examinees take a review course of some form. Classroom, audio, video, correspondence, and Internet courses are available. Classroom review courses are useful for a number of reasons. Courses provide several significant advantages over self-directed study, some of which may apply to you.

(1) A course structures and paces your review. It keeps you going forward without getting bogged down in one subject.

(2) A course focuses you on a limited amount of material. Without a course, you may not know what subjects to study.

(3) A course provides you with the problems you need to solve. You won't have to spend time looking for problems to solve.

(4) The course "spoon-feeds" you the material. You may not need to read the book!

(5) The course instructor can answer your questions when you get stuck.

You probably already know if any of these advantages apply to you. A review course will be less valuable if you are thorough, self-motivated, and highly disciplined.

HOW LONG SHOULD YOU STUDY?

We've all heard stories of the person who didn't crack a book until the week before the exam and still passed it with flying colors. Yes, these people really exist. But I'm not one of them, and you probably aren't either.

A thorough review takes approximately 300 hours. Most of this time is spent solving problems. Some of it may be spent in class; some is spent at home solving problems. Some examinees spread this time over a year; others cram it all into two months. Most classroom review courses last for three to four months. When you start studying will depend on how much time you can spend per week.

SHOULD YOU LOOK FOR OLD EXAMS?

The traditional approach for preparing for standardized tests includes working sample tests. However, NCEES does not release old tests or problems after they are used. Therefore, there are no official problems or tests available from legitimate sources. NCEES does publish a booklet of sample problems and solutions to illustrate the format of the exam. This publication is not a true "old exam," however.

ADDITIONAL REVIEW MATERIAL

In addition to your *Reference Manual* and the accompanying *Practice Problems*, PPI can provide you with a number of references and study aids for the electrical and computer engineering PE exam. For a complete listing of what is available, please go to PPI's website at **www.ppi2pass.com** and click on the "Catalog" button. The NCEES publication of sample problems, mentioned previously, is available from this catalog, along with other books, videos, calculator software, and other exam-related products.

DO YOU NEED A SCHEDULE?

It is important that you develop and adhere to a review outline and schedule. Once you have decided which subjects you are going to study, you can allocate the available time to those subjects in a manner that makes sense to you. If you are not taking a classroom review course (where the order of preparation is determined by the lectures), you should prepare your own Outline of Subjects for Self-Study to schedule your preparation.

HOW YOU CAN MAKE YOUR REVIEW REALISTIC

In the exam, you must be able to quickly recall solution procedures, formulas, and important data. You must remain sharp for eight hours or more. When you played a sport back in school, your coach probably tried to put you in game-related situations. Preparing for the PE exam isn't much different from preparing for a big game. Some part of your preparation should be realistic and representative of the examination environment.

There are several things that you can do to make your review more representative. For example, if you gather most of your review resources (i.e., books) in advance and try to use them exclusively during your review, you will become more familiar with them. (Of course, you can add to or change your references if you find inadequacies.)

Another good suggestion is to work all of your homework problems on grid paper. During the actual exam, your answers to the essay problems will be written on grid paper (lines drawn six to the inch). Some examinees find this grid paper difficult to get used to, so adjusting to it before the exam can be helpful.

Learning to use your time wisely is one of the most important things you can do during your review. You will undoubtedly encounter review problems that end up taking much longer than you expected. In some instances, you will cause your own delays by spending too much time looking through books for things you need (or just by looking for the books themselves!). Other times, the problems will just entail too much work. Learn to recognize these situations so that you can make intelligent decisions about such problems during the exam.

WHAT TO DO A FEW DAYS BEFORE THE EXAM

There are a few things that you should do a week or so before the exam. Make arrangements for child care and transportation. Since the exam does not always start or end at the designated time, make sure such arrangements are flexible.

If it is convenient, visit the exam site in order to find the building, parking areas, examination room, and restrooms.

If your spare calculator is not the same as your primary calculator, take a few minutes to familiarize yourself with its operation.

Remove the battery cover from your calculator and check to make sure you are bringing the correct replacement batteries. Put the cover back on and secure it with a piece of masking tape. Write your name on the tape to identify your calculator.

Second in importance to your scholastic preparation is the preparation of your two examination kits. The first kit consists of a bag or box (plastic milk crates work well, but handled suitcases with wheels are better) containing items to be brought with you into the examination room.

- [ ] letter admitting you to the examination
- [ ] photographic identification (e.g., driver's license)
- [ ] this book
- [ ] other textbooks and reference books
- [ ] regular dictionary
- [ ] scientific/engineering dictionary
- [ ] review course notes in a three-ring binder
- [ ] cardboard boxes or plastic milk crates to use as a bookcase

[ ] primary calculator

[ ] spare calculator

[ ] instruction booklets for your calculators

[ ] extra calculator batteries

[ ] erasers

[ ] straight-edge and rulers

[ ] compass

[ ] protractor

[ ] scissors

[ ] stapler

[ ] transparent tape

[ ] magnifying glass

[ ] unobtrusive snacks or candies (keep in your pocket)

[ ] travel pack of Kleenex™ (keep in your pocket)

[ ] handkerchief

[ ] headache remedy

[ ] personal medication

[ ] $3.00 in miscellaneous change

[ ] light, comfortable sweater

[ ] loose shoes or slippers

[ ] cushion for your chair

[ ] earplugs

[ ] wristwatch with alarm

[ ] wire coat hanger

[ ] small screwdriver (to repair eyeglasses and replace calculator batteries)

[ ] extra set of car keys

The second kit consists of the following items, which should be left in a separate bag or box in your car in case they are needed.

[ ] copy of your application

[ ] proof of delivery

[ ] light lunch

[ ] beverage in thermos and cans

[ ] sunglasses

[ ] extra pair of prescription glasses

[ ] raincoat, boots, gloves, hat, and umbrella

[ ] street map of the examination area

[ ] note to the parking patrol for your windshield

[ ] battery-powered desk lamp

[ ] your cellular phone

The following items *cannot* be used during the examination and should be left at home.

[ ] fountain pens

[ ] radio or tape player

[ ] battery charger

[ ] extension cords

[ ] scratch paper

[ ] note pads

Practice setting up your examination work environment. Carry your boxes to the kitchen table. Arrange your "bookcases" and supplies. Decide what stays on the floor in boxes and what gets an "honored position" on the tabletop.

WHAT TO DO THE DAY BEFORE THE EXAM

Take the day before the examination off from work to relax. Do not cram the last night. A good night's sleep is the best way to start the exam. If you live a considerable distance from the examination site, consider getting a hotel room in which to spend the night.

Make sure your exam kits are packed and ready to go.

Turn off the quarterly and hourly alerts on your wristwatch. Leave your pager at home. If you must bring it, change it to silent mode.

Calculate your wake-up time and set the alarms on two bedroom clocks.

Select and lay out your clothing items.

Select and lay out your breakfast items.

Make sure you have gas in your car and money in your wallet.

WHAT TO DO THE DAY OF THE EXAM

Bring or buy a morning newspaper.

Arrive at least 30 minutes before the examination starts. This will allow time to find a convenient parking place, bring your materials to the examination room, make room and seating changes, and calm down. Be prepared, though, to find that the examination room is not open or ready at the designated time.

Once you have arranged the materials around you on your table, take out your morning newspaper and look cool. (Only nervous people work crossword puzzles.)

WHAT TO DO DURING THE EXAM

All of the procedures typically associated with timed, proctored, computer-graded assessment tests will be in effect when you take the PE examination.

The proctors will distribute the examination booklets and answer sheets (or blank solution booklets) if they are not already on your table. However, you should not open the booklets until instructed to do so. You may read the information on the front and back covers, and you should write your name in the appropriate blank spaces.

Listen carefully to everything the proctors say. Do not ask your proctors any engineering questions. Even if they are knowledgeable in engineering, they will not be permitted to answer your questions.

Answers are recorded on an answer sheet contained in the test booklet. During the morning session, the proctors will guide you through the process of putting your name and other biographical information on this sheet, which will take approximately 15 minutes. You will be given the full four hours to work problems. Time to initialize the answer sheet is not part of your four hours.

The common suggestions to "use number 2 pencils, completely fill the bubbles, and erase completely" apply here. Each examinee will be provided with a mechanical pencil with HB lead at the exam site. Use of ballpoint pens and felt-tip markers is prohibited for several reasons.

If you finish the exam early and there are still more than 30 minutes remaining, you will be permitted to leave the room. If you finish less than 30 minutes before the end of the exam, you may be required to remain until the end, in order to be considerate of the people who are still working.

When you leave, you must return your exam booklet. You may not keep the exam booklet for later review.

HOW TO APPROACH THE MULTIPLE-CHOICE PROBLEMS

When solving multiple-choice problems, observe the following suggestions.

- Do not spend more than ten minutes or so per problem. If you have not finished a problem in that amount of time, make a note of it and move on.

- Set your wristwatch alarm for five minutes before the end of each four-hour session and use that remaining time to guess at all of the unsolved multiple-choice problems. You will be successful with about 25% of your guesses, and these points will more than make up for the few points that you might earn by working during the last five minutes.

- Make sure all of your responses on the answer sheet are dark and completely fill the bubbles.

HOW TO SOLVE PROBLEMS CAREFULLY

Many points are lost to carelessness. Keep the following items in mind when you are solving problems. Hopefully, these suggestions will be automatic to you in the exam.

[ ] Did you recheck all data obtained from other sources, tables, and figures?

[ ] Did you convert between radius and diameter?

[ ] Did you recheck all of your mathematical equations?

[ ] Did you convert to the requested units?

[ ] Do the units cancel out in your calculations?

SHOULD YOU TALK TO OTHER EXAMINEES AFTER THE EXAM?

The jury is out on this question. People react quite differently to the examination experience. Some people are energized. Most are exhausted. Some people need to unwind by talking with other examinees, describing every detail of their experience, and dissecting every examination question. Other people need lots of quiet space, and prefer just get into a hot tub to soak and sulk.

Since everyone who took the exam has seen it, you will not be violating your "oath of silence" if you talk about the details with other examinees. It's pretty difficult not to ask how someone else approached a problem that had you completely stumped. However, it is also very disquieting to think you did well on a problem, only to have someone tell you where you went wrong.

AFTER THE EXAM

Yes, there is something to do after the exam. Most people come home, throw their exam "kits" into the corner, and collapse. A week later, when they can bear to think about the experience again, they start integrating their exam kits back into their normal lives. The calculators go back into the desk, the books go back on the shelves, the $3.00 in change goes back into the piggy bank, and all of the miscellaneous stuff you brought with you to the exam is put back wherever it came from.

Here's what I suggest you do as soon as you get home, before you collapse.

[ ] Thank your spouse and children for helping you during your preparation.

[ ] Take any paperwork you received on exam day out of your pocket, purse, or wallet. Put this inside your *Electrical Engineering Reference Manual*.

[ ] Reflect on any statements regarding exam secrecy to which you signed your agreement in the exam.

[ ] Visit the PPI website and complete the after-exam survey to help PPI improve the quality of its service and products.

[ ] If you participated in a PPI Passing Zone, log on one last time to thank the instructors. (Passing Zones remain posted for a week after the exam.)

[ ] Call your employer and tell him/her that you need to take a mental health day off on Monday.

A few days later, when you can face the world again, do the following.

[ ] Make notes about anything you would do differently if you had to take the exam over again.

[ ] Consolidate all of your application paperwork, correspondence to/from your state, and any paperwork that you received on exam day.

[ ] If you took a live review course, call the instructor (or write a note) to say "Thanks."

[ ] Visit the Engineering Exam Forum on PPI's website and see what other people are saying about the exam you took.

[ ] Return any books you borrowed.

[ ] Write thank-you notes to all of the people who wrote letters of recommendation or reference for you.

[ ] Find and read the chapter in this book that covers ethics. There were no ethics questions on your PE exam, but it doesn't make any difference. Ethical behavior is expected of a PE in any case. Spend a few minutes reflecting on how your performance (obligations, attitude, presentation, behavior, appearance, etc.) might be about to change once you are licensed. Consider how you are going to be a role model for others around you.

[ ] Put all of your review books, binders, and notes someplace where they will be out of sight.

AND THEN THERE'S THE WAIT . . .

Waiting for the exam results is its own form of mental torture. There is no pattern to the release of results. None. The exam results are not released to all states simultaneously. They are not released by discipline. They are not released alphabetically by state or examinee name. The people who failed are not notified first, or last. Your coworker might receive his or her notification today, and you might wait another three weeks.

It all depends on when the work gets done. Some states have to approve the results at a board meeting. Some prepare the certificates before sending out notifications. Some states are more highly computerized than others. Some states have 50 examinees, while others have 5000. Some states are shut down by blizzards and hurricanes. Some states are understaffed and inadequately trained.

The size of the envelope you receive does not mean anything. Some states send a big congratulations package and certificate. Some states send a big package with a new application to repeat the exam. Some states send a postcard. Some states send a one-page letter. You just have to open it to find out.

If you are interested, you can check the Engineering Exam Forum on PPI's website (**www.ppi2pass.com**) periodically to find out which states have released their results. You will find many other anxious examinees there, as well as supportive PEs who know what the waiting is like.

Topic I: Mathematics

For the most current information about the exam, visit **www.ppi2pass.com** regularly.

1 Systems of Units

1. INTRODUCTION

The purpose of this chapter is to eliminate some of the confusion regarding the many units available for each engineering variable. In particular, an effort has been made to clarify the use of the so-called English systems, which for years have used the *pound* unit both for force and mass—a practice that has resulted in confusion even for those familiar with it.

2. COMMON UNITS OF MASS

The choice of a mass unit is the major factor in determining which system of units will be used in solving a problem. It is obvious that one will not easily end up with a force in pounds if the rest of the problem is stated in meters and kilograms. Actually, the choice of a mass unit determines more than whether a conversion factor will be necessary to convert from one system to another (e.g., between SI and English units). An inappropriate choice of a mass unit may actually require a conversion factor *within* the system of units.

The common units of mass are the gram, pound, kilogram, and slug.[1] There is nothing mysterious about these units. All represent different quantities of matter,

as Fig. 1.1 illustrates. In particular, note that the pound and slug do not represent the same quantity of matter.[2]

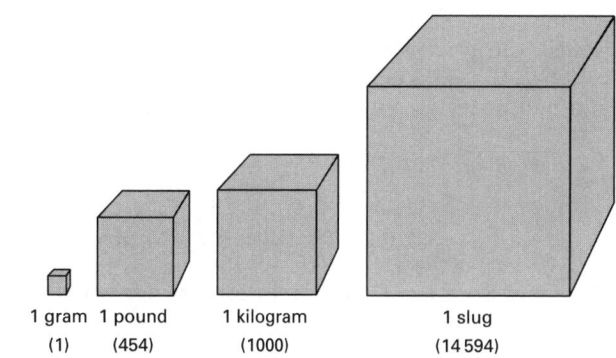

| 1 gram | 1 pound | 1 kilogram | 1 slug |
| (1) | (454) | (1000) | (14 594) |

Figure 1.1 *Common Units of Mass*

3. MASS AND WEIGHT

In SI, *kilograms* are used for mass and *newtons* for weight (force). The units are different, and there is no confusion between the variables. However, for years the term *pound* has been used for both mass and weight. This usage has obscured the distinction between the two: mass is a constant property of an object; weight varies with the gravitational field. Even the conventional use of the abbreviations *lbm* and *lbf* (to distinguish between pounds-mass and pounds-force) has not helped eliminate the confusion.

It is true that an object with a mass of one pound will have an earthly weight of one pound, but this is true only on the Earth. The weight of the same object will be much less on the moon. Therefore, care must be taken when working with mass and force in the same problem.

The relationship that converts mass to weight is familiar to every engineering student.

$$W = mg \qquad 1.1$$

Equation 1.1 illustrates that an object's weight will depend on the local acceleration of gravity as well as the object's mass. The mass will be constant, but gravity will depend on location. Mass and weight are not the same.

---

[1]Normally, one does not distinguish between a unit and a multiple of that unit, as is done here with the gram and the kilogram. However, these two units actually are bases for different consistent systems.

[2]A slug is equal to 32.1740 pounds-mass.

4. ACCELERATION OF GRAVITY

Gravitational acceleration on the Earth's surface is usually taken as 32.2 ft/sec^2 or 9.81 m/s^2. These values are rounded from the more exact standard values of 32.1740 ft/sec^2 and 9.8066 m/s^2. However, the need for greater accuracy must be evaluated on a problem-by-problem basis. Usually, three significant digits are adequate, since gravitational acceleration is not constant anyway but is affected by location (primarily latitude and altitude) and major geographical features.

The term *standard gravity*, g_0, is derived from the acceleration at essentially any point at sea level and approximately 45° N latitude. If additional accuracy is needed, the gravitational acceleration can be calculated from Eq. 1.2. This equation neglects the effects of large land and water masses. ϕ is the latitude in degrees.

$$g_{\text{surface}} = g'[1 + (5.305 \times 10^{-3})\sin^2 \phi \qquad \textit{1.2}$$
$$- (5.9 \times 10^{-6})\sin^2 2\phi]$$
$$g' = 32.0881 \text{ ft/sec}^2 = 9.78045 \text{ m/s}^2$$

If the effects of the Earth's rotation are neglected, the gravitational acceleration at an altitude h above the Earth's surface is given by Eq. 1.3. R_e is the Earth's radius.

$$g_h = g_{\text{surface}} \left(\frac{R_e}{R_e + h} \right)^2 \qquad \textit{1.3}$$

$$R_e = 3960 \text{ mi} = 6.37 \times 10^6 \text{ m}$$

5. CONSISTENT SYSTEMS OF UNITS

A set of units used in a calculation is said to be *consistent* if no conversion factors are needed.[3] For example, a moment is calculated as the product of a force and a lever arm length.

$$M = Fd \qquad \textit{1.4}$$

A calculation using Eq. 1.4 would be consistent if M was in newton-meters, F was in newtons, and d was in meters. The calculation would be inconsistent if M was in ft-kips, F was in kips, and d was in inches (because a conversion factor of 1/12 would be required). The concept of a consistent calculation can be extended to a system of units. A *consistent system of units* is one in which no conversion factors are needed for any calculation. For example, Newton's second law of motion can be written without conversion factors. Newton's second law simply states that the force required to accelerate an object is proportional to the acceleration of the object. The constant of proportionality is the object's mass.

$$F = ma \qquad \textit{1.5}$$

Notice that Eq. 1.5 is $F = ma$, not $F = Wa/g$ or $F = ma/g_c$. Equation 1.5 is consistent: it requires

no conversion factors. This means that in a consistent system where conversion factors are not used, once the units of m and a have been selected, the units of F are fixed. This has the effect of establishing units of work and energy, power, fluid properties, and so on.

It should be mentioned that the decision to work with a consistent set of units is desirable but unnecessary, depending on tradition and environment. Problems in fluid flow and thermodynamics are routinely solved in the United States with inconsistent units. This causes no more of a problem than working with inches and feet when calculating a moment. It is necessary only to use the proper conversion factors.

6. THE ENGLISH ENGINEERING SYSTEM

Through common and widespread use, pounds-mass (lbm) and pounds-force (lbf) have become the standard units for mass and force in the *English Engineering System*.

There are subjects in the United States where the practice of using pounds for mass is firmly entrenched. For example, most thermodynamics, fluid flow, and heat transfer problems have traditionally been solved using the units of lbm/ft^3 for density, Btu/lbm for enthalpy, and Btu/lbm-°F for specific heat. Unfortunately, some equations contain both lbm-related and lbf-related variables, as does the steady flow conservation of energy equation, which combines enthalpy in Btu/lbm with pressure in lbf/ft^2.

The units of pounds-mass and pounds-force are as different as the units of gallons and feet, and they cannot be canceled. A mass conversion factor, g_c, is needed to make the equations containing lbf and lbm dimensionally consistent. This factor is known as the *gravitational constant* and has a value of 32.1740 lbm-ft/lbf-sec^2. The numerical value is the same as the standard acceleration of gravity, but g_c is not the local gravitational acceleration, g.[4] g_c is a conversion constant, just as 12.0 is the conversion factor between feet and inches.

The English Engineering System is an inconsistent system as defined according to Newton's second law. $F = ma$ cannot be written if lbf, lbm, and ft/sec^2 are the units used. The g_c term must be included.

$$F \text{ in lbf} = \frac{(m \text{ in lbm})\left(a \text{ in } \dfrac{\text{ft}}{\text{sec}^2}\right)}{g_c \text{ in } \dfrac{\text{lbm-ft}}{\text{lbf-sec}^2}} \qquad \textit{1.6}$$

It is important to note in Eq. 1.6 that g_c does more than "fix the units." Since g_c has a numerical value of 32.174, it actually changes the calculation numerically. A force of 1.0 pound will not accelerate a 1.0-pound mass at the rate of 1.0 ft/sec^2.

---

[3]The terms *homogeneous* and *coherent* are also used to describe a consistent set of units.

[4]It is acceptable (and recommended) that g_c be rounded to the same number of significant digits as g. Therefore, a value of 32.2 for g_c would typically be used.

In the English Engineering System, work and energy are typically measured in ft-lbf (mechanical systems) or in British thermal units, Btu (thermal and fluid systems). One Btu is equal to 778.17 ft-lbf.

Example 1.1

Calculate the weight in lbf of a 1.00 lbm object in a gravitational field of 27.5 ft/sec^2.

Solution

From Eq. 1.6,

$$F = \frac{ma}{g_c} = \frac{(1.00 \text{ lbm}) \left(27.5 \frac{\text{ft}}{\text{sec}^2}\right)}{32.2 \frac{\text{lbm-ft}}{\text{lbf-sec}^2}} = 0.854 \text{ lbf}$$

7. FORMULAS AFFECTED BY INCONSISTENCY

It is not a significant burden to include g_c in a calculation, but it may be difficult to remember when g_c should be used. Knowing when to include the gravitational constant can be learned through repeated exposure to the formulas in which it is needed, but it is safer to carry the units along in every calculation.

The following is a representative (but not exhaustive) listing of formulas that require the g_c term. In all cases, it is assumed that the standard English Engineering System units will be used.

• kinetic energy

$$E = \frac{m\text{v}^2}{2g_c} \quad \text{[in ft-lbf]} \qquad 1.7$$

• potential energy

$$E = \frac{mgz}{g_c} \quad \text{[in ft-lbf]} \qquad 1.8$$

• pressure at a depth

$$p = \frac{\rho g h}{g_c} \quad \text{[in lbf/ft}^2\text{]} \qquad 1.9$$

Example 1.2

A rocket with a mass of 4000 lbm travels at 27,000 ft/sec. What is its kinetic energy in ft-lbf?

Solution

From Eq. 1.7,

$$E_k = \frac{m\text{v}^2}{2g_c} = \frac{(4000 \text{ lbm}) \left(27,000 \frac{\text{ft}}{\text{sec}}\right)^2}{(2) \left(32.2 \frac{\text{lbm-ft}}{\text{lbf-sec}^2}\right)}$$

$$= 4.53 \times 10^{10} \text{ ft-lbf}$$

8. WEIGHT AND WEIGHT DENSITY

Weight is a force exerted on an object due to its placement in a gravitational field. If a consistent set of units is used, Eq. 1.1 can be used to calculate the weight of a mass. In the English Engineering System, however, Eq. 1.10 must be used.

$$W = \frac{mg}{g_c} \qquad 1.10$$

Both sides of Eq. 1.10 can be divided by the volume of an object to derive the *weight density*, γ, of the object. Equation 1.11 illustrates that the weight density (in lbf/ft^3) can also be calculated by multiplying the mass density (in lbm/ft^3) by g/g_c. Since g and g_c usually have the same numerical values, the only effect of Eq. 1.12 is to change the units of density.

$$\frac{W}{V} = \left(\frac{m}{V}\right) \left(\frac{g}{g_c}\right) \qquad 1.11$$

$$\gamma = \frac{W}{V} = \left(\frac{m}{V}\right) \left(\frac{g}{g_c}\right) = \frac{\rho g}{g_c} \qquad 1.12$$

Weight does not occupy volume. Only mass has volume. The concept of weight density has evolved to simplify certain calculations, particularly fluid calculations. For example, pressure at a depth is calculated from Eq. 1.13. (Compare this with Eq. 1.9.)

$$p = \gamma h \qquad 1.13$$

9. THE ENGLISH GRAVITATIONAL SYSTEM

Not all English systems are inconsistent. Pounds can still be used as the unit of force as long as pounds are not used as the unit of mass. Such is the case with the consistent *English Gravitational System*.

If acceleration is given in ft/sec^2, the units of mass for a consistent system of units can be determined from Newton's second law. The combination of units in Eq. 1.14 is known as a *slug*. g_c is not needed at all since this system is consistent. It would be needed only to convert slugs to another mass unit.

$$\text{units of } m = \frac{\text{units of } F}{\text{units of } a}$$

$$= \frac{\text{lbf}}{\frac{\text{ft}}{\text{sec}^2}} = \frac{\text{lbf-sec}^2}{\text{ft}} \qquad 1.14$$

Slugs and pounds-mass are not the same, as Fig. 1.1 illustrates. However, both are units for the same quantity: mass. Equation 1.15 will convert between slugs and pounds-mass.

$$\text{no. of slugs} = \frac{\text{no. of lbm}}{g_c} \qquad 1.15$$

It is important to recognize that the number of slugs is not derived by dividing the number of pounds-mass by the local gravity. g_c is used regardless of the local gravity. The conversion between feet and inches is not dependent on local gravity; neither is the conversion between slugs and pounds-mass.

Since the English Gravitational System is consistent, Eq. 1.16 can be used to calculate weight. Notice that the local gravitational acceleration is used.

$$W \text{ in lbf} = (m \text{ in slugs}) \left(g \text{ in } \frac{\text{ft}}{\text{sec}^2} \right) \qquad \textit{1.16}$$

10. THE ABSOLUTE ENGLISH SYSTEM

The obscure *Absolute English System* takes the approach that mass must have units of pounds-mass (lbm) and the units of force can be derived from Newton's second law. The units for F cannot be simplified any more than they are in Eq. 1.17. This particular combination of units is known as a *poundal*.[5] A poundal is not the same as a pound.

$$\text{units of } F = (\text{units of } m)(\text{units of } a)$$
$$= (\text{lbm}) \left(\frac{\text{ft}}{\text{sec}^2} \right)$$
$$= \frac{\text{lbm-ft}}{\text{sec}^2} \qquad \textit{1.17}$$

Poundals have not seen widespread use in the United States. The English Gravitational System (using slugs for mass) has greatly eclipsed the Absolute English System in popularity. Both are consistent systems, but there seems to be little need for poundals in modern engineering.

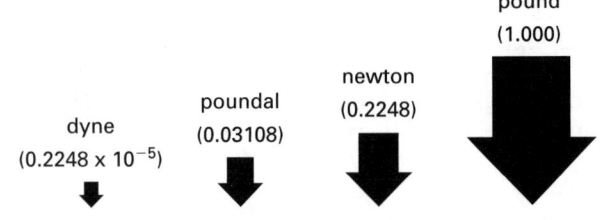

Figure 1.2 *Common Force Units*

11. METRIC SYSTEMS OF UNITS

Strictly speaking, a *metric system* is any system of units that is based on meters or parts of meters. This broad definition includes *mks systems* (based on meters, kilograms, and seconds) as well as *cgs systems* (based on centimeters, grams, and seconds).

Metric systems avoid the pounds-mass versus pounds-force ambiguity in two ways. First, a unit of weight is not established at all. All quantities of matter are specified as mass. Second, force and mass units do not share a common name.

The term *metric system* is not explicit enough to define which units are to be used for any given variable. For example, within the cgs system there is variation in how certain electrical and magnetic quantities are represented (resulting in the ESU and EMU systems). Also, within the mks system, it is common practice in some industries to use kilocalories as the unit of thermal energy, while the SI unit for thermal energy is the joule. Thus, there is a lack of uniformity even within the metricated engineering community.[6]

The SI (metric) system is used in this book, since it is the most developed and codified of the so-called metric systems.[7] There will be occasional variances with local engineering custom, but it is difficult to anticipate such variances within a book that must itself be consistent.[8]

12. THE cgs SYSTEM

The *cgs system* is used widely by chemists and physicists. It is named for the three primary units used to construct its derived variables: the centimeter, the gram, and the second.

When Newton's second law is written in the cgs system, the following combination of units results.

$$\text{units of force} = (m \text{ in g}) \left(a \text{ in } \frac{\text{cm}}{\text{s}^2} \right)$$
$$= \text{g·cm/s}^2 \qquad \textit{1.18}$$

This combination of units for force is known as a *dyne*. Energy variables in the cgs system have units of dyne·cm or, equivalently, g·cm^2/s^2. This combination is known as an *erg*. There is no uniformly accepted unit of power in the cgs system, although calories per second is frequently used.

The fundamental volume unit in the cgs system is the cubic centimeter (cc). Since this is the same volume as one thousandth of a liter, units of milliliters (mL) are also used.

---

[5] A poundal is equal to 0.03108 pounds-force.

[6] In the "field test" of the metric system conducted over the past 200 years, other conventions are to use kilograms-force (kgf) instead of newtons and kgf/cm^2 for pressure (instead of pascals).
[7] SI units are an outgrowth of the *General Conference of Weights and Measures*, an international treaty organization that established the *Système International d'Unités (International System of Units)* in 1960. The United States subscribed to this treaty in 1975.
[8] Conversion to pure SI units is essentially complete in Australia, Canada, New Zealand, and South Africa. The use of nonstandard metric units is more common among European engineers.

13. SI UNITS (THE mks SYSTEM)[9]

SI units comprise an *mks system* (so named because it uses the meter, kilogram, and second as base units). All other units are derived from the base units, which are completely listed in Table 1.1. This system is fully consistent, and there is only one recognized unit for each physical quantity (variable).

Table 1.1 *SI Base Units*

quantity	name	symbol
length	meter	m
mass	kilogram	kg
time	second	s
electric current	ampere	A
temperature	kelvin	K
amount of substance	mole	mol
luminous intensity	candela	cd

Two types of units are used: base units and derived units. The *base units* are dependent only on accepted standards or reproducible phenomena. The *derived units* (Tables 1.2 and 1.3) are made up of combinations of base units. The old *supplementary units* (Table 1.4) were classified as derived units in 1995.

Table 1.2 *Some SI Derived Units with Special Names*

quantity	name	symbol	expressed in terms of other units
frequency	hertz	Hz	$1/s$
force	newton	N	$kg \cdot m/s^2$
pressure, stress	pascal	Pa	N/m^2
energy, work, quantity of heat	joule	J	$N \cdot m$
power, radiant flux	watt	W	J/s
quantity of electricity, electric charge	coulomb	C	
electric potential, potential difference, electromotive force	volt	V	W/A
electric capacitance	farad	F	C/V
electric resistance	ohm	Ω	V/A
electric conductance	siemens	S	A/V
magnetic flux	weber	Wb	$V \cdot s$
magnetic flux density	tesla	T	Wb/m^2
inductance	henry	H	Wb/A
luminous flux	lumen	lm	
illuminance	lux	lx	lm/m^2

In addition, there is a set of non-SI units that may be used. This concession is primarily due to the significance and widespread acceptance of these units. Use of the non-SI units listed in Table 1.5 will usually create an inconsistent expression requiring conversion factors.

Table 1.3 *Some SI Derived Units*

quantity	description	symbol
area	square meter	m^2
volume	cubic meter	m^3
speed—linear	meter per second	m/s
—angular	radian per second	rad/s
acceleration—linear	meter per second squared	m/s^2
—angular	radian per second squared	rad/s^2
density, mass density	kilogram per cubic meter	kg/m^3
concentration (of amount of substance)	mole per cubic meter	mol/m^3
specific volume	cubic meter per kilogram	m^3/kg
luminance	candela per square meter	cd/m^2
absolute viscosity	pascal second	$Pa \cdot s$
kinematic viscosity	square meters per second	m^2/s
moment of force	newton meter	$N \cdot m$
surface tension	newton per meter	N/m
heat flux density, irradiance	watt per square meter	W/m^2
heat capacity, entropy	joule per kelvin	J/K
specific heat capacity, specific entropy	joule per kilogram kelvin	$J/kg \cdot K$
specific energy	joule per kilogram	J/kg
thermal conductivity	watt per meter kelvin	$W/m \cdot K$
energy density	joule per cubic meter	J/m^3
electric field strength	volt per meter	V/m
electric charge density	coulomb per cubic meter	C/m^3
surface density of charge, flux density	coulomb per square meter	C/m^2
permittivity	farad per meter	F/m
current density	ampere per square meter	A/m^2
magnetic field strength	ampere per meter	A/m
permeability	henry per meter	H/m
molar energy	joule per mole	J/mol
molar entropy, molar heat capacity	joule per mole kelvin	$J/mol \cdot K$
radiant intensity	watt per steradian	W/sr

Table 1.4 *SI Supplementary Units*[a]

quantity	name	symbol
plane angle	radian	rad
solid angle	steradian	sr

[a]classified as derived units in 1995.

[9]The mks system is used in mechanics and has the fundamental quantities of length, mass, and time, the units of which are the meter, kilogram, and second. When the fundamental quantity of current is added, with the associated unit of the ampere, the system is called the *mksa system* and is associated with the field of physics, or in this case, electrical engineering.

The SI unit of force can be derived from Newton's second law. This combination of units for force is known as a *newton*.

$$\text{units of force} = (m \text{ in kg}) \left(a \text{ in } \frac{\text{m}}{\text{s}^2}\right)$$

$$= \text{kg·m/s}^2 \qquad \qquad \textit{1.19}$$

Energy variables in SI units have units of N·m or, equivalently, kg·m^2/s^2. Both of these combinations are known as a *joule*. The units of power are joules per second, equivalent to a *watt*.

Example 1.3

A 10 kg block hangs from a cable. What is the tension in the cable? (Standard gravity equals 9.81 m/s^2.)

Solution

$$F = mg = (10 \text{ kg}) \left(9.81 \frac{\text{m}}{\text{s}^2}\right)$$

$$= 98.1 \text{ kg·m/s}^2 \text{ (98.1 N)}$$

Example 1.4

A 10 kg block is raised vertically 3 m. What is the change in potential energy?

Solution

$$\Delta E_p = mg\Delta h$$

$$= (10 \text{ kg}) \left(9.81 \frac{\text{m}}{\text{s}^2}\right) (3 \text{ m})$$

$$= 294 \text{ kg·m}^2/\text{s}^2 \text{ (294 J)}$$

14. RULES FOR USING SI UNITS

In addition to having standardized units, the set of SI units also has rigid syntax rules for writing the units and combinations of units. Each unit is abbreviated with a specific symbol. The following rules for writing and combining these symbols should be adhered to.

- The expressions for derived units in symbolic form are obtained by using the mathematical signs of multiplication and division. For example, units of velocity are m/s. Units of torque are N·m (not N-m or Nm).

- Scaling of most units is done in multiples of 1000.

- The symbols are always printed in roman type, regardless of the type used in the rest of the text. The only exception to this is in the use of the symbol for liter, where the use of the lower case "el" (l) may be confused with the numeral one (1). In this case, "liter" should be written out in full, or the script ℓ or L used. (L is used in this book.)

Table 1.5 *Acceptable Non-SI Units*

quantity	unit name	symbol or abbreviation	relationship to SI unit
area	hectare	ha	1 ha = 10 000 m^2
energy	kilowatt-hour	kW·h	1 kW·h = 3.6MJ
mass	metric ton^a	t	1 t= 1000 kg
plane angle	degree (of arc)	°	1°= 0.017 453 rad
speed of rotation	revolution per minute	r/min	1 r/min = 2π/60 rad/s
temperature interval	degree Celsius	°C	1°C = 1K
time	minute	min	1 min = 60 s
	hour	h	1 h = 3600 s
	day (mean solar)	d	1 d = 86 400 s
	year (calendar)	a	1 a = 31 536 000 s
velocity	kilometer per hour	km/h	1 km/h = 0.278 m/s
volume	liter^b	L	1 L = 0.001 m^3

^{a}The international name for metric ton is *tonne*. The metric ton is equal to the *megagram* (Mg).
^{b}The international symbol for liter is the lowercase l, which can be easily confused with the numeral 1. Several English-speaking countries have adopted the script ℓ or uppercase L (as does this book) as a symbol for liter in order to avoid any misinterpretation.

- Symbols are not pluralized: 1 kg, 45 kg (not 45 kgs).

- A period after a symbol is not used, except when the symbol occurs at the end of a sentence.

- When symbols consist of letters, there is always a full space between the quantity and the symbols: 45 kg (not 45kg). However, when the first character of a symbol is not a letter, no space is left: 32°C (not 32° C or 32 °C); or 42° 12' 45" (not 42 ° 12 ' 45 ").

- All symbols are written in lowercase, except when the unit is derived from a proper name: m for meter; s for second; A for ampere, Wb for weber, N for newton, W for watt.

- Prefixes are printed without spacing between the prefix and the unit symbol (e.g., km is the symbol for kilometer).

- In text, symbols should be used when associated with a number. However, when no number is involved, the unit should be spelled out. Examples: The area of the carpet is 16 m^2, not 16 square meters. Carpet is sold by the square meter, not by the m^2.

- Where a decimal fraction of a unit is used, a zero should always be placed before the decimal marker: 0.45 kg (not .45 kg). This practice draws attention to the decimal marker and helps avoid errors of scale.

Table 1.6 SI Prefixes[a]

prefix	symbol	value
exa	E	10^{18}
peta	P	10^{15}
tera	T	10^{12}
giga	G	10^{9}
mega	M	10^{6}
kilo	k	10^{3}
hecto	h	10^{2}
deka (or "deca")	da	10^{1}
deci	d	10^{-1}
centi	c	10^{-2}
milli	m	10^{-3}
micro	μ	10^{-6}
nano	n	10^{-9}
pico	p	10^{-12}
femto	f	10^{-15}
atto	a	10^{-18}

[a]There is no "B" (billion) prefix. In fact, the word "billion" means 10^9 in the United States but 10^{12} in most other countries. This unfortunate ambiguity is handled by avoiding the use of the term billion.

- A practice in some countries is to use a comma as a decimal marker, while the practice in North America, the United Kingdom, and some other countries is to use a period (or dot) as the decimal marker. Furthermore, in some countries that use the decimal comma, a dot is frequently used to divide long numbers into groups of three. Because of these differing practices, spaces must be used instead of commas to separate long lines of digits into easily readable blocks of three digits with respect to the decimal marker: 32 453.246 072 5. A space (half-space preferred) is optional with a four-digit number: 1 234 or 1234.

- Some confusion may arise with the word "tonne" (1000 kg). When this word occurs in French text of Canadian origin, the meaning may be a ton of 2000 pounds.

15. PRIMARY DIMENSIONS

Regardless of the system of units chosen, each variable representing a physical quantity will have the same *primary dimensions*. For example, velocity may be expressed in miles per hour (mph) or meters per second (m/s), but both units have dimensions of length per unit time. Length and time are two of the primary dimensions, as neither can be broken down into more basic dimensions. The concept of primary dimensions is useful when converting little-used variables between different systems of units, as well as in correlating experimental results (i.e., dimensional analysis).

There are three different sets of primary dimensions in use.[10] In the $ML\theta T$ system, the primary dimensions are mass (M), length (L), time (θ), and temperature (T). Notice that all symbols are uppercase. In order to avoid confusion between time and temperature, the Greek letter theta is used for time.[11]

All other physical quantities can be derived from these primary dimensions.[12] For example, work in SI units has units of N·m. Since a newton is a kg·m/s², the primary dimensions of work are ML^2/θ^2. The primary dimensions for many important engineering variables are shown in Table 1.7. If it is more convenient to stay with traditional English units, it may be more desirable to work in the $FML\theta TQ$ system (sometimes called the *engineering dimensional system*). This system adds the primary dimensions of force (F) and heat (Q). Thus, work (ft-lbf in the English system) has the primary dimensions of FL. (Compare this with the primary dimensions for work in the $ML\theta T$ system.) Thermodynamic variables are similarly simplified.

Dimensional analysis will be more conveniently carried out when one of the four-dimension systems ($ML\theta T$ or $FL\theta T$) is used. Whether the $ML\theta T$, $FL\theta T$, or $FML\theta TQ$ system is used depends on what is being derived and who will be using it, and whether or not a consistent set of variables is desired. Conversion constants such as g_c and J will almost certainly be required if the $ML\theta T$ system is used to generate variables for use in the English systems. It is also much more convenient to use the $FML\theta TQ$ system when working in the fields of thermodynamics, fluid flow, heat transfer, and so on.

16. DIMENSIONLESS GROUPS

A *dimensionless group* is derived as a ratio of two forces or other quantities. Considerable use of dimensionless groups is made in certain subjects, notably fluid mechanics and heat transfer. For example, the Reynolds number, Mach number, and Froude number are used to distinguish between distinctly different flow regimes in pipe flow, compressible flow, and open channel flow, respectively.

Table 1.8 contains information about the most common dimensionless groups used in fluid mechanics and heat transfer.

[10]One of these, the $FL\theta T$ system, is not discussed here but appears in Table 1.7.
[11]This is the most common usage. There is a lack of consistency in the engineering world about the symbols for the primary dimensions in dimensional analysis. Some writers use t for time instead of θ. Some use H for heat instead of Q. And, in the worst mix-up of all, some have reversed the use of T and θ.
[12]A *primary dimension* is the same as a *base unit* in the SI set of units. The SI units add several other base units, as shown in Table 1.1, to deal with variables that are difficult to derive in terms of the four primary base units.

Table 1.7 *Dimensions of Common Variables*

variable	dimensional system		
	$ML\theta T$	$FL\theta T$	$FML\theta TQ$
mass (m)	M	$F\theta^2/L$	M
force (F)	ML/θ^2	F	F
length (L)	L	L	L
time (θ)	θ	θ	θ
temperature (T)	T	T	T
work (W)	ML^2/θ^2	FL	FL
heat (Q)	ML^2/θ^2	FL	Q
acceleration (a)	L/θ^2	L/θ^2	L/θ^2
frequency (N)	$1/\theta$	$1/\theta$	$1/\theta$
area (A)	L^2	L^2	L^2
coefficient of thermal expansion (β)	$1/T$	$1/T$	$1/T$
density (ρ)	M/L^3	$F\theta^2/L^4$	M/L^3
dimensional constant (g_c)	1.0	1.0	$ML/\theta^2 F$
specific heat at constant pressure (c_p); at constant volume (c_v)	$L^2/\theta^2 T$	$L^2/\theta^2 T$	Q/MT
heat transfer coefficient (h); overall (U)	$M/\theta^3 T$	$F/\theta LT$	$Q/\theta L^2 T$
power (P)	ML^2/θ^3	FL/θ	FL/θ
heat flow rate ($\dot{Q}$)	ML^2/θ^3	FL/θ	Q/θ
kinematic viscosity (ν)	L^2/θ	L^2/θ	L^2/θ
mass flow rate ($\dot{m}$)	M/θ	$F\theta/L$	M/θ
mechanical equivalent of heat (J)	–	–	FL/Q
pressure (p)	$M/L\theta^2$	F/L^2	F/L^2
surface tension (σ)	M/θ^2	F/L	F/L
angular velocity (ω)	$1/\theta$	$1/\theta$	$1/\theta$
volumetric flow rate ($\dot{m}/\rho = \dot{V}$)	L^3/θ	L^3/θ	L^3/θ
conductivity (k)	$ML/\theta^3 T$	$F/\theta T$	$Q/L\theta T$
thermal diffusivity (α)	L^2/θ	L^2/θ	L^2/θ
velocity (v)	L/θ	L/θ	L/θ
viscosity, absolute (μ)	$M/L\theta$	$F\theta/L^2$	$F\theta/L^2$
volume (V)	L^3	L^3	L^3

17. LINEAL AND BOARD FOOT MEASUREMENTS

The term *lineal* is often mistaken as a typographical error for *linear*. Although "lineal" has its own specific meaning slightly different from "linear," the two are often used interchangeably by engineers.[13] The adjective *lineal* is often encountered in the building trade (e.g., 12 lineal feet of lumber), where the term is used to distinguish it from board feet measurement.

A *board foot* (abbreviated bd-ft) is not a measure of length. Rather, it is a measure of volume used with lumber. Specifically, a board foot is equal to 144 in^3 (2.36×10^{-3} m^3). The name is derived from the volume of a board 1 foot square and 1 inch thick. In that sense, it is parallel in concept to the acre-foot. Since lumber cost is directly related to lumber weight and volume, the board foot unit is used in determining the overall lumber cost.

18. DIMENSIONAL ANALYSIS

Dimensional analysis is a means of obtaining an equation that describes some phenomenon without understanding the mechanism of the phenomenon. The most serious limitation is the need to know beforehand which variables influence the phenomenon. Once these are known or assumed, dimensional analysis can be applied by a routine procedure.

The first step is to select a system of primary dimensions. (See Sec. 15.) Usually the $ML\theta T$ system is used, although this choice may require the use of g_c and J in the final results. The dimensional formulas and symbols for variables most frequently encountered are given in Table 1.7.

The second step is to write a functional relationship between the dependent variable and the independent variable, x_i.

$$y = f(x_1, x_2, \ldots, x_m) \qquad 1.20$$

This function can be expressed as an exponentiated series. The $C_1, a_i, b_i \ldots, z_i$ in Eq. 1.21 are unknown constants.

$$y = C_1 x_1^{a_1} x_2^{b_1} x_3^{c_1} \ldots x_m^{z_1} + C_2 x_1^{a_2} x_2^{b_2} x_3^{c_2} \ldots x_m^{z_2} + \cdots \quad 1.21$$

The key to solving Eq. 1.21 is that each term on the right-hand side must have the same dimensions as y. Simultaneous equations are used to determine some of the a_i, b_i, c_i, and z_i. Experimental data are required to determine the C_i and remaining exponents. In most analyses, it is assumed that the $C_i = 0$ for $i \geq 2$.

Since this method requires working with m different variables and n different independent dimensional quantities (such as M, L, θ, and T), an easier method is desirable. One simplification is to combine the m variables into dimensionless groups called *pi-groups*. (See Table 1.8.)

If these dimensionless groups are represented by $\pi_1, \pi_2, \pi_3, \ldots, \pi_k$, the equation expressing the relationship between the variables is given by the *Buckingham π-theorem*.

$$f(\pi_1, \pi_2, \pi_3 \ldots, \pi_k) = 0 \qquad 1.22$$

$$k = m - n \qquad 1.23$$

The dimensionless pi-groups are usually found from the m variables according to an intuitive process.

---

[13] *Lineal* is best used when discussing a line of succession (e.g., a lineal descendant of a particular person). *Linear* is best used when discussing length (e.g., a linear dimension of a room).

Table 1.8 Common Dimensionless Groups

name	symbol	formula	interpretation
Biot number	Bi	$\dfrac{hL}{k_s}$	$\dfrac{\text{surface conductance}}{\text{internal conduction of solid}}$
Cauchy number	Ca	$\dfrac{\text{v}^2}{\dfrac{B_s}{\rho}} = \dfrac{\text{v}^2}{a^2}$	$\dfrac{\text{inertia force}}{\text{compressive force}} = \text{Mach number}^2$
Eckert number	Ec	$\dfrac{\text{v}^2}{2c_p\,\Delta T}$	$\dfrac{\text{temperature rise due to energy conversion}}{\text{temperature difference}}$
Euler number	Eu	$\dfrac{\Delta p}{\rho V^2}$	$\dfrac{\text{pressure force}}{\text{inertia force}}$
Fourier number	Fo	$\dfrac{kt}{\rho c_p L^2} = \dfrac{\alpha t}{L^2}$	$\dfrac{\text{rate of conduction of heat}}{\text{rate of storage of energy}}$
Froude number[a]	Fr	$\dfrac{\text{v}^2}{gL}$	$\dfrac{\text{inertia force}}{\text{gravity force}}$
Graetz number[a]	Gz	$\left(\dfrac{D}{L}\right)\left(\dfrac{\text{v}\rho c_p D}{k}\right)$	$\dfrac{(\text{Re})(\text{Pr})}{L/D}$ $\dfrac{\text{heat transfer by convection in entrance region}}{\text{heat transfer by conduction}}$
Grashof number[a]	Gr	$\dfrac{g\beta\Delta T L^3}{\nu^2}$	$\dfrac{\text{buoyancy force}}{\text{viscous force}}$
Knudsen number	Kn	$\dfrac{\lambda}{L}$	$\dfrac{\text{mean free path of molecules}}{\text{characteristic length of object}}$
Lewis number[a]	Le	$\dfrac{\alpha}{D_c}$	$\dfrac{\text{thermal diffusivity}}{\text{molecular diffusivity}}$
Mach number	M	$\dfrac{\text{v}}{a}$	$\dfrac{\text{macroscopic velocity}}{\text{speed of sound}}$
Nusselt number	Nu	$\dfrac{hL}{k}$	$\dfrac{\text{temperature gradient at wall}}{\text{overall temperature difference}}$
Péclet number	Pé	$\dfrac{\text{v}\rho c_p D}{k}$	$\dfrac{(\text{Re})(\text{Pr})}{}$ $\dfrac{\text{heat transfer by convection}}{\text{heat transfer by conduction}}$
Prandtl number	Pr	$\dfrac{\mu c_p}{k} = \dfrac{\nu}{\alpha}$	$\dfrac{\text{diffusion of momentum}}{\text{diffusion of heat}}$
Reynolds number	Re	$\dfrac{\rho \text{v} L}{\mu} = \dfrac{\text{v}L}{\nu}$	$\dfrac{\text{inertia force}}{\text{viscous force}}$
Schmidt number	Sc	$\dfrac{\mu}{\rho D_c} = \dfrac{\nu}{D_c}$	$\dfrac{\text{diffusion of momentum}}{\text{diffusion of mass}}$
Sherwood number[a]	Sh	$\dfrac{k_D L}{D_c}$	$\dfrac{\text{mass diffusivity}}{\text{molecular diffusivity}}$
Stanton number	St	$\dfrac{h}{\text{v}\rho c_p} = \dfrac{h}{c_p G}$	$\dfrac{\text{heat transfer at wall}}{\text{energy transported by stream}}$
Stokes number	Sk	$\dfrac{\Delta p L}{\mu\,\text{v}}$	$\dfrac{\text{pressure force}}{\text{viscous force}}$
Strouhal number[a]	Sl	$\dfrac{L}{t\text{v}} = \dfrac{L\omega}{\text{v}}$	$\dfrac{\text{frequency of vibration}}{\text{characteristic frequency}}$
Weber number	We	$\dfrac{\rho \text{v}^2 L}{\sigma}$	$\dfrac{\text{inertia force}}{\text{surface tension force}}$

[a]Multiple definitions exist.

Example 1.5

A solid sphere rolls down a submerged incline. Find an equation for the velocity, v.

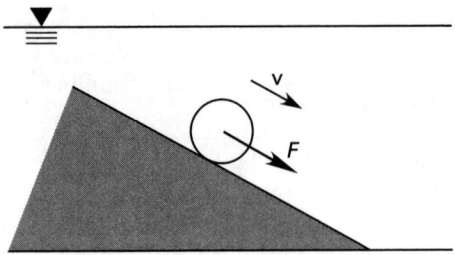

Solution

Assume that the velocity depends on the force, F, due to gravity, the diameter of the sphere, D, the density of the fluid, ρ, and the viscosity of the fluid, μ.

$$v = f(F, D, \rho, \mu) = CF^a D^b \rho^c \mu^d$$

This equation can be written in terms of the primary dimensions of the variables.

$$\frac{L}{\theta} = C \left(\frac{ML}{\theta^2} \right)^a L^b \left(\frac{M}{L^3} \right)^c \left(\frac{M}{L\theta} \right)^d$$

Since L on the left-hand side has an implied exponent of one, a necessary condition is

$$1 = a + b - 3c - d \qquad (L)$$

Similarly, the other necessary conditions are

$$-1 = -2a - d \qquad (\theta)$$
$$0 = a + c + d \qquad (M)$$

Solving simultaneously yields

$$b = -1$$
$$c = a - 1$$
$$d = 1 - 2a$$
$$v = CF^a D^{-1} \rho^{a-1} \mu^{1-2a}$$
$$= C \left(\frac{\mu}{D\rho} \right) \left(\frac{F\rho}{\mu^2} \right)^a$$

C and a must be determined experimentally.

2 Energy, Work, and Power

Nomenclature

c	specific heat	Btu/lbm-°F	J/kg·°C
C	molar specific heat	Btu/lbmol-°F	J/kmol·°C
E	energy	ft-lbf	J
F	force	lbf	N
g	gravitational acceleration	ft/sec²	m/s²
g_c	gravitational constant	lbm-ft/lbf-sec²	n.a.
h	height	ft	m
I	mass moment of inertia	lbm-ft²	kg·m²
J	Joule's constant	ft-lbf/Btu	n.a.
k	spring constant	lbf/ft	N/m
m	mass	lbm	kg
MW	molecular weight	lbm/lbmol	kg/kmol
p	pressure	lbf/ft²	Pa
P	power	ft-lbf/sec	W
q	heat	Btu/lbm	J/kg
Q	heat	Btu	J
r	radius	ft	m
s	distance	ft	m
t	time	sec	s
T	temperature	°F	°C
T	torque	ft-lbf	N·m
u	specific energy	ft-lbf/lbm	J/kg
U	internal energy	Btu	J
v	velocity	ft/sec	m/s
W	work	ft-lbf	J
x	displacement	ft	m

Symbols

δ	displacement	ft	m
η	efficiency	–	–
θ	angular position	rad	rad
ρ	mass density	lbm/ft³	kg/m³
υ	specific volume	ft³/lbm	m³/kg
ϕ	angle	deg	deg
ω	angular velocity	rad/sec	rad/s

Subscripts

f	frictional
p	constant pressure
v	constant volume

1. ENERGY OF A MASS

The *energy* of a mass represents the capacity of the mass to do work. Such energy can be stored and released. There are many forms that it can take, including mechanical, thermal, electrical, and magnetic energies. Energy is a positive, scalar quantity (although the change in energy can be either positive or negative).

The total energy of a body can be calculated from its mass, m, and its *specific energy* (i.e., the energy per unit mass).[1]

$$E = m \times \text{specific energy} \qquad 2.1$$

Typical units of mechanical energy are foot-pounds and joules. (A joule is equivalent to the units of N·m and kg·m²/s².) In traditional English-unit countries, the *British thermal unit* (Btu) is used for thermal energy, whereas the kilocalorie (kcal) is still used in some applications in SI countries. *Joule's constant* or the *Joule equivalent* (778.26 ft-lbf/Btu, usually shortened to 778, three significant digits) is used to convert between English mechanical and thermal energy units.

$$\text{energy in Btu} = \frac{\text{energy in ft-lbf}}{J} \qquad 2.2$$

Two other units of large amounts of energy are the therm and the quad. A *therm* is 10^5 Btu (1.055×10^8 J). A *quad* is equal to a quadrillion (10^{15}) Btu. This is 1.055×10^{18} J, roughly the energy contained in 200 million barrels of oil.

2. LAW OF CONSERVATION OF ENERGY

The *law of conservation of energy* says that energy cannot be created or destroyed. However, energy can be

---

[1] The use of symbols E and U for energy is not consistent in the engineering field.

converted into different forms. Therefore, the sum of all energy forms is constant.

$$\sum E = \text{constant} \qquad 2.3$$

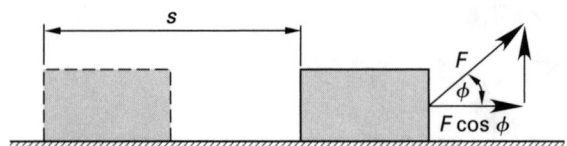

Figure 2.1 *Work of a Constant Force*

3. WORK

Work, W, is the act of changing the energy of a particle, body, or system. For a mechanical system, *external work* is work done by an external force, whereas *internal work* is done by an internal force. Work is a signed, scalar quantity. Typical units are inch-pounds, foot-pounds, and joules. Mechanical work is seldom expressed in British thermal units or kilocalories.

For a mechanical system, work is positive when a force acts in the direction of motion and helps a body move from one location to another. Work is negative when a force acts to oppose motion. (Friction, for example, always opposes the direction of motion and can do only negative work.) The work done on a body by more than one force can be found by superposition.

From a thermodynamic standpoint, work is positive if a particle or body does work on its surroundings. Work is negative if the surroundings do work on the object. (Thus, blowing up a balloon represents negative work to the balloon.) Although this may be a difficult concept, it is consistent with the conservation of energy, since the sum of negative work and the positive energy increase is zero (i.e., no net energy change in the system).[2]

The work performed by a variable force or torque is calculated from the dot products of Eqs. 2.4 and 2.5,

$$W_{\text{variable force}} = \int \mathbf{F} \cdot d\mathbf{s} \quad \text{[linear systems]} \qquad 2.4$$

$$W_{\text{variable torque}} = \int \mathbf{T} \cdot d\boldsymbol{\theta} \quad \text{[rotational systems]} \quad 2.5$$

The work done by a force or torque of constant magnitude is

$$W_{\text{constant force}} = \mathbf{F} \cdot \mathbf{s} = Fs\cos\phi \quad \text{[linear systems]} \qquad 2.6$$

$$W_{\text{constant torque}} = \mathbf{T} \cdot \boldsymbol{\theta} = Fr\theta\,\cos\phi$$
$$\text{[rotational systems]} \qquad 2.7$$

The nonvector forms, Eqs. 2.6 and 2.7, illustrate that only the component of force or torque in the direction of motion contributes to work.

Common applications of the work done by a constant force are frictional work and gravitational work. The work to move an object a distance s against a frictional force of F_f is

$$W_{\text{friction}} = F_f s \qquad 2.8$$

The work done by gravity when a mass m changes in elevation from h_1 to h_2 is

$$W_{\text{gravity}} = mg(h_2 - h_1) \qquad \text{[SI]} \qquad 2.9(a)$$

$$W_{\text{gravity}} = \left(\frac{mg}{g_c}\right)(h_2 - h_1) \quad \text{[U.S.]} \qquad 2.9(b)$$

The work done by or on a *linear spring* whose length or deflection changes from δ_1 to δ_2 is given by Eq. 2.10.[3] It does not make any difference whether the spring is a compression spring or an extension spring.

$$W_{\text{spring}} = \tfrac{1}{2}k\left(\delta_2^2 - \delta_1^2\right) \qquad 2.10$$

Example 2.1

A lawn mower engine is started by pulling a cord wrapped around a sheave. The sheave radius is 8.0 cm. The cord is wrapped around the sheave two times. If a constant tension of 90 N is maintained in the cord during starting, what work is done?

Solution

The starting torque on the engine is

$$T = Fr = (90 \text{ N})\left(\frac{8 \text{ cm}}{100 \dfrac{\text{cm}}{\text{m}}}\right)$$

$$= 7.2 \text{ N·m}$$

The cord wraps around the sheave $(2)(2\pi) = 12.6$ rad. From Eq. 2.7, the work done by a constant torque is

$$W = T\theta = (7.2 \text{ N·m})(12.6 \text{ rad})$$

$$= 90.7 \text{ J}$$

---

[2]This is just a partial statement of the *first law of thermodynamics*.

[3]A *linear spring* is one for which the linear relationship $F = kx$ is valid.

Example 2.2

A 200 lbm crate is pushed 25 ft at constant velocity across a warehouse floor. There is a frictional force of 60 lbf between the crate and floor. What work is done by the frictional force on the crate?

Solution

From Eq. 2.8,

$$W_{\text{friction}} = F_f s = (60 \text{ lbf})(25 \text{ ft})$$

$$= 1500 \text{ ft-lbf}$$

4. POTENTIAL ENERGY OF A MASS

Potential energy (gravitational energy) is a form of mechanical energy possessed by a body due to its relative position in a gravitational field. Potential energy is lost when the elevation of a body decreases. The lost potential energy usually is converted to kinetic energy or heat.

$$E_{\text{potential}} = mgh \quad \text{[SI]} \qquad \textit{2.11(a)}$$

$$E_{\text{potential}} = \frac{mgh}{g_c} \quad \text{[U.S.]} \qquad \textit{2.11(b)}$$

In the absence of friction and other nonconservative forces, the change in potential energy of a body is equal to the work required to change the elevation of the body.

$$W = \Delta E_{\text{potential}} \qquad \textit{2.12}$$

5. KINETIC ENERGY OF A MASS

Kinetic energy is a form of mechanical energy associated with a moving or rotating body. The kinetic energy of a body moving with instantaneous linear velocity v is

$$E_{\text{kinetic}} = \tfrac{1}{2}m\text{v}^2 \quad \text{[SI]} \qquad \textit{2.13(a)}$$

$$E_{\text{kinetic}} = \frac{m\text{v}^2}{2g_c} \quad \text{[U.S.]} \qquad \textit{2.13(b)}$$

According to the *work-energy principle* (see Sec. 2-9), the kinetic energy is equal to the work necessary to initially accelerate a stationary body, or to bring a moving body to rest.

$$W = \Delta E_{\text{kinetic}} \qquad \textit{2.14}$$

A body can also have rotational kinetic energy.

$$E_{\text{rotational}} = \tfrac{1}{2}I\omega^2 \quad \text{[SI]} \qquad \textit{2.15(a)}$$

$$E_{\text{rotational}} = \frac{I\omega^2}{2g_c} \quad \text{[U.S.]} \qquad \textit{2.15(b)}$$

Example 2.3

A solid disk flywheel ($I = 200 \text{ kg·m}^2$) is rotating with a speed of 900 rpm. What is its rotational kinetic energy?

Solution

The angular rotational velocity is

$$\omega = \frac{(900 \text{ rpm}) \left(2\pi \dfrac{\text{rad}}{\text{rev}}\right)}{60 \dfrac{\text{s}}{\text{min}}} = 94.25 \text{ rad/s}$$

From Eq. 2.15, the rotational kinetic energy is

$$E = \tfrac{1}{2}I\omega^2 = \left(\tfrac{1}{2}\right)(200 \text{ kg·m}^2)\left(94.25 \dfrac{\text{rad}}{\text{s}}\right)^2$$

$$= 888 \times 10^3 \text{ J} \quad (888 \text{ kJ})$$

6. SPRING ENERGY

A spring is an energy storage device, since the spring has the ability to perform work. In a perfect spring, the amount of energy stored is equal to the work required to compress the spring initially. The stored spring energy does not depend on the mass of the spring. Given a spring with spring constant (stiffness) k, the *spring energy* is

$$E_{\text{spring}} = \tfrac{1}{2}k\delta^2 \qquad \textit{2.16}$$

Example 2.4

A body of mass m falls from height h onto a massless, simply supported beam. The mass adheres to the beam. If the beam has a lateral stiffness k, what will be the deflection, δ, of the beam?

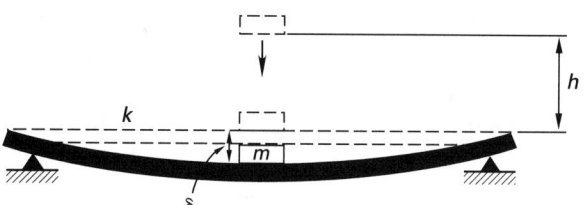

Solution

The initial energy of the system consists of only the potential energy of the body. Using consistent units, the change in potential energy is

$$E = mg(h + \delta) \qquad \text{[consistent units]}$$

All of this energy is stored in the spring. Therefore,

$$\tfrac{1}{2}k\delta^2 = mg(h + \delta)$$

Solving for the deflection,

$$\delta = \frac{mg \pm \sqrt{mg(2hk + mg)}}{k}$$

7. PRESSURE ENERGY OF A MASS

Since work is done in increasing the pressure of a system (i.e., it takes work to blow up a balloon), mechanical energy can be stored in pressure form. This is known as *pressure energy, static energy, flow energy, flow work,* and *p-V work (energy)*. For a system of pressurized mass m, the pressure energy is

$$E_{\text{flow}} = \frac{mp}{\rho} = mpv \qquad 2.17$$

8. INTERNAL ENERGY OF A MASS

The total internal energy, usually given the symbol U, of a body increases when the body's temperature increases.[4] In the absence of any work done on or by the body, the change in internal energy is equal to the heat flow, Q, into the body. Q is positive if the heat flow is into the body and negative otherwise.

$$U_2 - U_1 = Q \qquad 2.18$$

The property of internal energy is encountered primarily in thermodynamics problems. Typical units are British thermal units, joules, and kilocalories.

An increase in internal energy is needed to cause a rise in temperature. Different substances differ in the quantity of heat needed to produce a given temperature increase. The ratio of heat, Q, required to change the temperature of a mass, m, by an amount ΔT is called the *specific heat (heat capacity)* of the substance, c.

Because specific heats of solids and liquids are slightly temperature dependent, the mean specific heats are used when evaluating processes covering a large temperature range.

$$Q = mc\,\Delta T \qquad 2.19$$

$$c = \frac{Q}{m\Delta T} \qquad 2.20$$

The lowercase c implies that the units are Btu/lbm-°F or J/kg·°C. Typical values of specific heat are given in Table 2.1. The *molar specific heat*, designated by the symbol C, has units of Btu/lbmol − °F or J/kmol·°C.

$$C = (\text{MW}) \times c \qquad 2.21$$

[4]The *thermal energy*, represented by the body's enthalpy, is the sum of internal and pressure energies.

For gases, the specific heat depends on the type of process during which the heat exchange occurs. Specific heats for constant-volume and constant-pressure processes are designated by c_v and c_p, respectively.

$$Q = mc_v\Delta T \quad \text{[constant-volume process]} \qquad 2.22$$

$$Q = mc_p\Delta T \quad \text{[constant-pressure process]} \qquad 2.23$$

Approximate values of c_p and c_v for solids and liquids are essentially the same. However, the designation c_p is often encountered for solids and liquids.

9. WORK-ENERGY PRINCIPLE

Since energy can neither be created nor destroyed, external work performed on a conservative system goes into changing the system's total energy. This is known as the *work-energy principle* (or *principle of work and energy*).

$$W = \Delta E = E_2 - E_1 \qquad 2.24$$

Table 2.1 *Approximate Specific Heats of Selected Liquids and Solids*[a]

| | c_p | |
substance	Btu/lbm-°F	kJ/kg·K
aluminum, pure	0.23	0.96
aluminum, 2024-T4	0.2	0.84
ammonia	1.16	4.86
asbestos	0.20	0.84
benzene	0.41	1.72
brass, red	0.093	0.39
bronze	0.082	0.34
concrete	0.21	0.88
copper, pure	0.094	0.39
Freon-12	0.24	1.00
gasoline	0.53	2.20
glass	0.18	0.75
gold, pure	0.031	0.13
ice	0.49	2.05
iron, pure	0.11	0.46
iron, cast (4% C)	0.10	0.42
lead, pure	0.031	0.13
magnesium, pure	0.24	1.00
mercury	0.033	0.14
oil, light hydrocarbon	0.5	2.09
silver, pure	0.06	0.25
steel, 1010	0.10	0.42
steel, stainless 301	0.11	0.46
tin, pure	0.055	0.23
titanium, pure	0.13	0.54
tungsten, pure	0.032	0.13
water	1.0	4.19
wood (typical)	0.6	2.50
zinc, pure	0.088	0.37

(Multiply Btu/lbm-°F by 4.1868 to obtain kJ/kg·K.)
[a]Values in cal/g·°C are the same as Btu/lbm-°F.

Generally, the term *work-energy principle* is limited to use with mechanical energy problems (i.e., conversion of work into kinetic or potential energies). When energy is limited to kinetic energy, the work-energy principle is a direct consequence of Newton's second law but is valid for only inertial reference systems.

By directly relating forces, displacements, and velocities, the work-energy principle introduces some simplifications into many mechanical problems.

- It is not necessary to calculate or know the acceleration of a body to calculate the work performed on it.

- Forces that do not contribute to work (e.g., are normal to the direction of motion) are irrelevant.

- Only scalar quantities are involved.

- It is not necessary to individually analyze the particles or component parts in a complex system.

Example 2.5

A 4000 kg elevator starts from rest, accelerates uniformly to a constant speed of 2.0 m/s, and then decelerates uniformly to a stop 20 m above its initial position. Neglecting friction and other losses, what work was done on the elevator?

Solution

By the work-energy principle, the work done on the elevator is equal to the change in the elevator's energy. Since the initial and final kinetic energies are zero, the only mechanical energy change is the potential energy change.

Taking the initial elevation of the elevator as the reference (i.e., $h_1 = 0$),

$$W = E_{2,\,\text{potential}} - E_{1,\,\text{potential}} = mg(h_2 - h_1)$$

$$= (4000 \text{ kg})\left(9.81 \ \frac{\text{m}}{\text{s}^2}\right)(20 \text{ m})$$

$$= 785 \times 10^3 \text{ J } (785 \text{ kJ})$$

10. CONVERSION BETWEEN ENERGY FORMS

Conversion of one form of energy into another does not violate the conservation of energy law. However, most problems involving conversion of energy are really just special cases of the work-energy principle. For example, consider a falling body that is acted upon by a gravitational force. The conversion of potential energy into kinetic energy can be interpreted as equating the work done by the constant gravitational force to the change in kinetic energy.

In general terms, *Joule's law* states that one energy form can be converted without loss into another. There are two specific formulations of *Joule's Law*. As related to electricity, $P = I^2 R = V^2/R$ is the common formulation of Joule's law. As related to thermodynamics and ideal gases, Joule's law states that "the change in internal energy of an ideal gas is a function of the temperature change, not of the volume." This latter form can also be stated more formally as "at constant temperature, the internal energy of a gas approaches a finite value that is independent of the volume as the pressure goes to zero."

Example 2.6

A 2.0 lbm projectile is launched straight up with an initial velocity of 700 ft/sec. Neglecting air friction, calculate the (a) kinetic energy immediately after launch, (b) kinetic energy at maximum height, (c) potential energy at maximum height, (d) total energy at an elevation where the velocity has dropped to 300 ft/sec, and (e) maximum height attained.

Solution

(a) From Eq. 2.13, the kinetic energy is

$$E_{\text{kinetic}} = \frac{mv^2}{2g_c} = \frac{(2 \text{ lbm})\left(700 \ \dfrac{\text{ft}}{\text{sec}}\right)^2}{(2)\left(32.2 \ \dfrac{\text{lbm-ft}}{\text{sec}^2\text{-lbf}}\right)}$$

$$= 15{,}217 \text{ ft-lbf}$$

(b) The velocity is zero at the maximum height. Therefore, the kinetic energy is zero.

(c) At the maximum height, all of the kinetic energy has been converted into potential energy. Therefore, the potential energy is 15,217 ft-lbf.

(d) Although some of the kinetic energy has been transformed into potential energy, the total energy is still 15,217 ft-lbf.

(e) Since all of the kinetic energy has been converted into potential energy, the maximum height can be found from Eq. 2.11.

$$E_{\text{potential}} = \frac{mgh}{g_c}$$

$$15{,}217 \text{ ft-lbf} = \frac{(2 \text{ lbm})\left(32.2 \ \dfrac{\text{ft}}{\text{sec}^2}\right)(h)}{32.2 \ \dfrac{\text{lbm-ft}}{\text{sec}^2\text{-lbf}}}$$

$$h = 7609 \text{ ft}$$

Example 2.7

A 4500 kg ore car rolls down an incline and passes point A traveling at 1.2 m/s. The ore car is stopped by a spring bumper that compresses 0.6 m. A constant friction force of 220 N acts on the ore car at all times. What spring constant is required?

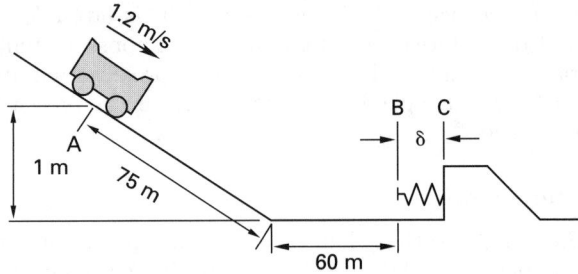

Solution

The car's total energy at point A is the sum of the kinetic and potential energies.

$$E_{\text{total,A}} = E_{\text{kinetic}} + E_{\text{potential}}$$

$$= \tfrac{1}{2}mv^2 + mgh$$

$$= \left(\tfrac{1}{2}\right)(4500 \text{ kg})\left(1.2 \,\frac{\text{m}}{\text{s}}\right)^2$$

$$+ (4500 \text{ kg})\left(9.81 \,\frac{\text{m}}{\text{s}^2}\right)(1 \text{ m})$$

$$= 47\,385 \text{ J}$$

At point B, the potential energy has been converted into additional kinetic energy. However, except for friction, the total energy is the same as at point A. Since the frictional force does negative work, the total energy remaining at point B is

$$E_{\text{total,B}} = E_{\text{total,A}} - W_{\text{friction}}$$

$$= 47\,385 \text{ J} - (220 \text{ N})(75 \text{ m} + 60 \text{ m})$$

$$= 17\,685 \text{ J}$$

At point C, the maximum compression point, the remaining energy has gone into compressing the spring a distance $\delta = 0.6$ m and performing a small amount of frictional work.

$$E_{\text{total,B}} = E_{\text{total,C}} = W_{\text{spring}} + W_{\text{friction}}$$

$$= \tfrac{1}{2}k\delta^2 + F_f\delta$$

$$17\,685 \text{ J} = \left(\tfrac{1}{2}\right)(k)(0.6 \text{ m})^2 + (220 \text{ N})(0.6 \text{ m})$$

The spring constant can be determined directly.

$$k = 97\,520 \text{ N/m} \quad (97.5 \text{ kN/m})$$

11. POWER

Power is the amount of work done per unit time. It is a scalar quantity. (Although power is calculated from two vectors, the vector dot-product operation is seldom needed.)

$$P = \frac{W}{\Delta t} \qquad 2.25$$

For a body acted upon by a force or torque, the instantaneous power can be calculated from the velocity.

$$P = F\text{v} \quad \text{[linear systems]} \qquad 2.26$$

$$P = T\omega \quad \text{[rotational systems]} \qquad 2.27$$

Typical basic units of power are ft-lbf/sec and watts (J/s), although *horsepower* is widely used. Table 2.2 can be used to convert units of power.

Table 2.2 *Useful Power Conversion Formulas*

1 hp = 550 ft-lbf/sec
= 33,000 ft-lbf/min
= 0.7457 kW
= 0.7068 Btu/sec
1 kW = 737.6 ft-lbf/sec
= 44,250 ft-lbf/min
= 1.341 hp
= 0.9483 Btu/sec
1 Btu/sec = 778.26 ft-lbf/sec
= 46,680 ft-lbf/min
= 1.415 hp

Example 2.8

When traveling at 100 km/h, a car supplies a constant horizontal force of 50 N to the hitch of a trailer. What tractive power (in horsepower) is required for the trailer alone?

Solution

From Eq. 2.26, the power being generated is

$$P = F\text{v} = \frac{(50 \text{ N})\left(100 \,\frac{\text{km}}{\text{h}}\right)\left(1000 \,\frac{\text{m}}{\text{km}}\right)}{\left(60 \,\frac{\text{s}}{\text{min}}\right)\left(60 \,\frac{\text{min}}{\text{h}}\right)\left(1000 \,\frac{\text{W}}{\text{kW}}\right)}$$

$$= 1.389 \text{ kW}$$

Using a conversion from Table 2.2, the horsepower is

$$P = \left(1.341 \; \frac{\text{hp}}{\text{kW}}\right)(1.389 \text{ kW})$$

$$= 1.86 \text{ hp}$$

12. EFFICIENCY

For energy-using systems (such as cars, electrical motors, elevators, etc.), the *energy-use efficiency*, η, of a system is the ratio of an ideal property to an actual property. The property used is commonly work, power, or, for thermodynamic problems, heat. When the rate of work is constant, either work or power can be used to calculate the efficiency. Otherwise, power should be used. Except in rare instances, the numerator and denominator of the ratio must have the same units.[5]

$$\eta = \frac{P_{\text{ideal}}}{P_{\text{actual}}} \quad [P_{\text{actual}} \geq P_{\text{ideal}}] \qquad 2.28$$

For energy-producing systems (such as electrical generators, prime movers, and hydroelectric plants), the *energy-production efficiency* is

$$\eta = \frac{P_{\text{actual}}}{P_{\text{ideal}}} \quad [P_{\text{ideal}} \geq P_{\text{actual}}] \qquad 2.29$$

The efficiency of an *ideal machine* is 1.0 (100%). However, all *real machines* have efficiencies of less than 1.0.

---

[5]The *energy-efficiency ratio* used to evaluate refrigerators, air conditioners, and heat pumps, for example, has units of Btu per watt-hour (Btu/W-hr).

3 Engineering Drawing Practice[1]

1. INTRODUCTION

This chapter is designed as a brief introduction to the reading of drawings and schematics to which the electrical engineer may be exposed. As many mechanical-electrical interfaces exist, sections 3.2 through 3.12 present the basic conventions utilized in such drawings. Section 3.13 is an overview of the electrical symbology used in schematics.[2] The various symbols are discussed throughout this book.

2. NORMAL VIEWS OF LINES AND PLANES

A *normal view* of a line is a perpendicular projection of the line onto a viewing plane parallel to the line. In the normal view, all points of the line are equidistant from the observer. Therefore, the true length of a line is viewed and can be measured.

Generally, however, a line will be viewed from an oblique position and will appear shorter than it actually is. The normal view can be constructed by drawing an auxiliary view (see Sec. 3-6) from the orthographic view.[3]

Similarly, a normal view of a plane figure is a perpendicular projection of the figure onto a viewing plane parallel to the plane of the figure. All points of the

plane are equidistant from the observer. Therefore, the true size and shape of any figure in the plane can be determined.

3. INTERSECTING AND PERPENDICULAR LINES

A single orthographic view is not sufficient to determine whether two lines intersect. However, if two or more views show the lines as having the same common point (i.e., crossing at the same position in space), then the lines intersect. In Fig. 3.1, the subscripts F and T refer to front and top views, respectively.

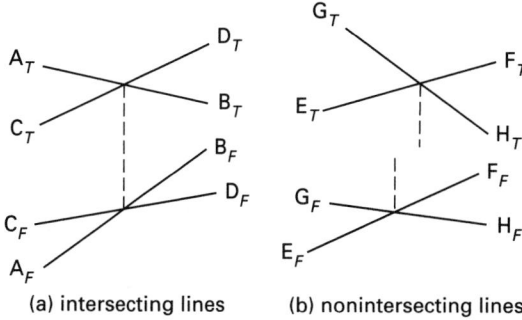

(a) intersecting lines   (b) nonintersecting lines

Figure 3.1 *Intersecting and Nonintersecting Lines*

According to the *perpendicular line principle*, two perpendicular lines appear perpendicular only in a normal view of either one or both of the lines. Conversely, if two lines appear perpendicular in any view, the lines are perpendicular only if the view is a normal view of one or both of the lines.

4. TYPES OF VIEWS

Objects can be illustrated in several different ways depending on the number of views, the angle of observation, and the degree of artistic latitude taken for the purpose of simplifying the drawing process.[4] Table 3.1 categorizes the types of views.

---

[1]This chapter is not meant to show "how to do it" as much as it is to present the conventions and symbols of engineering drawing.
[2]A drawing is the surface portrayal of a form or figure in line. A schematic is a presentation of the element-by-element relationship of all parts of a system.
[3]The technique for constructing a normal view is covered in engineering drafting texts.

[4]The omission of perspective from a drawing is an example of a step taken to simplify the drawing process.

Table 3.1 *Types of Views of Objects*

orthographic views
   principal views
   auxiliary views
   oblique views
      cavalier projection
      cabinet projection
      clinographic projection
      axonometric views
         isometric
         dimetric
         trimetric
perspective views
   parallel perspective
   angular perspective

The different types of views are easily distinguished by their *projectors* (i.e., projections of parallel lines on the object). For a cube, there are three sets of projectors corresponding to the three perpendicular axes. In an *orthographic (orthogonal) view*, the projectors are parallel. In a *perspective (central) view*, some or all of the projectors converge to a point.

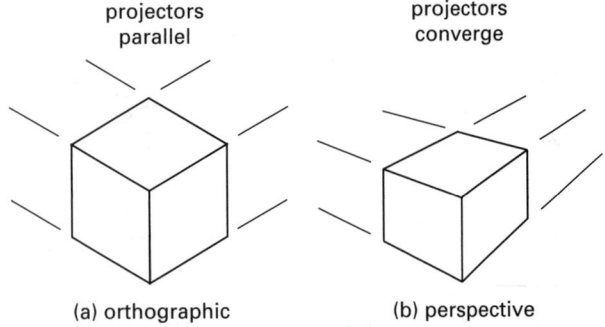

Figure 3.2 *Orthographic and Perspective Views of a Block*

5. PRINCIPAL (ORTHOGRAPHIC) VIEWS

In a *principal view* (also known as a *planar view*), one of the sets of projectors is normal to the view. That is, one of the planes of the object is seen in a normal view. The other two sets of projectors are orthogonal and are usually oriented horizontally and vertically on the paper. Because background details of an object may not be visible in a principal view, it is necessary to have at least three principal views to completely illustrate a symmetrical object. At most, six principal views will be needed to illustrate complex objects.

The relative positions of the six views have been standardized and are shown in Fig. 3.3, which also defines the *width* (also known as *depth*), *height*, and *length* of the object. The views that are not needed to illustrate features or provide dimensions (i.e., *redundant views*) can be omitted. The usual combination selected consists of the top, front, and right side views.

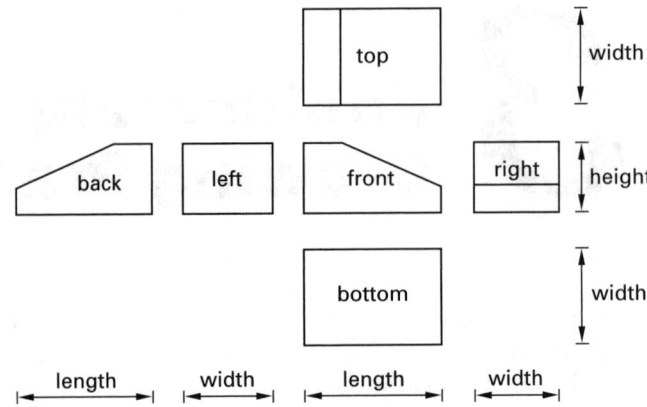

Figure 3.3 *Positions of Standard Orthographic Views*

It is common to refer to the front, side, and back views as *elevations* and to the top and bottom views as *plan views*. These terms are not absolute since any plane can be selected as the front.

6. AUXILIARY (ORTHOGRAPHIC) VIEWS

An *auxiliary view* is needed when an object has an inclined plane or curved feature or when there are more details than can be shown in the six principal views. As with the other orthographic views, the auxiliary view is a normal (face-on) view of the inclined plane.

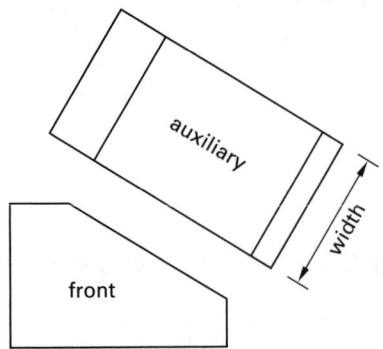

Figure 3.4 *Auxiliary View*

The projectors in an auxiliary view are perpendicular to only one of the directions in which a principal view is observed. Accordingly, only one of the three dimensions of width, height, and depth can be measured (scaled). In a *profile auxiliary view*, the object's width can be measured. In a *horizontal auxiliary view (auxiliary elevation)*, the object's height can be measured. In a *frontal auxiliary view*, the depth of the object can be measured.

7. OBLIQUE (ORTHOGRAPHIC) VIEWS

If the object is turned so that three principal planes are visible, it can be completely illustrated by a single *oblique view.*[5] In an oblique view, the direction from which the object is observed is not (necessarily) parallel to any of the directions from which principal and auxiliary views are observed.

In two common methods of oblique illustration, one of the view planes coincides with an orthographic view plane. Two of the drawing axes are at right angles to each other; one of these is vertical, and the other (the *oblique axis*) is oriented at 30° or 45° (originally chosen to coincide with standard drawing triangles). The ratio of scales used for the horizontal, vertical, and oblique axes can be 1:1:1 or 1:1:$\frac{1}{2}$. The latter ratio helps to overcome the visual distortion due to the absence of perspective in the oblique direction.

Cavalier (45° oblique axis and 1:1:1 scale ratio) and *cabinet* (45° oblique axis and 1:1:$\frac{1}{2}$ scale ratio) *projections* are the two common types of oblique views that incorporate one of the orthographic views. If an angle of 9.5° is used (as in illustrating crystalline lattice structures), the technique is known as *clinographic projection.*

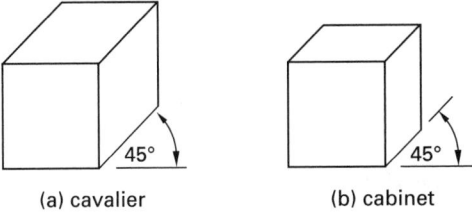

(a) cavalier   (b) cabinet

Figure 3.5 *Cavalier and Cabinet Oblique Drawings*

8. AXONOMETRIC (ORTHOGRAPHIC OBLIQUE) VIEWS

In axonometric views, the view plane is not parallel to any of the principal orthographic planes. Axonometric views and axonometric drawings are not the same. In a *view (projection)*, one or more of the face lengths is foreshortened. In a *drawing*, the lengths are drawn full length, resulting in a distorted illustration. Table 3.2 lists the proper ratios that should be observed.

Table 3.2 *Axonometric Foreshortening*

view	projector intersection angles	proper ratio of sides
isometric	120°,120°, 120°	0.82:0.82:0.82
dimetric	131°25′, 131°25′, 97°10′	1:1:$\frac{1}{2}$
	103°38′, 103°38′, 152°44′	$\frac{3}{4}$:$\frac{3}{4}$:1
trimetric	102°28′, 144°16′, 113°16′	1:$\frac{2}{3}$:$\frac{7}{8}$
	138°14′, 114°46′, 107°	1:$\frac{3}{4}$:$\frac{7}{8}$

[5]Oblique views are not unique in this capability—perspective drawings share it. Oblique and perspective drawings are known as *pictorial drawings* because they give depth to the object by illustrating it in three dimensions.

In an *isometric view*, the three projectors intersect at equal angles (120°) with the plane. This simplifies construction with standard 30° drawing triangles. All of the faces are foreshortened an equal amount, to $\sqrt{2/3}$, or approximately 81.6% of the true length. In a *dimetric view*, two of the projectors intersect at equal angles, and only two of the faces are equally reduced in length. In a *trimetric view*, all three intersection angles are different, and all three faces are reduced different amounts.

9. PERSPECTIVE VIEWS

In a *perspective view*, one or more sets of projectors converge to a fixed point known as the *center of vision*. In the *parallel perspective*, all vertical lines remain vertical in the picture; all horizontal frontal lines remain horizontal. Therefore, one face is parallel to the observer and only one set of projectors converges. In the *angular perspective*, two sets of projectors converge. In the little-used *oblique perspective*, all three sets of projectors converge.

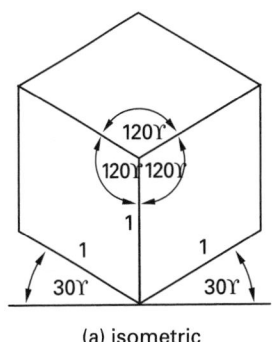

(a) isometric

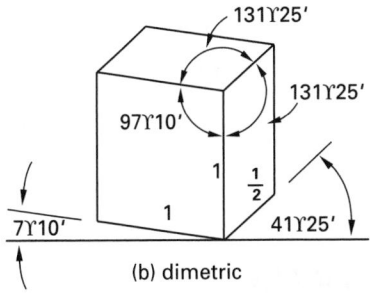

(b) dimetric

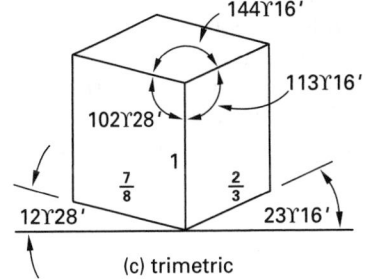

(c) trimetric

Figure 3.6 *Types of Axonometric Views*

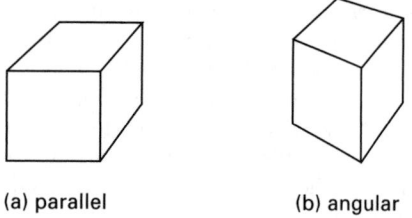

(a) parallel   (b) angular

Figure 3.7 *Types of Perspective Views*

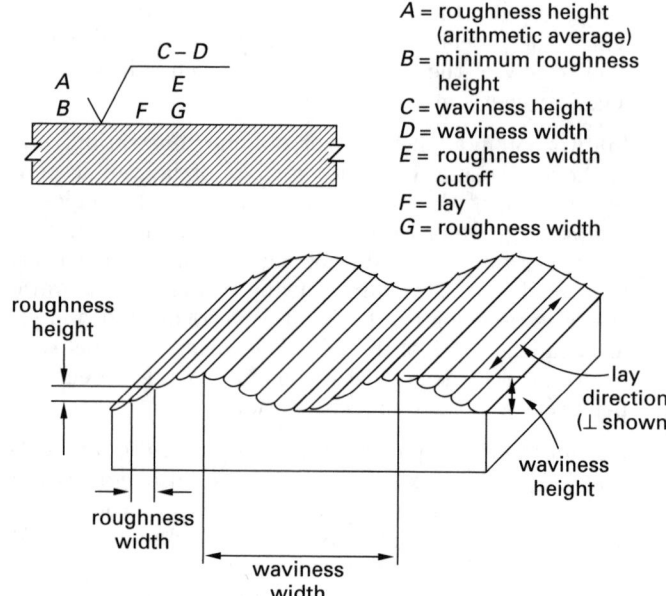

A = roughness height (arithmetic average)
B = minimum roughness height
C = waviness height
D = waviness width
E = roughness width cutoff
F = lay
G = roughness width

Figure 3.9 *Surface Finish Designations*

10. SECTIONS

The term *section* is an imaginary cut taken through an object to reveal the shape or interior construction.[6] Fig. 3.8 illustrates the standard symbol for a *sectioning cut* and the resulting sectional view. Section arrows are perpendicular to the cutting plane and indicate the viewing direction.

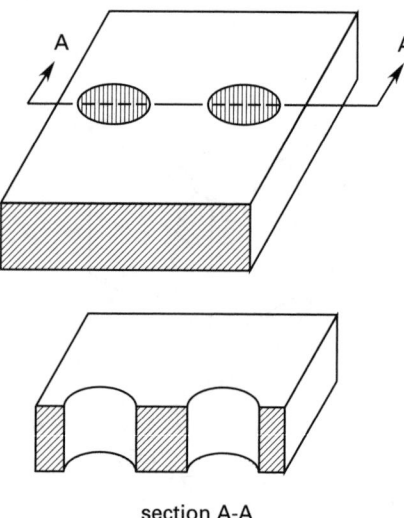

section A-A

Figure 3.8 *Sectioning Cut Symbol and Sectional View*

11. TOLERANCES

The *tolerance* for a dimension is the total permissible variation or difference between the acceptable limits. The tolerance for a dimension can be specified in two ways: either as a general rule in the title block (e.g., $\pm$ 0.001 in unless otherwise specified) or as specific limits given with each dimension (e.g., 2.575 in $\pm$ 0.005 in).

12. SURFACE FINISH

Surface finish is specified by a combination of parameters.[7] The basic symbol for designating these factors is shown in Fig. 3.9. In the symbol, A is the maximum *roughness height index*, B is the optional minimum *roughness height*, C is the peak-to-valley *waviness height*, D is the optional peak-to-valley *waviness spacing (width)* rating, E is the optional *roughness width cutoff (roughness sampling length)*, F is the lay, and G is the *roughness width*. Unless minimums are specified, all parameters are maximum allowable values, and all lesser values are permitted.

Since the roughness varies, the waviness height is an arithmetic average within a sampled square, and the designation A is known as the *roughness weight*, R_a.[8] Values are normally given in micrometers (μm) or microinches (μin) in SI or customary U.S. units, respectively. A value for the roughness width of 0.80 mm (32 μin) is assumed when D is not specified. The lay symbol, F, can be = (parallel to indicated surface), $\perp$ (perpendicular), C (circular), M (multidirectional), P (pitted), R (radial), or X (crosshatch).

If a small circle is placed at the A position, no machining is allowed and only cast, forged, die-cast, injection-molded, and other unfinished surfaces are acceptable.

---

[6]The term *section* is also used to mean a *cross section*—a slice of finite but negligible thickness that is taken from an object to show the cross section or interior construction at the plane of the slice.

[7]Specification does not indicate appearance (i.e., color, luster) or performance (i.e., hardness, corrosion resistance, microstructure).
[8]The symbol R_a is the same as the AA (arithmetic average) and CLA (centerline average) terms used in other (and earlier) standards.

13. ELECTRICAL SCHEMATICS

Although their use is no longer widespread, tubes are still used in certain applications. Basic vacuum tube symbology is shown in Table 3.3.

The basic symbology for semiconductor devices is displayed in Table 3.5.

Table 3.5 Semiconductor Symbology

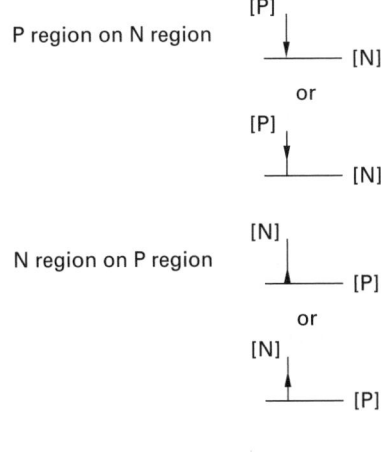

Table 3.3 Tube Symbology

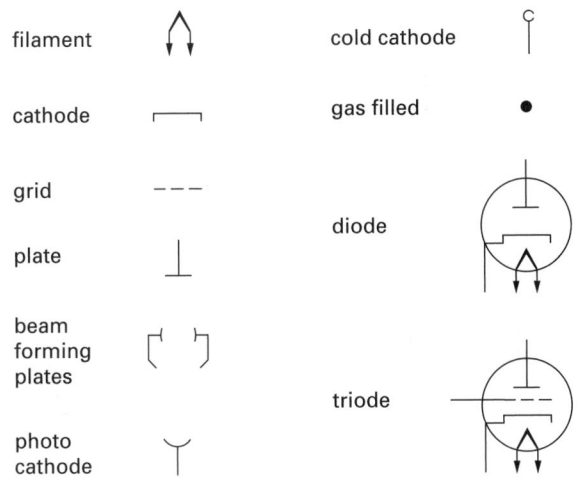

Basic circuit elements (i.e., resistors, inductors, and capacitors) are shown in Table 3.4.

Table 3.4 Basic Circuit Element Symbology

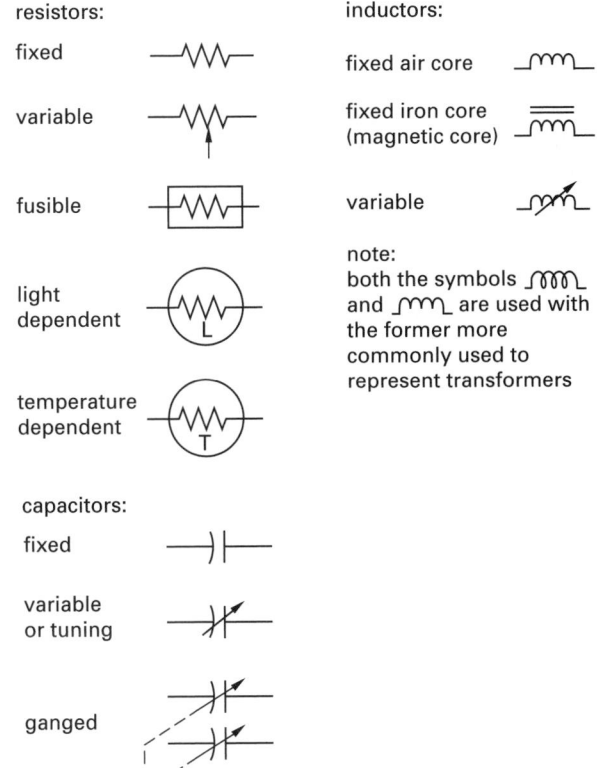

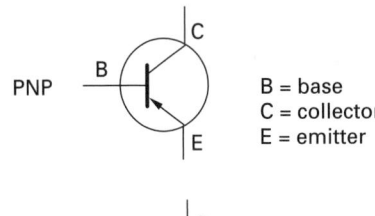

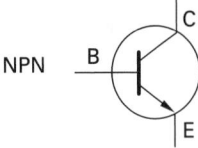

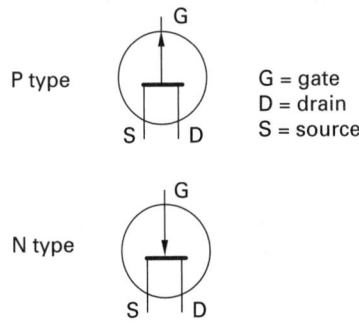

Table 3.5 *Semiconductor Symbology (continued)*

UJT (unijunction transistors):

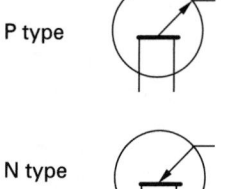

P type

N type

two-terminal devices:

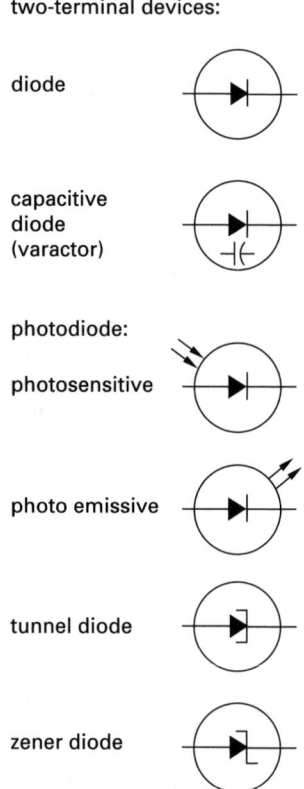

diode

capacitive
diode
(varactor)

photodiode:

photosensitive

photo emissive

tunnel diode

zener diode

miscellaneous devices:

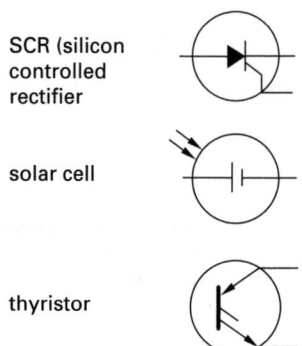

SCR (silicon
controlled
rectifier

solar cell

thyristor

One-line electrical diagrams, often associated with power equipment or transmission line representations, have a set of standard symbols. These symbols are present on electrical and electronic schematics, especially those for ground connections, connections in general, and electrical rotating machinery. Table 3.6 exhibits some of these symbols.

Table 3.6 *One-Line Electrical Diagram Symbols*

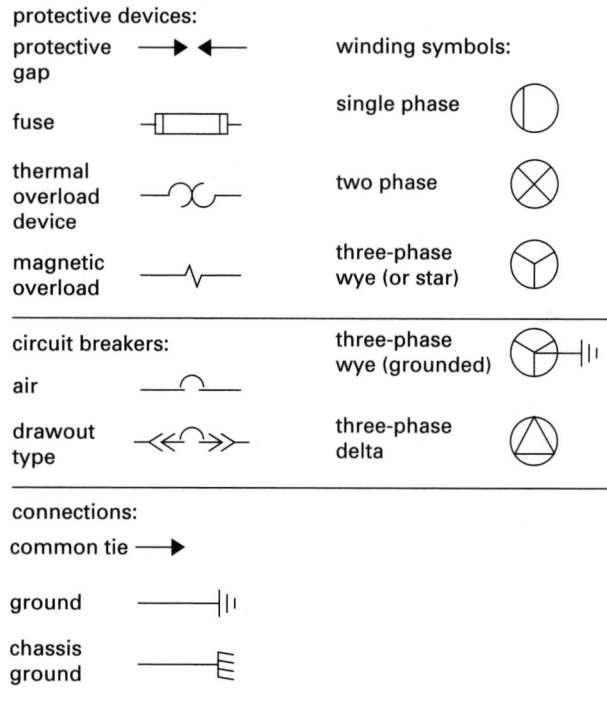

Always refer to the legend of a particular drawing or schematic to ensure understanding of the symbology utilized.

4 Algebra

1. INTRODUCTION

Engineers working in design and analysis encounter mathematical problems on a daily basis. Although algebra and simple trigonometry are often sufficient for routine calculations, there are many instances when certain advanced subjects are needed. This chapter and the others in the mathematics topic, in addition to supporting the calculations used in other chapters, consolidate the mathematical concepts most often needed by engineers.

2. SYMBOLS USED IN THIS BOOK

Many symbols, letters, and Greek characters are used to represent variables in the formulas used throughout this book. These symbols and characters are defined in the nomenclature section of each chapter. However, some of the other symbols in this book are listed in Table 4.2.

3. GREEK ALPHABET

Table 4.1 The Greek Alphabet

A	α	alpha	N	ν	nu
B	β	beta	Ξ	ξ	xi
Γ	γ	gamma	O	o	omicron
Δ	δ	delta	Π	π	pi
E	ϵ	epsilon	P	ρ	rho
Z	ζ	zeta	Σ	σ	sigma
H	η	eta	T	τ	tau
Θ	θ	theta	Υ	υ	upsilon
I	ι	iota	Φ	ϕ	phi
K	κ	kappa	X	χ	chi
Λ	λ	lambda	Ψ	ψ	psi
M	μ	mu	Ω	ω	omega

4. TYPES OF NUMBERS

The *numbering system* consists of three types of numbers: real, imaginary, and complex. *Real numbers*, in turn, consist of rational numbers and irrational numbers. *Rational real numbers* are numbers that can be written as the ratio of two integers (e.g., 4, $^2/_5$, and $^1/_3$).[1] *Irrational real numbers* are nonterminating, nonrepeating numbers that cannot be expressed as the ratio of two integers (e.g., π and $\sqrt{2}$). Real numbers can be positive or negative.

Imaginary numbers are square roots of negative numbers. The symbols i and j are both used to represent the square root of -1.[2] For example, $\sqrt{-5} = \sqrt{5}\sqrt{-1} = \sqrt{5}i$. *Complex numbers* consist of combinations of real and imaginary numbers (e.g., $3 - 7i$).

5. SIGNIFICANT DIGITS

The significant digits in a number include the leftmost, nonzero digits to the rightmost digit written. Final answers from computations should be rounded off to the number of decimal places justified by the data. The answer can be no more accurate than the least accurate number in the data. Of course, rounding should be done on final calculation results only. It should not be done on interim results.

---

[1] Notice that 0.3333333 is a nonterminating number, but as it can be expressed as a ratio of two integers (i.e., $^1/_3$), it is a rational number.

[2] The symbol j is used to represent the square root of -1 in electrical calculations to avoid confusion with the current variable, i.

Table 4.2 *Symbols Used in This Book*

symbol	name	use	example
$\sum$	sigma	series summation	$\sum_{i=1}^{3} x_i = x_1 + x_2 + x_3$
π	pi	3.1415927...	$p = \pi D$
e	base of natural logs	2.71828...	
$\prod$	pi	series multiplication	$\prod_{i=1}^{3} x_i = x_1 x_2 x_3$
Δ	delta	change in quantity	$\Delta h = h_2 - h_1$
$-$	over bar	average value	$\overline{x}$
$\cdot$	over dot	per unit time	$\dot{m} =$ mass flowing per second
!	factorial[a]		$x! = x(x-1)(x-2)\cdots(2)(1)$
$\|\ \|$	absolute value[b]		$\|-3\| = +3$
$\approx$	approximately equal to		$x \approx 1.5$
$\propto$	proportional to		$x \propto y$
∞	infinity		$x \to \infty$
log	base 10 logarithm		$\log(5.74)$
ln	natural logarithm		$\ln(5.74)$
exp	exponential power		$\exp(x) = e^x$
rms	root-mean-square	$\sqrt{\dfrac{1}{n}\sum_{i=1}^{n} x_i^2}$	V_{rms}
$\angle$	phasor or angle		$\angle 53°$

[a] *Zero factorial* (0!) is frequently encountered in the form of $(n-n)!$ when calculating permutations and combinations. Zero factorial is defined as 1.
[b] The notation abs (x) is also used to indicate the absolute value.

Table 4.3 *Examples of Significant Digits*

number as written	number of significant digits	implied range
341	3	340.5 to 341.5
34.1	3	34.05 to 34.15
0.00341	3	0.003405 to 0.003415
341×10^7	3	340.5×10^7 to 341.5×10^7
3.41×10^{-2}	3	3.405×10^{-2} to 3.415×10^{-2}
3410	4	3409.5 to 3410.5
341.0	4	340.95 to 341.05

There are two ways that significant digits can affect calculations. For the operations of multiplication and division, the final answer is rounded to the number of significant digits in the least significant multiplicand, divisor, or dividend. So, $2.0 \times 13.2 = 26$ since the first multiplicand (2.0) has two significant digits only.

For the operations of addition and subtraction, the final answer is rounded to the position of the least significant digit in the addenda, minuend, or subtrahend. So, $2.0 + 13.2 = 15.2$ because both addenda are significant to the tenth's position; but $2 + 13.4 = 15$ since the 2 is significant only in the ones' position.

The multiplication rule should not be used for addition or subtraction, as this can result in strange answers. For example, it would be incorrect to round $1700 + 0.1$ to 2000 simply because 0.1 has only one significant digit.

6. EQUATIONS

An *equation* is a mathematical statement of equality, such as $5 = 3 + 2$. *Algebraic equations* are written in terms of *variables*. In the equation $y = x^2 + 3$, the value of variable y depends on the value of variable x. Therefore, y is the *dependent variable* and x is the *independent variable*. The dependency of y on x is clearer when the equation is written in *functional form*: $y = f(x)$.

A *parametric equation* uses one or more independent variables (*parameters*) to describe a function.[3] For example, the parameter θ can be used to write the parametric equations of a unit circle.

$$x = \cos\theta \qquad \qquad 4.1$$

$$y = \sin\theta \qquad \qquad 4.2$$

---

[3] As used in this section, there is no difference between a parameter and an independent variable. However, the term *parameter* is also used as a descriptive measurement that determines or characterizes the form, size, or content of a function. For example, the radius is a parameter of a circle, and mean and variance are parameters of a probability distribution. Once these parameters are specified, the function is completely defined.

A unit circle can also be described by a *nonparametric equation*.[4]

$$x^2 + y^2 = 1 \qquad 4.3$$

7. FUNDAMENTAL ALGEBRAIC LAWS

Algebra provides the rules that allow complex mathematical relationships to be expanded or condensed. Algebraic laws may be applied to complex numbers, variables, and real numbers. The general rules for changing the form of a mathematical relationship are given as follows.

- commutative law for addition:
$$A + B = B + A \qquad 4.4$$

- commutative law for multiplication:
$$AB = BA \qquad 4.5$$

- associative law for addition:
$$A + (B + C) = (A + B) + C \qquad 4.6$$

- associative law for multiplication:
$$A(BC) = (AB)C \qquad 4.7$$

- distributive law:
$$A(B + C) = AB + AC \qquad 4.8$$

8. POLYNOMIALS

A *polynomial* is a rational expression—usually the sum of several variable terms known as *monomials*—that does not involve division. The *degree of the polynomial* is the highest power to which a variable in the expression is raised. The following *standard polynomial forms* are useful when trying to find the roots of an equation.

$$(a + b)(a - b) = a^2 - b^2 \qquad 4.9$$

$$(a \pm b)^2 = a^2 \pm 2ab + b^2 \qquad 4.10$$

$$(a \pm b)^3 = a^3 \pm 3a^2b + 3ab^2 \pm b^3 \qquad 4.11$$

$$(a^3 \pm b^3) = (a \pm b)(a^2 \mp ab + b^2) \qquad 4.12$$

$$(a^n - b^n) = (a - b)(a^{n-1} + a^{n-2}b + a^{n-3}b^2 + \cdots + b^{n-1}) \quad [n \text{ is any positive integer}] \quad 4.13$$

$$(a^n + b^n) = (a + b)(a^{n-1} - a^{n-2}b + a^{n-3}b^2 - \cdots + b^{n-1}) \quad [n \text{ is any positive odd integer}]$$
$$4.14$$

The *binomial theorem* defines a polynomial of the form $(a + b)^n$.

$$(a + b)^n = \underset{[i=0]}{a^n} + \underset{[i=1]}{na^{n-1}b} + \underset{[i=2]}{C_2 a^{n-2}b^2} + \cdots$$
$$+ C_i a^{n-i}b^i + \cdots + nab^{n-1} + b^n \qquad 4.15$$

$$C_i = \frac{n!}{i!(n-i)!} \qquad [i = 0,1,2,\ldots n] \qquad 4.16$$

---

[4]Since only the coordinate variables are used, this equation is also said to be in *Cartesian equation form*.

The coefficients of the expansion can be determined quickly from *Pascal's triangle*, Fig. 4.1. Notice that each entry is the sum of the two entries directly above it.

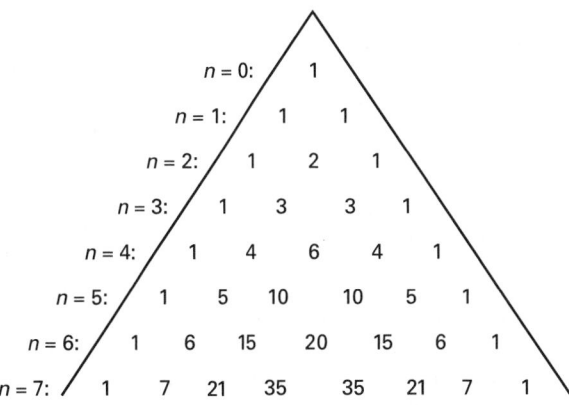

Figure 4.1 *Pascal's Triangle*

The values $r_1, r_2, \ldots, r_n$ of the independent variable x that satisfy a polynomial equation $f(x) = 0$ are known as *roots* or *zeros* of the polynomial. A polynomial of degree n with real coefficients will have at most n real roots, although they need not all be distinctly different.

9. ROOTS OF QUADRATIC EQUATIONS

A *quadratic equation* is an equation of the general form $ax^2 + bx + c = 0 \; [a \neq 0]$. The *roots*, x_1 and x_2, of the equation are the two values of x that satisfy it.

$$x_1, x_2 = \frac{-b \pm \sqrt{b^2 - 4ac}}{2a} \qquad 4.17$$

$$x_1 + x_2 = -\frac{b}{a} \qquad 4.18$$

$$x_1 x_2 = \frac{c}{a} \qquad 4.19$$

The types of roots of the equation can be determined from the *discriminant* (i.e., the quantity under the radical in Eq. 4.17).

- If $(b^2 - 4ac) > 0$, the roots are real and unequal.

- If $(b^2 - 4ac) = 0$, the roots are real and equal. This is known as a *double root*.

- If $(b^2 - 4ac) < 0$, the roots are complex and unequal.

10. ROOTS OF GENERAL POLYNOMIALS

It is more difficult to find roots of cubic and higher-degree polynomials because few general techniques exist.

- *inspection*: Finding roots by inspection is equivalent to making reasonable guesses about the roots and substituting into the polynomial.

- *graphing*: If the value of a polynomial $f(x)$ is calculated and plotted for different values of x, an approximate value of a root can be determined as the value of x at which the plot crosses the x-axis.

- *numerical methods*: If an approximate value of a root is known, numerical methods (bisection method, Newton's method, etc.) can be used to refine the value. The more efficient techniques are too complex to be performed by hand.

- *factoring*: If at least one root (say, $x = r$) of a polynomial $f(x)$ is known, the quantity $(x - r)$ can be factored out of $f(x)$ by long division. The resulting quotient will be lower by one degree, and the remaining roots may be easier to determine. This method is particularly applicable if the polynomial is in one of the standard forms presented in Sec. 4-8.

- *special cases*: Certain polynomial forms can be simplified by substitution or solved by standard formulas if they are recognized as being special cases. (The standard solution to the quadratic equation is such a special case.) For example, $ax^4 + bx^2 + c = 0$ can be reduced to a polynomial of degree 2 if the substitution $u = x^2$ is made.

11. EXTRANEOUS ROOTS

With simple equalities, it may appear possible to derive roots by basic algebraic manipulations.[5] However, multiplying each side of an equality by a power of a variable may introduce *extraneous roots*. Such roots do not satisfy the original equation even though they are derived according to the rules of algebra. Checking a calculated root is always a good idea, but is particularly necessary if the equation has been multiplied by one of its own variables.

Example 4.1

Use algebraic operations to determine a value that satisfies the following equation. Determine if the value is a valid or extraneous root.

$$\sqrt{x - 2} = \sqrt{x} + 2$$

Solution

Square both sides.

$$x - 2 = x + 4\sqrt{x} + 4$$

Subtract x from each side, and combine the constants.

$$4\sqrt{x} = -6$$

---

[5] In this sentence, *equality* means a combination of two expressions containing an equal sign. Any two expressions can be linked in this manner, even those that are not actually equal. For example, the expressions for two nonintersecting ellipses can be equated even though there is no intersection point. Finding extraneous roots is more likely when the underlying equality is false to begin with.

Solve for x.

$$x = \left(\frac{-6}{4}\right)^2 = \frac{9}{4}$$

Substitute $x = 9/4$ into the original equation.

$$\sqrt{\frac{9}{4} - 2} = \sqrt{\frac{9}{4}} + 2$$
$$\frac{1}{2} = \frac{7}{2}$$

Since the equality is not established, $x = \frac{9}{4}$ is an extraneous root.

12. DESCARTES' RULE OF SIGNS

Descartes' rule of signs determines the maximum number of positive (and negative) real roots that a polynomial will have by counting the number of sign reversals (i.e., changes in sign from one term to the next) in the polynomial. The polynomial $f(x)$ must have real coefficients and must be arranged in terms of descending powers of x.

- The number of positive roots of the polynomial equation $f(x) = 0$ will not exceed the number of sign reversals.

- The difference between the number of sign reversals and the number of positive roots is an even number.

- The number of negative roots of the polynomial equation $f(x) = 0$ will not exceed the number of sign reversals in the polynomial $f(-x)$.

- The difference between the number of sign reversals in $f(-x)$ and the number of negative roots is an even number.

Example 4.2

Determine the possible numbers of positive and negative roots that satisfy the following polynomial equation.

$$4x^5 - 5x^4 + 3x^3 - 8x^2 - 2x + 3 = 0$$

Solution

There are four sign reversals, so up to four positive roots exist. To keep the difference between the number of positive roots and the number of sign reversals an even number, the number of positive real roots is limited to zero, two, and four.

Substituting $-x$ for x in the polynomial results in

$$-4x^5 - 5x^4 - 3x^3 - 8x^2 + 2x + 3 = 0$$

There is only one sign reversal, so the number of negative roots cannot exceed one. There must be exactly one negative real root in order to keep the difference to an even number (zero in this case).

13. RULES FOR EXPONENTS AND RADICALS

In the expression $b^n = a$, b is known as the *base* and n is the *exponent* or *power*. In Eqs. 4.20 through 4.33, a, b, m, and n are any real numbers with limitations listed.

$$b^0 = 1 \qquad [b \neq 0] \qquad 4.20$$

$$b^1 = b \qquad 4.21$$

$$b^{-n} = \frac{1}{b^n} = \left(\frac{1}{b}\right)^n \qquad [b \neq 0] \qquad 4.22$$

$$\left(\frac{a}{b}\right)^n = \frac{a^n}{b^n} \qquad [b \neq 0] \qquad 4.23$$

$$(ab)^n = a^n b^n \qquad 4.24$$

$$b^{\frac{m}{n}} = \sqrt[n]{b^m} = \left(\sqrt[n]{b}\right)^m \qquad 4.25$$

$$(b^n)^m = b^{nm} \qquad 4.26$$

$$b^m b^n = b^{m+n} \qquad 4.27$$

$$\frac{b^m}{b^n} = b^{m-n} \qquad [b \neq 0] \qquad 4.28$$

$$\sqrt[n]{b} = b^{\frac{1}{n}} \qquad 4.29$$

$$\left(\sqrt[n]{b}\right)^n = \left(b^{\frac{1}{n}}\right)^n = b \qquad 4.30$$

$$\sqrt[n]{ab} = \sqrt[n]{a}\,\sqrt[n]{b} = a^{\frac{1}{n}} b^{\frac{1}{n}} = (ab)^{\frac{1}{n}} \qquad 4.31$$

$$\sqrt[n]{\frac{a}{b}} = \frac{\sqrt[n]{a}}{\sqrt[n]{b}} = \left(\frac{a}{b}\right)^{\frac{1}{n}} \qquad [b \neq 0] \qquad 4.32$$

$$\sqrt[m]{\sqrt[n]{b}} = \sqrt[mn]{b} = b^{\frac{1}{mn}} \qquad 4.33$$

14. LOGARITHMS

Logarithms can be considered to be exponents. For example, the exponent n in the expression $b^n = a$ is the logarithm of a to the base b. Therefore, the two expressions $\log_b a = n$ and $b^n = a$ are equivalent.

The base for *common logs* is 10. Usually, "log" will be written when common logs are desired, although "$\log_{10}$" appears occasionally. The base for *natural (Napierian) logs* is 2.71828..., a number which is given the symbol e. When natural logs are desired, usually "ln" will be written, although "$\log_e$" is also used.

Most logarithms will contain an integer part (the *characteristic*) and a decimal part (the *mantissa*). The logarithm of any number less than one is negative. If the number is greater than one, its logarithm is positive. Although the logarithm may be negative, the mantissa is always positive. For negative logarithms, the characteristic is found by expressing the logarithm as the sum of a negative characteristic and a positive mantissa.

For common logarithms of numbers greater than one, the characteristics will be positive and equal to one less than the number of digits in front of the decimal. If the number is less than one, the characteristic will be negative and equal to one more than the number of zeros immediately following the decimal point.

If a negative logarithm is to be used in a calculation, it must first be converted to *operational form* by adding the characteristic and mantissa. The operational form should be used in all calculations and is the form displayed by scientific calculators.

The logarithm of a negative number is a complex number.

Example 4.3

Use logarithm tables to determine the operational form of $\log_{10}(0.05)$.

Solution

Since the number is less than one and there is one leading zero, the characteristic is found by observation to be -2. From a book of logarithm tables, the mantissa of 5.0 is 0.699. Two ways of combining the mantissa and characteristic are possible.

method 1: $\overline{2}.699$

method 2: $8.699 - 10$

The operational form of this logarithm is $-2 + 0.699 = -1.301$.

15. LOGARITHM IDENTITIES

Prior to the widespread availability of calculating devices, logarithm identities were used to solve complex calculations by reducing the solution method to table look-up, addition, and subtraction. Logarithm identities are still useful in simplifying expressions containing exponentials and other logarithms. In Eqs. 4.34 through 4.45, $a \neq 1$, $b \neq 1$, $x > 0$, and $y > 0$.

$$\log_b(b) = 1 \qquad 4.34$$

$$\log_b(1) = 0 \qquad 4.35$$

$$\log_b(b^n) = n \qquad 4.36$$

$$\log(x^a) = a\log(x) \qquad 4.37$$

$$\log\left(\sqrt[n]{x}\right) = \log\left(x^{\frac{1}{n}}\right) = \frac{\log(x)}{n} \qquad 4.38$$

$$b^{n\log_b(x)} = x^n = \text{antilog}[n\log_b(x)] \qquad 4.39$$

$$b^{\frac{\log_b(x)}{n}} = x^{\frac{1}{n}} \qquad 4.40$$

$$\log(xy) = \log(x) + \log(y) \qquad 4.41$$

$$\log\left(\frac{x}{y}\right) = \log(x) - \log(y) \qquad 4.42$$

$$\log_a(x) = \log_b(x)\,\log_a(b) \qquad 4.43$$

$$\ln(x) = \log_{10}(x)\,\ln(10) \approx 2.3026\,\log_{10}(x) \quad 4.44$$

$$\log_{10}(x) = \ln(x)\,\log_{10}(e) \approx 0.4343\,\ln(x) \qquad 4.45$$

Example 4.4

The surviving fraction, x, of a radioactive isotope is given by $x = e^{-0.005t}$. For what value of t will the surviving percentage be 7%?

Solution

$$x = 0.07 = e^{-0.005t}$$

Take the natural log of both sides.

$$\ln(0.07) = \ln\!\left(e^{-0.005t}\right)$$

From Eq. 4.36, $\ln e^x = x$. Therefore,

$$-2.66 = -0.005t$$
$$t = 532$$

16. PARTIAL FRACTIONS

The method of *partial fractions* is used to transform a proper polynomial fraction of two polynomials into a sum of simpler expressions, a procedure known as *resolution*.[6,7] The technique can be considered to be the act of "unadding" a sum to obtain all of the addends.

Suppose $H(x)$ is a proper polynomial fraction of the form $P(x)/Q(x)$. The object of the resolution is to determine the partial fractions u_1/v_1, u_2/v_2, etc., such that

$$H(x) = \frac{P(x)}{Q(x)} = \frac{u_1}{v_1} + \frac{u_2}{v_2} + \frac{u_3}{v_3} + \cdots \qquad 4.46$$

The form of the denominator polynomial $Q(x)$ will be the main factor in determining the form of the partial fractions. The task of finding the u_i and v_i is simplified by categorizing the possible forms of $Q(x)$.

case 1: $Q(x)$ factors into n different linear terms.

$$Q(x) = (x - a_1)(x - a_2)\cdots(x - a_n) \quad 4.47$$

Then,

$$H(x) = \sum_{i=1}^{n} \frac{A_i}{x - a_i} \qquad 4.48$$

---

[6]To be a *proper polynomial fraction*, the degree of the numerator must be less than the degree of the denominator. If the polynomial fraction is improper, the denominator can be divided into the numerator to obtain whole and fractional polynomials. The method of partial fractions can then be used to reduce the fractional polynomial.
[7]This technique is particularly useful for calculating integrals and inverse Laplace transforms in subsequent chapters.

case 2: $Q(x)$ factors into n identical linear terms.

$$Q(x) = (x - a)(x - a)\cdots(x - a) \quad 4.49$$

Then,

$$H(x) = \sum_{i=1}^{n} \frac{A_i}{(x - a)^i} \qquad 4.50$$

case 3: $Q(x)$ factors into n different quadratic terms, $x^2 + p_i x + q_i$. Then,

$$H(x) = \sum_{i=1}^{n} \frac{A_i x + B_i}{x^2 + p_i x + q_i} \qquad 4.51$$

case 4: $Q(x)$ factors into n identical quadratic terms, $x^2 + px + q$. Then,

$$H(x) = \sum_{i=1}^{n} \frac{A_i x + B_i}{(x^2 + px + q)^i} \qquad 4.52$$

Once the general forms of the partial fractions have been determined from inspection, the *method of undetermined coefficients* is used. The partial fractions are all cross-multiplied to obtain $Q(x)$ as the denominator, and the coefficients are found by equating $P(x)$ and the cross-multiplied numerator.

Example 4.5

Resolve $H(x)$ into partial fractions.

$$H(x) = \frac{x^2 + 2x + 3}{x^4 + x^3 + 2x^2}$$

Solution

Here, $Q(x) = x^4 + x^3 + 2x^2$, which factors into $x^2(x^2 + x + 2)$. This is a combination of cases 2 and 3.

$$H(x) = \frac{A_1}{x} + \frac{A_2}{x^2} + \frac{A_3 + A_4 x}{x^2 + x + 2}$$

Cross-multiplying to obtain a common denominator yields

$$\frac{(A_1 + A_4)x^3 + (A_1 + A_2 + A_3)x^2 + (2A_1 + A_2)x + 2A_2}{x^4 + x^3 + 2x^2}$$

Since the original numerator is known, the following simultaneous equations result.

$$A_1 + A_4 = 0$$
$$A_1 + A_2 + A_3 = 1$$
$$2A_1 + A_2 = 2$$
$$2A_2 = 3$$

The solutions are $A_1 = 0.25$, $A_2 = 1.5$, $A_3 = -0.75$, and $A_4 = -0.25$.

$$H(x) = \frac{1}{4x} + \frac{3}{2x^2} - \frac{x+3}{(4)(x^2+x+2)}$$

17. SIMULTANEOUS LINEAR EQUATIONS

A *linear equation* with n variables is a polynomial of degree 1 describing a geometric shape in n-space. A *homogeneous linear equation* is one that has no constant term, and a *nonhomogeneous linear equation* has a constant term.

A solution to a set of simultaneous linear equations represents the intersection point of the geometric shapes in n-space. For example, if the equations are limited to two variables (e.g., $y = 4x - 5$), they describe straight lines. The solution to two simultaneous linear equations in 2-space is the point where the two lines intersect. The set of the two equations is said to be a *consistent system* when there is such an intersection.[8]

Simultaneous equations do not always have unique solutions, and some have none at all. In addition to crossing in 2-space, lines can be parallel or they can be the same line expressed in a different equation format (i.e., dependent equations). In some cases, parallelism and dependency can be determined by inspection. In most cases, however, matrix and other advanced methods must be used to determine whether a solution exists. A set of linear equations with no simultaneous solution is known as an *inconsistent system*.

Several methods exist for solving linear equations simultaneously by hand.[9]

- *graphing*: The equations are plotted and the intersection point is read from the graph. This method is possible only with two-dimensional problems.

- *substitution*: An equation is rearranged so that one variable is expressed as a combination of the other variables. The expression is then substituted into the remaining equations wherever the selected variable appears.

- *reduction*: All terms in the equations are multiplied by constants chosen to eliminate one or more variables when the equations are added or subtracted. The remaining sum can then be solved for the other variables. This method is also known as *eliminating the unknowns*.

- *Cramer's rule*: This method is covered in Ch. 5.

---

[8]A homogeneous system always has at least one solution: the *trivial solution*, in which all variables have a value of zero.
[9]Other methods exist, but they require a computer.

Example 4.6

Solve the following set of linear equations by (a) substitution and (b) reduction.

$$2x + 3y = 12 \quad \text{[Eq. 1]}$$
$$3x + 4y = 8 \quad \text{[Eq. 2]}$$

Solution

(a) From Eq. 1, solve for variable x.

$$x = 6 - 1.5y \quad \text{[Eq. 3]}$$

Substitute $6 - 1.5y$ into Eq. 2 wherever x appears.

$$(3)(6 - 1.5y) + 4y = 8$$
$$18 - 4.5y + 4y = 8$$
$$y = 20$$

Substitute 20 for y in Eq. 4.

$$x = 6 - (1.5)(20) = -24$$

The solution $(-24, 20)$ should be checked to verify that it satisfies both original equations.

(b) Eliminate variable x by multiplying Eq. 1 by 3 and Eq. 2 by 2.

$$3 \times \text{Eq. 1: } 6x + 9y = 36 \quad \text{[Eq. 1']}$$
$$2 \times \text{Eq. 2: } 6x + 8y = 16 \quad \text{[Eq. 2']}$$

Subtract Eq. 2' from Eq. 1'.

$$y = 20 \quad \text{[Eq. 1'- Eq. 2']}$$

Substitute $y = 20$ into Eq. 1'.

$$6x + (9)(20) = 36$$
$$x = -24$$

The solution $(-24, 20)$ should be checked to verify that it satisfies both original equations.

18. COMPLEX NUMBERS

A *complex number*, **Z**, is a combination of real and imaginary numbers. When expressed as a sum (e.g., $a + bi$), the complex number is said to be in *rectangular* or *trigonometric form*. The complex number can be plotted on the real-imaginary coordinate system known as the *complex plane*, as illustrated in Fig. 4.2.

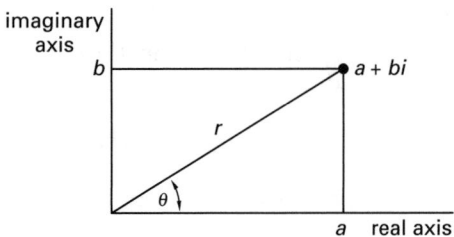

Figure 4.2 *A Complex Number in the Complex Plane*

The complex number $\mathbf{Z} = a + bi$ can also be expressed in *exponential form*.[10] The quantity r is known as the *modulus* of $\mathbf{Z}$; θ is the *argument*.

$$a + bi = re^{i\theta} \qquad 4.53$$

$$r = \text{mod}(\mathbf{Z}) = \sqrt{\mathbf{a^2 + b^2}} \qquad 4.54$$

$$\theta = \arg(\mathbf{Z}) = \arctan\left(\frac{b}{a}\right) \qquad 4.55$$

Similarly, the *phasor form* (also known as the *polar form*) is

$$\mathbf{Z} = r\angle\theta \qquad 4.56$$

The *rectangular form* can be determined from r and θ.

$$a = r\cos\theta \qquad 4.57$$

$$b = r\sin\theta \qquad 4.58$$

$$\mathbf{Z} = a + bi = r\cos\theta + ir\sin\theta$$
$$= r(\cos\theta + i\sin\theta) \qquad 4.59$$

The *cis form* is a shorthand method of writing a complex number in rectangular (trigonometric) form.

$$a + bi = r(\cos\theta + i\sin\theta) = r\operatorname{cis}\theta \qquad 4.60$$

Euler's equation (Eq. 4.61) expresses the equality of complex numbers in exponential and trigonometric form.

$$e^{i\theta} = \cos\theta + i\sin\theta \qquad 4.61$$

Related expressions are

$$e^{-i\theta} = \cos\theta - i\sin\theta \qquad 4.62$$

$$\cos\theta = \frac{e^{i\theta} + e^{-i\theta}}{2} \qquad 4.63$$

$$\sin\theta = \frac{e^{i\theta} - e^{-i\theta}}{2i} \qquad 4.64$$

---

[10]The terms *polar form*, *phasor form*, and *exponential form* are all used somewhat interchangeably.

Example 4.7

What is the exponential form of the complex number $\mathbf{Z} = 3 + 4i$?

Solution

$$r = \sqrt{a^2 + b^2} = \sqrt{3^2 + 4^2} = \sqrt{25} = 5$$

$$\theta = \arctan\left(\frac{b}{a}\right) = \arctan\left(\frac{4}{3}\right) = 0.927 \text{ rad}$$

$$\mathbf{Z} = re^{i\theta} = 5e^{i(0.927)}$$

19. OPERATIONS ON COMPLEX NUMBERS

Most algebraic operations (addition, multiplication, exponentiation, etc.) work with complex numbers, but notable exceptions are the inequality operators. The concept of one complex number being less than or greater than another complex number is meaningless.

When adding two complex numbers, real parts are added to real parts, and imaginary parts are added to imaginary parts.

$$(a_1 + ib_1) + (a_2 + ib_2) = (a_1 + a_2) + i(b_1 + b_2) \qquad 4.65$$

$$(a_1 + ib_1) - (a_2 + ib_2) = (a_1 - a_2) + i(b_1 - b_2) \qquad 4.66$$

Multiplication of two complex numbers in rectangular form is accomplished by the use of the algebraic distributive law, remembering that $i^2 = -1$.

Division of complex numbers in rectangular form requires use of the *complex conjugate*. The complex conjugate of the complex number $(a+bi)$ is $(a-bi)$. By multiplying the numerator and the denominator by the complex conjugate, the denominator will be converted to the real number $a^2 + b^2$. This technique is known as *rationalizing* the denominator and is illustrated in Ex. 4.8(c).

Multiplication and division are often more convenient when the complex numbers are in exponential or phasor forms, as Eqs. 4.67 and 4.68 show.

$$(r_1 e^{i\theta_1})(r_2 e^{i\theta_2}) = r_1 r_2 e^{i(\theta_1 + \theta_2)} \qquad 4.67$$

$$\frac{r_1 e^{i\theta_1}}{r_2 e^{i\theta_2}} = \left(\frac{r_1}{r_2}\right) e^{i(\theta_1 - \theta_2)} \qquad 4.68$$

Taking powers and roots of complex numbers requires *de Moivre's theorem*, Eqs. 4.69 and 4.70.

$$\mathbf{Z}^n = \left(re^{i\theta}\right)^n = r^n e^{in\theta} \qquad 4.69$$

$$\sqrt[n]{\mathbf{Z}} = \left(re^{i\theta}\right)^{\frac{1}{n}} = \sqrt[n]{r}\, e^{i\left(\frac{\theta + k(360^\circ)}{n}\right)}$$
$$[k = 0, 1, 2, \ldots n - 1] \qquad 4.70$$

Example 4.8

Perform the following complex arithmetic.

(a) $(3 + 4i) + (2 + i)$

(b) $(7 + 2i)(5 - 3i)$

(c) $\dfrac{2 + 3i}{4 - 5i}$

Solution

(a) $(3 + 4i) + (2 + i) = (3 + 2) + (4 + 1)i$
$$= 5 + 5i$$

(b) $(7 + 2i)(5 - 3i) = (7)(5) - (7)(3i) + (2i)(5)$
$$- (2i)(3i)$$
$$= 35 - 21i + 10i - 6i^2$$
$$= 35 - 21i + 10i - (6)(-1)$$
$$= 41 - 11i$$

(c) Multiply the numerator and denominator by the complex conjugate of the denominator.

$$\frac{2 + 3i}{4 - 5i} = \frac{(2 + 3i)(4 + 5i)}{(4 - 5i)(4 + 5i)}$$
$$= \frac{-7 + 22i}{(4)^2 + (5)^2}$$
$$= \left(\frac{-7}{41}\right) + i\left(\frac{22}{41}\right)$$

20. LIMITS

A *limit* (*limiting value*) is the value a function approaches when an independent variable approaches a target value. For example, suppose the value of $y = x^2$ is desired as x approaches 5. This could be written as

$$\lim_{x \to 5} x^2 \qquad 4.71$$

The power of limit theory is wasted on simple calculations such as this but is appreciated when the function is undefined at the target value. The object of limit theory is to determine the limit without having to evaluate the function at the target. The general case of a limit evaluated as x approaches the target value a (see Fig. 4.3(a)) is written as

$$\lim_{x \to a} f(x) \qquad 4.72$$

It is not necessary for the actual value $f(a)$ to exist for the limit to be calculated. The function $f(x)$ may be undefined at point a. However, it is necessary that $f(x)$

be defined on both sides of point a for the limit to exist. (See Fig. 4.3(b).) If $f(x)$ is undefined on one side, or if $f(x)$ is discontinuous at $x = a$ (as in Fig. 4.3(c) and 4.3(d)), the limit does not exist.

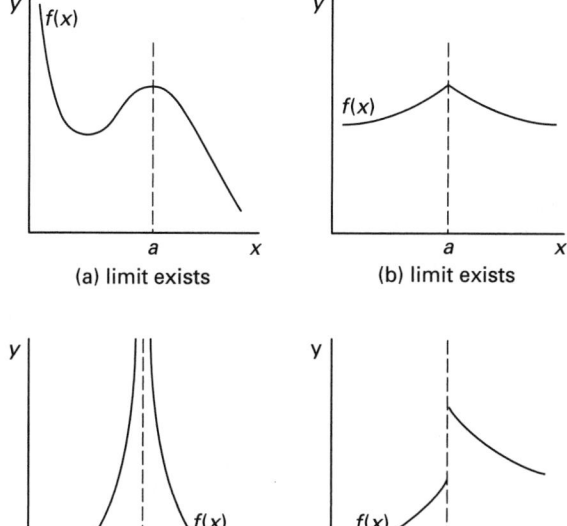

Figure 4.3 *Existence of Limits*

The following theorems can be used to simplify expressions when calculating limits.

$$\lim_{x \to a} x = a \qquad 4.73$$

$$\lim_{x \to a} (mx + b) = ma + b \qquad 4.74$$

$$\lim_{x \to a} b = b \qquad 4.75$$

$$\lim_{x \to a} (kF(x)) = k \lim_{x \to a} F(x) \qquad 4.76$$

$$\lim_{x \to a} \left(F_1(x) \begin{Bmatrix} + \\ - \\ \times \\ \div \end{Bmatrix} F_2(x) \right)$$
$$= \lim_{x \to a} (F_1(x)) \begin{Bmatrix} + \\ - \\ \times \\ \div \end{Bmatrix} \lim_{x \to a} (F_2(x)) \qquad 4.77$$

The following identities can be used to simplify limits of trigonometric expressions.

$$\lim_{x \to 0} \sin x = 0 \qquad 4.78$$

$$\lim_{x \to 0} \left(\frac{\sin x}{x} \right) = 1 \qquad 4.79$$

$$\lim_{x \to 0} \cos x = 1 \qquad 4.80$$

The following standard methods (tricks) can be used to determine limits.

- If the limit is taken to infinity, all terms can be divided by the largest power of x in the expression. This will leave at least one constant. Any quantity divided by a power of x vanishes as x approaches infinity.

- If the expression is a quotient of two expressions, any common factors should be eliminated from the numerator and denominator.

- *L'Hôpital's rule*, Eq. 4.81, should be used when the numerator and denominator of the expression both approach zero or both approach infinity.[11] $P^k(x)$ and $Q^k(x)$ are the kth derivatives of the functions $P(x)$ and $Q(x)$, respectively.[12] (L'Hôpital's rule can be applied repeatedly as required.)

$$\lim_{x \to a}\left(\frac{P(x)}{Q(x)}\right) = \lim_{x \to a}\left(\frac{P^k(x)}{Q^k(x)}\right) \qquad 4.81$$

Example 4.9

Evaluate the following limits.

(a) $\lim\limits_{x \to 3}\left(\dfrac{x^3 - 27}{x^2 - 9}\right)$

(b) $\lim\limits_{x \to \infty}\left(\dfrac{3x - 2}{4x + 3}\right)$

(c) $\lim\limits_{x \to 2}\left(\dfrac{x^2 + x - 6}{x^2 - 3x + 2}\right)$

Solution

(a) Factor the numerator and denominator. (L'Hôpital's rule can also be used.)

$$\lim_{x \to 3}\left(\frac{x^3 - 27}{x^2 - 9}\right) = \lim_{x \to 3}\left(\frac{(x-3)(x^2 + 3x + 9)}{(x-3)(x+3)}\right)$$

$$= \lim_{x \to 3}\left(\frac{x^2 + 3x + 9}{x + 3}\right) = \frac{(3)^2 + (3)(3) + 9}{3 + 3}$$

$$= {}^9\!/_2$$

---

[11] L'Hôpital's rule should not be used when only the denominator approaches zero. In that case, the limit approaches infinity regardless of the numerator.

[12] The calculation of derivatives is covered in Ch. 9.

(b) Divide through by the largest power of x. (L'Hôpital's rule can also be used.)

$$\lim_{x \to \infty}\left(\frac{3x - 2}{4x + 3}\right) = \lim_{x \to \infty}\left(\frac{3 - \dfrac{2}{x}}{4 + \dfrac{3}{x}}\right)$$

$$= \frac{3 - \dfrac{2}{\infty}}{4 + \dfrac{3}{\infty}}$$

$$= \frac{3 - 0}{4 + 0} = \frac{3}{4}$$

(c) Use L'Hôpital's rule. (Factoring can also be used.) Take the first derivative of the numerator and denominator.

$$\lim_{x \to 2}\left(\frac{x^2 + x - 6}{x^2 - 3x + 2}\right) = \lim_{x \to 2}\left(\frac{2x + 1}{2x - 3}\right)$$

$$= \frac{(2)(2) + 1}{(2)(2) - 3}$$

$$= \frac{5}{1} = 5$$

21. SEQUENCES AND PROGRESSIONS

A *sequence* is an ordered *progression* of numbers, a_i, such as $1, 4, 9, 16, 25, \ldots$ The *terms* in a sequence can be all positive, all negative, or of alternating signs. a_n is known as the *general term* of the sequence.

$$\{A\} = \{a_1, a_2, a_3, \ldots a_n\} \qquad 4.82$$

A sequence is said to *diverge* (i.e., be *divergent*) if the terms approach infinity or if the terms fail to approach any finite value, and is said to *converge* (i.e., be *convergent*) if the terms approach any finite value (including zero). That is, the sequence converges if the limit defined by Eq. 4.83 exists.

$$\lim_{n \to \infty} a_n \begin{cases} \text{converges if } L \text{ is finite} \\ \text{diverges if } L \text{ is infinite} \\ \text{or does not exist} \end{cases} \qquad 4.83$$

The main task associated with a sequence is determining the next (or the general) term. If several terms of a sequence are known, the next (unknown) term must usually be found by intuitively determining the pattern of the sequence. In some cases, though, the method of *Rth-order differences* can be used to determine the next term. This method consists of subtracting each term from the following term to obtain a set of differences. If the differences are not all equal, the next order of differences can be calculated.

Example 4.10

What is the general term of the sequence A?

$$\{A\} = \left\{3, \frac{9}{2}, \frac{27}{6}, \frac{81}{24}, \dots\right\}$$

Solution

The solution is purely intuitive. The numerator is recognized as a power series based on the number 3. The denominator is recognized as the factorial sequence. The general term is

$$a_n = \frac{3^n}{n!}$$

Example 4.11

Find the sixth term in the sequence $\{7, 16, 29, 46, 67, a_6\}$.

Solution

The sixth term is not intuitively obvious, so the method of Rth-order differences is tried. The pattern is not obvious from the first order differences, but the second order differences are all 4.

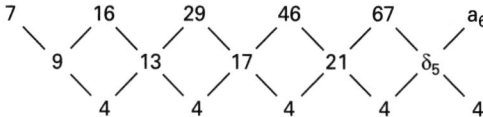

$$\delta_5 - 21 = 4$$
$$\delta_5 = 25$$
$$a_6 - 67 = \delta_5 = 25$$
$$a_6 = 92$$

Example 4.12

Does the sequence with general term e^n/n converge or diverge?

Solution

See if the limit exists.

$$\lim_{n \to \infty} \left(\frac{e^n}{n}\right) = \frac{\infty}{\infty}$$

Since ∞/∞ is inconclusive, apply L'Hôpital's rule. Take the derivative of both the numerator and the denominator with respect to n.

$$\lim_{n \to \infty} \left(\frac{e^n}{1}\right) = \frac{\infty}{1}$$
$$= \infty$$

The sequence diverges.

22. STANDARD SEQUENCES

There are four standard sequences: the geometric, arithmetic, harmonic, and p-sequences.

- *geometric sequence*: The geometric sequence converges for $-1 < r \le 1$ and diverges otherwise. a is known as the *first term*; r is known as the *common ratio*.

$$a_n = ar^{n-1} \qquad \begin{bmatrix} a \text{ is a constant} \\ n = 1, 2, 3, \dots, \infty \end{bmatrix} \qquad 4.84$$

  example: $\{1, 2, 4, 8, 16, 32\}$ $(a = 1, r = 2)$

- *arithmetic sequence*: The arithmetic sequence always diverges.

$$a_n = a + (n-1)d \qquad \begin{bmatrix} a \text{ and } d \text{ are constants} \\ n = 1, 2, 3, \dots, \infty \end{bmatrix} \qquad 4.85$$

  example: $\{2, 7, 12, 17, 22, 27\}$ $(a = 2, d = 5)$

- *harmonic sequence*: The harmonic sequence always converges.

$$a_n = \frac{1}{n} \qquad [n = 1, 2, 3, \dots, \infty] \qquad 4.86$$

  example: $\{1, 1/2, 1/3, 1/4, 1/5, 1/6\}$

- *p-sequence*: The p-sequence converges if $p \ge 0$ and diverges if $p < 0$. (Notice that this is different than the p-series.)

$$a_n = \frac{1}{n^p} \qquad [n = 1, 2, 3, \dots, \infty] \qquad 4.87$$

  example: $\{1, 1/4, 1/9, 1/16, 1/25, 1/36\}$ $(p = 2)$

23. SERIES

A *series* is the sum of terms in a sequence. There are two types of series. A *finite series* has a finite number of terms, and an *infinite series* has an infinite number of terms.[13] The main tasks associated with series are determining the sum of the terms and whether the series converges. A series is said to *converge* (be *convergent*) if the sum, S_n, of its term exists.[14] A finite series is always convergent.

The performance of a series based on standard sequences (defined in Sec. 4-22) is well known.

- *geometric series*:

$$S_n = \sum_{i=1}^{n} ar^{i-1} = \frac{a(1 - r^n)}{1 - r} \qquad [\text{finite}] \qquad 4.88$$

$$S_n = \sum_{i=1}^{\infty} ar^{i-1} = \frac{a}{1 - r} \qquad \begin{bmatrix} \text{infinite} \\ -1 < r < 1 \end{bmatrix} \qquad 4.89$$

---

[13]The term *infinite series* does not imply the sum is infinite.
[14]This is different from the definition of convergence for a sequence where only the last term was evaluated.

- *arithmetic series*: The infinite series diverges unless $a = d = 0$.

$$S_n = \sum_{i=1}^{n}(a + (i-1)d) = \frac{n(2a + (n-1)d)}{2}$$
$$\text{[finite]} \qquad \textbf{4.90}$$

- *harmonic series*: The infinite series diverges.

- *p-series*: The infinite series diverges if $p \leq 1$. The infinite series converges if $p > 1$. (Notice that this is different than the *p-sequence*.)

24. TESTS FOR SERIES CONVERGENCE

It is obvious that all *finite series* (i.e., series having a finite number of terms) converge. That is, the sum, S_n, defined by Eq. 4.91 exists.

$$S_n = \sum_{i=1}^{n} a_i \qquad \textbf{4.91}$$

Convergence of an infinite series can be determined by taking the limit of the sum. If the limit exists, the series converges; otherwise, it diverges.

$$\lim_{n \to \infty} S_n = \lim_{n \to \infty} \sum_{i=1}^{n} a_i \qquad \textbf{4.92}$$

In most cases, the expression for the general term a_n will be known, but there will be no simple expression for the sum S_n. Therefore, Eq. 4.92 cannot be used to determine convergence. It is helpful, but not conclusive, to look at the limit of the general term. If L, as defined in Eq. 4.93, is nonzero, the series diverges. If L equals zero, the series may either converge or diverge. Additional testing is needed in that case.

$$\lim_{n \to \infty} a_n \begin{cases} = 0 & \text{inconclusive} \\ \neq 0 & \text{diverges} \end{cases} \qquad \textbf{4.93}$$

Two tests can be used independently or after Eq. 4.93 has proven inconclusive: the ratio and comparison tests. The *ratio test* calculates the limit of the ratio of two consecutive terms.

$$\lim_{n \to \infty} \frac{a_{n+1}}{a_n} \begin{cases} < 1 & \text{converges} \\ = 1 & \text{inconclusive} \\ > 1 & \text{diverges} \end{cases} \qquad \textbf{4.94}$$

The *comparison test* is an indirect method of determining convergence of an unknown series. It compares a standard series (geometric and *p*-series are commonly used) against the unknown series. If all terms in a positive standard series are smaller than the terms in the unknown series and the standard series diverges, the unknown series must also diverge. Similarly, if all terms

in the standard series are larger than the terms in the unknown series and the standard series converges, then the unknown series also converges.

In mathematical terms, if A and B are both series of positive terms such that $a_n < b_n$ for all values of n, then (a) B diverges if A diverges, and (b) A converges if B converges.

Example 4.13

Does the infinite series A converge or diverge?

$$A = 3 + \frac{9}{2} + \frac{27}{6} + \frac{81}{24} + \cdots$$

Solution

The general term was found in Ex. 4.9 to be

$$a_n = \frac{3^n}{n!}$$

Since limits of factorials are not easily determined, use the ratio test.

$$\lim_{n \to \infty} \left(\frac{a_{n+1}}{a_n} \right) = \lim_{n \to \infty} \left(\frac{\dfrac{3^{n+1}}{(n+1)!}}{\dfrac{3^n}{n!}} \right) = \lim_{n \to \infty} \left(\frac{3}{n+1} \right)$$
$$= \frac{3}{\infty} = 0$$

Since the limit is less than 1, the infinite series converges.

Example 4.14

Does the infinite series A converge or diverge?

$$A = 2 + \frac{3}{4} + \frac{4}{9} + \frac{5}{16} + \cdots$$

Solution

By observation, the general term is

$$a_n = \frac{1+n}{n^2}$$

The general term can be expanded by partial fractions to

$$a_n = \frac{1}{n} + \frac{1}{n^2}$$

However, $1/n$ is the harmonic series. Since the harmonic series is divergent and this series is larger than the harmonic series (by the term $1/n^2$), this series also diverges.

25. SERIES OF ALTERNATING SIGNS[15]

Some series contain both positive and negative terms. The ratio and comparison tests can both be used to determine if a series with alternating signs converges. If a series containing all positive terms converges, then the same series with some negative terms also converges. Therefore, the all-positive series should be tested for convergence. If the all-positive series converges, the original series is said to be *absolutely convergent*. (If the all-positive series diverges and the original series converges, the original series is said to be *conditionally convergent*.)

Alternatively, the ratio test can be used with the absolute value of the ratio. The same criteria apply.

$$\lim_{n\to\infty} \left| \frac{a_{n+1}}{a_n} \right| \quad \begin{cases} < 1 & \text{converges} \\ = 1 & \text{inconclusive} \\ > 1 & \text{diverges} \end{cases} \qquad 4.95$$

---

[15]This terminology is commonly used even though it is not necessary that the signs strictly alternate.

5 Linear Algebra

1. MATRICES

A *matrix* is an ordered set of *entries* (*elements*) arranged rectangularly and set off by brackets.[1] The entries can be variables or numbers. A matrix by itself has no particular value—it is merely a convenient method of representing a set of numbers.

The size of a matrix is given by the number of rows and columns, and the nomenclature $m \times n$ is used for a matrix with m rows and n columns. For a *square matrix*, the number of rows and columns will be the same, a quantity known as the *order* of the matrix.

Bold uppercase letters are used to represent matrices, while lowercase letters represent the entries. For example, a_{23} would be the entry in the second row and third column of matrix $\mathbf{A}$.

$$\mathbf{A} = \begin{bmatrix} a_{11} & a_{12} & a_{13} \\ a_{21} & a_{22} & a_{23} \\ a_{31} & a_{32} & a_{33} \end{bmatrix}$$

A *submatrix* is the matrix that remains when selected rows or columns are removed from the original matrix.[2] For example, for matrix $\mathbf{A}$, the submatrix remaining after the second row and second column have been removed is

$$\begin{bmatrix} a_{11} & a_{13} \\ a_{31} & a_{33} \end{bmatrix}$$

---

[1] The term *array* is synonymous with *matrix*, although the former is more likely to be used in computer applications.
[2] By definition, a matrix is a submatrix of itself.

An *augmented matrix* results when the original matrix is extended by repeating one or more of its rows or columns or by adding rows and columns from another matrix. For example, for the matrix $\mathbf{A}$, the augmented matrix created by repeating the first and second columns is

$$\begin{bmatrix} a_{11} & a_{12} & a_{13} & \vdots & a_{11} & a_{12} \\ a_{21} & a_{22} & a_{23} & \vdots & a_{21} & a_{22} \\ a_{31} & a_{32} & a_{33} & \vdots & a_{31} & a_{32} \end{bmatrix}$$

2. SPECIAL TYPES OF MATRICES

Certain types of matrices are given special designations.

- *cofactor matrix*: the matrix that is formed when every entry is replaced by the cofactor (see Sec. 5-4) of that entry
- *column matrix*: a matrix with only one column
- *complex matrix*: a matrix with complex number entries
- *diagonal matrix*: a square matrix with all zero entries except for the a_{ij} for which $i = j$
- *echelon matrix*: a matrix in which the number of zeros preceding the first nonzero entry of a row increases row by row until only zero rows remain. A *row-reduced echelon matrix* is an echelon matrix in which the first nonzero entry in each row is a 1 and all other entries in the columns are zero.
- *identity matrix*: a diagonal (square) matrix with all nonzero entries equal to 1, usually designated as $\mathbf{I}$, having the property that $\mathbf{AI} = \mathbf{IA} = \mathbf{A}$
- *null matrix*: the same as a zero matrix
- *row matrix*: a matrix with only one row
- *scalar matrix*:[3] a diagonal (square) matrix with all diagonal entries equal to some scalar k
- *singular matrix*: a matrix whose determinant is zero (see Sec. 5-10)
- *skew symmetric matrix*: a square matrix whose transpose (see Sec. 5-9) is equal to the negative of itself (i.e., $\mathbf{A} = -\mathbf{A}^t$)

---

[3] Although the term *complex matrix* means a matrix with complex entries, the term *scalar matrix* means more than a matrix with scalar entries.

- *square matrix*: a matrix with the same number of rows and columns (i.e., $m = n$)

- *symmetric(al) matrix*: a square matrix whose transpose is equal to itself (i.e., $\mathbf{A}^t = \mathbf{A}$), which occurs only when $a_{ij} = a_{ji}$

- *triangular matrix*: a square matrix with zeros in all positions above or below the diagonal

- *unit matrix*: the same as the identity matrix

- *zero matrix*: a matrix with all zero entries

$$\begin{bmatrix} 9 & 0 & 0 & 0 \\ 0 & -6 & 0 & 0 \\ 0 & 0 & 1 & 0 \\ 0 & 0 & 0 & 5 \end{bmatrix} \qquad \begin{bmatrix} 2 & 18 & 2 & 18 \\ 0 & 0 & 1 & 9 \\ 0 & 0 & 0 & 9 \\ 0 & 0 & 0 & 0 \end{bmatrix} \qquad \begin{bmatrix} 1 & 9 & 0 & 0 \\ 0 & 0 & 1 & 0 \\ 0 & 0 & 0 & 1 \\ 0 & 0 & 0 & 0 \end{bmatrix}$$

(a) diagonal   (b) echelon   (c) row-reduced echelon

$$\begin{bmatrix} 1 & 0 & 0 & 0 \\ 0 & 1 & 0 & 0 \\ 0 & 0 & 1 & 0 \\ 0 & 0 & 0 & 1 \end{bmatrix} \qquad \begin{bmatrix} 2 & 0 & 0 & 0 \\ 7 & 6 & 0 & 0 \\ 9 & 1 & 1 & 0 \\ 8 & 0 & 4 & 5 \end{bmatrix} \qquad \begin{bmatrix} 3 & 0 & 0 & 0 \\ 0 & 3 & 0 & 0 \\ 0 & 0 & 3 & 0 \\ 0 & 0 & 0 & 3 \end{bmatrix}$$

(d) identity   (e) triangular   (f) scalar

Figure 5.1 *Examples of Special Matrices*

3. ROW EQUIVALENT MATRICES

A matrix $\mathbf{B}$ is said to be *row equivalent* to a matrix $\mathbf{A}$ if it is obtained by a finite sequence of *elementary row operations* on $\mathbf{A}$:

- interchanging the ith and jth rows

- multiplying the ith row by a nonzero scalar

- replacing the ith row by the sum of the original ith row and k times the jth row

However, two matrices that are row equivalent as defined do not have the same determinants. (See Sec. 5-5.)

Gauss-Jordan elimination is the process of using these elementary row operations to row-reduce a matrix to echelon or row-reduced echelon forms, as illustrated in Ex. 5.8. When a matrix has been converted to a row-reduced echelon matrix, it is said to be in *row canonical form*. Thus, the terms *row-reduced echelon form* and *row canonical form* are synonymous.

4. MINORS AND COFACTORS

Minors and cofactors are determinants of submatrices associated with particular entries in the original square matrix. The *minor* of entry a_{ij} is the determinant of a submatrix resulting from the elimination of the single

row i and the single column j. For example, the minor corresponding to entry a_{12} in a 3×3 matrix $\mathbf{A}$ is the determinant of the matrix created by eliminating row 1 and column 2.

$$\text{minor of } a_{12} = \begin{vmatrix} a_{21} & a_{23} \\ a_{31} & a_{33} \end{vmatrix} \qquad \textit{5.1}$$

The *cofactor* of entry a_{ij} is the minor of a_{ij} multiplied by either $+1$ or -1, depending on the position of the entry. (That is, the cofactor either exactly equals the minor or it differs only in sign.) The sign is determined according to the following positional matrix.[4]

$$\begin{bmatrix} +1 & -1 & +1 & \cdots \\ -1 & +1 & -1 & \cdots \\ +1 & -1 & +1 & \cdots \\ \vdots & \vdots & \vdots & \end{bmatrix}$$

For example, the cofactor of entry a_{12} in matrix $\mathbf{A}$ (described in Sec. 5-4) is

$$\text{cofactor of } a_{12} = -\begin{vmatrix} a_{21} & a_{23} \\ a_{31} & a_{33} \end{vmatrix} \qquad \textit{5.2}$$

Example 5.1

What is the cofactor corresponding to the -3 entry in the following matrix?

$$\mathbf{A} = \begin{bmatrix} 2 & 9 & 1 \\ -3 & 4 & 0 \\ 7 & 5 & 9 \end{bmatrix}$$

Solution

The minor's submatrix is created by eliminating the row and column of the -3 entry.

$$\mathbf{M} = \begin{bmatrix} 9 & 1 \\ 5 & 9 \end{bmatrix}$$

The minor is the determinant of $\mathbf{M}$.

$$|\mathbf{M}| = (9)(9) - (5)(1) = 76$$

The sign corresponding to the -3 position is negative. Therefore, the cofactor is -76.

---

[4]The sign of the cofactor a_{ij} is positive if $(i + j)$ is even and is negative if $(i + j)$ is odd.

5. DETERMINANTS

A *determinant* is a scalar calculated from a square matrix. The determinant of matrix $\mathbf{A}$ can be represented as $D\{\mathbf{A}\}$, Det $(\mathbf{A})$, $\Delta\mathbf{A}$, or $|\mathbf{A}|$.[5] The following rules can be used to simplify the calculation of determinants.

- If $\mathbf{A}$ has a row or column of zeros, the determinant is zero.

- If $\mathbf{A}$ has two identical rows or columns, the determinant is zero.

- If $\mathbf{B}$ is obtained from $\mathbf{A}$ by adding a multiple of a row (column) to another row (column) in $\mathbf{A}$, then $|\mathbf{B}| = |\mathbf{A}|$.

- If $\mathbf{A}$ is triangular, the determinant is equal to the product of the diagonal entries.

- If $\mathbf{B}$ is obtained from $\mathbf{A}$ by multiplying one row or column in $\mathbf{A}$ by a scalar k, then $|\mathbf{B}| = k|\mathbf{A}|$.

- If $\mathbf{B}$ is obtained from the $n \times n$ matrix $\mathbf{A}$ by multiplying by the scalar matrix k, then $|k\mathbf{A}| = k^n|\mathbf{A}|$.

- If $\mathbf{B}$ is obtained from $\mathbf{A}$ by switching two rows or columns in $\mathbf{A}$, then $|\mathbf{B}| = -|\mathbf{A}|$.

Calculation of determinants is laborious for all but the smallest or simplest of matrices. For a 2×2 matrix, the formula used to calculate the determinant is easy to remember.

$$\mathbf{A} = \begin{bmatrix} a & b \\ c & d \end{bmatrix}$$

$$|\mathbf{A}| = \begin{vmatrix} a & b \\ c & d \end{vmatrix} = ad - bc \qquad 5.3$$

Two methods are commonly used for calculating the determinant of 3×3 matrices by hand. The first uses an augmented matrix constructed from the original matrix and the first two columns (as presented in Sec. 5-1).[6] The determinant is calculated as the sum of the products in the left-to-right downward diagonals less the sum of the products in the left-to-right upward diagonals.

$$\mathbf{A} = \begin{bmatrix} a & b & c \\ d & e & f \\ g & h & i \end{bmatrix}$$

$$\text{augmented } \mathbf{A} = \begin{bmatrix} \overset{+}{a} & \overset{+}{b} & \overset{+}{c} & \overset{-}{a} & \overset{-}{b} \\ d & e & f & d & e \\ g & h & i & g & h \end{bmatrix} \qquad 5.4$$

$$|\mathbf{A}| = aei + bfg + cdh - gec - hfa - idb \qquad 5.5$$

The second method of calculating the determinant is somewhat slower than the first for a 3×3 matrix but illustrates the method that must be used to calculate determinants of 4×4 and larger matrices. This method is known as *expansion by cofactors*. One row (column) is selected as the base row (column). The selection is arbitrary, but the number of calculations required to obtain the determinant can be minimized by choosing the row (column) with the most zeros. The determinant is equal to the sum of the products of the entries in the base row (column) and their corresponding cofactors.

$$\mathbf{A} = \begin{bmatrix} a & b & c \\ d & e & f \\ g & h & i \end{bmatrix}$$

$$|\mathbf{A}| = a\begin{vmatrix} e & f \\ h & i \end{vmatrix} - d\begin{vmatrix} b & c \\ h & i \end{vmatrix} + g\begin{vmatrix} b & c \\ e & f \end{vmatrix} \qquad 5.6$$

Example 5.2

Calculate the determinant of matrix $\mathbf{A}$ (a) by cofactor expansion, and (b) by the augmented matrix method.

$$\mathbf{A} = \begin{bmatrix} 2 & 3 & -4 \\ 3 & -1 & -2 \\ 4 & -7 & -6 \end{bmatrix}$$

Solution

(a) Since there are no zero entries, it does not matter which row or column is chosen as the base. Choose the first column as the base.

$$|\mathbf{A}| = 2\begin{vmatrix} -1 & -2 \\ -7 & -6 \end{vmatrix} - 3\begin{vmatrix} 3 & -4 \\ -7 & -6 \end{vmatrix} + 4\begin{vmatrix} 3 & -4 \\ -1 & -2 \end{vmatrix}$$

$$= (2)(6 - 14) - (3)(-18 - 28) + (4)(-6 - 4)$$

$$= 82$$

(b)

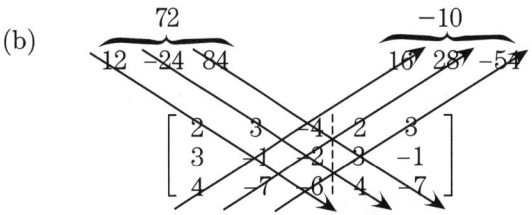

$$|\mathbf{A}| = 72 - (-10) = 82$$

6. MATRIX ALGEBRA[7]

Matrix algebra differs somewhat from standard algebra.

- *equality*: Two matrices, $\mathbf{A}$ and $\mathbf{B}$, are equal only if they have the same numbers of rows and columns *and* if all corresponding entries are equal.

- *inequality*: The $>$ and $<$ operators are not used in matrix algebra.

---

[5]The vertical bars should not be confused with the square brackets used to set off a matrix, nor with absolute value.
[6]It is not actually necessary to construct the augmented matrix, but doing so helps avoid errors.

[7]Since matrices are used to simplify the presentation and solution of sets of linear equations, matrix algebra is also known as *linear algebra*.

- *commutative law of addition*:

$$\mathbf{A} + \mathbf{B} = \mathbf{B} + \mathbf{A} \qquad 5.7$$

- *associative law of addition*:

$$\mathbf{A} + (\mathbf{B} + \mathbf{C}) = (\mathbf{A} + \mathbf{B}) + \mathbf{C} \qquad 5.8$$

- *associative law of multiplication*:

$$(\mathbf{AB})\mathbf{C} = \mathbf{A}(\mathbf{BC}) \qquad 5.9$$

- *left distributive law*:

$$\mathbf{A}(\mathbf{B} + \mathbf{C}) = \mathbf{AB} + \mathbf{AC} \qquad 5.10$$

- *right distributive law*:

$$(\mathbf{B} + \mathbf{C})\mathbf{A} = \mathbf{BA} + \mathbf{CA} \qquad 5.11$$

- *scalar multiplication*:

$$k(\mathbf{AB}) = (k\mathbf{A})\mathbf{B} = \mathbf{A}(k\mathbf{B}) \qquad 5.12$$

It is important to recognize that, except for trivial and special cases, matrix multiplication is not commutative. That is,

$$\mathbf{AB} \neq \mathbf{BA}$$

7. MATRIX ADDITION AND SUBTRACTION

Addition and subtraction of two matrices is possible only if both matrices have the same numbers of rows and columns (i.e., order). They are accomplished by adding or subtracting the corresponding entries of the two matrices.

8. MATRIX MULTIPLICATION

A matrix can be multiplied by a scalar, an operation known as *scalar multiplication*, in which case all entries of the matrix are multiplied by that scalar. For example, for the 2×2 matrix $\mathbf{A}$,

$$k\mathbf{A} = \begin{bmatrix} ka_{11} & ka_{12} \\ ka_{21} & ka_{22} \end{bmatrix}$$

A matrix can be multiplied by another matrix, but only if the left-hand matrix has the same number of columns as the right-hand matrix has rows. *Matrix multiplication* occurs by multiplying the elements in each left-hand matrix row by the entries in each right-hand matrix column, adding the products, and placing the sum at the intersection point of the participating row and column.

Matrix division can only be accomplished by multiplying by the inverse of the denominator matrix. There is no specific division operation in matrix algebra.

Example 5.3

Determine the product matrix $\mathbf{C}$.

$$\mathbf{C} = \begin{bmatrix} 1 & 4 & 3 \\ 5 & 2 & 6 \end{bmatrix} \begin{bmatrix} 7 & 12 \\ 11 & 8 \\ 9 & 10 \end{bmatrix}$$

Solution

The left-hand matrix has three columns, and the right-hand matrix has three rows. Therefore, the two matrices can be multiplied.

The first row of the left-hand matrix and the first column of the right-hand matrix are worked with first. The corresponding entries are multiplied and the products are summed.

$$c_{11} = (1)(7) + (4)(11) + (3)(9) = 78$$

The intersection of the top row and left column is the entry in the upper left-hand corner of the matrix $\mathbf{C}$.

The remaining entries are calculated similarly.

$$c_{12} = (1)(12) + (4)(8) + (3)(10) = 74$$
$$c_{21} = (5)(7) + (2)(11) + (6)(9) = 111$$
$$c_{22} = (5)(12) + (2)(8) + (6)(10) = 136$$

The product matrix is

$$\mathbf{C} = \begin{bmatrix} 78 & 74 \\ 111 & 136 \end{bmatrix}$$

9. TRANSPOSE

The *transpose*, $\mathbf{A}^t$, of an $m \times n$ matrix $\mathbf{A}$ is an $n \times m$ matrix constructed by taking the ith row and making it the ith column. The diagonal is unchanged. For example,

$$\mathbf{A} = \begin{bmatrix} 1 & 6 & 9 \\ 2 & 3 & 4 \\ 7 & 1 & 5 \end{bmatrix}$$

$$\mathbf{A}^t = \begin{bmatrix} 1 & 2 & 7 \\ 6 & 3 & 1 \\ 9 & 4 & 5 \end{bmatrix}$$

Transpose operations have the following characteristics.

$$(\mathbf{A}^t)^t = \mathbf{A} \qquad 5.13$$

$$(k\mathbf{A})^t = k(\mathbf{A}^t) \qquad 5.14$$

$$I^t = I \qquad 5.15$$

$$(AB)^t = B^t A^t \qquad 5.16$$

$$(A + B)^t = A^t + B^t \qquad 5.17$$

$$|A^t| = |A| \qquad 5.18$$

10. SINGULARITY AND RANK

A *singular matrix* is one whose determinant is zero. Similarly, a *nonsingular* matrix is one whose determinant is nonzero.

The *rank* of a matrix is the maximum number of linearly independent row or column vectors.[8] A matrix has rank r if it has at least one nonsingular square submatrix of order r but has no nonsingular square submatrix of order more than r. While the submatrix must be square (in order to calculate the determinant), the original matrix need not be.

The rank of an $m \times n$ matrix will be, at most, the smaller of m and n. The rank of a null matrix is zero. The ranks of a matrix and its transpose are the same. If a matrix is in echelon form, the rank will be equal to the number of rows containing at least one nonzero entry. For a 3×3 matrix, the rank can either be 3 (if it is nonsingular), 2 (if any one of its 2×2 submatrices is nonsingular), 1 (if it and all 2×2 submatrices are singular), or 0 (if it is null).

The determination of rank is laborious if done by hand. Either the matrix is reduced to echelon form by using elementary row operations, or exhaustive enumeration is used to create the submatrices and many determinants are calculated. If a matrix has more rows than columns and row-reduction is used, the work required to put the matrix in echelon form can be reduced by working with the transpose of the original matrix.

Example 5.4

What is the rank of matrix A?

$$A = \begin{bmatrix} 1 & -2 & -1 \\ -3 & 3 & 0 \\ 2 & 2 & 4 \end{bmatrix}$$

Solution

Matrix A is singular because $|A| = 0$. However, there is at least one 2×2 nonsingular submatrix:

$$\begin{vmatrix} 1 & -2 \\ -3 & 3 \end{vmatrix} = (1)(3) - (-3)(-2) = -3$$

Therefore, the rank is 2.

---

[8]The *row rank* and *column rank* are the same.

Example 5.5

Determine the rank of matrix A by reducing it to echelon form.

$$A = \begin{bmatrix} 7 & 4 & 9 & 1 \\ 0 & 2 & -5 & 3 \\ 0 & 4 & -10 & 6 \end{bmatrix}$$

Solution

By inspection, the matrix can be row-reduced by subtracting two times the second row from the third row. The matrix cannot be further reduced. Since there are two nonzero rows, the rank is 2.

$$\begin{bmatrix} 7 & 4 & 9 & 1 \\ 0 & 2 & -5 & 3 \\ 0 & 0 & 0 & 0 \end{bmatrix}$$

11. CLASSICAL ADJOINT

The *classical adjoint* is the transpose of the cofactor matrix. The resulting matrix can be designated as A_{adj}, adj$\{A\}$ or A^{adj}.

Example 5.6

What is the classical adjoint of matrix A?

$$A = \begin{bmatrix} 2 & 3 & -4 \\ 0 & -4 & 2 \\ 1 & -1 & 5 \end{bmatrix}$$

Solution

The matrix of cofactors is

$$\begin{bmatrix} -18 & 2 & 4 \\ -11 & 14 & 5 \\ -10 & -4 & -8 \end{bmatrix}$$

The transpose of the matrix of cofactors is

$$A_{adj} = \begin{bmatrix} -18 & -11 & -10 \\ 2 & 14 & -4 \\ 4 & 5 & -8 \end{bmatrix}$$

12. INVERSE

The product of a matrix A and its inverse, A^{-1}, is the identity matrix, I. Only square matrices have inverses, but not all square matrices are invertible. A matrix has an inverse if and only if it is nonsingular (i.e., its determinant is nonzero).

$$\mathbf{A}\mathbf{A}^{-1} = \mathbf{A}^{-1}\mathbf{A} = \mathbf{I} \qquad \textit{5.19}$$

$$(\mathbf{A}\mathbf{B})^{-1} = \mathbf{B}^{-1}\mathbf{A}^{-1} \qquad \textit{5.20}$$

The inverse of a 2×2 matrix is easily determined by formula.

$$\mathbf{A} = \begin{bmatrix} a & b \\ c & d \end{bmatrix}$$

$$\mathbf{A}^{-1} = \frac{\begin{bmatrix} d & -b \\ -c & a \end{bmatrix}}{|\mathbf{A}|} \qquad \textit{5.21}$$

For a 3×3 or larger matrix, the inverse is determined by dividing every entry in the classical adjoint by the determinant of the original matrix.

$$\mathbf{A}^{-1} = \frac{\mathbf{A}_{\text{adj}}}{|\mathbf{A}|} \qquad \textit{5.22}$$

Example 5.7

What is the inverse of matrix $\mathbf{A}$?

$$\mathbf{A} = \begin{bmatrix} 4 & 5 \\ 2 & 3 \end{bmatrix}$$

Solution

The determinant is calculated as

$$|\mathbf{A}| = (4)(3) - (2)(5) = 2$$

Using Eq. 5.22, the inverse is

$$\mathbf{A}^{-1} = \frac{\begin{bmatrix} 3 & -5 \\ -2 & 4 \end{bmatrix}}{2} = \begin{bmatrix} \frac{3}{2} & -\frac{5}{2} \\ -1 & 2 \end{bmatrix}$$

Check.

$$\mathbf{A}\mathbf{A}^{-1} = \begin{bmatrix} 4 & 5 \\ 2 & 3 \end{bmatrix} \begin{bmatrix} \frac{3}{2} & -\frac{5}{2} \\ -1 & 2 \end{bmatrix} = \begin{bmatrix} 6-5 & -10+10 \\ 3-3 & -5+6 \end{bmatrix}$$

$$= \begin{bmatrix} 1 & 0 \\ 0 & 1 \end{bmatrix} = \mathbf{I} \qquad [\text{ok}]$$

13. WRITING SIMULTANEOUS LINEAR EQUATIONS IN MATRIX FORM

Matrices are used to simplify the presentation and solution of sets of simultaneous linear equations. For example, the following three methods of presenting simultaneous linear equations are equivalent:

$$a_{11}x_1 + a_{12}x_2 = b_1$$

$$a_{21}x_1 + a_{22}x_2 = b_2$$

$$\begin{bmatrix} a_{11} & a_{12} \\ a_{21} & a_{22} \end{bmatrix} \begin{bmatrix} x_1 \\ x_2 \end{bmatrix} = \begin{bmatrix} b_1 \\ b_2 \end{bmatrix}$$

$$\mathbf{A}\mathbf{X} = \mathbf{B}$$

In the second and third representations, $\mathbf{A}$ is known as the *coefficient matrix*, $\mathbf{X}$ as the *variable matrix*, and $\mathbf{B}$ as the *constant matrix*.

Not all systems of simultaneous equations have solutions, and those that do may not have unique solutions. The existence of a solution can be determined by calculating the determinant of the coefficient matrix. These rules are summarized in Table 5.1.

- If the system of linear equations is homogeneous (i.e., $\mathbf{B}$ is a zero matrix) and $|\mathbf{A}|$ is zero, there are an infinite number of solutions.

- If the system is homogeneous and $|\mathbf{A}|$ is nonzero, only the trivial solution exists.

- If the system of linear equations is nonhomogeneous (i.e., $\mathbf{B}$ is not a zero matrix) and $|\mathbf{A}|$ is nonzero, there is a unique solution to the set of simultaneous equations.

- If $|\mathbf{A}|$ is zero, a nonhomogeneous system of simultaneous equations may still have a solution. The requirement is that the determinants of all substitutional matrices (see Sec. 5-14) are zero, in which case there will be an infinite number of solutions. Otherwise, no solution exists.

Table 5.1 *Solution Existence Rules for Simultaneous Equations*

	$\mathbf{B} = 0$	$\mathbf{B} \neq 0$
$|\mathbf{A}| = 0$	infinite number of solutions (linearly dependent equations)	either an infinite number of solutions or no solution at all
$|\mathbf{A}| \neq 0$	trivial solution only ($x_i = 0$)	unique nonzero solution

14. SOLVING SIMULTANEOUS LINEAR EQUATIONS

Gauss-Jordan elimination can be used to obtain the solution to a set of simultaneous linear equations. The coefficient matrix is augmented by the constant matrix. Then, elementary row operations are used to reduce the coefficient matrix to canonical form. All of the operations performed on the coefficient matrix are performed on the constant matrix. The variable values that satisfy the simultaneous equations will be the entries in the constant matrix when the coefficient matrix is in canonical form.

Determinants are used to calculate the solution to linear simultaneous equations through a procedure known as *Cramer's rule*.

The procedure is to calculate determinants of the original coefficient matrix $\mathbf{A}$ and of the n matrices resulting from the systematic replacement of a column in $\mathbf{A}$ by the constant matrix $\mathbf{B}$. For a system of three equations in three unknowns, there are three substitutional matrices, $\mathbf{A}_1$, $\mathbf{A}_2$, and $\mathbf{A}_3$, as well as the original coefficient matrix, for a total of four matrices whose determinants must be calculated.

The values of the unknowns that simultaneously satisfy all of the linear equations are

$$x_1 = \frac{|\mathbf{A}_1|}{|\mathbf{A}|} \qquad\qquad 5.23$$

$$x_2 = \frac{|\mathbf{A}_2|}{|\mathbf{A}|} \qquad\qquad 5.24$$

$$x_3 = \frac{|\mathbf{A}_3|}{|\mathbf{A}|} \qquad\qquad 5.25$$

Example 5.8

Use Gauss-Jordan elimination to solve the following system of simultaneous equations.

$$2x + 3y - 4z = 1$$
$$3x - y - 2z = 4$$
$$4x - 7y - 6z = -7$$

Solution

The augmented matrix is created by appending the constant matrix to the coefficient matrix.

$$\begin{bmatrix} 2 & 3 & -4 & \vdots & 1 \\ 3 & -1 & -2 & \vdots & 4 \\ 4 & -7 & -6 & \vdots & -7 \end{bmatrix}$$

Elementary row operations are used to reduce the coefficient matrix to canonical form. For example, two times the first row is subtracted from the third row. This step obtains the 0 needed in the a_{31} position.

$$\begin{bmatrix} 2 & 3 & -4 & \vdots & 1 \\ 3 & -1 & -2 & \vdots & 4 \\ 0 & -13 & 2 & \vdots & -9 \end{bmatrix}$$

This process continues until the following form is obtained.

$$\begin{bmatrix} 1 & 0 & 0 & \vdots & 3 \\ 0 & 1 & 0 & \vdots & 1 \\ 0 & 0 & 1 & \vdots & 2 \end{bmatrix}$$

$x = 3$, $y = 1$, and $z = 2$ satisfy this system of equations.

Example 5.9

Use Cramer's rule to solve the following system of simultaneous equations.

$$2x + 3y - 4z = 1$$
$$3x - y - 2z = 4$$
$$4x - 7y - 6z = -7$$

Solution

The determinant of the coefficient matrix is

$$|\mathbf{A}| = \begin{vmatrix} 2 & 3 & -4 \\ 3 & -1 & -2 \\ 4 & -7 & -6 \end{vmatrix} = 82$$

The determinants of the substitutional matrices are

$$|\mathbf{A}_1| = \begin{vmatrix} 1 & 3 & -4 \\ 4 & -1 & -2 \\ -7 & -7 & -6 \end{vmatrix} = 246$$

$$|\mathbf{A}_2| = \begin{vmatrix} 2 & 1 & -4 \\ 3 & 4 & -2 \\ 4 & -7 & -6 \end{vmatrix} = 82$$

$$|\mathbf{A}_3| = \begin{vmatrix} 2 & 3 & 1 \\ 3 & -1 & 4 \\ 4 & -7 & -7 \end{vmatrix} = 164$$

The values of x, y, and z that satisfy the linear equations are

$$x = \frac{246}{82} = 3$$

$$y = \frac{82}{82} = 1$$

$$z = \frac{164}{82} = 2$$

15. EIGENVALUES AND EIGENVECTORS

Eigenvalues and eigenvectors (also known as *characteristic values* and *characteristic vectors*) of a square matrix $\mathbf{A}$ are the scalars k and matrices $\mathbf{X}$ such that

$$\mathbf{AX} = k\mathbf{X} \qquad\qquad 5.26$$

The scalar k is an eigenvalue of $\mathbf{A}$ if and only if the matrix $(k\mathbf{I} - \mathbf{A})$ is singular; that is, if $|k\mathbf{I} - \mathbf{A}| = 0$. This equation is called the *characteristic equation* of the matrix $\mathbf{A}$. When expanded, the determinant is called the *characteristic polynomial*. The method of using the characteristic polynomial to find eigenvalues and eigenvectors is illustrated in Ex. 5.10.

If all of the eigenvalues are unique (i.e., nonrepeating), then Eq. 5.27 is valid.

$$[k\mathbf{I} - \mathbf{A}]\mathbf{X} = 0 \qquad \text{5.27}$$

Example 5.10

Find the eigenvalues and nonzero eigenvectors of the matrix $\mathbf{A}$.

$$\mathbf{A} = \begin{bmatrix} 2 & 4 \\ 6 & 4 \end{bmatrix}$$

Solution

$$k\mathbf{I} - \mathbf{A} = \begin{bmatrix} k & 0 \\ 0 & k \end{bmatrix} - \begin{bmatrix} 2 & 4 \\ 6 & 4 \end{bmatrix} = \begin{bmatrix} k-2 & -4 \\ -6 & k-4 \end{bmatrix}$$

The characteristic polynomial is found by setting the determinant $|k\mathbf{I} - \mathbf{A}|$ equal to zero.

$$(k-2)(k-4) - (-6)(-4) = 0$$
$$k^2 - 6k - 16 = (k-8)(k+2) = 0$$

The roots of the characteristic polynomial are $k = +8$ and $k = -2$. These are the eigenvalues of $\mathbf{A}$.

Substituting $k = 8$,

$$k\mathbf{I} - \mathbf{A} = \begin{bmatrix} 8-2 & -4 \\ -6 & 8-4 \end{bmatrix} = \begin{bmatrix} 6 & -4 \\ -6 & 4 \end{bmatrix}$$

This can be interpreted as the linear equation $6x_1 - 4x_2 = 0$. The values of x that satisfy this equation define the eigenvector. An eigenvector $\mathbf{X}$ associated with the eigenvalue $+8$ is

$$\mathbf{X} = \begin{bmatrix} x_1 \\ x_2 \end{bmatrix} = \begin{bmatrix} 4 \\ 6 \end{bmatrix}$$

All other eigenvectors for this eigenvalue are multiples of $\mathbf{X}$. Normally $\mathbf{X}$ is reduced to smallest integers.

$$\mathbf{X} = \begin{bmatrix} 2 \\ 3 \end{bmatrix}$$

Similarly, the eigenvector associated with the eigenvalue -2 is

$$\mathbf{X} = \begin{bmatrix} x_1 \\ x_2 \end{bmatrix} = \begin{bmatrix} +4 \\ -4 \end{bmatrix}$$

Reducing this to smallest integers gives

$$\mathbf{X} = \begin{bmatrix} +1 \\ -1 \end{bmatrix}$$

6 Vectors

1. INTRODUCTION

A physical property or quantity can be a scalar, vector, or tensor. A *scalar* has only magnitude. Knowing its value is sufficient to define a scalar. Mass, enthalpy, density, and speed are examples of scalars.

Force, momentum, displacement, and velocity are examples of vectors. A *vector* is a directed straight line with a specific magnitude. Thus, a vector is specified completely by its direction (consisting of the vector's *angular orientation* and its *sense*) and magnitude. A vector's *point of application* (*terminal point*) is not needed to define the vector.[1] Two vectors with the same direction and magnitude are said to be *equal vectors* even though their *lines of action* may be different.[2]

A vector can be designated by a boldface variable (as in this book) or as a combination of the variable and some other symbol. For example, the notations $\mathbf{V}$, $\bar{V}$, $\hat{V}$, $\vec{V}$, and $\underline{V}$ are used by different authorities to represent vectors. The magnitude of a vector can be designated by either $|\mathbf{V}|$ or V (not bold).

Stress, dielectric constant, and magnetic susceptibility are examples of tensors. A *tensor* has magnitude in a specific direction but the direction is not unique. Tensors are frequently associated with *anisotropic materials* that have different properties in different directions. A tensor in three-dimensional space is defined by nine components, compared with the three that are required

to define vectors. These components are written in matrix form. Stress, σ, at a point, for example, would be defined by the following tensor matrix.

$$\sigma = \begin{pmatrix} \sigma_{xx} & \sigma_{xy} & \sigma_{xz} \\ \sigma_{yx} & \sigma_{yy} & \sigma_{yz} \\ \sigma_{zx} & \sigma_{zy} & \sigma_{zz} \end{pmatrix}$$

2. VECTORS IN n-SPACE

In some cases, a vector, $\mathbf{V}$, will be designated by its two endpoints in n-dimensional vector space. A common example is three-dimensional force-space. Usually, one of the points will be the origin, in which case the vector is said to be "based at the origin," "origin-based," or "zero-based."[3] If one of the endpoints is the origin, specifying a terminal point P would represent a force directed from the origin to point P.

If a coordinate system is superimposed on the vector space, a vector can be specified in terms of the n coordinates of its two endpoints. The magnitude of the vector $\mathbf{V}$ is the distance in vector space between the two points, as given by Eq. 6.1. Similarly, the direction is defined by the angle the vector makes with one of the axes.

$$|\mathbf{V}| = \sqrt{(x_2 - x_1)^2 + (y_2 - y_1)^2} \qquad 6.1$$

$$\phi = \arctan\left(\frac{y_2 - y_1}{x_2 - x_1}\right) \qquad 6.2$$

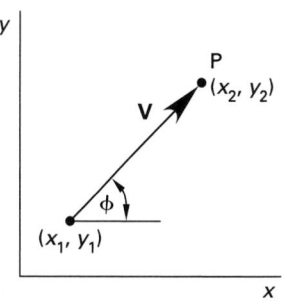

Figure 6.1 *Vector in Two-Dimensional Space*

---

[1]A vector that is constrained to act at or through a certain point is a *bound vector (fixed vector)*. A *sliding vector (transmissible vector)* can be applied anywhere along its line of action. A *free vector* is not constrained and can be applied at any point in space.
[2]A distinction is sometimes made between equal vectors and equivalent vectors. *Equivalent vectors* produce the same effect but are not necessarily equal.

[3]Any vector directed from P_1 to P_2 can be transformed into a zero-based vector by subtracting the coordinates of point P_1 from the coordinates of terminal point P_2. The transformed vector will be equivalent to the original vector.

The *components* of a vector are the projections of the vector on the coordinate axes. (For a zero-based vector, the components and the coordinates of the endpoint are the same.) Simple trigonometric principles are used to resolve a vector into its components. A vector constructed from its components is known as a *resultant vector.*

$$V_x = |\mathbf{V}| \cos \phi_x \qquad 6.3$$

$$V_y = |\mathbf{V}| \cos \phi_y \qquad 6.4$$

$$V_z = |\mathbf{V}| \cos \phi_z \qquad 6.5$$

$$|\mathbf{V}| = \sqrt{V_x^2 + V_y^2 + V_z^2} \qquad 6.6$$

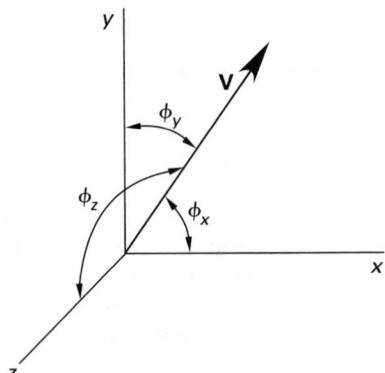

Figure 6.2 *Direction Angles of a Vector*

In Eqs. 6.3 through 6.5, ϕ_x, ϕ_y, and ϕ_z are the *direction angles*—the angles between the vector and the x-, y-, and z-axis, respectively. The cosines of these angles are known as *direction cosines*. The sum of the squares of the direction cosines is equal to 1.

$$\cos^2 \phi_x + \cos^2 \phi_y + \cos^2 \phi_z = 1 \qquad 6.7$$

3. UNIT VECTORS

Unit vectors are vectors with unit magnitudes (i.e., magnitudes of 1). They are represented in the same notation as other vectors. (Unit vectors in this book are written in boldface type.) Although they can have any direction, the standard unit vectors (the *Cartesian unit vectors* **i**, **j**, and **k**) have the directions of the x-, y-, and z-coordinate axes and constitute the *Cartesian triad.*

A vector **V** can be written in terms of unit vectors and its components.

$$\mathbf{V} = |\mathbf{V}|\mathbf{a} = V_x\mathbf{i} + V_y\mathbf{j} + V_z\mathbf{k} \qquad 6.8$$

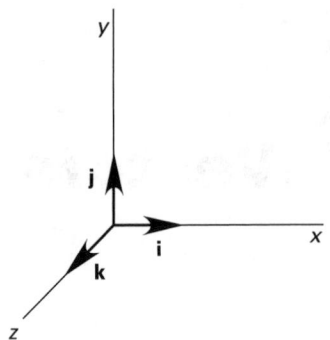

Figure 6.3 *Cartesian Unit Vectors*

The unit vector, **a**, has the same direction as the vector **V** but has a length of 1. This unit vector is calculated by dividing the original vector, **V**, by its magnitude, $|\mathbf{V}|$.

$$\mathbf{a} = \frac{\mathbf{V}}{|\mathbf{V}|} = \frac{V_x\mathbf{i} + V_y\mathbf{j} + V_z\mathbf{k}}{\sqrt{V_x^2 + V_y^2 + V_z^2}} \qquad 6.9$$

4. VECTOR REPRESENTATION

The most common method of representing a vector is by writing it in *rectangular form*—a vector sum of its orthogonal components. In rectangular form, each of the orthogonal components has the same units as the resultant vector.

$$\mathbf{A} \equiv A_x\mathbf{i} + A_y\mathbf{j} + A_z\mathbf{k} \quad \text{[three dimensions]}$$

However, the vector is also completely defined by its magnitude and associated angle. These two quantities can be written together in *phasor form*, sometimes referred to as *polar form*.

$$\mathbf{A} \equiv |\mathbf{A}|\angle\phi = A\angle\phi$$

5. CONVERSION BETWEEN SYSTEMS

The choice of the **ijk** triad may be convenient but is arbitrary. A vector can be expressed in terms of another set of unit vectors, **uvw**.

$$\mathbf{V} = V_x\mathbf{i} + V_y\mathbf{j} + V_z\mathbf{k} = V_x'\mathbf{u} + V_y'\mathbf{v} + V_z'\mathbf{w} \qquad 6.10$$

The two representations are related.

$$V_x' = \mathbf{V} \cdot \mathbf{u} = (\mathbf{i} \cdot \mathbf{u})V_x + (\mathbf{j} \cdot \mathbf{u})V_y + (\mathbf{k} \cdot \mathbf{u})V_z \qquad 6.11$$

$$V_y' = \mathbf{V} \cdot \mathbf{v} = (\mathbf{i} \cdot \mathbf{v})V_x + (\mathbf{j} \cdot \mathbf{v})V_y + (\mathbf{k} \cdot \mathbf{v})V_z \qquad 6.12$$

$$V_z' = \mathbf{V} \cdot \mathbf{w} = (\mathbf{i} \cdot \mathbf{w})V_x + (\mathbf{j} \cdot \mathbf{w})V_y + (\mathbf{k} \cdot \mathbf{w})V_z \qquad 6.13$$

Equations 6.11 through 6.13 can be expressed in matrix form. The dot products are known as the *coefficients of transformation*, and the matrix containing them is the *transformation matrix*.

$$\begin{pmatrix} V'_x \\ V'_y \\ V'_z \end{pmatrix} = \begin{pmatrix} \mathbf{i}\cdot\mathbf{u} & \mathbf{j}\cdot\mathbf{u} & \mathbf{k}\cdot\mathbf{u} \\ \mathbf{i}\cdot\mathbf{v} & \mathbf{j}\cdot\mathbf{v} & \mathbf{k}\cdot\mathbf{v} \\ \mathbf{i}\cdot\mathbf{w} & \mathbf{j}\cdot\mathbf{w} & \mathbf{k}\cdot\mathbf{w} \end{pmatrix} \begin{pmatrix} V_x \\ V_y \\ V_z \end{pmatrix} \quad 6.14$$

6. VECTOR ADDITION

Addition of two vectors by the *polygon method* is accomplished by placing the tail of the second vector at the head (tip) of the first. The sum (i.e., the *resultant vector*) is a vector extending from the tail of the first vector to the head of the second. Alternatively, the two vectors can be considered as the two sides of a parallelogram, while the sum represents the diagonal. This is known as addition by the *parallelogram method*.

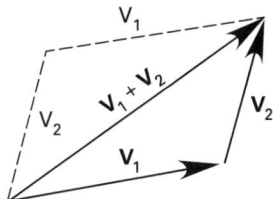

Figure 6.4 *Addition of Two Vectors*

The components of the resultant vector are the sums of the components of the added vectors ($V_{1x} + V_{2x}$, $V_{1y} + V_{2y}$, $V_{1z} + V_{2z}$).

Vector addition is both commutative and associative.

$$\mathbf{V}_1 + \mathbf{V}_2 = \mathbf{V}_2 + \mathbf{V}_1 \quad 6.15$$

$$\mathbf{V}_1 + (\mathbf{V}_2 + \mathbf{V}_3) = (\mathbf{V}_1 + \mathbf{V}_2) + \mathbf{V}_3 \quad 6.16$$

7. MULTIPLICATION BY A SCALAR

A vector, $\mathbf{V}$, can be multiplied by a scalar, c. If the original vector is represented by its components, each of the components is multiplied by c.

$$c\mathbf{V} = c|\mathbf{V}|\,\mathbf{a} = cV_x\mathbf{i} + cV_y\mathbf{j} + cV_z\mathbf{k} \quad 6.17$$

Scalar multiplication is distributive.

$$c(\mathbf{V}_1 + \mathbf{V}_2) = c\mathbf{V}_1 + c\mathbf{V}_2 \quad 6.18$$

8. VECTOR DOT PRODUCT

The *dot product (scalar product)*, $\mathbf{V}_1 \cdot \mathbf{V}_2$, of two vectors is a scalar that is proportional to the length of the projection of the first vector onto the second vector.[4]

---

[4]The dot product is also written in parentheses without a dot, that is, $(\mathbf{V}_1\mathbf{V}_2)$.

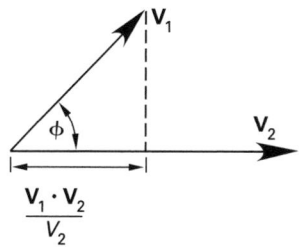

Figure 6.5 *Vector Dot Product*

The dot product is commutative and distributive.

$$\mathbf{V}_1 \cdot \mathbf{V}_2 = \mathbf{V}_2 \cdot \mathbf{V}_1 \quad 6.19$$

$$\mathbf{V}_1 \cdot (\mathbf{V}_2 + \mathbf{V}_3) = \mathbf{V}_1 \cdot \mathbf{V}_2 + \mathbf{V}_1 \cdot \mathbf{V}_3 \quad 6.20$$

The dot product can be calculated in two ways, as Eq. 6.21 indicates. ϕ is limited to 180° and is the angle between the two vectors.

$$\mathbf{V}_1 \cdot \mathbf{V}_2 = |\mathbf{V}_1||\mathbf{V}_2|\cos\phi$$
$$= V_{1x}V_{2x} + V_{1y}V_{2y} + V_{1z}V_{2z} \quad 6.21$$

When Eq. 6.21 is solved for the angle between the two vectors, ϕ, it is known as the *Cauchy-Schwartz theorem*.

$$\cos\phi = \frac{V_{1x}V_{2x} + V_{1y}V_{2y} + V_{1z}V_{2z}}{|\mathbf{V}_1||\mathbf{V}_2|} \quad 6.22$$

The dot product can be used to determine whether a vector is a unit vector and to show that two vectors are orthogonal (perpendicular). For two non-null orthogonal vectors,

$$\mathbf{V}_1 \cdot \mathbf{V}_2 = 0 \quad 6.23$$

For any unit vector, $\mathbf{u}$,

$$\mathbf{u} \cdot \mathbf{u} = 1 \quad 6.24$$

Equations 6.23 and 6.24 can be extended to the Cartesian unit vectors.

$$\mathbf{i} \cdot \mathbf{i} = 1 \quad 6.25$$
$$\mathbf{j} \cdot \mathbf{j} = 1 \quad 6.26$$
$$\mathbf{k} \cdot \mathbf{k} = 1 \quad 6.27$$
$$\mathbf{i} \cdot \mathbf{j} = 0 \quad 6.28$$
$$\mathbf{i} \cdot \mathbf{k} = 0 \quad 6.29$$
$$\mathbf{j} \cdot \mathbf{k} = 0 \quad 6.30$$

Example 6.1

What is the angle between the zero-based vectors $\mathbf{V}_1 = (-\sqrt{3}, 1)$ and $\mathbf{V}_2 = (2\sqrt{3}, 2)$?

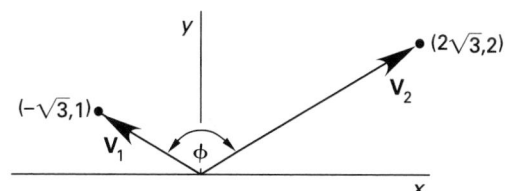

Solution

From Eq. 6.22,

$$\cos \phi = \frac{V_{1x}V_{2x} + V_{1y}V_{2y}}{|\mathbf{V}_1||\mathbf{V}_2|} = \frac{V_{1x}V_{2x} + V_{1y}V_{2y}}{\sqrt{V_{1x}^2 + V_{1y}^2}\sqrt{V_{2x}^2 + V_{2y}^2}}$$

$$= \frac{\left(-\sqrt{3}\right)\left(2\sqrt{3}\right) + (1)(2)}{\sqrt{\left(-\sqrt{3}\right)^2 + (1)^2}\sqrt{\left(2\sqrt{3}\right)^2 + (2)^2}}$$

$$= \frac{-4}{8} = -\frac{1}{2}$$

$$\phi = \arccos\left(-\tfrac{1}{2}\right) = 120°$$

9. VECTOR CROSS PRODUCT

The *cross product (vector product)*, $\mathbf{V}_1 \times \mathbf{V}_2$, of two vectors is a vector that is orthogonal (perpendicular) to the plane of the two vectors.[5] The unit vector representation of the cross product can be calculated as a third-order determinant.

$$\mathbf{V}_1 \times \mathbf{V}_2 = \begin{vmatrix} \mathbf{i} & V_{1x} & V_{2x} \\ \mathbf{j} & V_{1y} & V_{2y} \\ \mathbf{k} & V_{1z} & V_{2z} \end{vmatrix} \qquad 6.31$$

The direction of the cross-product vector corresponds to the direction a right-hand screw would progress if vectors $\mathbf{V}_1$ and $\mathbf{V}_2$ are placed tail-to-tail in the plane they define and $\mathbf{V}_1$ is rotated into $\mathbf{V}_2$. The direction can also be found from the *right-hand rule*.

The magnitude of the cross product can be determined from Eq. 6.32, in which ϕ is the angle between the two vectors and is limited to 180°. The magnitude corresponds to the area of a parallelogram that has $\mathbf{V}_1$ and $\mathbf{V}_2$ as two of its sides.

$$|\mathbf{V}_1 \times \mathbf{V}_2| = |\mathbf{V}_1||\mathbf{V}_2|\sin\phi \qquad 6.32$$

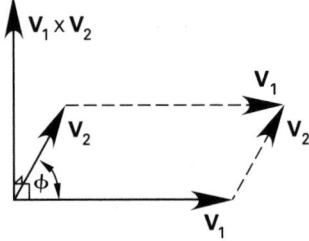

Figure 6.6 *Vector Cross Product*

Vector cross multiplication is distributive but not commutative.

$$\mathbf{V}_1 \times \mathbf{V}_2 = -(\mathbf{V}_2 \times \mathbf{V}_1) \qquad 6.33$$

$$c(\mathbf{V}_1 \times \mathbf{V}_2) = (c\mathbf{V}_1) \times \mathbf{V}_2 = \mathbf{V}_1 \times (c\mathbf{V}_2) \qquad 6.34$$

$$\mathbf{V}_1 \times (\mathbf{V}_2 + \mathbf{V}_3) = \mathbf{V}_1 \times \mathbf{V}_2 + \mathbf{V}_1 \times \mathbf{V}_3 \qquad 6.35$$

---

[5]The cross product is also written in square brackets without a cross, that is, $[\mathbf{V}_1\mathbf{V}_2]$.

If the two vectors are parallel, their cross product will be zero.

$$\mathbf{i} \times \mathbf{i} = \mathbf{j} \times \mathbf{j} = \mathbf{k} \times \mathbf{k} = 0 \qquad 6.36$$

Equation 6.36 can be extended to the unit vectors.

$$\mathbf{i} \times \mathbf{j} = -\mathbf{j} \times \mathbf{i} = \mathbf{k} \qquad 6.37$$

$$\mathbf{j} \times \mathbf{k} = -\mathbf{k} \times \mathbf{j} = \mathbf{i} \qquad 6.38$$

$$\mathbf{k} \times \mathbf{i} = -\mathbf{i} \times \mathbf{k} = \mathbf{j} \qquad 6.39$$

Example 6.2

Find a unit vector orthogonal to $\mathbf{V}_1 = \mathbf{i} - \mathbf{j} + 2\mathbf{k}$ and $\mathbf{V}_2 = 3\mathbf{j} - \mathbf{k}$.

Solution

The cross product is a vector orthogonal to $\mathbf{V}_1$ and $\mathbf{V}_2$.

$$\mathbf{V}_1 \times \mathbf{V}_2 = \begin{vmatrix} \mathbf{i} & 1 & 0 \\ \mathbf{j} & -1 & 3 \\ \mathbf{k} & 2 & -1 \end{vmatrix}$$

$$= -5\mathbf{i} + \mathbf{j} + 3\mathbf{k}$$

Check to see whether this is a unit vector.

$$|\mathbf{V}_1 \times \mathbf{V}_2| = \sqrt{(-5)^2 + (1)^2 + (3)^2} = \sqrt{35}$$

Since its length is $\sqrt{35}$, the vector must be divided by $\sqrt{35}$ to obtain a unit vector.

$$\mathbf{a} = \frac{-5\mathbf{i} + \mathbf{j} + 3\mathbf{k}}{\sqrt{35}}$$

10. MIXED TRIPLE PRODUCT

The *mixed triple product (triple scalar product* or just *triple product)* of three vectors is a scalar quantity representing the volume of a parallelepiped with the three vectors making up the sides. It is calculated as a determinant. Since Eq. 6.40 can be negative, the absolute value must be used to obtain the volume in that case.

$$\mathbf{V}_1 \cdot (\mathbf{V}_2 \times \mathbf{V}_3) = \begin{vmatrix} V_{1x} & V_{1y} & V_{1z} \\ V_{2x} & V_{2y} & V_{2z} \\ V_{3x} & V_{3y} & V_{3z} \end{vmatrix} \qquad 6.40$$

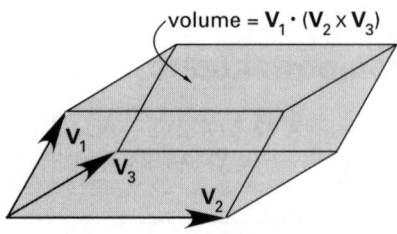

Figure 6.7 *Vector Mixed Triple Product*

The mixed triple product has the property of *circular permutation*, as defined by Eq. 6.41.

$$\mathbf{V}_1 \cdot (\mathbf{V}_2 \times \mathbf{V}_3) = (\mathbf{V}_1 \times \mathbf{V}_2) \cdot \mathbf{V}_3 \qquad 6.41$$

11. VECTOR TRIPLE PRODUCT

The *vector triple product* is a vector defined by Eq. 6.42. The quantities in parentheses on the right-hand side are scalars.

$$\mathbf{V}_1 \times (\mathbf{V}_2 \times \mathbf{V}_3) = (\mathbf{V}_1 \cdot \mathbf{V}_3)\mathbf{V}_2 - (\mathbf{V}_1 \cdot \mathbf{V}_2)\mathbf{V}_3 \qquad 6.42$$

12. VECTOR FUNCTIONS

A vector can be a function of another parameter. For example, a vector $\mathbf{V}$ is a function of variable t when its V_x, V_y, and V_z are functions of t.

$$\mathbf{V}(t) = (2t - 3)\mathbf{i} + (t^2 + 1)\mathbf{j} + (-7t + 5)\mathbf{k} \qquad 6.43$$

When the functions of t are differentiated (or integrated) with respect to t, the vector itself is differentiated (integrated).[6] (Chs. 9 and 10 cover differentiation and integration.)

$$\frac{d\mathbf{V}(t)}{dt} = \left(\frac{dV_x}{dt}\right)\mathbf{i} + \left(\frac{dV_y}{dt}\right)\mathbf{j} + \left(\frac{dV_z}{dt}\right)\mathbf{k} \qquad 6.44$$

Similarly, the integral of the vector is

$$\int \mathbf{V}(t)\,dt = \mathbf{i}\int V_x\,dt + \mathbf{j}\int V_y\,dt + \mathbf{k}\int V_z\,dt \qquad 6.45$$

---

[6]This is particularly valuable when converting among position, velocity, and acceleration vectors.

7 Trigonometry

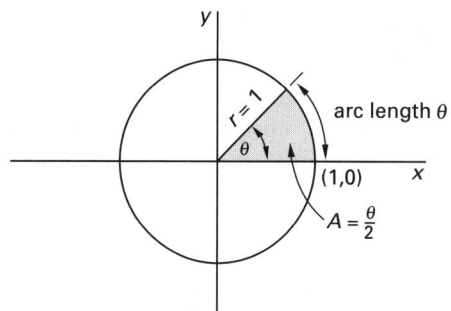

Figure 7.1 *Radians and Area of Unit Circle*

1. DEGREES AND RADIANS

Degrees and *radians* are two units for measuring angles. One complete circle is divided into 360 degrees (written 360°) or 2π radians (abbreviated *rad*).[1] The conversions between degrees and radians are

multiply	by	to obtain
radians	$\dfrac{180}{\pi}$	degrees
degrees	$\dfrac{\pi}{180}$	radians

The number of radians in an angle θ corresponds to the area within a circular sector with arc length θ and a radius of one. Alternatively, the area of a sector with central angle θ radians is $\theta/2$ for a *unit circle* (i.e., a circle with a radius of one unit).

2. PLANE ANGLES

A *plane angle* (usually referred to as just an *angle*) consists of two intersecting lines and an intersection point known as the *vertex*. The angle can be referred to by a capital letter representing the vertex (e.g., B in Fig. 7.2), a Greek letter representing the angular measure (e.g., β), or by three capital letters, where the middle letter is the vertex and the other two letters are two points on different lines, and either the symbol $\angle$ or $\sphericalangle$ (e.g., $\sphericalangle$ABC).

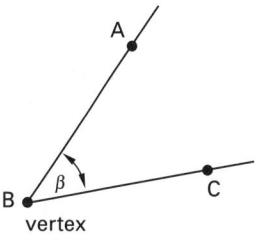

Figure 7.2 *Angle*

The angle between two intersecting lines generally is understood to be the smaller angle created.[2] Angles have been classified as follows.

- *acute angle*: an angle less than 90° ($\pi/2$ rad)

- *obtuse angle*: an angle more than 90° ($\pi/2$ rad) but less than 180° (π rad)

- *reflex angle*: an angle more than 180° (π rad) but less than 360° (2π rad)

---

[1]The abbreviation *rad* is also used to represent *radiation absorbed dose*, a measure of radiation exposure.

[2]In books on geometry, the term *ray* is used instead of *line*.

- *related angle*: an angle that differs from another by some multiple of 90° ($\pi/2$ rad)

- *right angle*: an angle equal to 90° ($\pi/2$ rad)

- *straight angle*: an angle equal to 180° (π rad), that is, a straight line

Complementary angles are two angles whose sum is 90° ($\pi/2$ rad). *Supplementary angles* are two angles whose sum is 180° (π rad). *Adjacent angles* share a common vertex and one (the interior) side. Adjacent angles are supplementary only if their exterior sides form a straight line.

Vertical angles are the two angles with a common vertex and with sides made up by two intersecting straight lines. Vertical angles are equal.

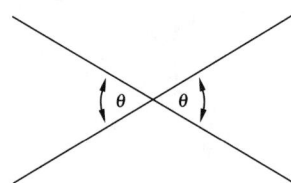

Figure 7.3 *Vertical Angles*

Angle of elevation and *angle of depression* are surveying terms referring to the angle above and below the horizontal plane of the observer, respectively.

3. TRIANGLES

A *triangle* is a three-sided closed polygon with three angles whose sum is 180° (π rad). Triangles are identified by their vertices and the symbol Δ (e.g., ΔABC in Fig. 7.4). A side is designated by its two endpoints (e.g., AB in Fig. 7.4) or by a lowercase letter corresponding to the capital letter of the opposite vertex (e.g., c).

In *similar triangles*, the corresponding angles are equal and the corresponding sides are in proportion. (Since there are only two independent angles in a triangle, showing that two angles of one triangle are equal to two angles of the other triangle is sufficient to show similarity.) The symbol for similarity is $\sim$. In Fig. 7.4, ΔABC $\sim$ ΔDEF (i.e., ΔABC is similar to ΔDEF).

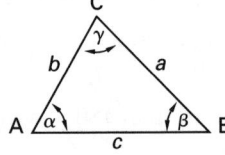 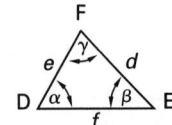

Figure 7.4 *Similar Triangles*

4. RIGHT TRIANGLES

A *right triangle* is a triangle in which one of the angles is 90° ($\pi/2$ rad). The remaining two angles are complementary. If one of the acute angles is chosen as the reference, the sides forming the right angle are known as the *adjacent side*, x, and the *opposite side*, y. The longest side is known as the *hypotenuse*, r. The *Pythagorean theorem* relates the lengths of these sides.

$$x^2 + y^2 = r^2 \qquad 7.1$$

In certain cases, the lengths of unknown sides of right triangles can be determined by inspection.[3] This occurs when the lengths of the sides are in the ratios of 3:4:5, 1:1:$\sqrt{2}$, 1:$\sqrt{3}$:2, and 5:12:13.

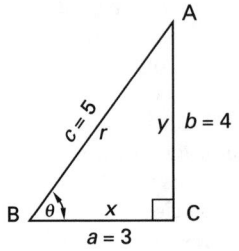

Figure 7.5 *3:4:5 Right Triangle*

5. CIRCULAR TRANSCENDENTAL FUNCTIONS

The *circular transcendental functions* (usually referred to as the *transcendental functions*, *trigonometric functions*, or *functions of an angle*) are calculated from the sides of a right triangle. Equations 7.2 through 6.7 refer to Fig. 7.5.

$$\text{sine: } \sin\theta = \frac{y}{r} = \frac{\text{opposite}}{\text{hypotenuse}} \qquad 7.2$$

$$\text{cosine: } \cos\theta = \frac{x}{r} = \frac{\text{adjacent}}{\text{hypotenuse}} \qquad 7.3$$

$$\text{tangent: } \tan\theta = \frac{y}{x} = \frac{\text{opposite}}{\text{adjacent}} \qquad 7.4$$

$$\text{cotangent: } \cot\theta = \frac{x}{y} = \frac{\text{adjacent}}{\text{opposite}} \qquad 7.5$$

$$\text{secant: } \sec\theta = \frac{r}{x} = \frac{\text{hypotenuse}}{\text{adjacent}} \qquad 7.6$$

$$\text{cosecant: } \csc\theta = \frac{r}{y} = \frac{\text{hypotenuse}}{\text{opposite}} \qquad 7.7$$

Three of the transcendental functions are reciprocals of the others. Notice that, while the tangent and cotangent functions are reciprocals of each other, the sine and cosine functions are not.

---

[3]These cases are almost always contrived examples. There is nothing intrinsic in nature to cause the formation of triangles with these proportions.

$$\cot\theta = \frac{1}{\tan\theta} \qquad 7.8$$

$$\sec\theta = \frac{1}{\cos\theta} \qquad 7.9$$

$$\csc\theta = \frac{1}{\sin\theta} \qquad 7.10$$

The trigonometric functions correspond to the lengths of various line segments in a right triangle with a unit hypotenuse. Figure 7.6 shows such a triangle inscribed in a unit circle.

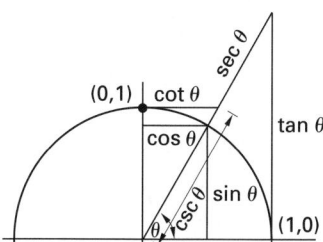

Figure 7.6 *Trigonometric Functions in a Unit Circle*

6. SMALL ANGLE APPROXIMATIONS

When an angle is very small, the hypotenuse and adjacent sides are essentially equal in length and certain approximations can be made. (The angle θ must be expressed in radians in Eqs. 7.11 and 7.12.)

$$\sin\theta \approx \tan\theta \approx \theta \Big|_{\theta<10° \ (0.175 \ \text{rad})} \qquad 7.11$$

$$\cos\theta \approx 1 \Big|_{\theta<5° \ (0.0873 \ \text{rad})} \qquad 7.12$$

7. GRAPHS OF THE FUNCTIONS

Figure 7.7 illustrates the periodicity of the sine, cosine, and tangent functions.[4]

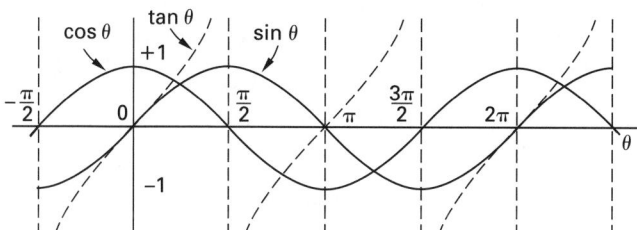

Figure 7.7 *Graphs of Sine, Cosine, and Tangent Functions*

[4]The remaining functions, being reciprocals of these three functions, are also periodic.

8. SIGNS OF THE FUNCTIONS

Table 7.1 shows how the sine, cosine, and tangent functions vary in sign with different values of θ. All three functions are positive for angles $0° \leq \theta \leq 90°$ ($0 \leq \theta \leq \pi/2$ rad), but only the sine is positive for angles $90° < \theta \leq 180°$ ($\pi/2$ rad $< \theta \leq \pi$ rad). The concept of quadrants is used to summarize the signs of the functions: angles up to $90°$ ($\pi/2$ rad) are in quadrant I, between $90°$ and $180°$ ($\pi/2$ and π rad) are in quadrant II, and so on.

Table 7.1 *Signs of the Functions by Quadrant*

function	quadrant			
	I	II	III	IV
sine	+	+	−	−
cosine	+	−	−	+
tangent	+	−	+	−

9. FUNCTIONS OF RELATED ANGLES

Figure 7.7 shows that the sine, cosine, and tangent curves are symmetrical with respect to the horizontal axis. Furthermore, portions of the curves are symmetrical with respect to a vertical axis. The values of the sine and cosine functions repeat every $360°$ (2π rad), and the absolute values repeat every $180°$ (π rad). This can be written as

$$\sin(\theta + 180°) = -\sin\theta \qquad 7.13$$

Similarly, the tangent function repeats every $180°$ (π rad), and its absolute value repeats every $90°$ ($\pi/2$ rad). Table 7.2 summarizes the functions of the related angles.

Table 7.2 *Functions of Related Angles*

$f(\theta)$	$-\theta$	$90° - \theta$	$90° + \theta$	$180° - \theta$	$180° + \theta$
sin	$-\sin\theta$	$\cos\theta$	$\cos\theta$	$\sin\theta$	$-\sin\theta$
cos	$\cos\theta$	$\sin\theta$	$-\sin\theta$	$-\cos\theta$	$-\cos\theta$
tan	$-\tan\theta$	$\cot\theta$	$-\cot\theta$	$-\tan\theta$	$\tan\theta$

10. TRIGONOMETRIC IDENTITIES

There are many relationships between trigonometric functions. For example, Eqs. 7.14 through 7.16 are well known.

$$\sin^2\theta + \cos^2\theta = 1 \qquad 7.14$$

$$1 + \tan^2\theta = \sec^2\theta \qquad 7.15$$

$$1 + \cot^2\theta = \csc^2\theta \qquad 7.16$$

Other relatively common identities are listed as follows.[5]

- *double-angle formulas:*

$$\sin 2\theta = 2\sin\theta\cos\theta = \frac{2\tan\theta}{1+\tan^2\theta} \qquad 7.17$$

$$\cos 2\theta = \cos^2\theta - \sin^2\theta = 1 - 2\sin^2\theta$$
$$= 2\cos^2\theta - 1 = \frac{1-\tan^2\theta}{1+\tan^2\theta} \qquad 7.18$$

$$\tan 2\theta = \frac{2\tan\theta}{1-\tan^2\theta} \qquad 7.19$$

$$\cot 2\theta = \frac{\cot^2\theta - 1}{2\cot\theta} \qquad 7.20$$

- *two-angle formulas:*

$$\sin(\theta \pm \phi) = \sin\theta\cos\phi \pm \cos\theta\sin\phi \qquad 7.21$$

$$\cos(\theta \pm \phi) = \cos\theta\cos\phi \mp \sin\theta\sin\phi \qquad 7.22$$

$$\tan(\theta \pm \phi) = \frac{\tan\theta \pm \tan\phi}{1 \mp \tan\theta\tan\phi} \qquad 7.23$$

$$\cot(\theta \pm \phi) = \frac{\cot\phi\cot\theta \mp 1}{\cot\phi \pm \cot\theta} \qquad 7.24$$

- *half-angle formulas* $(\theta < 180°)$:

$$\sin\frac{\theta}{2} = \sqrt{\frac{1-\cos\theta}{2}} \qquad 7.25$$

$$\cos\frac{\theta}{2} = \sqrt{\frac{1+\cos\theta}{2}} \qquad 7.26$$

$$\tan\frac{\theta}{2} = \sqrt{\frac{1-\cos\theta}{1+\cos\theta}} = \frac{\sin\theta}{1+\cos\theta} = \frac{1-\cos\theta}{\sin\theta} \quad 7.27$$

- *miscellaneous formulas* $(\theta < 90°)$:

$$\sin\theta = 2\sin\left(\frac{\theta}{2}\right)\cos\left(\frac{\theta}{2}\right) \qquad 7.28$$

$$\sin\theta = \sqrt{\frac{1-\cos 2\theta}{2}} \qquad 7.29$$

$$\cos\theta = \cos^2\left(\frac{\theta}{2}\right) - \sin^2\left(\frac{\theta}{2}\right) \qquad 7.30$$

$$\cos\theta = \sqrt{\frac{1+\cos 2\theta}{2}} \qquad 7.31$$

$$\tan\theta = \frac{2\tan\left(\frac{\theta}{2}\right)}{1-\tan^2\left(\frac{\theta}{2}\right)}$$
$$= \frac{2\sin\left(\frac{\theta}{2}\right)\cos\left(\frac{\theta}{2}\right)}{\cos^2\left(\frac{\theta}{2}\right) - \sin^2\left(\frac{\theta}{2}\right)} \qquad 7.32$$

$$\tan\theta = \sqrt{\frac{1-\cos 2\theta}{1+\cos 2\theta}}$$
$$= \frac{\sin 2\theta}{1+\cos 2\theta} = \frac{1-\cos 2\theta}{\sin 2\theta} \qquad 7.33$$

$$\cot\theta = \frac{\cot^2\left(\frac{\theta}{2}\right) - 1}{2\cot\left(\frac{\theta}{2}\right)}$$
$$= \frac{\cos^2\left(\frac{\theta}{2}\right) - \sin^2\left(\frac{\theta}{2}\right)}{2\sin\left(\frac{\theta}{2}\right)\cos\left(\frac{\theta}{2}\right)} \qquad 7.34$$

$$\cot\theta = \sqrt{\frac{1+\cos 2\theta}{1-\cos 2\theta}}$$
$$= \frac{1+\cos 2\theta}{\sin 2\theta} = \frac{\sin 2\theta}{1-\cos 2\theta} \qquad 7.35$$

11. INVERSE TRIGONOMETRIC FUNCTIONS

Finding an angle from a known trigonometric function is a common operation known as an *inverse trigonometric operation*. The inverse function can be designated by adding "inverse," "arc-," or the superscript -1 to the name of the function. For example,

$$\text{inverse}\sin(0.5) = \arcsin(0.5) = \sin^{-1}(0.5) = 30°$$

12. HYPERBOLIC TRANSCENDENTAL FUNCTIONS

Hyperbolic transcendental functions (normally referred to as *hyperbolic functions*) are specific equations containing combinations of the terms e^{θ} and $e^{-\theta}$. These combinations appear regularly in certain types of problems (e.g., analysis of cables and heat transfer from fins) and are given specific names and symbols to simplify presentation.[6]

$$\text{hyperbolic sine: } \sinh\theta = \frac{e^{\theta} - e^{-\theta}}{2} \qquad 7.36$$

$$\text{hyperbolic cosine: } \cosh\theta = \frac{e^{\theta} + e^{-\theta}}{2} \qquad 7.37$$

---

[5]It is an idiosyncrasy that these formulas are conventionally referred to as *formulas*, not *identities*.

[6]The hyperbolic sine and cosine functions are pronounced (by some) as "sinch" and "cosh," respectively.

hyperbolic tangent: $\tanh \theta = \dfrac{e^\theta - e^{-\theta}}{e^\theta + e^{-\theta}} = \dfrac{\sinh \theta}{\cosh \theta}$ *7.38*

hyperbolic cotangent: $\coth \theta = \dfrac{e^\theta + e^{-\theta}}{e^\theta - e^{-\theta}} = \dfrac{\cosh \theta}{\sinh \theta}$ *7.39*

hyperbolic secant: $\operatorname{sech} \theta = \dfrac{2}{e^\theta + e^{-\theta}} = \dfrac{1}{\cosh \theta}$ *7.40*

hyperbolic cosecant: $\operatorname{csch} \theta = \dfrac{2}{e^\theta - e^{-\theta}} = \dfrac{1}{\sinh \theta}$ *7.41*

Hyperbolic functions cannot be related to a right triangle, but they are related to a rectangular (equilateral) hyperbola, as shown in Fig. 7.8. The shaded area has a value of $\theta/2$ and is sometimes given the units of *hyperbolic radians*.

$$\sinh \theta = \frac{y}{a} \qquad\qquad 7.42$$

$$\cosh \theta = \frac{x}{a} \qquad\qquad 7.43$$

$$\tanh \theta = \frac{y}{x} \qquad\qquad 7.44$$

$$\coth \theta = \frac{x}{y} \qquad\qquad 7.45$$

$$\operatorname{sech} \theta = \frac{a}{x} \qquad\qquad 7.46$$

$$\operatorname{csch} \theta = \frac{a}{y} \qquad\qquad 7.47$$

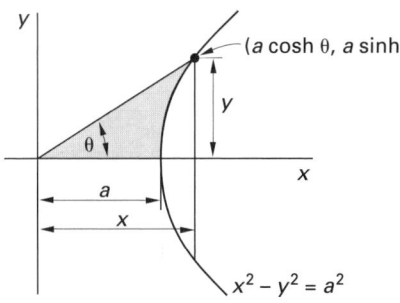

Figure 7.8 *Equilateral Hyperbola and Hyperbolic Functions*

13. HYPERBOLIC IDENTITIES

The hyperbolic identities are different from the standard trigonometric identities. Some of the most important identities are presented as follows.

$$\cosh^2 \theta - \sinh^2 \theta = 1 \qquad\qquad 7.48$$

$$1 - \tanh^2 \theta = \operatorname{sech}^2 \theta \qquad\qquad 7.49$$

$$1 - \coth^2 \theta = -\operatorname{csch}^2 \theta \qquad\qquad 7.50$$

$$\cosh \theta + \sinh \theta = e^\theta \qquad\qquad 7.51$$

$$\cosh \theta - \sinh \theta = e^{-\theta} \qquad\qquad 7.52$$

$$\sinh(\theta \pm \phi) = \sinh \theta \cosh \phi \pm \cosh \theta \sinh \phi \qquad 7.53$$

$$\cosh(\theta \pm \phi) = \cosh \theta \cosh \phi \pm \sinh \theta \sinh \phi \qquad 7.54$$

$$\tanh(\theta \pm \phi) = \frac{\tanh \theta \pm \tanh \phi}{1 \pm \tanh \theta \tanh \phi} \qquad\qquad 7.55$$

14. GENERAL TRIANGLES

A *general triangle* (also known as an *oblique triangle*) is one that is not specifically a right triangle, as shown in Fig. 7.9. Equation 7.56 calculates the area of a general triangle.[7]

$$\text{area} = \tfrac{1}{2}ab \sin C = \tfrac{1}{2}bc \sin A = \tfrac{1}{2}ca \sin B \qquad 7.56$$

The *law of sines* (Eq. 7.57) relates the sides and the sines of the angles.

$$\frac{\sin A}{a} = \frac{\sin B}{b} = \frac{\sin C}{c} \qquad\qquad 7.57$$

The *law of cosines* relates the cosine of an angle to an opposite side. (Equation 7.58 can be extended to the two remaining sides.)

$$a^2 = b^2 + c^2 - 2bc \cos A \qquad\qquad 7.58$$

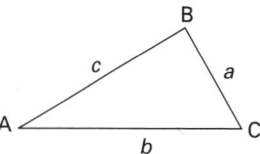

Figure 7.9 *General Triangle*

The *law of tangents* relates the sum and difference of two sides. (Equation 7.59 can be extended to the two remaining sides.)

$$\frac{a - b}{a + b} = \frac{\tan\left(\dfrac{A - B}{2}\right)}{\tan\left(\dfrac{A + B}{2}\right)} \qquad\qquad 7.59$$

15. SPHERICAL TRIGONOMETRY

A *spherical triangle* is a triangle that has been drawn on the surface of a sphere, as shown in Fig. 7.10. The *trihedral angle* O–ABC is formed when the vertices A, B, and C are joined to the center of the sphere. The *face angles* (BOC, COA, and AOB in Fig. 7.10) are

---

[7]Other methods of calculating the area of a general triangle are given in Ch. 8.

used to measure the sides (a, b, and c in Fig. 7.10). The *vertex angles* are A, B, and C. Thus, angles are used to measure both vertex angles and sides.

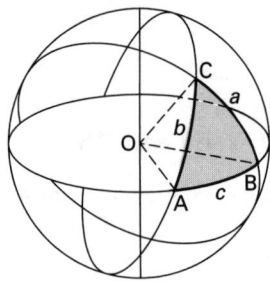

Figure 7.10 *Spherical Triangle ABC*

The following rules are valid for spherical triangles for which each side and angle is less than 180°.

- The sum of the three vertex angles is greater than 180° and less than 540°.

$$180° < A + B + C < 540° \qquad 7.60$$

- The sum of any two sides is greater than the third side.

- The sum of the three sides is less than 360°.

$$0° < a + b + c < 360° \qquad 7.61$$

- If the two sides are equal, the corresponding angles opposite are equal, and the converse is also true.

- If two sides are unequal, the corresponding angles opposite are unequal. The greater angle is opposite the greater side.

The *spherical excess*, ϵ, is the amount by which the sum of the vertex angles exceeds 180°. The *spherical defect*, d, is the amount by which the sum of the sides differs from 360°.

$$\epsilon = A + B + C - 180° \qquad 7.62$$

$$d = 360° - (a + b + c) \qquad 7.63$$

There are many trigonometric identities that define the relationships between angles in a spherical triangle. Some of the more common identities are presented as follows.

- *law of sines*:

$$\frac{\sin A}{\sin a} = \frac{\sin B}{\sin b} = \frac{\sin C}{\sin c} \qquad 7.64$$

- *first law of cosines*:

$$\cos a = \cos b \cos c + \sin b \sin c \cos A \qquad 7.65$$

- *second law of cosines*:

$$\cos A = -\cos B \cos C + \sin B \sin C \cos a \qquad 7.66$$

16. SOLID ANGLES

A *solid angle*, ω, is a measure of the angle subtended at the vertex of a cone. The solid angle has units of *steradians* (abbreviated *sr*). A steradian is the solid angle subtended at the center of a unit sphere (i.e., a sphere with a radius of one) by a unit area on its surface. Since the surface area of a sphere of radius r is r^2 times the surface area of a unit sphere, the solid angle is equal to the area cut out by the cone divided by r^2. (The surface area of a spherical segment is given in App. 8.B.)

$$\omega = \frac{\text{surface area}}{r^2} \qquad 7.67$$

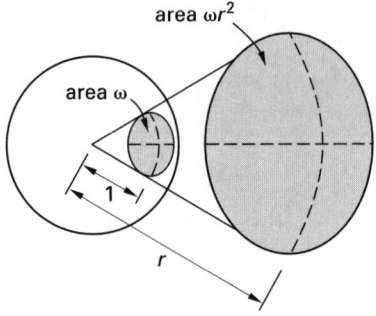

Figure 7.11 *Solid Angle*

8 Analytic Geometry

1. MENSURATION OF REGULAR SHAPES

The dimensions, perimeter, area, and other geometric properties constitute the *mensuration* (i.e., the measurements) of a geometric shape. Appendices 8.A and 8.B contain formulas and tables used to calculate these properties.

Example 8.1

In the study of open channel fluid flow, the hydraulic radius is defined as the ratio of flow area to wetted perimeter. What is the hydraulic radius of a 6 in diameter pipe filled to a depth of 2 in?

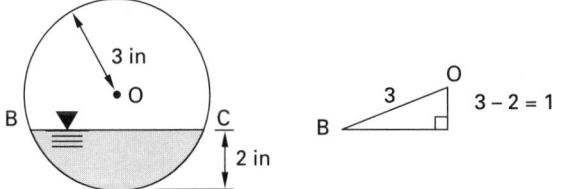

Solution

Points O, B, and C constitute a circular segment and are used to find the central angle of the circular segment.

$$\left(\tfrac{1}{2}\right)(\angle \text{BOC}) = \arccos\left(\tfrac{1}{3}\right) = 70.53°$$

$$\phi = \angle \text{BOC} = (2)(70.53°) = 141.06° = 2.462 \text{ rad}$$

From App. 8.A, the area in flow and arc length are

$$A = \tfrac{1}{2}r^2(\phi - \sin\phi)$$
$$= \left(\tfrac{1}{2}\right)(3 \text{ in})^2[2.462 \text{ rad} - \sin(2.462 \text{ rad})]$$
$$= 8.251 \text{ in}^2$$
$$s = r\phi$$
$$= (3 \text{ in})(2.462 \text{ rad}) = 7.386 \text{ in}$$

The hydraulic radius is

$$r_h = \frac{A}{s}$$
$$= \frac{8.251 \text{ in}^2}{7.386 \text{ in}} = 1.12 \text{ in}$$

2. AREAS WITH IRREGULAR BOUNDARIES

Areas of sections with irregular boundaries (such as creek banks) cannot be determined precisely, and approximation methods must be used. If the irregular side can be divided into a series of cells of equal width, either the trapezoidal rule or Simpson's rule can be used.

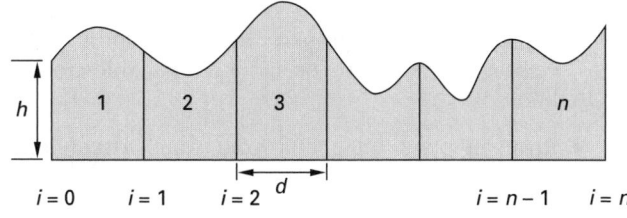

Figure 8.1 *Irregular Areas*

If the irregular side of each cell is fairly straight, the *trapezoidal rule* is appropriate.

$$A = \left(\frac{d}{2}\right)\left(h_0 + h_n + 2\sum_{i=1}^{n-1} h_i\right) \qquad 8.1$$

If the irregular side of each cell is curved (parabolic), *Simpson's rule* should be used. (n must be even to use Simpson's rule.)

$$A = \left(\frac{d}{3}\right)\left(h_0 + h_n + 4\sum_{\substack{i \text{ odd} \\ i=1}}^{n-1} h_i + 2\sum_{\substack{i \text{ even} \\ i=2}}^{n-2} h_i\right) \quad \text{8.2}$$

3. GEOMETRIC DEFINITIONS

The following terms are used in this book to describe the relationship or orientation of one geometric figure to another.

- *abscissa*: the horizontal coordinate, typically designated as x in a rectangular coordinate system

- *asymptote*: a straight line that is approached but not intersected by a curved line

- *asymptotic*: approaching the slope of another line; attaining the slope of another line in the limit

- *center*: a point equidistant from all other points

- *collinear*: falling on the same line

- *concave*: curved inward (in the direction indicated)[1]

- *convex*: curved outward (in the direction indicated)

- *convex hull*: a closed figure whose surface is convex everywhere

- *coplanar*: falling on the same plane

- *inflection point*: a point where the second derivative changes sign or the curve changes from concave to convex. (Also known as a *point of contraflexure*.)

- *locus of points*: a set or collection of points having some common property and being so infinitely close together as to be indistinguishable from a line

- *node*: a point on a line from which other lines enter or leave

- *normal*: rotated 90°; being at right angles

- *ordinate*: the vertical coordinate, typically designated as y in a rectangular coordinate system

- *orthogonal*: rotated 90°; being at right angles

---

[1]This is easily remembered since one must go inside to explore a cave.

- *saddle point*: a point in three-dimensional space where all adjacent points are higher in one direction (the direction of the saddle) and lower in an orthogonal direction (the direction of the sides)

- *tangent*: having equal slopes at a common point

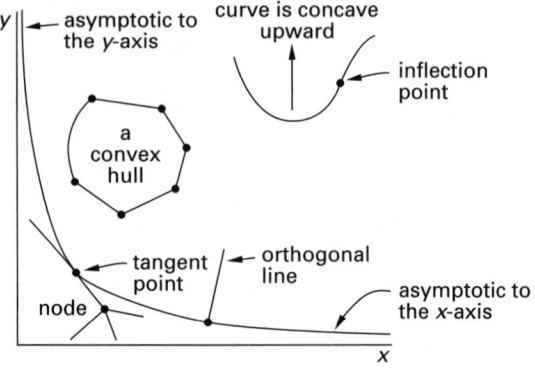

Figure 8.2 *Geometric Definitions*

4. CONCAVE CURVES

Concavity is a term that is applied to curved lines. A *concave up curve* is one whose function's first derivative increases continuously from negative to positive values. Straight lines drawn tangent to concave up curves are all below the curve. The graph of such a function may be thought of as being able to "hold water."

The first derivative of a *concave down curve* decreases continuously from positive to negative. A graph of a concave down function may be thought of as "spilling water." See Fig. 8.3.

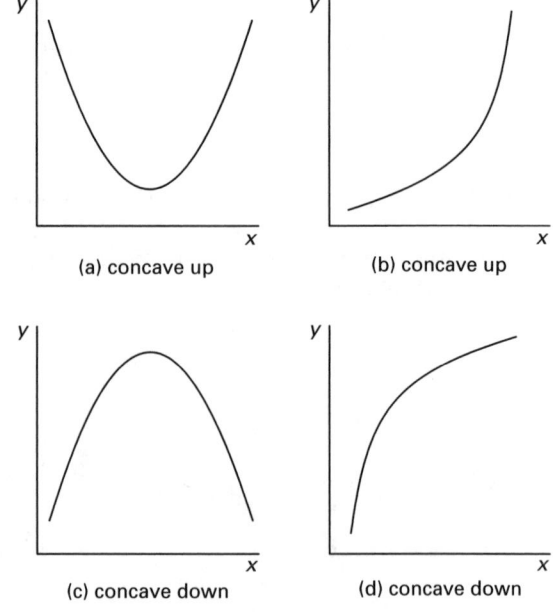

Figure 8.3 *Concave Curves*

5. CONVEX REGIONS

Convexity is a term that is applied to sets and regions.[2] It plays an important role in many mathematics subjects. A set or multidimensional region is *convex* if it contains the line segment joining any two of its points; that is, if a straight line is drawn connecting any two points in a convex region, that line will lie entirely within the region. For example, the interior of a parabola is a convex region, as is a solid sphere. The *void* or *null region* (i.e., an empty set of points), single points, and straight lines are convex sets. A convex region bounded by separate, connected line segments is known as a *convex hull*. (See Fig. 8.4.)

Within a convex region, a local maximum is also the global maximum. Similarly, a local minimum is also the global minimum. (See Sec. 8-3.) The intersection of two convex regions is also convex.

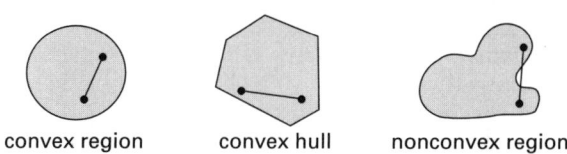

convex region   convex hull   nonconvex region

Figure 8.4 *Convexity*

6. CONGRUENCY

Congruence in geometric figures is analogous to *equality* in algebraic expressions. Congruent line segments are segments that have the same length. Congruent angles have the same angular measure. Congruent triangles have the same vertex angles and side lengths.

In general, *congruency*, indicated by the symbol $\cong$, means that there is one-to-one correspondence between all points on two objects. This correspondence is defined by the *mapping function* or *isometry*, which can be a translation, rotation, or reflection. Since the identity function is a valid mapping function, every geometric shape is congruent to itself.

Two congruent objects can be in different spaces. For example, a triangular area in three-dimensional space can be mapped into a triangle in two-dimensional space.

---

[2]It is tempting to define regions that fail the convexity test as being "concave." However, it is more proper to define such regions as "nonconvex." In any case, it is important to recognize that convexity depends on the reference point: An observer within a sphere will see the spherical boundary as convex; an observer outside the sphere may see the boundary as nonconvex.

7. COORDINATE SYSTEMS

The manner in which a geometric figure is described depends on the coordinate system that is used. The three-dimensional *rectangular coordinate system* (also known as the *Cartesian coordinate system*) with its x-, y-, and z-coordinates is the most commonly used in engineering. Table 8.1 summarizes the components needed to specify a point in the various coordinate systems. Figure 8.5 illustrates the use of and conversion between the coordinate systems.

Table 8.1 *Components of Coordinate Systems*

name	dimensions	components
rectangular	2	x, y
rectangular	3	x, y, z
polar	2	r, θ
cylindrical	3	r, θ, z
spherical	3	r, θ, ϕ

8. CURVES

A *curve* (commonly called a *line*) is a function over a finite or infinite range of the independent variable. When a curve is drawn in two- or three-dimensional space, it is known as a *graph of the curve*. It may or may not be possible to describe the curve mathematically. The *degree of a curve* is the highest exponent in the function. For example, Eq. 8.3 is a fourth-degree curve.

$$f(x) = 2x^4 + 7x^3 + 6x^2 + 3x + 9 = 0 \qquad 8.3$$

An *ordinary cycloid* (Fig. 8.6) is a curve traced out by a point on the rim of a wheel that rolls without slipping. It is described in parametric form by Eqs. 8.4 and 8.5 and in rectangular form by Eq. 8.6. In Eqs. 8.4 through 8.6, using the minus sign results in a cusp at the origin; using the plus sign results in a vertex (trough) at the origin.

$$x = r(\theta \pm \sin\theta) \qquad 8.4$$

$$y = r(1 \pm \cos\theta) \qquad 8.5$$

$$x = r\arccos\left(\frac{r-y}{r}\right) \pm \sqrt{2ry - y^2} \qquad 8.6$$

An *epicycloid* is a curve generated by a point on the rim of a wheel that rolls on the outside of a circle. A *hypocycloid* is a curve generated by a point on the rim of a wheel that rolls on the inside of a circle. The equation of a hypocycloid of four cusps is

$$x^{\frac{2}{3}} + y^{\frac{2}{3}} = r^{\frac{2}{3}} \qquad \text{[4 cusps]} \qquad 8.7$$

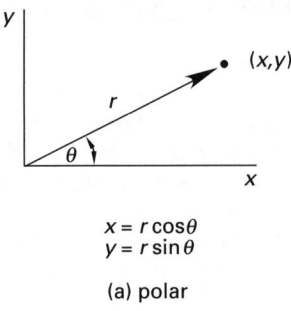

$$x = r\cos\theta$$
$$y = r\sin\theta$$

(a) polar

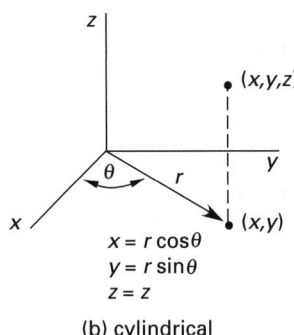

$$x = r\cos\theta$$
$$y = r\sin\theta$$
$$z = z$$

(b) cylindrical

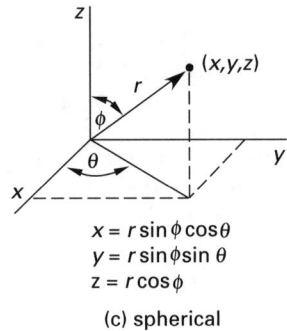

$$x = r\sin\phi\cos\theta$$
$$y = r\sin\phi\sin\theta$$
$$z = r\cos\phi$$

(c) spherical

Figure 8.5 *Different Coordinate Systems*

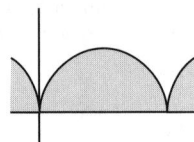

Figure 8.6 *Cycloid (cusp at origin)*

9. SYMMETRY OF CURVES

Two points, P and Q, are symmetrical with respect to a line if the line is a perpendicular bisector of the line segment PQ. If the graph of a curve is unchanged when y is replaced with $-y$, the curve is symmetrical with respect to the x-axis. If the curve is unchanged when x is replaced with $-x$, the curve is symmetrical with respect to the y-axis.

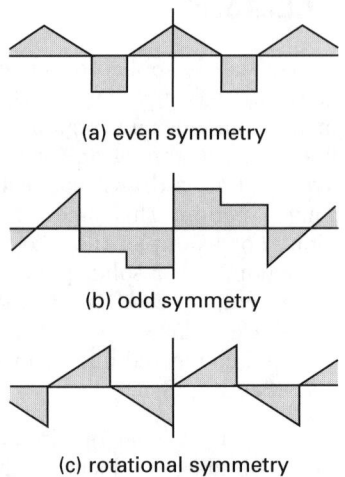

(a) even symmetry

(b) odd symmetry

(c) rotational symmetry

Figure 8.7 *Waveform Symmetry*

Repeating waveforms can be symmetrical with respect to the y-axis. A curve $f(x)$ is said to have *even symmetry* if $f(x) = f(-x)$. (Alternatively, $f(x)$ is said to be a *symmetrical function.*) With even symmetry, the function to the left of $x = 0$ is a reflection of the function to the right of $x = 0$. (In effect, the y-axis is a mirror.) The cosine curve is an example of a curve with even symmetry.

A curve is said to have *odd symmetry* if $f(x) = -f(-x)$. (Alternatively, $f(x)$ is said to be an *asymmetrical function.*)[3] The sine curve is an example of a curve with odd symmetry.

A curve is said to have *rotational symmetry (half-wave symmetry)* if $f(x) = -f(x + \pi)$.[4] Curves of this type are identical except for a sign reversal on alternate half-cycles. Table 8.2 describes the type of function resulting from the combination of two functions.

Table 8.2 *Combinations of Functions*

	operation			
	+	−	×	÷
$f_1(x)$ even, $f_2(x)$ even	even	even	even	even
$f_1(x)$ odd, $f_2(x)$ odd	odd	odd	even	even
$f_1(x)$ even, $f_2(x)$ odd	neither	neither	odd	odd

10. STRAIGHT LINES

Figure 8.8 illustrates a straight line in two-dimensional space. The *slope* of the line is m, the y-intercept is b, and the x-intercept is a. The equation of the line can be represented in several forms, and the procedure

---

[3]The semantics of "even symmetry" and "asymmetrical function" are contradictory.

[4]The symbol π represents half of a full cycle of the waveform, not the value $3.141\ldots$.

for finding the equation depends on the form chosen to represent the line. In general, the procedure involves substituting one or more known points on the line into the equation in order to determine the coefficients.

- *general form*:

$$Ax + By + C = 0 \qquad 8.8$$
$$A = -mB \qquad 8.9$$
$$B = \frac{-C}{b} \qquad 8.10$$
$$C = -aA = -bB \qquad 8.11$$

- *slope-intercept form*:

$$y = mx + b \qquad 8.12$$
$$m = \frac{-A}{B} = \tan\theta = \frac{y_2 - y_1}{x_2 - x_1} \qquad 8.13$$
$$b = \frac{-C}{B} \qquad 8.14$$
$$a = \frac{-C}{A} \qquad 8.15$$

- *point-slope form*:

$$y - y_1 = m(x - x_1) \qquad 8.16$$

- *intercept form*:

$$\frac{x}{a} + \frac{y}{b} = 1 \qquad 8.17$$

- *two-point form*:

$$\frac{y - y_1}{x - x_1} = \frac{y_2 - y_1}{x_2 - x_1} \qquad 8.18$$

- *normal form*:

$$x\cos\beta + y\sin\beta - d = 0 \qquad 8.19$$

(d and β are constants; x and y are variables.)

- *polar form*:

$$r = \frac{d}{\cos(\beta - \alpha)} \qquad 8.20$$

(d and β are constants; r and α are variables.)

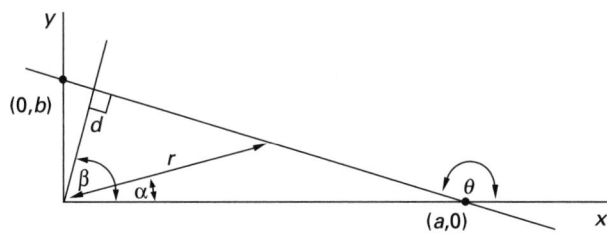

Figure 8.8 *Straight Line*

11. DIRECTION NUMBERS, ANGLES, AND COSINES

Given a directed line from (x_1, y_1, z_1) to (x_2, y_2, z_2), the *direction numbers* are

$$L = x_2 - x_1 \qquad 8.21$$
$$M = y_2 - y_1 \qquad 8.22$$
$$N = z_2 - z_1 \qquad 8.23$$

The distance between two points is

$$d = \sqrt{L^2 + M^2 + N^2} \qquad 8.24$$

The *direction cosines* are

$$\cos\alpha = \frac{L}{d} \qquad 8.25$$
$$\cos\beta = \frac{M}{d} \qquad 8.26$$
$$\cos\gamma = \frac{N}{d} \qquad 8.27$$

Note that

$$\cos^2\alpha + \cos^2\beta + \cos^2\gamma = 1 \qquad 8.28$$

The *direction angles* are the angles between the axes and the lines. They are found from the inverse functions of the direction cosines.

$$\alpha = \arccos\left(\frac{L}{d}\right) \qquad 8.29$$
$$\beta = \arccos\left(\frac{M}{d}\right) \qquad 8.30$$
$$\gamma = \arccos\left(\frac{N}{d}\right) \qquad 8.31$$

The direction cosines can be used to write the equation of the straight line in terms of the unit vectors. The line $\mathbf{R}$ would be defined as

$$\mathbf{R} = \mathbf{i}\cos\alpha + \mathbf{j}\cos\beta + \mathbf{k}\cos\gamma \qquad 8.32$$

Similarly, the line may be written in terms of its direction numbers.

$$\mathbf{R} = L\mathbf{i} + M\mathbf{j} + N\mathbf{k} \qquad 8.33$$

Example 8.2

A line passes through the points $(4, 7, 9)$ and $(0, 1, 6)$. Write the equation of the line in terms of its (a) direction numbers and (b) direction cosines.

Solution

(a) The direction numbers are

$$L = 4 - 0 = 4$$
$$M = 7 - 1 = 6$$
$$N = 9 - 6 = 3$$

Using Eq. 8.33,

$$\mathbf{R} = 4\mathbf{i} + 6\mathbf{j} + 3\mathbf{k}$$

(b) The distance between the two points is

$$d = \sqrt{(4)^2 + (6)^2 + (3)^2} = 7.81$$

The line in terms of its direction cosines is

$$\mathbf{R} = \frac{4\mathbf{i} + 6\mathbf{j} + 3\mathbf{k}}{7.81}$$
$$= 0.512\mathbf{i} + 0.768\mathbf{j} + 0.384\mathbf{k}$$

12. INTERSECTION OF TWO LINES

The intersection of two lines is a point. The location of the intersection point can be determined by setting the two equations equal and solving them in terms of a common variable. Alternatively, Eqs. 8.34 and 8.35 can be used to calculate the coordinates of the intersection point.

$$x = \frac{B_2 C_1 - B_1 C_2}{A_2 B_1 - A_1 B_2} \qquad 8.34$$

$$y = \frac{A_1 C_2 - A_2 C_1}{A_2 B_1 - A_1 B_2} \qquad 8.35$$

13. PLANES

A *plane* in three-dimensional space is completely determined by one of the following:

- three noncollinear points
- two nonparallel vectors $\mathbf{V}_1$ and $\mathbf{V}_2$ and their intersection point P_0
- a point P_0 and a vector, $\mathbf{N}$, normal to the plane (i.e., the *normal vector*)

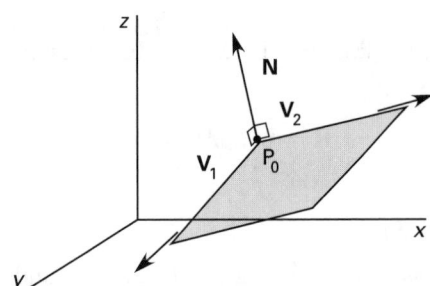

Figure 8.9 *Plane in Three-Dimensional Space*

The plane can be specified mathematically in one of two ways: in rectangular form or as a parametric equation. The general form is

$$A(x - x_0) + B(y - y_0) + C(z - z_0) = 0 \qquad 8.36$$

x_0, y_0, and z_0 are the coordinates of the intersection point of any two vectors in the plane. The coefficients A, B, and C are the same as the coefficients of the normal vector, $\mathbf{N}$.

$$\mathbf{N} = \mathbf{V}_1 \times \mathbf{V}_2 = A\mathbf{i} + B\mathbf{j} + C\mathbf{k} \qquad 8.37$$

Equation 8.36 can be simplified as follows.

$$Ax + By + Cz + D = 0 \qquad 8.38$$

$$D = -(Ax_0 + By_0 + Cz_0) \qquad 8.39$$

The following procedure can be used to determine the equation of a plane from three noncollinear points, P_1, P_2, and P_3, or from a normal vector and a single point.

step 1: (If the normal vector is known, go to step 3.) Determine the equations of the vectors $\mathbf{V}_1$ and $\mathbf{V}_2$ from two pairs of the points. For example, determine $\mathbf{V}_1$ from points P_1 and P_2, and determine $\mathbf{V}_2$ from P_1 and P_3. Express the vectors in the form $A\mathbf{i} + B\mathbf{j} + C\mathbf{k}$.

$$\mathbf{V}_1 = (x_2 - x_1)\mathbf{i} + (y_2 - y_1)\mathbf{j} + (z_2 - z_1)\mathbf{k} \qquad 8.40$$

$$\mathbf{V}_2 = (x_3 - x_1)\mathbf{i} + (y_3 - y_1)\mathbf{j} + (z_3 - z_1)\mathbf{k} \qquad 8.41$$

step 2: Find the normal vector, $\mathbf{N}$, as the cross product of the two vectors.

$$\mathbf{N} = \mathbf{V}_1 \times \mathbf{V}_2$$
$$= \begin{vmatrix} \mathbf{i} & (x_2 - x_1) & (x_3 - x_1) \\ \mathbf{j} & (y_2 - y_1) & (y_3 - y_1) \\ \mathbf{k} & (z_2 - z_1) & (z_3 - z_1) \end{vmatrix} \qquad 8.42$$

step 3: Write the general equation of the plane in rectangular form (Eq. 8.36) using the coefficients A, B, and C from the normal vector and any one of the three points as P_0.

The parametric equations of a plane also can be written as a linear combination of the components of two vectors in the plane. Referring to Fig. 8.9, the two known vectors are

$$\mathbf{V}_1 = V_{1x}\mathbf{i} + V_{1y}\mathbf{j} + V_{1z}\mathbf{k} \qquad 8.43$$

$$\mathbf{V}_2 = V_{2x}\mathbf{i} + V_{2y}\mathbf{j} + V_{2z}\mathbf{k} \qquad 8.44$$

If s and t are scalars, the coordinates of each point in the plane can be written as Eqs. 8.45 through 8.47. These are the parametric equations of the plane.

$$x = x_0 + sV_{1x} + tV_{2x} \qquad \textbf{8.45}$$

$$y = y_0 + sV_{1y} + tV_{2y} \qquad \textbf{8.46}$$

$$z = z_0 + sV_{1z} + tV_{2z} \qquad \textbf{8.47}$$

Example 8.3

The following points are coplanar.

$$P_1 = (2, 1, -4)$$
$$P_2 = (4, -2, -3)$$
$$P_3 = (2, 3, -8)$$

Determine the equation of the plane in (a) general form and (b) parametric form.

Solution

(a) Use the first two points to find a vector, $\mathbf{V}_1$.

$$\begin{aligned}\mathbf{V}_1 &= (x_2 - x_1)\mathbf{i} + (y_2 - y_1)\mathbf{j} + (z_2 - z_1)\mathbf{k} \\ &= (4-2)\mathbf{i} + (-2-1)\mathbf{j} + (-3-(-4))\mathbf{k} \\ &= 2\mathbf{i} - 3\mathbf{j} + 1\mathbf{k}\end{aligned}$$

Similarly, use the first and third points to find $\mathbf{V}_2$.

$$\begin{aligned}\mathbf{V}_2 &= (x_3 - x_1)\mathbf{i} + (y_3 - y_1)\mathbf{j} + (z_3 - z_1)\mathbf{k} \\ &= (2-2)\mathbf{i} + (3-1)\mathbf{j} + (-8-(-4))\mathbf{k} \\ &= 0\mathbf{i} + 2\mathbf{j} - 4\mathbf{k}\end{aligned}$$

From Eq. 8.42, determine the normal vector as a determinant.

$$\mathbf{N} = \begin{vmatrix} \mathbf{i} & 2 & 0 \\ \mathbf{j} & -3 & 2 \\ \mathbf{k} & 1 & -4 \end{vmatrix}$$

Expand the determinant across the top row.

$$\begin{aligned}\mathbf{N} &= \mathbf{i}(12 - 2) - 2(-4\mathbf{j} - 2\mathbf{k}) \\ &= 10\mathbf{i} + 8\mathbf{j} + 4\mathbf{k}\end{aligned}$$

The rectangular form of the equation of the plane uses the same constants as in the normal vector. Use the first point and write the equation of the plane in the form of Eq. 8.36.

$$(10)(x - 2) + (8)(y - 1) + (4)(z + 4) = 0$$

The three constant terms can be combined by using Eq. 8.39.

$$D = -[(10)(2) + (8)(1) + (4)(-4)] = -12$$

The equation of the plane is

$$10x + 8y + 4z - 12 = 0$$

(b) The parametric equations based on the first point and for any values of s and t are

$$x = 2 + 2s + 0t$$
$$y = 1 - 3s + 2t$$
$$z = -4 + 1s - 4t$$

The scalars s and t are not unique. Two of the three coordinates can also be chosen as the parameters. Dividing the rectangular form of the plane's equation by 4 to isolate z results in an alternate set of parametric equations.

$$x = x$$
$$y = y$$
$$z = 3 - 2.5x - 2y$$

14. DISTANCES BETWEEN GEOMETRIC FIGURES

The smallest distance, d, between various geometric figures is given by the following equations.

- between two points in (x, y, z) format:

$$d = \sqrt{(x_2 - x_1)^2 + (y_2 - y_1)^2 + (z_2 - z_1)^2} \quad \textbf{8.48}$$

- between a point (x_0, y_0) and a line $Ax + By + C = 0$:

$$d = \frac{|Ax_0 + By_0 + C|}{\sqrt{A^2 + B^2}} \qquad \textbf{8.49}$$

- between a point (x_0, y_0, z_0) and a plane $Ax + By + Cz + D = 0$:

$$d = \frac{|Ax_0 + By_0 + Cz_0 + D|}{\sqrt{A^2 + B^2 + C^2}} \qquad \textbf{8.50}$$

- between two parallel lines $Ax + By + C = 0$:

$$d = \left| \frac{|C_2|}{\sqrt{A_2^2 + B_2^2}} - \frac{|C_1|}{\sqrt{A_1^2 + B_1^2}} \right| \qquad \textbf{8.51}$$

Example 8.4

What is the minimum distance between the line $y = 2x + 3$ and the origin $(0, 0)$?

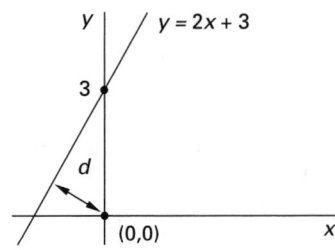

Solution

Put the equation in general form.

$$Ax + By + C = 2x - y + 3 = 0$$

Use Eq. 8.49 with $(x, y) = (0, 0)$.

$$d = \frac{|Ax + By + C|}{\sqrt{A^2 + B^2}} = \frac{(2)(0) + (-1)(0) + 3}{\sqrt{(2)^2 + (-1)^2}}$$
$$= \frac{3}{\sqrt{5}}$$

15. ANGLES BETWEEN GEOMETRIC FIGURES

The angle, ϕ, between various geometric figures is given by the following equations.

- between two lines in $Ax + By + C = 0$, $y = mx + b$, or direction angle formats:

$$\phi = \arctan\left(\frac{A_1 B_2 - A_2 B_1}{A_1 A_2 + B_1 B_2}\right) \qquad 8.52$$

$$= \arctan\left(\frac{m_2 - m_1}{1 + m_1 m_2}\right) \qquad 8.53$$

$$= |\arctan(m_1) - \arctan(m_2)| \qquad 8.54$$

$$= \arccos\left(\frac{L_1 L_2 + M_1 M_2 + N_1 N_2}{d_1 d_2}\right) \qquad 8.55$$

$$= \arccos(\cos\alpha_1 \cos\alpha_2 + \cos\beta_1 \cos\beta_2$$
$$+ \cos\gamma_1 \cos\gamma_2) \qquad 8.56$$

If the lines are parallel, then $\phi = 0$.

$$\frac{A_1}{A_2} = \frac{B_1}{B_2} \qquad 8.57$$

$$m_1 = m_2 \qquad 8.58$$

$$\alpha_1 = \alpha_2; \ \beta_1 = \beta_2; \ \gamma_1 = \gamma_2 \qquad 8.59$$

If the lines are perpendicular, then $\phi = 90°$.

$$A_1 A_2 = -B_1 B_2 \qquad 8.60$$

$$m_1 = \frac{-1}{m_2} \qquad 8.61$$

$$\alpha_1 + \alpha_2 = \beta_1 + \beta_2 = \gamma_1 + \gamma_2 = 90° \qquad 8.62$$

- between two planes in $A\mathbf{i} + B\mathbf{j} + C\mathbf{k} = 0$ format, the coefficients A, B, and C are the same as the coefficients for the normal vector. (See Eq. 8.37.) ϕ is equal to the angle between the two normal vectors.

$$\cos\phi = \frac{|A_1 A_2 + B_1 B_2 + C_1 C_2|}{\sqrt{A_1^2 + B_1^2 + C_1^2}\sqrt{A_2^2 + B_2^2 + C_2^2}} \qquad 8.63$$

Example 8.5

Use Eqs. 8.52, 8.53, and 8.54 to find the angle between the lines.

$$y = -0.577x + 2$$
$$y = +0.577x - 5$$

Solution

Write both equations in general form.

$$-0.577x - y + 2 = 0$$
$$0.577x - y - 5 = 0$$

(a) From Eq. 8.52,

$$\phi = \arctan\left(\frac{A_1 B_2 - A_2 B_1}{A_1 A_2 + B_1 B_2}\right)$$
$$= \arctan\left(\frac{(-0.577)(-1) - (0.577)(-1)}{(-0.577)(0.577) + (-1)(-1)}\right) = 60°$$

(b) Use Eq. 8.53.

$$\phi = \arctan\left(\frac{m_2 - m_1}{1 + m_1 m_2}\right)$$
$$= \arctan\left(\frac{0.577 - (-0.577)}{1 + (0.577)(-0.577)}\right) = 60°$$

(c) Use Eq. 8.54.

$$\phi = |\arctan(m_1) - \arctan(m_2)|$$
$$= |\arctan(-0.577) - \arctan(0.577)|$$
$$= |-30° - 30°| = 60°$$

Mathematics

16. CONIC SECTIONS

A *conic section* is any one of several curves produced by passing a plane through a cone as shown in Fig. 8.10. If α is the angle between the vertical axis and the cutting plane and β is the cone generating angle, Eq. 8.64 gives the *eccentricity*, ϵ, of the conic section. Values of the eccentricity are given in Fig. 8.10.

$$\epsilon = \frac{\cos\alpha}{\cos\beta} \qquad 8.64$$

All conic sections are described by second-degree polynomials (i.e., are *quadratic equations*) of the following form.[5]

$$Ax^2 + Bxy + Cy^2 + Dx + Ey + F = 0 \qquad 8.65$$

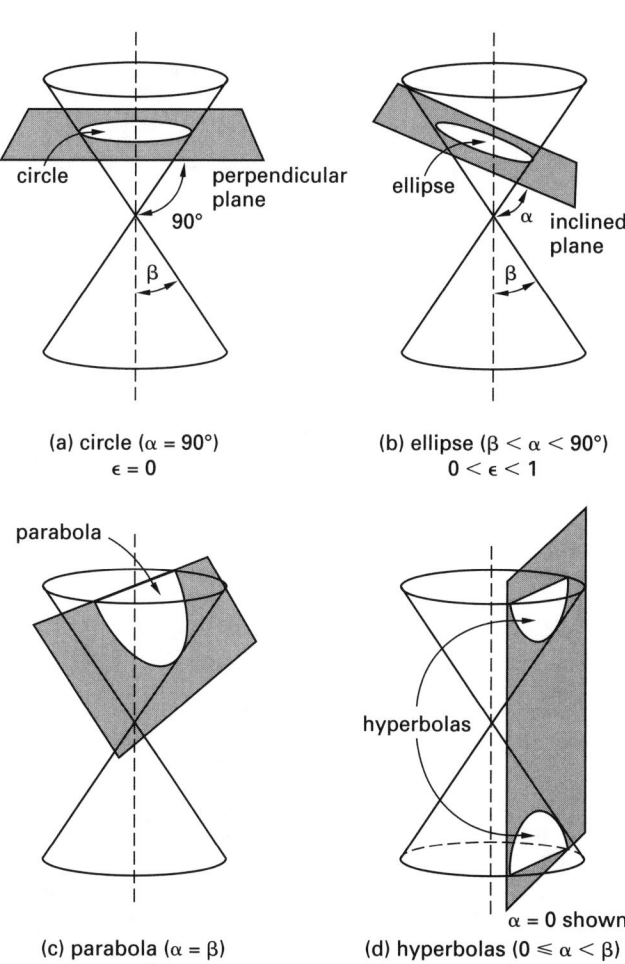

(a) circle ($\alpha = 90°$)
$\epsilon = 0$

(b) ellipse ($\beta < \alpha < 90°$)
$0 < \epsilon < 1$

(c) parabola ($\alpha = \beta$)
$\epsilon = 1$

(d) hyperbolas ($0 \leq \alpha < \beta$)
$\epsilon > 1$

Figure 8.10 *Conic Sections Produced by Cutting Planes*

This is the *general form*, which allows the figure axes to be at any angle relative to the coordinate axes. The

---

[5]One or more straight lines are produced when the cutting plane passes through the cone's vertex. Straight lines can be considered to be quadratic functions without second-degree terms.

standard forms presented in the following sections pertain to figures whose axes coincide with the coordinate axes, thereby eliminating certain terms of the general equation.

Figure 8.11 can be used to determine which conic section is described by the quadratic function. The quantity $B^2 - 4AC$ is known as the *discriminant*. Figure 8.11 determines only the type of conic section; it does not determine whether the conic section is degenerate (e.g., a circle with a negative radius).

Example 8.6

What geometric figures are described by the following equations?

(a) $4y^2 - 12y + 16x + 41 = 0$

(b) $x^2 - 10xy + y^2 + x + y + 1 = 0$

(c) $x^2 + 4y^2 + 2x - 8y + 1 = 0$

(d) $x^2 + y^2 - 6x + 8y + 20 = 0$

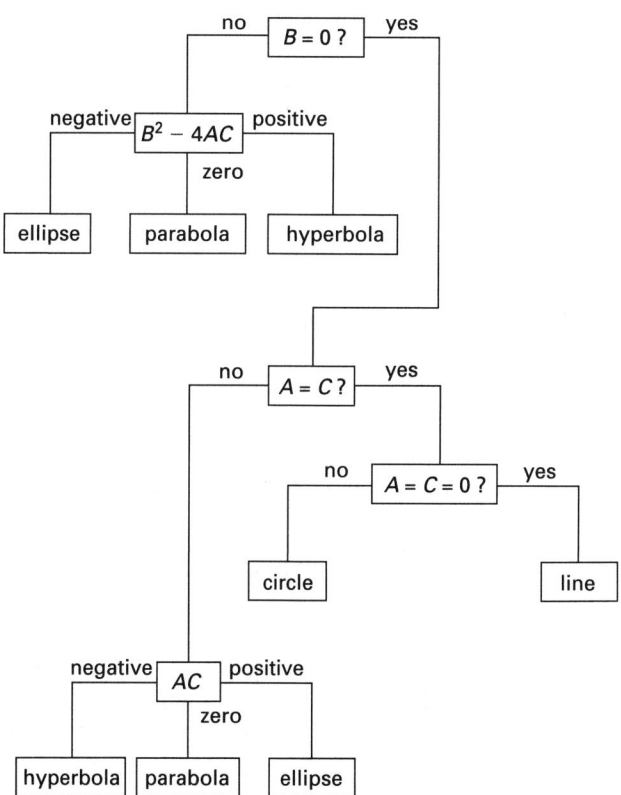

Figure 8.11 *Determining Conic Sections from Quadratic Equations*

Solution

(a) Referring to Fig. 8.11, $B = 0$ since there is no xy term, $A = 0$ since there is no x^2 term, and $AC = (0)(4) = 0$. This is a parabola.

(b) $B \neq 0$; $B^2 - 4AC = (-10)^2 - (4)(1)(1) = +96$. This is a hyperbola.

(c) $B = 0$; $A \neq C$; $AC = (1)(4) = +4$. This is an ellipse.

(d) $B = 0$; $A = C$; $A = C = 1$ $(\neq 0)$. This is a circle.

17. CIRCLE

The general form of the equation of a circle is

$$Ax^2 + Ay^2 + Dx + Ey + F = 0 \qquad 8.66$$

The *center-radius form* of the equation of a circle with radius r and center at (h, k) is

$$(x - h)^2 + (y - k)^2 = r^2 \qquad 8.67$$

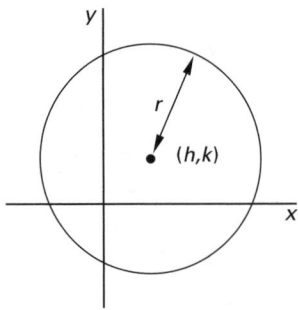

Figure 8.12 Circle

The two forms can be converted by use of Eqs. 8.68 through 8.70.

$$h = \frac{-D}{2A} \qquad 8.68$$

$$k = \frac{-E}{2A} \qquad 8.69$$

$$r^2 = \frac{D^2 + E^2 - 4AF}{4A^2} \qquad 8.70$$

If the right-hand side of Eq. 8.70 is positive, the figure is a circle. If it is zero, the circle shrinks to a point. If the right-hand side is negative, the figure is imaginary. A *degenerate circle* is one in which the right-hand side is less than or equal to zero.

18. PARABOLA

A *parabola* is the locus of points equidistant from the *focus* (point F in Fig. 8.13) and a line called the *directrix*. A parabola is symmetric with respect to its *parabolic axis*. The line normal to the parabolic axis and passing through the focus is known as the *latus rectum*. The eccentricity of a parabola is 1.

There are two common types of parabolas in the Cartesian plane—those that open right and left, and those

that open up and down. Equation 8.65 is the general form of the equation of a parabola. With Eq. 8.71, the parabola points to the right if $CD > 0$ and to the left if $CD < 0$. With Eq. 8.72, the parabola points up if $AE > 0$ and down if $AE < 0$.

$$Cy^2 + Dx + Ey + F = 0 \Big|_{\substack{C,D \neq 0 \\ \text{opens horizontally}}} \qquad 8.71$$

$$Ax^2 + Dx + Ey + F = 0 \Big|_{\substack{A,E \neq 0 \\ \text{opens vertically}}} \qquad 8.72$$

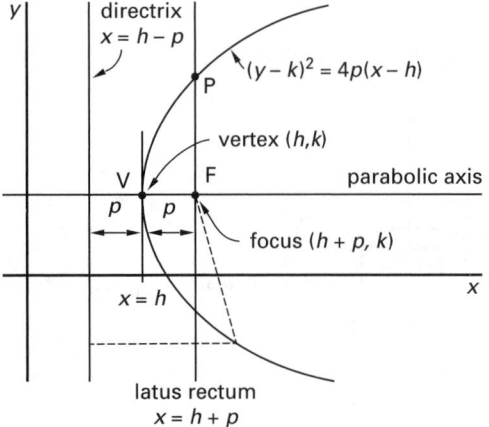

Figure 8.13 Parabola

The *standard form* of the equation of a parabola with vertex at (h, k), focus at $(h + p, k)$, and directrix at $x = h - p$ and that opens to the right or left is given by Eq. 8.73. The parabola opens to the right (points to the left) if $p > 0$ and opens to the left (points to the right) if $p < 0$.

$$(y - k)^2 = 4p(x - h) \Big|_{\text{opens horizontally}} \qquad 8.73$$

$$y^2 = 4px \Big|_{\substack{\text{vertex at origin} \\ h = k = 0}} \qquad 8.74$$

The *standard form* of the equation of a parabola with vertex at (h, k), focus at $(h, k + p)$, and directrix at $y = k - p$ and that opens up or down is given by Eq. 8.75. The parabola opens up (points down) if $p > 0$ and opens down (points up) if $p < 0$.

$$(x - h)^2 = 4p(y - k) \Big|_{\text{opens vertically}} \qquad 8.75$$

$$x^2 = 4py \Big|_{\text{vertex at origin}} \qquad 8.76$$

The general and vertex forms of the equations can be reconciled with Eqs. 8.77 through 8.79. Whether the first or second forms of these equations are used depends on whether the parabola opens horizontally or vertically (i.e., whether $A = 0$ or $C = 0$), respectively.

$$h = \begin{cases} \dfrac{E^2 - 4CF}{4CD} & \text{[opens horizontally]} \\[2ex] \dfrac{-D}{2A} & \text{[opens vertically]} \end{cases} \qquad 8.77$$

$$k = \begin{cases} \dfrac{-E}{2C} & \text{[opens horizontally]} \\[2ex] \dfrac{D^2 - 4AF}{4AE} & \text{[opens vertically]} \end{cases} \qquad 8.78$$

$$p = \begin{cases} \dfrac{-D}{4C} & \text{[opens horizontally]} \\[2ex] \dfrac{-E}{4A} & \text{[opens vertically]} \end{cases} \qquad 8.79$$

19. ELLIPSE

An *ellipse* has two foci separated along the *major axis* by a distance $2c$. The line perpendicular to the major axis passing through the center of the ellipse is the *minor axis*. The lines perpendicular to the major axis passing through the foci are the *latus recta*. The distance between the two vertices is $2a$. The ellipse is the locus of points such that the sum of the distances from the two foci is $2a$. Referring to Fig. 8.14,

$$F_1 P + P F_2 = 2a \qquad 8.80$$

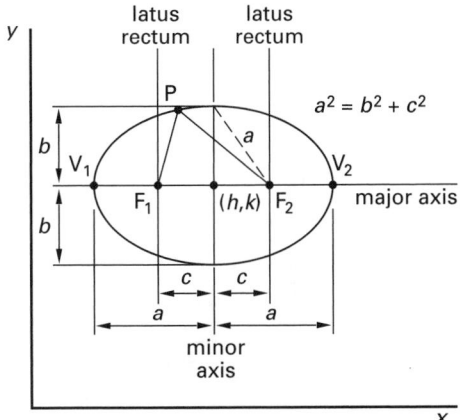

Figure 8.14 *Ellipse*

Equation 8.81 is the standard equation for an ellipse with axes parallel to the coordinate axes. (Equation 8.65 is the general form.) F is not independent of A, C, D, and E for the ellipse.

$$Ax^2 + Cy^2 + Dx + Ey + F = 0 \Big|_{\substack{AC > 0 \\ A \neq C}} \qquad 8.81$$

Equation 8.82 gives the standard form of the equation of an ellipse centered at (h, k). Distances a and b are known as the *semimajor distance* and *semiminor distance*, respectively.

$$\frac{(x - h)^2}{a^2} + \frac{(y - k)^2}{b^2} = 1 \qquad 8.82$$

The distance between the two foci is $2c$.

$$2c = 2\sqrt{a^2 - b^2} \qquad 8.83$$

The *aspect ratio* of the ellipse is

$$\text{aspect ratio} = \frac{a}{b} \qquad 8.84$$

The *eccentricity*, ϵ, of the ellipse is always less than 1. If the eccentricity is zero, the figure is a circle (another form of a *degenerative ellipse*).

$$\epsilon = \frac{\sqrt{a^2 - b^2}}{a} < 1 \qquad 8.85$$

The standard and center forms of the equations of an ellipse can be reconciled by using Eqs. 8.86 through 8.89.

$$h = \frac{-D}{2A} \qquad 8.86$$

$$k = \frac{-E}{2C} \qquad 8.87$$

$$a = \sqrt{C} \qquad 8.88$$

$$b = \sqrt{A} \qquad 8.89$$

20. HYPERBOLA

A *hyperbola* has two foci separated along the *transverse axis* by a distance $2c$. Lines perpendicular to the transverse axis passing through the foci are the *conjugate axes*. The distance between the two vertices is $2a$, and the distance along a conjugate axis between two points on the hyperbola is $2b$. The hyperbola is the locus of points such that the difference in distances from the two foci is $2a$. Referring to Fig. 8.15,

$$F_2 P - P F_1 = 2a \qquad 8.90$$

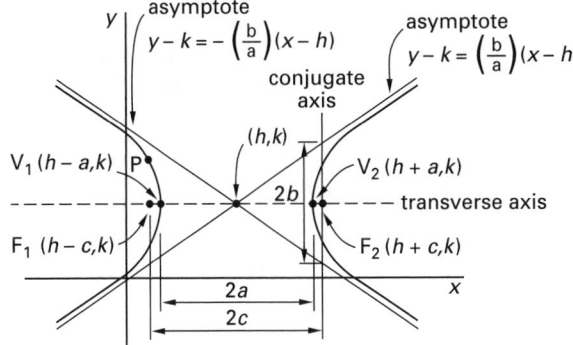

Figure 8.15 *Hyperbola*

Equation 8.91 is the standard equation of a hyperbola. Coefficients A and C have opposite signs.

$$Ax^2 + Cy^2 + Dx + Ey + F = 0\big|_{AC \, < \, 0} \qquad 8.91$$

Equation 8.92 gives the standard form of the equation of a hyperbola centered at (h, k) and opening to the left and right.

$$\frac{(x-h)^2}{a^2} - \frac{(y-k)^2}{b^2} = 1\bigg|_{\text{opens horizontally}} \qquad 8.92$$

Equation 8.93 gives the standard form of the equation of a hyperbola centered at (h, k) and opening up and down.

$$\frac{(y-k)^2}{a^2} - \frac{(x-h)^2}{b^2} = 1\bigg|_{\text{opens vertically}} \qquad 8.93$$

The distance between the two foci is $2c$.

$$2c = 2\sqrt{a^2 + b^2} \qquad 8.94$$

The *eccentricity*, ϵ, of the hyperbola is calculated from Eq. 8.95 and is always greater than 1.

$$\epsilon = \frac{c}{a} = \frac{\sqrt{a^2 + b^2}}{a} > 1 \qquad 8.95$$

The hyperbola is asymptotic to the lines given by Eqs. 8.96 and 8.97.

$$y = \pm\frac{b}{a}(x-h) + k\bigg|_{\text{opens horizontally}} \qquad 8.96$$

$$y = \pm\frac{a}{b}(x-h) + k\bigg|_{\text{opens vertically}} \qquad 8.97$$

For a *rectangular (equilateral) hyperbola*, the asymptotes are perpendicular, $a = b$, $c = \sqrt{2}a$, and the eccentricity is $\epsilon = \sqrt{2}$. If the hyperbola is centered at the origin (i.e., $h = k = 0$), then the equations are $x^2 - y^2 = a^2$ (opens horizontally) and $y^2 - x^2 = a^2$ (opens vertically).

If the asymptotes are the x- and y-axes, the equation of the hyperbola is simply

$$xy = \pm\frac{a^2}{2} \qquad 8.98$$

The general and center forms of the equations of a hyperbola can be reconciled by using Eqs. 8.99 through 8.103. Whether the hyperbola opens left and right or up and down depends on whether M/A or M/C is positive, respectively, where M is defined by Eq. 8.99.

$$M = \frac{D^2}{4A} + \frac{E^2}{4C} - F \qquad 8.99$$

$$h = \frac{-D}{2A} \qquad 8.100$$

$$k = \frac{-E}{2C} \qquad 8.101$$

$$a = \begin{cases} \sqrt{-C} & \text{[opens horizontally]} \\ \sqrt{-A} & \text{[opens vertically]} \end{cases} \qquad 8.102$$

$$b = \begin{cases} \sqrt{A} & \text{[opens horizontally]} \\ \sqrt{C} & \text{[opens vertically]} \end{cases} \qquad 8.103$$

21. SPHERE

Equation 8.104 is the general equation of a sphere. The coefficient A cannot be zero.

$$Ax^2 + Ay^2 + Az^2 + Bx + Cy + Dz + E = 0 \qquad 8.104$$

Equation 8.105 gives the standard form of the equation of a sphere centered at (h, k, l) with radius r.

$$(x-h)^2 + (y-k)^2 + (z-l)^2 = r^2 \qquad 8.105$$

The general and center forms of the equations of a sphere can be reconciled by using Eqs. 8.106 through 8.109.

$$h = \frac{-B}{2A} \qquad 8.106$$

$$k = \frac{-C}{2A} \qquad 8.107$$

$$l = \frac{-D}{2A} \qquad 8.108$$

$$r = \sqrt{\frac{B^2 + C^2 + D^2}{4A^2} - \frac{E}{A}} \qquad 8.109$$

Planes tangent to spheres and other solids are covered in Sec. 9-8.

22. HELIX

A *helix* is a curve generated by a point moving on, around, and along a cylinder such that the distance the point moves parallel to the cylindrical axis is proportional to the angle of rotation about that axis. For a cylinder of radius r, Eqs. 8.110 through 8.112 define the three-dimensional positions of points along the helix. The quantity $2\pi k$ is the *pitch* of the helix.

$$x = r \cos \theta \qquad 8.110$$

$$y = r \sin \theta \qquad 8.111$$

$$z = k\theta \qquad 8.112$$

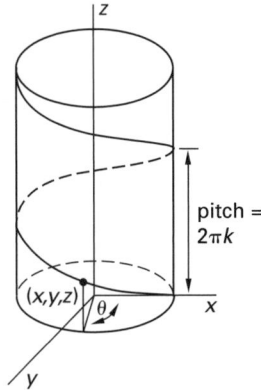

Figure 8.16 *Helix*

9 Differential Calculus

1. DERIVATIVE OF A FUNCTION

In most cases, it is possible to transform a continuous function, $f(x_1, x_2, x_3, \ldots)$, of one or more independent variables into a derivative function.[1] In simple cases, the *derivative* can be interpreted as the slope (tangent or rate of change) of the curve described by the original function. Since the slope of the curve depends on x, the derivative function will also depend on x. The derivative, $f'(x)$, of a function $f(x)$ is defined mathematically by Eq. 9.1. However, limit theory is seldom needed to actually calculate derivatives.

$$f'(x) = \lim_{\triangle x \to 0} \frac{\triangle f(x)}{\triangle x} \qquad 9.1$$

The derivative of a function $f(x)$, also known as the *first derivative*, is written in various ways, including

$$f'(x), \frac{df(x)}{dx}, \frac{df}{dx}, \mathbf{D}f(x), \mathbf{D}_x f(x), \dot{f}(x), sf(x)$$

A *second derivative* may exist if the derivative operation is performed on the first derivative—that is, a derivative is taken of a derivative function. This is written as

$$f''(x), \frac{d^2 f(x)}{dx^2}, \frac{d^2 f}{dx^2}, \mathbf{D}^2 f(x), \mathbf{D}_x^2 f(x), \ddot{f}(x), s^2 f(x)$$

---

[1] A function, $f(x)$, of one independent variable, x, is used in this section to simplify the discussion. Although the derivative is taken with respect to x, the independent variable can be anything.

A *regular* (*analytic* or *holomorphic*) *function* possesses a derivative. A point at which a function's derivative is undefined is called a *singular point*, as Fig. 9.1 illustrates.

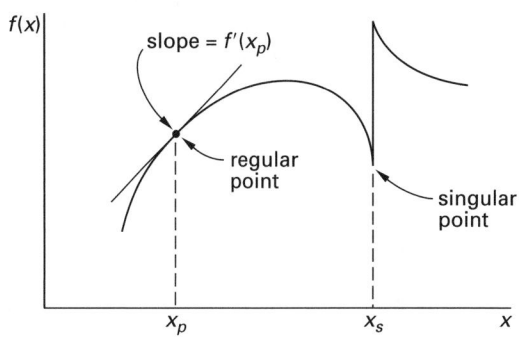

Figure 9.1 *Derivatives and Singular Points*

2. ELEMENTARY DERIVATIVE OPERATIONS

Equations 9.2 through 9.5 summarize the elementary derivative operations on polynomials and exponentials. Equations 9.2 and 9.3 are particularly useful. (a, n, and k represent constants. $f(x)$ and $g(x)$ are functions of x.)

$$\mathbf{D}k = 0 \qquad 9.2$$

$$\mathbf{D}x^n = nx^{n-1} \qquad 9.3$$

$$\mathbf{D}\ln x = \frac{1}{x} \qquad 9.4$$

$$\mathbf{D}e^{ax} = ae^{ax} \qquad 9.5$$

Equations 9.6 through 9.17 summarize the elementary derivative operations on transcendental (trigonometric) functions.

$$\mathbf{D}\sin x = \cos x \qquad 9.6$$

$$\mathbf{D}\cos x = -\sin x \qquad 9.7$$

$$\mathbf{D}\tan x = \sec^2 x \qquad 9.8$$

$$\mathbf{D}\cot x = -\csc^2 x \qquad 9.9$$

$$\mathbf{D}\sec x = \sec x \tan x \qquad 9.10$$

$$\mathbf{D}\csc x = -\csc x \cot x \qquad 9.11$$

$$\mathbf{D}\arcsin x = \frac{1}{\sqrt{1-x^2}} \qquad 9.12$$

$$\mathbf{D}\arccos x = -\mathbf{D}\arcsin x \qquad 9.13$$

$$\mathbf{D}\arctan x = \frac{1}{1+x^2} \qquad 9.14$$

$$\mathbf{D}\operatorname{arccot} x = -\mathbf{D}\arctan x \qquad 9.15$$

$$\mathbf{D}\operatorname{arcsec} x = \frac{1}{x\sqrt{x^2-1}} \qquad 9.16$$

$$\mathbf{D}\operatorname{arccsc} x = -\mathbf{D}\operatorname{arcsec} x \qquad 9.17$$

Equations 9.18 through 9.23 summarize the elementary derivative operations on hyperbolic transcendental functions. Derivatives of hyperbolic functions are not completely analogous to those of the regular transcendental functions.

$$\mathbf{D}\sinh x = \cosh x \qquad 9.18$$

$$\mathbf{D}\cosh x = \sinh x \qquad 9.19$$

$$\mathbf{D}\tanh x = \operatorname{sech}^2 x \qquad 9.20$$

$$\mathbf{D}\coth x = -\operatorname{csch}^2 x \qquad 9.21$$

$$\mathbf{D}\operatorname{sech} x = -\operatorname{sech} x \tanh x \qquad 9.22$$

$$\mathbf{D}\operatorname{csch} x = -\operatorname{csch} x \coth x \qquad 9.23$$

Equations 9.24 through 9.29 summarize the elementary derivative operations on functions and combinations of functions.

$$\mathbf{D}kf(x) = k\mathbf{D}f(x) \qquad 9.24$$

$$\mathbf{D}(f(x) \pm g(x)) = \mathbf{D}f(x) \pm \mathbf{D}g(x) \qquad 9.25$$

$$\mathbf{D}(f(x) \cdot g(x)) = f(x)\mathbf{D}g(x) + g(x)\mathbf{D}f(x) \qquad 9.26$$

$$\mathbf{D}\left(\frac{f(x)}{g(x)}\right) = \frac{g(x)\mathbf{D}f(x) - f(x)\mathbf{D}g(x)}{(g(x))^2} \qquad 9.27$$

$$\mathbf{D}\left(f(x)\right)^n = n\left(f(x)\right)^{n-1}\mathbf{D}f(x) \qquad 9.28$$

$$\mathbf{D}f(g(x)) = \mathbf{D}_g f(g)\mathbf{D}_x g(x) \qquad 9.29$$

Example 9.1

What is the slope at $x = 3$ of the curve $f(x) = x^3 - 2x$?

Solution

The derivative function found from Eq. 9.3 determines the slope.

$$f'(x) = 3x^2 - 2$$

The slope at $x = 3$ is

$$f'(3) = (3)(3)^2 - 2 = 27 - 2 = 25$$

Example 9.2

What are the derivatives of the following functions?

(a) $f(x) = 5\sqrt[3]{x^5}$

(b) $f(x) = \sin x \cos^2 x$

(c) $f(x) = \ln(\cos e^x)$

Solution

(a) Using Eqs. 9.3 and 9.24,

$$\begin{aligned}
f'(x) = 5\mathbf{D}\sqrt[3]{x^5} &= 5\mathbf{D}\left((x^5)^{\frac{1}{3}}\right) \\
&= (5)\left(\tfrac{1}{3}\right)(x^5)^{-\frac{2}{3}}\mathbf{D}x^5 \\
&= (5)\left(\tfrac{1}{3}\right)(x^5)^{-\frac{2}{3}}(5)(x^4) \\
&= \frac{25x^{\frac{2}{3}}}{3}
\end{aligned}$$

(b) Using Eq. 9.26,

$$\begin{aligned}
f'(x) &= \sin x\mathbf{D}\cos^2 x + \cos^2 x\mathbf{D}\sin x \\
&= (\sin x)(2\cos x)(\mathbf{D}\cos x) + \cos^2 x\cos x \\
&= (\sin x)(2\cos x)(-\sin x) + \cos^2 x\cos x \\
&= -2\sin^2 x\cos x + \cos^3 x
\end{aligned}$$

(c) Using Eq. 9.29,

$$\begin{aligned}
f'(x) &= \left(\frac{1}{\cos e^x}\right)\mathbf{D}\cos e^x \\
&= \left(\frac{1}{\cos e^x}\right)(-\sin e^x)\mathbf{D}e^x \\
&= \left(\frac{-\sin e^x}{\cos e^x}\right)e^x \\
&= -e^x\tan e^x
\end{aligned}$$

3. CRITICAL POINTS

Derivatives are used to locate the local *critical points* of functions of one variable—that is, *extreme points* (also known as *maximum* and *minimum* points) as well as the *inflection points* (*points of contraflexure*). The plurals *extrema*, *maxima*, and *minima* are used without the word "points." These points are illustrated in Fig. 9.2. There is usually an inflection point between two adjacent local extrema.

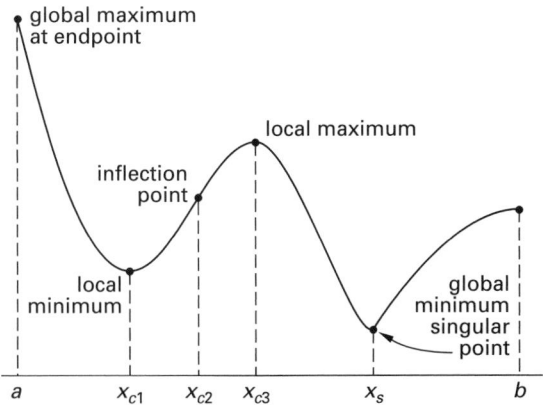

Figure 9.2 *Extreme and Inflection Points*

The first derivative is calculated to determine the locations of the critical points. The second derivative is calculated to determine whether a critical point is a local maximum, minimum, or inflection point, according to the following conditions. With this method, no distinction is made between local and global extrema. Therefore, the extrema should be compared with the function values at the endpoints of the interval, as illustrated in Ex. 9.3.[2] Note that $f'(x) \neq 0$ at an inflection point.

$$f'(x_c) = 0 \text{ at any extreme point, } x_c \qquad 9.30$$

$$f''(x_c) < 0 \text{ at a maximum point} \qquad 9.31$$

$$f''(x_c) > 0 \text{ at a minimum point} \qquad 9.32$$

$$f''(x_c) = 0 \text{ at an inflection point} \qquad 9.33$$

Example 9.3

Find the global extrema of the function $f(x)$ on the interval $[-2, +2]$.

$$f(x) = x^3 + x^2 - x + 1$$

Solution

The first derivative is

$$f'(x) = 3x^2 + 2x - 1$$

Since the first derivative is zero at extreme points, set $f'(x)$ equal to zero and solve for the roots of the quadratic equation.

$$3x^2 + 2x - 1 = (3x - 1)(x + 1) = 0$$

The roots are $x_1 = \frac{1}{3}$, $x_2 = -1$. These are the locations of the two extrema.

---

[2]It is also necessary to check the values of the function at singular points (i.e., points where the derivative does not exist).

The second derivative is

$$f''(x) = 6x + 2$$

Substituting x_1 and x_2 into $f''(x)$,

$$f''(x_1) = (6)\left(\tfrac{1}{3}\right) + 2 = 4$$

$$f''(x_2) = (6)(-1) + 2 = -4$$

Therefore, x_1 is a local minimum point (because $f''(x)$ is positive), and x_2 is a local maximum point (because $f''(x)$ is negative). The inflection point between these two extrema is found by setting $f''(x)$ equal to zero.

$$f''(x) = 6x + 2 = 0 \text{ or } x = -\tfrac{1}{3}$$

Since the question asked for the global extreme points, it is necessary to compare the values of $f(x)$ at the local extrema with the values at the endpoints.

$$f(-2) = -1$$

$$f(-1) = 2$$

$$f\left(\tfrac{1}{3}\right) = 22/27$$

$$f(2) = +11$$

Therefore, the actual global extrema are the endpoints.

4. DERIVATIVES OF PARAMETRIC EQUATIONS

The derivative of a function $f(x_1, x_2, \ldots x_n)$ can be calculated from the derivatives of the parametric equations $f_1(s), f_2(s), \ldots, f_n(s)$. The derivative will be expressed in terms of the parameter, s, unless the derivatives of the parametric equations can be expressed explicitly in terms of the independent variables.

Example 9.4

A circle is expressed parametrically by the equations

$$x = 5\cos\theta$$

$$y = 5\sin\theta$$

Express the derivative dy/dx (a) as a function of the parameter θ and (b) as a function of x and y.

Solution

(a) Taking the derivative of each parametric equation with respect to θ,

$$\frac{dx}{d\theta} = -5\sin\theta$$

$$\frac{dy}{d\theta} = 5\cos\theta$$

Then,

$$\frac{dy}{dx} = \frac{\frac{dy}{d\theta}}{\frac{dx}{d\theta}} = \frac{5\cos\theta}{-5\sin\theta} = -\cot\theta$$

(b) The derivatives of the parametric equations are closely related to the original parametric equations.

$$\frac{dx}{d\theta} = -5\sin\theta = -y$$

$$\frac{dy}{d\theta} = 5\cos\theta = x$$

$$\frac{dy}{dx} = \frac{\frac{dy}{d\theta}}{\frac{dx}{d\theta}} = \frac{-x}{y}$$

5. PARTIAL DIFFERENTIATION

Derivatives can be taken with respect to only one independent variable at a time. For example, $f'(x)$ is the derivative of $f(x)$ and is taken with respect to the independent variable x. If a function, $f(x_1, x_2, x_3, \ldots)$, has more than one independent variable, a *partial derivative* can be found, but only with respect to one of the independent variables. All other variables are treated as constants. Symbols for a partial derivative of f taken with respect to variable x are $\partial f/\partial x$ and $f_x(x,y)$.

The geometric interpretation of a partial derivative $\partial f/\partial x$ is the slope of a line tangent to the surface (a sphere, ellipsoid, etc.) described by the function when all variables except x are held constant. In three-dimensional space with a function described by $z = f(x,y)$, the partial derivative $\partial f/\partial x$ (equivalent to $\partial z/\partial x$) is the slope of the line tangent to the surface in a plane of constant y. Similarly, the partial derivative $\partial f/\partial y$ (equivalent to $\partial z/\partial y$) is the slope of the line tangent to the surface in a plane of constant x.

Example 9.5

What is the partial derivative $\partial z/\partial x$ of the following function?

$$z = 3x^2 - 6y^2 + xy + 5y - 9$$

Solution

The partial derivative with respect to x is found by considering all variables other than x to be constants.

$$\frac{\partial z}{\partial x} = 6x - 0 + y + 0 - 0 = 6x + y$$

Example 9.6

A surface has the equation $x^2 + y^2 + z^2 - 9 = 0$. What is the slope of a line that lies in a plane of constant y and is tangent to the surface at $(x,y,z) = (1,2,2)$?[3]

---
[3]Although only implied, it is required that the point actually be on the surface (i.e., it must satisfy the equation $f(x,y,z) = 0$).

Solution

Solve for the dependent variable. Then, consider variable y to be a constant.

$$z = \sqrt{9 - x^2 - y^2}$$

$$\frac{\partial z}{\partial x} = \frac{\partial(9 - x^2 - y^2)^{\frac{1}{2}}}{\partial x}$$

$$= \left(\tfrac{1}{2}\right)(9 - x^2 - y^2)^{-\frac{1}{2}}\left[\frac{\partial(9 - x^2 - y^2)}{\partial x}\right]$$

$$= \left(\tfrac{1}{2}\right)(9 - x^2 - y^2)^{-\frac{1}{2}}(-2x)$$

$$= \frac{-x}{\sqrt{9 - x^2 - y^2}}$$

At the point $(1,2,2)$, $x = 1$ and $y = 2$.

$$\left.\frac{\partial z}{\partial x}\right|_{(1,2,2)} = \frac{-1}{\sqrt{9 - (1)^2 - (2)^2}} = -\tfrac{1}{2}$$

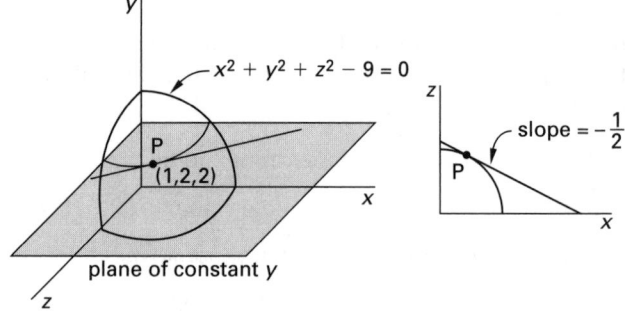

6. IMPLICIT DIFFERENTIATION

When a relationship between n variables cannot be manipulated to yield an explicit function of $n-1$ independent variables, that relationship implicitly defines the nth variable. Finding the derivative of the implicit variable with respect to any other independent variable is known as *implicit differentiation*.

An implicit derivative is the quotient of two partial derivatives. The two partial derivatives are chosen so that dividing one by the other eliminates a common differential. For example, if z cannot be explicitly extracted from $f(x,y,z) = 0$, the partial derivatives $\partial z/\partial x$ and $\partial z/\partial y$ can still be found as follows.

$$\frac{\partial z}{\partial x} = \frac{-\frac{\partial f}{\partial x}}{\frac{\partial f}{\partial z}} \qquad 9.34$$

$$\frac{\partial z}{\partial y} = \frac{-\frac{\partial f}{\partial y}}{\frac{\partial f}{\partial z}} \qquad 9.35$$

Example 9.7

Find the derivative dy/dx of

$$f(x,y) = x^2 + xy + y^3$$

Solution

Implicit differentiation is required because x cannot be extracted from $f(x,y)$.

$$\frac{\partial f}{\partial x} = 2x + y$$

$$\frac{\partial f}{\partial y} = x + 3y^2$$

$$\frac{dy}{dx} = \frac{-\dfrac{\partial f}{\partial x}}{\dfrac{\partial f}{\partial y}} = \frac{-(2x+y)}{x+3y^2}$$

Example 9.8

Solve Ex. 9.6 using implicit differentiation.

Solution

$$f(x,y,z) = x^2 + y^2 + z^2 - 9 = 0$$

$$\frac{\partial f}{\partial x} = 2x$$

$$\frac{\partial f}{\partial z} = 2z$$

$$\frac{\partial z}{\partial x} = \frac{-\dfrac{\partial f}{\partial x}}{\dfrac{\partial f}{\partial z}} = \frac{-2x}{2z} = -\frac{x}{z}$$

At the point $(1,2,2)$,

$$\frac{\partial z}{\partial x} = -\tfrac{1}{2}$$

7. TANGENT PLANE FUNCTION

Partial derivatives can be used to find the equation of a plane tangent to a three-dimensional surface defined by $f(x,y,z) = 0$ at some point, P_0.

$$T(x_0,y_0,z_0) = (x - x_0)\frac{\partial f(x,y,z)}{\partial x}\Big|_{P_0}$$

$$+ (y - y_0)\frac{\partial f(x,y,z)}{\partial y}\Big|_{P_0}$$

$$+ (z - z_0)\frac{\partial f(x,y,z)}{\partial z}\Big|_{P_0}$$

$$= 0 \qquad\qquad 9.36$$

The coefficients of x, y, and z are the same as the coefficients of $\mathbf{i}$, $\mathbf{j}$, and $\mathbf{k}$ of the normal vector at point P_0. (See Secs. 8.13 and 9.10.)

Example 9.9

What is the equation of the plane that is tangent to the surface defined by $f(x,y,z) = 4x^2 + y^2 - 16z = 0$ at the point $(2,4,2)$?

Solution

First, calculate the partial derivatives and substitute the coordinates of the point.

$$\frac{\partial f(x,y,z)}{\partial x}\Big|_{P_0} = 8x|_{(2,4,2)} = (8)(2) = 16$$

$$\frac{\partial f(x,y,z)}{\partial y}\Big|_{P_0} = 2y|_{(2,4,2)} = (2)(4) = 8$$

$$\frac{\partial f(x,y,z)}{\partial z}\Big|_{P_0} = -16|_{(2,4,2)} = -16$$

$$T(2,4,2) = (16)(x-2) + (8)(y-4) - (16)(z-2)$$

$$= 2x + y - 2z - 4$$

Next, substitute into Eq. 9.36.

$$2x + y - 2z - 4 = 0$$

8. GRADIENT VECTOR

The slope of a function is the change in one variable with respect to a distance in a chosen direction. Usually, the direction is parallel to a coordinate axis. However, the maximum slope at a point on a surface may not be in a direction parallel to one of the coordinate axes.

The *gradient vector function* $\nabla f(x,y,z)$ (pronounced "del f") gives the maximum rate of change of the function $f(x,y,z)$.

$$\nabla f(x,y,z) = \left(\frac{\partial f(x,y,z)}{\partial x}\right)\mathbf{i} + \left(\frac{\partial f(x,y,z)}{\partial y}\right)\mathbf{j}$$

$$+ \left(\frac{\partial f(x,y,z)}{\partial z}\right)\mathbf{k} \qquad 9.37$$

Example 9.10

A two-dimensional function is defined as

$$f(x,y) = 2x^2 - y^2 + 3x - y$$

(a) What is the gradient vector for this function? (b) What is the direction of the line passing through the point $(1,-2)$ that has a maximum slope? (c) What is the maximum slope at the point $(1,-2)$?

Solution

(a) It is necessary to calculate two partial derivatives in order to use Eq. 9.37.

$$\frac{\partial f(x,y)}{\partial x} = 4x + 3$$

$$\frac{\partial f(x,y)}{\partial y} = -2y - 1$$

$$\nabla f(x,y) = (4x+3)\mathbf{i} + (-2y-1)\mathbf{j}$$

(b) The direction of the line passing through $(1,-2)$ with maximum slope is found by inserting $x = 1$ and $y = -2$ into the gradient vector function.

$$\mathbf{V} = ((4)(1)+3)\mathbf{i} + ((-2)(-2)-1)\mathbf{j}$$

$$= 7\mathbf{i} + 3\mathbf{j}$$

(c) The magnitude of the slope is

$$|\mathbf{V}| = \sqrt{(7)^2 + (3)^2} = \sqrt{58} = 7.62$$

9. DIRECTIONAL DERIVATIVE

Unlike the gradient vector (covered in Sec. 9-8), which calculates the maximum rate of change of a function, the *directional derivative*, indicated by $\nabla_u f(x,y,z)$, $D_u f(x,y,z)$, or $f'_u(x,y,z)$, gives the rate of change in the direction of a given vector, $\mathbf{u}$ or $\mathbf{U}$. The subscript u implies that the direction vector is a unit vector, but it does not need to be, as only the direction cosines are calculated from it.

$$\nabla_u f(x,y,z) = \frac{\partial f(x,y,z)}{\partial x}\cos\alpha + \frac{\partial f(x,y,z)}{\partial y}\cos\beta$$

$$+ \frac{\partial f(x,y,z)}{\partial z}\cos\gamma \qquad 9.38$$

$$\mathbf{U} = U_x\mathbf{i} + U_y\mathbf{j} + U_z\mathbf{k} \qquad 9.39$$

$$\cos\alpha = \frac{U_x}{|\mathbf{U}|} = \frac{U_x}{\sqrt{U_x^2 + U_y^2 + U_z^2}} \qquad 9.40$$

$$\cos\beta = \frac{U_y}{|\mathbf{U}|} \qquad 9.41$$

$$\cos\gamma = \frac{U_z}{|\mathbf{U}|} \qquad 9.42$$

Example 9.11

What is the rate of change of $f(x,y) = 3x^2 + xy - 2y^2$ at the point $(1,-2)$ in the direction $4\mathbf{i} + 3\mathbf{j}$?

Solution

The direction cosines are given by Eqs. 9.40 and 9.41.

$$\cos\alpha = \frac{U_x}{|\mathbf{U}|} = \frac{4}{\sqrt{(4)^2 + (3)^2}} = 4/5$$

$$\cos\beta = \frac{U_y}{|\mathbf{U}|} = 3/5$$

The partial derivatives are

$$\frac{\partial f(x,y)}{\partial x} = 6x + y$$

$$\frac{\partial f(x,y)}{\partial y} = x - 4y$$

The directional derivative is given by Eq. 9.38.

$$\nabla_u f(x,y) = \left(\tfrac{4}{5}\right)(6x+y) + \left(\tfrac{3}{5}\right)(x-4y)$$

Substituting the given values of $x = 1$ and $y = -2$,

$$\nabla_u f(1,-2) = \left(\tfrac{4}{5}\right)((6)(1)-2) + \left(\tfrac{3}{5}\right)(1-(4)(-2))$$

$$= \tfrac{43}{5} = 8.6$$

10. NORMAL LINE VECTOR

Partial derivatives can be used to find the vector normal to a three-dimensional surface defined by $f(x,y,z) = 0$ at some point P_0. Notice that the coefficients of $\mathbf{i}$, $\mathbf{j}$, and $\mathbf{k}$ are the same as the coefficients of x, y, and z calculated for the equation of the tangent plane at point P_0. (See Secs. 8-13 and 9-7.)

$$\mathbf{N} = \frac{\partial f(x,y,z)}{\partial x}\bigg|_{P_0}\mathbf{i} + \frac{\partial f(x,y,z)}{\partial y}\bigg|_{P_0}\mathbf{j}$$

$$+ \frac{\partial f(x,y,z)}{\partial z}\bigg|_{P_0}\mathbf{k} \qquad 9.43$$

Example 9.12

What is the vector normal to the surface of $f(x,y,z) = 4x^2 + y^2 - 16z = 0$ at the point $(2,4,2)$?

Solution

The equation of the tangent plane at this point was calculated in Ex. 9.9 to be

$$T(2,4,2) = 2x + y - 2z - 4 = 0$$

A vector normal to the tangent plane through this point is

$$\mathbf{N} = 2\mathbf{i} + \mathbf{j} - 2\mathbf{k}$$

11. DIVERGENCE OF A VECTOR FIELD

The *divergence*, div $\mathbf{F}$, of a vector field $\mathbf{F}(x, y, z)$ is a scalar function defined by Eqs. 9.44 through 9.46.[4] The divergence of $\mathbf{F}$ can be interpreted as the *accumulation* of flux (i.e., a flowing substance) in a small region (i.e., at a point). One of the uses of the divergence is to determine whether flow (represented in direction and magnitude by $\mathbf{F}$) is compressible. Flow is incompressible if div $\mathbf{F} = 0$, since the substance is not accumulating.

$$\mathbf{F} = P(x, y, z)\mathbf{i} + Q(x, y, z)\mathbf{j} + R(x, y, z)\mathbf{k} \quad 9.44$$

$$\text{div } \mathbf{F} = \frac{\partial P}{\partial x} + \frac{\partial Q}{\partial y} + \frac{\partial R}{\partial z} \quad\quad 9.45$$

It may be easier to calculate the divergence from Eq. 9.46.

$$\text{div } \mathbf{F} = \mathbf{\nabla} \cdot \mathbf{F} \quad\quad 9.46$$

The vector del operator, $\mathbf{\nabla}$, is defined as

$$\mathbf{\nabla} = \frac{\partial}{\partial x}\mathbf{i} + \frac{\partial}{\partial y}\mathbf{j} + \frac{\partial}{\partial z}\mathbf{k} \quad\quad 9.47$$

If there is no divergence, then the dot product calculated in Eq. 9.46 is zero.

Example 9.13

Calculate the divergence of the following vector function.

$$\mathbf{F}(x, y, z) = xz\mathbf{i} + e^x y\mathbf{j} + 7x^3 y\mathbf{k}$$

Solution

From Eq. 9.45,

$$\text{div } \mathbf{F} = \frac{\partial}{\partial x}(xz) + \frac{\partial}{\partial y}(e^x y) + \frac{\partial}{\partial z}(7x^3 y)$$

$$= z + e^x + 0 = z + e^x$$

12. CURL OF A VECTOR FIELD

The *curl*, curl $\mathbf{F}$, of a vector field $\mathbf{F}(x, y, z)$ is a vector field defined by Eqs. 9.51 and 9.52. The curl $\mathbf{F}$ can be interpreted as the *vorticity* per unit area of flux (i.e., a flowing substance) in a small region (i.e., at a point). One of the uses of the curl is to determine whether flow (represented in direction and magnitude by $\mathbf{F}$) is rotational. Flow is irrotational if curl $\mathbf{F} = 0$.[5]

---

[4]Notice that a bold letter, $\mathbf{F}$, is used to indicate that the vector is a function of x, y, and z.

[5]If the velocity vector is $\mathbf{V}$, then the *vorticity* is

$$\boldsymbol{\omega} = \mathbf{\nabla} \times \mathbf{V} = \omega_x \mathbf{i} + \omega_y \mathbf{j} + \omega_z \mathbf{k} \quad\quad 9.49$$

The *circulation* is the line integral of the velocity $\mathbf{V}$ along a closed curve.

$$\Gamma = \oint \mathbf{V} \cdot d\mathbf{s} = \oint \boldsymbol{\omega} \cdot d\mathbf{A} \quad\quad 9.50$$

$$\mathbf{F} = P(x, y, z)\mathbf{i} + Q(x, y, z)\mathbf{j} + R(x, y, z)\mathbf{k} \quad 9.48$$

$$\text{curl } \mathbf{F} = \left(\frac{\partial R}{\partial y} - \frac{\partial Q}{\partial z}\right)\mathbf{i} + \left(\frac{\partial P}{\partial z} - \frac{\partial R}{\partial x}\right)\mathbf{j}$$

$$+ \left(\frac{\partial Q}{\partial x} - \frac{\partial P}{\partial y}\right)\mathbf{k} \quad\quad 9.51$$

It may be easier to calculate the curl from Eq. 9.52. (The vector del operator, $\mathbf{\nabla}$, was defined in Eq. 9.47.)

$$\text{curl } \mathbf{F} = \mathbf{\nabla} \times \mathbf{F}$$

$$= \begin{vmatrix} \mathbf{i} & \mathbf{j} & \mathbf{k} \\ \dfrac{\partial}{\partial x} & \dfrac{\partial}{\partial y} & \dfrac{\partial}{\partial z} \\ P(x, y, z) & Q(x, y, z) & R(x, y, z) \end{vmatrix} \quad 9.52$$

Example 9.14

Calculate the curl of the following vector function.

$$\mathbf{F}(x, y, z) = 3x^2\mathbf{i} + 7e^x y\mathbf{j}$$

Solution

Using Eq. 9.52,

$$\text{curl } \mathbf{F} = \begin{vmatrix} \mathbf{i} & \mathbf{j} & \mathbf{k} \\ \dfrac{\partial}{\partial x} & \dfrac{\partial}{\partial y} & \dfrac{\partial}{\partial z} \\ 3x^2 & 7e^x y & 0 \end{vmatrix}$$

Expand the determinant across the top row.

$$\mathbf{i}\left[\frac{\partial}{\partial y}(0) - \frac{\partial}{\partial z}(7e^x y)\right] - \mathbf{j}\left[\frac{\partial}{\partial x}(0) - \frac{\partial}{\partial z}(3x^2)\right]$$

$$+ \mathbf{k}\left[\frac{\partial}{\partial x}(7e^x y) - \frac{\partial}{\partial y}(3x^2)\right]$$

$$= \mathbf{i}(0 - 0) - \mathbf{j}(0 - 0) + \mathbf{k}(7e^x y - 0) = 7e^x y\mathbf{k}$$

13. TAYLOR'S FORMULA

Taylor's formula (series) can be used to expand a function around a point (i.e., approximate the function at one point based on the function's value at another point). The approximation consists of a series, each term composed of a derivative of the original function and a polynomial. Using Taylor's formula requires that the original function be continuous in the interval $[a, b]$ and have the required number of derivatives. To expand a function, $f(x)$, around a point, a, in order to obtain $f(b)$, Taylor's formula is[6]

$$f(b) = f(a) + \frac{f'(a)}{1!}(b - a) + \frac{f''(a)}{2!}(b - a)^2 + \cdots$$

$$+ \frac{f^n(a)}{n!}(b - a)^n + R_n(b) \qquad 9.53$$

In Eq. 9.53, the expression f^n designates the nth derivative of the function $f(x)$. To be a useful approximation, point a must satisfy two requirements: It must be relatively close to point b, and the function and its derivatives must be known or easy to calculate. The last term, $R_n(b)$, is the uncalculated remainder after n derivatives. It is the difference between the exact and approximate values. By using enough terms, the remainder can be made arbitrarily small. That is, $R_n(b)$ approaches zero as n approaches infinity.

It can be shown that the remainder term can be calculated from Eq. 9.54, where c is some number in the interval $[a, b]$. With certain functions, the constant c can be completely determined. In most cases, however, it is possible only to calculate an upper bound on the remainder from Eq. 9.55. M_n is the maximum (positive) value of $f^{(n+1)}(x)$ on the interval $[a, b]$.

$$R_n(b) = \frac{f^{n+1}(c)}{(n+1)!}(b - a)^{n+1} \qquad 9.54$$

$$|R_n(b)| \leq M_n \frac{|(b - a)^{n+1}|}{(n+1)!} \qquad 9.55$$

14. COMMON SERIES APPROXIMATIONS

Taylor's formulas can be used (by expanding about $a = 0$) to derive the following series approximations.

$$\sin x \approx x - \frac{x^3}{3!} + \frac{x^5}{5!} - \frac{x^7}{7!} + \cdots$$

$$+ (-1)^n \frac{x^{2n+1}}{(2n+1)!} \qquad 9.56$$

$$\cos x \approx 1 - \frac{x^2}{2!} + \frac{x^4}{4!} - \frac{x^6}{6!} + \cdots + (-1)^n \frac{x^{2n}}{(2n)!} \qquad 9.57$$

$$\sinh x \approx x + \frac{x^3}{3!} + \frac{x^5}{5!} + \frac{x^7}{7!} + \cdots + \frac{x^{2n+1}}{(2n+1)!} \qquad 9.58$$

$$\cosh x \approx 1 + \frac{x^2}{2!} + \frac{x^4}{4!} + \frac{x^6}{6!} + \cdots + \frac{x^{2n}}{(2n)!} \qquad 9.59$$

$$e^x \approx 1 + x + \frac{x^2}{2!} + \frac{x^3}{3!} + \cdots + \frac{x^n}{n!} \qquad 9.60$$

$$\ln(1 + x) \approx x - \frac{x^2}{2} + \frac{x^3}{3} - \frac{x^4}{4} + \cdots + (-1)^{n+1} \frac{x^n}{n} \qquad 9.61$$

$$\frac{1}{1 - x} \approx 1 + x + x^2 + x^3 + \cdots + x^n \qquad 9.62$$

---

[6]If $a = 0$, Eq. 9.53 is known as the *Maclaurin series*.

10

Integral Calculus

1. INTEGRATION

Integration is the inverse operation of differentiation. For that reason, *indefinite integrals* are sometimes referred to as *antiderivatives*. [1] Although expressions can be functions of several variables, integrals can only be taken with respect to one variable at a time. The *differential term* (dx in Eq. 10.1) indicates that variable. In Eq. 10.1, the function $f'(x)$ is the *integrand*, and x is the variable of integration.

$$\int f'(x)\,dx = f(x) + C \qquad 10.1$$

While most of a function, $f(x)$, can be "recovered" through integration of its derivative, $f'(x)$, a constant term will be lost. This is because the derivative of a constant term vanishes (i.e., is zero), leaving nothing to recover from. A *constant of integration*, C, is added to the integral to recognize the possibility of such a term.

2. ELEMENTARY OPERATIONS

Equations 10.2 through 10.8 summarize the elementary integration operations on polynomials and exponentials.[2] Equations 10.2 and 10.3 are particularly useful.

---

[1] The difference between an indefinite and definite integral (covered in Sec. 7) is simple: An *indefinite integral* is a function, while a *definite integral* is a number.

[2] More extensive listings, known as *tables of integrals*, are widely available.

(C and k represent constants. $f(x)$ and $g(x)$ are functions of x.)

$$\int k\,dx = kx + C \qquad 10.2$$

$$\int x^m\,dx = \frac{x^{m+1}}{m+1} + C \qquad [m \neq -1] \qquad 10.3$$

$$\int \frac{1}{x}\,dx = \ln|x| + C \qquad 10.4$$

$$\int e^{kx}\,dx = \frac{e^{kx}}{k} + C \qquad 10.5$$

$$\int xe^{kx}\,dx = \frac{e^{kx}(kx-1)}{k^2} + C \qquad 10.6$$

$$\int k^{ax}\,dx = \frac{k^{ax}}{a\,\ln k} + C \qquad 10.7$$

$$\int \ln x\,dx = x\,\ln x - x + C \qquad 10.8$$

Equations 10.9 through 10.20 summarize the elementary integration operations on transcendental functions.

$$\int \sin x\,dx = -\cos x + C \qquad 10.9$$

$$\int \cos x\,dx = \sin x + C \qquad 10.10$$

$$\int \tan x\,dx = \ln|\sec x| + C \qquad 10.11$$

$$\int \cot x\,dx = \ln|\sin x| + C \qquad 10.12$$

$$\int \sec x\,dx = \ln|(\sec x + \tan x)| + C \qquad 10.13$$

$$\int \csc x\,dx = \ln|(\csc x - \cot x)| + C \qquad 10.14$$

$$\int \frac{dx}{k^2 + x^2} = \frac{1}{k}\arctan\frac{x}{k} + C \qquad 10.15$$

$$\int \frac{dx}{\sqrt{k^2 - x^2}} = \arcsin\frac{x}{k} + C \qquad [k^2 > x^2] \qquad 10.16$$

$$\int \frac{dx}{x\sqrt{x^2 - k^2}} = \frac{1}{k}\operatorname{arcsec}\frac{x}{k} + C \qquad [x^2 > k^2] \qquad 10.17$$

$$\int \sin^2 x\,dx = \tfrac{1}{2}x - \tfrac{1}{4}\sin 2x + C \qquad 10.18$$

$$\int \cos^2 x\,dx = \tfrac{1}{2}x + \tfrac{1}{4}\sin 2x + C \qquad 10.19$$

$$\int \tan^2 x\,dx = \tan x - x + C \qquad 10.20$$

Equations 10.21 through 10.26 summarize the elementary integration operations on hyperbolic transcendental functions. Integrals of hyperbolic functions are not completely analogous to those of the regular transcendental functions.

$$\int \sinh x \, dx = \cosh x + C \qquad \text{10.21}$$

$$\int \cosh x \, dx = \sinh x + C \qquad \text{10.22}$$

$$\int \tanh x \, dx = \ln |\cosh x| + C \qquad \text{10.23}$$

$$\int \coth x \, dx = \ln |\sinh x| + C \qquad \text{10.24}$$

$$\int \operatorname{sech} x \, dx = \arctan(\sinh x) + C \qquad \text{10.25}$$

$$\int \operatorname{csch} x \, dx = \ln \left| \tanh \left(\frac{x}{2} \right) \right| + C \qquad \text{10.26}$$

Equations 10.27 through 10.31 summarize the elementary integration operations on functions and combinations of functions.

$$\int k f(x) dx = k \int f(x) dx \qquad \text{10.27}$$

$$\int (f(x) + g(x)) dx = \int f(x) dx + \int g(x) dx \qquad \text{10.28}$$

$$\int \frac{f'(x)}{f(x)} dx = \ln |f(x)| + C \qquad \text{10.29}$$

$$\int f(x) dg(x) = f(x) \int dg(x) - \int g(x) df(x) + C$$
$$= f(x) g(x) - \int g(x) df(x) + C \quad \text{10.30}$$

Example 10.1

Find the integral with respect to x of

$$3x^2 + \tfrac{1}{3} x - 7 = 0$$

Solution

This is a polynomial function, and Eq. 10.3 can be applied to each of the three terms.

$$\int \left(3x^2 + \tfrac{1}{3} x - 7\right) dx = x^3 + \tfrac{1}{6} x^2 - 7x + C$$

3. INTEGRATION BY PARTS

Equation 10.30, repeated here, is known as *integration by parts*. $f(x)$ and $g(x)$ are functions. The use of this method is illustrated by Ex. 10.2.

$$\int f(x) dg(x) = f(x) g(x) - \int g(x) df(x) + C \qquad \text{10.31}$$

Example 10.2

Find the following integral.

$$\int x^2 e^x \, dx$$

Solution

$x^2 e^x$ is factored into two parts so that integration by parts can be used.

$$f(x) = x^2$$
$$dg(x) = e^x \, dx$$
$$df(x) = 2x \, dx$$
$$g(x) = \int dg(x) = \int e^x \, dx = e^x$$

From Eq. 10.31, disregarding the constant of integration (which cannot be evaluated),

$$\int f(x) dg(x) = f(x) g(x) - \int g(x) df(x)$$
$$\int x^2 e^x \, dx = x^2 e^x - \int e^x (2x) dx$$

The second term is also factored into two parts, and integration by parts is used again. This time,

$$f(x) = x$$
$$dg(x) = e^x \, dx$$
$$df(x) = dx$$
$$g(x) = \int dg(x) = \int e^x \, dx = e^x$$

From Eq. 10.31,

$$\int 2x e^x \, dx = 2 \int x e^x \, dx$$
$$= (2) \left(x e^x - \int e^x \, dx \right)$$
$$= (2)(x e^x - e^x)$$

Then, the complete integral is

$$\int x^2 e^x \, dx = x^2 e^x - (2)(x e^x - e^x) + C$$
$$= e^x (x^2 - 2x + 2) + C$$

4. SEPARATION OF TERMS

Equation 10.28 shows that the integral of a sum of terms is equal to a sum of integrals. This technique is known as *separation of terms*. In many cases, terms are easily separated. In other cases, the technique of *partial fractions* (see Sec. 4-16) can be used to obtain individual terms. These techniques are illustrated by Exs. 10.3 and 10.4.

Example 10.3

Find the following integral.

$$\int \frac{(2x^2 + 3)^2}{x} dx$$

Solution

$$\int \frac{(2x^2 + 3)^2}{x} dx = \int \frac{4x^4 + 12x^2 + 9}{x} dx$$

$$= \int \left(4x^3 + 12x + \frac{9}{x} \right) dx$$

$$= x^4 + 6x^2 + 9 \ln |x| + C$$

Example 10.4

Find the following integral.

$$\int \frac{3x + 2}{3x - 2} dx$$

Solution

The integrand is larger than 1, so use long division to simplify it.

$$
\begin{array}{r}
1 \text{ rem } \dfrac{4}{3x-2} \\[4pt]
3x - 2 \overline{\smash{\big)}\ 3x+2} \\[2pt]
\underline{3x-2} \\[2pt]
4 \text{ remainder}
\end{array}
$$

$$\int \frac{3x + 2}{3x - 2} dx = \int \left(1 + \frac{4}{3x - 2} \right) dx$$

$$= \int dx + \int \frac{4}{3x - 2} dx$$

$$= x + \tfrac{4}{3} \ln |(3x - 2)| + C$$

5. DOUBLE AND HIGHER-ORDER INTEGRALS

A function can be successively integrated. (This is analogous to successive differentiation.) A function that is integrated twice is known as a *double integral*; if integrated three times, it is a *triple integral*; and so on. Double and triple integrals are used to calculate areas and volumes, respectively.

The successive integrations do not need to be with respect to the same variable. Variables not included in the integration are treated as constants.

There are several notations used for a multiple integral, particularly when the product of length differentials represents a differential area or volume. A double integral (i.e., two successive integrations) can be represented by one of the following notations.

$$\iint f(x,y)dx\, dy, \quad \int_{R^2} f(x,y)dx\, dy,$$

$$\text{or } \iint_{R^2} f(x,y)dA$$

A triple integral can be represented by one of the following notations.

$$\iiint f(x,y,z)dx\, dy\, dz,$$

$$\int_{R^3} f(x,y,z)dx\, dy\, dz,$$

$$\text{or } \iiint_{R^3} f(x,y,z)dV$$

Example 10.5

Find the following double integral.

$$\iint (x^2 + y^3 x)dx\, dy$$

Solution

$$\int (x^2 + y^3 x)dx = \tfrac{1}{3}x^3 + \tfrac{1}{2}y^3 x^2 + C_1$$

$$\int \left(\tfrac{1}{3}x^3 + \tfrac{1}{2}y^3 x^2 + C_1 \right) dy = \tfrac{1}{3}yx^3 + \tfrac{1}{8}y^4 x^2 + C_1 y + C_2$$

So,

$$\iint (x^2 + y^3 x)dx\, dy = \tfrac{1}{3}yx^3 + \tfrac{1}{8}y^4 x^2 + C_1 y + C_2$$

6. INITIAL VALUES

The constant of integration, C, can be found only if the value of the function $f(x)$ is known for some value of x_0. The value $f(x_0)$ is known as an *initial value*. To completely define a function, as many initial values, $f(x_0), f'(x_0), f''(x_0)$, and so on, as there are integrations are needed.

Example 10.6

It is known that $f(x) = 4$ when $x = 2$ (i.e., the initial condition is $f(2) = 4$). Find the original function.

$$\int (3x^3 - 7x)dx$$

Solution

The function is

$$f(x) = \int (3x^3 - 7x)dx = \tfrac{3}{4}x^4 - \tfrac{7}{2}x^2 + C$$

Substituting the initial value determines that C equals 6.

$$4 = \left(\tfrac{3}{4}\right)(2)^4 - \left(\tfrac{7}{2}\right)(2)^2 + C$$
$$4 = 12 - 14 + C$$
$$C = 6$$

The function is

$$f(x) = \tfrac{3}{4}x^4 - \tfrac{7}{2}x^2 + 6$$

7. DEFINITE INTEGRALS

A *definite integral* is restricted to a specific range of the independent variable. (Unrestricted integrals of the types shown in all preceding examples are known as *indefinite integrals*.) A definite integral restricted to the region bounded by *lower* and *upper limits* (also known as *bounds*), x_1 and x_2, is written as

$$\int_{x_1}^{x_2} f(x)dx$$

Equation 10.32 indicates how definite integrals are evaluated. It is known as the *fundamental theorem of calculus*.

$$\int_{x_1}^{x_2} f'(x)dx = f(x)\Big|_{x_1}^{x_2} = f(x_2) - f(x_1) \qquad 10.32$$

A common use of a definite integral is the calculation of work performed by a force, F, that moves from position x_1 to x_2.

$$W = \int_{x_1}^{x_2} F\, dx \qquad 10.33$$

Example 10.7

Evaluate the following definite integral.

$$\int_{\frac{\pi}{4}}^{\frac{\pi}{3}} \sin x\, dx$$

Solution

From Eq. 10.32,

$$\int_{\frac{\pi}{4}}^{\frac{\pi}{3}} \sin x\, dx = \left[-\cos x\right]_{\frac{\pi}{4}}^{\frac{\pi}{3}}$$
$$= -\cos \tfrac{\pi}{3} - \left(-\cos \tfrac{\pi}{4}\right)$$
$$= -0.5 - (-0.707) = 0.207$$

8. AVERAGE VALUE

The average value of a function $f(x)$ that is integrable over the interval $[a, b]$ is

$$\text{average value} = \frac{1}{b-a}\int_a^b f(x)dx \qquad 10.34$$

9. AREA

Equation 10.35 calculates the area, A, bounded by $x = a$, $x = b$, $f_1(x)$ above and $f_2(x)$ below. ($f_2(x) = 0$ if the area is bounded by the x-axis.) This is illustrated in Fig. 10.1.

$$A = \int_a^b (f_1(x) - f_2(x))dx \qquad 10.35$$

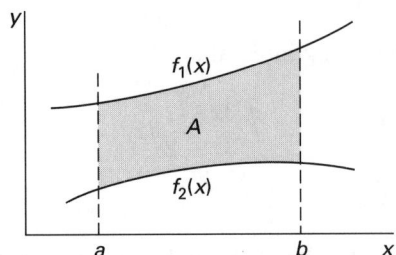

Figure 10.1 *Area Between Two Curves*

Example 10.8

Find the area between the x-axis and the parabola $y = x^2$ in the interval $[0,4]$.

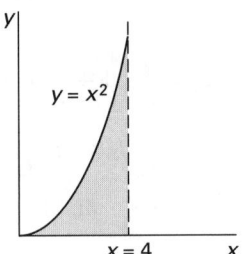

Solution

Referring to Eq. 10.35,

$$f_1(x) = x^2$$
$$f_2(x) = 0$$
$$A = \int_a^b (f_1(x) - f_2(x))dx = \int_0^4 x^2 dx$$
$$= \left[\frac{x^3}{3}\right]_0^4 = 64/3$$

10. ARC LENGTH

Equation 10.36 gives the length of a curve defined by $f(x)$ whose derivative exists in the interval $[a, b]$.

$$\text{length} = \int_a^b \sqrt{1 + (f'(x))^2}\, dx \qquad 10.36$$

11. PAPPUS' THEOREMS[3]

The first and second theorems of Pappus are:[4]

- *First Theorem*: Given a curve, C, that does not intersect the y-axis, the area of the *surface of revolution* generated by revolving C around the y-axis is equal to the product of the length of the curve and the circumference of the circle traced by the centroid of curve C.

$$A = \text{length} \times \text{circumference} = \text{length} \times 2\pi \times \text{radius} \qquad 10.37$$

- *Second Theorem*: Given a plane region, R, that does not intersect the y-axis, the *volume of revolution* generated by revolving R around the y-axis is equal to the product of the area and the circumference of the circle traced by the centroid of area R.

$$V = \text{area} \times \text{circumference} = \text{area} \times 2\pi \times \text{radius} \qquad 10.38$$

12. SURFACE OF REVOLUTION

The surface area obtained by rotating $f(x)$ about the x-axis is

$$A = 2\pi \int_{x=a}^{x=b} f(x)\sqrt{1 + (f'(x))^2}\, dx \qquad 10.39$$

The surface area obtained by rotating $f(y)$ about the y-axis is

$$A = 2\pi \int_{y=c}^{y=d} f(y)\sqrt{1 + (f'(y))^2}\, dy \qquad 10.40$$

Example 10.9

The curve $f(x) = \tfrac{1}{2}x$ over the region $x = [0, 4]$ is rotated about the x-axis. What is the surface of revolution?

---

[3]This section is an introduction to surfaces and volumes of revolution but does not involve integration.
[4]Some authorities call the first theorem the second and vice versa.

Solution

The surface of revolution is

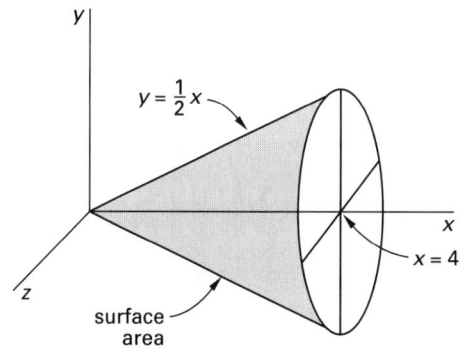

Since $f(x) = \tfrac{1}{2}x$, $f'(x) = \tfrac{1}{2}$. From Eq. 10.39, the area is

$$\begin{aligned}
A &= 2\pi \int_{x=a}^{x=b} f(x)\sqrt{1 + (f'(x))^2}\, dx \\
&= 2\pi \int_0^4 \tfrac{1}{2}x\sqrt{1 + \left(\tfrac{1}{2}\right)^2}\, dx \\
&= \frac{\sqrt{5}}{2}\pi \int_0^4 x\, dx \\
&= \frac{\sqrt{5}}{2}\pi \left[\frac{x^2}{2}\right]_0^4 \\
&= \frac{\sqrt{5}}{2}\pi \left(\frac{(4)^2 - (0)^2}{2}\right) \\
&= 4\sqrt{5}\pi
\end{aligned}$$

13. VOLUME OF REVOLUTION

The volume obtained by rotating $f(x)$ about the x-axis is given by Eq. 10.41. $f^2(x)$ is the square of the function, not the second derivative.

$$V = \pi \int_{x=a}^{x=b} f^2(x)\, dx \qquad 10.41$$

The volume obtained by rotating $f(x)$ about the y-axis is

$$V = 2\pi \int_{x=a}^{x=b} x f(x)\, dx \qquad 10.42$$

Example 10.10

The curve $f(x) = x^2$ over the region $x = [0, 4]$ is rotated about the x-axis. What is the volume of revolution?

Solution

The volume of revolution is

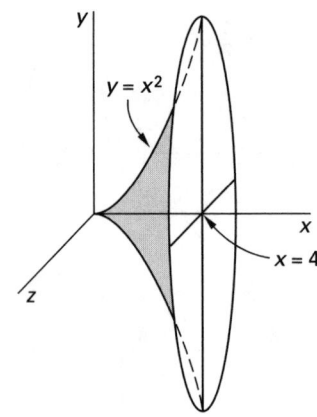

$$V = \pi \int_a^b f^2(x)dx = \pi \int_0^4 (x^2)^2 dx$$

$$= \pi \left[\frac{x^5}{5}\right]_0^4 = \pi \left(\frac{1024}{5} - 0\right) = 204.8\pi$$

14. MOMENTS OF A FUNCTION

The *first moment of a function* is a concept used in finding centroids and centers of gravity. Equations 10.43 and 10.44 are for one- and two-dimensional problems, respectively. It is the exponent of x (1 in this case) that gives the moment its name.

$$\text{first moment} = \int xf(x)dx \qquad 10.43$$

$$\text{first moment} = \iint xf(x,y)dx\,dy \qquad 10.44$$

The *second moment of a function* is a concept used in finding moments of inertia with respect to an axis. Equations 10.45 and 10.46 are for two- and three-dimensional problems, respectively. Second moments with respect to other axes are analogous.

$$(\text{second moment})_x = \iint y^2 f(x,y)dy\,dx \qquad 10.45$$

$$(\text{second moment})_x = \iiint (y^2 + z^2) f(x,y,z)dy\,dz\,dx$$
$$10.46$$

15. FOURIER SERIES

Any periodic waveform can be written as the sum of an infinite number of sinusoidal terms, known as *harmonic terms* (i.e., an infinite series). Such a sum of terms is known as a *Fourier series*, and the process of finding

the terms is *Fourier analysis*. (Extracting the original waveform from the series is known as *Fourier inversion*.) Since most series converge rapidly, it is possible to obtain a good approximation to the original waveform with a limited number of sinusoidal terms.

Fourier's theorem is Eq. 10.47.[5] The object of a Fourier analysis is to determine the coefficients a_n and b_n. The constant a_0 can often be determined by inspection since it is the average value of the waveform.

$$f(t) = \tfrac{1}{2}a_0 + a_1\cos\omega t + a_2\cos 2\omega t + \cdots$$
$$+ b_1\sin\omega t + b_2\sin 2\omega t + \cdots \qquad 10.47$$

ω is the *natural (fundamental) frequency* of the waveform. It depends on the actual waveform period, T.

$$\omega = \frac{2\pi}{T} \qquad 10.48$$

To simplify the analysis, the time domain can be normalized to the radian scale. The normalized scale is obtained by dividing all frequencies by ω. Then the Fourier series becomes

$$f(t) = \tfrac{1}{2}a_0 + a_1\cos t + a_2\cos 2t + \cdots$$
$$+ b_1\sin t + b_2\sin 2t + \cdots \qquad 10.49$$

The coefficients a_n and b_n are found from the following relationships.

$$a_0 = \frac{1}{\pi}\int_0^{2\pi} f(t)dt$$

$$= \frac{2}{T}\int_0^T f(t)dt \qquad 10.50$$

$$a_n = \frac{1}{\pi}\int_0^{2\pi} f(t)\cos nt\,dt$$

$$= \frac{2}{T}\int_0^T f(t)\cos nt\,dt \qquad [n \geq 1] \qquad 10.51$$

$$b_n = \frac{1}{\pi}\int_0^{2\pi} f(t)\sin nt\,dt$$

$$= \frac{2}{T}\int_0^T f(t)\sin nt\,dt \qquad [n \geq 1] \qquad 10.52$$

While Eqs. 10.51 and 10.52 are always valid, the work of integrating and finding a_n and b_n can be greatly simplified if the waveform is recognized as being symmetrical. Table 10.1 summarizes the simplifications.

---

[5]The independent variable used in this section is t, since Fourier analysis is most frequently used in the time domain.

Table 10.1 Fourier Analysis Simplifications for Symmetrical Waveforms

	even symmetry $f(-t) = f(t)$	odd symmetry $f(-t) = -f(t)$
	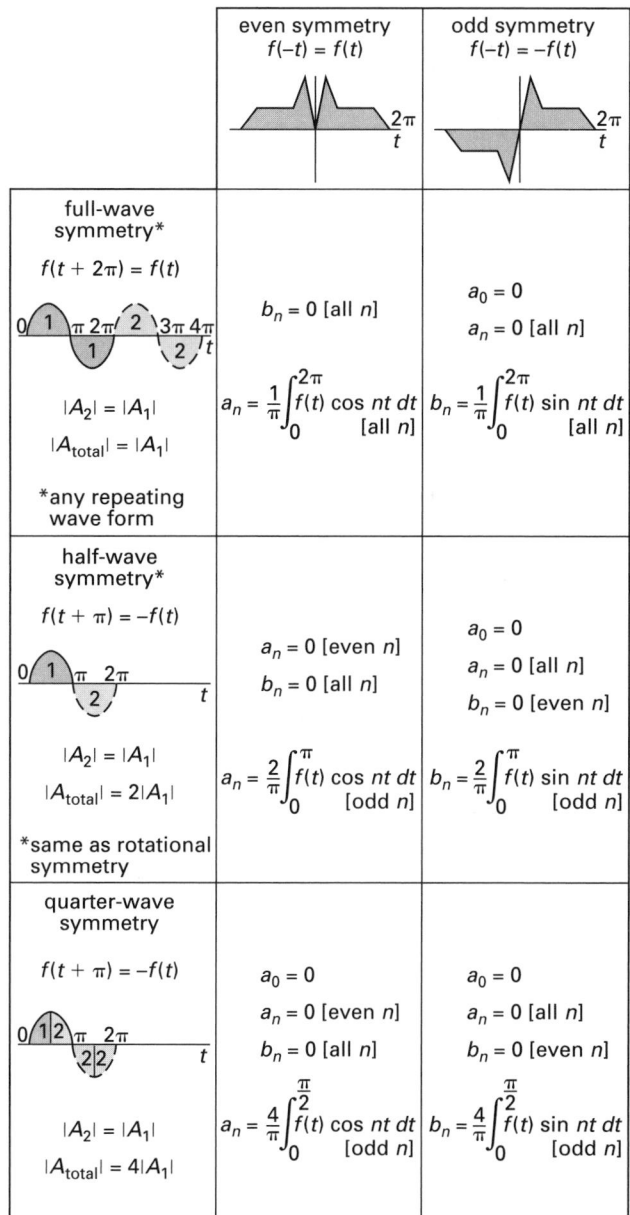	
full-wave symmetry* $f(t + 2\pi) = f(t)$ $\mid A_2 \mid = \mid A_1 \mid$ $\mid A_{total} \mid = \mid A_1 \mid$ *any repeating wave form	$b_n = 0$ [all n] $a_n = \frac{1}{\pi}\int_0^{2\pi} f(t)\cos nt \, dt$ [all n]	$a_0 = 0$ $a_n = 0$ [all n] $b_n = \frac{1}{\pi}\int_0^{2\pi} f(t)\sin nt \, dt$ [all n]
half-wave symmetry* $f(t + \pi) = -f(t)$ $\mid A_2 \mid = \mid A_1 \mid$ $\mid A_{total} \mid = 2\mid A_1 \mid$ *same as rotational symmetry	$a_n = 0$ [even n] $b_n = 0$ [all n] $a_n = \frac{2}{\pi}\int_0^{\pi} f(t)\cos nt \, dt$ [odd n]	$a_0 = 0$ $a_n = 0$ [all n] $b_n = 0$ [even n] $b_n = \frac{2}{\pi}\int_0^{\pi} f(t)\sin nt \, dt$ [odd n]
quarter-wave symmetry $f(t + \pi) = -f(t)$ $\mid A_2 \mid = \mid A_1 \mid$ $\mid A_{total} \mid = 4\mid A_1 \mid$	$a_0 = 0$ $a_n = 0$ [even n] $b_n = 0$ [all n] $a_n = \frac{4}{\pi}\int_0^{\frac{\pi}{2}} f(t)\cos nt \, dt$ [odd n]	$a_0 = 0$ $a_n = 0$ [all n] $b_n = 0$ [even n] $b_n = \frac{4}{\pi}\int_0^{\frac{\pi}{2}} f(t)\sin nt \, dt$ [odd n]

Solution

From Eq. 10.50,

$$a_0 = \frac{1}{\pi}\int_0^{\pi}(1)\,dt + \frac{1}{\pi}\int_{\pi}^{2\pi}(0)\,dt = 1$$

This value of $(1/2)a_0 = 1/2$ corresponds to the average value of $f(t)$. It could have been found by observation.

$$a_1 = \frac{1}{\pi}\int_0^{\pi}(1)\cos t \, dt + \frac{1}{\pi}\int_{\pi}^{2\pi}(0)\cos t \, dt$$

$$= \frac{1}{\pi}\Big[\sin t\Big]_0^{\pi} + 0 = 0$$

In general,

$$a_n = \frac{1}{\pi}\left[\frac{\sin nt}{n}\right]_0^{\pi} = 0$$

$$b_1 = \frac{1}{\pi}\int_0^{\pi}(1)\sin t \, dt + \frac{1}{\pi}\int_{\pi}^{2\pi}(0)\sin t \, dt$$

$$= \frac{1}{\pi}\Big[-\cos t\Big]_0^{\pi} = \frac{2}{\pi}$$

In general,

$$b_n = \frac{1}{\pi}\left[\frac{-\cos nt}{n}\right]_0^{\pi} = \begin{cases} 0 & \text{for } n \text{ even} \\ \dfrac{2}{\pi n} & \text{for } n \text{ odd} \end{cases}$$

The series is

$$f(t) = \frac{1}{2} + \frac{2}{\pi}\left[\sin t + \frac{1}{3}\sin 3t + \frac{1}{5}\sin 5t + \cdots\right]$$

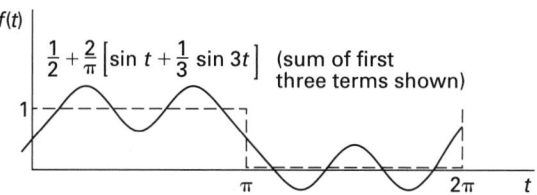

Example 10.11

Find the first four terms of a Fourier series that approximates the repetitive step function illustrated.

$$f(t) = \begin{cases} 1 & 0 < t < \pi \\ 0 & \pi < t < 2\pi \end{cases}$$

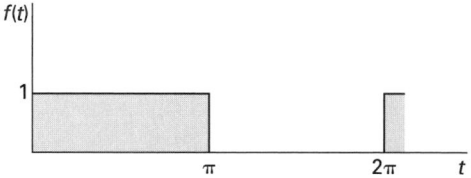

16. FAST FOURIER TRANSFORMS

Many mathematical operations are needed to implement a true Fourier transform. While the terms of a Fourier series might be slowly derived by integration, a faster method is needed to analyze real-time data. The *fast Fourier transform* (FFT) is a computer algorithm implemented in *spectrum analyzers (signal analyzers or FFT analyzers)* and replaces integration and multiplication operations with table look-ups and additions.[6]

---

[6] *Spectrum analysis,* also known as *frequency analysis, signature analysis,* and *time-series analysis,* develops a relationship (usually graphical) between some property (e.g., amplitude or phase shift) versus frequency.

Since the complexity of the transform is reduced, the transformation occurs more quickly, enabling efficient analysis of waveforms with little or no periodicity.[7]

Using a spectrum analyzer requires choosing the frequency band (e.g., 0 to 20 kHz) to be monitored. (This step automatically selects the sampling period. The lower the frequencies sampled, the longer the sampling period.) If they are not fixed by the analyzer, the numbers of time-dependent input variable samples (e.g., 1024) and frequency-dependent output variable values (e.g., 400) are chosen.[8] There are half as many frequency lines as data points because each line contains two pieces of information—real (amplitude) and imaginary (phase). The *resolution* of the resulting frequency analysis is

$$\text{resolution} = \frac{\text{frequency bandwidth}}{\text{no. of output variable values}} \qquad 10.53$$

17. INTEGRAL FUNCTIONS

Integrals that cannot be evaluated as finite combinations of elementary functions are called *integral functions*. These functions are evaluated by series expansion. Some of the more common functions are listed as follows.[9],[10]

- *integral sine function*:

$$\text{Si}(x) = \int_0^x \frac{\sin x}{x}dx$$

$$= x - \frac{x^3}{3\cdot 3!} + \frac{x^5}{5\cdot 5!} - \frac{x^7}{7\cdot 7!} + \cdots \qquad 10.54$$

- *integral cosine function*:

$$\text{Ci}(x) = \int_{-\infty}^x \frac{\cos x}{x}dx = -\int_x^\infty \frac{\cos x}{x}dx$$

$$= C_E + \ln\ x - \frac{x^2}{2\cdot 2!} + \frac{x^4}{4\cdot 4!} - \cdots \qquad 10.55$$

- *integral exponential function*:

$$\text{Ei}(x) = \int_{-\infty}^x \frac{e^x}{x}dx = -\int_{-x}^\infty \frac{e^{-x}}{x}dx$$

$$= C_E + \ln\ x + x + \frac{x^2}{2\cdot 2!} + \frac{x^3}{3\cdot 3!} + \cdots \qquad 10.56$$

- *error function*:

$$\text{erf}(x) = \frac{2}{\sqrt{\pi}}\int_0^x e^{-x^2}dx$$

$$= \left(\frac{2}{\sqrt{\pi}}\right)\left(\frac{x}{1\cdot 0!} - \frac{x^3}{3\cdot 1!} + \frac{x^5}{5\cdot 2!} - \frac{x^7}{7\cdot 3!} + \cdots\right)$$

$$10.57$$

- *Fresnel integrals*:

$$C(x) = \int_0^x \cos(t^2)dt$$

$$10.58$$

$$S(x) = \int_0^x \sin(t^2)dt$$

$$10.59$$

---

[7]Hours and days of manual computations are compressed into milliseconds.

[8]Two samples per time-dependent cycle (at the maximum frequency) is the lower theoretical limit for sampling, but the practical minimum rate is approximately 2.5 samples per cycle. This will ensure that *alias components* (i.e., low-level frequency signals) do not show up in the frequency band of interest.

[9]Other integral functions include the gamma function and elliptic integral.

[10]C_E in Eqs. 10.55 and 10.56 is *Euler's constant*.

$$C_E = \int_{+\infty}^0 e^{-x}\ln x\,dx$$

$$= \lim_{m\to\infty}\left(1 + \tfrac{1}{2} + \tfrac{1}{3} + \cdots + \frac{1}{m} - \ln m\right)$$

$$= 0.577215665$$

11 Differential Equations

1. TYPES OF DIFFERENTIAL EQUATIONS

A *differential equation* is a mathematical expression combining a function (e.g., $y = f(x)$) and one or more of its derivatives. The *order* of a differential equation is the highest derivative in it. *First-order differential equations* contain only first derivatives of the function, *second-order differential equations* contain second derivatives (and may contain first derivatives as well), and so on.

A *linear differential equation* can be written as a sum of products of multipliers of the function and its derivatives. If the multipliers are scalars, the differential equation is said to have *constant coefficients*. If the function or one of its derivatives is raised to some power (other than one) or is embedded in another function (e.g., y embedded in $\sin y$ or e^y), the equation is said to be *nonlinear*.

Each term of a *homogeneous differential equation* contains either the function (y) or one of its derivatives—that is, the sum of derivative terms is equal to zero. In a *nonhomogeneous differential equation*, the sum of derivative terms is equal to a nonzero *forcing function* of the independent variable (e.g., $g(x)$). In order to solve a nonhomogeneous equation, it is often necessary to solve the homogeneous equation first. The homogeneous equation corresponding to a nonhomogeneous equation is known as a *reduced equation* or *complementary equation*.

The following examples illustrate the types of differential equations.

- $y' - 7y = 0$ homogeneous, first-order linear, with constant coefficients

- $y'' - 2y' + 8y = \sin 2x$ nonhomogeneous, second-order linear, with constant coefficients

- $y'' - (x^2 - 1)y^2 = \sin 4x$ nonhomogeneous, second-order, nonlinear

An *auxiliary equation* (also called the *characteristic equation*) can be written for a homogeneous linear differential equation with constant coefficients, regardless of order. This auxiliary equation is simply the polynomial formed by replacing all derivatives with variables raised to the power of their respective derivatives.

The purpose of solving a differential equation is to derive an expression for the function in terms of the independent variable. The expression does not need to be explicit in the function, but there can be no derivatives in the expression. Since, in the simplest cases, solving a differential equation is equivalent to finding an indefinite integral, it is not surprising that *constants of integration* must be evaluated from knowledge of how the system behaves. Additional data are known as *initial* values, and any problem that includes them is known as an *initial value problem*.[1]

Most differential equations require lengthy solutions and are not efficiently solved by hand. However, several types are fairly simple and are presented in this chapter.

Example 11.1

Write the complementary differential equation for the following nonhomogeneous differential equation.

$$y'' + 6y' + 9y = e^{-14x} \sin 5x$$

Solution

The complementary equation is found by eliminating the forcing function, $e^{-14x} \sin 5x$.

$$y'' + 6y' + 9y = 0$$

Example 11.2

Write the auxiliary equation to the following differential equation.

$$y'' + 4y' + y = 0$$

Solution

Replacing each derivative with a polynomial term whose degree equals the original order, the auxiliary equation is

$$r^2 + 4r + 1 = 0$$

2. HOMOGENEOUS, FIRST-ORDER LINEAR DIFFERENTIAL EQUATIONS WITH CONSTANT COEFFICIENTS

A homogeneous, first-order linear differential equation with constant coefficients has the general form of Eq. 11.1.

$$y' + ky = 0 \qquad 11.1$$

The auxiliary equation is $r + k = 0$ and has a root of $r = -k$. Equation 11.2 is the solution.

$$y = Ae^{rx} = Ae^{-kx} \qquad 11.2$$

---

[1]The term *initial* implies that time is the independent variable. While this may explain the origin of the term, initial value problems are not limited to the time domain. A *boundary value problem* is similar, except that the data come from different points. For example, additional data in the form $y(x_0)$ and $y'(x_0)$ or $y(x_0)$ and $y'(x_1)$ that need to be simultaneously satisfied constitute an initial value problem. Data of the form $y(x_0)$ and $y(x_1)$ constitute a boundary value problem. Until solved, it is difficult to know whether a boundary value problem has no, one, or more than one solution.

If the initial condition is known to be $y(0) = y_0$, the solution is

$$y = y_0 e^{-kx} \qquad 11.3$$

3. FIRST-ORDER LINEAR DIFFERENTIAL EQUATIONS

A first-order linear differential equation has the general form of Eq. 11.4. $p(x)$ and $g(x)$ can be constants or any function of x (but not of y). However, if $p(x)$ is a constant and $g(x)$ is zero, it is easier to solve the equation as shown in Sec. 11-2.

$$y' + p(x)y = g(x) \qquad 11.4$$

The *integrating factor* (which is actually a function) to this differential equation is

$$u(x) = \exp\left[\int p(x)dx \right] \qquad 11.5$$

The closed-form solution to Eq. 11.4 is

$$y = \frac{1}{u(x)} \left[\int u(x)g(x)dx + C \right] \qquad 11.6$$

For the special case where $p(x)$ and $g(x)$ are both constants, Eq. 11.4 becomes

$$y' + ay = b \qquad 11.7$$

If the initial condition is $y(0) = y_0$, the solution to Eq. 11.7 is

$$y = \left(\frac{b}{a}\right)\left(1 - e^{-ax}\right) + y_0 e^{-ax} \qquad 11.8$$

Example 11.3

Find a solution to the following differential equation.

$$y' - y = 2xe^{2x} \qquad y(0) = 1$$

Solution

This is a first-order linear equation with $p(x) = -1$ and $g(x) = 2xe^{2x}$. The integrating factor is

$$u(x) = \exp\left[\int p(x)dx \right] = \exp\left[\int -1dx \right] = e^{-x}$$

The solution is given by Eq. 11.6.

$$
\begin{aligned}
y &= \frac{1}{u(x)}\left[\int u(x)g(x)dx + C\right] \\
&= \frac{1}{e^{-x}}\left[\int e^{-x}2xe^{2x}dx + C\right] \\
&= e^x\left[2\int xe^x dx + C\right] \\
&= e^x[2xe^x - 2e^x + C] \\
&= e^x[2e^x(x-1) + C]
\end{aligned}
$$

From the initial condition,

$$
\begin{aligned}
y(0) &= 1 \\
e^0[(2)(e^0)(0-1) + C] &= 1 \\
1[(2)(1)(-1) + C] &= 1
\end{aligned}
$$

Therefore, $C = 3$. The complete solution is

$$
y = e^x[2e^x(x-1) + 3]
$$

4. FIRST-ORDER SEPARABLE DIFFERENTIAL EQUATIONS

First-order separable differential equations can be placed in the form of Eq. 11.9. For clarity, y' is written as dy/dx.

$$
m(x) + n(y)\frac{dy}{dx} = 0 \qquad 11.9
$$

Equation 11.9 can be placed in the form of Eq. 11.10, both sides of which are easily integrated. An initial value will establish the constant of integration.

$$
m(x)dx = -n(y)dy \qquad 11.10
$$

5. FIRST-ORDER EXACT DIFFERENTIAL EQUATIONS

A *first-order exact differential equation* has the form

$$
f_x(x,y) + f_y(x,y)y' = 0 \qquad 11.11
$$

Notice that $f_x(x,y)$ is the exact derivative of $f(x,y)$ with respect to x, and $f_y(x,y)$ is the exact derivative of $f(x,y)$ with respect to y. The solution is

$$
f(x,y) - C = 0 \qquad 11.12
$$

6. HOMOGENEOUS, SECOND-ORDER LINEAR DIFFERENTIAL EQUATIONS WITH CONSTANT COEFFICIENTS

Homogeneous second-order linear differential equations with constant coefficients have the form of Eq. 11.13. They are most easily solved by finding the two roots of the auxiliary equation (Eq. 11.14).

$$
y'' + k_1y' + k_2y = 0 \qquad 11.13
$$

$$
r^2 + k_1r + k_2 = 0 \qquad 11.14
$$

There are three cases. If the two roots of Eq. 11.14 are real and different, the solution is

$$
y = A_1e^{r_1x} + A_2e^{r_2x} \qquad 11.15
$$

If the two roots are real and the same, the solution is

$$
y = A_1e^{rx} + A_2xe^{rx} \qquad 11.16
$$

$$
r = \frac{-k_1}{2} \qquad 11.17
$$

If the two roots are imaginary, they will be of the form $(\alpha + i\omega)$ and $(\alpha - i\omega)$, and the solution is

$$
y = A_1e^{\alpha x}\cos\omega x + A_2e^{\alpha x}\sin\omega x \qquad 11.18
$$

In all three cases, A_1 and A_2 must be found from the two initial conditions.

Example 11.4

Solve the following differential equation.

$$
y'' + 6y' + 9y = 0
$$

$$
y(0) = 0 \qquad y'(0) = 1
$$

Solution

The auxiliary equation is

$$
\begin{aligned}
r^2 + 6r + 9 &= 0 \\
(r+3)(r+3) &= 0
\end{aligned}
$$

The roots to the auxiliary equation are $r_1 = r_2 = -3$. Therefore, the solution has the form of Eq. 11.16.

$$
y = A_1e^{-3x} + A_2xe^{-3x}
$$

The first initial condition is

$$
\begin{aligned}
y(0) &= 0 \\
A_1e^0 + A_2(0)e^0 &= 0 \\
A_1 + 0 &= 0 \\
A_1 &= 0
\end{aligned}
$$

To use the second initial condition, the derivative of the equation is needed. Making use of the known fact that $A_1 = 0$,

$$
y' = \frac{d}{dx}\left(A_2xe^{-3x}\right) = -3A_2xe^{-3x} + A_2e^{-3x}
$$

Using the second initial condition,

$$y'(0) = 1$$
$$-3A_2(0)e^0 + A_2e^0 = 1$$
$$0 + A_2 = 1$$
$$A_2 = 1$$

The solution is

$$y = xe^{-3x}$$

7. NONHOMOGENEOUS DIFFERENTIAL EQUATIONS

A nonhomogeneous equation has the form of Eq. 11.19. $f(x)$ is known as the *forcing function*.

$$y'' + p(x)y' + q(x)y = f(x) \qquad \textbf{11.19}$$

The solution to Eq. 11.19 is the sum of two equations. The *complementary solution*, y_c, solves the complementary (i.e., homogeneous) problem. The *particular solution*, y_p, is any specific solution to the nonhomogeneous Eq. 11.19 that is known or can be found. Initial values are used to evaluate any unknown coefficients in the complementary solution *after* y_c and y_p have been combined. (The particular solution will not have any unknown coefficients.)

$$y = y_c + y_p \qquad \textbf{11.20}$$

Two methods are available for finding a particular solution. The *method of undetermined coefficients*, as presented here, can be used only when $p(x)$ and $q(x)$ are constant coefficients and $f(x)$ takes on one of the forms in Table 11.1.

The particular solution can be read from Table 11.1 if the forcing function is of one of the forms given. Of course, the coefficients A_i and B_i are not known—these are the *undetermined coefficients*. The exponent s is the smallest nonnegative number (and will be 0, 1, or 2), which ensures that no term in the particular solution, y_p, is also a solution to the complementary equation, y_c. s must be determined prior to proceeding with the solution procedure.

Table 11.1 *Particular Solutions**

form of $f(x)$	form of y_p
$P_n(x) = a_0x^n + a_1x^{n-1}$ $+ \cdots + a_n$	$x^s(A_0x^n + A_1x^{n-1} + \cdots + A_n)$
$P_n(x)e^{\alpha x}$	$x^s(A_0x^n + A_1x^{n-1} + \cdots + A_n)e^{\alpha x}$
$P_n(x)e^{\alpha x}\begin{Bmatrix} \sin\omega x \\ \cos\omega x \end{Bmatrix}$	$x^s[(A_0x^n + A_1x^{n-1} + \cdots + A_n)e^{\alpha x}\cos\omega x$ $+ (B_0x^n + B_1x^{n-1} + \cdots + B_n)e^{\alpha x}\sin\omega x]$

*$P_n(x)$ is a polynomial of degree n.

Once y_p (including s) is known, it is differentiated to obtain y_p' and y_p'', and all three functions are substituted into the original nonhomogeneous equation. The resulting equation is rearranged to match the forcing function, $f(x)$, and the unknown coefficients are determined, usually by solving simultaneous equations.

If the forcing function, $f(x)$, is more complex than the forms shown in Table 11.1, or if either $p(x)$ or $q(x)$ is a function of x, the method of *variation of parameters* should be used. This complex and time-consuming method is not covered in this book.

Example 11.5

Solve the following nonhomogeneous differential equation.

$$y'' + 2y' + y = e^x\cos x$$

Solution

step 1: Find the solution to the complementary (homogeneous) differential equation.

$$y'' + 2y' + y = 0$$

Since this is a differential equation with constant coefficients, write the auxiliary equation.

$$r^2 + 2r + 1 = 0$$

The auxiliary equation factors in $(r+1)^2 = 0$ with two identical roots at $r = -1$. Therefore, the solution to the homogeneous differential equation is

$$y_c(x) = C_1e^{-x} + C_2xe^{-x}$$

step 2: Use Table 11.1 to determine the form of a particular solution. Since the forcing function has the form $P_n(x)e^{\alpha x}\cos\omega x$ with $P_n(x) = 1$ (equivalent to $n = 0$), $\alpha = 1$, and $\omega = 1$, the particular solution has the form

$$y_p(x) = x^s(Ae^x\cos x + Be^x\sin x)$$

step 3: Determine the value of s. Check to see if any of the terms in $y_p(x)$ will themselves solve the homogeneous equation. Try $Ae^x\cos x$ first.

$$\frac{d}{dx}(Ae^x\cos x) = Ae^x\cos x - Ae^x\sin x$$

$$\frac{d^2}{dx^2}(Ae^x\cos x) = -2Ae^x\sin x$$

Substitute these quantities into the homogeneous equation.

$$y'' + 2y' + y = 0$$
$$-2Ae^x\sin x + 2Ae^x\cos x$$
$$- 2Ae^x\sin x + Ae^x\cos x = 0$$
$$3Ae^x\cos x - 4Ae^x\sin x = 0$$

Disregarding the trivial ($A = 0$) solution, $Ae^x \cos x$ does not solve the homogeneous equation.

Next, try $Be^x \sin x$.

$$\frac{d}{dx}\left(Be^x \sin x\right) = Be^x \cos x + Be^x \sin x$$

$$\frac{d^2}{dx^2}\left(Be^x \sin x\right) = 2Be^x \cos x$$

Substitute these quantities into the homogeneous equation.

$$y'' + 2y' + y = 0$$
$$2Be^x \cos x + 2Be^x \cos x$$
$$+2Be^x \sin x + Be^x \sin x = 0$$
$$3Be^x \sin x + 4Be^x \cos x = 0$$

Disregarding the trivial ($B = 0$) case, $Be^x \sin x$ does not solve the homogeneous equation.

Since none of the terms in $y_p(x)$ solve the homogeneous equation, $s = 0$, and a particular solution has the form

$$y_p(x) = Ae^x \cos x + Be^x \sin x$$

step 4: Use the method of unknown coefficients to determine A and B in the particular solution. Drawing on the previous steps, substitute the quantities derived from the particular solution into the nonhomogeneous equation.

$$y'' + 2y' + y = e^x \cos x$$
$$-2Ae^x \sin x + 2Be^x \cos x$$
$$+2Ae^x \cos x - 2Ae^x \sin x$$
$$+2Be^x \cos x + 2Be^x \sin x$$
$$+Ae^x \cos x + Be^x \sin x = e^x \cos x$$

Combining terms,

$$(-4A + 3B)e^x \sin x + (3A + 4B)e^x \cos x$$
$$= e^x \cos x$$

Equating the coefficients of like terms on either side of the equal sign results in the following simultaneous equations.

$$-4A + 3B = 0$$
$$3A + 4B = 1$$

The solution to these equations is

$$A = \tfrac{3}{25}$$
$$B = \tfrac{4}{25}$$

A particular solution is

$$y_p(x) = \left(\tfrac{3}{25}\right)\left(e^x \cos x\right) + \left(\tfrac{4}{25}\right)\left(e^x \sin x\right)$$

step 5: Write the general solution.

$$y(x) = y_c(x) + y_p(x)$$
$$= C_1 e^{-x} + C_2 x e^{-x} + \left(\tfrac{3}{25}\right)\left(e^x \cos x\right)$$
$$+ \left(\tfrac{4}{25}\right)\left(e^x \sin x\right)$$

The values of C_1 and C_2 would be determined at this time if initial conditions were known.

8. NAMED DIFFERENTIAL EQUATIONS

Some differential equations with specific forms are named after the individuals who developed solution techniques for them.

- *Bessel equation of order ν:*

$$x^2 y'' + xy' + (x^2 - \nu^2)y = 0 \qquad \textit{11.21}$$

- *Cauchy equation:*

$$a_0 x^n \frac{d^n y}{dx^n} + a_1 x^{n-1} \frac{d^{n-1} y}{dx^{n-1}} + \cdots$$
$$+ a_{n-1} x \frac{dy}{dx} + a_n y = f(x) \qquad \textit{11.22}$$

- *Euler's equation:*

$$x^2 y'' + \alpha xy' + \beta y = 0 \qquad \textit{11.23}$$

- *Gauss' hypergeometric equation:*

$$x(1-x)y'' + \big(c - (a+b+1)x\big)y' - aby = 0 \qquad \textit{11.24}$$

- *Legendre equation of order λ:*

$$(1 - x^2)y'' - 2xy' + \lambda(\lambda + 1)y = 0$$
$$[-1 < x < 1] \qquad \textit{11.25}$$

9. LAPLACE TRANSFORMS

Traditional methods of solving nonhomogeneous differential equations by hand are usually difficult and/or time consuming. *Laplace transforms* can be used to reduce many solution procedures to simple algebra.

Every mathematical function, $f(t)$, for which Eq. 11.26 exists has a Laplace transform, written as $\mathcal{L}(f)$ or $F(s)$. The transform is written in the s-domain, regardless of the independent variable in the original function.[2] (The variable s is equivalent to a derivative operator, although it may be handled in the equations as a simple variable.) Equation 11.26 converts a function into a Laplace transform.

$$\mathcal{L}\big(f(t)\big) = F(s) = \int_0^\infty e^{-st} f(t)\,dt \qquad 11.26$$

Equation 11.26 is not often needed because tables of transforms are readily available. (Appendix 11.A contains some of the most common transforms.)

Extracting a function from its transform is the *inverse Laplace transform* operation. Although other methods exist, this operation is almost always done by finding the transform in a set of tables.[3]

$$f(t) = \mathcal{L}^{-1}\big(F(s)\big) \qquad 11.27$$

Example 11.6

Find the Laplace transform of the following function.

$$f(t) = e^{at} \quad [s > a]$$

Solution

Applying Eq. 11.26,

$$\mathcal{L}(e^{at}) = \int_0^\infty e^{-st} e^{at}\,dt = \int_0^\infty e^{-(s-a)t}\,dt$$

$$= -\left[\frac{e^{-(s-a)t}}{s-a}\right]_0^\infty = \frac{1}{s-a} \quad [s > a]$$

10. STEP AND IMPULSE FUNCTIONS

Many forcing functions are sinusoidal or exponential in nature; others, however, can only be represented by a step or impulse function. A *unit step function*, u_t, is a function describing the disturbance of magnitude 1 that is not present before time t but is suddenly there after time t. A step of magnitude 5 at time $t = 3$ would be represented as $5u_3$. (The notation $5u(t-3)$ is used in some books.)

The *unit impulse function*, δ_t, is a function describing a disturbance of magnitude 1 that is applied and removed so quickly as to be instantaneous. An impulse of magnitude 5 at time 3 would be represented by $5\delta_3$. (The notation $5\delta(t-3)$ is used in some books.)

---

[2]It is traditional to write the original function as a function of the independent variable t rather than x. However, Laplace transforms are not limited to functions of time.

[3]Other methods include integration in the complex plane, convolution, and simplification by partial fractions.

Example 11.7

What is the notation for a forcing function of magnitude 6 that is applied at $t = 2$ and completely removed at $t = 7$?

Solution

The notation is $f(t) = 6(u_2 - u_7)$.

Example 11.8

Find the Laplace transform of u_0, a unit step at $t = 0$.

$$f(t) = 0 \text{ for } t < 0$$
$$f(t) = 1 \text{ for } t \geq 0$$

Solution

Since the Laplace transform is an integral that starts at $t = 0$, the value of $f(t)$ prior to $t = 0$ is irrelevant.

$$\mathcal{L}(u_0) = \int_0^\infty e^{-st}(1)\,dt = -\left[\frac{e^{-st}}{s}\right]_0^\infty$$

$$= 0 - \left(\frac{-1}{s}\right) = \frac{1}{s}$$

11. ALGEBRA OF LAPLACE TRANSFORMS

Equations containing Laplace transforms can be simplified by applying the following principles.

- *linearity theorem*: (c is a constant.)

$$\mathcal{L}\big(cf(t)\big) = c\mathcal{L}\big(f(t)\big) = cF(s) \qquad 11.28$$

- *superposition theorem*: ($f(t)$ and $g(t)$ are different functions.)

$$\mathcal{L}\big(f(t) \pm g(t)\big) = \mathcal{L}\big(f(t)\big) \pm \mathcal{L}\big(g(t)\big)$$
$$= F(s) \pm G(s) \qquad 11.29$$

- *time-shifting theorem (delay theorem)*:

$$\mathcal{L}\big(f(t-b)u_b\big) = e^{-bs}F(s) \qquad 11.30$$

- *Laplace transform of a derivative*:

$$\mathcal{L}\big(f^n(t)\big) = -f^{n-1}(0) - sf^{n-2}(0) - \cdots$$
$$- s^{n-1}f(0) + s^n F(s) \qquad 11.31$$

- *other properties*:

$$\mathcal{L}\left(\int_0^t f(u)\,du\right) = \left(\frac{1}{s}\right)F(s) \qquad 11.32$$

$$\mathcal{L}\big(tf(t)\big) = -\frac{dF}{ds} \qquad 11.33$$

$$\mathcal{L}\left(\frac{1}{t}f(t)\right) = \int_s^\infty F(u)\,du \qquad 11.34$$

12. CONVOLUTION INTEGRAL

A complex Laplace transform, $F(s)$, will often be recognized as the product of two other transforms, $F_1(s)$ and $F_2(s)$, whose corresponding functions $f_1(t)$ and $f_2(t)$ are known. Unfortunately, Laplace transforms cannot be computed with ordinary multiplication. That is, $f(t) \neq f_1(t)f_2(t)$ even though $F(s) = F_1(s)F_2(s)$.

However, it is possible to extract $f(t)$ from its *convolution*, $h(t)$, as calculated from either of the *convolution integrals* in Eq. 11.35. This process is demonstrated in Ex. 11.9. χ is a dummy variable.

$$f(t) = \mathcal{L}^{-1}\big(F_1(s)F_2(s)\big)$$

$$= \int_0^t f_1(t - \chi)f_2(\chi)d\chi$$

$$= \int_0^t f_1(\chi)f_2(t - \chi)d\chi \qquad \textit{11.35}$$

Example 11.9

Use the convolution integral to find the inverse transform of

$$F(s) = \frac{3}{s^2(s^2 + 9)}$$

Solution

$F(s)$ can be factored as

$$F_1(s)F_2(s) = \left(\frac{1}{s^2}\right)\left(\frac{3}{s^2 + 9}\right)$$

Since the inverse transforms of $F_1(s)$ and $F_2(s)$ are $f_1(t) = t$ and $f_2(t) = \sin 3t$, respectively, the convolution integral from Eq. 11.35 is

$$f(t) = \int_0^t (t - \chi)\sin 3\chi \, d\chi$$

$$= \int_0^t (t \, \sin 3\chi - \chi \, \sin 3\chi)d\chi$$

$$= t\int_0^t \sin 3\chi \, d\chi - \int_0^t \chi \, \sin 3\chi \, d\chi$$

Expand using integration by parts.

$$f(t) = \frac{3t - \sin 3t}{9}$$

13. USING LAPLACE TRANSFORMS

Any nonhomogeneous linear differential equation with constant coefficients can be solved with the following procedure, which reduces the solution to simple algebra.

A complete table of transforms simplifies or eliminates step 5.

step 1: Put the differential equation in standard form (i.e., isolate the y'' term).

$$y'' + k_1y' + k_2y = f(t) \qquad \textit{11.36}$$

step 2: Take the Laplace transform of both sides. Use the linearity and superposition theorems, Eqs. 11.28 and 11.29.

$$\mathcal{L}(y'') + k_1\mathcal{L}(y') + k_2\mathcal{L}(y) = \mathcal{L}\big(f(t)\big) \quad \textit{11.37}$$

step 3: Use Eqs. 11.38 and 11.39 to expand the equation. (These are specific forms of Eq. 11.31.) Use a table to evaluate the transform of the forcing function.

$$\mathcal{L}(y'') = s^2\mathcal{L}(y) - sy(0) - y'(0) \quad \textit{11.38}$$

$$\mathcal{L}(y') = s\mathcal{L}(y) - y(0) \qquad \textit{11.39}$$

step 4: Use algebra to solve for $\mathcal{L}(y)$.

step 5: If needed, use partial fractions (see Ch. 4) to simplify the expression for $\mathcal{L}(y)$.

step 6: Take the inverse transform to find $y(t)$.

$$y(t) = \mathcal{L}^{-1}\big(\mathcal{L}(y)\big) \qquad \textit{11.40}$$

Example 11.10

Find $y(t)$ for the following differential equation.

$$y'' + 2y' + 2y = \cos t$$
$$y(0) = 1 \qquad y'(0) = 0$$

Solution

step 1: The equation is already in standard form.

step 2: $\mathcal{L}(y'') + 2\mathcal{L}(y') + 2\mathcal{L}(y) = \mathcal{L}(\cos t)$

step 3: Use Eqs. 11.38 and 11.39. Use App. 11.A to find the transform of $\cos t$.

$$s^2\mathcal{L}(y) - sy(0) - y'(0)$$
$$+ 2s\mathcal{L}(y) - 2y(0) + 2\mathcal{L}(y) = \frac{s}{s^2 + 1}$$

But, $y(0) = 1$ and $y'(0) = 0$.

$$s^2\mathcal{L}(y) - s + 2s\mathcal{L}(y) - 2 + 2\mathcal{L}(y) = \frac{s}{s^2 + 1}$$

step 4: Combine terms and solve for $\mathcal{L}(y)$.

$$\mathcal{L}(y)(s^2 + 2s + 2) - s - 2 = \frac{s}{s^2 + 1}$$

$$\mathcal{L}(y) = \frac{\dfrac{s}{s^2 + 1} + s + 2}{s^2 + 2s + 2}$$

$$= \frac{s^3 + 2s^2 + 2s + 2}{(s^2 + 1)(s^2 + 2s + 2)}$$

step 5: Expand the expression for $\mathcal{L}(y)$ by partial fractions. (Refer to Ch. 4.)

$$\mathcal{L}(y) = \frac{s^3 + 2s^2 + 2s + 2}{(s^2 + 1)(s^2 + 2s + 2)}$$

$$= \frac{A_1 s + B_1}{s^2 + 1} + \frac{A_2 s + B_2}{s^2 + 2s + 2}$$

$$= \frac{\begin{pmatrix} s^3(A_1 + A_2) + s^2(2A_1 + B_1 + B_2) \\ + s(2A_1 + 2B_1 + A_2) + (2B_1 + B_2) \end{pmatrix}}{(s^2 + 1)(s^2 + 2s + 2)}$$

The following simultaneous equations result.

$$
\begin{array}{ccccccccc}
A_1 & + & A_2 & & & & & = & 1 \\
2A_1 & & & + & B_1 & + & B_2 & = & 2 \\
2A_1 & + & A_2 & + & 2B_1 & & & = & 2 \\
& & & & 2B_1 & + & B_2 & = & 2
\end{array}
$$

These equations have the solutions $A_1 = 1/5$, $A_2 = 4/5$, $B_1 = 2/5$, and $B_2 = 6/5$.

step 6: Refer to App. 11.A and take the inverse transforms.

$$y = \mathcal{L}^{-1}\big(\mathcal{L}(y)\big)$$

$$= \mathcal{L}^{-1}\left(\frac{\left(\tfrac{1}{5}\right)(s+2)}{s^2 + 1} + \frac{\left(\tfrac{1}{5}\right)(4s+6)}{s^2 + 2s + 2} \right)$$

$$= \left(\tfrac{1}{5}\right)\left(\mathcal{L}^{-1}\left(\frac{s}{s^2+1}\right) + 2\mathcal{L}^{-1}\left(\frac{1}{s^2+1}\right) \right.$$

$$+ 4\mathcal{L}^{-1}\left(\frac{s - (-1)}{\big(s-(-1)\big)^2 + 1} \right)$$

$$+ \left. 2\mathcal{L}^{-1}\left(\frac{1}{\big(s-(-1)\big)^2 + 1} \right) \right)$$

$$= \tfrac{1}{5}(\cos t + 2\sin t + 4e^{-t}\cos t + 2e^{-t}\sin t)$$

14. THIRD- AND HIGHER-ORDER LINEAR DIFFERENTIAL EQUATIONS WITH CONSTANT COEFFICIENTS

The solutions of third- and higher-order linear differential equations with constant coefficients are extensions of the solutions for second-order equations of this type. Specifically, if an equation is homogeneous, the auxiliary equation is written and its roots are found. If the equation is nonhomogeneous, Laplace transforms can be used to simplify the solution.

Consider the following homogeneous differential equation with constant coefficients.

$$y^n + k_1 y^{n-1} + \cdots + k_{n-1} y' + k_n y = 0 \qquad 11.41$$

The auxiliary equation to Eq. 11.41 is

$$r^n + k_1 r^{n-1} + \cdots + k_{n-1} r + k_n = 0 \qquad 11.42$$

For each real and distinct root r, the solution contains the term

$$y = A e^{rx} \qquad 11.43$$

For each real root r that repeats m times, the solution contains the term

$$y = \left(A_1 + A_2 x + A_3 x^2 + \cdots + A_m x^{m-1} \right) e^{rx} \qquad 11.44$$

For each pair of complex roots of the form $r = \alpha \pm i\omega$, the solution contains the terms

$$y = e^{\alpha x}(A_1 \sin \omega x + A_2 \cos \omega x) \qquad 11.45$$

15. APPLICATION: ENGINEERING SYSTEMS

There is a wide variety of engineering systems (mechanical, electrical, fluid flow, heat transfer, and so on) whose behavior is described by linear differential equations with constant coefficients.

16. APPLICATION: MIXING

A typical mixing problem involves a liquid-filled tank. The liquid may initially be pure or contain some solute. Liquid (either pure or as a solution) enters the tank at a known rate. A drain may be present to remove thoroughly mixed liquid. The concentration of the solution (or, equivalently, the amount of solute in the tank) at some given time is generally unknown.

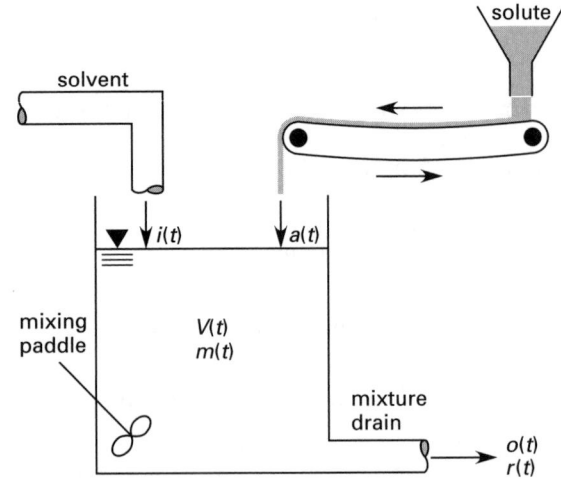

Figure 11.1 *Fluid Mixture Problem*

If $m(t)$ is the mass of solute in the tank at time t, the rate of solute change will be $m'(t)$. If the solute is being added at the rate of $a(t)$ and being removed at the rate of $r(t)$, the rate of change is

$$m'(t) = \text{rate of addition} - \text{rate of removal}$$

$$= a(t) - r(t) \qquad 11.46$$

The rate of solute addition $a(t)$ must be known and, in fact, may be constant. However, $r(t)$ depends on the concentration, $c(t)$, of the mixture and volumetric flow rates at time t. If $o(t)$ is the volumetric flow rate out of the tank, then

$$r(t) = c(t)o(t) \qquad \textit{11.47}$$

However, the concentration depends on the mass of solute in the tank at time t. Recognizing that the volume, $V(t)$, of the liquid in the tank may be changing with time,

$$c(t) = \frac{m(t)}{V(t)} \qquad \textit{11.48}$$

The differential equation describing this problem is

$$m'(t) = a(t) - \frac{m(t)o(t)}{V(t)} \qquad \textit{11.49}$$

Example 11.11

A tank contains 100 gal of pure water at the beginning of an experiment. Pure water flows into the tank at a rate of 1 gal/min. Brine containing $\frac{1}{4}$ lbm of salt per gallon enters the tank from a second source at a rate of 1 gal/min. A perfectly mixed solution drains from the tank at a rate of 2 gal/min. How much salt is in the tank 8 min after the experiment begins?

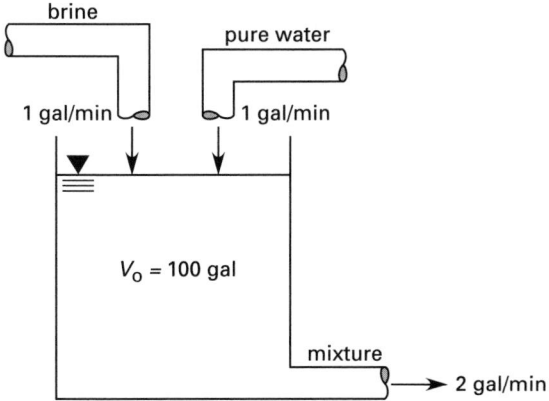

Solution

Let $m(t)$ be the mass of salt in the tank at time t. 0.25 lbm of salt enters the tank per minute (i.e., $a(t) = 0.25$ lbm/min). The salt removal rate depends on the concentration in the tank. That is,

$$r(t) = o(t)c(t)$$

$$= \left(2 \; \frac{\text{gal}}{\text{min}}\right)\left(\frac{m(t)}{100 \; \text{gal}}\right) = \left(0.02 \; \frac{1}{\text{min}}\right)m(t)$$

From Eq. 11.46, the rate of change of salt in the tank is

$$m'(t) = a(t) - r(t)$$

$$= 0.25 \; \frac{\text{lbm}}{\text{min}} - \left(0.02 \; \frac{1}{\text{min}}\right)m(t)$$

$$m'(t) + \left(0.02 \; \frac{1}{\text{min}}\right)m(t) = 0.25 \; \text{lbm/min}$$

This is a first-order linear differential equation of the form of Eq. 11.7. Since the initial condition is $m(0) = 0$, the solution is

$$m(t) = \left(\frac{0.25 \; \frac{\text{lbm}}{\text{min}}}{0.02 \; \frac{1}{\text{min}}}\right)\left(1 - e^{(0.02 \, \text{min}^{-1})t}\right)$$

$$= (12.5 \; \text{lbm})\left(1 - e^{(0.02 \, \text{min}^{-1})t}\right)$$

At $t = 8$,

$$m(t) = \left(12.5 \; \frac{1}{\text{min}}\right)\left(1 - e^{-(0.02 \, \text{min}^{-1})(8 \, \text{min})}\right)$$

$$= (12.5 \; \text{lbm})(1 - 0.852)$$

$$= 1.85 \; \text{lbm}$$

17. APPLICATION: EXPONENTIAL GROWTH AND DECAY

Equation 11.50 describes the behavior of a substance whose quantity, $m(t)$, changes at a rate proportional to the quantity present. The constant of proportionality, k, will be negative for decay (e.g., radioactive decay) and positive for growth (e.g., compound interest).

$$m'(t) = km(t) \qquad \textit{11.50}$$

$$m'(t) - km(t) = 0 \qquad \textit{11.51}$$

If the initial quantity of substance is $m(0) = m_0$, Eq. 11.51 has the solution

$$m(t) = m_0 e^{kt} \qquad \textit{11.52}$$

If $m(t)$ is known for some time t, the constant of proportionality is

$$k = \left(\frac{1}{t}\right)\ln\left(\frac{m(t)}{m_0}\right) \qquad \textit{11.53}$$

For the case of a decay, the *half-life*, $t_{1/2}$, is the time at which only half of the substance remains. The relationship between k and $t_{1/2}$ is

$$kt_{1/2} = \ln\left(\tfrac{1}{2}\right) = -0.693 \qquad \textit{11.54}$$

18. APPLICATION: EPIDEMICS

During an epidemic in a population of n people, the density of sick (contaminated, contagious, affected, etc.) individuals is $\rho_s(t) = s(t)/n$, where $s(t)$ is the number of sick individuals at time t. Similarly, the density of well (uncontaminated, unaffected, susceptible, etc.) individuals is $\rho_\omega(t) = \omega(t)/n$, where $\omega(t)$ is the number of well individuals. Assuming there is no quarantine, the population size is constant, individuals move about freely, and sickness does not limit the activities of individuals, the rate of contagion, $\rho_s'(t)$, will be $k\rho_s(t)\rho_\omega(t)$, where k is a proportionality constant.

$$\rho_s'(t) = k\rho_s(t)\rho_\omega(t) = k\rho_s(t)\big(1 - \rho_s(t)\big) \qquad 11.55$$

This is a separable differential equation with the solution

$$\rho_s(t) = \frac{\rho_s(0)}{\rho_s(0) + \big(1 - \rho_s(0)\big)e^{-kt}} \qquad 11.56$$

19. APPLICATION: SURFACE TEMPERATURE

Newton's law of cooling states that the surface temperature, T, of a cooling object changes at a rate proportional to the difference between the surface and ambient temperatures. The constant k is a positive number.

$$T'(t) = -k(T(t) - T_{\text{ambient}}) \quad [k>0] \qquad 11.57$$

$$T'(t) + kT(t) - kT_{\text{ambient}} = 0 \quad [k>0] \qquad 11.58$$

This first-order linear differential equation with constant coefficients has the following solution (from Eq. 11.8).

$$T(t) = T_{\text{ambient}} + (T(0) - T_{\text{ambient}})e^{-kt} \qquad 11.59$$

If the temperature is known at some time t, the constant k can be found from Eq. 11.60.

$$k = \left(\frac{-1}{t}\right) \ln\left(\frac{T(t) - T_{\text{ambient}}}{T(0) - T_{\text{ambient}}}\right) \qquad 11.60$$

20. APPLICATION: EVAPORATION

The mass of liquid evaporated from a liquid surface is proportional to the exposed surface area. Since quantity, mass, and remaining volume are all proportional, the differential equation is

$$\frac{dV}{dt} = -kA \qquad 11.61$$

For a spherical drop of radius r, Eq. 11.61 reduces to

$$\frac{dr}{dt} = -k \qquad 11.62$$

$$r(t) = r(0) - kt \qquad 11.63$$

For a cube with sides of length s, Eq. 11.61 reduces to

$$\frac{ds}{dt} = -2k \qquad 11.64$$

$$s(t) = s(0) - 2kt \qquad 11.65$$

12 Probability and Statistical Analysis of Data

1. SET THEORY

A *set* (usually designated by a capital letter) is a population or collection of individual items known as *elements* or *members*. The *null set*, ϕ, is empty (i.e., contains no members). If A and B are two sets, A is a *subset* of B if every member in A is also in B. A is a *proper subset* of B if B consists of more than the elements in A. These relationships are denoted

$$A \subseteq B \qquad \text{[subset]}$$
$$A \subset B \qquad \text{[proper subset]}$$

The *universal set, U,* is one from which other sets draw their members. If A is a subset of U, then A' (also designated as A^{-1}, $\tilde{A}$, $-A$, and $\overline{A}$) is the *complement* of A and consists of all elements in U that are not in A. This is illustrated in a *Venn diagram* in Fig. 12.1(a).

The *union of two sets*, denoted by $A \cup B$ and shown in Fig. 12.1(b), is the set of all elements that are either in A or B or both. The *intersection of two sets*, denoted

by $A \cap B$ and shown in Fig. 12.1(c), is the set of all elements that belong to both A and B. If $A \cap B = \phi$, A and B are said to be *disjoint sets*.

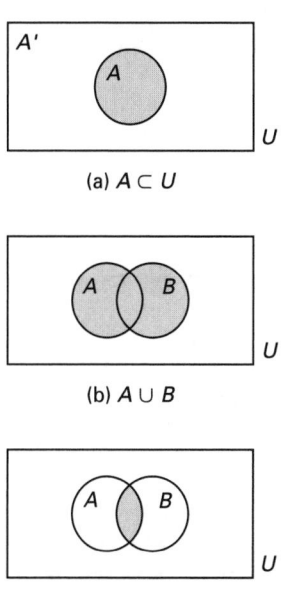

(a) $A \subset U$

(b) $A \cup B$

(c) $A \cap B$

Figure 12.1 Venn Diagrams

If A, B, and C are subsets of the universal set, the following laws apply.

- *identity laws:*

$$A \cup \phi = A \qquad\qquad 12.1$$
$$A \cup U = U \qquad\qquad 12.2$$
$$A \cap \phi = \phi \qquad\qquad 12.3$$
$$A \cap U = A \qquad\qquad 12.4$$

- *idempotent laws:*

$$A \cup A = A \qquad\qquad 12.5$$
$$A \cap A = A \qquad\qquad 12.6$$

- *complement laws:*

$$A \cup A' = U \qquad\qquad 12.7$$
$$(A')' = A \qquad\qquad 12.8$$
$$A \cap A' = \phi \qquad\qquad 12.9$$
$$U' = \phi \qquad\qquad 12.10$$

- *commutative laws*:

$$A \cup B = B \cup A \qquad\qquad 12.11$$
$$A \cap B = B \cap A \qquad\qquad 12.12$$

- *associative laws*:

$$(A \cup B) \cup C = A \cup (B \cup C) \qquad 12.13$$
$$(A \cap B) \cap C = A \cap (B \cap C) \qquad 12.14$$

- *distributive laws*:

$$A \cup (B \cap C) = (A \cup B) \cap (A \cup C) \qquad 12.15$$
$$A \cap (B \cup C) = (A \cap B) \cup (A \cap C) \qquad 12.16$$

- *de Morgan's laws*:

$$(A \cup B)' = A' \cap B' \qquad\qquad 12.17$$
$$(A \cap B)' = A' \cup B' \qquad\qquad 12.18$$

2. COMBINATIONS OF ELEMENTS

There are a finite number of ways in which n elements can be combined into distinctly different groups of r items. For example, suppose a farmer has a hen, a rooster, a duck, and a cage that holds only two birds. The possible *combinations* of three birds taken two at a time are (hen, rooster), (hen, duck), and (rooster, duck). The birds in the cage will not remain stationary, and the combination (rooster, hen) is not distinctly different from (hen, rooster). That is, the groups are not *order conscious*.

The number of combinations of n items taken r at a time is written $C(n,r)$, C_r^n, $_nC_r$, or $\binom{n}{r}$ (pronounced "n choose r") and given by Eq. 12.19. It is sometimes referred to as the *binomial coefficient*.

$$\binom{n}{r} = C(n,r) = \frac{n!}{(n-r)!r!} \qquad \text{[for } r \le n] \quad 12.19$$

Example 12.1

Six people are on a sinking yacht. There are four life jackets. How many combinations of survivors are there?

Solution

The groups are not order-conscious. From Eq. 12.19,

$$C(6,4) = \frac{6!}{(6-4)!4!} = \frac{6 \cdot 5 \cdot 4 \cdot 3 \cdot 2 \cdot 1}{(2 \cdot 1)(4 \cdot 3 \cdot 2 \cdot 1)}$$
$$= 15$$

3. PERMUTATIONS

An order-conscious subset of r items taken from a set of n items is the *permutation* $P(n,r)$, also written P_r^n and $_nP_r$. The permutation is order conscious because the arrangement of two items (say a_i and b_i) as $a_i b_i$ is different from the arrangement $b_i a_i$. The number of permutations is

$$P(n,r) = \frac{n!}{(n-r)!} \qquad \text{[for } r \le n] \quad 12.20$$

If groups of the entire set of n items are being enumerated, the number of permutations of n items taken n at a time is

$$P(n,n) = \frac{n!}{(n-n)!} = \frac{n!}{0!} = n! \qquad 12.21$$

A *ring permutation* is a special case of n items taken n at a time. There is no identifiable beginning or end, and the number of permutations is divided by n.

$$P_{\text{ring}}(n,n) = \frac{P(n,n)}{n} = (n-1)! \qquad 12.22$$

Example 12.2

A pianist knows four pieces but will have enough stage time to play only three of them. Pieces played in a different order constitute a different program. How many different programs can be arranged?

Solution

The groups are order conscious. From Eq. 12.20,

$$P(4,3) = \frac{4!}{(4-3)!} = \frac{4 \cdot 3 \cdot 2 \cdot 1}{1} = 24$$

Example 12.3

Seven diplomats from different countries enter a circular room. The only furnishings are seven chairs arranged around a circular table. How many ways are there of arranging the diplomats?

Solution

All seven diplomats must be seated, so the groups are permutations of seven objects taken seven at a time. Since there is no head chair, the groups are ring permutations. From Eq. 12.22,

$$P_{\text{ring}}(7,7) = (7-1)! = 6 \cdot 5 \cdot 4 \cdot 3 \cdot 2 \cdot 1 = 720$$

4. PROBABILITY THEORY

The act of conducting an experiment (trial) or taking a measurement is known as *sampling*. *Probability theory* determines the relative likelihood that a particular event will occur. An *event*, e, is one of the possible outcomes of the *trial*. Taken together, all of the possible events constitute a finite *sample space*, $E = [e_1, e_2, \ldots, e_n]$. The trial is drawn from the *population* or *universe*. Populations can be finite or infinite in size.

Events can be numerical or nonnumerical, discrete or continuous, and dependent or independent. An example of a nonnumerical event is getting tails on a coin toss. The roll of a die is a discrete numerical event. The measured diameter of a bolt produced from an automatic screw machine is a numerical event. Since the diameter can (within reasonable limits) take on any value, its measured value is a continuous numerical event.

An event is *independent* if its outcome is unaffected by previous outcomes (i.e., previous runs of the experiment) and *dependent* otherwise. Whether or not an event is independent depends on the population size and how the sampling is conducted. Sampling (a trial) from an infinite population is implicitly independent. When the population is finite, *sampling with replacement* produces independent events, while *sampling without replacement* changes the population and produces dependent events.

The terms *success* and *failure* are loosely used in probability theory to designate obtaining and not obtaining, respectively, the tested-for condition. "Failure" is not the same as a *null event* (i.e., one that has a zero probability of occurrence).

The *probability* of event e_1 occurring is designated as $p\{e_1\}$ and is calculated as the ratio of the total number of ways the event can occur to the total number of outcomes in the sample space.

Example 12.4

There are 380 students in a rural school—200 girls and 180 boys. One student is chosen at random and is checked for gender and height. (a) Define and categorize the population. (b) Define and categorize the sample space. (c) Define the trials. (d) Define and categorize the events. (e) In determining the probability that the student chosen is a boy, define success and failure. (f) What is the probability that the student is a boy?

Solution

(a) The population consists of 380 students and is finite.

(b) In determining the gender of the student, the sample space consists of the two outcomes $E = $ [girl, boy]. This sample space is nonnumerical and discrete. In determining the height, the sample space consists of a range of values and is numerical and continuous.

(c) The trial is the actual sampling (i.e., the determination of gender and height).

(d) The events are the outcomes of the trials (i.e., the gender and height of the student). These events are independent if each student returns to the population prior to the random selection of the next student; otherwise, the events are dependent.

(e) The event is a success if the student is a boy and is a failure otherwise.

(f) From the definition of probability,

$$p\{\text{boy}\} = \frac{\text{no. of boys}}{\text{no. of students}} = \frac{180}{380} = \frac{9}{19} = 0.47$$

5. JOINT PROBABILITY

Joint probability rules specify the probability of a combination of events. If n mutually exclusive events from the set E have probabilities $p\{e_i\}$, the probability of any one of these events occurring in a given trial is the sum of the individual probabilities. Notice that the events in Eq. 12.23 come from a single sample space and are linked by the word *or*.

$$p\{e_1 \text{ or } e_2 \text{ or } \ldots \text{ or } e_k\} = p\{e_1\} + p\{e_2\} + \ldots + p\{e_k\}$$
$$12.23$$

Given two independent sets of events, E and G, Eq. 12.24 gives the probability that either event e_i or g_i, or both, will occur. Notice that the events in Eq. 12.24 come from two different sample spaces and are linked by the word *or*.

$$p\{e_i \text{ or } g_i\} = p\{e_i\} + p\{g_i\} - p\{e_i\}p\{g_i\} \qquad 12.24$$

Given two independent sets of events, E and G, Eq. 12.25 gives the probability that events e_i and g_i will both occur. Notice that the events in Eq. 12.25 come from two different sample spaces and are linked by the word *and*.

$$p\{e_i \text{ and } g_i\} = p\{e_i\}p\{g_i\} \qquad 12.25$$

Example 12.5

A bowl contains five white balls, two red balls, and three green balls. What is the probability of getting either a white ball or a red ball in one draw from the bowl?

Solution

The two possible events are mutually exclusive and come from the same sample space, so Eq. 12.23 can be used.

$$p\{\text{white or red}\} = p\{\text{white}\} + p\{\text{red}\} = \frac{5}{10} + \frac{2}{10} = \frac{7}{10}$$

Example 12.6

One bowl contains five white balls, two red balls, and three green balls. Another bowl contains three yellow balls and seven black balls. What is the probability of getting a red ball from the first bowl and a yellow ball from the second bowl in one draw from each bowl?

Solution

The two trials are independent, so Eq. 12.25 can be used.

$$p\{\text{red and yellow}\} = p\{\text{red}\}p\{\text{yellow}\}$$
$$= \left(\frac{2}{10}\right)\left(\frac{3}{10}\right) = \frac{6}{100}$$

6. COMPLEMENTARY PROBABILITIES

The probability of an event occurring is equal to one minus the probability of the event not occurring. This is known as *complementary probability*.

$$p\{e_i\} = 1 - p\{\text{not } e_i\} \qquad 12.26$$

Equation 12.26 can be used to simplify some probability calculations. Specifically, calculation of the probability of numerical events being "greater than" or "less than" or quantities being "at least" a certain number can often be simplified by calculating the probability of complementary event.

Example 12.7

A fair coin is tossed five times.[1] What is the probability of getting at least one tail?

Solution

The probability of getting at least one tail in five tosses could be calculated as

$$p\{\text{at least 1 tail}\} = p\{1 \text{ tail}\} + p\{2 \text{ tails}\}$$
$$+ p\{3 \text{ tails}\} + p\{4 \text{ tails}\}$$
$$+ p\{5 \text{ tails}\}$$

However, it is easier to calculate the complementary probability of getting no tails (i.e., getting all heads).

$$p\{\text{at least 1 tail}\} = 1 - p\{0 \text{ tails}\}$$
$$= 1 - (0.5)^5 = 0.96875$$

7. CONDITIONAL PROBABILITY

Given two dependent sets of events, E and G, the probability that event e_k will occur given the fact that the dependent event g has already occurred is written as $p\{e_k|g\}$ and given by *Bayes' theorem*, Eq. 12.27.

$$p\{e_k|g\} = \frac{p\{e_k \text{ and } g\}}{p\{g\}} = \frac{p\{g|e_k\}p\{e_k\}}{\sum_{i=1}^{n} p\{g|e_i\}p\{e_i\}} \qquad 12.27$$

---

[1]It makes no difference whether one coin is tossed five times or five coins are each tossed once.

8. PROBABILITY DENSITY FUNCTIONS

A *density function* is a nonnegative function whose integral taken over the entire range of the independent variable is unity. A *probability density function* (PDF) is a mathematical formula that gives the probability of a discrete numerical event. A *numerical event* is an occurrence that can be described (usually) by an integer. For example, 27 cars passing through a bridge toll booth in an hour is a discrete numerical event.

A probability density function, $f(x)$, gives the probability that discrete event x will occur. That is, $p\{x\} = f(x)$. Important discrete probability density functions are the binomial, hypergeometric, and Poisson distributions.

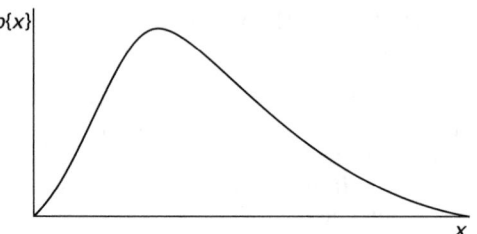

Figure 12.2 *Probability Density Function*

9. BINOMIAL DISTRIBUTION

The *binomial probability density function* is used when all outcomes can be categorized as either successes or failures. The probability of success in a single trial is $\hat{p}$, and the probability of failure is the complement, $\hat{q} = 1 - \hat{p}$. The population is assumed to be infinite in size so that sampling does not change the values of $\hat{p}$ and $\hat{q}$. (The binomial distribution can also be used with finite populations when sampling with replacement.)

Equation 12.28 gives the probability of x successes in n independent *successive trials*. The quantity $\binom{n}{x}$ is the binomial coefficient, identical to the number of combinations of n items taken x at a time.

$$p\{x\} = f(x) = \binom{n}{x}\hat{p}^x\hat{q}^{n-x} \qquad 12.28$$

$$\binom{n}{x} = \frac{n!}{(n-x)!x!} \qquad 12.29$$

Equation 12.28 is a true (discrete) distribution, taking on values for each integer value up to n. The mean, μ, and variance, σ^2 (see Secs. 12-18 and 12-19), of this distribution are

$$\mu = n\hat{p} \qquad 12.30$$
$$\sigma^2 = n\hat{p}\hat{q} \qquad 12.31$$

Example 12.8

Five percent of a large batch of high-strength steel bolts purchased for bridge construction are defective. (a) If seven bolts are randomly sampled, what is the probability that exactly three will be defective? (b) What is the probability that two or more bolts will be defective?

Solution

(a) The bolts are either defective or not, so the binomial distribution can be applied.

$$\hat{p} = 0.05$$
$$\hat{q} = 1 - 0.05 = 0.95$$

From Eq. 12.28,

$$p\{3\} = f(3) = \binom{7}{3} \hat{p}^3 \hat{q}^{7-3}$$
$$= \left(\frac{7 \cdot 6 \cdot 5 \cdot 4 \cdot 3 \cdot 2 \cdot 1}{4 \cdot 3 \cdot 2 \cdot 1 \cdot 3 \cdot 2 \cdot 1} \right) (0.05)^3 (0.95)^4$$
$$= 0.00356$$

(b) The probability that two or more bolts will be defective could be calculated as

$$p\{x \geq 2\} = p\{2\} + p\{3\} + p\{4\} + p\{5\} + p\{6\} + p\{7\}$$

This method would require six probability calculations. It is easier to use the complement of the desired probability.

$$p\{x \geq 2\} = 1 - p\{x \leq 1\} = 1 - (p\{0\} + p\{1\})$$
$$p\{0\} = \binom{7}{0} (0.05)^0 (0.95)^7 = (0.95)^7$$
$$p\{1\} = \binom{7}{1} (0.05)^1 (0.95)^6 = (7)(0.05)(0.95)^6$$
$$p\{x \geq 2\} = 1 - \left((0.95)^7 + (7)(0.05)(0.95)^6 \right)$$
$$= 1 - (0.6983 + 0.2573)$$
$$= 0.0444$$

10. HYPERGEOMETRIC DISTRIBUTION

Probabilities associated with sampling from a finite population without replacement are calculated from the *hypergeometric distribution*. If a population of finite size M contains K items with a given characteristic (e.g., red color, defective construction), then the probability of finding x items with that characteristic in a sample of n items is

$$p\{x\} = f(x) = \frac{\binom{K}{x} \binom{M-K}{n-x}}{\binom{M}{n}} \quad \text{[for } x \leq n \text{]} \quad \textit{12.32}$$

11. MULTIPLE HYPERGEOMETRIC DISTRIBUTION

Sampling without replacement from finite populations containing several different types of items is handled by the *multiple hypergeometric distribution*. If a population of finite size M contains K_i items of type i (such that $\sum K_i = M$), the probability of finding x_1 items of type 1, x_2 items of type 2, and so on, in a sample size n (such that $\sum x_i = n$) is

$$p\{x_1, x_2, x_3, \ldots\} = \frac{\binom{K_1}{x_1} \binom{K_2}{x_2} \binom{K_3}{x_3} \cdots}{\binom{M}{n}} \quad \textit{12.33}$$

12. POISSON DISTRIBUTION

Certain events occur relatively infrequently but at a relatively regular rate. The probability of such an event occurring is given by the *Poisson distribution*. Suppose an event occurs, on the average, λ times per period. The probability that the event will occur x times per period is

$$p\{x\} = f(x) = \frac{e^{-\lambda} \lambda^x}{x!} \quad [\lambda > 0] \quad \textit{12.34}$$

λ is both the mean and the variance of the Poisson distribution.

Example 12.9

The number of customers arriving at a hamburger stand in the next period is a Poisson distribution having a mean of eight. What is the probability that exactly six customers will arrive in the next period?

Solution

$\lambda = 8$ and $x = 6$. From Eq. 12.34,

$$p\{6\} = \frac{e^{-\lambda} \lambda^x}{x!} = \frac{e^{-8}(8)^6}{6!} = 0.122$$

13. CONTINUOUS DISTRIBUTION FUNCTIONS

Most numerical events are *continuously distributed* and are not constrained to discrete or integer values. For example, the resistance of a 10% 1 Ω resistor may be any value between 0.9 and 1.1 Ω. The probability of an exact numerical event is zero for continuously distributed variables. That is, there is no chance that a numerical event will be *exactly* x.[2] It is possible to determine only

---

[2] It is important to understand the rationale behind this statement. Since the variable can take on any value and has an infinite number of significant digits, we can infinitely continue to increase the precision of the value. For example, the probability is zero that a resistance will be exactly 1 Ω because the resistance is really 1.03 or 1.0260008 or 1.02600080005, and so on.

the probability that a numerical event will be less than x, greater than x, or between the values of x_1 and x_2, but not exactly equal to x.

While an expression, $f(x)$, for a probability density function can be written, it is used to derive the *continuous distribution function* (CDF), $F(x_0)$, which gives the probability of numerical event x_0 or less occurring.

$$p\{X < x_0\} = F(x_0) = \int_0^{x_0} f(x)dx \qquad 12.35$$

$$f(x) = \frac{dF(x)}{dx} \qquad 12.36$$

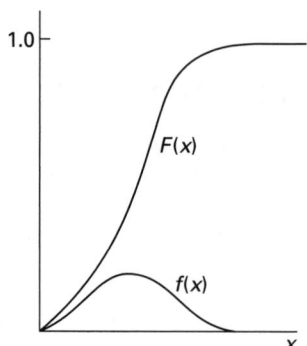

Figure 12.3 *Continuous Distribution Function*

14. EXPONENTIAL DISTRIBUTION

The continuous *exponential distribution* is given by its probability density and continuous distribution functions.

$$f(x) = \lambda e^{-\lambda x} \qquad 12.37$$

$$p\{X < x\} = F(x) = 1 - e^{-\lambda x} \qquad 12.38$$

The mean and variance of the exponential distribution are

$$\mu = \frac{1}{\lambda} \qquad 12.39$$

$$\sigma^2 = \frac{1}{\lambda^2} \qquad 12.40$$

15. NORMAL DISTRIBUTION

The *normal distribution (Gaussian distribution)* is a symmetrical distribution commonly referred to as the *bell-shaped curve*, which represents the distribution of outcomes of many experiments, processes, and phenomena. The probability density and continuous distribution functions for the normal distribution with mean μ and variance σ^2 are

$$f(x) = \frac{e^{-\frac{1}{2}\left(\frac{x-\mu}{\sigma}\right)^2}}{\sigma\sqrt{2\pi}} \qquad [-\infty < x < +\infty] \qquad 12.41$$

$$p\{\mu < X < x_0\} = F(x_0)$$
$$= \frac{1}{\sigma\sqrt{2\pi}} \int_0^{x_0} e^{\frac{1}{2}\left(\frac{x-\mu}{\sigma}\right)^2} dx \qquad 12.42$$

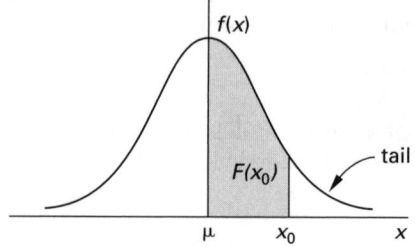

Figure 12.4 *Normal Distribution*

Since $f(x)$ is difficult to integrate, Eq. 12.41 is seldom used directly, and a *standard normal table* (see App. 12.A) is used instead. The standard normal table is based on a normal distribution with a mean of zero and a standard deviation of 1. Since the range of values from an experiment or phenomenon will not generally correspond to the standard normal table, a value, x_0, must be converted to a *standard normal value, z*. In Eq. 12.43, μ and σ are the mean and standard deviation, respectively, of the distribution from which x_0 comes. For all practical purposes, all normal distributions are completely bounded by $\mu \pm 3\sigma$.

$$z = \frac{x_0 - \mu}{\sigma} \qquad 12.43$$

Numbers in the standard normal table given by App. 12.A are the probabilities of the normalized x being between zero and z and represent the areas under the curve up to point z. When x is less than μ, z will be negative. However, the curve is symmetrical, so the table value corresponding to positive z can be used. The probability of x being greater than z is the complement of the table value. The curve area past point z is known as the *tail of the curve*.

The *error function*, erf(x), and its complement, erfc(x), are defined by Eqs. 12.44 and 12.45. The error function is used to determine the probable error of a measurement.

$$\text{erf}(x_0) = \frac{2}{\sqrt{\pi}} \int_0^{x_0} e^{-x^2} dx \qquad 12.44$$

$$\text{erfc}(x_0) = 1 - \text{erf}(x_0) \qquad 12.45$$

Example 12.10

The mass, m, of a particular hand-laid fiberglass (Fibreglas™) part is normally distributed with a mean of 66 kg and a standard deviation of 5 kg. (a) What percent of the parts will have a mass less than 72 kg? (b) What percent of the parts will have a mass in excess of 72 kg? (c) What percent of the parts will have a mass between 61 and 72 kg?

Solution

(a) The value of 72 kg must be normalized by using Eq. 12.43. The standard normal variable is

$$z = \frac{x - \mu}{\sigma} = \frac{72 \text{ kg} - 66 \text{ kg}}{5 \text{ kg}} = 1.2$$

Reading from App. 12.A, the area under the normal curve is 0.3849. This represents the probability of the mass, m, being between 66 kg and 72 kg (i.e., z being between 0 and 1.2). However, the probability of the mass being less than 66 kg is also needed. Since the curve is symmetrical, this probability is 0.5. Therefore,

$$p\{m < 72 \text{ kg}\} = p\{z < 1.2\} = 0.5 + 0.3849 = 0.8849$$

(b) The probability of the mass exceeding 72 kg is the area under the tail past point z.

$$p\{m > 72 \text{ kg}\} = p\{z > 1.2\} = 0.5 - 0.3849 = 0.1151$$

(c) The standard normal variable corresponding to $m = 61$ kg is

$$z = \frac{x - \mu}{\sigma} = \frac{61 \text{ kg} - 66 \text{ kg}}{5 \text{ kg}} = -1$$

Since the two masses are on opposite sides of the mean, the probability will have to be determined in two parts.

$$p\{61 < m < 72\} = p\{61 < m < 66\} + p\{66 < m < 72\}$$
$$= p\{-1 < z < 0\} + p\{0 < z < 1.2\}$$
$$= 0.3413 + 0.3849 = 0.7262$$

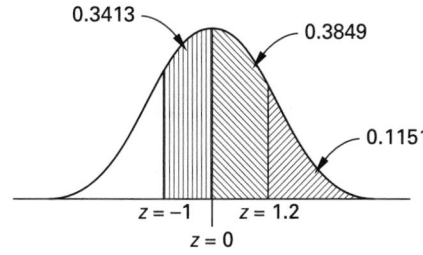

16. APPLICATION: RELIABILITY

Introduction

Reliability, $R\{t\}$, is the probability that an item will continue to operate satisfactorily up to time t. The *bathtub distribution*, Fig. 12.5, is often used to model the probability of failure of an item (or, the number of failures from a large population of items) as a function of time. Items initially fail at a high rate, a phenomenon

known as *infant mortality*. For the majority of the operating time, known as the *steady-state operation*, the failure rate is constant (i.e., is due to random causes). After a long period of time, the items begin to deteriorate and the failure rate increases. (No mathematical distribution describes all three of these phases simultaneously.)

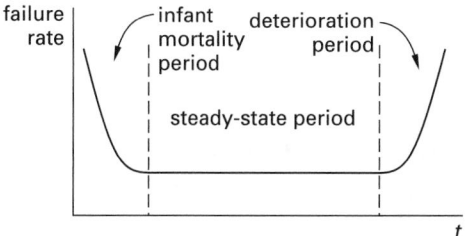

Figure 12.5 *Bathtub Reliability Curve*

The *hazard function*, $z\{t\}$, represents the *conditional probability of failure*—the probability of failure in the next time interval, given that no failure has occurred thus far.[3]

$$z\{t\} = \frac{f(t)}{R(t)} = \frac{\dfrac{dF(t)}{dt}}{1 - F(t)} \qquad \text{12.46}$$

Exponential Reliability

Steady-state reliability is often described by the *negative exponential distribution*. This assumption is appropriate whenever an item fails only by random causes and does not experience deterioration during its life. The parameter λ is related to the *mean time to failure* (MTTF) of the item.[4]

$$R\{t\} = e^{-\lambda t} = e^{\frac{-t}{\text{MTTF}}} \qquad \text{12.47}$$

$$\lambda = \frac{1}{\text{MTTF}} \qquad \text{12.48}$$

Equation 12.47 and the exponential continuous distribution function, Eq. 12.38, are complementary.

$$R\{t\} = 1 - F(t) = 1 - (1 - e^{-\lambda t}) = e^{-\lambda t} \qquad \text{12.49}$$

The hazard function for the negative exponential distribution is

$$z\{t\} = \lambda \qquad \text{12.50}$$

Thus, the hazard function for exponential reliability is constant and does not depend on t (i.e., on the age of the item). In other words, the expected future life of an item is independent of the previous history (length of operation). This lack of memory is consistent with the

---

[3]The symbol $z\{t\}$ is traditionally used for the hazard function and is not related to the standard normal variable.
[4]The term "mean time *between* failures" is improper. However, the term *mean time before failure* (MTBF) is acceptable.

assumption that only random causes contribute to failure during steady-state operations. And since random causes are unlikely discrete events, their probability of occurrence can be represented by a Poisson distribution with mean λ. That is, the probability of having x failures in any given period is

$$p\{x\} = \frac{e^{-\lambda}\lambda^x}{x!} \qquad 12.51$$

Serial System Reliability

In the analysis of system reliability, the binary variable X_i is defined as one if item i operates satisfactorily and zero otherwise. Similarly, the binary variable Φ is one only if the entire system operates satisfactorily. Thus, Φ will depend on a *performance function* containing the X_i.

A *serial system* is one for which all items must operate correctly for the system to operate. Each item has its own reliability, R_i. For a serial system of n items, the performance function is

$$\Phi = X_1 X_2 X_3 \cdots X_n = \min(X_i) \qquad 12.52$$

The probability of a serial system operating correctly is

$$p\{\Phi = 1\} = R_{\text{serial system}} = R_1 R_2 R_3 \cdots R_n \qquad 12.53$$

Parallel System Reliability

A *parallel system* with n items will fail only if all n items fail. Such a system is said to be *redundant* to the nth degree. Using redundancy, a highly reliable system can be produced from components with relatively low individual reliabilities.

The performance function of a redundant system is

$$\Phi = 1 - (1 - X_1)(1 - X_2)(1 - X_3) \cdots (1 - X_n)$$
$$= \max(X_i) \qquad 12.54$$

The reliability of the parallel system is

$$R = p\{\Phi = 1\} = 1 - (1 - R_1)(1 - R_2)(1 - R_3) \cdots (1 - R_n) \qquad 12.55$$

Example 12.11

The reliability of an item is exponentially distributed with mean time to failure (MTTF) of 1000 hr. What is the probability that the item will not have failed after 1200 hr of operation?

Solution

The probability of not having failed before time t is the reliability. From Eqs. 12.48 and 12.49,

$$\lambda = \frac{1}{\text{MTTF}} = \frac{1}{1000\,\text{hr}} = 0.001$$

$$R\{1200\} = e^{-\lambda t} = e^{(-0.001)(1200\ \text{hr})} = 0.3$$

Example 12.12

What are the reliabilities of the following systems?

(a)

(b)

Solution

(a) This is a serial system. From Eq. 12.53,

$$R = R_1 R_2 R_3 R_4 = (0.93)(0.98)(0.91)(0.87)$$
$$= 0.72$$

(b) This is a parallel system. From Eq. 12.55,

$$R = 1 - (1 - R_1)(1 - R_2)(1 - R_3)$$
$$= 1 - (1 - 0.76)(1 - 0.52)(1 - 0.39) = 0.93$$

17. ANALYSIS OF EXPERIMENTAL DATA

Experiments can take on many forms. An experiment might consist of measuring the mass of one cubic foot of concrete, or measuring the speed of a car on a roadway. Generally, such experiments are performed more than once to increase the precision and accuracy of the results.

Both systematic and random variations in the process being measured will cause the observations to vary, and the experiment would not be expected to yield the same result each time it was performed. Eventually, a collection of experimental outcomes (observations) will be available for analysis.

The *frequency distribution* is a systematic method for ordering the observations from small to large, according to some convenient numerical characteristic. The *step interval* should be chosen so that the data are presented in a meaningful manner. If there are too many intervals, many of them will have zero frequencies; if there are too few intervals, the frequency distribution will have little value. Generally, 10 to 15 intervals are used.

Once the frequency distribution is complete, it can be represented graphically as a *histogram*. The procedure in drawing a histogram is to mark off the interval limits (also known as *class limits*) on a number line and then draw contiguous bars with lengths that are proportional to the frequencies in the intervals and that are centered on the midpoints of their respective intervals. The continuous nature of the data can be depicted by a *frequency polygon*. The number or percentage of observations that occur up to and including some value can be shown in a *cumulative frequency table*.

Example 12.13

The number of cars that travel through an intersection between 12 noon and 1 p.m. is measured for 30 consecutive working days. The results of the 30 observations are

79, 66, 72, 70, 68, 66, 68, 76, 73, 71, 74, 70, 71, 69, 67, 74, 70, 68, 69, 64, 75, 70, 68, 69, 64, 69, 62, 63, 63, 61

(a) What are the frequency and cumulative distributions? (Use a distribution interval of two cars per hour.) (b) Draw the histogram. (Use a cell size of two cars per hour.) (c) Draw the frequency polygon. (d) Graph the cumulative frequency distribution.

Solution

(a)

cars per hour	frequency	cumulative frequency	cumulative percent
60–61	1	1	3
62–63	3	4	13
64–65	2	6	20
66–67	3	9	30
68–69	8	17	57
70–71	6	23	77
72–73	2	25	83
74–75	3	28	93
76–77	1	29	97
78–79	1	30	100

(b)

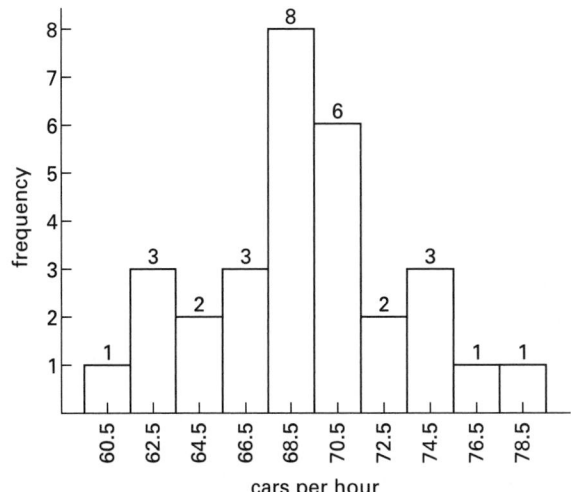

(c)

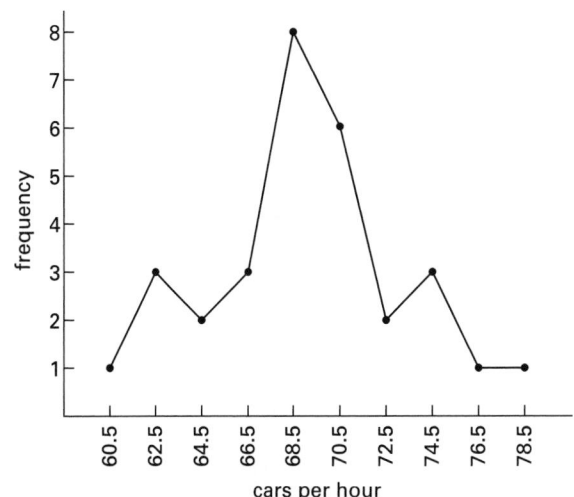

(d)

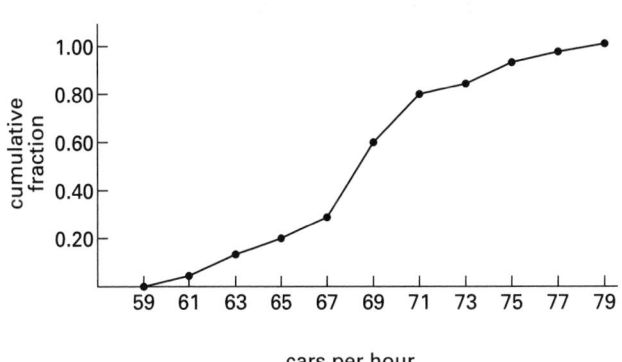

18. MEASURES OF CENTRAL TENDENCY

It is often unnecessary to present the experimental data in their entirety, either in tabular or graphical form. In such cases, the data and distribution can be represented by various parameters. One type of parameter is a measure of *central tendency*. Mode, median, and mean are measures of central tendency.

The *mode* is the observed value that occurs most frequently. The mode may vary greatly between series of observations. Therefore, its main use is as a quick measure of the central value since little or no computation is required to find it. Beyond this, the usefulness of the mode is limited.

The *median* is the point in the distribution that partitions the total set of observations into two parts containing equal numbers of observations. It is not influenced by the extremity of scores on either side of the distribution. The median is found by counting up (from either end of the frequency distribution) until half of the observations have been accounted for.

Similar in concept to the median are *percentiles (percentile ranks)*, *quartiles*, and *deciles*. The median could also have been called the *50th percentile* observation. Similarly, the 80th percentile would be the observed value (e.g., the number of cars per hour) for which the cumulative frequency was 80%. The quartile and decile points on the distribution divide the observations or distribution into segments of 25% and 10%, respectively.

The *arithmetic mean* is the arithmetic average of the observations. The sample mean, $\bar{x}$, can be used as an unbiased estimator of the population mean, μ. The *mean* may be found without ordering the data (as was necessary to find the mode and median). The mean can be found from the following formula.

$$\bar{x} = \left(\frac{1}{n}\right)(x_1 + x_2 + \cdots + x_n) = \frac{\sum x_i}{n} \qquad \textbf{12.56}$$

The *geometric mean* is used occasionally when it is necessary to average ratios. The geometric mean is calculated as

$$\text{geometric mean} = \sqrt[n]{x_1 x_2 x_3 \cdots x_n} \quad [x_i > 0] \qquad \textbf{12.57}$$

The *harmonic mean* is defined as

$$\text{harmonic mean} = \frac{n}{\dfrac{1}{x_1} + \dfrac{1}{x_2} + \cdots + \dfrac{1}{x_n}} \qquad \textbf{12.58}$$

The *root-mean-squared* (rms) *value* of a series of observations is defined as

$$x_{\text{rms}} = \sqrt{\frac{\sum x_i^2}{n}} \qquad \textbf{12.59}$$

Example 12.14

Find the mode, median, and arithmetic mean of the distribution represented by the data given in Ex. 12.13.

Solution

The mode is the interval 68–69, since this interval has the highest frequency. If 68.5 is taken as the interval center, then 68.5 would be the mode.

Since there are 30 observations, the median is the value that separates the observations into two groups of 15. From Ex. 12.13, the median occurs someplace within the 68–69 interval. Up through interval 66–67, there are nine observations, so six more are needed to make 15. Interval 68–69 has eight observations, so the median is found to be 6/8 or 3/4 of the way through the interval. Since the real limits of the interval are 67.5 and 69.5, the median is located at

$$67.5 + \left(\tfrac{3}{4}\right)(69.5 - 67.5) = 69$$

The mean can be found from the raw data or from the grouped data using the interval center as the assumed observation value. Using the raw data,

$$\bar{x} = \frac{\sum x}{n} = \frac{2069}{30} = 68.97$$

19. MEASURES OF DISPERSION

The simplest statistical parameter that describes the variability in observed data is the *range*. The range is found by subtracting the smallest value from the largest. Since the range is influenced by extreme (low probability) observations, its use as a measure of variability is limited.

The *standard deviation* is a better estimate of variability because it considers every observation. That is, N in Eq. 12.60 is the total population size, not the sample size, n.

$$\sigma = \sqrt{\frac{\sum (x_i - \mu)^2}{N}} = \sqrt{\frac{\sum x_i^2}{N} - \mu^2} \qquad \textbf{12.60}$$

The standard deviation of a sample (particularly a small sample) is a biased (i.e., is not a good) estimator of the population standard deviation. An *unbiased estimator* of the population standard deviation is the *sample standard deviation, s*.[5]

$$s = \sqrt{\frac{\sum (x_i - \bar{x})^2}{n-1}} = \sqrt{\frac{\sum x_i^2 - \dfrac{(\sum x_i)^2}{n}}{n-1}} \qquad \textbf{12.61}$$

---

[5]There is a subtle yet significant difference between *standard deviation of the sample*, σ (obtained from Eq. 12.60 for a finite sample drawn from a larger population) and the *sample standard deviation*, s (obtained from Eq. 12.61). While σ can be calculated, it has no significance or use as an estimator. It is true that the difference between σ and s approaches zero when the sample size, n, is large, but this convergence does nothing to legitimize the use of σ as an estimator of the true standard deviation. (Some people say "large" is 30, others say 50 or 100.)

If the sample standard deviation, s, is known, the standard deviation of the sample, σ_{sample}, can be calculated.

$$\sigma_{\text{sample}} = s\sqrt{\frac{n-1}{n}} \qquad \textit{12.62}$$

The *variance* is the square of the standard deviation. Since there are two standard deviations, there are two variances. The *variance of the sample* is σ^2, and the *sample variance* is s^2.

The *relative dispersion* is defined as a measure of dispersion divided by a measure of central tendency. The *coefficient of variation* is a relative dispersion calculated from the sample standard deviation and the mean.

$$\text{coefficient of variation} = \frac{s}{\overline{x}} \qquad \textit{12.63}$$

Skewness is a measure of a frequency distribution's lack of symmetry.

$$
\begin{aligned}
\text{skewness} &= \frac{\overline{x} - \text{mode}}{s} \\
&\approx \frac{(3)(\overline{x} - \text{median})}{s} \qquad \textit{12.64}
\end{aligned}
$$

Example 12.15

For the data given in Ex. 12.13, calculate (a) the sample range, (b) the standard deviation of the sample, (c) an unbiased estimator of the population standard deviation, (d) the variance of the sample, and (e) the sample variance.

Solution

$$\sum x_i = 2069$$
$$\left(\sum x_i\right)^2 = (2069)^2 = 4{,}280{,}761$$
$$\sum x_i^2 = 143{,}225$$
$$n = 30$$
$$\overline{x} = \frac{2069}{30} = 68.967$$

(a) $R = x_{\max} - x_{\min} = 79 - 61 = 18$

(b) $\sigma = \sqrt{\dfrac{\sum x_i^2}{n} - (\overline{x})^2} = \sqrt{\dfrac{143{,}225}{30} - \left(\dfrac{2069}{30}\right)^2}$

$\quad = \sqrt{17.766} = 4.215$

(c) $s = \sqrt{\dfrac{\sum x_i^2 - \dfrac{\left(\sum x_i\right)^2}{n}}{n-1}} = \sqrt{\dfrac{143{,}225 - \dfrac{4{,}280{,}761}{30}}{29}}$

$\quad = \sqrt{18.378} = 4.287$

(d) $\sigma^2 = 17.77$

(e) $s^2 = 18.38$

20. CENTRAL LIMIT THEOREM

Measuring a sample of n items from a population with mean μ and standard deviation σ is the general concept of an experiment. The sample mean, $\overline{x}$, is one of the parameters that can be derived from the experiment. This experiment can be repeated k times, yielding a set of averages $(\overline{x}_1, \overline{x}_2, \ldots, \overline{x}_k)$. The k numbers in the set themselves represent samples from distributions of averages. The average of averages, $\overline{\overline{x}}$, and sample standard deviation of averages, $s_{\overline{x}}$ (known as the *standard error of the mean*), can be calculated.

The *central limit theorem* defines the mean of the sample averages. The theorem can be stated in several ways, but the essential elements are the following points.

(1) The averages, $\overline{x}_i$, are normally distributed variables, even if the original data from which they are calculated are not normally distributed.

(2) The grand average, $\overline{\overline{x}}$ (i.e., the average of the averages), approaches and is an unbiased estimator of μ.

$$\mu \approx \overline{\overline{x}} \qquad \textit{12.65}$$

The standard deviation of the original distribution, σ, is much larger than the standard error of the mean.

$$\sigma \approx \sqrt{n}\, s_{\overline{x}} \qquad \textit{12.66}$$

21. CONFIDENCE LEVEL

The results of experiments are seldom correct 100% of the time. Recognizing this, researchers accept a certain probability of being wrong. In order to minimize this probability, the experiment is repeated several times. The number of repetitions depends on the desired level of confidence in the results.

If the results have a 5% probability of being wrong, the *confidence level*, C, is 95% that the results are correct, in which case the results are said to be *significant*. If the results have only a 1% probability of being wrong, the confidence level is 99%, and the results are said to be *highly significant*. Other confidence levels (90%, 99.5%, etc.) are used as appropriate.

22. APPLICATION: CONFIDENCE LIMITS

As a consequence of the central limit theorem, sample means of n items taken from a normal distribution with mean μ and standard deviation σ will be normally distributed with mean μ and variance σ^2/n. Thus, the probability that any given average, $\overline{x}$, exceeds some value, L, is

$$p\{\overline{x} > L\} = p\left\{ z > \left| \frac{L - \mu}{\frac{\sigma}{\sqrt{n}}} \right| \right\} \qquad \textit{12.67}$$

L is the *confidence limit* for the confidence level $1 - p\{\overline{x} > L\}$ (expressed as a percent). Values of z are read directly from the standard normal table. As an example, $z = 1.645$ for a 95% confidence level since only 5% of the curve is above that z in the upper tail. Similar values are given in Table 12.1. This is known as a *one-tail confidence limit* because all of the probability is given to one side of the variation.

Table 12.1 *Values of z for Various Confidence Levels*

confidence level, C	one-tail limit z	two-tail limit z
90%	1.28	1.645
95%	1.645	1.96
97.5%	1.96	2.17
99%	2.33	2.575
99.5%	2.575	2.81
99.75%	2.81	3.00

With *two-tail confidence limits*, the probability is split between the two sides of variation. There will be upper and lower confidence limits, UCL and LCL, respectively.

$$p\{\text{LCL} < \overline{x} < \text{UCL}\} = p\left\{ \frac{\text{LCL} - \mu}{\frac{\sigma}{\sqrt{n}}} < z < \frac{\text{UCL} - \mu}{\frac{\sigma}{\sqrt{n}}} \right\}$$

$$12.68$$

23. APPLICATION: BASIC HYPOTHESIS TESTING

A *hypothesis test* is a procedure that answers the question, "Did these data come from [a particular type of] distribution?" There are many types of tests, depending on the distribution and parameter being evaluated. The simplest hypothesis test determines whether an average value obtained from n repetitions of an experiment could have come from a population with known mean, μ, and standard deviation, σ. A practical application of this question is whether a manufacturing process has changed from what it used to be or should be. Of course, the answer (i.e., "yes" or "no") cannot be given with absolute certainty—there will be a confidence level associated with the answer.

The following procedure is used to determine whether the average of n measurements can be assumed (with a given confidence level) to have come from a known population.

step 1: Assume random sampling from a normal population.

step 2: Choose the desired confidence level, C.

step 3: Decide on a one-tail or two-tail test. If the hypothesis being tested is that the average has or has not *increased* or *decreased*, choose a one-tail test. If the hypothesis being tested is that the average has or has not *changed*, choose a two-tail test.

step 4: Use Table 12.1 or the standard normal table to determine the z-value corresponding to the confidence level and number of tails.

step 5: Calculate the actual standard normal variable, z'.

$$z' = \frac{\overline{x} - \mu}{\frac{\sigma}{\sqrt{n}}} \qquad 12.69$$

step 6: If $z' \geq z$, the average can be assumed (with confidence level C) to have come from a different distribution.

Example 12.16

When it is operating properly, a cement plant has a daily production rate that is normally distributed with a mean of 880 tons/day and a standard deviation of 21 tons/day. During an analysis period, the output is measured on 50 consecutive days, and the mean output is found to be 871 tons/day. With a 95% confidence level, determine whether the plant is operating properly.

Solution

step 1: Given.

step 2: $C = 0.95$ is given.

step 3: Since a specific direction in the variation is not given (i.e., the example does not ask whether the average has decreased), use a two-tail hypothesis test.

step 4: From Table 12.1, $z = 1.96$.

step 5: From Eq. 12.69,

$$z' = \left| \frac{\overline{x} - \mu}{\frac{\sigma}{\sqrt{n}}} \right| = \left| \frac{871 - 880}{\frac{21}{\sqrt{50}}} \right| = 3.03$$

Since $3.03 > 1.96$, the distributions are not the same. There is at least a 95% probability that the plant is not operating correctly.

24. APPLICATION: STATISTICAL PROCESS CONTROL

All manufacturing processes contain variation due to random and nonrandom causes. Random variation cannot be eliminated. *Statistical process control* (SPC) is the act of monitoring and adjusting the performance of a process to detect and eliminate nonrandom variation.

Mathematics

Statistical process control is based on taking regular (hourly, daily, etc.) samples of n items and calculating the mean, $\overline{x}$, and range, R, of the sample. To simplify the calculations, the range is used as a measure of the dispersion. These two parameters are graphed on their respective x-bar and *R-control charts*.[6] Confidence limits are drawn at $\pm 3\sigma/\sqrt{n}$. From a statistical standpoint, the control chart tests a hypothesis each time a point is plotted. When a point falls outside these limits, there is a 99.75% probability that the process is out of control. Until a point exceeds the control limits, no action is taken.[7]

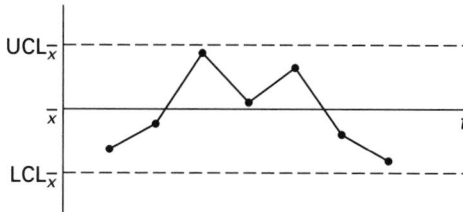

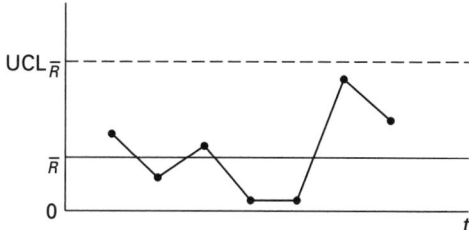

Figure 12.6 *Typical Statistical Process Control Charts*

25. MEASURES OF EXPERIMENTAL ADEQUACY

An experiment is said to be *accurate* if it is unaffected by experimental error. In this case, *error* is not synonymous with *mistake*, but rather includes all variations not within the experimenter's control.

For example, suppose a gun is aimed at a point on a target and five shots are fired. The mean distance from the point of impact to the sight in point is a measure of the alignment accuracy between the barrel and sights. The difference between the actual value and the experimental value is known as *bias*.

Precision is not synonymous with accuracy. Precision is concerned with the repeatability of the experimental results. If an experiment is repeated with identical results, the experiment is said to be precise.

The average distance of each impact from the centroid of the impact group is a measure of the precision of the

experiment. Thus, it is possible to have a highly precise experiment with a large bias.

Most of the techniques applied to experiments in order to improve the accuracy of the experimental results (e.g., repeating the experiment, refining the experimental methods, or reducing variability) actually increase the precision.

Sometimes the word *reliability* is used with regard to the precision of an experiment. Thus, a "reliable estimate" is used in the same sense as a "precise estimate."

Stability and *insensitivity* are synonymous terms. A stable experiment will be insensitive to minor changes in the experimental parameters. For example, suppose the centroid of a bullet group is 2.1 in from the target point at 65°F and 2.3 in away at 80°F. The sensitivity of the experiment to temperature change would be

$$\text{sensitivity} = \frac{\Delta x}{\Delta T} = \frac{2.3 \text{ in} - 2.1 \text{ in}}{80°\text{F} - 65°\text{F}} = 0.0133 \text{ in/}°\text{F}$$

26. LINEAR REGRESSION

If it is necessary to draw a straight line ($y = mx + b$) through n data points $(x_1, y_1), (x_2, y_2), \ldots, (x_n, y_n)$, the following method based on the *method of least squares* can be used.

step 1: Calculate the following nine quantities.

$$\sum x_i \quad \sum x_i^2 \quad \left(\sum x_i\right)^2 \quad \overline{x} = \frac{\sum x_i}{n} \quad \sum x_i y_i$$

$$\sum y_i \quad \sum y_i^2 \quad \left(\sum y_i\right)^2 \quad \overline{y} = \frac{\sum y_i}{n}$$

step 2: Calculate the slope, m, of the line.

$$m = \frac{n \sum (x_i y_i) - (\sum x_i)(\sum y_i)}{n \sum x_i^2 - (\sum x_i)^2} \qquad 12.70$$

step 3: Calculate the y-intercept, b.

$$b = \overline{y} - m\overline{x} \qquad 12.71$$

step 4: To determine the goodness of fit, calculate the correlation coefficient, r.

$$r = \frac{n \sum (x_i y_i) - (\sum x_i)(\sum y_i)}{\sqrt{\left(n \sum x_i^2 - (\sum x_i)^2\right)\left(n \sum y_i^2 - (\sum y_i)^2\right)}} \qquad 12.72$$

If m is positive, r will be positive; if m is negative, r will be negative. As a general rule, if the absolute value of r exceeds 0.85, the fit is good; otherwise, the fit is poor. r equals 1.0 if the fit is a perfect straight line.

---

[6]Other charts (e.g., the *sigma chart*, *p-chart*, and *c-chart*) are less common but are used as required.

[7]Other indications that a correction may be required are seven measurements on one side of the average and seven consecutively increasing measurements. Rules such as these detect shifts and trends.

A low value of r does not eliminate the possibility of a nonlinear relationship existing between x and y. It is possible that the data describe a parabolic, logarithmic, or other nonlinear relationship. (Usually this will be apparent if the data are graphed.) It may be necessary to convert one or both variables to new variables by taking squares, square roots, cubes, or logarithms, to name a few of the possibilities, in order to obtain a linear relationship. The apparent shape of the line through the data will give a clue to the type of variable transformation that is required. The curves in Fig. 12.7 may be used as guides to some of the simpler variable transformations.

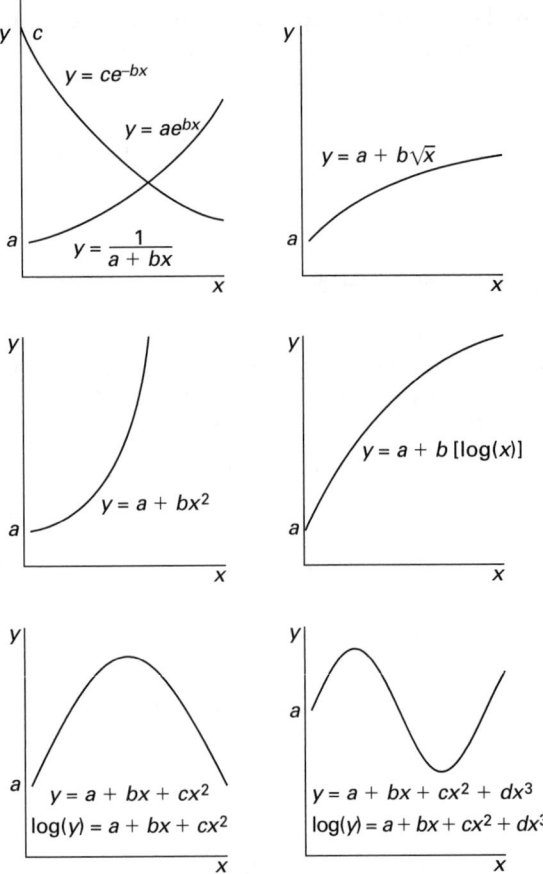

Figure 12.7 *Nonlinear Data Curves*

Figure 12.8 illustrates several common problems encountered in trying to fit and evaluate curves from experimental data. Figure 12.8(a) shows a graph of clustered data with several extreme points. There will be moderate correlation due to the weighting of the extreme points, although there is little actual correlation at low values of the variables. The extreme data should be excluded, or the range should be extended by obtaining more data.

Figure 12.8(b) shows that good correlation exists in general, but extreme points are missed and the overall correlation is moderate. If the results within the small linear range can be used, the extreme points should be

excluded. Otherwise, additional data points are needed, and curvilinear relationships should be investigated.

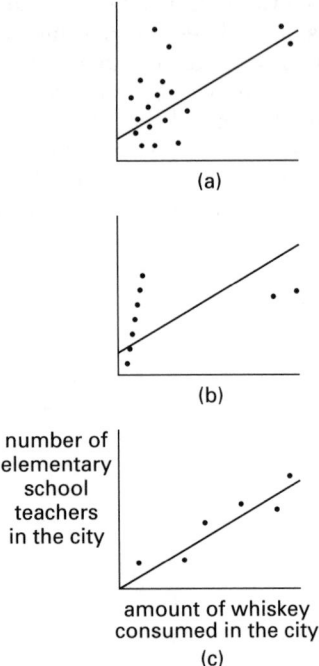

Figure 12.8 *Common Regression Difficulties*

Figure 12.8(c) illustrates the problem of drawing conclusions of cause and effect. There may be a predictable relationship between variables, but that does not imply a cause and effect relationship. In the case shown, both variables are functions of a third variable, the city population. But there is no direct relationship between the plotted variables.

Example 12.17

An experiment is performed in which the dependent variable y is measured against the independent variable x. The results are as follows.

x	y
1.2	0.602
4.7	5.107
8.3	6.984
20.9	10.031

(a) What is the least squares straight line equation that represents this data? (b) What is the correlation coefficient?

Solution

(a)
$$\sum x_i = 35.1$$

$$\sum y_i = 22.72$$

$$\sum x_i^2 = 529.23$$

$$\sum y_i^2 = 175.84$$

$$\left(\sum x_i\right)^2 = 1232.01$$

$$\left(\sum y_i\right)^2 = 516.38$$

$$\bar{x} = 8.775$$

$$\bar{y} = 5.681$$

$$\sum x_i y_i = 292.34$$

$$n = 4$$

From Eq. 12.70, the slope is

$$m = \frac{(4)(292.34) - (35.1)(22.72)}{(4)(529.23) - (35.1)^2} = 0.42$$

From Eq. 12.71, the y-intercept is

$$b = 5.681 - (0.42)(8.775) = 2.0$$

The equation of the line is

$$y = 0.42x + 2.0$$

(b) From Eq. 12.72, the correlation coefficient is

$$r = \frac{(4)(292.34) - (35.1)(22.72)}{\sqrt{\begin{array}{c}((4)(529.23) - 1232.01) \\ \times((4)(175.84) - 516.38)\end{array}}} = 0.914$$

Example 12.18

Repeat Ex. 12.17 assuming the relationship between the variables is nonlinear.

Solution

The first step is to graph the data. Since the graph has the appearance of the fourth case, it can be assumed that the relationship between the variables has the form of $y = a + b[\log(x)]$. Therefore, the variable change $z = \log(x)$ is made, resulting in the following set of data.

z	y
0.0792	0.602
0.672	5.107
0.919	6.984
1.32	10.031

If the regression analysis is performed on this set of data, the resulting equation and correlation coefficient are

$$y = -0.036 + 7.65z$$

$$r = 0.999$$

This is a very good fit. The relationship between the variable x and y is approximately

$$y = -0.036 + 7.65\log(x)$$

13 Computer Mathematics

1. POSITIONAL NUMBERING SYSTEMS

A *base-b number*, N_b, is made up of individual *digits*. In a *positional numbering system*, the position of a digit in the number determines that digit's contribution to the total value of the number. Specifically, the position of the digit determines the power to which the *base* (also known as the *radix*), b, is raised. For *decimal numbers*, the radix is 10, hence the description *base-10 numbers*.

$$(a_n a_{n-1} \cdots a_2 a_1 a_0)_b = a_n b^n + a_{n-1} b^{n-1} + \cdots$$
$$+ a_2 b^2 + a_1 b + a_0 \quad \textit{13.1}$$

The leftmost digit, a_n, contributes the greatest to the number's magnitude and is known as the *most significant digit* (MSD). The rightmost digit, a_0, contributes the least and is known as the *least significant digit* (LSD).

2. CONVERTING BASE-b NUMBERS TO BASE-10

Equation 13.1 converts base-b numbers to base-10 numbers. The calculation of the right-hand side of Eq. 13.1 is performed in the base-10 arithmetic and is known as the *expansion method*.[1]

---

[1]Equation 13.1 works with any base number. The *double-dabble (double and add) method* is a specialized method of converting from base-2 to base-10 numbers.

Converting base-b fractions to base-10 is similar to converting whole numbers and is accomplished by Eq. 13.2.

$$(0.a_1 a_2 \cdots a_m)_b = a_1 b^{-1} + a_2 b^{-2} + \cdots + a_m b^{-m} \quad \textit{13.2}$$

3. CONVERTING BASE-10 NUMBERS TO BASE-b

The *remainder method* is used to convert base-10 numbers to base-b numbers. This method consists of successive divisions by the base, b, until the quotient is zero. The base-b number is found by taking the remainders in the reverse order from which they were found. This method is illustrated in Exs. 13.1 and 13.3.

Converting a base-10 fraction to base-b requires multiplication of the base-10 fraction and subsequent fractional parts by the base. The base-b fraction is formed from the integer parts of the products taken in the same order in which they were determined. This is illustrated in Ex. 13.2(d).

4. BINARY NUMBER SYSTEM

There are only two *binary digits (bits)* in the *binary number system*: zero and one.[2] Thus, all binary numbers consist of strings of bits (i.e., zeros and ones). The leftmost bit is known as the *most significant bit* (MSB), and rightmost bit is the *least significant bit* (LSB).

As with digits from other numbering systems, bits can be added, subtracted, multiplied, and divided, although only digits 0 and 1 are allowed in the results. The rules of bit addition are

$$0 + 0 = 0$$
$$0 + 1 = 1$$
$$1 + 0 = 1$$
$$1 + 1 = 0 \text{ carry } 1$$

Example 13.1

(a) Convert $(1011)_2$ to base-10. (b) Convert $(75)_{10}$ to base-2.

---

[2]Alternatively, the binary states may be called *true* and *false*, *on* and *off*, *high* and *low*, or *positive* and *negative*.

Solution

(a) Using Eq. 13.1 with $b = 2$,

$$(1)(2)^3 + (0)(2)^2 + (1)(2)^1 + 1 = 11$$

(b) Use the remainder method (see Sec. 13-3).

$$75 \div 2 = 37 \text{ remainder } 1$$
$$37 \div 2 = 18 \text{ remainder } 1$$
$$18 \div 2 = 9 \text{ remainder } 0$$
$$9 \div 2 = 4 \text{ remainder } 1$$
$$4 \div 2 = 2 \text{ remainder } 0$$
$$2 \div 2 = 1 \text{ remainder } 0$$
$$1 \div 2 = 0 \text{ remainder } 1$$

The binary representation of $(75)_{10}$ is $(1001011)_2$.

5. OCTAL NUMBER SYSTEM

The *octal (base-8) system* is one of the alternatives to working with long binary numbers. Only the digits 0 through 7 are used. The rules for addition in the octal system are the same as for the decimal system except that the digits 8 and 9 do not exist. For example,

$$7 + 1 = 6 + 2 = 5 + 3 = (10)_8$$
$$7 + 2 = 6 + 3 = 5 + 4 = (11)_8$$
$$7 + 3 = 6 + 4 = 5 + 5 = (12)_8$$

Example 13.2

Perform the following operations.

(a) $(2)_8 + (5)_8$

(b) $(7)_8 + (6)_8$

(c) Convert $(75)_{10}$ to base-8.

(d) Convert $(0.14)_{10}$ to base-8.

(e) Convert $(13)_8$ to base-10.

(f) Convert $(27.52)_8$ to base-10.

Solution

(a) The sum of 2 and 5 in base-10 is 7, which is less than 8 and, therefore, is a valid number in the octal system. The answer is $(7)_8$.

(b) The sum of 7 and 6 in base-10 is 13, which is greater than 8 (and, therefore, needs to be converted). Using the remainder method (Sec. 13-3),

$$13 \div 8 = 1 \text{ remainder } 5$$
$$1 \div 8 = 0 \text{ remainder } 1$$

The answer is $(15)_8$.

(c) Use the remainder method (see Sec. 13-3).

$$75 \div 8 = 9 \text{ remainder } 3$$
$$9 \div 8 = 1 \text{ remainder } 1$$
$$1 \div 8 = 0 \text{ remainder } 1$$

The answer is $(113)_8$.

(d) Refer to Sec. 13-3.

$$0.14 \times 8 = 1.12$$
$$0.12 \times 8 = 0.96$$
$$0.96 \times 8 = 7.68$$
$$0.68 \times 8 = 5.44$$
$$0.44 \times 8 = \text{etc.}$$

The answer, $(0.1075\cdots)_8$, is constructed from the integer parts of the products.

(e) Use Eq. 13.1.

$$(1)(8) + 3 = (11)_{10}$$

(f) Use Eqs. 13.1 and 13.2.

$$(2)(8)^1 + (7)(8)^0 + (5)(8)^{-1} + (2)(8)^{-2} = 16 + 7 + \frac{5}{8} + \frac{2}{64}$$
$$= (23.656)_{10}$$

6. HEXADECIMAL NUMBER SYSTEM

The *hexadecimal (base-16) system* is a shorthand method of representing the value of four binary digits at a time.[3] Since 16 distinctly different characters are needed, the capital letters A through F are used to represent the decimal numbers 10 through 15. The progression of hexadecimal numbers is illustrated in Table 13.1.

Example 13.3

(a) Convert $(4D3)_{16}$ to base-10. (b) Convert $(1475)_{10}$ to base-16. (c) Convert $(0.8)_{10}$ to base-16.

Solution

(a) The hexadecimal number D is 13 in base-10. Using Eq. 13.1,

$$(4)(16)^2 + (13)(16)^1 + 3 = (1235)_{10}$$

(b) Use the remainder method (see Sec. 13-3).

$$1475 \div 16 = 92 \text{ remainder } 3$$
$$92 \div 16 = 5 \text{ remainder } 12$$
$$5 \div 16 = 0 \text{ remainder } 5$$

Since $(12)_{10}$ is $(C)_{16}$, (or hex C), the answer is $(5C3)_{16}$.

---

[3]The term *hex number* is often heard.

(c) Refer to Sec. 13-3.

$$0.8 \times 16 = 12.8$$
$$0.8 \times 16 = 12.8$$
$$0.8 \times 16 = \text{etc.}$$

Since $(12)_{10} = (C)_{16}$, the answer is $(0.CCCCC\cdots)_{16}$.

7. CONVERSIONS AMONG BINARY, OCTAL, AND HEXADECIMAL NUMBERS

The octal system is closely related to the binary system since $(2)^3 = 8$. Conversion from a binary to an octal number is accomplished directly by starting at the LSB (right-hand bit) and grouping the bits in threes. Each group of three bits corresponds to an octal digit. Similarly, each digit in an octal number generates three bits in the equivalent binary number.

Conversion from a binary to a hexadecimal number starts by grouping the bits (starting at the LSB) into fours. Each group of four bits corresponds to a hexadecimal digit. Similarly, each digit in a hexadecimal number generates four bits in the equivalent binary number.

Table 13.1 *Binary, Octal, Decimal, and Hexadecimal Equivalents*

binary	octal	decimal	hexadecimal
0	0	0	0
1	1	1	1
10	2	2	2
11	3	3	3
100	4	4	4
101	5	5	5
110	6	6	6
111	7	7	7
1000	10	8	8
1001	11	9	9
1010	12	10	A
1011	13	11	B
1100	14	12	C
1101	15	13	D
1110	16	14	E
1111	17	15	F
10000	20	16	10

Conversion between octal and hexadecimal numbers is easiest when the number is first converted to a binary number.

Example 13.4

(a) Convert $(5431)_8$ to base-2. (b) Convert $(1001011)_2$ to base-8. (c) Convert $(1011111101111001)_2$ to base-16.

Solution

(a) Convert each octal digit to binary digits.

$$(5)_8 = (101)_2$$
$$(4)_8 = (100)_2$$
$$(3)_8 = (011)_2$$
$$(1)_8 = (001)_2$$

The answer is $(101100011001)_2$.

(b) Group the bits into threes starting at the LSB.

$$1 \quad 001 \quad 011$$

Convert these groups into their octal equivalents.

$$(1)_2 = (1)_8$$
$$(001)_2 = (1)_8$$
$$(011)_2 = (3)_8$$

The answer is $(113)_8$.

(c) Group the bits into fours starting at the LSB.

$$1011 \quad 1111 \quad 0111 \quad 1001$$

Convert these groups into their hexadecimal equivalents.

$$(1011)_2 = (B)_{16}$$
$$(1111)_2 = (F)_{16}$$
$$(0111)_2 = (7)_{16}$$
$$(1001)_2 = (9)_{16}$$

The answer is $(BF79)_{16}$.

8. COMPLEMENT OF A NUMBER

The *complement*, N^*, of a number, N, depends on the machine (computer, calculator, etc.) being used. Assuming that the machine has a maximum number, n, of digits per integer number stored, the b's and $(b-1)$'s complements are

$$N_b^* = b^n - N \qquad 13.3$$
$$N_{b-1}^* = N_b^* - 1 \qquad 13.4$$

For a machine that works in base-10 arithmetic, the *tens* and *nines* complements are

$$N_{10}^* = 10^n - N \qquad 13.5$$
$$N_9^* = N_{10}^* - 1 \qquad 13.6$$

For a machine that works in base-2 arithmetic, the *twos* and *ones complements* are

$$N_2^* = 2^n - N \qquad 13.7$$
$$N_1^* = N_2^* - 1 \qquad 13.8$$

9. APPLICATION OF COMPLEMENTS TO COMPUTER ARITHMETIC

Equations 13.9 and 13.10 are the practical applications of complements to computer arithmetic.

$$(N^*)^* = N \qquad 13.9$$
$$M - N = M + N^* \qquad 13.10$$

The binary ones complement is easily found by switching all of the ones and zeros to zeros and ones, respectively. It can be combined with a technique known as *end-around carry* to perform subtraction. End-around carry is the addition of the *overflow bit* to the sum of N and its ones complement.

Example 13.5

(a) Simulate the operation of a base-10 machine with a capacity of four digits per number and calculate the difference $(18)_{10} - (6)_{10}$ with tens complements.

(b) Simulate the operation of a base-2 machine with a capacity of five digits per number and calculate the difference $(01101)_2 - (01010)_2$ with twos complements.

(c) Solve part (b) with a ones complement and end-around carry.

Solution

(a) The tens complement of 6 is

$$(6)_{10}^* = (10)^4 - 6 = 10{,}000 - 6 = 9994$$

Using Eq. 13.10,

$$18 - 6 = 18 + (6)_{10}^* = 18 + 9994 = 10{,}012$$

However, the machine has a maximum capacity of four digits. Therefore, the leading 1 is dropped, leaving 0012 as the answer.

(b) The twos complement of $(01010)_2$ is

$$N_2^* = (2)^5 - N = (32)_{10} - N$$
$$= (100000)_2 - (01010)_2$$
$$= (10110)_2$$

From Eq. 13.10,

$$(01101)_2 - (01010)_2 = (01101)_2 + (10110)_2$$
$$= (100011)_2$$

Since the machine has a capacity of only five bits, the leftmost bit is dropped, leaving $(00011)_2$ as the difference.

(c) The ones complement is found by reversing all the digits.

$$(01010)_1^* = (10101)_2$$

Adding the ones complement,

$$(01101)_2 + (10101)_2 = (100010)_2$$

The leading bit is the overflow bit, which is removed and added to give the difference.

$$(00010)_2 + (1)_2 = (00011)_2$$

10. COMPUTER REPRESENTATION OF NEGATIVE NUMBERS

On paper, a minus sign indicates a negative number. This representation is not possible in a machine. Hence, one of the n digits, usually the MSB, is reserved for sign representation. (This reduces the machine's capacity to represent numbers to $n - 1$ bits per number.) It is arbitrary whether the sign bit is 1 or 0 for negative numbers as long as the MSB is different for positive and negative numbers.

The ones complement is ideal for forming a negative number since it automatically reverses the MSB. For example, $(00011)_2$ is a five-bit representation of decimal 3. The ones complement is $(11100)_2$, which is recognized as a negative number because the MSB is 1.

Example 13.6

Simulate the operation of a six-digit binary machine that uses ones complements for negative numbers.

(a) What is the machine representation of $(-27)_{10}$?
(b) What is the decimal equivalent of the twos complement of $(-27)_{10}$? (c) What is the decimal equivalent of the ones complement of $(-27)_{10}$?

Solution

(a) $(27)_{10} = (011011)_2$. The negative of this number is the same as the ones complement: $(100100)_2$.

(b) The twos complement is one more than the ones complement. (See Eqs. 13.7 and 13.8.) Therefore, the twos complement is

$$(100100)_2 + 1 = (100101)_2$$

This represents $(-26)_{10}$.

(c) From Eq. 13.9, the complement of a complement of a number is the original number. Therefore, the decimal equivalent is 27.

11. BOOLEAN ALGEBRA FUNDAMENTAL POSTULATES

Boolean Algegra is a system of mathematics that deals with the algebraic treatment of logic. When the variables of the system are limited to only one of two values, the system is called two-valued. *Switching algebra* is the term often used to describe two-valued Boolean algebra. The basic postulates of switching algebra follow.

- Postulate 1: A Boolean variable x has two possible exclusive values, 0 and 1. That is,

$$\text{if } x = 0 \text{ then } x \neq 1, \text{ and}$$
$$\text{if } x = 1 \text{ then } x \neq 0$$

- Postulate 2: The NOT operation, "$\overline{x}$" or x', is defined as

$$\overline{0} = 1 \quad \overline{1} = 0$$

- Postulate 3: The logical operations AND ($\wedge$ or $\cdot$), and OR ($\vee$ or $+$) are defined as

$$0 \cdot 0 = 0 \qquad 0 + 0 = 0$$
$$0 \cdot 1 = 0 \qquad 0 + 1 = 1$$
$$1 \cdot 0 = 0 \qquad 1 + 0 = 1$$
$$1 \cdot 1 = 1 \qquad 1 + 1 = 1$$

Note: Sometimes the symbology for the AND operation is dropped completely and is represented as $x_1 x_2$ for $x_1 \cdot x_2$.

12. BOOLEAN ALGEBRA LAWS

The process of reducing logical expressions to simpler forms is aided by the use of the following relationships, or laws, of switching algebra.[4] In the following, x can represent a single variable or a general logical function.

- Special properties of 0 and 1

$$0 + x = x \qquad 0 \cdot x = 0$$
$$1 + x = 1 \qquad 1 \cdot x = x$$

- Idempotence Laws

$$x + x = x \qquad x \cdot x = x$$

- Complementation Laws

$$x + \overline{x} = 1 \qquad x \cdot \overline{x} = 0$$

- Involution

$$\overline{(\overline{x})} = x$$

---

[4]The reduction of such expressions is desired to minimize circuit complexity and costs when trying to realize a given expression with electronic logic circuits.

- Commutative Laws

$$x + y = y + x \qquad x \cdot y = y \cdot x$$

- Associative Laws

$$x + (y + z) = (x + y) + z$$
$$x \cdot (y \cdot z) = (x \cdot y) \cdot z$$

- Distributive Laws

$$x \cdot (y + z) = (x \cdot y) + (x \cdot z)$$
$$x + (y \cdot z) = (x + y) \cdot (x + z)$$

- Absorption Laws

$$x + (x \cdot y) = x \qquad x \cdot (x + y) = x$$
$$x + (\overline{x} \cdot y) = x + y \qquad x \cdot (\overline{x} + y) = x \cdot y$$

13. BOOLEAN ALGEBRA THEOREMS

The simplification of logical expressions is further enhanced by a set of relationships that exist as theorems, although one takes the name of a law. These theorems follow.

- Simplification Theorems

$$xy + x\overline{y} = x \qquad (x + y)(x + \overline{y}) = x$$

Note: The AND symbol is often assumed between parenthesis.

$$x + xy = x \qquad x(x + y) = x$$
$$(x + \overline{y})y = xy \qquad x\overline{y} + y = x + y$$

- De Morgan's Laws

$$\overline{(x + y + z + \cdots)} = \overline{x}\, \overline{y}\, \overline{z} \cdots$$
$$\overline{(xyz \cdots)} = \overline{x} + \overline{y} + \overline{z} \cdots$$
$$\overline{[f(x_1, x_2, x_3, \cdots, x_n, 0, 1, +, \cdot)]}$$
$$= f(\overline{x_1}, \overline{x_2}, \overline{x_3}, \cdots, \overline{x_n}, 1, 0, \cdot, +)$$

- Duality

$$(x + y + z + \cdots)^{\mathrm{D}} = xyz \cdots$$
$$(xyz \cdots)^{\mathrm{D}} = x + y + z + \cdots$$
$$[f(x_1, x_2, x_3, \cdots x_n, 0, 1, +, \cdot)]^{\mathrm{D}}$$
$$= f(x_1, x_2, x_3, \cdots, x_n, 1, 0, \cdot, +)]$$

- Consensus Theorem

$$xy + yz + \overline{x}z = xy + \overline{x}z$$
$$(x + y)(y + z)(\overline{x} + z) = (x + y)(\overline{x} + z)$$
$$(x + y)(\overline{x} + z) = xz + \overline{x}y$$

14 Numerical Analysis

1. NUMERICAL METHODS

Although the roots of second-degree polynomials are easily found by a variety of methods (by factoring, completing the square, or using the quadratic equation), easy methods of solving cubic and higher-order equations exist only for specialized cases. However, cubic and higher-order equations occur frequently in engineering, and they are difficult to factor. Trial and error solutions, including graphing, are usually satisfactory for finding only the general region in which the root occurs.

Numerical analysis is a general subject that covers, among other things, iterative methods for evaluating roots to equations. The most efficient numerical methods are too complex to present and, in any case, work by hand. However, some of the simpler methods are presented here. Except in critical problems that must be solved in real time, a few extra calculator or computer iterations will make no difference.[1]

2. FINDING ROOTS: BISECTION METHOD

The *bisection method* is an iterative method that "brackets" (also known as "straddles") an interval containing the *root* or *zero* of a particular equation.[2] The size of the interval is halved after each iteration. As the method's name suggests, the best estimate of the root after any iteration is the midpoint of the interval. The maximum error is half the interval length. The procedure continues until the size of the maximum error is "acceptable."[3]

---

[1]Most advanced hand-held calculators now have "root finder" functions that use numerical methods to iteratively solve equations.

[2]The equation does not have to be a pure polynomial. The bisection method requires only that the equation be defined and determinable at all points in the interval.

[3]The bisection method is not a closed method. Unless the root actually falls on the midpoint of one iteration's interval, the method continues indefinitely. Eventually, the magnitude of the maximum error is small enough not to matter.

The disadvantages of the bisection method are (a) the slowness in converging to the root, (b) the need to know the interval containing the root before starting, and (c) the inability to determine the existence of or find other real roots in the starting interval.

The bisection method starts with two values of the independent variable, $x = L_0$ and $x = R_0$, which straddle a root. Since the function passes through zero at a root, $f(L_0)$ and $f(R_0)$ will have opposite signs. The following algorithm describes the remainder of the bisection method.

Let n be the iteration number. Then, for $n = 0, 1, 2, \ldots$, perform the following steps until sufficient accuracy is attained.

step 1: Set $m = \frac{1}{2}(L_n + R_n)$.

step 2: Calculate $f(m)$.

step 3: If $f(L_n)f(m) \leq 0$, set $L_{n+1} = L_n$ and $R_{n+1} = m$. Otherwise, set $L_{n+1} = m$ and $R_{n+1} = R_n$.

step 4: $f(x)$ has at least one root in the interval (L_{n+1}, R_{n+1}). The estimated value of that root, x^*, is

$$x^* \approx \tfrac{1}{2}(L_{n+1} + R_{n+1})$$

The maximum error is $\frac{1}{2}(R_{n+1} - L_{n+1})$.

Example 14.1

Use two iterations of the bisection method to find a root of

$$f(x) = x^3 - 2x - 7$$

Solution

The first step is to find L_0 and R_0, which are the values of x that straddle a root and have opposite signs. A table can be made and values of $f(x)$ calculated for random values of x.

x	-2	-1	0	$+1$	$+2$	$+3$
$f(x)$	-11	-6	-7	-8	-3	$+14$

Since $f(x)$ changes sign between $x = 2$ and $x = 3$, $L_0 = 2$ and $R_0 = 3$. First iteration, $n = 0$:

$$m = \tfrac{1}{2}(2 + 3) = 2.5$$
$$f(2.5) = (2.5)^3 - (2)(2.5) - 7 = 3.625$$

Since $f(2.5)$ is positive, a root must exist in the interval $(2, 2.5)$. Therefore, $L_1 = 2$ and $R_1 = 2.5$. At this point, the best estimate of the root is

$$x^* \approx \tfrac{1}{2}(2 + 2.5) = 2.25$$

The maximum error is $\tfrac{1}{2}(2.5 - 2) = 0.25$.

second iteration, $n = 1$:

$$m = \tfrac{1}{2}(2 + 2.5) = 2.25$$
$$f(2.25) = (2.25)^3 - (2)(2.25) - 7 = -0.1094$$

Since $f(2.25)$ is negative, a root must exist in the interval $(2.25, 2.5)$. Therefore, $L_2 = 2.25$ and $R_2 = 2.5$. The best estimate of the root is

$$x^* \approx \tfrac{1}{2}(2.25 + 2.5) = 2.375$$

The maximum error is $\tfrac{1}{2}(2.5 - 2.25) = 0.125$.

3. FINDING ROOTS: NEWTON'S METHOD

Many other methods have been developed to overcome one or more of the disadvantages of the bisection method. These methods have their own disadvantages.[4]

Newton's method is a particular form of *fixed-point iteration*. In this sense, "fixed point" is often used as a synonym for "root" or "zero." However, fixed-functions with the characteristic property $x = g(x)$ such that the limit of $g(x)$ is the fixed point (i.e., is the root).

All fixed-point techniques require a starting point. Preferably, the starting point will be close to the actual root.[5] And, while Newton's method converges quickly, it requires the function to be continuously differentiable.

Newton's method algorithm is simple. At each iteration ($n = 0, 1, 2$, etc.), Eq. 14.1 estimates the root. The maximum error is determined by looking at how much the estimate changes after each iteration. If the change between the previous and current estimates (representing the magnitude of error in the estimate) is too large, the current estimate is used as the independent variable for the subsequent iteration.[6]

$$x_{n+1} = g(x_n) = x_n - \frac{f(x_n)}{f'(x_n)} \qquad 14.1$$

---

[4]The *regula falsi (false position) method* converges faster than the bisection method but is unable to specify a small interval containing the root. The *secant method* is prone to round-off errors and gives no indication of the remaining distance to the root.
[5]Theoretically, the only penalty for choosing a starting point too far away from the root will be a slower convergence to the root.
[6]Actually, the theory defining the maximum error is more definite than this. For example, for a large enough value of n, the error decreases approximately linearly. Therefore, the consecutive values of x_n converge linearly to the root as well.

Example 14.2

Solve Ex. 14.1 using two iterations of Newton's method. Use $x_0 = 2$.

Solution

The function and its first derivative are

$$f(x) = x^3 - 2x - 7$$
$$f'(x) = 3x^2 - 2$$

first iteration, $n = 0$:

$$x_0 = 2$$
$$f(x_0) = f(2) = (2)^3 - (2)(2) - 7 = -3$$
$$f'(x_0) = f'(2) = (3)(2)^2 - 2 = 10$$
$$x_1 = x_0 - \frac{f(x_0)}{f'(x_0)} = 2 - \frac{-3}{10} = 2.3$$

second iteration, $n = 1$:

$$x_1 = 2.3$$
$$f(x_1) = (2.3)^3 - (2)(2.3) - 7 = 0.567$$
$$f'(x_1) = (3)(2.3)^2 - 2 = 13.87$$
$$x_2 = x_1 - \frac{f(x_1)}{f'(x_1)} = 2.3 - \frac{0.567}{13.87} = 2.259$$

4. NONLINEAR INTERPOLATION: LAGRANGIAN INTERPOLATING POLYNOMIAL

Interpolating between two points of known data is common in engineering. Primarily due to its simplicity and speed, straight-line interpolation is used most often. Even if more than two points on the curve are explicitly known, they are not used. Since straight-line interpolation ignores all but two of the points on the curve, it ignores the effects of curvature.

A more powerful technique that accounts for the curvature is the *Lagrangian interpolating polynomial*. This method uses an nth degree parabola (polynomial) as the interpolating curve.[7] This method requires that $f(x)$ be continuous and real-valued on the interval $[x_0, x_n]$ and that $n + 1$ values of $f(x)$ are known corresponding to $x_0, x_1, x_2, \ldots, x_n$.

The procedure for calculating $f(x)$ at some intermediate point x^* starts by calculating the Lagrangian interpolating polynomial for each known point.

$$L_k(x^*) = \prod_{\substack{i=0 \\ i \neq k}}^{n} \frac{x^* - x_i}{x_k - x_i} \qquad 14.2$$

---

[7]The Lagrangian interpolating polynomial reduces to straight-line interpolation if only two points are used.

The value of $f(x)$ at x^* is calculated from Eq. 14.3.

$$f(x^*) = \sum_{k=0}^{n} f(x_k) L_k(x^*) \qquad 14.3$$

The Lagrangian interpolating polynomial has two primary disadvantages. The first is that a large number of additions and multiplications are needed.[8] The second is that the method does not indicate how many interpolating points should be (or should have been) used. Other interpolating methods have been developed that overcome these disadvantages.[9]

Example 14.3

A real-valued function has the following values.

$$f(1) = 3.5709$$
$$f(4) = 3.5727$$
$$f(6) = 3.5751$$

Use the Lagrangian interpolating polynomial to determine the value of the function at 3.5.

Solution

The procedure for applying Eq. 14.2, the Lagrangian interpolating polynomial, is illustrated in tabular form. Notice that the term corresponding to $i = k$ is omitted from the product.

$$\begin{array}{ccc} i=0 & i=1 & i=2 \end{array}$$

$$k = 0: \quad L_0(3.5) = \left(\frac{3.5 - 1}{1 - 1}\right)\left(\frac{3.5 - 4}{1 - 4}\right)\left(\frac{3.5 - 6}{1 - 6}\right)$$
$$= 0.08333$$

$$k = 1: \quad L_1(3.5) = \left(\frac{3.5 - 1}{4 - 1}\right)\left(\frac{3.5 - 4}{4 - 4}\right)\left(\frac{3.5 - 6}{4 - 6}\right)$$
$$= 1.04167$$

$$k = 2: \quad L_2(3.5) = \left(\frac{3.5 - 1}{6 - 1}\right)\left(\frac{3.5 - 4}{6 - 4}\right)\left(\frac{3.5 - 6}{6 - 6}\right)$$
$$= -0.12500$$

Equation 14.3 is used to calculate the estimate.

$$f(3.5) = (3.5709)(0.08333) + (3.5727)(1.04167)$$
$$+ (3.5751)(-0.12500)$$
$$= 3.57225$$

---

[8]As with the numerical methods for finding roots previously discussed, the number of calculations probably will not be an issue if the work is performed by a calculator or computer.
[9]Other common methods for performing interpolation include the *Newton form* and *divided difference table*.

5. NONLINEAR INTERPOLATION: NEWTON'S INTERPOLATING POLYNOMIAL

Newton's form of the interpolating polynomial is more efficient than the Lagrangian method of interpolating between known points.[10] Given $n + 1$ known points for $f(x)$, the *Newton form of the interpolating polynomial* is

$$f(x^*) = \sum_{i=0}^{n} \left(f[x_0, x_1, \ldots, x_i] \prod_{j=0}^{i-1} (x^* - x_j) \right) \qquad 14.4$$

$f[x_0, x_1, \ldots, x_i]$ is known as the ith *divided difference*.

$$f[x_0, x_1, \ldots, x_i] = \sum_{k=0}^{i} \left(\frac{f(x_k)}{(x_k - x_0) \cdots (x_k - x_{k-1})} \times (x_k - x_{k+1}) \cdots (x_k - x_i) \right) \qquad 14.5$$

It is necessary to define the following two terms.

$$f[x_0] = f(x_0) \qquad 14.6$$

$$\prod (x^* - x_j) = 1 \qquad [i = 0] \qquad 14.7$$

Example 14.4

Repeat Ex. 14.3 using Newton's form of the interpolating polynomial.

Solution

Since there are $n + 1 = 3$ data points, $n = 2$. Evaluate the terms for $i = 0$ to 2.

$i = 0$:

$$f[x_0] \prod_{j=0}^{-1} (x^* - x_j) = f(x_0)(1) = f(x_0)$$

$i = 1$:

$$f[x_0, x_1] \prod_{j=0}^{0} (x^* - x_j) = f[x_0, x_1](x^* - x_0)$$

$$f[x_0, x_1] = \frac{f(x_0)}{x_0 - x_1} + \frac{f(x_1)}{x_1 - x_0}$$

$i = 2$:

$$f[x_0, x_1, x_0] \prod_{j=0}^{1} (x^* - x_j) = f[x_0, x_1, x_0](x^* - x_0)$$
$$\times (x^* - x_1)$$

$$f[x_0, x_1, x_2] = \frac{f(x_0)}{(x_0 - x_1)(x_0 - x_2)}$$
$$+ \frac{f(x_1)}{(x_1 - x_0)(x_1 - x_2)}$$
$$+ \frac{f(x_2)}{(x_2 - x_0)(x_2 - x_1)}$$

---

[10]In this case, "efficiency" relates to the ease in adding new known points without having to repeat all previous calculations.

Substitute known values.

$$f(3.5) = 3.5709 + \left(\frac{3.5709}{1-4} + \frac{3.5727}{4-1} \right)(3.5-1)$$
$$+ \left[\frac{3.5709}{(1-4)(1-6)} + \frac{3.5727}{(4-1)(4-6)} \right.$$
$$\left. + \frac{3.5751}{(6-1)(6-4)} \right]$$
$$\times (3.5-1)(3.5-4)$$
$$= 3.57225$$

This answer is the same as that determined in Ex. 14.3.

15 Advanced Engineering Mathematics

1. OVERVIEW

This chapter emphasizes advanced engineering mathematics not covered in earlier chapters. A grasp of such material is necessary in order to understand various aspects of electrical engineering, especially at the design and theoretical levels. The province of the engineer in mathematics, however, resides in the recognition of the form of, and restrictions on, the solution. To that end, this chapter is devoted.

2. POWER SERIES METHOD: HOMOGENEOUS LINEAR DIFFERENTIAL EQUATIONS

When the coefficients of a homogeneous linear differential equation are a function of x, the methods explained in Ch. 11 are no longer applicable.[1] Further, the solutions may be *nonelementary functions*.[2] Equations of this type include Bessel's equation and Legendre's equation.[3]

- Bessel Equation of Order ν

$$x^2 y'' + xy' + \left(x^2 - \nu^2\right) y = 0 \qquad \textbf{15.1}$$

- Legendre Equation of Order λ

$$\left(1 - x^2\right) y'' - 2xy' + \lambda\left(\lambda + 1\right) y = 0$$
$$[-1 < x < 1] \qquad \textbf{15.2}$$

The solutions of such equations are power series; the method of obtaining them is the *power series method*. A power series is of the following form.[4]

$$\sum_{m=0}^{\infty} c_m \left(x - a\right)^m$$
$$= c_0 + c_1 \left(x - a\right) + c_2 \left(x - a\right)^2 + \ldots \quad \textbf{15.3}$$

The constants $c_0, c_1, \ldots$ are the *coefficients* of the series. The constant a is called the *center*. The variable is x.

A common power series is the Maclaurin series.

$$\frac{1}{1 - x} = \sum_{m=0}^{\infty} x^m$$
$$= 1 + x + x^2 + \ldots$$
$$\left[\begin{array}{l}\text{If } |x| < 1, \text{ then this power series} \\ \text{is known as a geometric series.}\end{array}\right] \quad \textbf{15.4}$$

---

[1]Equations discussed in this chapter are *ordinary differential equations*, or ODEs, as are most equations in Ch. 11. The term "ordinary" distinguishes these equations from *partial differential equations*, or PDEs.
[2]*Elementary functions* are those that can be formed by a finite number of mathematical operations (addition, subtraction, multiplication, and division) from algebraic functions and exponential, logarithmic, and trigonometric functions. *Nonelementary functions* are those that cannot be so formed. Do not confuse nonelementary functions with transcendental functions. *Transcendental functions* are nonalgebraic functions, for example, exponential, logarithmic, and trigonometric.
[3]Bessel and Legendre equations and their associated polynomials occur in many engineering applications. As a single example, they are used in approximating the response of filter circuits.
[4]Note that a power series does not include negative or fractional powers of x.

$$e^x = \sum_{m=0}^{\infty} \frac{x^m}{m!}$$

$$= 1 + x + \frac{x^2}{2!} + \frac{x^3}{3!} + \dots \qquad \textbf{15.5}$$

$$\cos x = \sum_{m=0}^{\infty} \frac{(-1)^m x^{2m}}{2m!}$$

$$= 1 - \frac{x^2}{2!} + \frac{x^4}{4!} - \dots \qquad \textbf{15.6}$$

$$\sin x = \sum_{m=0}^{\infty} \frac{(-1)^m x^{2m+1}}{(2m+1)!}$$

$$= x - \frac{x^3}{3!} + \frac{x^5}{5!} - \dots \qquad \textbf{15.7}$$

The steps to the power series method are as follows.

step 1: Given a differential equation, represent the given functions in the equation by a power series in powers of x (or $x - a$, if solutions of this type are desired). If the functions are polynomials, as is often the case in practical applications, nothing need be done in this first step.

step 2: Assume a solution in the form of a power series.

step 3: Substitute this series and the series obtained by termwise differentiation into the original equation.

step 4: Collect like powers of x and equate the sum of their coefficients to zero.

step 5: Determine the unknown coefficients in the assumed power series solution successively. The result is the solution.

Example 15.1

Solve the following homogeneous, first-order linear ordinary differential equation with constant coefficients.[5]

$$y' + ky = 0$$

Solution

step 1: Represent the given functions in the equation by a power series in powers of x. Since the coefficients are constant (i.e., no functions are used in the equation), this step is not required.

step 2: Assume a solution in the form of a power series.

$$y(x) = c_0 + c_1 x + c_2 x^2 + \dots = \sum_{m=0}^{\infty} c_m x^m$$

---
[5]The solution to this differential equation was found using other methods in Sec. 2 of Ch. 11. The power series method is used on this simple differential equation to clarify the method.

step 3: Substitute this series and the series obtained by termwise differentiation into the original equation.

Differentiation of $y(x)$ termwise results in the following equation.

$$y'(x) = 0 + c_1 + 2c_2 x + 3c_3 x^2 + \dots$$

Substituting into the original equation gives

$$y' + ky = 0$$
$$\left(0 + c_1 + 2c_2 x + 3c_3 x^2 + \dots\right)$$
$$+ k\left(c_0 + c_1 x + c_2 x^2 + \dots\right) = 0$$

step 4: Collect like powers of x and equate the sum of their coefficients to zero.

$$(c_1 + kc_0) + (2c_2 + kc_1) x$$
$$+ (3c_3 + kc_2) x^2 + \dots = 0$$

For this equation to be true, the sum of the coefficients of each power of x must be zero. Thus,

$$c_1 + kc_0 = 0$$
$$2c_2 + kc_1 = 0$$
$$3c_3 + kc_2 = 0 \dots$$

step 5: Determine the unknown coefficients in the assumed power series successively. That is, using the results of step 4, determine the coefficients in terms of c_0, which remains arbitrary.

$$c_1 = -kc_0$$
$$c_2 = \frac{-k}{2}c_1 = \frac{k^2}{2!}c_0$$
$$c_3 = \frac{-k}{3}c_2 = \frac{-k^3}{3!}c_0 \dots$$

Substitute these values into the assumed power series solution from step 2.

$$y(x) = c_0 + c_1 x + c_2 x^2 + \dots = \sum_{m=0}^{\infty} c_m x^m$$

$$= c_0 + (-kc_0) x + \left(\frac{k^2}{2!}c_0\right) x^2$$
$$+ \left(\frac{-k^3}{3!}c_0\right) x^3 + \dots$$

$$= c_0 \left(1 - kx + \frac{k^2}{2!}x^2 - \frac{k^3}{3!}x^3 + \dots\right)$$

For clarification, let $-kx = z$.

$$y(z) = c_0 \left(1 + z + \frac{z^2}{2!} + \frac{z^3}{3!} + \ldots\right)$$

The terms in parentheses represent the Maclaurin series for e^z. The solution is

$$y(z) = c_0 e^z$$

or

$$y(x) = c_0 e^{-kx}$$

The initial conditions can be used to determine the value of the constant c_0.

3. EXTENDED POWER SERIES METHOD: HOMOGENEOUS LINEAR DIFFERENTIAL EQUATIONS

If a negative power of x exists in an equation—in other words, if the equation is not analytic at $x = 0$—the *extended power series method* is used. (In simpler terms, with a negative power of x, the x is in the denominator; thus, the equation cannot be defined as $x = 0$ since this involves division by zero. This is also referenced as a *singularity*.) The extended power series method can be used on any differential equation of the following form.

$$y'' + \left(\frac{a(x)}{x}\right) y' + \left(\frac{b(x)}{x^2}\right) y = 0 \qquad 15.8$$

The Cauchy or Euler equation, Eq. 15.9, is of this form.[6]

$$x^2 y'' + axy' + by = 0 \qquad 15.9$$

Rearranging to show that the equation is not analytic at $x = 0$ (i.e., dividing by x^2 and allowing a and b to be functions of x) changes Eq. 15.9 into the form of Eq. 15.8.

Gauss's hypergeometric equation, Eq. 15.10, can also be solved using this method.[7]

$$x(1-x) y'' + \big(c - (a+b+1) x\big) y' - aby = 0 \qquad 15.10$$

In Eq. 15.8, it is assumed that $a(x)$ and $b(x)$ are analytic at $x = 0$; therefore, at least one solution can be represented in the form[8]

$$y(x) = x^r \sum_{m=0}^{\infty} c_m x^m$$

$$= x^r \left(c_0 + c_1 x + c_2 x^2 + \ldots\right) \qquad 15.11$$

---

[6]This form is equivalent to that given in Ch. 11. One should become accustomed to seeing various formats for the same-named equation.
[7]This equation finds application in probability and statistics.
[8]Using the form of the solution shown, the method is called the *Frobenius' method*.

The exponent r may be any real or complex number and is chosen so that $c_0 \neq 0$.[9] Expanding $a(x)$ and $b(x)$ into a power series, differentiating Eq. 15.11, substituting into Eq. 15.8, and equating the sum of the coefficients of each power of x to zero yields the *indicial equation*, from the coefficients associated with x^r.

$$r^2 + (a_0 - 1) r + b_0 = 0 \qquad 15.12$$

Equation 15.12 will yield a basis of solutions. One of the solutions will be of the form of Eq. 15.11; the other form depends on the roots of the indicial equation. There are three possible cases depending on these roots.

case 1: distinct roots r_1 and r_2, which do not differ by an integer[10]

$$y_1(x) = x^{r_1} \left(c_0 + c_1 x + c_2 x^2 + \ldots\right)$$

$$y_2(x) = x^{r_2} \left(d_0 + d_1 x + d_2 x^2 + \ldots\right)$$

case 2: double root, r

$$y_1(x) = x^r \left(c_0 + c_1 x + c_2 x^2 + \ldots\right)$$

$$r = \frac{1 - a_0}{2}$$

$$y_2(x) = y_1(x) \ln x + x^r \sum_{m=1}^{\infty} A_m x^m$$

Note that the a_0 in the equation for $y_1(x)$ is from the indicial equation, Eq. 15.12.

case 3: distinct roots, r_1 and r_2, that differ by an integer, for example, 1, 2, 3, ...

$$y_1(x) = x^{r_1} \left(c_0 + c_1 x + c_2 x^2 + \ldots\right)$$

$$y_2(x) = k_p y_1(x) \ln x + x^{r_2} \sum_{m=0}^{\infty} c_m x^m \quad [x > 0]$$

4. GENERAL METHOD FOR SOLVING NONHOMOGENEOUS DIFFERENTIAL EQUATIONS

In Ch. 11, the method of undetermined coefficients was used to find a particular solution to nonhomogeneous equations. It was limited to simpler differential equations in which the coefficients $p(x)$ and $q(x)$ are constants and the forcing function $f(x)$ was of a particular form. The method of variation of parameters is completely general but more complicated, and it removes all such restrictions.

---

[9]Restraining r to be a nonnegative integer would make the entire expression a power series as covered in the Power Series Method section of this chapter.
[10]This includes complex conjugate roots.

Consider the nonhomogeneous second-order linear differential equation, Eq. 15.13.[11]

$$y'' + p(x)y' + q(x)y = f(x) \qquad 15.13$$

The homogeneous portion of Eq. 15.13 has a solution of the following form.

$$y_h(x) = c_1 y_1(x) + c_2 y_2(x) \qquad 15.14$$

The coefficients c_1 and c_2 are selected to be functions $u(x)$ and $v(x)$ that make the following a particular solution to Eq. 15.13.[12]

$$y_p(x) = u(x)y_1(x) + v(x)y_2(x) \qquad 15.15$$

The parameters $u(x)$ and $v(x)$ are then determined in terms of $y_1(x)$ and $y_2(x)$ by rearranging the particular solution, Eq. 15.15. The result is substituted into the homogeneous solution, Eq. 15.14. The associated first and second derivatives are determined. The results are substituted into Eq. 15.13. The result is a particular solution of the following completely general form.

$$y_p(x) = -y_1(x)\int \frac{y_2(x)f(x)}{W}dx$$
$$+ y_2(x)\int \frac{y_1(x)f(x)}{W}dx \qquad 15.16$$

The term W is called the Wronskian of y_1 and y_2 and is given by

$$W = y_1(x)y_2'(x) - y_1'(x)y_2(x) \qquad 15.17$$

The parameters $u(x)$ and $v(x)$ are given by

$$u(x) = -\int \frac{y_2(x)f(x)}{W}dx \qquad 15.18$$

$$v(x) = \int \frac{y_1(x)f(x)}{W}dx \qquad 15.19$$

The overall result is that the combination of Eq. 15.14 and Eq. 15.16 constitute a general solution to Eq. 15.13, that is, to any nonhomogeneous differential equation.

Example 15.2

Solve the following differential equation.

$$y'' + y = \sec x$$

---

[11] The functions p, q, and f are continuous on an open interval I.
[12] This is defined on the same open interval I mentioned in the previous footnote. Also, recall that $j = 1\angle 90°$, thus $j^2 = 1\angle 180° = -1$, making $j = \sqrt{-1}$.

Solution

The homogeneous portion of the equation is

$$y'' + y = 0$$

The characteristic equation is[13]

$$r^2 + 0r + 1 = 0$$
$$r^2 = -1$$
$$r = \pm\sqrt{-1} = \pm j$$

The homogeneous solution is

$$y_h(x) = c_1 \cos x + c_2 \sin x \qquad [\text{I}]$$

Thus,

$$y_1 = \cos x \text{ and } y_2 = \sin x$$

From Eq. 15.17, the Wronskian is

$$W(y_1(x), y_2(x)) = y_1(x)y_2'(x) - y_1'(x)y_2(x)$$
$$= \cos x \cos x - (-\sin x)\sin x$$
$$= \cos^2 x + \sin^2 x = 1$$

Substitute into Eq. 15.16 to determine the particular solution.

$$y_p(x) = -y_1(x)\int \frac{y_2(x)f(x)}{W}dx$$
$$+ y_2(x)\int \frac{y_1(x)f(x)}{W}dx$$
$$= -\cos x \int \frac{\sin x \sec x}{1}dx$$
$$+ \sin x \int \frac{\cos x \sec x}{1}dx$$

Substitute the definition of the secant and then integrate.

$$y_p(x) = -\cos x \int \sin x \left(\frac{1}{\cos x}\right)dx$$
$$+ \sin x \int \cos x \left(\frac{1}{\cos x}\right)dx$$
$$= -\cos x \int \tan x\,dx + \sin x \int dx$$
$$= -\cos x \left(-\ln|\cos x|\right) + (\sin x)(x)$$
$$= \cos x \ln|\cos x| + x \sin x \qquad [\text{II}]$$

---

[13] See Ch. 11.

Combine the homogeneous (or complimentary) solution I with the particular solution II to obtain the general solution.

$$y(x) = y_h(x) + y_p(x)$$

$$= c_1 \cos x + c_2 \sin x + \cos x \ln|\cos x| + x \sin x$$

$$= (c_1 + \ln|\cos x|)\cos x + (c_2 + x)\sin x$$

5. ORTHOGONALITY OF FUNCTIONS

Let $g_m(x)$ and $g_n(x)$ be two real-valued functions defined on an interval $a \le x \le b$. Further, define the integral of the two as

$$(g_m, g_n) = \int_a^b g_m(x)g_n(x)dx \qquad \textbf{15.20}$$

The functions g_m and g_n are *orthogonal* on the interval $a \le x \le b$ if the integral in Eq. 15.20 equals zero.[14] That is,

$$(g_m, g_n) = \int_a^b g_m(x)g_n(x)dx$$

$$= 0 \quad [m \ne n] \qquad \textbf{15.21}$$

The nonnegative square root of (g_m, g_m) is called the *norm* of g_m.

$$\|g_m\| = \sqrt{(g_m, g_m)} = \sqrt{\int_a^b g_m^2(x)dx} \qquad \textbf{15.22}$$

6. LAPLACE TRANSFORM: INITIAL AND FINAL VALUE THEOREMS

Laplace transforms simplify the solution of nonhomogeneous differential equations by reducing many steps in the process to algebra. Additionally, the initial conditions are automatically included in the analysis.[15] Nevertheless, obtaining the inverse transform (i.e., moving from the s domain back into the time domain) may not be an easy task.

$$f(t) = L^{-1}\{F(s)\}$$

$$= \frac{1}{2\pi j}\int_{c-j\infty}^{c+j\infty} F(s)e^{st}ds \qquad \textbf{15.23}$$

Eq. 15.23 can be used for this task, although the use of transform tables is more common. Further, the exact

inverse transform may not be necessary. It may be more fundamental to understanding how the equation, that is, the circuit, reacts at $t = 0$ and $t = \infty$. To that end, the *initial value theorem*, Eq. 15.24, can be used.[16]

$$f(0) = \lim_{s \to \infty} sF(s) \qquad \textbf{15.24}$$

The *final value theorem*, Eq. 15.25, can be used to determine the value of the function in the time domain as time approaches infinity.[17]

$$f(\infty) = \lim_{s \to 0} sF(s) \qquad \textbf{15.25}$$

Example 15.3

The Laplace transform of a given voltage is shown. Evaluate the voltage at $t = 0$.

$$V(s) = \frac{4s}{(s^2 + 9)(s^2 + 1)}$$

Expand the denominator and divide the numerator and denominator by s^2.

$$V(s) = \frac{4s}{s^4 + 10s^2 + 9} = \frac{\dfrac{4}{s}}{s^2 + 10 + \dfrac{9}{s^2}}$$

Apply the initial value theorem, Eq. 15.24.

$$v(0) = \lim_{s \to \infty} sV(s)$$

$$= \lim_{s \to \infty} s\left(\frac{\dfrac{4}{s}}{s^2 + 10 + \dfrac{9}{s^2}}\right)$$

$$= \lim_{s \to \infty}\left(\frac{4}{s^2 + 10 + \dfrac{9}{s^2}}\right)$$

$$= 0$$

Note that the voltage function associated with the given voltage is

$$v(t) = \sin t \sin 2t$$

The value of $v(t)$ is indeed zero at $t = 0$ as $\sin(0) = 0$, thus confirming the answer.

Example 15.4

Given the voltage in the frequency domain (or s domain), as shown, what is the value of the voltage at $t = \infty$?[18]

$$V(s) = \frac{1}{s(s + 1)^2}$$

---

[14]Both Legendre polynomials and Bessel functions are orthogonal. Important orthogonal sets of functions arise as solutions to linear second-order differential equations.

[15]For example, the values for $f(0)$ and $f'(0)$ are required when transforming a second-order differential, thus immediately placing the initial conditions into use.

[16]$f(0)$ is more technically $f(0^+)$, that is, the value at zero when approached from the positive side. This is based on the normally used Laplace transform being defined from zero to infinity.

[17]Technically, Eq. 15.25 should be $\lim_{t \to \infty} f(t) = \lim_{s \to 0} sF(s)$.

[18]It should be noted that this is the equivalent of asking for the steady-state value.

Expand the denominator.

$$V(s) = \frac{1}{s(s+1)^2} = \frac{1}{s(s+1)(s+1)}$$

$$= \frac{1}{s(s^2 + 2s + 1)} = \frac{1}{s^3 + 2s^2 + s}$$

Apply the final value theorem, Eq. 15.25.

$$v(\infty) = \lim_{s \to 0} sV(s)$$

$$= \lim_{s \to 0} s\left(\frac{1}{s^3 + 2s^2 + s}\right)$$

$$= \lim_{s \to 0} \left(\frac{s}{s^3 + 2s^2 + s}\right)$$

$$= \lim_{s \to 0} \left(\frac{1}{s^2 + 2s + 1}\right) = 1$$

Note that the voltage function associated with the given voltage is

$$v(t) = 1 - e^{-t} - te^{-t}$$

The value of $v(t)$ as $t \to \infty$ can be seen to be equal to 1 from this equation, confirming the solution.[19]

7. HALF-RANGE EXPANSIONS

In some engineering problems a function $f(t)$ defined over a finite interval may be more readily manipulated and a solution more easily obtained if the function is represented as a Fourier series, as will be shown in the Partial Differential Equations section of this chapter.

Given a function $f(t)$ defined on some interval $0 \le t \le l$ or $0 \le t \le T/2$ where T is the period,[20] the *even periodic extension* of $f(t)$ of period $T = 2l$ is given by the Fourier cosine series.

$$f(t) = a_0 + \sum a_n \cos \frac{n\pi t}{l} \qquad 15.26$$

The coefficients of the series are given by the following.

$$a_0 = \frac{1}{l} \int_0^l f(t)dt \qquad 15.27$$

$$a_n = \frac{2}{l} \int_0^l f(t) \cos\left(\frac{n\pi t}{l}\right) dt$$

$$[n = 1, 2, 3, \ldots] \qquad 15.28$$

---

[19]As t approaches ∞, the value of s approaches 0. In other words, as the function, or circuit, approaches a steady-state condition, the number of oscillations (frequency) approaches zero. The physical model represented by the mathematics matches what one would expect in the real world.

[20]This means the period is $T/2 = l$ or $T = 2l$.

The *odd periodic extension* of $f(t)$ on the same interval is given by the Fourier sine series.

$$f(t) = \sum_{n=1}^{\infty} b_n \sin\left(\frac{n\pi t}{l}\right) \qquad 15.29$$

The coefficients are

$$b_n = \frac{2}{l} \int_0^l f(t) \sin\left(\frac{n\pi t}{l}\right) dt$$

$$[n = 1, 2, 3, \ldots] \qquad 15.30$$

The series represented by Eqs. 15.26 and 15.29 along with their associated coefficients are called the *half-range expansions* of the given function $f(t)$. Figure 15.1 shows a given function $f(t)$ and its half-range expansions.

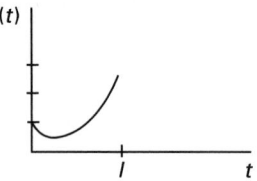

(a) given function $f(t)$

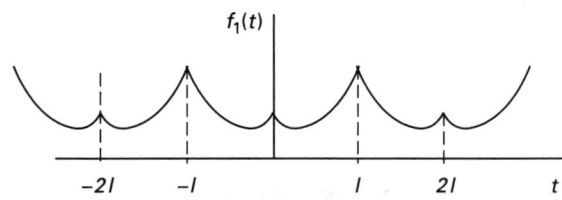

(b) even periodic extension of $f(t)$

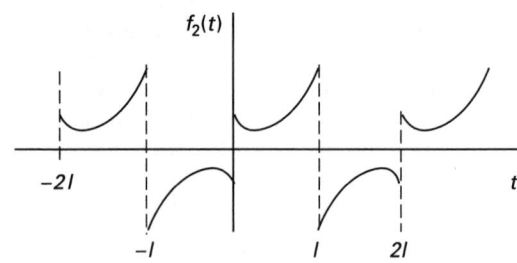

(c) odd periodic extension of $f(t)$

Note in the half-range expansion, $f_1(t) = f(t)$ on the interval $0 \le t \le l$. Also, $f_2(t) = f(t)$ on this same interval.

Figure 15.1 *Periodic Extensions*

Note that mentally summing the infinite series from the graphs in Fig. 15.1 leaves the original function $f(t)$ in place as was given mathematically in Eq. 15.29. This is especially apparent in part (c) of the figure, providing a clearer view for what is occurring mathematically.

8. PARTIAL DIFFERENTIAL EQUATIONS: SEPARATION OF VARIABLES

A *partial differential equation* involves one or more partial derivatives of an unknown function of two or more variables. As before, the order of the equation is the order of the highest derivative. The equation is *linear* if the dependent variable and its partial derivatives are of the first degree, that is, not raised to a power greater than one. The equation is defined as *homogeneous* if each term consists of a dependent variable or one of its derivatives.

Some important second-order partial differential equations follow. (Note that in the following equations, c is a constant, t is time, and u represents the unknown function. All are shown in Cartesian coordinates. Additionally, the constant c is shown as c^2 to indicate a positive quantity in the physical world.)

$$\frac{\partial^2 u}{\partial t^2} = c^2 \frac{\partial^2 u}{\partial x^2}$$

$$\begin{bmatrix} \text{one-dimensional} \\ \text{wave equation}^{21} \end{bmatrix} \qquad 15.31$$

$$\frac{\partial u}{\partial t} = c^2 \frac{\partial^2 u}{\partial x^2}$$

$$\begin{bmatrix} \text{one-dimensional} \\ \text{heat equation} \end{bmatrix} \qquad 15.32$$

$$\frac{\partial^2 u}{\partial x^2} + \frac{\partial^2 u}{\partial y^2} = 0$$

$$\begin{bmatrix} \text{two-dimensional} \\ \text{Laplace equation} \end{bmatrix} \qquad 15.33$$

$$\frac{\partial^2 u}{\partial x^2} + \frac{\partial^2 u}{\partial y^2} = f(x, y)$$

$$\begin{bmatrix} \text{two-dimensional} \\ \text{Poisson equation} \end{bmatrix} \qquad 15.34$$

$$\frac{\partial^2 u}{\partial x^2} + \frac{\partial^2 u}{\partial y^2} + \frac{\partial^2 u}{\partial z^2} = 0$$

$$\begin{bmatrix} \text{three-dimensional} \\ \text{Laplace equation}^{22} \end{bmatrix} \qquad 15.35$$

The equations are solved based on the conditions at a boundary, that is, the *boundary conditions*. Those equations having time as one of the variables may already have the conditions at $t = 0$ (the *initial conditions*) given or known, and they are solved as constrained by those conditions.

An important aspect of the solutions to linear partial differential equations is the following fundamental theorem. If u_1 and u_2 are any of the solutions of a linear homogeneous partial differential equation in some region, then another solution is

$$u = c_1 u_1 + c_2 u_2 \qquad 15.36$$

The terms c_1 and c_2 represent constants.

The following method of solving such problems is termed the *separation of variables*.[23]

step 1:   Apply the separation of variables to obtain ordinary differential equations.[24]

step 2:   Determine solutions that satisfy the boundary conditions.

step 3:   Compose the solutions such that the initial conditions are satisfied.

Example 15.5

Solve the one-dimensional wave equation with the given boundary and initial conditions.

$$\frac{\partial^2 u}{\partial t^2} = c^2 \left(\frac{\partial^2 u}{\partial x^2} \right)$$

The boundary conditions are that the function $u(x, t)$ equals zero at $x = 0$ and $x = l$.[25] This can be expressed in mathematical terms as

$$u(0, t) = 0 \qquad \text{[I]}$$

$$u(l, t) = 0 \qquad \text{[II]}$$

Further, this function must take on the values in conditions I and II for all t. Let the initial deflection be $f(x)$ and the initial velocity be $g(x)$. Thus,

$$u(x, 0) = f(x) \qquad \text{[III]}$$

$$\left. \frac{\partial u}{\partial t} \right|_{t=0} = g(x) \qquad \text{[IV]}$$

---

[21] A solution to this one-dimensional case will be shown. Solutions to the two-dimensional case in Cartesian coordinates involve double Fourier series. Solutions to the two-dimensional case in polar coordinates involve Bessel functions.
[22] The theory of solutions to this equation is called *potential theory*.

---

[23] This method is also called the *product method*.
[24] Recall that ordinary differential equations are those that do not involve partial differentiation.
[25] A vibrating string secured at each end would provide such conditions, as would a confined electromagnetic wave.

Solution

step 1: Apply the separation of variables to obtain ordinary differential equations. Separating variables yields an unknown function of the following form.

$$u(x,t) = F(x)G(t)$$

Thus,

$$\frac{\partial^2 u(x,t)}{\partial t^2} = F(x)\ddot{G}(t)$$

$$\frac{\partial^2 u(x,t)}{\partial x^2} = F''(x)G(t)$$

Note that the dots represent differentiation with respect to t. The primes represent differentiation with respect to x. Insert these results into the original wave equation.

$$\frac{\partial^2 u}{\partial t^2} = c^2\left(\frac{\partial^2 u}{\partial x^2}\right)$$

$$F(x)\ddot{G}(t) = c^2 F''(x)G(t)$$

$$\frac{F''(x)}{F(x)} = \frac{\ddot{G}(t)}{c^2 G(t)}$$

The variables t and x are now separated. Further, the result must be equal to a constant, say k. This can be deduced from the fact that changing the variable t changes only one side of the equation, and vice versa for the variable x. Thus,

$$\frac{F''(x)}{F(x)} = \frac{\ddot{G}(t)}{c^2 G(t)} = k$$

This leads to two linear ordinary differential equations, ODEs.

$$F''(x) - kF(x) = 0$$

$$\ddot{G}(t) - kc^2 G(t) = 0$$

step 2: Determine solutions based on the boundary conditions. Starting with the function $F(x)$, the boundary conditions in I and II were

$$u(0,t) = 0 = F(0)G(t)$$

$$u(l,t) = 0 = F(l)G(t)$$

Letting $G(t) \equiv 0$ gives $u(x,t) \equiv 0$, which is trivial. Thus, $G(t) \neq 0$. This means that

$$F(0) = 0 \text{ and } F(l) = 0$$

If $k = 0$, the general solution of the ODE for the function $F(x)$ is of the form $F(x) = ax + b$.[26] Since $F(0) = 0$, b must be zero. Since $F(l) = 0$, a must be zero. Hence, $F(x) \equiv 0$ giving $u(x,t) \equiv 0$, all of which results in a trivial solution.

Let k be some positive value, $k = \mu^2$. The solution is now of the following form.

$$F(x) = Ae^{\mu x} + Be^{-\mu x}$$

But again, applying the boundary conditions makes $A = 0$ and $B = 0$. Therefore, $F(x) \equiv 0$ and $u(x,t) \equiv 0$.

Thus, to have a useful solution, k must be negative, say $k = -p^2$. This provides a solution of the following form.

$$F(x) = A\cos px + B\sin px$$

Now, $F(0) = 0$ makes $A = 0$. The second boundary condition yields the following.

$$F(l) = 0 = B\sin pl$$

Since $B = 0$ would give $F(x) \equiv 0$ and $u(x,t) \equiv 0$, $\sin pl$ must be equal to zero. That is,

$$pl = n\pi \text{ or } p = \frac{n\pi}{l} \quad [n = 0, 1, 2, \ldots]$$

Letting $B = 1$, one obtains infinitely many solutions of the following form.

$$F_n(x) = \sin\left(\left(\frac{n\pi}{l}\right)x\right) \quad [n = 1, 2, 3, \ldots]$$

Since k is restricted to negative values, that is, $k = -p^2$, the ODE for the function containing the t variable is

$$\ddot{G}(t) - kc^2 G(t) = 0$$

$$\ddot{G}(t) + p^2 c^2 G(t) = 0$$

$$\ddot{G}(t) + \left(\frac{n\pi}{l}\right)^2 c^2 G(t) = 0$$

$$\ddot{G}(t) + \lambda_n^2 G(t) = 0 \quad [\lambda_n = \left(\frac{n\pi}{l}\right)c]$$

The solution is therefore[27]

$$G_n(t) = B_n\cos\lambda_n t + B_n^*\sin\lambda_n t$$

---

[26] The characteristic equation results in a double root of zero. See Ch. 11 for the method used on homogeneous, second-order linear differential equations with constant coefficients.

[27] The symbols B_n and B_n^* are commonly used as b_n in representations of Fourier sine series, as will be shown momentarily. Do not confuse the symbology B_n^* with that of a complex number.

Writing the function $u_n(x,t) = F_n(t)G_n(t)$ gives

$$u(x,t) = \sin\left(\left(\frac{n\pi}{l}\right)x\right)$$
$$\times \left(B_n \cos\lambda_n t + B_n^* \sin\lambda_n t\right)$$
$$[n = 1, 2, 3, \ldots] \qquad \text{[V]}$$

This function satisfies the boundary conditions given in Eqs. I and II. The functions are *eigenfunctions*, or characteristic functions. The values for λ_n are *eigenvalues*, or characteristic values. The function u_n represents harmonic motion of a frequency $\lambda_n/2\pi = cn/2l$ cycles per unit time. The *fundamental mode* occurs when $n = 1$. Each mode n has $n - 1$ nodes, that is, points that do not move.[28] Some fundamental nodes are shown in the following illustration.

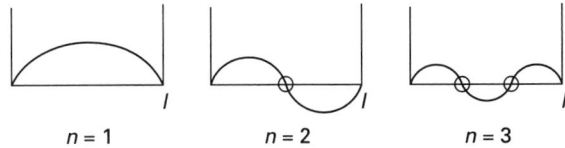

$n = 1$ $n = 2$ $n = 3$

step 3: Compose the solution such that the initial conditions are satisfied.

A single solution of the form of Eq. V will not, in general, satisfy both initial conditions III and IV. Based on the fundamental theorem given by Eq. 15.36, the sum of the solutions is a solution. Therefore, the following combination of solutions to Eq. V provides a solution to the one-dimensional wave equation.

$$u(x,t) = \sum_{n=1}^{\infty} u_n(x,t)$$
$$= \sum \sin\left(\left(\frac{n\pi}{l}\right)x\right)$$
$$(B_n \cos\lambda_n t + B_n^* \sin\lambda_n t) \qquad \text{[VI]}$$

Applying the initial condition III to Eq. VI gives

$$u(x,0) = \sum_{n=1}^{\infty} B_n \sin\left(\left(\frac{n\pi}{l}\right)x\right) = f(x) \quad \text{[VII]}$$

Thus, to make Eq. VI satisfy initial condition III, B_n must be chosen as the half-range

expansion of $f(x)$, that is, the Fourier sine series of $f(x)$.[29]

$$B_n = \frac{2}{l}\int_0^l f(x)\sin\left(\left(\frac{n\pi}{l}\right)x\right)dx$$
$$[n = 1, 2, 3, \ldots] \qquad \text{[VIII]}$$

By making this choice for B_n, Eq. VII becomes valid and Eq. VI satisfies initial condition III.

One item now remains: make Eq. VI satisfy initial condition IV. Start by taking the derivative of Eq. VI with respect to time and applying condition IV. This gives the following equation.

$$\left.\frac{\partial u}{\partial t}\right|_{t=0}$$

$$= \left[\sum_{n=1}^{\infty} \sin\left(\left(\frac{n\pi}{l}\right)x\right)\left(\begin{array}{c}-B_n\lambda_n \sin\lambda_n t \\ + B_n^*\lambda_n \cos\lambda_n t\end{array}\right)\right]\Bigg|_{t=0}$$

$$= \sum_{n=1}^{\infty} B_n^*\lambda_n \sin\left(\left(\frac{n\pi}{l}\right)x\right) = g(x) \qquad \text{[IX]}$$

Using the same logic, and recalling that $\lambda_n = cn\pi/l$ gives

$$B_n^*\lambda_n = \frac{2}{l}\int_0^l g(x)\sin\left(\left(\frac{n\pi}{l}\right)x\right)dx$$

$$B_n^* = \frac{2}{cn\pi}\int_0^l g(x)\sin\left(\left(\frac{n\pi}{l}\right)x\right)dx \qquad \text{[X]}$$

Thus, with B_n^* equal to the condition given in Eq. X, Eq. VI satisfies initial condition IV. All the conditions are now met.

Eq. VI is a solution to the one-dimensional wave equation with the coefficients B_n and B_n^* defined by Eqs. VIII and IX.[30]

---

[28]This is not counting the constrained endpoints.

[29]The Fourier sine series results in an odd extension of $f(x)$. In other words, the original function $f(x)$ is extended in the coefficient B_n which, when summed, leaves only $f(x)$. See the graphs in the Half-Range Expansions section of this chapter for a physical picture of what is occurring.

[30]Also, the series represented in Eq. VI must converge, and the series obtained by differentiating Eq. VI twice (termwise) with respect to x and t must converge and sum to $\partial^2 u/\partial x^2$ and $\partial^2 u/\partial t^2$. Other solution methods are possible. One is called the *D'Alembert* solution. Another is using a variation of parameters and Laplace transforms. Also, Laplace transforms can change the partial differential equation into an ordinary differential equation that can be solved using standard methods.

9. SPECIAL FUNCTIONS

Several different special functions of use occur as *named equations* with different forms and symbology. Some of particular use to the electrical engineer are discussed in this section.

Unit Step Function

Symbology: u_t; $S_k(t)$; $H(t-k)$

This function could be used to represent what occurs when a switch is turned on.

$$u_t = \begin{cases} 0 & 0 < t < k \\ 1 & t > k \end{cases} \qquad \text{15.37}$$

The above symbology mathematically represents what is shown in Fig. 15.2. The symbol $H(t-k)$ is called the *Heaviside step function.*[31]

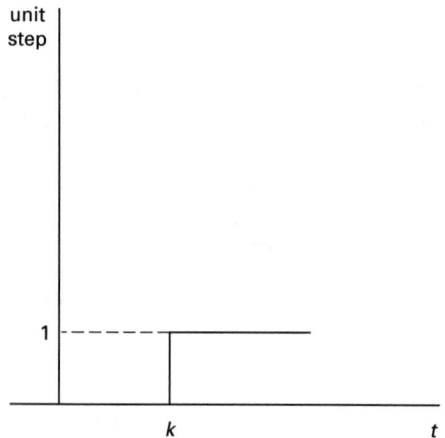

Figure 15.2 *Unit Step Function*

Unit Finite Impulse Function

$$I(h, t-t_0) = H(t-t_0) - H\big(t-(t_0+h)\big) \qquad \text{15.38}$$

$$I(h, t-t_0) = \begin{cases} 0 & t < t_0 \text{ and } t > (t_0+h) \\ 1 & t_0 < t < (t_0+h) \end{cases} \qquad \text{15.39}$$

The *unit finite impulse function* is shown graphically in Fig. 15.3.

Of special importance in Fig. 15.3 part (c) is that the unit finite impulse function, that is, the shaded region, represents an area equal to 1 for any $h > 0$. This area represents a rectangle with the base equal to $t_0 - (t_0 + h) = h$ and the height equal to $1/h$; therefore, the area is $h(1/h) = 1$. The impulse is the integral of force $(1/h)$ over the interval of time (h).

---

[31]Oliver Heaviside was an English electrical engineer whose treatment of the derivative process (represented by D) in algebraic terms was the spur to the development of the Laplace transform.

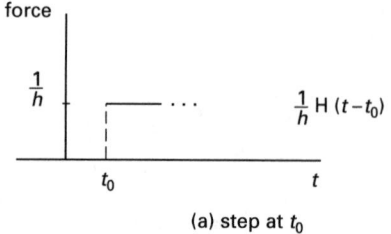

(a) step at t_0

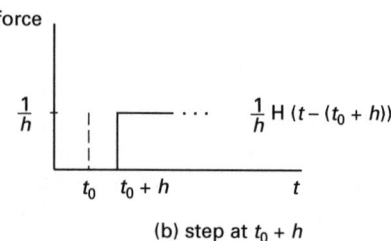

(b) step at $t_0 + h$

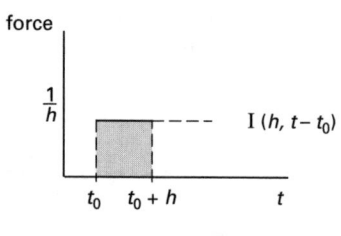

(c) unit finite impulse

Figure 15.3 *Unit Finite Impulse Function*

Delta Function; Dirac Function; Impulse Function; Unit Impulse Function

Symbology: δt; $\delta(t - t_0)$

If the limit as $h \to 0$ is taken on the unit finite impulse function, the result is the *delta*, or *Dirac function.*[32] This function is sometimes called the *impulse function* or *unit impulse function* and not always differentiated from the unit finite impulse function.

$$\lim_{h \to 0} I(h, t-t_0) \equiv \delta(t-t_0) = \begin{cases} 0 & t \neq t_0 \\ \infty & t = t_0 \end{cases} \qquad \text{15.40}$$

This is shown graphically in Fig. 15.4.[33]

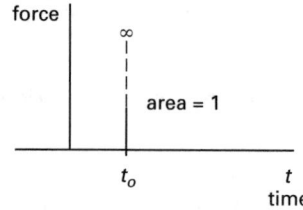

Figure 15.4 *Delta Function*

---

[32]Dirac was an English physicist whose theory of negative energy holes predicted the existence of positrons.
[33]There is much more to the delta function. The mathematics shown is not strictly true. The delta function is not a proper function and is therefore shown as equivalent to, $\equiv$, the limit.

The delta function is strictly defined only in terms of integration. The delta function displays a property called *sifting* that is given by Eq. 15.41.[34]

$$\int\limits_{-\infty}^{+\infty} \delta(t - t_0) f(t)\, dt = f(t_0) \qquad \textbf{15.41}$$

Gamma Function

Symbology: $\Gamma(\alpha); (\alpha - 1)!$

Also known as the *factorial function*, the *gamma function* is defined by the following integral.[35]

$$\Gamma(\alpha) = \int\limits_{0}^{\infty} e^{-t} t^{(\alpha-1)}\, dt \qquad [\alpha > 0] \qquad \textbf{15.42}$$

The gamma function is shown graphically in Fig. 15.5.

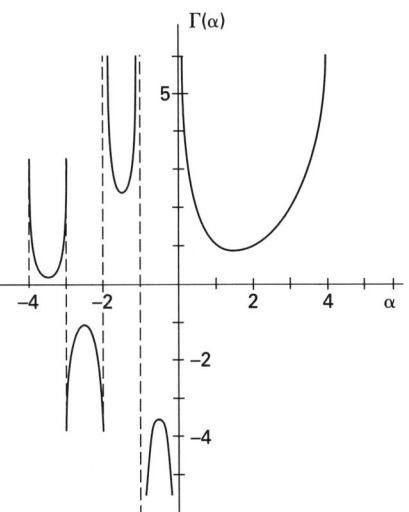

Figure 15.5 *Gamma Function*

The important functional relation of the gamma function is

$$\Gamma(\alpha + 1) = \alpha\Gamma(\alpha) \qquad \textbf{15.43}$$

---

[34]The impulse function is a powerful analytical tool. For example, electrical signals can be represented as weighted sums of shifted impulses—essentially this can be the interpretation of Eq. 15.41. Using this interpretation, the response of a *linear time invariant* (LTI) system can be analyzed for any arbitrary input, including *discrete* inputs, that is, those inputs that are not continuous. (Recall that the term "linear" indicates that superposition can be used; "time invariant" means a time shift in the input signal causes an identical time shift in the output signal.) Using the principle of superposition, one can then obtain the output of an electrical circuit to any input. Using the unit finite impulse function, one can do the same for continuous time systems.

[35]Also defined in terms of the gamma function are the *beta function* and *incomplete gamma functions*.

Example 15.6

Find $\Gamma(2)$.

Solution

Apply the definition given by Eq. 15.42.

$$\Gamma(\alpha) = \int\limits_{0}^{\infty} e^{-t} t^{(\alpha-1)}\, dt \qquad [\alpha > 1]$$

$$\Gamma(2) = \int\limits_{0}^{\infty} e^{-t} t^{(2-1)}\, dt$$

$$= \int\limits_{0}^{\infty} e^{-t} t\, dt$$

From a mathematical handbook of integrals, the result of the previous equation is

$$\Gamma(2) = \left[\left(\frac{e^{-t}}{(-1)^2} \right)(-t - 1) \right]\Big|_{0}^{\infty}$$

$$= \left[e^{-t}(-t - 1) \right]\Big|_{0}^{\infty}$$

$$= \frac{-(t + 1)}{e^t}\Big|_{0}^{\infty}$$

Use L'Hôpital's rule to determine the value at infinity. The result is

$$\frac{-(t + 1)}{e^t}\Big|_{0}^{\infty} = 0 - \left(\frac{-(0 + 1)}{e^0} \right) = 1$$

Note that this result could also have been accomplished using the functional relation of Eq. 15.43 as shown.

$$\Gamma(\alpha + 1) = \alpha\Gamma(\alpha)$$

$$\Gamma(2) = \Gamma(1 + 1) = 1\Gamma(1) = 1$$

Finally, the factorial symbology could have been used.[36]

$$\Gamma(\alpha) = (\alpha - 1)!$$

$$\Gamma(2) = (2 - 1)! = 1! = 1$$

A table of gamma function values is given in App. 15.A.

Error Function and Complementary Error Function

The error function, also called the *probability integral*, was defined in Ch. 10 as

$$\text{erf}(x) = \frac{2}{\sqrt{\pi}} \int\limits_{0}^{x} e^{-t^2}\, dt \qquad \textbf{15.44}$$

---

[36]An important value of gamma is $\Gamma(1/2) = \sqrt{\pi}$.

The $2/\sqrt{\pi}$ normalizes the function so that $\mathrm{erf}(\infty) = 1$. By contrast, the *complimentary error function* is defined as

$$\mathrm{erfc}(x) = 1 - \mathrm{erf}(x) = \frac{2}{\sqrt{\pi}} \int_x^\infty e^{-t^2} dt \qquad 15.45$$

The error function is shown graphically in Fig. 15.6.

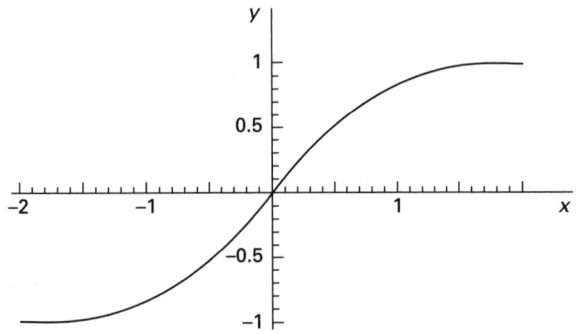

Figure 15.6 *Error Function*

Bessel Functions of the First Kind

A Bessel differential equation of order ν is given by

$$x^2 y'' + xy' + \left(x^2 - \nu^2\right) y = 0 \qquad 15.46$$

A particular solution is given by the Bessel function of the first kind of order ν. (Note that this result is obtained using the power series method.)

$$J_\nu(x) = x^\nu \sum_{m=0}^\infty \frac{(-1)^m x^{2m}}{2^{2m+\nu} m! \Gamma(\nu + m + 1)} \qquad 15.47$$

Integer values of ν are often denoted by n. The functions J_0 and J_1 are important in many engineering problems. These functions are shown graphically in Fig. 15.7.

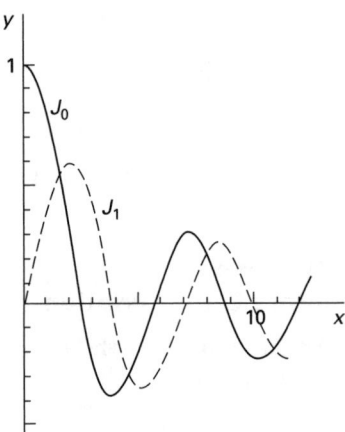

Figure 15.7 *Bessel Functions of the First Kind*

A table of values for $J_0(x)$ and $J_1(x)$ is given in App. 15.B.

Bessel Functions of the Second Kind

If the value of ν or n is zero in a Bessel differential equation and one divides by the variable x, the result is

$$xy'' + y' + xy = 0 \qquad 15.48$$

The standard particular solution is known as the Bessel function of the second kind of order zero and is given by the following equation. This is also called *Neumann's function* of order zero.

$$Y_0(x) = \left(\frac{2}{\pi}\right) \left(J_0(x) \left(\ln \frac{x}{2} + \gamma\right)\right.$$
$$\left. + \sum \left(\frac{(-1)^{m-1} h_m}{2^{2m} (m!)^2}\right) x^{2m}\right) \qquad 15.49$$

The constant γ is Euler's constant (also given as C_E, C, and $\ln \gamma$) which is defined as

$$\gamma = \lim_{s \to \infty} \left(1 + \frac{1}{2} + \ldots + \frac{1}{s} - \ln s\right)$$
$$\approx 0.577 \qquad 15.50$$

The variable h_m is defined as

$$h_m = 1 + \frac{1}{2} + \ldots + \frac{1}{m}$$
$$[m = 1, 2, 3 \ldots] \qquad 15.51$$

The second standard solution to Eq. 15.48 for all ν is given by

$$Y_\nu(x) = \left(\frac{1}{\sin \nu\pi}\right)$$
$$\times \left(J_\nu(x) \cos \nu\pi - J_{-\nu}(x)\right) \qquad 15.52$$

$$Y_n(x) = \lim_{\nu \to n} Y_\nu(x) \qquad 15.53$$

Eq. 15.52 and 15.53 are known as Bessel functions of the second kind of order ν. These are also called *Neuman's functions* of order ν and are sometimes given as $N_\nu(x)$. They are also called *Weber's functions* in some texts. These functions are shown graphically in Fig. 15.8.

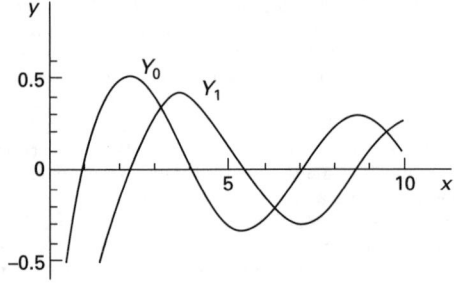

Figure 15.8 *Bessel Functions of the Second Kind*

Mathematics

Bessel Functions of the Third Kind

A general solution to Bessel's equation for all values of ν is given by the following formula.

$$y(x) = c_1 J_\nu(x) + c_2 Y_\nu(x) \qquad 15.54$$

Sometimes a solution needs to be complex for real values of x. In such cases, the solutions take on the following form.

$$H_\nu^{(1)}(x) = J_\nu(x) + jY_\nu(x) \qquad 15.55$$

$$H_\nu^{(2)}(x) = J_\nu(x) - jY_\nu(x) \qquad 15.56$$

These linearly independent functions are called Bessel functions of the third kind of order ν. They are also called the *first and second Hankel functions* of order ν, hence the nomenclature.

Hyperbolic Bessel Functions

Hyperbolic Bessel functions are given by Eq. 15.57.

$$I_m(x) = j^{-m} J_m(jx) \qquad 15.57$$

10. CONTINUOUS-TIME SYSTEMS: FOURIER SERIES AND TRANSFORMS

Fourier series are used to analyze periodic signals. Fourier integrals, also called Fourier transforms, are used to analyze aperiodic signals. The power of Fourier analysis is that if the input signal can be expressed in terms of periodic complex exponentials or sinusoids, then the output can be expressed in the same form, thus simplifying analysis. Further, representing otherwise intractable equations as a periodic Fourier series allows for analysis in ordinary and partial differential equations. This important concept is shown in the following example.

Example 15.7

Show that the response of a linear time-invariant system (LTI) to a complex exponential input is the same complex exponential with a change in amplitude.[37] That is, show[38]

$$e^{st} \rightarrow H(s)e^{st}$$

The term $H(s)$ is the complex amplitude factor and in general will be a function of the complex variable s.

---

[37] "Linear" indicates that superposition applies. "Time invariant" indicates that a shift in the input signal $x(t)$ to $x(t - \tau)$ causes a shift in the output signal $y(t)$ to $y(t - \tau)$. Note that when such a shift occurs in an equation, the equation is referred to as a *difference equation* and, as will be shown, represents a discrete-time system.

[38] This is equivalent to showing the same thing for sinusoids, since from Euler's relation $e^{j\omega t} = \cos \omega t + j \sin \omega t$.

Solution

Assume an input signal of the form $x(t) = e^{st}$ provided to an LTI circuit with impulse response $h(t)$.[39]

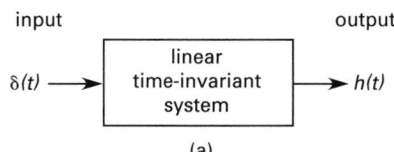

(a)
LTI system with unit impulse response *h(t)*

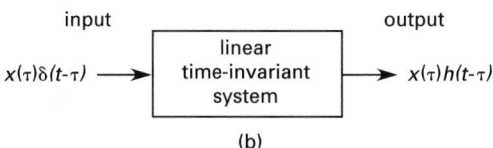

(b)
LTI system response to an impulse at time τ with amplitude *x*(τ)

Writing the output $y(t)$ in terms of the convolution integral gives the following equation. (Recall that the convolution integral takes two functions defined in the s domain, in this case the impulse and the input signal $x(t)$, and gives the output in the time domain t directly without determining the inverse Laplace transform.)

$$y(t) = \int_{-\infty}^{+\infty} h(t - \tau)x(\tau)d\tau = \int_{-\infty}^{+\infty} h(\tau)e^{-s\tau}e^{st}d\tau$$

Since the term e^{st} does not affect the integral, it can be moved outside the integral.

$$y(t) = e^{st}\int_{-\infty}^{+\infty} h(\tau)e^{-s\tau}d\tau$$

This can be rewritten as

$$y(t) = e^{st}H(s) = H(s)e^{st}$$

$$H(s) = \int_{-\infty}^{+\infty} h(\tau)e^{-s\tau}d\tau$$

Thus, any complex exponential is an eigenfunction of an LTI system. The constant $H(s)$ for a specified value s is the eigenvalue of the eigenfunction e^{st}.[40] In practical terms, what this means can best be illustrated as follows.

---

[39] The impulse response $h(t)$ occurs due to an input $\delta(t)$. This is simply a notation, which can vary considerably.

[40] The value $H(s)$ is determined by the electrical circuit and represents the circuit's response to an impulse $\delta(t - \tau)$. As will be seen later in this chapter, $H(s)$, with $s = j\omega$, is the Fourier transform of the response of the circuit to an impulse signal. Generally written as $H(\omega)$, it is termed the *frequency response* of the system.

Consider an input signal of the form

$$x(t) = a_1 e^{s_1 t} + a_2 e^{s_2 t} + a_3 e^{s_3 t}$$

The response to each term separately is

$$a_1 e^{s_1 t} \rightarrow a_1 H(s_1) e^{s_1 t}$$

$$a_2 e^{s_2 t} \rightarrow a_2 H(s_2) e^{s_2 t}$$

$$a_3 e^{s_3 t} \rightarrow a_3 H(s_3) e^{s_3 t}$$

Because, by the definition of "linear," superposition applies, the output $y(t)$ is

$$y(t) = a_1 H(s_1) e^{s_1 t} + a_2 H(s_2) e^{s_2 t} + a_3 H(s_3) e^{s_3 t}$$

In general terms,

$$\text{input} \rightarrow \text{output}$$

$$\sum_k a_k e^{s_k t} \rightarrow \sum_k a_k H(s_k) e^{s_k t}$$

One final note before considering how this applies to electrical circuits in specific cases: In general, s is complex and of the form $s = a + j\omega$. If s is restricted to be equal to $j\omega$, that is, $s = j\omega$, then the exponentials are of the form $e^{j\omega t}$ and Fourier transforms are used. If s is allowed to be of the more general form $s = a + j\omega$, then Laplace transforms are used.

11. CONTINUOUS-TIME SYSTEMS: FOURIER SERIES REPRESENTATION OF A PERIODIC SIGNAL

A *continuous-time signal* is one in which the independent variable is continuous, that is, the signal is defined for a continuum of values. Fig. 15.9 shows an example of such a signal.

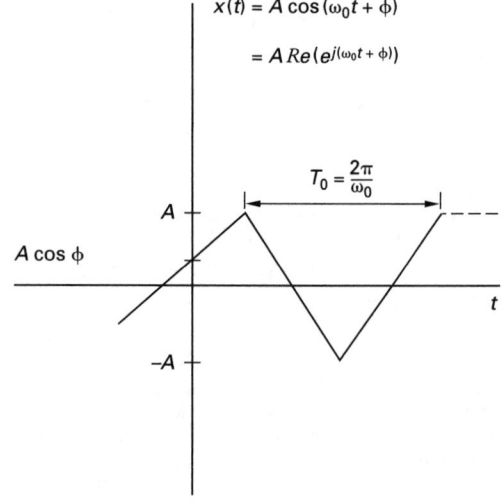

Figure 15.9 *Continuous-Time Periodic Signal*

In Fig. 15.9, the values T_0 and ω_0 represent the *fundamental period* and *fundamental frequency*, respectively. That is, they represent the minimum positive, nonzero value for which $x(t) = x(t + T)$ is true for a periodic function.

The Fourier series representation for any continuous-time periodic signal is

$$x(t) = \sum_{k=-\infty}^{+\infty} a_k e^{jk\omega_0 t} \qquad 15.58$$

$$a_k = \frac{1}{T_0} \int_{T_0} x(t) e^{-jk\omega_0 t} dt \qquad 15.59$$

There are several important items to note:

(1) There are other representations of a Fourier series. (See Sec. 10-15 for an example.) This complex exponential form is very useful in electrical analysis. It is called both the *complex Fourier series* and the *exponential Fourier series*.

(2) The input term $x(t)$ is real and has the form of a cosine function.[41]

(3) The value of $x(t)$ when $k = 0$ represents the DC or constant value.

(4) Fourier analysis is the determination of coefficients using the *analysis equation*, Eq. 15.59, to obtain the result.

(5) Item (4) restated is the crux of this concept. The output $y(t)$ for a circuit is the input $x(t)$ represented as a Fourier series (see Eqs. 15.58 and 15.59) multiplied by an amplitude factor $H(s)$. The amplitude factor $H(s)$ is determined by the circuit through which the given input (i.e., the signal) is processed.[42] The amplitude factor $H(s)$ is the circuit's response, $h(t - \tau)$, to an impulse, $\delta(t - \delta)$.[43] Some properties of the Fourier series for periodic signals are given in App. 15.C.

---

[41] This is because the summation is from $-\infty$ to $+\infty$ and $a_k = a_{-k}$ (sometimes written $a_k = a_{-k}^*$), thus the imaginary terms cancel. Hint: Use Euler's relation and trigonometric identities on the first few terms as a proof.

[42] This may seem rather circular. In different terms, if one represents the input as a Fourier series, that is, as Eq. 15.58, and then replaces the a_k with the Fourier coefficients (also called the *spectral coefficients*) as given by Eq. 15.59, the output is obtained by multiplying by the Fourier transform of the impulse input signal (given by $H(s)$ earlier in this section). Note that the s is restricted to $s = j\omega$ when using the Fourier transform.

[43] When using the Fourier transform, the s is limited to $j\omega$ and $H(s)$ is the Fourier transform of $h(t - \tau)$. If s has both real and imaginary parts, that is, $s = a + j\omega$, $H(s)$ is the Laplace transform of $h(t - \tau)$.

Example 15.8

Represent $x(t)$ from Fig. 15.9 as a Fourier series.

Solution

One could apply the analysis equation given in Eq. 15.59. However, in this simple case, it is easier to expand the equation in terms of complex exponentials and identify the Fourier series coefficients by inspection. Specifically, the $\cos(\omega_0 t + \phi)$ function can be expressed as

$$\cos(\omega_0 t + \phi) = \tfrac{1}{2}e^{j(\omega_0 t + \phi)} + \tfrac{1}{2}e^{-j(\omega_0 t + \phi)}$$

In the first term $k = 1$, therefore

$$a_k = a_1 = \tfrac{1}{2}$$

In the second term $k = -1$, therefore

$$a_k = a_{-1} = \tfrac{1}{2}$$

For all other values of k, $a_k = 0$. Thus, the function in terms of a Fourier series can be represented as

$$x(t) = \sum_{-\infty}^{+\infty} a_k e^{jk(\omega_0 t + \phi)} = \tfrac{1}{2}e^{j(\omega_0 t + \phi)} + \tfrac{1}{2}e^{-j(\omega_0 t + \phi)}$$

Note the shift of $+\phi$ is the equivalent of multiplying by $e^{j\phi}$.

To confirm the result, use Euler's relation to expand the Fourier series and obtain the original equation. That is,

$$x(t) = \tfrac{1}{2}e^{j(\omega_0 t + \phi)} + \tfrac{1}{2}e^{-j(\omega_0 t + \phi)}$$
$$= \tfrac{1}{2}\big(\cos(\omega_0 t + \phi) + j\sin(\omega_0 t + \phi) \big)$$
$$+ \tfrac{1}{2}\begin{pmatrix} \cos\big(-(\omega_0 t + \phi)\big) \\ + j\sin\big(-(\omega_0 t + \phi)\big) \end{pmatrix}$$

Applying the trigonometric identities $\sin(-\alpha) = -\sin\alpha$ and $\cos(-\alpha) = \cos\alpha$ gives

$$x(t) = \tfrac{1}{2}\big(\cos(\omega_0 t + \phi) + j\sin(\omega_0 t + \phi) \big)$$
$$+ \tfrac{1}{2}\big(\cos(\omega_0 t + \phi) - j\sin(\omega_0 t + \phi) \big)$$
$$= \cos(\omega_0 t + \phi)$$

12. CONTINUOUS-TIME SYSTEMS: FOURIER TRANSFORM OF AN APERIODIC SIGNAL

Many signals of interest to an electrical engineer are not periodic, that is, they are *aperiodic*. The Fourier series representation can still be utilized if one considers the aperiodic signal as the limit of a periodic signal as the period becomes arbitrarily large.[44] The result is[45]

$$x(t) = \frac{1}{2\pi} \int_{-\infty}^{+\infty} X(\omega) e^{j\omega t}\, d\omega \qquad \text{15.60}$$

$$X(\omega) = \int_{-\infty}^{+\infty} x(t) e^{-j\omega t}\, dt \qquad \text{15.61}$$

Equation 15.60 and 15.61 are referred to as a *Fourier transform pair*. Equation 15.61 is the *Fourier transform* or *Fourier integral*. Equation 15.60 is the *inverse Fourier transform* that is the *synthesis equation* of the pair.[46]

Some properties of the Fourier transform for aperiodic signals are given in App. 15.D. Some Fourier transform pairs are given in App. 15.E.

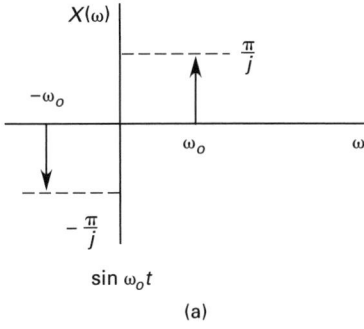

(a)

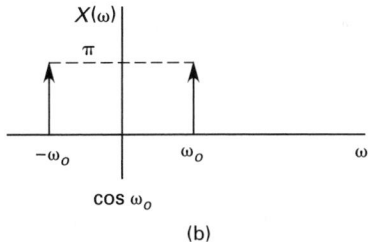

(b)

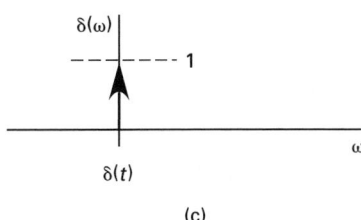

(c)

Figure 15.10 Fourier Transformations

[44]This is equivalent to saying $\omega_0 \to 0$ since $\omega_0 = 2\pi/T_0$. In other words, no periodicity exists.

[45]The derivation of most of the formulas used in this book will not be given unless they enhance the understanding of a fundamental principle. The electrical engineer is, after all, concerned with the results and how to utilize these results.

[46]This is because the inverse Fourier transform decomposes a signal into a linear combination of complex exponentials similar to the function of Eq. 15.58.

The arrangement for aperiodic signals is analogous to that for periodic functions. Eq. 15.60 correlates with Eq. 15.58. The complex exponentials formed by the synthesis equations have amplitudes represented by the coefficients a_k given by Eq. 15.59 in the periodic case—representing a discrete set of harmonically related frequencies $k\omega_0$ with $k = 0, \pm 1, \pm 2, \ldots$. For aperiodic signals, Eq. 15.60 represents the linear combination of complex exponentials with amplitudes $(X(\omega)/2\pi)\,d\omega$, with $X(\omega)$ given by Eq. 15.61. Examples of Fourier transformations are shown in Fig. 15.10. (Obviously, the sine and cosine functions shown in the figure are not aperiodic, but rather periodic. The Fourier transform of a periodic signal with Fourier coefficients can be thought of as a train of impulses occurring at harmonically related frequencies for which the area of the impulse at the kth harmonic frequency ($k\omega_0$) is 2π times the kth Fourier series coefficient a_k. All the calculations in the aperiodic section apply.)

Fourier transforms are a useful tool for the analysis of continuous-time systems that are linear time-invariant. (The broad range of circuitry using resistors, capacitors and inductors are linear time-invariant.) The output signals of such circuits are related proportionally to the input signals by an amplitude factor.

In mathematical terms, an input signal $x(t)$ can be represented by an infinite number of pulses, summed as shown in Eq. 15.62.

$$x(t) = \int_{-\infty}^{+\infty} x(\tau)\delta(t - \delta)d\tau \qquad \textbf{\textit{15.62}}$$

The output signal is then given by the following equation.[47]

$$y(t) = \int_{-\infty}^{+\infty} x(\tau)h(t - \tau)d\tau \qquad \textbf{\textit{15.63}}$$

Equation 15.63 is of the form of a convolution integral (see Sec. 15-13). Using data from App. 15.D, Eq. 15.63 can be written as[48]

$$y(t) = \int_{-\infty}^{+\infty} x(\tau)h(t - \tau)d\tau$$

$$= \mathcal{F}^{-1}\big(X(\omega)H(\omega)\big) \qquad \textbf{\textit{15.64}}$$

Herein lies the power of the Fourier transform: the often difficult convolution can be completely avoided. Convert the input signal $x(t)$ into the frequency domain as

$X(\omega)$. Convert the impulse response $h(t)$ into $H(\omega)$. The convolution of $x(t)$ and $h(t)$ is simply the product of $X(\omega)$ and $H(\omega)$, and the output $y(t)$ is realized by applying the inverse Fourier transform to the result. In other words, a convolution in the time domain is multiplication in the frequency domain.[49]

13. CONVOLUTION

The convolution property of a given transform T is one that gives the inverse transform $T^{-1}\{X(\tau)H(\tau)\}$ directly in terms of the original functions $x(t)$ and $h(t)$.[50] This was evident in Eq. 15.64 where the inverse transform $T^{-1}\{X(\tau)H(\tau)\}$, that is, $y(t)$, was given in terms of the original functions $x(t)$ and $h(t)$.[51] A generalized form of the *convolution integral* is

$$f(t) * g(t) = \int_{a}^{b} f(\tau)g(t - \tau)d\tau$$

$$= T^{-1}\big\{F(s)G(s)\big\} \qquad \textbf{\textit{15.65}}$$

When the convolution property is invoked for a Fourier transform, the limits of integration a and b become $-\infty$ and $+\infty$, respectively. Also, s becomes ω (or $j\omega$ in some texts) and $G(\omega)$ represents $H(\omega)$. ($H(\omega)$ is the Fourier transform of the system impulse response, or the frequency response, for LTI circuits as discussed in previous sections of this chapter.) When invoked for a Laplace transformation, a and b become 0 and $+\infty$, respectively, and s is generalized to real and imaginary components, $s = \sigma + j\omega$. (This essentially gives the properties of magnitude and phase.)

Convolution properties follow. The variable t has been dropped in the symbology of Eq. 15.66 through 15.69 as convolution can be used with more than one variable.

$$f * g = g * f$$
$$\text{[commutative law]} \qquad \textbf{\textit{15.66}}$$

$$f * (g_1 + g_2) = f * g_1 + f * g_2$$
$$\text{[distributive law]} \qquad \textbf{\textit{15.67}}$$

$$(f * g) * \nu = f * (g * \nu)$$
$$\text{[associative law]} \qquad \textbf{\textit{15.68}}$$

$$f * 0 = 0 * f = 0 \qquad \textbf{\textit{15.69}}$$

---

[47]Note that here no change in the input signal representation occurs. When using the Fourier series of $x(t)$ in the process of obtaining $y(t)$, such a change does occur. Though the transform method is more straightforward, the Fourier series representation has numerous applications and was therefore presented in depth in Sec. 11.

[48]Some texts write $X(\omega)$ and $H(\omega)$ as $X(j\omega)$ and $H(j\omega)$. The j is normally understood to be present.

---

[49]For reference, the transformation is from the domain of k-space, which is referred to as the *frequency domain* when ω replaces k.

[50]The dummy variable τ is used to represent the transform and can be any transform: Fourier, Fourier sine, Fourier cosine, Laplace and others.

[51]It is often easier to find the result using convolution rather than applying the inverse transform to $X(\omega)H(\omega)$.

14. LAPLACE TRANSFORMS

The Laplace transform, whose properties and uses are given in the Ch. 11, is repeated here for convenience.

$$\mathcal{L}\big(f(t)\big) = F(s) = \int_0^\infty e^{-st} f(t) dt \qquad 15.70$$

Equation 15.70 is more specifically termed the *unilateral Laplace transformation*. The primary use of this transform is in the solution of linear constant-coefficient differential equations with initial conditions. (See Sec. 11-13 for an example.)

By contrast, the *bilateral Laplace transform* of a general signal $f(t)$ is given by Eq. 15.71.

$$\mathcal{L}\big(f(t)\big) = F(s) = \int_{-\infty}^{+\infty} e^{-st} f(t) dt \qquad 15.71$$

The unilateral transform is more commonly encountered. The term unilateral is not often used. The differentiation between the two types is simply shown in the limits of integration. From the bilateral Laplace transform, the relationship between it and the Fourier transform is easily seen.

$$F(s)\big|_{s=j\omega} = \mathcal{F}\big(f(t)\big) \qquad 15.72$$

If s is not purely imaginary, the relationship is given mathematically as

$$F(\sigma + j\omega) = \int_{-\infty}^{+\infty} \big(f(t)e^{-\sigma t}\big) e^{-j\omega t} dt \qquad 15.73$$

In other words, the Laplace transform of a function with $s = \sigma + j\omega$ is the Fourier transform of $f(t)$ multiplied by the real exponential $e^{-\sigma t}$. Properties of the Laplace transform as well as some common transforms themselves are given in App. 15.F and 15.G. These appendices include the *region of convergence* (ROC or R in some texts), that is, the restrictions necessary to make the transform valid, which will become important when the frequency response of electrical circuits is considered.

15. DISCRETE-TIME SYSTEMS: FOURIER SERIES AND TRANSFORMS[52]

A *discrete-time signal* is one in which the independent variable is defined only at discrete times. Figure 15.11 shows an example of such a signal.

---

[52]These systems, and their associated Fourier series and transforms, are sometimes called *finite* vice discrete.

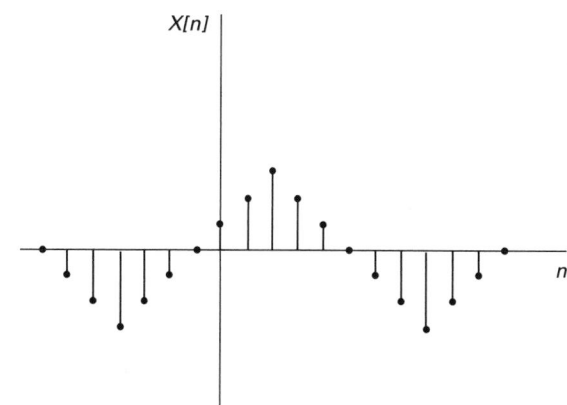

X[n]

Figure 15.11 *Discrete-Time Periodic Signal*

Note that n replaces t as the variable. This is to distinguish between the continuous-time and discrete-time cases. Additionally, brackets are used, $[n]$, to further distinguish discrete-time systems.

The response of an LTI system to inputs that can be represented as complex exponential sequences is analogous to continuous-time systems.[53] That is, in general terms,

$$x[n] = \sum_k a_k z_k^n \rightarrow y[n] = \sum_k a_k H[z_k] z_k^n$$

$$\text{input} \rightarrow \text{output} \qquad 15.74$$

Here $H[z_k]$, instead of being the Fourier or Laplace transform of the system impulse response $h(t)$, is the Fourier or the z-transform of the discrete time system impulse response $h[n]$. When $z = e^{j\Omega}$ with Ω real, $H[z_k]$ is the Fourier transform. When $z = e^{\sigma + j\Omega}$, $H[z_k]$ is the z-transform.[54]

16. DISCRETE-TIME SYSTEMS: FOURIER SERIES REPRESENTATION OF A PERIODIC SIGNAL

Similar to the continuous-time case, the Fourier series representation of a discrete-time periodic signal, of period N, is given by the following synthesis and analysis equations. (The symbol $\langle N \rangle$ indicates an interval of length N.)

$$k[n] = \sum_{k=\langle N \rangle} a_k e^{jk\left(\frac{2\pi}{N}\right)n} \qquad 15.75$$

$$a_k = \frac{1}{N} \sum_{n=\langle N \rangle} x[n] e^{-jk\left(\frac{2\pi}{N}\right)n} \qquad 15.76$$

---

[53]See the example in Sec. 15-10 for the form of the proof, as only the results will be presented here.

[54]This discrete transform, the z-transform, is discussed in Sec. 15-18. It is the discrete-time counterpart to the Laplace transform for continuous-time cases.

Note the important difference in the Fourier series representation in this discrete-time case as compared to the continuous-time case—it is finite (over the interval N).[55]

17. DISCRETE-TIME SYSTEMS: FOURIER TRANSFORM OF AN APERIODIC SIGNAL

Applying an analogous procedure to that used for continuous-time systems, that is, constructing a periodic signal from the aperiodic signal and then taking the limit as the period approaches infinity, gives the following.

$$x[n] = \frac{1}{2\pi} \int_{2\pi} x[\Omega] e^{j\Omega n} d\Omega \qquad 15.77$$

$$X[\Omega] = \sum_{n=-\infty}^{+\infty} x[n] e^{-j\Omega n} \qquad 15.78$$

The function given by Eq. 15.78 is called the *discrete-time Fourier transform* and the two equations together, Eq. 15.77 and 15.78, are termed the *Fourier transform pair*. As before, Eq. 15.77 is the synthesis equation and Eq. 15.78 is the analysis equation. The properties of the discrete-time Fourier series (used with periodic signals) are given in App. 15.H. Discrete-time Fourier transforms (used with aperiodic signals) are given in App. 15.I.

18. Z-TRANSFORMS

The z-transform is the discrete-time counterpart to the Laplace transform. Also, it represents the generalization of the Fourier transform for discrete-time systems; that is, it is the transform to use when the value of s in a discrete-time system is $s = \sigma + j\Omega$.[56] The z-transform of a sequence is defined as[57]

$$X[z] = \sum_{n=-\infty}^{+\infty} x[n] z^{-n} \qquad 15.79$$

The variable z is complex. The relationship between the z-transform and the discrete-time Fourier transform is given by the following.

$$X[z]\big|_{z=e^{j\Omega}} = \mathcal{F}\big[x[n]\big] \qquad 15.80$$

Some properties of the z-transform are given in App. 15.J. Some common z-transforms are given in App. 15.K.

19. TRANSFORMATION OF INTEGRALS

A *transform* takes on the following form.[58]

$$T\{f\} = \int_a^b f(k)k(x,\tau)dx = f(\tau) \qquad 15.81$$

Essentially one integrates out a variable, x in this case, and places the function in another *space* or *domain*, that of τ in this case. The symbology $T\{f\}$ indicates the transform of the function and $k(x,\tau)$ is referred to as the *kernal*. For the Fourier transform the kernal is $e^{j\omega_0 \tau}$. For the Laplace transform, the kernal is $e^{-s\tau}$. Many others exist. A transform places the function in another *function space*, for example, k, ω, s, and so on.

By contrast, the *transformation of integrals* changes the way a physical system is viewed and has important consequences in electrical engineering. Such transformations follow.

Gauss's Divergence Theorem[59]

$$\oiint_{A(V)} \mathbf{F} \cdot d\mathbf{A} = \iiint_{V(A)} \nabla \cdot \mathbf{F} dV \qquad 15.82$$

An arbitrary vector field is represented by $\mathbf{F}$. The symbology $A(V)$ indicates the surface area over the volume of concern. The $d\mathbf{A}$ indicates the normal unit differential surface area, and $V(A)$ is the volume bounded by the surface A. Equation 15.82 says the net flux through a closed surface equals the summation of the scaler sources inside; that is, it equals the divergence of the field integrated over the volume. Note that the symbols

$$\oiint_{A(V)} \text{ and } \iiint_{V(A)}$$

are sometimes represented as

$$\int_S \text{ or } \oint_S \text{ or } \iint_S \text{ or } \oiint_S$$

and

$$\int_V \text{ or } \iiint_V$$

---

[55]This finiteness is important due to its application in the FFT or Fast Fourier Transform. The FFT algorithm is well-suited to computers allowing efficient analysis of signals using discrete methods, which prior to computers had been too time consuming.

[56]This means that s has both real and imaginary parts, that is, a magnitude and a phase.

[57]This is the *bilateral z-transform*. The *unilateral z-transform* is summed from $n = 0$ to $n = \infty$ much as the unilateral Laplace transform commonly used. Also, again similar to the unilateral Laplace transform, the unilateral z-transform is useful in solving for the output on circuits with a response given by constant coefficient difference equations with nonzero initial conditions.

[58]This is a general linear integral transformation; that is, superposition is applicable.

[59]More will be said on the electrical consequences and applications of this and other theorems later in this section.

Stokes' Theorem

Stokes' theorem is analogous to Gauss's theorem. It relates the curl of a vector field **F** inside a closed line, or contour, to the circulation—that is, the curl of **F** along that contour. See Fig. 15.12 for a physical picture of the theorem.[60]

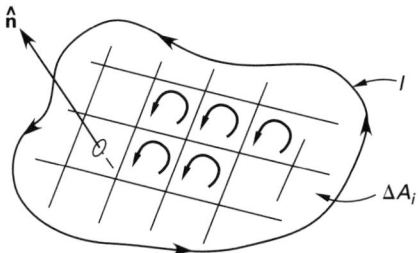

Figure 15.12 *Curl and Circulation of Vector Field*

In mathematical terms, Stokes' theorem is

$$\oint_{l(A)} \mathbf{F} \cdot d\mathbf{l} = \iint_{A(l)} \boldsymbol{\nabla} \times \mathbf{F} \cdot d\mathbf{A} \qquad 15.83$$

Note that the symbols

$$\oint_{l(A)} \quad \text{and} \quad \iint_{A(l)}$$

are sometimes represented as

$$\int_{C} \quad \text{and} \quad \int_{S}$$

where C indicates a *contour*, or *line*, *integral* and S indicates a *surface integral*.

Green's Theorem

Green's theorem is actually a special case of the Stokes' theorem with the vector **F** in the same plane as the surface A. If $g(x,y)$ and $f(x,y)$ are functions defined in this plane, say x and y, then Green's theorem in mathematical terms is

$$\oint_{l(A)} \big(f(x,y)dx + g(x,y)dy \big)$$

$$= \iint_{A(l)} \left(\frac{\delta g}{\delta x} - \frac{\delta f}{\delta y} \right) dx\, dy \qquad 15.84$$

This theorem transforms double integrals, that is, surface integrals, into line integrals. (This is similar to

Gauss's theorem that transformed volume integrals, that is, triple integrals, into area integrals (double integrals).) Green's theorem is sometimes useful in simplifying the calculation of the integral.

20. LINE INTEGRALS

A line integral is a generalization of the concept of a definite integral. For example, a definite integral is given by the following.

$$\int_{a}^{b} f(x)dx \qquad 15.85$$

Let this integral apply to the curve in Fig. 15.13.

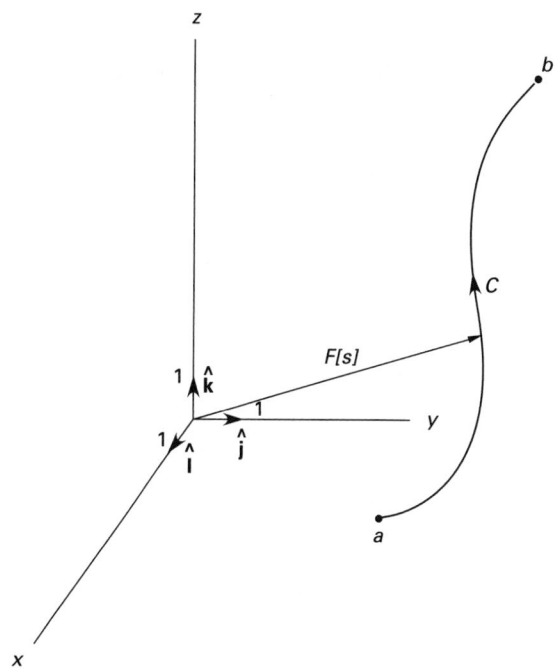

Figure 15.13 *Simple Curve*

Since curve C is *simple*, that is, it has no double points, it may be represented in Cartesian coordinates as[61]

$$a \leq s \leq b$$
$$\mathbf{r}(s) = x(s)\mathbf{i} + y(s)\mathbf{j} + z(s)\mathbf{k} \qquad 15.86$$

The *line integral*[62] of f along C from a to b is

$$\int_{C} f(x,y,z)ds \qquad 15.87$$

---

[60] A paddle wheel's rate of rotation can be thought of as measuring the curl of a turbulent fluid of velocity **F** with the curl in the direction of the paddle wheel axis, $\mathbf{n}dA = d\mathbf{A}$.

[61] Another way of defining a smooth curve is to say it has unique tangents at all points within the curve of line.
[62] Probably a better name would be *curve integral* rather than line integral, but this term is not commonly used.

The curve C is called the *path of integration*.

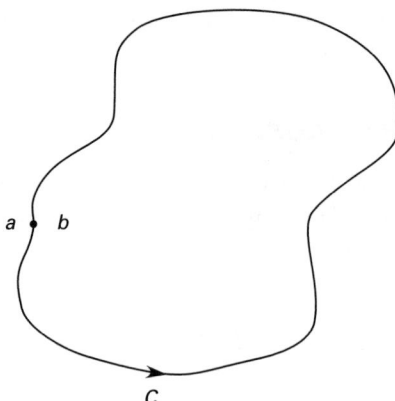

Figure 15.14 *Closed Path*

If the line integral is over a closed path as shown in Fig. 15.14, the integral would be written as

$$\oint_C f(x, y, z)ds \qquad 15.88$$

Line integrals are evaluated by changing them into definite integrals. The key is to represent the path of integration C in the desired form. One possible form is

$$\int_C f(x, y, z)ds$$

$$= \int_a^b f\big(x(s), y(s), z(s)\big)ds \qquad 15.89$$

The value of $\mathbf{r}$ may also be represented as

$$\mathbf{r}(t) = x(t)\mathbf{i} + y(t)\mathbf{j} + z(t)\mathbf{k}$$

$$t_o \le t \le t_1 \qquad 15.90$$

The variable t can be any parameter, not just time.

In this case, the line integral can be calculated using the following.

$$\int_C f(x, y, z)ds$$

$$= \int_{t_o}^{t_i} f\big(x(t), y(t), z(t)\big)\frac{ds}{dt}dt \qquad 15.91$$

Note that

$$\frac{ds}{dt} = \sqrt{\dot{\mathbf{r}}\cdot\dot{\mathbf{r}}}$$

$$= \sqrt{\dot{x}^2 + \dot{y}^2 + \dot{z}^2} \qquad 15.92$$

Example 15.9

Integrate the function $f(x, y) = 3xy^2$ over the portion of the circle from a to b as shown. The direction of C determines the positive direction and the integration path.

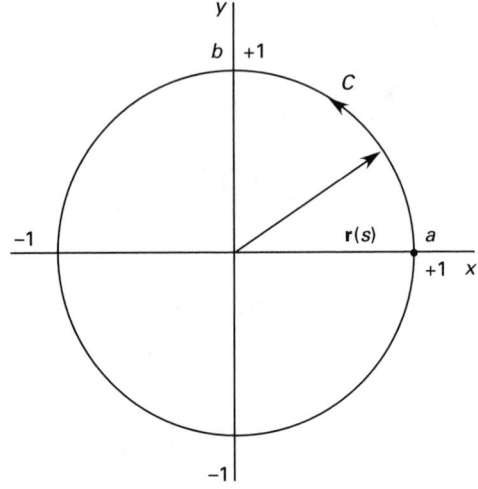

Solution

The value of $\mathbf{r}(s)$ can be seen as

$$\mathbf{r}(s) = \cos s\mathbf{i} + \sin s\mathbf{j}$$

Using Eq. 15.87 gives

$$\int_C f(x, y)ds = \int_a^b f\big(x(s), y(s)\big)ds$$

$$\int_C 3xy^2 ds = \int_0^{\frac{\pi}{2}} 3\big(\cos(s)\big)\big(\sin(s)\big)^2 ds$$

$$= 3\int_0^{\frac{\pi}{2}} \cos(s)\sin^2(s)ds$$

Using a change of variables, let $u = \sin s$. The integration becomes[63]

$$3\int_0^{\frac{\pi}{2}} \cos(s)\sin^2 s\, ds = 3\int_0^1 u^2 du$$

$$= 3\left[\frac{u^3}{3}\right]\Big|_0^1 = 3\left[\frac{1}{3} - \frac{0}{3}\right]$$

$$= 3\left[\frac{1}{3}\right] = 1$$

---

[63] At this point, one could choose to determine the integral from standard math tables.

Example 15.10

Consider a particle upon which a variable force **p** acts along the curve C. The work done by the force **p** is given by the line integral

$$w = \int_C \mathbf{p} \cdot d\mathbf{r}$$

Show that this equals the gain in kinetic energy.

Solution

First, note that

$$d\mathbf{r} = \frac{d\mathbf{r}}{dt} dt = \mathbf{v} dt$$

Substituting gives

$$w = \int_{t_0}^{t_1} \mathbf{p} \cdot \mathbf{v} dt$$

The values t_0 and t_1 are the initial and final values of dt. Newton's second law states that

$$\mathbf{p} = m \frac{d^2\mathbf{r}}{dt^2} = m \frac{d\mathbf{v}}{dt}$$

The variable m represents the mass. Substituting a second time gives

$$w = \int_{t_o}^{t_1} \mathbf{p} \cdot \mathbf{v} dt = \int_{t_0}^{t_1} \left(m \frac{d\mathbf{v}}{dt} \right) \cdot \mathbf{v} dt$$

$$= \int_{t_0}^{t_1} \frac{d}{dt} [m\mathbf{v} \cdot \mathbf{v}] dt = \int_{t_0}^{t_1} dm |\mathbf{v}|^2$$

$$= \frac{m}{2} |\mathbf{v}|^2 \Big|_{t_0}^{t_1}$$

This is the desired result.

An important final note: The value of a line integral of a given function depends not only on the endpoints but also on the shape of the integration path.

21. COMPLEX NUMBERS

Two additional important items not covered in Ch. 4 regarding complex numbers are now considered.

Let any complex number z be represented by

$$z = x + jy \qquad \text{15.93}$$

Note that the z is not bolded as in Ch. 4 to avoid confusion with vector nomenclature. Although it is somewhat similar, it is a point and not a direction. Further, j is used instead of i to represent $\sqrt{-1}$ to avoid confusion with the nomenclature for current.

The complex conjugate is

$$\overline{z} = z^* = x - jy \qquad \text{15.94}$$

The real portion (Re or $\mathcal{R}e$) of any complex number is given by

$$\mathcal{R}e\, z = x = \frac{1}{2}(z + z^*) \qquad \text{15.95}$$

The imaginary part (Im or $\mathcal{I}m$) of any complex number is given by

$$\mathcal{I}m\, z = y = \frac{1}{2j}(z - z^*) \qquad \text{15.96}$$

Also, the familiar formulas for real trigonometric functions hold for complex values.

Example 15.11

Express the Euler formula for complex values in terms of real functions. The Euler formula is

$$e^{j\theta} = \cos\theta + j\sin\theta$$

Solution

If θ is complex, and representing θ as z (which is standard in electrical engineering symbology), gives

$$e^{jz} = \cos z + j\sin z$$

Using a trigonometric identity, the individual terms are

$$\cos z = \cos(x + jy)$$
$$= \cos(x)\cos(jy) - \sin(x)\sin(jy)$$

$$\sin z = \sin(x + jy)$$
$$= \sin(x)\cos(jy) + \cos(x)\sin(jy)$$

Since the familiar formulas apply, the hyperbolic cosine and sine for a complex number are

$$\cos z = \frac{1}{2}\left(e^{jz} + e^{-jz}\right)$$

$$\sin z = \frac{1}{2j}\left(e^{jz} - e^{-jz}\right)$$

This allows the following representation.

$$\cos jy = \frac{1}{2}\left(e^{j(jy)} + e^{-j(jy)}\right)$$

$$= \frac{1}{2}\left(e^{-y} + e^{+y}\right) = \cosh y$$

Also,

$$\sin jy = \frac{1}{2j} \left(e^{j(jy)} - e^{-j(jy)} \right)$$

$$= \frac{1}{2j} \left(e^{-y} - e^{+y} \right) = j \sinh y$$

Substituting gives

$$\cos z = \cos(x + jy)$$
$$= \cos x \cosh y - j \sin x \sinh y$$

$$\sin z = \sin(x + jy)$$
$$= \sin x \cosh y + j \cos x \sinh y$$

It is this real representation of the terms that is useful in the actual calculation of $\cos z$ and $\sin z$.

The theory of *complex analytic functions* can be useful in the study of potential theory in electrostatics. The development of such functions will not be presented but is mentioned to place them in the context of electrical engineering. The results of such calculations are shown as force field lines, or potential lines, in field theory.

Topic II: Theory

Theory

For the most current information about the exam, visit **www.ppi2pass.com** regularly.

16 Electromagnetic Theory

Nomenclature

A	area	m^2
B	magnetic flux density	T or Wb/m^2
C	capacitance	F
d	distance	m
D	electric flux density	C/m^2
e^-	electron	C
esu	electrostatic unit, 1.6022×10^{-19}	C
E	electric field strength	V/m
F	force	N
g	gravitational constant	$N{\cdot}m^2/kg^2$
G	universal gravitation constant, 6.673×10^{-11}	$N{\cdot}m^2/kg^2$
G	conductance	S or $1/\Omega$ or A/V
h	Planck's constant, 6.6256×10^{-34}	J·s
H	magnetic field strength	A/m
i, I	current	A
J	current density	A/m^2
k	electrostatic proportionality constant, 8.9887×10^9	$N{\cdot}m^2/C^2$
ℓ	azimuthal quantum number	
l	length	m
L	inductance	H
m	magnetic moment	$A{\cdot}m^2$ or J/T
m	mass	kg
m	spatial quantum number	
m_s or s	spin quantum number	
M	magnetization	A/m
M	mutual induction	H
n	principal quantum number	
N	number of turns	–
p	pole strength	Wb or N/T
p^+	proton	C
q, Q	charge	C
r	radius	m
R	ratio	–
R	resistance	Ω
$\mathcal{R}$	reluctance	A/Wb
s	surface area	m^2
S	elastance	1/F
t	time	s
u	energy density	J/m^3
U	energy	J
v	velocity[1]	m/s
v, V	voltage	V
V	volume	m^3

Symbols

Γ	reciprocal inductance	1/H
ϵ	permittivity	$C^2/N{\cdot}m^2$ or F/m

---

[1]The term "velocity" is often used in text where it would be more technically correct to use the term "speed." Velocity is a vector quantity with speed as its magnitude.

ϵ_0	free-space permittivity, 8.854×10^{-12}	$C^2/N\cdot m^2$
Λ	flux linkage	Wb
μ	mobility	$m^2/V\cdot s$
μ	permeability	N/A^2 or H/m
μ_0	free-space permeability, 1.2566×10^{-6}	N/A^2
ρ	charge density	C/m^3
σ	conductivity	$1/\Omega\cdot m$
υ	volume	m^3
ϕ, Φ, Ψ	magnetic or electric flux	Wb or C
χ	susceptibility	–
ω	angular velocity	rad/s

Subscripts

1–2	from 1 to 2; or effect on 2 due to the field of 1
ave	average
ci	cast iron
d	displacement drift
e	electric
m	magnetic
m	mean
r	relative
V	volume

Equivalent Units

symbol	equivalent units				
A	C/s	W/V	V/Ω	$J/s\cdot V$	$N/T\cdot m$	Wb/H
C	$A\cdot s$	J/V	$N\cdot m/V$	$V\cdot F$		
F	C/V	C^2/J	$C^2/N\cdot m$	$A^2\cdot s^4/kg\cdot m^2$		
F/m	$C/V\cdot m$	$C^2/J\cdot m$	$C^2/N\cdot m^2$	$A^2\cdot s^4/kg\cdot m^3$		
H	Wb/A	$V\cdot s/A$	$T\cdot m^2/A$	$kg\cdot m^2/A^2\cdot s^2$	$\Omega\cdot s$	
Hz	1/s					
J	$N\cdot m$	$V\cdot C$	$W\cdot s$	C^2/F	$kg\cdot m^2/s^2$	
m^2/s^2	$V\cdot C/kg$					
N	J/m	$V\cdot C/m$	$kg\cdot m/s^2$			
N/A^2	$Wb^2/N\cdot m^2$					
Pa	N/m^2	$kg/m\cdot s^2$				
Ω	V/A	$kg\cdot m^2/A^2\cdot s^3$				
S	A/V	$A^2\cdot s^3/kg\cdot m^2$	$1/\Omega$			
T	Wb/m^2	$N/A\cdot m$	$N\cdot s/C\cdot m$	$kg/A\cdot s^2$		
V	W/A	C/F	J/C	$kg\cdot m^2/A\cdot s^3$		
V/m	N/C					
W	J/s	$V\cdot A$	$kg\cdot m^2/s^3$			
Wb	$V\cdot s$	$H\cdot A$	$T\cdot m^2$	$kg\cdot m^2/A\cdot s^2$		

Part 1: Fundamentals

As a general rule, electrical effects are caused by or emanate from static charges, while magnetic effects are caused by or emanate from charges in motion. To be more precise, electricity is any manifestation of energy conversion of charge carriers that results in forces in the direction of motion of those charge carriers.[2] Magnetism is any manifestation of the kinetic energy of charge carriers that arises from forces or produces forces in the direction perpendicular to the motion of those charge carriers. The two were long thought to be separate phenomena, but were united by Maxwell's equations.[3] Electromagnetism deals with control of the average movement of charges, in generally linear elements, and usually at considerable power levels. For comparison, see the description of electronics in Sec. 17-1.

---

[2]The forces produce displacement, velocity, or acceleration.

[3]Special Relativity also confirms the link between electricity and magnetism in that it shows that electric and magnetic fields appear or vanish depending upon the motion of the observer. Further, it was the Lorentz transformation that required invariance in Maxwell's equations; that is, electric and magnetic fields must remain in the same form for stationary and moving systems. This requirement is necessary because, without it, electrical and optical phenomena would differ depending on one's motion. It is important to understand that, while the experimental results do differ depending on the motion of the observer, the fundamental phenomena remain unchanged. For example, Maxwell's equations describing electromagnetism remain in the same form. The Lorentz transformation was the genesis of Einstein's Special Theory of Relativity.

Table 16.2 *Common Conductors*[a]

material	gauge[b]	diameter (in)	area[c] (circular mils)	nominal DC resistance (ohms/1000 ft at 68°F (20°C))
aluminum[d]	AWG 8	0.1285	16,510	1.030
copper[e]	AWG 8	0.1285	16,510	0.6533
steel[f]	BWG 8	0.165	27,239[g]	3.28[h]

[a] The English Engineering System is used in this table, as it represents the system most commonly used in this area of electrical engineering.

[b] AWG stands for American Wire Gauge, the usual standard for nonferrous wires, rods, and plates. BWG stands for Birmingham Wire Gauge, the usual standard for galvanized iron and steel wire.

[c] One circular mil is a unit having the area corresponding to the area of a circle with a diameter of 0.001 in.

[d] This information is for bare, solid, all-aluminum hard-drawn wire and is taken from ASTM Standard B230-55T.

[e] This value is considered a trade maximum. It is dependent upon the specific process used when manufacturing the wire. ASTM (American Society of Testing Materials) requirements do exist for the resistance of copper for various processes.

[f] The type of steel referenced here is used for telephone and telegraph wire.

[g] The area in circular mils varies slightly depending upon the construction, that is, the number of strands and arrangement.

[h] This number is referenced to 1000 ft here for convenience in comparison with the other conductors. Transmission cables and telephone or telegraph wires are often referenced to longer lengths such as in ohms/mile.

The charges of interest are the electron and the proton, designated as e^- and p^+. The mass, charge, and charge-to-mass ratios for the proton and electron are given in Table 16.1. The SI unit of charge is the coulomb (C). The charge of one electron is sometimes referred to as one *electrostatic unit* (esu). One C is approximately equal to 6.24×10^{18} esu. One mole of electrons is referred to as a *faraday*. One faraday is approximately equal to 96 500 C.

Table 16.1 *Properties of the Electron and Proton*

	electron	proton
mass at rest, kg	9.1096×10^{-31}	1.6726×10^{-27}
charge, C	-1.6022×10^{-19}	$+1.6022 \times 10^{-19}$
charge-to-mass ratio, C/kg	1.7588×10^{11}	9.5791×10^7

The *extranuclear structure* of an atom, that is, its electrical or electronic structure, determines the defining characteristics of that atom. Orbiting electrons are arranged in shells designated *K, L, M, N, O, P,* and *Q*. The inner shell electrons are tightly bound and interact only with high-energy particles, such as gamma rays. The outer shells in complex atoms—those with numerous neutrons and protons, and thus numerous electrons—tend to be loosely bound. It is these outer shells that determine the electrical and chemical properties of the elements. The energy of the orbiting electron is determined by four quantum numbers: the *principle quantum number* (n), the *azimuthal quantum number* (ℓ), the *spatial quantum number* (m), and the *spin quantum number* $(m_s$ or $s)$.[4,5]

---

[4]The azimuthal quantum number, ℓ, specifies the angular orbital momentum. Electrons whose value of ℓ is 0, 1, 2, and 3 are referred to as the *s, p, d,* and *f* electrons (for historical reasons related to the "picture" these electrons produced during experiments).

[5]The spin has two values, $\pm 1/2$, and is equal to $h/2\pi$ where h is Planck's constant.

When electrons are in close proximity, such as in a crystalline solid, nearby atoms affect their behavior and the electron's energy is no longer uniquely determined. Indeed, the single energy level of an electron in a free atom is spread into a band, or range, of energy levels.[6] The *conduction band* is the range of energy states in a solid in which electrons can move freely (that is, they can effect transitions between energy levels). In other words, the band is not full of *conduction electrons*. When every energy level is full, the material is known as an *insulator* or a *dielectric*.

Part 2: Electrical and Magnetic Phenomena

1. ELECTROMAGNETIC EFFECTS

Electromagnetic effects are caused by the dynamic behavior of elementary particles, that is, the electron and the proton, due to their mass and charge. These effects can differ greatly from free-space effects, depending upon the material in which the particles find themselves. There are three major classes of electromagnetic materials.

- *Conductors or Semiconductors* are materials through which charges flow more or less easily. Some common conductors are given in Table 16.2. The properties of widely used semiconductors are given in Table 16.3. Good conductors have a conductivity range of 1×10^3 S/m or $1/\Omega \cdot$m to 6×10^3 S/m or $\Omega \cdot$m.

---

[6]This band of energy levels allows what is referred to in some texts as an *electron cloud* to exist. The theory is by no means universal. Consult electrical conduction theory texts for more information.

- *Dielectrics or Insulators* are materials that inhibit the passages of charges. Some common insulators are given in Table 16.4. Insulators have a conductivity range of 10^{-18} S/m or $1/\Omega \cdot$m up to approximately 10^{-4} S/m or $1/\Omega \cdot$m. The determining factors for the conductivity of insulators are the application and the voltage stress (that is, the magnitude of the voltage being insulated).

- *Magnetic materials* are materials in which the motion of the charges produces perpendicular effects, that is, enhanced transverse effects. A very small portion of the elements in the periodic table exhibit such effects. These materials are given in Table 16.5.

Table 16.3 Common Semiconductors

semiconductor	symbol	periodic table group
silicon	Si	IV
germanium	Ge	IV
gallium arsenide	GaAs	III and V
diamond and graphite	–	–
selenium	Se	VI
silicon carbide	SiC	IV
indium antimonide	InSb	III and V

Table 16.4 Common Insulators

insulating material
polyvinyl chloride (PVC)
butyl rubber
neoprene
plastics
ceramics
insulating gasses and liquids

Table 16.5 Magnetic Materials

magnetic material	symbol	atomic number
iron	Fe	26
cobalt	Co	27
nickel	Ni	28

2. CONDUCTION EFFECTS

Conduction effects occur in systems having mobile charges when an electric force is applied. The charges move in the direction of the applied force. As mentioned in Sec. 16-1, conduction effects result from the loosely bound outer shell electrons of atoms. In very strong conductors, electrons can be released and move under the application of small chemical, thermal, or other forces.

3. DIELECTRIC PHENOMENA

When a system contains bound charges, the electrons still move under the application of an electric force, but only to a limited extent. This process is called the *dielectric phenomenon*. It results in a current termed the *displacement current*, which is related to the *electric flux density*, **D**, in the dielectric material. In turn, the electric flux density, **D**, is proportional to the *electric field strength* or *intensity*, **E**.[7] Heat losses occur in dielectrics due to hysteresis effects at very high frequencies. Hysteresis effects occur in magnetic materials at much lower frequencies.

Dielectrics can sustain only a certain field strength before they break down and conduction occurs. This maximum field strength is called the *dielectric strength*, and its unit of measure is volts per meter of thickness of the material tested. The results vary, depending upon thickness and the environmental conditions of the test. For the same material, thin dielectrics break down at lower field strengths than thick dielectrics.

4. MAGNETIC PHENOMENA

Magnetic phenomena are caused by the directed motion of charges. The effects occur perpendicular to this motion. In magnetic materials, the phenomena are due to the orientation of the orbiting electrons and, to some extent, the spin of the electrons. Magnetic phenomena occur in a relatively small number of materials whose outer shell orbits fill prior to the inner shells. This occurs because the outer shell orbit configuration is actually a lower energy configuration than the inner shell.

Several types of magnetism exist.

- *Ferromagnetism* is produced by the exchange of forces between atomic moments. This type of magnetism produces strongly magnetic materials with high permeability. Large clusters of atoms group together to form *magnetic domains*, each of which has the same atomic moment alignment within the domain. Ferromagnetism, named for iron, is the common magnetism one associates with horseshoes or toy magnets. See Fig. 16.1 for an example of a crystalline structure with domains.

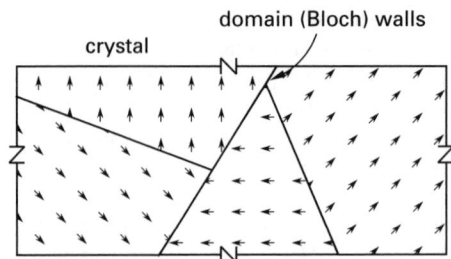

Figure 16.1 Magnetic Domains

---

[7]Both of these are discussed more fully in Sec. 16-11.

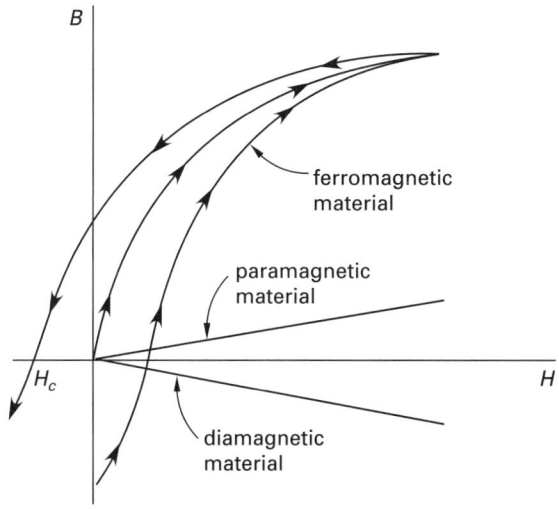

Figure 16.2 *Magnetization Curves*

- *Paramagnetism* is produced by the orbital or spin moments of the electrons, or both. The atomic alignment is minimal, that is, there is little or no domain formation, and thus paramagnetic materials are not strongly magnetic.

- *Antiferromagnetism* is produced by the exchange of forces between atomic moments. Antiferromagnetic materials have an antiparallel arrangement of equal spins. Their magnetism is of a similar strength to that of paramagnetic materials.

- *Diamagnetism* is produced by electron spins in antiparallel pairs in closed electron shells. Diamagnetic materials are weakly repulsive to external magnetic fields. These materials possess an internal magnetic field that opposes any externally applied magnetic field.

- *Ferrimagnetism* is produced by the moment that results from the combination of two antiferromagnetic lattices. Ferrimagnetic materials contain two kinds of magnetic ions, arranged antiparallel but with unequal spins.

5. THERMOELECTRIC PHENOMENA

There are three significant relations between electricity and heat of importance to the electrical engineer. These are the Seebeck effect, the Thomson effect, and the Peltier effect. Each will be explained in terms of its applications. For now, the important point is that heat is capable of providing the energy to release and transport electrons.

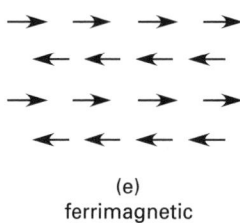

(a)
ferromagnetic

(b)
paramagnetic

(c)
antiferromagnetic

(d)
diamagnetic

Note: The circle represents the internal magnetic field.

(e)
ferrimagnetic

Figure 16.3 *Magnetic Spin Arrangements*

Part 3: Electrostatics

As the name implies, electrostatics is the field of electrical engineering that deals with the electrical effects of stationary charges. An important concept in this field is that all electrical phenomena are expressible in terms of the masses and charges of the elementary particles involved.

6. ELECTRIC CHARGES

An electrically neutral material has equal numbers of positive and negative charge carriers, and thus the following is true for bodies in an unelectrified state.

$$Q^+ = Q^- = 0 \qquad 16.1$$

Only integer numbers of charge carriers exist. Therefore, the transport of an integer number n charge carriers to a body from another body means that

$$Q^+ = np^+ \text{ and } Q^- = ne^- \qquad 16.2$$

Like charges exhibit an electric repulsion. Unlike charges exhibit an electric attraction. Note that an attraction also exists between the masses of any two charged particles. The force associated with the gravitational attraction is very small in comparison to the force of the electrical attraction. Therefore, the gravitational force can be ignored except when the charges are accelerated.

Static electricity formed by positive charges, such as that produced by rubbing silk on a glass rod, is called *vitreous static electricity*. When the static electricity is formed by negative charges, such as that produced by rubbing fur on a rubber, amber, or plastic rod, it is termed *resinous static electricity*.

Example 16.1

What is the gravitational force of attraction between two electrons located 1 cm apart?

Solution

From Newton's Law of Gravitation the following formula is applicable.

$$F = \frac{Gm_1 m_2}{r^2}$$

Newton's universal constant or *gravitational constant*, G, is equal to $6.672 \times 10^{-11} \text{N·m}^2/\text{kg}^2$. The mass of the electrons (taken from Table 16.1) and the 1 cm distance between the electrons are substituted into the equation, giving the following.

$$F = \frac{Gm_1 m_2}{r^2} = \frac{Gm_e^2}{r^2}$$

$$= \frac{\left(6.672 \times 10^{-11} \frac{\text{N·m}^2}{\text{kg}^2}\right)\left(9.1096 \times 10^{-31} \text{ kg}\right)^2}{(1 \text{ cm})^2 \left(\frac{1 \text{ m}}{100 \text{ cm}}\right)^2}$$

$$= 5.537 \times 10^{-67} \text{ N}$$

7. COULOMB'S LAW

Coulomb's law states that the force between charges at rest in an isotropic medium is proportional to the product of the charges and inversely proportional to both the square of the distance between them and the dielectric coefficient of the medium. The force acts in a straight line between the centers of the charges.[8] Equation 16.3 describes Coulomb's law in mathematical terms.

$$\mathbf{F}_{1\text{-}2} = \left(\frac{Q_1 Q_2}{4\pi \epsilon r_{1\text{-}2}^2}\right) \mathbf{a}_{r_{1\text{-}2}} \qquad 16.3$$

The vector $\mathbf{a}_{r_{1\text{-}2}}$ is a unit vector having a direction along a straight line connecting Q_1 and Q_2. The term ϵ represents the *dielectric coefficient*, more commonly called the *permittivity*. The 4π term occurs in many electric field calculations and is thus included in this *rationalized* form of Coulomb's law in order to cancel said term and make calculations easier. This is perhaps more readily seen when Coulomb's law is written in terms of the electric field $\mathbf{E}$.

$$\mathbf{F}_{1\text{-}2} = \mathbf{F}_2 = Q_2 \mathbf{E}_1 \qquad 16.4$$

Coulomb's law is also stated in its Gaussian SI form as follows.

$$\mathbf{F}_{1\text{-}2} = \frac{kQ_1 Q_2}{\epsilon_r r^2} \mathbf{a}_{1\text{-}2} \qquad 16.5$$

The term ϵ_r represents the *relative permittivity* of the dielectric. The constant k has an approximate value of 8.987×10^9 N·m^2/C^2. (Note that $\epsilon_r = 1.0$ for a vacuum.) In Eqs. 16.3 through 16.5 the force is measured in newtons, the distance in meters, and the charge in coulombs.

Example 16.2

What is the magnitude of the electrostatic force between two electrons 1 cm apart in a vacuum?

Solution

Take the magnitude of Eq. 16.5 and substitute the values for the relative permittivity and k given in this section. Note that the relative permittivity has no units. As the term "relative" suggests, it is a comparison. This

---

[8]In Eq. 16.3, the unit vector, and thus the force, is *from 1 to 2*. Don't be confused by this notation. The force direction is the direction charge 2 will move due to the electric field of charge 1. See Eq. 16.4. The notation often varies. The unit vector $\mathbf{a}_{r_{1\text{-}2}}$ (Eq. 16.3) or $a_{1\text{-}2}$ (Eq. 16.4) is sometimes written $\mathbf{a}_r$, $\mathbf{a}_{12}$, or even $\mathbf{a}_{21}$. The $\mathbf{a}_{21}$ notation is used as a reminder that, although the vector is *from 1 to 2*, it is determined mathematically by subtracting the position elements of 2 from 1; for example, $x_2 - x_1$, $y_2 - y_1$, and $z_2 - z_1$. (The unit vector is then determined by dividing the distance between the two points.) At times, the notation 1-2 merely indicates the connection between the points and one must determine the correct direction from other information.

is discussed more fully in Sec. 16-9.

$$\mathbf{F}_{1\text{-}2} = k\left(\frac{Q_1 Q_2}{\epsilon_r r^2}\right)\mathbf{a}_{1\text{-}2}$$

$$|\mathbf{F}_{1\text{-}2}| = F = k\left(\frac{Q_1 Q_2}{\epsilon_r r^2}\right)$$

$$= \frac{\left(8.987 \times 10^9 \, \frac{\text{N·m}^2}{\text{C}^2}\right)(-1.6022 \times 10^{-19} \, \text{C})^2}{(1)(1 \, \text{cm})^2 \left(\frac{1 \, \text{m}}{100 \, \text{cm}}\right)^2}$$

$$= 2.307 \times 10^{-24} \, \text{N}$$

Example 16.3

What is the ratio of the electrostatic force to the gravitational force between two electrons 1 cm apart in a vacuum?

Solution

From Ex. 16.2, the electrostatic repulsion is 2.307×10^{-24} N. From Ex. 16.1, the gravitational attraction is 5.537×10^{-67} N. The ratio of electrostatic force to gravitational force is

$$R = \frac{2.307 \times 10^{-24} \, \text{N}}{5.537 \times 10^{-67} \, \text{N}}$$

$$= 4.166 \times 10^{42}$$

Thus, the electrostatic force that these elementary particles experience is 42 powers of ten greater than the gravitational force of attraction between them. Clearly, the gravitational attraction between such particles can be ignored in all but the most exacting calculations.

Coulomb's law implies that the principle of superposition applies for electric charges. This means that a system of charges is a linear system. The total resultant effect at any point within the system can be determined from the vector sum of the individual components.

Example 16.4

Three point charges in a vacuum are arranged in a straight line. Find the force on point charge A.

$$
\begin{array}{ccccc}
\text{A} & \xleftarrow{3\,\text{m}} & \text{B} & \xleftarrow{4\,\text{m}} & \text{C} \\
400\,\mu\text{C} & & -200\,\mu\text{C} & & 800\,\mu\text{C}
\end{array}
$$

Solution

Since all three charges are in line, vector analysis is not needed. The force on point charge A is the sum of the individual forces from the other two point charges. Use the Gaussian form of Coulomb's law.

From Eq. 16.5,

$$F_{\text{A-B\&C}} = F_{\text{B\&C-A}}$$

$$= F_{\text{A-B}} + F_{\text{A-C}} = \frac{kQ_A Q_B}{r_{\text{A-B}}^2} + \frac{kQ_A Q_C}{r_{\text{A-C}}^2}$$

$$= 8.987 \times 10^9 \, \frac{\text{N·m}^2}{\text{C}^2}$$

$$\times \left(\frac{(400 \times 10^{-6}\,\text{C})(-200 \times 10^{-6} \, \text{C})}{(3\,\text{m})^2}\right.$$

$$\left. + \frac{(400 \times 10^{-6} \, \text{C})(800 \times 10^{-6} \, \text{C})}{(7\,\text{m})^2}\right)$$

$$= -79.9 \, \text{N} + 58.7 \, \text{N}$$

$$= -21.2 \, \text{N}$$

Since the sign is negative, the force on A is toward B and C.

Example 16.5

Determine the magnitude and direction of the force on point charge A due to point charges B and C.

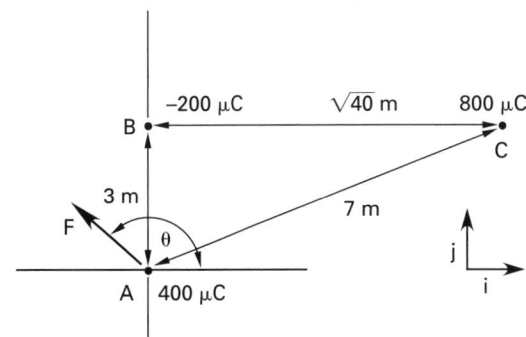

Solution

The charges and distances are the same as in Ex. 16.4.

$$F_{\text{B-A}} = F_{\text{A-B}} = -79.9 \, \text{N} \quad [\text{attractive}]$$

$$F_{\text{C-A}} = F_{\text{A-C}} = 58.7 \, \text{N} \quad [\text{repulsive}]$$

Determine the unit vectors for these forces. Let point A correspond to the origin. The unit vectors ($\mathbf{a}_{\text{B-A}}$ and $\mathbf{a}_{\text{C-A}}$) from B and C to A are determined as follows:

$$\mathbf{a}_{\text{B-A}} = \frac{(x_A - x_B)\mathbf{i} + (y_A - y_B)\mathbf{j}}{\sqrt{(x_A - x_B)^2 + (y_A - y_B)^2}}$$

$$= \frac{(0-0)\mathbf{i} + (0-3)\mathbf{j}}{\sqrt{(0-0)^2 + (0-3)^2}} = -\mathbf{j}$$

$$\mathbf{a}_{\text{C-A}} = \frac{(0-\sqrt{40})\mathbf{i} + (0-3)\mathbf{j}}{\sqrt{(0-\sqrt{40})^2 + (0-3)^2}}$$

$$= -0.904\mathbf{i} - 0.429\mathbf{j}$$

The total force is

$$F_{B-A}\mathbf{a}_{B-A} + F_{C-A}\mathbf{a}_{C-A}$$

$$= (-79.9 \text{ N})(-\mathbf{j}) + (58.7 \text{ N})(-0.904\mathbf{i} - 0.429\mathbf{j})$$

$$= -53.1\mathbf{i} + 54.7\mathbf{j}$$

The magnitude of the force is

$$|\mathbf{F}_{A-B\&C}| = \sqrt{(-53.1)^2 + (54.7)^2} = 76.2 \text{ N}$$

The direction (counterclockwise angle with respect to the horizontal) is

$$\theta = 180° - \arctan\left(\frac{54.7}{53.1}\right) = 180 - 45.9 = 134.1°$$

8. ELECTRIC FIELDS

The *electric field*, $\mathbf{E}$, with units of N/C (or V/m), is the region in which an electric charge exerts a measurable force on another charge, above and beyond the gravitational force that the two charges exert on each other. It is a *vector force field* because a test charge placed in the field will experience a force in a given direction and with a specific magnitude. The direction of the field is along flux lines. These lines, ideally, represent the direction of the force on an infinitely small, isolated positive test charge.[9] The actual direction is represented by the unit vector $\mathbf{a}$. Equation 16.6 represents the *electric field strength*, or *electric field intensity*, in a substance with permittivity ϵ at a distance r from a point charge Q.

$$\mathbf{E} = \frac{Q}{4\pi\epsilon r^2}\mathbf{a} \qquad 16.6$$

Coulomb forces obey the principle of linear superposition in free space and in many dielectrics. Therefore, the electric field strength or intensity acting on a charge Q located at point 1 as a result of charges $Q_1, Q_2, \ldots Q_i$ at distances r_{i1} from point 1 can be represented as the following sum. The vector a_{ri1} is the unit vector with direction from Q_i to point 1.

$$\mathbf{E}_1 = \sum_{i=1}^{i} \frac{Q_i}{4\pi\epsilon r_{i1}^2}\mathbf{a}_{ri1} \qquad 16.7$$

Additionally, the electric field intensity at any point p produced by charges $Q_1, Q_2, \ldots Q_n$ at distances $r_1, r_2, \ldots r_n$ from the point is the vector sum of the field strengths produced by the charges individually.

If one is dealing with charges distributed uniformly throughout space, the electric field strength at a distance r that is large when compared to dv is given by Eq. 16.8.

$$\mathbf{E}_1 = \int_V \frac{\rho dV}{4\pi\epsilon r_{i1}^2}\mathbf{a}_{ri1} \qquad 16.8$$

The unit vector a_{ri1} is directed from dv to point 1, at which $\mathbf{E}$ is evaluated. The unit vector, the distance r, and the charge density ρ are usually functions of the variable of integration. The total charge Q in the volume V that has the charge density ρ is

$$Q = \int_V \rho dV \qquad 16.9$$

The electric fields discussed to this point have been radial in nature. Not all fields are radial. As seen in Eq. 16.8, the electric field depends on the orientation of the surface or volume that contains the charge. A *uniform electric field*, commonly used in sensing instruments and other important applications, can be created between two flat plates, as shown in Fig. 16.4.

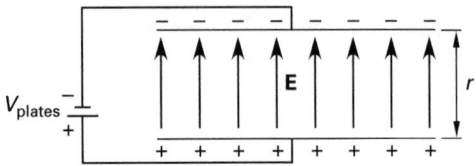

Figure 16.4 *Uniform Electric Field*

With a potential difference of V volts between the flat plates separated by distance r, the electric field strength is

$$E_{\text{uniform}} = \frac{V_{\text{plates}}}{r} \qquad 16.10$$

The field strengths, or intensities, for several configurations are given in Table 16.6.

An electric field is an abstraction that allows one to visualize how charges in different locations within the field will experience a force. A *line of electric force* is a curve drawn so that it has the direction of the force acting on the test charge, (i.e., it is in the direction of the electric field). A *tube of force* is created by drawing lines of force through the boundary of any closed curve. The lines form a tubular surface not cut by any other lines of force.

9. PERMITTIVITY AND SUSCEPTIBILITY

The forces between electric charges depend upon the environment in which they are located. The maximum coulombic force occurs in free space. In all other materials, the coulombic force is reduced. Another way of viewing this phenomenon is to say that the electric flux does not pass equally through all materials.[10] In fact, it cannot pass through conductive material at all and is diminished to varying degrees by various dielectrics.

---

[9]The positive test charge is arbitrary, but universal. The charge would ideally have to be infinitely small in order to prevent its own electric field from distorting the original field.

[10]The concept of flux is another abstraction used to help one visualize electric interactions. It will be discussed more fully in the Electric Flux section in this chapter.

Table 16.6 *Electric Fields and Capacitance for Various Configurations*

isolated point charge	infinite coaxial cylinder
$E = \dfrac{Q}{4\pi\epsilon r^2}$ $C = 0$	$E = \dfrac{\rho_l}{2\pi\epsilon r}$ $(a < r < b)$ $\dfrac{C}{L} = \dfrac{2\pi\epsilon}{\ln\left(\dfrac{b}{a}\right)}$
isolated sphere	infinite line distribution inside an infinite cylinder
$E = \dfrac{Q}{4\pi\epsilon(r+R)^2}$ $(r > 0)$ $C = 4\pi\epsilon R$	$E = \dfrac{\rho_l}{2\pi\epsilon r}$
concentric spheres	infinite sheet distribution
$E = \dfrac{Q}{4\pi\epsilon r^2}$ $(a < r < b)$ $C = \dfrac{4\pi\epsilon ab}{b-a}$	$E = \dfrac{\rho_s}{2\epsilon}$ $\rho_s = \dfrac{Q}{A}$
infinite line distribution	infinite parallel plates
$E = \dfrac{\rho_l}{2\pi\epsilon r}$	$E = \dfrac{\rho_s}{\epsilon} = \dfrac{V}{r}$ $\dfrac{C}{A} = \dfrac{\epsilon}{r}$
infinite isolated cylinder	dipole (doublet)
$E = \dfrac{\rho_l}{2\pi\epsilon(r+R)}$ $(r > 0)$	$E = \dfrac{Qd}{4\pi\epsilon r^3}(2\cos\theta\, a_r + \sin\theta\, a_\theta)$

Reprinted with permission from Professional Publications, Inc., *EIT Reference Manual*, eighth edition, by Michael R. Lindeburg, copyright © 1998, by Professional Publications, Inc.

The *permittivity of free space*, that is, a vacuum, ϵ_0, may be considered to be representative of the dielectric properties of free space. In actuality, it is a factor that provides consistent units for Coulomb's law when units of force, charge, and distance are arbitrary. The product of the permittivity of free space and the *relative permittivity* of a medium gives the *permittivity* of that medium.

$$\epsilon = \epsilon_0\epsilon_r \qquad 16.11$$

In general, the relative permittivity is a complex number. Except for very high frequencies, the imaginary component, or quadrature component, can be ignored. The real component is often termed the *dielectric constant*. Since the value of ϵ_r depends on frequency as well as other factors, the term *dielectric coefficient* is more appropriate. The relative permittivity can also be viewed in terms of capacitance, as shown in Eq. 16.12.

$$\epsilon_r = \frac{C_{\text{with dielectric}}}{C_{\text{vacuum}}} \qquad 16.12$$

Typical relative permittivities are given in Table 16.7.

The permittivity can also be expressed as follows.

$$\epsilon = \epsilon_0(1 + \chi_e) \qquad 16.13$$

The term χ_e represents the electric susceptibility of a given dielectric. It is a numeric measure of the polarization or displacement of electrons in the atoms or molecules of the dielectric.

The units for permittivity are $C^2/\text{N}\cdot\text{m}^2$ or F/m. The permittivity of free space, or vacuum, ϵ_0, in SI units is 8.854×10^{-12} $C^2/\text{N}\cdot\text{m}^2$.

Table 16.7 *Typical Relative Permittivities (20°C and 1 atmosphere)*

material	ϵ_r	material	ϵ_r
acetone	21.3	mineral oil	2.24
air	1.00059	mylar	2.8–3.5
alcohol	16–31	olive oil	3.11
amber	2.9	paper	2.0–2.6
asbestos paper	2.7	paper (kraft)	3.5
asphalt	2.7	paraffin	1.9–2.5
bakelite	3.5–10	polyethylene	2.25
benzene	2.284	polystyrene	2.6
carbon dioxide	1.001	porcelain	5.7–6.8
carbon tetrachloride	2.238	quartz	5
castor oil	4.7	rock	≈ 5
diamond	16.5	rubber	2.3–5.0
glass	5–10	shellac	2.7–3.7
glycerine	56.2	silicon oil	2.2–2.7
hydrogen	1.003	slate	6.6–7.4
lucite	3.4	sulfur	3.6–4.2
marble	8.3	teflon	2.0–2.2
methanol	22	vacuum	1.000
mica	2.5–8	water	80.37
		wood	2.5–7.7

Reprinted with permission from Professional Publications, Inc., *EIT Reference Manual*, eighth edition, by Michael R. Lindeburg, copyright © 1998, by Professional Publications, Inc.

Example 16.6

A certain capacitor is constructed of two square parallel plates with air as the dielectric. If a teflon insert is used in place of the air, by what factor does the capacitance increase?

Solution

The capacitance is directly proportional to the permittivity and is given by the following formula.

$$C = \frac{\epsilon A}{r} = \frac{\epsilon_0 \epsilon_r A}{r}$$

Since what was asked for was the factor by which the capacitance increases, not the actual value of the capacitance, one need merely compare the permittivity of teflon to that of air. Using data from Table 16.7,

$$\frac{\epsilon_{r,\text{teflon}}}{\epsilon_{r,\text{air}}} = \frac{2.0}{1.00059} = 2.0$$

One could use the upper bound of the permittivity for the teflon, in which case the factor by which the capacitance increases would be approximately 2.2.

10. ELECTRIC FLUX

Flux is defined as the electric or magnetic lines of force in a region. The *electric flux*, Ψ, is a scalar field representing these imaginary lines of force. Flux lines leave or enter any surface at right angles to that surface, and the orientations of the electric field lines and the electric flux lines always coincide. By convention, the flux lines are directed outward from the positive charge and inward toward the negative charge, or to infinity if no negative charge exists in the area, as shown in Fig. 16.5. Note that in Fig. 16.5 only a few representative lines of flux are shown. Also, if the charge on a hollow sphere is all the same sign, the flux outside this sphere is the same as the flux that would exist from a point charge at the center. Therefore, for the purposes of drawing flux lines and calculating forces, a hollow charged sphere can be replaced by a point charge at the center.

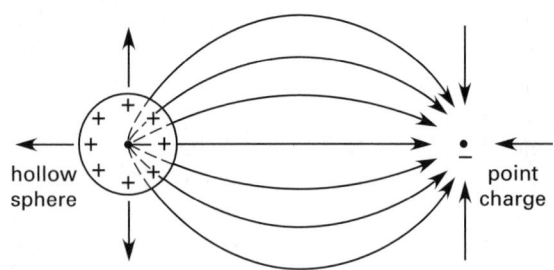

Figure 16.5 *Flux Lines*

The flux Ψ is numerically equal to the charge. That is, by definition, one coulomb of electric charge gives rise to one coulomb of electric flux.

$$\Psi = Q \qquad \qquad 16.14$$

The electric flux is also called the *displacement flux*. This is because the flux is a quantity associated with the amount of bound charge, Q, displaced in a dielectric material that is subject to an electric field. This terminology can be understood by referring to Fig. 16.6.

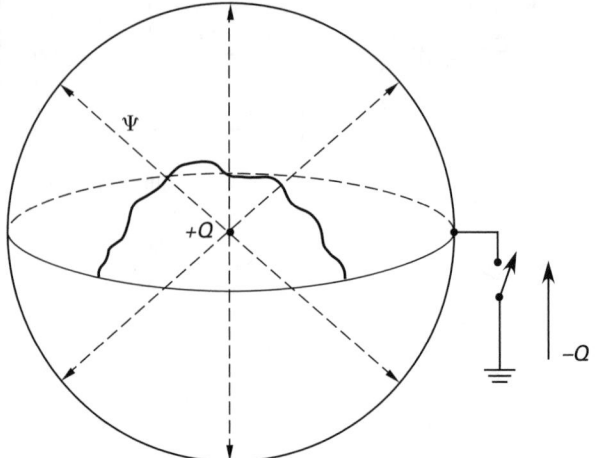

Figure 16.6 *Displacement Flux Experiment*

Early experimenters enclosed a fixed positive charge inside a spherical conducting shell as shown in Fig. 16.6. When the switch was closed, a charge of $-Q$, equal in magnitude to the enclosed positive charge, was found on the shell and was believed to be due to the transient flow of negative charge induced by the flux from the $+Q$. In other words, the positive charge *displaced* the $-Q$ charge and forced it onto the surface. The scalar electric flux field, Ψ, is not directly measurable as is the electric field, $\mathbf{E}$. Instead it is inferred from such experiments.

11. ELECTRIC FLUX DENSITY

The *electric flux density*, $\mathbf{D}$, is also referred to as the *displacement density* or simply the *displacement*. It is also loosely referred to as the *vector flux density*. It is a vector quantity that, like the electric flux, Ψ, cannot be directly measured. It is given in Eq. 16.15. Note that, unlike the electric field strength $\mathbf{E}$, the electric flux density or displacement $\mathbf{D}$ is independent of the medium.[11]

$$\mathbf{D} = \epsilon \mathbf{E} = \epsilon_0 \epsilon_r \mathbf{E} \qquad \qquad 16.15$$

The magnitude of $\mathbf{D}$, or D, is also referred to as the *flux density*. It is the number of flux lines per unit area perpendicular to the flux. It is a scalar quantity with units of C/m^2.

$$D = |\mathbf{D}| = \frac{\psi}{A} = \frac{Q}{A} = \sigma \qquad \qquad 16.16$$

---

[11] This is because $\mathbf{E}$ is multiplied by the permittivity. See Eq. 16.6 by way of explanation, noting the permittivity in the denominator.

When the medium is nonisotropic, the permittivity ϵ is a tensor and the displacement **D** and electric field strength **E** are not necessarily in the same direction.

Many terms with differing units are used to refer to "flux density," such as ρ_l in units of C/m and ρ_V in units of C/m³. This confusion is avoided in texts by referring to such terms as ρ_l and ρ_V as charge densities.[12] Care must be exercised when interpreting of the phrase "flux density."

12. GAUSS'S LAW FOR ELECTROSTATICS

Gauss's law for electrostatics is a necessary consequence of Coulomb's law. Gauss's law states that the amount of electric flux, Ψ, passing through any closed surface is proportional to (and in our rationalized system of SI units, equal to) the total charge, Q, contained within the surface. The surface is called a *Gaussian surface*. Gauss's law in a variety of forms is given in Eqs. 16.17 through 16.19. Note that ds is an element of the surface area and θ is the angle between the normal to the surface area and **D**.

$$\Psi = \oiint D dA = \oiint \sigma dA$$
$$= Q \quad [d\mathbf{A} \text{ parallel to } \mathbf{D}] \qquad 16.17$$

$$\Psi = \oiint \mathbf{D} \cdot d\mathbf{s} = \oiint D \cos \theta ds$$
$$= Q \quad [\text{arbitrary surface}] \qquad 16.18$$

$$\Psi = \iiint \rho dv = Q \quad [\text{arbitrary volume}] \qquad 16.19$$

Since the electric field strength or intensity **E** is related to the displacement or electric flux density **D** by the permittivity, Gauss's law as given in Eq. 16.17 can also be stated mathematically as

$$\Psi = \oiint \epsilon E dA = Q \quad [d\mathbf{A} \text{ parallel to } \mathbf{E}] \qquad 16.20$$

Special Gaussian surfaces are used in highly symmetrical charge configurations to simplify calculations with Gauss's law. Such surfaces meet the following criteria.

(1) The surface is closed.

(2) At each point on the chosen surface, **D** is either normal or tangential to the surface.

(3) The flux density, or displacement, magnitude D is sectionally constant over the portion of the surface where **D** is normal.

---

[12]Note that this is consistent with the electric flux, Ψ, being numerically equal to the charge. Another way of considering this is to realize that a rationalized set of units (the SI system) that makes the flux equal to the charge is being used. If the system were not rationalized, the flux would merely be proportional to the charge.

Requirement (1) is a general requirement of Gauss's law. Requirement (2) completely eliminates integrals and makes the calculation simple. Requirement (3) allows D to be removed from the integral.

Example 16.7

Given a point charge of constant value Q, what is the flux density D at a distance r from the charge?

Solution

First, arbitrarily construct a sphere around the point charge as shown.

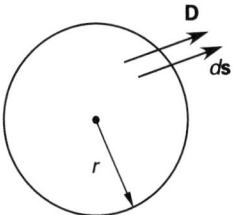

A sphere surrounding a point charge meets all the requirements of a special Gaussian surface. Using Eq. 16.18 gives the following.

$$\Psi = \oiint \mathbf{D} \cdot d\mathbf{s} = \oiint D \cos \theta ds = Q$$

Since the lines of flux emanate radially outward from the point charge, **D** and ds are parallel and $\cos \theta = 1$. Additionally, the magnitude of the flux density, D, is constant at a given r. This results in the following manipulations.

$$Q = \oint D \cos \theta ds$$
$$= \oint D(1)ds = D \oint ds$$

The surface area of a sphere is given by $4\pi r^2$. Thus,

$$Q = D \oint ds = D 4\pi r^2$$

$$D = \frac{Q}{4\pi r^2}$$

Using the same logic as in Ex. 16.7 on a uniform line charge of density ρ_l of length l at a distance of r gives the following equation for displacement magnitude.

$$D = \frac{\rho_l}{2\pi r} \qquad 16.21$$

13. CAPACITANCE AND ELASTANCE

If charges Q^+ and Q^- exist on conductors separated by free space or by a dielectric, the mutual attraction "stores" the charges, holding them in place. The charges produce a stress in the dielectric measured by the electric field strength $\mathbf{E}$, and a potential difference exists between the plates. *Capacitance* is the property of a system of conductors and dielectrics that permits this storage of charges.

If a potential difference is applied across two conductors, charges will build up, creating an electric field between the conductors. The amount of the charge, Q, is proportional to the applied voltage. The proportionality constant is called the *capacitance* and has units of *farads*, F. This is shown in Eq. 16.22. A farad is a large capacitance and very few capacitors are constructed with a value of even 1.0 F. Most circuit elements have capacitances in the range of μF (10^{-6}) or pF (10^{-12}).

$$Q = CV \qquad \text{16.22(a)}$$

$$C = \frac{Q}{V} \qquad \text{16.22(b)}$$

The *elastance*, S, is the reciprocal of the capacitance.

$$S = \frac{1}{C} = \frac{V}{Q} = \frac{\dfrac{U}{Q}}{Q} \qquad \text{16.23}$$

As shown in Eq. 16.23, the elastance can also be defined as the amount of energy, U, required per charge Q to transfer a unit charge between two conductors separated by a dielectric.

14. CAPACITORS

A *capacitor* (once called a *condenser*) is a device that stores electric charge. It consists of conductors separated by a dielectric. A common type of capacitor consists of two parallel plates of equal area A separated by a distance r. This type of capacitor is called a *parallel plate capacitor* and its capacitance is determined by Eq. 16.24.

$$C = \frac{\epsilon A}{r} \qquad \text{16.24}$$

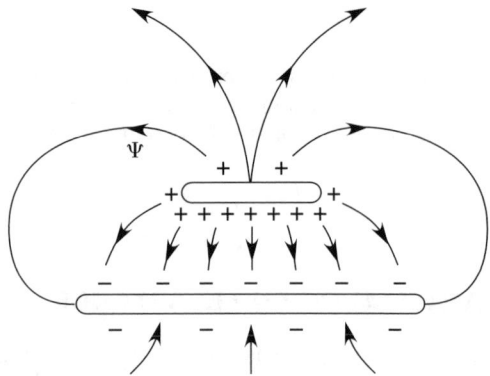

Figure 16.7 *Fringing*

Equation 16.24 ignores the effects of *fringing*, as do many calculations involving capacitors. An example of fringing is shown in Fig. 16.7. The formula for the capacitance of other configurations is given in Table 16.6.

Assuming the plates in a capacitor begin with the same potential ($V = 0$), the total energy, U (in J), in the electric field of a capacitor with capacitance C charged to a potential V is given by Eq. 16.25.

$$U = \frac{1}{2}CV^2 = \frac{1}{2}VQ = \left(\frac{1}{2}\right)\left(\frac{Q^2}{C}\right) \qquad \text{16.25}$$

For n capacitors in parallel, the total capacitance is

$$C_{\text{total}} = C_1 + C_2 + C_3 + \cdots + C_n \qquad \text{16.26}$$

For n capacitors in series, the total capacitance is

$$\frac{1}{C_{\text{total}}} = \frac{1}{C_1} + \frac{1}{C_2} + \frac{1}{C_3} + \cdots + \frac{1}{C_n} \qquad \text{16.27}$$

15. ENERGY DENSITY IN AN ELECTRIC FIELD

In a rectangular dielectric solid of surface area A and thickness t with a potential difference of V applied to either surface, the average energy (in J) is given by

$$U_{\text{ave}} = \frac{1}{2}V_{\text{voltage}}Q = \left(\frac{1}{2}\right)(Et)(DA) = \frac{1}{2}EDV_{\text{volume}}$$

$$= \frac{1}{2}\epsilon E^2 V_{\text{volume}} = \left(\frac{1}{2}\right)\left(\frac{D^2}{\epsilon}\right)V_{\text{volume}} \qquad \text{16.28}$$

Equation 16.28 in all its forms assumes that $\mathbf{D}$ and $\mathbf{E}$ are in the same direction. If not, the dot product must be used to determine the energy.

The average energy density at any point in this electric field is given in Eq. 16.29.

$$u_{\text{ave}} = \frac{U_{\text{ave}}}{V_{\text{volume}}} = \frac{1}{2}\epsilon E^2 = \left(\frac{1}{2}\right)\left(\frac{D^2}{\epsilon}\right) \qquad \text{16.29}$$

Part 4: Electrokinetics

16. SPEED AND MOBILITY OF CHARGE CARRIERS

A positive free charge moving in an electric field through a vacuum is shown in Fig. 16.8. Such a particle would experience a force given by the following.[13]

$$\mathbf{F} = Q\mathbf{E} = m\mathbf{a} \qquad \text{16.30}$$

---

[13]This is identical to Eq. 16.4 without the subscripts. Nominally, subscripts need only be used to avoid confusion when multiple fields or charges are involved.

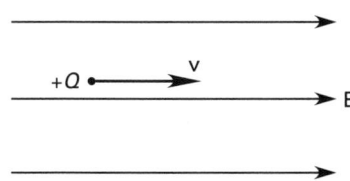

Figure 16.8 *Free Charge in the Electric Field of a Vacuum*

The force the particle experiences is unopposed and results in the constant acceleration of the particle as long as it remains in the electric field.

In metals, according to *electron-gas theory*, the electrons instead reach an average *drift velocity*, $\mathbf{v}_d$, due to the collisions they undergo as they move through the electric field. These collisions are between the electrons and the atoms within the crystalline structure of the metal and are due to the thermal vibrations of the individual atoms.[14] The average drift velocity is

$$\mathbf{v}_d = \int \mathbf{a}\,dt = \int \frac{\mathbf{F}}{m}\,dt$$

$$= \frac{Q}{m} t_m \mathbf{E} = \mu \mathbf{E} \qquad \text{16.31}$$

The term t_m is the mean free time between collisions and m is the mass of the particle.[15] The mobility of the moving charge is given by μ. Since the particles in the metal undergo an increase in vibratory motion as the temperature increases, the mobility varies with the temperature—and with the crystalline structure of the material. As the temperature increases, the mobility decreases, resulting in a lower drift velocity and thus a lower current. In circuit analysis this phenomenon is called *resistivity*. The resistivity is the reciprocal of the conductivity.

17. CURRENT

An *electric current* is the movement of charges past a particular reference point or through a specified surface. The symbol I is generally used for constant currents while i is used for time-varying currents. Current is measured in *amperes* (A), with 1 A = 1 C/s. By convention, the current moves in the direction a positive charge would move (from the positive terminal to the negative terminal), that is, opposite to the flow of electrons. *Ohm's law* relates current to voltage and resistance as shown in Eq. 16.32.

$$V = IR \qquad \text{16.32}$$

---

[14]The thermal vibrations can be treated in a quantum mechanical fashion as discrete particles called *phonons*.
[15]The variation in the overall field within the metal due to the periodic variations established by the lattice atoms, which can result in collisions, is accounted for in some formulas by an *effective electron mass*, m_e^*. Thus, only the thermal vibrations must be determined when such a mass is used.

For simple DC circuits, this relationship is adequate and the current is defined by Eq. 16.33.[16]

$$I = \frac{dQ}{dt} \qquad \text{16.33}$$

However, when the charges exist in a liquid or a gas, as in the field of electronic engineering, or when the charges are both positive and negative and have different characteristics, as in semiconductors, Eq. 16.33 is inadequate and one must be more specific in defining current. As a result, the *current density vector*, $\mathbf{J}$, with units of A/m^2 is more frequently utilized in electromagnetic theory.

The *total current* is given by the sum of the convection current, the displacement current, and the conduction current. The convection and conduction currents are considered true currents and the displacement current a virtual current. Unless otherwise stated, the "current" will normally be the conduction current, but care must be used when determining what "current" is being referenced.

18. CONVECTION CURRENT

In a material containing a volume V of mobile charges with a charge density of ρ moving across a surface area A with an average speed of v_{ave}, the *convection current* will be given by Eq. 16.34.

$$I = \rho A v_{ave} = \left(\frac{Q}{V}\right) A v_{ave}$$

$$= \left(\frac{Q}{Al}\right) A \left(\frac{l}{t}\right) = \frac{Q}{t} \qquad \text{16.34}$$

In terms of the current density vector, the convection current is given by Eq. 16.35 and can be visualized as shown in Fig. 16.9.

$$\mathbf{J} = \rho \mathbf{v}_d \qquad \text{16.35}$$

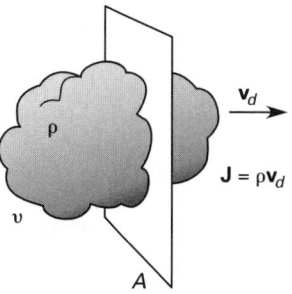

Figure 16.9 *Convection Current*

---

[16]The use of the capital I indicates that the speed of the charges referred to is constant. If the speed is not constant then the current varies with time and i would be used as the appropriate symbol. In practical terms, this means that currents in DC circuits (unidirectional current flow) in which the current magnitude varies could be labeled i.

Convection, as the name implies, involves the diffusion of charges in a medium, as opposed to the movement of charges under the influence of an electric field. Although not of general use in electrical engineering, the concept of a convection current is sometimes useful in electromagnetic field theory.

19. DISPLACEMENT CURRENT

When the electric field varies with time, a current can be made to flow in a dielectric material. That is, charges can be displaced within a dielectric by the electric fields of the charges moving outside the borders of the dielectric. This is because the electric fields of the charged particles extend well beyond the dimensions of the charges themselves. The current so created is called the *displacement current*, i_d, and is given by Eq. 16.36.

$$i_d = \frac{d\psi}{dt} \qquad \qquad 16.36$$

The displacement current in terms of the current density vector, $\mathbf{J}$, is given by Eq. 16.37.

$$\mathbf{J}_d = \frac{\delta \mathbf{D}}{\delta t} \qquad \qquad 16.37$$

The displacement current, i_d, through any given surface can be obtained by integration of the normal component of the displacement current density vector, $\mathbf{J}_d$, over the surface as in Eq. 16.38.

$$i_d = \frac{d}{d_t} \int_A \mathbf{D} \cdot d\mathbf{A} \qquad \qquad 16.38$$

The displacement current is equal to zero if the electric flux is constant with respect to time. That is, to cause a current to "flow" through a dielectric without the benefit of a direct connection to a conductor (as in the case of a capacitor), the electric flux must be changing with respect to time.[17]

20. CONDUCTION CURRENT

A *conduction current* occurs in the presence of an electric field within a conductor. The magnitude of this current is given by the number of mobile charges transferred per unit time within the conductor. The charges may be electrons (in metals and semiconductors), holes (in semiconductors), or ions (in gases and electrolytes). Conduction current is given by Eq. 16.39.

$$I_c = \sigma \left(\frac{A}{l} \right) V = GV \qquad \qquad 16.39$$

The symbol σ indicates the conductivity of the material and is a property of a unit volume of the material. The area of the conductor is given by A and the potential difference by V. The length represented by l is the length over which the potential is applied. The quantity G is the conductance of the specified conductor and depends on the dimensions of the material used.

The conduction current in terms of the conduction current density vector, $\mathbf{J}_c$, is given by Eq. 16.40.

$$\mathbf{J}_c = \rho \mathbf{v}_d \qquad \qquad 16.40$$

This is identical to Eq. 16.35. Applying Eq. 16.31, specifically $\mathbf{v}_d = \mu \mathbf{E}$, puts Eq. 16.40 in the more widely recognized form shown in Eq. 16.41.

$$\mathbf{J} = \sigma \mathbf{E} \qquad \qquad 16.41$$

Equation 16.41 is called the *point form of Ohm's law*.

Part 5: Magnetostatics

21. MAGNETIC POLES

Mobile charges set in motion by an electric field carry their own electric fields with them. The subsequent three-dimensional field can be resolved into longitudinal and transverse components, referenced to the direction of motion. The transverse fields produce magnetic forces between moving charges. This force also acts between the conductors in which the charges move. It is this force that constitutes the basis for motor and generator theory in electric power engineering. Moving charges represented by the current, i, and the magnetic field established by their motion are shown in Fig. 16.10(a). Another completely valid view of the same situation is shown in Fig. 16.10(b). If the charges move with a uniform motion, the field is referred to as *magnetostatic*. If the charges move nonuniformly, the field is referred to as *magnetokinetic*.

A magnetic field can exist only with two equal and opposite poles, referred to as the *north pole* and the *south pole*.[18] The combination of the north and south pole linked together is referred to as a *dipole*. The magnetic *dipole moment*, m, depends on the pole strength, p, and the distance between the poles.

$$m = pd \qquad \qquad 16.42$$

---

[17]It is important to note that the charges do not flow through the dielectric but are instead displaced within the dielectric. Also, this displacement ceases if the electric flux within the dielectric no longer changes.

[18]An electric charge can exist as a single charged object. Magnetic poles cannot exist as single north or south poles. (Mathematically there is no reason for the nonexistence of such a *magnetic monopole*, but none has yet been found.)

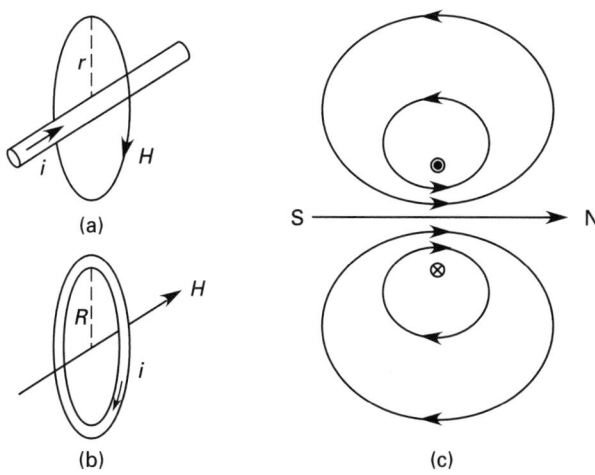

(a)

(b)

(c)

Note: The "x" indicates the current flows in to the page. The "dot" in it indicates flow out of the page. This convention is meant to represent the tail of an arrow (that is, the "x") and the head of the arrow (that is, the "dot.") Assume positive current flow. This correlates with the *right-hand rule* that states that placing the thumb in the direction of the current flow and curling the fingers inward gives the direction of the magenetic field.

Figure 16.10 *Magnetic Fields of Moving Charges*

The magnetic moment is a mathematical construct. Since the poles cannot be separated for measurement, the distance, d, is not measurable. Nevertheless, the magnetic moment is an important and fundamental quantity, and it can be measured. Magnetic moments are measured in terms of *Bohr magnetons*. In ferromagnetic material, a Bohr magneton is the moment produced by one unpaired electron that is equivalent to 9.24×10^{-27} A·m².

Figure 16.10(c) represents a loop of current creating a magnetic field as shown. As the diameter of the current loop becomes infinitely small, the field approaches that of a magnetic dipole. A magnetic dipole field is shown in Fig. 16.11.

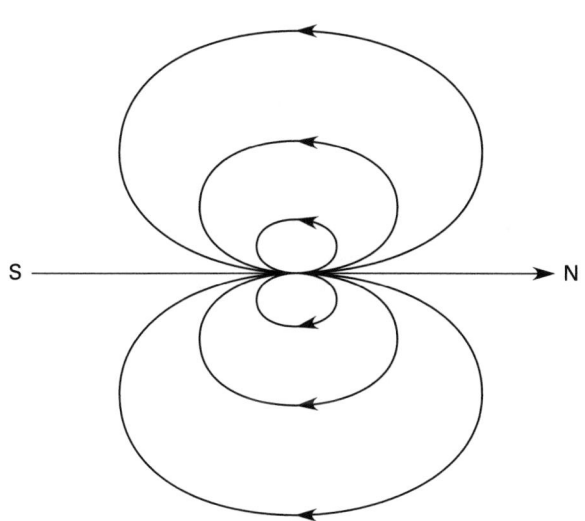

Figure 16.11 *Magnetic Dipole Field*

The infinitely small loop of current can be viewed as a single electron moving in an orbit. Hence, the magnetic moments of individual atoms and the usefulness of the Bohr magneton.[19]

For ease of presentation, and because in magnetic materials many groups of atoms tend to align themselves into domains, individual electrons or loops of current are not usually shown. Instead they are more often shown as in Fig. 16.12, which illustrates the standard convention that lines of *magnetic flux* are directed outward from the north pole (that is, the *magnetic source*), and inward at the south pole (that is, the *magnetic sink*).

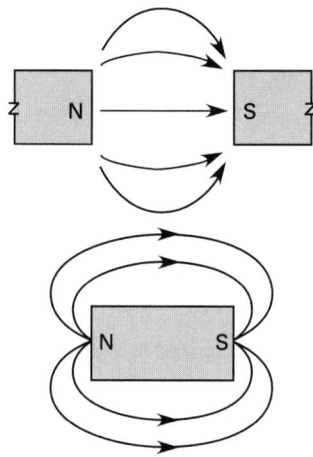

Figure 16.12 *Typical Magnetic Field*

22. BIOT-SAVART LAW

The Biot-Savart law is basic to magnetostatics and magnetokinetics just as Coulomb's law is basic to electrostatics and electrokinetics. The law, in conjunction with Ampere's law, relates the force $\mathbf{F}_{1\text{-}2}$ between two current elements as in Eq. 16.43.

$$\mathbf{F}_{1\text{-}2} = \left(\frac{\mu}{4\pi}\right)\left(\frac{(I_2 d\mathbf{l}_2) \times (I_1 d\mathbf{l}_1) \times (\mathbf{r}_{1\text{-}2})}{r_{1\text{-}2}^2}\right) \quad \textit{16.43}$$

The term μ is the magnetic permeability and $\mathbf{r}_{1\text{-}2}$ is a unit vector from the point represented by differential current element $d\mathbf{l}_1$ to $d\mathbf{l}_2$.[20] Both current elements are in the magnetic field of the other. For parallel currents Eq. 16.43 becomes

$$F_{1\text{-}2} = \frac{\mu}{4\pi r^2} I_1 dl_1 I_2 dl_2 \quad \textit{16.44}$$

Since differential elements of current cannot be isolated, Eqs. 16.43 and 16.44 are of use only when integrated

[19]It was Ampère who suggested over a hundred years ago that the magnetism observed in materials was due to "molecular currents."
[20]The magnetic permeability, μ, can only be used where one is operating over a linear region of the BH curve. If not, the magnetic permeability of free space, μ_0, should be used in its place. For hard ferromagnetic materials, $\mathbf{B}$ and $\mathbf{M}$ are directly related and μ need not be used.

over the entire path of the currents I_1 and I_2. The situation described by Eqs. 16.43 and 16.44 is shown in Fig. 16.13.

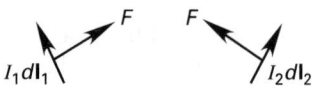

Figure 16.13 *Magnetic Force Between Current Elements*

The magnetic force between two moving charges is given by Eq. 16.45. A specific case is shown in Fig. 16.13.

$$\mathbf{F}_{1\text{-}2} = \left(\frac{\mu}{4\pi}\right)\left(\frac{Q_1Q_2}{r_{1\text{-}2}^2}\right)(\mathbf{v}_2)$$
$$\times (\mathbf{v}_1 \times \mathbf{r}_{1\text{-}2}) \qquad 16.45$$

If the charges are moving parallel to one another, the maximum magnetic force is

$$F = \frac{\mu}{4\pi r^2}Q_1Q_2\mathbf{v}^2 \qquad 16.46$$

Comparison of Magnetic and Electric Force

The maximum coulombic force between moving charges in free space can be deduced from Eq. 16.3 to be

$$F_e = \frac{Q_1Q_2}{4\pi\epsilon_0 r^2} \qquad 16.47$$

The maximum magnetic force between moving charges in free spaces can be deduced from Eq. 16.45 to be

$$F_m = \frac{\mu_0 Q_1Q_2\mathbf{v}^2}{4\pi r^2} \qquad 16.48$$

Comparing the maximum magnetic force F_m to the maximum electric force F_e gives

$$\frac{F_m}{F_e} = \epsilon_0\mu_0\mathbf{v}^2 \qquad 16.49$$

Consider Eq. 16.49. Since in electrical engineering practice the velocities of electrons are much less than the velocity of light, the magnetic force is much weaker than the electric force. Additionally, the speed of light squared is fundamentally related to magnetic permeability and the electric permittivity as follows.

$$c^2 = \frac{1}{\mu_0\epsilon_0} \qquad 16.50$$

Finally, the magnetic force can be viewed as the electric force multiplied by the factor $(\mathbf{v}/c)^2$. That is, the magnetic force is a result of charges in relative motion.

$$F_m = F_e\left(\frac{\mathbf{v}}{c}\right)^2 \qquad 16.51$$

The magnetic and electric forces on two reference charges moving in parallel are shown in Fig. 16.14.

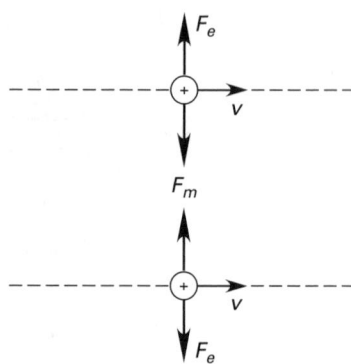

Figure 16.14 *Magnetic and Electric Forces on Moving Charges*

Magnetic Superposition

Magnetic forces occur as a result of the interaction of a pair of current-carrying conductors or the equivalent. One sets up the magnetic field and the other exhibits a force reaction to this same magnetic field. In free space and most nonmagnetic materials, the magnitude of the magnetic field is proportional to the total current that sets up the field as long as the permeability of the medium is constant. However, this is not true for iron and other ferromagnetic materials. Therefore, in general, the principle of linear superposition does not apply to current-carrying conductors in the vicinity of ferromagnetic materials.

23. MAGNETIC FIELDS

A magnetic field is a region in which a moving electric charge exerts a measurable force on another moving charge. It is a vector force field because a moving test charge placed in the field will experience a force in a given direction and with a specific magnitude. The direction of the field is along the flux lines, which are perpendicular to the current (that is, the moving charges creating the magnetic field). These lines, ideally, represent the direction of the force exerted on an infinitely small, isolated positive test charge moving with speed v. The actual direction is represented by the unit vector **a**. Equation 16.52 represents the *strength of the B-field*, or the *magnetic flux density*, in a substance with permeability μ at a distance r from current element dl_1 that gives the direction of one turn, N, of the total current I.[21] The denominator, $4\pi r^2$, represents the surface area of a sphere centered on the current element dl whose surface just touches the point in space where the B-field is being evaluated. Specific configurations may be evaluated with Eq. 16.52 by changing **B** to $d\mathbf{B}$ and l to $d\mathbf{l}$ and integrating.

$$\mathbf{B} = \frac{\mu Il}{4\pi r^2}\mathbf{a} \qquad 16.52$$

---

[21] The **B** field has two sources. One is the current, and the other is the magnetization, **M**.

B should not be called the magnetic field strength. Magnetic field strength is designated by **H**.[22] Although **H**, which will be defined later in this chapter in the section titled "Magnetic Field Strength," is directly related to the current I (as **E** is directly related to the charge), the fundamental field is **B**, as **B** enters into the force relations (as **E** enters into the force relations in electrokinetics). Thus, **B** is analogous with **E**, and **H** is analogous with **D**. This is true though a review of the units in each case would seem to indicate otherwise.

The amount of *magnetic flux* in a magnetic field is symbolized by Φ, measured in webers (Wb). The magnetic flux density, **B**, is measured in teslas (T). One T is equivalent to 1 Wb/m^2. **B** is also known as the *magnetic induction*, the *intensity of magnetization*, and the *dipole moment per unit volume*.[23] **B** is given in terms of the flux in Eqs. 16.53 and 16.54.

$$B = |\mathbf{B}| = \frac{\phi}{A} \qquad 16.53$$

$$\mathbf{B} = \frac{\phi}{A}\mathbf{a} \qquad 16.54$$

24. PERMEABILITY AND SUSCEPTIBILITY

The behavior and magnitude of magnetic forces between moving charges depend on the environment in which the moving charges are located. The largest magnetic forces occur in ferromagnetic materials. Another way of viewing this phenomenon is to say that magnetic flux does not pass equally through all materials. In fact, diamagnetic material repels magnets.

The *permeability of free space*, that is, a vacuum, μ_0, may be considered to be representative of the magnetic properties of free space. However, it is better regarded as a constant relating the units in mechanical and electromagnetic systems. The value for μ_0 in the SI system is $1.2566 \times 10^{-6}\,\text{N/A}^2 = 4\pi \times 10^{-7}\,\text{N/A}^2$. The product of the permeability of free space and the *relative permeability* of a medium gives the *permeability* of the medium.

$$\mu = \mu_0 \mu_r \qquad 16.55$$

Permeability is often defined in terms of the magnetic flux density and the magnetic field strength necessary to create it, as in Eq. 16.56.

$$\mu = \frac{B}{H} \qquad 16.56$$

When **B** and **H** are not parallel, the permeability, μ, is a tensor. In general, when losses occur within a material, the permeability becomes a complex number and curves that show the variations in the real and imaginary terms are called the *magnetic spectrum* or *permeability spectrum* of the material.

The permeability can also be expressed as follows.

$$\mu = \mu_0 (1 + \chi_m) \qquad 16.57$$

The term χ_m is a dimensionless quantity called the *magnetic susceptibility*. Magnetic susceptibility is a measure of the alignment in a magnetic material. It is also the ratio of the magnetization, M, to the external applied magnetic field, H.

$$\chi = \frac{M}{H} \qquad 16.58$$

For diamagnetic and paramagnetic materials, the susceptibility is essentially constant. For ferromagnetic materials, the susceptibility is nonlinear with respect to the applied field, as shown in Fig. 16.2.

25. MAGNETIC FLUX

The magnetic flux, Ψ_m, is defined as the surface integral of the normal component of the magnetic flux density, **B**, over an area, A, and is given by Eq. 16.59.

$$\Psi_m = \iint \mathbf{B} \cdot d\mathbf{A} \qquad 16.59$$

It is also useful to define the flux, ϕ, as the magnetic lines of force in a region.[24] In this case, the flux is given by Eq. 16.60, with N representing the number of complete turns of closely spaced conductors of length l carrying current I. The term NI in Eq. 16.60 is sometimes referred to as the *magnetomotive force* (mmf or preferably F_m).

$$\phi = \mu NIl = BA \qquad 16.60$$

The orientation of the magnetic field lines and magnetic flux lines depends on the properties of the material, specifically the permeability. By convention, the flux lines exit the north pole as given by the right-hand rule.[25] The magnetic field lines always form closed loops. They do not terminate on anything but themselves (see Figs. 16.10 and 16.11). Since no magnetic charges exist, the magnetic flux on a closed sphere (or any closed surface) is equal to zero.

---

[22]Due to historical reasons, the concept of the **B** field in terms of magnetic flux density developed first, prior to the concept of a field or field strength **H**. The names thus are opposites of those used when discussing electric quantities. Nevertheless, **B** is analogous to **E**, and **H** is analogous to **D**.

[23]The word "induction" stems from an earlier era. When an unmagnetized piece of iron was brought near a magnet, magnetic poles were said to be "induced" in the iron.

[24]The symbology for the flux, Ψ_m, does not necessarily have to change. It is shown here as commonly used, ϕ.

[25]In the case of straight wire, the thumb indicates the current direction and the fingers curl in the direction of the magnetic field. For a coil of wire, the fingers curl in the direction of the positive current flow and the thumb points in the direction of the magnetic field direction. The thumb points toward the north pole.

26. MAGNETIC FIELD STRENGTH

Table 16.8 *Magnetic Field and Inductance for Various Configurations*

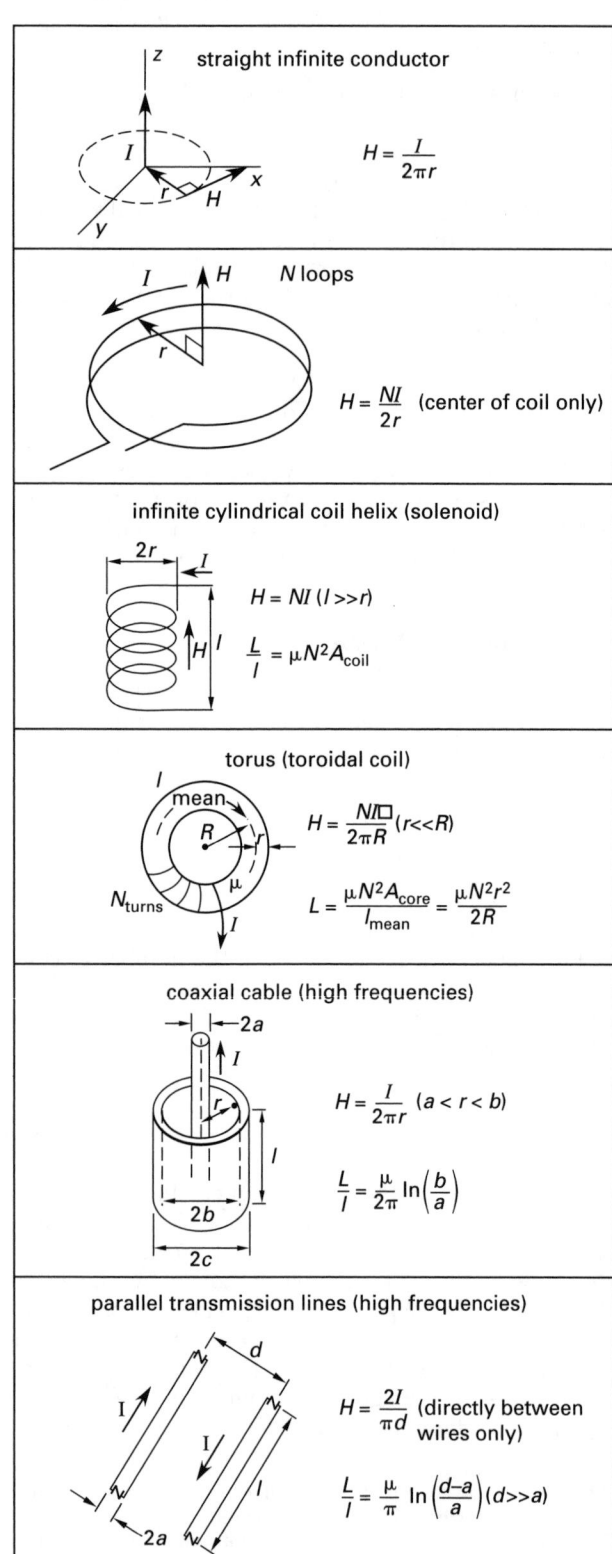

The *magnetic field strength*, **H**, in units of A/m, is derived from the magnetic flux density.

$$\mathbf{H} = \frac{1}{\mu}\mathbf{B} \qquad 16.61$$

The magnetic field strength is a vector point function whose curl is the current density. Because it is proportional to the magnetic flux density in regions where no magnetized substances exist, and because the *cgs* system units define **H** in terms of lines of flux per unit area, **H** can be referred to as the magnetic flux density and is done so in texts on magnetic theory. This explains its correlation with the electric flux density **D**.[26] When using the SI system of units, **H** should be referred to only as the magnetic field strength.

The magnetic field and inductance for various configurations are given in Table 16.8.

Example 16.8

A long wire with a radius of 0.001 m carries 30 A of current. What are the magnetic field strength and the magnetic flux density 0.01 m from the surface of the wire? (This is roughly equivalent to asking what are the magnetic field strength and flux density at the inside surface of a conduit containing a 10 AWG copper wire carrying the maximum current allowed by the National Electrical Code.)

Solution

From Table 16.8, the magnetic field strength is

$$H = \frac{I}{2\pi r} = \frac{30 \text{ A}}{2\pi(0.001 \text{ m} + 0.01 \text{ m})} = 434.06 \text{ A/m}$$

From Eq. 16.56, using the permeability of free space, the flux density is

$$B = \mu H = \left(4\pi \times 10^{-7}\,\frac{\text{Wb}}{\text{A·m}}\right)\left(434.06\,\frac{\text{A}}{\text{m}}\right)$$
$$= 5.45 \times 10^{-4} \text{ T}$$

27. GAUSS'S LAW FOR MAGNETIC FLUX

A law similar to Gauss's law for electric flux can be written for magnetic flux. (See section titled "Gauss's Law for Electrostatics" earlier in this chapter.) Such a law states that the amount of flux passing through any closed surface is equal to zero.

$$\psi_m = \oiint \mathbf{B} \cdot d\mathbf{A} = 0 \qquad 16.62$$

---

[26]There are other reasons that **H** is analogous to **D**. For instance, **D** is used to avoid direct reference to polarization charges and the polarization **P**. Thus, it is then easier to work with **E** and **D**. **H** is used to avoid direct reference to the magnetization **M**. Thus, it is then easier to work with **B** and **H**.

The physical significance of Eq. 16.62 is that magnetic flux lines form continuous closed loops. Additionally, this indicates that the divergence of any magnetic field is zero. In other words, no magnetic charge exists. The law refers to magnetic flux vice magnetostatics as it is also applicable to time-varying magnetic fields (that is, magnetokinetics).

28. INDUCTANCE AND RECIPROCAL INDUCTANCE

When charges are in motion, a current is said to be flowing and a magnetic field is established. When the current varies with time, the generated magnetic field opposes the current change and in doing so "stores" energy in the magnetic field. *Inductance* is the property of a system of conductors and circuits that permits this storage.

If a time-varying current is applied to a conductor, usually in the form of a coil of wire, the magnetic field builds up, producing a potential difference across the ends of the conductor. The amount of potential, V, is proportional to the magnitude of the rate of change of current. The proportionality constant is called the *inductance* and is measured in *henrys*, H. This is shown in Eq. 16.63.[27]

$$\mathrm{v} = L\frac{di}{dt} \qquad 16.63$$

The *reciprocal inductance*, Γ, a seldom used term, is the reciprocal of the inductance.

$$\Gamma = \frac{1}{L} \qquad 16.64$$

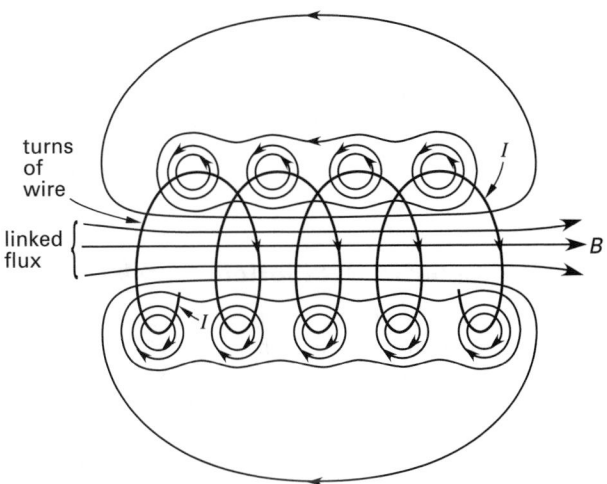

Figure 16.15 *Magnetic Flux Links in a Solenoid*

---

[27]This equation technically belongs in the Magnetokinetics section. It is often stated as the definition of inductance and is therefore used here.

The type of inductance just described is also called *self-inductance*. It can also be described in terms of *flux linkage*, Λ (measured in Wb). Flux linkage is defined as the lines of flux that link the entire magnetic circuit. For example, in Fig. 16.15 the "linked flux" is the flux through the center of the solenoid that links the current flows of all the wire turns (the small flux loops around each turn of wire are ignored). Using this definition, the inductance would be given by Eq. 16.65.

$$L = \frac{\Lambda}{I} = \frac{N\phi}{I} \qquad 16.65$$

Mutual inductance, M, vital to transformer theory, is the term used to describe the fact that a changing current in one circuit induces an electromotive force in another. It is given by Eq. 16.66 where Λ represents the flux links in one winding and I is the current in the other winding.

$$M = \frac{\delta\Lambda}{\delta I} \qquad 16.66$$

29. INDUCTORS

An *inductor* is a device that stores magnetic energy. It typically consists of coils of wire with a core material in the center, which can be air. A common type of inductor consists of N turns of wire with radius r and an iron core center with permeability μ. This type of inductor is called a *solenoid*, and its inductance is determined by Eq. 16.67.

$$L = \mu\frac{N^2 A_{\mathrm{coil}}}{l} \qquad 16.67$$

The formula for the inductance of other configurations is given in Table 16.8.

The total energy, U (in J), in the magnetic field of an inductor with inductance L and current I is

$$U = \frac{1}{2}LI^2 \qquad 16.68$$

For inductors in series, the total inductance is

$$L_{\mathrm{total}} = L_1 + L_2 + L_3 + \ldots + L_n \qquad 16.69$$

The reciprocal of the total inductance for inductors in parallel is

$$\frac{1}{L_{\mathrm{total}}} = \frac{1}{L_1} + \frac{1}{L_2} + \frac{1}{L_3} + \ldots + \frac{1}{L_n} \qquad 16.70$$

30. ENERGY DENSITY IN A MAGNETIC FIELD

The average energy (in J) in a magnetic field is

$$U_{\mathrm{ave}} = \frac{1}{2}\phi N I \qquad 16.71$$

The average energy density at any point in this magnetic field is

$$u_{\text{ave}} = \frac{U_{\text{ave}}}{V_{\text{volume}}} = \frac{1}{2}BH = \frac{1}{2}\mu H^2 = \frac{1}{2}\frac{B^2}{\mu} \qquad 16.72$$

The average energy stored in inductors that are mutually coupled is given by Eq. 16.73. Here M is the magnetization (with units of henrys) and I_1 and I_2 represent the currents in the two inductors.

$$U_{\text{ave}} = MI_1I_2 \qquad 16.73$$

Part 6: Magnetokinetics

31. SPEED AND DIRECTION OF CHARGE CARRIERS

A positive free charge moving with velocity, $\mathbf{v}$, in a magnetic field experiences a force given by Eq. 16.74.

$$\mathbf{F} = Q\mathbf{v} \times \mathbf{B} \qquad 16.74$$

The magnetic force can change a charge's direction but not its kinetic energy. This is in contrast to the force the charge experiences in an electric field, which does work on the charge and therefore changes its kinetic energy.[28]

If there are many charges flowing through a conductor in a magnetic field, that is, if a current-carrying conductor exists in an external magnetic field, the differential force equation is

$$d\mathbf{F} = (dq)(\mathbf{v} \times \mathbf{B}) = (Idt)(\mathbf{v} \times \mathbf{B})$$
$$= I(d\mathbf{l} \times \mathbf{B}) \qquad 16.75$$

In Eq. 16.75, $d\mathbf{l}$ is a unit length in the direction of conventional (i.e., positive) current flow I. If the conductor is straight and the magnetic field is constant along the length of the conductor, a version of *Ampère's law* results.

$$F = NIBl \sin\theta \qquad 16.76$$

N represents the number of turns of current-carrying conductor in the magnetic field and θ is the angle between the conductor (i.e., the current element $d\mathbf{l}$) and the magnetic field, $\mathbf{B}$. If these are at right angles, which is often the case, the force is represented by

$$F = NIBl \qquad 16.77$$

The situations in Eqs. 16.75 through 16.77 are illustrated in Fig. 16.16.

Care must be taken to properly interpret directions in the figure. If an external force is applied in the direction shown, the conductor has a velocity in the same

---

[28]Consider that if a charge is not in motion with respect to a magnetic field, it cannot "see" the magnetic field and therefore experiences no change in kinetic energy. A charge stationary with respect to an electric field can "see" the field and therefore experiences a force that changes its kinetic energy.

direction. Applying Eq. 16.74 then gives the direction of force on the charge carriers within the conductor—in this case, in the direction of the current shown. This is the basis of generator theory. If instead a current is applied to a loop of wire, of which the conductor represents one section, the velocity in Eq. 16.74 is in the direction of the charge carriers, that is, the direction of the current flow shown. In this case, the force given by Eq. 16.74 is in a direction opposite to the force arrow shown. This is the basis of motor theory.

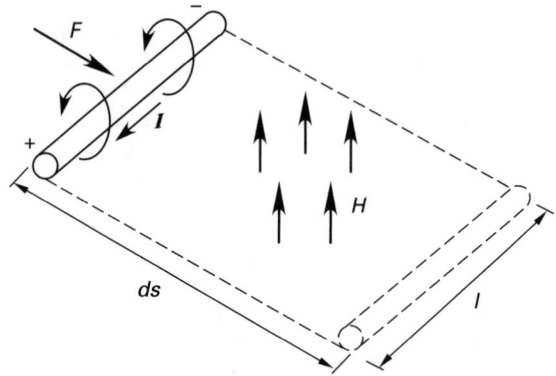

Figure 16.16 *Conductor in a Magnetic Field*

If a magnetic field, $\mathbf{B}$, is uniform within a region and a charge has an initial velocity perpendicular to this field, the resulting path of the particle is a circle of radius r. The magnitude of the force is determined by Eq. 16.74. Such a force would be directed toward the center of the circle. The centripetal acceleration is of a magnitude $\omega^2 r = \mathrm{v}^2/r$. Applying Newton's second law gives the following for the radius.

$$|Q|\,\mathrm{v}B = m\frac{\mathrm{v}^2}{r} \qquad 16.78$$

Thus,

$$r = \frac{m\mathrm{v}}{|Q|\,B} \qquad 16.79$$

When both electric and magnetic fields are in a given region at the same time, the force is given by the *Lorentz force equation*, Eq. 16.80.

$$\mathbf{F} = Q\,(\mathbf{E} + \mathbf{v} \times \mathbf{B}) \qquad 16.80$$

Electromagnetic Oscillations and Waves

A time-varying electric field produces time-varying currents that generate magnetic fields. If the electromagnetic variations are restricted to a small region, such as within a circuit, no spatial variations of the field occur and electrical oscillations are produced. If the electromagnetic variations are generated over large areas of space, compared to the wavelength of the oscillation, electromagnetic waves are generated that propagate through space with a speed given by

$$\mathrm{v} = \frac{1}{\sqrt{\epsilon\mu}} \qquad 16.81$$

When a conduction current generates the electromagnetic field, the magnetic field strength $\mathbf{H}$ is proportional to the electric field strength $\mathbf{E}$ and an induction field is set up, keeping the energy within the circuit. If the electromagnetic field is generated by a displacement current, the magnetic field strength $\mathbf{H}$ is proportional to the rate of change of the electric field strength, $d\mathbf{E}/dt$. A radiation field is thus set up and energy can leave the system, (i.e., be radiated into space).

Electric and Magnetic Energy Conversion

Applying the law of conservation of energy to a lossless, energy-storing system in which charges move in directed nonuniform motion means the sum of the instantaneous values of electric and magnetic energy must remain constant. Thus, changes in the electric energy necessarily mean changes in the magnetic energy as given in Eq. 16.82.

$$du_e = -du_m \qquad 16.82$$

32. VOLTAGE AND THE MAGNETIC CIRCUIT

An *electric potential* or *voltage* is the work done on a unit charge to bring it from some specified reference point to another point. The symbol V is generally used for constant voltages and v is used for time-varying systems. The unit is the *volt* with $1\ \text{V} = 1\ \text{J/C}$. By convention, current flows from the positive terminal on a voltage source to the negative outside the source and from negative to positive inside the source. Also by convention, the current flows from the positive terminal to the negative terminal through a resistor.[29] This situation is illustrated in Fig. 16.17 along with an analogous *magnetic circuit*.

In a magnetic circuit, the flux density is proportional to the *magnetomotive force* (mmf), F_m, given in Eq. 16.83.

$$F_m = IN \qquad 16.83$$

The magnetomotive force is in units of amp-turns. Turns, like revolutions or radians, are for clarification only and can be disregarded in equation manipulations.

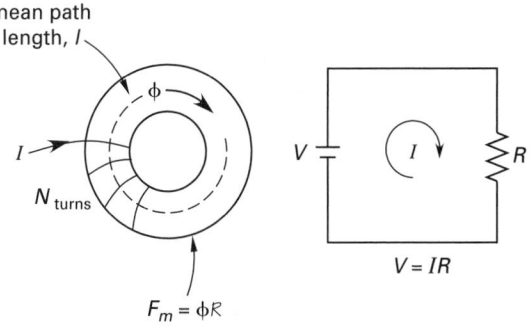

Figure 16.17 *Magnetic-Electric Circuit Analogy*

---

[29]These directions are based on conventional current flow and are arbitrary. As long as one remains consistent when choosing reference directions for both voltage and current, the correct results will be obtained.

The *reluctance*, $\mathcal{R}$, measured in A/Wb, is analogous to electrical resistance and is given by

$$\mathcal{R} = \frac{l}{\mu A} \qquad 16.84$$

The permeability depends on the flux density. Therefore, iterations between Eqs. 16.84 and 16.83 may be needed if either the reluctance or flux density is an unknown.

The magnetic equation that correlates with Ohm's law in electric circuits is

$$F_m = Hl = \phi\mathcal{R} \qquad 16.85$$

Example 16.9

A 0.001 m slice is taken from a cast-iron toroidal coil ($\mu_r = 2000$) with a 0.35 m mean toroidal diameter and a 0.07 m core diameter. A steady unknown current flows through 300 turns of wire wrapped around the coil, producing a constant flux density in the core and air gap. What current is required to establish a flux of 10 mWb across the air gap?

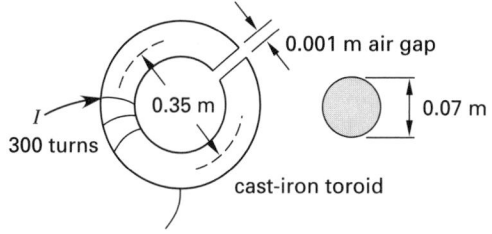

Solution

The mean path length through the cast iron is

$$l_{\text{ci}} = \pi(0.35\ \text{m}) - 0.001\ \text{m} = 1.0986\ \text{m}$$

The cross-sectional area of the flux path is

$$A = \frac{\pi}{4}d^2 = \frac{\pi}{4}(0.07\ \text{m})^2 = 0.003848\ \text{m}^2$$

The total reluctance is the sum of the reluctances of the cast iron and air paths. From Eq. 16.84,

$$\mathcal{R} = \sum \frac{l}{\mu A} = \frac{1}{\mu_0 A}\sum \frac{l}{\mu_r}$$

$$= \frac{1}{\left(4\pi \times 10^{-7}\ \dfrac{\text{Wb}}{\text{A·m}}\right)(0.003848\ \text{m}^2)}$$

$$\times \left(\frac{1.0986\ \text{m}}{2000} + \frac{0.001\ \text{m}}{1}\right)$$

$$= 3.204 \times 10^5\ \text{A/Wb}$$

The flux, ϕ, is given as 10^{-2} Wb. From Eq. 16.83, the magnetomotive force is

$$F_m = IN = \mathrm{I}\,300$$

From Eqs. 16.83 and 16.85 the current is

$$I = \frac{\phi\mathcal{R}}{N}$$

$$= \frac{(10^{-2}\ \text{Wb})\left(3.203 \times 10^5\ \dfrac{\text{A}}{\text{Wb}}\right)}{300\ \text{turns}} = 10.68\ \text{A}$$

33. MAGNETIC FIELD-INDUCED VOLTAGE

When a conductor "cuts" lines of magnetic flux at a given rate, a voltage called the *induced electromotance*, or *electromotive force* (emf) is produced in the conductor, as illustrated in Fig. 16.16. The magnitude of such an *electromagnetic induction* is given in Eq. 16.86, which is the mathematical statement of *Faraday's law*. Faraday's law states that the induced voltage is proportional to the time rate of change of the magnetic flux linked with the circuit. The minus sign is mandated by *Lenz's law*, which states that the current flow in a conductor due to an induced voltage moves in a direction that creates a magnetic flux that opposes the magnetic flux inducing the voltage. More succinctly, the direction of the induced emf is such as to oppose any change in current. This is a consequence of the law of conservation of energy. (The minus sign indicates this opposing nature of the induced voltage. It is not shown in all equations. Instead, the polarity of the voltage in the circuit indicates the opposition.)

$$V = -N\frac{d\phi}{dt} \qquad \text{16.86}$$

Since magnetic flux linkage is given by $\Lambda = N\phi$, and the induced voltage is the time rate of change of these linkages, the induced voltage can be written in terms of the velocity of the conductors.

$$V = N\frac{d\phi}{dt} = NBl\frac{ds}{dt} = NBlv \qquad \text{16.87}$$

In both Eqs. 16.86 and 16.87, N is the number of conductors, l is the length of the conductor in the magnetic field, and ds is the differential distance traveled by the conductor through the field. This is shown in Fig. 16.16. Equation 16.86 is the *flux-changing method* of generating a voltage. Equation 16.87 is the *flux-cutting method* of generating a voltage.

34. EDDY CURRENTS

Eddy currents, also known as *Foucault currents*, occur as a result of induced voltages in a conducting medium by a varying magnetic field. Since the currents so induced cannot leave the conducting medium, they "deflect" off the edges to form loops, or circulating currents. Such currents result in Joule heating of the conducting medium and can result in considerable losses if not adequately designed against.

35. DISPLACEMENT CURRENT MAGNETIC EFFECT

The displacement current explained in the section titled "Displacement Current" earlier in this chapter occurs only in dielectrics and insulators. The conduction electrons in these materials are tightly bound and cannot move freely through the material. Nevertheless, the displacement current creates a magnetic field intensity $\mathbf{H}$ proportional to the time rate of change of the electric field intensity $\mathbf{E}$ producing the displacement current. (By contrast, in a conduction current, $\mathbf{H}$ and $\mathbf{E}$ are in time phase.) Since $\mathbf{H}$ is proportional to $d\mathbf{E}/dt$, and $\mathbf{E}$ is proportional to $d\mathbf{H}/dt$, it is possible to generate electromagnetic waves in nonconducting material.

36. MAGNETIC HYSTERESIS

Plotting the values of the magnetic flux density $\mathbf{B}$ for various values of the magnetic field strength, or magnetizing force, $\mathbf{H}$, for a ferromagnetic material reveals a nonlinear relationship. This is illustrated in Fig. 16.18. The phenomenon causing values of $\mathbf{B}$ to lag behind $\mathbf{H}$ so that the fields differ in magnitude is called *hysteresis*. The loop traces a complete cycle of **BH** values and is termed the *hysteresis loop*.

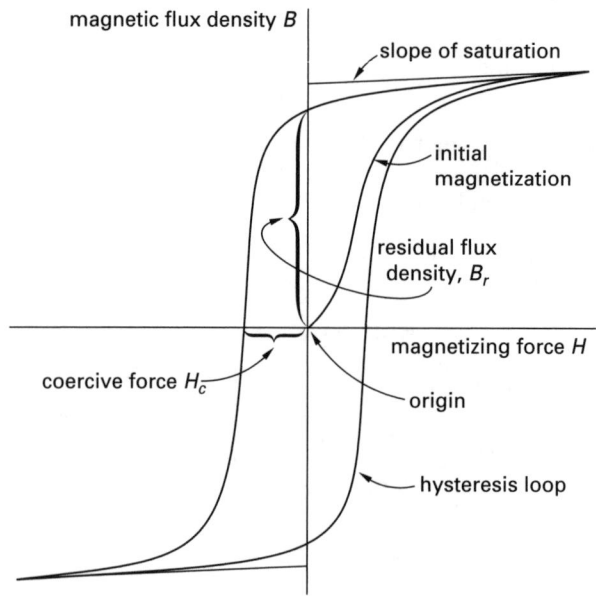

Figure 16.18 *Magnetic Hysteresis*

17 Electronic Theory

Nomenclature

A_0	constant	–
B	magnetic flux density	T or Wb/m^2
C	constant	–
C	capacitance	F
D	diffusion constant	m^2/s
e^-	electron	C
E	electric field strength	V/m or N/C
E	energy	J or eV
E_a	acceptor ionization energy	J or eV
E_c	conduction-band energy edge	J or eV
E_d	donor ionization energy	J or eV

Nomenclature *(continued)*

E_v	valence band energy edge	J or eV
f	frequency	Hz or s^{-1} or cycles/s
F	force	N
h	Planck's constant 6.6256×10^{-34}	J·s
i, I	current	A
J	current density	A/m^2
k	Boltzmann's constant 1.38×10^{-23}	J/K
m	mass	kg
m_e	electron mass	kg
m_0	rest mass	kg
M	magnetization	A/m
n	electron carrier density	m^{-3}
N	number of turns	–
N_c	effective conduction-band density state	cm^{-3}

Equivalent Units

symbol	equivalent units					
A	C/s	W/V	V/Ω	J/s·V	N/T·m	Wb/H
C	A·s	J/V	N·m/V	V·F		
F	C/V	C^2/J	$C^2/N{\cdot}m$	$A^2{\cdot}s^4/kg{\cdot}m^2$		
F/m	C/V·m	$C^2/J{\cdot}m$	$C^2/N{\cdot}m^2$	$A^2{\cdot}s^4/kg{\cdot}m^3$		
H	Wb/A	V·s/A	$T{\cdot}m^2/A$	$kg{\cdot}m^2/A^2{\cdot}s^2$	Ω·s	
Hz	1/s					
J	N·m	V·C	W·s	C^2/F	$kg{\cdot}m^2/s^2$	
m^2/s^2	V·C/kg					
N	J/m	V·C/m	$kg{\cdot}m/s^2$			
N/A^2	$Wb^2/N{\cdot}m^2$					
Pa	N/m^2	$kg/m{\cdot}s^2$				
Ω	V/A	$kg{\cdot}m^2/A^2{\cdot}s^3$				
S	A/V	$A^2{\cdot}s^3/kg{\cdot}m^2$	1/Ω			
T	Wb/m^2	N/A·m	N·s/C·m	$kg/A{\cdot}s^2$		
V	W/A	C/F	J/C	$kg{\cdot}m^2/A{\cdot}s^3$		
V/m	N/C					
W	J/s	V·A	$kg{\cdot}m^2/s^3$			
Wb	V·s	H·A	$T{\cdot}m^2$	$kg{\cdot}m^2/A{\cdot}s^2$		

Nomenclature *(continued)*

N_v	effective valence-band density state	cm^{-3}
p	hole carrier density	m^{-3}
p	power	J/s
p^+	proton	C
q	unit of charge 1.6022×10^{-19}	C
q, Q	charge	C
r	radius	m
t	time	s
T	temperature	K
U	energy	J
v	velocity[1]	m/s
v, V	voltage	V
W	work	J

Symbols

ϵ_0	free-space permittivity, 8.854×10^{-12}	$C^2/N{\cdot}m^2$
μ	mobility	$m^2/V{\cdot}s$
ν	frequency	rad/s
ρ	charge density	C/m^3
ρ_e	electron charge density	electrons/m^3
σ	conductivity	S/m
τ	mean lifetime	s
φ	work function[2]	J
ω	angular velocity	rad/s

Subscripts

0	initial or rest
d	drift
e	electron
g	gap
h	hole
i	intrinsic
n	electrons
p	holes
T	temperature
$\|$	parallel
$\perp$	perpendicular

1. FUNDAMENTALS

Electronic theory is the study of subatomic particles, such as electrons, in materials other than metals. It is concerned with the release, transport, control, and collection of charge carriers in a vacuum, in gases, and in semiconductors. While the charge carriers are usually electrons, they can be holes (a concept used to describe the absence of an electron), positive ions, or negative ions.

---

[1]The term "velocity" is often used in text where it would be more technically correct to use the term "speed." Velocity is a vector quantity with speed as its magnitude.
[2]The units for the work function are often electron volts, eV. One eV is equivalent to 1.602×10^{-19} J.

The fundamentals of electronic engineering are identical to those for electricity and magnetism. Because of this, many of the formulas used in this chapter will be repetitious. Their form and symbology may differ slightly, as often occurs in electrical engineering texts; this is purposefully done to expose the reader to such variations as may be encountered. The principles and results are consistent. Thus, all the electromagnetic theory presented in Ch. 16 is applicable to electronic theory as well.

The primary difference between electronics and electromagnetism is in their applications. Electronic engineering allows for much greater control over the instantaneous, rather than the average, movement of charge carriers. Additionally, the control of these carriers can be extremely rapid. Electronic devices are generally nonlinear. That is, the output terminals can provide an amplified version of the current, voltage, or power supplied to the input terminals. The devices tend to be smaller than their electromagnetic counterparts and use considerably less energy. The field of electronic engineering is focused on *active electron devices*, those devices requiring an external power source in order to maintain their terminals at suitable operating voltages and currents. Electronics is generally more dependent on the material properties of a medium, such as composition and electronic structure, as well as the mode of electric conduction.

2. CHARGES IN A VACUUM

A charge Q in a vacuum experiencing an electric field intensity $\mathbf{E}$ (with units of V/m) will undergo an acceleration in the direction of the field. The force is

$$\mathbf{F}_{\|} = Q\mathbf{E} \qquad 17.1$$

If the charge is in a magnetic field of flux density $\mathbf{B}$ (with units of T) and velocity $\mathbf{v}$ in m/s, it will experience a transverse force of

$$\mathbf{F}_{\perp} = Q\mathbf{v} \times \mathbf{B} \qquad 17.2$$

The direction of motion of the particle will be a circle if the magnetic field is uniform and the particle enters it at right angles. The radius of the circle is

$$r = \frac{m\mathrm{v}}{QB} \qquad 17.3$$

If the charge is in an electromagnetic field, the total force is

$$\mathbf{F} = \mathbf{F}_{\|} + \mathbf{F}_{\perp} = Q\left(\mathbf{E} + (\mathbf{v} \times \mathbf{B})\right) \qquad 17.4$$

The energy of a charge Q in a vacuum, moving in an electric field $\mathbf{E}$ from point 1 to point 2, will experience a change in energy U (in units of J)[3] of

$$U = Q \int_{x_1}^{x_2} \mathbf{E} \cdot d\mathbf{x} = Q\,(V_1 - V_2) \qquad 17.5$$

Vacuum Tubes

A *vacuum tube* is a device in which electrons move through a vacuum or gaseous medium in a gastight envelope. In a basic vacuum tube, called a *triode*, the electrons are emitted from a heated *cathode* and move within the tube to the *plate* by an intervening *grid* that is used for amplification.

Tubes dominated electronic technology in the first half of the twentieth century as semiconductors do today. The release of electrons from the surface of a metal, which is the basis of the vacuum tube, remains an important area of study in electronic applications in space and in the construction of metal-semiconductor junctions.

Work Function

The work function, φ, is a measure of the energy required to remove an electron from the surface of a metal. A unit of energy (J), it is usually given in terms of electron volts, eV. It is specifically the minimum energy necessary to move an electron from the *Fermi level* of a metal (an energy level below which all available energy levels are filled, and above which all are empty, at 0K) to infinity, that is, the vacuum level. This energy level must be reached in order to move electrons from semiconductor devices into the metal conductors that constitute the remainder of an electrical circuit. It is also the energy level of importance in the design of optical electronic devices.

In photosensitive electronic devices an incoming photon provides the energy to release an electron and make it available to the circuit, that is, free it so that it may move under the influence of an electric field. This phenomenon whereby a short wavelength photon interacts with an atom and releases an electron is called the *photoelectric effect*. The photon energy required to release an electron is

$$h\nu = \varphi + \tfrac{1}{2}mv^2 \qquad 17.6$$

The term h is Planck's constant and ν is the *threshold frequency*. If the photon does not have the minimum energy required, no electron is released.

---

[3]The magnetic field acts transverse to the direction of motion and thus can only change the direction, not the energy.

3. CHARGES IN LIQUIDS AND GASES

In any substance, such as a liquid or a gas, where the motion of a charged particle is hindered by surrounding particles, an average velocity called the *drift velocity*, $\mathbf{v}_d$, is established. The drift velocity is dependent on the electric field intensity, $\mathbf{E}$, and the mobility, μ, of the charge carrier as shown in Eq. 17.7.

$$\mathbf{v}_d = \mu \mathbf{E} \qquad 17.7$$

Mobility is measured in $m^2/V \cdot s$. The concept of drift velocity in liquids and gases is illustrated in Fig. 17.1.

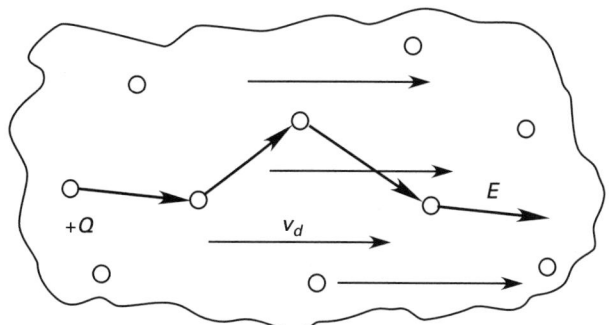

Figure 17.1 *Drift Velocity in Liquids and Gases*

Conductivity and Resistivity

The mobility of the charge carriers is of special importance in electronic engineering, more so than in electrical engineering, although the principles are identical. The mobility is proportional to the conductivity.

$$\sigma = \rho \mu \qquad 17.8$$

The conductivity, σ, is measured in siemens per meter, S/m. The charge density is in units of C/m^3. Metals have approximately 10^{28} atoms/m^3 with one to two electrons free to move, while silicon semiconductors have 10^{16} intrinsic charge carriers per cubic meter. Concentrations in gases vary widely with the type of gas and the conditions under which the gas is held.

In metallic conductors, the conductivity is often defined by the point form of Ohm's law and given by

$$\sigma = \frac{|\mathbf{J}|}{|\mathbf{E}|} = \frac{J}{E} \qquad 17.9$$

Conductivity is more complex for liquids and gases. In general, both positive and negative ions may be present, some singly charged and some doubly charged, possibly with different masses. The conductivity must account for all these factors. If one assumes that the negative and positive ions are alike, the formula for conductivity in liquids and gases has only two terms. See Fig. 17.2 for the formula and a comparison to the formulas for conductors and semiconductors.

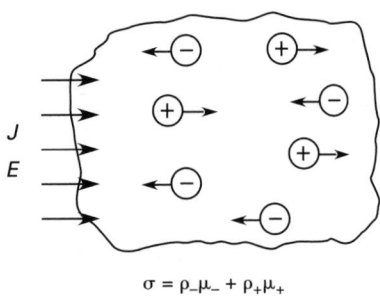

$$\sigma = \rho_-\mu_- + \rho_+\mu_+$$

(a) liquid or gas

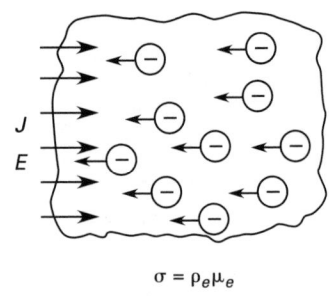

$$\sigma = \rho_e\mu_e$$

(b) conductor

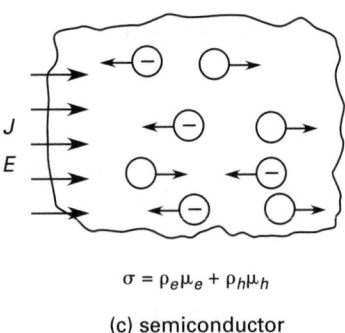

$$\sigma = \rho_e\mu_e + \rho_h\mu_h$$

(c) semiconductor

Figure 17.2 *Conductivity Formulas*

Example 17.1

A certain 10 AWG copper wire has a conductivity of 58×10^6 S/m. Assuming 10^{28} conduction electrons per cubic meter, what is the mobility of the electrons in the copper?

Solution

From Eq. 17.8, the conductivity is

$$\sigma = \rho\mu$$

Rearranging gives

$$\mu = \frac{\sigma}{\rho}$$

The conductivity is given. The charge density is

$$\rho = \left(10^{28}\,\frac{\text{electron}}{\text{m}^3}\right)\left(\frac{-1.6022 \times 10^{-19}\ \text{C}}{\text{electron}}\right)$$

$$= -1.6022 \times 10^9\ \text{C/m}^3$$

Substituting gives

$$\mu = \frac{\sigma}{\rho} = \frac{58 \times 10^6\,\dfrac{\text{S}}{\text{m}}}{-1.6022 \times 10^9\,\dfrac{\text{C}}{\text{m}^3}}$$

$$= \left(-3.6 \times 10^{-2}\,\frac{\text{S·m}^2}{\text{C}}\right)\left(1\frac{\frac{\text{A}}{\text{V}}}{\text{S}}\right)\left(1\frac{\text{C}}{\text{A·s}}\right)$$

$$= -3.6 \times 10^{-2}\ \text{m}^2/\text{V·s}$$

Note that in metallic conductors, the charge carriers are electrons that move in a direction opposite that of the electric field. Thus, the charge density, ρ, and the mobility, μ, are both negative, leaving the conductivity positive. By convention, the electrons moving in one direction are treated as positive charges moving in the opposite direction, and ρ and μ are reported as positive.

Plasmas

Plasma, often viewed as a fourth state of matter, can be considered a highly ionized gas. Plasmas are highly conductive, with high concentrations of roughly equal numbers of positive and negative ions and a low voltage gradient. The negative charge carriers tend to be electrons that vibrate longitudinally about their mean position with a frequency of

$$f = \left(\frac{1}{2\pi}\right)\left(\frac{\rho_e Q^2}{\epsilon_0 m_e}\right)^{\frac{1}{2}} \qquad 17.10$$

Let λ be the wavelength of oscillation of the frequency given by Eq. 17.10. Then the velocity of the electrons is

$$\text{v} = \lambda\left(\frac{\rho_e Q^2}{\epsilon_0 m_e}\right)^{\frac{1}{2}} \qquad 17.11$$

4. CHARGES IN SEMICONDUCTORS

The conductivity is proportional to the concentration of free charge carriers (see Eq. 17.8). In a good conductor, this concentration is very large ($\approx 10^{28}$ electrons/m^3). In an insulator, it is very small ($\approx 10^7$ electrons/m^3). Semiconductor charge carrier concentrations vary between these approximate values. The semiconducting nature of these types of materials originates not so much from the number of carriers but from the arrangement of allowed energy states into bands. Figure 17.3 illustrates the general arrangement of energy levels for metals, semiconductors, and insulators. The energy bands and gaps are not to scale and vary according to the element or compound utilized. The importance of the figure can be summed up by the following statement: an electric field cannot impart energy to an electron in a completely filled band. The electron must be able to absorb energy and move into a new allowable energy level in order to move under the influence of the field or of a potential difference.

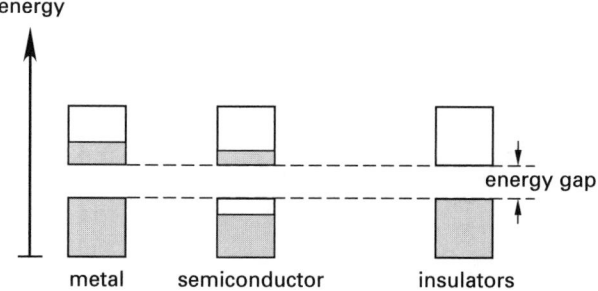

Figure 17.3 *Energy Band Diagrams*

Energy-Band Theory[4]

Electrons exist in *core shells* (i.e., completely filled allowed energy levels) and in *valence shells* (i.e., allowed energy levels, some of which remain vacant). For example, the carbon atom has a core shell that is completely filled, the K shell (in electron configuration parlance, $1s^2$). The carbon atom's valence shell, the L shell, has only four of the possible eight electrons it could hold (in electron configuration parlance, $2s^2 2p^2$). When individual carbon atoms are brought together in a crystalline structure, the valence electrons are exposed to the potential energy changes of nearby atoms as illustrated in Fig. 17.4.

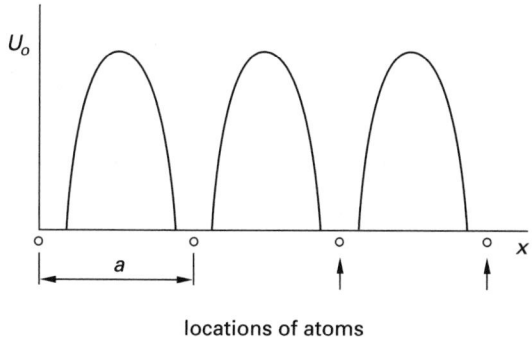

Figure 17.4 *Electron Potential Energy Variation in a One-Dimensional Crystal*

Because of this periodic potential, each energy level in the outermost shell splits into a number of discrete

levels.[5] This is illustrated in Fig. 17.5. The periodic potential is usually accounted for in electronic equations by the use of an *effective mass* for electrons and holes, m_e^* and m_h^*.

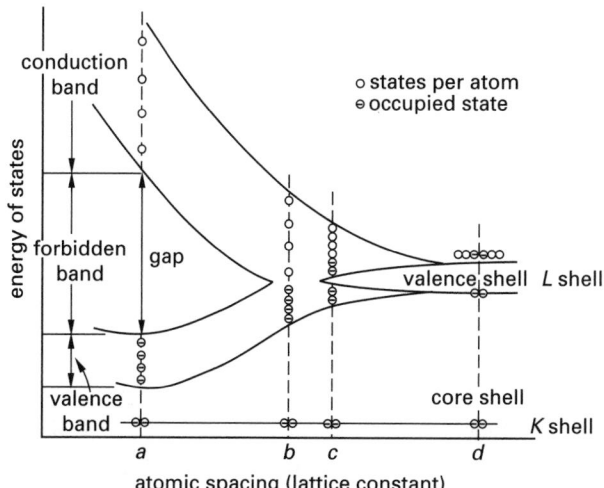

Figure 17.5 *Energy Bands of Carbon as a Function of Atomic Spacing*

At the atomic spacing d, atoms behave as individual atoms with discrete energy levels centered around the K and L shells. At atomic spacing c, the valence band splits, with two states in a lower band and six in an upper. At atomic spacing b, the bands overlap. Discrete energy levels still exist, but now an electron can move from low in the band to high in the band in small discrete steps of energy called *quanta*.[6] At atomic spacing a, the original L valence shell has been split into two distinct bands of energy. The lower retains the name *valence band* and the upper is called the *conduction band*. These bands are normally shown as in Fig. 17.6.

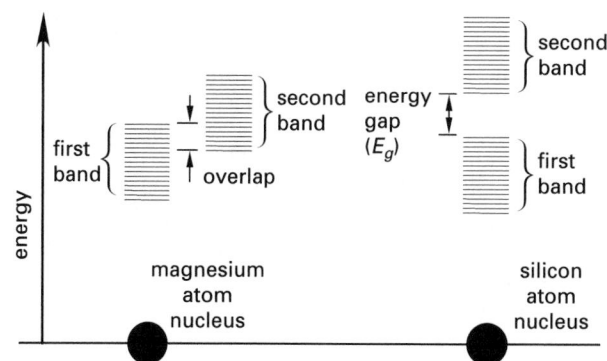

Figure 17.6 *Multiple Energy Bands*

---

[4]Quantum theory plays the dominant role in explaining energy-band theory. Electrical engineering is primarily concerned with the results.

[5]In quantum mechanical terms, this occurs because the eigenfunctions of each atom obtained from a solution of Schrödinger's equation overlap those of neighboring atoms. This phenomenon is similar to the resonant double peak that occurs when two LC tuning circuits with identical resonant frequencies are brought close together.

[6]At atomic spacing c and d, the electrons can also move in discrete energy levels, but of several quanta.

Intrinsic and Extrinsic Semiconductors

Silicon and germanium are the bases of most semiconductor devices. They each possess four valence electrons, leaving four unfilled electron positions. Their crystalline lattice structure uses covalent bonding, that is, a sharing of the electrons, to "complete" the outer shell of eight electrons. (The sharing of electrons does not actually fill the remaining positions but does bond the individual atoms into the crystalline structure.) This situation is illustrated symbolically in Fig. 17.7. Note in the figure that only silicon atoms are present. When a semiconductor is pure, it is called an *intrinsic semiconductor*.

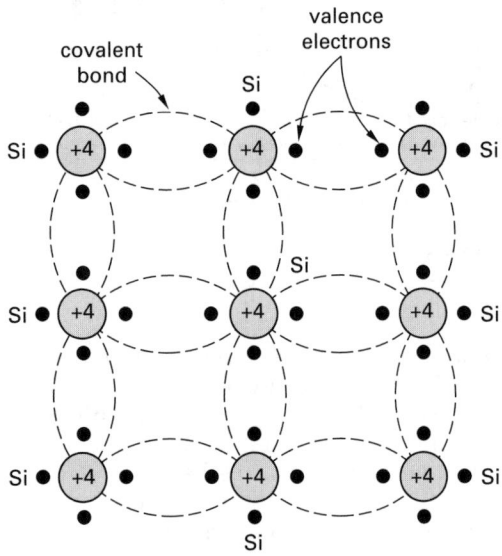

Figure 17.7 *Crystal Structure of Silicon*

The energy the structure possesses simply due to its temperature is sometimes adequate to break a covalent bond and free an electron, that is, give it the necessary energy to overcome the energy gap, E_g, and rise to the conduction band. In this manner, a *hole* is created in the valence band and a free electron in the conduction band, as illustrated in Fig. 17.8. The hole and the free electron are both *carriers*. That is, they both contribute significantly to the current upon the application of a voltage. This type of carrier generation is termed *thermal carrier generation*. The current so generated is the *intrinsic conduction*. The intrinsic carrier density, n_i, is given by Eq. 17.12.

$$n_i^2 = Ce^{-\frac{E_g}{kT}} \qquad\qquad 17.12$$

As Eq. 17.12 illustrates, the intrinsic concentration is a function of temperature. The constant, C, takes into account the effective masses of electrons and holes and is actually constant only at thermal equilibrium.[7] If temperature varies, the constant C can be replaced with

---

[7]The constant C is the product of the *effective density states* N_c and N_v (see App. 17.A).

$A_0 T^3$, where A_0 is constant and T is the absolute temperature. Boltzmann's constant, k, is given in units consistent with the energy gap units, 8.62×10^{-5} eV/K. The energy gap in silicon is approximately 1.1 eV, and in germanium is approximately 0.8 eV.

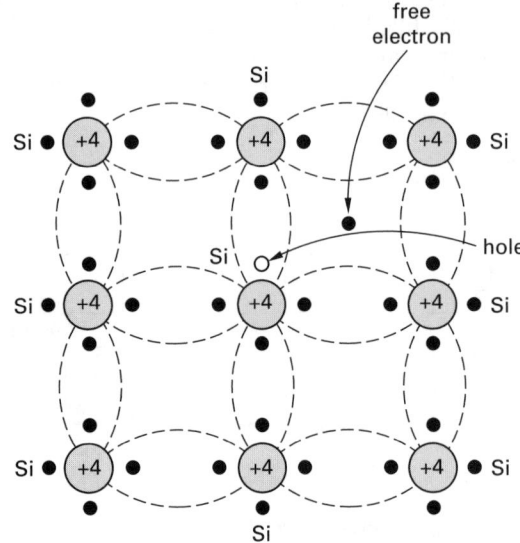

Figure 17.8 *Silicon Crystal with Broken Covalent Bond*

Silicon and germanium atoms are tetravalent (i.e., they have four electrons in the valence band). The conductivity of semiconductors composed of these elements can be increased by *doping*, that is, adding impurities to the mix. When a semiconductor is doped, it is called an *extrinsic semiconductor*.

The doping of silicon with pentavalent substitutional impurities such as phosphorus, arsenic, or antimony results in a higher electron density. This is because the fifth electron on the impurity atom is easily ionized, since its energy level is very close in energy to the edge of the conduction-band energy, E_c. This situation is illustrated in Fig. 17.9(a). Such *dopants* are called *donors*. Any such donor mix is called *n-type*, because the *majority carriers* are electrons. Regions so doped are sometimes labeled n^+. The *minority carriers* in such materials are the holes.

The addition of trivalent impurity atoms such as boron, aluminum, and gallium attracts electrons and thereby creates vacancies (i.e., a lack of electrons—in essence, holes). This results in a higher hole density, since the energy level necessary to attract the electron (and create the hole) is very close to that at the edge of the valence band, E_v. This situation is illustrated in Fig. 17.9(b). Such dopants are called *acceptors*. Any such acceptor mix is called *p-type*, because the majority carriers are holes. Regions so doped are sometimes labeled p^+. The minority carriers in such materials are electrons.

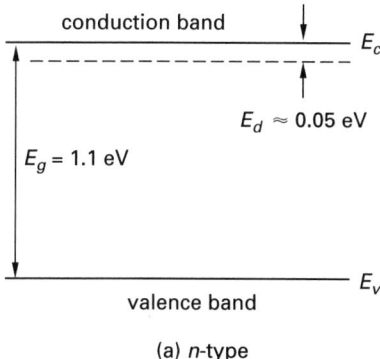

(a) *n*-type

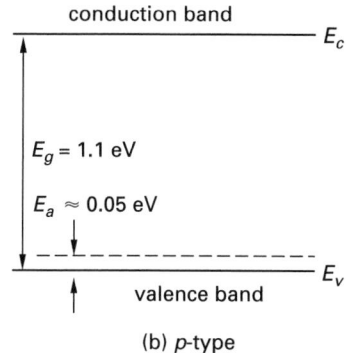

(b) *p*-type

Figure 17.9 *Energy Band Diagrams of n-type and p-type Semiconductor*

Example 17.2

Given a constant, C, of 7.62×10^{38} cm^{-6}, what is the intrinsic carrier concentration in silicon at room temperature (300K)?

Solution

The intrinsic carrier concentration is given by Eq. 17.12.

$$n_i^2 = Ce^{-\frac{E_g}{kT}}$$

$$= \left(7.62 \times 10^{38} \text{ cm}^{-6}\right)$$

$$\exp\left(\frac{-1.1 \text{ eV}}{\left(8.62 \times 10^{-5} \dfrac{\text{eV}}{\text{K}}\right)(300 \text{ K})}\right)$$

$$= 2.56 \times 10^{20} \text{ cm}^{-6}$$

$$n_i = \sqrt{2.56 \times 10^{20} \text{ cm}^{-6}}$$

$$= 1.6 \times 10^{10} \text{ cm}^{-3}$$

Generation and Recombination

Thermal agitation continually generates new electron-hole pairs within a semiconductor.[8] However, electron-hole pairs disappear as a result of *recombination*. This is the process whereby free electrons fall into empty covalent bonds, filling the holes and reducing the number of charge carriers. The average electron or hole lives a certain mean time, τ_n for electrons and τ_p for holes.[9] These are important parameters in the design of semiconductor devices.

Law of Mass Action

Adding *n*-type impurities decreases the number of holes below the number existing in an intrinsic semiconductor; conversely, adding *p*-type impurities decreases the number of free electrons. Theoretical analysis indicates that, under thermal equilibrium, the product of free negative and positive (hole) concentrations remains constant regardless of the amount of doping. This is called the *law of mass action* and is given by

$$np = n_i^2 \qquad 17.13$$

The result of this law is that the doping of an intrinsic semiconductor increases the conductivity and results in a conductor that utilizes either predominately electrons (*n*-type) or predominately holes (*p*-type) as carriers. The intrinsic concentration is a function of temperature (see Eq. 17.12).

Diffusion

Whenever a concentration gradient of any kind exists, a flow will occur in an attempt to equalize the concentration. This phenomenon is called *diffusion* and is governed by a *diffusion coefficient, D*, measured in units of m^2/s. Thus, in a semiconductor, when an *n*-type region is placed in contact with a *p*-type region, a *diffusion current* flows in an attempt to equalize the concentration of electrons and holes. This diffusion current has no analog in metals. The magnitude of the diffusion current density is

$$J = -qD\frac{d\rho}{dx} \qquad 17.14$$

The minus sign indicates that the current flow is in a direction opposite the concentration gradient. This is shown in Fig. 17.10.

---

[8]Any form of energy that can be absorbed by the semiconductor, such as illumination, radiation, and acoustic energy, can create electron-hole pairs.

[9]Do not confuse these terms with relaxation time that occurs for conduction and uses identical symbology.

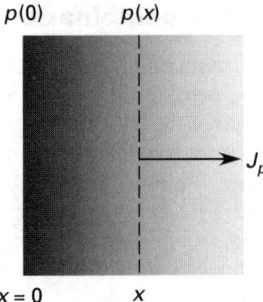

Figure 17.10 *Diffusion Current Flow*

The diffusion and mobility of the electron and hole carriers are governed by the *Einstein equation*, Eq. 17.15.

$$V_T = \frac{D_p}{\mu_p} = \frac{D_n}{\mu_n} \qquad 17.15$$

The term V_T is the *volt-equivalent of temperature* and is defined by Eq. 17.16.

$$V_T = \frac{kT}{q} \qquad 17.16$$

Boltzmann's constant in Eq. 17.16 must be in units of J/K and the temperature, T, is in units of K. The unit of charge, q, is 1.602×10^{-19} C.

The diffusion current will continue to flow until an electric field is established that opposes the flow. When this occurs, a region is created between the two semiconductor types that is devoid of free charge carriers. This region is called the *space-charge region*, because the region is space containing a charge but no free carriers. This region is also called the *depletion region* or *transition region*. The space-charge region is illustrated in Fig. 17.11 and forms the basis for the operation and control of all semiconductor devices.

A partial listing of the properties of silicon, germanium, and gallium arsenide is given in App. 17.A.

Example 17.3

What is the volt-equivalent of temperature at 300K?

Solution

From Eq. 17.16, the volt-equivalent of temperature is

$$V_T = \frac{kT}{q}$$

$$= \frac{\left(1.38 \times 10^{-23} \, \frac{\text{J}}{\text{K}}\right)(300\text{K})}{1.602 \times 10^{-19} \, \text{C} \left(1 \frac{\text{V}\cdot\text{C}}{\text{J}}\right)}$$

$$= 2.58 \times 10^{-2} \text{ V}$$

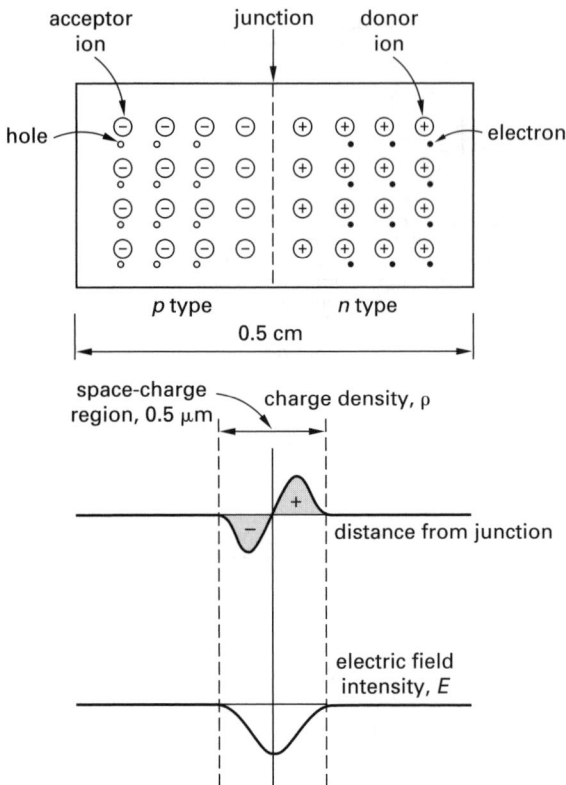

Figure 17.11 *pn Junction Properties*

Example 17.4

If the mobility of electrons in silicon is known to be 1.5 cm^2/V·s at room temperature (300 K), what is the value of the diffusion coefficient, D?

Solution

From Eq. 17.15, the Einstein relationship gives

$$V_T = \frac{D_n}{\mu_n}$$

Rearranging gives

$$D_n = V_T \mu_n$$

$$= (0.0258 \text{ V}) \left(1.5 \, \frac{\text{cm}^2}{\text{V}\cdot\text{s}}\right) \left(\frac{1 \text{ m}}{100 \text{ cm}}\right)^2$$

$$= 3.87 \times 10^{-6} \text{ m}^2/\text{s}$$

Summary of Fundamental Semiconductor Principles

The principles of semiconductor behavior are summarized as follows.

- Two types of charge carriers exist: electrons and holes.

- A semiconductor may have donor impurities and thus mobile charge carriers that are primarily electrons.

- A semiconductor may have acceptor impurities and thus mobile charge carriers that are primarily holes.

- The intrinsic concentration of carriers is a function of temperature.

- At room temperature, essentially all donor and acceptor impurities are ionized.

- Current occurs as a result of two phenomena: (1) diffusion current due to a concentration gradient and (2) drift current due to an external electric field (identical to the conduction current in metals).

- Charge carriers are continuously generated, due to thermal generation, and are simultaneously disappearing, due to recombination.

- Across an open circuited pn junction, a potential exists.

18 Communication Theory

Nomenclature

A	alphabet or message size	–
BW	bandwidth	Hz
$c(t)$	carrier signal	–
C	communication channel	–
E	encoder	–
f	frequency	Hz
$f(t)$	time domain function	–
$F(\omega)$	frequency domain function	–
H	information entropy	–
i, I	current	A
$\{I\}$	information set	–
$\{I_E\}$	encoded information	–
$I(x_i)$	information content of x_i	–
k	Boltzmann's constant, 1.38×10^{-23}	J/K
n	number of carriers	–
$p(x_i)$	probability of element x_i	–
q	charge per carrier	C
r	ratio	dB
rms	root mean square	–
R	resistance	Ω
R	receiver or destination	–
S	source or transmitter	–
t	time	s
T	temperature	K
T_S	sampling period	s
v, V	voltage	V

Symbols

ΔT	rise time or time duration	s
$\Delta \Omega$	equivalent rectangular bandwidth	Hz
τ_d	delay time	s
ω	angular frequency	rad/s

Subscripts

0	initial or center
c	center or carrier
I	information (signal)
m	message (signal)
n	noise
P	power
R	received or receiver

s	signal
S	sampling

1. FUNDAMENTALS

The purpose of any communication system is to deliver *information* represented as a set of data, $\{I\}$, over a given *channel*, C, to the desired *destination* or *receiver*, R. The information comes from a *source* and is usually transmitted as electric signals controlled by the sender. This is illustrated in Fig. 18.1.

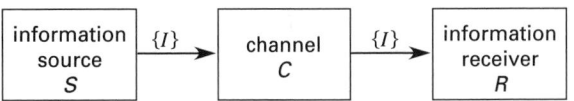

Figure 18.1 *Communication System*

Ideally, the information received would match the information sent. However, due to channel limitations and noise, this may not be the case. Instead, a corrupted version of the information, $\{I^*\}$, could be received. The design objectives in any communication system are to minimize channel distortion and noise and to maximize the size and rate of the information sent. As in most engineering problems, a tradeoff exists. Sending more information at a greater rate results in increased distortion and error.

Some forms of information are better suited for transmission over given channels than others. Therefore, encoders are used at the source to change the information set, $\{I\}$, into a more suitable form of information, $\{I_E\}$.[1] This is illustrated by Fig. 18.2.

Analog and Digital Signals

An *analog signal* operates with variables that are continuously measurable. An analog signal has an infinite number of values, although a given instrument may not have the resolution necessary to determine them. An example of such a signal is the human voice. The receiver, the human ear, does not always hear or resolve the entire signal, but the continuously varying values are in the voice waveform.

---

[1]The information is encoded in the sense that its format is changed to that best suited to the medium of the channel, not for security purposes. Encoding for security purposes is termed *encrypting*.

Figure 18.2 *Communication System with Encoding and Decoding*

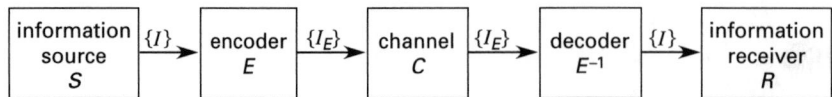

A *digital signal* operates with nominally discontinuous, or discrete, variables that can vary in frequency, amplitude, or polarity. Computer inputs and computer information processing utilize digital signals.

It is often necessary in electrical engineering to change an analog signal into a digital signal. To do so, the analog signal must be sampled. Fig. 18.3 illustrates the sampling of a portion of an analog signal, part (a), in the time domain. The sampling is accomplished with impulse pulses, part (b), and results in a digital signal whose individual pulses vary in amplitude, part (c).

The *sampling theorem* is the basis for conversion of an analog signal to a digital signal. A signal that is sampled the minimum number of times while retaining all the properties of the original in the sampled signal is called an *ideal sampled signal*. The rate that achieves this ideal sampling is called the *Nyquist rate*, f_S, and is given by Eq. 18.1.[2]

$$f_S = \frac{1}{T_S} = 2f_I \qquad \textbf{18.1}$$

The term f_I is the frequency of the original information signal (sometimes called the message signal). The *Nyquist interval*, T_S, is the sampling period corresponding to the Nyquist rate and is given by

$$T_S = \frac{1}{2f_I} \qquad \textbf{18.2}$$

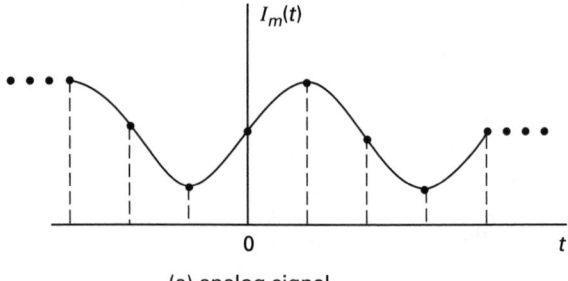

(a) analog signal

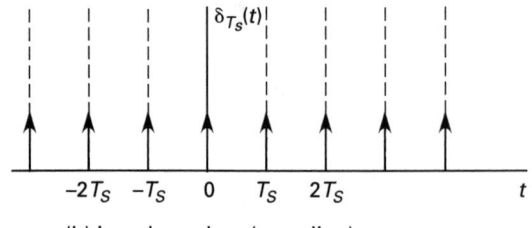

(b) impulse pulses (sampling)

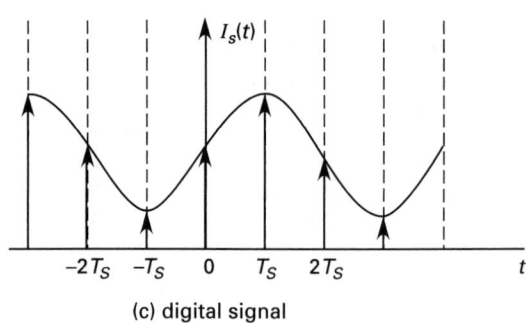

(c) digital signal

Figure 18.3 *Sampling*

Example 18.1

In general, telephone line voice signals are limited to about 3.3 kHz. What is the minimum sampling rate required in order to retain the original signal?

Solution

The minimum sampling rate is given by the Nyquist rate, Eq. 18.1. Thus,

$$f_S = 2f_I$$

$$= (2)(3.3 \text{ kHz}) = 6.6 \text{ kHz}$$

Example 18.2

What is the Nyquist interval for a wideband channel with a bandwidth of 4 kHz?

Solution

The Nyquist interval is the period of sampling corresponding to the Nyquist rate. That is, it is the period corresponding to twice the information frequency. Using Eq. 18.2 and substituting gives

$$T_S = \frac{1}{2f_I} = \left(\frac{1}{(2)(4 \text{ kHz})} \right) \left(\frac{1 \text{ kHz}}{1000 \frac{\text{cycles}}{\text{s}}} \right)$$

$$= \frac{1}{8000 \frac{\text{cycles}}{\text{s}}} = 1.250 \times 10^{-4} \text{ s}$$

---

[2]In actual practice the sampling rate is slightly higher than the Nyquist criteria to avoid *aliasing*, or *foldover*, which distorts the original signal. Also, sampling rates higher than the Nyquist rate are required to properly sample a sinusoidal signal because if one happened to sample when the sinusoid was equal to zero, all subsequent samples would be zero.

Basic Signal Theory

Consider the arbitrary nonnegative function, $f(t)$, shown in Fig. 18.4(a), which is the impulse response of a particular linear circuit. The time that $f(t)$ is appreciably different from zero is the *duration* of $f(t)$. When used in terms of the system, the duration is called the *rise time* or *response time* and is given the symbol ΔT. The term τ_d is called the *delay time* of the system. The integral of the pulse represents the step-function response to the impulse $f(t)$ (see Fig. 18.4).

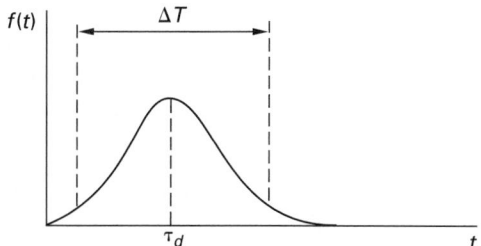

(a) impulse response of a linear system

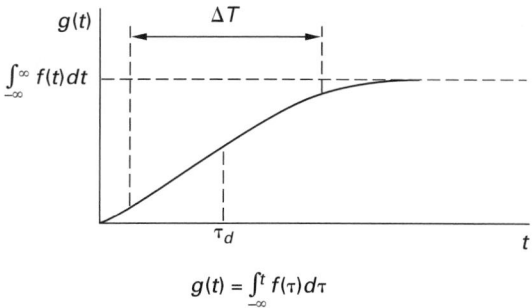

$$g(t) = \int_{-\infty}^{t} f(\tau)\, d\tau$$

(b) integral of the impulse response

Figure 18.4 *Signal Duration and Delay*

Since the function $f(t)$ is nonnegative, a new nonnegative function with unit area can be defined. This function is

$$\frac{f(t)}{\int\limits_{-\infty}^{+\infty} f(t)\, dt} \qquad\qquad 18.3$$

Eq. 18.3 and its frequency domain counterpart, that is, its Fourier transform counterpart, are used to define an equivalent rectangular signal of bandwidth, $\Delta\Omega$, and height, $F(\omega)$. This is illustrated by the dashed lines

in Fig. 18.5. The same process is followed in the time domain and yields an equivalent rectangular duration, ΔT.

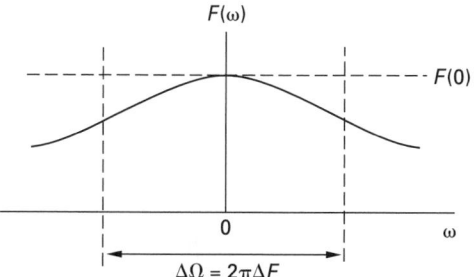

Figure 18.5 *Equivalent Rectangular Bandwidth*

The *bandwidth* is defined by the frequency range over which $F(\omega)$ is appreciably greater than zero. The duration in the time domain is directly related to the bandwidth in the frequency domain. Furthermore, the two are related in a fundamental way by various *uncertainty relationships*. The primary uncertainty relationship is

$$\Delta T \Delta\Omega = 2\pi \qquad\qquad 18.4$$

The second moment uncertainty relationship, which relates the energy, is

$$\Delta T_2 \Delta\Omega_2 \geq \frac{1}{2} \qquad\qquad 18.5$$

Many other uncertainty relationships exist.

Signals, such as the function $f(t)$, are modified prior to being placed in a communication channel for transmission. Signals containing the desired information are placed on, or mixed with, a carrier signal and then transmitted to the receiver. The situation is illustrated in mathematical terms in Fig. 18.6.

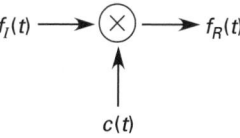

Figure 18.6 *Basic Signal Processing: Functional Representation*

The processing indicated in the figure takes place in the information source, S, of Fig. 18.1 or in the encoder, E, of Fig. 18.2. Consider a specific signal. Let the information signal, $f_I(t)$, be given by Fig. 18.7(a). When mixed with a carrier signal, $c(t)$, of Fig. 18.7(b), the resultant signal, $f_R(t)$, is given as Fig. 18.7(c).[3]

---

[3]The specific signal shown in Fig. 18.7(c) represents $f_R(t) \approx f(t)\cos\omega_c t$ and is called a *suppressed-carrier amplitude modulation system*.

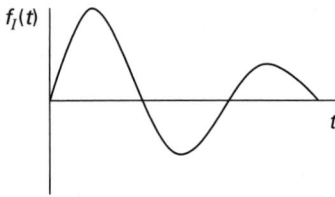

(a) information signal

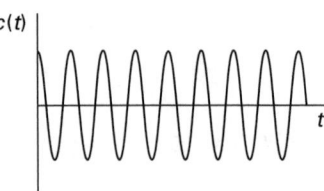

(b) carrier signal

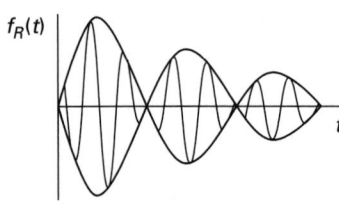

(c) resultant signal

Figure 18.7 *Basis Signal Processing: Graphical Representation*

If the amplitude of the carrier signal is sufficient, the resultant signal, $f_R(t)$, does not cross the x-axis and is represented as shown in Fig. 18.8.

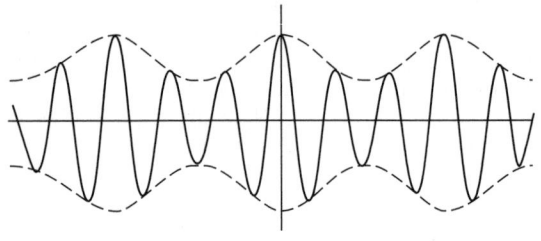

(a) double sideband AM signal

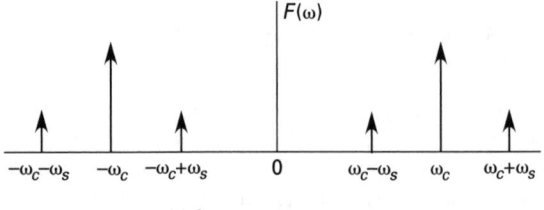

(b) frequency spectrum

Figure 18.8 *Double Sideband AM Signal and Frequency Spectrum*

The carrier is centered at zero on the frequency axis. One *sideband* is centered at ω_c and the other at $-\omega_c$.

In each sideband, the frequencies above the center frequency $|\omega_c|$ are termed the *upper sideband* and those below constitute the *lower sideband*. If the information signal is not sinusoidal, the frequency spectrum spreads as shown in Fig. 18.9.

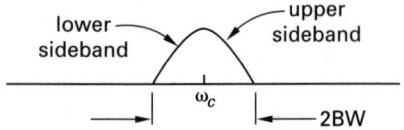

Figure 18.9 *Spreading of Frequency Spectrum*

Information signals, or message sets, are modified in order to transmit information over a given channel with less error. Additionally, a carrier makes it easier to demodulate the signal at the receiver. However, the information desired, that is, the original signal, is completely contained within the upper or lower sideband of a single sideband. The transmitted carrier simply absorbs power and the second sideband unnecessarily expands the bandwidth. The former lowers the efficiency. The latter makes poor use of the frequency spectrum, meaning less information can be passed in a given bandwidth. Nevertheless, such a system is simple to construct and less expensive than comparable systems. In fact, this is the type of signal used for AM radio.

Example 18.3

If the bandwidth of a narrowband, or telegraph-grade, communication system is 300 Hz, what is the response time?

Solution

Assuming the bandwidth given is the equivalent rectangular bandwidth, the applicable formula is given by Eq. 18.4. Thus,

$$\Delta T \Delta \Omega = 2\pi$$

$$\Delta T = \frac{2\pi}{\Delta \Omega} = \frac{2\pi}{(300 \text{ Hz}) \left(1 \frac{\text{s}^{-1}}{\text{Hz}} \right)} = 0.02 \text{ s}$$

Information Entropy

Information theory quantifies the communication process. To model a communication system, the information must be predictable. It must be possible for each piece of information to be represented by a *probability of occurrence*, $p(x_i)$. The more probable a particular piece of information, the less *self-information*, $I(x_i)$, a message contains. The information content is defined by Eq. 18.6.

$$I(x_i) = \log \left(\frac{1}{p(x_i)} \right) = -\log p(x_i) \qquad \textbf{18.6}$$

The choice of the base for the logarithm is arbitrary, but is normally chosen as 2. In this case the unit of self-information, $I(x_i)$, is the common *binary digit* or *bit*. If base e is used the unit is called the *nat*. (Base e is convenient due to its analytical properties when integration and differentiation are realized with electric circuitry.) Base 10, rarely used, would result in the unit called a *Hartley*.

Example 18.4

What is the information content of a message with only two possible values, zero and one, of equal probability?

Solution

Assume base 2. Using Eq. 18.6, and noting that equal probability means a 0.5 chance of a zero and a 0.5 chance of a one, gives

$$I(x_i) = \log_2\left(\frac{1}{p(x_i)}\right) = \log_2\left(\frac{1}{0.5}\right) = 1 \text{ bit}$$

Note: Unless otherwise told, assume base 2. Also, the bit is not a unit in the true sense.

Example 18.5

The probability of occurrence of a given message in a computer system is 1.5259×10^{-5}. What is the information content of a message?

Solution

Use Eq. 18.6 and substitute the given information.

$$I(x_i) = \log_2\left(\frac{1}{p(x_i)}\right)$$

$$= \log_2\left(\frac{1}{1.529 \times 10^{-5}}\right) = 16 \text{ bits}$$

One of the more important statistical properties of the information variable $I(x_i)$ is its mean or average value. In information theory, the average value of a message is called its *entropy* and is given the symbol $H(X)$. The entropy is defined by Eq. 18.7.

$$H(X) = \sum_{i=1}^{A} p(x_i)I(x_i)$$

$$= -\sum_{i=1}^{A} p(x_i)\log_2 p(x_i) \qquad \textbf{18.7}$$

The term A is the alphabet size, that is, the total number of elements in the message set. Properly interpreted, Eq. 18.7 is identical to entropy in statistical mechanics or thermodynamics. In statistical mechanics, entropy is a measure of the disorder of the system.

Similarly, in information theory, entropy is a measure of the uncertainty associated with a message. The units of entropy are the *bit*, the *nat*, and the *Hartley* and are sometimes given as bits/message, nats/message, or Hartleys/message.

The following properties emerge for a binary system.

- $H(X) \geq 0$.
- $H(X) = 0$ only for $p = 0$ or 1.
- $H(X)$ is a maximum at $\frac{1}{2}$.

In the general case, for an alphabet size of A, the entropy has the following properties.

- $H(X) \geq 0$.
- $H(X) = 0$ if and only if all the probabilities are zero except for one, which is unity.
- $H(X) \leq \log_b A$.
- $H(X) = \log_b A$ if and only if all the probabilities are equal so that $p(x_i) = 1/A$ for all i.

Example 18.6

What is the entropy of a binary system with probabilities p and $1 - p$?

Solution

The entropy is given by Eq. 18.7. The alphabet size is two. That is, the value of the elements can be either a zero or a one in a binary system. Thus,

$$H(X) = -\sum_{i=1}^{A} p(x_i)\log_2 p(x_i)$$

$$= -\sum_{i=1}^{2} p(x_i)\log_2 p(x_i)$$

$$= -\big(p\log_2 p + (1-p)\log_2(1-p)\big)$$

$$= -\big((0.5)\log_2(0.5) + (1-0.5)\log_2(1-0.5)\big)$$

$$= 1 \text{ bit/message}$$

More enlightening than the mathematical answer in Ex. 18.6 is the plot of $H(X)$ versus the probability.

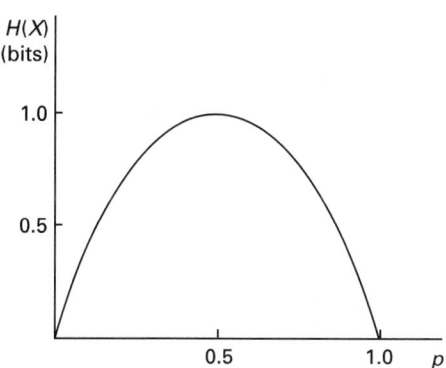

Example 18.7

A certain communication system consists of three possible messages with probabilities of ½, ¼, and ¼. What is the entropy of the message?

Solution

The information content of the probability ½ message is as follows.

From Eq. 18.6,

$$I(x_i) = \log_2\left(\frac{1}{p(x_i)}\right) = \log_2\left(\frac{1}{0.5}\right) = 1 \text{ bit}$$

The information content of the probability ¼ messages is as follows.

From Eq. 18.6,

$$I(x_i) = \log_2\left(\frac{1}{p(x_i)}\right) = \log_2\left(\frac{1}{0.25}\right) = 2 \text{ bits}$$

From Eq. 18.7, the entropy is

$$H(X) = -\sum_{i=1}^{A} p(x_i) \log_2 p(x_i)$$

$$= \sum_{i=1}^{3} p(x_i) I(x_i)$$

$$= \left(\tfrac{1}{2}\right)\left(1 \text{ bit}\right) + \left(\tfrac{1}{4}\right)\left(2 \text{ bits}\right) + \left(\tfrac{1}{4}\right)\left(2 \text{ bits}\right)$$

$$= 1.5 \text{ bits/message}$$

Decibels

The use of a unit known as the *decibel*, dB, is common in communications theory and electrical engineering in general. The decibel is a unit that describes the ratio of two powers or intensities. In acoustics, it is also the ratio of a power to a reference power. In mathematical terms, the decibel is

$$r = 10 \log_{10}\left(\frac{P_2}{P_1}\right) \qquad \text{18.8}$$

When the ratio is negative, a power loss is indicated. Since the decibel is a log function, decibel gain or loss in the successive stages of a communication system can be added algebraically.

When the ratio is taken with respect to 1 watt, that is, when $P_1 = 1$ W, the ratio is commonly called the *decibel watt* and abbreviated dBW. When the ratio is taken with respect to 1 milliwatt, that is, when $P_1 = 1 \times 10^{-3}$ W, the ratio is termed the *decibels above 1*

milliwatt and abbreviated dBm. Several other decibel ratios are used in electrical engineering and care must be taken to ensure that the correct reference power level is used.

The decibel is also used as a ratio of currents and voltages, though technically this should only be done if the impedances of the measured quantities are equal. This is often ignored, and the decibel for currents or voltages is given by Eq. 18.9.

$$r = 10 \log\left(\frac{I_2}{I_1}\right)^2 = 10 \log\left(\frac{V_2}{V_1}\right)^2$$

$$= 20 \log\left(\frac{I_2}{I_1}\right) = 20 \log\left(\frac{V_2}{V_1}\right) \qquad \text{18.9}$$

2. CHANNELS AND BANDS

As shown in Fig. 18.1, all communication systems use a channel. The *channel* is the medium that connects the information source to the destination, or receiver. The medium can be metallic conductors such as copper, specialized cables such as coaxial cable or fiber optic cable, waveguides (devices that guide the propagation of electromagnetic waves using the physical construction of the guide), the atmosphere or, in the case of satellites, space itself.

Channels are classified according to bandwidth. Narrowband channels use bandwidths of 300 Hz or less. Voice-grade channels use 300 Hz to 4 kHz (the audible frequency range of the human voice is approximately 300 Hz to 3400 Hz). Wideband channels use 4 kHz and greater bandwidths.[4]

Bands represent a somewhat arbitrary range of electromagnetic-wave frequencies between definite limits. Though somewhat arbitrary, the properties of each band vary, creating different uses for each of the bands. For example, VLF (very low frequency) transmissions form surface waves that travel great distances and are able to penetrate water to a certain degree, making them useful for submarine communications. Standard AM radio uses the MF band while FM radio uses the VHF band. Television utilizes the VHF and UHF bands. Satellites, with their need for high data rates and minimal environmental interference, as well as their ability to utilize line-of-sight communication from Earth, function in the UHF and SHF bands. The bands are shown in Table 18.1. The electromagnetic spectrum is shown in App. 18.A.

---

[4]The channel capacity is determined by the type of channel and is given by the *Shannon coding theorem*. The results of this theorem are one of the most important in information theory and communications engineering.

Table 18.1 Frequency Bands

symbol	frequency range	frequency band
VLF	3–30 kHz	very low
LF	30–300 kHz	low
MF	300–3000 kHz	medium
HF	3–30 MHz	high
VHF	30–300 MHz	very high
UHF	300–3000 MHz	ultra high
SHF	3–30 GHz	super high
EHF	30–300 GHz	extremely high

3. CODING

In designing a communication system, one attempts to minimize the probability of information error, or bit error, and at the same time minimize the length of the message. Coding theorems guarantee the existence of codes that allow transmission within a channel's capacity at an arbitrarily small error rate. However, to do so, especially in the presence of noise, may require excessively long messages—that is, a low data rate— or raising the signal power above acceptable rates. To overcome this difficulty, *error-correcting codes* have been developed to correct some of the errors that occur during the transmission of an information signal through a noisy channel. A simple example is a parity check code. An example of an *even parity* check code for detecting single errors is shown in Table 18.2. This code determines the total number of ones or zeros in the message and assigns a one if the total is odd and a zero if the total is even.

Table 18.2 Even Parity Check Code

information	check digit	transmitted signal
000	0	0000
001	1	1001
010	1	1010
011	0	0011
100	1	1100
101	0	0101
110	0	0110
111	1	1111

4. NOISE

Noise or *interference* is any unwanted input to a communication system. Such input causes meaningless or erroneous information that must be ignored or removed.

Noise results from random processes and can be classified in mathematical terms, but it is more useful to classify random processes as either *natural noise* or *artificial noise*. However, one mathematical classification is worthy of mention. *White noise* or *additive white Gaussian noise* is defined as a waveform in which all observations are jointly Gaussian and mutually independent. The power spectral density is constant with respect to frequency variation. That is, the noise is random and unaffected by changes in frequency. (Such noise cannot technically exist, but the concept is a useful one when applied to a system with a constant power spectral density over a frequency range wider than the bandwidth.)

Natural Sources of Noise

Thermal noise, or *Johnson noise*, is generated by the random motion of electrons in conductors and semiconductors. The mean squared value of the thermal noise voltage is given by Eq. 18.10.

$$\overline{v_n^2} = 4kTR(\text{BW}) \qquad \textit{18.10}$$

The bandwidth, BW, is the frequency bandwidth of the monitoring equipment. The resistance of the circuit or element is given by R in ohms. The temperature, T, is in kelvins. Because the voltage is squared, Eq. 18.10 is proportional to the average power generated by the noise. Thermal noise is constant from DC values to very high frequencies and is thus considered white noise.

Shot noise, sometimes called *Schottky noise*, is caused by electrons moving across a potential barrier in a random way. It was originally defined for vacuum tubes as the voltage developed because of random variations in the number and velocity of electrons emitted by a heated cathode. Current can be defined as

$$I = q\frac{dn}{dt} \qquad \textit{18.11}$$

q represents the charge per carrier and dn/dt is the number of carriers passing through the potential per second. Because of the random thermal velocities of the electrons, the value of dn/dt fluctuates. If the average current is given by I and the total current by $i(t)$, then the noise current is

$$i_n(t) = i(t) - I \qquad \textit{18.12}$$

Assuming the process is completely random, all frequencies are equally likely. The mean squared value of the shot noise current is

$$\overline{i_n^2} = 2qI(\text{BW}) \qquad \textit{18.13}$$

This noise causes sputtering or popping sounds in radio receivers and snow effects in televisions.

Defect noise describes a number of related phenomena exhibited by noise voltages across terminals of devices when DC currents pass through them. The magnitude depends on the surface imperfections of the terminals. Such noise is also called *flicker noise, contact noise, excess noise, current noise,* and *1/f noise.*

Example 18.8

What is the noise voltage of a 10 kΩ resistor operating in a 4 kHz frequency bandwidth at room temperature?

Solution

Use Eq. 18.10 and note that room temperature is 300K.

$$\overline{v_n^2} = 4kTR(\text{BW})$$

$$= (4)\left(1.38 \times 10^{-23}\,\frac{\text{J}}{\text{K}}\right)$$

$$\times (300\text{K})\left(10 \times 10^3\,\Omega\right)(4000\,\text{Hz})$$

$$= \left(6.62 \times 10^{-13}\,\frac{\text{J}\cdot\Omega}{\text{s}}\right)\left(1\frac{\frac{\text{V}}{\text{A}}}{\Omega}\right)\left(1\frac{\text{A}}{\frac{\text{C}}{\text{s}}}\right)\left(1\frac{\text{V}}{\frac{\text{J}}{\text{C}}}\right)$$

$$= 6.62 \times 10^{-13}\,\text{V}^2$$

Solve for the voltage directly.

$$\overline{v_n^2} = 6.62 \times 10^{-13}\,\text{V}^2$$

$$v_n = \sqrt{6.62 \times 10^{-13}\,\text{V}^2}$$

$$= 8.14 \times 10^{-7}\,\text{V}$$

Artificial Noise

The natural noise explained in the section titled "Natural Sources of Noise" is fundamental in nature in that the proximate cause of the noise is the random, noncontinuous nature of electric charge under the influence of a potential. When a noise is generated by electrical or electronic equipment, which is in principle under human control, the noise is classified as *artificial* or *environmental noise.* Artificial noise may be grouped into three main categories.

Interference noise results when a radio station or television channel interferes with another. This may be due to poor antenna design or unexpected atmospheric effects. This also includes *crosstalk* between communication links due to multipath propagation or reflection.

Hum is a periodic signal arising from power lines. It occurs at the power supply frequency or its harmonics (60 Hz or 120 Hz in the U.S.). In transformer iron cores with loose laminations, the frequency of the hum is twice the supply frequency and is related to the alignment of magnetic domains in the material. Hum

can be eliminated by proper construction, filtering, and shielding.

Impulse noise describes numerous phenomena. Impulse noises are often modeled as low-frequency shot processes, that is, as the superposition of a small number of large impulses. Sources can include such things as ignition noise from automobiles, corona noise from high-voltage transmission lines, switching noise, and lightning. Impulse noises are usually best handled with limiters or blankers that protect the system during the occurrence of the noise and return it to normal when the noisy period has passed.

5. MODULATION

Modulation is the process whereby some parameter of one wave is altered by some parameter of another wave. A particular case of modulation is illustrated in Figs. 18.6 and 18.7. Modulation is necessary because in many instances it may be inefficient or impossible to transmit a given spectrum through the required channel. In addition, modulation is utilized for frequency translation of a signal in order to maximize the efficiency of an antenna or move the signal to an assigned frequency band. Modulation makes signal processing, particularly the recovery of the signal information, easier. It can also be used to enhance *multiplexing,* a process whereby multiple signals are transmitted within the same channel. Fig. 18.2 shows the block diagram for a communication system with the encoder replaced by a modulator. In fact, modulation is a form of encoding. Further, the encoding covered in Sec. 18-3 would still occur alongside any modulation.

Signals can be modulated in amplitude or frequency. (AM stands for amplitude modulation and FM for frequency modulation.) Both sidebands or a single sideband can be transmitted. The original carrier wave may or may not be present.

19 Acoustic and Transducer Theory

Nomenclature

B	bulk modulus	Pa
c	specific heat	J/kg·K
C	capacitance	F
E	modulus of elasticity	Pa
f	frequency	Hz
I	intensity	W/m²
L	inductance	H
L_p	sound pressure level	dB
m	mass per area	kg/m
M	total mass	kg
MW	molecular weight (mass)	kg/kmol[1]
p	pressure	Pa
R	resistance	Ω
R	specific gas constant	kJ/kg·K
R^*	universal gas constant	kJ/kmol·K
T	period	s
T	temperature	K
v	velocity	m/s
V	voltage	V

Symbols

β	sound intensity level	dB
γ	ratio of specific heats	–
γ	isentropic exponent	–
η	energy density	J/m³
λ	wavelength	m
ρ	density	kg/m³

Subscripts

0	reference
a	acoustic
e	electrical
m	mechanical
p	pressure
s	sound
v	volume

---

[1]This is also called the *kilogram molecular weight*. It is truly a mass, not a weight. The gram molecular weight, measured in g/mol·K, is also common. Care should be exercised when applying the units.

1. INTRODUCTION

Sound is an alteration of properties, such as pressure, particle displacement or velocity, or density, in an elastic medium. Sound is a wave phenomenon, like electromagnetic radiation, but with two important fundamental differences. First, sound is transmitted as a *longitudinal wave*, also known as a *compression wave*, that causes alternating compression and expansion (*rarefaction*), while electromagnetic waves are transverse waves. Second, sound waves require a medium through which to travel. Electromagnetic waves do not; they can travel through a vacuum.

For visualization purposes, sound waves are often represented as transverse waves as shown in Fig. 19.1. The *wavelength* is the distance between successive compressions or expansions.

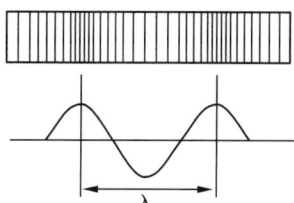

Figure 19.1 *Longitudinal Waves*

The usual relationships among wavelength, propagation velocity, frequency, and period remain valid.

$$v_s = f\lambda \qquad\qquad 19.1$$

$$T = \frac{1}{f} \qquad\qquad 19.2$$

Sound waves traveling through a medium change the energy level of the medium. This energy is called *acoustic energy* and represents the difference between the total energy of the system and the system energy without the sound waves present. Electrical engineering is concerned with the manipulation and use of this energy. The basic interface between acoustic energy and electrical engineering is illustrated in Fig. 19.2.

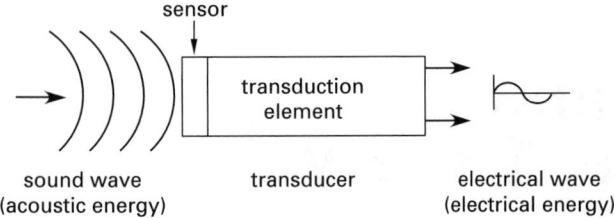

Figure 19.2 *Acoustic/Electrical Interface*

A *transducer* provides the interface and is the device that provides a usable output in response to a specified *measurand*. The measurand is a physical quantity, property, or condition to be measured.[2] The transducer can be *self-generating*, in which case no external excitation power is required. Such transduction elements are termed *active elements*, whereas those that require excitation power are *passive elements*.

2. SOUND WAVE PROPAGATION VELOCITY

The propagation velocity of sound, v_s, is commonly called the *speed of sound* and *sonic velocity*. The sound velocity depends on the compressibility of the medium through which the sound wave is traveling. This occurs because the sound wave physically moves atoms in the medium, that is, places them closer together (compression) or farther apart (rarefaction), as it propagates. In solids, the compressibility is accounted for by the *elastic modulus*, E. In liquids, the elastic modulus is referred to as the *bulk modulus*, B. The speed of sound in solids and liquids is given by

$$v_s = \sqrt{\frac{E}{\rho}} \qquad 19.3$$

For ideal gases, the propagation velocity is

$$v_s = \sqrt{\frac{\gamma R^* T}{MW}} \qquad 19.4$$

The term γ is the ratio of the specific heat at a constant pressure to the specific heat at a constant volume, c_p/c_v, and is called the *isentropic exponent* in thermodynamics.

The speed of sound in a few common materials is given in Table 19.1. Various values of the universal gas constant are given in Table 19.2. When the universal gas constant is divided by the molecular weight, that is, R^*/MW, it becomes the *specific gas constant*, R.

---

[2]These definitions are paraphrased from the American National Standard ANSI MC6.1-1982.

Table 19.1 *Approximate Speeds of Sound (at one atmospheric pressure $\approx 10^5$ Pa)*

material	speed of sound (m/s)
air	330 @ 0°C
aluminum	4990
carbon dioxide	260 @ 0°C
hydrogen	970 @ 0°C
steel	5150
water	1490 @ 20°C

Table 19.2 *Values of the Universal Gas Constant, R^* units in SI and metric systems*

8.3143 kJ/kmol·K
8314.3 J/kmol·K
0.08206 atm·L/mol·K
1.986 cal/mol·K
8.314 J/mol·K
82.06 atm·cm³/mol·K
0.08206 atm·m³/kmol·K
8314.3 kg·m²/s²·kmol·K
8314.3 m³·Pa/kmol·K
8.314 × 10⁷ erg/mol·K

Example 19.1

What is the speed of sound in air at 293K given an isentropic exponent of 1.4?

Solution

Use Eq. 19.4.

$$v_s = \sqrt{\frac{\gamma R^* T}{MW}}$$

The universal gas constant is 8.314 J/mol·K. The molecular weight of air is approximately 29 g/mol. (This assumes 78% N_2, 21% O_2, and 1% Ar.) Substituting gives

$$v_s = \sqrt{\frac{\gamma R^* T}{MW}}$$

$$= \sqrt{\frac{(1.4)\left(8.314 \ \dfrac{J}{mol \cdot K}\right) \times (293K)\left(1 \dfrac{\frac{kg \cdot m^2}{s^2}}{J}\right)}{\left(29 \ \dfrac{g}{mol}\right)\left(\dfrac{1 \ kg}{10^3 \ g}\right)}}$$

$$= 343 \text{ m/s}$$

It is not necessary to do the unit conversions, but doing so acts as a check on the solution.

3. ENERGY AND INTENSITY OF SOUND WAVES

A sound wave, or acoustic energy, manifests itself in the displacement and motion of the medium through which it travels. One could measure the energy of the medium in terms of this displacement or motion. However, the most widely used and measurable property of sound waves is *sound pressure*, the fluctuation above and below the equilibrium pressure (or reference pressure), p_0, of the medium in which the sound wave travels. The energy density, η, is

$$\eta = \tfrac{1}{2}\frac{p_0^2}{\rho v^2}\qquad \textbf{\textit{19.5}}$$

Like many other items in electrical engineering, the *intensity* of the sound wave, which is the power delivered per unit of wave front, is the value of concern. The intensity is

$$I = \eta v = \tfrac{1}{2}\frac{p_0^2}{\rho v}\qquad \textbf{\textit{19.6}}$$

The human ear can sense sound wave intensities from approximately 10^{-12} W/m^2 up to the pain threshold of 1 W/m^2. Since this encompasses a 12-decade range, the intensities are normally measured in decibels, dB, compared to a reference intensity, as shown in Eq. 19.7.

$$\beta = 10 \log\left(\frac{I}{I_0}\right)\qquad \textbf{\textit{19.7}}$$

Since the pressures that sound waves create are the normally measured quantities, the *sound-pressure level*, L_p, as defined in Eq. 19.8 is normally used in lieu of Eq. 19.7.

$$L_p = 20 \log\left(\frac{p}{p_0}\right)\qquad \textbf{\textit{19.8}}$$

The defined reference pressure for sounds in air is 20 μPa. This correlates with 0 dB, which is the threshold of human hearing.

Example 19.2

The density of air is approximately 1.3 kg/m^3. What pressure is expected at a transducer's sensor for a sound intensity level of 1 W/m^2 in air at 273K?

Solution

Use Eq. 19.6.

$$I = \tfrac{1}{2}\frac{p_0^2}{\rho v}$$

From Table 19.1, the velocity (speed) of sound in air at 273K (0°C) is 330 m/s. The density of air is given. Rearranging to solve for the equilibrium pressure and substituting gives

$$I = \tfrac{1}{2}\frac{p_0^2}{\rho v}$$

$$p_0 = \sqrt{2I\rho v} = \sqrt{\left(1\,\frac{\text{W}}{\text{m}^2}\right)(2)\left(1.3\,\frac{\text{kg}}{\text{m}^3}\right)\left(330\,\frac{\text{m}}{\text{s}}\right)}$$

$$= 29.3\ \text{Pa}$$

Note that this small change in pressure due to a sound wave is superimposed on normal atmospheric pressure of approximately 10^5 Pa. Though small, this pressure results in a sensation of pain for most people. Thus, any audio system design should use this pressure as an absolute maximum.

4. TRANSDUCTION PRINCIPLES

The transduction principle of a given transducer determines nearly all its other characteristics. There are three *self-generating* transduction types: photovoltaic, piezoelectric, and electromagnetic. All other types require the use of an external excitation power source.

In *photovoltaic transduction*, light is directed onto the junction of two dissimilar metals, generating a voltage. This type of transduction is used primarily in optical sensors. It can also be used with the measurand controlling a mechanical-displacement shutter that varies the intensity of the built-in light source. *Piezoelectric transduction* occurs because certain crystals generate an electrostatic charge or potential when placed in compression or tension or by bending forces. In *electromagnetic transduction*, the measurand is converted into a voltage by a change in magnetic flux that occurs when magnetic material moves relative to a coil with a ferrous core. These self-generating types of transduction are illustrated in Fig. 19.3.

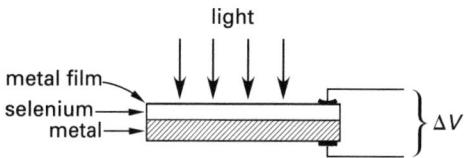

(a) photovoltaic transduction

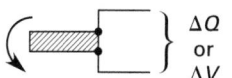

(b) piezoelectric transduction

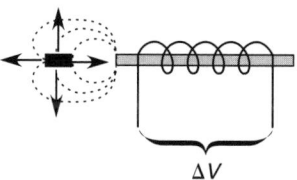

(c) electromagnetic transduction

Figure 19.3 *Self-Generating Transducers*

In *capacitive transduction*, the measurand is converted into a change in capacitance. This can be realized by moving the electrodes of a given capacitor or changing the position of the dielectric. *Inductive transduction* is the change in self-inductance of a single coil. *Photoconductive transduction* converts the measurand into a conductance change within a semiconductor material due to a change in the incident illumination. This is similar to photovoltaic transduction (see Fig. 19.4 to compare).

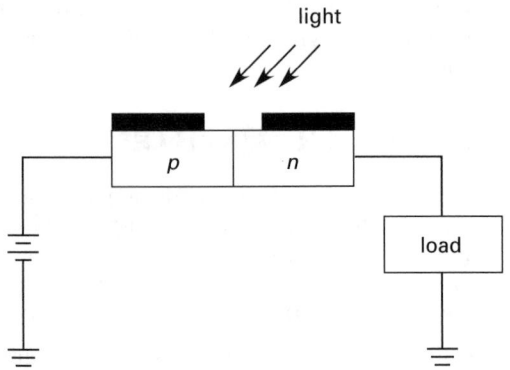

Figure 19.4 *Photoconductive Transduction*

In *reluctive transduction*, the measurand is converted to an AC voltage change due to a change in the reluctance path between two or more coils. Reluctance in a magnetic circuit is analogous to resistance in an electrical circuit. This principle also applies to differential transformers and inductance bridges. *Potentiometric transduction* converts the measurand into a position change of a moveable contact on a resistance element. This changes the end-to-end resistance relative to the wiper-arm position controlled by the measurand. *Resistive transduction* is similar to potentiometric transduction except that the change in resistance is brought about by heating or cooling, mechanical stresses, or other external forces. *Strain-gauge transduction* is a special case of resistive transduction generated by a resistance change, due to a strain, in two of the four arms of a Wheatstone bridge. Here, however, the output is always given by the bridge output voltage.

5. EQUIVALENT CIRCUIT ELEMENTS

In electrical engineering, analogs have developed for the basic elements of acoustic and mechanical systems. These analogs are illustrated in Fig. 19.5.

Resistance in acoustic systems can be defined as air friction or viscosity effects in narrow slots. In mechanical systems, resistance is analogous to friction. Inductance in acoustic systems is mass per unit area in a constriction or narrow opening, while in a mechanical system it is the mass itself. Capacitance in an acoustic system is the inverse stiffness of an enclosed volume of air when subjected to piston-like action. It is proportional to the volume enclosed. In mechanical systems, a spring is analogous to capacitance.

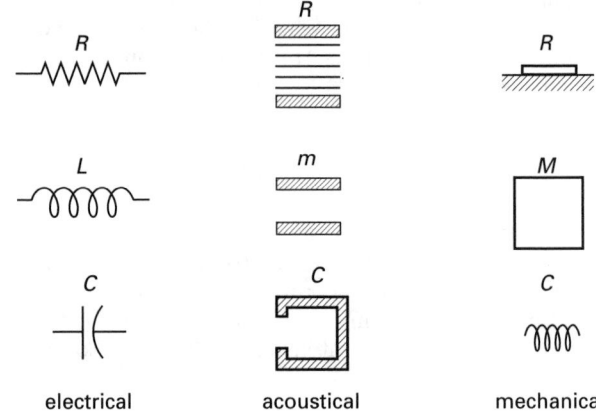

Figure 19.5 *Basic Elements of Electrical, Acoustic, and Mechanical Systems*

Topic III: Fields

Fields

For the most current information about the exam, visit **www.ppi2pass.com** regularly.

20 Electrostatics

Nomenclature

d	distance	m
D	electric flux density	C/m^2
E	electric field strength	N/C (V/m)
F	force	N
l, L	length	m
p	dipole moment	C·m
q, Q[1]	charge	C
ΔQ	infinitesimal test charge	C
r	radius or distance	m
s	surface area	m^2
T	torque	N·m
U	energy	J
v, V	voltage	V
ν, V	volume	m^3
ΔV	potential difference	V
W	work	J

Symbols

ϵ	permittivity	$C^2/N·m^2$ or F/m
ϵ_0	free-space permittivity	$C^2/N·m^2$
ρ	charge density	C/m^3
ρ_l	line charge density	C/m
ρ_s	surface charge density	C/m^2
ρ_ν	volume charge density	C/m^3
Ψ	electric flux	C

Subscripts

1–2	from 1 to 2; or effect on 2 due to the field of 1
A–B	at B with respect to A or from A to B[2]

e	electric
l	line
n	normal
r	radial
s	surface
v	volume
x	x component
y	y component
z	z component

Fields

Part 1: Electrostatic Fields

A *field* is an entity that acts as an intermediary in interactions between particles. It is distributed over part or all of space and is a function of space coordinates and time (if the field is not static). An electric field is fundamental in nature, causing a charged body to be attracted or repelled by another charged body. The electric field, **E**, represents the force and direction a positive test charge would experience if placed in the field. It is measured in N/C, which is equivalent to V/m. The electric field strength is given by[3]

$$\mathbf{E}_1 = \frac{\mathbf{F}_{1-2}}{\Delta Q_2} = \left(\frac{Q_1}{4\pi\epsilon r^2} \right) \mathbf{a} \qquad 20.1$$

The unit vector, **a**, points in the direction the positive test charge will move at the point of concern within the field, that is, along the lines of flux. The electric field is a vector force field in that a charge placed within the field experiences a force in a specific direction. The force field for a single charged particle is illustrated in Fig. 20.1.

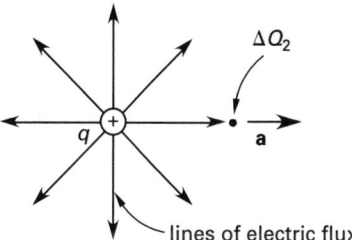

Figure 20.1 *Electric Field of a Positive Charge*

---

[1]In some formulas the small q represents the absolute value of a single charge, that is, 1.602×10^{-19} C.

[2]The A-B notation is by no means consistent from one text to another. Care should be taken to determine the actual convention used. Many authors avoid any convention at all and instead use figures to illustrate the physical principle.

---

[3]The electric field is defined in terms of the force per unit positive test charge. The rightmost equation is actually the realization of this definition for a point charge.

By convention, lines of flux are said to originate at positive charges and terminate at negative charges. Within an electric field, like charges repel and unlike charges attract. Therefore, a positive sign on vector $\mathbf{E}$ indicates repulsion and a negative sign indicates attraction.

The force fields of multiple charges interact and the principle of superposition is applicable. The interaction of two unlike charges and of two like charges is illustrated in Fig. 20.2.

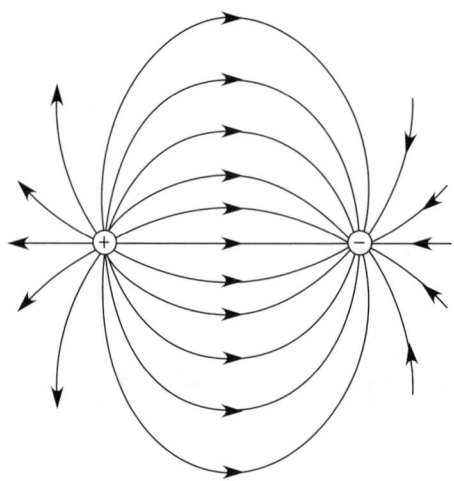

(a) unlike charges

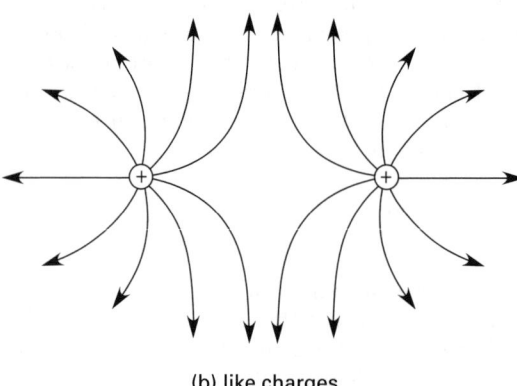

(b) like charges

Figure 20.2 *Electric Field Interaction*

Equation 20.1 was written in terms of the test charge. Normally an electric field is understood to be the field as it would exist without the test charge present. Further, subscripts are unnecessary in simple configurations. Therefore, in general, the force on a charged object (or distribution of charged objects) is

$$\mathbf{F} = Q\mathbf{E} \qquad 20.2$$

The lines of electric flux comprise a scalar field normally given the symbol Ψ. The flux is equal in magnitude to the charge enclosed. Thus, by definition, one coulomb of electric charge gives rise to one coulomb of electric flux. The electric flux density, $\mathbf{D}$, is a vector field with units of flux per unit area or C/m^2. This is not a measurable

field. It is used in electrical engineering to remove the effect of the medium, as can be seen by its defining equation.

$$\mathbf{D} = \epsilon\mathbf{E} \qquad 20.3$$

1. POINT CHARGE

The electric field of a point charge was given in Eq. 20.1. It is repeated here with the unit vector specifically showing the radial spreading of the electric field. The electric field of a point charge follows the *inverse-square law* and is normally written in spherical coordinates.

$$\mathbf{E} = \left(\frac{Q}{4\pi\epsilon r^2}\right)\mathbf{a}_r \qquad 20.4$$

The potential, V, associated with a point charge is given by Eq. 20.5. The potential is a *work field* and is always a scalar. It represents the energy required to transport a test charge from one point to another in an electric field. The unit of potential is the volt (V or v), which is equivalent to J/C .

$$V = \frac{Q}{4\pi\epsilon r} \qquad 20.5$$

Example 20.1

A positive charge is located at the origin of a Cartesian coordinate system. What is the electric field at point (2,2,3) (distance in units of meters) with no dielectric present?

Solution

The situation is as shown.

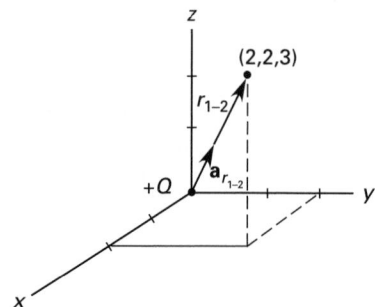

Using Eq. 20.1 and noting that point 1 of the equation is the origin and point 2 is the point (2,2,3) gives

$$\mathbf{E}_1 = \left(\frac{Q_1}{4\pi\epsilon r_{1-2}^2}\right)\mathbf{a}_{r_{1-2}}$$

Since no dielectric is present, $\epsilon = \epsilon_0 = 8.854 \times 10^{-12}$ $C^2/$ N·m². The distance r_{1-2} is

$$r_{1-2} = \sqrt{(2-0)^2 + (2-0)^2 + (3-0)^2} = 4.1$$

The vector $\mathbf{r}_{1-2} = 2\mathbf{a}_x + 2\mathbf{a}_y + 3\mathbf{a}_z$. Therefore the unit vector in the direction of $\mathbf{r}_{1-2}$ is given by

$$\mathbf{a}_{r_{1-2}} = \frac{\mathbf{r}_{1-2}}{r_{1-2}} = \frac{2\mathbf{a}_x + 2\mathbf{a}_y + 3\mathbf{a}_z}{4.1}$$

$$= 0.49\mathbf{a}_x + 0.49\mathbf{a}_y + 0.73\mathbf{a}_z$$

The magnitude of the electric field is

$$E_1 = \frac{Q_1}{4\pi\epsilon_0 r_{1-2}^2}$$

$$= \frac{1.602 \times 10^{-19} \text{ C}}{4\pi \left(8.854 \times 10^{-12} \dfrac{\text{C}^2}{\text{N·m}^2}\right)(4.1 \text{ m})^2}$$

$$= 8.6 \times 10^{-11} \text{ N/C}$$

Therefore, the electric field is

$$\mathbf{E}_1 = E_1\mathbf{a}_{r_{1-2}}$$

$$= \left(8.6 \times 10^{-11} \frac{\text{N}}{\text{C}}\right)(0.49\mathbf{a}_x + 0.49\mathbf{a}_y + 0.73\mathbf{a}_z)$$

2. DIPOLE CHARGE

A *dipole*, or *doublet*, consists of equal and opposite point charges separated by a small distance d. (Substances whose molecules are dipoles are referred to as *polar substances*. Water is the most significant of these substances. The two hydrogen ions are separated by 105° with the oxygen molecule at the vertex.) A dipole is illustrated in Fig. 20.3.

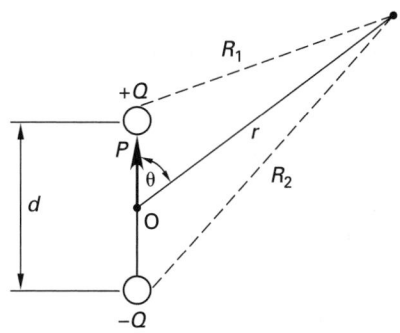

Figure 20.3 *Dipole*

The dipole moment is defined by Eq. 20.6.[4]

$$\mathbf{p} = q\mathbf{d} \qquad \textit{20.6}$$

---

[4]The charge, q, used in the dipole moment equation is equal to the unit positive charge, often called simply the positive charge. This is done because a unit positive charge is the arbitrary reference. In any work (energy) equation involving a dipole, a 2 multiplies the result to account for both charges.

The electric field of a dipole, given in spherical coordinates, is

$$\mathbf{E} = \left(\frac{Qd}{4\pi\epsilon r^3}\right)(2\cos\theta\mathbf{a}_r + \sin\theta\mathbf{a}_\theta) \qquad \textit{20.7}$$

The potential arising from two charges at a distance r from the dipole's center of symmetry is[5]

$$V = \frac{Q(R_2 - R_1)}{4\pi\epsilon R_1 R_2} \approx \frac{Qd\cos\theta}{4\pi\epsilon r^2} \qquad \textit{20.8}$$

A dipole in an electric field will experience a torque that tends to align the dipole moment, $\mathbf{p}$, with the field, $\mathbf{E}$. This is shown in Fig. 20.4. The torque is given by Eqs. 20.9 and 20.10.

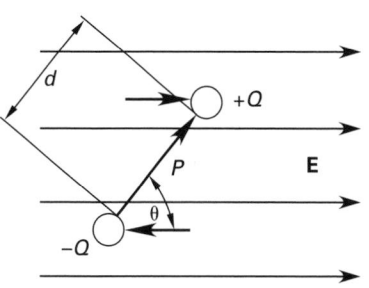

Figure 20.4 *Dipole in an Electric Field*

$$\mathbf{T} = \mathbf{p} \times \mathbf{E} \qquad \textit{20.9}$$

$$T = dF\sin\theta = dQE\sin\theta \qquad \textit{20.10}$$

The minimum and maximum energy configurations for a dipole in the presence of an electric field are shown in Fig. 20.5.

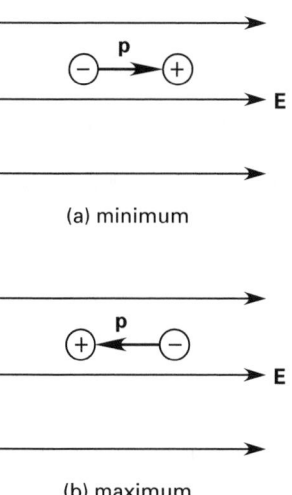

Figure 20.5 *Dipole Energy Configurations*

---

[5]When $r \gg d$, the distances R_1 and R_2 from the two dipole charges to the point (r, θ) are essentially the same.

3. COULOMB'S LAW

So far, the equations have described the effect of the two charges in the dipole on a third test charge within their electric field. The force between the two charges in the dipole, or between any two charges, is described by Coulomb's law and is defined by

$$\mathbf{F} = \left(\frac{Q_1 Q_2}{4 \pi \epsilon r^2} \right) \mathbf{a} \qquad 20.11$$

4. LINE CHARGE

Another useful standard charge configuration is that of an infinite, straight-line distribution of charges. This distribution is illustrated in Fig. 20.6.

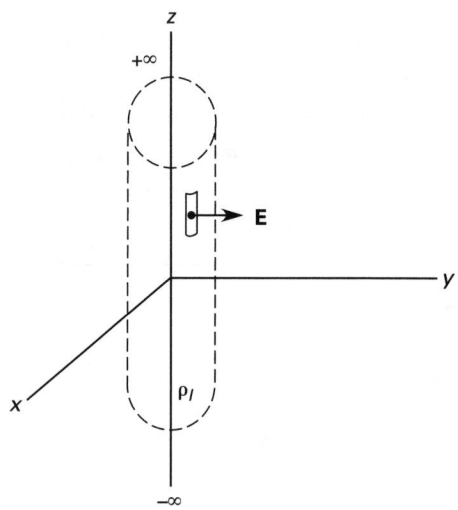

Figure 20.6 *Infinite Line Charge*

The electric field of such a distribution is given in cylindrical coordinates as

$$\mathbf{E} = \left(\frac{\rho_l}{2 \pi \epsilon r} \right) \mathbf{a}_r \qquad 20.12$$

If the line is not an infinite line charge but instead exists over some finite length, L, then a line integral must be utilized. The electric field for a finite line charge density is

$$\mathbf{E} = \int_L \left(\frac{\rho_l \mathbf{a}_r}{4 \pi \epsilon r^2} \right) dl \qquad 20.13$$

Example 20.2

An infinite uniform line charge in free space has a charge density, ρ_l, of 10 nC/m and lies along the x-axis of an arbitrary Cartesian coordinate system using meters as the unit of distance. What is the electric field strength, $\mathbf{E}$, at point (2,2,3)?

Solution

The situation is as shown.

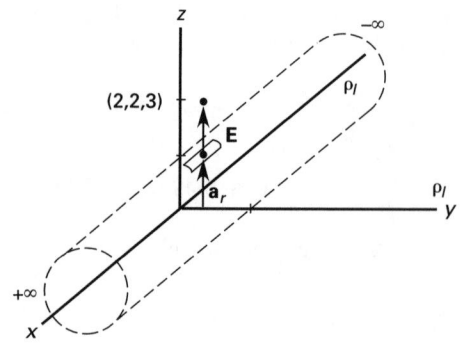

The field is constant with respect to x. Thus, charges in the y-z plane must be determined in cylindrical coordinates.

$$r = \sqrt{(2 \text{ m})^2 + (3 \text{ m})^2} = 3.6 \text{ m}$$

The electric field is given by

$$\mathbf{E} = \left(\frac{\rho_l}{2 \pi \epsilon_0 r} \right) \mathbf{a}_r$$

$$= \left(\frac{10 \times 10^{-9} \, \frac{\text{C}}{\text{m}}}{2 \pi \left(8.854 \times 10^{-12} \, \frac{\text{C}^2}{\text{N·m}^2} \right) (3.6 \text{ m})} \right) \mathbf{a}_r$$

$$= \left(49.9 \, \frac{\text{N}}{\text{C}} \right) \mathbf{a}_r$$

$$= (49.9 \text{ V/m}) \, \mathbf{a}_r$$

5. PLANE CHARGE

Consider electric charges distributed uniformly over an infinite plane with density ρ_s as shown in Fig. 20.7. Such a charge density produces an electric field given by Eq. 20.14. The term $\mathbf{a}_n$ represents the unit normal vector to the surface, S.

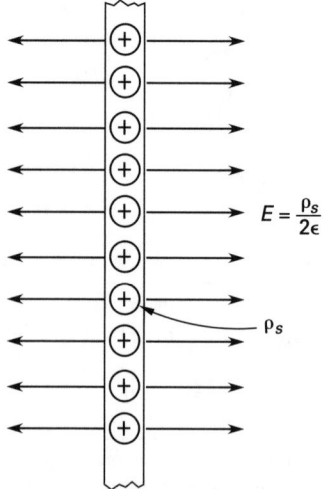

Figure 20.7 *Infinite Surface Charge*

$$\mathbf{E} = \left(\frac{\rho_s}{2\epsilon}\right)\mathbf{a}_n \qquad 20.14$$

This type of charge configuration is also known as a *surface charge* or *sheet charge*. For a finite surface charge that exists in more than one plane as shown in Fig. 20.8, the electric field is given by the surface integral shown in Eq. 20.15.

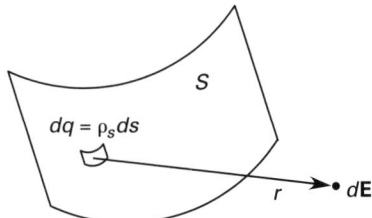

Figure 20.8 *Sheet Charge*

$$\mathbf{E} = \int_S \left(\frac{\rho_s}{4\pi\epsilon r^2}\right) ds \qquad 20.15$$

Part 2: Divergence

The main indicators of a changing vector field are the *divergence* and the *curl*. The curl is best studied in terms of magnetic fields. The divergence is a scalar that bears a similarity to the derivative of a function. In terms of any vector field, **A**, the *divergence theorem*, or *Gauss's divergence theorem*, is

$$\oint_S \mathbf{A}\cdot d\mathbf{s} = \int_V (\mathbf{\nabla}\cdot\mathbf{A})\, dv \qquad 20.16$$

If the vector field is the electric flux density, **D**, then the divergence theorem relates the flux emitted over a closed surface to the divergence of the flux density vector from the same volume, that is, the outward flow of the flux per unit volume. The term $\mathbf{\nabla}\cdot\mathbf{D}$ is defined as the divergence of **D**. The term $\mathbf{\nabla}$ is called the *del operator*.[6]

$$\text{div } \mathbf{D} = \mathbf{\nabla}\cdot\mathbf{D}$$
$$= \left(\frac{\delta}{\delta x}\right)\mathbf{x} + \left(\frac{\delta}{\delta y}\right)\mathbf{y} + \left(\frac{\delta}{\delta z}\right)\mathbf{z}\cdot(D_x + D_y + D_z)$$
$$= \frac{\delta D_x}{\delta x} + \frac{\delta D_y}{\delta y} + \frac{\delta D_z}{\delta z} \qquad 20.17$$

---

[6]The del operator is defined only in the Cartesian coordinate system. It is used in the cylindrical and spherical coordinate systems, but care must be taken to ensure correct results.

From Gauss's law for electrostatics, the divergence of the electric flux density is

$$\text{div } \mathbf{D} = \mathbf{\nabla}\cdot\mathbf{D} = \rho \qquad 20.18$$

Equation 20.18 is the differential form of Gauss's Law given in integral form in Sec. 16-12.

Example 20.3

In the space charge region on the *n*-side of a *pn* junction, the electric flux density is

$$\mathbf{D} = 8x^2\mathbf{x} + 2\mathbf{y} + 2\mathbf{z}$$

What is the charge density in this region?

Solution

The charge density can be determined from Eq. 20.18. Thus,

$$\rho = \mathbf{\nabla}\cdot\mathbf{D} = \left(\frac{\delta D_x}{\delta x}\right)\mathbf{x} + \left(\frac{\delta D_y}{\delta D_y}\right)\mathbf{y} + \left(\frac{\delta D_z}{\delta D_z}\right)\mathbf{z}$$
$$= \left(\frac{\delta(8x^2)}{\delta x}\right)\mathbf{x} + \left(\frac{\delta(2)}{\delta y}\right)\mathbf{y} + \left(\frac{\delta(2)}{\delta z}\right)\mathbf{z}$$
$$= 16x + 0 + 0$$
$$\rho = 16x \text{ C/m}^3$$

Part 3: Potential

The term "potential" was defined in Sec. 20-1 as the work that must be done against an electric field to bring a unit charge from an arbitrary reference point to the point in question. The reference point is normally infinity. For practical purposes, the surface of the earth or another large conductor is the reference point. The potential at a given point does not depend on the charge placed at the point, but rather on the charge or charge distribution that creates the electric field. The units of potential are volts, or J/C. One volt thus represents one joule of work expended moving one coulomb of charge from one point to another. The potential is given in terms of the electric field as

$$\mathbf{E} = -\mathbf{\nabla}V \qquad 20.19$$

Thus, the negative gradient of the potential is equal to the electric field. The minus sign is required, since positive charges moving in the direction of the electric field move from higher to lower potentials. That is, the potential decreases as the distance from the source of the electric field increases.

Since the potential is the work done per unit positive test charge in moving the charge from, say, point B to point A, the potential can be represented in integral form as follows.

$$V_{BA} = \frac{W}{\Delta Q} = -\int_{B}^{A} \mathbf{E} \cdot d\mathbf{l} \qquad 20.20$$

Note that the potential is at point A with respect to point B and is the work done in moving the positive test charge from B to A. The potential difference between two points is often represented as ΔV (not to be confused with the gradient of the potential, ∇V).

It is often useful to represent the potential in yet another way. Gauss's law was stated in Eq. 20.18 in terms of the electric flux density, $\mathbf{D}$. Restating this equation in terms of the electric field, $\mathbf{E}$, and assuming the permittivity is a constant, Gauss's law becomes[7]

$$\text{div } \mathbf{E} = \nabla \cdot \mathbf{E} = \frac{\rho}{\epsilon} \qquad 20.21$$

Substituting Eq. 20.19 into Eq. 20.21 gives the following useful result and introduces the Laplacian operator, ∇^2.

$$\nabla^2 v = -\frac{\rho}{\epsilon} \qquad 20.22$$

Equation 20.22 is called *Poisson's equation*. When there are no charges present, the right-hand side of Eq. 20.22 is zero and the equation is called *Laplace's equation*. The Laplacian operator for Cartesian coordinates is

$$\nabla^2 \equiv \frac{\delta^2}{\delta x^2} + \frac{\delta^2}{\delta y^2} + \frac{\delta^2}{\delta z^2} \qquad 20.23$$

Table 20.1 shows the relationship between the electric field, the potential, and the distance, r, from the charge or charge distribution.

Table 20.1 *Electric Field and Potential versus Distance*

charge distribution	electric field	potential
point charge	$\propto \dfrac{1}{r^2}$	$\propto \dfrac{1}{r}$
line charge	$\propto \dfrac{1}{r}$	$\propto \ln r$
surface charge	$\propto$ constant	$\propto r$

---

[7]The permittivity assumption is valid for linear, homogeneous, and isotropic mediums—a wide class of useful materials.

Example 20.4

Determine the equation for the potential of a point charge.

Solution

The potential is defined by Eq. 20.20 as

$$V_{BA} = -\int_{B}^{A} \mathbf{E} \cdot d\mathbf{l}$$

Since a point charge's electric field is radial at all points, the term $\mathbf{E} \cdot d\mathbf{l}$ can be replaced with $E_r dr$.

$$V_{BA} = -\int_{B}^{A} E_r dr$$

$$= -\int_{B}^{A} \left(\frac{Q}{4\pi\epsilon r^2} \right) dr = \frac{-Q}{4\pi\epsilon} \int_{B}^{A} \frac{1}{r^2} dr$$

$$= \left(\frac{Q}{4\pi\epsilon} \right) \left(\frac{1}{r_A} - \frac{1}{r_B} \right)$$

The normal reference point is infinity. Thus,

$$V_{A\text{-}\infty} = \left(\frac{Q}{4\pi\epsilon} \right) \left(\frac{1}{r_A} - \frac{1}{\infty} \right) = \frac{Q}{4\pi\epsilon r_A}$$

The subscripts are normally not shown, and free space is assumed unless otherwise indicated. Therefore,

$$V = \frac{Q}{4\pi\epsilon_0 r}$$

Note that Eq. 20.19 could have been used as the starting point as well.

Part 4: Work and Energy

Work is force acting over a distance. In an electric field, a positive charge will experience a force tending to move it in the direction of the electric field. Work, dW, is accomplished only by the force in the direction of motion. When a force is applied in a direction other than the direction of motion, the charge will change direction but no energy will be transferred to the charge. This is why magnetic fields change the direction of moving charges but transfer no energy.

Let the direction of motion be given by dl. Let θ be the angle between the direction of motion, dl (the actual path of the charge), and the direction of the electric field, **E**. The situation for an arbitrary path is shown in Fig. 20.9.[8]

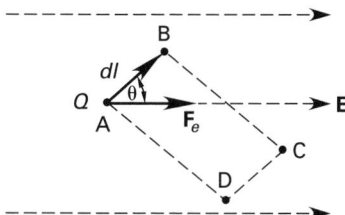

Figure 20.9 *Positive Charge in an Electric Field (arbitrary path)*

The differential work done by the electric field is

$$dw = -Q\mathbf{E}{\cdot}dl = -QE\cos\theta dl \qquad 20.24$$

The total work done in moving a charge from point A to point B in an electric field is

$$W_{AB} = -\int_A^B Q\mathbf{E} \cdot dl$$

$$= -\int_A^B QE\cos\theta dl \qquad 20.25$$

The work is positive if an external force is required to move the charge (as in bringing two repulsive charges closer together or moving a charge against the force of the electric field). Work is negative if the electric field does the work, for example, allowing attracting charges to come together or allowing a positive charge to move in the direction of the electric field. Thus, positive work indicates that the system requires an energy input and negative work indicates work output. No work is done in moving a charge perpendicular to a field, that is, across the lines of electric flux or anywhere along the lines of an equipotential surface.

When Eq. 20.25 is calculated for all the charges in a given charge configuration creating an electrostatic field, the result is the total energy, U, stored in the field. The energy is "stored" in the volume of the electrostatic field and is given by the following expressions.

$$U = \frac{1}{2}\int \rho V dv = \frac{1}{2}\int \mathbf{D}{\cdot}\mathbf{E} dv$$

$$= \frac{1}{2}\int \epsilon E^2 dv = \frac{1}{2}\int \frac{D^2}{\epsilon} dv \qquad 20.26$$

An electrostatic field is a *conservative field*. No work is done in moving a charge around a closed path within an electrostatic field, such as path ABCD in Fig. 20.9.

---

[8]External forces not shown would be required to make the positive charge follow the indicated path.

Further, the work done in moving a charge between any two points within the field is independent of the path. This means the integral of the electric field around a closed path equals zero as shown in Eq. 20.26.

$$W = -\oint Q\mathbf{E}{\cdot}dl = 0 \qquad 20.27$$

The result given by Eq. 20.26 is true for electrostatic fields and must be modified for time varying fields.

Example 20.5

Neglecting fringing, what is the total energy stored in a parallel plate capacitor with a capacitance of 50 μF across 120 V?

Solution

Neglecting fringing, the capacitor's electric field is identical to one for infinite parallel plates and is uniform. That is, from Table 16.6,

$$E = \frac{V}{r}$$

Substituting and noting that the field is constant and thus can be removed from inside the integral gives

$$U = \frac{1}{2}\int \epsilon E^2 dv$$

$$U = \left(\frac{1}{2}\right)\epsilon E^2 \int dv = \left(\frac{1}{2}\right)\epsilon\left(\frac{V}{r}\right)^2 \int dv$$

$$= \left(\frac{1}{2}\right)\frac{\epsilon V^2}{r^2}\int dv$$

The capacitance is given by

$$C = \frac{\epsilon A}{r}$$

Substituting,

$$U = \left(\frac{1}{2}\right)\frac{\epsilon V^2}{r^2}\int dv = \left(\frac{1}{2}\right)\frac{CV^2}{rA}\int dv$$

The volume equals the area of the plates, A, multiplied by the distance between them, r. Thus,

$$U = \left(\frac{1}{2}\right)\left(\frac{CV^2}{rA}\right)(Ar) = \left(\frac{1}{2}\right)CV^2$$

Substituting the given information gives

$$U = \left(\frac{1}{2}\right)\left(50\times10^{-6}\text{ F}\right)\left(120\text{ V}\right)^2$$

$$= 0.36\text{ J}$$

21 Electrostatic Fields

Nomenclature

A	area	m^2
d	distance	m
D	electric flux density	C/m^2
E	electric field strength	N/C (V/m)
f	frequency	Hz
J	current density	A/m^2
p	dipole moment	C·m
P	polarization	C/m^2
q, Q	charge[1]	C
v, V	volume	m^3
W	work	J

Symbols

ε	permittivity	$C^2/N \cdot m^2$ or F/m
ε_0	free-space permittivity	$C^2/N \cdot m^2$
ρ_{sp}	polarization surface charge density	C/m^2
σ	conductivity	S/m
χ	susceptibility	–
ω	angular frequency	rad/s

Subscripts

0	free space
c	conduction
d	displacement
e	electric
i	induced or internal
r	relative
sp	surface polarization

1. POLARIZATION

When an electric field is present in a dielectric, the bound electric charges tend to align themselves into the lowest energy configuration possible. In doing so, they increase the flux density, **D**. The alignment is referred to as the *polarization* of the dielectric and is given the symbol **P**. The polarization is the total dipole moment, **p**, per unit volume, v.

$$\mathbf{P} = \frac{\mathbf{p}}{v} \qquad 21.1$$

---

[1] In some formulas the small q represents the absolute value of a single charge, that is, 1.602×10^{-19} C.

The flux density is

$$\mathbf{D} = \varepsilon_0 \mathbf{E} + \mathbf{P} \qquad 21.2$$

Polarization thus increases the flux density above the level that would exist under free-space conditions with the same electric field intensity (the polarization of free space equals zero). This occurs even when permanent dipole molecules are not initially present in the dielectric. This can be understood by considering that when an atom is placed within the influence of an electric field, the positive protons tend to move toward the direction of the electric field and the electrons tend to move toward the direction opposite the electric field. In this manner, a dipole is set up within the atom. See Fig. 21.1 for an illustration of this concept. The charge in such induced dipoles is usually not a complete unit charge (1.602×10^{-19} C) but is instead some fraction thereof, because the charges are not completely separate. The dipole is analogous to spring energy in mechanical systems. The work done in the distortion of the atom is recoverable upon removal of the electric field.

(a) neutral atom or molecule

(b) application of an external
electric field

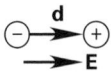

(c) equivalent dipole
representation

Figure 21.1 *Induced Dipole Moment*

The electric field and the polarization can have different directions, though this is often not the case. In isotropic, linear material **E** and **P** are parallel at each point and thus

$$\mathbf{P} = \chi_e \varepsilon_0 \mathbf{E} \qquad 21.3$$

The term *isotropic* indicates that the medium's properties are independent of direction. The term *linear* indicates that the permittivity, ε, is constant regardless of the magnitude of the charge present. The term χ_e is the electric susceptibility. This dimensionless term measures the ease of polarization in a material. Thus, for an isotropic linear material, Eq. 21.2 can be rewritten as[2]

$$\mathbf{D} = \varepsilon_0 \left(1 + \chi_e\right) \mathbf{E} \qquad 21.4$$

The relative permittivity can thus be represented by Eq. 21.5.

$$\varepsilon_r = 1 + \chi_e \qquad 21.5$$

In free space, then, since no dielectrics are present, no polarization occurs. The electric flux density, $\mathbf{D}$, is

$$\mathbf{D} = \varepsilon_0 \mathbf{E} \qquad 21.6$$

If a dielectric is present, the flux density is increased and is given by Eq. 21.2. Importantly, Eq. 21.2 is often written without direct reference to polarization, but rather in terms of the relative permittivity. The relationship is shown explicitly in Eq. 21.7.

$$\mathbf{D} = \varepsilon \mathbf{E} = \varepsilon_0 \varepsilon_r \mathbf{E} = \varepsilon_0 \mathbf{E} + \mathbf{P} \qquad 21.7$$

Example 21.1

Water in vapor form has a fractional charge of approximately 5×10^{-20} C and a dipole distance of approximately 10^{-10} m. What is the maximum possible polarization for one mole of water molecules at standard temperature and pressure (STP)?

Solution

The maximum polarization is

$$P = \frac{p}{v}$$

Determine the dipole moment.

$$p = qd$$
$$= \left(5 \times 10^{-20} \text{ C}\right) \left(10^{-10} \text{ m}\right) = 5 \times 10^{-30} \text{ C·m}$$

At STP, the volume of 1 mol of an ideal gas is 22.4 L or 0.0224 m³. The maximum polarization is

$$P = \frac{p}{v}$$
$$= \frac{5 \times 10^{-30} \text{ C·m}}{0.0224 \text{ m}^3}$$
$$= 2.2 \times 10^{-28} \text{ C/m}^2$$

---

[2]If the material is anisotropic or nonlinear, or both, χ_e is a tensor.

Example 21.2

What are the magnitudes of $\mathbf{P}$ and $\mathbf{D}$ for a dielectric in an electric field with a field strength of $\mathbf{E} = 38.8 \times 10^3$ V/m with a relative permittivity of 4.1?

Solution

First,

$$P = \chi_e \varepsilon_0 E$$

The susceptibility is found from

$$\varepsilon_r = 1 + \chi_e$$
$$\chi_e = \varepsilon_r - 1 = 4.1 - 1 = 3.1$$

Thus,

$$P = \chi_e \varepsilon_0 E$$
$$= (3.1) \left(8.854 \times 10^{-12} \frac{\text{C}^2}{\text{N·m}^2}\right) \left(38.8 \times 10^3 \frac{\text{V}}{\text{m}}\right)$$
$$= 1.06 \times 10^{-6} \text{ C/m}^2$$

Second,

$$D = \varepsilon_0 \varepsilon_r E$$
$$= \left(8.854 \times 10^{-12} \frac{\text{C}^2}{\text{N·m}^2}\right) (4.1) \left(38.8 \times 10^3 \frac{\text{V}}{\text{m}}\right)$$
$$= 1.41 \times 10^{-6} \text{ C/m}^2$$

Example 21.3

A given electric field strength is measured as 0.20 MV/m. If design requirements are such that the electric flux density must be 8 μC/m², what is the minimum susceptibility required of the dielectric?

Solution

The flux density is given. Thus,

$$D = \varepsilon_0 \left(1 + \chi_e\right) E$$
$$\chi_e = \frac{D}{\varepsilon_0 E} - 1$$
$$= \frac{8 \times 10^{-6} \dfrac{\text{C}}{\text{m}^2}}{\left(8.854 \times 10^{-12} \dfrac{\text{C}^2}{\text{N·m}^2}\right) \left(0.20 \times 10^6 \dfrac{\text{V}}{\text{m}}\right)} - 1$$
$$= 3.5$$

2. POLARIZED FIELDS

If a dipole is present, it can be aligned either partially or fully with the electric field. An example of such an alignment is illustrated in Fig. 21.2.

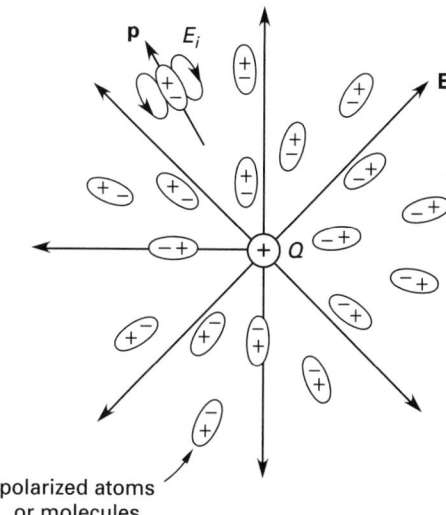

Figure 21.2 *Dipole Alignment About a Free Positive Charge*

If a molecule contains a permanent dipole moment, it is called a *polar molecule*. Two common examples are water and salt (NaCl). If the molecule has no dipole moment in the absence of an electric field, it is called a *nonpolar molecule*. In this case, the dipole moment can be induced as explained earlier in this chapter and shown in Fig. 21.1. Yet a third type of dielectric exists that exhibits permanent polarization in the absence of an electric field but does so only after an induced polarization has taken place. Some waxes, for instance, are polarized when molten and exposed to an electric field. After cooling some of the induced polarization remains. This special case of aligned polar molecules occurs below a temperature called the *Curie point*. Such materials are analogous to permanent magnets and are called ferroelectrics. Ferroelectrics exhibit the properties of electrical hysteresis.

A dielectric with its dipole moments aligned produces an internal electric field, E_i, that opposes the original electric field, E_0, that created it. The energy expended in the alignment results in a dielectric, or dipole, electric field, E_d. This situation is illustrated in Fig. 21.3(a). The net result is that a surface charge appears at either end of the dielectric as shown in Fig. 21.3(b). Thus, the electric flux density is increased above the level that would exist in free space by an amount equal to the polarization as indicated by Eq. 21.2. This can be inferred from the units used to measure electric flux density, C/m^2. The charges per unit area have been increased above the level that exists in free space. An important point to remember is that the net charge in the dielectric remains zero. Also, the polarization surface charge is a real accumulation of charge. It does contribute to the internal and external electric fields. It does not,

however, enter into boundary conditions when solving electrical engineering problems. Furthermore, the divergence of **D** depends only on the free charge density.

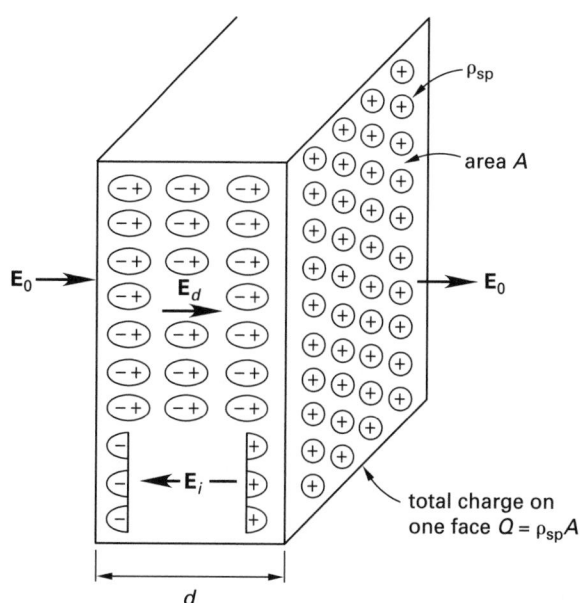

(a) dielectric electric field, $\mathbf{E}_d$

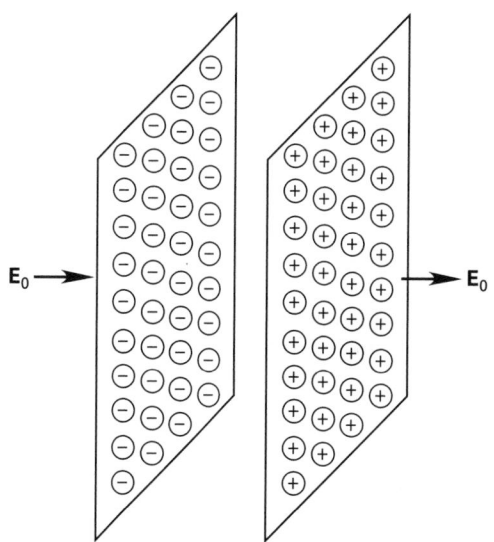

(b) dielectric surface charge

Figure 21.3 *Dielectric Electric Fields and Surface Charge*

3. ELECTRIC DISPLACEMENT

The electric flux density, **D**, is also called the displacement density, for reasons now clear. It represents the displacement of charges within a dielectric. The motion of the charges within the dielectric is not translational, as it is for conduction current. Instead, the dipole moments rotate under the influence of the electric field (see Fig. 20.4). If the angle $\pi/2$ is chosen as the arbitrary zero energy reference, the work of rotating a dipole is

$$W = -\mathbf{p}\cdot\mathbf{E} \qquad\qquad 21.8$$

This motion of charges is the *displacement current*. In terms of the current density vector, $\mathbf{J}$, the displacement current is defined by Eq. 21.9.

$$\mathbf{J}_D = \frac{\delta\mathbf{D}}{\delta t} \qquad\qquad 21.9$$

Equation 21.9 indicates that a time-varying electric field, and subsequently a time-varying electric flux density, results in current flow in a dielectric material that contains no free charges. In this manner, current is made to "flow" from one side of a capacitor to another. For a poor conductor, or *lossy dielectric*, the ratio of the conduction current density to the displacement current is

$$\frac{J_c}{J_d} = \frac{\sigma}{\omega\varepsilon} \qquad\qquad 21.10$$

Displacement and displacement current define the electrical phenomena in a dielectric. While polarization is necessary for an understanding of the situation, it complicates calculations. Thus, in practice with simple dielectrics, work is done directly with $\mathbf{D}$ and not $\mathbf{P}$ by using the relative permittivity as shown in Eq. 21.7. The relative permittivity is easily determined (for isotropic, linear dielectrics) and thus can be used to account for all polarization effects.

Example 21.4

What is the total energy required to "flip" a dipole of unit positive charge a dipole distance of 10^{-10} m in an electric field of 0.20 MV/m?

Solution

The work done in rotating a dipole is

$$W = -\mathbf{p}\cdot\mathbf{E}$$

$$= -pE\cos\theta$$

The zero reference point is $\theta = \pi/2$. Thus, the total energy required to rotate the dipole 90° is

$$|W_{90°}| = pE$$

To flip the dipole (i.e., to rotate 180°) requires double that amount of energy.

$$|W_{180°}| = 2pE$$

Substituting,

$$W_{\text{total}} = 2pE = 2qdE$$

$$= (2)\left(1.602\times10^{-19}\text{ C}\right)\left(10^{-10}\text{ m}\right)$$

$$\times\left(0.20\times10^6\,\frac{\text{V}}{\text{m}}\right)$$

$$= 6.4\times10^{-24}\text{ J}$$

Example 21.5

What is the ratio of conduction current to displacement current in a conductor operating at 60 Hz with a conductance of 30 MS/m and a ε_r of 1?

Solution

The applicable formula is

$$\frac{J_c}{J_d} = \frac{\sigma}{\omega\varepsilon}$$

Expanding and substituting,

$$\frac{J_c}{J_d} = \frac{\sigma}{\omega\varepsilon} = \frac{\sigma}{2\pi f\varepsilon_0\varepsilon_r}$$

$$= \frac{30\times10^6\,\dfrac{\text{S}}{\text{m}}}{2\pi(60\text{ Hz})\left(8.854\times10^{-12}\,\dfrac{\text{C}^2}{\text{N}\cdot\text{m}^2}\right)(1)}$$

$$= 9.0\times10^{15}$$

The displacement current is thus a minor effect in conductors. Only at high frequencies need it be considered.

4. POLARIZATION IN DIELECTRICS VERSUS CONDUCTORS

In a metal conductor, free conduction charges are mobile and move to the surface of the metal under the influence of an electrostatic field. Because the charges are plentiful and mobile, the surface charge builds up until the induced electric field cancels the applied electric field. Consequently, the electric field on the interior of a metal is zero. That is, the polarization of a conductor by the external electric field is complete.

For dielectrics, on the other hand, the charges are bound and can only move a small fraction of an atomic diameter. The induced electric field, E_i, partially cancels the applied electric field. The result is an internal electric field in the dielectric that points in the same direction as the applied field (for isotropic, linear dielectrics). Furthermore, the induced charges on the surface of the dielectric are not equal in magnitude to the induced charges on metallic surfaces. (This could occur only if ε_r were infinite.) In short, a dielectric in an external electric field becomes polarized, but not to such an extent that the interior electric field goes to zero.

22 Magnetostatics

Nomenclature

$\mathbf{A}$	vector magnetic potential	Wb/m (T·m)
A	area	m^2
B	magnetic flux density	T (Wb/m^2)
c	speed of light (vacuum)	3.00×10^8 m/s
d	distance	m
D	electric flux density	C/m^2
e^-	electron	C
E	electric field strength	V/m
F	force	N
F_V	force per unit volume	N/m^3
h	Planck's constant	J·s
H	magnetic field strength	A/m
i, I	current	A
J	current density	A/m^2
K	sheet current	A
l	length	m
m	magnetic moment	A·m^2 (J/T)
m	mass	kg
M	magnetization	A/m
N	number of turns	–
p	electric dipole moment	C·m
p	pole strength	N/(A·m) or J/A or Wb
p_m	magnetic dipole moment	A·m^2 (J/T)
q, Q	charge	C
q_m	magnetic charge	A·m or N/T
r	radius	m
T	torque	N·m
U	magnetic potential	A
v	velocity[1]	m/s
v, V	volume	m^3
V	effective or DC voltage	V
V_m	magnetomotance	A
W	work	J

Symbols

μ	permeability	N/A^2 or H/m
μ_0	free-space permeability	N/A^2 or H/m
ρ	charge density	C/m^3
Φ, ϕ	magnetic flux	Wb or N/(A·m) or J/A
ω	angular frequency	rad/s

Subscripts

0	initial
1–2	from 1 to 2; or effect on 2 due to the field of 1
f	final
e	electric or electron
m	magnetic
p	pole
r	relative
V	volume

Part 1: The Magnetic Field

A *field* is an entity that acts as an intermediary in interactions between moving charges. It is distributed over part or all of space and is a function of space coordinates and time (if the field is not static). A *magnetic field* is elementary in nature, causing charges moving through the field to experience a force. The strength of the **B**-field is the electromotive force (V or J/C) produced when one turn of a loop of area A is linked in 1 sec.[2] It is measured in Wb/m^2, equivalent to V· s/m^2. The strength of the **B**-field is

$$\mathbf{B} = \left(\frac{\mu I l}{4\pi r^2} \right) \mathbf{a} \qquad 22.1$$

The unit vector, **a**, points in the direction of the **B**-field, along the lines of flux. This is perpendicular to

---

[1]The term "velocity" is often used in text where it would be more technically correct to use the term "speed." Velocity is a vector quantity with speed as its magnitude.
[2]The term *linked* indicates that the magnetic flux is passed through the area of concern—in this case, the area of the loop.

the direction of the force experienced by the positive moving charges, that is, the current. This occurs because a magnetic force always acts in a direction transverse to the movement of charges. The denominator, $4\pi r^2$, represents the surface area of a sphere centered on the current element dl whose surface just touches the point in space where the **B**-field is being evaluated. Specific configurations may be evaluated with Eq. 22.1 by changing **B** to $d\mathbf{B}$ and l to dl and integrating. The magnetic field is a vector force field in that a moving charge within the field experiences a force in a specific direction. The force field for a single line of moving charges is illustrated in Fig. 22.1

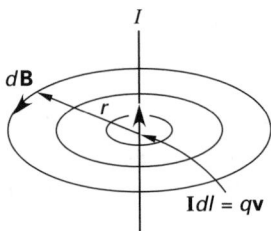

Figure 22.1 *Magnetic Field of a Positive Current*

Lines of flux form continuous loops. By convention, these lines are said to emanate from the north pole of a magnetic dipole and enter the south pole. Like poles repel, and unlike poles attract.

The magnetic force fields of multiple currents interact, and the principle of superposition applies as long as the permeability of the medium is constant. Unfortunately, constant permeability is not exhibited by iron and other ferromagnetic materials, which are widely used around magnetic fields. Therefore, in general, the principle of linear superposition does not apply to current-carrying conductors in the vicinity of ferromagnetic materials.

Equation 22.1 is normally written in terms of the magnetic flux density as shown in Eq. 22.2. Both equations use identical units, Wb/m^2, which in the SI system is a tesla, T. The term **B** is most often called the magnetic flux density.[3]

$$B = \frac{\Phi}{A} \qquad 22.2$$

The force on a current element in a magnetic field is given by the Biot-Savart law in Eq. 16.43. This law can also be viewed in terms of the magnetic field produced by the current elements as explained in Sec. 22-2.

Magnetic flux, Φ, is defined as the surface integral of the normal component of the flux density, **B**, over the area.

---

[3]The magnetic flux density, **B**, is the fundamental field in magnetics—it enters into the force equations. Though the name is reminiscent of the electric flux density, **D**, this is a result of the historical development in the study of magnetics. **B** and **E** are the fundamental fields. **H** and **D** are constructs to aid in the solution of electrical engineering problems.

$$\Phi = \int_A \mathbf{B} \cdot d\mathbf{A} \qquad 22.3$$

It is often more helpful to visualize the magnetic flux as the total quantity of lines of magnetic force set up around a current-carrying conductor. If there are N conductors of length l, the flux is

$$\Phi = \mu N I l = BA \qquad 22.4$$

Example 22.1

A wire carrying a current of 6 A is shown. The current is in the plane of the paper. Determine the magnetic flux density 0.5 m away from a 1 cm segment of the wire at the point indicated.

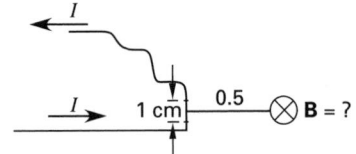

Solution

The magnetic flux density due to the small segment, dl, 1 cm long is

$$\mathbf{B} = \left(\frac{\mu I l}{4\pi r^2}\right)\mathbf{a}$$

Since the segment is small compared to the distance, r, integration is avoided. (When to integrate, that is, when to use $d\mathbf{B}$ and dl, depends on the amount of error tolerable in the solution.) Substituting and assuming free-space conditions gives

$$\mathbf{B} = \left(\frac{\mu I l}{4\pi r^2}\right)\mathbf{a}$$

$$= \frac{\left(1.2566 \times 10^{-6}\ \frac{N}{A^2}\right)(6\ A)(1\ cm)\left(\frac{0.01\ m}{1\ cm}\right)\mathbf{a}}{4\pi(0.5\ m)^2}$$

$$= \left(2.4 \times 10^{-8}\ T\right)\mathbf{a}$$

Using the right-hand rule, the direction of the **B**-field is into the paper as shown.

Example 22.2

A loop of wire in an electric motor approximates a square with sides of length 0.15 m. What is the total flux of the loop if the motor is rated to carry 15 A?

Solution

The applicable equation is

$$\Phi = \mu N I l = BA$$

There are four conductors, each of length 0.15 m. Since no other condition was specified, free space is assumed. Substituting known and given values results in the following.

$$\Phi = \mu NIl = \left(1.2566 \times 10^{-6} \; \frac{\mathrm{H}}{\mathrm{m}}\right)(4)(15 \; \mathrm{A})(0.15 \; \mathrm{m})$$

$$= 1.1 \times 10^{-5} \; \mathrm{Wb}$$

1. BIOT-SAVART LAW

The Biot-Savart law is normally given as relating the force between two current elements as in Sec. 16-5. This is done to emphasize the similarity between electrostatics and magnetostatics and to exhibit the relationship between the Biot-Savart law and Coulomb's law in electromagnetics. When considering $\mathbf{B}$ as the flux density, the Biot-Savart law is

$$d\mathbf{B} = \left(\frac{\mu_0}{4\pi}\right)\left(\frac{\mathbf{I} \times \mathbf{r}}{r^2}\right) dl \qquad 22.5$$

The term $\mathbf{r}$ is the unit vector pointing from the current element $\mathbf{I}dl$ to $d\mathbf{B}$.[4] The magnitude of Eq. 22.5 is given in Eq. 22.6.

$$dB = \left(\frac{\mu_0}{4\pi}\right)\left(\frac{Idl}{r^2}\right)\sin\theta \qquad 22.6$$

Comparing Eq. 22.6 with the equation for an electric field, Eq. 16.6, the magnetic charge (if one existed) would have the form $Q_m = Idl$. Since the current element $\mathbf{I}dl$ is equivalent to the moving point charge $q\mathbf{v}$, the Biot-Savart law can be written as

$$\mathbf{B} = \left(\frac{\mu_0}{4\pi}\right)\left(\frac{q\mathbf{v} \times \mathbf{r}}{r^2}\right) \qquad 22.7$$

The cross-product in Eqs. 22.5 and 22.7 is accounted for when determining the direction of the magnetic field by using the *right-hand rule*. The right-hand rule states that when the thumb is pointed in the direction of the positive current flow, the fingers curled inward toward the palm give the direction of the magnetic field. The rule can be verified by using the directions shown in Fig. 22.1.

2. FORCE ON A MOVING CHARGED PARTICLE

A magnetic field has no effect on stationary charged particles. If, however, the charge Q is moving with velocity $\mathbf{v}$ in an external magnetic field with flux density $\mathbf{B}$, the force on the charge is

$$\mathbf{F} = Q\mathbf{v} \times \mathbf{B} \qquad 22.8$$

---

[4]Importantly, the direction of $\mathbf{I}dl$ can be ascribed to the current and written as shown, or be ascribed to the differential length and be written as $I d\mathbf{l}$ or $\mathbf{I}dl$.

The magnetic field referred to in Eq. 22.8 is an external field, not the field created by the moving charged particle. The acceleration of the particle is given by Newton's second law and, due to the cross-product, is at an angle perpendicular to both the velocity v and the external magnetic field $\mathbf{B}$. A charged particle in a uniform magnetic field will travel in a circular path with radius and angular velocity of

$$r = \frac{m\mathrm{v}}{QB} \qquad 22.9$$

$$\omega = \frac{QB}{m} \qquad 22.10$$

Equation 22.8 is sometimes used to define the magnetic field. In doing so, the magnitude of Eq. 22.8 is written as

$$B = \frac{F}{q\mathrm{v}} \qquad 22.11$$

Comparing Eq. 22.11 to Eq. 16.4 shows that just as the electric field can be defined in terms of the force per unit charge that is exerted at a point, the magnetic field can be defined as the force per unit moving charge that is exerted at a point.

Example 22.3

What is the maximum magnitude of the force on an electron moving at $0.1c$ in a uniform magnetic field of 5.0×10^{-5} T?

Solution

One-tenth the speed of light is the approximate maximum speed at which relativistic effects can be ignored. The electron is moving at $0.1c = 0.3 \times 10^8$ m/s. (The magnetic field strength is that of the Earth's.) The applicable equation is

$$\mathbf{F} = Q\mathbf{v} \times \mathbf{B}$$

Since the maximum magnitude is to be determined, the following adjustments can be made.

$$F = |Q|\,\mathrm{v}B\sin\theta = |Q|\,\mathrm{v}B(1) = |Q|\,\mathrm{v}B$$

Substitute the given values.

$$F = |Q|\,\mathrm{v}B$$

$$= \left(1.6022 \times 10^{-19} \; \mathrm{C}\right)\left(0.3 \times 10^8 \; \frac{\mathrm{m}}{\mathrm{s}}\right)\left(5 \times 10^{-5} \; \mathrm{T}\right)$$

$$= 2.4 \times 10^{-16} \; \mathrm{N}$$

3. FORCE ON CURRENT ELEMENTS

A current element $\mathbf{I}dl$ within a magnetic field $\mathbf{B}$ experiences a force of

$$d\mathbf{F} = \mathbf{I}dl \times \mathbf{B} \qquad 22.12$$

Since the term $\mathbf{I}dl$ is a mathematical construct and cannot exist alone but must instead be part of a complete loop or circuit, the force on the entire loop can be obtained from integrating around the loop as shown in Eq. 22.13.

$$\mathbf{F} = \oint \mathbf{I}dl \times \mathbf{B} \qquad 22.13$$

For a straight conductor of length l, the total force on a current-carrying conductor is

$$\mathbf{F} = l\mathbf{I} \times \mathbf{B} \qquad 22.14$$

In terms of magnitude and assuming a uniform magnetic field, Eq. 22.14 can be written as

$$F = lIB \sin \theta \qquad 22.15$$

The direction of the magnetic force can be ascertained by the left-hand *FBI rule* as shown in Fig. 22.2.

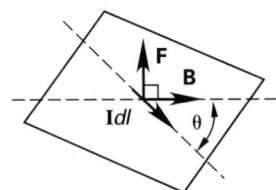

(a) magnetic force direction in terms of $\mathbf{I}dl$

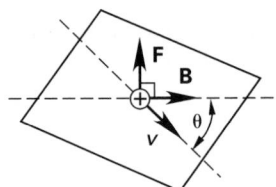

(b) magnetic force direction in terms of individual charge velocities

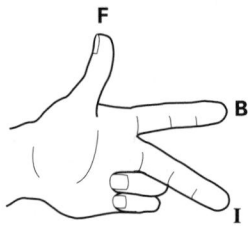

(c) FBI rule

Figure 22.2 *Magnetic Force Direction*

For a straight, infinitely long conductor, the magnitude of the magnetic field is

$$B = \frac{\mu I}{2\pi r} \qquad 22.16$$

Two straight, infinitely long conductors situated within the magnetic fields of one another experience a force of

$$\frac{\mathbf{F}}{l} = \mathbf{I} \times \mathbf{B} = \left(\frac{\mu I_1 I_2}{2\pi r}\right)\mathbf{a} \qquad 22.17$$

The relationship given by Eq. 22.17 is useful in defining current in terms of force and distance, enabling the measurement of electrical current in terms of familiar mechanical variables. The distance, r, is perpendicular to the wires.

Example 22.4

The force on two parallel wires, approximated as infinite, is measured to be 30 mN/m. The wires carry equal currents and repel one another. What are the magnitude and direction of the current in the two wires if the wires are 1 cm apart?

Solution

The force per unit length is given. Assuming free space, the applicable equation is

$$\frac{\mathbf{F}}{l} = \mathbf{I} \times \mathbf{B} = \left(\frac{\mu I_1 I_2}{2\pi r}\right)\mathbf{a}$$

Determine the magnitude of the current by rearranging the equation to

$$I^2 = \frac{2\pi \left(\frac{F}{l}\right) r}{\mu_0} = \frac{2\pi \left(30 \times 10^{-3}\ \frac{\text{N}}{\text{m}}\right)(0.01\ \text{m})}{1.2566 \times 10^{-6}\ \frac{\text{N}}{\text{A}^2}}$$

$$= 1.5 \times 10^3\ \text{A}^2$$

$$I = \sqrt{1.5 \times 10^3\ \text{A}^2} = 38.7\ \text{A}$$

The direction can be determined a number of ways, such as by using the right-hand rule or by determining if the magnetic fields aid or oppose each other (that is, if they "link"). Using the equation, and knowing that the force must be in a direction that causes repulsion, the currents must be as shown.

$$\mathbf{F_1} = l\mathbf{I_1} \times \mathbf{B_2} \qquad \qquad \mathbf{F_2} = l\mathbf{I_2} \times \mathbf{B_1}$$

The magnitude of the current is 38.7 A flowing in opposite directions in the two wires.

4. FORCE ON DISTRIBUTED CURRENT ELEMENTS

Equations 22.8 and 22.12 can be generalized to the case of force per unit volume.[5] In doing so, the force can be written as

$$\mathbf{F}_V = \mathbf{J} \times \mathbf{B} \qquad 22.18$$

---

[5]This is useful for understanding, among other things, plasmas and convection or diffusion currents in semiconductors.

Part 2:
Lorentz Force Law

When an electric field and a magnetic field are present at the same time, the force on a charge is given by the general force law, commonly called the *Lorentz force law*. The Lorentz force combines the electrical and magnetic force and is given in Eq. 22.19.

$$\mathbf{F} = \mathbf{F}_e + \mathbf{F}_m = Q\left(\mathbf{E} + \mathbf{v} \times \mathbf{B}\right) \qquad 22.19$$

Similarly, if a charge cloud of density ρ at each point exists in a volume, the Lorentz force is

$$d\mathbf{F} = \rho\left(\mathbf{E} + \mathbf{v} \times \mathbf{B}\right)dV \qquad 22.20$$

Part 3:
Traditional Magnetism

Traditional magnetism was based on concepts similar to those in electrostatics. As a result, the concept of a magnetic charge, q_m, was used and magnets were described in terms of pole strength, p. The isolated magnetic charge is called a monopole, which does not exist in nature. Nevertheless, using the concept of an isolated magnetic charge at the end of a bar magnet is useful in understanding the effects of magnetism. The concept is illustrated for a bar magnet (magnetic dipole) in an external magnetic field in Fig. 22.3. The force on a magnetic charge is defined as the charge multiplied by the strength of the **B**-field as shown in Eq. 22.21. A magnetic field may therefore be defined in the same manner as an electric field, that is, the field represents the force exerted on an infinitesimal magnetic test charge at a point in space ($B = F/q_m$).

$$\mathbf{F} = q_m\mathbf{B} \qquad 22.21$$

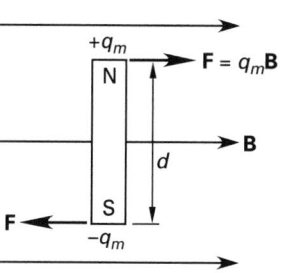

Figure 22.3 *Magnetic Charge in a Magnetic Field*
(Dipole in a Magnetic Field)

The torque exerted on the magnetic charge in an external magnetic field is

$$\mathbf{T} = q_m\mathbf{d} \times \mathbf{B} \qquad 22.22$$

The magnetic flux density, **B**, is also called the *intensity of magnetization* and the *dipole moment per unit volume*. For a bar magnet of pole strength p, length l, and cross-sectional area A, the magnetic flux density is

$$B = \frac{pl}{lA} = \frac{p}{A} = \frac{m}{V} \qquad 22.23$$

The cgs system of units was the first to be used in magnetism. This resulted in equations whose forms differ significantly from those in the SI system. Care should be exercised when using equations in systems other than the SI system to ensure consistency. A comparison of the cgs and SI system units for magnetism is given in Table 22.1.

5. MAGNETIC POLE

The *pole strength*, not to be confused with the magnetic field strength, is defined in order to provide a method of calculating the force between two permanent magnetic

Table 22.1 *Magnetic Units*

quantity	symbol	cgs units	SI units	conversion: cgs to SI
pole strength or flux	p or Φ or ϕ^a	maxwells[b]	Wb	10^8 maxwells = 1 Wb
flux density	B	gauss[c]	T[d]	10^4 gauss = 1 T
field strength (intensity)	H	oersted	A/m[e]	$4\pi \times 10^{-3}$ oersted = 1 A/m
magnetization[f]	M	oersted	A/m[e]	$4\pi \times 10^{-3}$ oersted = 1 A/M
permeability	μ	gauss/oersted	H/m[g]	$\mu_{\text{cgs}}\left(4\pi \times 10^{-7}\right) = \mu_{\text{mks}}$[h]

[a] The SI symbol for magnetic flux is Φ. The symbol p is used when referring to magnetic poles. The symbol ϕ is a generic flux symbol used in many texts.
[b] A *maxwell* is a line of force.
[c] This is equivalent to lines/cm² or maxwells/cm².
[d] This is equivalent to Wb/m².
[e] This is equivalent to N/Wb.
[f] The units used in this row assume the following equation for flux density: $B = \mu_0 H + \mu_0 M$. Other forms of this equation exist. Care must be used in determining the correct units or in comparing values from different references.
[g] This is equivalent to Wb/A·m and Ω·s/m.
[h] The value of μ_{cgs} in the cgs system is 1.

poles analogous to Coulomb's law. The number of *unit-poles* or just *poles*, p, is numerically equal to the flux, ϕ, emanating from the end of the magnet. (In the cgs system, the flux is numerically equal to $4\pi p$.) The area of the pole, A_p, and the flux density from the pole, B_p, are related to the pole strength by Eq. 22.24.

$$p = \phi = B_p A_p \qquad 22.24$$

The pole strength, as used here, and the magnetic charge used in Sec. 22-3 are related by Eq. 22.25.

$$q_m = \frac{p}{\mu} \qquad 22.25$$

In the cgs system the value of the permeability of free space is 1. Thus, in free space, the magnetic charge and the pole strength can be used interchangeably.[6] In the SI system, this is not the case and care must be exercised to ensure consistent units and terminology. In some SI texts, the pole strength referred to is q_m.

The moving charge of modern magnetism in Eq. 22.11 is the magnetic charge of traditional magnetism in Eq. 22.21. Modern electrical engineering teaches that magnetism is caused by circulating currents (called *amperian currents*) and electron spins. The magnetic charge in such a context is equal to $I dl$. Equation 22.26 relates these concepts.

$$q_m = qv = I dl \qquad 22.26$$

6. MAGNETIC DIPOLE

Individual magnetic poles do not exist. Magnetic dipoles do exist and are the result of circulating currents. A small loop of circulating current has an associated *magnetic moment*, **m**, whose magnitude depends on the current flow and the area encompassed by that flow as given in Eq. 22.27. The direction of the magnetic moment is perpendicular to the plane of the current loop and is shown as the unit vector **n**.

$$\mathbf{m} = I A \mathbf{n} \qquad 22.27$$

Two views of the magnetic moment are illustrated in Fig. 22.4. The first is associated with a loop of wire in a motor or generator (Fig. 22.4(a) and (b)). The second might be associated with a circulating electron (Fig. 22.4(c)).

The torque experienced by the magnetic moment is given by Eq. 22.28.

$$\mathbf{T} = \mathbf{m} \times \mathbf{B} \qquad 22.28$$

---

[6]The permeability, μ, in the cgs system is equivalent to the relative permeability, μ_r, in the mks system.

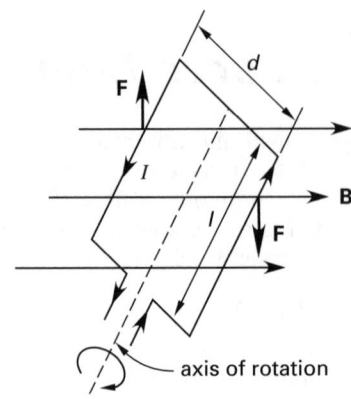

(a) loop of wire carrying current I and encompassing area $A = ld$

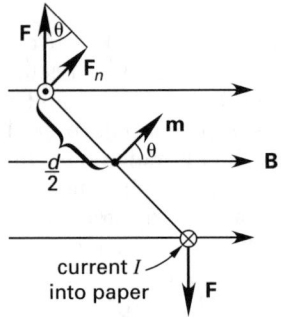

(b) cross section of loop showing the magnetic moment, **m**

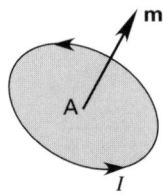

(c) generic view of a magnetic moment

Figure 22.4 *Magnetic Moment*

In traditional magnetism, the magnetic dipole was viewed as a bar magnet with a magnetic dipole moment $p_m = q_m d$. A comparison of the electric dipole moment, the modern view of the magnetic moment, and the traditional magnetic dipole moment is given in Fig. 22.5.

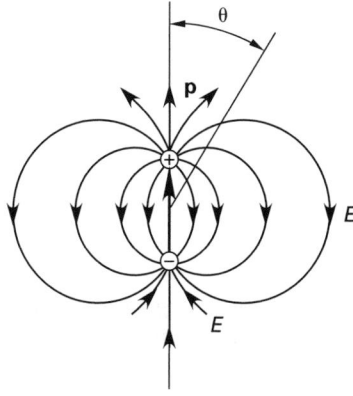

(a) electric dipole moment

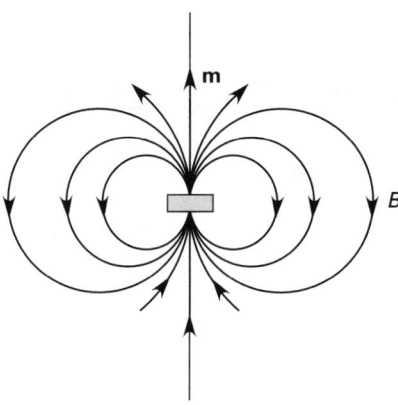

(b) magnetic dipole moment

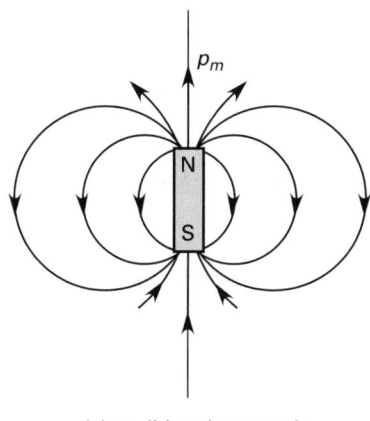

(c) traditional magnetic
dipole moment

Figure 22.5 *Dipole Moment Comparison*

The magnetic moment in terms of current loops, magnetic charge, and pole strength is

$$m = IA = q_m d = \left(\frac{p}{\mu}\right) l \qquad 22.29$$

Example 22.5

Determine the magnitude of the magnetic moment of a single electron at the first *Bohr radius* of 5.292×10^{-11} m if the electron moving in the orbit is equivalent to approximately 1.05 mA.

Solution

The magnitude of the magnetic moment, m, is

$$m = IA$$

The area correlating to the Bohr radius is

$$A = \pi r^2 = \pi \left(5.292 \times 10^{-11} \text{ m}\right)^2$$
$$= 8.798 \times 10^{-21} \text{ m}^2$$

Substituting gives

$$m = IA = \left(1.05 \times 10^{-3} \text{ A}\right)\left(8.798 \times 10^{-21} \text{ m}^2\right)$$
$$= 9.24 \times 10^{-24} \text{ A·m}^2$$

This is the approximate equivalent of one Bohr magneton, which equals $q_e h/4\pi m_e$.

7. COULOMB'S LAW EQUIVALENT: FORCE BETWEEN MAGNETIC POLES

The force between the pole ends of two magnets separated by a distance r, written similar to Coulomb's law, is given by Eq. 22.30. The orientation of the magnets is not important. Vector addition is required if more than two poles are present.

$$\mathbf{F}_{1-2} = p_2 \mathbf{H}_1 = \left(\frac{p_1 p_2}{4\pi \mu r^2}\right) \mathbf{a} \qquad 22.30$$

8. MAGNETIC POTENTIAL

The traditional *magnetic potential* U_m (measured in amps) possessed by a magnet of pole strength p_A is

$$U_m = \frac{U_{\text{potential}}}{p_A} = \frac{-p_B}{4\pi \mu r} \qquad 22.31$$

The *magnetic potential gradient* is

$$H = \frac{-\Delta U_m}{\Delta r} \qquad 22.32$$

In modern magnetism, the magnetic potential gradient, called the *magnetic field strength*, is

$$H = \frac{dI}{dl} \qquad 22.33$$

9. MAGNETIC POTENTIAL ENERGY

The *work* required to change the separation distance, r, between two magnetic poles is

$$W = \int_{r_0}^{r_f} \mathbf{F} \cdot d\mathbf{r} = \left(\frac{-p_A p_B}{4\pi\mu} \right) \left(\frac{1}{r_f} - \frac{1}{r_0} \right) \quad 22.34$$

The total *magnetic potential energy* in a system of two magnetic poles separated by a distance r is

$$U_{\text{potential}} = W_{\infty - r} = \int_{\infty}^{r} \mathbf{F} \cdot d\mathbf{r} = \frac{-p_A p_B}{4\pi\mu r} \quad 22.35$$

Part 4: Curl

In Electrostatics, Sec. 20-3, the divergence of an electric field was found to equal the charge enclosed within. The applicable equation is repeated here for convenience. Equation 22.36 assumes the permittivity is constant (i.e., the medium is isotropic and linear).

$$\nabla \cdot \mathbf{E} = \frac{\rho}{\varepsilon} \quad 22.36$$

The curl of an electrostatic field is equal to zero.

$$\nabla \times \mathbf{E} = 0 \quad 22.37$$

Equations 22.36 and 22.37 can be understood as follows: Any point in space surrounded by a closed surface with net flux emanating from the volume must contain a charge, Q, within that volume. The charge is the cause of the net flux and is called the *scalar source* of the flux. No sideways variation in the field is present. That is, the field varies in the direction of the source. The electric field is an *irrotational field*. (An irrotational field is any force field in which $\oint \mathbf{F} \cdot dl = 0$.)

In magnetostatics, no magnetic charge exists. The divergence of the **B**-field is zero.

$$\nabla \cdot \mathbf{B} = 0 \quad 22.38$$

The curl of the **B**-field in free space is

$$\nabla \times \mathbf{B} = \mu_0 \mathbf{J} \quad 22.39$$

Equations 22.38 and 22.39 can be understood as follows: The current is the source of the flux in a magnetic field. The net flux from any volume, however, will be zero, since the lines of flux close upon themselves (no magnetic charges are ever enclosed). A sidewise variation exists and can be used to measure the source as given

by Eq. 22.39. The current is a vector source. The magnetic field is called a *solenoidal field*. (A solenoidal field is any field in which the divergence is zero.)

In terms of the magnetic field strength, Eq. 22.39 is

$$\nabla \times \mathbf{H} = \mathbf{J} \quad 22.40$$

In integral form, the results of Eqs. 22.39 and 22.40 are known as *Ampere's law*. Ampere's law for steady currents states that the line integral of the magnetic field strength around a closed loop is equal to the current enclosed by the path. Ampere's law is

$$\oint \mathbf{H} \cdot d\mathbf{l} = \int_{\text{surface}} \mathbf{J} \cdot d\mathbf{s} = I_{\text{enclosed}} \quad 22.41$$

Part 5: Magnetic Potential

The difference in the magnetic potential between any two points in a magnetic field is given by the product of the magnetic field strength and the length of the path.

$$\int_{0}^{f} \mathbf{H} \cdot d\mathbf{l} = \mathbf{U}_f - \mathbf{U}_0 \quad 22.42$$

If the path is closed and no current is contained within it, no potential difference exists, as shown by Eq. 22.43.

$$\oint \mathbf{H} \cdot d\mathbf{l} = 0 \quad 22.43$$

In a closed path that encloses N turns carrying current I, the potential is

$$\oint \mathbf{H} \cdot d\mathbf{l} = NI = V_m \quad 22.44$$

The term NI, given the symbol V_m, is called the *magnetomotance*. The magnetic potential and the magnetomotance are specified in terms of amperes or ampere-turns.

Part 6: Magnetic Vector Potential

The electric field intensity, **E**, was first obtained from known charge configurations. The concept of a potential was then developed and **E** was found, determined from the negative gradient of the potential. As a starting point in electrical engineering problems, the potentials on the boundary conductors are known. Then

Laplace's equations are used to determine the potential V. Once V is known, $\mathbf{E}$ and $\mathbf{D}$ can be determined. Similarly, a *vector magnetic potential*, $\mathbf{A}$, can be defined that functions in the same manner as potential in electric circuits.

To ensure the vector magnetic potential satisfies known conditions in magnetic fields, the following conditions must be met.

$$\nabla \times \mathbf{A} = \mathbf{B} \qquad 22.45$$

$$\nabla \cdot \mathbf{A} = 0 \qquad 22.46$$

The condition required by Eq. 22.46 is called the *gauge condition*, known in magnetostatics as the *Coulomb gauge*. Magnetic vector potential is measured in Wb/m or T·m.

The vector magnetic potential for three standard current configurations, that of a current filament (Eq. 22.47), that of a sheet current (Eq. 22.48), and that of a volume current (Eq. 22.49), follow.

$$\mathbf{A} = \oint \frac{\mu I d\mathbf{l}}{4\pi r} \qquad 22.47$$

$$\mathbf{A} = \int_s \frac{\mu \mathbf{K} ds}{4\pi r} \qquad 22.48$$

$$\mathbf{A} = \int_V \frac{\mu \mathbf{J} dV}{4\pi r} \qquad 22.49$$

The magnetic vector potential is assumed to be zero at infinity. Equations 22.47 through 22.49 cannot be applied if the current distributions are infinite.

Part 7: Work and Energy

Magnetic forces on moving charge particles and current-carrying conductors result from the magnetic field. Any forces used to counter these must be applied ($\mathbf{F}_a$) from a source external to the magnetic field and be equal and opposite to the force of the field. If motion of the charged particle or current-carrying conductor occurs, the work is given by

$$W = \int_0^f \mathbf{F}_a \cdot d\mathbf{l} \qquad 22.50$$

A positive result from the initial position on the path (l_0) to the final position (l_f) indicates that work was done by an external force on the system; that is, energy has been expended against the field. Because the magnetic force is generally not a conservative force field, the entire path of integration must be specified.

Fields

23 Magnetostatic Fields

Nomenclature

A	area	m^2
B	magnetic flux density	T (Wb/m^2)
H	magnetic field strength	A/m
i, I	current	A
l	length	m
m	magnetic moment	A·m^2 (J/T)
M	magnetization	A/m
N	number of turns	–
p_m	magnetic dipole moment	A·m^2 (J/T)
P	polarization	C/m^2
q, Q	charge	C
q_m	magnetic charge	A·m or N/T
r	radius	m
T_c	Curie temperature	K
v, V	volume	m^3

Symbols

μ	permeability	N/A^2 or H/m
μ_0	free-space permeability 1.2566×10^{-6}	N/A^2 or H/m
χ	susceptibility	–

Subscripts

0	initial or free space
C	Curie
m	magnetic
r	relative

1. MAGNETIZATION

When a magnetic field is present in a magnetizable material, the magnetic moments tend to align themselves into the lowest energy configuration possible. In doing so, they increase the flux density, **B**. The alignment is referred to as *magnetization* of the material and is given the symbol **M**. The magnetization is the total magnetic moment, **m**, per unit volume, v.

$$\mathbf{M} = \frac{\mathbf{m}}{v} \qquad 23.1$$

The flux density is then given by the following general formula.[1]

$$\mathbf{B} = \mu_0 \mathbf{H} + \mathbf{M} \qquad 23.2$$

Magnetization is technically defined as the magnetic polarization divided by the magnetic constant of the system used. In the SI system, the magnetic constant is the permeability of free space. Therefore, the equation becomes

$$\mathbf{B} = \mu_0 (\mathbf{H} + \mathbf{M}) = \mu_0 \mathbf{H} + \mu_0 \mathbf{M} \qquad 23.3$$

The magnetic **B**-field has two sources. The first is the current, I; and the other is the magnetization, **M**. The current is the source of the magnetic field strength, **H**. The magnetic dipoles of individual atoms or molecules (which make up the larger domains) are the source of the magnetization, **M**.

Magnetization thus increases the flux density in magnetizable materials above the level that would exist under free-space conditions with the same magnetic field strength.[2] This occurs even when the magnetic moments of a material are not initially aligned. When an atom is placed within the influence of a magnetic field, its magnetic moment tends to align with the external magnetic field. The extent to which this occurs depends on the material. This is analogous to spring energy in mechanical systems. The work done in rotating the magnetic moment is recoverable upon removal of the magnetic field. The extent of the recovery again depends on the material.

The magnetic field and the magnetization can have different directions, though this is often not the case on a macroscopic level. In isotropic, linear material, **H** and **M** are parallel and related by

$$\mathbf{M} = \chi_m \mathbf{H} \qquad 23.4$$

---

[1]Equation 23.2 is of a form similar to that used in electrostatic fields but it cannot be used in the SI system as written. The study of magnetics developed mainly with the cgs system. In the cgs system $\mu_0 = 1$, but cgs was not a rationalized system so $\mathbf{B} = \mathbf{H} + 4\pi\mathbf{M}$. (The terms **B**, **H**, and **M** are often parallel and the equation is therefore written as $B = H + 4\pi M$.) The SI system is rationalized, but $\mu_0 = 4\pi \times 10^{-7}$ H/m.

[2]The magnetization, **M**, of free space is zero. A diamagnetic material's magnetization is negative and opposes the applied field. Diamagnetic materials are thus an exception to the statement in the text.

The term *isotropic* indicates that the medium's properties are independent of direction. The term *linear* indicates that the permeability, μ, is constant regardless of the magnitude of the current present. The term χ_m is the magnetic susceptibility. This dimensionless term measures the ease of magnetization in a material. Thus, for an isotropic, linear material, Eq. 23.3 can be rewritten as

$$\mathbf{B} = \mu_0 \left(1 + \chi_m\right) \mathbf{H} \qquad 23.5$$

Importantly, for ferromagnetic materials (those materials with large permeabilities), Eqs. 23.4 and 23.5 do not hold true. (By contrast, Eq. 23.3 is always valid.) Ferromagnetic materials are nonlinear and exhibit the property of hysteresis. This nonlinearity is easily seen on the *BH* curve in Fig. 23.1, which displays typical hysteresis loops.[3] Because ferromagnetic materials are so important in engineering applications, many permeabilities are defined in the various linear portions of the *BH* curve—and utilized only in those linear regions.

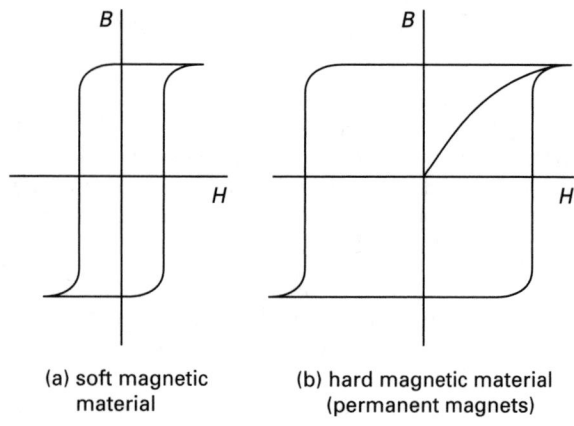

(a) soft magnetic material

(b) hard magnetic material (permanent magnets)

Figure 23.1 *Hysteresis Loops*

For those magnetic materials that are linear, and for ferromagnetic materials on linear portions of the *BH* curve, the relative permeability is

$$\mu_r = 1 + \chi_m \qquad 23.6$$

In free space, no magnetization occurs because no magnetic materials are present, and the magnetic flux density, $\mathbf{B}$, is given as

$$\mathbf{B} = \mu_0 \mathbf{H} \qquad 23.7$$

If a magnetizable material is present, the flux density is increased and is given by Eq. 23.3. Importantly, Eq. 23.3 is often written without direct reference to the magnetization but rather in terms of the relative permeability. The relationship is shown explicitly in Eq. 23.8.

$$\mathbf{B} = \mu \mathbf{H} = \mu_0 \mu_r \mathbf{H} = \mu_0 \mathbf{H} + \mu_0 \mathbf{M} \qquad 23.8$$

---

[3]*MH* curves are not commonly plotted, though they are of identical shape to *BH* curves.

Example 23.1

The dimensions of a small magnet used in instrumentation are 1 cm × 1 cm × 0.3 cm. The volume is uniformly magnetized and possesses a magnetic moment of 0.01 A·m² and a measured flux density of 3×10^{-2} T. What is the magnetization, $\mathbf{M}$, of the magnet?

Solution

The magnetization is determined from Eq. 23.1.

$$\mathbf{M} = \frac{\mathbf{m}}{v}$$

The volume is determined from the dimensions

$$v = (1 \text{ cm})(1 \text{ cm})(0.3 \text{ cm}) = 0.3 \text{ cm}^3 = 0.3 \times 10^{-6} \text{ m}^3$$

Substituting gives

$$\mathbf{M} = \frac{\mathbf{m}}{v} = \frac{0.01 \text{ A·m}^2}{0.3 \times 10^{-6} \text{ m}^3} = 3.3 \times 10^4 \text{ A/m}$$

Example 23.2

What is the magnetic field strength in the magnet described in Ex. 23.1?

Solution

The magnetic field strength, $\mathbf{H}$, is related to the flux density and magnetization in all cases as follows.

$$\mathbf{B} = \mu_0 \left(\mathbf{H} + \mathbf{M}\right) = \mu_0 \mathbf{H} + \mu_0 \mathbf{M}$$

Rearranging to solve for the desired quantity and substituting the given and calculated information gives

$$\mathbf{B} = \mu_0 \mathbf{H} + \mu_0 \mathbf{M}$$

Inside a magnet, the $\mathbf{B}$-field and magnetization $\mathbf{M}$ are parallel, while the $\mathbf{H}$-field is in opposition. Knowing this, one can change from vector quantities to magnitudes.

$$B = \mu_0 H + \mu_0 M$$

$$H = \frac{B - \mu_0 M}{\mu_0}$$

$$= \frac{3 \times 10^{-2} \text{ T} - \left(1.2566 \times 10^{-6} \frac{\text{H}}{\text{m}}\right)\left(3.3 \times 10^4 \frac{\text{A}}{\text{m}}\right)}{1.2566 \times 10^{-6} \frac{\text{H}}{\text{m}}}$$

$$= -9.1 \times 10^3 \text{ A/m}$$

Consider the following.[4]

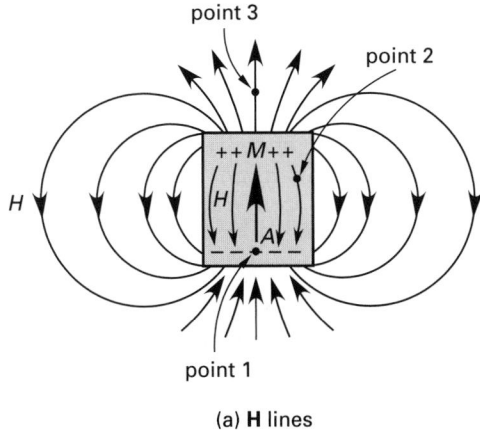

(a) **H** lines

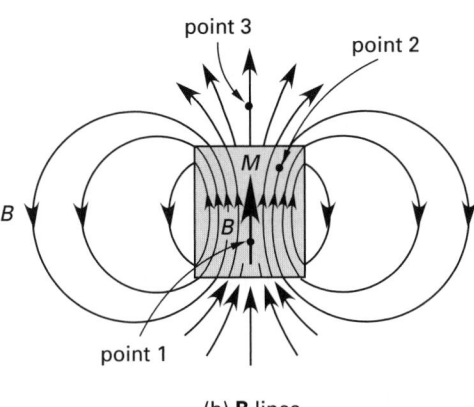

(b) **B** lines

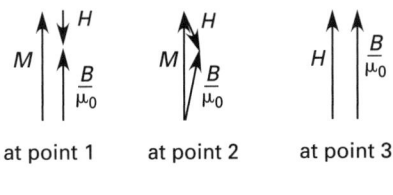

(c) relationships

Outside the magnet, $\mathbf{B} = \mu_0\mathbf{H}$ and $\mathbf{M}$ equals zero. Thus, $\mathbf{B}$ and $\mathbf{H}$ are parallel outside the magnet.

2. MAGNETIC FIELDS

All materials of concern to electronic engineers contain electrons in motion. In specific orbits, these electrons can be viewed as circulating currents. All materials then contain magnetic dipoles. Such dipoles can fully or partially align with a magnetic field. Parts (a) and (c) of Fig. 23.2 illustrate such an alignment.

---

[4]Considerably more explanation is possible but unnecessary for the majority of magnetic applications in electrical engineering.

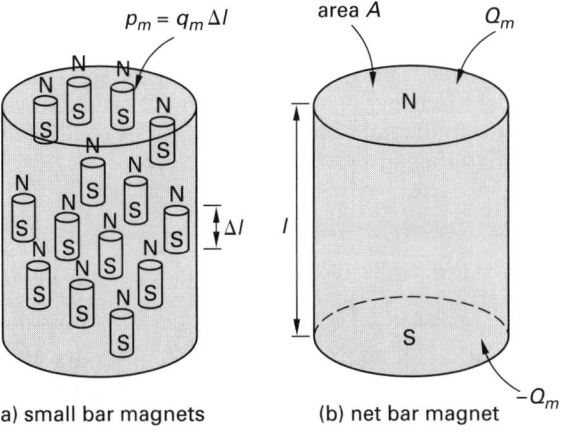

(a) small bar magnets   (b) net bar magnet

traditional view

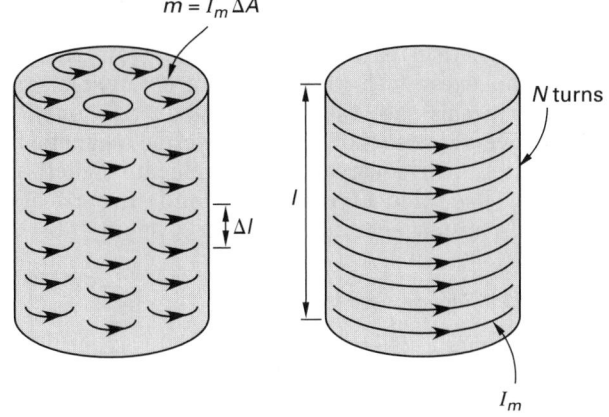

(c) small circulating currents   (d) net circulating currents

modern view

Figure 23.2 *Permanent Magnet: Two Theoretical Views*

In addition to the movement of the electron in its orbit, the electron exhibits a property known as *spin*. The spin is the equivalent of the rotation of the electron about its axis. This rotation can be viewed as the equivalent of a circulating current and the source of an additional magnetic moment.

The magnetic moment of an atom may take two forms. First, the magnetic moments of all the electrons may be so oriented that they cancel, so that the magnet as a whole has no net magnetic moment. This condition leads to diamagnetism. Second, the cancellation of the electron moments, orbital and spin, may be partial, so that the atom is left with a net magnetic moment. Such an atom is referred to as a *magnetic atom*. This condition leads to several types of magnetism, including para-, ferro-, antiferro-, and ferrimagnetism.

Diamagnetism is exhibited by all materials and is due to the orbital motion of the electrons. When an external magnetic field is applied, the magnetic moments of orbital pairs of electrons are unbalanced. Currents are induced in the circulating electrons in such a way as to create an internal magnetic field that opposes the external magnetic fields.[5] This is known as *antimagnetism*. No permanent magnetization, **M**, occurs. The magnetic field induced is too weak to be of use. The values of permeability and susceptibility for several diamagnetic materials are given in Table 23.1.

Table 23.1 *Diamagnetic Materials*

substance	relative permeability $\mu_r = 1 + \chi_m$	susceptibility χ_m
bismuth	0.99983	-1.66×10^{-4}
mercury	0.999968	-3.20×10^{-5}
gold	0.999964	-3.60×10^{-5}
silver	0.99998	-2.60×10^{-5}
lead	0.999983	-1.70×10^{-5}
copper	0.999991	-0.98×10^{-5}
water	0.999991	-0.88×10^{-5}

Paramagnetism is caused by the partial alignment of the spin axes of electrons with an external magnetic field. In this case the induced magnetization, **M**, is parallel and proportional to the external magnetic field. The internal field is weak. No permanent magnetization occurs. The values of permeability and susceptibility for several paramagnetic materials are given in Table 23.2.

Table 23.2 *Paramagnetic Materials*

substance	relative permeability $\mu_r = 1 + \chi_m$	susceptibility χ_m
air	1.00000036	3.6×10^{-7}
aluminum	1.000021	2.5×10^{-5}
palladium	1.00082	8.2×10^{-4}

The most important type of magnetism is ferromagnetism. Ferromagnetism is primarily caused by unpaired electron magnetic moments. The magnetic moments of individual atoms or molecules align themselves within small volumes called *domains*.[6] This effect occurs in iron, nickel, cobalt, and their alloys; in some manganese compounds; and in some rare earth elements. The domains can be completely aligned with weak external magnetic fields, resulting in very large magnetization, **M**. This effect is shown in Fig. 23.2. In part (a) of this figure, the individual magnetic moments (or groups of atoms in domains) are viewed as small bar magnets that have aligned with the external magnetic field (not shown). The net result is shown in part (b). Parts (a) and (b) illustrate the traditional view of magnetism. In the modern view, ferromagnetism is caused by circulating currents (both orbital and spin) as shown in part (c). Because of the cancellation of interior currents, the net

---

[5]This reaction is best known as *Lenz's law* and is the *counter-electromotive force* (emf) in motors and generators.
[6]Both paramagnetism and ferromagnetism occur as a result of unpaired electron moments. The difference is that ferromagnets exhibit domain formation due to the quantum mechanical effect of *exchange forces*.

result is a sheet of current on the surface of the cylinder as shown in part (d). The magnetization is significant enough in ferromagnetic materials that for all practical purposes **H** ≪ **M**, so

$$\mathbf{B} \approx \mu\mathbf{M} \qquad 23.9$$

The internal magnetic field is thus very strong. Permanent magnetization can occur. The values of permeability and susceptibility for several ferromagnetic materials are given in Table 23.3.

Table 23.3 *Ferromagnetic Materials*

substance	relative permeability $\mu_r = 1 + \chi_m$
cobalt	250
nickel	600
commercial iron (0.2 impurity)	6000
high-purity iron (0.05 impurity)	2×10^5

Note: The relative permeabilities are nonlinear for ferromagnetic materials. The value, however, will not change regardless of the system of units used. Nevertheless, the numbers are for comparison of materials only. To have meaning, the *B* or *H* value at which the permeability was measured would need to be specified.

Both paramagnetic and ferromagnetic materials are temperature sensitive. When ferromagnetic materials are heated above the *Curie temperature*, T_C, they become paramagnetic. The magnetization and temperature dependence of these three classes of magnetic materials are shown in Fig. 23.3.

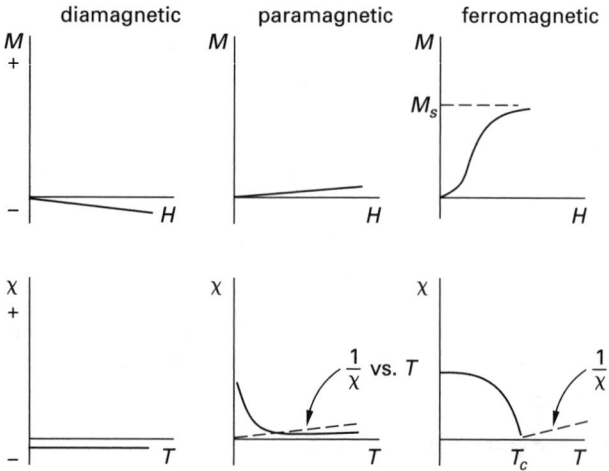

Figure 23.3 *Magnetization Curves and Temperature Dependence*

3. AUXILIARY FIELD H

The vector **H**, known as the magnetic field strength, is used to separate the sources of the **B**-field into those caused by external currents and those caused by the magnetization of the material itself. The magnetic field strength, **H**, represents the portion of the flux caused by external currents. The magnetization, **M**, represents the portion of the flux caused by the movement of magnetic dipoles within the material itself.

The magnetic field strength specifies the amount of force per unit flux produced by a conductor of N turns and length l carrying current I. After some manipulation, the magnitude of the magnetic field strength is given by

$$H = \frac{NI}{l} \qquad 23.10$$

Example 23.3

What is the magnetic field strength in a circular electrical device of radius 2.5 cm with 10 turns of wire carrying 30 A?

Solution

The magnetic field strength, H, which does not depend on the material present, is

$$H = \frac{NI}{l}$$

The path length is the circumference of a circle with a radius of 2.5 cm. That is,

$$l = 2\pi r = 2\pi\,(0.025\ \text{m}) = 0.16\ \text{m}$$

Substituting the given and the calculated gives

$$H = \frac{NI}{l} = \frac{(10)\,(30\ \text{A})}{0.16\ \text{m}} = 1.9 \times 10^3\ \text{A/m}$$

4. MAGNETIZATION IN NONMAGNETIC AND MAGNETIC MATERIALS

The use of the magnetic field strength, **H**, allows material media to be categorized into three groups. The magnetization vector, **M**, is avoided if possible (as is the polarization vector, **P**) to simplify calculations.

The first group consists of free space and nonmagnetic materials. In this group, the current is the only source of the magnetic field. The magnetization is zero. Thus,

$$\mathbf{B} = \mu_0\mathbf{H} \qquad 23.11$$

The second group consists of *soft ferromagnetic materials*. Such materials are suitable for linearization. That is, the current can be restricted to operate over a limited portion of the BH curve so that μ is constant. In this case, the governing equation is

$$\mathbf{B} = \mu\mathbf{H} = \mu_0\mu_r\mathbf{H} \qquad 23.12$$

The magnetization is embedded in the relative permeability term.

The third group consists of the *hard ferromagnetic materials*. These materials are very nonlinear and are suitable for use as permanent magnets. Due to the nonlinearity, the relative permeability term cannot be used. Instead, B and H are given by graphs that are experimentally determined and are typically as shown in Fig. 23.1. Since M is significantly larger than H in such materials, B is related directly to M as indicated in the magnetization curves of Fig. 23.3.

Example 23.4

What is the magnitude of the magnetic flux density in air associated with a magnetic field strength of 50 A/m?

Solution

The relative permeability, μ_r, of air (1.00000036) very closely approximates that of free space (1). Thus, when air is specified, μ_r can be assumed to be equal to one. Use Eq. 23.11.

$$B = \mu_0 H = \left(1.2566 \times 10^{-6}\ \frac{\text{H}}{\text{m}}\right)\left(50\ \frac{\text{A}}{\text{m}}\right)$$
$$= 6.3 \times 10^{-5}\ \text{T}$$

Example 23.5

What is the flux density if commercial iron is substituted for air in Ex. 23.4?

Solution

Commercial iron has a relative permeability of 6000. From Ex. 23.4, $\mu_0 H = 6.3 \times 10^{-5}$ T. Using these values and Eq. 23.12 gives

$$B = \mu_0 H \mu_r = \left(6.3 \times 10^{-5}\ \text{T}\right)(6000) = 0.38\ \text{T}$$

24 Electrodynamics

Nomenclature

A	area	m^2
AW	atomic weight	g/mol
B	magnetic flux density	T (Wb/m^2)
c	speed of light	m/s
e^-	electron	–
E	electric field strength	V/m
F	faraday	C/mol
F	force	N
H	magnetic field strength	A/m
i, I	current	A
J	current density	A/m^2
l	length	m
L	self-inductance	H
m_g	mass in grams	g
N	number of turns	–
P	active or real power	W
P_a	apparent power	V·A
$\mathcal{P}_{ave}$	average Poynting vector	W/m^2
Re	real part	–
s	surface area	m^2
$\mathbf{S}$	Poynting's vector	W/m^2
t	time	s
v	velocity	m/s
v, V	voltage	V
V	volume	m^3
V_m	magnetic potential	A or A-turns
W_e	equivalent weight	–
W	work	J
z	valence change	–

Symbols

ε	permittivity	C^2/N·m or F/m
$\mathcal{E}$	electromotive force	V
μ	permeability	N/A^2 or H/m
μ_0	free-space permeability	
	1.2566×10^{-6}	N/A^2
		or H/m
σ	conductivity	S/m
ϕ	magnetic flux	Wb

Subscripts

ave	average
e	equivalent
g	grams
m	magnetic

1. ELECTROMOTIVE FORCE

It takes force to drive a charge through a conducting medium. The source of the force can be electromagnetic, chemical, or mechanical. Whatever the source, the net effect of the force in any closed circuit is called the *electromotive force*, emf, or *electromotance*, and is given the symbol $\mathcal{E}$. It can be found from the line integral of the force over the path of the circuit.

$$\mathcal{E} = \oint \mathbf{F} \cdot d\mathbf{l} = \oint \frac{\mathbf{J}}{\sigma} \cdot d\mathbf{l}$$

$$= \oint \frac{I}{A\sigma} dl = I \oint \frac{1}{A\sigma} dl = IR \qquad 24.1$$

The term R represents the total resistance in the external circuit. The force that moves the charges is the electric field, $\mathbf{E}$. The electromotive force is measured in volts. In actuality, it is not a force, but the integral of the force per unit charge. The emf can be viewed as energy per unit charge, especially when considering that the unit for emf is the volt. The term electromotive force and the symbol $\mathcal{E}$ are reserved for the sources within an electric circuit, while v and V are used for the remaining components.[1]

Electromotive force can be generated by chemical means as in a battery, mechanically as in a piezoelectric crystal, by heat as in a thermocouple, or by many other methods. The two methods of considerable concern in electrical engineering are dynamic in nature.

The first is the time-varying magnetic flux method. The time-varying flux induces a voltage in any conductor present that depends on the number of turns, N, of the conductor and the *flux linkage*, $N\phi$ (the total number of flux lines cut). This is given by Eq. 24.2.

$$v = -N\frac{d\phi}{dt} \qquad 24.2$$

---

[1]Though this is generally true, in many instances the term *voltage* and its associated symbols (V, v) are used to indicate sources. It is understood that the voltage in those cases represents an electromotive force. Except where confusion may arise, this text will use v and V.

Fields

The minus sign is used to indicate that any current that flows due to an induced electromotive force does so in a direction that generates a magnetic field opposing the flux that induced the emf. This is known as *Lenz's law*. The minus sign is not always shown. In circuit analysis, it is considered understood, though it is still used in equations written to obtain a numeric result. Equation 24.2 is often called *Faraday's law*, though in actuality it is the result of the law for time-varying fields.

Equation 24.2 holds even when the time-varying magnetic flux in a given circuit element is changing because of the current in the element itself. Therefore, Eq. 24.2 can be written as follows.

$$v = -N \frac{d\phi}{di} \frac{di}{dt} = -L \frac{di}{dt} \qquad 24.3$$

The term L is known as *self-inductance*.

The second dynamic method of inducing an electromotive force is to maintain the flux constant and move the conductors. This is called the flux-cutting method. It results in an induced voltage, or electromotive force, of

$$V = -NBl\text{v} \qquad 24.4$$

2. FARADAY'S LAW OF ELECTROMAGNETIC INDUCTION

Faraday's law of electromagnetic induction states that the induced voltage is proportional to the rate of change of magnetic flux linked with the circuit. If the rate of change of magnetic flux linked is due to a time rate of change of the magnetic flux, that is, if the circuit (conductor) is stationary, then Faraday's law is

$$v = \oint \mathbf{E} \cdot d\mathbf{l} = -\frac{d}{dt} \int_s \mathbf{B} \cdot d\mathbf{s}$$

$$= -\int_s \frac{\partial \mathbf{B}}{\partial t} \cdot d\mathbf{s} = -\frac{d\phi}{dt} \qquad 24.5$$

If the rate of change of the magnetic flux linked is due to the motion of the circuit (conductor) and the magnetic field is constant with respect to time, Faraday's law is

$$v = \oint \mathbf{E} \cdot d\mathbf{l} = \oint (\text{v} \times \mathbf{B}) \cdot d\mathbf{l} \qquad 24.6$$

The general case of Faraday's law when the magnetic flux is changing and the circuit (conductor) is in motion is

$$v = \oint \mathbf{E} \cdot d\mathbf{l} = \oint (\text{v} \times \mathbf{B}) \cdot d\mathbf{l}$$

$$= -\int_s \frac{\partial \mathbf{B}}{\partial t} \cdot d\mathbf{s} \qquad 24.7$$

Faraday's law applies to changing magnetic fields. In electrical engineering, the formulas of magnetostatics (Ampere's law, the Biot-Savart law, and others) are used to calculate the fields. While they give results that are only approximately correct, the errors are negligible. Even in the case where the current through a wire is interrupted by a fast-acting switch, or is simply cut, Ampere's law still holds. Magnetostatic formulas can be used in conjunction with Faraday's law in *quasistatic conditions*. The magnetostatic equations must be abandoned only when the frequencies are extremely high, as in electromagnetic waves and radiation.

3. FARADAY'S LAW OF ELECTROLYSIS

Faraday's law of electrolysis states that the mass of substance deposited at one electrode or liberated at the other varies as in Eq. 24.8.

$$m_g = \frac{ItW_e}{1F} \qquad 24.8$$

The term W_e is the equivalent weight of an element, the atomic weight in grams divided by the valence change that occurs during electrolysis, z.

$$W_e = \frac{\text{AW}}{z} \qquad 24.9$$

The atomic weight in the case of Faraday's law is given in grams. This is because one *faraday* of electricity will produce one gram of equivalent weight.[2]

Example 24.1

The following electrolysis reaction can occur to copper.

$$Cu^{+2} + 2e^- \rightarrow Cu$$

What current is required to produce five grams of metallic copper (AW = 63.5) from a copper sulfate solution in 1 hr?

Solution

Since the change in the valence of copper is 2, the equivalent weight is

$$W_e = \frac{\text{AW}}{z} = \frac{63.5}{2} = 31.8$$

---

[2]An electron has a charge of 1.602×10^{-19} C. A *faraday* is the charge associated with one mole of electrons. Therefore, a faraday is

$$F = \left(1.602 \times 10^{-19} \frac{\text{C}}{\text{electron}}\right) \left(6.022 \times 10^{23} \text{ electrons}\right)$$

$$= 96{,}485 \text{ C}$$

Rearranging Eq. 24.8 to determine the current required gives

$$m_g = \frac{ItW_e}{1F}$$

$$I = \frac{m_g F}{tW_e}$$

$$= \frac{(5 \text{ g}) \left(96{,}485 \dfrac{\text{C}}{\text{g-equivalent}}\right)}{(1 \text{ hr}) \left(\dfrac{3600 \text{ sec}}{1 \text{ hr}}\right)(31.8)}$$

$$= 4.2 \text{ A}$$

4. POTENTIAL

The electrostatic potential, or voltage, V, is a steady-state phenomenon that is well defined in space and is associated with a conservative electric field. The induced voltage, or potential, of Faraday's law, v, is time dependent and a multivalued function of position. It is associated with a nonconservative field, that of electromotive force.[3] When the potential is associated with a source of energy, the potential is called the electromotive force, or emf.

In magnetic circuits, the potential is called the *magnetomotive force*, mmf, or *magnetomotance*. Like its electric counterpart, mmf is actually not a force, but the line integral of the force around a path. In the magnetic case, it is the force moving the flux around the path. Hence, the mmf, or magnetic potential, is

$$\oint \mathbf{H} \cdot d\mathbf{l} = NI = V_m \qquad \textit{24.10}$$

5. ENERGY AND MOMENTUM: POYNTING'S VECTOR

Momentum conservation in electrodynamics occurs because the electric and magnetic fields themselves carry momentum. The conservation of energy in electrodynamics is given by *Poynting's theorem* as

$$\frac{dW}{dt} = -\frac{d}{dt} \int\limits_V \frac{1}{2} \left(\varepsilon E^2 + \mu H^2\right) dv$$

$$\qquad - \oint\limits_A (\mathbf{E} \times \mathbf{H}) \cdot d\mathbf{A} \qquad \textit{24.11}$$

The first integral represents the stored energy in the fields. The second integral represents the rate at which energy is transferred out of volume V across the surface boundary A. Poynting's theorem says the work done

on charges by the electromotive force is equal to the decrease in the energy stored in the field less the energy that flows out of the surface.

The cross product of $\mathbf{E}$ and $\mathbf{H}$ is Poynting's vector, $\mathbf{S}$. It represents the instantaneous power density (in W/m^2) carried by an electromagnetic wave. The direction of $\mathbf{S}$ is perpendicular to both $\mathbf{E}$ and $\mathbf{H}$, which is normal to the wave front as shown in Fig. 24.1. (In Fig. 24.1 $\mathbf{E}$ and $\mathbf{B}$ are shown as the primary fields, consistent with these two vectors representing the fundamental fields.) Poynting's vector is given in Eq. 24.12. In Eq. 24.12, $\mathbf{a}$ is a unit vector in the direction of propagation of the wave, that is, normal to the wave front.

$$\mathbf{S} = c\varepsilon E^2 \mathbf{a} = c\mu H^2 \mathbf{a} = \mathbf{E} \times \mathbf{H} \qquad \textit{24.12}$$

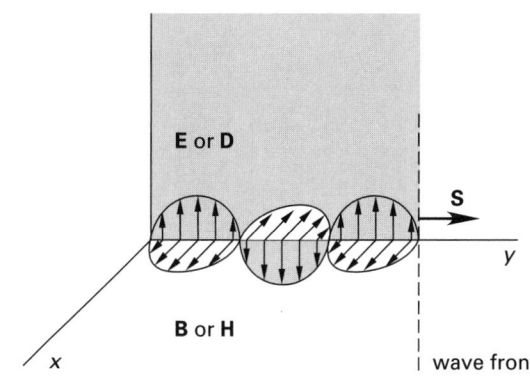

Figure 24.1 *Poynting's Vector*

Poynting's vector can be used to solve any problem involving the flow of electric or magnetic energy. The only restriction is that the electric and magnetic fields be related—from the same source. Though Eq. 24.12 represents the instantaneous value of the power density, allowing $\mathbf{E}$ and $\mathbf{H}$ to vary dynamically, with a time dependence of $e^{j\omega t}$, leads to a time-average of Poynting's vector, $\mathcal{P}_{\text{ave}}$, given by

$$\mathcal{P}_{\text{ave}} = \tfrac{1}{2}(\text{Re})\left(\mathbf{E} \times \mathbf{H}^*\right) \qquad \textit{24.13}$$

The complex conjugate of $\mathbf{H}$ is represented by $\mathbf{H}^*$. This follows the results of complex power analysis in which the apparent power is $\mathbf{P}_a = \frac{1}{2}\mathbf{VI}^*$, where the active power is the real part of the apparent power, that is, $\mathcal{P} = \frac{1}{2}(\text{Re})\mathbf{VI}^*$.

Poynting's vector, $\mathbf{S}$, describes the flow of energy just as the current density, $\mathbf{J}$, describes the flow of current.

6. ELECTROMAGNETIC WAVES

Stationary charges produce an electric field. If the charges are moving with a uniform velocity, a second effect occurs called *magnetism*. If the charges are accelerating, they produce a radiating electromagnetic field capable of transporting energy. The radiating electromagnetic field can be treated as a wave.[4]

---

[3]A conservative field is one where the work done in moving a charge depends only on its initial and final positions and not its path.

[4]Electromagnetic waves undergo all of the processes commonly affecting waves, that is, reflection, refraction, diffraction, wave interference, dispersion, and scattering.

The treatment of electrical phenomena as waves dominates electrical engineering fields where the transmission of electromagnetic energy takes place over long distances (often without metallic conductors), such as communications, power transmission, and related areas. In other electrical engineering fields, it is usually more convenient to work with the results, that is, potentials, currents, and resistances.

25 Maxwell's Equations

Subscripts

0	initial or free space
c	conduction
e	electric
m	magnetic

Fields

Nomenclature

A	area	m^2
B	magnetic flux density	T (Wb/m^2)
D	electric flux density	C/m^2
E	electric field strength	V/m
F	force	N
F_m	magnetomotive force	A or Amp-turns
G	conductance	S or $1/\Omega$
H	magnetic field strength	A/m
i, I	current	A
J	current density	A/m^2
l	length	m
L	inductance	H
M	magnetization	A/m
P	polarization	C/m^2
P_m	permeance	H
Q	charge	C
R	resistance	Ω
s	surface area	m^2
t	time	s
T	torque	N·m
U	energy density	J/m^3
v	velocity	m/s
v, V	voltage	V
V	volume	m^3
V_m	magnetic potential	A or A-turns
W	energy	J

Symbols

$\mathcal{E}$	electromotive force	V
ε	permittivity	C^2/N·m^2 or F/m
ε_0	free-space permittivity 8.854×10^{-12}	C^2/N·m^2
μ	permeability	N/A^2 or H/m
μ_0	free-space permeability 1.2566×10^{-6}	N/A^2 or H/m
ρ	charge density	C/m^3
ρ	resistivity	Ω·m
$\mathcal{R}$	reluctance	1/H
σ	conductivity	S/m
ϕ	magnetic flux	Wb
χ	susceptibility	–

1. MAXWELL'S EQUATIONS

Maxwell's equations are Gauss's, Coulomb's, Faraday's, and Ampere's laws extended by the concept of a displacement current and the notion of a field. Though seldom used in the exact form presented in Table 25.1, they govern all of the electromagnetic phenomena in a medium that is stationary with respect to the coordinate system used.[1] They are valid for linear and nonlinear, isotropic and nonisotropic, homogeneous and nonhomogeneous media for frequencies ranging from zero to microwave frequencies, including many phenomena of light. The general set of Maxwell's equations is given in Table 25.1.

In free space, no charges exist ($\rho = 0$) and no conduction currents exist ($\mathbf{J}_c = 0$). Maxwell's equations then take on the forms given in Table 25.2.

The integral forms of Maxwell's equations are more fundamental in that they display the underlying physical laws. Maxwell's equations describe electromagnetic phenomena on a macroscopic, that is, a nonquantum, level. Thus, regions used in the integrals should be larger than atomic dimensions and time intervals should be long enough to average atomic fluctuations.

2. ELECTROMAGNETIC FIELD VECTORS

The following vectors are electromagnetic field vectors: $\mathbf{E}$, $\mathbf{D}$, $\mathbf{B}$, $\mathbf{H}$, and $\mathbf{J}$. These vectors define the properties of any medium in terms of the auxiliary equations summarized in Table 25.3.[2]

---

[1]If the medium is moving, the concepts of relativity must be invoked.

[2]These are auxiliary equations in the sense that they refine terms given in Maxwell's equations.

Table 25.1 *Maxwell's Equations*

integral form	point form	remarks
$\oint_s \mathbf{D} \cdot ds = \int_V \rho \, dv$	$\nabla \cdot \mathbf{D} = \rho$	Gauss's law
$\oint_s \mathbf{B} \cdot ds = 0$	$\nabla \cdot \mathbf{B} = 0$	nonexistence of magnetic monopoles
$\oint \mathbf{E} \cdot dl = \int_s \left(\dfrac{-\partial \mathbf{B}}{\partial t} \right) \cdot ds$	$\nabla \times \mathbf{E} = -\dfrac{\partial \mathbf{B}}{\partial t}$	Faraday's law
$\oint \mathbf{H} \cdot dl = \int_s \left(\mathbf{J}_c + \dfrac{\partial \mathbf{D}}{\partial t} \right) \cdot ds$	$\nabla \times \mathbf{H} = \mathbf{J}_c + \dfrac{\partial \mathbf{D}}{dt}$	Ampere's law

Table 25.2 *Maxwell's Equations: Free-Space Form*

integral form	point form
$\oint_s \mathbf{D} \cdot ds = 0$	$\nabla \cdot \mathbf{D} = 0$
$\oint_s \mathbf{B} \cdot ds = 0$	$\nabla \cdot \mathbf{B} = 0$
$\oint \mathbf{E} \cdot dl = \int_s \left(\dfrac{-\partial \mathbf{B}}{\partial t} \right) \cdot ds$	$\nabla \times \mathbf{E} = -\dfrac{\partial \mathbf{B}}{\partial t}$
$\oint \mathbf{H} \cdot dl = \int_s \left(\dfrac{\partial \mathbf{D}}{\partial t} \right) \cdot ds$	$\nabla \times \mathbf{H} = \dfrac{\partial \mathbf{D}}{dt}$

Table 25.3 *Electromagnetic Field Vector Equations*

$$\mathbf{D} = \varepsilon \mathbf{E} = \varepsilon_0 \mathbf{E} + \mathbf{P} = \varepsilon_0 \left(1 + \chi_e \right) \mathbf{E}$$
$$\mathbf{B} = \mu \mathbf{H} = \mu_0 \mathbf{H} + \mu_0 \mathbf{M} = \mu_0 \left(1 + \chi_m \right) \mathbf{H}$$
$$\mathbf{J} = \sigma \mathbf{E} = \rho \mathbf{v}$$

Even though the continuity equation is embedded in Maxwell's equation (as Gauss's and Ampere's laws), for completeness one should list it separately along with the Lorentz force equation. The continuity equation is

$$\nabla \cdot \mathbf{J} = -\frac{\partial \rho}{\partial t} \qquad 25.1$$

The Lorentz force equation is repeated here for convenience.

$$\mathbf{F} = Q \left(\mathbf{E} + \mathbf{v} \times \mathbf{B} \right) \qquad 25.2$$

The electromagnetic field vector equations define the electromagnetic properties of the medium. The field vector equations along with the continuity equation and the Lorentz force equation are used as the practical applications of Maxwell's equations to the relationships between $\mathbf{E}$, $\mathbf{D}$, $\mathbf{B}$, $\mathbf{H}$, and $\mathbf{J}$.

3. COMPARISON OF ELECTRIC AND MAGNETIC EQUATIONS

A comparison of electric and magnetic phenomena in terms of circuits is given in Table 25.4.

Table 25.4 *Electric and Magnetic Circuit Analogies*

electric	magnetic
$\text{emf} = V = IR$	$\text{mmf} = V_m = \phi \mathcal{R}$
current I	flux ϕ
emf $\mathcal{E}$ or V	mmf V_m
resistance $R = \rho l / A = l / \sigma A$	reluctance $\mathcal{R} = l / \mu A$
resistivity ρ	reluctivity $1 / \mu$
conductance $G = 1/R$	permeance $P_m = \mu A / l$
conductivity $\sigma = 1/\rho$	permeability μ

A more extensive comparison of electric and magnetic equations and concepts is given in App. 25.A.

Topic IV: Circuit Theory

Circuit Theory

PROFESSIONAL PUBLICATIONS, INC.

26 DC Circuit Fundamentals

Nomenclature

A	area	m^2
d	diameter	m
E	electromotive force	V
G	conductance	S
i, I	current	A
l	length	m
N	number	–
P	power	W
Q	charge	C
R	resistance	Ω
S	surface area	m^2
t	time	s
T	temperature	°C
v, V	voltage	V
W	work	J

Symbols

α	thermal coefficient of resistivity	$1/°C$
ρ	resistivity	$\Omega \cdot cm$ or $\Omega \cdot m$
σ	conductivity	S/m

Subscripts

0	initial
cmils	circular mils
Cu	copper
e	equivalent
fl	full load
nl	no load
N	Norton
oc	open circuit
s	shunt or source
sc	short circuit
Th	Thevenin

1. VOLTAGE

Voltage is the energy, or work, per unit charge exerted in moving a charge from one position to another. One of the positions is a reference position and is given an arbitrary value of zero. Voltage is also referred to as the *potential difference* to distinguish it from the unit of potential difference, the volt. One volt, V, is the potential difference if one joule of energy moves one coulomb of charge from the reference position to another position. Thus 1 V = 1 J/C. Additional equivalent units are W/A, C/F, A/S, and Wb/s.

In practical terms, a potential of one volt is defined as the potential existing between two points of a conducting wire carrying a constant current, I, of one ampere when the power dissipated between these two points is one watt. When the voltage refers to a source of electrical energy, it is termed the *electromotive force* (emf) and sometimes given the symbol E. In a *direct-current* (DC) *circuit*, the voltage, v, may vary in amplitude but not polarity. In many applications, the voltage magnitude, V, is constant as well. That is, it does not vary with time.

2. CURRENT

Current is the amount of charge transported past a given point per unit time. When current is constant or time-invariant, it is given the symbol I.[1] The unit of

---

[1] The symbol for current is derived from the French word intensité.

measure for current is the ampere (A), which is equivalent to coulombs per second (C/s). That is, one ampere is equal to the flow of one coulomb of charge past a plane surface, S, in one second. Though current is now known to be electron movement, it was originally viewed as flowing positive charges. This frame-of-reference current is called *conventional current* and is used in this and most other texts. When current refers to the actual flow of electrons, it is termed *electron current*. These concepts are illustrated in Fig. 26.1.

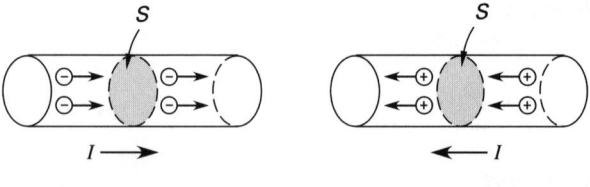

(a) electron current   (b) conventional current

Figure 26.1 *Current Flow*

Current does not "flow," though this term is often used in conjunction with current. Additionally, due to the conservation of charge, all circuit elements are electrically neutral, meaning that no net positive or negative charge accumulates on any circuit element.

Example 26.1

A conductor has a constant current of 10 A. How many electrons pass a fixed plane in the conductor in 1 min?

Solution

First, using amperes, determine the charge flowing per minute.

$$10 \text{ A} = \left(10 \; \frac{\text{C}}{\text{s}}\right)\left(60 \; \frac{\text{s}}{\text{min}}\right) = 600 \text{ C/min}$$

Using the charge of an electron, compute the number of electrons flowing.

$$\frac{600 \; \dfrac{\text{C}}{\text{min}}}{1.602 \times 10^{-19} \; \dfrac{\text{C}}{\text{electron}}} = 3.75 \times 10^{21} \text{ electrons/min}$$

3. RESISTANCE

Resistance, R, is the property of a circuit element that impedes current flow.[2] It is measured in ohms, Ω. A circuit with zero resistance is a *short circuit*, while a circuit with infinite resistance is an *open circuit*.

Resistors are often constructed from carbon compounds, ceramics, oxides, or coiled wire. Adjustable resistors are known as *potentiometers* and *rheostats*.

The resistance of any substance depends on its *resistivity* (ρ), its length, and its cross-sectional area. Assuming resistivity and cross-sectional area are constant, the resistance of a substance is

---

[2]Resistance is the real part of *impedance*, which opposes changes in current flow.

$$R = \frac{\rho l}{A} \qquad \text{26.1}$$

Though SI system units are the standard, the units of area, A, are often in the English Engineering System, especially in many tabulated values in the National Electrical Code. The area is given and tabulated for various conductors in *circular mils*, abbreviated cmil. One cmil is the area of a 0.001 in diameter circle. The concept of area in circular mils is represented by Eqs. 26.2, 26.3, and 26.4.

$$A_{\text{cmil}} = \left(\frac{d_{\text{inches}}}{0.001}\right)^2 \qquad \text{26.2}$$

$$A_{\text{in}^2} = 7.854 \times 10^{-7} \times A_{\text{cmil}} \qquad \text{26.3}$$

$$A_{\text{cm}^2} = 5.067 \times 10^{-6} \times A_{\text{cmil}} \qquad \text{26.4}$$

Resistivity is dependent on temperature. In most conductors, it increases with temperature, since at higher temperatures electron movement through the lattice structure becomes increasingly difficult. The variation of resistivity with temperature is specified by a *thermal coefficient of resistivity*, α, with typical units of $1/^\circ\text{C}$. The actual resistance or resistivity for a given temperature is calculated with Eqs. 26.5 and 26.6.

$$R = R_0 \left(1 + \alpha \Delta T\right) \qquad \text{26.5}$$

$$\rho = \rho_0 \left(1 + \alpha \Delta T\right) \qquad \text{26.6}$$

The thermal coefficients for several common conducting materials are given in Table 26.1.

Table 26.1 *Approximate Temperature Coefficients of Resistance[a] and Percent Conductivities*

material	α, $1/^\circ\text{C}$	% conductivity
aluminum, 99.5% pure	0.00423	63.0
aluminum, 97.5% pure	0.00435	59.8
constantan[b]	0.00001	3.1
copper, IACS (annealed)	0.00402	100.0
copper, pure annealed	0.00428	102.1
copper, hard drawn	0.00402	97.8
gold, 99.9% pure	0.00377	72.6
iron, pure	0.00625	17.5
iron wire, EBB[c]	0.00463	16.2
iron wire, BB[d]	0.00463	13.5
manganin[e]	0.00000	3.41
nickel	0.00622	12.9
platinum, pure	0.00367	14.6
silver, pure annealed	0.00400	108.8
steel wire	0.00463	11.6
tin, pure	0.00440	12.2
zinc, very pure	0.00406	27.7

(Multiply $1/^\circ\text{C}$ by 0.5556 to obtain $1/^\circ\text{F}$.)

[a]between 0°C and 100°C
[b]58% Cu, 41% Ni, 1% Mn
[c]Extra Best Best grade
[d]Best Best grade
[e]84% Cu, 4% Ni, 12% Mn

Example 26.2

What is the resistance of 305 m of 10 AWG (with an area of approximately 10,000 cmil) copper conductor with a resistivity of 2.82×10^{-6} Ω·cm?

Solution

From Eq. 26.1, and using the conversion of Eq. 26.4, the resistance is

$$R = \frac{\rho l}{A}$$

$$= \frac{(2.82 \times 10^{-6} \ \Omega\text{·cm})(305 \text{ m})\left(\dfrac{100 \text{ cm}}{1 \text{ m}}\right)}{(10{,}000 \text{ cmil})\left(\dfrac{5.067 \times 10^{-6} \text{ cm}^2}{1 \text{ cmil}}\right)}$$

$$= 1.70 \ \Omega$$

4. CONDUCTANCE

The reciprocal of resistivity is *conductivity*, σ, measured in siemens per meter (S/m). The reciprocal of resistance is *conductance*, G, measured in siemens (S). Siemens is the metric unit for the older unit known as the *mho* ($1/\Omega$).

$$\sigma = \frac{1}{\rho} \qquad 26.7$$

$$G = \frac{1}{R} \qquad 26.8$$

The *percent conductivity* is the ratio of a given substance's conductivity to the conductivity of standard IACS (International Annealed Copper Standard) copper, simply called *standard copper*. Alternatively, the percent conductivity is the ratio of standard copper's resistivity to the substance's resistivity.

$$\% \text{ conductivity} = \frac{\sigma}{\sigma_{\text{Cu}}} \times 100\%$$

$$= \frac{\rho_{\text{Cu}}}{\rho} \times 100\% \qquad 26.9$$

The standard resistivity of copper at 20°C is approximately

$$\rho_{\text{Cu,20°C}} = 1.7241 \times 10^{-6} \ \Omega\text{·cm}$$

$$= 0.3403 \ \Omega\text{·cmil/cm} \qquad 26.10$$

The percent conductivities for various substances are given in Table 26.1.

5. OHM'S LAW

Voltage, current, and resistance are related by Ohm's law.[3] The numerical result of Ohm's law is called the *voltage drop* or the *IR* drop.[4]

$$V = IR \qquad 26.11$$

Ohm's law presupposes a *linear circuit*, that is, a circuit consisting of linear elements and linear sources. In mathematical terms, this means that voltage plotted against current will result in a straight line with the slope represented by resistance. A *linear element* is a passive element, such as a resistor, whose performance can be portrayed by a linear voltage-current relationship. A *linear source* is one whose output is proportional to the first power of the voltage or current in the circuit. Many elements are linear or can be represented by equivalent linear circuits over some portion of their operation. Most sources, though not linear, can be represented as ideal sources with resistors in series or parallel to account for the nonlinearity.

6. POWER

If a steady current and voltage produce work, W, in time interval t, the electric power (energy conversion rate) is

$$P = \frac{W}{t} = \frac{VIt}{t} = VI = \frac{V^2}{R} \qquad 26.12$$

If the voltage and current vary with time, Eq. 26.12 still applies with these terms expressed as v and i. Equation 26.12 then represents the instantaneous power. The same is true for Eq. 26.13.

Equation 26.13 represents a form of power sometimes referred to as *I squared R* (I^2R) *losses*. Equation 26.13 is also the mathematical statement of *Joule's law of heating effect*.

$$P = I^2 R \qquad 26.13$$

7. DECIBELS

Power changes in many circuits range over decades, that is, over several orders of magnitude. As a result, decibels are adopted to express such changes. The use of

---

[3]This book uses the convention where uppercase letters represent fixed, maximum, effective values or direct current (DC) values, and lowercase letters represent values that change with time, such as alternating current (AC) values. Direct current (DC) values can change in amplitude over time (but not polarity or direction), and for this reason are sometimes shown as lowercase letters.
[4]It is sometimes helpful to consider electrical problems and theories in terms of their mechanical analogs. The mechanical analogy to Ohm's law in terms of fluid flow is as follows: Voltage is the pressure, current is the flow, and resistance is the head loss caused by friction and restrictions within the system.

decibels also has an advantage in that the power gain (or loss) of cascaded stages in series is the sum of the individual stage decibel gains (or losses), making the mathematics easier. Strictly speaking, decibels refer to a power ratio with the denominator arbitrarily chosen as a particular value, P_0. The reference value changes, depending upon the specific area of electrical engineering usage. For example, communications and acoustics use different references.[5] Decibels have come into common usage for referring to voltage and current ratios as well, though an accurate comparison between circuits requires equivalent resistances between the two terminals being compared.

$$\text{ratio (in dB)} = 10\log_{10}\left(\frac{P}{P_0}\right)$$

$$\approx 20\log_{10}\left(\frac{V_2}{V_1}\right)$$

$$= 20\log_{10}\left(\frac{I_2}{I_1}\right) \qquad 26.14$$

Time-varying quantities can also be in the ratio. Care must be taken to compare only instantaneous values or effective (rms) or time-averaged values to one another.

8. ENERGY SOURCES

Sources of electrical energy include friction between dissimilar substances, contact between dissimilar substances, thermoelectric action (for example, the Thomson, Peltier, and Seebeck effects), the Hall effect, electromagnetic induction, the photoelectric effect, and chemical action. These sources of energy manifest themselves by the potential, V, they generate. When sources of energy are processed through certain electronic circuits, a current source can be created.

An *ideal voltage source* supplies power at a constant voltage, regardless of the current drawn. An *ideal current source* supplies power in terms of a constant current, regardless of the voltage between its terminals. However, real sources have internal resistances that, at higher currents, reduce the available voltage. Consequently, a *real voltage source* cannot maintain a constant voltage when the currents are large. A *real current source* cannot maintain a constant current completely independent of the voltage at its terminals. Real and ideal voltage and current sources are shown in electrical schematic form in Fig. 26.2.

---

[5]Communications uses many such reference values. One references all power changes to 1 kW. In audio acoustics, the reference is the power necessary to cause the minimum sound pressure level audible to the human ear at 2000 Hz, that is, a pressure of 20 μPa.

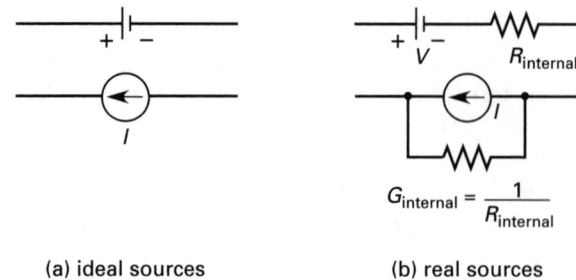

(a) ideal sources   (b) real sources

Figure 26.2 *Ideal and Real Energy Sources*

The change in the output of a voltage source is measured as its *regulation*. The formula is given by

$$\text{regulation} = \frac{V_{nl} - V_{fl}}{V_{fl}} \times 100\% \qquad 26.15$$

Independent sources deliver voltage and current at their rated values regardless of circuit parameters. *Dependent sources* deliver voltage and current at levels determined by a voltage or current somewhere else in the circuit.

9. VOLTAGE SOURCES IN SERIES AND PARALLEL

Voltage sources connected in series, as in Fig. 26.3(a), can be reduced to an equivalent circuit with a single voltage source and a single resistance, as in Fig. 26.3(b). The equivalent voltage and resistance are calculated per Eqs. 26.16 and 26.17.

$$V_e = \sum V_i \qquad 26.16$$

$$R_e = \sum R_i \qquad 26.17$$

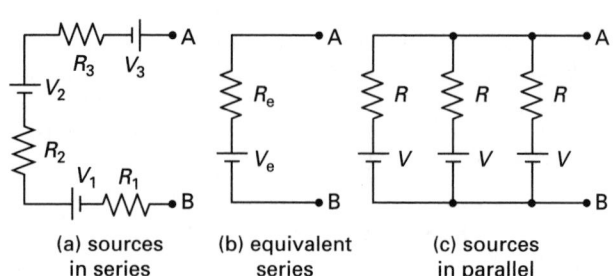

(a) sources   (b) equivalent   (c) sources
in series   series   in parallel

Figure 26.3 *Voltage Sources in Series and Parallel*

Millman's theorem states that N identical voltage sources of voltage V and resistance R connected in parallel, as in Fig. 26.3(c), can be reduced to an equivalent circuit with a single voltage source and a single resistance, as in Fig. 26.3(b). The equivalent voltage and resistance are calculated per Eqs. 26.18 and 28.19.

$$V_e = V \qquad 26.18$$

$$R_e = \frac{R}{N} \qquad 26.19$$

Nonidentical sources can be connected in parallel, but the lower-voltage sources may be "charged" by the higher-voltage sources. That is, the current may enter the positive terminal of the source. A loop current analysis is needed to determine the current direction and magnitude through the voltage sources.

10. CURRENT SOURCES IN SERIES AND PARALLEL

Currents sources may be placed in parallel. The circuit is then analyzed using any applicable method to determine the power delivered to the various elements. Current sources of differing magnitude cannot be placed in series without damaging one of them.

11. SOURCE TRANSFORMATIONS

A voltage source of V_s volts with an internal series resistance of R_s ohms can be represented by an equivalent circuit with a current source of I_s amps with an internal parallel resistance of the same R_s, and vice versa. Consequently, if Eq. 26.20 is valid, the circuits in Fig. 26.4 are equivalent.

$$V_s = I_s R_s \qquad 26.20$$

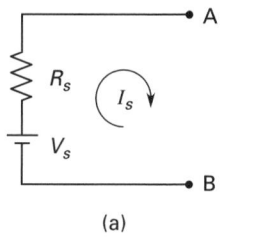 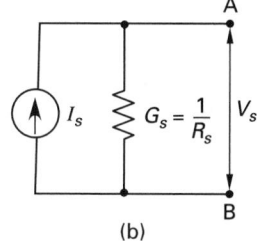

(a)   (b)

Figure 26.4 *Equivalent Sources*

12. MAXIMUM ENERGY TRANSFER

The maximum energy transfer, that is, the maximum power transfer, from a voltage source is attained when the series source resistance, R_s, of Fig. 26.4(a) is reduced to the minimum possible value, with zero being the ideal case. This assumes the load resistance is fixed and the source resistance can be changed. Though this is the ideal situation, it is not often the case. Where the load resistance varies and the source resistance is fixed, maximum power transfer occurs when the load and source resistances are equal.

Similarly, the maximum power transfer from a current source occurs when the parallel source resistance, R_s, of Fig. 26.4(b) is increased to the maximum possible value. This assumes the load resistance is fixed and the source resistance can be changed. In the more common case, where the load resistance varies and the source

resistance is fixed, the maximum power transfer occurs when the load and source resistances are equal.

The maximum power transfer occurs in any circuit when the load resistance equals the Norton or Thevenin equivalent resistance.

13. VOLTAGE AND CURRENT DIVIDERS

At times, a source voltage will not be at the required value for the operation of a given circuit. For example, the standard automobile battery voltage is 12 V, while some electronic circuitry uses 8 V for DC biasing. One method of obtaining the required voltage is by use of a circuit referred to as a *voltage divider*. A voltage divider is illustrated in Fig. 26.5(a). The voltage across resistor 2 is

$$V_2 = V_s \left(\frac{R_2}{R_1 + R_2} \right) \qquad 26.21$$

If the fraction of the source voltage, V_s, is found as a function of V_2, the result is known as the *gain* or the *voltage-ratio transfer function*. Rearranging Eq. 26.21 to show the specific relationship between R_1 and R_2 gives

$$\frac{V_2}{V_s} = \frac{1}{1 + \dfrac{R_1}{R_2}} \qquad 26.22$$

An analogous circuit called a *current divider* can be used to produce a specific current. A current divider circuit is shown in Fig. 26.5(b). The current through resistor 2 is

$$I_2 = I_s \left(\frac{R_1}{R_1 + R_2} \right) = I_s \left(\frac{G_2}{G_1 + G_2} \right) \qquad 26.23$$

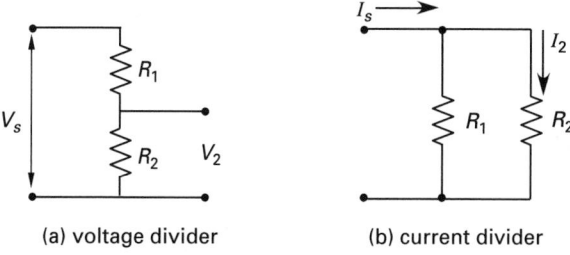

(a) voltage divider   (b) current divider

Figure 26.5 *Divider Circuits*

14. KIRCHHOFF'S LAWS

A fundamental principle in the design and analysis of electrical circuits is that the dimensions of the circuit are small. That is, the circuit's dimensions are small when compared to the wavelengths of the electromagnetic quantities that pass within them. Thus, only time variations need be considered, not spatial variations.

Consequently, Maxwell's equations, which are partial integrodifferential equations, become ordinary integro-differential equations that vary only with time.

The net result is that Gauss's law, in integral form, reduces to Kirchhoff's current law. This occurs because the net charge enclosed in any particular volume of the circuit is zero and thus the sum of the currents at any point must be zero.

Another result is that Faraday's law of electromagnetic induction reduces to Kirchhoff's voltage law. This is because, with a "small" circuit, the surface integral of magnetic flux density is zero; thus the sum of the voltages around any closed loop must be zero.

15. KIRCHHOFF'S VOLTAGE LAW

Kirchhoff's voltage law (KVL) states that the algebraic sum of the voltages around any closed path within a circuit or network is zero. Some of the voltages may be sources, while others will be voltages due to current in passive elements. Voltages through the passive elements are often called *voltage drops*. A *voltage rise* may also appear to occur across passive elements in the initial stages of the application of Kirchhoff's voltage law. For resistors, this is only because the direction of the current is arbitrarily chosen. For inductors and capacitors, a voltage rise indicates the release of magnetic or electrical energy. Stated in terms of the voltage rises and drops, KVL is

$$\sum_{\text{loop}} \text{voltage rises} = \sum_{\text{loop}} \text{voltage drops} \qquad 26.24$$

Kirchhoff's voltage law can also be stated as follows: The sum of the voltages around a closed loop must equal zero. The reference directions for voltage rises and drops used in this book are shown in Fig. 26.6.[6]

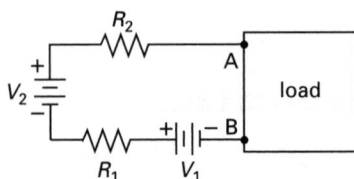

Figure 26.6 *Voltage Reference Directions*

Example 26.3

Consider the following circuit. Determine the expression for V_{AB} across the load in the circuit in the figure.

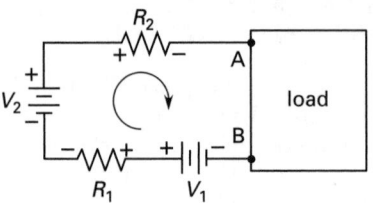

Solution

For consistency in all problems, apply KVL in a clockwise direction as shown. The unknown voltage is V_{AB}, which indicates that the voltage is arbitrarily referenced from A to B. That is, terminal A is assumed to be the high potential (+). Assign polarities to the resistors based on the assumed direction.

The KVL expression for the loop, starting at terminal B, is

$$V_1 - V_{R_1} + V_2 - V_{R_2} - V_{AB} = 0$$

Rearrange and solve for V_{AB}.

$$V_{AB} = V_1 - V_{R_1} + V_2 - V_{R_2}$$

16. KIRCHHOFF'S CURRENT LAW

Kirchhoff's current law (KCL) states that the algebraic sum of the currents at a node is zero. A *node* is a connection of two or more circuit elements. When the connection is between two elements, the node is a *simple node*. When the connection is between three or more elements, the node is referred to as a *principal node*. Stated in terms of the currents directed into and out of the node, KCL is

$$\sum_{\text{node}} \text{currents in} = \sum_{\text{node}} \text{currents out} \qquad 26.25$$

Kirchhoff's current law can also be stated as follows: The sum of the currents flowing out of a node must equal zero.[7] The reference directions for positive currents (currents "out") and negative currents (currents "in") used in this book are illustrated in Fig. 26.7.

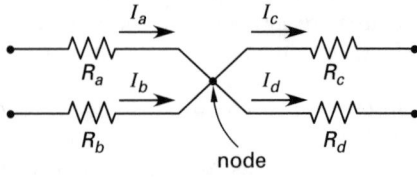

(a) I_a, I_b: negative currents or currents in
(b) I_c, I_d: positive currents or currents out

Figure 26.7 *Current Reference Directions*

---

[6]This alternative statement of Kirchhoff's voltage law and the arbitrary directions chosen are sometimes useful in simplifying the writing of the mathematical equations used in applying KVL.

[7]This alternative statement of Kirchhoff's current law and the arbitrary directions chosen are sometimes useful in simplifying the writing of the mathematical equations used in applying KCL.

17. SERIES CIRCUITS

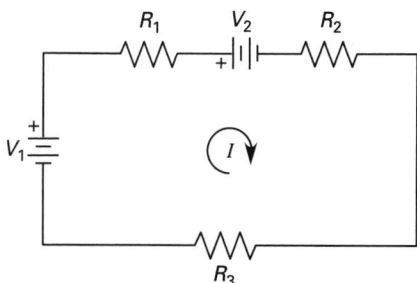

Figure 26.8 *Series Circuit*

A simple series circuit, as shown in Fig. 26.8, has the following properties (generalized to any number of voltage sources and series resistances).

- Current is the same through all the circuit elements.

$$I = I_{R_1} = I_{R_2} = I_{R_3} \cdots = I_{R_N} \qquad 26.26$$

- The equivalent resistance is the sum of the individual resistances.

$$R_e = R_1 + R_2 + R_3 \cdots + R_N \qquad 26.27$$

- The equivalent applied voltage is the sum of all the voltage sources, with the polarity considered.

$$V_e = \pm V_1 \pm V_2 \cdots \pm V_N \qquad 26.28$$

- The sum of the voltage drops across all circuit elements is equal to the equivalent applied voltage (KVL).

$$V_e = I R_e \qquad 26.29$$

18. PARALLEL CIRCUITS

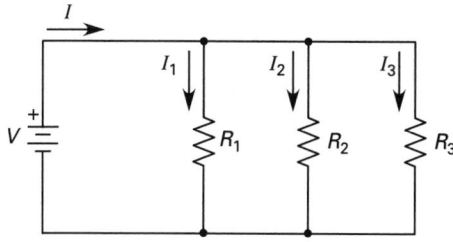

Figure 26.9 *Parallel Circuit*

A simple parallel circuit with only one active source, as shown in Fig. 26.9, has the following properties (generalized to any number of resistors).

- The voltage across all legs of the circuit is the same.

$$V = V_1 = V_2 = V_3 \cdots = V_N$$

$$= I_1 R_1 = I_2 R_2 = I_3 R_3 \cdots = V_N \qquad 26.30$$

- The reciprocal equivalent resistance is the sum of the reciprocals of the individual resistances. The equivalent conductance is the sum of the individual conductances.

$$\frac{1}{R_e} = \frac{1}{R_1} + \frac{1}{R_2} + \frac{1}{R_3} \cdots + \frac{1}{R_N} \qquad 26.31(a)$$

$$G_e = G_1 + G_2 + G_3 \cdots + G_N \qquad 26.31(b)$$

- The total current is the sum of the currents in the individual legs of the circuit (KCL).

$$I = I_1 + I_2 + I_3 \cdots + I_N$$

$$= \frac{V}{R_1} + \frac{V}{R_2} + \frac{V}{R_3} \cdots + \frac{V}{R_N}$$

$$= V \left(G_1 + G_2 + G_3 \cdots + G_N \right) \qquad 26.32$$

19. ANALYSIS OF COMPLICATED RESISTIVE NETWORKS

The following is a general method that can be used to determine the current flow and voltage drops within a complex resistive circuit.

step 1: If the circuit is three-dimensional, draw a two-dimensional representation.

step 2: Combine series voltage sources.

step 3: Combine parallel current sources.

step 4: Combine series resistances.

step 5: Combine parallel resistances.

step 6: Repeat steps 2–5 as needed to obtain the current and/or voltage at the desired point, junction, or branch of the circuit. (Do not simplify the circuit beyond what is required to determine the desired quantities.)

step 7: If applicable, utilize the delta-wye transformation of Sec. 26-20.

20. DELTA-WYE TRANSFORMATIONS

Electrical resistances arranged in the shape of the Greek letter delta or the English letter Y (wye) are known as *delta-wye configurations*. They are also called *pi-T configurations*. Two such circuits are shown in Fig. 26.10.

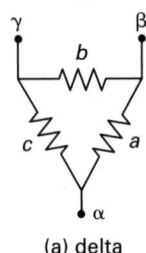

(a) delta

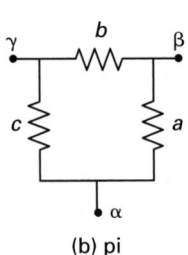

(b) pi

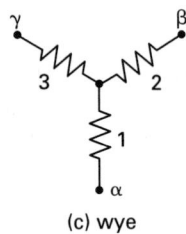

(c) wye

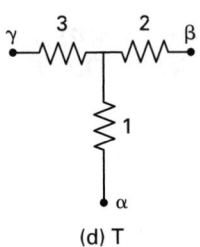

(d) T

Figure 26.10 *Delta (Pi)-Wye (T) Configurations*

The equivalent resistances, which allow transformation between configurations, are

$$R_a = \frac{R_1 R_2 + R_1 R_3 + R_2 R_3}{R_3} \qquad 26.33$$

$$R_b = \frac{R_1 R_2 + R_1 R_3 + R_2 R_3}{R_1} \qquad 26.34$$

$$R_c = \frac{R_1 R_2 + R_1 R_3 + R_2 R_3}{R_2} \qquad 26.35$$

$$R_1 = \frac{R_a R_c}{R_a + R_b + R_c} \qquad 26.36$$

$$R_2 = \frac{R_a R_b}{R_a + R_b + R_c} \qquad 26.37$$

$$R_3 = \frac{R_b R_c}{R_a + R_b + R_c} \qquad 26.38$$

Example 26.4

Simplify the circuit and determine the total current.

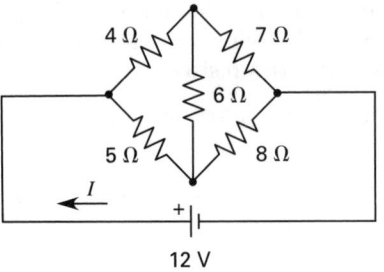

Solution

Convert the 4 Ω-5 Ω-6 Ω delta connection to wye form.

$$R_1 = \frac{R_a R_c}{R_a + R_b + R_c} = \frac{(5\ \Omega)(4\ \Omega)}{5\ \Omega + 6\ \Omega + 4\ \Omega}$$

$$= \frac{20}{15} = 1.33\ \Omega$$

$$R_2 = \frac{(5)(6)}{15\ \Omega} = 2\ \Omega$$

$$R_3 = \frac{(6)(4)}{15} = 1.6\ \Omega$$

The transformed circuit is

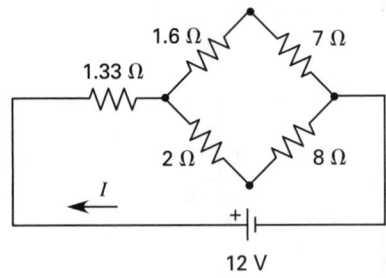

The total equivalent resistance is

$$R_e = 1.33 + \cfrac{1}{\cfrac{1}{1.6 + 7} + \cfrac{1}{2 + 8}} = 5.95\ \Omega$$

The current is

$$I = \frac{V}{R_e} = \frac{12 \text{ V}}{5.95 \text{ } \Omega} = 2.02 \text{ A}$$

21. SUBSTITUTION THEOREM

The *substitution theorem*, also known as the *compensation theorem*, states that any branch in a circuit can be replaced by a substitute branch, as long as the branch voltage and current remain the same, without affecting voltages and currents in any other portion of the circuit.

22. RECIPROCITY THEOREM

In any linear, time-independent circuit with independent current and voltage sources, the ratio of the current in a short circuit in one part of the network to the output of a voltage source in another part is constant—even when the positions of the voltage source and the short-circuit positions are interchanged. This principle of *reciprocity* is also applicable to the ratio of the current from a current source and the voltage across an open circuit. Reciprocity between a voltage source and short circuit is illustrated in Fig. 26.11.

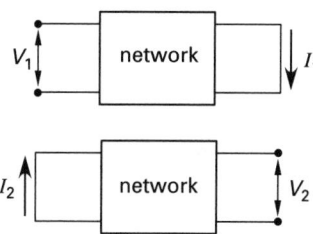

Figure 26.11 *Reciprocal Measurements*

Alternatively, if the network is comprised of linear resistors, the ratio of the applied voltage to the current measured at any point is constant—even when the positions of the source and meter are interchanged. The ratio in both cases is

$$R_{\text{transfer}} = \frac{V_1}{I_1} = \frac{V_2}{I_2} \qquad \textit{26.39}$$

The ratio is called the *transfer resistance*. The term *transfer* is used when a given response is determined at a point in a network other than where the driving force is applied. In this case, the voltage is the driving force on one side of the circuit and the current is measured on the other side.

23. SUPERPOSITION THEOREM

The principle of *superposition* is that the response (that is, the voltage across or current through) of a linear circuit element in a network with multiple independent

sources is equal to the response obtained if each source is considered individually and the results are summed. The steps involved in determining the desired quantity are as follows.

step 1: Replace all sources except one with their internal resistances.[8] Replace ideal current sources with open circuits. Replace ideal voltage sources with short circuits.

step 2: Compute the desired quantity, either voltage or current, for the element in question attributable to the single source.

step 3: Repeat steps 1 and 2 for each of the sources in turn.

step 4: Sum the calculated values obtained for the current or voltage obtained in step 2. The result is the actual value of the current or voltage in the element for the complete circuit.

Example 26.5

Determine the current through the center leg.

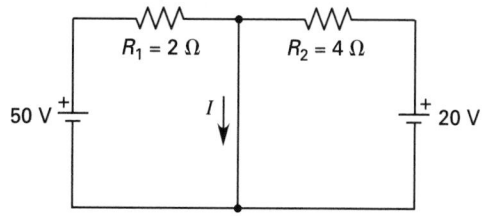

Solution

First, work with the left (50 V) battery. Short out the right (20 V) battery.

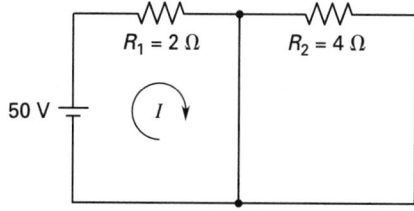

No current will flow through R_2. The equivalent resistance and current are

$$R_e = 2 \text{ } \Omega$$

$$I = \frac{V}{R_e} = \frac{50 \text{ V}}{2 \text{ } \Omega} = 25 \text{ A}$$

Next, work with the right (20 V) battery. Short out the left (50 V) battery.

---

[8]As in all the DC theorems presented in this chapter, this principle also applies to AC circuits or networks. In the case of AC circuits or networks, the term *impedance* would be applicable instead of *resistance*.

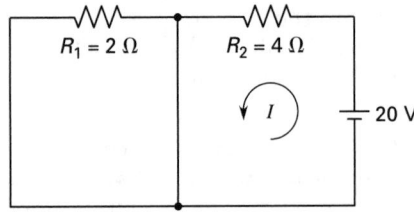

No current will flow through R_1. The equivalent resistance and current are

$$R_e = 4 \ \Omega$$

$$I = \frac{V}{R_e} = \frac{20 \text{ V}}{4} = 5 \text{ A}$$

Considering both batteries, the total current flowing is $25 \text{ A} + 5 \text{ A} = 30 \text{ A}$.

24. THEVENIN'S THEOREM

Thevenin's theorem states that, insofar as the behavior of a linear circuit at its terminals is concerned, any such circuit can be replaced by a single voltage source, V_{Th}, in series with a single resistance, R_{Th}. The method for determining and utilizing the *Thevenin equivalent circuit*, with designations referring to Fig. 26.12, follows.

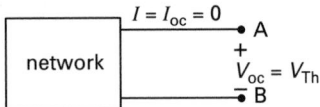

(a) open-circuit measurements

(b) short-circuit measurements

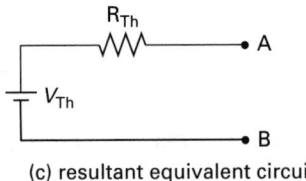

(c) resultant equivalent circuit

Figure 26.12 *Thevenin Equivalent Circuit*

step 1: Separate the network that is to be changed into a Thevenin equivalent circuit from its load at two terminals, say A and B.

step 2: Determine the open-circuit voltage, V_{oc}, at terminals A and B.

step 3: Short circuit terminals A and B and determine the current, I_{sc}.

step 4: Calculate the Thevenin equivalent voltage and resistance from the following equations.

$$V_{\text{Th}} = V_{\text{oc}} \qquad \qquad 26.40$$

$$R_{\text{Th}} = \frac{V_{\text{oc}}}{I_{\text{sc}}} \qquad \qquad 26.41$$

step 5: Using the values calculated in Eqs. 26.40 and 26.41, replace the network with the Thevenin equivalent. Reconnect the load at terminals A and B. Determine the desired electrical parameters in the load.

Steps 3 and 4 can be altered by using the following shortcut. Determine the Thevenin equivalent resistance by looking into terminals A and B toward the network with all the power sources altered. Specifically, change independent voltage sources into short circuits and independent current sources into open circuits, then calculate the resistance of the altered network. The resulting resistance is the Thevenin equivalent resistance.

Example 26.6

Determine the Thevenin equivalent circuit that the load resistor, R_{load}, sees.

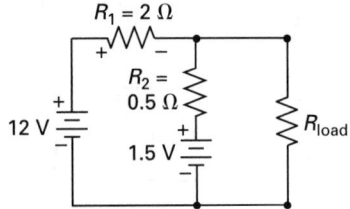

Solution

step 1: Separate the load resistor from the portion of the network to be changed.

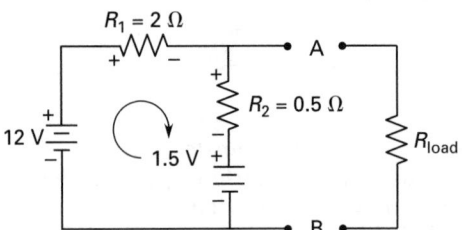

step 2: Apply Kirchhoff's voltage law to find the open-circuit voltage. Starting at terminal A, the equation is

$$12 \text{ V} - V_{R_1} - V_{R_2} - 1.5 \text{ V} = 0$$

$$V_{R_1} + V_{R_2} = 10.5 \text{ V}$$

For clarification, the circuit is redrawn using the result calculated.

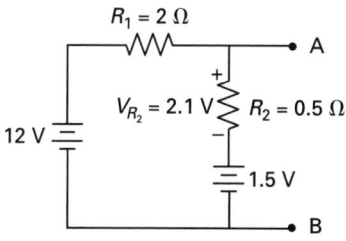

The voltage across R_2 is found using the voltage-divider concept.

$$V_{R_2} = (10.5 \text{ V}) \left(\frac{R_2}{R_1 + R_2} \right)$$

$$= (10.5 \text{ V}) \left(\frac{0.5 \text{ }\Omega}{2 \text{ }\Omega + 0.5 \text{ }\Omega} \right) = 2.1 \text{ V}$$

The circuit branch from A to B can now be illustrated as follows.

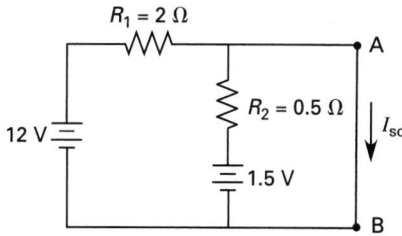

The open-circuit voltage is the voltage across R_2 and the 1.5 V source, that is $V_{oc} = 3.6$ V. (This can be found by using KVL around the loop containing the terminals A and B.)

step 3: Short terminals A and B to determine the short-circuit current.

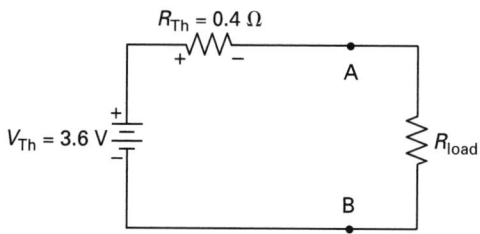

Using either Ohm's law or KVL around the outer loop, determine the short-circuit current contribution from the 12 V source.

$$I_{sc} = \frac{12 \text{ V}}{2 \text{ }\Omega} = 6 \text{ A}$$

Using either Ohm's law or KVL around the loop containing the 1.5 V source, R_2,

and terminals A and B, determine the short-circuit current contribution from the 1.5 V source.

$$I_{sc} = \frac{1.5 \text{ V}}{0.5 \text{ }\Omega} = 3 \text{ A}$$

The total short-circuit current, using the principle of superposition, is 6 A+3 A = 9 A.

step 4: Using Eqs. 26.40 and 26.41, calculate the Thevenin equivalent voltage and resistance.

$$V_{Th} = V_{oc} = 3.6 \text{ V}$$

$$R_{Th} = \frac{V_{oc}}{I_{sc}} = \frac{3.6 \text{ V}}{9 \text{ A}} = 0.4 \text{ }\Omega$$

step 5: Replace the network with the Thevenin equivalent.

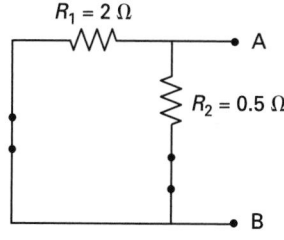

The method outlined is systematic. The shortcut for finding the Thevenin equivalent circuit mentioned earlier in this section would have simplified obtaining the solution. For example, short the voltage sources and determine the Thevenin equivalent resistance.

$$R_{Th} = \frac{R_1 R_2}{R_1 + R_2} = \frac{(2 \text{ }\Omega)(0.5 \text{ }\Omega)}{2 \text{ }\Omega + 0.5 \text{ }\Omega} = 0.4 \text{ }\Omega$$

The short-circuit current, found using superposition, is 9 A. The Thevenin equivalent voltage can then be found by combining Eqs. 26.29 and 26.41.

$$V_{Th} = I_{Th} R_{Th}$$

Circuit Theory

Substituting the short-circuit current for I_{sc} and the calculated value for R_{Th} results in an identical answer, as required by Thevenin's theorem.

25. NORTON'S THEOREM

Norton's theorem states that, insofar as the behavior of a linear circuit at its terminals is concerned, any such circuit can be replaced by a single current source, I_N, in parallel with a single resistance, R_N. The method for determining and utilizing the *Norton equivalent circuit*, with designations referring to Fig. 26.13, follows.

step 1: Separate the network that is to be changed into a Norton equivalent circuit from its load at two terminals, say A and B.

step 2: Determine the open-circuit voltage, V_{oc}, at terminals A and B.

step 3: Short circuit terminals A and B and determine the current, I_{sc}.

step 4: Calculate the Norton equivalent current and resistance from the following equations.

$$I_N = I_{sc} \qquad \textbf{26.42}$$

$$R_N = \frac{V_{oc}}{I_{sc}} \qquad \textbf{26.43}$$

step 5: Using the values calculated in Eqs. 26.42 and 26.43, replace the network with the Norton equivalent. Reconnect the load at terminals A and B. Determine the desired electrical parameters in the load.

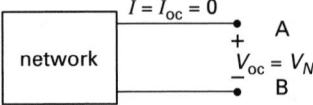

(a) open-circuit measurements

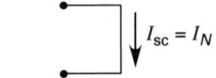

(b) short-circuit measurements

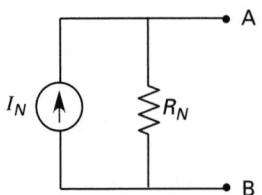

(c) resultant equivalent circuit

Figure 26.13 *Norton Equivalent Circuit*

Steps 3 and 4 can be altered by using the following shortcut. Determine the Norton equivalent resistance by looking into terminals A and B toward the network with all the power sources altered. Specifically, change independent voltage sources into short circuits and independent current sources into open circuits, then calculate the resistance of the altered network. The resulting resistance is the Norton equivalent resistance.

The Norton equivalent resistance equals the Thevenin equivalent resistance for identical networks.

26. LOOP-CURRENT METHOD

The *loop-current method* is a systematic network-analysis procedure that uses currents as the unknowns. It is also called *mesh analysis* or the *Maxwell loop-current method*. The method uses Kirchhoff's voltage law and is performed on planar networks. It requires $n-1$ simultaneous equations for an n-loop system. The method's steps are as follows.

step 1: Select $n-1$ loops, that is, one loop less than the total number of possible loops.

step 2: Assume current directions for the selected loops. (While this is arbitrary, clockwise directions will always be chosen in this text for consistency. Any incorrectly chosen current direction will result in a negative result when the simultaneous equations are solved in step 4.) Show the direction of the current with an arrow.

step 3: Write Kirchhoff's voltage law for each of the selected loops. Assign polarities based on the direction of the loop current. The voltage of a source is positive when the current flows out of the positive terminal, that is, from the negative terminal to the positive terminal inside the source. The selected direction of the loop current always results in a voltage drop in the resistors of the loop (see Fig. 26.6(c)). Where two loop currents flow through an element, they are summed to determine the voltage drop in that element, using the direction of the current in the loop for which the equation is being written as the positive (i.e., correct) direction. (Any incorrect direction for the loop current will be indicated by a negative sign in the solution in step 4.)

step 4: Solve the $n-1$ equations (from step 3) for the unknown currents.

step 5: If required, determine the actual current in an element by summing the loop currents flowing through the element. (Sum the absolute values to obtain the correct magnitude. The correct direction is given by the loop current with the positive value.)

27. NODE-VOLTAGE METHOD

The *node-voltage method* is a systematic network-analysis procedure that uses voltages as the unknowns. The method uses Kirchhoff's current law. It requires $n-1$ equations for an n-principal node system (equations are not necessary at simple nodes, that is, nodes connecting only two circuit elements). The method's steps are as follows.

step 1: Simplify the circuit, if possible, by combining resistors in series or parallel or by combining current sources in parallel. Identify all nodes. (The minimum total number of equations required will be $n-1$, where n represents the number of principal nodes.)

step 2: Choose one node as the reference node, that is, the node that will be assumed to have ground potential (0 V). (To minimize the number of terms in the equations, pick the node with the largest number of circuit elements as the reference node.)

step 3: Write Kirchhoff's current law for each principal node except the reference node, which is assumed to have a zero potential.

step 4: Solve the $n-1$ equations (from step 3) to determine the unknown voltages.

step 5: If required, use the calculated node voltages to determine any branch current desired.

28. DETERMINATION OF METHOD

When analyzing electrical networks, the method used depends on the circuit elements and their configurations. The loop-current method using Kirchhoff's voltage law is used in circuits without current sources. The node-voltage method using Kirchhoff's current law is used in circuits without voltage sources. When both types of sources are present, one of the following two methods may be used.

method 1: Use each of Kirchhoff's laws, assigning voltages and currents as needed, and substitute any known quantity into the equations as written.

method 2: Use source transformation or source shifting to change the appearance of the circuit so that it contains only the desired sources, that is, voltage sources when using KVL and current sources when using KCL. (Source shifting is manipulating the circuit so that each voltage source has a resistor in series and each current source has a resistor in parallel.)

Use the method that results in the least number of equations. The loop-current method produces $n-1$ equations, where n is the total number of loops. The node-voltage method produces $n-1$ equations, where n is the number of principal nodes. Count the number of loops and the number of principal nodes prior to writing the equations; whichever is least determines the method used.

Additional methods exist, some of which require less work. The advantage of the loop-current and node-voltage methods is that they are systematic and thus guarantee a solution.

29. PRACTICAL APPLICATION: BATTERIES

In general, a battery is defined as a direct-current voltage source made up of one or more units that convert chemical, thermal, nuclear, or solar energy into electrical energy. The most widely used battery type is one that converts the chemical energy contained in its active materials directly into electrical energy by means of an oxidation-reduction reaction.[9] A battery consists of two dissimilar metals, an anode and a cathode, immersed in an electrolyte. The anode is the component that gives up electrons, that is, it is oxidized during the reaction. The anode is labeled as the positive terminal since by definition this is the terminal through which current enters the battery. (The current is conventional current flow. Therefore, positive "charges" enter the anode and it is thus the source of the electrons to the external circuit.) The cathode is reduced during the reaction. The transfer of charge is completed within the electrolyte by the flow of ions. This is shown for a single cell in Fig. 26.14.

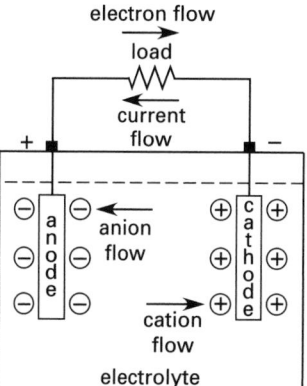

Figure 26.14 *Electrochemical Battery*

The battery terminals have no absolute voltage value. They have a value relative only to each other, and thus the terminals and the battery are said to *float*. Usually, one of the terminals is assigned as the reference and given the value of 0 V. This terminal is then the *reference* or *datum* and is said to be *grounded*. If the battery is allowed to float and the circuit is grounded elsewhere, all voltages are assumed to be measured with respect to this ground. The earth is generally regarded as being

---

[9]In *nonelectrochemical reactions*, the transfer of electrons takes place directly and only heat is involved.

at 0 V and any circuit tied to earth by an electrical wire is grounded. Various symbols for grounds are shown in Fig. 26.15.

Figure 26.15 *Ground Symbols*

The voltage of a battery is determined by the number of cells used. The voltage of the cell is determined by the materials used, as this determines the half-cell oxidation potentials. The capacity of the battery is determined by the amount of materials used and is measured in ampere-hours (A·h). One gram-equivalent weight of material supplies approximately 96,480 C or 26.805 A·h of electric charge.

A *primary battery* is one that uses an electrochemical reaction that is not efficiently reversible. An example is the common flashlight battery. This type of battery is also called a *dry cell*, as the electrolyte is a moist paste instead of a liquid solution. A *secondary battery* is rechargeable and has a much higher energy density. An example is a lead-acid storage battery used in automobiles. Such batteries are treated as ideal voltage sources with a series resistance representing internal resistance. Using the specified voltage and resistance, any network containing a battery is analyzed using the techniques described in this chapter.

27 AC Circuit Fundamentals

Nomenclature

B	susceptance	S, Ω^{-1}, or mho
C	capacitance	F
f	function	–
f	frequency	Hz, s^{-1}, or cycles/s
G	conductance	S, Ω^{-1}, or mho
i	instantaneous current	A
I	effective or DC current	A
Im	imaginary	–
L	inductance	H
p	instantaneous power	W
pf	power factor	–
P	power	W
Q	reactive power	var (volt-amps reactive or VAR)
R	resistance	Ω
Re	real portion[1]	–
S	apparent power	voltampere (volt-amps or VA)
t	time	s
T	period	s
v, V	voltage	V
X	reactance	Ω
Y	admittance	S, Ω^{-1}, or mho
$\mathbf{Z}$	complex number	–
Z	impedance	Ω

Symbols

θ	phase angle	rad
ϕ	phase difference angle	rad
ϕ	impedance angle	rad
φ	current angle $(\theta \pm \phi)$	rad
Ψ	power factor angle	rad
ω	angular frequency	rad/s

Subscripts

0	initial
ave	average
C	capacitor
e	equivalent
eff	effective
i	imaginary
I	current
L	inductor
m	maximum
p	peak
r	real
rms	root-mean-square
R	resistor
s	source
thr	threshold
V	voltage
Z	impedance

1. FUNDAMENTALS

Alternating waveforms have currents and voltages that vary with time in a symmetrical manner. Possible waveforms include square, sawtooth, and triangular along with many variations on these themes. However, for the most part, the variations are sinusoidal in time for many applications in electrical engineering. In this book, unless otherwise specified, currents and voltages referred to are sinusoidal.[2] When sinusoidal, the waveform is nearly always referred to as AC, that is, *alternating current*, indicating that the current is produced by the application of a sinusoidal voltage. This means that the flow of electrons changes directions, unlike DC circuits, where the flow of electrons is unidirectional (though the

---

[1]Do not confuse this symbology, Re, for that of the Reynolds number.

[2]Nearly all periodic functions can be represented as sinusoidal functions. The superposition theorem allows the effects of individual sinusoids to be summed to obtain an overall effect, which simplifies the mathematics necessary to obtain a result.

Circuit Theory

magnitude can change in time). A circuit is said to be in a *steady-state* condition if the current and voltage time variation is purely constant (DC) or purely sinusoidal (AC).[3] In this book, unless otherwise specified, the circuits referred to are in a steady-state condition. Importantly, all the methods, basic definitions, and equations presented in Ch. 26 involving DC circuits are applicable to AC circuits. AC electrical parameters have both magnitudes and angles. Nevertheless, following common practice, phasor notation will be used only in cases where not doing so would cause confusion.

2. VOLTAGE

Sinusoidal variables can be expressed in terms of sines or cosines without any loss of generality.[4] A sine waveform is often the standard. If this is the case, Eq. 27.1 gives the value of the instantaneous voltage as a function of time.

$$v(t) = V_m \sin(\omega t + \theta) \qquad 27.1$$

The *maximum value* of the sinusoid is given the symbol V_m and is also known as the *amplitude*. If $v(t)$ is not zero at $t = 0$, the sinusoid is *shifted* and a *phase angle*, θ, must be used, as shown in Fig. 27.1.[5] Also shown in Fig. 27.1 is the *period*, T, which is the time that elapses in one cycle of the sinusoid. The *cycle* is the smallest portion of the sinusoid that repeats.

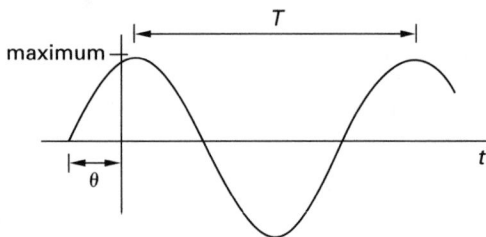

Figure 27.1 *Sinusoidal Waveform with Phase Angle*

Since the horizontal axis of the voltage in Fig. 27.1 corresponds to time, not distance, the waveform does not have a wavelength. The frequency, f, of the sinusoid is the reciprocal of the period in hertz (Hz). The angular frequency, ω, in rad/s can also be used.

$$f = \frac{1}{T} = \frac{\omega}{2\pi} \qquad 27.2$$

$$\omega = 2\pi f = \frac{2\pi}{T} \qquad 27.3$$

---

[3]Steady-state AC may have a *DC offset*.
[4]The point at which time begins, that is, where $t = 0$, is of no consequence in steady-state AC circuit problems, as the signal repeats itself every cycle. Therefore, it makes no difference whether a sine or cosine waveform is used (though, if one exists, care must be taken to keep the phase angle correct).
[5]The term *phase* is not the same as *phase difference*, which is the difference between corresponding points on two sinusoids of the same frequency.

An AC voltage waveform without a phase angle is plotted as a function of differing variables in Fig. 27.2.

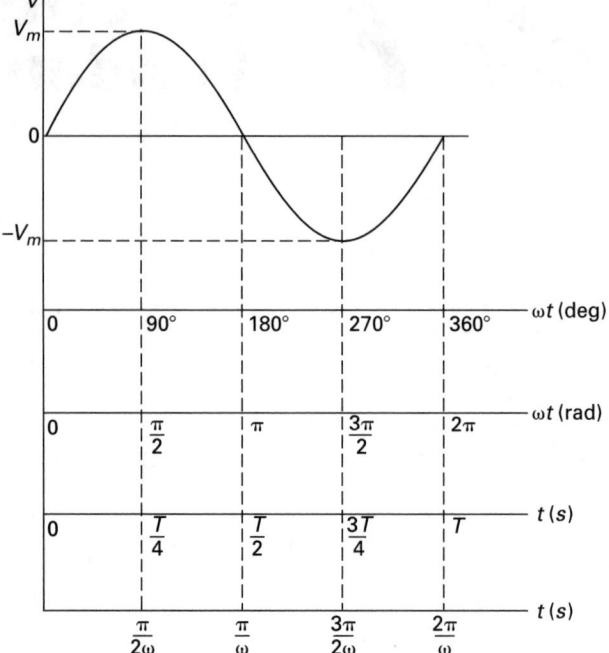

Figure 27.2 *Sine Wave Plots*

Exponentials, $e^{j\theta}$ and $e^{-j\theta}$, can be combined to produce $\sin\theta$ and $\cos\theta$ terms. As a result, sinusoids can be represented in the following equivalent forms.

- trigonometric: $V_m \sin(\omega t + \theta)$
- exponential: $V_m e^{j\theta}$
- polar or phasor: $V_m \angle \theta$
- rectangular: $V_r + jV_i$

The use of complex exponentials and phasor analysis allows sinusoidal functions to be more easily manipulated mathematically, especially when dealing with derivatives. Exponents are manipulated algebraically, and the resulting sinusoid is then recovered using Euler's relation. This avoids complicated trigonometric mathematics. The angles used in exponentials and with angular frequency, ω, must be in radians. From Euler's relation,

$$e^{j\theta} = \cos\theta + j\sin\theta \qquad 27.4$$

$$j = \sqrt{-1} \qquad 27.5$$

Therefore, to regain the sinusoid from the exponential, use Euler's equation, keeping in mind two factors. First, either the real or the imaginary part contains the sinusoid desired, not both.[6] Second, the exponential should

---

[6]The cosine function could be used as the reference sinusoid, and is in some texts. In this text, unless otherwise specified, assume the sine function is the desired form. The desired form is determined by the representation of the voltage source waveform as either a cosine or a sine function (see Eq. 27.1). Either way, only the phase angle changes.

not be factored out of any equation until all the derivatives, or integrals, have been taken.

3. CURRENT

Current is the net transfer of electric charge per unit time. When a circuit's driving force, the voltage, is sinusoidal, the resulting current is also (though it may differ by an amount called the *phase angle difference*). The equations and representations in Sec. 27-2, as well as any other representations of sinusoidal waveforms provided in this chapter, apply to current as well as to voltage.

4. IMPEDANCE

Electrical impedance, also known as *complex impedance*, is the total opposition a circuit presents to alternating current. It is equal to the ratio of the complex voltage to the complex current. Impedance, then, is a ratio of phasor quantities and is not itself a function of time. Relating voltage and current in this manner is analogous to Ohm's law, which for AC analysis is referred to as *extended Ohm's law*. Impedance is given the symbol **Z** and is measured in ohms. The three passive circuit elements (resistors, capacitors, and inductors), when used in an AC circuit, are assigned an angle, θ, known as the *impedance angle*. This angle corresponds to the phase difference angle produced when a sinusoidal voltage is applied across that element alone.

Impedance is a complex quantity with a magnitude and associated angle. It can be written in *phasor form*—also known as *polar form*—for example, $\mathbf{Z}\angle\theta$, or in *rectangular form* as the complex sum of its resistive (R) and reactive (X) components.

$$\mathbf{Z} \equiv R + jX \qquad \text{27.6}$$

$$R = Z\cos\phi \quad \begin{bmatrix} \text{resistive or} \\ \text{real part} \end{bmatrix} \qquad \text{27.7}$$

$$X = Z\sin\phi \quad \begin{bmatrix} \text{reactive or} \\ \text{imaginary part} \end{bmatrix} \qquad \text{27.8}$$

The resistive and reactive components can be combined to form an *impedance triangle*. Such a triangle is shown in Fig. 27.3.

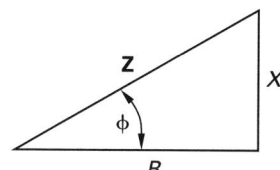

Figure 27.3 *Impedance Triangle*

The characteristics of the passive elements in AC circuits, including impedance, are given in Table 27.1.

Table 27.1 *Characteristics of Resistors, Capacitors, and Inductors*

	resistor	capacitor	inductor
value	R (Ω)	C (F)	L (H)
reactance, X	0	$\dfrac{-1}{\omega C}$	ωL
rectangular impedance, **Z**	$R + j0$	$0 - \dfrac{j}{\omega C}$	$0 + j\omega L$
phasor impedance, **Z**	$R\angle 0°$	$\dfrac{1}{\omega C}\angle{-90°}$	$\omega L\angle 90°$
phase	in-phase	leading	lagging
rectangular admittance, **Y**	$\dfrac{1}{R} + j0$	$0 + j\omega C$	$0 - \dfrac{j}{\omega L}$
phasor admittance, **Y**	$\dfrac{1}{R}\angle 0°$	$\omega C\angle 90°$	$\dfrac{1}{\omega L}\angle{-90°}$

5. ADMITTANCE

The reciprocal of impedance is called *admittance*, **Y**. Admittance is useful in analyzing parallel circuits, as admittances can be added directly. The defining equation is

$$\mathbf{Y} = \frac{1}{\mathbf{Z}} = \frac{1}{Z}\angle{-\phi} \qquad \text{27.9}$$

The reciprocal of the resistive portion of an impedance is known as *conductance*, G. The reciprocal of the reactive part is termed the *susceptance*, B.

$$G = \frac{1}{R} \qquad \text{27.10}$$

$$B = \frac{1}{X} \qquad \text{27.11}$$

Using these definitions and multiplying by a complex conjugate, admittance can be written in terms of resistance and reactance. Using the same method, impedance can be written in terms of conductance and susceptance. Equations 27.12 and 27.13 show this and can be used for conversion between admittance and impedance and vice versa.

$$\mathbf{Y} = G + jB = \frac{R}{R^2 + X^2} - j\left(\frac{X}{R^2 + X^2}\right) \qquad \text{27.12}$$

$$\mathbf{Z} = R + jX = \frac{G}{G^2 + B^2} - j\left(\frac{B}{G^2 + B^2}\right) \qquad \text{27.13}$$

Circuit Theory

6. VOLTAGE SOURCES

The energy for AC voltage sources comes primarily from electromagnetic induction. The concepts of ideal and real sources, as well as regulation, apply to AC sources (see Sec. 26-8). Independent sources deliver voltage and current at their rated values regardless of circuit parameters. Dependent sources, often termed *controlled sources*, deliver voltage and current at levels determined by a voltage or current somewhere else in the circuit. These types of sources occur in electronic circuitry and are also used to model electronic elements, such as transistors. The symbols used for AC voltage sources are shown in Fig. 27.4.

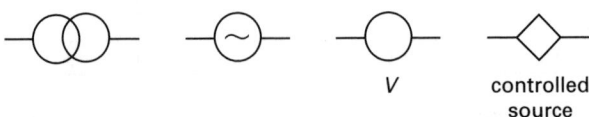

Figure 27.4 *AC Voltage Sources Symbology*

7. AVERAGE VALUE

In purely mathematical terms, the average value of a periodic waveform is the first term of a Fourier series representing the function, that is, it is the zero frequency or DC value. If a function, $f(t)$, repeats itself in a time period T, then the average value of the function is given by Eq. 27.14. In Eq. 27.14, t_1 is any convenient time for evaluating the integral, that is, the time that simplifies the integration. The integral itself is computed over the period.

$$f_{\text{ave}} = \frac{1}{T} \int_{t_1}^{t_1+T} f(t)dt \qquad 27.14$$

Integration can be interpreted as the area under the curve of the function. Equation 27.14 divides the net area of the waveform by the period T. This concept is illustrated in Fig. 27.5 and stated in mathematical terms by Eq. 27.15.

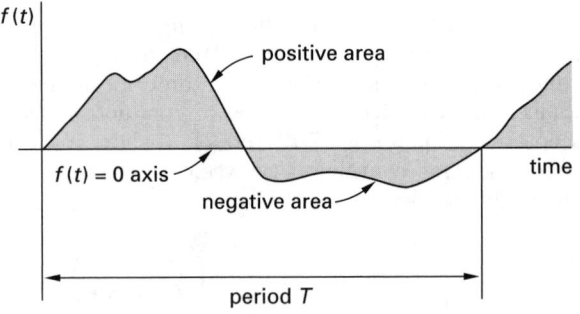

Figure 27.5 *Average Value Areas Defined*

$$f_{\text{ave}} = \frac{\text{positive area} - \text{negative area}}{T} \qquad 27.15$$

For the function shown in Fig. 27.5, the area above the axis is called the positive area and the area below is called the negative area. The average value is the net area remaining after the negative area is subtracted from the positive area and the result is divided by the period.

For any periodic voltage, Eq. 27.16 calculates the average value.

$$V_{\text{ave}} = \frac{1}{2\pi} \int_0^{2\pi} v(\theta)d\theta = \frac{1}{T} \int_0^T v(t)dt \qquad 27.16$$

Any waveform that is symmetrical with respect to the horizontal axis will result in a value of zero for Eq. 27.16. While mathematically correct, the electrical effects of such a voltage occur on both the positive and negative half-cycles. Therefore, the average is instead taken over only half a cycle. This is equivalent to determining the average of a *rectified waveform*, that is, the absolute value of the waveform. The average voltage for a rectified sinusoid is

$$V_{\text{ave}} = \frac{1}{\pi} \int_0^{\pi} v(\theta)d\theta$$

$$= \frac{2V_m}{\pi} \quad \text{[rectified sinusoid]} \qquad 27.17$$

A DC current equal in value to the average value of a rectified AC current produces the same electrolytic effects, such as capacitor charging, plating operations, and ion formation. Nevertheless, not all AC effects can be accounted for using the average value. For instance, in a typical DC meter, the average DC current determines the response of the needle. An AC current sinusoid of equal average magnitude will not result in the same effect since the torque on each half-cycle is in opposite directions, resulting in a net zero effect. Unless the AC signal is rectified, the reading will be zero.

Example 27.1

What is the average value of the pure sinusoid shown?

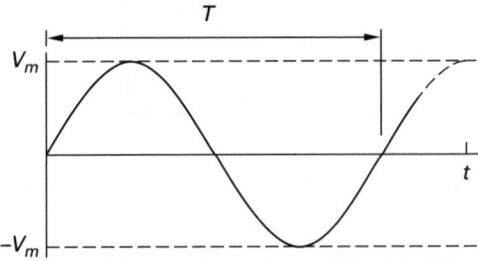

Solution

The sinusoid shown is a sine wave. Substituting into Eq. 27.14 with $t_1 = 0$ gives

$$f_{ave} = \frac{1}{T} \int_{t_1}^{t_1+T} f(t)dt = \frac{1}{T} \int_0^T V_m \sin t\, dt$$

$$= \left(\frac{V_m}{T}\right)\left(\cos t \Big|_0^T\right) = \left(\frac{V_m}{T}\right)\left(\cos T - \cos(0)\right)$$

$$= \left(\frac{V_m}{T}\right)(1-1) = 0$$

This result is as expected, since a pure sinusoid has equal positive and negative areas.

Example 27.2

Diodes are used to rectify AC waveforms. Real diodes will not pass current until a threshold voltage is reached. As a result, the output is the difference between the sinusoid and the threshold value. This is illustrated in the following figures.

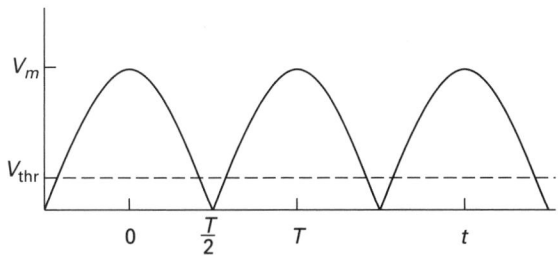

(a) ideal rectified waveform

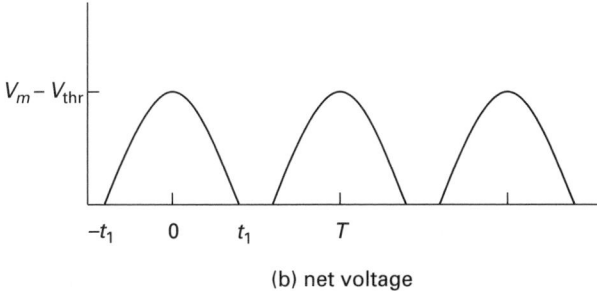

(b) net voltage

What is the expression for the average voltage on the output of a real rectifier diode with a sinusoidal voltage input?

Solution

Because of the symmetry, the average value can be found using half the average determined from Eq. 27.16 by integrating from zero to $T/2$, simplifying the calculation. The waveform is represented by

$$v(t) = \begin{cases} V_m \cos\left(\left(\frac{\pi}{T}\right)t\right) - V_{thr} & \text{[for } 0 < t < t_1] \\ 0 & \text{[for } t_1 < t < {}^T\!/_2] \end{cases}$$

At $t = t_1$, the following condition exists.

$$V_m \cos\left(\left(\frac{\pi}{T}\right)t_1\right) = V_{thr}$$

The equation for one-half the average is

$$\frac{1}{2}v_{ave} = \frac{1}{T} \int_0^{t_1} \left(V_m \cos\left(\left(\frac{\pi}{T}\right)t\right) - V_{thr}\right)dt$$

$$= \left(\frac{1}{T}\right)\left(\frac{T}{\pi}\right)V_m \sin\left(\left(\frac{\pi}{T}\right)t_1\right) - V_{thr}\left(\frac{t_1}{T}\right)$$

Using the equation for the condition at $t = t_1$ and the trigonometric identity $\sin^2 x + \cos^2 x = 1$ results in the following equations.

$$\sin\left(\left(\frac{\pi}{T}\right)t_1\right) = \sqrt{1 - \left(\frac{V_{thr}}{V_m}\right)^2}$$

$$\frac{t_1}{T} = \frac{1}{\pi}\arccos\left(\frac{V_{thr}}{V_m}\right)$$

Substituting into the equation for one-half the average and rearranging gives the following final result.

$$v_{ave} = \left(\frac{2}{\pi}\right)V_m\sqrt{1 - \left(\frac{V_{thr}}{V_m}\right)^2}$$
$$- \left(\frac{2}{\pi}\right)V_{thr}\arccos\left(\frac{V_{thr}}{V_m}\right)$$

The arccos must be expressed in radians. The conclusion drawn is that real diode average values are more complex than values for ideal diode full-wave rectification. However, if ideal diodes are assumed, that is, if the threshold voltage is considered negligible, the following errors are generated.

- 1% error for $V_{thr}/V_m = 0.0064$
- 2% error for $V_{thr}/V_m = 0.0128$
- 5% error for $V_{thr}/V_m = 0.0321$
- 10% error for $V_{thr}/V_m = 0.0650$

For most practical applications, the ideal diode assumption results in an error of less than 10%.

8. ROOT-MEAN-SQUARE VALUE

In purely mathematical terms, the *effective value* (also known as the *root-mean-square* (rms) *value*) of a periodic waveform represented by a function, $f(t)$, which repeats itself in a time period is given by Eq. 27.18. In Eq. 27.18, t_1 is any convenient time for evaluating the integral, that is, the time that simplifies the integration. The integral itself is computed over the period.

$$f_{rms}^2 = \frac{1}{T} \int_{t_1}^{t_1+T} f^2(t)dt \qquad \textbf{27.18}$$

For any periodic voltage, Eq. 27.19 calculates the effective or rms value.

$$V = V_{eff} = V_{rms} = \sqrt{\frac{1}{2\pi} \int_0^{2\pi} v^2(\theta)\,d\theta}$$

$$= \sqrt{\frac{1}{T} \int_0^T v^2(t)\,dt} \qquad 27.19$$

Normally when a voltage or current variable, V or I, is left unsubscripted, it represents the effective (rms) value.[7] Rectification of the waveform is not necessary to calculate the effective value. (The squaring of the waveform ensures the net result of integration is something other than zero for a sinusoid.) For a sinusoidal waveform, $V = V_m/\sqrt{2} \approx 0.707 V_m$. A DC current of magnitude I produces the same heating effect as an AC current of magnitude I_{eff}.

Table 27.2 gives the characteristics of various commonly encountered alternating waveforms. In this table, two additional terms are introduced. The *form factor* is given by

$$FF = \frac{V_{eff}}{V_{ave}} \qquad 27.20$$

The *crest factor*, CF, also known as the *peak factor* or *amplitude factor*, is

$$CF = \frac{V_m}{V_{eff}} \qquad 27.21$$

Nearly all periodic functions can be represented by a Fourier series. A Fourier series can be written as

$$f(t) = \tfrac{1}{2}a_0 + \sum_{n=1}^{\infty} a_n \cos\left(\left(\frac{2\pi n}{T}\right)t\right)$$

$$+ \sum_{n=1}^{\infty} b_n \sin\left(\left(\frac{2\pi n}{T}\right)t\right) \qquad 27.22$$

The average value is the first term, as mentioned in Sec. 27-7. The rms value is

$$f_{rms} = \sqrt{\left(\tfrac{1}{2}a_0\right)^2 + \tfrac{1}{2}\sum_{n=1}^{\infty}\left(a_n^2 + b_n^2\right)} \qquad 27.23$$

[7]The value of the standard voltage in the United States, reported as 115–120 V, is an effective value.

Table 27.2 *Characteristics of Alternating Waveforms*

waveform	$\dfrac{V_{ave}}{V_m}$	$\dfrac{V_{rms}}{V_m}$	FF	CF
sinusoid	0	$\dfrac{1}{\sqrt{2}}$	–	$\sqrt{2}$
full-wave rectified sinusoid	$\dfrac{2}{\pi}$	$\dfrac{1}{\sqrt{2}}$	$\dfrac{\pi}{2\sqrt{2}}$	$\sqrt{2}$
half-wave rectified sinusoid	$\dfrac{1}{\pi}$	$\dfrac{1}{2}$	$\dfrac{\pi}{2}$	2
symmetrical square wave	0	1	–	1
unsymmetrical square wave	$\dfrac{t}{T}$	$\sqrt{\dfrac{t}{T}}$	$\sqrt{\dfrac{T}{t}}$	$\sqrt{\dfrac{T}{t}}$
sawtooth and symmetrical triangular	0	$\dfrac{1}{\sqrt{3}}$	–	$\sqrt{3}$
sawtooth and symmetrical triangular	$\dfrac{1}{2}$	$\dfrac{1}{\sqrt{3}}$	$\dfrac{2}{\sqrt{3}}$	$\sqrt{3}$

Example 27.3

A peak sinusoidal voltage, V_p, of 170 V is connected across a 240 Ω resistor in a light bulb. What is the power dissipated by the bulb?

Solution

From Table 27.2, the effective voltage is

$$V = \frac{V_p}{\sqrt{2}} = \frac{170\text{ V}}{\sqrt{2}} = 120.21\text{ V}$$

The power dissipated is

$$P = \frac{V^2}{R} = \frac{(120.21 \text{ V})^2}{240 \text{ }\Omega} = 60.21 \text{ W}$$

Example 27.4

What is the rms value of a constant 3 V signal?

Solution

Using Eq. 27.18, the rms value is

$$v_{\text{rms}}^2 = \frac{1}{T} \int_0^T (3)^2 dt = \left(\frac{1}{T}\right)(9)(T - 0)$$

$$v_{\text{rms}} = \sqrt{9} = 3 \text{ V}$$

9. PHASE ANGLES

AC circuit elements inductors and capacitors have the ability to store energy in magnetic and electric fields, respectively. Consequently, the voltage and current waveforms, while the same shape, differ by an amount called the *phase angle difference*, ϕ. Ordinarily, the voltage and current sinusoids do not peak at the same time. In a *leading circuit*, the phase angle difference is positive and the current peaks before the voltage. A leading circuit is termed a *capacitive circuit*. In a *lagging circuit*, the phase angle difference is negative and the current peaks after the voltage. A lagging circuit is termed an *inductive circuit*. These cases are represented mathematically as

$$v(t) = V_m \sin(\omega t + \theta) \quad \text{[reference]} \qquad \textbf{27.24}$$

$$i(t) = I_m \sin(\omega t + \theta + \phi) \quad \text{[leading]} \qquad \textbf{27.25}$$

$$i(t) = I_m \sin(\omega t + \theta - \phi) \quad \text{[lagging]} \qquad \textbf{27.26}$$

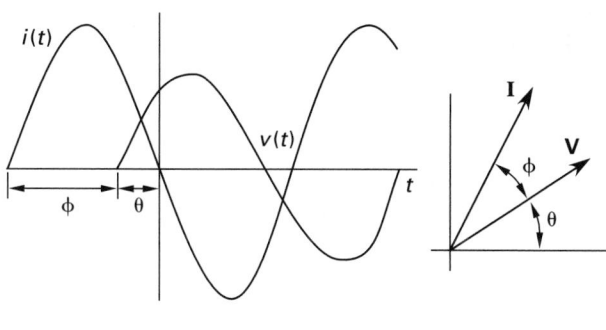

(a) time response   (b) phasor response

Figure 27.6 *Leading Phase Angle Difference*

10. SINUSOID

The most important waveform in electrical engineering is the sine function, or a waveform that is sinusoidal in value with respect to time. A sinusoidal waveform has an amplitude, also known as a magnitude or peak value, that remains constant, such as V_m. The waveform, however, repeats or goes through cycles. From Eq. 27.1, the sinusoid can be characterized by three quantities: magnitude, frequency, and phase angle.[8] The major properties of the sinusoid are covered in Sec. 27-2.

A circuit processes or changes waveforms. This is called *signal processing* or *waveform processing*. *Signal analysis* is the determination of these waveforms. The sinusoid can be represented as a phasor and used to analyze electrical circuits only if the circuit is in a steady-state condition. The terms "AC" and "DC" imply steady-state conditions. For AC circuits, "sinusoidal steady-state" indicates that all voltages and currents within the circuit are sinusoids of the same frequency as the excitation, that is, the driving voltage.

Example 27.5

During an experiment, a reference voltage of $v(t) = 170 \sin \omega t$ is used. A measurement of a second voltage, $v_2(t)$ occurs. The second voltage reaches its peak 2.5 ms before the reference voltage and has the same peak value 20 ms later. The peak value is 1.8 times the reference peak. What is the expression for $v_2(t)$?

Solution

Any sinusoidal voltage can be represented as in Eq. 27.1.

$$v(t) = V_m \sin(\omega t + \theta)$$

The peak value (equivalent to V_m) is 1.8 times the reference, or $1.8 \times 170 \text{ V} = 306 \text{ V}$. Calculate the angular frequency using Eq. 27.3 and the given period of 20 ms.

$$\omega = \frac{2\pi}{T} = \frac{2\pi}{20 \text{ ms}} = 314 \text{ rad/s}$$

The phase angle is determined by changing the time, 2.5 ms, into its corresponding angle, θ, as follows.

$$\theta = \omega t = \left(314 \, \frac{\text{rad}}{\text{s}}\right)\left(2.5 \times 10^{-3} \text{ s}\right) = 0.79 \text{ rad}$$

Converting 0.79 rad gives a result of 45°. Substituting the calculated values gives the result.

$$v_2(t) = V_m \sin(\omega t + \theta) = 306 \sin(314t + 45°)$$

Note that the result mixes the use of radians (314) and degrees (45). This is often done for clarity. Ensure one or the other is converted prior to calculations. It is often best to deal with all angles in radians.

---

[8]Note that a phase angle of $\pm 90°$ changes the sine function into a cosine function. Also, frequency refers in this case to the angular frequency, ω, in radians per second. The term frequency also refers to the term f, measured in cycles per second or hertz (Hz).

11. PHASORS

The addition of voltages and currents of the same frequency is simplified mathematically by treating them as phasors.[9] Using a method called the *Steinmetz algorithm*, the following two quantities, introduced in Sec. 27-2, are considered analogs.

$$v(t) = V_m \sin(\omega t + \theta) \text{ and } V = V_m \angle \theta \qquad \text{27.27}$$

Both forms show the magnitude and the phase, but the *phasor*, $V_m \angle \theta$, does not show frequency. In the phasor form, the frequency is implied. The phasor form of Eq. 27.27 is shown in Fig. 27.7.

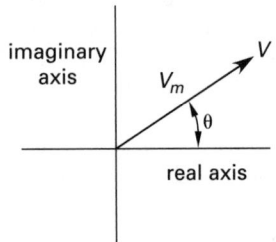

Figure 27.7 *Phasor of Magnitude V_m and Angle θ*

The phasor is actually a point, but is represented by an arrow of magnitude V_m at an angle θ with respect to a reference, normally taken to be $\theta = 0$. A *reference phasor* is one of known value, usually the driving voltage of a circuit (i.e., the voltage phasor). The reference phasor would be shown in the position of the real axis with θ equal to zero. Phasors are summed using phasor addition, commonly referred to as *vector addition*.[10] The Steinmetz algorithm is illustrated in Fig. 27.8.

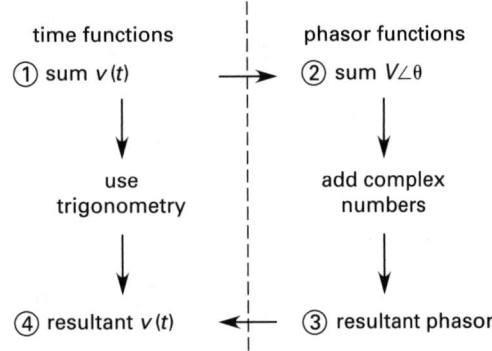

Figure 27.8 *Steinmetz Algorithm Steps*

In electric circuits with sinusoidal waveforms, an alternate phasor representation is often used. The alternate representation is called the *effective value phasor notation*. In this notation, the rms values of the voltage and current are used instead of the peak values. The

---

[9]For all circuit elements at the same frequency, the circuit must be in a steady-state condition.
[10]The term "vector addition" is not technically correct since phasors, with the exception of impedance and admittance phasors, rotate with time. The methods, however, are identical.

angles are also given in degrees. The phasor notation of Eq. 27.27 is then modified to

$$V = V_{\text{rms}} \angle \theta = \frac{V_m}{\sqrt{2}} \angle \theta \qquad \text{27.28}$$

Ensure that angles substituted into exponential forms are in radians or errors in mathematical calculations will result.

12. COMPLEX REPRESENTATION

Phasors are plotted in the complex plane. Figure 27.9 shows the relationships among the complex quantities introduced.

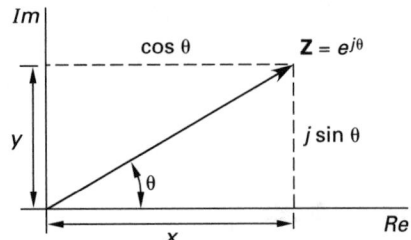

Figure 27.9 *Complex Quantities*

In Fig. 27.9, **Z** represents a complex number, not the impedance. Equation 27.1 is repeated here for convenience.

$$v(t) = V_m \sin(\omega t + \theta) \qquad \text{27.29}$$

Using Euler's relation, Eq. 27.4, the voltage $v(t)$ can be represented as

$$v(t) = V_m e^{j(\omega t + \theta)} \qquad \text{27.30}$$

Consequently, $v(t)$ is the imaginary part of Eq. 27.30. If the cosine function is used, $v(t)$ is the real part of Eq. 27.30. The fixed portion of Eq. 27.30 $(e^{j\theta})$ can be separated from the time-variable portion $(e^{j\omega t})$ of the function. This is shown in Fig. 27.10(a) and (b). The magnitude of $e^{j\omega t}$ remains equal to one, but the angle increases (rotates counterclockwise) linearly with time. All functions with an $e^{j\omega t}$ term are assumed to rotate counterclockwise with an angular velocity of ω in the complex plane. The voltage can thus be represented as in Fig. 27.10(c) and

$$v(t) = V_m e^{j\omega t} e^{j\theta} \qquad \text{27.31}$$

If the rotating portion of Eq. 27.31 is assumed to exist, the voltage can be written

$$v(\theta) = V_m e^{j\theta} \qquad \text{27.32}$$

Changing Eq. 27.32 to the phasor form yields

$$\mathbf{V} = V_m \angle \theta \qquad \text{27.33}$$

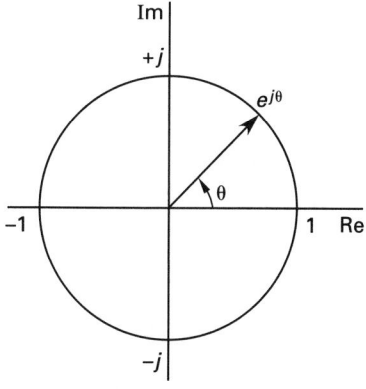

(a) $e^{j\theta}$ for all time

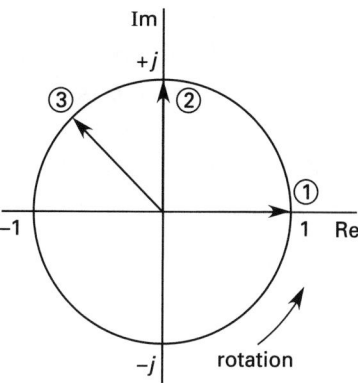

(b) $e^{j(\omega t)}$ for

① $\omega t = 0$
② $\omega t = \frac{\pi}{2}$
③ $\omega t = \frac{3\pi}{4}$

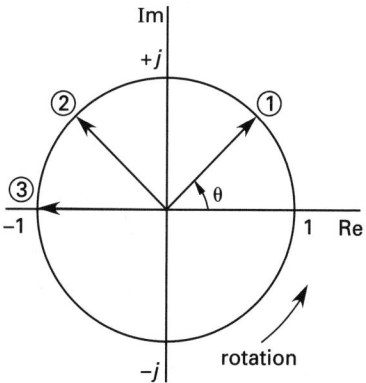

(c) $e^{j(\omega t + \theta)}$ for

① $\omega t = 0$
② $\omega t = \frac{\pi}{2}$
③ $\omega t = \frac{3\pi}{4}$

Figure 27.10 *Phasor Rotation in the Complex Plane*

The properties of complex numbers, which can represent voltage, current, or impedance, are summarized in Table 27.3. The designations used in the table are illustrated in Fig. 27.9.

13. RESISTORS

Resistors oppose the movement of electrons. In an *ideal* or *pure resistor*, no inductance or capacitance exists. The magnitude of the impedance is the resistance, R, with units of ohms and an impedance angle of zero. Therefore, voltage and current are in phase in a purely resistive circuit.

$$\mathbf{Z}_R = R\angle 0 = R + j0 \qquad 27.34$$

14. CAPACITORS

Capacitors oppose the movement of electrons by storing energy in an electric field and using this energy to resist changes in voltage over time. Unlike a resistor, an *ideal* or *perfect capacitor* consumes no energy. Equation 27.35 gives the impedance of an ideal capacitor with capacitance C. The magnitude of the impedance is termed the *capacitive reactance*, X_C, with units of ohms and an impedance angle of $-\pi/2(-90°)$. Consequently, the current leads the voltage by 90° in a purely capacitive circuit.[11]

$$\mathbf{Z}_C = X_C\angle{-90°} = 0 + jX_C \qquad 27.35$$

$$X_C = \frac{-1}{\omega C} = \frac{-1}{2\pi f C} \qquad 27.36$$

15. INDUCTORS

Inductors oppose the movement of electrons by storing energy in a magnetic field and using this energy to resist changes in current over time. Unlike a resistor, an *ideal* or *perfect inductor* consumes no energy. Equation 27.37 gives the impedance of an ideal inductor with inductance L. The magnitude of the impedance is termed the *inductive reactance*, X_L, with units of ohms and an impedance angle of $+\pi/2$ $(+90°)$. Consequently, the current lags the voltage by 90° in a purely inductive circuit.[12]

$$\mathbf{Z}_L = X_L\angle{+90°} = 0 + jX_L \qquad 27.37$$

$$X_L = \omega L = 2\pi f L \qquad 27.38$$

---

[11]The impedance angle for a capacitor is negative, but the current phase angle difference is positive—hence the term "leading." This occurs mathematically since the current is obtained by dividing the voltage by the impedance: $\mathbf{I} = \mathbf{V}/\mathbf{Z}$.

[12]The impedance angle for an inductor is positive, but the current phase angle difference is negative—hence the term "lagging." This occurs mathematically since the current is obtained by dividing the voltage by the impedance: $\mathbf{I} = \mathbf{V}/\mathbf{Z}$.

Circuit Theory

Table 27.3 *Properties of Complex Numbers*

	rectangular form	polar/exponential form
	$\mathbf{Z} = x + jy$	$\mathbf{Z} = \lvert\mathbf{Z}\rvert\angle\theta$ $\mathbf{Z} = \lvert\mathbf{Z}\rvert e^{j\theta} = \lvert\mathbf{Z}\rvert\cos\theta + j\lvert\mathbf{Z}\rvert\sin\theta$
relationship between forms	$x = \lvert\mathbf{Z}\rvert\cos\theta$ $y = \lvert\mathbf{Z}\rvert\sin\theta$	$\lvert\mathbf{Z}\rvert = \sqrt{x^2 + y^2}$ $\theta = \tan^{-1}\left(\dfrac{y}{x}\right)$
complex conjugate	$\mathbf{Z}^* = x - jy$ $\mathbf{Z}\mathbf{Z}^* = (x^2 + y^2) = \lvert z\rvert^2$	$\mathbf{Z}^* = \lvert\mathbf{Z}\rvert e^{-j\theta} = \lvert\mathbf{Z}\rvert\angle{-\theta}$ $\mathbf{Z}\mathbf{Z}^* = (\lvert\mathbf{Z}\rvert e^{j\theta})(\lvert\mathbf{Z}\rvert e^{-j\theta}) = \lvert\mathbf{Z}\rvert^2$
addition	$\mathbf{Z}_1 + \mathbf{Z}_2 = (x_1 + x_2) + j(y_1 + y_2)$	$\mathbf{Z}_1 + \mathbf{Z}_2 = (\lvert\mathbf{Z}_1\rvert\cos\theta_1 + \lvert\mathbf{Z}_2\rvert\cos\theta_2)$ $\quad + j\,(\lvert\mathbf{Z}_1\rvert\sin\theta_1 + \lvert\mathbf{Z}_2\rvert\sin\theta_2)$
multiplication	$\mathbf{Z}_1\mathbf{Z}_2 = (x_1 x_2 - y_1 y_2) + j(x_1 y_2 + x_2 y_1)$	$\mathbf{Z}_1\mathbf{Z}_2 = \lvert\mathbf{Z}_1\rvert\lvert\mathbf{Z}_2\rvert\angle\theta_1 + \theta_2$
division	$\dfrac{\mathbf{Z}_1}{\mathbf{Z}_2} = \dfrac{(x_1 x_2 + y_1 y_2) + j(x_2 y_1 - x_1 y_2)}{\lvert\mathbf{Z}_2\rvert^2}$	$\dfrac{z_1}{z_2} = \dfrac{\lvert\mathbf{Z}_1\rvert}{\lvert\mathbf{Z}_2\rvert}\angle\theta_1 - \theta_2$

16. COMBINING IMPEDANCES

Impedances in combination are like resistors: Impedances in series are added, while the reciprocals of impedances in parallel are added. For series circuits, the resistive and reactive parts of each impedance element are calculated separately and summed. For parallel circuits, the conductance and susceptance of each element are summed. The total impedance is found by a complex addition of the resistive (conductive) and reactive (susceptive) parts. It is convenient to perform the addition in rectangular form.

$$Z_e = \sum Z$$

$$= \sqrt{\left(\sum R\right)^2 + \left(\sum X_L - \sum X_C\right)^2} \quad \text{[series]}$$

$$27.39$$

$$\frac{1}{Z_e} = \sum \frac{1}{Z} = Y_e$$

$$= \sqrt{\left(\sum G\right)^2 + \left(\sum B_L - \sum B_C\right)^2} \quad \text{[parallel]}$$

$$27.40$$

The impedance of various series-connected circuit elements is given in App. 27.A. The impedance of various parallel-connected circuit elements is given in App. 27.B.

Example 27.6

Determine the impedance and admittance of the following circuits.

(a)

(b)

Solution

(a) From Eq. 27.39,

$$Z = \sqrt{R^2 + Z_C^2} = \sqrt{(2\ \Omega)^2 + (4\ \Omega)^2} = 4.47\ \Omega$$

$$\phi = \arctan\left(\frac{X_C}{R}\right) = \arctan\left(\frac{-4}{2}\right) = -63.4°$$

$$\mathbf{Z} = 4.47\angle{-63.4°}\,\Omega$$

From Eq. 27.9,

$$\mathbf{Y} = \frac{1}{\mathbf{Z}} = \frac{1}{4.47\angle-63.4°\,\Omega} = 0.224\angle63.4°\text{ S}$$

(b) Since this is a parallel circuit, work with the admittances.

$$G = \frac{1}{R} = \frac{1}{0.5\ \Omega} = 2\text{ S}$$

$$B_C = \frac{1}{X_C} = \frac{1}{0.25\ \Omega} = 4\text{ S}$$

$$Y = \sqrt{G^2 + B_C^2} = \sqrt{(2\text{ S})^2 + (4\text{ S})^2} = 4.47\text{ S}$$

$$\phi = \arctan\left(\frac{B_C}{G}\right) = \arctan\left(\frac{4}{2}\right) = 63.4°$$

$$\mathbf{Y} = 4.47\angle63.4°\text{ S}$$

$$\mathbf{Z} = \frac{1}{\mathbf{Y}} = \frac{1}{4.47\angle63.4°\text{ S}} = 0.224\angle-63.4°\ \Omega$$

17. OHM'S LAW

Ohm's law for AC circuits with linear circuit elements is similar to Ohm's law for DC circuits.[13] The difference is that all the terms are represented as phasors.[14]

$$\mathbf{V} = \mathbf{IZ} \qquad\qquad 27.41$$

$$V\angle\theta_V = (I\angle\theta_I)(Z\angle\phi_Z) \qquad\qquad 27.42$$

18. POWER

The instantaneous power, p, in a purely resistive circuit is given by

$$p_R(t) = i(t)v(t) = (I_m \sin\omega t)(V_m \sin\omega t)$$

$$= I_m V_m \sin^2\omega t$$

$$= \tfrac{1}{2}I_m V_m - \tfrac{1}{2}I_m V_m \cos 2\omega t \qquad 27.43$$

The second term of Eq. 27.43 integrates to zero over the period. Therefore, the *average power* dissipated is

$$P_R = \tfrac{1}{2}I_m V_m = \frac{V_m^2}{2R} = I_{\text{rms}}V_{\text{rms}} = IV \qquad 27.44$$

---

[13]Ohm's law can be used on *nonlinear devices* (NLD) if the region analyzed is restricted to be approximately linear. When this condition is applied, the analysis is termed *small signal analysis*.
[14]In general, phasors are considered to rotate with time. A secondary definition of a phasor is any quantity that is a complex number. As a result, the impedance is also considered a phasor even though it does not change with time.

Equation 27.28 was used to change the maximum, or peak, values into the more usable rms, or effective, values.

In a purely capacitive circuit, the current leads the voltage by 90°. This allows the current term of Eq. 27.43 to be replaced by a cosine (since the sine and cosine differ by a 90° phase). The instantaneous power in a purely capacitive circuit is

$$p_C(t) = i(t)v(t)$$

$$= (I_m \cos\omega t)(V_m \sin\omega t)$$

$$= I_m V_m \sin\omega t \cos\omega t$$

$$= \tfrac{1}{2}I_m V_m \sin 2\omega t \qquad\qquad 27.45$$

Since the $\sin 2\omega t$ term integrates to zero over the period, the average power is zero. Nevertheless, power is stored during the charging process and returned to the circuit during the discharging process.

$$P_C = 0 \qquad\qquad 27.46$$

Similarly, the instantaneous power in an inductor is

$$p_L(t) = -\tfrac{1}{2}I_m V_m \sin 2\omega t \qquad\qquad 27.47$$

Again, the $\sin 2\omega t$ integrates to zero over the period, and the average power is zero. Nevertheless, power is stored during the expansion of the magnetic field and returned to the circuit during the contraction of the magnetic field.

$$P_L = 0 \qquad\qquad 27.48$$

19. REAL POWER AND THE POWER FACTOR

In a circuit that contains all three circuit elements (resistors, capacitors, and inductors) or the effects of all three, the average power is calculated from Eq. 27.14 as

$$P_{\text{ave}} = \frac{1}{T}\int_0^T i(t)v(t)\,dt \qquad\qquad 27.49$$

Let the generic voltage and current waveforms be represented by Eqs. 27.50 and 27.51. The current angle φ is equal to $\theta \pm \phi$.[15]

$$i(t) = I_m \sin(\omega t + \varphi) \qquad\qquad 27.50$$

$$v(t) = V_m \sin(\omega t + \theta) \qquad\qquad 27.51$$

---

[15]Using a current angle simplifies the derivational mathematics (not shown) and clarifies the definition of the power factor angle given in this section.

PROFESSIONAL PUBLICATIONS, INC.

Circuit Theory

Substituting Eqs. 27.50 and 27.51 into Eq. 27.49 gives

$$P_{\text{ave}} = \left(\frac{I_m V_m}{2}\right) \cos \Psi = I_{\text{rms}} V_{\text{rms}} \cos \Psi \qquad 27.52$$

The angle $\Psi = \theta - \varphi$ is the *power factor angle*. It represents the difference between the voltage and current angles. This difference is the impedance angle, $\pm\phi$. Since the $\cos(-\phi) = \cos(+\phi)$, only the absolute value of the impedance angle is used. Because the absolute value is used in the equation, the terms "leading" (for a capacitive circuit) and "lagging" (for an inductive circuit) must be used when describing the power factor. The power factor of a purely resistive circuit equals one; the power factor of a purely reactive circuit equals zero.

Equation 27.52 determines the magnitude of the product of the current and voltage in phase with one another, that is, of a resistive nature. Consequently, Eq. 27.52 determines the power consumed by the resistive elements of a circuit. This average power is called the *real power* or *true power*, and sometimes the *active power*. The term $\cos \Psi$, or $\cos |\phi|$, is called the *power factor* or *phase factor* and given the symbol pf. The power factor is also equal to the ratio of the real power to the apparent power (see Sec. 27-22). Often, no subscript is used on P when it represents the real power, nor are subscripts used on rms or effective values. The real power is then given by

$$P = IV \cos \Psi = IV \, \text{pf} \qquad 27.53$$

It is important to realize that the real power cannot be obtained by multiplying phasors. Doing so results in the addition of the voltage and current angles when the difference is required.

$$P \neq I \angle \varphi V \angle \theta = V I \angle \theta + \varphi \qquad 27.54$$

If phasor power is used, the difficulty can be alleviated by using Eq. 27.55.

$$P = \text{Re}\{\mathbf{VI}^*\} \qquad 27.55$$

Equation 27.55 is equivalent to Eq. 27.53.

20. REACTIVE POWER

The *reactive power*, Q, measured in units of VAR, is given by

$$Q = IV \sin \Psi \qquad 27.56$$

The reactive power, sometimes called the *wattless power*, is the product of the rms values of the current and voltage multiplied by the *quadrature* of the current. $\text{Sin}\Psi$ is called the *reactive factor*.[16] The reactive power represents the energy stored in the inductive and capacitive elements of a circuit.

21. APPARENT POWER

The apparent power, S, measured in voltamperes, is given by

$$S = IV \qquad 27.57$$

The apparent power is the product of the rms values of the voltage and current without regard to the angular relationship between them. The apparent power is representative of the combination of real and reactive power (see Sec. 22). As such, not all of the apparent power is dissipated or consumed. Nevertheless, electrical engineers must design systems to adequately handle this power as it exists in the system.

22. COMPLEX POWER: THE POWER TRIANGLE

The real, reactive, and apparent powers can be related to one another as vectors. The *complex power vector*, $\mathbf{S}$, is the vector sum of the *reactive power vector*, $\mathbf{Q}$, and the *real power vector*, $\mathbf{P}$. The magnitude of $\mathbf{S}$ is given by Eq. 27.57. The magnitude of $\mathbf{Q}$ is given by Eq. 27.56. The magnitude of $\mathbf{P}$ is given by Eq. 27.53. In general, $\mathbf{S} = \mathbf{I}^*\mathbf{V}$ where $\mathbf{I}^*$ is the complex conjugate of the current, that is, the current with the phase difference angle reversed. This convention is arbitrary, but emphasizes that the *power angle*, ϕ, associated with $\mathbf{S}$ is the same as the overall impedance angle, ϕ, whose magnitude equals the power factor angle, Ψ.

The relationship between the magnitudes of these powers is

$$S^2 = P^2 + Q^2 \qquad 27.58$$

A drawing of the power vectors in a complex plane is shown in Fig. 27.11 for leading and lagging conditions. The following relationships are determined from the drawing.

$$P = S \cos \phi \qquad 27.59$$

$$Q = S \sin \phi \qquad 27.60$$

---

[16]The power factor angle, Ψ, is used instead of the absolute value of the impedance angle (as in the power factor) since the sine function results in positive and negative values depending upon the difference between the voltage and current angles. If Ψ were not used, $\pm\phi$ would have to be used.

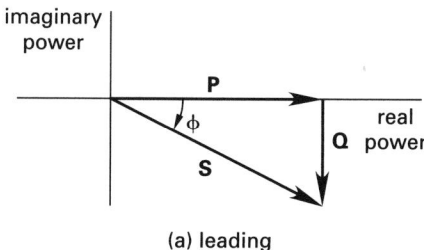

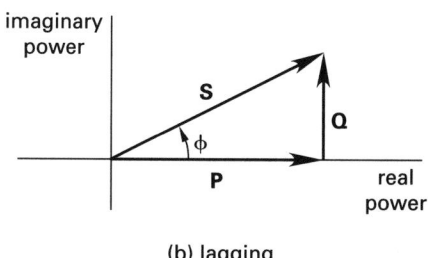

Figure 27.11 Power Triangle

Example 27.7

For the circuit shown, find the (a) apparent power, (b) real power, and (c) reactive power; and (d) draw the power triangle.

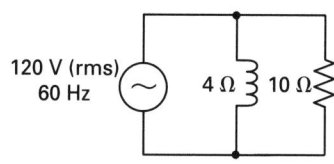

Solution

The equivalent impedance is

$$\frac{1}{Z} = \frac{1}{j4} + \frac{1}{10} = -j0.25 + 0.10 \text{ S}$$

$$Z = 3.714 \ \Omega$$

$$\phi_Z = \arctan\left(\frac{0.25}{0.10}\right) = 68.2°$$

The total current is

$$I = \frac{V}{Z} = \frac{120\angle 0° \text{ V}}{3.714\angle 68.2° \ \Omega} = 32.31\angle -68.2° \text{ A}$$

(a) The apparent power is

$$S = I^*V = (32.31\angle 68.2° \text{ A})(120\angle 0° \text{ V})$$

$$= 3877\angle 68.2° \text{ VA}$$

(The angle of apparent power is usually not reported.)

(b) The real power is

$$P = \frac{V_R^2}{R} = \frac{(120\angle 0° \text{ V})^2}{10 \ \Omega} = 1440 \text{ W}$$

Alternatively, the real power can be calculated from Eq. 27.59.

$$P = S\cos\phi = (3877)\cos(68.2°) = 1440 \text{ W}$$

(c) The reactive power is

$$Q = \frac{V_L^2}{X_L} = \frac{(120 \text{ V})^2}{4 \ \Omega} = 3600 \text{ VAR}$$

Alternatively, the reactive power can be calculated from Eq. 27.60.

$$Q = S\sin\phi = (3877)(\sin 68.2°) = 3600 \text{ VAR}$$

(d) The real power is represented by the vector (in rectangular form) $1440 + j0$. The reactive power is represented by the vector $0 + j3600$. The apparent power is represented (in phasor form) by the vector $3877\angle 68.2°$. The power triangle is

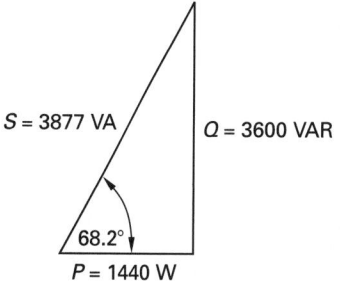

23. MAXIMUM POWER TRANSFER

Assuming a fixed primary impedance, the maximum power condition in an AC circuit is similar to that in a DC circuit. That is, the maximum power is transferred from the source to the load when the impedances match. This occurs when the circuit is in resonance. Resonance is discussed more fully in Ch. 29. The conditions for maximum power transfer, and thus resonance, are given by

$$R_{\text{load}} = R_s \qquad\qquad 27.61$$

$$X_{\text{load}} = -X_s \qquad\qquad 27.62$$

24. AC CIRCUIT ANALYSIS

All of the methods and equations presented in Ch. 26, such as Ohm's and Kirchhoff's laws and loop-current and node-voltage methods can be used to analyze AC circuits as long as complex arithmetic is utilized.

28 Transformers

Nomenclature

a	turns ratio	–
B	magnetic flux density	T
f	frequency	Hz, or s[1], or cycles/s
H	magnetic field strength	A/m
i	instantaneous current	A
I	effective or DC current	A
k	coupling coefficient	–
L	self-inductance	H
M	mutual inductance	H
N	number	–
N	number of turns	–
O	origin	–
t	time	s
v	instantaneous voltage	V
V	effective or DC voltage	V
X	reactance	Ω
Z	impedance	Ω

Symbols

μ	permeability	H/m
Φ	magnetic flux	Wb
ω	angular frequency	rad/s

Subscripts

0	initial
c	coercive
c	core
C	capacitor
ep	effective primary
E	exciting
L	inductor
m	magnetizing
m	maximum
m	mutual
p	primary
r	residual
ref	reflected
s	secondary

1. FUNDAMENTALS

A *transformer* is an electrical device consisting of two or more multiturn coils of wire in close proximity to link the magnetic field of one to the other. This linkage allows the transfer of electrical energy from one or more alternating circuits to one or more other alternating circuits by the process of magnetic induction. The transformer can be used to raise or lower the value of a capacitor, inductor, or resistor. It enables the efficient transmission of electrical energy at high voltage over great distances, and then is used to lower the voltage to safe values for industrial, commercial, and household use. The principle of the transformer is also the basic principle of induction motors, alternators, and synchronous motors. The transformer and these other devices are based on laws of magnetic induction.

2. MAGNETIC COUPLING

Any coil of wire, energized by a sinusoidal waveform, produces a magnetic flux, Φ, that is also sinusoidal.[1]

$$\Phi(t) = \Phi_m \sin \omega t \qquad 28.1$$

Faraday's law gives the voltage induced when a portion of the magnetic flux produced in one coil of wire with N_p turns passes through a second coil of wire with N_s turns. The minus sign is a result of Lenz's law, which states that the induced voltage, and thus the induced flux, always acts to oppose the voltage (flux) that created it.

$$v_s(t) = -N_s \left(\frac{d\Phi(t)}{dt} \right) = -N_s \omega \Phi_m \cos \omega t \qquad 28.2$$

The effective voltage is obtained by dividing Eq. 28.2 by the square root of two. The net result is[2]

$$V_s(t) = \frac{-N_s (2\pi f) \Phi_m \cos \omega t}{\sqrt{2}}$$

$$= -4.44 N_s f \Phi_m \cos \omega t \qquad 28.3$$

The magnetic flux that passes through both coils is called the *mutual flux*. The magnetic flux that couples only one coil is called the *leakage flux*. The term

---

[1]This ignores harmonic effects in the core of the transformer stemming from variable permeability. These harmonics must be accounted for if inductive interference with communications circuits is a concern.

[2]The capital V_s is used in Eq. 28.3 to indicate the effective value, even though the voltage remains a function of time.

"leakage" is used because any flux that does not link the coils will not contribute to inducing a voltage in the secondary of the transformer. The fluxes are illustrated for a *core transformer* in Fig. 28.1. The magnetic flux Φ_{12} is the flux generated by coil number one that links coil two. The magnetic flux Φ_{11} is the flux generated by coil one that links only coil one. Similarly, the magnetic flux Φ_{21} is the flux generated by coil two that links coil one, and the magnetic flux Φ_{22} is the flux generated by coil two that links only coil two. Magnetic fluxes Φ_{11} and Φ_{22} result in the *self-inductance* of coils one (L_p) and two (L_s), normally called the *primary winding* and *secondary winding*, respectively.

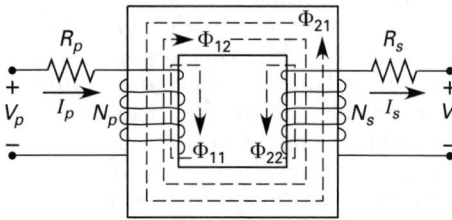

Figure 28.1 *Fluxes in a Magnetically Coupled Circuit*

The *mutual inductance*, M, resulting from magnetic fluxes Φ_{12} and Φ_{21} is defined as the ratio of the electromotive force induced in one coil (circuit) to the rate of change of current in the other coil (circuit). It is the proportionality constant between the induced voltage and the rate of current change.

$$v_p(t) = M\left(\frac{di_s(t)}{dt}\right) \qquad 28.4$$

$$v_s(t) = M\left(\frac{di_p(t)}{dt}\right) \qquad 28.5$$

Just as leakage flux is associated with self-inductance, mutual flux is associated with a *mutual reactance* given by

$$X_m = \omega M \qquad 28.6$$

The induced voltage and generated flux associated with mutual reactance always oppose the flux and current that created them. For example, Φ_{12} in Fig. 28.1, generated by the current in the primary winding, results in a voltage in the secondary winding. This secondary voltage causes a current to flow that generates Φ_{21}, which opposes the original flux and generates a voltage drop of $-(I_s)j\omega M$ in the *primary*. By similar reasoning, a voltage drop of $-I_p j\omega M$ is generated in the *secondary* of the transformer as well. This is illustrated in Fig. 28.2(a), which is equivalent to Fig. 28.1, and in Eqs. 28.7 and 28.8, which are obtained by applying KVL around the indicated loops.

$$V_p = I_p(R_p + j\omega L_p) - I_s j\omega M$$
$$= I_p(R_p + jX_p) - I_s jX_m$$
$$= I_p Z_p - I_s jX_m \qquad 28.7$$

$$V_s = I_s(R_s + j\omega L_s) - I_p j\omega M$$
$$= I_s(R_s + jX_s) - I_p jX_m$$
$$= I_s Z_s - I_p jX_m \qquad 28.8$$

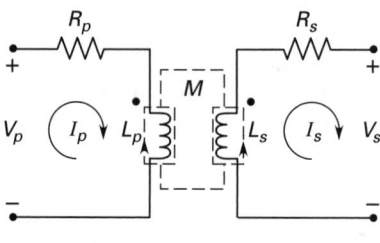

(a) flux linked circuit with mutual inductance

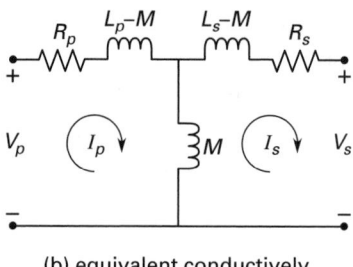

(b) equivalent conductively coupled circuit

Figure 28.2 *Transformer Models*

The dots shown in Fig. 28.2(a) are used to determine the sign of the mutual inductance *when using assumed directions of current flow* in analysis. When analyzing magnetically coupled circuits, the winding sense, or direction, can be determined using the right-hand rule (see Fig. 28.3). First, select a current direction on the primary side and place a dot where the current enters the winding (see Fig. 28.3(a)). Then use the right-hand rule to determine the current direction necessary for an opposing flux to be generated (see Fig. 28.3(b)). Place a dot where the current *leaves* this winding.[3] The two dotted terminals are positive simultaneously. With the instantaneous polarity determined, the need for the pictorial representation of the windings shown in Fig. 28.1 is eliminated. The coupled coils can be shown as in Fig. 28.3(c).

---

[3]This is equivalent to considering the primary of the transformer to be a load on the source and the secondary of the transformer to be the source for the secondary portion of the circuit.

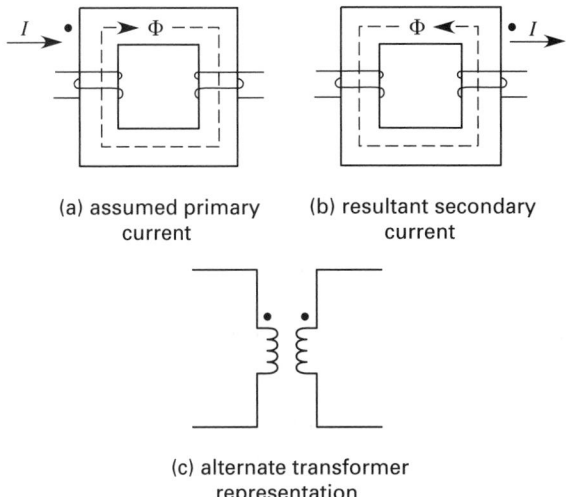

(a) assumed primary current

(b) resultant secondary current

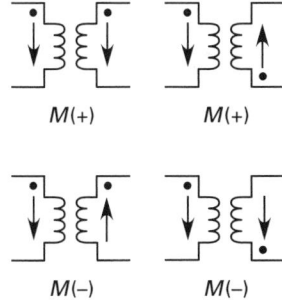

(c) alternate transformer representation

Figure 28.3 *Dot Rule Determination*

The *dot rule* is stated as follows.

- When the assumed currents both enter or both leave a pair of coupled coils by the dotted terminals, the sign of the mutual reactance, or M-terms, is the same as the L-terms in Eqs. 28.7 and 28.8.

- When one current enters by a dotted terminal and the other current leaves by a dotted terminal, the sign of the M-terms is opposite the sign of the L-terms in Eqs. 28.7 and 28.8.

This is shown in Fig. 28.4. The rule is used so that in the application of analysis laws, such as KVL, the resulting equations (see Eqs. 28.7 and 28.8) will be consistent and the mathematical result will be valid for the circuit.

$M(+)$ $\qquad$ $M(+)$

$M(-)$ $\qquad$ $M(-)$

Figure 28.4 *Positive and Negative Mutual Inductances*

The amount of magnetic coupling is measured by the *coupling coefficient (coefficient of coupling)*, k, which is the fraction of the total flux that links both coils. The value of k varies from near zero for radio coils to near 1.0 for iron-core transformers.[4]

$$k = \frac{M}{\sqrt{L_p L_s}} = \frac{X_m}{\sqrt{X_p X_s}} \qquad \textbf{28.9}$$

---

[4]In terms of the fluxes in Fig. 28.1, the coupling coefficient $k = 1 - \Phi_{11}/(\Phi_{11} + \Phi_{12})$. The high value of the coupling coefficient in iron cores is the primary reason for their use in transformers.

3. IDEAL TRANSFORMERS

Transformers, which are magnetically coupled circuits, are used to change voltages, match impedances, and isolate circuits. A primary winding current produces a magnetic flux, which induces a current in the secondary coil. To minimize flux leakage and improve efficiency, transformer coils are wound on magnetic permeable cores that improve the mutual inductance, M (see Sec. 28-2).[5] Two common designs, *core transformers* and *shell transformers*, are shown in Fig. 28.5.

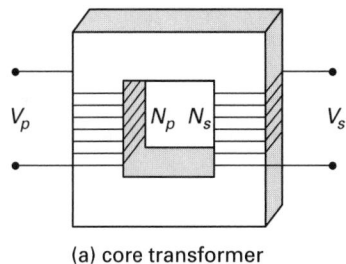

(a) core transformer

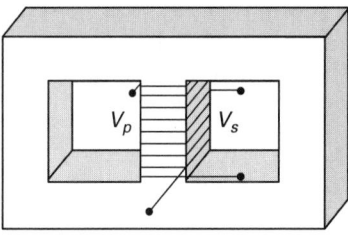

(b) shell transformer

Figure 28.5 *Core and Shell Transformers*

An *ideal transformer* has the following properties.

- The coupling coefficient is unity ($k = 1$), that is, all the flux passes through each coil. There is no leakage flux. Thus, $|M| = |L_p| = |L_s|$, and $|X_m| = |X_p| = |X_s|$.

- The coils have no resistance, that is, $R_p = R_s = 0$. Combined with the first condition, this means that $Z_p = Z_s$ (see Fig. 28.2(b)).

- All the flux is contained inside a path of constant area. This allows the dimensions of the core to be used.

- Core *iron losses* are zero. This indicates that no heating losses exist due to eddy currents or hysteresis loop effects, that is $R_c = 0$ (see Secs. 28-5, 28-6, and 28-7).

- The core is infinitely permeable. Thus, no *magnetizing current* (also called the *exciting current*) is required to create the flux in the core (see Sec. 28-5).

---

[5]An iron core transformer is indicated by one or more lines drawn between the coil symbols on a circuit diagram.

The last two conditions are technically the conditions for an *ideal core*. The transformer and the core can be treated and modeled separately if necessary. The model for an ideal transformer with an ideal core is shown in Fig. 28.6.

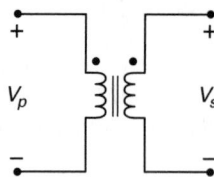

Figure 28.6 *Ideal Transformer and Core*

The ratio of the number of turns of primary to secondary windings is called the *turns ratio (ratio of transformation)*, a.[6]

$$a = \frac{N_p}{N_s} \qquad 28.10$$

In a lossless transformer, that is, a transformer with 100% efficiency, the power transfer from the primary side equals the power to the secondary side. The relationship in Eq. 28.11 is thus valid.[7]

$$I_p V_p = I_s V_s \qquad 28.11$$

$$a = \frac{N_p}{N_s} = \frac{V_p}{V_s} = \frac{I_s}{I_p} = \sqrt{\frac{Z_p}{Z_s}} \qquad 28.12$$

If the turns ratio is greater than unity, the transformer decreases the voltage and is termed a *step-down transformer*. If the turns ratio is less than unity, the transformer increases the voltage and is termed a *step-up transformer*. The assumptions made for ideal transformers are very good approximations, especially for large iron-core transformers.

Example 28.1

An ideal step-down transformer has 200 primary turns and 50 secondary turns in its coils. 440 V are applied across the primary. The secondary load resistance is 5 Ω. What is the secondary voltage and current?

Solution

The turns ratio is

$$a = \frac{N_p}{N_s} = \frac{200}{50} = 4$$

---

[6]The turns ratio, a, is alternatively defined as the ratio of secondary to primary turns, in which case the relationships of Eq. 28.12 are inverted. Unless specified otherwise, assume the ratio is defined as in Eq. 28.10.

[7]Phasor notation is not used, as all the elements are in phase in an ideal transformer. Further, this is not the formula for power. The 1/2 has been dropped as has the complex conjugate of the current (see Ch. 27).

The secondary voltage and current are

$$V_s = \frac{V_p}{a} = \frac{440 \text{ V}}{4} = 110 \text{ V}$$

$$I_s = \frac{V_s}{R_s} = \frac{110 \text{ V}}{5 \text{ }\Omega} = 22 \text{ A}$$

4. IMPEDANCE MATCHING

Matching of impedances between source and load is accomplished to effect the maximum power transfer to the load. The transformer can impedance match. In Fig. 28.7, the impedance in the secondary is

$$Z_s = \frac{V_s}{I_s} \qquad 28.13$$

The impedance "seen" by the primary side, Z_{ep}, is termed the *effective primary impedance* or the *reflected impedance*, Z_{ref}. The reflected impedance is

$$Z_{\text{ep}} = Z_{\text{ref}} = \frac{V_p}{I_p} = Z_p + a^2 Z_s \qquad 28.14$$

The turns ratio can be found from any of the relationships in Eq. 28.12, or specifically from

$$a = \sqrt{\frac{Z_p}{Z_s}} \qquad 28.15$$

For a fixed primary impedance, the maximum power consumption in the load, that is, the maximum power transfer from the source, takes place when the secondary impedance is the conjugate of the primary impedance. Consequently, the conditions necessary to impedance match are

$$R_p = a^2 R_s \qquad 28.16$$

$$X_p = -a^2 X_s \qquad 28.17$$

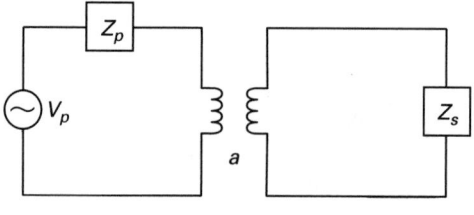

(a) circuit

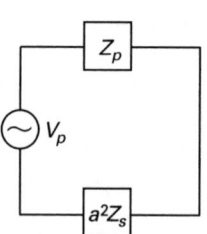

(b) equivalent circuit

Figure 28.7 *Impedance Matching*

Example 28.2

A 15 Ω speaker must be impedance matched to a 10 kΩ amplifier in order to produce the highest quality sound. What is the required turns ratio?

Solution

From Eq. 28.15,

$$a = \sqrt{\frac{Z_p}{Z_s}} = \sqrt{\frac{10 \times 10^3 \ \Omega}{15 \ \Omega}} = 25.8$$

Example 28.3

The primary of an ideal transformer with a turns ratio of 5 contains a 10 Ω resistor and an 80 mH inductor in series. What is the secondary reactance required to maximize the power transfer in this 60 Hz circuit?

Solution

The relationship between the reactances is given by Eq. 28.17.

$$X_p = -a^2 X_s$$

$$X_s = \frac{X_p}{-a^2}$$

The primary reactance is found from

$$X_p = X_L = \omega L = 2\pi f L$$

$$= 2\pi \left(60 \ \mathrm{s}^{-1}\right) (80 \ \mathrm{mH}) = 30.16 \ \Omega$$

Substituting the known and calculated values gives

$$X_s = \frac{X_p}{-a^2} = \frac{30.16 \ \Omega}{-(5)^2} = -1.21 \ \Omega$$

Example 28.4

For the reactance in Ex. 28.3, what is the value of the capacitance required?

Solution

From Ex. 28.3 and the definition of capacitive reactance,

$$X_s = X_c = -\frac{1}{\omega C}$$

$$C = -\frac{1}{2\pi f X_c} = -\frac{1}{(2\pi) \left(60 \ \mathrm{s}^{-1}\right) (-1.21 \ \Omega)}$$

$$= 2.19 \times 10^{-3} \ \mathrm{F}$$

5. REAL TRANSFORMERS

Real transformers have flux leakage, resistance in the coils, and resistance and reactance associated with moving the domains within the core. A model of a real transformer is shown in Fig. 28.8.

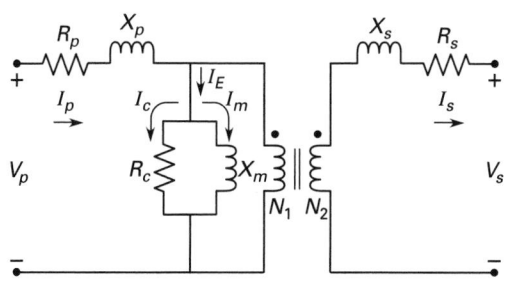

Figure 28.8 *Real Transformer Equivalent Circuit*

In the model of Fig. 28.8, all the core terms are shown on the primary side. This is arbitrary. Many other equivalent models are possible, though they will all contain these components. Components not yet defined follow.

- *Exciting current*, I_E: the total current needed to produce the magnetic flux in the core.

- *Mutual reactance*, X_m, and *magnetizing current*, I_m: the magnetizing current is the current necessary to generate the flux in the core. It is 90° out of phase with the core current.

- *Core resistance*, R_c, and *core current*, I_c: the core resistance represents iron losses (see Secs. 28-6 and 28-7). The core current is the current necessary to overcome those losses. This current flows in the primary windings but occurs due to resistive losses physically occurring in the core.

Each leakage flux corresponds to a "loss" of induced voltage in its coil. This is considered to have been caused by an equivalent reactance, X_p or X_s. In terms of the flux, these reactances are

$$X_p = \omega L_p = \frac{V_{X_p}}{I_p} = \frac{4.44 f \Phi_p N_p}{I_p} \qquad 28.18$$

$$X_s = \omega L_s = \frac{V_{X_s}}{I_s} = \frac{4.44 f \Phi_s N_s}{I_s} \qquad 28.19$$

The turns ratio is related to the leakage inductances by the following.

$$a = \sqrt{\frac{L_p}{L_s}} \qquad 28.20$$

6. MAGNETIC HYSTERESIS: BH CURVES

The core of a transformer, especially when specifically designed to increase the coupling, is made from iron. Iron is a ferromagnetic material. When a magnetic field, in this case produced by the coil of the transformer, is applied to a ferromagnetic material, a phenomenon known as *magnetic hysteresis* takes place. This phenomenon is shown in Fig. 28.9.

Circuit Theory

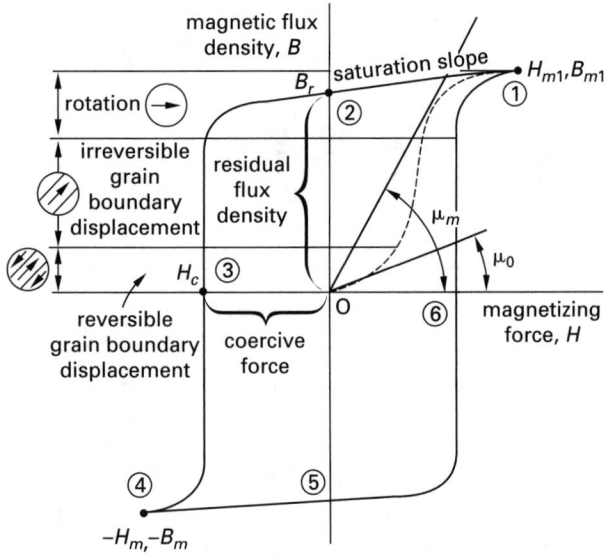

Figure 28.9 *Magnetic Hysteresis Loop*

Figure 28.9 plots the magnetic flux density, B, produced for various values of the magnetizing force, H. The relationship is decidedly nonlinear.[8] When a material is first magnetized, the initial BH values occur along the curve from the origin, O, to point 1. This part of the curve, O1, is known as the *initial* or *normal magnetization curve*. The initial slope of O1 is called the *initial permeability*, μ_o (not to be confused with the permeability of free space). The slope of the line from the origin to a point tangent to the normal magnetization curve is called the *maximum permeability*, μ_m. The maximum value of B reached at point 1 is called the *saturation flux density*. If the magnetic field is then decreased, the flux density decreases as well, but not in proportion to the decrease in magnetizing field strength. When H equals zero, at point 2, the amount of magnetic flux density remaining in the core is called the *residual flux density*, B_r. In order to reduce the magnetic field in the core to zero, the magnetizing force must increase in reverse direction of that initially applied. The magnitude of the magnetizing force where B again becomes zero, at point 3 is called the *coercive force*, H_c.

As the magnetizing force continues to increase in the negative direction, the flux density again increases, but in opposite polarity, eventually reaching saturation in this direction at point 4. As the magnetic field is again removed, the magnitude of the magnetic flux density decreases to point 5. A positive value of H (equal to H_c) is required to drive the flux density to zero, at point 6.

As the magnetizing force is increased further, the curve from point 6 to point 1 is followed, completing the cycle but not returning to the origin. The domain, or grain boundary movement, that is the ultimate cause of this curve is shown for the positive portion of the cycle. The phenomenon whereby the values of B lag behind the

values of H is called *hysteresis*. The entire loop is called the *hysteresis loop*. The area within the loop represents the energy loss that occurs due to heating from the movement of the magnetic domains.[9]

7. EDDY CURRENTS

Any conducting mass exposed to variations in magnetic flux contains *eddy*, or *Foucault*, *currents*. These currents are a result of magnetic induction caused by the relative motion due either to variations in the magnetic field or the movement of the mass itself. Such currents result in Joule heating in the mass—in this case, in the core of the transformer. Such an energy loss, in a transformer or in the iron portions of electric machinery, is minimized by dividing the magnetically conducting material into small segments called *laminations*, coated with insulating material. This reduces the length of the path in which current exists.

8. CORE LOSSES

Core losses are defined as the rate of energy conversion into heat in a magnetic material due to the presence of an alternating or pulsing magnetic field. This type of loss is also called the *excitation loss* or *iron loss* (see Sec. 28-5). Core losses are the combination of magnetic hysteresis and eddy current losses.

---

[8]To this point, all the relationships and equations have assumed linearity in the transformers.

[9]The portion of the energy used to create the flux comes from I_m (see Sec. 28-5). The portion of the energy that is lost due to heating comes from I_c (see Sec. 28-5). The current I_c also supplies any eddy current losses (see Sec. 28-7).

29 Linear Circuit Analysis

Nomenclature

A	area	m^2
b	y-intercept	–
C	arbitrary constant	–
C	capacitance	F
f	frequency	Hz, s^{-1}, or cycles/s
i	instantaneous current	A
I	effective or DC current	A
l	length	m
L	self-inductance	H
m	slope	–
M	mutual inductance	H
n	integer	–
N	number of turns	–
Q	charge	C
r	distance	m
R	resistance	Ω
t	time	s
T	temperature	$°C$
U	energy	J
v	instantaneous voltage	V
V	effective or DC voltage	V
Z	impedance	Ω

Symbols

α	thermal coefficient of resistance	$1/°C$
ε	permittivity	F/m
ε_0	free-space permittivity	8.854×10^{-12} F/m
ε_r	relative permittivity	–
κ	arbitrary constant	–
μ	permeability	H/m
μ_0	free-space permeability	1.2566×10^{-6} H/m
ρ	resistivity	$\Omega \cdot cm$
Ψ	magnetic flux	Wb
ω	angular frequency	rad/s
$\mho$	conductance	S

Subscripts

0	initial or free space (vacuum)
C	capacitor
e	equivalent
fl	full load
int	internal or source
l	load
L	inductor
m	magnetizing
nl	no load
N	Norton
oc	open circuit
s	source
sc	short circuit
Th	Thevenin

1. FUNDAMENTALS

Circuit analysis is fundamental to the practice of electrical engineering. Such analysis is based on Kirchhoff's two circuit laws and the values, as well as the fluctuations, of the *circuit variables*, that is, the voltages between terminals and the currents within the network. Circuit theory is nominally divided into determination of *equivalent circuits* and *mathematical analysis* of those equivalent circuits.[1] Equivalent circuits are composed of *lumped elements*. A lumped element is a

---

[1]In digital systems, the final equivalent circuit is a logic diagram.

Circuit Theory

single element representing a particular electrical property in the entire circuit, or in some portion of the circuit.[2] Equivalent circuit theory is sometimes called *network theory*. The term "network," while often used synonymously with "circuit," is usually reserved for more complicated arrangements of elements. In fact, "circuit" often designates a single loop of a network. Networks are often modeled using ideal elements.

2. IDEAL INDEPENDENT VOLTAGE SOURCES

An ideal independent voltage source maintains the voltage at its terminals regardless of the current flowing through the terminals. The symbology for an ideal independent voltage source and its characteristics is shown in Fig. 29.1. The subscript s generally indicates an independent source. The source value can vary with time, but its effective value does not change. That is, the magnitude of its variations in time is without regard to the current. The polarity assigned is arbitrary, but it is often assigned in the direction of positive power flow out of the source. This type of notation is called *source notation*.

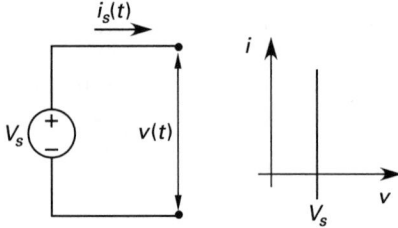

Figure 29.1 *Ideal Voltage Source*

A *real voltage source* cannot maintain the voltage at its terminals when the current becomes too large (in either the positive or the negative direction). The equivalent circuit of a real voltage source is shown in Fig. 29.2. The decrease in voltage as the current increases is measured by the *voltage regulation* given in Eq. 29.1.

$$\text{regulation} = \frac{V_{nl} - V_{fl}}{V_{fl}} \times 100\% \qquad 29.1$$

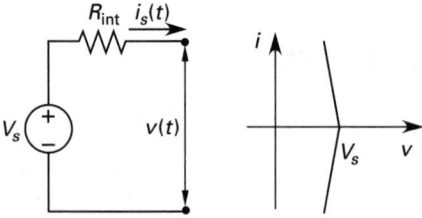

Figure 29.2 *Real Voltage Source*

---

[2]Electrical properties are distributed throughout elements, though they are concentrated in certain areas. Analyzing such *distributed elements* is difficult and they are thus represented by lumped parameters. This method is important in the analysis of electrical energy transmission.

3. IDEAL INDEPENDENT CURRENT SOURCES

An *ideal independent current source* maintains the current regardless of the voltage at its terminals. The symbology for an ideal independent current source and its characteristics is shown in Fig. 29.3, the subscript s indicating the independent source. The source value can vary with time, but its effective value does not change. That is, the magnitude of its variations in time is without regard to the voltage. The polarity is again arbitrary but is assigned using the source notation.

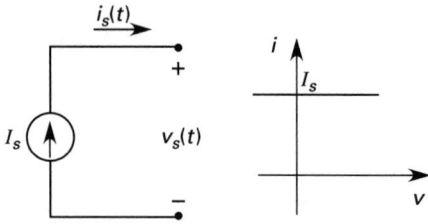

Figure 29.3 *Ideal Current Source*

A *real current source* cannot maintain the current at its terminals when the voltage becomes too large (in either the positive or the negative direction). The equivalent circuit of a real current source is shown in Fig. 29.4.

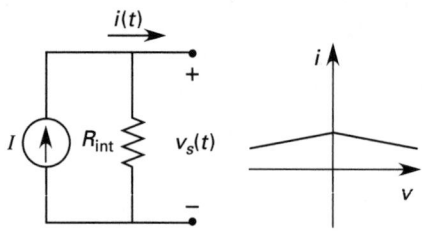

Figure 29.4 *Real Current Source*

4. IDEAL RESISTORS

An *ideal resistor* maintains the same resistance at its terminals regardless of the power consumed. That is, it is a linear element whose voltage/current relationship is given by Ohm's law.

$$v = iR \qquad 29.2$$

The symbology for an ideal resistor is identical to that for a real resistor and is shown in Fig. 29.5, along with the characteristics of the ideal resistor. The polarity is assigned so as to make Eq. 29.2, Ohm's law, valid as written. This is sometimes called the *sink notation*, as it corresponds to positive power flow into the resistor.

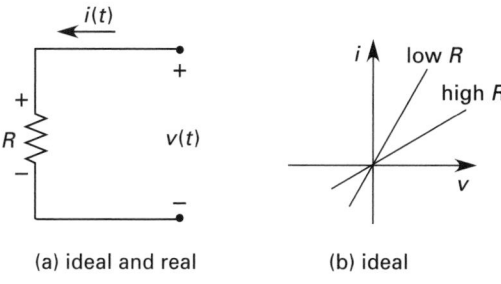

(a) ideal and real   (b) ideal

Figure 29.5 *Resistor*

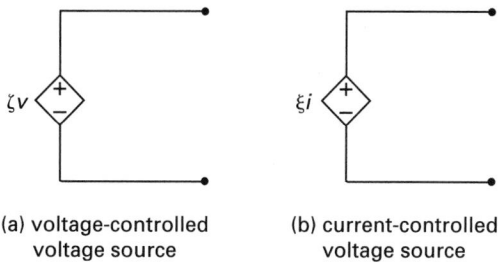

(a) voltage-controlled
voltage source

(b) current-controlled
voltage source

Figure 29.6 *Dependent Voltage Sources*

A real resistor undergoes changes in resistance as the temperature changes. The resistance of metals and most alloys increases with temperature while that of carbon and electrolytes decreases. The resistance change is accounted for in circuit design by ensuring the design allows for proper circuit functioning over the expected temperature range (see Sec. 26-3).

5. DEPENDENT SOURCES

Any circuit that is to deliver more power to the load than is available at the signal frequency must use an *active device*. An active device is defined as a component capable of amplifying the current or voltage, such as a transistor. An active device uses a source of power other than the input signal in order to accomplish the modification. In order to analyze circuits containing active devices, the devices are replaced by an equivalent circuit of *passive linear elements*, that is, elements that are not sources of energy (such as resistors, inductors, and capacitors), and one or more dependent sources. Such sources may be treated as standard sources using all the linear circuit analysis techniques, with the exception of superposition. Superposition can only be used if the device is limited to operation over a linear region.

6. DEPENDENT VOLTAGE SOURCES

A *dependent voltage source* is one in which the output is controlled by a variable elsewhere in the circuit. For this reason, dependent voltage sources are sometimes called *controlled voltage sources*. The controlling variable can be another voltage, a current, or any other physical quantity, such as light intensity or temperature. The symbology for dependent voltage sources is shown in Fig. 29.6. A *voltage-controlled voltage source* is controlled by a voltage, v, present somewhere else in the circuit. A *current-controlled voltage source* is controlled by a current, i, present somewhere else in the circuit. The terms ζ and ξ are constants.

7. DEPENDENT CURRENT SOURCES

A *dependent current source* is one in which the output is controlled by a variable elsewhere in the circuit. For this reason, dependent current sources are sometimes called *controlled current sources*. The controlling variable can be another voltage, a current, or any other physical quantity, such as light intensity or temperature. The symbology for dependent current sources is shown in Fig. 29.7. A *voltage-controlled current source* is controlled by a voltage, v, present somewhere else in the circuit. A *current-controlled current source* is controlled by a current, i, present somewhere else in the circuit. The terms ζ and ξ are constants.

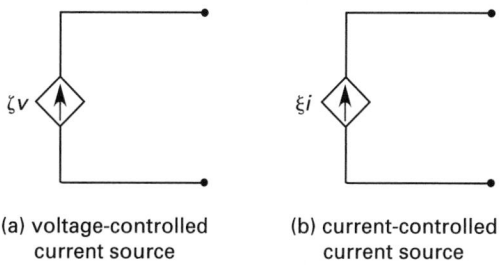

(a) voltage-controlled
current source

(b) current-controlled
current source

Figure 29.7 *Dependent Current Sources*

8. LINEAR CIRCUIT ELEMENTS

The basic circuit elements are resistance, capacitance, and inductance. Related to the basic element of inductance are mutual inductance and ideal (linear) transformers. A *linear network* is one in which the resistance, capacitance, and inductance are constant with respect to current and voltage, and in which the current or voltage of sources is independent or directly proportional to the other currents and voltages, or their derivatives, in the network. That is, a linear network is one that consists of linear elements and linear sources.[3] Circuit elements or sources are linear if the superposition principle is valid for them, and vice versa. A summary of linear circuit element parameters is given in Table 29.1. The time domain and frequency domain behavior of linear circuit elements, along with their defining equations, are shown in Table 29.2.

---

[3] A linear source is one in which the output is proportional to the first power of a voltage or current in the circuit.

9. RESISTANCE

Resistance is the opposition to current flow in a conductor. Technically, it is the opposition that a device or material offers to the flow of direct current. In AC circuits, it is the real part of the complex impedance. The resistance of a conductor depends on the resistivity, ρ, of the conducting material. The resistivity is a function of temperature that varies approximately linearly in accordance with Eq. 29.3. A slightly different version of this equation was given in Sec. 26-3. Here, the resistivity is referenced to 20°C and is written ρ_{20}. For copper, ρ_{20} is approximately 1.8×10^{-8} $\Omega \cdot$m. The thermal coefficient of resistivity, also called the *temperature coefficient of resistivity*, is referenced to 20°C and written α_{20}. For copper, α_{20} is approximately 3.9×10^{-3} C^{-1}. The equation for resistivity at any temperature, T (in °C), then becomes[4]

$$\rho = \rho_{20}\left(1 + \alpha_{20}\left(T - 20\right)\right) \qquad 29.3$$

The resistance of a conductor of length l and cross-sectional area A is

$$R = \frac{\rho l}{A} \qquad 29.4$$

The defining equation for resistance is Ohm's law, or the extended Ohm's law for AC circuits, which relates the linear relationship between the voltage and the current.

$$V = IR \quad \text{or} \quad \mathbf{V} = \mathbf{I}\mathbf{Z}_R \qquad 29.5$$

The power dissipated by the resistor is[5]

$$P = IV = I^2 R = \frac{V^2}{R} \qquad 29.6$$

The equivalent resistance of resistors in series is

$$R_e = R_1 + R_2 + \ldots R_n \qquad 29.7$$

The equivalent resistance of resistors in parallel is

$$\frac{1}{R_e} = \frac{1}{R_1} + \frac{1}{R_2} + \ldots \frac{1}{R_n} \qquad 29.8$$

10. CAPACITANCE

Capacitance is a measure of the ability to store charge. Technically, it is the opposition a device or material offers to changing voltage. In a DC circuit, that is, a

circuit in which the voltage is constant, the capacitor appears as an open circuit (after any transient condition has subsided). In a DC circuit, if the voltage is zero, the capacitor appears as a short circuit. In AC circuits, it is the constant of proportionality between the changing voltage and current.

$$i = C\left(\frac{dv}{dt}\right) \qquad 29.9$$

Capacitance is a function of the insulating material's *permittivity*, ε, between the conducting portions of the capacitor. The permittivity is the product of the *free-space permittivity*, ε_0, and the *relative permittivity* or *dielectric constant*, ε_r. For free space, that is, a vacuum, the permittivity is 8.854×10^{-12} F/m. The dielectric constant varies, and for many linear materials it has values between 1 and 10.

$$\varepsilon = \varepsilon_r \varepsilon_0 \qquad 29.10$$

The capacitance of a simple parallel plate capacitor is given by Eq. 29.11. The term r is the distance between the two parallel plates of equal area, A, and ε is the permittivity of the material between the plates.

$$C = \frac{\varepsilon A}{r} \qquad 29.11$$

The defining equation for capacitance, which relates the amount of charge stored to the voltage across the capacitor terminals, is

$$C = \frac{Q}{V} \qquad 29.12$$

The (average) power dissipated by a capacitor is zero. The (average) energy stored in the electric field of a capacitor is

$$U = \tfrac{1}{2}CV^2 = \tfrac{1}{2}QV = \tfrac{1}{2}\left(\frac{Q^2}{C}\right) \qquad 29.13$$

The equivalent capacitance of capacitors in series is

$$\frac{1}{C_e} = \frac{1}{C_1} + \frac{1}{C_2} + \ldots + \frac{1}{C_n} \qquad 29.14$$

The equivalent capacitance of capacitors in parallel is

$$C_e = C_1 + C_2 + \ldots + C_n \qquad 29.15$$

From Eq. 29.9, the voltage cannot change instantaneously, as this would make the current and the power infinite, which clearly cannot be the case. Using this fact, the transient behavior of capacitors can be analyzed. Using this principle is the equivalent of treating a capacitor's arbitrary constant κ in Table 29.1 as the initial voltage, V_0.

---

[4]Copper with a conductivity of 100% will experience a resistance change of approximately 10% for a 25°C (77°F) temperature change.

[5]When the term "power" is used, the average power is the referenced quantity. The instantaneous power depends on the time within the given cycle and is written p.

Example 29.1

In the circuit shown, the switch has been open for an "extended time," which in such circuits is considered to be a minimum of five time constants. The capacitor has no charge on it at $t = 0$ when the switch is closed. What is the current through the capacitor at the instant the switch is closed?

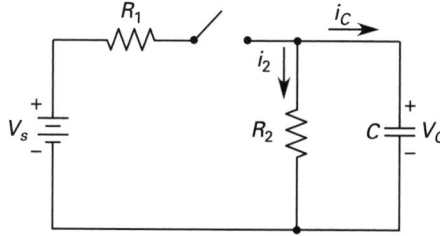

Solution

Since the voltage across the capacitor cannot change instantaneously, $v(0^-) = v(0^+)$, and the capacitor remains uncharged. The capacitor thus initially acts as a short circuit. All the current flows through the capacitor, bypassing R_2. The current is given by

$$i_c(0^+) = \frac{V_s}{R_1}$$

The initial voltage across the capacitor, $V_c(0^+)$, is zero.

Example 29.2

Using the circuit given in Ex. 29.1, what is the steady-state voltage across the capacitor?

Solution

The term "steady-state" is another way of saying "an extended period of time," which in such circuits is considered to be a minimum of five time constants. The capacitor will be fully charged and will act as an open circuit. Consequently, $dv/dt = 0$ and $i_c = 0$. All the current flows through R_2 and is given by

$$I_{R_2} = \frac{V_s}{R_1 + R_2}$$

Since they are in parallel, the voltage across the capacitor is the same as the voltage across R_2. Thus,

$$V_c(\infty) = V_{R_2} = I_2 R_2$$

$$= \left(\frac{V_s}{R_1 + R_2}\right) R_2 = V_s\left(\frac{R_2}{R_1 + R_2}\right)$$

The steady-state current through the capacitor, $I_c(\infty)$, is zero.

11. INDUCTANCE

Inductance is a measure of the ability to store magnetic energy. Technically, it is the opposition a device or material offers to changing current. In a DC circuit, that is, a circuit in which the current is constant, the inductor appears as a short circuit (after any transient condition has subsided). In a DC circuit, if the current is zero, the inductor appears as an open circuit. In AC circuits, it is the constant of proportionality between the changing current and voltage.

$$v = L\left(\frac{di}{dt}\right) \qquad 29.16$$

Inductance is a function of a medium's *permeability*, μ, between the magnetically linked portions of the inductor. The permeability is the product of the *free-space permeability*, μ_0, and the *relative permeability* or *permeability ratio*, μ_r. For free space, that is, a vacuum, the permeability is 1.2566×10^{-6} H/m. The permeability ratio varies. Depending on the material and the flux density, it can have values in the hundreds and sometimes thousands.

$$\mu = \mu_r \mu_0 \qquad 29.17$$

The inductance of a simple toroid shown in Fig. 29.8 is given by Eq. 29.18. The term l is the average flux path distance around the toroid of equal area, A, and μ is the permeability of the medium between the coil turns, N.

$$L = \frac{\mu N^2 A}{l} \qquad 29.18$$

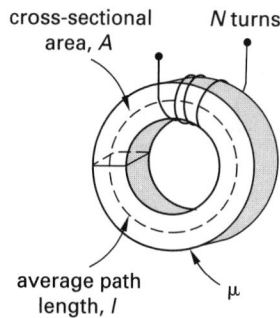

Figure 29.8 *Toroid Inductance Terms*

The defining equation for inductance, which relates the amount of flux present to the current flowing through the inductor terminals, is

$$L = \frac{\Psi}{I} \qquad 29.19$$

The (average) power dissipated by an inductor is zero. The (average) energy stored in the magnetic field of an inductor is given by Eq. 29.20. Positive energy indicates energy stored in the inductor. Negative energy indicates energy supplied to the circuit by the inductor.

$$U = \tfrac{1}{2}LI^2 = \tfrac{1}{2}\Psi I = \tfrac{1}{2}\left(\frac{\Psi^2}{L}\right) \qquad 29.20$$

The equivalent inductance of inductors in series is

$$L_e = L_1 + L_2 + \ldots + L_n \qquad 29.21$$

The equivalent inductance of inductors in parallel is

$$\frac{1}{L_e} = \frac{1}{L_1} + \frac{1}{L_2} + \ldots + \frac{1}{L_n} \qquad 29.22$$

From Eq. 29.16, the current cannot change instantaneously, as this would make the voltage and the power infinite, which clearly cannot be the case. Using this fact, the transient behavior of inductors can be analyzed. Using this principle is the equivalent of treating an inductor's arbitrary constant κ in Table 29.1 as the initial current, I_0.

Example 29.3

In the circuit shown, the switch has been open for an extended time, that is, greater than five time constants. The inductor has no associated magnetic field on it at $t = 0$ when the switch is closed. What is the current through the inductor at the instant the switch is closed?

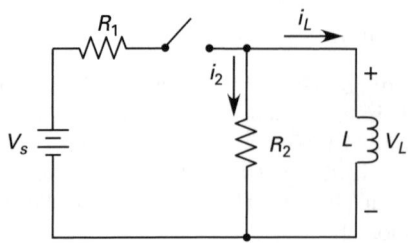

Solution

Since the current across the inductor cannot change instantaneously, $i(0^-) = i(0^+)$ and the inductor remains without a magnetic field. The inductor thus initially acts as an open circuit. All the current flows through R_2.

$$i_L(0^+) = 0$$

Example 29.4

Using the circuit given in Ex. 29.3, what is the initial voltage across the inductor?

Solution

All the current flows through R_2 and is given by

$$I_{R_2} = \frac{V_s}{R_1 + R_2}$$

Since they are in parallel, the voltage across the inductor is the same as the voltage across R_2. Thus,

$$V_L(0^+) = V_{R_2} = I_2 R_2$$
$$= \left(\frac{V_s}{R_1 + R_2}\right) R_2 = V_s \left(\frac{R_2}{R_1 + R_2}\right)$$

Table 29.1 *Linear Circuit Element Parameters*

circuit element	voltage	current	instantaneous power	average power	average energy stored
	$v = iR$	$i = \dfrac{v}{R}$	$p = iv = i^2R = \dfrac{v^2}{R}$	$P = IV = I^2R$ $= \dfrac{V^2}{R}$	—
	$v = \dfrac{1}{C}\displaystyle\int idt + \kappa$	$i = C\left(\dfrac{dv}{dt}\right)$	$p = iv = Cv\left(\dfrac{dv}{dt}\right)$	0	$U = \tfrac{1}{2}CV^2$
	$v = L\left(\dfrac{di}{dt}\right)$	$i = \dfrac{1}{L}\displaystyle\int vdt + \kappa$	$p = iv = Li\left(\dfrac{di}{dt}\right)$	0	$U = \tfrac{1}{2}LI^2$

Example 29.5

Using the circuit given in Ex. 29.3, what is the steady-state value of the current through the inductor?

Solution

The term "steady-state" is another way of saying "an extended period of time," which in such circuits is considered to be a minimum of five time constants. The inductor's magnetic field will be at a maximum and the inductor acts as a short circuit. Consequently, $di/dt = 0$, and $v_L = 0$. All the current thus flows through the inductor, bypassing R_2, and, using Ohm's law, is

$$i_L(\infty) = \frac{V_s}{R_1}$$

12. MUTUAL INDUCTANCE

Mutual inductance is the ratio of the electromotive force induced in one circuit to the rate of change of current in the other circuit. Applicable equations are given in Table 29.2. The concept is explained in Ch. 28. The total energy stored in a system involving mutual inductance is

$$U = \tfrac{1}{2}L_1 I_1^2 + \tfrac{1}{2}L_2 I_2^2 + M I_1 I_2 \qquad \textbf{29.23}$$

13. LINEAR SOURCE MODELS

Electric power sources tend to be nonlinear. In order to analyze such sources using circuit analysis, they are modeled as linear sources in combination with linear impedance. Such a source is classified as an *independent source*.

Dependent sources, whose output depends on a parameter elsewhere in the circuit, are associated with active devices, such as a transistor. These are not actually sources of power but can also be modeled as linear sources in combination with an impedance. The range of operation over which the model is valid must be selected to be approximately linear. Circuit analysis of this type is called *small-signal analysis*.

A third source type is the *transducer* that produces a current or voltage output from a nonelectric input. *Thermocouples* are transducers that produce voltages as functions of temperature. *Photodiodes* are transducers that produce currents as functions of incident illumination level. *Microphones* are transducers that produce voltages as functions of incident sound intensity.

All these sources can be modeled by linear approximation to measured values. Specific models for the electronic components are given in Ch. 43.

Circuit Theory

Table 29.2 *Linear Circuit Parameters, Time and Frequency Domain Representation*

parameter	defining equation	time domain	frequency domain[a]
resistance	$R = \dfrac{v}{i}$	$v = iR$	$V = IR$
capacitance	$C = \dfrac{Q}{V}$	$i = C\left(\dfrac{dv}{dt}\right)$	$I = j\omega C V$
self-inductance	$L = \dfrac{\Psi}{I}$	$v = L\left(\dfrac{di}{dt}\right)$	$V = j\omega L I$
mutual inductance[b]	$M_{12} = M_{21} = M = \dfrac{\Psi_{12}}{I_2}$ $= \dfrac{\Psi_{21}}{I_1}$	$v_1 = L_1\left(\dfrac{di_1}{dt}\right) + M\left(\dfrac{di_2}{dt}\right)$ $v_2 = L_2\left(\dfrac{di_2}{dt}\right) + M\left(\dfrac{di_1}{dt}\right)$	$V_1 = j\omega(L_1 I_1 + M I_2)$ $V_2 = j\omega(L_2 I_2 + M I_1)$

[a]The voltages and currents in the frequency domain column are not shown as vectors; for example, **I** or **V**. They are, however, shown in phasor form. Across any single circuit element (or parameter), the phase angle is determined by the impedance and embodied by the j. If the $j\omega$ (C or L or M) were not shown and instead given as **Z**, the current and voltage would be shown as **I** and **V**, respectively.
[b]Currents I_1 and I_2 (in the mutual inductance row) are in phase. The angle between either I_1 or I_2 and V_1 or V_2 is determined by the impedance angle embodied by the j. Even in a real transformer, this relationship holds, since the equivalent circuit accounts for any phase difference with a magnetizing current, I_m (see Sec. 28-5).

Example 29.6

The accompanying figure shows a representative plot of battery voltage versus load current. (a) What is the linear expression of the battery for the range of operation from 8 to 12 A? (b) Show the associated model.

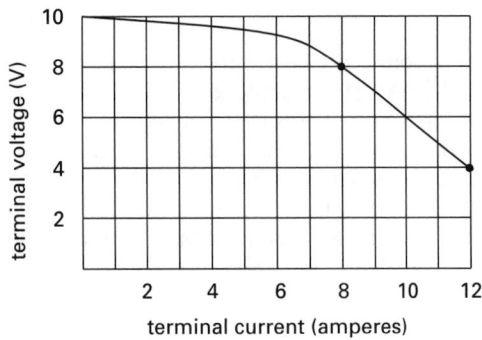

Solution

(a) Using the *two-point method*, equivalent to determining the parameters of a straight line in the formula $y = mx + b$, for the points (8,8) and (4,12) gives

$$\frac{I-8}{V-8} = \frac{12-8}{4-8} = -1$$

(b) Either the voltage or the current may be selected as the dependent variable.

$$V = -I + 16 \quad [I]$$

$$I = -V + 16 \quad [II]$$

Equation I is a loop voltage equation with the terminal voltage determined by two sources, one independent (the 16 V) and the other dependent on the current (the $-1\ \Omega$). Equation II is a node current equation with the terminal current determined by two sources, one independent (the 16 A) and the other dependent on the voltage (the $-1\ \Omega^{-1}$). The resulting models are

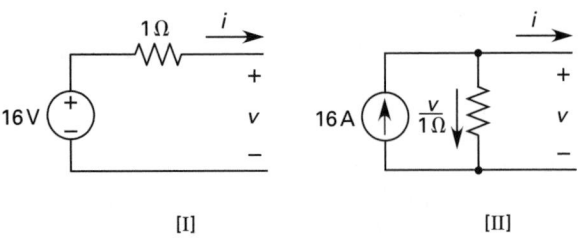

14. SOURCE TRANSFORMATIONS

An electrical source can be modeled as either a voltage source in series with an impedance or a current source in parallel with an impedance. The two models are equivalent in that they have the same terminal voltage/current relationships. Changing from one model to

the other is called *source transformation* and is a *circuit reduction* technique. The models' parameters are determined by

$$v_{\text{Th}} = i_N Z_{\text{Th}} \qquad 29.24$$

$$i_N = \frac{v_{\text{Th}}}{Z_{\text{Th}}} \qquad 29.25$$

Using Eqs. 29.24 and 29.25, it is possible to convert from one model to the other. The models are shown in Fig. 29.9.

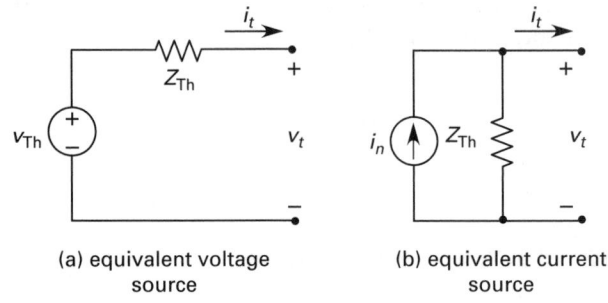

(a) equivalent voltage source (b) equivalent current source

Figure 29.9 *Source Transformation Models*

The source transformation is not defined for $Z = 0$ or $Z = \infty$. Also, both the open-circuit voltage and the short-circuit current direction must be the same in each model (see Sec. 26-11).

15. SERIES AND PARALLEL SOURCE COMBINATION RULES

One aspect of circuit reduction is the removal of sources that are not required. Section 29-14 referred to ideal current and voltage sources. These can be manipulated, combined, or eliminated as indicated by the following rules.

- An ideal voltage source cannot be placed in parallel with a second ideal voltage source, since the resulting voltage is indeterminate.
- An ideal current source cannot be placed in series with a second ideal current source, since the resulting current is indeterminate.
- An ideal voltage source in parallel with an ideal current source makes the current source redundant. The voltage is set by the voltage source, which absorbs all the current from the current source.
- An ideal current source in series with an ideal voltage source makes the voltage source redundant. The current is set by the current source, which is able to overcome any voltage of the voltage source.
- Ideal voltage sources in series can be combined algebraically using Kirchhoff's voltage law.
- Ideal current sources in parallel can be combined algebraically using Kirchhoff's current law.

The rules above can be used in an iterative manner to reduce a network to a single equivalent source and impedance. This technique is useful during phasor analysis or where a capacitance or inductance is present in a complicated network and the transient current or voltage is of interest. The iterative technique is also useful in the analysis of networks containing nonlinear or active devices.

16. REDUNDANT IMPEDANCES

Circuit reduction occurs when unnecessary impedances are removed according to the following rules.

- Impedance in parallel with an ideal voltage source may be removed. Theoretically, the ideal voltage supplies whatever energy is required to maintain the voltage, so the current flowing through the parallel impedance has no impact on the remainder of the circuit. Care must be exercised when calculating the source current to take into account the current through the parallel impedance.

- Impedance in series with an ideal current source may be removed. Theoretically, the ideal current source supplies whatever energy is required to maintain the current, so the energy loss in the series impedance has no impact on the remainder of the circuit. The voltage across the current source or current source/series impedance combination is determined by the remainder of the circuit.

17. DELTA-WYE TRANSFORMATIONS

Electrical impedances arranged in the shape of the Greek letter delta (Δ) or the English letter Y (wye) are known as *delta-wye configurations*, or *pi-T configurations*. A delta-wye configuration can be useful in reducing circuits in power networks.

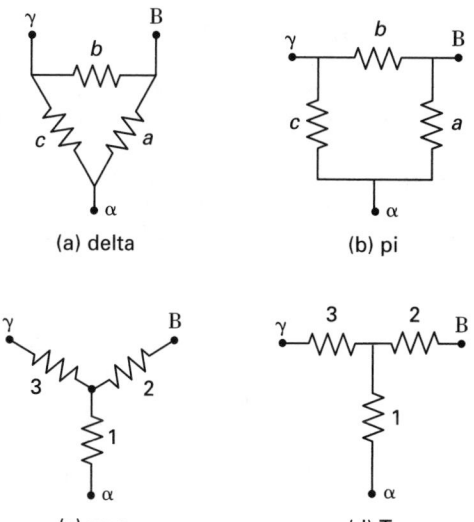

(a) delta (b) pi

(c) wye (d) T

Figure 29.10 *Delta (Pi)-Wye (T) Configurations*

The equivalent impedances, which allow transformation between configurations, are

$$Z_a = \frac{Z_1 Z_2 + Z_1 Z_3 + Z_2 Z_3}{Z_3} \qquad 29.26$$

$$Z_b = \frac{Z_1 Z_2 + Z_1 Z_3 + Z_2 Z_3}{Z_1} \qquad 29.27$$

$$Z_c = \frac{Z_1 Z_2 + Z_1 Z_3 + Z_2 Z_3}{Z_2} \qquad 29.28$$

$$Z_1 = \frac{Z_a Z_c}{Z_a + Z_b + Z_c} \qquad 29.29$$

$$Z_2 = \frac{Z_a Z_b}{Z_a + Z_b + Z_c} \qquad 29.30$$

$$Z_3 = \frac{Z_b Z_c}{Z_a + Z_b + Z_c} \qquad 29.31$$

18. THEVENIN'S THEOREM

Thevenin's theorem states that, insofar as the behavior of a linear circuit at its terminals is concerned, any such circuit can be replaced by a single voltage source, $\mathbf{V}_{\text{Th}}$, in series with a single impedance, $\mathbf{Z}_{\text{Th}}$. The method for determining and utilizing the *Thevenin equivalent circuit*, with designations referring to Fig. 29.11, follows.

step 1: Separate the network that is to be changed into a Thevenin equivalent circuit from its load at two terminals, say A and B.

step 2: Determine the open-circuit voltage, $\mathbf{V}_{\text{oc}}$, at terminals A and B.

step 3: Short-circuit terminals A and B and determine the current, $\mathbf{I}_{\text{sc}}$.

step 4: Calculate the Thevenin equivalent voltage and resistance from the following equations.

$$\mathbf{V}_{\text{Th}} = \mathbf{V}_{\text{oc}} \qquad 29.32$$

$$\mathbf{Z}_{\text{Th}} = \frac{\mathbf{V}_{\text{oc}}}{\mathbf{I}_{\text{sc}}} \qquad 29.33$$

step 5: Using the values calculated in Eqs. 29.32 and 29.33, replace the network with the Thevenin equivalent. Reconnect the load at terminals A and B. Determine the desired electrical parameters in the load.

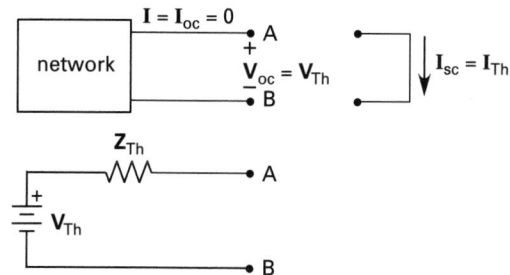

Figure 29.11 *Thevenin Equivalent Circuit*

Steps 3 and 4 can be altered by using the following shortcut. Determine the Thevenin equivalent impedance by looking into terminals A and B toward the network with all the power sources altered. Specifically, change independent voltage sources into short circuits and independent current sources into open circuits. Then calculate the resistance of the altered network. The resulting resistance is the Thevenin equivalent impedance.

19. NORTON'S THEOREM

Norton's theorem states that, insofar as the behavior of a linear circuit at its terminals is concerned, any such circuit can be replaced by a single current source, $\mathbf{I}_N$, in parallel with a single impedance, $\mathbf{Z}_N$. The method for determining and utilizing the *Norton equivalent circuit*, with designations referring to Fig. 29.12, follows.

step 1: Separate the network that is to be changed into a Norton equivalent circuit from its load at two terminals, say A and B.

step 2: Determine the open-circuit voltage, $\mathbf{V}_{oc}$, at terminals A and B.

step 3: Short-circuit terminals A and B and determine the current, $\mathbf{I}_{sc}$.

step 4: Calculate the Norton equivalent current and resistance from the following equations.

$$\mathbf{I}_N = \mathbf{I}_{sc} \qquad \text{29.34}$$

$$\mathbf{Z}_N = \frac{\mathbf{V}_{oc}}{\mathbf{I}_{sc}} \qquad \text{29.35}$$

step 5: Using the values calculated in Eqs. 29.34 and 29.35, replace the network with the Norton equivalent. Reconnect the load at terminals A and B. Determine the desired electrical parameters in the load.

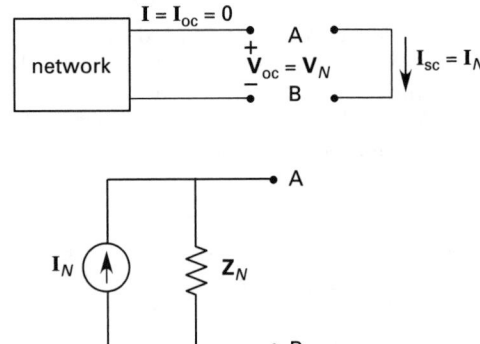

Figure 29.12 *Norton Equivalent Circuit*

Steps 2 and 4 can be altered by using the following shortcut. Determine the Norton equivalent impedance

by looking into terminals A and B toward the network with all the power sources altered. Specifically, change independent voltage sources into short circuits and independent current sources into open circuits. Then calculate the resistance of the altered network. The resulting impedance is the Norton equivalent impedance.

The Norton equivalent impedance equals the Thevenin equivalent impedance for identical networks.

20. MAXIMUM POWER TRANSFER THEOREM

The maximum energy, that is, maximum power transfer from a voltage source, occurs when the series source impedance, $\mathbf{Z}_s$, is reduced to the minimum possible value, with zero being the ideal case. This assumes the load impedance is fixed and the source impedance can be changed. Though this is the ideal situation, it is not often the case. Where the load impedance varies and the source impedance is fixed, maximum power transfer occurs when the load and source impedances are complex conjugates. That is, $\mathbf{Z}_l = \mathbf{Z}_s^*$ or $R_{\text{load}} + jX_{\text{load}} = R_s - jX_s$.

Similarly, the maximum power transfer from a current source occurs when the parallel source impedance, $\mathbf{Z}_s$, is increased to the maximum possible value. This assumes the load impedance is fixed and the source impedance can be changed. In the more common case where the load impedance varies and the source impedance is fixed, the maximum power transfer occurs when the load and source impedances are complex conjugates.

The maximum power transfer in any circuit occurs when the load impedance equals the complex conjugate of the Norton or Thevenin equivalent impedance.

21. SUPERPOSITION THEOREM

The principle of *superposition* is that the response of (that is, the voltage across or current through) a linear circuit element in a network with multiple independent sources is equal to the response obtained if each source is considered individually and the results summed. The steps involved in determining the desired quantity follow.

step 1: Replace all sources except one by their internal resistances. Ideal current sources are replaced by open circuits. Ideal voltage sources are replaced by short circuits.

step 2: Compute the desired quantity, either voltage or current, for the element in question due to the single source.

step 3: Repeat steps 1 and 2 for each of the sources in turn.

step 4: Sum the calculated values obtained for the current or voltage obtained in step 2. The result is the actual value of the current or voltage in the element for the complete circuit.

Superposition is not valid for circuits in which the following conditions exist.

- The capacitors have an initial charge (i.e., an initial voltage) not equal to zero. This principle can be used if the charge is treated as a separate voltage source and the equivalent circuit analyzed.

- The inductors have an initial magnetic field (i.e., an initial current) not equal to zero. This principle can be used if the energy in the magnetic field is treated as a separate current source and this equivalent circuit analyzed.

- Dependent sources are used.

22. MILLER'S THEOREM

A standard two-port network is shown in Fig. 29.13(a). *Miller's theorem* states that if an admittance, Y, is connected between the input and output terminals of a two-port network, as shown in Fig. 29.13(b), the output voltage is linearly dependent on the input voltage. Further, the circuit can be transformed as shown in Fig. 29.13(c). The transformation equations, with A as a constant that must be determined by independent means, are

$$Y_1 = Y(1 - A) \qquad 29.36$$

$$Y_2 = Y\left(1 - \frac{1}{A}\right) \qquad 29.37$$

Similar formulas can be derived from Eqs. 29.36 and 29.37 for impedances Z_1 and Z_2, though the theorem is not often used in such a manner. Miller's theorem is useful in transistor high-frequency amplifiers, among other applications.

23. KIRCHHOFF'S LAWS

Once an equivalent circuit is determined and reduced to its simplest form, Kirchhoff's voltage and current laws are used to determine the circuit behavior. If the KVL and KCL equations are written in terms of instantaneous quantities, $v(t)$ and $i(t)$, a differential equation results with an order equal to the number of independent energy-storing devices in the circuit. The solution of this equation results in a steady-state component and a transient component, which decays with time. If the equation is written in terms of phasor quantities, $\mathbf{V}$ and $\mathbf{I}$, an algebraic equation results. The solution of this equation alone results in a steady-state component. For any equation with an order higher than two, the most efficient solution method becomes a computer-aided design software package.

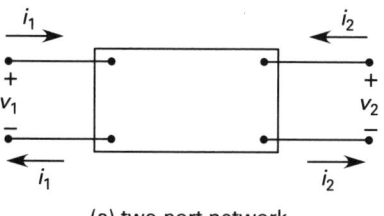

(a) two-port network

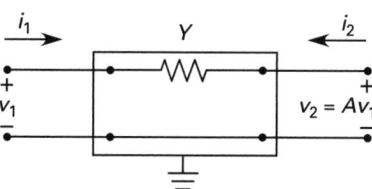

(b) connecting admittance

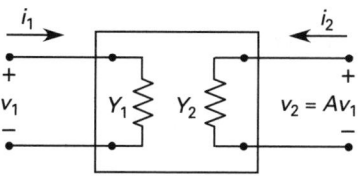

(c) Miller's transformation

Figure 29.13 Miller's Theorem

24. KIRCHHOFF'S VOLTAGE LAW

Kirchhoff's voltage law (KVL) states that the algebraic sum of the phasor voltages around any closed path within a circuit or network is zero. Stated in terms of the voltage rises and drops, the sum of the phasor voltage rises equals the sum of the phasor voltage drops around any closed path within a circuit or network. The method for applying KVL follows.

step 1: Identify the loop.

step 2: Pick a loop direction.[6]

step 3: Assign the loop current in the direction picked in step 2.

step 4: Assign voltage polarities consistent with the loop current direction in step 3.

step 5: Apply KVL to the loop using Ohm's law to express the voltages across each circuit element.

step 6: Solve the equation for the desired quantity.

---

[6]Clockwise is the direction chosen as the standard in this text. Any direction is allowed. The resulting mathematics determines the correct direction of current flow and voltage polarities. Consistency is the key to preventing calculation errors.

25. KIRCHHOFF'S CURRENT LAW

Kirchhoff's current law (KCL) states that the algebraic sum of the phasor currents at any node is zero. Stated in terms of the currents out of and into a node, the sum of the currents directed out of a node must equal the sum of the currents directed into the same node. The method for applying KCL follows.

step 1: Identify the nodes and pick a reference or datum node.

step 2: Label the node-to-datum voltage for each unknown node.

step 3: Pick a current direction for each path at every node.[7]

step 4: Apply KCL to the nodes using Ohm's law to express the currents through each circuit branch.

step 5: Solve the equations for the desired quantity.

26. LOOP ANALYSIS

The *loop-current* method is a systematic network-analysis procedure that uses currents as the unknowns. It is also called *mesh analysis* or the *Maxwell loop-current* method. The method uses Kirchhoff's voltage law and is performed on planar networks. It requires $n - 1$ simultaneous equations for an n-loop system. The method's steps are as follows.

step 1: Select $n - 1$ loops, that is, one loop less than the total number of possible loops.

step 2: Assign current directions for the selected loops. (While this is arbitrary, clockwise directions will always be chosen in this text for consistency.) Any incorrectly chosen current direction will cause a negative result when the simultaneous equations are solved in step 4. Show the direction of the current with an arrow.

step 3: Write Kirchhoff's voltage law for each of the selected loops. Assign polarities based on the direction of the loop current. The voltage of a source is positive when the current flows out of the positive terminal, that is, from the negative terminal to the positive terminal inside the source. The selected direction of the loop current always results in a voltage drop in the resistors of the loop (see Fig. 26.6(c)). Where two loop currents flow through an element, they are summed to determine the voltage drop in that element, using the direction of the current in the loop for which the equation is being written as

the positive (i.e., correct) direction. Any incorrect direction for the loop current will be indicated by a negative sign in the solution in step 4.

step 4: Solve the $n - 1$ equations from step 3 for the unknown currents.

step 5: If required, determine the actual current in an element by summing the loop currents flowing through the element. Sum the absolute values to obtain the correct magnitude. The correct direction is given by the loop current with the positive value.

27. NODE ANALYSIS

The *node-voltage* method is a systematic network-analysis procedure that uses voltages as the unknowns. The method uses Kirchhoff's current law. It requires $n - 1$ equations for an n-principal node system (equations are not necessary at simple nodes, that is, nodes connecting only two circuit elements). The method's steps are as follows.

step 1: Simplify the circuit, if possible, by combining resistors in series or parallel or by combining current sources in parallel. Identify all nodes. The minimum number of equations required will be $n - 1$ where n represents the number of principal nodes.

step 2: Choose one node as the reference node, that is, the node that will be assumed to have ground potential (0 V). To minimize the number of terms in the equations, select the node with the largest number of circuit elements to serve as the reference node.

step 3: Write Kirchhoff's current law for each principal node except the reference node, which is assumed to have a zero potential.

step 4: Solve the $n - 1$ equations from step 3 to determine the unknown voltages.

step 5: If required, use the calculated node voltages to determine any branch current desired.

28. DETERMINATION OF METHOD

The method used to analyze an electrical network depends on the circuit elements and their configurations. The loop-current method employing Kirchhoff's voltage law is used in circuits without current sources. The node-voltage method employing Kirchhoff's current law is used in circuits without voltage sources. When both types of sources are present, one of two methods may be used.

method 1: Use each of Kirchhoff's laws, assigning voltages and currents as needed, and substitute any known quantity into the equations as written.

---

[7]Current flow out of the node is the direction chosen as the standard for positive current in this text. Any direction is allowed. The resulting mathematics determines the correct direction of current flow and voltage polarities. Consistency is the key to preventing calculation errors.

method 2: Use source transformation or source shifting to change the appearance of the circuit so that it contains only the desired sources, that is, voltage sources when using KVL and current sources when using KCL. *Source shifting* is manipulating the circuit so that each voltage source has a resistor in series and each current source has a resistor in parallel.

Use the method that results in the least number of equations. The loop-current method produces $n - 1$ equations where n is the total number of loops. The node-voltage method produces $n - 1$ equations where n is the number of principal nodes. Count the number of loops and the number of principal nodes prior to writing the equations. Whichever is least determines the method used.

Additional methods exist, some of which require less work. The advantage of the loop-current and node-voltage methods is that they are systematic and thus guarantee a solution.

29. VOLTAGE AND CURRENT DIVIDERS

At times, a source voltage will not be at the required value for the operation of a given circuit. For example, the household voltage of 120 V is too high to properly bias electronic circuitry. One method of obtaining the required voltage without using a transformer is to use a *voltage divider*. A voltage divider is illustrated in Fig. 29.14(a). The voltage across impedance 2 is

$$\mathbf{V}_2 = \mathbf{V}_s \left(\frac{\mathbf{Z}_2}{\mathbf{Z}_1 + \mathbf{Z}_2} \right) \qquad 29.38$$

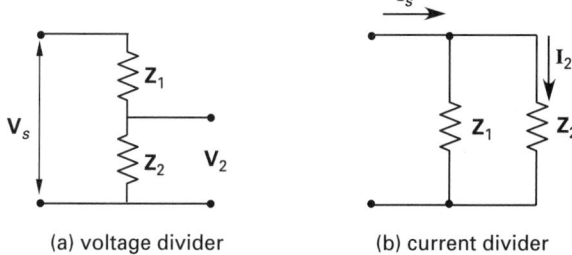

(a) voltage divider   (b) current divider

Figure 29.14 *Divider Circuits*

If the fraction of the source voltage, $\mathbf{V}_s$, is found as a function of $\mathbf{V}_2$, the result is known as the *gain* or the *voltage-ratio transfer function*. Rearranging Eq. 29.38 to show the specific relationship between $\mathbf{Z}_1$ and $\mathbf{Z}_2$ gives

$$\frac{\mathbf{V}_2}{\mathbf{V}_s} = \frac{1}{1 + \dfrac{\mathbf{Z}_1}{\mathbf{Z}_2}} \qquad 29.39$$

A specific current can be obtained through an analogous circuit called a *current divider*, shown in Fig. 29.14(b). The current through impedance 2 is

$$\mathbf{I}_2 = \mathbf{I}_s \left(\frac{\mathbf{Z}_1}{\mathbf{Z}_1 + \mathbf{Z}_2} \right) = \mathbf{I}_s \left(\frac{\mathbf{G}_2}{\mathbf{G}_1 + \mathbf{G}_2} \right) \qquad 29.40$$

30. STEADY-STATE AND TRANSIENT IMPEDANCE ANALYSIS

When the electrical parameters of a circuit do not change with time, the circuit is said to be in a *steady-state* condition. In DC circuits, this condition exists when the magnitude of the parameter is constant. In AC circuits, this condition exists when the frequency is constant. A *transient* is a temporary phenomenon occurring prior to a network reaching steady state.

DC steady-state impedance analysis is based on constant voltages and currents and, therefore, time derivatives of zero. The DC impedances are

- Resistance:

$$Z_R|_{\mathrm{DC}} = R \qquad 29.41$$

- Inductance:

$$v = L \left(\frac{di}{dt} \right) = L(0) = 0 \qquad 29.42$$

$$Z = \frac{v}{i} = \frac{0}{i} = 0 \qquad 29.43$$

$$Z_L|_{\mathrm{DC}} = 0 \quad \text{[short circuit]} \qquad 29.44$$

- Capacitance:

$$i = C \left(\frac{dv}{dt} \right) = C(0) = 0 \qquad 29.45$$

$$Z = \frac{v}{i} = \frac{v}{0} \to \infty \qquad 29.46$$

$$Z_C|_{\mathrm{DC}} = \infty \quad \text{[open circuit]} \qquad 29.47$$

AC steady-state impedance analysis is based on the phasor form where $df(t)/dt = j\omega f(t)$. The AC steady-state impedances are

- Resistance:

$$Z_R|_{\mathrm{AC}} = R \qquad 29.48$$

- Inductance:

$$v = L \left(\frac{di}{dt} \right) = Lj\omega i \qquad 29.49$$

$$Z = \left(\frac{v}{i} \right) = \frac{Lj\omega i}{i} = j\omega L \qquad 29.50$$

$$Z_L|_{\mathrm{AC}} = j\omega L \qquad 29.51$$

Circuit Theory

- Capacitance:

$$i = C\left(\frac{dv}{dt}\right) = Cj\omega v \qquad 29.52$$

$$Z = \frac{v}{i} = \frac{v}{Cj\omega v} = \frac{1}{j\omega C} \qquad 29.53$$

$$Z_C|_{\text{AC}} = \frac{1}{j\omega C} \qquad 29.54$$

Transient impedance analysis is based on the phasor form with the complex variable s substituted for $j\omega$. The variable $s = \sigma + j\omega$ and is the same as the Laplace transform variable. The derivative is $df(t)/dt = sf(t)$. The transient impedances are

- Resistance:

$$Z_R = R \qquad 29.55$$

- Inductance:

$$Z_L = sL \qquad 29.56$$

- Capacitance:

$$Z_C = \frac{1}{sC} \qquad 29.57$$

Transient impedances are useful in the analysis of stability as well as during transients.

31. TWO-PORT NETWORKS

An electric circuit or network is often used to connect a source to a load, modifying the source energy or information in a given manner as required or desired by the load. If the circuit is such that the current flow into one terminal is equal to the current flow out of a second terminal, the terminal pair is called a *port*. Any number of ports is possible for a given network, the most common being the *two-port network* shown in Fig. 29.15. Four variables exist in this representation: v_1, v_2, i_1, and i_2. The port using the subscript 1 is the *input port*, and the port using the subscript 2 is the *output port*.

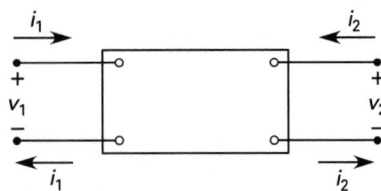

Figure 29.15 *Two-Port Network*

Two-port networks can represent either three- or four-terminal devices, such as transistors or transformers

(see Fig. 29.16). The electrical properties of the device within the network are described by a set of parameters identified by double subscripts, the first representing the row and the second the column of a matrix. The *parameter type* is determined by the selection of independent variables for which to solve.

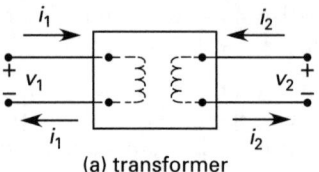

(a) transformer

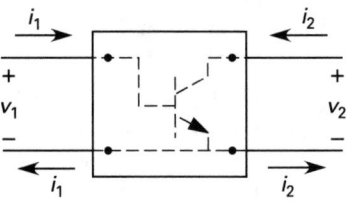

(b) transistor (common emitter configuration)

Figure 29.16 *Transformer and Transistor Two-Port Networks*

Open-circuit impedance parameters, or z parameters, occur when the two currents are selected as the independent variables. The z parameter model and *deriving equations* for the individual parameters are shown in Fig. 29.17. The applicable matrix and resulting equations follow.

$$\begin{bmatrix} v_1 \\ v_2 \end{bmatrix} = \begin{bmatrix} z_{11} & z_{12} \\ z_{21} & z_{22} \end{bmatrix} \begin{bmatrix} i_1 \\ i_2 \end{bmatrix} \qquad 29.58$$

$$v_1 = z_{11}i_1 + z_{12}i_2 \qquad 29.59$$

$$v_2 = z_{21}i_1 + z_{22}i_2 \qquad 29.60$$

Short-circuit admittance parameters, or y parameters, occur when the two voltages are selected as the independent variables. The y parameter model and *deriving equations* for the individual parameters are shown in Fig. 29.18. The applicable matrix and resulting equations follow.

$$\begin{bmatrix} i_1 \\ i_2 \end{bmatrix} = \begin{bmatrix} y_{11} & y_{12} \\ y_{21} & y_{22} \end{bmatrix} \begin{bmatrix} v_1 \\ v_2 \end{bmatrix} \qquad 29.61$$

$$i_1 = y_{11}v_1 + y_{12}v_2 \qquad 29.62$$

$$i_2 = y_{21}v_1 + y_{22}v_2 \qquad 29.63$$

Hybrid parameters, or h parameters, occur when one voltage and one current are picked as the independent variables. The h parameter model and *deriving equations* for the individual parameters are shown in

Fig. 29.19. The applicable matrix and resulting equations follow.

$$\begin{bmatrix} v_1 \\ i_2 \end{bmatrix} = \begin{bmatrix} h_{11} & h_{12} \\ h_{21} & h_{22} \end{bmatrix} \begin{bmatrix} i_1 \\ v_2 \end{bmatrix} \qquad \text{29.64}$$

$$v_1 = h_{11}i_1 + h_{12}v_2 \qquad \text{29.65}$$

$$i_2 = h_{21}i_1 + h_{22}v_2 \qquad \text{29.66}$$

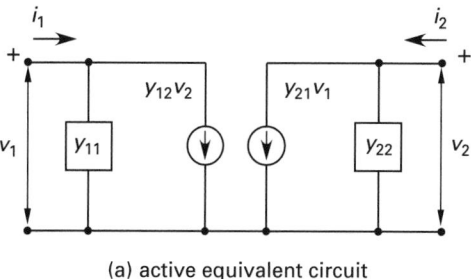

(a) active equivalent circuit

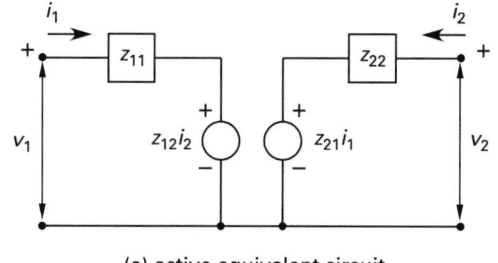

(a) active equivalent circuit

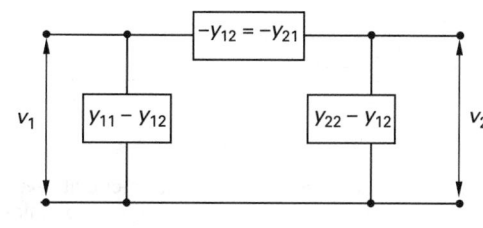

(b) passive π-model equivalent circuit

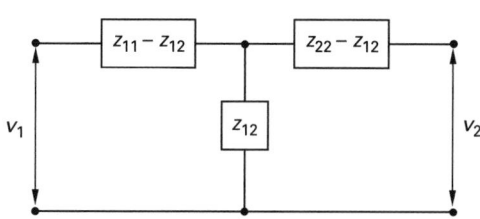

(b) passive T-model equivalent circuit

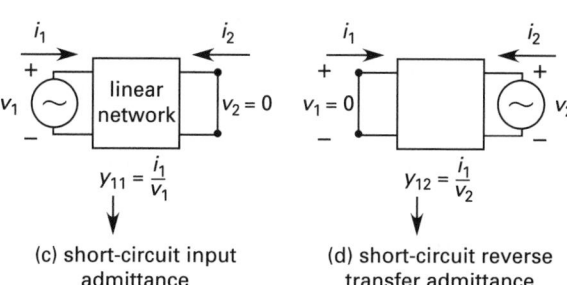

(c) short-circuit input admittance

(d) short-circuit reverse transfer admittance

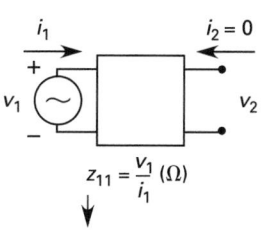

(c) open-circuit input impedance

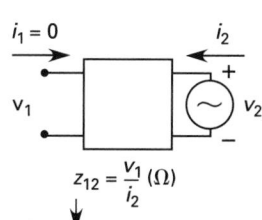

(d) open-circuit reverse transfer impedance

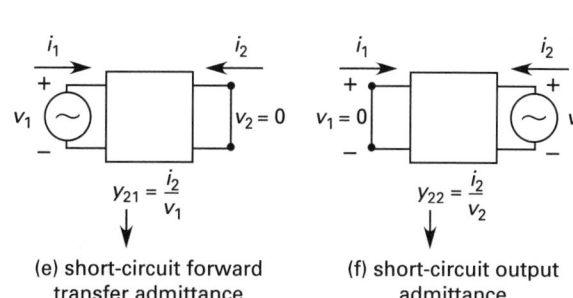

(e) short-circuit forward transfer admittance

(f) short-circuit output admittance

Figure 29.18 *Admittance Model Parameters*

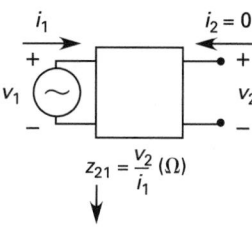

(e) open-circuit forward transfer impedance

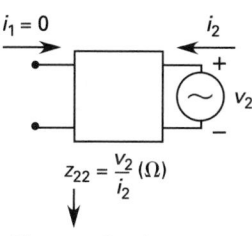

(f) open-circuit output impedance

Figure 29.17 *Impedance Model Parameters*

Inverse hybrid parameters, or *g* parameters, also occur when one voltage and one current are picked as the independent variables. The applicable matrix and resulting equations follow.

$$\begin{bmatrix} i_1 \\ v_2 \end{bmatrix} = \begin{bmatrix} g_{11} & g_{12} \\ g_{21} & g_{22} \end{bmatrix} \begin{bmatrix} v_1 \\ i_2 \end{bmatrix} \qquad \text{29.67}$$

$$i_1 = g_{11}v_1 + g_{12}i_2 \qquad \text{29.68}$$

$$v_2 = g_{21}v_1 + g_{22}i_2 \qquad \text{29.69}$$

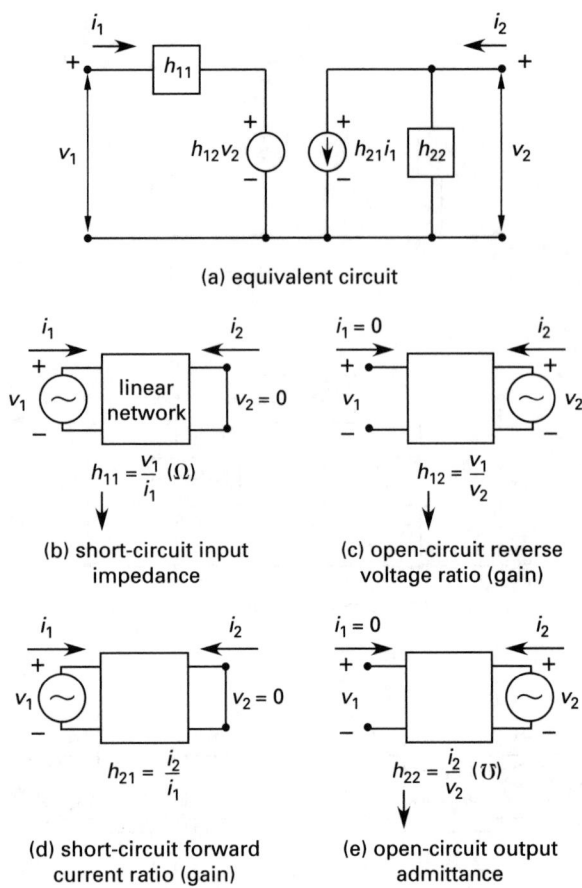

(a) equivalent circuit

$h_{11} = \dfrac{V_1}{i_1}$ (Ω)

(b) short-circuit input impedance

$h_{12} = \dfrac{V_1}{v_2}$

(c) open-circuit reverse voltage ratio (gain)

$h_{21} = \dfrac{i_2}{i_1}$

(d) short-circuit forward current ratio (gain)

$h_{22} = \dfrac{i_2}{v_2}$ $(\mho)$

(e) open-circuit output admittance

Figure 29.19 *Hybrid Model Parameters*

Other parameter representations are possible. Some examples include the *transmission line* or *chain parameters*, and the *inverse transmission line parameters*. The choice of parameter type depends on numerous factors, including whether or not all the parameters exist or are defined for a given network, the mathematical convenience of using a certain set of parameters, and the accuracy of sensitivity of the parameters when considered as part of the overall circuit to which it is connected. The parameters and their deriving equations are summarized in Table 29.3. The formulas for conversion between the parameter types are given in App. 29.A.

Table 29.3 *Two-Port Network Parameters*

representation	deriving equations			
impedance $\begin{bmatrix} z_{11} & z_{12} \\ z_{21} & z_{22} \end{bmatrix}$	$z_{11} = \dfrac{V_1}{I_1}$ $I_2 = 0$	$z_{12} = \dfrac{V_1}{I_2}$ $I_1 = 0$	$z_{21} = \dfrac{V_2}{I_1}$ $I_2 = 0$	$z_{22} = \dfrac{V_2}{I_2}$ $I_1 = 0$
admittance $\begin{bmatrix} y_{11} & y_{12} \\ y_{21} & y_{22} \end{bmatrix}$	$y_{11} = \dfrac{I_1}{V_1}$ $V_2 = 0$	$y_{12} = \dfrac{I_1}{V_2}$ $V_1 = 0$	$y_{21} = \dfrac{I_2}{V_1}$ $V_2 = 0$	$y_{22} = \dfrac{I_2}{V_2}$ $V_1 = 0$
hybrid $\begin{bmatrix} h_{11} & h_{12} \\ h_{21} & h_{22} \end{bmatrix}$	$h_{11} = \dfrac{V_1}{I_1}$ $V_2 = 0$	$h_{12} = \dfrac{V_1}{V_2}$ $I_1 = 0$	$h_{21} = \dfrac{I_2}{I_1}$ $V_2 = 0$	$h_{22} = \dfrac{I_2}{V_2}$ $I_1 = 0$
inverse hybrid $\begin{bmatrix} g_{11} & g_{12} \\ g_{21} & g_{22} \end{bmatrix}$	$g_{11} = \dfrac{I_1}{V_1}$ $I_2 = 0$	$g_{12} = \dfrac{I_1}{I_2}$ $V_1 = 0$	$g_{21} = \dfrac{V_2}{V_1}$ $I_2 = 0$	$g_{22} = \dfrac{V_2}{I_2}$ $V_1 = 0$
transmission or chain $\begin{bmatrix} A & B \\ C & D \end{bmatrix}$	$A = \dfrac{V_1}{V_2}$ $I_2 = 0$	$B = \dfrac{-V_1}{I_2}$ $V_2 = 0$	$C = \dfrac{I_1}{V_2}$ $I_2 = 0$	$D = \dfrac{-I_1}{I_2}$ $V_2 = 0$
inverse transmission $\begin{bmatrix} \alpha & \beta \\ \gamma & \delta \end{bmatrix}$	$\alpha = \dfrac{v_2}{v_1}$ $i_1 = 0$	$\beta = \dfrac{v_2}{i_1}$ $v_1 = 0$	$\gamma = \dfrac{-i_2}{v_1}$ $i_1 = 0$	$\delta = \dfrac{-i_2}{i_1}$ $v_i = 0$

30 Transient Analysis

Nomenclature

a	constant	–
A	natural response coefficient	–
b	constant	–
c	constant	–
C	capacitance	F
f	function	–
G	conductance	S
i, I	current	A
L	self-inductance	H
pf	power factor	–
P	power	W
q	instantaneous charge	C
Q	quality factor	–
R	resistance	Ω
s	Laplace transform variable	–
t	time	s
v, V	voltage	V
x	variable	–
X	reactance	Ω
y	variable	–
Y	admittance	S

Symbols

κ	integration constant	–
σ	damping, real part of s	–
τ	time constant	s
ϕ	angle	rad
ϕ	phase difference angle	rad
ω_0	zero resistance frequency	rad/s

Subscripts

0	initial
bat	battery
C	capacitor
d	delay
L	inductor
m	maximum
R	resistor
s	source
ss	steady-state
tr	transient

1. FUNDAMENTALS

Whenever a network or circuit undergoes a change, the currents and voltages experience a transitional period during which their properties shift from their former values to their new steady-state values. This time period is called a *transient*. The determination of a circuit's behavior during this period is *transient analysis*. There are three classes of transient problems, all based on the nature of the energy source. These are *DC switching transients*, *AC switching transients*, and *pulse transients*. *Switching transients* occur as a result of a change in topology of the circuit, that is, the physical elements within the circuit change (become connected or disconnected) as a result of the operation of a switch. The switch can be a physical switch or an electrical one, such as a transistor operating in the switching mode. A circuit's response to a switching transient is known as a *step response*.

Pulse transients involve a change in the current or voltage waveform, not in the topology of a circuit. A circuit's response to a pulse transient is known as an *impulse response*.

The energy storage elements in electric circuits are the capacitor and the inductor. When one of these is present, the mathematical representation of the circuit is a first-order differential equation of the form

$$f(t) = b \left(\frac{dx}{dt} \right) + cx \qquad \textit{30.1}$$

Either the current or the voltage can be represented as the dependent variable, that is, $f(t)$, in Eq. 30.1. For capacitors, the voltage is used because the energy storage is a function of voltage and a true differential equation results (the capacitor current is an integral equation). For inductors, similar reasoning holds and the current is considered the dependent variable, $f(t)$. The solution to Eq. 30.1 is of the general form

$$x(t) = \kappa + Ae^{-\frac{t}{\tau}} \qquad \textit{30.2}$$

See Table 30.1 for the applicable equations for both the inductor and capacitor.

The term A in Eq. 30.2 is the natural response coefficient and is set to match the conditions at some known time in the transient, typically at the onset. The term κ is value-dependent on the forcing function, $f(t)$. The term τ is called the *time constant* and is equal to

$$\tau = \frac{b}{c} \qquad \textit{30.3}$$

The time constant for an RC circuit is

$$\tau = RC \qquad \textit{30.4}$$

The time constant for an RL circuit is

$$\tau = \frac{L}{R} \qquad \textit{30.5}$$

There are two parts to Eq. 30.2, or any solution to a first-order differential equation: the homogeneous (or complementary) solution and the particular solution. The homogeneous solution, that is, the solution where $f(t) = 0$, is called the *natural response* of the circuit and is represented by the exponential term in Eq. 30.2. The decay behavior of a general exponential is shown in Fig. 30.1. The particular solution is called the *forced response*, since it depends on the forcing function, $f(t)$, and is represented by κ in Eq. 30.2.

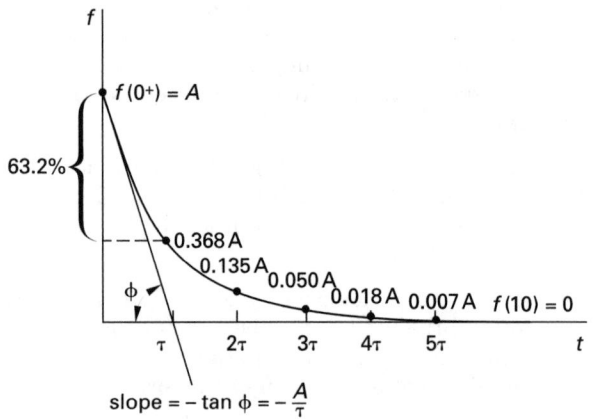

Figure 30.1 *Exponential Behavior*

When two independent energy storage elements are present, the mathematical representation of the circuit is a second-order differential equation of the form

$$f(t) = a\left(\frac{d^2x}{dt^2}\right) + b\left(\frac{dx}{dt}\right) + cx \qquad \textit{30.6}$$

There are three forms of the solution, depending on the magnitudes of the constants a, b, and c (see Ch. 31).

2. RESISTOR-CAPACITOR CIRCUITS: NATURAL RESPONSE

Consider the generic resistor-capacitor *source-free circuit* shown in Fig. 30.2. When the charged capacitor, C, is connected to the resistor, R, via a complete electrical path, the capacitor will discharge in an attempt to resist the change in voltage. In the process, the capacitor's stored electrical energy is dissipated in the resistor until no further energy remains and no current flows. This gradual decrease is the transient. The voltage follows the form of Fig. 30.1.

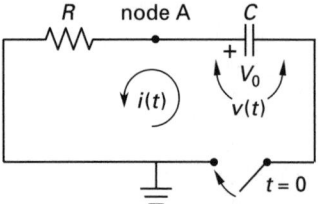

Figure 30.2 *Resistor-Capacitor Circuit: Natural Response*

Example 30.1

Determine the formula for the voltage in the circuit of Fig. 30.2 from $t = 0$ onward.

Solution

Write Kirchhoff's current law (KCL) for the simple node A between the resistor and the capacitor. The reference node is located at the switch, opposite node A. In keeping with the convention of this book, both currents are assumed to flow out of the node, giving

$$i_C + i_R = 0$$

Substitute the expression for current flow through a capacitor and the Ohm's law expression for current flow in a resistor.

$$C\left(\frac{dv}{dt}\right) + \frac{v}{R} = 0$$

$$\frac{dv}{dt} + \left(\frac{1}{RC}\right)v = 0$$

This is a homogeneous first-order linear differential equation. The solution could be written directly from the information in the mathematics chapters or by knowing that only an exponential form of v can be linearly combined with its derivative. In order to show the process, separate the variables and integrate both sides.

$$\frac{dv}{v} = -\left(\frac{1}{RC}\right)dt$$

$$\int \frac{1}{v}dv = \int -\left(\frac{1}{RC}\right)dt$$

$$\ln v = -\frac{1}{RC}t + \kappa \quad [\kappa \equiv \text{constant}]$$

$$v(t) = e^{-\left(\frac{1}{RC}\right)t+\kappa} = Ae^{-\frac{t}{RC}} = Ae^{-\frac{t}{\tau}}$$

This is the natural response of the system. To determine the constant A, the value of $v(t)$ at some time must be known. Since the initial voltage is V_0, the value of the constant A is

$$v(0) = V_0 = Ae^{-\frac{0}{\tau}} = A$$

The final solution is

$$v(t) = V_0 e^{-\frac{t}{RC}}$$

3. RESISTOR-CAPACITOR CIRCUITS: FORCED RESPONSE

Consider the generic resistor-capacitor circuit shown in Fig. 30.3. When the charged capacitor, C, is connected to the resistor, R, via a complete electrical path, the capacitor will discharge or charge, depending on the magnitude of V_{bat}, in an attempt to resist the change in voltage. In the process, the capacitor's stored electrical energy is either dissipated or enhanced until no further energy change occurs and no current flows. This gradual decrease or increase is the transient. After the passage of time equal to five time constants, 5τ, the final value is within less than 1% of its steady-state value and the transient is considered complete. The voltage follows the form of Fig. 30.1, or its inverse, with the exception that the final steady-state voltage is the voltage of the driving force.

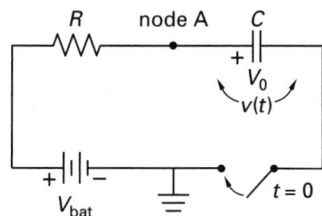

Figure 30.3 Resistor-Capacitor Circuit: Forced Response

Example 30.2

Determine the formula for the voltage in the circuit of Fig. 30.3 from $t = 0$ onward.

Solution

Write KCL for the simple node A between the resistor and the capacitor. The node between the negative terminal of the battery and the capacitor, opposite node A, is the reference node. Both currents are assumed to flow out of the node in keeping with the convention of this text, giving

$$i_C + i_R = 0$$

Substitute the expression for current flow through a capacitor and the Ohm's law expression for current flow in a resistor.

$$C\left(\frac{dv}{dt}\right) + \frac{v - V_{\text{bat}}}{R} = 0$$

$$\frac{dv}{dt} + \left(\frac{1}{RC}\right)(v - V_{\text{bat}}) = 0$$

$$\frac{dv}{dt} + \left(\frac{1}{RC}\right)v = \left(\frac{1}{RC}\right)V_{\text{bat}}$$

This is a nonhomogeneous first-order linear differential equation. The known constant to the right of the equal sign is called the *forcing* or *driving function*. Such an equation is solved in two steps. The first step is to determine the homogeneous, or complementary, solution, that is, the solution with the forcing function equal to zero. In Ex. 30.1 this was determined to be

$$v_{\text{tr}}(t) = Ae^{-\frac{t}{RC}}$$

This is the natural response of the system and represents the transient portion of the solution. The next step is to determine the particular solution, or forced response, of the system. Since the particular solution is the steady-state solution, the expectation is that the solution would be of the same form as the forcing function. Thus, assuming a constant, κ, as the particular solution and substituting this into the KCL expression for v gives

$$v_{\text{ss}} = \kappa$$

$$\frac{dv}{dt} + \left(\frac{1}{RC}\right)v = \left(\frac{1}{RC}\right)V_{\text{bat}}$$

$$\frac{dv_{\text{ss}}}{dt} + \left(\frac{1}{RC}\right)v_{\text{ss}} = \left(\frac{1}{RC}\right)V_{\text{bat}}$$

$$\frac{d\kappa}{dt} + \left(\frac{1}{RC}\right)\kappa = \left(\frac{1}{RC}\right)V_{\text{bat}}$$

$$0 + \left(\frac{1}{RC}\right)\kappa = \left(\frac{1}{RC}\right)V_{\text{bat}}$$

$$\kappa = V_{\text{bat}}$$

The particular solution is

$$v(\infty) = v_{\text{ss}} = V_{\text{bat}}$$

PROFESSIONAL PUBLICATIONS, INC.

Circuit Theory

The total solution is the combination of the complementary (homogeneous) and particular (nonhomogeneous) solutions, that is, the combination of the transient and steady-state solutions. The total solution is

$$v(t) = V_{\text{bat}} + Ae^{-\frac{t}{RC}}$$

The order of the transient and steady-state solutions is unimportant. The exponential term (the transient term) is often listed to the right in an equation.

The natural response coefficient A is again determined by the initial condition $v(0) = V_0$. Thus,

$$v(0) = V_0 = V_{\text{bat}} + Ae^{-\frac{0}{RC}} = V_{\text{bat}} + A$$

$$A = V_0 - V_{\text{bat}}$$

The final solution is

$$v(t) = V_{\text{bat}} + (V_0 - V_{\text{bat}})\,e^{-\frac{t}{RC}}$$

The voltage $v(t)$ is the voltage at node A, that is, the voltage across the capacitor.

4. RESISTOR-INDUCTOR CIRCUITS: NATURAL RESPONSE

Consider the generic resistor-inductor source-free circuit shown in Fig. 30.4. When the inductor, L, whose magnetic field is at a maximum due to initial current, I_0, is connected to the resistor, R, via a complete electrical path, the inductor will discharge in an attempt to resist the change in current. In the process, the inductor's stored magnetic energy is dissipated in the resistor until no further energy remains and no current flows. This gradual decrease is the transient. After the passage of time equal to five time constants, 5τ, the final value is within less than 1% of its steady-state value and the transient is considered complete. The current follows the form of Fig. 30.1.

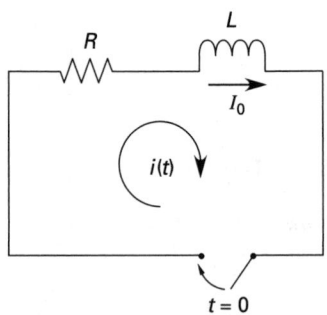

Figure 30.4 *Resistor-Inductor Circuit: Natural Response*

Example 30.3

Determine the formula for the current in the circuit of Fig. 30.4 from $t = 0$ onward.

Solution

Write Kirchhoff's voltage law (KVL) for the loop containing the resistor and the inductor. In keeping with the convention of this book, the reference direction is clockwise, giving

$$v_L + v_R = 0$$

Substitute the expression for voltage across an inductor and the Ohm's law expression for voltage drop in a resistor.

$$L\left(\frac{di}{dt}\right) + iR = 0$$

$$\frac{di}{dt} + \left(\frac{R}{L}\right)i = 0$$

This is a homogeneous first-order linear differential equation. The solution could be written directly from the information in the mathematics chapters or by knowing that only an exponential form of i can be linearly combined with its derivative. The process is similar to that in Ex. 30.1.

$$\frac{di}{i} = -\left(\frac{R}{L}\right)dt$$

$$\int\left(\frac{1}{i}\right)di = \int -\left(\frac{R}{L}\right)dt$$

$$\ln i = -\left(\frac{R}{L}\right)t + \kappa \quad [\kappa \equiv \text{constant}]$$

$$i(t) = e^{-\left(\frac{R}{L}\right)t + \kappa} = Ae^{-\frac{t}{L/R}} = Ae^{-\frac{t}{\tau}}$$

This is the natural response of the system. To determine the constant A, the value of $i(t)$ at some time must be known. Since the initial current is I_0, the value of the constant A is

$$i(0) = I_0 = Ae^{-\frac{0}{\tau}} = A$$

The final solution is

$$i(t) = I_0 e^{-\frac{t}{L/R}}$$

5. RESISTOR-INDUCTOR CIRCUITS: FORCED RESPONSE

Consider the generic resistor-inductor circuit shown in Fig. 30.5. When the inductor, L, whose magnetic field is at a maximum due to initial current, I_0, is connected to the resistor, R, via a complete electrical path, the inductor's magnetic field will collapse or expand in an attempt to resist the change in current. In the process, the inductor's stored magnetic energy is either dissipated or enhanced until no further energy change occurs and no current flows. This gradual decrease or increase is the transient. After the passage of time equal to five time constants, 5τ, the final value is within less than 1% of its steady-state value and the transient is considered complete. The voltage follows the form of Fig. 30.1 or its inverse, with the exception that the final steady-state current is the current resulting from the driving force and is determined by the remainder of the circuit.

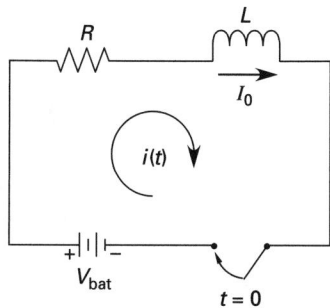

Figure 30.5 *Resistor-Inductor Circuit: Forced Response*

Example 30.4

Determine the formula for the current in the circuit of Fig. 30.5 from $t = 0$ onward.

Solution

Write KVL for the loop containing the resistor and the inductor. In keeping with the convention of this book, the reference direction is clockwise, giving

$$v_L + v_R - V_{bat} = 0$$

Substitute the expression for voltage across an inductor and the Ohm's law expression for voltage drop in a resistor.

$$L\left(\frac{di}{dt}\right) + iR - V_{bat} = 0$$

$$\frac{di}{dt} + \left(\frac{R}{L}\right)i = \left(\frac{1}{L}\right)V_{bat}$$

$$\frac{di}{dt} + \left(\frac{1}{\frac{L}{R}}\right)i = \left(\frac{1}{L}\right)V_{bat}$$

This is a nonhomogeneous first-order linear differential equation. The known constant to the right of the equal sign is called the *forcing* or *driving function*. Such an equation is solved in two steps. The first step is to determine the particular solution, or forced response, of the system. Since the particular solution is the steady-state solution, the expectation is that the solution would be of the same form as the forcing function. Thus, assuming a constant, κ, as the particular solution and substituting this into the KVL expression for i gives

$$i_{ss} = \kappa$$

$$\frac{di}{dt} + \left(\frac{1}{\frac{L}{R}}\right)i = \left(\frac{1}{L}\right)V_{bat}$$

$$\frac{di_{ss}}{dt} + \left(\frac{1}{\frac{L}{R}}\right)i_{ss} = \left(\frac{1}{L}\right)V_{bat}$$

$$\frac{d\kappa}{dt} + \left(\frac{1}{\frac{L}{R}}\right)\kappa = \left(\frac{1}{L}\right)V_{bat}$$

$$0 + \left(\frac{1}{\frac{L}{R}}\right)\kappa = \left(\frac{1}{L}\right)V_{bat}$$

$$\kappa = \left(\frac{1}{R}\right)V_{bat}$$

The particular solution is

$$i(\infty) = i_{ss} = \left(\frac{1}{R}\right)V_{bat}$$

The next step is to determine the homogeneous, or complementary, solution, that is, the solution with the forcing function equal to zero. In Ex. 30.3 this was determined to be

$$i_{tr}(t) = Ae^{-\frac{t}{L/R}}$$

This is the natural response of the system and represents the transient portion of the solution. The total solution is the combination of the particular (nonhomogeneous) and the complementary (homogeneous) solution, that is, the combination of the steady-state and transient solutions. The total solution is

$$i(t) = \left(\frac{1}{R}\right)V_{bat} + Ae^{-\frac{t}{L/R}}$$

The natural response coefficient A is again determined by the initial condition $i(0) = I_0$. Thus,

$$i(0) = I_0 = \left(\frac{1}{R}\right)V_{bat} + Ae^{-\frac{0}{L/R}} = \left(\frac{1}{R}\right)V_{bat} + A$$

$$A = I_0 - \left(\frac{1}{R}\right)V_{bat}$$

The final solution is

$$i(t) = \left(\frac{1}{R}\right)V_{bat} + \left(I_0 - \left(\frac{1}{R}\right)V_{bat}\right)e^{-\frac{t}{L/R}}$$

The current $i(t)$ is the current through the inductor.

Circuit Theory

6. RC AND RL CIRCUITS: SOLUTION METHOD

The method presented in Secs. 30-2 through 30-5 is valid for any circuit that can be represented by a single equivalent capacitor or inductor and a single equivalent resistor. The solutions were obtained for DC transients. Table 30.1 is a summary of DC transient responses.[1] AC transients are handled in the same manner except that the forcing function, that is, the source, is represented as a sinusoid. Phasor methods, or sinusoidal methods, are then used to solve the resulting differential equation. If the switch closure were instead treated as a pulse from the source and represented by the unit step function, $u(t)$, the voltage applied at $t = 0^+$ would have been $V_{\text{bat}}u(t)$ and the equations handled in the same manner. The steps for solving single resistor and single capacitor/inductor circuits follow.

step 1: Find the steady-state response from the particular solution of the nonhomogeneous equation. The solution will be of the same form as the forcing function.[2] Regardless of transient type, the result will be a steady-state value, κ.

step 2: Determine the transient response from the homogeneous equation or from the known parameters of the time constant. The transient response will be of the form $Ae^{-\frac{t}{\tau}}$.

step 3: Sum the steady-state and transient solutions, that is, the particular and complementary solutions, to obtain the total solution. This will be of the form $\kappa + Ae^{-\frac{t}{\tau}}$.

step 4: Determine the initial value from the circuit initial conditions. Use this information to calculate the constant A.

step 5: Write the final solution.

This method is used for resistor-capacitor circuits which are primarily used in electronics. Inductors tend to be large and change value with both temperature and time. They do not lend themselves well to miniaturization. Capacitors are more easily manufactured on integrated circuits. Inductors are important in power circuits that handle large amounts of current or high voltages and in transformers.

Table 30.1 Transient Response

type of circuit	response
series RC, charging $\tau = RC$ $e^{-N} = e^{-\frac{t}{\tau}} = e^{-\frac{t}{RC}}$	$V_{\text{bat}} = v_R(t) + v_C(t)$ $i(t) = \left(\dfrac{V_{\text{bat}} - V_0}{R}\right)e^{-N}$ $v_R(t) = i(t)R$ $\quad = (V_{\text{bat}} - V_0)e^{-N}$ $v_C(t) = V_0 + (V_{\text{bat}} - V_0)$ $\quad \times (1 - e^{-N})$ $Q_C(t) = C\big(V_0 + (V_{\text{bat}} - V_0)$ $\quad \times (1 - e^{-N})\big)$
series RC, discharging $\tau = RC$ $e^{-N} = e^{-\frac{t}{\tau}} = e^{-\frac{t}{RC}}$	$0 = v_R(t) + v_C(t)$ $i(t) = \left(\dfrac{V_0}{R}\right)e^{-N}$ $v_R(t) = -V_0 e^{-N}$ $v_C(t) = V_0 e^{-N}$ $Q_C(t) = CV_0 e^{-N}$
series RL, charging $\tau = \dfrac{L}{R}$ $e^{-N} = e^{-\frac{t}{\tau}} = e^{-\frac{t}{L/R}}$	$V_{\text{bat}} = v_R(t) + v_L(t)$ $i(t) = I_0 e^{-N}$ $\quad + \left(\dfrac{V_{\text{bat}}}{R}\right)\left(1 - e^{-N}\right)$ $v_R(t) = i(t)R$ $\quad = I_0 R e^{-N}$ $\quad + V_{\text{bat}}(1 - e^{-N})$ $v_L(t) = (V_{\text{bat}} - I_0 R)e^{-N}$
series RL, discharging $\tau = \dfrac{L}{R}$ $e^{-N} = e^{-\frac{t}{\tau}} = e^{-\frac{t}{L/R}}$	$0 = v_R(t) + v_L(t)$ $i(t) = I_0 e^{-N}$ $v_R(t) = I_0 R e^{-N}$ $v_L(t) = -I_0 R e^{-N}$

---

[1] The final expression of the response given in the table appears to differ from that given in step 3 of the solution method. The table uses the common form $1 - e^{-t/\tau}$. The forms are equivalent and can be interchanged. Both contain a steady-state and transient component. The steady-state term κ can be found in the $1 - e^{-t/\tau}$ form by letting $t \to \infty$.

[2] If the excitation is constant, the form of the particular solution is a constant and can be determined by simply open-circuiting the terminals of the capacitor and determining the voltage present. This voltage will be the steady-state voltage to which the capacitor will be driven.

7. RISE TIME

When the input of a capacitive or inductive circuit like the one shown in Fig. 30.6(a) undergoes a step change, the circuit response is called a *step response*. The step response is illustrated in Fig. 30.6(b). The speed of the response may be quantified in several ways. The time constant, τ, is one possibility. The shorter the time constant, the less time it takes for the circuit to reach a specified value. The value of the time constant is dependent on the circuit. A second possibility is the *rise time*, which is a *defined quantity* applicable to responses in general regardless of the circuit. Let t_{10} be the time a response has reached 10% of its final steady-state value. Let t_{90} be the time a response has reached 90% of its final steady-state value. The rise time, t_r, is then defined as

$$t_r = t_{90} - t_{10} \qquad \textit{30.7}$$

The rise time is related to the time constant by Eq. 30.8.

$$t_r = 2.2\tau \qquad \textit{30.8}$$

Another defined quantity is the *time delay*, t_d, which is defined as the time for a response to reach 50% of its final steady-state value. The rise and delay times and their relationship to the time constant are shown graphically in Fig. 30.6(b).

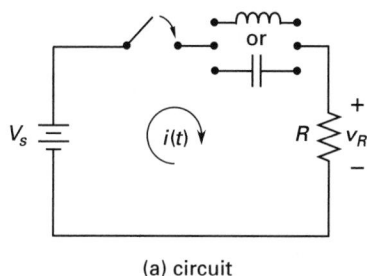

(a) circuit

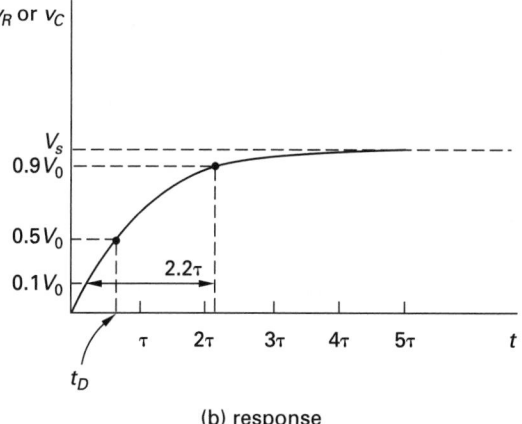

(b) response

Figure 30.6 *Rise Time*

8. DAMPED OSCILLATIONS: RINGING

Some circuits when energized undergo voltage or current oscillations that decrease in magnitude over time. Such oscillation is called *ringing*. An example of such a circuit and its response to a switch closure at $t = 0$ is shown in Fig. 30.7. The capacitor is initially charged to V_0 and the inductor has no initial magnetic field. Writing KVL for the circuit yields

$$\frac{1}{C}\int_0^t i\,dt + iR + L\left(\frac{di}{dt}\right) = V_0 \qquad \textit{30.9}$$

Note that the current is the rate of change of the charge on the capacitor, that is, $i = dq/dt$. Rearranging mathematically gives

$$\frac{d^2q}{dt^2} + \left(\frac{R}{L}\right)\left(\frac{dq}{dt}\right) + \left(\frac{1}{LC}\right)q = \frac{q_0}{LC} \qquad \textit{30.10}$$

Representing this nonhomogeneous second-order differential equation in the s domain, appropriate for Laplace transform analysis, results in the following characteristic equation.[3]

$$s^2 + \left(\frac{R}{L}\right)s + \frac{1}{LC} = 0 \qquad \textit{30.11}$$

The roots of Eq. 30.11, found from the quadratic equation, take one of three possible forms: (1) real and unequal, (2) real and equal, or (3) complex conjugates.

Ringing occurs for case 3 with roots given by

$$s_1, s_2 = -\sigma_1 \pm j\omega_1 \qquad \textit{30.12}$$

The solution to Eq. 30.10 for the ringing circuit is

$$v_C(t) = V_0 e^{-\sigma_1 t}\cos\omega_1 t \qquad \textit{30.13}$$

The term σ_1 is called the *damping*. The exponential, $e^{-\sigma_1 t}$, is termed the *envelope* of the waveform. The frequency of the oscillation is given by ω_1. These quantities are illustrated in Fig. 30.7(b). The damping and frequency are found in terms of the circuit parameters from Eqs. 30.14 and 30.15.

$$\sigma_1 = \frac{R}{2L} \qquad \textit{30.14}$$

$$\omega_1 = \frac{1}{LC} - \left(\frac{R}{2L}\right)^2 \qquad \textit{30.15}$$

The damping and frequency are also defined in terms of the *quality factor*, Q, of the inductor coil. If ω_0 is the frequency of the circuit when $R = 0$, the quality factor is

$$Q = \frac{\omega_0 L}{R} = \frac{1}{\omega_0 C R} = \frac{X_L}{R} = \frac{X_C}{R} \qquad \textit{30.16}$$

The damping and frequency can be defined as

$$\sigma_1 = \frac{\omega_0}{2Q} \qquad \textit{30.17}$$

$$\omega_1 = \omega_0\sqrt{1 - \left(\frac{1}{2Q}\right)^2} \qquad \textit{30.18}$$

---

[3]A second-order equation should be expected since the circuit contains two independent energy storage devices.

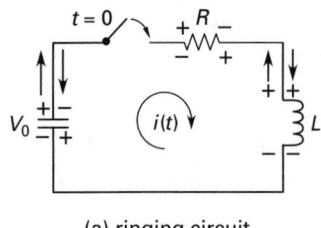

(a) ringing circuit

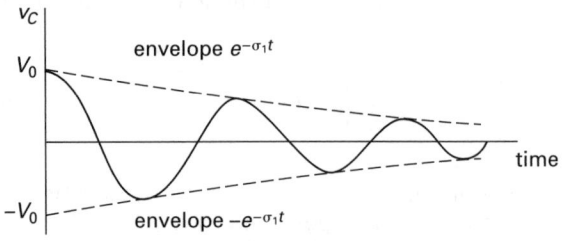

(b) ringing response

Figure 30.7 *Ringing Circuit*

9. SUSTAINED OSCILLATIONS: RESONANCE

Capacitors and inductors within the same circuit exchange energy. In the ideal, that is, lossless, circuit shown in Fig. 30.8, this exchange can continue indefinitely and is known as a *sustained oscillation*.[4]

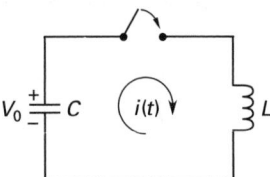

Figure 30.8 *Resonant Circuit*

The characteristic equation for the circuit in Fig. 30.8 is

$$s^2 + \frac{1}{LC} = 0 \qquad \textit{30.19}$$

The solution is

$$s_1, s_2 = \pm j\sqrt{\frac{1}{LC}} = \pm j\omega_0 \qquad \textit{30.20}$$

No damping term, σ_1, exists and the oscillations can continue unfettered. This unimpeded flow of energy naturally occurs at the frequency ω_0 and is called *resonance*.

---

[4]This type of circuit is more than academic and can be realized, or nearly so, with superconducting inductors.

In a real circuit, resistive components are present and will thus yield a damping term, σ_1, in the solution to the circuit equation. The result is that an external source of energy is required to make up the loss. Resonance is then defined as a phenomenon of an AC circuit whereby relatively large currents occur near certain frequencies, and a relatively unimpeded oscillation of energy from potential to kinetic occurs. The frequency at which this occurs is called the *resonant frequency*. At the resonant frequency the capacitive reactance equals the inductive reactance, that is, $X_C = X_L$, the current phase angle difference is zero ($\phi = 0$), and the power factor equals one (pf $= 1$).

10. RESONANT CIRCUITS

A *resonant circuit* has a zero current phase angle difference. This is equivalent to saying the circuit is purely resistive (i.e., the power factor is equal to one) in its response to an AC voltage. The frequency at which the circuit becomes purely resistive is the *resonant frequency*.

For frequencies below the resonant frequency, a series RLC circuit will be capacitive (leading) in nature; above the resonant frequency, the circuit will be inductive (lagging) in nature.

For frequencies below the resonant frequency, a parallel GLC circuit will be inductive (lagging) in nature; above the resonant frequency, the circuit will be capacitive (leading) in nature.

Circuits can become resonant in two ways. If the frequency of the applied voltage is fixed, the elements must be adjusted so that the capacitive reactance cancels the inductive reactance (i.e., $X_L - X_C = 0$). If the circuit elements are fixed, the frequency must be adjusted.

As Figs. 30.9 and 30.10 illustrate, a circuit approaches resonant behavior gradually. ω_1 and ω_2 are the *half-power points* (*70 percent points* or *3 dB points*) because at those frequencies, the power dissipated in the resistor is half of the power dissipated at the resonant frequency.

$$Z_{\omega 1} = \sqrt{2}R \qquad \textit{30.21}$$

$$I_{\omega 1} = \frac{V}{Z_{\omega 1}} = \frac{V}{\sqrt{2}R} = \frac{I_0}{\sqrt{2}} \qquad \textit{30.22}$$

$$P_{\omega 1} = I^2 R = \left(\frac{I_0}{\sqrt{2}}\right)^2 R = \tfrac{1}{2}P_0 \qquad \textit{30.23}$$

The frequency difference between the half-power points is the *bandwidth*, BW, a measure of circuit selectivity. The smaller the bandwidth, the more selective the circuit.

$$\text{BW} = f_2 - f_1 \qquad \textit{30.24}$$

The *quality factor*, Q, for a circuit is a dimensionless ratio that compares the reactive energy stored in an inductor each cycle to the resistive energy dissipated.[5] Figure 30.9 illustrates the effect the quality factor has on the frequency characteristic.

$$Q = 2\pi \left(\frac{\text{maximum energy stored per cycle}}{\text{energy dissipated per cycle}} \right)$$

$$= \frac{f_0}{(\text{BW})_{\text{Hz}}} = \frac{\omega_0}{(\text{BW})_{\text{rad/s}}}$$

$$= \frac{f_0}{f_2 - f_1} = \frac{\omega_0}{\omega_2 - \omega_1} \quad \begin{bmatrix} \text{parallel} \\ \text{or series} \end{bmatrix} \qquad 30.25$$

Then, the energy stored in the inductor of a series RLC circuit each cycle is

$$U = \frac{I_m^2 L}{2} = I^2 L = Q \left(\frac{I^2 R}{2\pi f_0} \right) \qquad 30.26$$

The relationships between the half-power points and quality factor are

$$f_1, f_2 = f_0 \left(\sqrt{1 + \frac{1}{4Q^2}} \mp \frac{1}{2Q} \right)$$

$$\approx f_0 \mp \frac{f_0}{2Q} = f_0 \mp \frac{\text{BW}}{2} \qquad 30.27$$

Various resonant circuit formulas are tabulated in App. 30.A.

11. SERIES RESONANCE

In a resonant series RLC circuit,

- impedance is minimum
- impedance equals resistance
- current and voltage are in phase
- current is maximum
- power dissipation is maximum

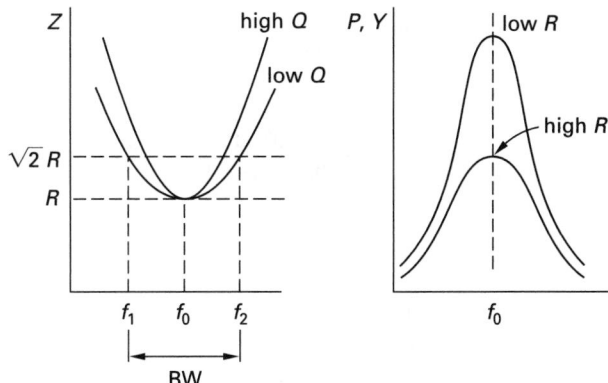

Figure 30.9 Series Resonance (Band-Pass Filter)

---

[5]The name *figure of merit* refers to the quality factor calculated from the inductance and internal resistance of a coil.

The total impedance (in rectangular form) of a series RLC circuit is $R + j(X_L - X_C)$. At the resonant frequency, $\omega_0 = 2\pi f_0$,

$$X_L = X_C \quad [\text{at resonance}] \qquad 30.28$$

$$\omega_0 L = \frac{1}{\omega_0 C} \qquad 30.29$$

$$\omega_0 = 2\pi f_0 = \frac{1}{\sqrt{LC}} \qquad 30.30$$

The power dissipation in the resistor is

$$P = \tfrac{1}{2} I_m^2 R = \frac{V_m^2}{2R}$$

$$= I^2 R$$

$$= \frac{V^2}{R} \qquad 30.31$$

The quality factor for a series RLC circuit is

$$Q = \frac{X}{R} = \frac{\omega_0 L}{R}$$

$$= \frac{1}{\omega_0 RC} = \frac{1}{R}\sqrt{\frac{L}{C}}$$

$$= \frac{\omega_0}{(\text{BW})_{\text{rad/s}}} = \frac{f_0}{(\text{BW})_{\text{Hz}}}$$

$$= G\omega_0 L = \frac{G}{\omega_0 C} \qquad 30.32$$

Example 30.5

A series RLC circuit is connected across a sinusoidal voltage with peak of 20 V. (a) What is the resonant frequency in rad/s? (b) What are the half-power points in rad/s? (c) What is the peak current at resonance? (d) What is the peak voltage across each component at resonance?

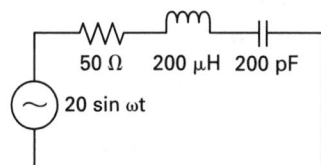

Solution

(a) Equation 30.30 gives the resonant frequency.

$$\omega_0 = \frac{1}{\sqrt{LC}} = \frac{1}{\sqrt{(200 \times 10^{-6}\text{ H})(200 \times 10^{-12}\text{ F})}}$$

$$= 5 \times 10^6 \text{ rad/s}$$

(b) The half-power points are given by

$$\omega_1, \omega_2 = \omega_0 \mp \frac{\text{BW}}{2} = \omega_0 \mp \frac{\omega_0}{2Q} = \omega_0 \mp \frac{R}{2L}$$

$$= 5 \times 10^6 \mp \frac{50\ \Omega}{(2)(200 \times 10^{-6}\text{ H})}$$

$$= 5.125 \times 10^6,\ 4.875 \times 10^6 \text{ rad/s}$$

(c) The total impedance is the resistance at resonance. The peak resonant current is

$$I_0 = \frac{V_m}{Z_0} = \frac{V_m}{R}$$

$$= \frac{20\angle 0° \, \text{V}}{50 \, \Omega}$$

$$= 0.4\angle 0° \, \text{A}$$

(d) The peak voltages across the components are

$$V_R = I_0 R = (0.4\angle 0° \text{A})(50 \, \Omega) = 20\angle 0° \, \text{V}$$

$$V_L = I_0 X_L = I_0 j \omega_0 L$$

$$= j(0.4\angle 0° \text{A}) \left(5 \times 10^6 \, \frac{\text{rad}}{\text{s}}\right)(200 \times 10^{-6} \, \text{H})$$

$$= 400\angle 90° \, \text{V}$$

$$V_C = I_0 X_C = \frac{I_0}{j \omega_0 C}$$

$$= \frac{0.4\angle 0° \, \text{A}}{j \left(5 \times 10^6 \, \frac{\text{rad}}{\text{s}}\right)(200 \times 10^{-12} \, \text{F})}$$

$$= 400\angle -90° \, \text{V}$$

12. PARALLEL RESONANCE

In a resonant parallel GLC circuit,

- impedance is maximum
- impedance equals resistance
- current and voltage are in phase
- current is minimum
- power dissipation is minimum

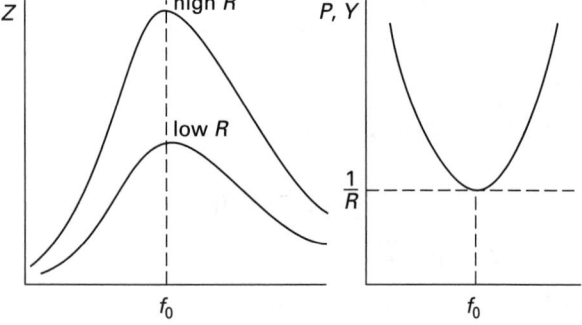

Figure 30.10 *Parallel Resonance (Band-Reject Filter)*

The total admittance (in rectangular form) of a parallel GLC circuit is $G + j(B_C - B_L)$. At resonance,

$$X_L = X_C \qquad\qquad 30.33$$

$$\omega_0 L = \frac{1}{\omega_0 C} \qquad\qquad 30.34$$

$$\omega_0 = 2\pi f_0 = \frac{1}{\sqrt{LC}} \qquad\qquad 30.35$$

The power dissipation in the resistor is

$$P = \tfrac{1}{2} I_m^2 R = \frac{V_m^2}{2R}$$

$$= I^2 R$$

$$= \frac{V^2}{R} \qquad\qquad 30.36$$

The quality factor for a parallel RLC circuit is

$$Q = \frac{R}{X} = \omega_0 RC = \frac{R}{\omega_0 L}$$

$$= R\sqrt{\frac{C}{L}} = \frac{\omega_0}{(\text{BW})_{\text{rad/s}}} = \frac{f_0}{(\text{BW})_{\text{Hz}}}$$

$$= \frac{\omega_0 C}{G} = \frac{1}{G\omega_0 L} \qquad\qquad 30.37$$

Example 30.6

A parallel RLC circuit containing a 10 Ω resistor has a resonant frequency of 1 MHz and a bandwidth of 10 kHz. To what should the resistor be changed in order to increase the bandwidth to 20 kHz without changing the resonant frequency?

Solution

From Eqs. 30.25 and 30.37,

$$Q_{\text{old}} = \frac{f_0}{\text{BW}} = 2\pi f_0 RC$$

$$C = \frac{1}{2\pi R(\text{BW})} = \frac{1}{(2\pi)(10 \, \Omega)(10 \times 10^3 \, \text{Hz})}$$

$$= \frac{1 \times 10^{-5}}{2\pi} \, \text{F}$$

The new quality factor is

$$Q_{\text{new}} = \frac{f_0}{\text{BW}}$$

$$= \frac{10^6 \, \text{Hz}}{20 \times 10^3 \, \text{Hz}}$$

$$= 50$$

From Eq. 30.37, the required resistance is

$$R = \frac{Q}{2\pi f_0 C}$$

$$= \frac{50}{(2\pi)(10^6 \, \text{Hz})\left(\dfrac{1 \times 10^{-5}}{2\pi} \, \text{F}\right)}$$

$$= 5 \, \Omega$$

31 Time Response

Nomenclature

A	natural response coefficient	–
C	capacitance	F
f	function	–
h	impulse response	–
i, I	current	A
Im	imaginary portion	–
k	Fourier transform variable	–
$\mathcal{L}$	Laplace transform	–
L	self-inductance	H
Q	quality factor	–
Q	charge	C
R	resistance	Ω
$\mathbf{s}$	complex frequency	s^{-1}
s	Laplace transform variable	–
t	time	s
v, V	voltage	V
x	variable	–
Z	impedance	Ω

Symbols

α	damping	s^{-1}
β	oscillation frequency	rad/s
ζ	damping ratio	–
θ	phase angle	rad
κ	steady-state constant	–
σ	neper frequency or napier frequency	Np/s or s^{-1}
τ	time constant	s
ϕ	angle	rad
ω	angular frequency	rad/s
ω_0	resonance frequency	rad/s

Subscripts

0	initial
C	capacitor
DC	direct current
e	exponential
i	current
L	inductor
L/C	inductor or capacitor
n	natural
N	Norton
P	peak
R	resistor
s	source
so	switch open
ss	steady-state
tr	transient
Th	Thevenin
v	voltage

1. FUNDAMENTALS

Time domain analysis is analysis that takes place with time as the governing variable. *Sinusoidal analysis* is analysis that occurs when the current and voltage functions are manipulated in their trigonometric form. *Frequency domain analysis* deals with Fourier or Laplace transforms of time functions, which are then functions of frequency. The Fourier transform occurs in k-space which, when k is replaced by ω, is called the frequency domain. The Laplace transform occurs in s-space, also considered the frequency domain. The variable $\mathbf{s}$ is called the *complex frequency* and is given by Eq. 31.1. (The variable $\mathbf{s}$ is shown in bold to emphasize its vector properties, important during pole-zero analysis. It is commonly shown as unbolded and will be shown as such in the remainder of this book.)

$$\mathbf{s} = \sigma + j\omega \qquad \textit{31.1}$$

The variable σ is termed the *neper frequency* and is assigned units of Np/s or s^{-1}. The variable σ represents the ratio of two currents, voltages, or other analogous quantities. The number of nepers is the natural log of this ratio. Thus, the term "napier" is sometimes used in place of neper.[1] The Laplace transform and the complex frequency s are used to represent exponential and damped sinusoidal motion in the same manner that ordinary frequency expresses simple harmonic motion.

---
[1] This occurs since the natural logarithm is alternately called the *Naperian logarithm*.

That is, using the variable s allows constants, exponential decay, sinusoids, and damped sinusoids (sinusoids with an exponential envelope) to be represented in the same form, Ae^{st} (see Sec. 31-11). *Sinusoidal analysis* is also considered any analysis using the assumption that the forcing function, the source, is sinusoidal in nature. Sinusoidal analysis in Chs. 27 through 30 assumed a single frequency. When the frequencies are allowed to vary, the circuit output is called the *frequency analysis* or *frequency response*.

2. FIRST-ORDER ANALYSIS

First-order analysis is performed upon circuits with one energy storage element—that is, one equivalent capacitor or inductor. In circuits containing numerous sources and resistors, the Thevenin or Norton equivalent is taken as seen from the terminals of the single capacitor or inductor. This allows the methods of Ch. 30 to be applied, modified so that the Thevenin voltage, v_{Th}, is used in place of the source voltage and the Thevenin current, i_{Th}, is used as the current through the capacitor and inductor. Figure 31.1(a) represents a generic circuit with a single capacitor or inductor. The source can be either AC or DC. Figure 31.1(b) shows the Thevenin equivalent of the circuit with the switch closed. Figure 31.1(c) shows the new Thevenin equivalent with the switch open. Part (c) of this figure is a convenient way of showing that if the source is eliminated (or changed to a DC source), the resulting currents and voltages will be exponentials (the natural response) and will share a new time constant and possibly new initial currents and voltages.

The differential equation for first-order circuits takes on the general form

$$f(t) = b\frac{dx}{dt} + cx \qquad 31.2$$

The solutions are of the form

$$x(t) = \kappa + Ae^{-\frac{t}{\tau}} \qquad 31.3$$

The time constant $\tau = b/c$. κ is a constant, the steady-state component of the solution. The method for determining the complete solution is as follows.

step 1: Determine the Thevenin or Norton equivalent of the circuit as seen from the terminals of the capacitor or inductor.

step 2: Find the steady-state response from the particular solution of the nonhomogeneous equation. The solution will be of the same form as the forcing function.[2] Regardless of transient type, the result will be a steady-state value, κ.

step 3: Determine the transient response from the homogeneous equation or from the known parameters of the time constant. The transient response will be of the form $Ae^{-\frac{t}{\tau}}$.

step 4: Sum the steady-state and transient solutions, that is, the complementary and particular solutions, to obtain the total solution.[3] This will be of the form $\kappa + Ae^{-\frac{t}{\tau}}$.

step 5: Determine the initial value from the circuit initial conditions. Use this information to calculate the constant A.

step 6: Write the final solution.

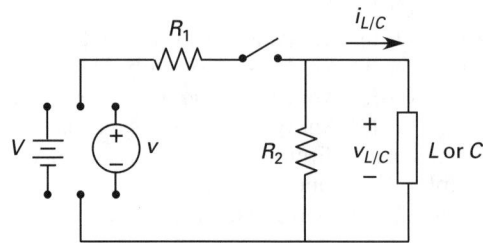

(a) first-order circuit: DC or AC

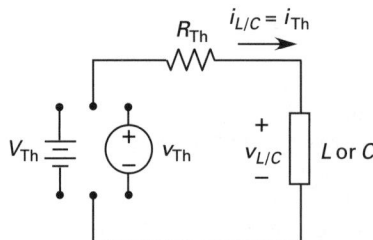

(b) Thevenin equivalent: DC or AC, switch closed

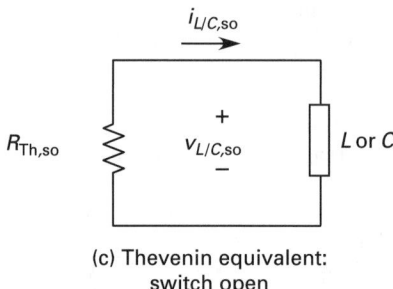

(c) Thevenin equivalent: switch open

Figure 31.1 *First-Order Circuit*

A capacitive first-order circuit generally uses voltage as the dependent variable. Applying KVL to Fig. 31.1(b) gives

$$v_{\text{Th}} = i_{\text{Th}}R_{\text{Th}} + v_C \qquad 31.4$$

---

[2]If the excitation is constant, the form of the particular solution is a constant and can be determined by simply open-circuiting the terminals of the capacitor and determining the voltage present. This voltage will be the steady-state voltage to which the capacitor will be driven.

[3]The final expression often varies between texts, with some using $1 - e^{-\frac{t}{\tau}}$. The forms are equivalent and can be interchanged. Both contain a steady-state and transient component. The steady-state term κ can be found in the $1 - e^{-\frac{t}{\tau}}$ form by letting $t \to \infty$.

The Thevenin current is the current flowing through the capacitor, $i_{Th} = i_C = C(dv_C/dt)$. Substituting gives

$$v_{Th} = R_{Th}C\left(\frac{dv_C}{dt}\right) + v_C \qquad \textbf{\textit{31.5}}$$

Equation 31.5 is in the general form of a first-order differential equation shown in Eq. 31.2. The solution is

$$v_C(t) = V_{C,\text{ss}} + Ae^{-\frac{t}{\tau}} \qquad \textbf{\textit{31.6}}$$

An inductive first-order circuit generally uses current as the dependent variable. Applying KVL to Fig. 31.1(b) gives

$$v_{Th} = i_{Th}R_{Th} + v_L \qquad \textbf{\textit{31.7}}$$

The Thevenin current is flowing through the inductor, $i_{Th} = i_L$. Substituting for the inductor voltage and rearranging gives

$$v_{Th} = L\left(\frac{di_L}{dt}\right) + R_{Th}i_L \qquad \textbf{\textit{31.8}}$$

Equation 31.8 is in the general form of a first-order differential equation shown in Eq. 31.2. The solution is

$$i_L(t) = I_{L,\text{ss}} + Ae^{-\frac{t}{\tau}} \qquad \textbf{\textit{31.9}}$$

3. FIRST-ORDER ANALYSIS: SWITCHING TRANSIENTS

First-order switching transients result from the operation of a switch. The switching mechanism can be an actual switch, relay contacts, or a nonlinear device such as a transistor operating in the saturation or cutoff region. A sudden change in the source results in sudden changes in the voltages and currents within a circuit. A jump in the capacitor voltage requires an impulse current. A jump in the inductor current requires an impulse voltage. Since the conditions in a capacitor or inductor cannot change instantaneously, the post-switched conditions of capacitance, C, or self-inductance, L, are derived from the pre-switched conditions.

Example 31.1[4]

For the network shown, let $R_1 = 20\ \Omega$, $R_2 = 10\ \Omega$, $V_{DC} = 12$ V, and $L = 0.20$ H. The switch has been open for a long time and is closed at $t = 0$. Determine the steady-state response of the network.

---

[4]Since transient conditions, and a possible AC source, are used in some examples, lowercase letters will be used to designate the voltage and current.

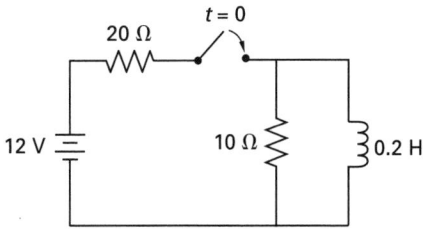

Solution

Using the voltage divider concept, the Thevenin voltage is 4 V. The Thevenin resistance is calculated as seen from the terminals of the inductor, as shown in Fig. 31.1(b). Thus,

$$R_{Th} = R_1 \| R_2 = \frac{R_1 R_2}{R_1 + R_2} = \frac{(20\ \Omega)(10\ \Omega)}{20\ \Omega + 10\ \Omega}$$

$$= 6.67\ \Omega$$

The steady-state condition is determined from Eq. 31.8, noting that steady-state current is constant $(di/dt = 0)$.

$$v_{Th} = L\left(\frac{di_L}{dt}\right) + R_{Th}i_L$$

$$= (0.2\ \text{H})(0) + (6.67\ \Omega)i_L = 4\ \text{V}$$

$$i_{L,\text{ss}} = \frac{4\ \text{V}}{6.67\ \Omega} = 0.60\ \text{A}$$

The inductor is a short circuit for steady-state DC conditions, as expected.

Example 31.2

Determine the transient response of the circuit in Ex. 31.1.

Solution

The transient response is of the form $Ae^{-\frac{t}{\tau}}$. The time constant is

$$\tau = \frac{L}{R_{Th}} = \frac{0.20\ \text{H}}{6.67\ \Omega} = 30\ \text{ms}$$

The steady-state current was found to be 0.60 A in Ex. 31.1. Using Eq. 31.9, the total current is

$$i_L(t) = I_{L,\text{ss}} + Ae^{-\frac{t}{\tau}} = 0.60 + Ae^{-\frac{t}{30 \times 10^{-3}\ \text{s}}}$$

Since the switch was open for a long time, steady-state conditions are assumed to exist and the initial current in the inductor is zero. That is, $i(0^-) = i(0^+) = 0$. Substituting this information to find the constant A gives

$$i_L(0) = 0.60 + Ae^{-\frac{0}{30 \times 10^{-3}\ \text{s}}} = 0.60 + A = 0$$

$$A = -0.60\ \text{A}$$

The transient response is

$$i_{tr}(t) = -0.60e^{-\frac{t}{30\times10^{-3}\text{ s}}}$$

Example 31.3

What is the current through the inductor in Ex. 31.1 120 ms after the switch closes?

Solution

The total solution is the combination of the steady-state and transient responses. From Exs. 31.1 and 31.2,

$$i_L(t) = 0.60 - 0.60e^{-\frac{t}{30\times10^{-3}\text{ s}}}$$

At 120 ms, or 4τ, after the switch closes, the current is

$$i_L(120\text{ ms}) = 0.60 - 0.60e^{\frac{-120\times10^{-3}}{30\times10^{-3}\text{ s}}}$$

$$= 0.60 - 0.60e^{-4} = 0.59\text{ A}$$

Example 31.4

For the network shown, let $R_1 = 10\ \Omega$, $R_2 = 10\ \Omega$, $V_{DC} = 10$ V, and C = 50 μF. The switch has been open for an extended time and is closed at $t = 0$. Determine the steady-state capacitance response.

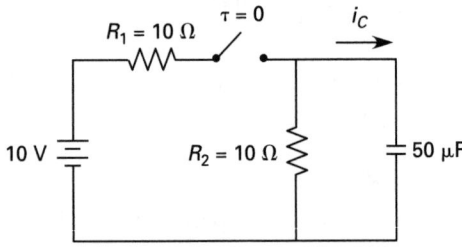

Solution

Using the voltage divider concept, the Thevenin voltage is 5 V. The Thevenin resistance is calculated as seen from the terminals of the capacitor. Thus,

$$R_{Th} = R_1 \| R_2 = \frac{R_1 R_2}{R_1 + R_2} = \frac{(10\ \Omega)(10\ \Omega)}{10\ \Omega + 10\ \Omega}$$

$$= 5\ \Omega$$

The steady-state condition is determined from Eq. 31.5, noting that the voltage is constant ($dv/dt = 0$).

$$v_{Th} = R_{Th}C\left(\frac{dv_C}{dt}\right) + v_C$$

$$= (5\ \Omega)(50\times10^{-6}\text{ F})(0) + v_C$$

$$= 5\text{ V}$$

$$v_{C,ss} = 5\text{ V}$$

The capacitor is an open circuit for steady-state DC conditions, as expected.

Example 31.5

For the network in Ex. 31.4, the switch opens three time constants into the transient. Determine the response of the network from this time onward.

Solution

After the switch opens, $v_{Th} = 0$ (see Fig. 31.1(c)). Therefore, the steady-state voltage will be zero. The transient response must be of the form $Ae^{-\frac{t}{\tau}}$. Thus,

$$v_C(t) = Ae^{-\frac{t}{\tau}} = Ae^{-\frac{t}{R_{Th}C}}$$

Once the switch is open, the Thevenin resistance is

$$R_{Th} = R_2 = 10\ \Omega$$

The time constant is

$$\tau = R_{Th}C = (10\ \Omega)(50\times10^{-6}\text{ F}) = 500\ \mu s$$

The total solution is the combination of the steady-state and transient responses, as given in Eq. 31.6.

$$v_C(t) = V_{C,ss} + Ae^{-\frac{t}{\tau}} = 0 + Ae^{-\frac{t}{500\times10^{-6}\text{ s}}}$$

In order to determine the constant A, the voltage condition must be known at a certain time. Since the capacitor voltage cannot change instantaneously, $v_C(0^-) = v_C(0^+)$. The switch was opened three time constants into the transient determined by the conditions in Ex. 31.4. The transient response of the circuit in Ex. 31.4 is

$$v_C(t) = Ae^{-\frac{t}{R_{Th}C}} = Ae^{-\frac{t}{(5\ \Omega)(50\times10^{-6}\text{F})}} = Ae^{-\frac{t}{250\ \mu s}}$$

The total solution for the circuit in Ex. 31.4, as given by Eq. 31.6, and noting that the initial voltage is zero, is

$$v_C(t) = V_{C,ss} + Ae^{-\frac{t}{\tau}} = 5 + Ae^{-\frac{t}{250\times10^{-6}\text{ s}}}$$

$$v_C(0) = 5 + Ae^{-\frac{0}{250\times5^{-6}\text{ s}}} = 5 + A = 0$$

$$A = -5$$

$$v_C(t) = 5 - 5e^{-\frac{t}{250\times10^{-6}\text{ s}}}$$

At three time constants into the initial transient,

$$v_C(3\tau) = 5 - 5e^{\frac{-(3)\left(250\times10^{-6}\text{ s}\right)}{250\times10^{-6}\text{ s}}} = 4.75\text{ V}$$

Use 4.75 V as the initial voltage for the network in Ex. 31.5 to determine the value of A.

$$v_C(t) = V_{C,\text{ss}} + Ae^{-\frac{t}{\tau}} = 0 + Ae^{-\frac{t}{500 \times 10^{-6}\ \text{s}}}$$

$$v_C(0^-) = v_C(0^+) = 4.75\ \text{V}$$

$$4.75\ \text{V} = 0 + Ae^{-\frac{0}{500 \times 10^{-6}\ \text{s}}} = A$$

Thus, the response of the network from the time the switch opens is

$$v_C(t) = V_{C,\text{ss}} + Ae^{-\frac{t}{\tau}} = 0 + 4.75e^{-\frac{t}{500 \times 10^{-6}\ \text{s}}}$$

$$= 4.75e^{-\frac{t}{500 \times 10^{-6}\ \text{s}}}\ \text{V}$$

Example 31.6

For the network shown, $R_1 = 10\ \Omega$, $R_2 = 10\ \Omega$, $v = 170\sin(377t + 30°)$, and $L = 0.15$ H. Determine the steady-state current with the switch closed.

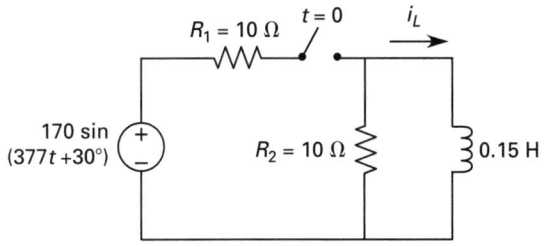

Solution

Using Eq. 31.8 and noting that during steady-state condition $di/dt = 0$ gives Ohm's law,

$$v_{\text{Th}} = L\left(\frac{di_L}{dt}\right) + R_{\text{Th}}i_L = L(0) + R_{\text{Th}}i_L$$

$$i_L = \frac{v_{\text{Th}}}{R_{\text{Th}}}$$

The source is given as sinusoidal. Analysis is the same for sinusoidal sources, but the mathematics is more cumbersome. Instead, use the phasor representation of the sinusoid source and Eq. 31.8.

$$i_L = \frac{v_{\text{Th}}}{R_{\text{Th}}}$$

$$\mathbf{I}_L = \frac{\mathbf{V}_{\text{Th}}}{\mathbf{Z}_{\text{Th}}} = \frac{\mathbf{V}_{\text{Th}}}{\mathbf{Z}_R + \mathbf{Z}_L}$$

Represent the source voltage in phasor form as $120\angle 30°$ and note that, using the voltage-divider concept, the Thevenin voltage is one-half this value. The impedances of the resistor and inductor are

$$\mathbf{Z}_{R_{\text{Th}}} = 5\ \Omega \angle 0°$$

$$\mathbf{Z}_L = j\omega L = j\left(377\ \frac{\text{rad}}{\text{s}}\right)(0.15\ \text{H}) = 56.55\ \Omega \angle 90°$$

$$\mathbf{Z}_{\text{Th}} = \mathbf{Z}_{R_{\text{Th}}} + \mathbf{Z}_L = (5 + j0) + (0 + j56.55)$$

$$= 5 + j56.55 = 56.77\ \Omega \angle 84.9°$$

Substitute to determine the steady-state current.

$$\mathbf{I}_L = \frac{\mathbf{V}_{\text{Th}}}{\mathbf{Z}_{\text{Th}}} = \frac{\mathbf{V}_{\text{Th}}}{\mathbf{Z}_R + \mathbf{Z}_L} = \frac{60\angle 30°}{56.77\angle 84.9°} = 1.06\ \angle -55°$$

Convert to the time (sinusoidal) form.

$$\mathbf{I}_L = 1.06\ \angle -55° = 1.5\sin(377t - 55°)$$

Effective or rms values are used with phasor quantities. Peak values are used with sinusoidal quantities.[5] Radians and degrees are mixed in the equations strictly for clarity. Radians are necessary for calculations. Transient calculation using sinusoids follows a similar pattern, with phasor analysis being an efficient calculational method. During transients using sinusoids, the exact time of the transient must be specified.

Example 31.7

For the network shown, $R_1 = 10\ \Omega$, $R_2 = 10\ \Omega$, $v = 636\cos(2513t + 60°)$, and $C = 1\ \mu\text{F}$. Determine the steady-state capacitance voltage with the switch closed.

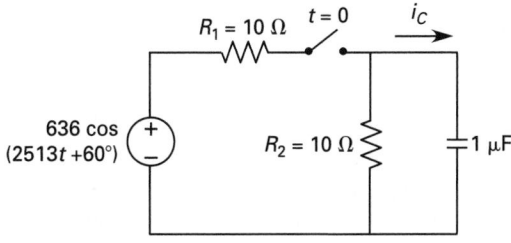

Solution

Using Eq. 31.5 gives

$$v_{\text{Th}} = R_{\text{Th}}C\left(\frac{dv_C}{dt}\right) + v_C$$

Represent the source voltage in phasor form as $450\angle 60°$ and note that, using the voltage divider concept, the Thevenin voltage is one-half this value. Recall that, in

---

[5]The representation of phasors as rms quantities and sinusoidal variables as peak quantities is standard in this text. Care should be taken to determine how phasors are being represented in a given document.

Circuit Theory

the frequency domain, multiplying by $j\omega$ is equivalent to taking the derivative. Combining the two gives

$$v_{\text{Th}} = R_{\text{Th}}C\left(\frac{dv_C}{dt}\right) + v_C$$

$$\mathbf{V}_{\text{Th}} = \mathbf{Z}_{R_{\text{Th}}}Cj\omega\mathbf{V}_C + \mathbf{V}_C = (j\omega\mathbf{Z}_{R_{\text{Th}}}C + 1)\,\mathbf{V}_C$$

$$\mathbf{V}_C = \frac{\mathbf{V}_{\text{Th}}}{j\omega\mathbf{Z}_{R_{\text{Th}}}C + 1}$$

$$= \frac{225 \text{ V}\angle 60°}{j\left(2513\,\dfrac{\text{rad}}{\text{s}}\right)(5\ \Omega)(1 \times 10^{-6}\text{ F}) + 1}$$

$$= \frac{225 \text{ V}\angle 60°}{1.01\ \Omega\angle 0.72°} = 225 \text{ V}\angle 59.28°$$

Transforming into the time domain gives

$$v_{C,\text{ss}}(t) = 318\cos(2513t + 59°)$$

At the angular frequency given, equivalent to 400 Hz, the capacitance is essentially an open circuit and has little impact on the network. The voltage source in this example was given as the cosine function to emphasize that the choice is arbitrary and does not affect the analysis.

4. FIRST-ORDER ANALYSIS: PULSE TRANSIENTS

A *pulse* is generally defined as a variation in a quantity that is normally constant. When the variation occurs at a single point in time, the pulse is called a *step pulse*. If the magnitude of the step pulse equals one, it is termed a *unit step*. A pulse has a finite duration that is normally brief compared to the time scale of interest. When the duration of the pulse is so short that it can be thought of as infinitesimal, it is called an *impulse*. Figure 31.2 shows a *unit pulse* of each type.

The unit step is defined by Eq. 31.10. The unit impulse is the derivative of the unit step function. The relationships between various forcing functions are shown in Table 31.1.

$$u(t - t_1) = 0 \quad [t < t_1]$$
$$u(t - t_1) = 1 \quad [t > t_1] \qquad \qquad \textbf{31.10}$$

Table 31.1 Operations on Forcing Functions

function	function when differentiated	function when integrated
unit impulse	–	unit step
unit step	unit impulse	unit ramp
unit ramp	unit step	unit parabola
unit parabola	unit ramp	(third degree)
unit exponential	unit exponential	unit exponential
unit sinusoid	unit sinusoid	unit sinusoid

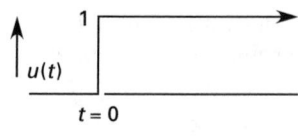

(a) unit step $u(t)$

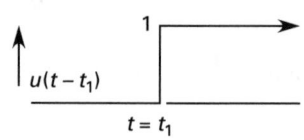

(b) unit step $u(t - t_1)$

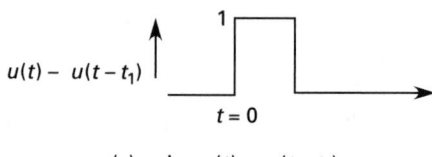

(c) pulse $u(t) - u(t - t_1)$

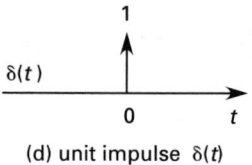

(d) unit impulse $\delta(t)$

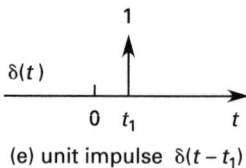

(e) unit impulse $\delta(t - t_1)$

Figure 31.2 Pulse Transients

The response of a circuit to a unit step is called a *step response*. The source voltage magnitude, in this case the Thevenin voltage magnitude, is one for a unit step. For a first-order linear circuit like the one shown in Fig. 31.3, the response is determined in terms of the *state variables* of the energy storage device, that is, inductive current or capacitive voltage.[6] If other electrical quantities are required, they can be found from any of the circuit relations. Thus, the response equations are

$$v_C(t) = u(t)\left(1 - e^{-\frac{t}{\tau}}\right) \qquad \qquad \textbf{31.11}$$

$$i_L(t) = u(t)\left(\frac{1}{R_{\text{Th}}}\right)\left(1 - e^{-\frac{t}{\tau}}\right) \qquad \qquad \textbf{31.12}$$

---

[6]A state variable is a variable sufficient to specify the condition of the network and define its future behavior.

The time constant, τ, for the capacitive circuit is $R_{\mathrm{Th}}C$ and for the inductive circuit is L/R_{Th}.

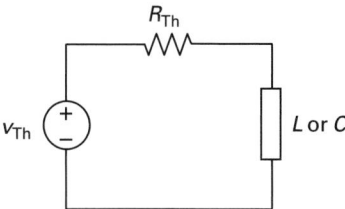

Figure 31.3 *First-Order Circuit*

The step and impulse responses for RC and RL circuits are shown in Table 31.2. Of importance, the magnitude of the voltage and current sources for a unit step response is one.[7]

Example 31.8

Determine the capacitance voltage in the circuit shown.

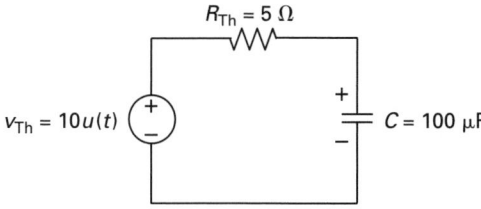

Solution

From Eq. 31.11 or Table 31.2, the form of the capacitance voltage is

$$v_C(t) = u(t)\left(1 - e^{-\frac{t}{\tau}}\right)$$

The time constant is

$$\tau = R_{\mathrm{Th}}C = (5\ \Omega)\left(100 \times 10^{-6}\ \mathrm{F}\right) = 500 \times 10^{-6}\ \mathrm{s}$$

The unit step has a magnitude of one. Thus, a factor of 10 is required to scale the unit step to the actual value of the Thevenin voltage. The final solution is

$$v_C(t) = 10u(t)\left(1 - e^{-\frac{t}{500 \times 10^{-6}\ \mathrm{s}}}\right)$$

Example 31.9

If the source in Ex. 31.8 outputs two 10 V pulses of 10 ms duration 20 ms apart, what is the capacitance voltage?

Solution

From Eq. 31.10 and Fig. 31.2, the voltage source is represented as follows. The brackets enclose each pulse for clarification.

---

[7]The magnitude of one for a unit step is the reason for the apparent nonappearance of the current and voltage in the equations.

$$v_{\mathrm{Th}}(t) = \left(10u(t) - 10u(t - 0.010\ \mathrm{s})\right)$$
$$+ \left(10u(t - 0.030\ \mathrm{s}) - 10u(t - 0.040\ \mathrm{s})\right)$$

Using Eq. 31.11 or Table 31.2, scaling by a factor of 10, and applying the principle of superposition gives the total response.

$$v_C(t) = 10u(t)\left(1 - e^{-\frac{t}{500 \times 10^{-6}\ \mathrm{s}}}\right)$$
$$- 10u(t - 0.010\ \mathrm{s})\left(1 - e^{\frac{-t - 0.010\ \mathrm{s}}{500 \times 10^{-6}\ \mathrm{s}}}\right)$$
$$+ 10u(t - 0.030\ \mathrm{s})\left(1 - e^{\frac{-t - 0.030\ \mathrm{s}}{500 \times 10^{-6}\ \mathrm{s}}}\right)$$
$$- 10u(t - 0.040\ \mathrm{s})\left(1 - e^{\frac{-t - 0.040\ \mathrm{s}}{500 \times 10^{-6}\ \mathrm{s}}}\right)$$

5. SECOND-ORDER ANALYSIS

Second-order analysis involves circuits with one equivalent capacitor and one equivalent inductor, that is, two energy storage elements. As in first-order analysis, the method for determining the desired quantity is mathematical analysis in the time domain. Frequency domain analysis, which is also performed in first-order networks, significantly simplifies second-order analysis and is accomplished by transforming either the applicable differential equation or the circuit itself into the s domain.

Laplace transform analysis is then used to determine the desired quantity. (Fourier transform analysis could also be used, but the Laplace transform has the advantage of representing constants, sinusoids, and exponential functions in a single variable, s.) Figure 31.4 represents a generic circuit with a single capacitor and inductor.

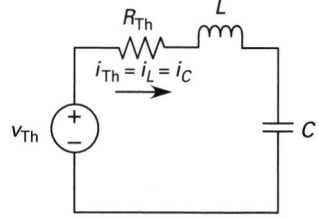

(a) series *LC* circuit

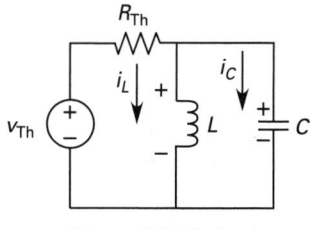

(b) parallel *LC* circuit

Figure 31.4 *Second-Order Circuits*

The differential equation for second-order circuits takes on the general form

$$f(t) = a\frac{d^2x}{dt^2} + b\frac{dx}{dt} + cx \qquad 31.13$$

Table 31.2 *Step and Impulse Responses for First-Order Circuits*

circuit	unit step response	unit impulse response
	$v_{\mathrm{Th}}(t) = u(t)$ $v_C(t) = u(t)\left(1 - e^{-\frac{t}{R_{\mathrm{Th}}C}}\right)$ $i_C(t) = u(t)\left(\frac{1}{R_{\mathrm{Th}}}\right)\left(e^{-\frac{t}{R_{\mathrm{Th}}C}}\right)$	$v_{\mathrm{Th}}(t) = \delta(t)$ $h_v(t) = u(t)\left(\frac{1}{R_{\mathrm{Th}}C}\right)\left(e^{-\frac{t}{R_{\mathrm{Th}}C}}\right)$ $h_i(t) = u(t)\left(-\frac{1}{R_{\mathrm{Th}}^2 C}\right)\left(e^{-\frac{t}{R_{\mathrm{Th}}C}}\right)$ $+ \delta(t)\left(\frac{1}{R_{\mathrm{Th}}}\right)$
	$i_{\mathrm{Th}}(t) = u(t)$ $v_C(t) = u(t)R_{\mathrm{Th}}\left(1 - e^{-\frac{t}{R_{\mathrm{Th}}C}}\right)$ $i_C(t) = u(t) = u(t)\left(e^{-\frac{t}{R_{\mathrm{Th}}C}}\right)$	$i_{\mathrm{Th}}(t) = \delta(t)$ $h_v(t) = u(t)\left(\frac{1}{C}\right)\left(e^{-\frac{t}{R_{\mathrm{Th}}C}}\right)$ $h_i(t) = u(t)\left(-\frac{1}{R_{\mathrm{Th}}C}\right)\left(e^{-\frac{t}{R_{\mathrm{Th}}C}}\right) + \delta(t)$
	$v_{\mathrm{Th}}(t) = u(t)$ $v_L(t) = u(t)\left(e^{-\frac{t}{L/R_{\mathrm{Th}}}}\right)$ $i_L(t) = u(t)\left(\frac{1}{R_{\mathrm{Th}}}\right)\left(1 - e^{-\frac{t}{L/R_{\mathrm{Th}}}}\right)$	$v_{\mathrm{Th}}(t) = \delta(t)$ $h_v(t) = u(t)\left(\frac{R_{\mathrm{Th}}}{L}\right)\left(e^{-\frac{t}{L/R_{\mathrm{Th}}}}\right) + \delta(t)$ $h_i(t) = u(t)\left(\frac{1}{L}\right)\left(e^{-\frac{t}{L/R_{\mathrm{Th}}}}\right)$
	$i_{\mathrm{Th}}(t) = u(t)$ $v_L(t) = u(t)R\left(e^{-\frac{t}{L/R_{\mathrm{Th}}}}\right)$ $i_L(t) = u(t)\left(1 - e^{-\frac{t}{L/R_{\mathrm{Th}}}}\right)$	$i_{\mathrm{Th}}(t) = \delta(t)$ $h_v(t) = u(t)\left(-\frac{R_{\mathrm{Th}}^2}{L}\right)\left(e^{-\frac{t}{L/R_{\mathrm{Th}}}}\right) + R\delta(t)$ $h_i(t) = u(t)\left(\frac{R_{\mathrm{Th}}}{L}\right)\left(e^{-\frac{t}{L/R_{\mathrm{Th}}}}\right)$

The associated characteristic equation, written with the roots shown in the *s* domain, is

$$as^2 + bs + c = 0 \qquad \textbf{31.14}$$

Equation 31.13 can be written in terms of any circuit variable. It is normally written in terms of capacitance voltage or inductance current for reasons identical to those in first-order circuits. The solutions take on one of three forms, depending upon the magnitude of the constants (see Secs. 31-5 through 31-7). The general response of the three forms is illustrated in Fig. 31.5.

Overdamping in a linear system is defined as damping-over that is required for critical damping. *Critical damping* in a linear system is damping on the threshold between oscillatory and exponential behavior. *Underdamping* is damping that results in oscillatory behavior.

The method for determining the complete solution to second-order networks follows.

step 1: Determine the second-order equation for the circuit to be analyzed by using electrical engineering relationships and laws such as Ohm's law, KVL, KCL, and others.

step 2: Write the characteristic equation associated with the second-order equation. Using the constants of this equation, a, b, and c, determine the form of the solution.

step 3: Write the solution form.

step 4: If required, determine the initial values from the circuit's initial conditions.[8] Determine the steady-state values, x_{ss} and x'_{ss}.

step 5: Calculate the roots of the characteristic equation.

step 6: Use the information in steps 4 and 5 to calculate the constants in the final solution, A and B. (This involves the solution form in step 3 and its derivative or the direct use of the generic equations for the constants.)

step 7: Write the final solution from the information found in steps 4, 5, and 6.

Steps 4, 5, and 6 are equivalent to performing the following.

step 4′: Find the steady-state response from the particular solution of the nonhomogeneous equation. The solution will be of the same form as the forcing function. Regardless of transient type, the result will be a steady-state value, κ.

step 5′: Determine the transient response from the homogeneous equation. Regardless of the transient type, the transient response will be of the form $Ae^{s_1 t} + Be^{s_2 t}$. For the critical damped case $s_1 = s_2$ and for the underdamped or oscillatory case the variable s will take on the values $\alpha \pm j\beta$.

step 6′: Sum the steady-state and transient solutions, that is, the complementary and particular solutions, to obtain the total solution. This will be of the form $\kappa + Ae^{s_1 t} + Be^{s_2 t}$. Calculate the constants in the final solution, A and B.

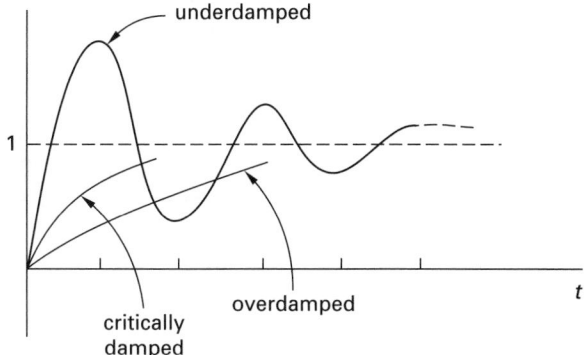

Figure 31.5 *Second-Order Circuit Response to a Step Change*

---

[8]Initial conditions in a circuit are the initial values of the state variable function and its derivative. Two known conditions are required for the solution of a second-order equation.

6. SECOND-ORDER ANALYSIS: OVERDAMPED

If the two roots of Eq. 31.14 are real and different from one another, equivalent to the condition $b^2 > 4ac$, the solution is of the form

$$x(t) = \kappa + Ae^{s_1 t} + Be^{s_2 t} \qquad \text{31.15}$$

The first term, κ, is a constant, that is, the steady-state component of the solution. This is the particular solution and results from the effect of an energy source or forcing function. The exponential terms represent the natural response of the circuit.[9] The roots of the characteristic equation are

$$s_1 = \frac{-b + \sqrt{b^2 - 4ac}}{2a} \qquad \text{31.16}$$

$$s_2 = \frac{-b - \sqrt{b^2 - 4ac}}{2a} \qquad \text{31.17}$$

The coefficients A and B in Eq. 31.15 are found from the initial and steady-state values of x and dx/dt. Since capacitor voltage and inductor current cannot change instantaneously, knowing the conditions immediately prior to any switching transient means knowing the conditions immediately after. By isolating the two energy storage elements and determining the Thevenin equivalent, any second-order circuit can be represented as in Fig. 31.4. In the series RLC circuits of Fig. 31.4(a), the results of applying the initial conditions are

$$v_C(0^+) = V_0 \qquad \text{31.18}$$

$$\frac{dv_C(0^+)}{dt} = \frac{I_0}{C} \qquad \text{31.19}$$

These conditions are a result of the common current path and $i_C = C(dv_C/dt)$.

For the parallel RLC circuits in Fig. 31.4(b), the results of applying the initial conditions are

$$i_L(0^+) = I_0 \qquad \text{31.20}$$

$$\frac{di_L(0^+)}{dt} = \frac{v_C(0^+)}{L} \qquad \text{31.21}$$

These conditions are a result of the common voltage and $v_L = L(di_L/dt)$.

---

[9]The natural response of an electric circuit is the response that would occur without a forcing function, that is, without an energy source. This portion of the solution goes by numerous names including natural, zero energy, homogeneous, and complementary.

The constants A and B are determined by substituting Eqs. 31.18 through 31.21 into Eq. 31.15 and its derivative (multiplying by s is the equivalent of taking the derivative) and solving for the specific circuit. Equations 31.18 through 31.21 can be used if network initial conditions are given and circuit element parameters, for example, L and C, are known. The results of such a substitution, in generic terms, are

$$x(0^+) = A + B + x_{\text{ss}} \qquad \textit{31.22}$$

$$x'(0^+) = s_1 A + s_2 B + x'_{\text{ss}} \qquad \textit{31.23}$$

Solving Eqs. 31.22 and 31.23 simultaneously gives the values of the constants.

$$A = \left(\frac{1}{2}\right)\left(1 + \frac{b}{\sqrt{b^2 - 4ac}}\right)\left(x(0^+) - x_{\text{ss}}\right)$$
$$+ \left(\frac{a}{\sqrt{b^2 - 4ac}}\right)\left(x'(0^+) - x'_{\text{ss}}\right) \qquad \textit{31.24}$$

$$B = \left(\frac{1}{2}\right)\left(1 - \frac{b}{\sqrt{b^2 - 4ac}}\right)\left(x(0^+) - x_{\text{ss}}\right)$$
$$- \left(\frac{a}{\sqrt{b^2 - 4ac}}\right)\left(x'(0^+) - x'_{\text{ss}}\right) \qquad \textit{31.25}$$

Example 31.10

Determine the second-order equation for the circuit in Fig. 31.4(a).

Solution

Using KVL and rearranging the result gives

$$v_{\text{Th}} = iR_{\text{Th}} + L\left(\frac{di}{dt}\right) + \frac{1}{C}\int i\,dt$$

Differentiating gives

$$\frac{dv_{\text{Th}}}{dt} = R_{\text{Th}}\left(\frac{di}{dt}\right) + L\left(\frac{d^2 i}{dt^2}\right) + \left(\frac{1}{C}\right)i$$

Rearrange to put into the form of Eq. 31.13.

$$\frac{dv_{\text{Th}}}{dt} = L\left(\frac{d^2 i}{dt^2}\right) + R_{\text{Th}}\left(\frac{di}{dt}\right) + \left(\frac{1}{C}\right)i$$

Example 31.11

A certain network is described by the following differential equation.[10]

$$24 = 2\left(\frac{d^2 v}{dt^2}\right) + 5\left(\frac{dv}{dt}\right) + 2v$$

---

[10]Dividing the equation by two shows that 12 is the value of the forcing function. Further, the 12 actually represents a step change of the source (Thevenin) voltage at $t = 0$. After the transient ends, the final value of the voltage across the capacitor will be 12 V.

The initial conditions are $v = 0$ and $dv/dt = 0$. Determine the expression for the voltage for $t \geq 0$.

Solution

Use the method for solving second-order networks outlined in Sec. 31-5.

step 1: Since the second-order equation is given, step 1 is complete.

step 2: The characteristic equation using constants a, b, and c from the second-order equation is

$$2s^2 + 5s + 2 = 0$$

Thus, $a = 2$, $b = 5$, $c = 2$, and $b^2 > 4ac$. The network is overdamped.

step 3: The solution form is

$$x(t) = \kappa + Ae^{s_1 t} + Be^{s_2 t}$$

step 4: The initial conditions are given. The forcing function is a constant 24 V. At steady-state, dv/dt and $d^2 v/dt^2$ equal zero. The steady-state voltages are[11]

$$24 = 2\left(\frac{d^2 v_{\text{ss}}}{dt^2}\right) + 5\left(\frac{dv_{\text{ss}}}{dt}\right) + 2v_{\text{ss}}$$

$$= 0 + 0 + 2v_{\text{ss}}$$

$$v_{\text{ss}} = \frac{24}{2} = 12 \text{ V}$$

The steady-state voltage in this case is not a function of time. Therefore,

$$v'_{\text{ss}} = 0$$

step 5: Using Eqs. 31.16 and 31.17, the roots of the characteristic equation are

$$s_1 = \frac{-b + \sqrt{b^2 - 4ac}}{2a}$$
$$= \frac{-5 + \sqrt{(5)^2 - (4)(2)(2)}}{(2)(2)} = -\frac{1}{2}$$

$$s_2 = \frac{-b - \sqrt{b^2 - 4ac}}{2a}$$
$$= \frac{-(5) - \sqrt{5^2 - (4)(2)(2)}}{(2)(2)} = -2$$

---

[11]Units are not shown where they might cause confusion with the variables used, such as the voltage v and volts V. Units should be shown in all final solutions.

step 6: Use Eqs. 31.24 and 31.25 to determine the constants A and B.

$$A = \left(\frac{1}{2}\right)\left(1 + \frac{b}{\sqrt{b^2 - 4ac}}\right)\left(v(0^+) - v_{ss}\right)$$

$$+ \left(\frac{a}{\sqrt{b^2 - 4ac}}\right)\left(v'(0^+) - v'_{ss}\right)$$

$$= \left(\frac{1}{2}\right)\left(1 + \frac{5}{\sqrt{(5)^2 - (4)(2)(2)}}\right)(0 - 12)$$

$$+ \left(\frac{2}{\sqrt{(5)^2 - (4)(2)(2)}}\right)(0 - 0)$$

$$= -16$$

$$B = \left(\frac{1}{2}\right)\left(1 - \frac{b}{\sqrt{b^2 - 4ac}}\right)\left(v(0^+) - v_{ss}\right)$$

$$- \left(\frac{a}{\sqrt{b^2 - 4ac}}\right)\left(v'(0^+) - v'_{ss}\right)$$

$$= \left(\frac{1}{2}\right)\left(1 - \frac{5}{\sqrt{(5)^2 - (4)(2)(2)}}\right)(0 - 12)$$

$$- \left(\frac{2}{\sqrt{(5)^2 - (4)(2)(2)}}\right)(0 - 0)$$

$$= 4$$

step 7: Write the final solution.

$$v(t) = \kappa + Ae^{s_1 t} + Be^{s_2 t}$$

$$= 12 - 16e^{-\frac{1}{2}t} + 4e^{-2t}$$

7. SECOND-ORDER ANALYSIS: CRITICALLY DAMPED

If the two roots of Eq. 31.14 are real and the same, equivalent to the condition $b^2 = 4ac$, the solution is of the form

$$x(t) = \kappa + Ae^{st} + Bte^{st} \qquad \text{31.26}$$

The first term, κ, is a constant, that is, the steady-state component of the solution. The exponential terms represent the natural response of the circuit. The root of the characteristic equation is

$$s = \frac{-b}{2a} \qquad \text{31.27}$$

The coefficients A and B in Eq. 31.26 are found from the initial and steady-state values of x and dx/dt. By the same reasoning used in Sec. 31-6, the initial conditions for a series RLC circuit remain unchanged and are repeated here for convenience.

$$v_C(0^+) = V_0 \qquad \text{31.28}$$

$$\frac{dv_C(0^+)}{dt} = \frac{I_0}{C} \qquad \text{31.29}$$

Similarly, the initial conditions for parallel RLC circuits are

$$i_L(0^+) = I_0 \qquad \text{31.30}$$

$$\frac{di_L(0^+)}{dt} = \frac{v_C(0^+)}{L} \qquad \text{31.31}$$

The constants A and B are determined by substituting Eqs. 31.28 through 31.31 into Eq. 31.26 and its derivative (multiplying by s is the equivalent of taking the derivative) and solving for the specific circuit. Equations 31.28 through 31.31 can be used if network initial conditions are given and circuit element parameters such as L and C are known. The results of such a substitution, in generic terms, are

$$x(0^+) = A + x_{ss} \qquad \text{31.32}$$

$$x'(0^+) = sA + B + x'_{ss} \qquad \text{31.33}$$

Solving Eqs. 31.32 and 31.33 simultaneously gives the values of the constants.

$$A = x(0^+) - x_{ss} \qquad \text{31.34}$$

$$B = x'(0^+) - x'_{ss} + \left(\frac{b}{2a}\right)A \qquad \text{31.35}$$

Example 31.12

Consider the series RLC circuit shown, with $v_{Th} = 10$ V, $R = 80$ Ω, $L = 80$ mH, and $C = 50$ μF. The capacitor is initially charged to 5 V and the initial current in the circuit is 187.5 mA. Determine the behavior of the circuit, in terms of the capacitance voltage, for $t \geq 0$.

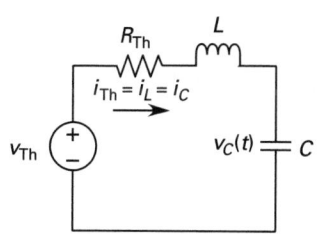

Circuit Theory

Solution

Using KVL and writing the resulting equation in terms of the capacitance voltage,

$$10 = iR_{\text{Th}} + L\left(\frac{di}{dt}\right) + v_C$$

Substituting $i = C(dv/dt)$,

$$10 = C\left(\frac{dv_C}{dt}\right)R_{\text{Th}} + LC\left(\frac{d^2v_C}{dt^2}\right) + v_C$$

$$10 = LC\left(\frac{d^2v_C}{dt^2}\right) + R_{\text{Th}}C\left(\frac{dv_C}{dt}\right) + v_C$$

The characteristic equation follows.

$$LCs^2 + RCs + 1 = 0$$

Determine the solution form from the relationship between the coefficients of the characteristic equation.

$$b^2 = (RC)^2 = \left((80)\left(50 \times 10^{-6}\right)\right)^2 = 16 \times 10^{-6}$$

$$4ac = (4)(LC)(1)$$

$$= (4)\left(80 \times 10^{-3}\right)\left(50 \times 10^{-6}\right)(1) = 16 \times 10^{-6}$$

$$b^2 = 4ac$$

The circuit is critically damped. The solution is of the form given by Eq. 31.26.

$$v_C(t) = \kappa + Ae^{st} + Bte^{st}$$

The initial conditions are given. The steady-state conditions are found from the results of the KVL analysis.

$$10 = LC\left(\frac{d^2v_C}{dt^2}\right) + R_{\text{Th}}C\left(\frac{dv_C}{dt}\right) + v_C$$

$$= LC(0) + R_{\text{Th}}C(0) + v_C$$

$$v_{C,\text{ss}} = 10$$

$$v'_{C,\text{ss}} = 0$$

The roots of the characteristic equation are found from Eq. 31.27 or actual calculation to be[12]

$$s = \frac{-b}{2a} = \frac{-R_{\text{Th}}C}{2LC}$$

$$= \frac{-R_{\text{Th}}}{2L} = \frac{-(80)}{(2)\left(80 \times 10^{-3}\right)} = -500$$

---

[12]Units are not commonly shown for s. They are inverse seconds.

Since the initial conditions and circuit parameters are known, instead of using the generic equations (Eqs. 31.32 through 31.35), use the solution form directly with the initial conditions from Eqs. 31.28 and 31.29.

$$v_C(t) = \kappa + Ae^{st} + Bte^{st}$$

$$v_C(0^+) = 10 + Ae^{s(0)} + B(0)e^{s(0)}$$

$$5 = 10 + A$$

$$A = -5$$

The second initial condition results in the following value for B.

$$v_C(t) = k + Ae^{st} + Bte^{st}$$

$$\frac{dv_C(0^+)}{dt} = \frac{I_0}{C} = 0 + Ase^{st} + B(e^{st} + tse^{st})$$

$$\frac{dv_C(0^+)}{dt} = \frac{I_0}{C} = 0 + Ase^{st} + B(e^{st} + tse^{st})$$

$$\frac{187.5 \times 10^{-3}}{50 \times 10^{-6}} = 0 + (-5)(-500)\left(e^{(-500)(0)}\right)$$

$$+ B\left(\begin{array}{c} e^{(-500)(0)} + (0)(-500) \\ \times \left(e^{(-500)(0)}\right) \end{array}\right)$$

$$3750 = 2500 + B$$

$$B = 1250$$

The final solution is

$$v_C(t) = \kappa + Ae^{st} + Bte^{st}$$

$$= 10 - 5e^{-500t} + 1250te^{-500t}$$

8. SECOND-ORDER ANALYSIS: UNDERDAMPED

If the two roots of Eq. 31.14 are complex conjugates, equivalent to the condition $b^2 < 4ac$, the solution is of the form[13]

$$x(t) = \kappa + e^{\alpha t}\left(A_e e^{j\beta t} + B_e e^{-j\beta t}\right) \qquad \textit{31.36}$$

This can also be represented in the sinusoidal form as

$$x(t) = \kappa + e^{\alpha t}\left(A\cos\beta t + B\sin\beta t\right) \qquad \textit{31.37}$$

The first term, κ, is a constant, that is, the steady-state component of the solution. The exponential terms represent the natural response of the circuit. Equation 31.36 is the ringing response, or damped oscillation, shown in Fig. 30.7(b). The sinusoidal response is enveloped by the exponential, since α is negative.

---

[13]The subscript e on the constant is a reminder that the constants in Eqs. 31.36 and 31.37 differ.

The variable s used in the overdamped and critically damped cases has been expanded to clarify this behavior and is represented by Eqs. 31.38 and 31.39.[14]

$$s_1 = \alpha + j\beta \qquad \text{31.38}$$

$$s_2 = \alpha - j\beta \qquad \text{31.39}$$

The roots of the characteristic equation are

$$\alpha = \frac{-b}{2a} \qquad \text{31.40}$$

$$\beta = \frac{\sqrt{4ac - b^2}}{2a} \qquad \text{31.41}$$

The coefficients A_e and B_e in Eq. 31.36, or A and B in Eq. 31.37, are found from the initial and steady-state values of x and dx/dt. By the same reasoning used in Sec. 31-6, the initial conditions for a series RLC circuit remain unchanged and are repeated here for convenience.

$$v_C(0^+) = V_0 \qquad \text{31.42}$$

$$\frac{dv_C(0^+)}{dt} = \frac{I_0}{C} \qquad \text{31.43}$$

Similarly, the initial conditions for parallel RLC circuits are

$$i_L(0^+) = I_0 \qquad \text{31.44}$$

$$\frac{di_L(0^+)}{dt} = \frac{v_C(0^+)}{L} \qquad \text{31.45}$$

The constants are determined by substituting Eqs. 31.42 through 31.45 into Eqs. 31.36 or 31.37 and their derivatives (multiplying by $j\beta$ is the equivalent of taking the derivative) and solving for the specific circuit. Equations 31.42 through 31.45 can be used if network initial conditions are given and circuit element parameters such as L and C are known. The results of such a substitution, in generic terms, are

$$x(0^+) = A + x_{\text{ss}} \qquad \text{31.46}$$

$$x'(0^+) = -\alpha A + \beta B + x'_{\text{ss}} \qquad \text{31.47}$$

Solving Eqs. 31.46 and 31.47 simultaneously gives the values of the constants.

$$A = x(0^+) - x_{\text{ss}} \qquad \text{31.48}$$

$$B = \left(\frac{\alpha}{\beta}\right) A + \left(\frac{1}{\beta}\right)\left(x'(0^+) - x'_{\text{ss}}\right) \qquad \text{31.49}$$

---

[14]The terminology of α and β is often used in transient analysis. The variable s was split for the underdamped case to clearly show the sinusoids enveloped by the exponentials.

9. SECOND-ORDER ANALYSIS: PULSE TRANSIENTS

The output of a second-order circuit to a step change is illustrated in Fig. 31.5. The electrical parameters defining the response vary depending upon the type of analysis required (see Sec. 31-10). The response of a second-order network with zero initial conditions, that is, $x(0) = 0$ and $x'(0) = 0$, to a step change has a wide range of applications. Since a unit step change can be scaled to any specific pulse, it forms the baseline input. Consider Eq. 31.13 with zero initial conditions and the unit step as the input, that is, $f(t) = 1$. The steady-state value of the state variable is

$$x_{\text{ss}} = \frac{f(t)}{c} = \frac{1}{c} \qquad \text{31.50}$$

The equations in Secs. 31-6 through 31-8 are then modified using Eq. 31.50 and the zero initial conditions requirement as follows.[15]

For an overdamped circuit ($b^2 > 4ac$)

$$x(t) = \frac{1}{c} + Ae^{s_1 t} + Be^{s_2 t} \qquad \text{31.51}$$

$$s_1 = \frac{-b + \sqrt{b^2 - 4ac}}{2a} \qquad \text{31.52}$$

$$s_2 = \frac{-b - \sqrt{b^2 - 4ac}}{2a} \qquad \text{31.53}$$

$$A = -\left(\frac{1}{2c}\right)\left(1 + \frac{b}{\sqrt{b^2 - 4ac}}\right) \qquad \text{31.54}$$

$$B = -\left(\frac{1}{2c}\right)\left(1 - \frac{b}{\sqrt{b^2 - 4ac}}\right) \qquad \text{31.55}$$

For a critically damped circuit ($b^2 = 4ac$)

$$x(t) = \frac{1}{c} + Ae^{st} + Bte^{st} \qquad \text{31.56}$$

$$s = \frac{-b}{2a} \qquad \text{31.57}$$

$$A = -\frac{1}{c} \qquad \text{31.58}$$

$$B = -\frac{b}{2ac} \qquad \text{31.59}$$

For an underdamped circuit ($b^2 < 4ac$),

$$x(t) = \frac{1}{c} + e^{\alpha t}(A \cos \beta t + B \sin \beta t) \qquad \text{31.60}$$

$$s_1 = \alpha + j\beta \qquad \text{31.61}$$

---

[15]Since x_{ss} is a constant then $x'_{\text{ss}} = 0$. This condition is also incorporated.

$$s_2 = \alpha - j\beta \qquad 31.62$$

$$\alpha = \frac{-b}{2a} \qquad 31.63$$

$$\beta = \frac{\sqrt{4ac - b^2}}{2a} \qquad 31.64$$

$$A = -\frac{1}{c} \qquad 31.65$$

$$B = -\frac{\alpha}{\beta c} \qquad 31.66$$

10. HIGHER-ORDER CIRCUITS

A circuit with two or more independent energy storage elements is considered a *higher-order circuit*. Second-order systems dominate the behavior of third- and higher-order systems and can be used to approximate the behavior of such networks. Higher-order analysis takes place primarily in the s domain, using the concept of a complex frequency, s. The characteristic equation derived from the network differential equation is described with different terminology depending upon the type of electrical analysis performed. In purely mathematical terms, the characteristic equation is

$$as^2 + bs + c \qquad 31.67$$

Control systems analysis uses the form given by Eq. 31.68. The *natural frequency, ω_n*, is sometimes written simply as ω. The term ζ is called the *damping ratio*.

$$s^2 + 2\zeta\omega_n s + \omega_n^2 \qquad 31.68$$

Resonance analysis uses the form given by Eq. 31.69. The resonant frequency is symbolized by ω_0 and the quality factor by Q.

$$s^2 + \left(\frac{\omega_0}{Q}\right)s + \omega_0^2 \qquad 31.69$$

Transient analysis uses the form given by Eq. 31.70. The *damping* is given the symbol α while the *frequency of oscillation* is β.

$$(s + \alpha)^2 + \beta^2 \qquad 31.70$$

In the transient analysis of Secs. 31-6 through 31-8, the three cases could have been represented as $\alpha > \omega_0$ (overdamped), $\alpha = \omega_0$ (critically damped), and $\alpha < \omega_0$ (underdamped). The parameters are related as follows.

$$\omega_0^2 = \omega_n^2 = \alpha^2 + \beta^2 \qquad 31.71$$

$$\zeta = \frac{1}{2Q} \qquad 31.72$$

$$\alpha = \frac{\omega_0}{2Q} \qquad 31.73$$

These characteristic equations are the denominators of transfer functions, which are ratios from a network's output point to a network's input point. Their roots, called *poles*, will aid in determining the frequency response of a network without solving the original differential equation. The roots of the characteristic equation determine the natural response of the network, that is, the response without a forcing function (without input). Thus, the transfer function and the characteristic equation describe the response of the network independent of input. Any input can then be attached and the output derived by knowing the circuit's transfer function and natural response.[16]

11. COMPLEX FREQUENCY

Electrical circuits can respond to input with a resulting constant output, A, and exponential output, $e^{\alpha t}$, or a sinusoid, $\sin \omega t$.[17] The complex frequency, s, is used to combine the three possibilities and represent each in terms of a single exponential, e^{st}.[18] The complex frequency is

$$\mathbf{s} = \sigma + j\omega \qquad 31.74$$

Consider a signal containing all three possibilities represented by Eq. 31.75, including a phase shift. In such a case, the peak value can be represented by the constant A. (The peak value is used to allow the constant to remain unchanged when in the exponential or sinusoidal form. In the sinusoidal form the peak value must be used to properly represent the function. The form $A\angle\theta$ normally uses an rms value for A, while the exponential form uses a peak value. Care must be taken to ensure the desired form is used.) Such a signal is given by

$$A_p e^{\sigma t} \sin(\omega t + \theta) \qquad 31.75$$

Representing the sinusoid in exponential terms, using the sine term, and dropping the Im prefix as is common practice, allows Eq. 31.75 to be represented as

$$A_p e^{j\theta} e^{(\sigma + j\omega)t} = A_p e^{j\theta} e^{st} \qquad 31.76$$

The e^{st} term will be common to all elements in a network, just as $e^{j\omega t}$ was common to all elements. Only the magnitude A and the angle θ of a given signal need to be carried in calculations. If desired, the e^{st} term is recovered upon completion of analysis, giving the exact form of the response. The variable σ is termed the *neper frequency* and assigned units of Np/s or s^{-1}. The neper

---

[16]The transfer function gives the steady-state response. The poles provide the transient response.
[17]Even aperiodic signals can be represented in terms of sinusoids. Consequently, these three possibilities represent the full range.
[18]It is common practice to write the complex frequency **s** simply as s, regardless of its actual value, which can be real or complex. This book will follow this practice.

frequency σ is always zero or negative in an actual circuit. In Eq. 31.75, for example, if σ were positive, the exponential term would increase without bound, resulting in an infinite response. Clearly, this is not possible. Nevertheless, the σ is considered to carry its sign (in other words, the negative is internal), though the sign is sometimes shown explicitly to clarify the decaying exponential aspect. The variable ω is the angular frequency and is assigned units of rad/s or s^{-1}. A negative value of ω merely implies that the sinusoid carries a negative sign, since $\sin(-\omega t) = -\sin(\omega t)$.

If $\sigma = 0$ and $\omega = 0$, the signal or input is a constant. If $\sigma < 0$ and $\omega = 0$, the signal represented is an exponential decay. If $\sigma < 0$ and $\omega \neq 0$, the signal is a damped sinusoid.[19] If $\sigma = 0$ and $\omega \neq 0$, the signal is a sinusoid. Table 31.3 shows examples of each of the signal response types based on Eq. 31.76 along with their values in the s domain. Table 31.4 gives the generalized impedances in the s domain. Table 31.4 also exhibits the similarity of form between the frequency domain and the s domain. Working in the s domain is always possible and simplifies the mathematics. If only a sinusoidal response is desired, upon completion of circuit manipulation, s can be replaced with $j\omega$ and the result will be in the correct form for any computations. Figure 31.6 illustrates how signal response terms plot.

Table 31.3 Example Signal Representation in the s Domain

time function	$A_p e^{j\theta}$ or $A_p \angle \theta^a$	s^b
50	$50\angle 0°$	$0 + j0$
$50e^{-100t}$	$50\angle 0°$	$-100 + j0$ Np/s
$50\sin(200t - 60°)$	$50\angle -60°$	$0 + j200$ rad/s
$50e^{-100t}\sin(-200t + 60°)$	$50\angle 60°$	$-100 - j200$ s^{-1}

[a]The format $A\angle\theta$ used in this text normally indicates rms values. Here the peak value is used. If the rms format is used, the values in this column would be divided by $\sqrt{2}$.
[b]A negative frequency $(-\omega)$ value merely implies that the sinusoid carries a negative sign, since $\sin(-\omega t) = -\sin(\omega t)$ and s would then have a negative j term.

Table 31.4 Generalized Impedance in the s Domain

impedance type	frequency domain value	s domain value
Z_R	R	R
Z_L	$j\omega L$	sL
Z_C	$\dfrac{1}{j\omega C}$	$\dfrac{1}{sC}$

---

[19]The condition for a damped sinusoid is sometimes given as $\sigma \neq 0$ and $\omega \neq 0$. This is true as long as the value of σ is restricted to negative values. (In real circuits, σ is either zero or negative.) If the negative sign is not carried with the σ but is instead shown in the exponential as $e^{-(\sigma+j\omega)t}$, the condition $\sigma > 0$ and $\omega \neq 0$ represents the damped sinusoid.

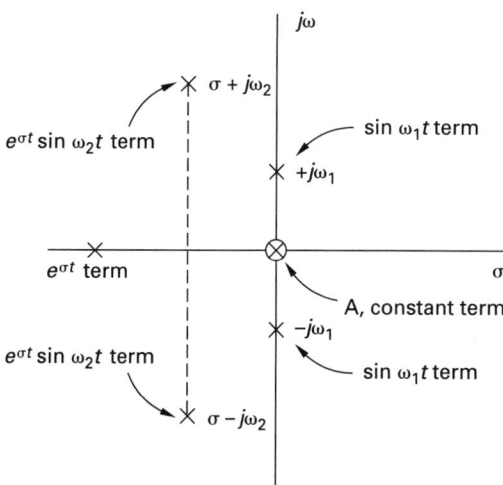

Figure 31.6 Signal Response in the s Domain

12. LAPLACE TRANSFORM ANALYSIS

The Laplace transform method of analysis moves the state variables, that is, the capacitive voltage and the inductive current, from the time domain into the s domain. In the s domain, many forms of excitation, such as the constant, the exponential, and the sinusoid, can be united and easily handled mathematically. The Laplace transform is given by[20]

$$\mathcal{L}\{f(t)\} = \int_{0^+}^{\infty} f(t)e^{-st}dt \qquad 31.77$$

The calculation of Eq. 31.77 seldom needs to be accomplished. Numerous tables of Laplace transforms exist (see App. 11.A). Using the data in a Laplace transform table, the properties given by Eqs. 31.78 and 31.79 are used to transform a differential equation representing the behavior of an electrical network.

$$\mathcal{L}\{f(t)\} = F(s) \qquad 31.78$$

$$\mathcal{L}\left\{\frac{d^2 f(t)}{dt^2}\right\} = s^2 F(s) - sf(0^+) - f'(0^+) \qquad 31.79$$

$$\mathcal{L}\left\{\frac{df(t)}{dt}\right\} = sF(s) - f(0^+) \qquad 31.80$$

Since the energy in a capacitor or inductor cannot change instantaneously, the conditions at $t = 0$ are

---

[20]The term s is the complex frequency or the *complex frequency variable* and could be shown as $\mathbf{s}$. Also, taking the Laplace transform of a physical quantity introduces an additional unit of time. For example, $I(s)$ has units of A·s or C. Since the inverse transform removes the extra time unit, such units will be omitted in the s domain and $I(s)$ is still referred to as the current. The same is done for the voltage, $V(s)$.

those at $t = 0^+$. Once the transformation is complete, the inverse Laplace transform can be taken using Eq. 31.81.

$$\mathcal{L}^{-1}\{F(s)\} = f(t) = \frac{1}{2\pi j}\int_{\sigma-j\infty}^{\sigma+j\infty} F(s)e^{st}dt \quad \textbf{31.81}$$

The calculation of Eq. 31.81 seldom needs to be accomplished. The same tables used for Laplace transforms contain the inverses. The path of integration of Eq. 31.81 is a straight line parallel to the $j\omega$ axis. All the poles of $F(s)$ lie to the left of this path.

The *classical method* for analyzing electrical networks using Laplace transform analysis follows.

step 1: Write the differential equation describing the circuit behavior from known circuit relationships and laws.

step 2: Transform the differential equation, term by term, into the s domain using Laplace transform tables (e.g., App. 11.A).

step 3: Solve for the transformed variable, $V(s)$ or $I(s)$, in terms of s.

step 4: Use partial fraction expansion, or any other desired method, to rearrange the terms into those recognizable in Laplace transform tables.

step 5: Equate the coefficients of the powers of s. Solve for the constants.

step 6: Obtain the inverse Laplace transform.

The classical method is useful when the desired quantity is one of the state variables, capacitor voltage or inductor current. If the desired variables are other than the state variables, the *circuit transformation method* is the most direct.

The circuit transformation method follows.

step 1: Transform each individual circuit element into the s domain. Initial condition sources will exist for energy storage elements. For the capacitor, an initial condition voltage source will be added in series. For the inductor, an initial condition current source will be added in parallel.

step 2: If the desired quantity is one of the variables of an energy storage device (i.e., a state variable), obtain the equivalent Thevenin or Norton form of the circuit as seen from the terminals of the capacitor or inductor.

step 3: Write the differential equation, describing the circuit behavior from known circuit relationships and laws.

step 4: Solve for the desired variable in terms of s.

step 5: Use partial fraction expansion, or any other desired method, to rearrange the terms into those recognizable in Laplace transform tables.

step 6: Equate the coefficients of the powers of s. Solve for the constants.

step 7: Obtain the inverse Laplace transform.

Regardless of the method used, the transformed equations obtained in step 3 are arranged using partial fraction expansion in order to be inversely transformed. This is necessary for transient analysis of first- and second-order systems, because the initial conditions impact the transient. Since the initial conditions do not impact the steady-state value, in sinusoidal analysis, usually called frequency analysis or frequency response, the equations of step 3 reduce to phasor form, so steps 4 through 7 can be skipped and the response analyzed using graphical means (see Ch. 32).

13. CAPACITANCE IN THE s DOMAIN

Capacitors oppose rates of change in voltage. In capacitive circuits, voltage is the state variable and is preferred in system equations. Since the capacitor voltage cannot change instantaneously, $v_C(0+) = v_C(0)$. With this information, the capacitive voltage, $v_C(t)$, and the capacitive current, $C(dv_C(t)/dt)$, are transformed by Eqs. 31.82 and 31.83.

$$\mathcal{L}\{v_C(t)\} = V_C(s) \quad \textbf{31.82}$$

$$\mathcal{L}\{i_C(t)\} = \mathcal{L}\left\{C\frac{dv_C(t)}{dt}\right\}$$
$$= sCV_C(s) - Cv_C(0) \quad \textbf{31.83}$$

The transformed capacitance must exhibit the properties at its terminals given by Eq. 31.83. Such a model is shown in Fig. 31.7. Using the extended Ohm's law on the current source $sCV_C(s)$, that is, taking the voltage across the source's terminals, $V_C(s)$, and dividing by the current, $sCV_C(s)$, gives the model of Fig. 31.8. Taking the Thevenin equivalent of the model shown in Fig. 31.8 results in a third model given in Fig. 31.9 that shows the initial voltage explicitly. Any of the models is valid; Fig. 31.9 is perhaps the most intuitive.

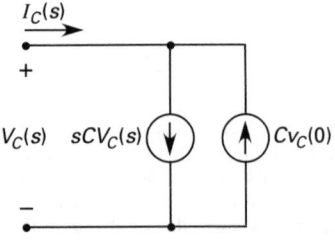

Figure 31.7 *Capacitor Two-Source Model*

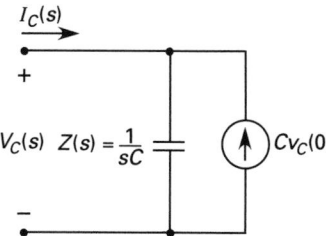

Figure 31.8 *Capacitor Impedance and Initial Current Source Model*

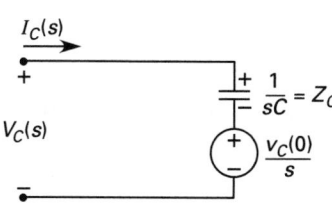

Figure 31.9 *Capacitor Impedance and Initial Voltage Source Model*

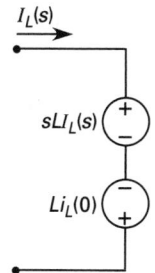

Figure 31.10 *Inductor Two-Source Model*

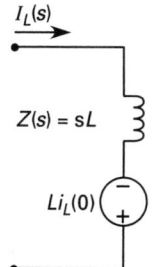

Figure 31.11 *Inductor Impedance and Initial Voltage Source Model*

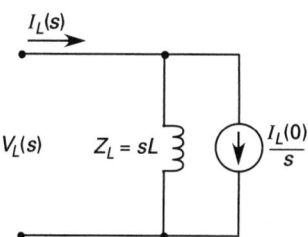

Figure 31.12 *Inductor Impedance and Initial Current Source Model*

14. INDUCTANCE IN THE s DOMAIN

Inductors oppose rates of change in current. In inductive circuits, current is the state variable and is preferred in system equations. Since the inductor current cannot change instantaneously, $i_L(0+) = i_L(0)$. With this information, the inductive current, $i_L(t)$, and the inductive voltage, $L(di_L(t)/dt)$, are transformed by Eqs. 31.84 and 31.85.

$$\mathcal{L}\left\{i_L(t)\right\} = I_L(s) \qquad 31.84$$

$$\mathcal{L}\left\{v_L(t)\right\} = \mathcal{L}\left\{L\frac{di_L(t)}{dt}\right\}$$

$$= sLI_L(s) - Li_L(0) \qquad 31.85$$

The transformed inductance must exhibit the properties at its terminals given by Eq. 31.85. Such a model is shown in Fig. 31.10. Using the extended Ohm's law on the voltage source $sLI_L(s)$, that is, dividing by the current through the source's terminals, $I_L(s)$, gives the model shown in Fig. 31.11. Taking the Norton equivalent of the model in Fig. 31.11 results in a third model given in Fig. 31.12, which shows the initial current explicitly. Any of the models is valid; Fig. 31.12 is perhaps the most intuitive.

15. LAPLACE TRANSFORM ANALYSIS: FIRST- AND SECOND-ORDER SYSTEMS

Either of the methods in Sec. 31-12 may be applied. If variables other than the capacitor voltage or inductor current are required, the circuit transformation method is the most direct method. The Thevenin form is preferred for capacitive circuits because the final capacitive voltage must equal the Thevenin voltage, thereby acting as a check on the solution. The Norton equivalent is preferred for inductive circuits, because the final inductive current must equal the Norton equivalent current, thereby acting as a check on the solution. Any of the network analysis techniques is applicable when working in the s domain.

Example 31.13

In the circuit shown, the capacitor has an initial charge of 4.5 mC. The switch closes at $t = 0$, applying a steady

120 V source. Determine the current using the Laplace transform method.

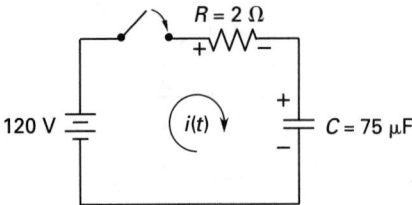

Solution

The classical approach will be used. Using KVL in the direction shown gives

$$V_s - i(t)R - \frac{1}{C}\int i(t)dt = 0$$

Though not stated explicitly, the behavior of the circuit from $t = 0$ onward is the desired result. Thus,

$$V_s = i(t)R + v_C(0^+) + \frac{1}{C}\int_0^\infty i(t)dt$$

The initial capacitor voltage can be determined from the initial charge, using $C = Q/V$. The constant-source voltage can be considered a unit step at $t = 0$. The initial capacitor voltage is a constant as well. Both transform to the s domain using $1/s$ (see App. 11.A). Integration equates to multiplication by $1/s$. Substituting and transforming gives

$$V_s = \left(i(t)\right)R + \frac{Q_0}{C} + \frac{1}{C}\int_0^\infty i(t)dt$$

$$\frac{120}{s} = \left(I(s)\right)(2) + \frac{4.5 \times 10^{-3}}{\left(75 \times 10^{-6}\right)(s)} + \frac{I(s)}{\left(75 \times 10^{-6}\right)(s)}$$

$$I(s) = \frac{30}{s + 6.65 \times 10^3}$$

Performing the inverse Laplace transform on the result gives

$$i(t) = 30e^{-6.65 \times 10^3 t}$$

32 Frequency Response

Nomenclature

D	denominator magnitude	–
$f(t), F(s)$	forcing or input function	–
G	gain	dB
i	instantaneous current	A
I	effective or DC current	A
K	constant	–
N	numerator magnitude	–
$\mathbf{p}, p$	pole	s^{-1}
$r(t), R(s)$	response function	–
R	resistance	Ω
$\mathbf{s}, s$	complex frequency	s^{-1}
s	Laplace transform variable	–
$t(t), T(s)$	transfer function	–
v	instantaneous voltage	V
V	effective or DC voltage	V
x	poles at s domain origin	–
y	zeros at s domain origin	–
$\mathbf{z}, z$	zero	s^{-1}
Z	impedance	Ω

Symbols

α	angle	rad
β	angle	rad
Π	multiplication of terms	–
σ	neper frequency or napier frequency	Np/s or s^{-1}
σ_0	break or corner frequency	Np/s or s^{-1}
Σ	sum of terms	–
ω	angular frequency	rad/s

Subscripts

d	number or denominator
d	delay
DC	direct current
e	exponential
e	equivalent
n	number or numerator
net	network
p	pole
x	input
y	output
z	zero

1. FUNDAMENTALS

In a network with capacitors or inductors, or both, transients occur.[1] Transients are represented by mathematical circuit models consisting of differential equations. The solutions to these equations depend on the circuit initial conditions.[2] The methods outlined in Ch. 31 can be used to obtain the solutions, with the Laplace transform method changing differential equation manipulation to algebraic manipulation and the phasor method simplifying sinusoidal mathematics.

Transient analysis requires the complete analysis of a circuit, since both the transient and steady-state components are relevant. The transient portion (natural response) of any output eventually settles on the steady-state (force response) value determined by the forcing function, that is, the *signal*.[3] Steady-state analysis of a circuit is thus independent of the circuit initial conditions. Solutions to problems in steady-state analysis can be determined from the circuit characteristics alone. These characteristics are contained within a *transfer function*, given in Eq. 32.1.

$$T(s) = \mathcal{L}\left\{\frac{r(t)}{f(t)}\right\} \qquad \textit{32.1}$$

The circuit's input signal or forcing function is $f(t)$, and the response to this function is $r(t)$. The transfer function is given the symbol $T(s)$.[4] The transfer function is evaluated in the frequency domain rather than in the time domain. This type of analysis is called *frequency response analysis*. When the signal is sinusoidal, that is, $s = j\omega$, the term *sinusoidal steady-state analysis* is used. Unlike phasor analysis where a fixed frequency is assumed, frequency response analysis allows the frequency behavior over a range of values to be examined. This allows the effects of *variable frequency inputs* and *spectral inputs*, which have multiple frequencies, to be analyzed.

---

[1]In purely resistive circuits, the response is considered instantaneous.
[2]The arbitrary constant occurring in integration of electrical parameters represents the initial condition.
[3]The natural response is determined from the homogenous solution (i.e., the forcing function is equal to zero) of the differential equation. The forced response is from the particular solution whose form must be that of the forcing function.
[4]The term $T(s)$ is actually the Laplace transform of the transform function. The symbolism for each of the terms varies widely, but Laplace transforms are normally shown as uppercase while time functions are shown as lowercase.

If the time response to a sinusoidal signal (that is, $s = j\omega$) is required, the signal magnitude is multiplied by the transfer function magnitude, $|F(s)|\,|T(s)|$, which gives the magnitude of the output or response, $|R(s)|$. The angle is determined by adding the signal angle to the transfer function angle, $\angle F(s) + \angle T(s)$, which gives the angle of the response, $\angle R(s)$. The result is a phasor that can be changed into the sinusoidal form using the known value of $j\omega$. The transfer function can thus be used to determine the output for a variety of inputs since it contains the circuit's characteristics.

Frequency response analysis is used to simplify calculations and increase understanding compared to equivalent analysis in the time domain. The main principle is that a sinusoidal input produces a sinusoidal output.

2. TRANSFER FUNCTION[5]

The term "transfer function" refers to the relationship of one electrical parameter in a network to a second electrical parameter elsewhere in the network.[6] The *network function*, $\mathbf{T}_{\text{net}}(\mathbf{s})$, refines this definition to be the ratio of the complex amplitude of an exponential output, $\mathbf{Y}(\mathbf{s})$, to the complex amplitude of an exponential input, $\mathbf{X}(\mathbf{s})$. If $x(t) = Xe^{st}$ and $y(t) = Ye^{st}$, and if the linear circuit is made up of lumped elements, the network function can be derived from an input-output differential equation as a rational function of $\mathbf{s}$ and can be written in the following general form.[7]

$$\mathbf{T}_{\text{net}}(\mathbf{s}) = \frac{\mathbf{Y}(\mathbf{s})}{\mathbf{X}(\mathbf{s})}$$

$$= A\left(\frac{(\mathbf{s}-\mathbf{z}_1)(\mathbf{s}-\mathbf{z}_2)\dots(\mathbf{s}-\mathbf{z}_n)}{(\mathbf{s}-\mathbf{p}_1)(\mathbf{s}-\mathbf{p}_2)\dots(\mathbf{s}-\mathbf{p}_d)}\right) \quad \textit{32.2}$$

The complex constants represented as $\mathbf{z}_y$ ($y = 1, 2, \dots, n$) are the *zeros* of $\mathbf{T}_{\text{net}}(\mathbf{s})$ and are plotted in the $\mathbf{s}$ domain as $\bigcirc$'s. The complex constants represented as $\mathbf{p}_x$ ($x = 1, 2, \dots, d$) are the *poles* of $\mathbf{T}_{\text{net}}(\mathbf{s})$ and are plotted in the $\mathbf{s}$ domain as X's. Equation 32.2 can be interpreted as the ratio of the response in one part of the $\mathbf{s}$ domain network to the excitation in another part of the network. When $\mathbf{s} = \mathbf{z}_y$, the response of the network will be zero regardless of the excitation. When $\mathbf{s} = \mathbf{p}_x$, the response will be infinite regardless of the excitation. Equation 32.2 is independent of the forcing function. That is, the equation contains the characteristics of the network itself. The response to some source

---

[5]Functions of $\mathbf{s}$ are not normally shown in bold, though they are complex, to avoid confusion with phasors. They are shown as complex in this section to clarify several points regarding their properties.
[6]When the function is unitless—that is, when it is the ratio of voltages, the ratio of currents, and so on—it is called the *transfer ratio*.
[7]This equation is often shown with positive signs. In this case, the zeros are the $-\mathbf{z}$ values. The minus signs are shown here to clarify how the $\mathbf{s}$ and $\mathbf{z}$ values would be combined in the $\mathbf{s}$ domain (see Sec. 32-3).

(input) is found by substituting the value of $\mathbf{s} = \sigma_s + j\omega_s$ for the source into Eq. 32.2. The properties of the circuit elements themselves are held in the complex zeros and poles, $\mathbf{z}$ or $\mathbf{p}$.

Example 32.1

Determine the network function of the circuit shown in terms of a current output, $\mathbf{I}(\mathbf{s})$, for a given voltage input, $\mathbf{V}(\mathbf{s})$.

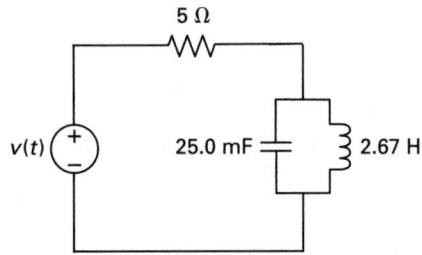

Solution

The network function is

$$\mathbf{T}_{\text{net}}(\mathbf{s}) = \frac{\mathbf{I}(\mathbf{s})}{\mathbf{V}(\mathbf{s})} = \frac{1}{\mathbf{Z}(\mathbf{s})}$$

Since the calculation will be accomplished in the $\mathbf{s}$ domain, transform the circuit. The result is

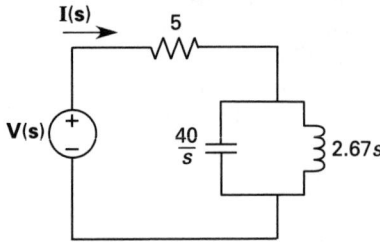

The impedance consists of a resistor in series with the parallel combination of a capacitor and an inductor. Thus,

$$\mathbf{Z}(\mathbf{s}) = 5 + \frac{\left(\dfrac{40}{\mathbf{s}}\right)(2.67\mathbf{s})}{\dfrac{40}{\mathbf{s}} + 2.67\mathbf{s}}$$

After manipulating mathematically, the result is

$$\mathbf{Z}(\mathbf{s}) = \frac{(5)(\mathbf{s}^2 + 8\mathbf{s} + 15)}{\mathbf{s}^2 + 15} = \frac{(5)(\mathbf{s}+3)(\mathbf{s}+5)}{\mathbf{s}^2 + 15}$$

The network function is

$$\mathbf{T}_{\text{net}}(\mathbf{s}) = \frac{1}{\mathbf{Z}(\mathbf{s})} = (0.2)\left(\frac{\mathbf{s}^2 + 15}{(\mathbf{s}+3)(\mathbf{s}+5)}\right)$$

Example 32.2

Plot zeros and poles of the network function in Ex. 32.1.

Solution

The zeros are determined by the roots in the numerator as $s = \pm j\sqrt{15}$, complex conjugate pairs. The poles are determined by the roots in the denominator as $s = -3$ and $s = -5$. Plotting gives the following result.

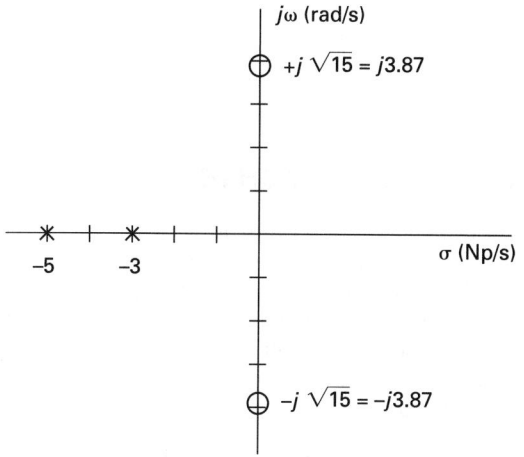

The plot is a complete representation of the network function.

3. STEADY-STATE RESPONSE[8]

The network function is a ratio of phasor quantities, such as $\mathbf{V}_2/\mathbf{V}_1$, $\mathbf{I}_2/\mathbf{I}_1$, $\mathbf{V}_2/\mathbf{I}_1$, and so on. The variables can also be impedances or admittances. As for any phasor quantity, the network function can be expressed as a magnitude at an angle, N or $D\angle\theta$. Each of the zero terms in the numerator and pole terms in the denominator in Eq. 32.2 can be represented as in Eqs. 32.3 and 32.4, respectively.

$$(\mathbf{s} - \mathbf{z}_n) = N_n\angle\alpha_n \qquad 32.3$$

$$(\mathbf{s} - \mathbf{p}_d) = D_d\angle\beta_d \qquad 32.4$$

Substituting Eqs. 32.3 and 32.4 into the generic network equation, Eq. 32.2, provides a mathematical picture of the steady-state response phasors.

$$\mathbf{T}_{\text{net}}(\mathbf{s}) = A\frac{(N_1\angle\alpha_1)(N_2\angle\alpha_2)\cdots(N_n\angle\alpha_n)}{(D_1\angle\beta_1)(D_2\angle\beta_2)\cdots(D_d\angle\beta_d)}$$

$$= A\left(\frac{N_1 N_2 \cdots N_n}{D_1 D_2 \dots D_d}\right)\angle(\alpha_1 + \alpha_2 + \cdots + \alpha_n)$$

$$- (\beta_1 + \beta_2 + \cdots \beta_d) \qquad 32.5$$

---

[8]Functions of s are shown as complex in this section to clarify several points regarding their properties, such as the phasor addition that takes place in the example.

Equations 32.3 and 32.4 considered together indicate that the steady-state response of a network to a forcing function input of $\mathbf{s} = \sigma + j\omega$ is determined by the lengths of the vectors from the zeros to the $\mathbf{s}$ and the vectors from the poles to the $\mathbf{s}$. In addition, the angle is determined by the angles these vectors make with the σ axis.[9] The angles α and β are always between $0°$ and $90°$. Equation 32.5 is a geometric means of determining the network function from the locations of the zeros and poles, without knowing the analytical expression. If the poles and zeros are known, the network function can be written to within a factor of A.

Example 32.3

For the circuit in Ex. 32.1 and the response given in Ex. 32.2, determine the current output for a voltage excitation input of $v = 1e^{\mathbf{s}t}$ where $\mathbf{s} = \sqrt{15}$ Np/s.

Solution

Place the input on the pole-zero plot. Draw the vectors from the zeros to the input $\mathbf{s}$. Draw the vectors from the poles to the input $\mathbf{s}$.

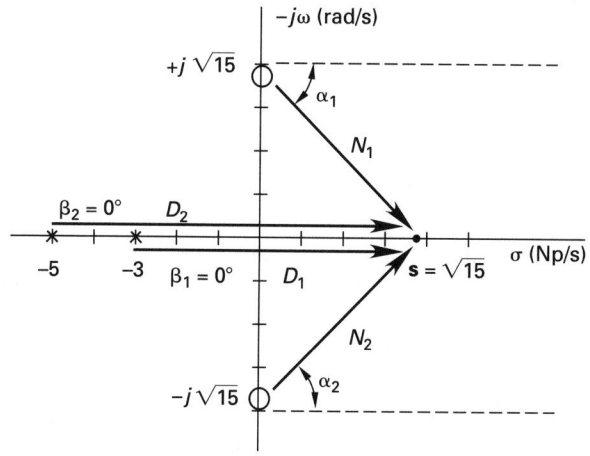

The magnitudes and angles are then calculated. For the zeros,

$$N_1 = N_2 = \sqrt{\left(\sqrt{15}\right)^2 + \left(\sqrt{15}\right)^2} = \sqrt{30}$$

$$\alpha_1 = \tan^{-1}\left(\frac{-\sqrt{15}}{\sqrt{15}}\right) = -45°$$

$$\alpha_2 = \tan^{-1}\left(\frac{\sqrt{15}}{\sqrt{15}}\right) = 45°$$

$$\alpha_1 + \alpha_2 = 0°$$

---

[9]The vectors represent what the circuit must do to respond to the forcing function. Recall that the steady-state output of a circuit always takes the form of the forcing or input function.

For the poles,

$$D_1 = 3 + \sqrt{15} = 6.87$$

$$D_2 = 5 + \sqrt{15} = 8.87$$

$$\beta_1 = 0°$$

$$\beta_2 = 0°$$

$$\beta_1 + \beta_2 = 0°$$

The factor A was determined in Ex. 32.1. The network function is calculated as

$$\mathbf{T}_{\text{net}}(\mathbf{s}) = A \left(\frac{N_1 N_2 \ldots N_n}{D_1 D_2 \ldots D_d} \right) \angle (\alpha_1 + \alpha_2 + \cdots \alpha_n)$$
$$- (\beta_1 + \beta_2 + \cdots \beta_d)$$

$$\mathbf{T}_{\text{net}} \left(\sqrt{15} \right) = (0.2) \left(\frac{(\sqrt{30})(\sqrt{30})}{(6.87)(8.87)} \right) \angle (0° + 0°)$$

$$= 98.46 \times 10^{-3} \angle 0°$$

The response for the given input can now be determined.

$$\mathbf{T}_{\text{net}}(\mathbf{s}) = \frac{\mathbf{I}(\mathbf{s})}{\mathbf{V}(\mathbf{s})}$$

$$i(t) = t_{\text{net}}(t)v(t)$$

$$= 98.46 \times 10^{-3} e^{\sqrt{15}t}$$

The result indicates that as time increases the current becomes infinite, clearly an impossibility. This is consistent with the condition in Sec. 31-10 that σ takes on zero or negative values only.

Example 32.4

Write the expression for the transfer function in the pole-zero plot shown.

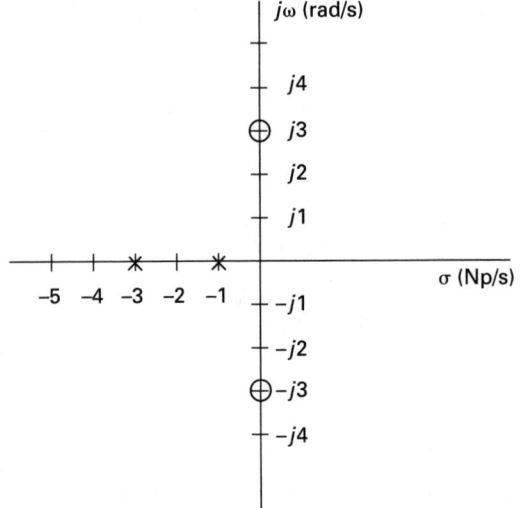

Solution

The zeros are located at $\pm j3$. Consequently, the numerator is given by

$$\mathbf{s}^2 + 9$$

The poles are located at -1 and -3. Consequently, the denominator is

$$(\mathbf{s} + 3)(\mathbf{s} + 1)$$

The transfer function is

$$\mathbf{T}_{\text{net}}(\mathbf{s}) = A \left(\frac{\mathbf{s}^2 + 9}{(\mathbf{s} + 3)(\mathbf{s} + 1)} \right)$$

4. TRANSIENT RESPONSE

The network function uses the complex frequency in such a way as to determine the steady-state response. Nevertheless, the circuit characteristics, including the natural response, are embedded in the input equation in the denominator of the network function. Specifically, the transient response is characterized by the poles of the network function.

The real portion of the complex number representing a pole contains the neper frequency, σ, of the circuit for that pole. The *neper frequency*, also called the napier frequency, determines the exponential terms in a solution.[10] The exponential terms represent the transient response of a circuit (see Ch. 31). Near a pole, then, the natural response of the circuit elements impacts the overall output of a network, while far away from a pole little effect occurs.

Example 32.5

What is the mathematical form of the transient response for the network function in Ex. 32.4?

Solution

The network function in Ex. 32.4 was

$$T_{\text{net}}(s) = A \left(\frac{s^2 + 9}{(s + 3)(s + 1)} \right)$$

In keeping with common practice, the s is no longer indicated as a complex number, though it still is. Since each term is $(s - p)$, the poles occur at $(s - (-p))$. That is, the poles are at -3 and -1. The expanded denominator is

$$s^2 + 4s + 3$$

The coefficients $a = 1$, $b = 4$, and $c = 3$ indicate that $b^2 > 4ac$. The transient response is of the form

$$\text{transient response} = Ae^{s_1 t} + Be^{s_2 t}$$

---

[10]The term "napier" is most commonly known for its association with natural logarithms. It is used here associated with the natural response.

The transient response could be either the current or voltage response, depending upon the ratio represented by the network function. The initial conditions, not included in the network function, would be required to determine the constants A and B.

5. MAGNITUDE AND PHASE PLOTS

A plot of the magnitude and phase of a transfer function in the **s** domain provides information regarding the response of that function as frequency varies.[11] Magnitude and phase plots are fundamental to constructing Bode plots.

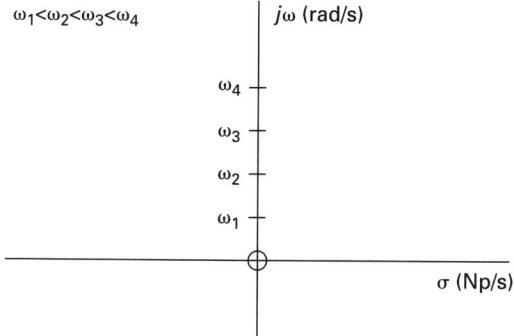

(a) zero at origin, frequency increasing

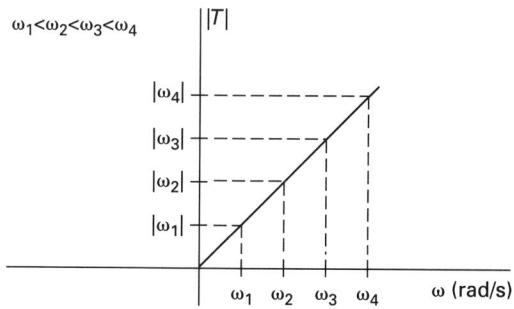

(b) transfer function magnitude, frequency increasing

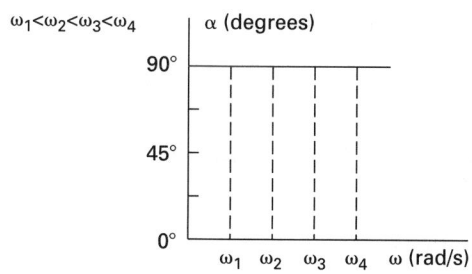

(c) transfer function angle, frequency increasing

Figure 32.1 Single Zero at Origin

Consider a transfer function with a single zero at the origin. As only the steady-state response to a sinusoid is of concern, let $s = j\omega$. The transfer function is

$$T(s) = s = s + 0 \qquad 32.6$$

$$T(j\omega) = j\omega \qquad 32.7$$

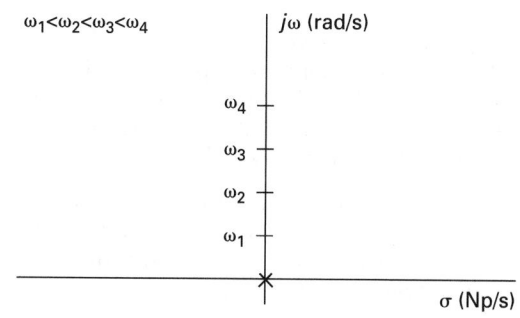

(a) pole at origin, frequency increasing

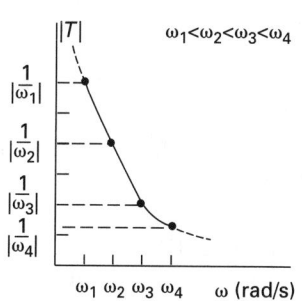

(b) transfer function magnitude, frequency increasing

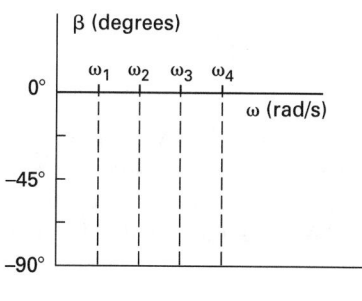

(c) transfer function angle, frequency increasing

Figure 32.2 Single Pole at Origin

The frequency is allowed to vary as illustrated in Fig. 32.1(a). The magnitude is plotted in Fig. 32.1(b) from $|T(jw)| = \omega$. The phase angle variation with frequency is shown in Fig. 32.1(c) from $\alpha = \tan^{-1}(w/0)$ or directly from j, which is $1\angle 90°$.

[11]Since the methods apply to all functions in the **s** domain, not merely network functions, the term "transfer function" and the symbol $T(s)$ will be used instead of $\mathbf{T}_{net}(\mathbf{s})$.

Consider a transfer function with a single pole at the origin. As only the steady-state response to a sinusoid is of concern, let $s = j\omega$. The transfer function is

$$T(s) = \frac{1}{s} = \frac{1}{s+0} \qquad 32.8$$

$$T(j\omega) = \frac{1}{j\omega} \qquad 32.9$$

The frequency is allowed to vary as illustrated in Fig. 32.2(a). The magnitude is plotted in Fig. 32.2(b) from $|T(j\omega)| = 1/|\omega|$. The phase angle variation with frequency is shown in Fig. 32.2(c) from $\beta = -\tan^{-1}(\omega/0)$ or directly from $1/j$, which is $-j$ or $1\angle-90°$.

Consider a transfer function with a single zero but located on the negative real axis at $-\sigma_n$.[12] Again, only the steady-state response to a sinusoid is of concern, so $s = j\omega$. The transfer function is

$$T(s) = s + \sigma_n \qquad 32.10$$

$$T(j\omega) = j\omega + \sigma_n \qquad 32.11$$

The frequency is allowed to vary as illustrated in Fig. 32.3(a).[13] The magnitude is plotted in Fig. 32.3(b) from $|T(jw)| = \sqrt{\omega^2 + \sigma_n^2}$. The phase angle variation with frequency is shown in Fig. 32.3(c) from $\alpha = \tan^{-1}(\omega/\sigma_n)$.

Consider a transfer function with a single pole but located on the negative real axis at $-\sigma_n$. Again, only the steady-state response to a sinusoid is of concern, so $s = j\omega$. The transfer function is

$$T(s) = \frac{1}{s + \sigma_d} \qquad 32.12$$

$$T(j\omega) = \frac{1}{j\omega + \sigma_d} \qquad 32.13$$

The frequency is allowed to vary as illustrated in Fig. 32.4(a).[14] The magnitude is plotted in Fig. 32.4(b) from $|T(jw)| = 1/\sqrt{\omega^2 + \sigma_d^2}$. The phase angle variation with frequency is shown in Fig. 32.4(c) from $\beta = -\tan^{-1}(\omega/\sigma_d)$.

[12]All practical devices will have zeros to the left of the imaginary axis. The restriction of being on the negative real axis will be removed for Bode plots but is used here to clarify the concept of a *break* or *corner frequency* that occurs at zeros and poles, that is, where the real portion of the zero or pole is σ_n or σ_p. The neper frequency, σ, represents the exponential solution or the natural response of the circuit elements that impacts the steady-state response in the area of a zero or a pole.
[13]The figure is not to scale. The relationships between drawings are approximate, but individual drawings show their true shapes.
[14]The figure is not to scale. The relationships between drawings are approximate, but individual drawings show their true shapes.

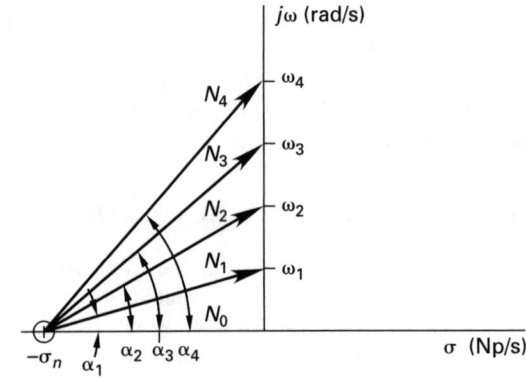

(a) zero on negative real axis, frequency increasing

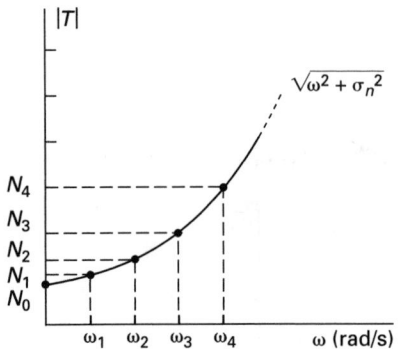

(b) transfer function magnitude, frequency increasing

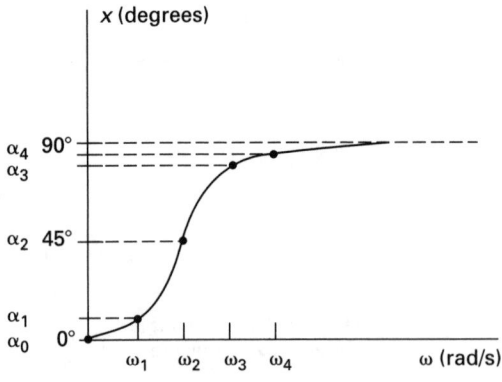

(c) transfer function angle, frequency increasing

Figure 32.3 *Single Zero on Negative Real Axis* ($-\sigma_n$)

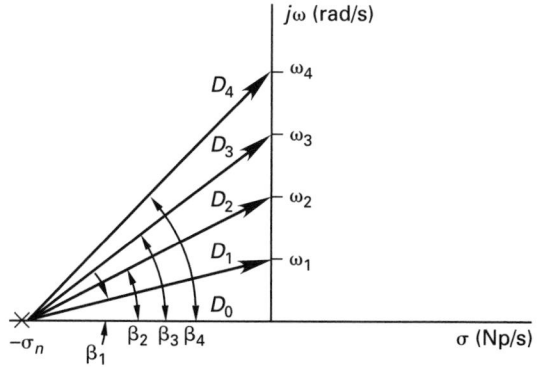

(a) pole on negative real axis,
frequency increasing

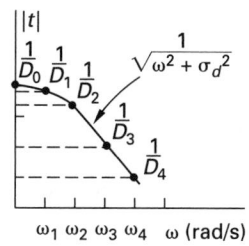

(b) transfer function magnitude,
frequency increasing

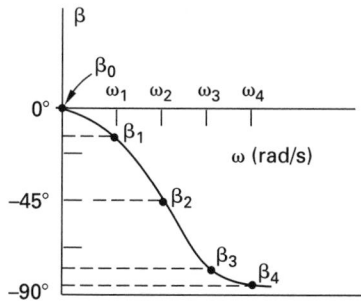

(c) transfer function angle,
frequency increasing

Figure 32.4 *Single Pole on Negative Real Axis $(-\sigma_p)$ Plot*

When $\sigma = \omega$, the magnitude of the transfer function for a zero is $\sqrt{\omega^2 + \sigma_n^2} = \sqrt{2}\sigma_n$ and the phase angle is $\alpha = +45°$. When $\sigma = \omega$, the magnitude of the transfer function for a pole is $1/(\sqrt{2}\sigma_d)$ and the angle is $\beta = -45°$. When the transfer function contains numerous zeros and poles, the magnitude is obtained by multiplying the magnitudes of the individual transfer functions, and the phase is obtained by summing the zero angles and subtracting the sum of the pole angles. Equation 32.5 is such a combination, with each $N\angle\alpha$ and $D\angle\beta$ representing the magnitude of one transfer function. The multiplication and division can be replaced by addition and subtraction using logarithmic quantities, guided by Eqs. 32.14 and 32.15.

$$\log(xy) = \log x + \log y \qquad 32.14$$

$$\log\left(\frac{x}{y}\right) = \log x - \log y \qquad 32.15$$

The magnitude of the transfer function, Eq. 32.5, can be written as

$$\log|T(s)| = \log A + \log N_1 + \log N_2$$
$$+ \cdots + \log N_n - \log D_1$$
$$- \log D_2 - \cdots - \log D_d \qquad 32.16$$

6. BODE PLOT PRINCIPLES: MAGNITUDE PLOT

A *Bode plot* or *Bode diagram* is a plot of the gain or phase of an electrical device or network against the frequency. The plot is normally of magnitude in decibels versus the logarithm of the frequency, and is utilized because of the algebraic manipulation possible with logarithms (see Eqs. 32.14 and 32.15) and the wide frequency range of response. The *decibel*, in units dB, is defined as 10 times the logarithmic ratio of two powers. Nevertheless, it is used for transfer functions in general. Since the I^2 and V^2 are proportional to power, the magnitude of transfer functions containing the voltage and current is[15]

$$G = 10\log\left|\left(\frac{I_2}{I_1}\right)\right|^2 = 20\log\left|\left(\frac{I_2}{I_1}\right)\right| \qquad 32.17$$

$$G = 10\log\left|\left(\frac{V_2}{V_1}\right)\right|^2 = 20\log\left|\left(\frac{V_2}{V_1}\right)\right| \qquad 32.18$$

In general terms, any transfer function's magnitude, called the *gain* (G), is

$$G = 20\log|T(s)| \qquad 32.19$$

The transfer function for a single zero at the origin is given by Eq. 32.6. Normalizing this with the factor σ_0 gives

$$T(s) = \sigma_0\left(\frac{s}{\sigma_0}\right) \qquad 32.20$$

Using Eqs. 32.19 and 32.20, $s = j\omega$, and the properties of logarithms from Eqs. 32.14 and 32.15 gives

$$G = 20\log\left|\sigma_0\left(\frac{j\omega}{\sigma_0}\right)\right|$$
$$= 20\log\sigma_0 + 20\log\left(\frac{\omega}{\sigma_0}\right) \qquad 32.21$$

---

[15]To be used in the ratio, the currents or voltages should act on the same impedance. Standard practice ignores this technicality.

The resulting Bode plot is shown in Fig. 32.5. The normalizing factor is the frequency where the gain (magnitude) has a value of 0 dB, which gives an actual normalizing factor of one. A zero at the origin results in a line of positive slope passing through $\omega = 1$ or $\log \omega = 0$.[16] The slope is 6 dB/octave or 20 dB/decade. An *octave* is an interval between any two frequencies with a ratio of two to one; for example, $\omega_2 = 2\omega_1$. On the log scale the magnitude change between octaves is

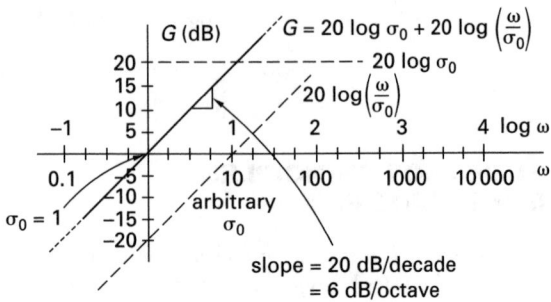

Figure 32.5 *Bode Plot: Single Zero at Origin*

$$\Delta G = 20 \log \omega_2 - 20 \log \omega_1$$

$$= 20 \log \left(\frac{\omega_2}{\omega_1} \right)$$

$$= 20 \log \left(\frac{2\omega_1}{\omega_1} \right)$$

$$= 20 \log(2)$$

$$= 6.02 \approx 6 \text{ dB/octave} \qquad 32.22$$

A decade of frequency change alters the magnitude by

$$\Delta G = 20 \log \omega_2 - 20 \log \omega_1$$

$$= 20 \log \left(\frac{\omega_2}{\omega_1} \right)$$

$$= 20 \log \left(\frac{10\omega_1}{\omega_1} \right)$$

$$= 20 \log(10)$$

$$= 20 \text{ dB/decade} \qquad 32.23$$

The two slopes are equivalent. The constant $20 \log \sigma_0$ does not change the shape of the plot. The constant can be accounted for in one of two ways. First, it can shift the plot upward, as was done in this case. Second, the value of 0 dB could be changed to a new value equal to the constant, after plotting the other factors of the gain. The angle for a zero at the origin remains a constant 90°.

The transfer function for a single pole at the origin is given by Eq. 32.8. Normalizing this with the factor σ_0 gives

$$T(s) = \frac{1}{\sigma_0 \left(\dfrac{s}{\sigma_0} \right)} \qquad 32.24$$

With $s = j\omega$, the gain is

$$G = 20 \log \left| \frac{1}{\sigma_0 \left(\dfrac{j\omega}{\sigma_0} \right)} \right|$$

$$= -20 \log \sigma_0 - 20 \log \left(\frac{\omega}{\sigma_0} \right) \qquad 32.25$$

The resulting Bode plot is shown in Fig. 32.6. The normalizing factor is the frequency where the gain (magnitude) has a value of 0 dB, which gives an actual normalizing factor of one. A pole at the origin results in a line of negative slope passing through $\omega = 1$ or $\log \omega = 0$.[17] The slope is -6 dB per octave or -20 dB per decade. The constant $-20 \log \sigma_0$ does not change the shape of the plot. In this case, the constant is used to shift the plot downward. Another option is to change the gain coordinates so that the zero value correlates with the value of the constant, after plotting the other terms of the gain. The angle for a pole at the origin remains a constant $-90°$.

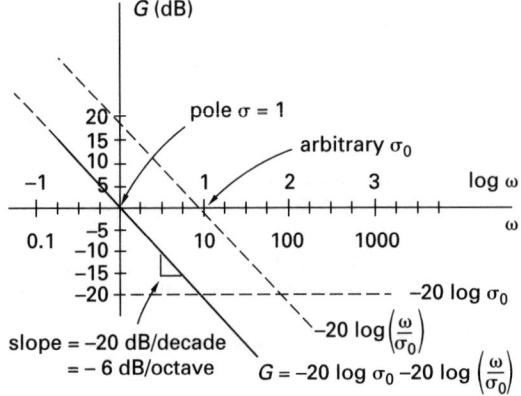

Figure 32.6 *Bode Plot: Single Pole at Origin*

The transfer function for a single zero on the negative real axis at $-\sigma_0$ in the s plane is given by Eq. 32.10. Normalizing this equation gives

$$T(s) = \sigma_0 \left(1 + \frac{s}{\sigma_0} \right) \qquad 32.26$$

---

[16]Considering Eq. 32.21 and Fig. 32.5, regardless of where the example normalizing factor, σ_0, is located (in this case, it was at $\sigma_0 = 10$), the final result remains unchanged: a +20 dB/decade line through $\omega = 1$ or $\log \omega = 0$ with an actual σ_0 of 1.

[17]Considering Eq. 32.25 and Fig. 32.6, regardless of where the example normalizing factor, σ_0, is located (in this case, it was at $\sigma_0 = 10$), the final result remains unchanged: a -20 dB per decade line through $\omega = 1$ or $\log \omega = 0$ with an actual σ_0 of 1.

With $s = j\omega$, the gain is

$$G = 20 \log \left| \sigma_0 \left(1 + \frac{j\omega}{\sigma_0} \right) \right|$$

$$= 20 \log \sigma_0 + 20 \log \left| \left(1 + \frac{j\omega}{\sigma_0} \right) \right| \qquad \textbf{32.27}$$

The resulting Bode plot is shown in Fig. 32.7. The constant $20 \log \sigma_0$ does not change the shape of the plot but rather shifts it upward or changes the value of the 0 dB to that of the constant (not shown). The normalizing factor is the frequency where the gain (magnitude) has a value of 0 dB. To determine the gain on either side of the normalizing factor, which will now be called the *break frequency* or *corner frequency*, consider the behavior of the asymptotes.[18] At low frequencies, ignoring the constant term of Eq. 32.27 since it only changes the plot's position, the gain is

$$G \approx 20 \log \left| \left(1 + \frac{0}{\sigma_0} \right) \right| = 0 \text{ dB} \quad [\omega << \sigma_0] \quad \textbf{32.28}$$

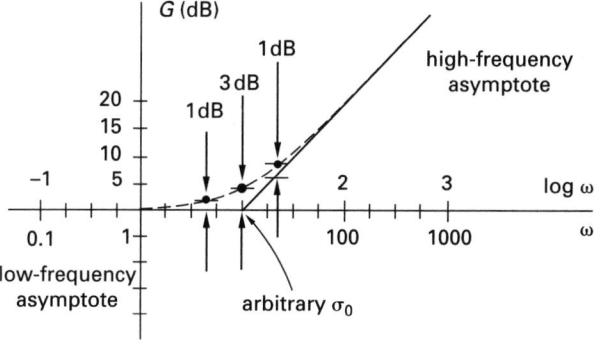

Figure 32.7 *Bode Plot: Single Zero on Negative Real Axis*

The low-frequency asymptote is approximated by drawing a line from the break frequency horizontally to the left. At high frequencies, ignoring the constant term of Eq. 32.27, the gain is

$$G \approx 20 \log \left| \left(\frac{j\omega}{\sigma_0} \right) \right|$$

$$= 20 \log \left(\frac{\omega}{\sigma_0} \right) \quad [\omega >> \sigma_0] \quad \textbf{32.29}$$

The high-frequency asymptote is approximated by a straight line with a slope of +20 dB/decade. That is, the curve "breaks upward" at the break frequency. At the break frequency $\omega = \sigma_0$, the magnitude of the gain is

---

[18]The break frequency or corner frequency corresponds to the zero or pole location.

$$G_0 = 20 \log \left| \left(1 + \frac{j\omega}{\sigma_0} \right) \right|$$

$$= 20 \log \left(\sqrt{(1)^2 + (1)^2} \right)$$

$$= 10 \log(2) = 3 \text{ dB} \qquad \textbf{32.30}$$

At $\frac{1}{2}\sigma_0$ and $2\sigma_0$, the gain differs by 1 dB from the approximating asymptotes. The exact value of the frequency for a given gain, G_ω, can be determined by rearranging Eq. 32.27 (ignoring the constant factor). The result is

$$\omega = \sigma_0 \sqrt{(10)^{\frac{G_\omega}{10}} - 1} \qquad \textbf{32.31}$$

While Eq. 32.31 could be used to determine exact values, the asymptotes provide reasonable accuracy and adequate information for most electrical engineering applications. Use of the asymptotic approximations results in an *idealized Bode plot*. Computer programs are available to accurately plot the response of a given circuit should such accuracy become necessary. The angle for a zero on the negative real axis of the s domain is determined from Eq. 32.26 to be $\tan^{-1}(\omega/\sigma_0)$. At the break frequency, this angle is 45°.

The transfer function for a single pole on the negative real axis at $-\sigma_0$ in the s plane is given by Eq. 32.12. Normalizing the equation and letting $s = j\omega$ results in a gain of

$$G = 20 \log \left| \left(\frac{1}{\sigma_0 \left(1 + \frac{j\omega}{\sigma_0} \right)} \right) \right|$$

$$= -20 \log \sigma_0 - 20 \log \left| \left(1 + \frac{j\omega}{\sigma_0} \right) \right| \qquad \textbf{32.32}$$

The plot for the pole is identical to that for a zero, with two exceptions: it has a negative slope and the plot breaks downward at the break frequency as shown in Fig. 32.8.

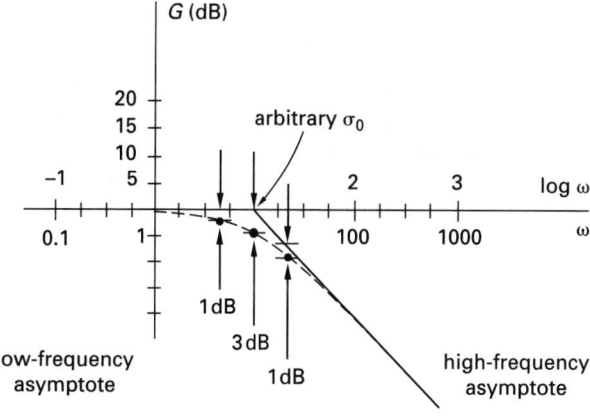

Figure 32.8 *Bode Plot: Single Pole on Negative Real Axis*

The transfer function for a circuit with numerous zeros and poles can be generalized as

$$T(s) = A \left(\frac{(s + \sigma_{z1})(s + \sigma_{z2})}{\cdots \times (s + \sigma_{zn})s^y} \Big/ \frac{(s + \sigma_{p1})(s + \sigma_{p2})}{\cdots \times (s + \sigma_{pd})s^x} \right) \qquad 32.33$$

The term s^y represents zeros at the origin, with y equal to the number of zeros. Zeros at the origin of the s domain have a normalizing factor of one. The term s^x represents poles at the origin, with x equal to the number of poles. Poles at the origin of the s domain also have a normalizing factor of one. Changing the form to that useful for Bode plotting,

$$G = 20 \log |T(s)|$$
$$= 20 \log |T(j\omega)|$$
$$= 20 \log \left(A \left(\frac{\sigma_{z1}\sigma_{z2} \cdots \sigma_{zn}}{\sigma_{p1}\sigma_{p2} \cdots \sigma_{pd}} \right) \right)$$
$$+ 20 \log \left| \left(1 + \frac{j\omega}{\sigma_{z1}} \right) \right| + 20 \log \left| \left(1 + \frac{j\omega}{\sigma_{z2}} \right) \right| + \cdots$$
$$+ 20 \log \left| \left(1 + \frac{j\omega}{\sigma_{zn}} \right) \right| + y 20 \log |j\omega|$$
$$- 20 \log \left| \left(1 + \frac{j\omega}{\sigma_{p1}} \right) \right| - 20 \log \left| \left(1 + \frac{j\omega}{\sigma_{p2}} \right) \right| - \cdots$$
$$- 20 \log \left| \left(1 + \frac{j\omega}{\sigma_{pd}} \right) \right| - x 20 \log |j\omega| \qquad 32.34$$

The normalization used in Eq. 32.34 results in each of the terms being plotted from the break frequency on the $G = 0$ dB line. The break frequency (normalizing factor) for the single zeros and poles at the origin is $\omega = 1$. (The zeros and poles plot as straight lines and do not actually "break" up or down at the origin. The zero or pole at zero frequency does not get plotted on a Bode plot, because the value is $-\infty$ and $+\infty$.)

The entire plotting process can be reversed. For example, if the output of a circuit is determined experimentally, the resulting Bode plot can be estimated and the transfer function derived from the plot. Then, using the value of the asymptotic plot at some point, the value of A in Eq. 32.33 can be determined.

Example 32.6

The figure shows a certain circuit's output, measured and plotted as a dashed line on a Bode plot. An asymptotic response is estimated and plotted as a solid line. Determine the transfer function.

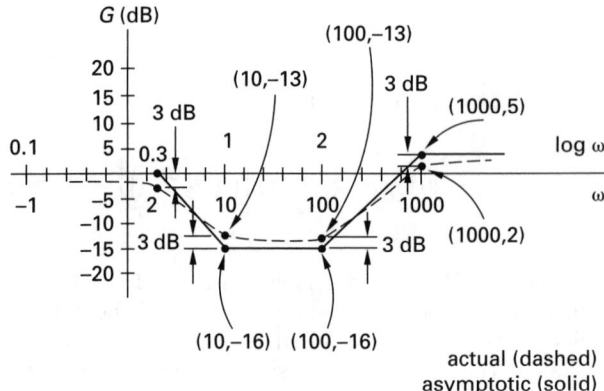

actual (dashed)
asymptotic (solid)

Solution

The transfer function must be in the form of Eq. 32.33. From the plot, the zeros and poles are the points where $G = 0$ dB on the frequency axis where the breaks occur. Consequently, the breaks upward at $\omega = 10$ and $\omega = 100$ indicate zeros. The breaks downward at $\omega = 2$ and $\omega = 1000$ indicate poles. The transfer function is

$$T(s) = A \left(\frac{(s + 10)(s + 100)}{(s + 2)(s + 1000)} \right)$$

Determine A from the value of the asymptotic plot when $s = 0$. At $s = 0$, the plot indicates a magnitude of 0 dB. Thus,

$$G = 20 \log |T(0)| = 0 \text{ dB}$$
$$|T(0)| = \text{antilog}(0) = 1$$

Substituting $T(0) = 1$ and $s = 0$ gives

$$T(s) = A \left(\frac{(s + 10)(s + 100)}{(s + 2)(s + 1000)} \right)$$
$$1 = A \left(\frac{(10)(100)}{(2)(1000)} \right)$$
$$A = 2$$

The final solution is

$$T(s) = (2) \left(\frac{(s + 10)(s + 100)}{(s + 2)(s + 1000)} \right)$$

7. BODE PLOT PRINCIPLES: PHASE PLOT

The phase of a zero at the origin, such as $s + 0$ in the numerator of Eq. 32.33, provides an angle of $+90°$. The phase of a pole at the origin, such as $s + 0$ in the denominator of Eq. 32.33, provides an angle of $-90°$. For zeros or poles on the negative real axis of the s domain, with $s = j\omega$, the angles for each term in Eq. 32.33 are determined from Eqs. 32.35 and 32.36, respectively.

$$\theta_z = \tan^{-1} \left(\frac{\omega}{\sigma_{zn}} \right) \quad \text{[radians]} \qquad 32.35$$

$$\theta_p = -\tan^{-1} \left(\frac{\omega}{\sigma_{pd}} \right) \quad \text{[radians]} \qquad 32.36$$

At the break frequency, $\omega = \sigma$ for either zeros or poles, and the phase angle is $+45°$ for zeros and $-45°$ for poles. Equations 32.35 and 32.36 could be used directly to plot the phase angle as a function of frequency. An *idealized Bode phase plot* similar to that used for magnitude can be plotted with approximations. Consider Eqs. 32.35 and 32.36 at a frequency $\omega = 0.1\sigma$. The absolute value of the zero or pole angle at this frequency is

$$|\theta| = \left|\pm \tan^{-1}\left(\frac{0.1\sigma}{\sigma}\right)\right| = 0.09 \text{ rad}$$

$$= 5° \approx 0 \qquad\qquad 32.37$$

This is analogous to the low-frequency asymptote for the magnitude of Eq. 32.28. Consider a frequency one decade higher than the break frequency, $\omega = 10\sigma$, for either a zero or a pole. The absolute value of the zero or pole angle at this frequency is

$$|\theta| = \left|\pm \tan^{-1}\left(\frac{10\sigma}{\sigma}\right)\right|$$

$$= 1.47 \text{ rad} = 84.3° \approx 90° \qquad 32.38$$

This is analogous to the high-frequency asymptote for the magnitude of Eq. 32.29. At the break frequency, $\omega = \sigma$ and the angle is $45°$ for either a zero or a pole. The result is that, from one decade below the break frequency to one decade above, the phase angle for a zero or pole changes from an absolute value of $0°$ to $90°$ at a rate of $45°$ per decade. This approximation is illustrated in Fig. 32.9 for an arbitrary break frequency, that is, arbitrary zero or pole. The approximations used in the idealized phase plot are accurate to within $6°$. The phase angles for the entire transfer function must be added as in Eq. 32.5 to obtain the resultant angle. This is done on the Bode plot in a manner analogous to adding the magnitudes.

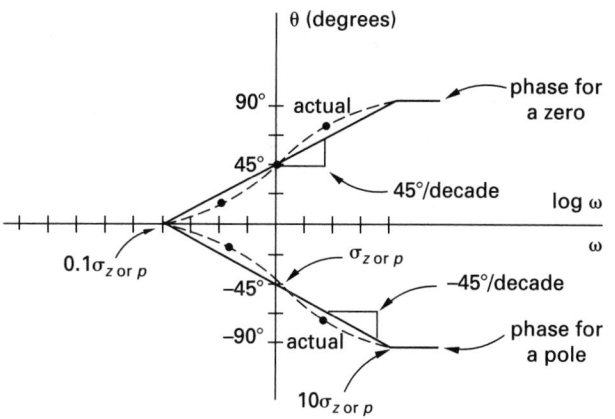

Figure 32.9 *Bode Phase Plot for a Zero or Pole*

8. BODE PLOT METHODS

The general form of the transfer function of a circuit is[19]

$$T(s) = A\left(\frac{\begin{array}{c}(s+z_1)(s+z_2)\\ \cdots \times (s+z_n)(s)^y\end{array}}{\begin{array}{c}(s+p_1)(s+p_2)\\ \cdots \times (s+p_d)(s)^x\end{array}}\right) \qquad 32.39$$

The break frequencies in Eq. 32.39 are the zeros and the poles.[20] The break frequency, better termed the *normalizing factor*, for the zeros at the origin (s^y) and poles at the origin (s^x) is $\omega = 1$. As the sinusoidal response is of interest, $s = j\omega$. Using the normalized notation changes Eq. 32.39 to Eq. 32.40.

$$T(j\omega) = A\left(\frac{z_1 z_2 \cdots z_n}{p_1 p_2 \cdots p_d}\right)$$

$$\times \left(\frac{\begin{array}{c}\left(1+\dfrac{j\omega}{z_1}\right)\left(1+\dfrac{j\omega}{z_2}\right)\\ \cdots \times \left(1+\dfrac{j\omega}{z_n}\right)(j\omega)^y\end{array}}{\begin{array}{c}\left(1+\dfrac{j\omega}{p_1}\right)\left(1+\dfrac{j\omega}{p_2}\right)\\ \cdots \times \left(1+\dfrac{j\omega}{p_d}\right)(j\omega)^x\end{array}}\right) \qquad 32.40$$

Combining the constant term A with the values of the zeros and poles gives a new constant, K.[21]

$$T(j\omega) = K\left(\frac{\begin{array}{c}\left(1+\dfrac{j\omega}{z_1}\right)\left(1+\dfrac{j\omega}{z_2}\right)\\ \cdots \times \left(1+\dfrac{j\omega}{z_n}\right)(j\omega)^y\end{array}}{\begin{array}{c}\left(1+\dfrac{j\omega}{p_1}\right)\left(1+\dfrac{j\omega}{p_2}\right)\\ \cdots \times \left(1+\dfrac{j\omega}{p_d}\right)(j\omega)^x\end{array}}\right) \qquad 32.41$$

The constant K is

$$K = A\frac{\displaystyle\prod_1^n z_n}{\displaystyle\prod_1^d p_d} \qquad 32.42$$

[19]Transfer functions are often called the *gain* of a circuit. This chapter uses gain to indicate $20\log|T(s)|$.
[20]Using the zero and pole notation removes the restriction of the break frequency occurring only on the negative real axis of the s domain. The zeros and poles are located at $-z$ and $-p$, respectively. This is the reason the positive sign is used.
[21]The constant K is a function of $j\omega$ because of the zeros and poles included.

The gain as used in Bode plots is derived from Eq. 32.41 to be

$$G = 20 \log K + \sum_{n=1}^{n} 20 \log \left| \left(1 + \frac{j\omega}{z_n} \right) \right| + 20y \log \omega$$

$$- \sum_{d=1}^{d} 20 \log \left| \left(1 + \frac{j\omega}{p_d} \right) \right| - 20x \log \omega \qquad \textbf{32.43}$$

The angle of the gain, θ_G, in degrees is

$$\theta_G = \sum_{n=1}^{n} \tan^{-1} \left(\frac{\omega}{z_n} \right) + 90y$$

$$- \sum_{d=1}^{d} \tan^{-1} \left(\frac{\omega}{p_d} \right) - 90x \qquad \textbf{32.44}$$

The overall method for determining the frequency response of a circuit utilizing Bode plots follows.

step 1: Determine the transfer function for the circuit of interest and arrange it in the form of Eq. 32.39 or 32.41. Note the break frequencies for each term, that is, the zeros and the poles. Recall that the break frequency for a zero or pole at the origin of the s domain is $\omega = 1$.

step 2: Plot the break frequency points on the axis of a log frequency scale. That is, plot the points where the magnitude, in decibels, of the transfer function is zero.

step 3: For each zero at the s domain origin (s^y terms), plot a straight line with a slope of $+20$ dB per decade through the origin of the Bode plot axis, that is, through $\omega = 1$ ($\log \omega = 0$) and $20 \log |T(j\omega)| = 0$.

step 4: For each pole at the s domain origin (s^x terms), plot a straight line with a slope of -20 dB per decade through the origin of the Bode plot axis, that is, through $\omega = 1$ ($\log \omega = 0$) and $20 \log |T(j\omega)| = 0$.

step 5: For all other zeros, draw two asymptotic straight lines from the associated break frequency. The first is of zero slope and extends to lower frequencies. The other has a slope of $+20$ dB per decade (or 6 dB per octave) and extends to higher frequencies.

step 6: For all other poles, draw two asymptotic straight lines from the associated break frequency. The first is of zero slope and extends to lower frequencies. The other has a slope of -20 dB per decade (or -6 dB per octave) and extends to higher frequencies.

step 7: Sum the asymptotic lines drawn. The result is the idealized Bode plot for the transfer function.

step 8: If greater accuracy is desired, at each break frequency plot a point displaced 3 dB from the intersection of the asymptotic lines. The point is $+3$ dB from lines that break upward and -3 dB from lines that break downward. At differences of plus or minus one octave from the break frequency, plot a point $+1$ dB from lines that break upward and -1 dB from lines that break downward. Additional points may be plotted using Eq. 32.31. Join the points in a smooth curve with the ends asymptotic to the idealized Bode plot lines. This smooth curve is the actual plot.

step 9: Determine the value of K from the value of the asymptotic or actual curve at some frequency or by using Eq. 32.42.[22] Using this value of $20 \log K$, either shift the plot upward or downward by the amount calculated, or change the 0 dB point on the axis to the value of $20 \log K$. In either case, the shape of the Bode plot remains unchanged.

The overall method for determining the phase response of a circuit utilizing Bode plots follows.

step 1: Determine the transfer function for the circuit of interest and arrange it in the form of Eq. 32.39 or 32.41. Note the break frequencies for each term, that is, the zeros and the poles. Recall that the break frequency for a zero or pole at the origin of the s domain is $\omega = 1$.

step 2: Plot the break frequency points on the axis of a log frequency scale. That is, plot the points where the angle $\theta = 0°$.

step 3: For each zero at the s domain origin (s^y terms), plot a constant angle of $+90°$.

step 4: For each pole at the s domain origin (s^x terms), plot a constant angle of $-90°$.

step 5: For all other zeros, plot a point at the break frequency corresponding to $+45°$. At one decade below the break frequency, plot a point corresponding to $0°$. At one decade above the break frequency, plot a point corresponding to $+90°$. Connect the points. The slope will be $+45°$ per decade.

step 6: For all other poles, plot a point at the break frequency corresponding to $-45°$. At one decade below the break frequency, plot a point corresponding to $0°$. At one decade above the break frequency, plot a point corresponding to $-90°$. Connect the points. The slope will be $-45°$ per decade.

step 7: Sum the asymptotic lines drawn. The result is an idealized Bode phase plot.

----

[22]A pole or zero at the origin in the s domain has a pole or zero value in Eq. 32.42 of 1.

step 8: If greater accuracy is required, additional points may be drawn using Eq. 32.35 for zeros and Eq. 32.36 for poles. Connecting the points with a smooth curve results in the actual plot.

Example 32.7

A certain circuit has the following transfer function.

$$T(s) = \frac{10}{s(s+10)}$$

Draw the associated Bode plot.

Solution

Unless otherwise specified, an idealized Bode plot is assumed. Following the steps in Sec. 32-8 results in the following plot.

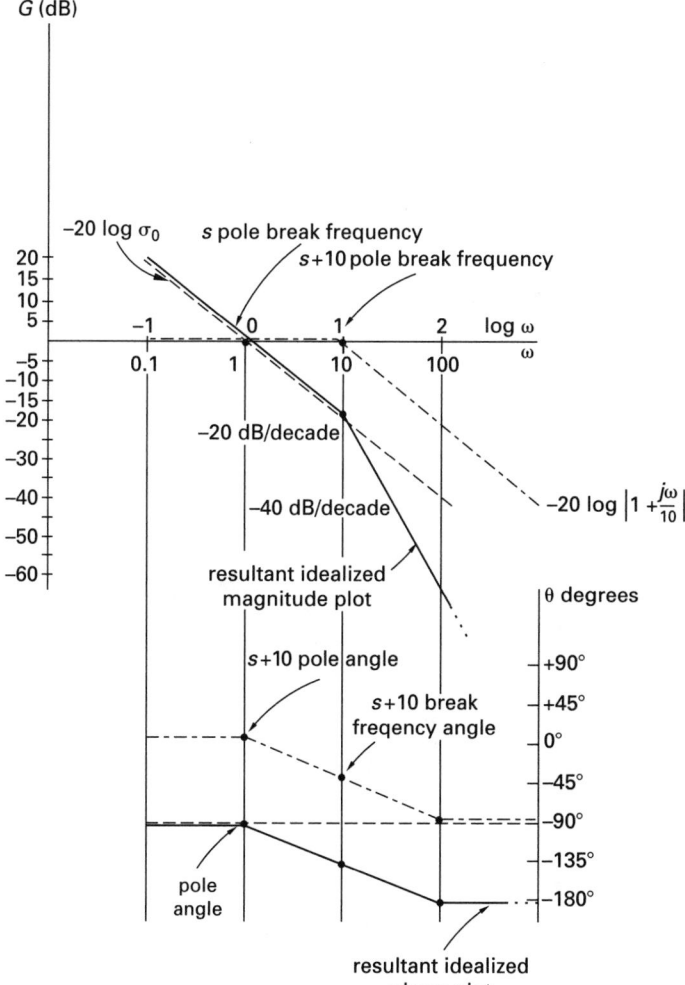

Circuit Theory

Topic V: Power—Generation

Chapter

Power–
Generation

For the most current information about the exam, visit **www.ppi2pass.com** regularly.

33

Generation Systems

Nomenclature

B	magnetic flux density	T
c	speed of light	m/s
E	energy	J
f	frequency	Hz, s^{-1}, or cycles/s
f_{droop}	frequency droop	Hz/kW
h	enthalpy	kJ/kg
I	effective or DC current	A
m	mass	kg
N	synchronous speed	r/min or min^{-1}
P	poles	–
Q	heat	J
s	specific entropy	kJ/kg·K
T	temperature	°C or K
V_{droop}	voltage droop	V/kVAR
W	work	kJ

Symbols

ω	armature angular speed	rad/s

Subscripts

nl	no load
s	synchronous
sys	system

1. FOSSIL FUEL PLANTS

The electric utility industry is one of the largest users of fossil fuels. Fossil fuels are those composed of hydrocarbons, such as petroleum, coal, and natural gas. Typical combustion of the fuel produces high-pressure, high-temperature steam.[1] The pressures range from 16.5–24 MPa (2400–3500 psig). A common steam temperature is 540°C (1000°F). The basic thermodynamic cycle is the Rankine cycle shown in Fig. 33.1, along with typical plant components. Turbines used to drive the electrical generators spin at 3600 rpm. The efficiency

---

[1]Steam is the most common substance used, but the principles discussed are applicable to vapor cycles in general.

of the cycle is proportional to the temperature difference between the temperature at the *source*, that is, the steam generator, and the *sink*, that is, the cooling medium of the condenser.

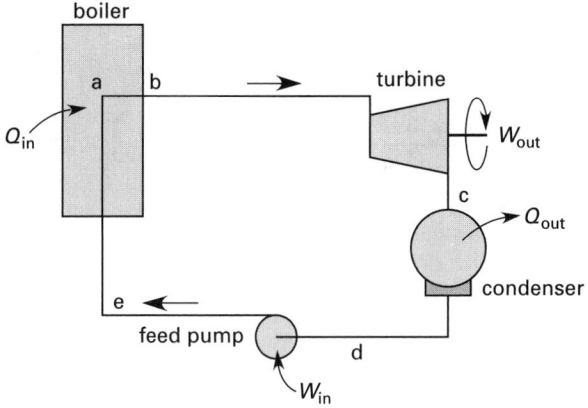

(a) generation components

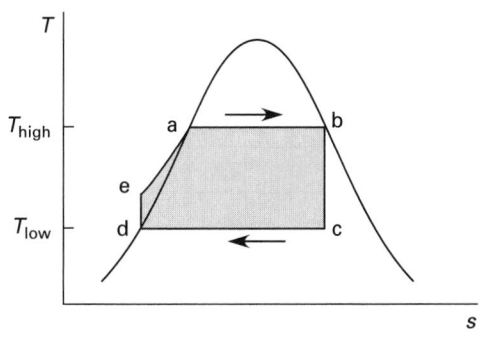

(b) thermodynamic properties

Figure 33.1 *Basic Rankine Cycle*

Efficiency improves when the source temperature is raised. Additionally, turbine blade wear is reduced as the moisture content in the expanding steam is reduced. Both are accomplished by adding heat to the steam beyond that required to maintain the steam in the vapor phase, that is, raising the steam temperature above the saturation temperature for the associated pressure, a concept known as *superheat*. The maximum practical metallurgical limit on superheat is approximately 625°C (1150°F). In some situations, moisture still forms in the low-pressure turbine stages. This problem is overcome by reheating the steam after partial isentropic expansion in the turbine, then allowing the remaining expansion to take place, a concept known as *reheat*. A generation plant operating with these modifications is said to use the *reheat cycle*. The steam flow through the plant and the resulting thermodynamic cycle are shown in Fig. 33.2.

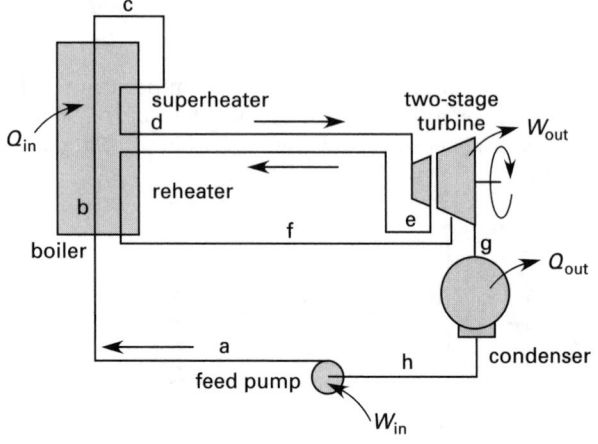

(a) generation components

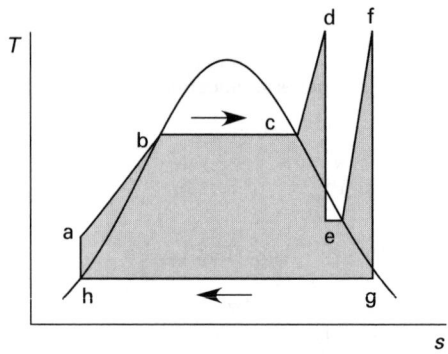

(b) thermodynamic properties

Figure 33.2 Basic Reheat Cycle

Efficiency can be further increased by minimizing the irreversibilities in the Rankine cycle associated with heating the compressed water to saturation by a finite temperature difference occurring between points e and a in Fig. 33.1 and points a and b in Fig. 33.2. This is accomplished by using heat sources elsewhere in the cycle that have temperatures slightly above that of the compressed liquid. The process is known as *regeneration* and is illustrated in Fig. 33.3. Superheat, reheat, and regeneration are all used to raise the mean effective temperature at which heat is added. Superheat raises the temperature prior to turbine input, reheat raises the temperature prior to low-pressure turbine input, and regeneration raises the temperature of the compressed liquid returning to the boiler, thus minimizing the heat addition required in the boiler.

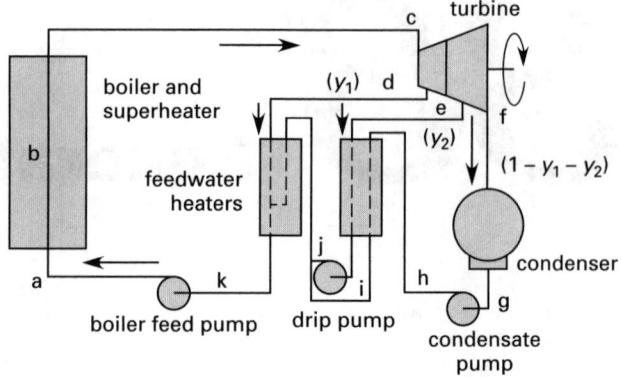

(a) generation components

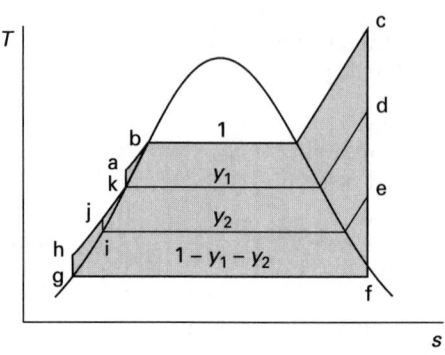

(b) thermodynamic properties

Figure 33.3 Basic Regenerative Cycle

Fossil fuels are broadly categorized into solid, liquid, and gaseous types. They may be further classified as natural, manufactured, or by-product. Coal is arguably the most important fossil fuel, since it is widely used and has well-known reserves. The most commonly used types of coal are anthracite, bituminous, subbituminous, and lignite. Environmental concerns involving the burning of coal include the emission of nitrogen oxides, sulfur oxides, particulate matter, and ash. The nitrogen oxides are minimized by control of the fuel-air mixture during combustion and postcombustion reduction using chemical reagents. Sulfur oxides are minimized using a variety of flue-gas desulfurization (FGD) systems. Particulate matter is minimized using electrostatic precipitators. Ash is controlled by a variety of means including coal selection, combustion techniques, and various filtration means.

Residual fuel oil is used as the source fuel for combustion in electric generating plants. Residual oil is what remains after lighter hydrocarbons, such as gasoline, have been removed from crude oil. The plant operating principles are similar to those for coal, with the significant difference being the design of the boiler. Environmental concerns include gaseous emissions, removal of oil from environmentally sensitive areas, and oil transportation safety.

Natural gas is highly valuable for various chemical and space-heating uses. Consequently, its use in large-scale electric power plants is minimal.

2. NUCLEAR POWER PLANTS

The energy process in nuclear power plants is fundamentally different from that of fossil fuel plants. While fossil fuels are burned or undergo combustion, nuclear fuels, normally uranium, change form and in the process release energy. The amount of energy released is given by Einstein's equation.

$$E = mc^2 \hspace{4em} 33.1$$

During *fission* a uranium atom is split by the absorption of a neutron into two or more elements of lower atomic mass, plus additional neutrons. When the total combined mass of the products is compared to the mass of the reactants, a *mass defect* of approximately 0.215 amu is noted. Substituting this mass defect into Eq. 33.1 results in an energy of approximately 200 MeV/fission (3.2×10^{-11} W·s/fission). *Fusion* combines lighter elements to create an element of higher mass, but it also results in a mass defect and subsequent release of energy. All current commercial electric generation plants are based on the fission process.

Nuclear power plants operate on the same thermodynamic cycles as fossil fuel plants. *Pressurized water reactors* (PWR) are constructed of two loops, a primary and a secondary, as shown in Fig. 33.4. The primary system operates in the range of 13.8–17.3 MPa (2000–2500 psia). The temperature of the primary, at approximately 300°C (572°F), is well below the saturation temperature for this pressure range, to prevent boiling and other adverse effects. As a result, the steam side operates at lower temperatures and pressures than in a fossil fuel plant—approximately 8.4 MPa (1230 psia), correlating with the 300°C (572°F) operating temperature on the primary side.

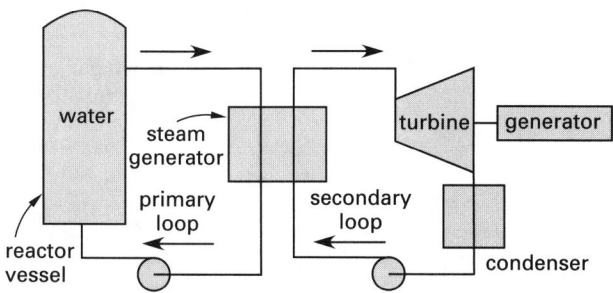

Figure 33.4 *Pressurized Water Reactor*

A *boiling water reactor* (BWR) generates steam directly in the core. Such a reactor operates at lower pressures than the typical PWR, usually 6.9 MPa (1000 psia). The steam is passed through various separators and dryers to minimize the moisture content prior to entering the turbine.

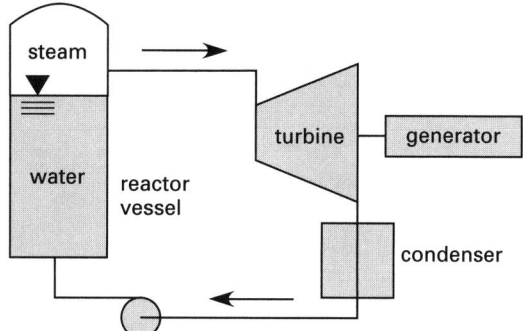

Figure 33.5 *Boiling Water Reactor*

Additional fundamental differences in nuclear power plants compared to fossil fuel plants include a significantly higher thermal energy density (0.80 kW/l versus 0.20 kW/l), heat generation after shutdown due to radioactive decay, and minimal environmental emissions. Fossil fuel plants emit carbon dioxide (CO_2), nitrogen dioxide (NO_2), sulfur trioxide (SO_3), particulate matter and, in the case of coal, ash. Nuclear power plants emit small quantities of gaseous products and, when no longer capable of generating power, solid radioactive waste products from the fuel.

3. HYDROELECTRIC POWER

The flow of water from higher elevations to lower elevations changes potential energy to kinetic energy that can be used to drive electric generators directly without the depletion of any fuel. While this eliminates the plant components that deliver heated steam to the turbine and condensed water back to the generator, it does require a source of water large enough to generate a steady output and the ability to control that source. Environmental considerations include the multiple effects of creating large reservoirs of water. Additionally, the multiple uses of the water—for navigation, irrigation, recreation, and flood control—compete with the production of power and significantly impact costs.

4. COGENERATION PLANTS

Cogeneration is the simultaneous on-site generation of electric energy and process steam or heat from the same plant. All power cycles discard a large portion of incoming energy as heat energy. Typical plant efficiencies are on the order of 30%. The other 70% of the energy is rejected to the environment as heat. If this heat is recovered and used for *space heating*, also called *district heating*, or cooling (using an *absorption system*), the process is called a *cogeneration cycle*. If the recovered heat instead vaporizes water in a steam power cycle, the process is called a *combined cycle*. In cogeneration, the recovered heat is used as heat and not converted into mechanical or electrical energy. Two terms are used to measure this recovered energy, neither of which is the same as thermal efficiency. The *fuel utilization* given in Eq. 33.2 is the ratio of useful energy to energy input. The *power-to-heat ratio* relates turbine work to the energy recovered, Eq. 33.3.

$$\text{fuel utilization} = \frac{W_{\text{turbine}} + Q_{\text{recovered}}}{Q_{\text{in}}} \quad \textit{33.2}$$

$$\text{power-to-heat ratio} = \frac{W_{\text{turbine}}}{Q_{\text{recovered}}} \quad \textit{33.3}$$

5. PRIME MOVERS

Steam prime movers are of two types, reciprocating and turbine. *Reciprocating prime movers* are used primarily in low-speed (100–400 rpm), high-efficiency applications requiring high starting torque. As such, they are used for auxiliary applications throughout a power plant. *Steam turbines* dominate as the mover of choice for electric power generation. They are variable-speed (1800–25,000 rpm), relatively highly efficient (85% in

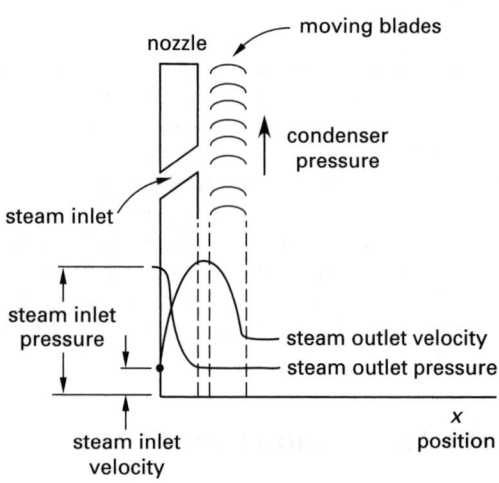

(a) simple impulse turbine

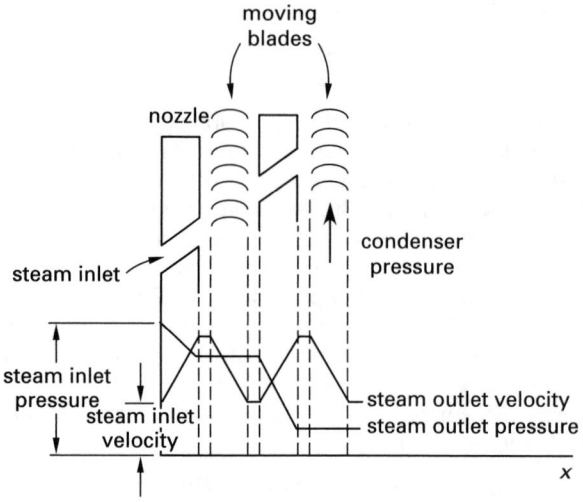

(b) Rateau impulse (pressure-staged impulse)

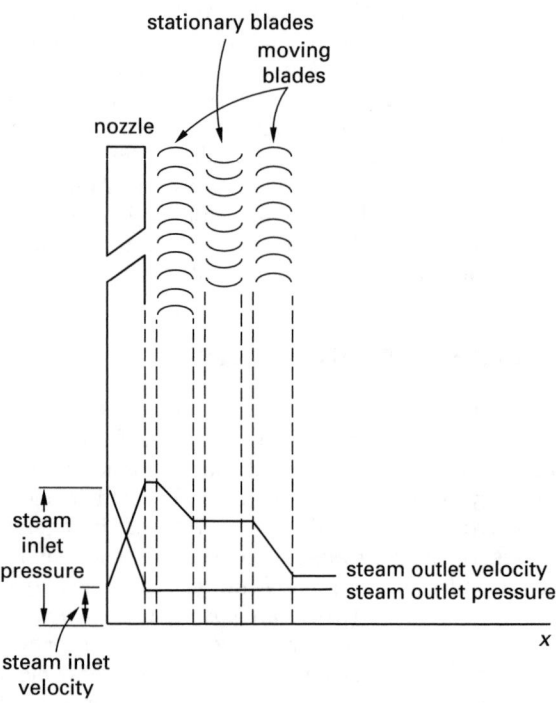

(c) Curtis impulse turbine (velocity-staged impulse turbine)

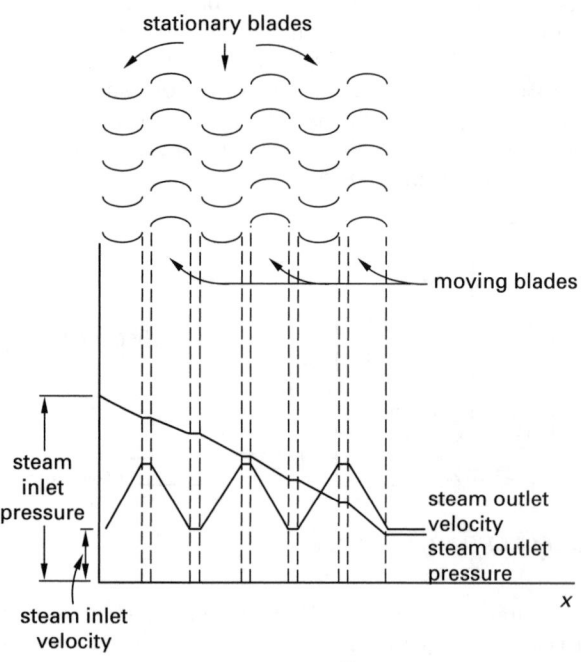

(d) Parsons turbine (reaction turbine)

Figure 33.6 *Steam Turbine Types*

large machines), require no internal lubrication, and operate at steam pressures ranging from 34.5 MPa (5000 psig) and 565°C (1050°F) to 1.7 MPa (0.5 in Hg absolute). Most important for electric generation, they can be built in capacities over 1,000,000 kW—more than any other prime mover.

Steam turbines are a series of nozzles in which heat energy is changed to kinetic energy that is then transferred to a rotating wheel or drum and ultimately to an output shaft. *Impulse turbines* use stationary nozzles that drop the pressure with the kinetic energy absorbed in rotating blades operating at approximately constant pressure. *Reaction turbines* drop pressure in both the stationary and rotating portions. The steam turbine types and their associated pressure-velocity diagrams are shown in Fig. 33.6.

All large turbines use multiple stages to improve efficiency. Reaction turbines generally have more stages than impulse designs. A large turbine's first stage is an impulse stage since no pressure drop occurs across the moving blades. This allows for partial-arc admission of the steam. To maximize efficiency, the simple impulse turbine (single row) is the first stage, also called the *control stage*, on heavily loaded reheat turbines. The two-row Rateau impulse turbine is the control stage on small and medium-sized nonreheat turbines, in order to maintain efficiency over a wider range of operating loads. Marine turbines and mechanical drives operating at high revolutions per minute use single-row impulse first stages in the forward directions but use a single-stage Curtis in the reverse direction to provide the required torque while minimizing windage loss in the forward direction. The Curtis stage absorbs approximately four times the energy of an impulse stage and eight times the energy of a reaction stage.

Losses in a steam turbine include *clearance leakage*, *nozzle leakage*, *rotation* or *windage*, *carryover*, *leaving*, *partial arc*, and *supersaturation* and *moisture loss*. The overall efficiency of a turbine is given by[2]

$$\eta_{\text{turbine}} = \frac{W_{\text{real}}}{W_{\text{ideal}}} = \frac{h_{\text{in}} - h_{\text{condenser}}}{h_{\text{in}} - h_{\text{out,ideal}}} \qquad \textbf{\textit{33.4}}$$

6. ALTERNATING CURRENT GENERATORS

A conductor moving relative to a magnetic field experiences an induced electromotive force or voltage in accordance with Faraday's law. This is called *generator action*. The conductor in which the electromotive force

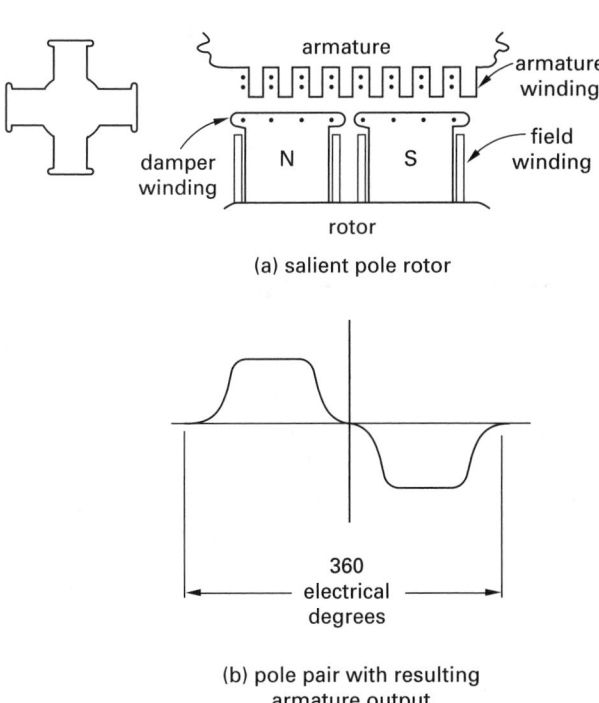

(a) salient pole rotor

(b) pole pair with resulting armature output

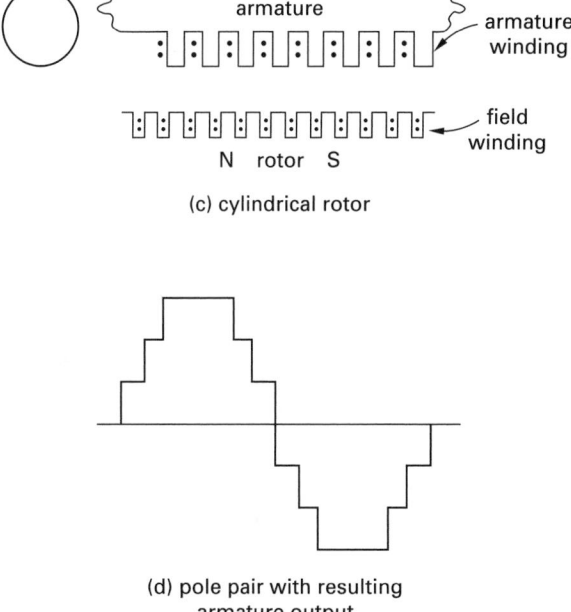

(c) cylindrical rotor

(d) pole pair with resulting armature output

Figure 33.7 *Rotor Construction and AC Generator Output*

---

[2]The ideal outlet enthalpy is determined assuming a constant entropy from the inlet pressure of the turbine to the condenser pressure, that is, a straight line on the Mollier diagram.

is induced is called the *armature*. For alternating current generators, the armature is physically located on the *stator*, that is, the stationary portion of the generator. The *field* produces the magnetic flux that reacts with the armature. For alternating current generators, the field is located on the *rotor*, that is, the rotating portion of the generator, and is supplied by a direct current that maintains the electromagnetic pole strength, and thus the output voltage, at the desired value. This arrangement—field on the rotor, armature on the stator—is used primarily because the field current is smaller, easing design requirements for the electric connection, which is commonly made through slip rings and brushes or via brushless exciters. This allows the high-current armature output connections to be made on the stationary portion of the generator. The rotors are either *salient pole* or *cylindrical* as shown in Fig. 33.7. Salient pole rotors are utilized in low rpm applications (approximately 300 rpm) due to mechanical restrictions. The generation of an AC output voltage using a permanent magnet field is shown in Fig. 33.8.

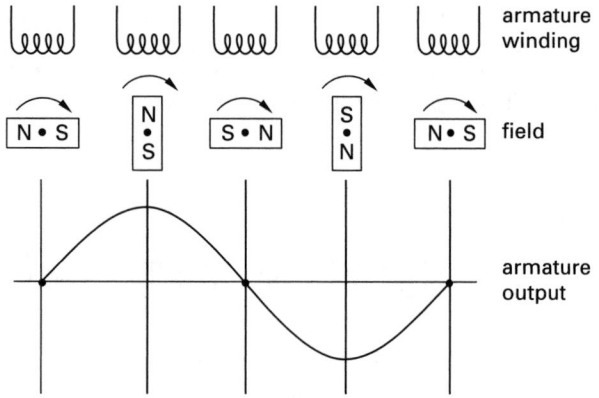

Figure 33.8 *AC Output Generation*

The speed of rotation of the magnetic field in a synchronous machine is called the *synchronous speed*. Since two poles must pass a given point on the armature in order to complete one cycle (360 electrical degrees), the synchronous speed is given by

$$N_s = \frac{120f}{P} \qquad 33.5$$

The term N is the synchronous speed in revolutions per minute, f is frequency in Hertz, and P is the number of poles.[3] The mechanical distance around the periphery of an electrical machine is often measured in electrical degrees, with 360° the distance between a pole pair.

---

[3]The magnetic field is moving on the armature but the armature itself is stationary. Also, a pole is either a single north or a single south. This formula is sometimes written for pole pairs, in which case the 120 would have to be 60.

Since one cycle of a sine wave is 360°, the mechanical and electrical degrees are related by Eq. 33.6.

$$\text{mechanical degrees} = \frac{\text{electrical degrees}}{\text{number of pole pairs}} \qquad 33.6$$

Losses in an AC electrical generator include *windage* and *friction*, *core*, *armature copper*, *field copper*, and *stray* or *load loss*.

Example 33.1

At what speed does the rotor of a four-pole AC generator turn?

Solution

The rotor in an AC generator contains the field, which rotates at synchronous speed. Since the frequency was not stated, assume 60 Hz. Substituting into Eq. 33.5 gives

$$N_s = \frac{120f}{P} = \frac{(120)(60 \text{ Hz})}{4} = 1800 \text{ rpm}$$

Example 33.2

What is the mechanical spacing, in degrees, between the poles of a 12-pole AC generator?

Solution

The mechanical spacing is determined by

$$\text{mechanical degrees} = \frac{\text{electrical degrees}}{\text{number of pole pairs}}$$
$$= \frac{360°}{6} = 60°$$

7. PARALLEL OPERATION

Parallel operation of generators occurs for myriad reasons, including shifting of the load to shut down a generator for maintenance, to increase the capacity available, and so on. Prior to paralleling two synchronous generators, the following conditions must be met.

- The phase sequence must be identical. The *phase sequence* is the order in which the phase voltages successively reach their maximum positive values. The phase sequence, normally a-b-c, is determined during construction and does not routinely require verification.

- Frequency must be matched. That is, the rotation speeds are matched. This minimizes current surges due to the shifting of real load when the oncoming generator is connected.

- Voltages must be matched. This minimizes current surges due to the shifting of reactive load when the oncoming generator is connected.

The division of real load among generators operating in parallel is determined by the speed of the generators and the characteristics of the prime mover speed-governing system as shown in Fig. 33.9. The slope of the *speed characteristic line*, also called the *generator* or *frequency droop* (f_{droop}), is determined by the speed-governing system and is constant. The total load is also constant, assuming no loads are started or secured. If the generator's speed is manually changed, that is, if the no-load frequency is adjusted, the associated characteristic line moves up or down and the load is shifted to or from the generator. The resulting frequency of the system (f_{sys}) would thus rise or fall as well. If a load is added to or removed from the system without a change in the generator's no-load frequency, the additional load results in a drop in the system frequency determined by Eq. 33.7.

$$P = \frac{f_{\text{sys}} - f_{\text{nl}}}{f_{\text{droop}}} \qquad 33.7$$

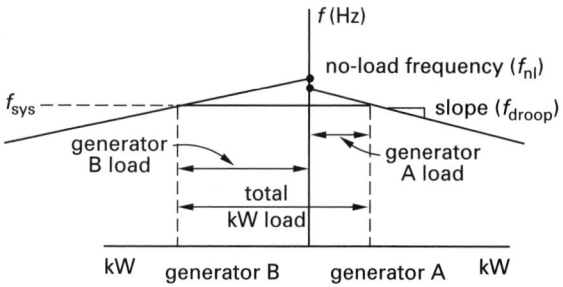

Figure 33.9 *Real Load Sharing*

P is the total real power carried by the generator, f_{sys} is the operating frequency of the system, and f_{nl} is the no-load frequency. The frequency droop, f_{droop}, is the slope of the speed characteristic line in units of Hz/kW. Although negative, the frequency droop is often stated as a positive value. If the droop is given or used as a positive value, Eq. 33.7 becomes

$$P = \frac{f_{\text{nl}} - f_{\text{sys}}}{f_{\text{droop}}} \qquad 33.8$$

The division of reactive load takes place in the same manner, but is controlled by the voltage regulators. Reactive load sharing is illustrated in Fig. 33.10. The reactive power carried by a generator is determined by Eq. 33.9.

$$Q = \frac{V_{\text{sys}} - V_{\text{nl}}}{V_{\text{droop}}} \qquad 33.9$$

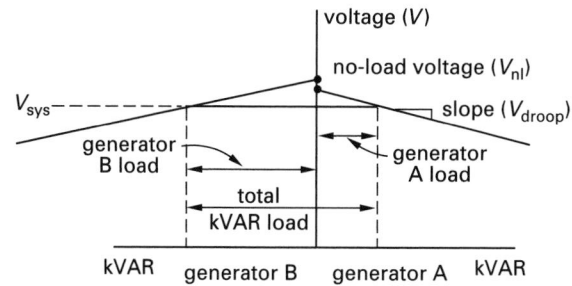

Figure 33.10 *Reactive Load Sharing*

If the droop is given or used as a positive value, Eq. 33.9 becomes

$$Q = \frac{V_{\text{nl}} - V_{\text{sys}}}{V_{\text{droop}}} \qquad 33.10$$

The slopes of the speed and voltage characteristic lines in Figs. 33.9 and 33.10 are normally given in percent change of rated frequency or voltage from a no-load to a full-load condition. These are called the *speed regulation* and *voltage regulation* of a generator, respectively.

Example 33.3

A 1000 MW generator has a speed droop of 1%. Rated frequency is 60 Hz. If the generator carries 500 MW at 60 Hz, what is the no-load frequency setpoint of the speed-governing system?

Solution

A speed droop of 1% indicates a change in frequency of 0.01×60 Hz from 0 kW to 1000 MW, that is, from no-load to full load. Thus, the slope of the speed characteristic line, or curve, is 0.6 Hz/1000 MW. As the droop is stated as a positive value, Eq. 33.8 is used.

$$P = \frac{f_{\text{nl}} - f_{\text{sys}}}{f_{\text{droop}}}$$

$$f_{\text{nl}} = P f_{\text{droop}} + f_{\text{sys}}$$

$$= (500 \text{ MW}) \left(\frac{0.6 \text{ Hz}}{1000 \text{ MW}} \right) + 60 \text{ Hz}$$

$$= 60.3 \text{ Hz}$$

8. DIRECT CURRENT GENERATORS

Direct current generators operate under the same principles as AC current generators, except the output is manipulated to provide the desired DC. The definitions of armature and stator remain the same as for AC generators, but their locations are reversed. The field of a DC generator is located on the stator, or stationary portion of the generator. The armature is located on the rotor, or rotating portion of the generator. This arrangement is used primarily because of the need to change the armature output to DC, which is accomplished by the *commutator*. *Commutation* is the process of current reversal in the armature windings that provides direct current to the brushes. Simplified commutation is shown in Fig. 33.11. The physical location of the windings and the electrical connections for a four-coil two-pole generator is shown in Fig. 33.12. The rotors are normally cylindrical. The DC output voltage is a rectified version of that shown in Fig. 33.8.

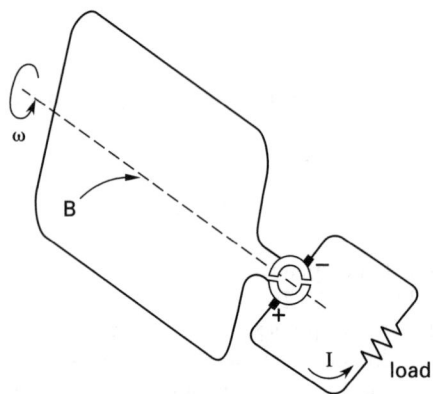

Figure 33.11 *Commutator Action*

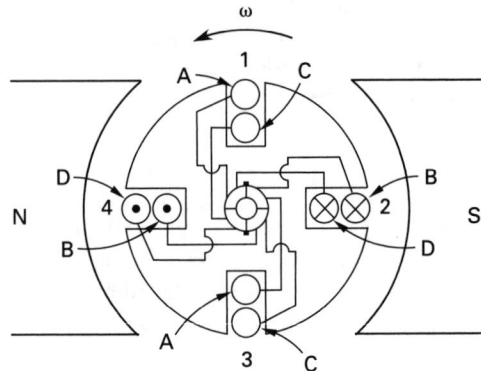

(a) four-coil winding: physical diagram

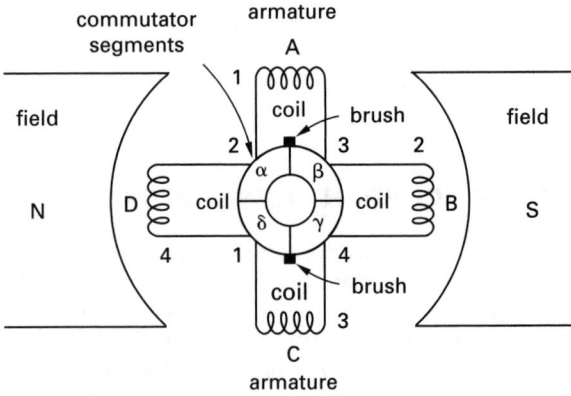

(b) four-coil winding: schematic diagram

Figure 33.12 *DC Machine Windings*

Armature reaction is the interaction between the magnetic flux produced by the armature current and the magnetic flux produced by the field current. Armature reaction occurs in both AC and DC generators, lowering the output voltage. The reaction manifests itself by shifting the *neutral plane* as shown in Fig. 33.13.[4]

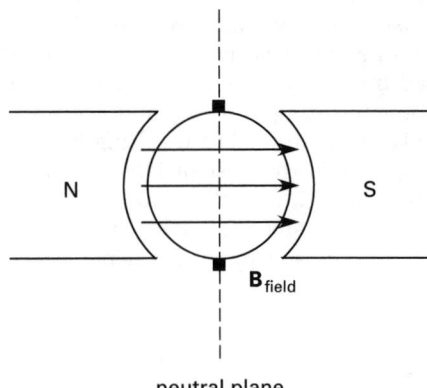

(a) stator magnetic field

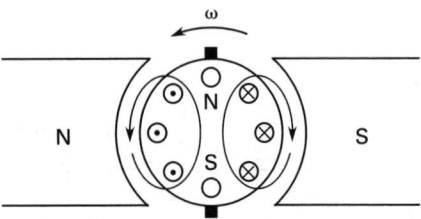

(b) armature magnetic field

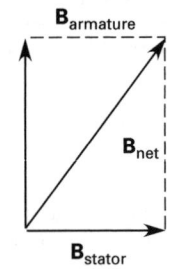

(c) net magnetic field

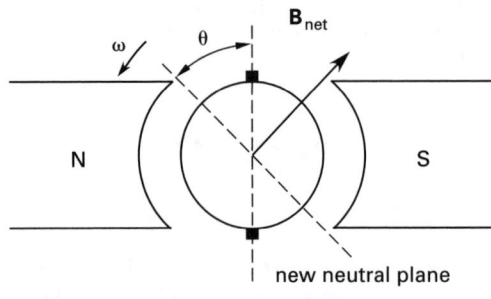

(d) neutral plane shift

Figure 33.13 *DC Generator Armature Flux*

---

[4]In generators, armature reaction shifts the neutral plane in the direction of angular motion. The effect occurs in DC motors as well, but the neutral plane shifts opposite the direction of angular motion.

The neutral plane, also called the *neutral zone*, is the plane where the surface of the armature experiences a magnetic flux density of zero. In DC generators, the shifting of the *neutral plane* results in commutation occurring on coils that have a net voltage. Normally commutation, which shorts the coil connection, occurs when the net voltage on the coil is zero. With a voltage present in the coil, arcing at the brush commutator interface occurs as the coil is shorted. Such arcing lowers brush life and damages the commutator. The problem is solved in most DC machines by using *commutating poles* placed between the main field poles. The commutating poles oppose the armature magnetic field in the vicinity of the brushes. In large DC machines, the armature magnetic field is negated by *compensating windings* electrically in series with the armature windings but physically located in the field pole pieces. These solutions are illustrated in Fig. 33.14.

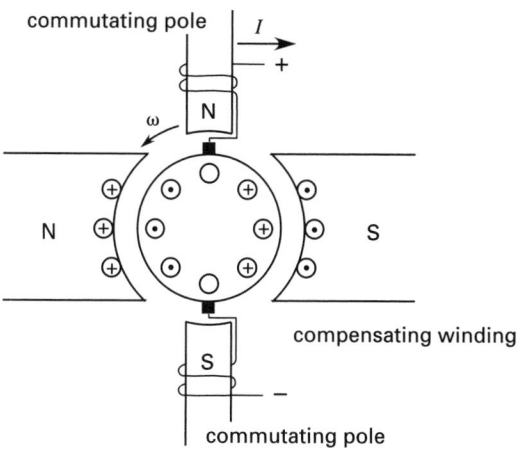

Figure 33.14 *DC Machine Commutating Poles and Compensating Windings*

Losses in a DC electrical generator include windage and friction; core; I^2R or copper losses in the armature, field, compensating, and commutating windings; and *shunt field*, load, and *brush I^2R* and *friction losses*.

9. ENERGY MANAGEMENT

To ensure the safety, security, and reliability, an electric power network must be managed properly. The system designed to accomplish these goals is called the *energy management system* (EMS). The EMS accomplishes four main tasks: generation control, generation scheduling, network analysis, and operator training. The subsystems within the EMS are supervisory control and data acquisition (SCADA), information management, applications, and communications.

The SCADA subsystem gathers data from throughout the network, allows authorized supervisors to control the network, and displays alarms and controls. The information management subsystem stores and controls

access to the data required by any portion of the system. The applications subsystem contains the software programs and packages used to accomplish the tasks of the EMS. The communications subsystem generates the connectivity between all the subsystems and the necessary interfaces with operators and outside concerns.

10. POWER QUALITY

Power quality is defined by the equipment supplied. Each piece of electrical equipment has certain power supply requirements, which may vary considerably depending upon purpose and usage. The major areas of power quality follow and are illustrated in Fig. 33.15.

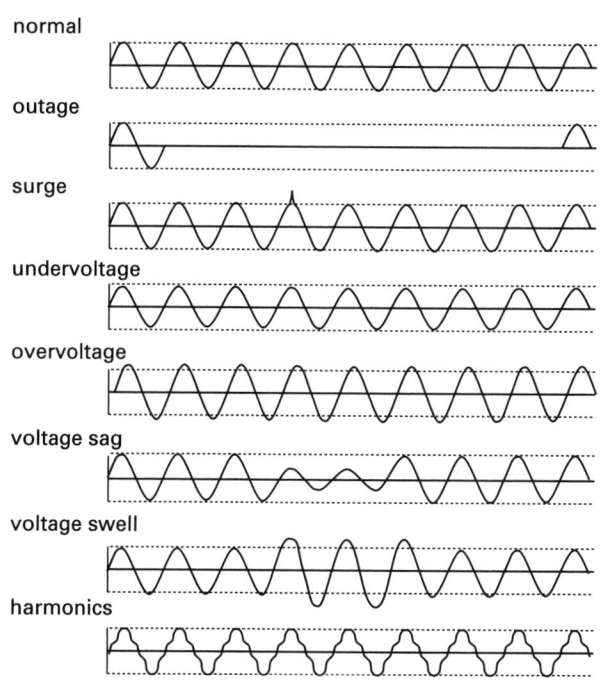

Figure 33.15 *Power Disturbances*

- *Outages*: a complete loss of electrical power. A *blackout* is a complete loss of voltage lasting from one cycle to several days. An outage condition results when faults cause protective devices to function, for example, circuit breakers to open or fuses to blow.

- *Surge*: a transient with a short duration and high magnitude. The term *spike* is used for peak voltages of 6000 V or more lasting approximately 100 μs to one-half of a cycle. The term *transient* describes a peak voltage of 20,000 or more lasting 10 to 100 μs. Surges are caused by switching operations or lightning strikes.

- *Undervoltage*: voltage below the rated voltage for a long duration, that is, for several cycles.[5] If the low-voltage condition lasts for an extended time, it is referred to as a *brownout*. Undervoltage conditions result from any number of factors, including loading beyond system capacity, failure of some sources on the grid, and ground faults in the system.

- *Overvoltage*: an increase in the steady-state voltage that lasts for an extended time. It is also called *chronic overvoltage*.[6] Overvoltages are usually due to improper regulation or voltage regulator failure.

- *Voltage sag*: a drop in voltage. If a low-voltage condition is 80–85% of its rated value for several cycles, the term *sag* or *dip* is used. Sags can result from faults or large starting currents.

- *Voltage swell*: a condition whereby a steady-state rise in voltage occurs for several seconds to approximately one minute.

- *Harmonics*: nonfundamental frequency components of the standard 60 Hz waveform, also called *line noise*. This can occur due to feedback from equipment connected to the line, radio frequency interference, and other electromagnetic sources.

---

[5]The ANSI standard for service rms voltage low limits is 110–114 V.
[6]The ANSI standard for service rms overvoltage is 126 V.

34

Three-Phase Electricity and Power

Nomenclature

a	ratio of transformation	–
I	current	A
N	number of turns	–
P	power	W
pf	power factor	–
Q	reactive power	VAR
R	resistance	Ω
S	apparent power	VA
t	time	s
T	transformer	–
V	voltage	V
X	reactance	Ω
Z	impedance	Ω

Symbols

ϕ	impedance angle	rad
Ψ	power factor angle	rad
ω	armature angular speed	rad/s

Subscripts

l	line
L	inductor
p	phase
pri	primary
pu	per unit
sec	secondary
t	total

1. BENEFITS OF THREE-PHASE POWER

Three-phase energy distribution systems use fewer and smaller conductors and, therefore, are more efficient than multiple single-phase systems providing the same power. Three-phase motors provide a uniform torque, not a pulsating torque as do single-phase motors. Three-phase induction motors do not require additional starting windings or associated switches. When rectified, three-phase voltage has a smoother waveform and less ripple to be filtered out.

2. GENERATION OF THREE-PHASE POTENTIAL

The symbolic representation of an AC generator that produces three equal sinusoidal voltages is shown in Fig. 34.1(a). The generated voltage is known as the *phase voltage*, V_p, or *coil voltage*. (Three-phase voltages are almost always stated as effective values.) Due primarily to the location of the windings, the three sinusoids are 120° apart in phase as shown in Fig. 34.1(b). If $\mathbf{V}_a$ is chosen as the reference voltage, then Eqs. 34.1 through 34.3 represent the phasor forms of the three sinusoids. At any moment, the vector sum of these three voltages is zero.

$$\mathbf{V}_a = V_p \angle 0° \qquad 34.1$$

$$\mathbf{V}_b = V_p \angle -120° \qquad 34.2$$

$$\mathbf{V}_c = V_p \angle -240° \qquad 34.3$$

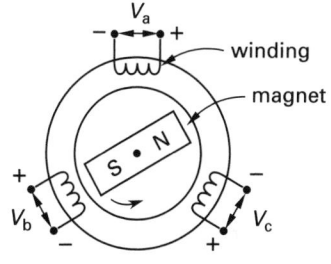

(a) alternator

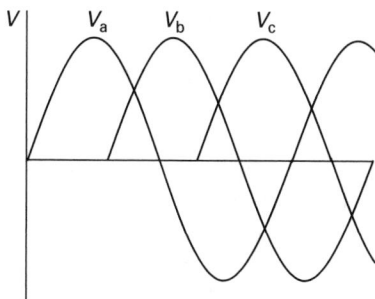

(b) ABC (positive) sequence

Figure 34.1 *Three-Phase Voltage*

Equations 34.1 through 34.3 define an ABC or *positive sequence*. That is, $\mathbf{V}_a$ reaches its peak before $\mathbf{V}_b$, and $\mathbf{V}_b$ peaks before $\mathbf{V}_c$. With a CBA (also written as

ACB) or *negative sequence*, obtained by rotating the *field magnet* in the opposite direction, the order of the delivered sinusoids is reversed (i.e., $\mathbf{V}_c$, $\mathbf{V}_b$, $\mathbf{V}_a$).

Although a six-conductor transmission line could be used to transmit the three voltages, it is more efficient to interconnect the windings. The two methods are commonly referred to as *delta (mesh)* and *wye (star)* connections.

Figure 34.2(a) illustrates delta source connections. The voltage across any two of the lines is known as the *line voltage (system voltage)* and is equal to the phase voltage. Any of the coils can be selected as the reference as long as the sequence is maintained. For a positive (ABC) sequence,

$$\mathbf{V}_{CA} = V_p \angle 0° \qquad 34.4$$

$$\mathbf{V}_{AB} = V_p \angle -120° \qquad 34.5$$

$$\mathbf{V}_{BC} = V_p \angle -240° \qquad 34.6$$

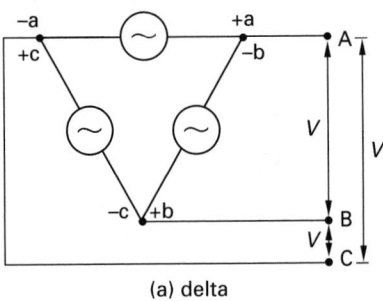

(a) delta

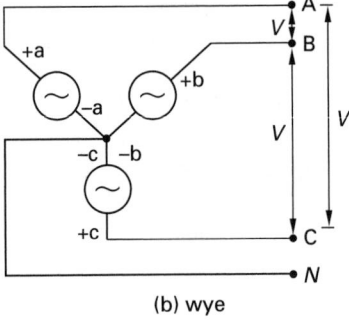

(b) wye

Figure 34.2 *Delta and Wye Source Connections*

Wye-connected sources are illustrated in Fig. 34.2(b). While the *line-to-neutral voltages* are equal to the phase voltage, the line voltages are greater—$\sqrt{3}$ times the phase voltage. The *grounded wire (neutral)* is needed to carry current only if the system is unbalanced.[1] (See Sec. 34-3.) For an ABC sequence, the line voltages are

$$\mathbf{V}_{AB} = \sqrt{3}V_p \angle 30° \qquad 34.7$$

$$\mathbf{V}_{BC} = \sqrt{3}V_p \angle -90° \qquad 34.8$$

$$\mathbf{V}_{CA} = \sqrt{3}V_p \angle -210° \qquad 34.9$$

---
[1] The neutral wire is usually kept to provide for minor imbalance.

Although the magnitude of the line voltage depends on whether the generator coils are delta- or wye-connected, each connection results in three equal sinusoidal voltages, each 120° out of phase with one another.

3. DISTRIBUTION SYSTEMS

Three-phase power is delivered by three-wire and four-wire systems. A *four-wire system* consists of three power conductors and a neutral conductor. A *three-wire system* contains only the three power conductors.

Utility power distribution starts with generation. The generator is connected through step-up *subtransmission transformers* that supply *transmission lines*. The actual transmission line voltage depends on the distance between the subtransmission transformers and the user. Distribution *substation transformers* reduce the voltage from the transmission line level to approximately 35 kV. The *primary distribution system* delivers power to *distribution transformers* that reduce voltage still further, to between 120 and 600 V.

4. BALANCED LOADS

Three impedances are required to fully load a three-phase voltage source. The impedances in a three-phase system are *balanced* when they are identical in magnitude and angle. The voltages and line currents, as well as the real, apparent (kVA), and reactive powers, are all identical in a balanced system. Also, the power factor is the same for each phase. Therefore, balanced systems can be analyzed on a per-phase basis. Such calculations are known as *one-line analyses*.

Figure 34.3 illustrates the vector diagram for a balanced delta three-phase system. The phase voltages, $\mathbf{V}$, are separated by 120° phase angles, as are the phase currents, $\mathbf{I}$. The phase difference angle, ϕ, between a phase voltage and its respective phase current depends on the phase impedance. With delta-connected balanced loads, the phase and line currents differ in phase by 30°.

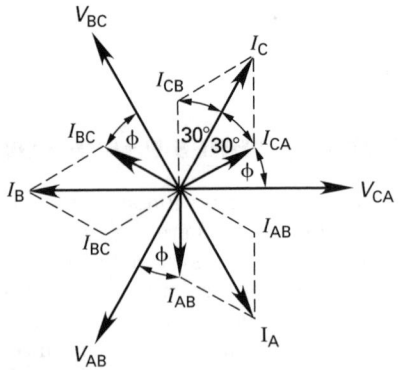

Figure 34.3 *Positive ABC Balanced Delta Load Vector Diagram*

5. DELTA-CONNECTED LOADS

Figure 34.4 illustrates delta-connected loads.

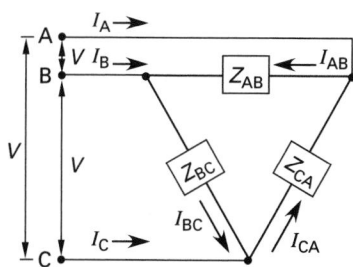

Figure 34.4 *Delta-Connected Loads*

The *phase currents* for a balanced system are calculated from the line voltage (same as the phase voltage). For a positive (ABC) sequence,

$$\mathbf{I}_{AB} = \frac{\mathbf{V}_{AB}}{\mathbf{Z}_{AB}} = \frac{V \angle -120°}{Z \angle \phi}$$

$$= \frac{V}{Z} \angle -120° - \phi \qquad \textbf{34.10}$$

$$\mathbf{I}_{BC} = \frac{\mathbf{V}_{BC}}{\mathbf{Z}_{BC}} = \frac{V}{Z} \angle -240° - \phi \qquad \textbf{34.11}$$

$$\mathbf{I}_{CA} = \frac{\mathbf{V}_{CA}}{\mathbf{Z}_{CA}} = \frac{V}{Z} \angle -\phi \qquad \textbf{34.12}$$

The *line currents* are not the same as the phase currents but are $\sqrt{3}$ times the phase current and displaced $-30°$ in phase from the phase current.

$$|\mathbf{I}_A| = |\mathbf{I}_{AB} - \mathbf{I}_{CA}| = \sqrt{3}I_{AB} \qquad \textbf{34.13}$$

$$|\mathbf{I}_B| = |\mathbf{I}_{BC} - \mathbf{I}_{AB}| = \sqrt{3}I_{BC} \qquad \textbf{34.14}$$

$$|\mathbf{I}_C| = |\mathbf{I}_{CA} - \mathbf{I}_{BC}| = \sqrt{3}I_{CA} \qquad \textbf{34.15}$$

$$I_A = I_B = I_C \quad \text{[balanced]} \qquad \textbf{34.16}$$

Each impedance in a balanced system dissipates the same real *phase power*, P_p. The total power dissipated is three times the phase power. (This is the same as for wye-connected loads.)

$$\begin{aligned}P_t &= 3P_p = 3V_pI_p \cos|\phi| \\ &= 3V_pI_p\text{pf} \\ &= \sqrt{3}VI\cos\Psi \\ &= \sqrt{3}VI\text{pf} \qquad \textbf{34.17}\end{aligned}$$

Example 34.1

Three identical impedances are connected in delta across a three-phase system with 240 V (rms) line voltages in an ABC sequence. Find the (a) phase current $\mathbf{I}_{AB}$, (b) phase real power, P_p, (c) line current $\mathbf{I}_B$, and (d) total real power.

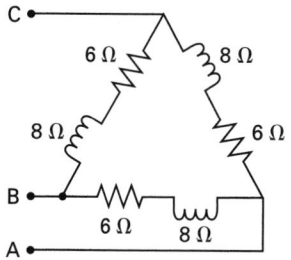

Solution

(a) The phase impedance is

$$\begin{aligned}\mathbf{Z}_p &= \sqrt{R^2 + X_L^2} \angle \arctan\left(\frac{X}{R}\right) \\ &= \sqrt{(6)^2 + (8)^2} \angle \arctan\left(\frac{8}{6}\right) \\ &= 10 \angle 53.13° \ \Omega\end{aligned}$$

The phase current is

$$\mathbf{I}_{AB} = \frac{\mathbf{V}_{AB}}{\mathbf{Z}_{AB}} = \frac{240\angle -120° \text{ V}}{10\angle 53.13° \ \Omega} = 24\angle -173.13° \text{ A}$$

(b) The average phase power is

$$P_p = I_p^2 R = (24)^2(6) = 3456 \text{ W}$$

(c) The phase current I_{BC} contributes to the line current.

$$\mathbf{I}_{BC} = \frac{\mathbf{V}_{BC}}{\mathbf{Z}_{BC}} = \frac{240\angle -240° \text{ V}}{10\angle 53.13° \ \Omega} = 24\angle -293.13$$

$$\mathbf{I}_B = \mathbf{I}_{BC} - \mathbf{I}_{AB} = 41.57\angle 36.87°$$

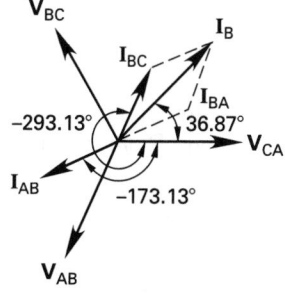

(d) The total power is three times the phase power.

$$P_t = 3P_p = (3)(3456 \text{ W}) = 10{,}368 \text{ W}$$

6. WYE-CONNECTED LOADS

Figure 34.5 illustrates three equal impedances connected in wye configuration. The line and phase currents are equal. However, the phase voltage is less than the line voltage since more than a single phase is connected across two lines. The line and phase currents are

$$\mathbf{I}_A = \mathbf{I}_{AN} = \frac{\mathbf{V}_{AN}}{\mathbf{Z}_{AN}} = \frac{\mathbf{V}_{AB}}{\sqrt{3}\mathbf{Z}_{AN}} \qquad 34.18$$

$$\mathbf{I}_B = \mathbf{I}_{BN} = \frac{\mathbf{V}_{BN}}{\mathbf{Z}_{BN}} = \frac{\mathbf{V}_{BC}}{\sqrt{3}\mathbf{Z}_{BN}} \qquad 34.19$$

$$\mathbf{I}_C = \mathbf{I}_{CN} = \frac{\mathbf{V}_{CN}}{\mathbf{Z}_{CN}} = \frac{\mathbf{V}_{CA}}{\sqrt{3}\mathbf{Z}_{CN}} \qquad 34.20$$

$$\mathbf{I}_N = 0 \quad \text{[balanced]} \qquad 34.21$$

The total power dissipated in a balanced wye-connected system is three times the phase power. (This is the same as for the delta connection.)

$$P_t = 3P_p = 3V_pI_p \cos\phi = \sqrt{3}VI \cos\phi \qquad 34.22$$

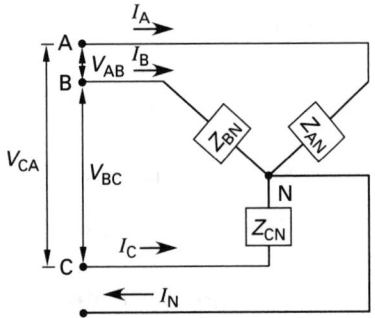

Figure 34.5 *Wye-Connected Loads*

Example 34.2

A three-phase, 480 V, 250 hp motor has an efficiency of 94% and a power factor of 90%. What is the line current?

Solution

It does not matter whether the motor's windings are delta- or wye-connected. From Eq. 34.22, the line current is

$$I = \frac{P_t}{\eta\sqrt{3}V \cos\phi} = \frac{(250 \text{ hp})\left(745.7 \frac{\text{W}}{\text{hp}}\right)}{(0.94)(\sqrt{3})(480 \text{ V})(0.9)}$$

$$= 265 \text{ A}$$

7. DELTA-WYE CONVERSIONS

It is occasionally convenient to convert a delta system to a wye system and vice versa. The equations in Sec. 26-20, used to convert between delta and wye resistor networks, are applicable if impedances are substituted for resistances.

8. PER-UNIT CALCULATIONS

Power systems use a multitude of voltages ranging from very high (35 kV) to very low (120 V). To simplify analysis, a *per-unit system* is employed that scales the sections of the distribution system in accordance with the voltage ratios of the transformers. The per-unit system is given mathematically by

$$\text{per unit} = \frac{\text{actual}}{\text{base}} = \frac{\text{percent}}{100} \qquad 34.23$$

For a three-phase system, the usual bases are the line voltage (in kV) and the total (three-phase) apparent power (in kVA) ratings. From these two bases, the base current and impedance can be calculated. Line and phase values may differ in three-phase systems. The relationship between the phase and line quantities for the bases is

$$V_p = \frac{V_l}{\sqrt{3}} \qquad 34.24$$

$$S_p = (VA)_p = \frac{S_t}{3} = \frac{(VA)_t}{3} \qquad 34.25$$

The line-to-phase voltage conversion is necessary in a wye connection only. In a delta connection, the line and phase voltages are equal. Care should be taken when using phase and line quantities due to the variation in symbology. Phase values are represented here with the subscript p but are also given as ϕ and ln (line-to-neutral). Line values are given with the subscript t (total) but are also given as ll (line-to-line), l (line), 3ϕ, or with no subscript. The per-unit system is represented by Eqs. 34.26 through 34.31 and is presented elsewhere as phase bases for use with single-phase systems. Conversion to a three-phase base, which uses line and total quantities, is accomplished with Eqs. 34.24 and 34.25 and is indicated by the subscript 3ϕ in Eqs. 34.26 through 34.31.

$$S_{\text{base}} = S_p = \left(\frac{S}{3}\right)_{3\phi} \qquad 34.26$$

$$V_{\text{base}} = V_p = \left(\frac{V}{\sqrt{3}}\right)_{3\phi} \qquad 34.27$$

$$I_{\text{base}} = \frac{S_{\text{base}}}{V_{\text{base}}} = \frac{S_p}{V_p} = \left(\frac{S}{\sqrt{3}V}\right)_{3\phi} \qquad 34.28$$

$$Z_{\text{base}} = \frac{V_{\text{base}}}{I_{\text{base}}} = \frac{V_p^2}{S_p} = \left(\frac{V^2}{S}\right)_{3\phi} \qquad 34.29$$

$$P_{\text{base}} = S_p = \left(\frac{S}{3}\right)_{3\phi} \qquad 34.30$$

$$Q_{\text{base}} = S_p = \left(\frac{S}{3}\right)_{3\phi} \qquad 34.31$$

The per-unit values are

$$I_{\text{pu}} = \frac{I_{\text{actual}}}{I_{\text{base}}} \qquad 34.32$$

$$V_{\text{pu}} = \frac{V_{\text{actual}}}{V_{\text{base}}} \qquad 34.33$$

$$Z_{\text{pu}} = \frac{Z_{\text{actual}}}{Z_{\text{base}}} \qquad 34.34$$

$$P_{\text{pu}} = \frac{P_{\text{actual}}}{P_{\text{base}}} \qquad 34.35$$

$$Q_{\text{pu}} = \frac{Q_{\text{actual}}}{Q_{\text{base}}} \qquad 34.36$$

Ohm's law and other circuit analysis methods can be used with the per-unit quantities.

$$V_{\text{pu}} = I_{\text{pu}} Z_{\text{pu}} \qquad 34.37$$

All values in a given portion of a power system must be expressed using the same base. The general method for converting from one per-unit value, call it χ, to another in a different base is

$$\chi_{\text{pu,new}} = \chi_{\text{pu,old}} \left(\frac{\chi_{\text{base,old}}}{\chi_{\text{base,new}}} \right) \qquad 34.38$$

The impedance per-unit value conversion is

$$Z_{\text{pu,new}} = Z_{\text{pu,old}} \left(\frac{V_{\text{base,old}}}{V_{\text{base,new}}} \right)^2 \left(\frac{S_{\text{base,new}}}{S_{\text{base,old}}} \right) \qquad 34.39$$

Example 34.3

A wye-connected three-phase device is rated to draw a total of 300 kVA when connected to a line-to-line voltage of 15 kV. The device's per-unit impedance is $0.1414 + j0.9900$. What is the actual impedance?

Solution

The bases are

$$V_{\text{base}} = V_p = \frac{V}{\sqrt{3}} = \frac{15 \text{ kV}}{\sqrt{3}} = 8.660 \text{ kV}$$

$$S_{\text{base}} = S_{\text{phase}} = \frac{300 \text{ kVA}}{3} = 100 \text{ kVA}$$

The base current is

$$I_{\text{base}} = \frac{S_{\text{base}}}{V_{\text{base}}} = \frac{100 \text{ kVA}}{8.660 \text{ kV}} = 11.55 \text{ A}$$

The base impedance is

$$Z_{\text{base}} = \frac{V_{\text{base}}}{I_{\text{base}}} = \frac{8660 \text{ V}}{11.55 \text{ A}} = 750 \ \Omega$$

From Eq. 34.34, the actual impedance is

$$Z_{\text{actual}} = Z_{\text{pu}} Z_{\text{base}} = (0.1414 + j0.9900)(750 \ \Omega)$$

$$= 106.1 + j742.5 \ \Omega$$

9. UNBALANCED LOADS

The three-phase loads are unequal in an unbalanced system. A fourth conductor, the *neutral conductor*, is required for the line voltages to be constant. Such a system is known as a *four-wire system*. Without the neutral conductor (i.e., in a *three-wire system*), the common point of the load connections is not at zero potential and the voltages across the three impedances vary from the line-to-neutral voltage. The voltage at the common point is known as the *displacement neutral voltage*.

Regardless of whether the line voltages are equal, the line currents are not the same nor do they have a 120° phase difference. Unbalanced systems can be evaluated by computing the phase currents and then applying Kirchhoff's current law (in vector form) to obtain the line currents. The *neutral current* will be

$$\mathbf{I}_N = -(\mathbf{I}_A + \mathbf{I}_B + \mathbf{I}_C) \qquad 34.40$$

Example 34.4

Unequal impedances are connected to a 208 V (rms) three-phase system as shown. The sequence is ABC with $\angle V_{\text{CA}} = 0°$. (a) What is the line current $\mathbf{I}_A$? (b) What is the total real power dissipated?

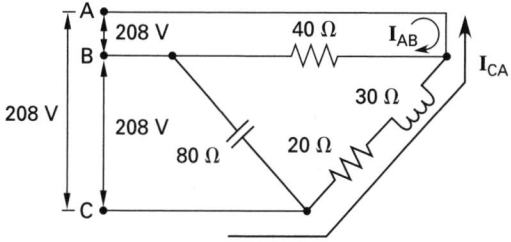

Solution

(a) First, calculate the phase current $\mathbf{I}_{AB}$.

$$\mathbf{I}_{AB} = \frac{\mathbf{V}_{AB}}{\mathbf{Z}_{AB}} = \frac{208 \angle -120° \text{ V}}{40 \angle 0° \ \Omega} = 5.20 \angle -120° \text{ A}$$

The phase impedance $\mathbf{Z}_{CA}$ is

$$\mathbf{Z}_{CA} = \sqrt{R^2 + X_L^2} \ \angle \arctan \left(\frac{X_L}{R} \right)$$

$$= \sqrt{(20)^2 + (30)^2} \ \angle \arctan \left(\frac{30}{20} \right)$$

$$= 36.06 \angle 56.31° \ \Omega$$

Next, calculate the phase current $\mathbf{I}_{CA}$.

$$\mathbf{I}_{CA} = \frac{\mathbf{V}_{CA}}{\mathbf{Z}_{CA}} = \frac{208 \angle 0° \text{ V}}{36.06 \angle 56.31° \ \Omega} = 5.77 \angle -56.31$$

The line current is

$$\mathbf{I}_A = \mathbf{I}_{AB} - \mathbf{I}_{CA} = \mathbf{I}_{AB} + \mathbf{I}_{AC}$$
$$= 5.8\angle -183°$$

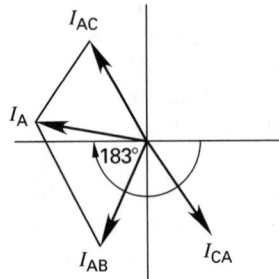

(b) The total power dissipated is

$$P = \sum I^2 R = (5.2\text{ A})^2(40\ \Omega) + (5.77)^2(20)$$
$$= 1747\text{ W}$$

10. THREE-PHASE TRANSFORMERS

Transformer banks for three-phase systems have three primary and three secondary windings. Each primary-secondary set is referred to as a *transformer*, as distinguished from the *bank*. Each side can be connected in delta or wye configuration, making a total of four possible transformer configurations (delta-delta, delta-wye, wye-delta, and wye-wye) as shown in Fig. 34.6. The turns ratio, a, is the same for each winding. Each transformer provides one-third of the total kVA rating, regardless of connection configuration. However, as the figure shows, the secondary voltages and currents depend on the configuration.

$$a = \frac{N_{\text{pri}}}{N_{\text{sec}}} \qquad\qquad 34.41$$

Example 34.5

A 240 V (rms) three-phase system drawing 1200 kVA is supplied by a 2400 V (primary-side) transformer bank. Each transformer is connected in a wye-delta (primary-secondary) configuration. What are the (a) ratio of transformation, (b) high-side and low-side winding voltages, (c) high-side and low-side winding currents, and (d) kVA rating?

Solution

This is case (c) in Fig. 34.6.

(a) The primary line voltage is 2400 V.

$$240\text{ V} = \frac{V}{\sqrt{3}a} = \frac{2400\text{ V}}{\sqrt{3}a}$$
$$a = \frac{2400}{(240)(\sqrt{3})} = 5.774$$

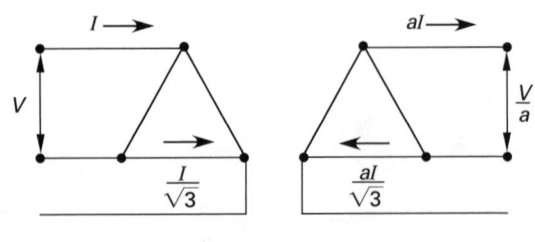

(a) delta-delta

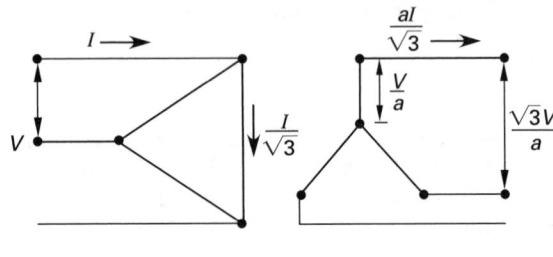

(b) delta-wye

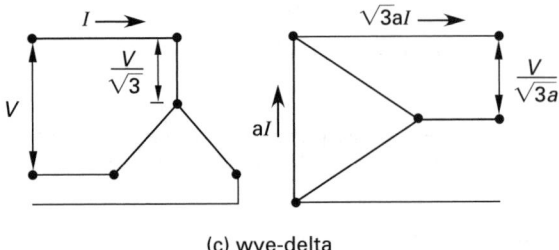

(c) wye-delta

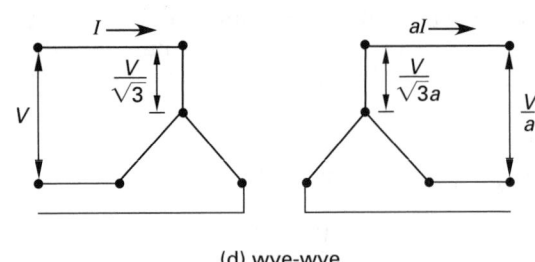

(d) wye-wye

Figure 34.6 *Three-Phase Transformer Configurations*

(b) The winding (phase) voltages are

$$V_{\text{high}} = \frac{V}{\sqrt{3}} = \frac{2400\text{ V}}{\sqrt{3}} = 1386\text{ V}$$
$$V_{\text{low}} = 240\text{ V} \qquad \text{[given]}$$

(c) The winding (phase) currents are

$$I_{\text{high}} = \frac{S_p}{V_p} = \frac{\sqrt{3}S_t}{3V}$$
$$= \frac{(\sqrt{3})(1200\times 10^3\text{ VA})}{(3)(2400\text{ V})} = 288.7\text{ A}$$
$$I_{\text{low}} = aI_{\text{high}} = (5.774)(288.7\text{ A})$$
$$= 1667\text{ A}$$

(d) The transformer winding kVA rating (per phase) is

$$S_{\text{high}} = I_{\text{high}} V_{\text{high}} = (288.7 \text{ A})(1386 \text{ V})$$
$$= (4 \times 10^5 \text{ VA})$$
$$= 400 \text{ kVA}$$

11. TWO-WATTMETER METHOD

Regardless of how the load impedances are connected, two *wattmeters* can be used to determine the total real power in a three-wire system when connected in the *two-wattmeter configuration* shown in Fig. 34.7. Since one or both of the power readings can be negative, the additions in this section must be performed algebraically.

$$P_t = P_1 + P_2 \quad\quad 34.42$$

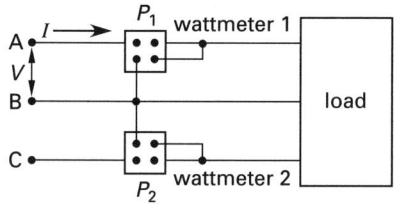

Figure 34.7 *Connections for the Two-Wattmeter Method*

The system does not need to be balanced in order for Eq. 34.42 to be used. However, if it is, Eq. 34.43 can be used. ϕ is the power factor angle of one phase of the load.

$$P_t = IV \cos(\phi - 30°) + IV \cos(\phi + 30°) \quad\quad 34.43$$

The two-wattmeter readings also determine the reactive and apparent power.

$$Q = \sqrt{3} IV \sin \phi = \sqrt{3}(P_1 - P_2) \quad\quad 34.44$$

$$S^2 = P^2 + Q^2 \quad\quad 34.45$$

12. FAULTS AND FAULT CURRENT

A *fault* is an unwanted connection (i.e., a short circuit) between a line and ground or another line. Although the *fault current* is usually very high before circuit breakers trip, it is not infinite because the transformers and transmission line have finite impedance up to the fault point. If the line impedance is known, the fault current can be found by Ohm's law.[2]

$$V = I_{\text{fault}} Z \qu\quad 34.46$$

---

[2]Equation 34.46 ignores the transient (DC) current component and the relatively small current flowing in the wire before the fault occurs. Although the fault current has a transient component, it dies out so quickly as to be insignificant.

Topic VI: Power—Transmission

Power–
Transmission

For the most current information about the exam, visit **www.ppi2pass.com** regularly.

35 Power Distribution

Nomenclature

C	capacitance	F
E	electric field strength	V/m
I	effective or DC current	A
l	length	m
L	inductance	H
pf	power factor	–
pu	per unit	–
P	pressure	Pa
r	radius	m
v	wind velocity	km/hr
V	effective or DC voltage	V
X	reactance	Ω
Z	impedance	Ω

Symbols

ε	permittivity	F/m
ε_0	free-space permittivity	8.854×10^{-12} F/m
μ_0	free-space permeability	1.2566×10^{-6} H/m
ξ	ratio of radii	–

Subscripts

0	zero sequence
1	positive sequence
2	negative sequence
d	direct
g	generator
l	line
m	motor
n	neutral
p	phase
q	quadrature
t	terminal

1. FUNDAMENTALS

The *electric distribution system* is the circuitry, high-voltage switchgear, transformers, and related equipment that receives high voltage from the source and delivers it at lower voltages. Its function is to receive power from large bulk sources, that is, generation sources, and distribute it to users at the voltage level and degree of reliability required. A hypothetical distribution system one-line diagram is shown in Fig. 35.1. The electric utility may consider the distribution system as that portion of Fig. 35.1 from the distribution substation to the consumer. The symbols used are explained in Fig. 35.2.

One-line diagrams can be used for balanced three-phase systems since the per-phase values are equal. If a neutral line exists, it is not shown. If the diagram is used for *load studies*, the circuit breaker, fuse, and other switching device locations are not shown since they are of no concern. If the diagram is used for *stability studies* or *fault analysis*, the locations and characteristics of circuit breakers, fuses, relays, and other protective devices are shown.

At the *generation level* in Fig. 35.1, the voltage is approximately 6.9 kV. The synchronous generators are shown with their neutrals connected through impedances designed to limit surges in case of a fault in the generator circuit. One or more generators may be attached to a given power bus.

At the *transmission level*, the transmission lines connect the various generators to one another and to the subtransmission lines.

At the *subtransmission level*, the voltage range is 12.5 to 245 kV. The most commonly used voltages in order of usage are 115 kV, 69 kV, 138 kV, and 230 kV. The subtransmission lines are usually in *grid form*. The grid form allows connections between input busses and various paths to each distribution substation, thus increasing reliability. A minimum of two switchable inputs to each substation input bus is normally used.

At the *primary distribution level*, the voltage range is 4.2 to 34.5 kV. The most commonly used voltages are 12.5 kV, 25 kV, and 34.5 kV. The actual voltage level is controlled with taps on the substation transformers. The distribution substation supplies several distribution transformers connected in a radial, tree, loop, or grid system. The distribution transformers are mounted on poles, grade-level pads, or underground near the user (a substation may be dedicated to a single large user.) Secondary mains operate at a voltage range of 120 to 600 V.[1]

---

[1] The ANSI standard for Voltage Range A is 114/228 to 126/252 V at the service entrance and 110/220 to 126/252 V at the utilization point.

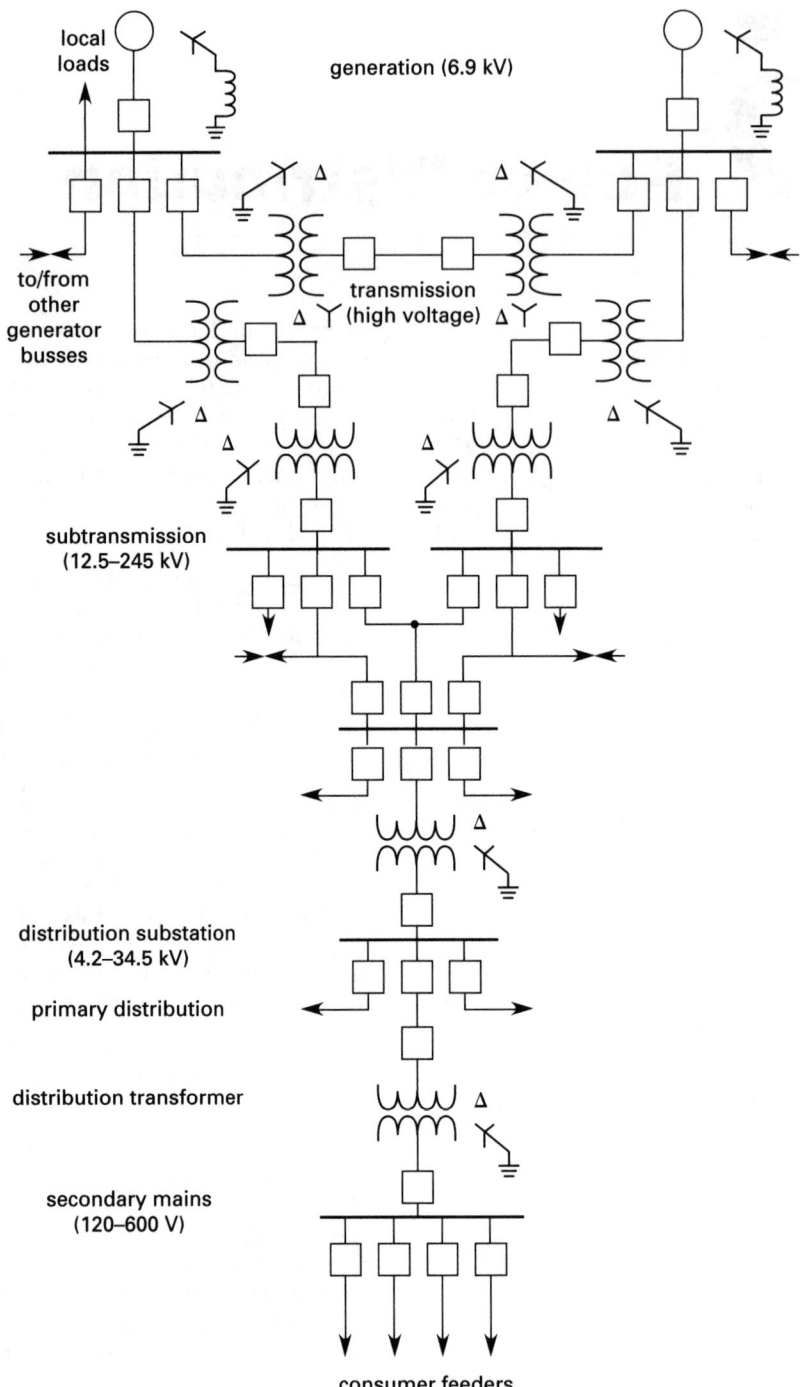

Figure 35.1 *Distribution System*

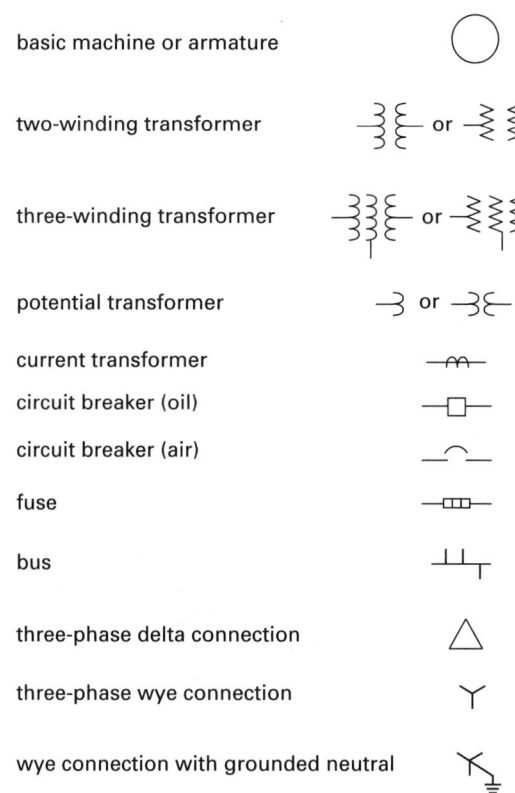

Figure 35.2 *Common Distribution Symbols*

basic machine or armature	
two-winding transformer	
three-winding transformer	
potential transformer	
current transformer	
circuit breaker (oil)	
circuit breaker (air)	
fuse	
bus	
three-phase delta connection	
three-phase wye connection	
wye connection with grounded neutral	

2. CLASSIFICATION OF DISTRIBUTION SYSTEMS

Distribution systems are classified in a number of ways, including

- current (AC or DC)
- voltage (120 V, 12.5 kV, 15 kV, 34.5 kV, etc.)
- type of load (residential, commercial, lighting, power, etc.)
- number of conductors (two-wire, three-wire, etc.)
- construction type (overhead or underground)
- connection type

Connection types include radial (seldom used), loop (see Fig. 35.3), network, multiple (see Fig. 35.4), and series (used primarily for street lighting).

3. COMMON-NEUTRAL SYSTEM

In practice, due to economic and operating advantages, almost all distribution systems are three-phase four-wire, common-neutral primary systems like the one shown in Fig. 35.5. The fourth wire acts as a neutral for both the primary and secondary systems (see Fig. 35.1) although one side of the transformer is delta. (The delta connection is used because the third-harmonic voltages generated by the nonlinear nature of the transformer circulate within the delta and do not affect the line

currents or voltages.) The neutral is grounded in numerous locations, including at each distribution transformer and to metallic water piping or grounding devices at each service entrance.[2] The fourth wire carries a portion of the unbalanced currents that may exist, while the remainder flows through the earth. Some of the advantages of such a system follow.

- Unbalanced currents flow through the neutral or the earth, thus ensuring approximately balanced three-phase current flow through the generator. This evens the countertorque, minimizes vibration, and extends bearing life.

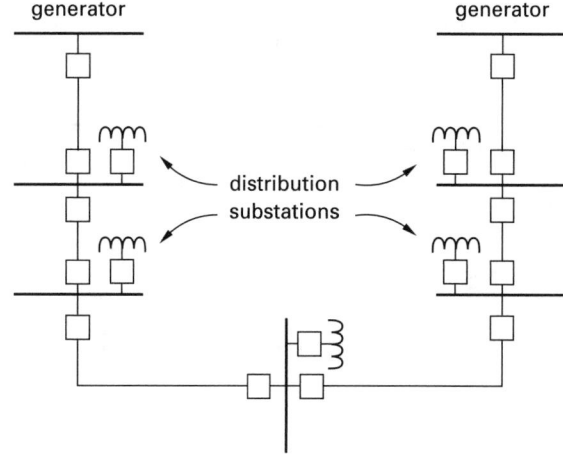

Figure 35.3 *Loop Distribution Pattern*

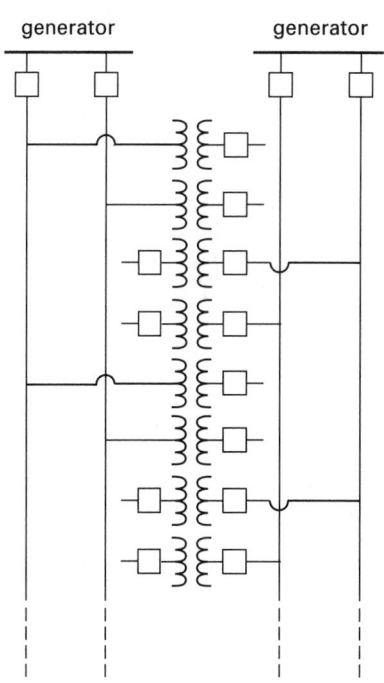

Figure 35.4 *Multiple Distribution Pattern*

---

[2]The use of nonmetallic water piping requires the use of other grounding techniques described in the National Electrical Code®.

- Single-phase circuits require only one insulated conductor and the uninsulated neutral, thus lowering wiring costs.

- Protective devices need to be placed on only one wire of a single-phase circuit to provide adequate protection, significantly lowering costs.

Other possible systems include the *three-phase three-wire system*, which is not widely used for public power distribution except in California. Such a system finds application in marine systems that operate ungrounded to improve reliability. *Two-phase systems* are rarely used. *Direct-current systems*, though they have some advantages, have been replaced by AC systems. *Series systems* are used primarily for street lighting but have been largely replaced by multiple systems.

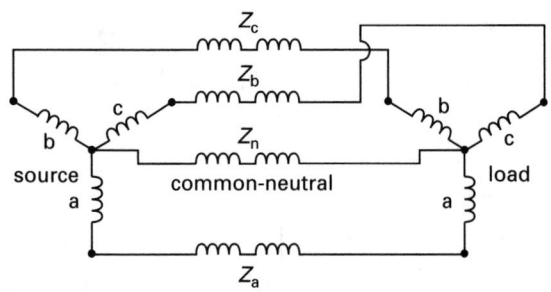

Figure 35.5 *Common-Neutral (Three-Phase Four-Wire System)*

4. OVERCURRENT PROTECTION

Overcurrent is current in excess of the rated current of the equipment or the ampacity of the conductors. It may be caused by an overload, short circuit, or ground. *Overload* is any condition beyond the rating of the electrical device supplied that if continued for an extended time will damage the equipment or cause overheating. A short circuit or ground fault is not an overload. The short circuit or ground fault conditions can result in extremely high currents for very short time periods. Overload conditions are longer term. Protective devices include relayed circuit breakers, reclosers, sectionalizers, and fuses.

A *circuit breaker* is an electromagnetic device used to open and close a circuit. The operation of a circuit breaker can be automatic or manual. For overcurrent protection, relays are installed that sense the current and create a corresponding magnetic field whose force is used to trip the breaker if necessary. The trip characteristics of some circuit breakers are adjustable. An *instantaneous trip* is one that occurs without delay. An *inverse time trip* is one in which a delay is deliberately instituted. As the current magnitude increases, the time to trip decreases. The National Electrical Code®, 1999, specifies that a circuit breaker is considered to provide adequate short-circuit protection if its rating is no more than six times the ampacity of the conductors.

A *recloser* is a device that opens a circuit instantaneously when a fault occurs but recloses after a short period. Reclosers typically have a continuous current rating of 560 A and an interrupting capacity of 16,000 A or less. A circuit breaker will typically have a continuous current rating of 1200 A and an interrupting capacity of 40,000 A under short-circuit conditions. A recloser normally operates at twice its current rating. This is called the *minimum pickup current*. Reclosers are used to protect the distribution system while minimizing outages.

A *sectionalizer* is a device that counts the consecutive number of times a recloser operates and opens the circuit at a predetermined number of counts, usually two or three. The sectionalizer has no interrupting capacity. It operates to keep the recloser open during the time that the recloser is open.

A *fuse* is an expendable device that opens a circuit when the current becomes excessive. The fuse is a sealed container that contains a conductor surrounded by quartz-sand filler. The conductor melts after the current exceeds a given value. Though normally used for overload conditions, special types of fuses called *current-limiting fuses* are designed to melt nearly instantaneously. The National Electrical Code®, 1999, specifies that such fuses can be used as short-circuit protection if their rating does not exceed three times the ampacity of the conductors.

5. POLE LINES

The poles that support overhead electric utility lines are constructed of wood, concrete, steel, and aluminum. Poles are subject to *vertical loading* due to the weight of the conducting cable. Normal *transverse loading* occurs when a pole is located at a corner, that is, a point where the geographic direction of the conducting cable changes. Additional vertical and transverse forces must be accounted for due to *wind loading* and *ice loading*. This is accomplished by first calculating the forces using standard mechanical engineering techniques. Then the expected pressures on flat surfaces due to the wind is determined from

$$P = 71.43 \text{v}^2 \qquad \qquad 35.1$$

The term v represents the wind velocity normal to the flat surface in km/hr. The pressure is in units of Pa. For curved surfaces, such as a standard pole, the pressure is determined from

$$P = 44.64 \text{v}^2 \qquad \qquad 35.2$$

Finally, a percentage of this force is added to the result of the mechanical and wind calculations to increase the amount of force that the pole must be designed to withstand. (The force calculations become moment calculations when the pole height is taken into consideration).

The percentage to be added is determined by the extreme wind and icing conditions expected in the area where the pole is to be located. The percentage may vary from 0.0 to 0.30.

Copper conductor used for overhead lines where spans are approximately 61 m (200 ft) or more is the hard-drawn type due to its greater strength. For shorter spans, medium hard-drawn or annealed copper is utilized. For very long spans, aluminum stranded around a steel core is used. This type of cable is called aluminum conductor steel-reinforced (ACSR). High-strength aluminum alloys are also used. One type is called aluminum conductor alloy-reinforced (ACAR). Another is the all-aluminum-alloy conductor (AAAC).

6. UNDERGROUND DISTRIBUTION

Underground installations have increased in popularity, especially in residential districts, primarily for aesthetic reasons, though there are indications that the frequency of faults is lower compared to overhead systems. When faults do occur, however, they are more difficult to access and repair. In a residential district, such a system is called an *underground residential distribution* (URD). Aluminum conductor is used almost exclusively for underground installations due to the development of synthetic insulation such as polyethylene.[3] In copper conductors, standard soft copper is used to improve flexibility. Unlike overhead lines, underground cables must be insulated to ensure the conductor does not contact ground or the cable's external sheath. The insulation also protects against mechanical, chemical, and other effects peculiar to an underground environment.

The insulating material surrounding a conductor is a dielectric. Figure 35.6(a) shows such an arrangement. The insulation thickness must be such that the electric field strength, E, at the surface of the conductor does not break down the insulation. If the cable is too large, it becomes difficult to handle and expensive. If the cable is too small, the dielectric loss becomes large, resulting in overheating of the cable. The optimal thickness is determined by the ratio in Eq. 35.3.

$$\frac{r_2}{r_1} = e = 2.718 \qquad \textit{35.3}$$

Figure 35.6(a) illustrates a cable with a single dielectric. When more than one dielectric is present, they are arranged to minimize the difference between the maximum and minimum electric field strength across the cable. This arrangement is known as *grading*. *Capacitance grading* is illustrated in Fig. 35.6(b). Given the restriction of a maximum electric field, E_{max}, the operating voltage for a capacitance-graded cable is given by

$$V = E_{max} \left(r_1 \ln \left(\frac{r_2}{r_1} \right) + r_2 \ln \left(\frac{r_3}{r_2} \right) \right) \qquad \textit{35.4}$$

---

[3]Aluminum conductor connections require special attention to ensure adequate contact and avoid corrosion.

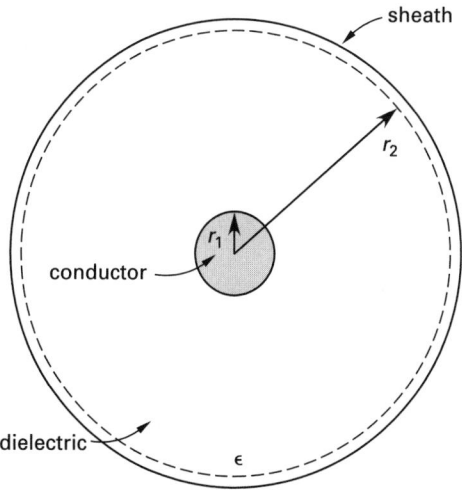

(a) single dielectric

(b) multiple dielectric: capacitance grading

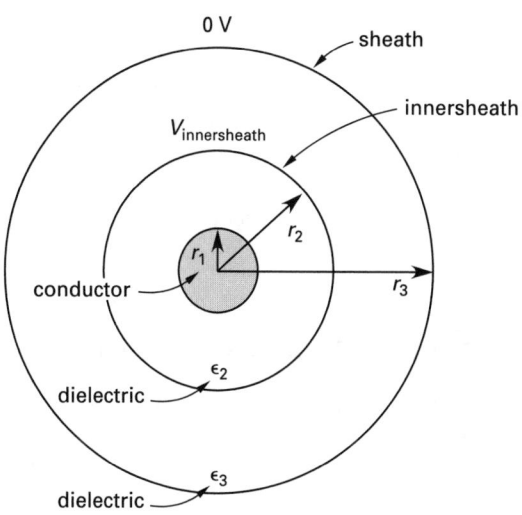

(c) multiple dielectric: innersheath grading

Figure 35.6 *Insulated Conductors*

When the dielectric material is separated by coaxial metallic sheaths maintained at a constant voltage level, the grading is called *innersheath grading* (see Fig. 35.6(c)). The sheathing is used to minimize the electric field. Let the ratio of the radii be

$$\xi = \frac{r_3}{r_2} = \frac{r_2}{r_1} \qquad 35.5$$

The maximum electric field at the surface of the conductor is

$$E_{1,\text{max}} = \frac{V - V_{\text{innersheath}}}{r_1 \ln\left(\dfrac{r_2}{r_1}\right)}$$

$$= \frac{V - V_{\text{innersheath}}}{r_1 \ln \xi} \qquad 35.6$$

The maximum electric field at the surface of the innersheath is

$$E_{2,\text{max}} = \frac{V_{\text{innersheath}}}{r_2 \ln\left(\dfrac{r_3}{r_2}\right)} = \frac{V_{\text{innersheath}}}{r_2 \ln \xi} \qquad 35.7$$

The permittivities are selected so that the maximum electric fields are the same. This condition results in the relationship given in Eq. 35.8 between the innersheath voltage and the operating voltage.

$$V_{\text{innersheath}} = \left(\frac{\xi}{1 + \xi}\right) V \qquad 35.8$$

The capacitance per unit length, C_l, for a single-conductor cable is

$$C_l = \frac{Q}{V} = \frac{2\pi\varepsilon}{\ln \xi} \quad \text{[farads per meter]} \qquad 35.9$$

The inductance per unit length, L_l, for a single-conductor cable is

$$L_l = \frac{\mu_0}{2\pi} \ln \xi \quad \text{[Henries per meter]} \qquad 35.10$$

7. FAULT ANALYSIS: SYMMETRICAL

A *fault* is any defect in a circuit, such as an open circuit, short circuit, or ground. Short-circuit faults, called *shunt faults*, are shown in Fig. 35.7. Open-circuit faults are called *series faults*. Any fault that connects a circuit to ground is termed a *ground fault*. The balanced three-phase short circuit shown in Fig. 35.7(a) is the least likely, yet most severe, fault and thus determines the ratings of the supplying circuit breaker.

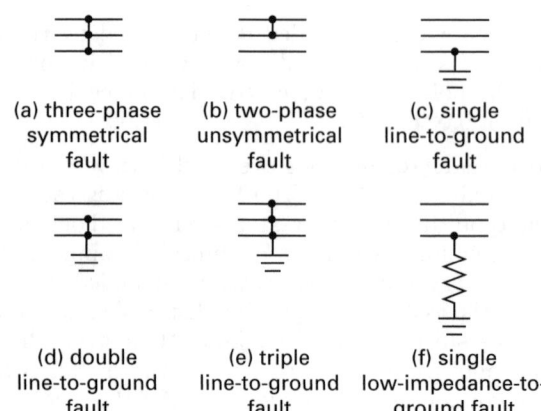

| (a) three-phase symmetrical fault | (b) two-phase unsymmetrical fault | (c) single line-to-ground fault |
| (d) double line-to-ground fault | (e) triple line-to-ground fault | (f) single low-impedance-to-ground fault |

Figure 35.7 *Fault Types*

A three-phase *symmetrical fault*, such as that in Fig. 35.7(a), has three specific time periods of concern as shown in Fig. 35.8. During the *subtransient period*, which lasts only for a few cycles, the current rapidly decreases and the synchronous reactance, X_s, changes to the subtransient reactance, X_d''.[4] The model for a synchronous generator changes from that in Fig. 35.9(a) to that in 35.9(b). The sudden change in armature current results in a lower armature reactance, but the current through the leakage reactance must remain the same for continuity of energy. Thus, for proper modeling, it is necessary to change the voltage, E_g, to E_g''. The voltage E_g'' is calculated for the subtransient interval *just prior to the initiation of the fault* using[5]

$$E_g'' = V_t + jI_L X_d'' \qquad 35.11$$

During the *transient period*, a similar situation exists and the model for the synchronous generator changes from that in Fig. 35.9(b) to that in Fig. 35.9(c). The correct generator voltage for this period is calculated *just prior to the initiation of the fault* using

$$E_g' = V_t + jI_L X_d' \qquad 35.12$$

Both E_g'' and E_g' depend on the impedance of the load and the resulting current prior to fault initiation. Typical reactances of three-phase synchronous machines are given in Table 35.1. When a synchronous motor is part of a system, it is considered a synchronous generator for fault analysis.[6]

---

[4]There are actually two reactances: the *direct reactance* that lags the generator voltage by $90°$, and a *quadrature reactance* (X_q) that is in phase with the generator voltage. During faults, the power factor is low and the quadrature reactance can be ignored.
[5]The general form for the synchronous generator voltage is $E_g'' = V_t + jIX$.
[6]The motor actually becomes a generator when a fault occurs. The spinning motor changes the mechanical energy it possesses into electrical energy, supplying the fault and slowing in the process.

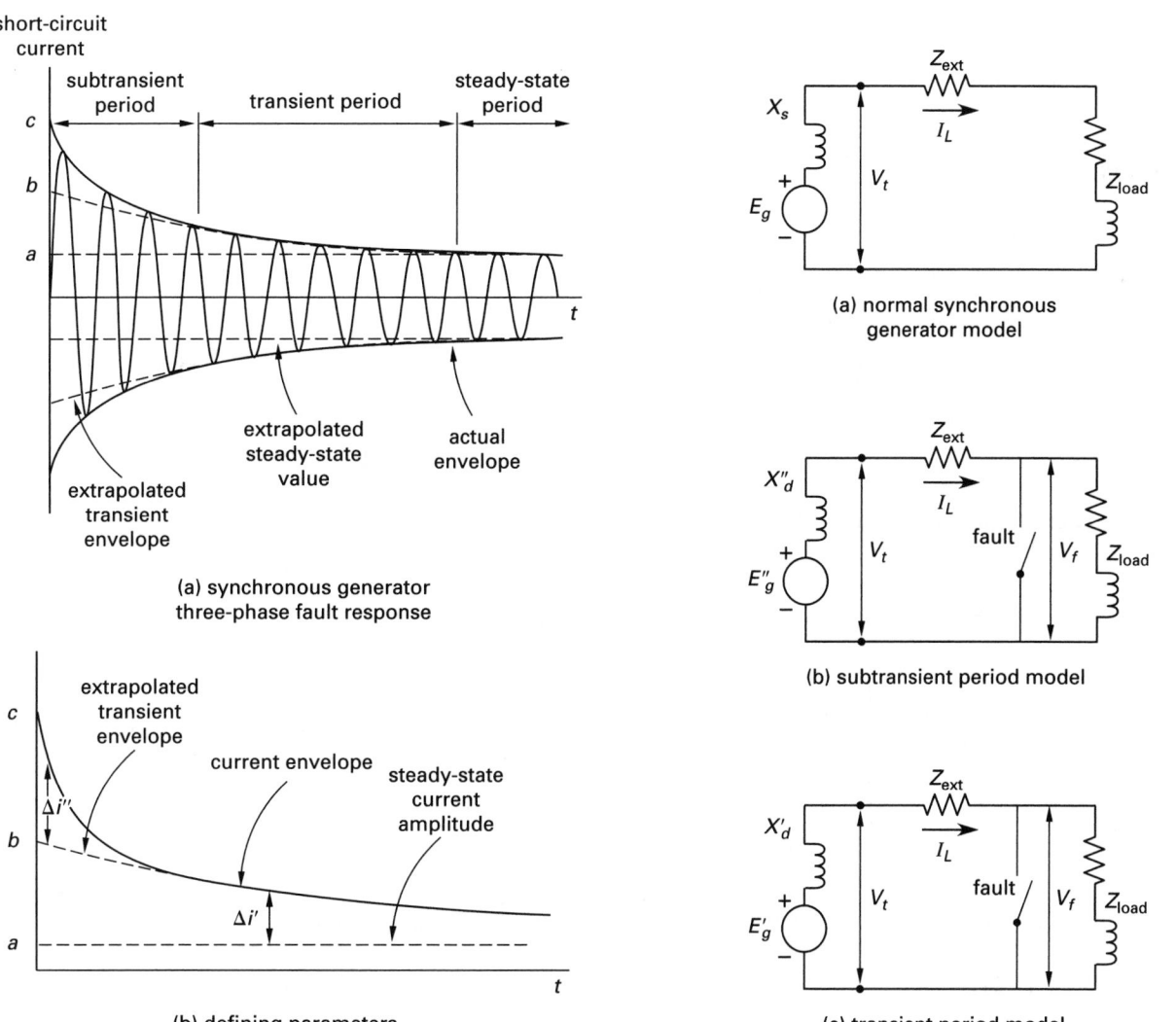

(a) synchronous generator three-phase fault response

(b) defining parameters

Figure 35.8 *Symmetrical Fault Terminology*

(a) normal synchronous generator model

(b) subtransient period model

(c) transient period model

Figure 35.9 *Synchronous Generator Fault Models*

Table 35.1 *Typical Reactances of Three-Phase Synchronous Machines*
(Typical values are given above the bars. The ranges are given below the bars.)

	X_d (unsaturated)	X_q (rated current)	X'_d (rated voltage)	X''_d (rated voltage)
two-pole turbine generators	1.20 / 0.95−1.45	1.16 / 0.92−1.42	0.15 / 0.12−0.21	0.09 / 0.07−0.14
four-pole turbine generators	1.20 / 1.00−1.45	1.16 / 0.92−1.42	0.23 / 0.20−0.28	0.14 / 0.12−0.17
salient-pole generators and motors (with dampers)	1.25 / 0.60−1.50	0.70 / 0.40−0.80	0.30 [a] / 0.20−0.50	0.20 [a] / 0.13−0.32
salient-pole generators (without dampers)	1.25 / 0.60−1.50	0.70 / 0.40−0.80	0.30 [a] / 0.20−0.50	0.30 [a] / 0.20−0.50
capacitors (air-cooled)	1.85 / 1.25−2.20	1.15 / 0.95−1.30	0.40 / 0.36−0.50	0.27 / 0.19−0.30
capacitors (hydrogen-cooled at $^1/_2$ psi)	2.20 / 1.50−2.65	1.35 / 1.10−1.55	0.48 / 0.36−0.60	0.32 / 0.23−0.36

[a]High-speed units tend to have low reactance, and low-speed units tend to have high reactance.

Used with permission from *Electrical Transmission and Distribution Reference Book*, by permission of the Westinghouse Electric Corporation.

Circuit breakers must be able to interrupt fault currents at the rated voltage. The usual fault ratings are in kVA or MVA. The fault rating is the product of the interrupt current capacity and the line-to-line kV rating.

Example 35.1

A synchronous generator and motor are rated for 15 MVA, 13.9 kV, and 25% subtransient reactance. The line impedance connecting the generator and motor is $0.3 + j0.4$ pu, given with the ratings as the base. The motor is operating at 12 MVA, 13 kV, with a 0.8 pf lagging. Determine the per-unit equivalent generated voltage for the motor, $E''_{g,m}$, with the expectation of a fault at the motor terminals.

Solution

The one-line diagram illustrating the situation is

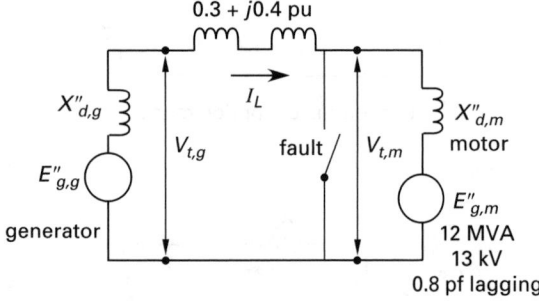

Determine the base quantities for this three-phase system.

$$S_{\text{base}} = S_t = 15 \text{ MVA}$$

$$V_{\text{base}} = V_l = 13.9 \text{ kV}$$

$$I_{\text{base}} = \frac{S_{\text{base}}}{\sqrt{3}V_{\text{base}}} = \frac{15 \text{ MVA}}{\sqrt{3}(13.9 \text{ kV})} = 0.62 \text{ kA}$$

$$Z_{\text{base}} = \frac{V_{\text{base}}}{I_{\text{base}}} = \frac{13.9 \text{ kV}}{0.62 \text{ kA}} = 22.4 \ \Omega$$

Determine the per-unit quantities of interest, that is, the voltage at the fault site (the terminals of the motor), and the current flowing.

$$V_{\text{pu}} = \frac{V_{\text{actual}}}{V_{\text{base}}} = \frac{13 \text{ kV}}{13.9 \text{ kV}} = 0.935 \text{ pu}$$

$$I_{\text{pu}} = \frac{I_{\text{actual}}}{I_{\text{base}}} = \frac{\dfrac{12 \text{ MVA}}{\sqrt{3}(13 \text{ kV})}}{0.62 \text{ kA}} = 0.860 \text{ pu}$$

The power factor is 0.8 lagging, thus

$$\text{lagging pf} = -\cos\theta = -0.8$$

$$\theta = -\cos^{-1} 0.8 = -36.9°$$

The line current, I_l, is

$$I_l = I_{\text{pu}}\angle\theta = 0.860 \text{ pu}\angle -36.9°$$

$$= 0.688 - j0.516 \text{ pu}$$

The equivalent generated voltage for the motor can now be found. Writing the equation for the voltage from the motor terminals to the motor armature gives

$$E''_{g,m} = V_{t,m} - jX''_d I_l$$

$$= 0.935 - j(0.25)(0.688 - j0.516)$$

$$= 0.806 - j0.172$$

$$= 0.824\angle -12° \text{ pu}$$

Example 35.2

Determine the per-unit equivalent generated voltage for the generator, $E''_{g,g}$, for the system in Ex. 35.1.

Solution

The motor terminal voltage is known and the line current is known. An equation for the equivalent generated voltage, written from the known quantities at the motor terminals, is

$$E''_{g,g} = V_{t,m} + I_l Z$$

The impedance is from the motor terminals back to the generator, which is the only unknown. Thus,

$$Z = Z_l + Z_g$$

$$= (0.3 + j0.4) + j0.25$$

$$= 0.3 + j0.65$$

$$= 0.715\angle 65° \text{ pu}$$

Substituting gives

$$E''_{g,g} = V_{t,m} + I_l Z$$

$$= 0.935 + (0.860\angle -36.9°)(0.715\angle 65°)$$

$$= 1.5\angle 11.1° \text{ pu}$$

Example 35.3

Determine the per-unit subtransient period current for the motor in Ex. 35.1.

Solution

A fault at the motor terminals results in a current given by

$$I_m = \frac{E''_{g,m}}{jX''_d} = \frac{0.824\angle -12°}{0.25\angle 90°} = 3.3\angle -102° \text{ pu}$$

Example 35.4

Determine the per-unit subtransient period current for the generator in Ex. 35.1.

Solution

The generator current that flows to the fault is given by

$$I_g = \frac{E''_{g,g}}{Z} = \frac{1.5\angle 11.1°}{0.715\angle 65°} = 2.09\angle -53.9° \text{ pu}$$

Example 35.5

What is the actual fault current for the circuit in Ex. 35.1?

Solution

The actual fault current is the total current through the fault from the generator and the motor. The synchronous motor acts as a generator to the fault as it gradually loses speed. The subtransient period is too short for the motor to lose a significant amount of speed. Thus, it can be treated as a generator. The total current is

$$I = I_m + I_g$$

$$= 3.3\angle -102° \text{ pu} + 2.09\angle -53.9° \text{ pu}$$

$$= 4.95\angle -83.7° \text{ pu}$$

The actual current is

$$I_{\text{actual}} = I_{\text{pu}} I_{\text{base}}$$

$$= (4.95\angle -83.7)(0.62 \text{ kA})$$

$$= 3.07\angle -83.7° \text{ kA}$$

Example 35.6

Determine the minimum rating necessary for the motor and generator circuitbreakers to protect against this fault.

Solution

The minimum rating is determined by the per-unit current drawn during the fault and the base apparent power, that is, the base VA. In other words, the rating is the base apparent power (VA_{base}) multiplied by a factor representing the current flowing above the base current.

$$\text{VA rating} = I_{\text{pu,fault}} VA_{\text{base}}$$

For the motor circuitbreaker,

$$\text{VA rating} = I_{\text{pu,fault}} VA_{\text{base}}$$

$$= (3.3)(15 \text{ MVA})$$

$$= 49.5 \text{ MVA}$$

For the generator circuitbreaker,

$$\text{VA rating} = I_{\text{pu,fault}} VA_{\text{base}}$$

$$= (2.09)(15 \text{ MVA})$$

$$= 31.4 \text{ MVA}$$

8. FAULT ANALYSIS: UNSYMMETRICAL

Unsymmetrical faults, also called *asymmetrical faults*, are any faults other than a three-phase short. Examples include the line-to-line and line-to-ground faults shown in Figs. 35.7(b) through 35.7(f). Such faults are more common than three-phase shorts, but they are more difficult to analyze because they result in uneven phase voltages and currents. Nevertheless, unsymmetrical faults can be analyzed by separating unbalanced phasor components into three sets of symmetrical components as shown in Fig. 35.10.

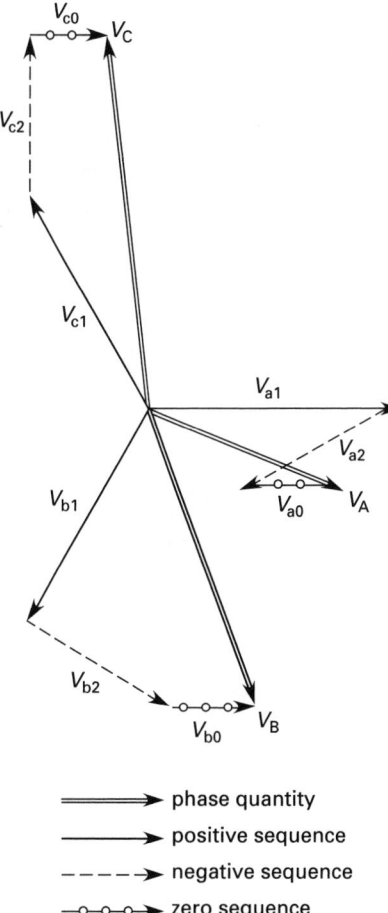

Figure 35.10 Phasor Diagram: Symmetrical Components of Unbalanced Phasors

The three sets of symmetrical components shown in Fig. 35.10 are referred to as the *positive-sequence*, *negative-sequence*, and *zero-sequence* components of the unsymmetrical phasors. The positive-sequence phasors, shown again in Fig. 35.11(a), are three equal-magnitude phasors rotating counterclockwise in sequence a, b, c. Positive-sequence phasors are 120 electrical degrees apart and sum to zero. Positive-sequence components are the ones normally used in electrical engineering. Such components represent balanced three-phase generators, motors, and transformers. The subscript 1 is normally used to indicate the positive sequence.

The negative-sequence phasors, shown again in Fig. 35.11(b), are three equal-magnitude phasors rotating counterclockwise in sequence a, c, b. They can also be represented as a mirror image of the positive-sequence phasors rotating in the clockwise direction. Negative-sequence phasors are 120 electrical degrees apart and sum to zero. The negative-sequence reactance is the average of the subtransient direct and quadrature reactances, as given in Eq. 35.13. The subscript 2 is normally used to indicate the negative sequence.

$$X_2 = \frac{1}{2}\left(X_d'' + X_q''\right) \qquad 35.13$$

The zero-sequence phasors, shown again in Fig. 35.11(c), are three equal-magnitude phasors coincident in phase sequence and rotating counterclockwise. The subscript 0 is normally used to indicate the zero sequence.

The unsymmetrical phasors of Fig. 35.10 are represented in terms of their symmetrical components by the following equations.

$$V_A = V_{a0} + V_{a1} + V_{a2} \qquad 35.14$$

$$V_B = V_{b0} + V_{b1} + V_{b2} \qquad 35.15$$

$$V_C = V_{c0} + V_{c1} + V_{c2} \qquad 35.16$$

Consider an operator **a** defined as a unit vector with a magnitude of one and an angle of 120°, $1\angle 120°$. (This is similar to the operator **j** of $1\angle 90°$.) The properties of such an operator are[7]

$$a = 1\angle 120° = 1 \times e^{j120°}$$

$$= -0.5 + j0.866 \qquad 35.17$$

$$a^2 = 1\angle 240° = -0.5 - j0.866 = a^* \quad 35.18$$

$$a^3 = 1\angle 360° = 1\angle 0° \qquad 35.19$$

$$a^4 = a \qquad 35.20$$

$$a^5 = a^2 \qquad 35.21$$

$$a^6 = a^3 \qquad 35.22$$

$$1 + a + a^2 = 0 \qquad 35.23$$

---

[7] Just as **j** is not normally written in vector notation, the operator **a** is normally written as "a."

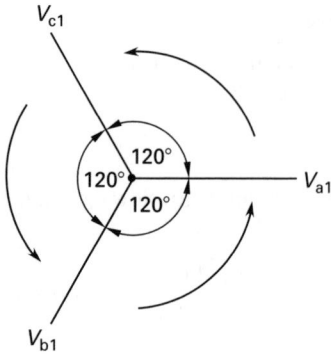

(a) positive sequence

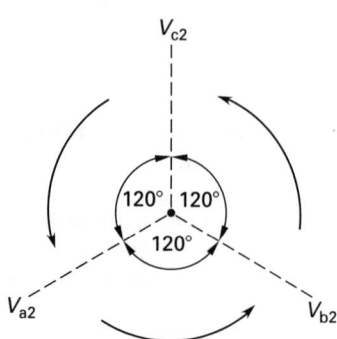

(b) negative sequence

$$V_{a0} = V_{b0} = V_{c0}$$

(c) zero sequence

Figure 35.11 *Components of Unsymmetrical Phasors*

Using Eqs. 35.14 through 35.16 and the concept of the phasor **a**, the unsymmetrical components can be represented in terms of a single phase. For example, using phase A gives

$$V_A = V_{a0} + V_{a1} + V_{a2} \qquad 35.24$$

$$V_B = V_{a0} + a^2 V_{a1} + a V_{a2} \qquad 35.25$$

$$V_C = V_{a0} + a V_{a1} + a^2 V_{a2} \qquad 35.26$$

Solving for the sequence components from Eqs. 35.24 through 35.26 gives

$$V_{a0} = \frac{1}{3}\left(V_A + V_B + V_C\right) \qquad 35.27$$

$$V_{a1} = \frac{1}{3}\left(V_A + a V_B + a^2 V_C\right) \qquad 35.28$$

$$V_{a2} = \frac{1}{3}\left(V_A + a^2 V_B + a V_C\right) \qquad 35.29$$

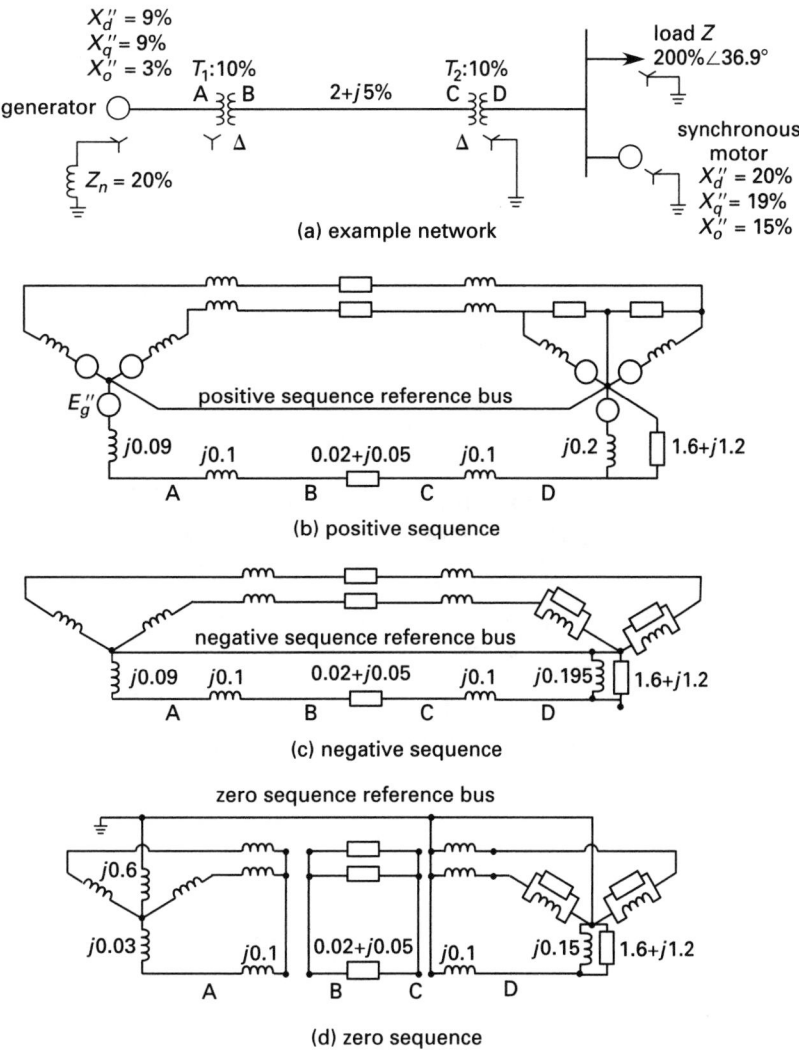

Figure 35.12 *Sample Sequence Networks*

Equations similar to Eqs. 35.24 through 35.29 exist for currents as well. Thus, sequence impedances may also be defined. Impedance through which only positive-sequence currents flow is called a *positive-sequence impedance*. Positive-sequence networks are normally used in electrical engineering and an example is given in Fig. 35.12(b). *Negative-sequence impedance* is similar to positive-sequence impedance with the reactance given by Eq. 35.13. In addition, a negative-sequence network omits all positive-sequence generators. A sample negative-sequence network is shown in Fig. 35.12(c). *Zero-sequence impedance* is significantly different from positive-sequence impedance. The only machine impedance seen by the zero-sequence impedance is the leakage reactance, X_0. Series reactance is greater than positive-sequence reactance by a factor of 2 to 3.5. A sample zero-sequence network is shown in Fig. 35.12(d). Only a wye-connected load with a grounded neutral permits zero-sequence currents. Only a delta-connected transformer secondary permits zero-sequence currents. Figure 35.13 shows zero-sequence impedances for various configurations. The use of sequence networks simplifies fault calculations.

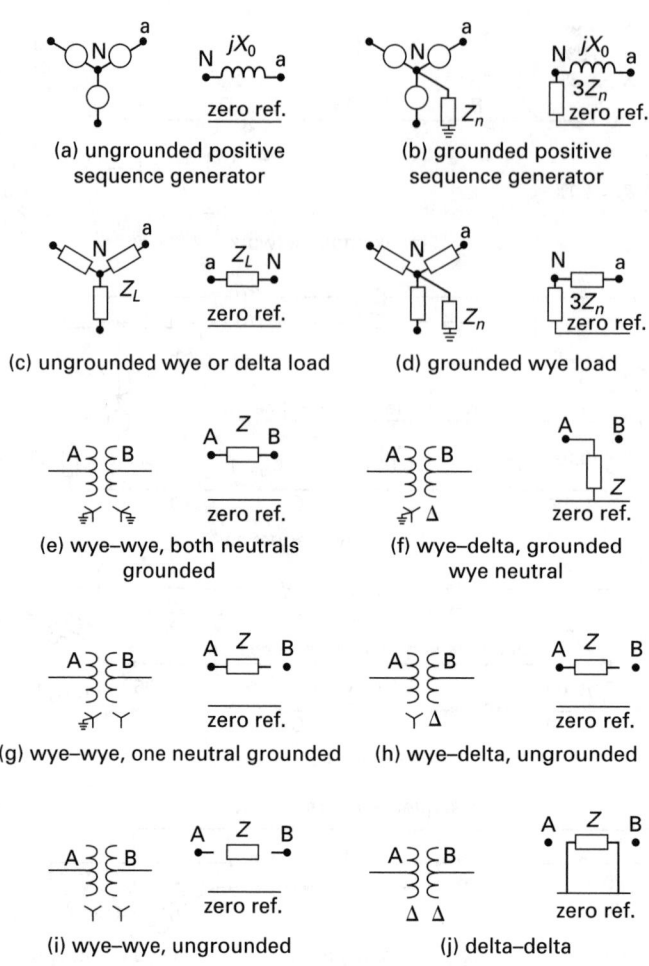

Figure 35.13 *Zero Sequence Impedances*

36

Power Transformers

Nomenclature

a	turns ratio	–
A	ABCD parameter	–
B	ABCD parameter	–
B	magnetic flux density	T
B	susceptance	S, Ω^{-1}, or mho
C	ABCD parameter	–
D	ABCD parameter	–
f	frequency	Hz, s^{-1}, or cycles/s
G	conductance	S, Ω^{-1}, or mho
I	effective or DC current	A
k	coupling coefficient	–
L	inductance	H
m	mass	kg
n	Steinmetz exponent	–
N	number of turns	–
P	power	W
R	resistance	Ω
S	apparent power	kVA
V	effective or DC voltage	V
VR	voltage regulation	–
X	reactance	Ω
Y	admittance	S, Ω^{-1}, or mho
Z	impedance	Ω

Subscripts

1	primary
2	secondary
ϕ	phase
c	core
Cu	copper
e	eddy current
fl	full load
h	hysteresis
l	line
m	maximum
nl	no load
oc	open circuit
O	origin
p	primary
ps	primary to secondary
pu	per unit
s	secondary
sc	short circuit

1. THEORY

A *transformer* is an electrical device consisting of two or more multiturn coils of wire placed in close proximity in order to link the magnetic field of one to the other. This linkage allows the transfer of electrical energy from one or more alternating circuits to one or more other alternating circuits through the process of magnetic induction. The transformer can be used to raise or lower the value of a capacitor, inductor, or resistor. It enables the efficient transmission of electrical energy at high voltage over great distances, and then is used to lower the voltage to safe values for industrial, commercial, and household use. A transformer model with the core parameters referred to the primary side is illustrated in Fig. 36.1.

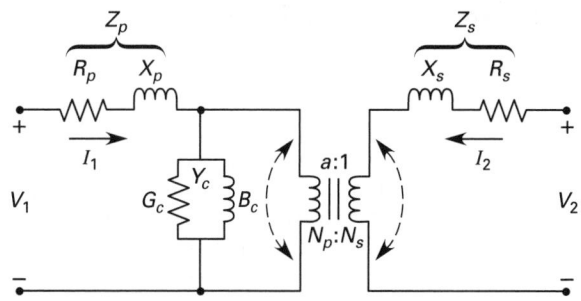

Figure 36.1 *Exact Transformer Model*

The parameters are defined as follows.

- R_p is the primary winding resistance.

- X_p is the primary winding reactance. This is also called the *primary winding leakage reactance*. The *primary winding leakage inductance*, L_p, is sometimes used in the model as well. All the terms represent the inductance not mutually coupled with the secondary winding.

- Z_p is the primary impedance, that is, $R_p + jX_p$.

- G_c is the *core conductance*. The reciprocal is sometimes used here and is referred to as the *core resistance*, R_c.

PROFESSIONAL PUBLICATIONS, INC.

Power–Transmission

- B_c is the *core susceptance*. The reciprocal is sometimes used here and is referred to as the *core reactance*, X_c. The reciprocal is also called the *core magnetizing reactance*, X_m. The *core inductance*, L_c, is sometimes used in the model as well.

- Y_c is the core admittance, that is, $G_c + jB_c$. The reciprocal is the core impedance, Z_c, that is, $R_c + jX_c$.

- The turns ratio N_p/N_s is given the symbol a, primarily for ease of use in equations. The relationships based on the turns ratio given in Ch. 28, $a = N_p/N_s = V_p/V_s = I_s/I_p = \sqrt{Z_p/Z_s}$, are valid only for ideal transformers, though they are an excellent approximation for large transformers. The relationships are valid for the real transformer in Fig. 36.1, but only if the electrical parameters are referenced to the ideal portion of the real model, shown by the dashed lines.[1]

- X_s is the *secondary winding reactance*. This is also called the *secondary winding leakage reactance*. The *secondary winding leakage inductance*, L_s, is sometimes used in the model as well. All the terms represent the inductance not mutually coupled with the primary winding.

- R_s is the secondary winding resistance.

- Z_s is the secondary impedance, that is, $R_s + jX_s$.

2. TRANSFORMER RATING

Transformers are rated according to winding voltages, frequency, and kVA rating. Continuous operation at rated values is possible without excessive heat build-up or malfunction of the transformer. The apparent power (measured in kVA) rather than the real power (measured in W) is used to rate the transformer since the total heating effect depends on the square of the actual current flowing, including that moving in the system between reactive components.

The two main power losses in transformers are *core losses* (*iron losses*) and *copper losses* (*winding losses*). Core losses, represented in Fig. 36.1 by G_c, consist of *eddy current losses* and *hysteresis losses*. Eddy current losses, P_e, are the result of microscopic circulating currents in the iron caused by the magnetic flux passing through. Hysteresis losses, P_h, are due to the cyclic changes in the magnetic state of the iron. Both these losses are constant and independent of transformer load.

That is, they do not vary from no-load to full-load conditions, but they do depend on the mass of iron, m, of the transformer core. Equations 36.1 and 36.2 can be used to calculate the losses. The maximum flux density, B_m, is calculated from Faraday's law. The exponent n is called the *Steinmetz exponent* and varies from 1.5 to 2.5 with a common value of 1.6. The coupling coefficient, k, depends on the transformer design.

$$P_e = k_e B_m^2 f^2 m \qquad \text{36.1}$$

$$P_h = k_h B_m^n f m \qquad \text{36.2}$$

The core losses, ignoring the small line loss caused by R_p, may be approximated from the transformer model in Fig. 36.1 using the following formula.

$$P_c = \frac{V_1^2}{R_c} \qquad \text{36.3}$$

The term R_c is the reciprocal of the core conductance, G_c. Copper losses, P_{Cu}, are due to the total wire resistance and can be calculated from

$$P_{\text{Cu}} = I^2 R = I_1^2 R_p + I_2^2 R_s \qquad \text{36.4}$$

The *transformer efficiency* is the ratio of the output power to the input power and is at a maximum when the eddy current losses equal the core losses. The *all-day efficiency* is the ratio of energy delivered by a transformer in a 24-hour period to the energy input during the same period.

$$\eta = \frac{P_{\text{out}}}{P_{\text{in}}} = \frac{P_{\text{in}} - \sum P_{\text{losses}}}{P_{\text{in}}}$$

$$= \frac{P_{\text{out}}}{P_{\text{out}} + P_c + P_{\text{Cu}}} \qquad \text{36.5}$$

Example 36.1

Determine the hysteresis power loss for a 500 kVA-rated 200 kg iron core transformer with a coupling coefficient of 4×10^{-4} in a 1.4 T peak magnetic field.

Solution

The hysteresis loss is determined from Eq. 36.2. The frequency is not given; thus, it is assumed to be 60 Hz. The Steinmetz exponent is not given; thus, it is assumed to be 1.6.

$$P_h = k_h B_m^n f m$$

$$= \left(4 \times 10^{-4}\right) \left(1.4 \text{ T}\right)^{1.6} \left(60 \text{ s}^{-1}\right) \left(200 \text{ kg}\right)$$

$$= 8.22 \text{ W}$$

---

[1]The ideal transformer model is not used for power transformers because the efficiency, regulation, and power factor must be determined. In this book, the subscripts p and s used for ideal transformers are not used on the current and voltage in the real transformer. They are reserved for electrical parameters of the ideal transformer, that is, the parameters referring to the area of the dashed lines in Fig. 36.1.

Example 36.2

The transformer in Ex. 36.1 uses a primary voltage of 345 kV. If the core resistance is 200 MΩ, estimate the total core power loss.

Solution

The total core losses can be estimated using Eq. 36.3.

$$P_c = \frac{V_1^2}{R_c} = \frac{\left(345 \times 10^3 \text{ V}\right)^2}{200 \times 10^6 \text{ }\Omega} = 595 \text{ W}$$

3. VOLTAGE REGULATION

Voltage regulation is the change in the output voltage from the no-load to the full-load condition.

$$\text{VR} = \frac{V_{\text{nl}} - V_{\text{fl}}}{V_{\text{fl}}} \qquad \textit{36.6}$$

The voltage regulation can be expressed in terms of rated quantities. At no load, reflecting the primary voltage to the secondary side gives V_p/a. The rated conditions of the secondary represent full load. The voltage regulation is then given by

$$\text{VR} = \frac{\dfrac{V_p}{a} - V_{s,\text{rated}}}{V_{s,\text{rated}}} \qquad \textit{36.7}$$

The primary voltage, due to the large generation source, can be considered constant. That is, the primary side is considered an infinite bus. This assumption is embedded in any regulation equation based on rated values and also when using per-unit values to calculate the regulation. With output voltage considered to have a 1.0 pu value, the difference between the primary voltage and the output voltage of the transformer secondary in per-unit values is the regulation. That is, to supply the load current and secondary voltage at rated values, the primary voltage must be greater than the secondary. The excess is termed the regulation and is found by using Eq. 36.7 with per-unit values instead of rated values.

Example 36.3

A single-phase transformer rated for 50 kVA operates at a rated secondary voltage of 440 V and 0.8 pf lagging. The primary winding impedance is given by

$$Z_p = R_p + jX_p = 0.014 + j0.026 \text{ pu}$$

What is the voltage regulation in percent?

Solution

To determine the regulation, the primary voltage is required. Since per-unit impedance is given, the per-unit method will be used. The output voltage, (i.e., the secondary voltage) is 1.0 pu at rated conditions. The

output current is 1.0 pu as well. Since the power factor is 0.8 lagging, the current is $1.0\angle{-36.9°}$ per unit. The impedance in polar form is

$$Z_p = 0.014 + j0.026 \text{ pu} = 0.030\angle61.7° \text{ pu}$$

Using the model given in Fig. 36.1 and noting that the per-unit current in the secondary must also flow in the primary, the phase voltage at the primary of the transformer (the voltage at the dashed lines) is

$$V_{\text{primary phase}} = I_{\text{pu}}Z_p$$
$$= \left(1\angle{-36.9°}\right)\left(0.030\angle61.7°\right)$$
$$= 0.030\angle24.8° \text{ pu}$$

Reflecting this voltage to the secondary side does not require the turns ratio since per-unit values are in use.[2] Given that the secondary voltage is 1.0 pu, at an angle of 0° since it is the reference, the primary voltage is

$$V_{\text{primary reflected}} = V_{\text{primary phase,pu}} + V_{s,\text{pu}}$$
$$= 0.030\angle24.8° + 1.0\angle0°$$
$$= 1.027\angle0.7° \text{ pu}$$

The voltage regulation is not concerned with the resulting angle. The regulation is given by

$$\text{VR} = V_{\text{primary reflected,pu}} - V_{\text{rated,pu}}$$
$$= 1.027 - 1.0$$
$$= 0.027 \text{ or } 2.7\%$$

4. CONNECTIONS

Three-phase transformers can be connected as either delta or wye. When voltage or current ratios are given for three-phase transformers, they are assumed to specify the line conditions, regardless of the connection. If only the turns ratio, a, is given, the line quantities must be calculated. For delta connections, the line and phase quantities are related as in Eqs. 36.8 and 36.9.

$$V_l = V_\phi \quad \text{[delta]} \qquad \textit{36.8}$$

$$I_l = \sqrt{3}I_\phi \quad \text{[delta]} \qquad \textit{36.9}$$

For wye connections, the line and phase quantities are related as in Eqs. 36.10 and 36.11.

$$V_l = \sqrt{3}V_\phi \quad \text{[wye]} \qquad \textit{36.10}$$

$$I_l = I_\phi \quad \text{[wye]} \qquad \textit{36.11}$$

---

[2]The advantage of the per-unit system is that by choosing base primary and secondary voltages related by the turns ratio, which occurs in this example by letting the secondary voltage be the base and relating the primary voltage to it, the transformer is no longer required in the electrical model.

Because either the voltage or the current contains a factor of $\sqrt{3}$, the power for either connection type is given by

$$P = \sqrt{3} I_l V_l \text{ pf} = 3 I_\phi V_\phi \text{ pf} \qquad \textbf{36.12}$$

The line-phase relationships are illustrated in Fig. 36.2 for various connection types.[3] Figure 36.3 shows how the connections would be displayed in an electrical schematic.

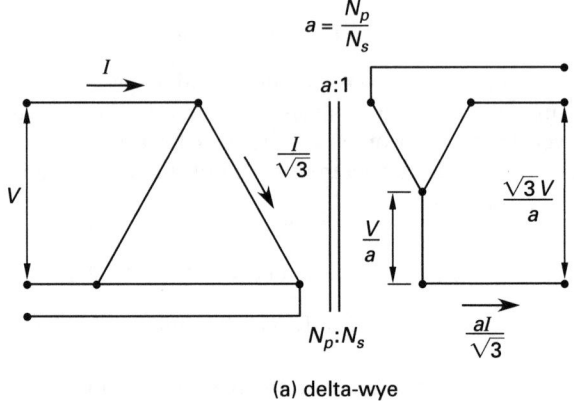

(a) delta-wye

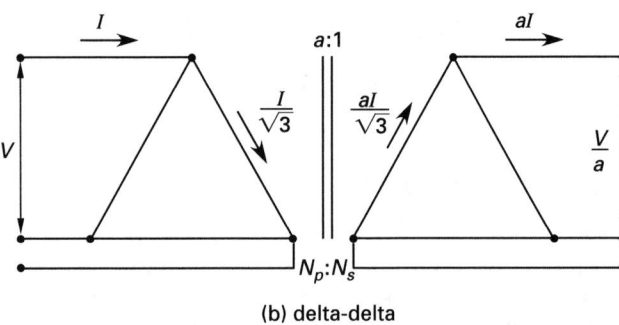

(b) delta-delta

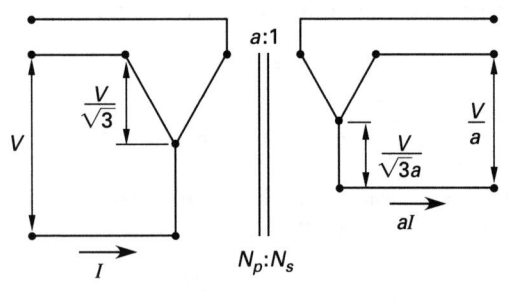

(c) wye-wye

Figure 36.2 Transformer Connections

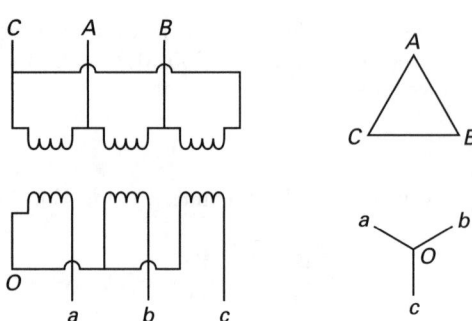

Figure 36.3 Standard Transformer Schematic

Wye connections, with a common neutral, offer economic and operating advantages (see Sec. 35-3). Delta connections allow third-harmonic voltages common to all transformers to circulate within the delta so that line electrical parameters are unaffected. For this reason, transformers normally have at least one delta-connected winding. Additionally, delta connections can suffer a failure of a single phase and still provide 57.7% of their rated load.[4]

5. TRANSFORMER TESTING

Transformer testing is done to determine the parameters for a real transformer, a model of which is shown in Fig. 36.1. Based on this model, testing determines the rating, efficiency, and equivalent circuit values. Two standardized tests are performed: open-circuit and short-circuit. Open-circuit and short-circuit conditions are established on the low-voltage side of the transformer, which can be either the primary or the secondary side.[5] The primary voltage and current, the secondary voltage and current, and the power are measured in each test. From these values, the transformer's properties are derived. The open-circuit test determines the core parameters and the turns ratio. The short-circuit test determines the winding impedances and verifies the turns ratio. The admittance parameter, Y, accounts for the power loss in the core. The susceptance parameter, B, accounts for energy storage in the core. The resistance parameter, R, accounts for power loss in the windings. The reactance parameter, X, accounts for the leakage (self-inductance) of the primary and secondary windings.

6. OPEN-CIRCUIT TEST

The open-circuit test determines the core parameters and the turns ratio. An open-circuit test is performed

---

[3]The turns ratio, a, is sometimes given as the ratio of the secondary to the primary turns. If this is the case, switch the a from the numerator (denominator) to the denominator (numerator) to obtain the correct relationship.

[4]When only two phases of a delta connection are purposefully used, it is called an *open delta* or *vee transformer*.

[5]Because of this, the terms "high-voltage" (HV) and "low-voltage" (LV) are sometimes used instead of the terms "primary" and "secondary." The terms "high-tension" and "low-tension," synonymous with high- and low-voltage, respectively, are also sometimes used.

by opening the secondary terminals of the transformer. Actually, a voltmeter measures the secondary voltage, V_{2oc}, but the meter has such high resistance it can be considered an open circuit. The rated voltage, V_{1oc}, is applied to the primary terminals. (The rated voltage is used because the core losses are dependent on the magnetic flux density, B_m.) The input current, I_{1oc}, is measured as well as the input power, P_{oc}. Because the current flow I_{1oc} is small, the voltage drop across Z_p is negligible. The open-circuit power, P_{oc}, represents the core power loss. The input open-circuit current, I_{1oc}, is the exciting current that maintains the flux in the core. The open-circuit model is shown in Fig. 36.4. The admittance is given by

$$Y_c = G_c + jB_c = \frac{I_{1oc}}{V_{1oc}} \qquad \textbf{36.13}$$

The conductance is

$$G_c = \frac{P_{oc}}{(V_{1oc})^2} \qquad \textbf{36.14}$$

The susceptance is

$$B_c = \frac{1}{X_c} = \frac{-1}{\omega L_c} = -\sqrt{Y_c^2 - G_c^2}$$

$$= \frac{-\sqrt{I_{1oc}^2 V_{1oc}^2 - P_{oc}^2}}{V_{1oc}^2} \qquad \textbf{36.15}$$

The susceptance is negative $(-1/\omega L)$ for a lagging condition.

The turns ratio is[6]

$$a_{\text{ps}} = \frac{V_{1oc}}{V_{2oc}} \qquad \textbf{36.16}$$

The power is

$$P_{oc} = V_{oc}^2 G_c \qquad \textbf{36.17}$$

The reactive power is

$$Q_{oc} = V_{1oc}^2 B_c \qquad \textbf{36.18}$$

The apparent power is

$$S_{oc} = V_{1oc}^2 Y_c = V_{1oc} I_{1oc} = \sqrt{P_{oc}^2 + Q_{oc}^2} \qquad \textbf{36.19}$$

---

[6]The subscript *ps* is added to clarify the turns ratio as being from the primary to the secondary, as is standard for this text. The turns ratio is sometimes used to indicate the secondary to primary ratio. Care should be taken to determine the definition used in a particular situation.

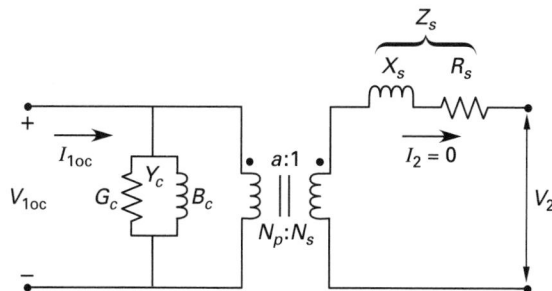

Figure 36.4 *Transformer Open-Circuit Test Model*

Example 36.4

An open-circuit test is conducted on a 120 V transformer winding. The results are $V_{1oc} = 120$ V, $V_{2oc} = 240$ V, $I_{1oc} = 0.25$ A, and $P_{oc} = 20$ W. Determine the transformer equivalent circuit element parameters given by this data.

Solution

The admittance is given by

$$Y_c = G_c + jB_c = \frac{I_{1oc}}{V_{1oc}} = \frac{0.25 \text{ A}}{120 \text{ V}}$$

$$= 2.08 \times 10^{-3} \text{ S}$$

The core conductance is given by

$$G_c = \frac{P_{oc}}{V_{1oc}^2} = \frac{20 \text{ W}}{(120 \text{ V})^2}$$

$$= 1.39 \times 10^{-3} \text{ S}$$

The core susceptance is given by

$$B_c = -\sqrt{Y_c^2 - G_c^2}$$

$$= -\sqrt{(2.08 \times 10^{-3} \text{ S})^2 - (1.39 \times 10^{-3} \text{ S})^2}$$

$$= -1.55 \times 10^{-3} \text{ S}$$

The turns ratio is

$$a_{\text{ps}} = \frac{V_{1oc}}{V_{2oc}} = \frac{120 \text{ V}}{240 \text{ V}} = 0.5$$

7. SHORT-CIRCUIT TEST

The short-circuit test determines the winding impedances and verifies the turns ratio. A short-circuit test is performed by shorting the secondary terminals of the transformer. Actually, an ammeter measures the secondary current, I_{2sc}, but the meter has such low resistance it can be ignored in the analysis of the circuit.

A voltage, $V_{\rm sc}$, is applied to the primary such that the rated current, that is, the volt-amp rating divided by the voltage rating, flows. The voltage will be low because the secondary winding has minimal impedance. The voltage is normally less than 5% of the rated value. Because of this minimal secondary impedance, the core admittance, Y_c, is considered shorted. The effective circuit then consists of the primary impedance in series with the reflected secondary impedance. The input current, $I_{\rm 1sc}$, and output current, $I_{\rm 2sc}$, are measured as well as the input power, $P_{\rm sc}$. The short-circuit power, $P_{\rm sc}$, represents the copper loss or $I^2 R$ losses in the windings. The short-circuit test model is shown in Fig. 36.5. The total impedance, Z, is

$$Z = R_p + jX_p + a_{ps}^2 \left(R_s + jX_s \right) = \frac{V_{\rm 1sc}}{I_{\rm 1sc}} \qquad 36.20$$

The total resistance, R, is

$$R = R_p + a_{ps}^2 R_s = \frac{P_{\rm sc}}{I_{\rm sc}^2} \qquad 36.21$$

To maximize efficiency, transformers are normally designed with R_p equal to $a_{ps}^2 R_s$. Thus,

$$R_p = a_{ps}^2 R_s = \frac{P_{\rm sc}}{2I_{\rm sc}^2} \qquad 36.22$$

The total reactance, X, is

$$X = X_p + a_{ps}^2 X_s = \frac{\sqrt{I_{\rm sc}^2 V_{\rm sc}^2 - P_{\rm sc}^2}}{I_{\rm sc}^2} \qquad 36.23$$

To maximize efficiency, transformers are normally designed with X_p equal to $a_{ps}^2 X_s$. Thus,

$$X_p = a_{ps}^2 X_s = \frac{Q_{\rm sc}}{2I_{\rm sc}^2} \qquad 36.24$$

The turns ratio is

$$a_{ps} = \frac{I_{\rm 2sc}}{I_{\rm 1sc}} \qquad 36.25$$

The power is

$$P_{\rm sc} = I_{\rm sc}^2 R = I_{\rm sc}^2 \left(R_p + a_{ps}^2 R_s \right) \qquad 36.26$$

The reactive power is

$$Q_{\rm sc} = I_{\rm sc}^2 X = I_{\rm sc}^2 \left(X_p + a_{ps}^2 X_s \right) \qquad 36.27$$

The apparent power is

$$S_{\rm sc} = I_{\rm 1sc}^2 Z_{\rm sc} = V_{\rm 1sc} I_{\rm 1sc} = \sqrt{P_{\rm sc}^2 + Q_{\rm sc}^2} \qquad 36.28$$

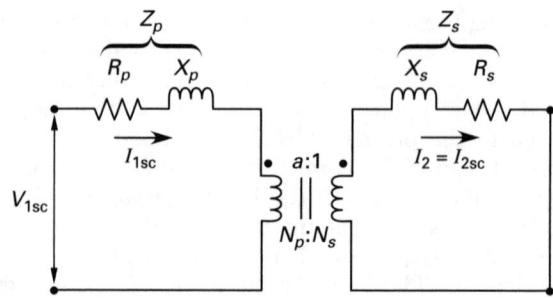

Figure 36.5 *Transformer Short-Circuit Test Model*

Example 36.5

A short-circuit test is conducted on a transformer rated at 15 kVA and 1300 primary volts. What is the value of the input current, $I_{\rm 1sc}$?

Solution

Short-circuit tests are conducted at the rated current, that is, $I_{\rm 1sc} = I_{\rm rated}$. The rated current is found from

$$I_{\rm rated} = \frac{S_{\rm rated}}{V_{\rm rated}} = \frac{15 \times 10^3 \text{ VA}}{1300 \text{ V}} = 11.5 \text{ A}$$

8. ABCD PARAMETERS

ABCD parameters, also known as *transfer* or *chain parameters*, are analytic tools for transmission and distribution problem solving. To use them on transformers, the secondary impedance is reflected to the primary as shown in Fig. 36.6. The result is a two-port network (see Sec. 29-31). The ABCD parameters for any two-port network are given by

$$V_{\rm in} = A V_{\rm out} - B I_{\rm out} \qquad 36.29$$

$$I_{\rm in} = C V_{\rm out} - D I_{\rm out} \qquad 36.30$$

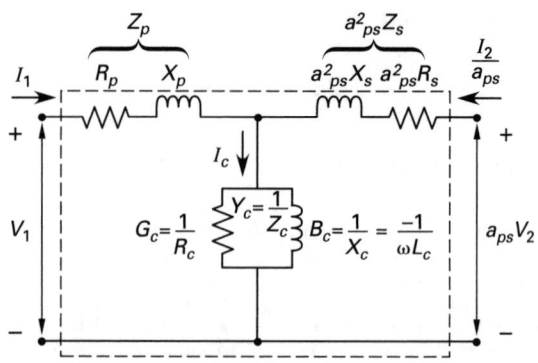

Figure 36.6 *Transformer Two-Port Network*

Circuit analysis of the transformer yields Eqs. 36.31 through 36.35, which are used to determine the ABCD parameters.[7]

$$V_1 = a_{ps}V_2\left(1 + Z_pY_c\right)$$
$$- \left(\frac{I_2}{a_{ps}}\right)\left(Z_p + a_{ps}^2 Z_s\left(1 + Z_pY_c\right)\right) \quad \textbf{36.31}$$

$$I_1 = a_{ps}V_2Y_c - \left(\frac{I_2}{a_{ps}}\right)\left(1 + a_{ps}^2 Z_sY_c\right) \quad \textbf{36.32}$$

$$Z_p = R_p + jX_p \quad \textbf{36.33}$$

$$Z_s = R_s + jX_s \quad \textbf{36.34}$$

$$Y_c = G_c + jB_c \quad \textbf{36.35}$$

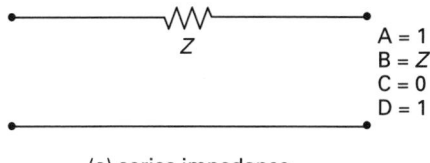

(a) series impedance

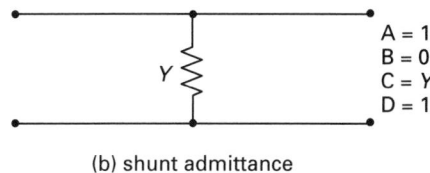

(b) shunt admittance

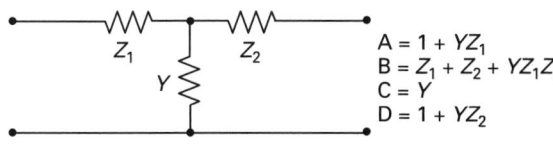

(c) unbalanced tee network

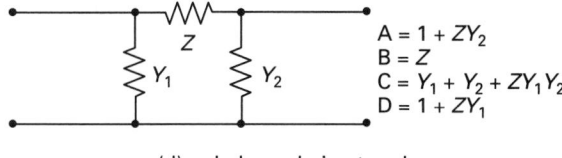

(d) unbalanced pi network

Figure 36.7 *Two-Port Network ABCD Parameters*

---

[7]Care should be used when dealing with the susceptance in Eq. 36.35. Susceptance is negative for an inductive element. Both the susceptance, B_c, and the associated reactance, X_c, carry any negative sign internally.

To maximize efficiency, transformers are normally designed with Z_p equal to $a_{ps}^2 Z_s$. In this case, the ABCD parameters are given by Eqs. 36.36 through 36.39.

$$A = 1 + Z_pY_c \quad \textbf{36.36}$$

$$B = Z_p\left(2 + Z_pY_c\right) \quad \textbf{36.37}$$

$$C = Y_c \quad \textbf{36.38}$$

$$D = A \quad \textbf{36.39}$$

ABCD parameters can be used for two-port networks that are chained together, such as power distribution systems and cascaded amplifiers. For two such cascaded networks, the ABCD parameters are given by Eqs. 36.40 through 36.43.

$$A = A_1A_2 + B_1C_2 \quad \textbf{36.40}$$

$$B = A_1B_2 + B_1D_2 \quad \textbf{36.41}$$

$$C = A_2C_1 + C_2D_1 \quad \textbf{36.42}$$

$$D = B_2C_1 + D_1D_2 \quad \textbf{36.43}$$

The ABCD parameters for common two-port network types are given in Fig. 36.7.

9. TRANSFORMER TYPES

Transformer principles and the analytic methods in this chapter and Ch. 28 are applicable to all types of transformers. The model used changes based on the transformer rating and the relative magnitudes of the electrical parameters, some of which are small enough to ignore in certain situations.

Power transformers are generally those transformers rated at approximately 500 kVA or greater that transfer energy between the generator and primary distribution circuits. In such large transformers, the total leakage reactance, X, is five times the value of the total resistance, R. As a result, the transformer can be accurately represented by a reactance only as shown in Fig. 36.8(a).

Distribution transformers are generally those transformers rated at approximately 500 kVA or less that transfer electrical energy from a primary distribution circuit to a secondary distribution circuit. Because the transformer primaries are connected to a large, essentially infinite bus, the voltage and power factors can be considered constant. The load on an individual distribution transformer does not change the primary voltage. Consequently, the distribution transformer can be represented by the series portion of the equivalent impedance only, that is, by the total leakage reactance and the total winding resistance as shown in Fig. 36.8(b).

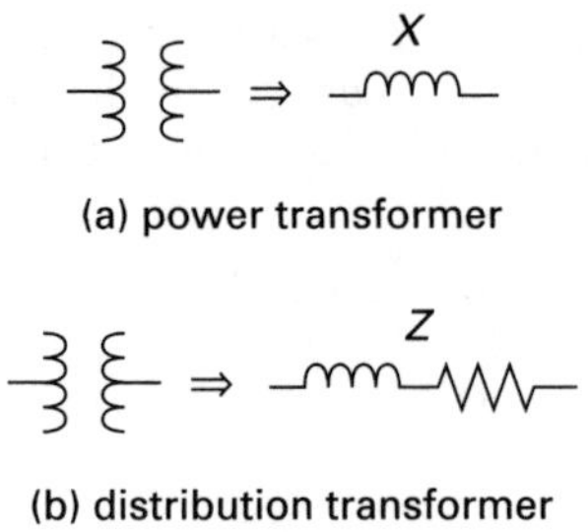

(a) power transformer

(b) distribution transformer

Figure 36.8 *Special Transformer Models*

Typical per-unit values for transformer parameters are given in Table 36.1.

Table 36.1 *Transformer Typical Values*

parameter (see Fig. 36.6)	typical per-unit values 3–250 kVA	1–100 MVA
R_p or R_s	0.009–0.005	0.005–0.002
X_p or X_s	0.008–0.025	0.030–0.060
R_c	20–50	100–500
X_c	20–30	30–50
I_c	0.05–0.03	0.03–0.02

Furnace transformers supply electric furnaces (welding machines) of the induction, resistance, open-arc, or submerged-arc types. The secondary voltage is in the range of a few hundred volts with currents as high as 100,000 A. *Grounding transformers* act as a neutral point for grounding purposes. *Instrument transformers* adjust current and voltage to levels that metering devices utilize while isolating the metering circuit from the system measured.

37 Power Transmission Lines

Nomenclature

a	phase	–
A	ABCD parameter	–
A	area	m^2
b	phase	–
B	ABCD parameter	–
B	susceptance	S, Ω^{-1}, or $\mho$
c	phase	–
c	speed of light	3.00×10^8 m/s
C	ABCD parameter	–
C	capacitance	F
D	ABCD parameter	–
D	distance	m
f	frequency	Hz, s^{-1}, or cycles/s
G	conductance	S, Ω^{-1}, or $\mho$
GMD	geometric mean distance	m
GMR	geometric mean radius	m
I	effective or DC current	A
I	rms phasor current	A
K	skin effect ratio	–
K	correction factor	–
l	length	m
L	inductance	H
M	mutual inductance	H
P	power	W
r	radius	m
R	resistance	Ω
SWR	standing wave ratio	–
T	temperature	°C or K
v	velocity (speed)	m/s
V	effective or DC voltage	V
V	rms phasor voltage	V
x	variable	–

X	reactance	Ω
y	admittance per unit length	S/m, 1/Ω·m, or $\mho$/m
Y	admittance	S, Ω^{-1}, or $\mho$
z	impedance per unit length	Ω/m
Z	impedance	Ω
Z_0	characteristic impedance	Ω

Symbols

α	attenuation constant	Np/m
α	thermal coefficient of resistance	1/°C
β	phase constant	rad/m
γ	propagation constant	rad/m
Γ	reflection coefficient	–
δ	skin depth	m
Δ	change, final to initial	–
ε	permittivity	F/m
ε_0	free-space permittivity	8.854×10^{-12} F/m
ε_r	relative permittivity	–
η	efficiency	–
θ	phase angle	rad
μ	permeability	H/m
μ_0	free-space permeability	1.2566×10^{-6} H/m
μ_r	relative permeability	–
ρ	resistivity	Ω·m
σ	conductivity	S/m

Subscripts

0	initial
0	free space (vacuum)
0	characteristic
ab	a to b
AC	alternating current
bc	b to c
c a	c to a
C	capacitor
Cu	copper
DC	direct current
e	equivalent
eff	effective
ext	external
fl	full load
int	internal
l	per unit length
ll	line-to-line
L	inductor

Power–Transmission

m	mutual
nl	no load
R	receiving end
R	resistance
S	sending end
w	wave

1. FUNDAMENTALS

A *transmission line* is a system of conductors, such as wires, waveguides, or coaxial cables, that conducts power or transmits signals from one terminal to another.[1] Specifically, a transmission line guides *transverse electromagnetic waves* (TEM). In transverse electromagnetic waves, the electric and magnetic fields are perpendicular to the direction of propagation. At low frequencies, where the dimensions of the line are small compared to the wavelength, lumped parameters provide an adequate model of the circuit.[2] This is because the displacement current is small and can be ignored. That is, the storage of electromagnetic energy in the space surrounding the line can be ignored. At high frequencies, the dimensions of the line are large compared to the wavelengths propagated. The displacement current becomes significant. That is, electric and magnetic energy is stored in the space surrounding the line (see Chs. 24 and 25).

To avoid using field theory to analyze transmission lines, distributed parameters are used along with currents and voltages associated with the electric and magnetic fields. *Distributed parameters* are circuit elements that exist along the entire line but are represented per unit length, which allows the use of common electrical analytic techniques. A transmission line represented by distributed parameters is illustrated in Fig. 37.1.

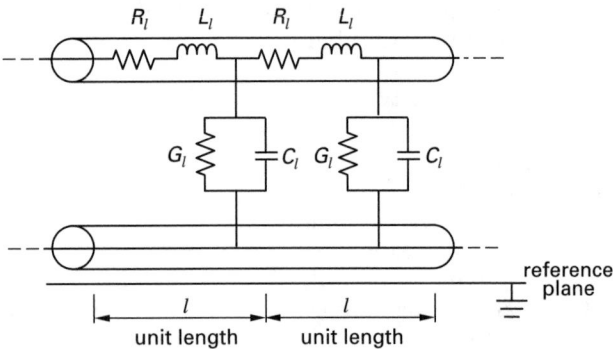

Figure 37.1 *Distributed Parameters*

---

[1]The transmission of signals is normally considered a communications function. Communications systems thus use transmission lines as well. The lines tend to be high frequency with different properties than power transmission lines. Waveguides are used in high-frequency transmission when the wavelength is less than 10 cm. High-frequency transmission lines are covered in the Communications topic of this book, Chs. 56 through 61.

[2]Lumped parameters are single-valued circuit elements that are equivalent to the value in the circuit as a whole, such as a resistor, inductor, or capacitor in a specific position in the circuit that accounts for the line electrical properties as well.

The distributed parameters in Fig. 37.1 are the resistance per unit length, R_l, inductance per unit length, L_l, capacitance per unit length, C_l, and *shunt conductance* per unit length, G_l. The resistance depends on the frequency because of a phenomenon known as the *skin effect*: as frequency rises, current flows closer to the surface of the conductor, thus raising the resistance. (The AC resistance at 60 Hz is 5 to 10% higher than the DC resistance.) The *temperature effect* on resistance must also be accounted for in transmission line design. The inductance consists of two terms, the internal inductance, L_{int}, and the external inductance, L_{ext}. The internal inductance depends on the skin effect, decreasing as the frequency increases. The external inductance depends on the geometric arrangement of the conductors. The capacitance also depends on the geometric arrangement of the conductors. The shunt conductance, also called the *line-to-line conductance*, is usually insignificant.

Additional parameters of concern include the characteristic impedance, the standing wave ratio, and the reflection coefficient.[3] The *characteristic impedance* is determined by

$$Z_0 = \sqrt{\frac{Z_l}{Y_l}} \quad \text{[in } \Omega\text{]} \qquad 37.1$$

If the transmission line is terminated with the characteristic impedance, no power is reflected back to the source. If the termination impedance is other than Z_0, then the power (signal) from the generator (source) will be partially reflected back to the generator. In wave terminology, the generator's waves will then combine in the transmission line with the reflected waves to form *standing waves*. The *standing wave ratio*, SWR, is the ratio of the maximum to the minimum voltages (currents) encountered along the transmission line. Typically, the SWR is greater than unity. If the terminating impedance matches the characteristic impedance, all the input power provided by the generator will be absorbed by the load and the SWR will equal one. The standing wave ratios are given by

$$\text{SWR} = \text{VSWR} = \text{CSWR}$$

$$= \max \begin{cases} Z_{\text{load}}/Z_0 \\ Z_0/Z_{\text{load}} \end{cases} \qquad 37.2$$

The *reflection coefficient*, Γ, is the ratio of the reflected voltage (current) to the incident voltage (current). The reflection coefficient is given by

$$\Gamma = \frac{V_{\text{reflected}}}{V_{\text{incident}}} = \frac{I_{\text{reflected}}}{I_{\text{incident}}}$$

$$= \left| \frac{Z_{\text{load}} - Z_0}{Z_{\text{load}} + Z_0} \right| \qquad 37.3$$

---

[3]The Smith chart is a convenient way to analyze these three parameters for low-loss lines and high-frequency lines.

The fraction of incident power that is reflected back to the source from the load is Γ^2. The relationship between the reflection coefficient and the standing wave ratio is given by Eq. 37.4.

$$\Gamma = \frac{\text{SWR} - 1}{\text{SWR} + 1} \qquad 37.4$$

The *velocity of propagation*, v_w, is the velocity at which a wave propagates along a transmission line and is given by

$$v_w = \frac{1}{\sqrt{L_l C_l}} \qquad 37.5$$

Transmission lines are considered in three lengths.[4] The *short transmission line* is from 0 to 80 km (0 to 50 mi). In a short line, the shunt parameters, G_l and C_l, are insignificant and ignored in the analysis. The *medium transmission line* is from 80 to 240 km (50 to 150 mi). In a medium line, the shunt capacitances are generally lumped in predetermined locations along the line. The *long transmission line* is greater than 240 km (150 mi). A long transmission line is also called a *uniform transmission line* since it has constant distributed parameters along the line from the sending end to the receiving end.

Many transmission line characteristics and formulas are tabulated in English Engineering System units. For this reason, a mixture of metric and English Engineering System units is utilized in this chapter. Additionally, all AC quantities have a magnitude and an angle. However, as is common practice, these items will be shown in phasor notation, or complex notation, only when possible confusion could occur.[5]

2. DC RESISTANCE

Resistance generates the I^2R power loss in transmission lines. Additionally, the IR drop, or voltage drop, affects the voltage regulation. The DC resistance, often called the *characteristic resistance*, R_0, of a conductor of length l, cross-sectional area A, and resistivity ρ is normally given as $\rho l/A$. Since the distributed parameters are given as per-unit-length quantities, the DC resistance per unit length is

$$R_{l,\text{DC}} = R_0 = \frac{\rho}{A} \quad [\text{in } \Omega/\text{m}] \qquad 37.6$$

The resistance sensitivity to temperature is calculated from

$$R_{\text{final}} = R_{\text{initial}}(1 + \alpha\Delta T) \qquad 37.7$$

---

[4]The distances given assume a 60 Hz signal. The engineering methods in each type of transmission line differ. As the frequency increases, the distances of each definition decrease, with the concurrent adjustment in the engineering method used.
[5]Phasor or complex notation usually presents the parameter in bold or with an arrow over the top ($\mathbf{V}$ or $\vec{V}$).

Voltage drops are often tabulated for various conductor sizes as shown in Table 37.1.

Table 37.1 *Voltage Drop in Two-Wire DC Circuits (Loop)[a]*

conductor size (AWG)		voltage drop per 100,000 A·m
copper	approximately equivalent aluminum	
6	4	31.3
4	2	19.7
2	1/0	12.4
1/0	3/0	7.80
2/0	4/0	6.19

[a]Values are calculated for 32°C (90°F) conductors.

Example 37.1

A 50 kW DC device is located 1.6 km from a motor-generator with a maximum output voltage of 600 V_{DC}. If the lowest voltage the DC device can properly utilize is 575 V, what is the minimum conductor size required?

Solution

The maximum allowable voltage drop is 25 V. To use Table 37.1, the total ampere-meters for this system is required. The maximum current is given by

$$P = IV$$
$$I = \frac{P}{V} = \frac{50 \times 10^3 \text{ W}}{575 \text{ V}} = 86.96 \text{ A}$$

The total cable distance to reach the site and provide the return path to the generator is

$$D = (2)(1.6 \times 10^3 \text{ m}) = 3.2 \times 10^3 \text{ m}$$

The total A·m is

$$(86.96 \text{ A})(3.2 \times 10^3 \text{ m}) = 278.27 \times 10^3 \text{ A·m}$$

The voltage drop entry point for the table, x, is

$$\text{allowable voltage drop} = (\text{A·m}_{\text{total}})$$
$$\times \left(\frac{\text{voltage drop}}{100,000 \text{ A·m}}\right)$$
$$25 \text{ V} = (278.27 \times 10^3 \text{ A·m})$$
$$\times \left(\frac{x}{100,000 \text{ A·m}}\right)$$
$$x = 8.98 \text{ V}$$

From Table 37.1, the conductor size with an 8.98 voltage drop, or less, is 1/0 for copper or 3/0 for aluminum.

3. SKIN EFFECT

The *skin effect* is the tendency of AC currents to flow near the surface of a conductor. This results in the AC current being restricted to less than the total cross-sectional area of the conductor, resulting in an increased resistance and a decreased internal inductance. The skin effect can be thought of as an electromagnetic phenomenon related to the time required for an external field to penetrate a conductor.[6] It occurs because the surface reactance is smaller than the reactance of other possible current paths within the conductor. The *skin depth*, δ, is the depth beneath the surface of the conductor that is carrying the current at a given frequency due to electromagnetic waves incident to the surface.[7] Specifically, it is the depth at which the current density is $1/e$, the value of the surface current density. The skin depth for a flat conducting plate is

$$\delta = \frac{1}{\sqrt{\dfrac{\pi f \mu}{\rho}}} = \frac{1}{\sqrt{\pi f \mu \sigma}} \qquad 37.8$$

The AC resistance per unit length for a flat conducting plate of unit width w is

$$R_{l,\text{AC}} = \frac{\rho}{\delta w} \quad [\text{in } \Omega/\text{m}] \qquad 37.9$$

Conductors are normally round. Equation 37.9 would be applicable only if the skin depth were much less than the radius, r, so that the wire approximated a flat surface. For example, a ratio of $r/\delta > 5.5$ results in less than a 10% error. The actual skin depth for copper conductors is

$$\delta_{\text{Cu}} = \frac{0.066}{\sqrt{f}} \quad [\text{in } \text{m}] \qquad 37.10$$

Example 37.2

Determine the copper wire radius in mils where the skin depth equals the radius for the following frequencies: 60 Hz, 10 kHz, 1 MHz.

Solution

First, convert Eq. 37.10 into an equivalent equation based on mils rather than meters.

---

[6]It is this principle that allows metal shielding to be used to prevent electromagnetic wave penetration of equipment. Conversely, such shielding also prevents the escape of radiated electromagnetic waves.

[7]In generators, the armature current is induced from an electromagnetic wave incident to the surface. Thus, this principle applies for generators and any conductor cabling attached to the generator.

$$\delta_{\text{Cu}} = \left(\frac{0.066}{\sqrt{f}}\right)\left(\frac{1 \text{ mil}}{0.001 \text{ in}}\right)\left(\frac{1 \text{ in}}{2.54 \text{ cm}}\right)\left(\frac{100 \text{ cm}}{1 \text{ m}}\right)$$

$$= \frac{2.60 \times 10^3}{\sqrt{f}} \text{ mil}$$

Substitute the given frequencies to determine the skin depth.

$$\delta_{\text{Cu}} = \frac{2.60 \times 10^3}{\sqrt{f}} \text{ mil} = \frac{2.60 \times 10^3}{\sqrt{60 \text{ s}^{-1}}} \text{ mil} = 336 \text{ mil}$$

$$\delta_{\text{Cu}} = \frac{2.60 \times 10^3}{\sqrt{f}} \text{ mil} = \frac{2.60 \times 10^3}{\sqrt{10 \times 10^3 \text{ s}^{-1}}} \text{ mil} = 26 \text{ mil}$$

$$\delta_{\text{Cu}} = \frac{2.60 \times 10^3}{\sqrt{f}} \text{ mil} = \frac{2.60 \times 10^3}{\sqrt{1 \times 10^6 \text{ s}^{-1}}} \text{ mil} = 2.6 \text{ mil}$$

The largest solid copper wire is AWG 4/0, that is, AWG 0000.[8] The radius is 230 mil, so the diameter is 460 mil. This indicates that at 60 Hz the electromagnetic wave does not completely penetrate the conductor. That is, some skin effect is occurring, as the skin depth is less than the diameter. By contrast, AWG 1/0 has a diameter of approximately 330 mil and would be completely penetrated by a 60 Hz wave. Thus, skin effect occurs at 60 Hz but only for the largest conductors, such as those used for commercial power applications. For most other applications, the 10 kHz frequency mark is often used as the point where skin effect must be accounted for and AC resistance calculated.

4. AC RESISTANCE

Skin effect and the resulting increase in resistance and decrease in internal inductance are often neglected in initial calculations or approximations. In the case of large copper wires at commercial power frequencies, exact calculations must consider the effect.[9] The effect on aluminum conductors is less because aluminum has a higher resistivity. Tables of skin effect ratios are available when the exact value of the AC resistance is required.[10] Table 37.2 gives sample ratios. The skin effect ratio for the AC resistance, K_R, is used to determine the effective resistance from

$$R_{\text{AC}} = K_R R_{\text{DC}} \qquad 37.11$$

---

[8]Any larger size consists of stranded conductors.

[9]The effect should be accounted for at communications system frequencies as well.

[10]Skin effect ratios are also called *AC/DC resistance ratios*.

The table is entered using a value of x determined from

$$x = 2\pi r \sqrt{\frac{2f\mu}{\rho}} = \frac{2r\sqrt{2\pi}}{\delta} \qquad 37.12$$

Table 37.2 Skin Effect Ratios

x	$K_R{}^a$	$K_{L,\text{int}}{}^b$
0.0	1.00000	1.00000
0.5	1.00032	0.99984
1.0	1.00519	0.99741
2.0	1.07816	0.96113
4.0	1.67787	0.68632
8.0	3.09445	0.35107
16.0	5.91509	0.17649
32.0	11.56785	0.08835
50.0	17.93032	0.05656
100.0	35.60666	0.02828
∞	∞	0

$^a K_R$ is the skin effect ratio for AC resistance.
$^b K_{L,\text{int}}$ is the skin effect ratio for internal inductance.

The use of tables is the most accurate practical method of determining the AC resistance. Curves normalized to the DC resistance are sometimes used for intermediate frequencies. The AC resistance can also be found by calculating the effective area of the conductor, that is, the area actually conducting current, and comparing this area to that used by the DC resistance. The DC resistance is given by Eq. 37.6 as ρ/A. Since the resistivity doesn't change, comparing an AC resistance to a DC resistance on the same wire gives

$$\frac{R_{l,\text{AC}}}{R_{l,\text{DC}}} = \frac{A}{A_{\text{eff}}} \qquad 37.13$$

The effective area is

$$A_{\text{eff}} = \pi r^2 - \pi (r - \delta)^2 \qquad 37.14$$

Example 37.3

An AWG #20 copper conductor has a diameter of 32 mil with a DC resistance of approximately 33 Ω/km at 25°C. Calculate the resistance per kilometer at 10 kHz with no temperature change.

Solution

Equation 37.13 contains the relationship desired. The skin depth, δ, from Ex. 37.2 is 26 mil. The radius is 16 mil. Rearranging Eq. 37.13 and substituting Eq. 37.14 gives

$$\frac{R_{l,\text{AC}}}{R_{l,\text{DC}}} = \frac{A}{A_{\text{eff}}}$$

$$R_{l,\text{AC}} = R_{l,\text{DC}} \left(\frac{A}{A_{\text{eff}}} \right)$$

$$= R_{l,\text{DC}} \left(\frac{\pi r^2}{\pi r^2 - \pi (r - \delta)^2} \right)$$

$$= R_{l,\text{DC}} \left(\frac{\pi r^2}{\pi r^2 \left(1 - \left(1 - \frac{\delta}{r} \right)^2 \right)} \right)$$

$$= \left(33 \ \frac{\Omega}{\text{km}} \right) \left(\frac{1}{1 - \left(1 - \frac{26 \ \text{mil}}{16 \ \text{mil}} \right)^2} \right)$$

$$= 54 \ \Omega/\text{km}$$

5. INTERNAL INDUCTANCE

The *internal inductance*, also called the *characteristic inductance*, is that inductance associated with the interior of a solid conductor. That is, the fluxes in the space surrounding the conductor are disregarded. For nonmagnetic materials, the relative permeability, μ_r, is approximately one. Thus, the magnetic permeability of a conductor is approximately that of free space, that is, $\mu \approx \mu_0$. The internal inductance per unit length for a conductor is given by

$$L_{l,\text{int}} = \frac{\mu_0}{8\pi} = 0.5 \times 10^{-7} \ \text{H/m} \qquad 37.15$$

For a single-phase system, two conductors are present, with current flow in opposite directions, and the total internal inductance is twice the value given in Eq. 37.15. The *single-phase internal inductance* is

$$L_{l,\text{int}} = \frac{\mu_0}{4\pi} = 1.0 \times 10^{-7} \ \text{H/m} \qquad 37.16$$

6. EXTERNAL INDUCTANCE

The *external inductance* is that inductance associated with the exterior of a solid conductor. That is, the fluxes in the space surrounding the conductor store the inductive energy. For air, the relative permeability, μ_r, is approximately one. Thus, the magnetic permeability of air is approximately that of free space, that is, $\mu \approx \mu_0$. The internal inductance per unit length for a single conductor is given by

$$L_{l,\text{ext}} = \left(\frac{\mu_0}{2\pi} \right) \ln \left(\frac{D}{r} \right)$$

$$= 2 \times 10^{-7} \ln \left(\frac{D}{r} \right) \quad [\text{in H/m}] \qquad 37.17$$

The term D is the distance between the centers of the conductors. The term r is the radius of the conductor. For a single-phase system, two conductors are present, with current flow in opposite directions, and the total external inductance is twice the value given in Eq. 37.17. The *single-phase external inductance* is

$$L_{l,\text{ext}} = \left(\frac{\mu_0}{\pi}\right) \ln\left(\frac{D}{r}\right)$$

$$= 4 \times 10^{-7} \ln\left(\frac{D}{r}\right) \quad \text{[in H/m]} \qquad 37.18$$

7. SINGLE-PHASE INDUCTANCE

The total inductance of a single-phase system, which consists of two conductors, is the sum of the internal and external inductances.[11]

$$L_l = L_{l,\text{int}} + L_{l,\text{ext}}$$

$$= \left(\frac{\mu_0}{4\pi}\right)\left(1 + 4\ln\left(\frac{D}{r}\right)\right) \quad \text{[in H/m]} \qquad 37.19$$

The first term of Eq. 37.19 represents the internal inductance of a solid conductor. The second term represents the inductance due to external fluxes. Equation 37.19 is simplified using the concept of the *geometric mean radius* (GMR), defined as[12]

$$\text{GMR} = re^{-\frac{1}{4}} \qquad 37.20$$

Substituting the geometric mean radius for the radius in Eq. 37.19 gives the following simplified form for the total single-phase inductance per unit length.

$$L_l = \left(\frac{\mu_0}{\pi}\right) \ln\left(\frac{D}{\text{GMR}}\right)$$

$$= 4 \times 10^{-7} \ln\left(\frac{D}{\text{GMR}}\right) \quad \text{[in H/m]} \qquad 37.21$$

Equation 37.21 is equivalent to Eq. 37.19 but represents a different model of the conductor. Specifically, Eq. 37.21 represents an equivalent thin-walled, hollow conductor of radius GMR with no internal flux linkage and thus no internal inductance.

Example 37.4

The wire in Ex. 37.3 is used in a single-phase system operating at 10 kHz with the lines spaced 0.5 m apart. Ignoring shunt effects, what is the impedance per unit length of line?

---

[11]The equations used in this text assume that $D \gg r$ and that r is identical in the two conductors, which is generally the case in power transmission lines.
[12]The geometric mean radius is sometimes called the *self geometric mean distance* (GMD).

Solution

Ignoring the shunt conductance and capacitance gives impedance per unit length of

$$Z_l = R_l + jX_{L,l}$$

The resistance per unit length was determined in Ex. 37.3 to be 54 Ω/km or 54×10^{-3} Ω/m. The inductive reactance per unit length is

$$X_{L,l} = 2\pi f L_l$$

Thus, the only unknown is the inductance. Using Eqs. 37.19 and 37.20, and noting that a radius of 16 mil is 4.064×10^{-4} m, gives

$$L_l = 4 \times 10^{-7} \ln\left(\frac{D}{\text{GMR}}\right)$$

$$= 4 \times 10^{-7} \ln\left(\frac{D}{re^{-\frac{1}{4}}}\right)$$

$$= 4 \times 10^{-7} \ln\left(\frac{0.5 \text{ m}}{(4.064 \times 10^{-4} \text{ m})\, e^{-\frac{1}{4}}}\right)$$

$$= 2.95 \times 10^{-6} \text{ H/m}$$

The reactance per unit length is

$$X_{L,l} = 2\pi f L_l$$

$$= (2\pi)(10 \times 10^3 \text{ Hz})(2.95 \times 10^{-6} \text{ H/m})$$

$$= 1.85 \times 10^{-1} \text{ } \Omega/\text{m}$$

The impedance per unit length is

$$Z_l = R_l + jX_{L,l} = 54 \times 10^{-3} + j1.85 \times 10^{-1} \text{ } \Omega/\text{m}$$

Note that the inductive reactance has an effect approximately four times greater than the resistance.

8. SINGLE-PHASE CAPACITANCE

The shunt capacitance is the capacitance between the solid conductors. Power transmission line conductors are normally separated by air. For air, the relative permittivity, ε_r, is approximately one. Thus, the permittivity of air is approximately that of free space, that is, $\varepsilon \approx \varepsilon_0$. The capacitance of a single-phase system, which consists of two conductors, is

$$C_l = \frac{\pi\varepsilon_0}{\ln\left(\dfrac{D}{r}\right)} = \frac{2.78 \times 10^{-11}}{\ln\left(\dfrac{D}{r}\right)} \quad \text{[in F/m]} \qquad 37.22$$

The term D represents the distance between the centers of the conductors. The term r is the radius of the conductors. For a single-phase system, two conductors are present, with current flow in opposite directions.

Example 37.5

Copper wire, AWG #20, is used in a single-phase system at 10 kHz, with the conductors 0.5 m apart. What is the capacitive reactance per unit length?

Solution

The capacitive reactance per unit length is given by

$$X_{C,l} = \frac{1}{2\pi f C_l}$$

From Ex. 37.4, AWG #20 has a radius of 16 mil or 4.064×10^{-4} m. The capacitance per unit length is

$$C_l = \frac{\pi \varepsilon_0}{\ln\left(\dfrac{D}{r}\right)}$$

$$= \frac{\pi \left(8.854 \times 10^{-12} \; \dfrac{\text{F}}{\text{m}}\right)}{\ln\left(\dfrac{0.5 \text{ m}}{4.064 \times 10^{-4} \text{ m}}\right)}$$

$$= 3.91 \times 10^{-12} \text{ F/m}$$

The capacitive reactance is

$$X_C = \frac{1}{2\pi f C}$$

$$= \frac{1}{(2\pi)(10 \times 10^3 \text{ s}^{-1})(3.91 \times 10^{-12} \text{ F})}$$

$$= 4.07 \times 10^6 \; \Omega$$

$$X_{C,l} = 4.0 \times 10^6 \; \Omega/\text{m}$$

Note that the capacitive reactance has a much larger effect on the per-unit reactance of the line than the inductive reactance.

9. THREE-PHASE TRANSMISSION

In three-phase, three-wire, or four-wire balanced transmission, the sum of the instantaneous currents is zero, and the sum of the instantaneous voltages is zero.[13] The conductors are assumed to be equilaterally spaced at distance D. When the conductors are not symmetrically arranged, as is often the case, formulas used for the inductance and capacitance are still valid if the equivalent distance, D_e, is substituted for the distance (see Fig. 37.2(a) and (b)).[14]

$$D_e = \sqrt[3]{D_{ab}D_{bc}D_{ca}} \qquad 37.23$$

---

[13]The three-wire or four-wire balanced assumption allows simplifications in the derivations of the equations in this section.
[14]The equivalent spacing is sometimes called the *mutual geometric mean distance* (D_m or GMD).

Nonsymmetrical spacing causes electrostatic and electromagnetic imbalance among the lines, resulting in unequal phase voltages and currents. In large commercial power transmission systems, the effect is minimal because of the balancing effect of the rotating loads. The effect can be minimized by *transposition* of the lines as illustrated in Fig. 37.2(c). Transposing the lines improves reliability by minimizing the current induced in the event of a fault, reduces power losses, and reduces interference with nearby telecommunications lines.

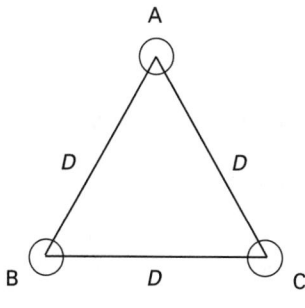

(a) symmetrical spacing

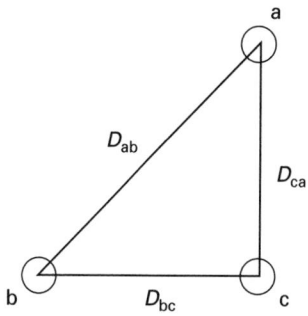

(b) unsymmetrical spacing

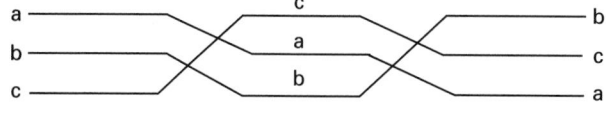

(c) transposition

Figure 37.2 *Transmission Line Spacing*

The per-phase inductance per unit length of a three-phase transmission line is

$$L_l = \frac{\mu_0}{2\pi} \ln\left(\frac{D_e}{\text{GMR}}\right)$$

$$= 2 \times 10^{-7} \ln\left(\frac{D_e}{\text{GMR}}\right) \quad \text{[in H/m]} \qquad 37.24$$

Mutual inductance, M, also exists, but for symmetrically spaced conductors the total is zero. This equation is similar to Eq. 37.21 for the total inductance of a single-phase system. Specifically, it is one-half the single-phase value. Per-phase values are equivalent to phase-to-neutral values.

The per-phase capacitance per unit length of a three-phase transmission line is

$$C_l = \frac{2\pi\varepsilon_0}{\ln\left(\dfrac{D_e}{r}\right)} = \frac{5.56 \times 10^{-11}}{\ln\left(\dfrac{D_e}{r}\right)} \quad \text{[in F/m]} \quad 37.25$$

This equation is similar to Eq. 37.22 for the total capacitance of a single-phase system. Specifically, it is twice the single-phase value. Per-phase values are equivalent to phase-to-phase values.

10. POWER TRANSMISSION LINES

Power transmission lines normally consist of aluminum strands with a steel core or all-aluminum alloy wires. The lines consist of either three or four conductors. The voltage drop in such lines can be calculated using the principles in Secs. 37-3 through 37-9 and the basic formula

$$\mathbf{V}_l = \mathbf{I}R_l + j\mathbf{I}X_l = \mathbf{I}\mathbf{Z}_l \qquad 37.26$$

The actual calculation is complicated. Frequency and spacing affect the resistance, inductance, and capacitance. The geometric mean radius given in Eq. 37.20 must be adjusted for stranded conductors, taking into account both the total number and the layering geometry.

Such calculations are unnecessary, because the properties of standard conductors have been tabulated and the conductors given code names for ease of reference. A sample table is given in App. 37.A, which uses English Engineering System units and a 1 ft symmetrical spacing between conductors. The formula for inductive reactance per mile used in App. 37.A is

$$X_L = 2.022 \times 10^{-3} f \ln\left(\frac{D}{\text{GMR}}\right) \quad \text{[in } \Omega/\text{mi]} \quad 37.27$$

For line-to-line spacings other than one foot, the correction factor given by Eq. 37.28 must be applied.[15]

$$K_L = 1 + \frac{\ln D}{\ln\left(\dfrac{1}{\text{GMR}}\right)} \qquad 37.28$$

The formula for capacitive reactance per mile used in App. 37.A is approximately

$$X_C = \left(\frac{1.781 \times 10^6}{f}\right) \ln\left(\frac{D}{r}\right) \quad \text{[in } \Omega/\text{mi]} \quad 37.29$$

---

[15]The GMR used in Eq. 37.27, not Eq. 37.20, is the one tabulated. Equation 37.20 is a good approximation of the tabulated value.

For line-to-line spacings other than one foot, the correction factor given by Eq. 37.30 must be applied.

$$K_C = 1 + \frac{\ln D}{\ln\left(\dfrac{1}{r}\right)} \qquad 37.30$$

The values of D, GMR, and r are in feet when used in Eqs. 37.27 through 37.30. Appendix 37.A can be used for single-phase systems as well. The values for inductive and capacitive reactance, Eqs. 37.27 and 37.29, double for single-phase systems.

Example 37.6

A single-phase system uses merlin conductors spaced 10 ft apart from center to center. What is the total shunt reactance for 10 mi of the system?

Solution

From App. 37.A, the capacitive reactance for a one-foot spacing is 0.1055 MΩ-mi. The diameter is taken from App. 37.A as 0.684 in, thus the radius is 0.342 in, or 0.0285 ft. The correction factor for the 10 ft spacing is

$$K_C = 1 + \frac{\ln D}{\ln\left(\dfrac{1}{r}\right)} = 1 + \frac{\ln(10)}{\ln\left(\dfrac{1}{0.0285}\right)} = 1.647$$

The capacitive reactance for a single conductor is

$$\begin{aligned} X_{C,\text{actual}} &= X_{C,\text{table}} K_C \\ &= (0.1055 \text{ M}\Omega\text{-mi})(1.647) \\ &= 0.1738 \text{ M}\Omega\text{-mi} \end{aligned}$$

This is a single-phase system. That is, two conductors are present. The total capacitive reactance is

$$X_{C,\text{total}} = (2)(0.1738 \text{ M}\Omega\text{-mi}) = 0.3475 \text{ M}\Omega\text{-mi}$$

For a 10 mi portion of the system, the capacitive reactance is

$$X_C = \frac{0.3475 \text{ M}\Omega\text{-mi}}{10 \text{ mi}} = 0.03475 \text{ M}\Omega$$

11. TRANSMISSION LINE REPRESENTATION

Analysis of transmission lines is accomplished by approximating the line with distributed parameters. The line is represented as a series-parallel combination of electrical components applicable to the length of the transmission system.[16] Three-phase transmission systems are operated in as balanced a configuration as possible, but are seldom equilaterally spaced. Transposition may alleviate this imbalance, but it is seldom used.

---

[16]The shunt conductance is normally insignificant and is excluded from the models used.

Table 37.3 *Per-Phase ABCD Constants for Transmission Lines*

transmission line length	equivalent circuit	A	B	C	D
short <80 km	series impedance Fig. 37.4	1	Z	0	1
medium 80−240 km	nominal-T Fig. 37.5(a)	$1 + \frac{1}{2}YZ$	$Z\left(1 + \frac{1}{4}YZ\right)$	Y	$1 + \frac{1}{2}YZ$
medium 80−240 km	nominal-Π Fig. 37.5(b)	$1 + \frac{1}{2}YZ$	Z	$Y\left(1 + \frac{1}{4}YZ\right)$	$1 + \frac{1}{2}YZ$
long >240 km	distributed parameters Fig. 37.6	$\cosh \gamma l$	$Z_0 \sinh \gamma l$	$\dfrac{\sinh \gamma l}{Z_0}$	$\cosh \gamma l$

Nevertheless, calculations assume equilateral spacing and transposition. Calculated values closely approximate actual values if the equivalent spacing, D_e, of Eq. 37.23 is used in place of the actual spacing. Given these assumptions, balanced three-phase systems can be analyzed on a per-phase basis using a hypothetical neutral, which contributes no resistance, inductance, or capacitance. Lines with significant imbalances must be analyzed using symmetrical components.

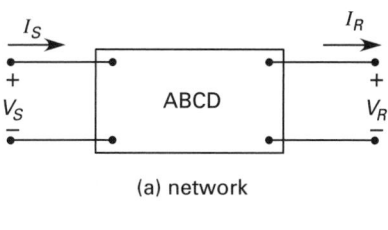

(a) network

$$V_S = AV_R + BI_R$$
$$I_S = CV_R + DI_R$$

(b) equations

$$\begin{bmatrix} V_S \\ I_S \end{bmatrix} = \begin{bmatrix} A & B \\ C & D \end{bmatrix} \begin{bmatrix} V_R \\ I_R \end{bmatrix}$$

(c) matrix form of equations

Figure 37.3 *Transmission Line Two-Port Network*

Design of all transmission lines, regardless of length, must account for the voltage regulation and efficiency of power transmission, η_P. In terms of a *sending end* (subscript S), or source, and a *receiving end* (subscript R), or load, these quantities are defined by Eqs. 37.31 and 37.32.

$$\text{VR} = \frac{|V_{R,\text{nl}}| - |V_{R,\text{fl}}|}{|V_{R,\text{fl}}|} \qquad 37.31$$

$$\eta_P = \frac{P_R}{P_S} \qquad 37.32$$

Any transmission line may be represented as a two-port network as shown in Fig. 37.3. The ABCD parameters are sometimes called *generalized circuit constants* and, in general, are complex. (Though complex, common practice is to not show them in complex notation. That is, they are not shown in bold.) The ABCD parameters for the various transmission lines are given in Table 37.3.

12. SHORT TRANSMISSION LINES

Short transmission lines are 60 Hz lines that are less than 80 km (50 mi) long. The shunt reactance is excluded from the model because it is much greater than most load impedances and significantly greater than the line impedance. Only the series resistance and inductance are significant as shown in Fig. 37.4.

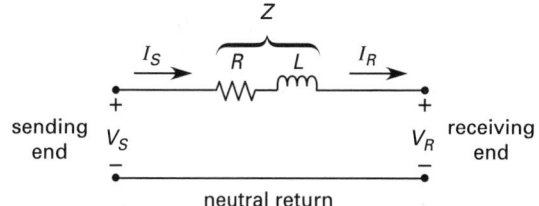

Figure 37.4 *Short Transmission Line Model*

The impedance is given by[17]

$$Z = R + jX_L \qquad 37.33$$

---

[17]The quantities represent total amounts but are calculated per unit length, corrected for conductor spacing, and multiplied by the length of the transmission line. It should also be noted that the impedance is a per-phase quantity and needs to be multiplied by three to get the line-to-line impedance in a delta-connected load.

Using the ABCD parameters in Table 37.3 results in Eqs. 37.34 and 37.35.

$$V_S = V_R + I_R Z \qquad \textit{37.34}$$

$$I_S = I_R \qquad \textit{37.35}$$

Example 37.7

A 30 mi hawk three-phase 60 Hz transmission line has a 10 ft spacing between conductors. The sending-end voltage is 11 kV per phase. The load draws 200 A per phase at 0.8 pf lagging. The entire system is assumed to operate at 20°C. What is the voltage regulation in percent?

Solution

This is a short transmission line. The voltage regulation is given by Eq. 37.31. Thus, the voltage on the receiving end needs to be determined. The current on the receiving end and its angle (via the power factor) are given. Thus, to determine the voltage on the receiving end, the impedance must be calculated. From App. 37.A, the uncorrected resistance and reactance are

$$R_{l,\text{AC}} = 0.1931 \ \Omega/\text{mi}$$

$$X_{L,l} = 0.430 \ \Omega/\text{mi}$$

The correction factor for a 10 ft spacing, using the GMR value in App. 37.A, is

$$K_L = 1 + \frac{\ln D}{\ln\left(\dfrac{1}{\text{GMR}}\right)} = 1 + \frac{\ln(10)}{\ln\left(\dfrac{1}{0.0289}\right)} = 1.6497$$

The corrected reactance is

$$\begin{aligned} X_{L,l(\text{corrected})} &= K_L X_{L,l} \\ &= (1.6497)(0.430) \\ &= 0.71 \ \Omega/\text{mi} \end{aligned}$$

The impedance for 30 mi of line is

$$\begin{aligned} Z &= (R_{l,\text{AC}} + jX_{L,l})\,(30 \text{ mi}) \\ &= \left(0.1931 \ \frac{\Omega}{\text{mi}} + j0.71 \ \frac{\Omega}{\text{mi}}\right)(30 \text{ mi}) \\ &= 5.79 + j21.3 \ \Omega \\ &= 22.1\angle 74.8° \ \Omega \end{aligned}$$

The power factor is given as 0.8 lagging, thus the current lags the receiving-end voltage by 36.8°. Since this is referenced to the receiving-end voltage, take V_R to be the reference at 0°. The current is

$$I_R = 200\angle{-36.8°} \text{ A}$$

The voltage drop across the line at full load is

$$\begin{aligned} V_{\text{drop}} &= I_R Z \\ &= (200\angle{-36.8°} \text{ A})\,(22.1\angle 74.8° \ \Omega) \\ &= 4.42\angle 38.0° \text{ kV} \end{aligned}$$

Using the relationship for the sending and receiving ends given by Eq. 37.34 gives

$$V_S = V_R + I_R Z$$

$$V_R = V_S - I_R Z$$

$$= 11\angle\theta \text{ kV} - 4.42\angle 38° \text{ kV}$$

$$|V_R|\angle 0° \text{ kV} = |V_R| + j0$$

$$= (11\cos\theta + j11\sin\theta) - (3.48 + j2.72)$$

The angle θ is the angle of the sending-end voltage with respect to the receiving-end voltage, which was selected as the reference. Since no imaginary portion exists on the left-hand side of the equation, the following must be true.

$$11\sin\theta - 2.72 = 0$$

$$\sin\theta = \frac{2.72}{11} = 0.247$$

Using this information to determine the angle and the term $\cos\theta$ gives

$$\theta = \sin^{-1}(0.247) = 14.3°$$

$$\cos\theta = \cos 14.3° = 0.969$$

Substituting this information gives the receiving-end voltage at full load.

$$|V_R|\angle 0° \text{ kV} = |V_R| + j0$$

$$= (11\cos\theta + j11\sin\theta) - (3.48 + j2.72)$$

$$= (11)(0.969) + j(11)(0.247) - 3.48 - j2.72$$

$$= 7.18\angle 0° \text{ kV}$$

At a no-load condition, $I_R = 0$ in Eq. 37.34, and the receiving-end voltage equals the sending-end voltage. That is, $V_{R,\text{nl}} = 11$ kV. The regulation, from Eq. 37.31, is

$$\begin{aligned} \text{VR} &= \frac{|V_{R,\text{nl}}| - |V_{R,\text{fl}}|}{|V_{R,\text{fl}}|} \\ &= \frac{11 \text{ kV} - 7.18 \text{ kV}}{7.18 \text{ kV}} \\ &= 0.532 \quad (53.2\%) \end{aligned}$$

13. MEDIUM-LENGTH TRANSMISSION LINES

Medium-length transmission lines are 60 Hz lines between 80 and 240 km (50 and 150 mi) long. The shunt reactance is significant enough to be included. The medium-length line is modeled in one of two ways as shown in Fig. 37.5.

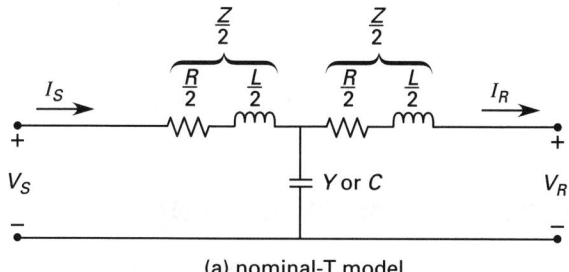

(a) nominal-T model

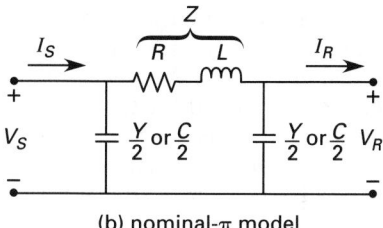

(b) nominal-π model

Figure 37.5 *Medium-Length Transmission Line Models*

The impedance is given by[18]

$$Z = R + jX_L \qquad 37.36$$

The admittance is given by

$$Y = jB_C = \frac{j}{X_C} = -\frac{1}{jX_C} \qquad 37.37$$

Using the ABCD parameters in Table 37.3 gives Eqs. 37.38 and 37.39 for the T approximation.

$$V_S = \left(1 + \tfrac{1}{2}YZ\right)V_R$$
$$+ \left(Z\left(1 + \tfrac{1}{4}YZ\right)\right)I_R \qquad 37.38$$

$$I_S = YV_R + \left(1 + \tfrac{1}{2}YZ\right)I_R \qquad 37.39$$

Using the ABCD parameters in Table 37.3 gives Eqs. 37.40 and 37.41 for the Π approximation.[19]

$$V_S = \left(1 + \tfrac{1}{2}YZ\right)V_R + ZI_R \qquad 37.40$$

$$I_S = \left(Y\left(1 + \tfrac{1}{4}YZ\right)\right)V_R$$
$$+ \left(1 + \tfrac{1}{2}YZ\right)I_R \qquad 37.41$$

---

[18]The quantities represent total amounts but are calculated per unit length, corrected for conductor spacing, and multiplied by the length of the transmission line. It should also be noted that the impedance is a per-phase quantity and needs to be multiplied by three to get the line-to-line impedance in a delta-connected load.

[19]The Π approximation results in simpler regulation calculations.

14. LONG TRANSMISSION LINES

Long transmission lines are 60 Hz lines greater than 240 km (150 mi) long. All the parameters must be represented as distributed parameters, that is, per unit length, as shown in Fig. 37.6. The shunt conductance is shown for completeness only and is excluded from most calculations, as will be done here.

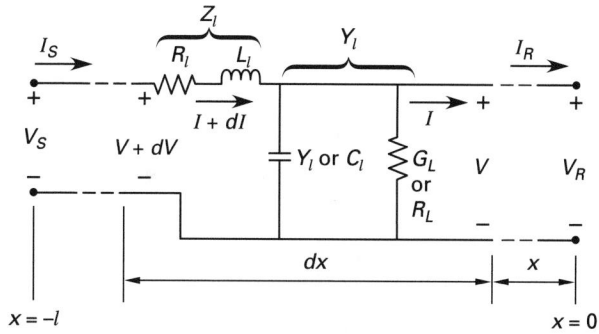

Figure 37.6 *Long Transmission Line Model*

The impedance per unit length is[20]

$$Z_l = R_l + jX_{L,l} \qquad 37.42$$

The admittance per unit length is

$$Y_l = jB_{C,l} = \frac{j}{X_{C,l}} = -\frac{1}{jX_{C,l}}$$
$$= j\omega C \qquad 37.43$$

The voltage at any point along the line of length l in Fig. 37.6 is

$$\frac{\partial^2 V}{\partial x^2} = \gamma^2 V \qquad 37.44$$

The term γ is called the *propagation constant*, measured in radians per meter (mile), and is given by

$$\gamma = \sqrt{Y_l Z_l} = \sqrt{-B_{C,l}X_{L,l} + jR_l B_{C,l}}$$
$$= \alpha + j\beta \qquad 37.45$$

The magnitude of the propagation constant is

$$|\gamma| = \sqrt{B_{C,l}}\sqrt[4]{R_l^2 + X_{L,l}^2} \qquad 37.46$$

The term α is called the *attenuation constant*, measured in nepers per meter (mile), and is given by

$$\alpha = |\gamma|\cos\left(\tfrac{1}{2}\arctan\left(\frac{-R_l}{X_{L,l}}\right)\right) \qquad 37.47$$

---

[20]The per unit length impedance and admittance are sometimes written with small letters, z and y, to emphasize their per-unit qualities.

Power–Transmission

The term β is called the *phase constant*, measured in radians per meter (mile), and is given by

$$\beta = |\gamma| \sin\left(\frac{1}{2}\arctan\left(\frac{-R_l}{X_{L,l}}\right)\right) \qquad \textit{37.48}$$

A solution to Eq. 37.44 giving the voltage at any point along the line is[21]

$$V = \frac{1}{2}V_R\left(e^{\gamma x} + e^{-\gamma x}\right)$$
$$+ \frac{1}{2}I_R Z_0 \left(e^{\gamma x} - e^{-\gamma x}\right)$$
$$= V_R \cosh\gamma x + I_R Z_0 \sinh\gamma x \qquad \textit{37.49}$$

Similarly, the current at any point along the line is given by

$$I = \frac{1}{2}I_R\left(e^{\gamma x} + e^{-\gamma x}\right)$$
$$+ \frac{1}{2}\left(\frac{V_R}{Z_0}\right)\left(e^{\gamma x} - e^{-\gamma x}\right)$$
$$= I_R \cosh\gamma x + \left(\frac{V_R}{Z_0}\right)\sinh\gamma x \qquad \textit{37.50}$$

The term Z_0 is called the *characteristic impedance* and is given by

$$Z_0 = \sqrt{\frac{Z_l}{Y_l}} = \sqrt{\frac{R_l + jX_{L,l}}{jB_{C,l}}}$$
$$= \sqrt{X_{L,l}X_{C,l} - jX_{C,l}R_l} \quad [\text{in } \Omega] \qquad \textit{37.51}$$

The magnitude of the characteristic impedance is

$$|Z_0| = \sqrt{X_{C,l}}\sqrt[4]{R_l^2 + X_l^2} \qquad \textit{37.52}$$

The angle of the characteristic impedance is

$$\angle Z_0 = \frac{1}{2}\arctan\left(\frac{-R_l}{X_{L,l}}\right) \qquad \textit{37.53}$$

Equations 37.49 and 37.50 represent *traveling waves.* Specifically, the terms with $e^{-\gamma x}$ represent waves traveling in the positive x-direction. (The distance x is defined as positive to the right, the direction toward the receiving end, that is, toward the direction power flows.) Terms with $e^{\gamma x}$ represent waves traveling in the negative x-direction. The two equations are the value of the two waves superimposed at that point in the line. Since $V = V_S$ (see Eq. 37.49) at $x = -l$ (see Fig. 37.6), and

---

[21]The solution assumes a sinusoidal steady-state condition. That is, transmission lines are generally AC steady-state networks.

consistent with the ABCD parameters in Table 37.3, the voltage at the sending end is

$$V_S = V_R \cosh\gamma l + I_R Z_0 \sinh\gamma l \qquad \textit{37.54}$$

Since $I = I_S$ (see Eq. 37.50) at $x = -l$ (see Fig. 37.6), and consistent with the ABCD parameters in Table 37.3, the current at the sending end is

$$I_S = I_R \cosh\gamma l + \left(\frac{V_R}{Z_0}\right)\sinh\gamma l \qquad \textit{37.55}$$

The term l in Eqs. 37.54 and 37.55 is the transmission line length.

15. REFLECTION COEFFICIENT

Energy propagated along a transmission line can be thought of as traveling electromagnetic waves (see Sec. 37-14). Loads interact with these waves in a manner that absorbs some of the energy and reflects the remainder. If the transmission line is terminated with the characteristic impedance, Z_0, no power is reflected back to the source or sending end. If the termination impedance is other than Z_0, the power (signal) from the generator (source) will be partially reflected back to the generator. Assuming a steady-state sinusoidal source, the generator's waves will then combine in the transmission line with the reflected waves to form *standing waves*. The *standing wave ratio*, SWR, is the ratio of the maximum to the minimum voltages (currents) encountered along the transmission line. Typically, the SWR is greater than unity. If the terminating impedance matches the characteristic impedance, all the input power provided by the generator is absorbed by the load and the SWR equals one. The standing wave ratios are given by

$$\text{VSWR} = \frac{V_{\max}}{V_{\min}} \qquad \textit{37.56}$$

$$\text{ISWR} = \frac{I_{\max}}{I_{\min}} \qquad \textit{37.57}$$

The *reflection coefficient*, Γ, is the ratio of the reflected to the incident electric parameter. The voltage reflection coefficient for the load is given by

$$\Gamma_L = \frac{V_{\text{reflected}}}{V_{\text{incident}}} = \frac{Z_{\text{load}} - Z_0}{Z_{\text{load}} + Z_0} \qquad \textit{37.58}$$

The current reflection coefficient is the negative of the reflection coefficient for the voltage. That is,

$$\Gamma_L = \frac{I_{\text{reflected}}}{I_{\text{incident}}} = \frac{Z_0 - Z_{\text{load}}}{Z_0 + Z_{\text{load}}} \qquad \textit{37.59}$$

The fraction of incident power that is reflected back to the source from the load is Γ^2. The relationship between the reflection coefficient and the standing wave ratio is

$$\Gamma = \frac{\text{SWR} - 1}{\text{SWR} + 1} \qquad \textit{37.60}$$

38 Batteries and Fuel Cells

Nomenclature

Ah	ampere-hours	A·h
$\mathcal{E}$	electromotive force	V
I	effective or DC current	A
P	power	W
Q	overcharge	A·h
R	resistance	Ω
V	effective or DC voltage	V
V	volume	L

Subscripts

bat	battery
cell	per cell
H_2	hydrogen
int	internal
s	source
t	terminal

1. BATTERY FUNDAMENTALS

A *battery* is a direct-current voltage source made up of one or more units that converts chemical, thermal, nuclear, or solar energy into electrical energy. The chemical conversion type is responsible for the majority of the batteries in use and will be the focus of this section. A *cell* is a single unit of a battery.[1] A *primary cell* undergoes chemical change in such a manner that one or both electrodes are unusable following discharge. A *secondary cell* undergoes chemical change in a reversible manner and thus can be recharged following operation. *Shelf life* is the time a battery can be stored without use before its capacity drops to a value that makes it uneconomical to store further. The *cycle life* is the number of times a battery can be discharged or charged before the output voltage drops to a prescribed value. A *cycle* is a predetermined number of amp-hours (A·h) of discharge or charge. The advantages of a battery as a source of electrical energy are its portability, storage capability, and nonpolluting operation. The disadvantages are its general energy inefficiency and low power capability, both of which are changing as battery technology advances.[2]

---

[1] The first storage cell was invented by Italian professor Alessandro Volta in 1800.

[2] *Energy efficiency* is the ratio of the discharging energy (in J) to the charging energy (in J). Values between 50 and 70% are common.

2. BATTERY THEORY

An *electrolyte* is a compound that, when dissolved in certain solvents (usually water), will conduct electricity. When placed in solution, an electrolyte dissociates into ions as shown in Fig. 38.1(a). Placing two dissimilar electrodes in the solution results in a migration of the ions due to each metal's affinity. The result is a potential difference between the two electrodes as shown in Fig. 38.1(b). The magnitude of the potential difference depends on the materials used for the electrodes and is determined by their specific electronic structures. Some typical combinations are shown in Table 38.1.

Table 38.1 *Cell Voltages*

battery type/ name	anode (negative electrode)[a]	cathode (positive electrode)[b]	nominal cell voltage (V)
primary/ Leclanche	Zn	MnO_2	1.2
primary/ mercury—zinc	Zn	HgO	1.2
primary/ silver—zinc	Zn	AgO	1.6
secondary/ lead—acid	Pb	PbO_2	1.8
secondary/ nickel—cadmium	Cd	NiOOH	1.2
secondary/ silver—zinc	Zn	AgO	1.5
secondary/ nickle—metal hydride	H_2·(metals)	NiOOH	1.2

[a] The anode is positive with respect to the battery, which results in a negatively charged electrode.

[b] The cathode is negative with respect to the battery, which results in a positively charged electrode.

When an external load is attached to the terminals, a current flows from the positive electrode (cathode) to the negative electrode (anode) in the external circuit as shown in Fig. 38.1(c). In the internal circuit, current flows from the anode to the cathode. Positive ions (cations) migrate toward the cathode (positive electrode), where they capture incoming electrons and become neutral atoms, transforming the material on the cathode. Negative ions (anions) migrate toward the anode where they lose their electrons by combining with the active material of the anode. When the material on either plate is consumed, the chemical reaction stops and the current ceases as shown in Fig. 38.1(d).

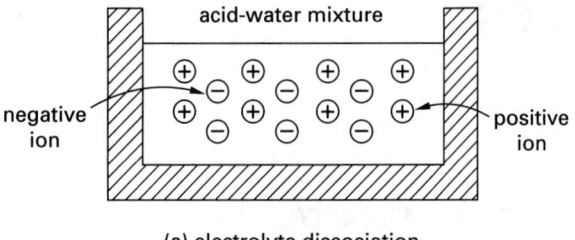

(a) electrolyte dissociation

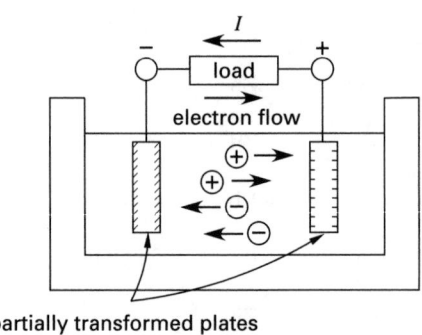

(b) potential due to electrode affinity

(c) current flow

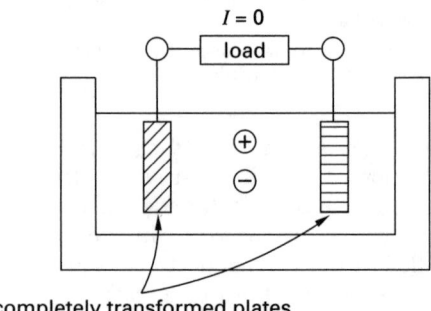

(d) cell discharged (electrode chemical change complete)

Figure 38.1 Battery Theory

Care should be taken when referring to positive and negative potential on a battery. The anode is positive with respect to the internal circuit, that is, the battery itself. The cathode is negative with respect to the internal circuit, that is, the battery itself. In electrical circuit schematics showing a battery, the power flows out of the positively charged electrode, that is, the cathode, during discharge, since it is positive with respect to the external circuit. The positive, or conventional, current flows from the anode to the cathode within the battery and from the cathode to the anode in the external circuit, that is, from the negative electrode to the positive electrode within the battery and from the positive to the negative in the external circuit. The cathode (positive electrode) is the one labeled as positive on an actual battery terminal. The anode (negative electrode) is alternatively defined as the terminal at which conventional current enters a battery.

All batteries possess an internal resistance to current flow that results in a lower output voltage as the current increases. The equivalent circuit for a battery is shown in Fig. 38.2.

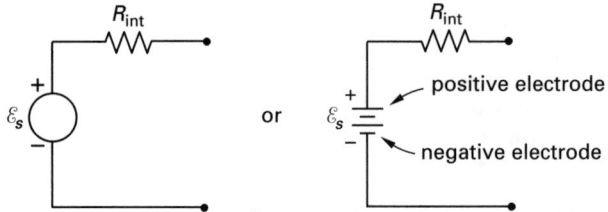

Figure 38.2 Battery Equivalent Circuit

Example 38.1

A 1.5 V battery with an internal resistance of 0.3 Ω is connected to a 1 Ω load. What is the load current?

Solution

The circuit is that of Fig. 38.2 with a 1 Ω resistor attached to the terminals. Using Ohm's law, the current is

$$V = IR$$

$$I = \frac{V}{R} = \frac{V}{R_{\text{int}} + R_{\text{load}}}$$

$$= \frac{1.5 \text{ V}}{0.3 \text{ } \Omega + 1 \text{ } \Omega} = 1.15 \text{ A}$$

Example 38.2

For the battery and load in Ex. 38.1, what is the power consumed by the internal resistance?

Solution

The power is determined by

$$P = I^2 R$$

$$= (1.15 \text{ A})^2 (0.3 \text{ } \Omega) = 0.397 \text{ W}$$

Example 38.3

For the battery and load in Ex. 38.1, what is the terminal voltage?

Solution

Using KVL around the loop of Fig. 38.2 gives

$$\mathcal{E}_s - IR_{\text{int}} - V_t = 0$$

Using the current from Ex. 38.1 and solving for the terminal voltage gives

$$V_t = \mathcal{E}_s - IR_{\text{int}}$$

$$= 1.5 \text{ V} - (1.15 \text{ A})(0.3 \text{ } \Omega) = 1.15 \text{ V}$$

3. BATTERY TYPES AND CAPACITY

Primary batteries are those that cannot be recharged. Primary batteries that contain no free or liquid electrolyte are called *dry cell batteries*. A *reserve battery* is one in which an essential component is withheld, making the battery inert and permitting long-term storage. A *secondary battery* is one designed so that it can be recharged electrically following discharge. Charging is accomplished by passing a current through the battery in the direction opposite that of the discharge current.

Batteries are also typed according to cell sizes established by the American National Standards Institute (ANSI). Some examples of typical sizes are given in Table 38.2.

Table 38.2 Cell Size

cell size	starting drain current (mA)	service capacity (h)
AAA	2	290
	10	45
	20	17
AA	3	350
	15	40
	30	15
B	5	420
	25	65
	50	25
C	5	430
	25	100
	50	40
D	10	500
	50	105
	100	45

As seen in Table 38.2, service capacity of a battery varies. As the current withdrawn is increased, the capacity drops. This occurs primarily because of the battery's internal resistance and the effect of *polarization*. Polarization is the phenomenon whereby, when a cell discharges current, hydrogen is given off at one of the terminals. The hydrogen bubbles act to insulate the electrode, lowering the chemical reaction rate and the current.

Hydrogen production also occurs during battery charging. For a battery that is fully charged, the approximate hydrogen production is given by

$$V_{\text{H}_2} = (0.42) \left(\frac{V_{\text{bat}} Q}{V_{\text{cell}}} \right) \qquad \text{38.1}$$

The term V_{H_2} is the volume of hydrogen produced in liters. Each liter of hydrogen produced correlates to the electrolysis of 0.8 mL of water. The term V_{bat} is the total battery voltage while V_{cell} is the individual cell voltage. The term Q is the amount of overcharge in amp-hours (A·h).

The service capacity combined with the current withdrawn determines the *battery capacity*. Since the capacity varies with current, it is normally measured in amp-hours (A·h) over a standard period. Standard time periods include one, five, eight, and ten hours.

Example 38.4

A 12 V lead-acid battery with six cells is left on a charger for one day following completion of the charge. Assuming a charging current of 3 A, how much hydrogen is produced?

Solution

The volume of hydrogen produced is given by

$$V_{\text{H}_2} = (0.42) \left(\frac{V_{\text{bat}} Q}{V_{\text{cell}}} \right)$$

The battery voltage is given. The cell voltage is 2 V. The amount of overcharge, Q, is

$$Q = (3 \text{ A})(24 \text{ h}) = 72 \text{ A·h}$$

Substituting gives

$$V_{\text{H}_2} = (0.42) \left(\frac{V_{\text{bat}} Q}{V_{\text{cell}}} \right) = (0.42) \left(\frac{(12 \text{ V})(72 \text{ A·h})}{2 \text{ V}} \right)$$

$$= 181 \text{ L}$$

Example 38.5

By what factor does the capacity of a D-cell battery drop when the drain current is increased from 10 to 100 mA?

Solution

From Table 38.2, the capacity in A·h at 10 mA is

$$\text{A·h} = (10 \times 10^{-3} \text{ A})(500 \text{ h}) = 5000 \times 10^{-3} \text{ A·h}$$

From Table 38.2, the capacity in A·h at 100 mA is

$$\text{A·h} = (100 \times 10^{-3} \text{ A})(45 \text{ h}) = 4500 \times 10^{-3} \text{ A·h}$$

The factor by which the capacity decreases is

$$\text{A·h}_{\text{factor}} = 1 - \frac{4500 \times 10^{-3} \text{ A·h}}{5000 \times 10^{-3} \text{ A·h}} = 0.1 \quad (10\%)$$

4. FUEL CELLS

A *fuel cell* is a device that converts chemical energy directly into electrical energy. A fuel cell differs from a battery in that the fuel and oxidant must be continuously supplied and, at least in theory, no heat is produced as a result of the reaction. A fuel cell uses both liquid and gaseous fuels, such as hydrogen, hydrazine, and coal gas. The oxidant is oxygen or air. A hydrogen-oxygen fuel cell is shown in Fig. 38.3.

A fuel, in this case hydrogen, is supplied continuously to an electrode, where it undergoes a catalytic reaction producing electrons and a fuel ion ($2H^+$). The electrons travel through the external circuit, supplying power to the load. The fuel ion migrates through the electrolyte and gas separation barrier to the oxygen electrode. The incoming oxygen reacts with the cathode to produce negative oxygen ions that diffuse in the electrolyte. At the oxidizing electrode's surface, the oxygen, hydrogen ions, and electrons combine through a catalytic reaction to form water. The net result is hydrogen and oxygen combining to form water and produce electrical energy.

Fuel cells are of interest because of their high efficiency, minimal environmental impact, and ability to be constructed in a modular fashion.

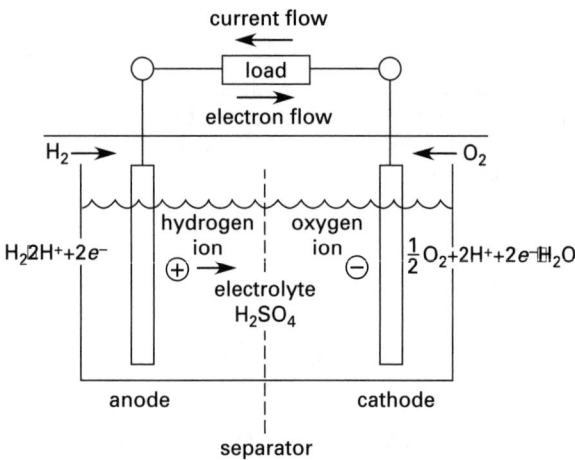

Figure 38.3 *Hydrogen-Oxygen Fuel Cell*

Topic VII: Power—Machinery

**Power–
Machinery**

For the most current information about the exam, visit **www.ppi2pass.com** regularly.

PROFESSIONAL PUBLICATIONS, INC.

Topic VIII: Power—Machinery

39 Rotating DC Machinery

Nomenclature

a	number of parallel armature paths	–
a	ratio of transformation	–
A	area	m^2
B	magnetic flux density	T
B	susceptance	S
E	generated emf	V
f	electrical frequency	Hz
G	conductance	S
I	current	A
k	constant	various
n	rotational speed	rpm
N	number of series armature paths	–
p	total number of poles	–
P	power	W
q	number of loops	–
r	radius	m
R	resistance	Ω
s	slip	–
S	apparent power	VA
SR	speed regulation	–
t	time	s
T	period	s
T	torque	N·m
V	line voltage	V
VR	voltage regulation	–
X	reactance	Ω
Y	admittance	S
z	total number of conductors	–
Z	impedance	Ω

Symbols

δ	torque angle	rad
η	efficiency	–
θ	phase difference angle	rad
Φ	magnetic flux	Wb
ω	electrical frequency	rad/s
Ω	armature rotational speed	rad/s

Subscripts

a	armature
adj	adjusted
b	blocked rotor
Cu	copper
E	emf
f	field
fl	full load
g	generator
L	load
m	maximum
nl	no load
p	per phase
r	rotor
rpm	revolutions per minute
s	synchronous
st	stator
t	terminal
t	total
T	torque

1. ROTATING MACHINES

Rotating machines are broadly categorized as AC and DC machines. Both categories include machines that use power (i.e., motors) and those that generate power (alternators and generators).

2. TORQUE AND POWER

Torque and power are operating parameters. It takes power to turn an alternator or generator. A motor converts electrical power into mechanical power. In the SI system, power is given in kilowatts (kW). One horsepower is equivalent to 745.7 W. The relationship between torque and power is

$$T_{\text{ft-lbf}} = \frac{5252 P_{\text{horsepower}}}{n_{\text{rpm}}} \qquad 39.1$$

$$T_{\text{N·m}} = \frac{1000 P_{\text{kW}}}{\Omega} = \frac{9549 P_{\text{kW}}}{n_{\text{rpm}}} \qquad 39.2$$

Power–Machinery

There are many important torque parameters for motors. The *starting torque* (also known as *static torque,* *breakaway torque,* and *locked-rotor torque*) is the turning effort exerted in starting a load from rest. *Pull-up torque* (*acceleration torque*) is the minimum torque developed during the period of acceleration from rest to full speed. The *steady-state torque* must be provided to the load on a continuous basis. It establishes the temperature increase that the motor can withstand without deterioration. The *rated torque* is developed at rated speed and rated horsepower. *Breakdown torque* is the maximum torque the motor can develop without stalling (i.e., without coming rapidly to a complete stop).

Equation 39.3 is the general torque expression for a rotating machine with N coils of cross-sectional area A, each carrying current I through a magnetic field of strength B.

$$T = NBAI \cos \omega t \qquad 39.3$$

3. SERVICE FACTOR

The horsepower and torque ratings listed on the nameplate of a motor can be provided on a continuous basis without overheating. Motors can be operated at slightly higher loads without exceeding a safe temperature rise, but the higher temperature has a deteriorating effect on the winding insulation. (A general rule of thumb is that a motor loses two or three hours of useful life for each hour run at the factored load.) The ratio of the safe to standard loads is the *service factor*, usually expressed as a decimal. Service factors vary from 1.15 to 1.4, with the lower values going to larger, more efficient motors.

$$\text{service factor} = \frac{\text{safe load}}{\text{nameplate load}} \qquad 39.4$$

4. MOTOR CLASSIFICATIONS

The National Electrical Manufacturers Association (NEMA) has categorized motors in several ways: *speed classification* (constant-, adjustable-, multi-, varying-speed, etc.); *service classification* (general, definite, and special purpose); and motor class. *Motor class* is a primary indicator of the maximum motor operating temperature, which, in turn, depends on the type of insulation used on the conductors. The classes are: Class A, 105°C; Class B, 130°C; Class F, 155°C; and Class H, 180°C.

5. POWER LOSSES[1]

The losses for all rotating machines can be divided into four categories. *Copper losses,* P_{Cu}, are real power

---

[1]The subject of transformer losses covered in Sec. 36-3 is equivalent in concept.

losses due to wire and winding resistance. In a DC machine, copper losses occur in the armature and field windings as well as from the brush contact resistance. In an AC machine, copper losses occur in the armature and exciter field windings. There are no brush losses in an induction machine.

$$P_{\text{Cu}} = \sum I^2 R \qquad 39.5$$

Core losses, including hysteresis and eddy current losses, are constant losses that are independent of the load and, for that reason, are also known as *open-circuit* and *no-load losses.* In DC and synchronous AC machines, core losses occur in the armature iron. In induction machines, core losses occur in the stator iron.

Mechanical losses (also known as *rotational losses*) include brush and bearing friction and *windage* (air friction). (Windage is a no-load loss but is not an electrical core loss.) Mechanical losses are determined by measuring the power input at the rated speed with no load.

Stray losses are due to nonuniform current distribution in the conductors. Stray losses are approximately 1% for DC machines and zero for AC machines.

Real power only is used to compute the *efficiency* of a rotating machine.

$$\eta = \frac{\text{output}}{\text{input}} = \frac{\text{output}}{\text{output} + \text{losses}}$$
$$= \frac{\text{input} - \text{losses}}{\text{input}} \qquad 39.6$$

6. REGULATION

The *voltage regulation*, VR, is

$$\text{VR} = \frac{\text{no-load voltage} - \text{full-load voltage}}{\text{full-load voltage}} \times 100\% \qquad 39.7$$

The *speed regulation*, SR, is

$$\text{SR} = \frac{\text{no-load speed} - \text{full-load speed}}{\text{full-load speed}} \times 100\% \qquad 39.8$$

7. NO-LOAD CONDITIONS

The meaning of the term *no load* is different for generators and motors. For unloaded shunt-wired alternators and generators (see Sec. 39-10), there is no electrical load connected across the output terminals, so although the field current flows, the line current, I, is zero. For unloaded shunt-wired motors, the work performed is zero, but line current is still drawn to keep the motor turning. All of the current is field current, however, and (neglecting mechanical losses) the armature current, I_a, is zero.

$$I = 0; \quad I_f \neq 0; \quad I_a = I_f \quad \text{[generator]} \qquad 39.9$$

$$I \neq 0; \quad I_f = I; \quad I_a = 0 \quad \text{[motor]} \qquad 39.10$$

8. PRODUCTION OF DC POTENTIAL

A *generator* is a device that produces DC potential. The actual voltage induced is sinusoidal (i.e., AC). However, brushes on *split-ring commutators* make the connection to the rotating armature and rectify the AC potential.[2] The more coils there are, the smoother the DC voltage. Two commutator segments are needed for each coil, and each coil produces its own sine wave. Since mechanical commutation is difficult unless the emf is produced in a rotating armature, DC generators are based on the simple design shown in Fig. 39.1.

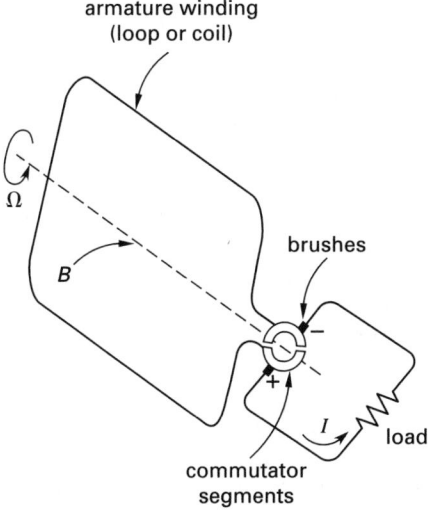

armature winding
(loop or coil)

brushes

B

Ω

load

commutator
segments

Figure 39.1 *Commutator Action*

DC generators suffer from several limitations. The rotating coils must be well insulated to prevent shorting with the high voltages that are induced. Structural bracing is required to counteract the large centrifugal force that results from rotating many coils of wire. It is difficult to make efficient high-voltage, high-power connections through slip rings.

The armature in a simple DC generator consists of a single coil with several turns (loops) of wire. The two ends of the coil terminate at the commutator, which consists of a single ring split into two halves known as *segments* as is shown in Fig. 39.1. The brushes slide on the commutator and make contact with the adjacent segment every half-rotation of the coil. This produces

---

[2] *Sparking* is one of the problems associated with commutators and occurs at points of low resistance. Brush resistance is non-linear and drops as current increases.

a rectified (though not constant) potential, as shown in Fig. 39.2.

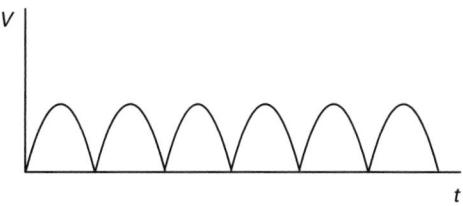

V

t

Figure 39.2 *Rectified DC Voltage Induced in a Single Cell*

Modern DC generators contain multiple coils connected in series, an arrangement known as *closed-coil winding*. The coils are spaced uniformly around the armature core. The single-ring commutator is divided into as many pairs of segments as there are coils. There are only two brushes, however, located on opposite sides of the commutator. Since there are many coils, as the armature rotates, the brushes always make contact with two segments of the commutator that are in nearly the same positions relative to the magnetic field. The average induced emf, $\overline{E}$, for a finite number of coils approaches the voltage induced with an infinite number of coils. N is the total number of series turns in all of the coils.

$$\overline{E} \approx E_\infty = (2)\left(\frac{n}{60}\right)NAB \qquad 39.11$$

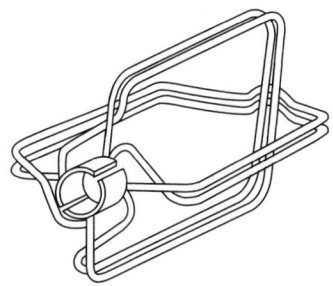

Figure 39.3 *Two-Coil, Four-Segment Closed-Coil Armature*

Since the coils are connected in series (in the modern closed-coil winding arrangement), the emf induced is the sum of the emfs induced in the individual coils. The voltage induced in each coil of a DC generator with multiple coils is still sinusoidal, but the terminal output is nearly constant, not a (rectified) sinusoid. The slight variations in the voltage are known as *ripple*. Since the output is nearly constant, the concept of electrical frequency has no meaning, and distinctions between maximum, effective, and average voltages are not made.

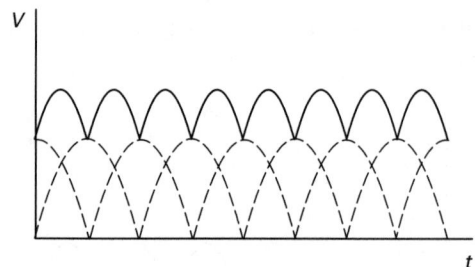

Figure 39.4 *Rectified DC Voltage from a Two-Coil, Four-Segment Generator*

It is also possible to use an *open-coil connection*, though this is seldom done. Lower voltages are produced, as the average induced voltage is the maximum voltage from a single coil.

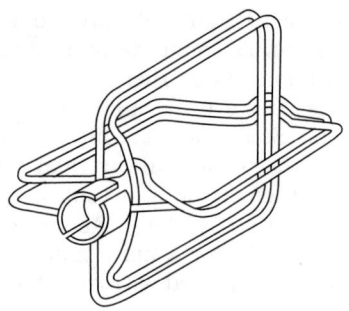

Figure 39.5 *Two-Coil, Four-Segment Open-Coil Armature*

The average emf, E, for a DC machine (motor or generator) is given by Eq. 39.12. Φ is the flux per pole. For a generator, the emf is greater than the armature voltage ($E > V_a$). For a motor, the armature voltage is greater than the back emf ($V_a > E$).

$$E = \frac{Np\Phi n}{60} = \frac{zp\Phi n}{60a} = k_E \Phi n \qquad 39.12$$

$$k_E = \frac{zp}{60a} \qquad 39.13$$

9. SERIES-WIRED DC MACHINES

The equivalent circuit of a *series-wired DC machine* is shown in Fig. 39.6. The only components in the circuit are the field and armature resistances in series (from which the name is derived). The *brush resistance* is also considered to be in series but is often included in the armature resistance specification. The governing equations are given by Eqs. 39.14 through 39.17. I_a is positive for a motor and negative for a generator. The magnetic flux varies with the armature current.

$$E = k'_E n\phi = k_E n I_a = V - I_a(R_a + R_f) \qquad 39.14$$

$$V = E + I_a(R_a + R_f) \qquad 39.15$$

The speed, torque, and current are related by

$$\frac{T_1}{T_2} = \left(\frac{I_{a,1}}{I_{a,2}}\right)^2 \approx \frac{n_1}{n_2} \qquad 39.16$$

The torque is

$$T = k'_T \Phi I_a = k_T I_a^2$$

$$= k_T \left(\frac{V}{k_E n + R_a + R_f}\right)^2 \qquad 39.17$$

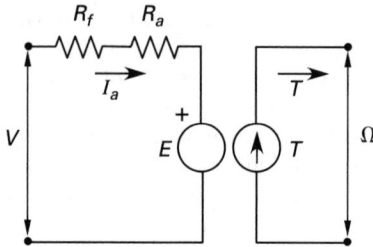

Figure 39.6 *Series-Wired DC Motor Equivalent Circuit*

For a motor, a reduction in load (torque) causes a corresponding reduction in armature current. However, since the field and armature currents are identical, the flux is also reduced and the speed increases to maintain Eq. 39.15. Therefore, a series motor is not a constant-speed device. A load should never be completely removed from a running DC motor, and gears (not belts, which can slip) are the preferred method of connecting DC motors to their loads.

The back emf, E, is zero when the motor starts from rest. Therefore, the armature current, I_a, must be excessively high in order to keep Eq. 39.15 valid. Thus, reduced voltages are required when starting, and the field resistance, R_f, is often a rheostat or switchable resistor bank.

At high speeds, the back emf, E, counteracts the applied voltage, increasing as the rotational speed increases. The *stall speed* is the speed at which Eq. 39.18 is valid.

$$E = I_a(R_a + R_f) = \frac{V}{2} \qquad \text{[stall]} \qquad 39.18$$

10. SHUNT-WIRED DC MACHINES

Shunt-wired DC machines have a constant (but adjustable) magnetic field since the field current is constant. For shunt-wired motors, this results in a relatively constant speed. The magnetic coil is fed from the same line as the armature (as it is in Fig. 39.7) in a *self-excited machine*; in a *separately excited machine* the field coil is fed from another source. In Eqs. 39.19 through 39.25, I_a is positive for a motor and negative for a generator.

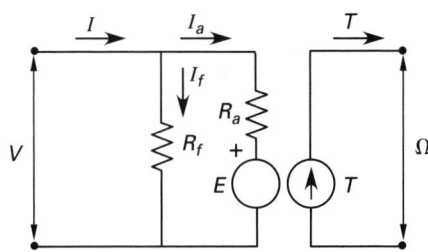

Figure 39.7 *Shunt-Wired DC Motor Equivalent Circuit*

$$E = k_E n \Phi \qquad 39.19$$

$$V = E + I_a R_a = I_f R_f \qquad 39.20$$

$$I = I_a + I_f \quad \text{[motor]} \qquad 39.21$$

$$I = I_a - I_f \quad \text{[generator]} \qquad 39.22$$

The speed-current relationship is

$$n = n_{\text{nl}} - k_n T = \frac{V - I_a R_a}{k_E \Phi} \qquad 39.23$$

The torque is

$$T = k_T \Phi I_a \qquad 39.24$$

The torque and current are proportional.

$$\frac{T_1}{T_2} = \frac{I_{a,1}}{I_{a,2}} \qquad 39.25$$

11. COMPOUND DC MACHINES

Compound DC machines have both series and shunt windings. Their performance is between those of shunt and series machines.

Example 39.1

A two-pole DC generator with a lap-wound armature is turned at 1800 rpm. There are 100 conductors between the brushes. The average magnetic flux density in the air gap between the pole faces and armature is 1.2 T. The pole faces have an area of 0.03 m². What is the no-load terminal voltage?

Solution

The flux per pole is

$$\Phi = BA = (1.2 \text{ T})(0.03 \text{ m}^2) = 0.036 \text{ Wb}$$

From Sec. 39-7, the term "no load" for a generator means the line current is zero. Also, $N = z/a = z/p$ for a lap-wound armature.

$$E = \frac{zp\Phi n}{60a} = \frac{Np\Phi n}{60}$$

$$= \frac{(100)(2)(0.036 \text{ Wb})\left(1800 \dfrac{\text{rev}}{\text{min}}\right)}{60 \dfrac{\text{s}}{\text{min}}} = 216 \text{ V}$$

12. VOLTAGE-CURRENT CHARACTERISTICS FOR DC GENERATORS

Figure 39.8 illustrates the voltage-current characteristics for a DC generator.

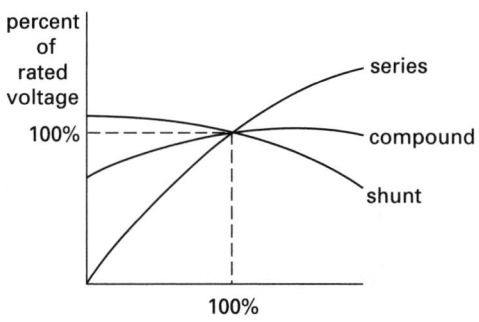

Figure 39.8 *DC Generator Voltage-Current Characteristics*

13. TORQUE CHARACTERISTICS FOR DC MOTORS

The torque produced by a DC motor is illustrated by Fig. 39.9.

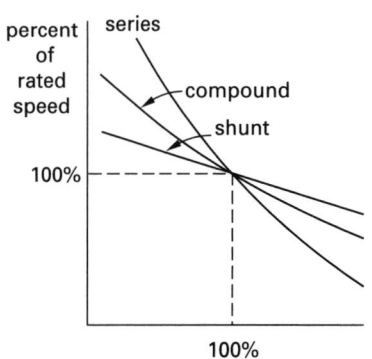

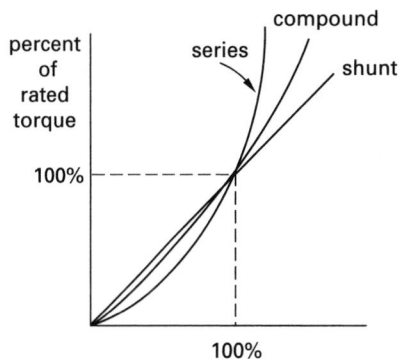

Figure 39.9 *DC Motor Torque Characteristics*

Power–Machinery

14. STARTING DC MOTORS

DC motors have very low armature resistance. At rest, there is no back emf, E. If connected across the full line voltage, the high current could damage the motor. Such motors are almost always started with a resistance in series with the armature winding.[3] An initial resistance is chosen that limits the starting current to approximately 150% of the full-load current. As the motor builds up speed, the back emf opposes the line voltage, reducing the current. The starting resistance can then be gradually reduced and removed.

Since torque depends on the current, starting torque is limited to approximately 225%, 175%, and 150% of the full-load torque for series, compound, and shunt motors, respectively, when the starting current is 150% of full-load current.

15. SPEED CONTROL FOR DC MOTORS

The speed of a DC motor can be controlled by changing the armature conditions, field conditions, or both. Changes can be made manually or automatically. Automatic starters may switch between positions based on time, current, voltage, or magnetic field.

Armature control techniques include (a) placing a variable resistance in series or parallel with the armature and (b) changing the voltage across it. The field voltage is held constant. For a constant torque load, the speed varies in approximate proportion to the voltage impressed on the armature.

Field control (*field weakening*) techniques include (a) changing the resistance of the field winding (series or shunt) and (b) changing the voltage across it. The armature voltage is held constant. Reducing the field current reduces the flux and increases the motor speed.

Electronic control of DC motors is achieved by supplying either the armature or field (or both) through electronic rectifiers (silicon-controlled rectifiers—SCRs). Through a feedback element or other control mechanism, rectifier output is automatically adjusted to maintain the requirements of the machine or process. Constant speed can be obtained in this manner, as well as great control over a widely varying range of speeds and torques.

---

[3]Small fractional horsepower ($^1/_4$ hp or smaller) motors are an exception.

40 Rotating AC Machinery

Nomenclature

a	number of parallel armature paths	–
a	ratio of transformation	–
A	area	m^2
B	magnetic flux density	T
B	susceptance	S
E	generated emf	V
f	electrical frequency	Hz
G	conductance	S
I	current	A
k	constant	various
n	rotational speed	rpm
N	number of series armature paths	–
p	total number of poles	–
P	power	W
q	number of loops	–
Q	reactive power	VAR
r	radius	m
R	resistance	Ω
s	slip	–
S	apparent power	VA
SR	speed regulation	–
t	time	s
T	period	s
T	torque	N·m
V	line voltage	V
VR	voltage regulation	–
X	reactance	Ω
Y	admittance	S
z	total number of conductors	–
Z	impedance	Ω

Symbols

δ	torque angle	rad
η	efficiency	–
θ	phase difference angle	rad
Φ	magnetic flux	Wb
ω	electrical frequency	rad/s
Ω	armature rotational speed	rad/s

Subscripts

a	armature
adj	adjusted
b	blocked rotor
Cu	copper
E	emf
f	field
fl	full load
g	generated
l	line
L	load
m	maximum
n	neutral
nl	no load
p	phase
r	rotor
s	synchronous
st	stator
t	total
T	torque

1. ROTATING MACHINES

Rotating machines are broadly categorized as AC and DC machines. Both categories include machines that use power (i.e., motors) and those that generate power (alternators and generators). Most machines can be constructed in either single-phase or polyphase configurations, although single-phase machines may be outclassed in terms of economics and efficiency.

Types of small AC motors include split-phase, repulsion-induction, universal, capacitor, and series motors. Large AC motors are almost always three-phase, but it is necessary to analyze only one phase of the motor. Torque and power are divided evenly among the three phases. Machines can be wye- or delta-wired or both.

Power–Machinery

Wye connections have several benefits:

- A neutral (ground) wire is intrinsically part of the circuit.

- Higher-order (harmonic) terms are not shorted out.

- Starting current is lower.

Nevertheless, high horsepower motors are usually run in delta. To avoid large starting currents, the motor can be started in wye and switched over to delta.

It is common to refer to line-to-line voltage as the *terminal voltage*, V.

$$V_p = \begin{cases} V & \text{[delta-wired]} \\ \dfrac{V}{\sqrt{3}} & \text{[wye-wired]} \end{cases} \qquad 40.1$$

$$T_p = \frac{T_t}{3} \qquad 40.2$$

$$P_p = \frac{P_t}{3} \qquad 40.3$$

$$S_p = \frac{S_t}{3} \qquad 40.4$$

2. TORQUE AND POWER

Torque and power are operating parameters. It takes power to turn an alternator or generator. A motor converts electrical power into mechanical power. In the SI system, power is given in kilowatts (kW). One horsepower is equivalent to 745.7 watts. The relationship between torque and power is

$$T_{\text{ft-lbf}} = \frac{5252\,P_{\text{horsepower}}}{n_{\text{rpm}}} \qquad 40.5$$

$$T_{\text{N·m}} = \frac{1000\,P_{\text{kW}}}{\Omega} = \frac{9549\,P_{\text{kW}}}{n_{\text{rpm}}} \qquad 40.6$$

There are many important torque parameters for motors. The *starting torque* (also known as *static torque, breakaway torque,* and *locked-rotor torque*) is the turning effort exerted in starting a load from rest. *Pull-up torque (acceleration torque)* is the minimum torque developed during the period of acceleration from rest to full speed. *Pull-in torque* (as developed in synchronous motors) is the maximum torque that brings the motor back to synchronous speed. (Nominal pull-in torque is the torque that is developed at 95% of synchronous speed.) The *steady-state torque* must be provided to the load on a continuous basis. It establishes the temperature increase that the motor must be able to withstand without deterioration. The *rated torque* is developed at rated speed and rated horsepower. The maximum torque a motor can develop at its synchronous speed

is the *pull-out torque. Breakdown torque* is the maximum torque the motor can develop without stalling (i.e., without coming rapidly to a complete stop).

Equation 40.7 is the general torque expression for a rotating machine with N coils of cross-sectional area A each carrying current I through a magnetic field of strength B.

$$T = NBAI\cos\omega t \qquad 40.7$$

3. SERVICE FACTOR

The horsepower and torque ratings listed on the nameplate of a motor can be provided on a continuous basis without overheating. Motors can be operated at slightly higher loads without exceeding a safe temperature rise, but the higher temperature has a deteriorating effect on the winding insulation. A general rule of thumb is that a motor loses two or three hours of useful life for each hour run at the factored load. The ratio of the safe to standard loads is the *service factor*, usually expressed as a decimal. Service factors vary from 1.15 to 1.4, with the lower values going to larger, more efficient motors.

$$\text{service factor} = \frac{\text{safe load}}{\text{nameplate load}} \qquad 40.8$$

4. MOTOR CLASSIFICATIONS

The National Electrical Manufacturers Association (NEMA) has categorized motors in several ways: *speed classification* (constant-, adjustable-, multi-, varying-speed, etc.); *service classification* (general, definite, and special purpose); and motor class. *Motor class* is a primary indicator of the maximum motor operating temperature, which, in turn, depends on the type of insulation used on the conductors. The classes are: Class A, 105°C; Class B, 130°C; Class F, 155°C; and Class H, 180°C.

5. POWER LOSSES[1]

The losses for all rotating machines can be divided into four categories. *Copper losses*, P_{Cu}, are real power losses due to wire and winding resistance. In a DC machine, copper losses occur in the armature and field windings as well as from the brush contact resistance. In an AC machine, copper losses occur in the armature and exciter field windings. There are no brush losses in an induction machine.

$$P_{\text{Cu}} = \sum I^2 R \qquad 40.9$$

Core losses, including hysteresis and eddy current losses, are constant losses that are independent of the load

---

[1]The subject of transformer losses covered in Sec. 36-3 is equivalent in concept.

and, for that reason, are also known as *open-circuit* and *no-load losses*. In DC and synchronous AC machines, core losses occur in the armature iron. In induction machines, core losses occur in the stator iron.

Mechanical losses (also known as *rotational losses*) include brush and bearing friction and *windage* (air friction). (Windage is a no-load loss but is not an electrical core loss.) Mechanical losses are determined by measuring the power input at rated speed and no load.

Stray losses are due to nonuniform current distribution in the conductors. Stray losses are approximately 1% for DC machines and zero for AC machines.

Real power only is used to compute the *efficiency* of a rotating machine.

$$\eta = \frac{\text{output}}{\text{input}}$$
$$= \frac{\text{output}}{\text{output} + \text{losses}}$$
$$= \frac{\text{input} - \text{losses}}{\text{input}} \qquad 40.10$$

6. REGULATION

The *voltage regulation*, VR, is

$$\text{VR} = \frac{\text{no-load voltage} - \text{full-load voltage}}{\text{full-load voltage}} \times 100\% \qquad 40.11$$

The *speed regulation*, SR, is

$$\text{SR} = \frac{\text{no-load speed} - \text{full-load speed}}{\text{full-load speed}} \times 100\% \qquad 40.12$$

7. NO-LOAD CONDITIONS

The meaning of the term *no load* is different for generators and motors. For unloaded shunt-wired alternators and generators, there is no electrical load connected across the output terminals, so although the field current flows, the line current, I, is zero. For unloaded shunt-wired motors, the work performed is zero, but line current is still drawn to keep the motor turning. All of the current is field current, however, and (neglecting mechanical losses) the armature current, I_a, is zero.

$$I = 0; \quad I_f \neq 0; \quad I_a = I_f \qquad \text{[generator]} \qquad 40.13$$
$$I \neq 0; \quad I_f = I; \quad I_a = 0 \qquad \text{[motor]} \qquad 40.14$$

8. PRODUCTION OF AC POTENTIAL

A potential of alternating polarity is produced by an *alternator (AC generator)*. A permanent magnet or DC electromagnet produces a constant magnetic field. Figure 40.1 illustrates how several loops of wire can be combined into a rotating *induction coil* or *armature* to produce a continuously varying potential in a *dynamo* (*coil dynamo*). Three-phase alternators have three sets of independent windings. A single-phase AC alternator has only one set of windings.

The induced voltage, E, is commonly called *electromotive force* (emf). In an elementary alternator, emf is the desired end result and is picked off by stationary brushes making contact with *slip rings* on the rotating shaft.[2] In a motor, emf is also produced but is referred to as *back emf* (*counter emf*) since it opposes the input current.

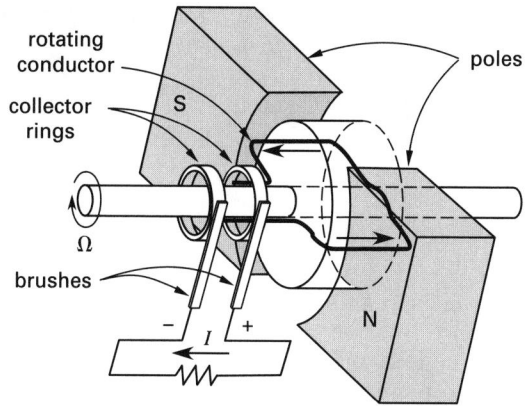

Figure 40.1 *Elementary Two-Pole, Single-Coil Dynamo*

Assuming the magnetic field flux density under each pole face, B, is uniform, the maximum flux linked by a coil with N turns and area A is NAB. Since the coil rotates, the flux linkage is a function of the projected coil area. The instantaneous induced voltage is predicted by Faraday's law. Care must be taken to distinguish between the armature speed, n (in rpm), the angular armature speed, Ω (in rad/s), and the linear and angular voltage frequencies, f and ω (in Hz and rad/s, respectively). Note the distinction in Eq. 40.15 between the rotational speeds of the armature ($\Omega = 2\pi n/60$) and the electrical waveform ($\omega = 2\pi f$).

$$V(t) = V_m \sin \omega t = \omega NAB \sin \omega t$$
$$= \left(\frac{p}{2}\right) \Omega NAB \sin\left(\left(\frac{p\Omega}{2}\right) t\right) \qquad 40.15$$
$$= \frac{\pi n p NAB}{60} \sin\left(\left(\frac{p\Omega}{2}\right) t\right)$$

---

[2]Practical alternators, described in Sec. 40-10, use slip rings to feed the field, not to transfer the generated voltage.

$$V = \frac{V_m}{\sqrt{2}} = \frac{\omega NAB}{\sqrt{2}} = \frac{p\Omega NAB}{2\sqrt{2}}$$

$$= \frac{\pi np NAB}{60\sqrt{2}} \quad \text{[effective]} \qquad 40.16$$

Table 40.1 summarizes the most frequently used formulas for a single-phase AC alternator.

Alternators are characterized by the number of magnetic poles, p. (Both north and south poles are counted to distinguish the quantity from the number of *pole pairs*.) The *coil pitch* (*pole pitch*) is the angle between the poles, or 360° divided by p. A two-pole alternator produces one complete sinusoidal cycle per revolution. An alternator with p poles produces $(^1/_2)p$ cycles per revolution. Since the armature normally turns at a constant speed known as the *synchronous speed*, n_s, the *electrical frequency*, f, of the generated potential is given by Eq. 40.18. The actual rotational speed, n, is known as the *mechanical frequency*.

$$n_s = \frac{120f}{p} = \frac{60\Omega}{2\pi} = \frac{60\omega}{\pi p} \quad \left[\begin{array}{c}\text{synchronous}\\\text{speed}\end{array}\right] \quad 40.17$$

$$f = \frac{1}{T} = \frac{\omega}{2\pi} = \frac{pn_s}{120} \qquad 40.18$$

Since the coils in an alternator have inductance as well as resistance (see Sec. 40-14), the rated capacity of an AC machine is stated as apparent power at some rated voltage and power factor.

Example 40.1

A four-pole alternator produces a 60 Hz potential. What is the (a) mechanical speed of the armature, (b) angular velocity of the potential, and (c) angular velocity of the armature?

Solution

(a) From Eq. 40.17, the rotational speed is

$$n = n_s = \frac{120f}{p} = \frac{\left(120\ \frac{\text{pole·s}}{\text{min}}\right)(60\ \text{Hz})}{4\ \text{poles}}$$

$$= 1800\ \text{rpm}$$

(b) The angular velocity of the 60 Hz potential is

$$\omega = 2\pi f = \left(2\pi\ \frac{\text{rad}}{\text{cycle}}\right)(60\ \text{Hz}) = 377\ \text{rad/s}$$

(c) From Eq. 40.15, the angular velocity of the armature is

$$\Omega = \frac{2\omega}{p} = \frac{2\pi n}{60} = \frac{\left(2\pi\ \frac{\text{rad}}{\text{rev}}\right)\left(1800\ \frac{\text{rev}}{\text{min}}\right)}{60\ \frac{\text{s}}{\text{min}}}$$

$$= 188.5\ \text{rad/s}$$

Table 40.1 *Single-Phase AC Alternator Formulas*

	in terms of n	in terms of Ω	in terms of f	in terms of ω
n (rpm)	–	$\dfrac{30\Omega}{\pi}$	$\dfrac{120f}{p}$	$\dfrac{60\omega}{\pi p}$
Ω (mechanical rad/s)	$\dfrac{\pi n}{30}$	–	$\dfrac{4\pi f}{p}$	$\dfrac{2\omega}{p}$
f (Hz)	$\dfrac{pn}{120}$	$\dfrac{p\Omega}{4\pi}$	–	$\dfrac{\omega}{2\pi}$
ω (electrical rad/s)	$\dfrac{\pi pn}{60}$	$\dfrac{p\Omega}{2}$	$2\pi f$	–
single-phase V_m (V)	$\dfrac{\pi pn NAB}{60}$	$\dfrac{\Omega p NAB}{2}$	$2\pi f NAB$	ωNAB
single-phase V_{eff} (V)	$\dfrac{\pi pn NAB}{60\sqrt{2}}$	$\dfrac{\Omega p NAB}{2\sqrt{2}}$	$\dfrac{2\pi f NAB}{\sqrt{2}}$	$\dfrac{\omega NAB}{\sqrt{2}}$
single-phase V_{ave} (V)	$\dfrac{pn NAB}{30}$	$\dfrac{\Omega p NAB}{\pi}$	$4f NAB$	$\dfrac{2\omega NAB}{\pi}$

Example 40.2

The rotor of a single-phase, four-pole alternator rotates at 1200 rpm. The effective diameter and length of the 20-turn (loop) coil are 0.12 m and 0.24 m, respectively. The magnetic flux density is 1.2 T. What is the effective voltage produced?

Solution

The coil area is

$$A = 2rl = (0.12 \text{ m})(0.24 \text{ m}) = 0.0288 \text{ m}^2$$

From Eq. 40.16, the effective voltage is

$$V = \frac{p\pi n NAB}{60\sqrt{2}}$$

$$= \frac{(4)(1200 \text{ rpm})\left(2\pi \dfrac{\text{rad}}{\text{rev}}\right)}{(2\sqrt{2})\left(60\dfrac{\text{s}}{\text{min}}\right)}$$

$$= 122.8 \text{ V}$$

9. ARMATURE WINDINGS

An armature winding consists of several *coils* of continuous wire formed into *q loops*. Each loop in an armature contributes two *conductors*, also known as *bars*. The number of conductors, z, is

$$z = 2q \qquad\qquad 40.19$$

Voltage is induced only in conductors and inductors that are parallel to the armature shaft (i.e., that cut magnetic flux lines). The number of *series paths*, N, between positive and negative brush sets depends on how the coils are wound and connected to the commutator.

$$N = \frac{z}{a} = \frac{2q}{a} \qquad\qquad 40.20$$

The number of parallel armature paths between each pair of brushes, a, equals the number of poles, p, in a *lap-wound armature* (*simplex-lap armature* or *multiple-drum armature*).[3] (The number of poles is also equal to the number of brushes.) The coil connections are made to adjacent commutator segments. This type of winding is commonly used in DC machines and induction motors and, to a lesser extent, in AC generators.

---

[3]Auxiliary windings are used in *shaded pole motors; split-phase motors* have two separate windings; *capacitor motors* have capacitors in series with an auxiliary winding.

In a *wave-wound armature* (*two-circuit* or *series-drum armature*), $a = 2$ and the coil connections are on commutator segments on opposite sides of the armature regardless of the number of poles. Only two brushes are needed, although more can be used. This configuration is commonly used in DC armatures requiring the generation or use of higher voltages than lap windings can tolerate.

A *multiplex winding* has two or more distinct coils connected in parallel. The number of parallel current paths, a, in the armature can be determined by the type of winding and the number of poles, p, as indicated in Table 40.2.

Table 40.2 *Multiplex Winding Current Paths*

type	lap winding	wave winding
simplex	$a = p$	$a = 2$
duplex	$a = 2p$	$a = 4$
triplex	$a = 3p$	$a = 6$
quadriplex	$a = 4p$	$a = 8$

10. PRACTICAL ALTERNATORS

There are several reasons why large alternators are not designed with a stationary magnetic field and rotating coil as shown in Fig. 40.1.

- The rotating coils must be well insulated to prevent shorting with the high voltages that are induced.

- Structural bracing is required to counteract the large centrifugal force that results from rotating many coils of wire.

- It is difficult to make efficient high-voltage, high-power connections through slip rings.

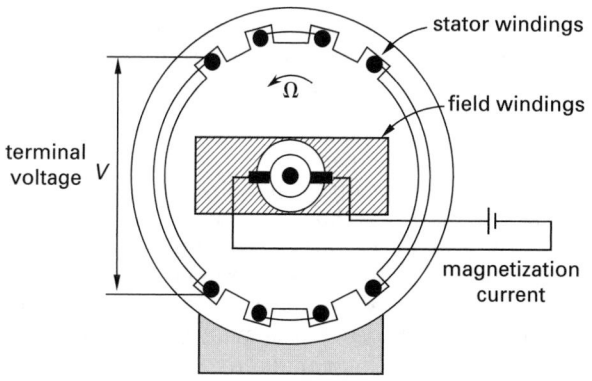

Figure 40.2 *Practical Alternator*

For these reasons, practical alternators reverse the locations of the field and induction coils so that the low-voltage field revolves and the high-voltage field is induced in stationary coils. The magnetic field is produced by *field windings* whose DC *magnetization current* is supplied through brushes and slip rings. Such an

Power–Machinery

armature containing field coils is known as a *rotor*. The stationary induction coils are placed in slots a small distance from the rotor and are known as the *stator*. The closer the stator and rotor are, the smaller the magnetization current required.

With *cylindrical rotors* (*wound rotors*), the field windings are embedded in an externally smooth-surfaced cylindrical rotor. This makes them suitable for high-speed (i.e., greater than 1800 rpm) operation since windage losses are reduced. The poles of *salient-pole rotors*, on the other hand, resemble exposed wound electromagnets. This is adequate for 1800 rpm and less.

11. SYNCHRONOUS MOTORS

Synchronous motors are essentially dynamo alternators operating in reverse. Alternating current is supplied to the stationary stator windings. DC current is applied to the field windings in the rotor through brushes and slip rings as in an alternator.[4] The field current interacts with the stator field, causing the armature to turn. Since the stator field frequency is fixed, the motor runs only at a single *synchronous speed* (see Eq. 40.17).

Important features of synchronous motors are as follows.

- They turn at constant speeds, regardless of the load. (The speed may momentarily change when the load is changed.) Stalling occurs when a motor's countertorque is exceeded.

- The power factor can be adjusted manually without losing synchronization by varying the field current. A unity power factor occurs with *normal excitation* current. The power factor is leading (lagging) when the current is more (less) than normal, and this is known as being *over-* (*under-*) *excited*.

- They can draw leading currents and be used for power factor correction, in which case they are known as *synchronous capacitors* (*synchronous condensers*).

Since the starting torque is zero, it is necessary to bring a synchronous motor up to speed by some other means. The most common method is to include auxiliary windings in the pole faces so that the motor can be started as an induction motor. At synchronous speed, the auxiliary windings draw no current. If the motor speed becomes nonsynchronous, the auxiliary windings draw power to resynchronize the rotor.

Since synchronous motors run at only one speed, curves of horsepower (or other characteristics) versus speed are not used. Instead, motors are categorized on the basis of speed, torque, and horsepower at a specific power factor.

---

[4]The magnetization energy can also be generated through induction. In that case, the rotor includes diodes to rectify the induced AC potential.

12. SYNCHRONOUS MACHINE EQUIVALENT CIRCUIT

Figure 40.3 illustrates a simple equivalent circuit for a synchronous machine. The vector voltage relationship is defined by Eqs. 40.21 and 40.22, which introduce the equivalent *synchronous inductance*, X_s, of each phase. V_p is the phase voltage. The series armature resistance, R_a, is small and normally disregarded.

$$\mathbf{E} = \mathbf{V}_p + (R_a + jX_s)\mathbf{I}_a$$
$$\approx \mathbf{V}_p + jX_s\mathbf{I}_a \quad \text{[alternator]} \qquad 40.21$$
$$\mathbf{V}_p = \mathbf{E} + (R_a + jX_s)\mathbf{I}_a$$
$$\approx \mathbf{E} + jX_s\mathbf{I}_s \quad \text{[motor]} \qquad 40.22$$

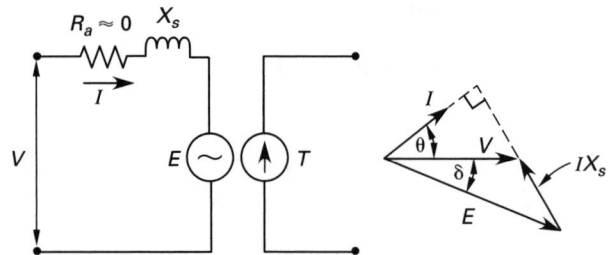

Figure 40.3 *Synchronous Motor Equivalent Circuit*

Equation 40.23 gives the real power generated per phase. For a motor, the power factor is determined by E and I_f. For an alternator, the power factor determines E and I_f. (Equation 40.23 mixes variables with different units.)

$$P_p = T_p\Omega = VI\cos\theta = \left(\frac{VE}{X_s}\right)\sin\delta \qquad 40.23$$

Equation 40.24 gives the torque produced per phase. Φ_r and Φ_{st} are the internal rotor and stator fluxes, respectively. The *torque angle*, δ (also known as the *power angle* and *displacement angle*), is the phase difference angle between the applied voltage, V, and the generated emf, E. It is positive for an alternator and negative for a motor. Torque is maximum when $\delta = 90°$, a condition equivalent to a unity power factor and known as *pull-out torque*. The *pull-out power* is found by setting $\delta = 90°$ in Eq. 40.23.

$$T_p = k_T\Phi_r\Phi_{st}\sin\delta \qquad 40.24$$

The apparent power per phase is

$$S = VI = \frac{P_p}{\cos\theta} \qquad 40.25$$

Example 40.3

A six-pole motor is connected to a three-phase, 240 V (rms), 60 Hz line. Its stator windings are connected in a wye configuration. The motor has a synchronous reactance of 3 Ω per phase and is rated at 10 kVA at its synchronous speed and 100% power factor. What are the (a) synchronous speed, (b) phase voltage, (c) line current, (d) voltage drop across the synchronous reactance, (e) generated back emf, and (f) torque angle?

Solution

(a) Equation 40.17 gives the synchronous speed.

$$n_s = \frac{120f}{p} = \frac{(120)(60 \text{ Hz})}{6} = 1200 \text{ rpm}$$

(b) Since the windings are in a wye configuration, the phase voltage is

$$V_p = \frac{V}{\sqrt{3}} = \frac{240 \text{ V}}{\sqrt{3}} = 138.6 \text{ V}$$

(c) The line current is the same as the phase current, which can be calculated from the apparent power. Each phase draws one-third of the kVA. The real power is

$$P_p = \left(\frac{S}{3}\right) \cos\theta$$

$$= \frac{(10{,}000 \text{ VA})\left(1 \dfrac{\text{W}}{\text{VA}}\right)}{3} = 3333 \text{ W}$$

$$I_p = \frac{P_p}{V_p} = \frac{3333 \text{ W}}{138.6 \text{ V}} = 24.05 \text{ A}$$

(d) The voltage drop across each winding is

$$V_p = I_p X_p = (24.05 \text{ A})(3 \text{ }\Omega) = 72.15 \text{ V}$$

(e) The back emf is

$$\mathbf{E} = \mathbf{V}_p - jIX = 138.6 - j72.15$$

$$= 156.3\angle{-27.50°}$$

(f) The torque angle was found in part (e) to be $\delta = -27.50°$.

13. INDUCTION MOTORS

Induction motors are essentially constant-speed devices that receive power through induction—there are no brushes or slip rings. A motor can be considered as a rotating transformer secondary (the rotor) with a stationary primary (the stator). The stator field rotates at the synchronous speed given in Eq. 40.17. An emf is induced as the stator field moves past the rotor conductors. Since the rotor windings have reactance, the rotor field lags the induced emf.

In order to have a change in flux linkage, the rotor must turn at less than the synchronous speed. The difference in speed is small but essential. *Percent slip, s,* typically 2 to 5%, is the percentage difference in speed between the rotor and stator fields. Slip can be expressed as either a decimal or a fraction (e.g., 0.05 slip or 5% slip). Percent slip and *percent synchronism* are complements (i.e., add to 100%). Slip in rpm is the difference between actual and synchronous speeds.

$$s = \frac{n_s - n}{n_s} = \frac{\Omega_s - \Omega}{\Omega_s} \qquad 40.26$$

The stator is identical to that in an alternator or synchronous motor. In a *wound rotor,* the rotor is similar to an armature winding in a dynamo.[5] However, in a *squirrel-cage rotor* there are no wire windings at all.[6] The squirrel-cage rotor consists of copper or aluminum bars embedded in slots in the cylindrical iron core of the rotor. The ends of the bars are shorted by conductive rings, as illustrated in Fig. 40.4, to form loops. The rotor and its components, connections, and lines constitute the *rotor circuit* or *secondary circuit.*

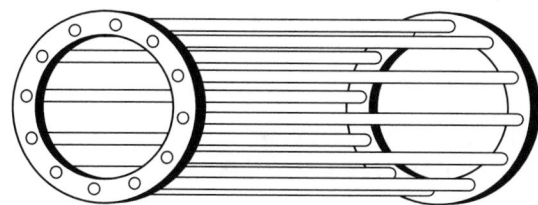

Figure 40.4 *Squirrel-Cage Rotor Induction Motor*

Example 40.4

An induction motor developing 10 hp is connected to a three-phase, 240 V (rms), 60 Hz power line. The stator windings are connected in a wye configuration. The synchronous speed is 1800 rpm, but the motor turns at 1738 rpm when loaded. Its energy efficiency is 80% and power factor (pf) is 70%. Calculate the (a) slip, (b) number of poles, (c) line current drawn, and (d) phase voltage.

Solution

(a) From Eq. 40.26, the slip is

$$s = \frac{n_s - n}{n_s} = \frac{1800 \text{ rpm} - 1738 \text{ rpm}}{1800 \text{ rpm}} = 0.03444$$

(b) The number of poles is calculated from the synchronous speed.

$$p = \frac{120f}{n_s} = \frac{(120)(60)}{1800} = 4$$

---

[5]Windings can be incorporated in the rotor to obtain better speed control or high torque.
[6]If the slip rings of a wound rotor are shorted, the motor behaves like a squirrel-cage motor.

(c) The total real power input is

$$P = \frac{(10 \text{ hp}) \left(745.7 \ \dfrac{\text{W}}{\text{hp}} \right)}{0.8} = 9321 \text{ W}$$

The power per phase is

$$P_p = \frac{P}{3} = \frac{9321 \text{ W}}{3} = 3107 \text{ W}$$

Since the phase and line currents are identical in wye-connected loads (see Sec. 49-6), the line current is

$$I = I_p = \frac{P_p}{V_p(\text{pf})} = \frac{3107 \text{ W}}{\left(\dfrac{240 \text{ V}}{\sqrt{3}} \right) (0.70)} = 32.03 \text{ A}$$

(d) The voltage across each phase is

$$V_p = \frac{V}{\sqrt{3}} = \frac{240 \text{ V}}{\sqrt{3}} = 138.6 \text{ V}$$

14. INDUCTION MOTOR EQUIVALENT CIRCUIT

Figure 40.5(a) illustrates the equivalent circuit for an induction motor.[7] It is very similar to the equivalent circuit for a transformer. R_1 and X_1 represent the equivalent stator resistance and reactance, respectively. G_{nl} and B_{nl} represent the stator core loss and susceptance, respectively, as determined from no-load testing. The dashed line represents the air gap across which energy is transferred to the rotor. R_2 and X_2 are the equivalent rotor resistance and reactance, respectively. There is no element to model the rotor core loss, which is negligible. Rotational losses are included in the stator core loss element, G. Any equivalent load resistance is included in the rotor resistance. The ratio of transformation, a, is taken as 1.0 for a squirrel-cage motor.[8]

Using an adjusted voltage, V_{adj}, simplifies the model, as shown in Fig. 40.5(b). The relationship between the applied terminal voltage, V_1, and the adjusted voltage is

$$\mathbf{V}_{\text{adj}} = \mathbf{V}_1 - \mathbf{I}_{nl}(R_1 + jX_1) \qquad 40.27$$

$$V_{\text{adj}} \approx V_1 - I_{nl}\sqrt{R_1^2 + X_1^2} \qquad 40.28$$

---

[7]This equivalent circuit cannot be used for a double squirrel-cage motor.
[8]With a phase-wound rotor, the brushes can be lifted from the slip rings. The ratio of transformation, a, can be determined as the ratio of applied voltage to voltage across the slip rings. This cannot be done with a squirrel-cage motor, so the equivalent circuit parameters are redefined with the ratio of transformation.

The total series resistance per phase, R, is

$$R = R_1 + \frac{R_2}{s} \qquad 40.29$$

Equation 40.30 gives the torque-speed relationship predicted by this model.

$$T_p = \frac{I_2^2 R_2}{s \omega_s} = \frac{V_{\text{adj}}^2 R_2}{s \omega_s \left(\left(R_1 + \dfrac{R_2}{s} \right)^2 + X^2 \right)} \qquad 40.30$$

$$T_t = 3 T_p \qquad 40.31$$

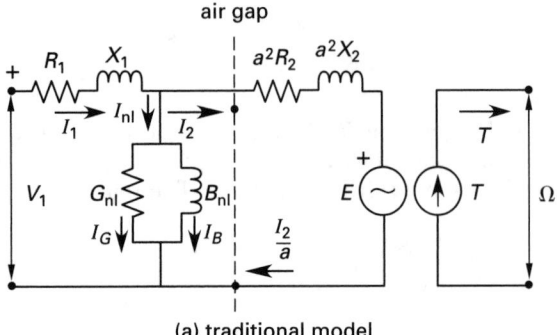

(a) traditional model

$$R = R_1 + \frac{R_2}{s} \quad X = X_1 + X_2 = 2X_1$$

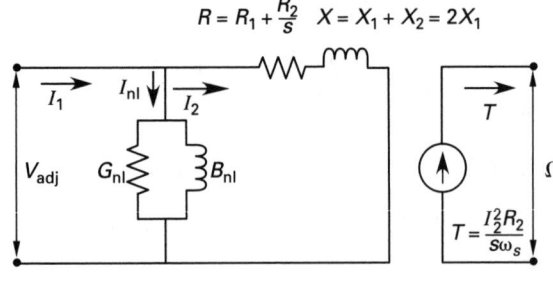

(b) simplified model ($a = 1$)

Figure 40.5 *Equivalent Circuits of an Induction Motor*

15. OPERATING CHARACTERISTICS OF INDUCTION MOTORS

Under normal operating conditions, slip is small (less than 0.05) and Eq. 40.32 predicts that the torque will be directly proportional to slip, s, and inversely proportional to the rotor resistance, R_2. At low speeds, the reactive term is larger than the resistive term in Eq. 40.33.

$$T_p \approx \frac{V_{\text{adj}}^2 s}{\omega_s R_2} \quad \text{[high speed, per phase]} \qquad 40.32$$

$$\approx \frac{V_{\text{adj}}^2 R_2}{s \omega_s X^2} \quad \text{[low speed, per phase]} \qquad 40.33$$

The starting torque per phase is proportional to rotor winding resistance and terminal voltage and can be found by setting s equal to one in Eq. 40.33.

$$T_{\text{starting}} \approx \frac{V_{\text{adj}}^2 R_2}{\omega_s X^2} \quad \text{[starting, per phase]} \quad \textbf{40.34}$$

The maximum torque (known as *breakdown torque*) is independent of the rotor circuit resistance, but the rotor resistance does affect the speed at which the maximum torque occurs. A large rotor circuit resistance only causes the maximum torque to occur at a larger slip. Maximum torque varies directly with the square of the stator voltage.

$$T_{\text{max}} = \frac{V_{\text{adj}}^2}{2\omega_s(R_1 + \sqrt{R_1^2 + X^2})} \quad \text{[per phase]} \quad \textbf{40.35}$$

$$s_{\text{max } T} = \frac{R_2}{\sqrt{R_1^2 + X^2}} \quad \textbf{40.36}$$

Figure 40.6 illustrates typical characteristic curves of an induction motor. Such curves are conventionally provided for the motor running at its optimum efficiency and power factor. Curves can be provided for polyphase or single-phase operation.

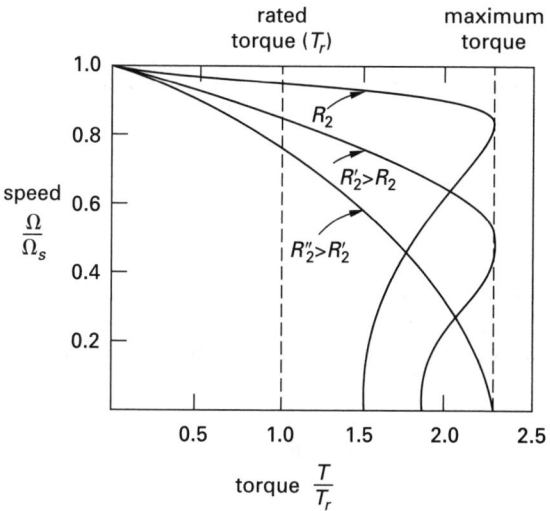

Figure 40.6 *Characteristic Curves for an Induction Motor*

16. TESTING INDUCTION MOTORS

Performance of induction motors is evaluated in a manner analogous to transformer testing. The *no-load motor test* (*running-light test*) corresponds to an open-circuit transformer test. This test determines the values of B and G in the equivalent circuit. The *blocked-rotor test* corresponds to a closed-circuit transformer test and determines the values of R and X.

In a *no-load test*, the motor is run at the rated voltage without load. The line voltage, V, line current, I_{nl}, and power per phase, P_{nl}, are measured.

$$\text{no-load power factor} = \cos\theta_{\text{nl}} = \frac{P_{\text{nl}}}{V_{\text{adj}} I_{\text{nl}}} \quad \textbf{40.37}$$

Referring to Fig. 40.5, the magnetization (quadrature) current, I_B, and the in-phase component of the exciting current, I_G, are

$$I_B = I_{\text{nl}} \sin\theta_{\text{nl}} \quad \textbf{40.38}$$

$$I_G = I_{\text{nl}} \cos\theta_{\text{nl}} \quad \textbf{40.39}$$

The stator parameters are

$$G_{\text{nl}} = \frac{P_{\text{nl}}}{V_{\text{adj}}^2} \quad \textbf{40.40}$$

$$B_{\text{nl}} = -\sqrt{Y_{\text{nl}}^2 - G_{\text{nl}}^2} = -\sqrt{\left(\frac{I_{\text{nl}}}{V_{\text{nl}}}\right)^2 - G_{\text{nl}}^2}$$

$$\approx \frac{I_B}{V_{\text{adj}}} \quad \textbf{40.41}$$

In a *blocked-rotor test*, the rotor is blocked. A (low) voltage is applied and measured so that the rated current flows. The phase current, I_b, power per phase, P_b, and phase voltage, V_b, are measured.

The phase current lags the phase voltage by the following angle.

$$\cos\theta_b = \frac{P_b}{I_b V_b} \quad \textbf{40.42}$$

The stator reactance, X_1, is usually assumed to be one-half of X, the remainder being the rotor reactance. However, since performance depends greatly on R_2, the division of R must be less arbitrary. One approach is to measure the DC resistance between the terminals and use this as an approximate value of $2R_1$.[9] (This assumes wye-connected motor windings.) The remainder of R is given to R_2.

$$X = X_1 + X_2 = 2X_1 = \frac{V_b}{I_b \sin\theta_b}$$

$$= \sqrt{Z^2 - R^2} \quad \textbf{40.43}$$

$$R = R_1 + R_2 = \frac{P_b}{I_b^2} \quad \textbf{40.44}$$

$$Z = \frac{V_b}{I_b} \quad \textbf{40.45}$$

---

[9]This value is corrected empirically, but the correction is beyond the scope of this book.

17. STARTING INDUCTION MOTORS

Induction motors draw their maximum current when starting (i.e., when slip, s, is one). The starting torque for a polyphase motor varies directly with the square of the stator voltage but also depends on the rotor resistance and reactance. (Starting torque for a single-phase induction motor is zero. Therefore, single-phase induction motors must be brought up to speed by some other means.)[10] For a given stator voltage, there is a particular rotor resistance that maximizes starting torque. Increasing or decreasing resistance from this optimum value decreases the starting torque. Starting current can be calculated from Eq. 40.46 by setting s equal to one.

$$I_{\text{starting}} = I_1 = I_{\text{nl}} + I_2$$

$$= I_{\text{nl}} + \frac{V_{\text{adj}}}{R_1 + \dfrac{R_2}{s} + jX} \qquad \textbf{40.46}$$

The *blocked- (locked-) rotor current* is the current drawn by the motor with the rotor held stationary. It is a worst-case starting current since the rotor begins to move immediately upon starting, reducing the current drawn. *Free-rotor starting current* is approximately 75% of the blocked-rotor current.

18. SPEED CONTROL FOR INDUCTION MOTORS

The rotational speed of an induction motor depends on the number of poles, line voltage, supply frequency, and rotor circuit resistance. For a given machine, the number of poles cannot be varied without excessive complexity in winding switching and increased manufacturing cost. Since the breakdown torque is proportional to the square of the voltage, reducing the voltage may stall the motor. As a result, voltage speed control is rarely used. Changing the supply frequency is also impractical in most instances. Introducing a resistance in series with the rotor decreases the motor speed but is possible only in wound-rotor motors.

Speed control is commonly accomplished by introducing a foreign voltage in the secondary (rotor) circuit. If the foreign voltage opposes the voltage induced in the secondary circuit, the motor speed will be reduced and vice versa.

If two induction motors are available, one can be used to control the other by connecting them in *cascade*. The shafts are rigidly connected and the rotor and stator windings are interconnected. The two armatures can also be constructed on a single shaft.

---

[10]Polyphase induction motors will run but will not start on a single phase. There is a danger of current overload if synchronization is lost during single-phase operation. Protective elements (e.g., circuit breakers) are used to limit current.

19. POWER TRANSFER IN INDUCTION MOTORS

Figure 40.7 illustrates the power transfer in an induction motor. Equations 40.47 through 40.52 are per phase.

$$\text{input power} = V_1 I_1 \cos\theta \qquad \textbf{40.47}$$

$$\text{stator copper losses} = I_1^2 R_1 \qquad \textbf{40.48}$$

$$\text{rotor input power} = \frac{I_2^2 R_2}{s} \qquad \textbf{40.49}$$

$$\text{rotor copper losses} = I_2^2 R_2 \qquad \textbf{40.50}$$

$$\text{electrical power delivered} = I_2^2 R_2 \left(\frac{1-s}{s}\right) \qquad \textbf{40.51}$$

$$\text{shaft output power} = T\Omega \qquad \textbf{40.52}$$

Core losses are constant (see Sec. 40-5). Copper losses are proportional to the square of the delivered power.

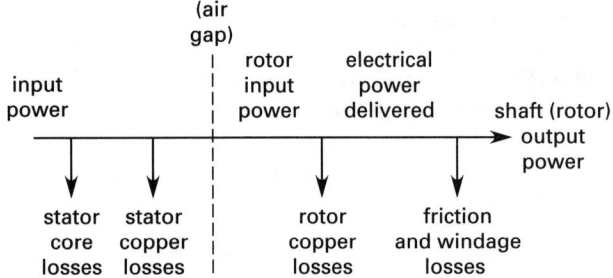

Figure 40.7 *Induction Motor Power Transfer*

Example 40.5

A 15 hp induction motor with six poles operates at 80% efficiency on a three-phase, 240 V (rms), 60 Hz line. The following losses are observed for full-load operation:

stator copper loss	540 W
friction/windage loss	975 W
core loss	675 W

What are the (a) speed and (b) torque when the motor delivers half power?

Solution

(a) The full-load output power is

$$P = (15 \text{ hp})\left(745.7 \, \frac{\text{W}}{\text{hp}}\right) = 11{,}186 \text{ W}$$

The input power is

$$P_{\text{in}} = \frac{P_{\text{out}}}{\eta} = \frac{11{,}186 \text{ W}}{0.80} = 13{,}982 \text{ W}$$

The full-load rotor copper loss is

$$\text{rotor copper loss} = 13{,}982 \text{ W} - 11{,}186 \text{ W}$$
$$- 540 \text{ W} - 975 \text{ W} - 675 \text{ W}$$
$$= 606 \text{ W} \qquad \text{[full power]}$$

The output power at half load is

$$P = \left(\tfrac{1}{2}\right)(15 \text{ hp})\left(745.7 \frac{\text{W}}{\text{hp}}\right) = 5593 \text{ W}$$

At half load, the friction and core losses are unchanged. From Eqs. 40.50 and 40.51, the ratio of actual to full-load copper losses is equal to the square of the ratio of actual to full-load output power.

$$\text{rotor copper loss} = \left(\tfrac{1}{2}\right)^2 (606 \text{ W}) = 152 \text{ W}$$
$$\text{stator copper loss} = \left(\tfrac{1}{2}\right)^2 (540 \text{ W}) = 135 \text{ W}$$

The rotor input power is

$$P_{\text{in}} = 5593 \text{ W} + 975 \text{ W} + 152 \text{ W} = 6720 \text{ W}$$

The synchronous speed is

$$n_s = \frac{120f}{p} = \frac{(120)(60)}{6} = 1200 \text{ rpm}$$

The slip at half power is found from Eq. 40.49.

$$s = \frac{\text{rotor copper loss}}{\text{rotor input power}} = \frac{152 \text{ W}}{6720 \text{ W}} = 0.0226$$

The speed at half load is

$$n = n_s(1 - s) = (1200 \text{ rpm})(1 - 0.0226)$$
$$= 1173 \text{ rpm}$$

(b) The torque is found from Eq. 40.5.

$$T = \frac{5252P}{n} = \frac{(5252)\left(\tfrac{1}{2}\right)(15 \text{ hp})}{1173 \text{ rpm}} = 33.6 \text{ ft-lbf}$$

Topic VIII: Power—Lightning

Chapter

    41. Lightning Protection and Grounding

For the most current information about the exam, visit **www.ppi2pass.com** regularly.

41 Lightning Protection and Grounding

Nomenclature

f	frequency	Hz, s^{-1}, or cycles/s
l	length	m
PM	protective margin	–
PQI	protection quality index	–
V	effective or DC voltage	V

Symbols

δ	skin depth	m
μ	permeability	H/m
σ	conductivity	S/m

Subscripts

p	protection
r	reseal
w	withstand

1. FUNDAMENTALS

Lightning is an atmospheric electrical discharge resulting from the creation and separation of electric charges in cumulonimbus clouds. The exact mechanism causing the charge separation is unknown, with several theories explaining some, but not all, of the effects. Among these theories are the *precipitation theory*, the *convection theory*, and the theory of *graupel-ice crystal interaction with charge-reversal temperature*. The combination of all three is nominally a tripolar-electrified thundercloud as shown in Fig. 41.1.

Lightning occurs when the charge accumulation is such that the electric field between the cloud and the earth is high enough to break down the dielectric between them (i.e., the air). During the first phase of the process, local ionization occurs and discharges called *pilot streamers* create a path for the *stepped leaders* that follow.[1] The luminous stepped leaders move at 15 to 20% the speed of light at approximately 50 m at a time (see Fig. 41.1). Each step is separated by a few microseconds. The second phase occurs when the stepped leader reaches the earth or a building or power line on the earth's surface. When this occurs, an extremely luminous high-intensity discharge occurs between the charge centers

[1]Stepped leaders may originate from the earth and move skyward.

of the earth and cloud. The discharge moves at 10 to 50% of the speed of light and results in a current of from several thousand amperes to 200,000 A peaking in from 1 to 10 μs. This discharge is called the *return stroke* or the *lightning stroke*.[2] In the final phase, strokes may occur between the charge centers within the cloud or to the ground through the established conducting channel because of the shifting potentials within the cloud. Approximately one-third of all lightning is a single stroke. When multiple strokes occur, the mean number is three. A *stroke* is any one of a series of repeated discharges that comprise a single discharge, which is called a *lightning flash* or *strike*.

The voltage of a lightning stroke varies from 10 to 1000 MV. The voltage of concern to the electrical engineer, however, is that which appears on the equipment. The overvoltage on the equipment depends on the impedance and the current. Lightning strokes are considered current sources at the point of the strike. The current varies from 1 to 200 kA.

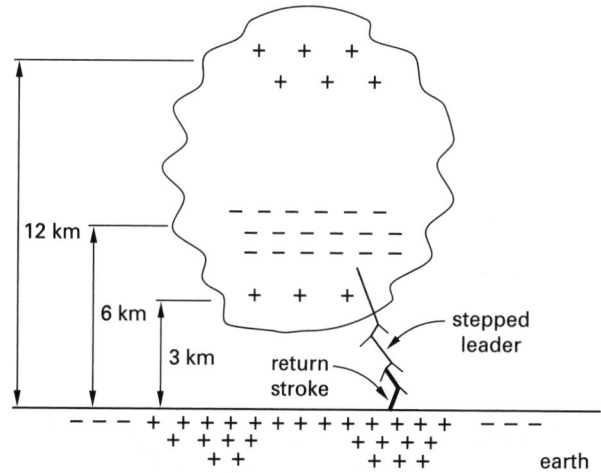

Figure 41.1 *Lightning Development*

2. CONCEPTS AND DEFINITIONS

A power system must use a coordinated design to protect from internally generated surges, usually due to switching, and externally generated surges, usually due

[2]The strokes mentioned are called *ground flashes*. Other strokes occur between clouds but are not of concern in power electrical engineering.

to lightning. Coordinated design includes shielding, grounding, surge arresters, switching resistors, breaker timing, and surge capacitors. The first three are primarily used to protect against lightning strikes and the last three are used to protect against switching transients. Protection devices include spark gaps and surge arresters. The main purpose of the protection is to prevent insulation breakdown, damage to equipment, and outages.

Power system protection terminology has been standardized.[3] Some of the more important terms follow.

- *Withstand voltage* is the voltage that electrical equipment can handle without failure or disruptive discharge.

- *Transient insulation level* (TIL) is an insulation level specified in terms of the crest value of the withstand voltage, for a specified waveform.

- *Lightning impulse insulation level* is an insulation level specified for the crest value of a lightning impulse withstand voltage.

- *Basic lightning impulse insulation level* (BIL) is an insulation level specified in terms of the crest value of the standard lightning impulse.

- *Standard lightning impulse* is a full impulse with a *front time* of 1.2 μs and a half-value time of 50 μs, normally termed a 1.2/50 impulse.[4] See Fig. 41.2.

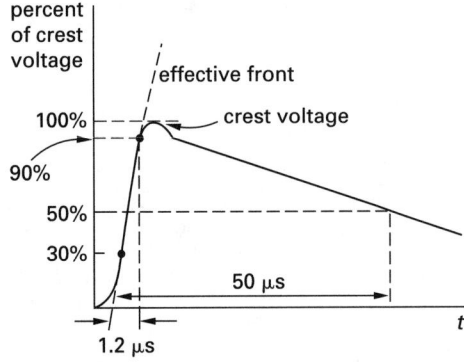

Figure 41.2 *Standard Lightning Impulse*

Systems of approximately 230 kV or less cannot be insulated against direct lightning strikes because of the high voltage of the lightning, on the order of several million volts. To protect such systems a combination of shielding and grounding is used as shown in Fig. 41.3.[5]

---

[3]See the American National Standards Institute (ANSI) Std. C92.1-1971.
[4]This is taken from American National Standard Measurement of Voltage in Dielectric Tests, C68.1-1968.
[5]The physical grounding of systems is covered in Ch. 65.

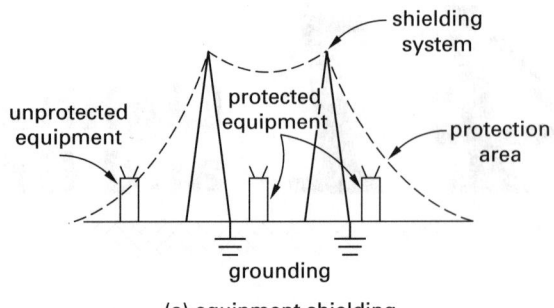

(a) equipment shielding

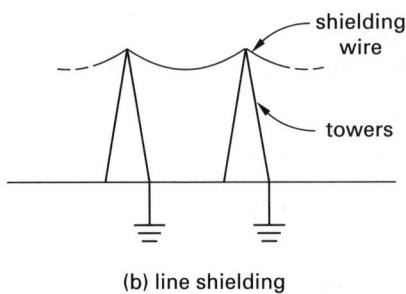

(b) line shielding

Figure 41.3 *Shielding System*

Indirect lightning strikes result in voltage surges due to conductive coupling through the soil or due to inductive and capacitive coupling to the lines. *Induced voltage surges* are typically less than 400 kV and can damage systems of 35 kV or less. Systems of 69 kV and more have sufficient insulation to withstand such surges. To protect such systems, surge arresters are used.

3. METHODS OF ANALYSIS

The first step in analysis is the determination of the models to use for the various components of a distribution system. The models can be lumped parameters, distributed parameters, nonlinear equations, time varying equations, or logical equations. Once the models have been determined, three primary methods of analysis are used. *Graphical analysis* is based on the representation of voltages and currents as traveling waves and the use of a *Bewley diagram* to determine the waves that exist at any given point. *Analytical analysis* consists of a systematic approach to obtaining the solution of the differential equations that are used to represent a system, usually with Laplace transforms. *Numerical analysis* transforms the differential equations into discrete time equations, which are algebraic. The algebraic equations can be interpreted as *resistive companion circuits* and analyzed using standard techniques, such as KVL or KCL. Additionally, Monte Carlo techniques are often applied to the many uncertainties in the analysis of lightning overvoltages.

4. GROUNDING MODELS

Two basic grounding models are used: low frequency and high frequency. Low-frequency models consider grounds to be purely resistive and require DC analysis.[6] High-frequency models require a complete electromagnetic analysis.[7] A general rule for determining which model to use is to compare the largest dimension of the grounding system, say length l, to the skin depth, δ, and then apply the rule of thumb given by Eq. 41.1.

$$\frac{l}{\delta} = \frac{l}{\dfrac{1}{\sqrt{\pi f \mu \sigma}}} \begin{cases} < 0.1 \text{ DC analysis} \\ > 0.1 \text{ electromagnetic analysis} \end{cases} \quad 41.1$$

Grounding for protection against lightning overvoltages is in addition to individual equipment grounding and power system grounding. The relationship among the different grounding systems is shown in Fig. 41.4.

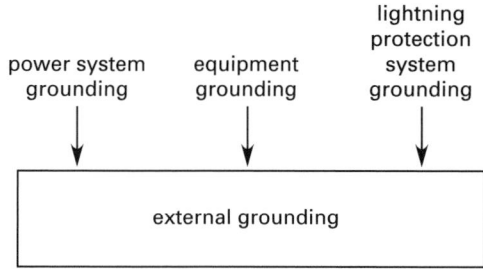

Figure 41.4 *Grounding Systems*

5. PROTECTIVE DEVICES

When overvoltages occur, the transient is minimized and the system is protected by connecting the overvoltage condition to ground until the energy is dissipated. This connection must be temporary and ideally will not allow current flow at normal operating voltages. This is accomplished by the use of strategically placed *arresters*, also called *surge arresters*. The quality of surge arresters is roughly measured by a *protection quality index* (PQI) given by

$$\text{PQI} = \frac{V_r}{V_p} \quad 41.2$$

The denominator term is the voltage *protection level*, the maximum voltage allowed by the arrester. The numerator term is the *reseal level*, the voltage level at which minimal current flow through the arrester occurs. The ideal protection quality index is one.

Basic protection is provided by *air gaps*, also known as *spark gaps*. These devices are composed of two electrodes, the shape of which varies depending upon the

---

[6]DC analysis is based on the *method of moments* or the *relaxation method*.

[7]Electromagnetic analysis is based on the *method of moments* or *finite-element analysis*.

application, separated by a distance. When the voltage across the air gap exceeds a given value, an arc is initiated. *Gapped surge arresters* use a series of spark gaps in conjunction with a nonlinear resistor as shown in Fig. 41.5. The nonlinear resistors have low resistance at high voltages, thus allowing the surge current. As the voltage of the lightning approaches that of the system, the resistance of the nonlinear resistors increases (typically as the fourth power of the voltage), thus extinguishing the arc without tripping protective devices.

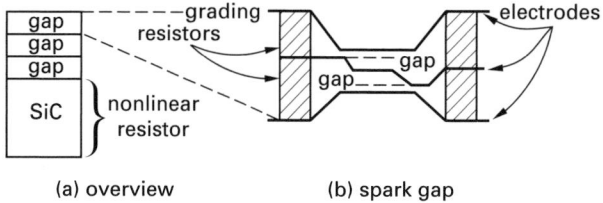

Figure 41.5 *Gapped Surge Arrester*

Improved protection is provided by *metal-oxide varistor arresters* (MOVs). Such an arrester is shown in Fig. 41.6. The device is manufactured in such a manner as to approximate an ideal protective device, that is, a device that clips the voltage level at the maximum allowed and reseals at the same level. An MOV functions like back-to-back diodes. The MOV conducts current at the voltages normally experienced by the electrical system. Thus, heating effects that could deteriorate the device must be accounted for in the design. The performance of MOVs at high discharge currents is improved by adding a shunt gap much like the construction in Fig. 41.5(a). When constructed in such a manner, they are called *shunt-gap MOVs*.

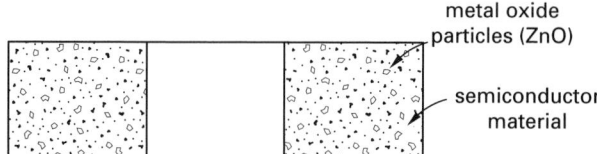

Figure 41.6 *Metal-Oxide Varistor (MOV) Arrester*

The effectiveness of surge arresters for a particular system is measured by the *protective margin* (PM) provided. The protective margin is determined by

$$\text{PM} = \frac{V_w - V_p}{V_p} \quad 41.3$$

The term V_w is the voltage a given device can withstand before damage or insulation breakdown occurs. The term V_p is the voltage protection level provided by the arrester.[8]

---

[8]The protection level provided is obtained from manufacturer's nameplate data on the arrester.

Topic IX:
Measurement and Instrumentation

Chapter

42. Measurement and Instrumentation

For the most current information about the exam, visit **www.ppi2pass.com** regularly.

PROFESSIONAL PUBLICATIONS, INC.

42 Measurement and Instrumentation

Nomenclature

a	Fourier series coefficient	–
B	magnetic flux density	T
BW	bandwidth	Hz
C	compensating winding	–
$f(t)$	function of time	–
f	frequency	Hz, s^{-1}, or cycles/s
H	magnetic field strength	A/m
I	effective or DC current	A
k	coupling coefficient	–
N	number of turns	–
pf	power factor	–
P	power	W
Q	quality factor	–
rms	root-mean-square	–
R	resistance	Ω
S	apparent power	voltampere (volt-amps or VA)
t	time	s
T	time of one cycle	s
T	temperature	K
T	torque	N·m
v	instantaneous voltage	V
V	effective or DC voltage	V

Symbols

Δ	change, high–low	–
θ	angular displacement	rad
κ	Boltzmann constant	1.3807×10^{-23} J/K
ϕ	angle	rad

Subscripts

0	initial
ave	average
ext	external
fs	full scale
H	higher
L	lower
n	number
x	unknown

1. FUNDAMENTALS

An *instrument* is a device for measuring, and sometimes for controlling or recording, a value under observation. *Instrumentation* is the use of such devices for measurement, detection, observation, computation, communication, or control. Instrument outputs are given in terms of some reference value. A *reference standard* is an accepted measurement base.[1] *Precision* is the range of values of a set of measurements. An instrument's precision is a measure of the reproducibility of observations. The *accuracy* is a statement of the limits that bound the departure of the measured value from the true quantity. Accuracy includes the imprecision of the instrument and any accumulated errors from the national reference standards in the measurement itself. The accuracy of meter movements is given in terms of the percentage of error relative to the full-scale value, regardless of where the error occurs. That is, the error may occur in the low end of the indicating range, but the accuracy is nevertheless given as a percentage of the highest indicated value.

Errors can be separated into two general categories: systematic and random. A *systematic error* is an error that is reproducible and the result of some physical effect, such as a weakened magnetic field, altered resistance value, repeatable signal distortion, and the like. A *random error* is a deviation in the output due to chance. Random errors are accidental, inconsistent in magnitude and sign, and not reproducible. The effects of errors are analyzed using statistical mathematics.

The *resolution* is the smallest increment that can be distinguished by the instrumentation system. The resolution is determined by the amount of deflection per unit input and is limited by the amount of signal beyond the noise. *Noise* is any portion of a signal that does not convey useful information, such as unwanted currents and voltages in an electrical device or system. Random disturbances due to thermal agitation of electrons are a significant contributor to noise in electrical

---

[1]United States national reference standards are maintained by the National Institute of Standards and Technology, formerly the National Bureau of Standards (NBS).

systems and are called *Johnson noise*. A measure of the Johnson noise in resistors is given by[2]

$$V_{\text{noise}} = \sqrt{4\kappa T R \Delta f} = \sqrt{4\kappa T R(\text{BW})} \qquad 42.1$$

The term V_{noise} is the rms equivalent voltage generated by the thermal agitation. Boltzmann's constant is κ, the absolute temperature is T, the resistance is R, and the bandwidth over which the noise occurs is Δf or BW.

Environmental factors influence the accuracy of instrumentation systems. Stray magnetic fields can combine with instruments' magnetic fields to alter the output. Ambient temperature differences between readings can result in differences due to mechanical changes in the instrumentation or changes in resistance in the circuitry. Electrostatic charges influence the position of indicating needles. For example, the rubbing of an instrument face can move a needle with constant sensing input. Mechanical shock, vibration, and even mounting position influence the output of instrumentation and must be accounted for or protected against in order to obtain accurate indications.

Most measurement circuits are not exposed to the full voltage or current of the system monitored. The measurement circuits are isolated from the system by *instrument transformers*. An instrument transformer extends the range measured by a given meter, isolates the meter from the line, and allows the instrument circuit to be grounded. When the primary winding is in series with the line, the instrument transformer is called a *current transformer*. When the primary winding is in parallel with the line, the instrument transformer is called a *potential transformer*. Instrument transformers are tested in the same manner as power transformers (see Ch. 36).

Example 42.1

What is the expected thermal agitation noise at 20°C in a digital voltmeter using a 10 kΩ resistor and designed to measure from 0 to 1000 Hz?

Solution

Ignoring any other resistance in the voltmeter, which will be small compared to the 10 kΩ resistance, and noting that 20°C is equal to 293K, the Johnson noise voltage is

$$V_{\text{noise}} = \sqrt{4\kappa T R \Delta f}$$

$$= \sqrt{\begin{array}{c}(4)\left(1.3807 \times 10^{-23}\ \dfrac{\text{J}}{\text{K}}\right)(293\ \text{K}) \\ \times (10 \times 10^3\ \Omega)(1000\ \text{Hz})\end{array}}$$

$$= 4.02 \times 10^{-7}\ \text{V}$$

---

[2]Random disturbances in mechanical systems are explained by the mechanism of *Brownian motion*. In magnetic systems, the *Barkhausen effect* explains such disturbances.

Such a small voltage level is of significance in only the most sensitive of measurements.

2. SIGNAL REPRESENTATION

DC signals are represented by their constant values. AC signals can be represented in a number of ways. The average value of a generic AC signal is given by

$$f_{\text{ave}} = \left(\frac{1}{T}\right) \int_{t_1}^{t_1+T} f(t)\,dt \qquad 42.2$$

The average value of a sinusoidal function is zero, so a separate method of measuring AC sinusoidal signals is required in order to use their values in electrical calculations. The root-mean-square (rms) is that method. The rms or effective value of a generic function is given by

$$f_{\text{rms}} = \sqrt{\left(\frac{1}{T}\right) \int_{t_1}^{t_1+T} f^2(t)\,dt} \qquad 42.3$$

Since nearly all periodic functions can be represented by a Fourier series, many important nonsinusoidal signals are changed into such series for further manipulation. A Fourier series is given by

$$f(t) = \tfrac{1}{2}a_0 + \sum_{n=1}^{\infty} a_n \cos\left(\frac{2\pi n}{T}\right)t$$

$$+ \sum_{n=1}^{\infty} b_n \sin\left(\frac{2\pi n}{T}\right)t \qquad 42.4$$

The average value of a function given by a Fourier series is

$$f_{\text{ave}} = \tfrac{1}{2}a_0 \qquad 42.5$$

The rms value of a function given by a Fourier series is

$$f_{\text{rms}} = \sqrt{\left(\tfrac{1}{2}a_0\right)^2 + \tfrac{1}{2}\sum_{n=1}^{\infty}\left(a_n^2 + b_n^2\right)} \qquad 42.6$$

For a sinusoid, the average and rms values are related as given in Eqs. 42.7 through 42.9. (Equation 42.7 is the average of a rectified sinusoid.)

$$V_{\text{ave}} = \left(\frac{2}{\pi}\right)V_p \qquad 42.7$$

$$V_{\text{rms}} = \left(\frac{1}{\sqrt{2}}\right)V_p \qquad 42.8$$

$$V_{\text{rms}} = 1.11 V_{\text{ave}} \qquad 42.9$$

Example 42.2

A voltage waveform is represented by the truncated Fourier series given as

$$v(t) = 5 + 3\cos 100t + 0.8\cos 300t$$
$$+ 4\sin 100t + 0.5\sin 200t$$

Determine the average value of this voltage.

Solution

The average value is one-half of the zero frequency term as determined by Eq. 42.5.

$$f_{ave} = \tfrac{1}{2}a_0 = 5 \text{ V}$$
$$a_0 = (2)(5 \text{ V}) = 10 \text{ V}$$
$$f_{ave} = 5 \text{ V}$$

Example 42.3

For the voltage waveform given in Ex. 42.2, what is the rms voltage?

Solution

The rms value is given by Eq. 42.6.

$$f_{rms} = \sqrt{\left(\tfrac{1}{2}a_0\right)^2 + \tfrac{1}{2}\sum_{n=1}^{\infty}\left(a_n^2 + b_n^2\right)}$$
$$= \sqrt{(5.0)^2 + \left(\frac{1}{2}\right)\left((3)^2 + (0.8)^2 + (4)^2 + (0.5)^2\right)}$$
$$= 6.16 \text{ V}$$

3. MEASUREMENT CIRCUIT TYPES

Electrical parameters can be measured with *electromechanical circuits*. Such instruments use the measured parameter to generate a magnetic field that interacts with a separate magnetic field to produce the rotation of a restrained needle. Because of mechanical inertia, the instruments do not function for AC in the range of 5 to 20 Hz. Due to inductive effects, the instruments do not function well above a maximum of 100 to 2500 Hz, depending on the design. Because of these restrictions, the electromechanical instruments are primarily used in DC systems.

Substitution circuits compare an unknown electrical quantity with a reference. *Bridge circuits* are a simple means of determining impedance by such substitution. A *Wheatstone bridge* is shown in Fig. 42.1.

Balance exists in the bridge when no current flows through the ammeter. When balanced, the relationship between the resistors is

$$\frac{R_x}{R_3} = \frac{R_2}{R_4} \qquad\qquad 42.10$$

Typically, R_3 and R_4 are fixed resistors that can be selected to vary their ratio from 10^{-3} to 10^3. Resistor R_2 is adjustable in steps and used to balance the bridge with the unknown resistor R_x. When balance is achieved, the unknown resistance is given by Eq. 42.10. Many other types of bridges are used to measure such things as inductance, capacitance, frequency, angles, and the like.[3]

Analog circuits use the mathematical relationships among electrical parameters to determine the desired parameters. Designs vary, but the instruments are used to determine such parameters as impedance, quality factor of coils (Q), the coefficient of coupling (k), magnetic parameters $(H$ and $B)$, frequency (f), and the like.

Digital circuits use digital processing of analog signals and analog-to-digital converters to display the results.[4] Digital instruments offer ease of use, quick response, and accurate results.

Transducer circuits are used in conjunction with transducers themselves to provide a signal of sufficient strength for use in follow-on systems.[5]

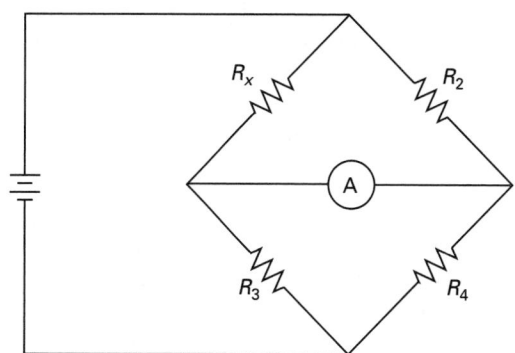

Figure 42.1 *Wheatstone Bridge*

Example 42.4

A Wheatstone bridge with an R_3/R_4 ratio of 500 is balanced when the adjustable resistor has a value of 10 Ω. What is the unknown resistance?

Solution

At a balanced condition, the resistance is given by Eq. 42.10. Thus,

$$\frac{R_x}{R_3} = \frac{R_2}{R_4}$$
$$R_x = \left(\frac{R_3}{R_4}\right)R_2 = (500)(10 \ \Omega) = 5 \text{ k}\Omega$$

---

[3]Some of the named bridges are the Maxwell-Wien, Anderson, Owen, Hay, Campbell, and Schering.
[4]The digital processing is often accomplished with the use of operational amplifiers, which are covered in Ch. 44.
[5]Chapter 19 provides background on transducer principles.

4. PERMANENT MAGNET MOVING-COIL INSTRUMENTS

A permanent magnet moving-coil instrument is shown in Fig. 42.2. Such an instrument is also called a *d'Arsonval movement*. A current is passed through the coil on the rotor. The rotor is located within a magnetic field created by the permanent magnets. Since a current-carrying conductor within a magnetic field will experience a torque, the coil rotates. The torque produced is proportional to the current through the coil and the number of turns of wire in the coil ($T_{\text{coil}} \propto IN$). The spring is restrained in such a manner as to provide an opposing torque proportional to the angular displacement ($T_{\text{spring}} \propto \theta$). The pointer comes to rest when the torques are equal in magnitude.

The current necessary to cause full-scale deflection of a specific instrument is given the symbol I_{fs}. The usual range of currents is 50 μA to 10 mA. In ammeter applications, the voltage at which full-scale current flow occurs is usually in the range of voltages from 50 to 100 mV, with a nominal value of 50 mV used as the standard. Since this is usually more than required by the coil for I_{fs} to flow, a *swamping resistor* is placed in series with the coil.

Because the permanent magnet's field is unidirectional, a change in the direction of the current in the coil reverses the torque, so the unmodified instrument is useful for detecting DC quantities only.[6] It can be made to detect AC quantities with the addition of a rectifier circuit on the coil input. When measuring AC rectified quantities, the d'Arsonval movement responds to the average voltage or current. In view of Eq. 42.9, adjusting the scale by a factor of 1.11 allows the rms value to be read directly.

The d'Arsonval movement can be designed to detect AC quantities with the use of a thermocouple. The hot junction of a thermocouple is placed in thermal, but not electrical, contact with a resistance carrying the current to be measured. The cold junction is connected to a heat sink that maintains ambient temperature. The measured current heats the wire via the resistance and subsequently heats one end of the thermocouple. A potential difference is developed between the ends of the dissimilar metals of the thermocouple proportional to the temperature difference ($I \propto \Delta T$), which is fed as input to the d'Arsonval meter movement. Because the heating effect is proportional to the square of the current, using a square-root factor on the scale output allows the rms value to be read directly. Such a *thermoelement instrument* can function up to several hundred megahertz.

---

[6]Specially designed d'Arsonval movements can respond to currents up to 100 Hz.

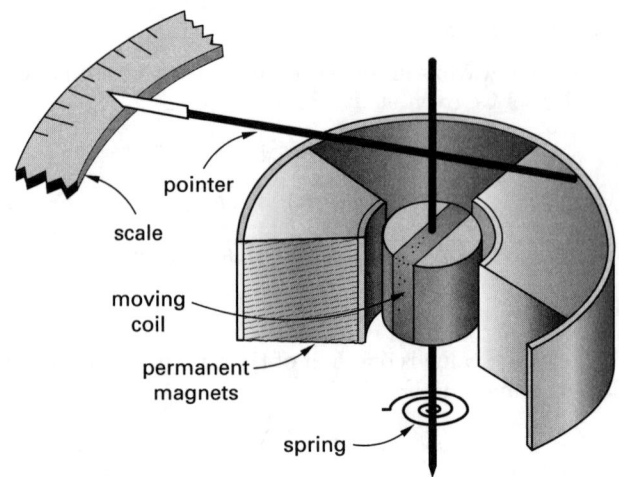

Figure 42.2 *Permanent Magnet Moving-Coil Mechanism (d'Arsonval)*

5. DC VOLTMETERS

A d'Arsonval meter movement configured to perform as a DC voltmeter is shown in Fig. 42.3.

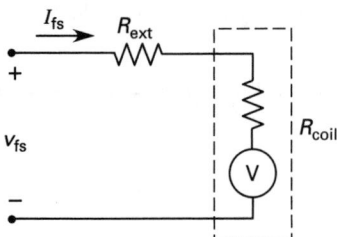

Figure 42.3 *DC Voltmeter*

The external resistance is used to limit the current to the full-scale value, I_{fs}, at the desired full-scale voltage, V_{fs}. The electrical relationships in the voltmeter are given by Eq. 42.11.

$$\frac{1}{I_{\text{fs}}} = \frac{R_{\text{ext}} + R_{\text{coil}}}{V_{\text{fs}}} \qquad 42.11$$

The quantity $1/I_{\text{fs}}$ is fixed for a given instrument and is called the *sensitivity*. The sensitivity is measured in ohms per volt (Ω/V).

Example 42.5

Common DC voltmeters have a sensitivity of 1000 Ω/V. Given such a meter with a coil resistance of 200 Ω and a 1% accuracy, determine the external resistance necessary to use a full-scale voltage of 10 V.

Solution

The external resistance is found from Eq. 42.11 by noting that the sensitivity is $1/I_{\text{fs}}$.

$$\frac{1}{I_{\text{fs}}} = \frac{R_{\text{ext}} + R_{\text{coil}}}{V_{\text{fs}}} = 1000 \ \Omega/\text{V}$$

$$R_{\text{ext}} = V_{\text{fs}}\left(1000 \ \frac{\Omega}{\text{V}}\right) - R_{\text{coil}}$$

$$= (10 \ \text{V})\left(1000 \ \frac{\Omega}{\text{V}}\right) - 200 \ \Omega$$

$$= 9.8 \ \text{k}\Omega$$

Example 42.6

An overall accuracy of 2% is desired for the meter in Ex. 42.5. What is the maximum error possible for the external resistance?

Solution

The maximum error introduced by the external resistance will occur at full-scale voltage. The maximum error from the movement itself is given at the full-scale voltage as 1%, regardless of where the error actually occurs. Thus, the external resistance must be a 9.8 kΩ resistor with a 1% accuracy.

Example 42.7

The manufacturer's data for a DC voltmeter lists the sensitivity as 20,000 $\Omega/$V. What current is required to deflect the movement to the 100% value?

Solution

The 100% value is the full-scale deflection. The full-scale deflection current is found from the definition of the sensitivity.

$$\text{sensitivity} = \frac{1}{I_{\text{fs}}}$$

$$I_{\text{fs}} = \frac{1}{20,000 \ \dfrac{\Omega}{\text{V}}} = 50 \ \mu\text{A}$$

6. DC AMMETERS

A d'Arsonval meter movement configured to perform as a DC ammeter is shown in Fig. 42.4.

The swamping resistance is used to limit the current to the full-scale value, I_{fs}, at the desired full-scale voltage, V_{fs}. For ammeters, the standard full-scale voltage is

50 mV.[7] The electrical relationships in the ammeter are given by Eq. 42.12.

$$I_{\text{design}} = \frac{V_{\text{fs}}}{R_{\text{shunt}}} + I_{\text{fs}} \qquad 42.12$$

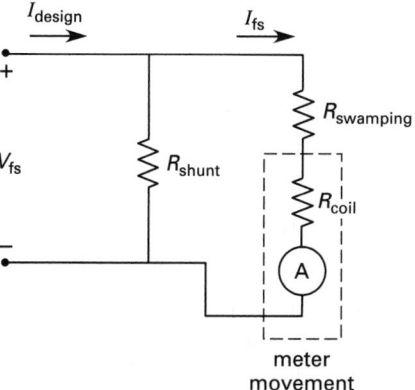

Figure 42.4 *DC Ammeter*

Example 42.8

A d'Arsonval movement is set up as an ammeter with coil and swamping resistors totaling 200 Ω and $I_{\text{fs}} = 0.5 \ \mu\text{A}$. If no shunt resistor is used, what percentage of full-scale current will flow through the meter movement?

Solution

Without a shunt resistor and noting that the standard full-scale standard voltage is 50 mV, the full-scale current is given by

$$I_{\text{fs}} = \frac{V_{\text{fs}}}{R_{\text{swamping}} + R_{\text{coil}}}$$

$$= \frac{50 \times 10^{-3} \ \text{V}}{200 \ \Omega} = 2.5 \times 10^{-4} \ \text{A}$$

The actual full-scale current should be 0.5 μA. Thus, the percentage of full-scale current flowing without the shunt resistor is

$$\%I = \left(\frac{2.5 \times 10^{-4} \ \text{A}}{0.5 \times 10^{-6} \ \text{A}}\right) \times 100 = 50,000\%$$

It should be noted that if the full-scale current had not been specified as 0.5 μA, the calculated current of 2.5×10^{-4} A would be the correct value for proper operation.

Example 42.9

What value of shunt resistance is required for the ammeter in Ex. 42.8 to function properly with a 1 A full-scale deflection?

---

[7]The standard ammeter is designed for a 50 mV voltage across the shunt with I_{fs} flowing in the movement at the desired design current flow.

Solution

The proper relationship is given by Eq. 42.12.

$$I_{\text{design}} = \frac{V_{\text{fs}}}{R_{\text{shunt}}} + I_{\text{fs}}$$

$$R_{\text{shunt}} = \frac{V_{\text{fs}}}{I_{\text{design}} - I_{\text{fs}}} = \frac{50 \times 10^{-3} \text{ V}}{1 \text{ A} - 0.5 \times 10^{-6} \text{ A}} = 0.05 \ \Omega$$

7. MOVING-IRON INSTRUMENTS

A *moving-iron instrument* is shown in Fig. 42.5. The instrument shown is called a *radial-vane mechanism*. A current is passed through the coil. A stationary vane and a moving vane are located within the resulting magnetic field and are similarly magnetized. Since the induced magnetic field is the same for both vanes, they repel one another. The moving vane is attached to a pointer and restraining spring. The angular displacement of the moving vane is proportional to the square of the current and the number of turns of wire in the coil ($T_{\text{coil}} \propto I^2 N$). The spring is restrained in such a manner as to provide an opposing torque proportional to the angular displacement ($T_{\text{spring}} \propto \theta$). The pointer comes to rest when the torques are equal in magnitude.

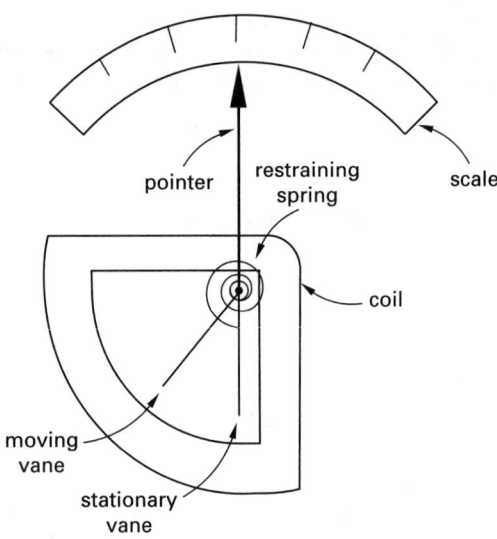

Figure 42.5 *Moving-Iron Mechanism*

The vanes repel one another regardless of the polarity of the current or its changing value. As a result, the instrument can be used to measure AC quantities. Because the torque is proportional to the square of the current, using a square-root factor on the scale output allows the rms value to be read directly. The instrument can measure AC quantities up to approximately 500 Hz. With special compensation for eddy currents and skin effects, this range can be extended to 10 kHz.

Most moving-iron instruments can also measure DC quantities. They are less useful for DC quantities because significant remnant magnetization occurs over time.

Moving-iron instruments can be used for voltmeters or ammeters using setups similar to those of d'Arsonval instruments. The sensitivity and power consumption of moving-iron instruments is between that of d'Arsonval movements and electrodynamometer movements.

8. ELECTRODYNAMOMETER INSTRUMENTS

An *electrodynamometer instrument* is shown in Fig. 42.6(a). A current is passed through a moving coil and two stationary coils. The magnetic field of the moving coil interacts with the magnetic fields of the stationary coils to produce a torque. The torque produced is proportional to the current and the number of turns of wire in the moving coil as well as the current and number of turns in the stationary coils ($T \propto I^2 N$). The spring is restrained in a manner that provides an opposing torque proportional to the angular displacement ($T_{\text{spring}} \propto \theta$). The pointer comes to rest when the torques are equal in magnitude.

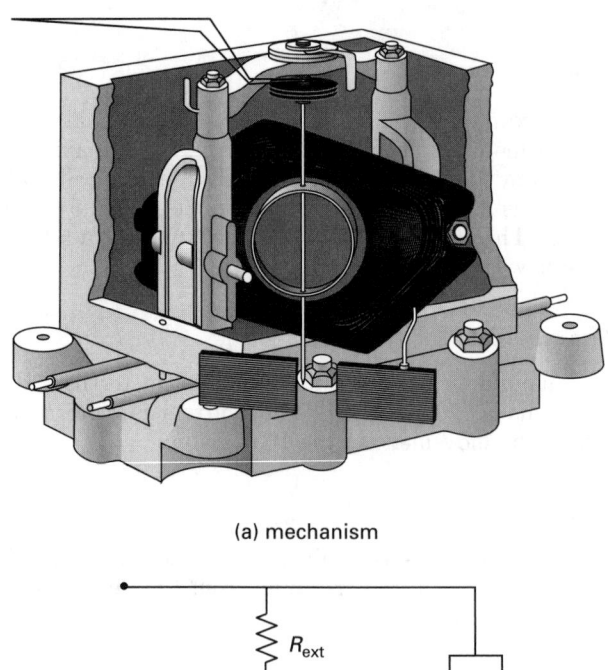

(a) mechanism

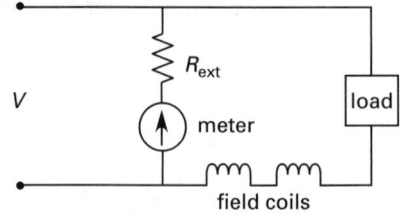

(b) configured to measure power

Figure 42.6 *Electrodynamometer*

Because the torque is proportional to the square of the current, using a square-root factor on the scale output allows the rms value to be read directly.[8] The instrument can measure AC quantities in the range of 25 to 2500 Hz. The lower limit is due to the instrument's

---

[8]This is true when the same current is flowing in all coils. That is, the device is configured so that the coils are in series.

attempt to follow the instantaneous values of the AC quantities. The upper limit is due to the significance of inductive reactance.

Though currents and voltages can be measured, the dynamometer is most commonly used to measure power. When the moving coil is placed across a load, the current in the moving coil will be proportional to the voltage across the load. When the stationary coils are placed in series with the load, the current in the stationary coils will be proportional to the current in the load. The combined magnetic field is then proportional to the instantaneous product of the voltage and the current—that is, the power. One possible configuration for measuring power is shown in Fig. 42.6(b). Corrections for the power consumption of the coils are required to ensure an accurate result.

The electrodynamometer can also be used to measure DC quantities since average DC values are equivalent to rms AC values.

The instruments can be used for voltmeters or ammeters using setups similar to those of d'Arsonval instruments. The sensitivity is higher than either d'Arsonval or moving-iron instruments, but the power consumption is higher as well.

9. POWER MEASUREMENTS

Many different types of instruments are used to measure power. All require compensation for the power they consume, inductance of the measuring coils, and phase angle corrections for any instrument transformers used. Three-phase systems require a minimum of two wattmeters to measure power (see Sec. 34-11). The power meter most commonly used in residential applications is the induction-type watthour meter shown in Fig. 42.7.

The parallel coil V produces a magnetic flux at its pole tip that lags the applied voltage by nearly 90°, due to its high inductance and the large number of turns used in the coil. The fluxes in the series coils, I_1 and I_2, are nearly in phase with the current. All three poles induce currents in a circular disk located between the poles.[9] A torque, proportional to the power, is generated in the disk and causes it to rotate.[10] The disk speed is made proportional to the driving torque by a braking torque that is also proportional to the disk speed. This braking torque (countertorque) arises from the interaction of the rotating disk with a permanent magnetic field (not shown) that induces eddy currents in the disk. The compensating winding, C, with adjustable resistor, R, is used to ensure that the flux from the parallel pole, V,

is exactly 90° out of phase. The net result is a rotating disk speed proportional to the power delivered to the load.

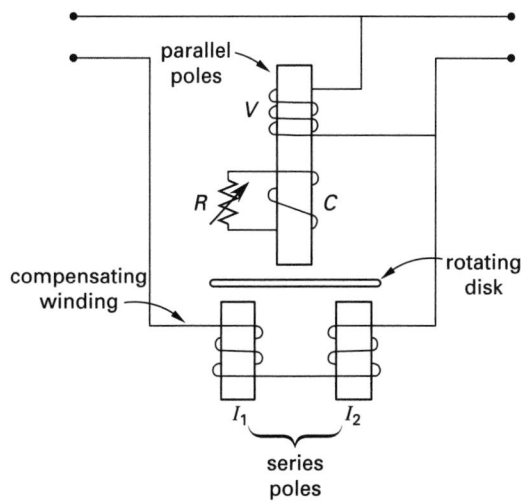

Figure 42.7 *Induction-Type Watthour Meter*

10. POWER FACTOR MEASUREMENTS

The power factor, pf, is given by the ratio of the true power to the apparent power.

$$\mathrm{pf} = \frac{P}{S} \qquad \text{42.13}$$

For a balanced three-phase, three-wire system, results obtained using the two-wattmeter method (see Sec. 34-11) can be used to determine the angle between the current and the voltage, and thus the power factor. Let P_H be the larger power and P_L the power measured by the other wattmeter. The desired angle is determined from

$$\tan \phi = \frac{\sqrt{3}\,(P_H - P_L)}{P_H + P_L} \qquad \text{42.14}$$

Power factor meters can also be constructed from electrodynamometers with two moving coils. Both coils carry the series current of the load, but one does so in series with a resistor and the other does so in series with an inductor. Consequently, the two coils' currents are nearly in quadrature, that is, 90° apart. At unity power factor, the torque on the coil in series with the resistor will be at a maximum. At zero power factor, the torque on the coil in series with the inductor will be at a maximum. The angular displacement of the needle measures the power factor of the load.

Example 42.10

Using the two-wattmeter method, the power of a three-phase circuit is measured as 1074 kW and 426 kW. What is the power factor?

---

[9]This is the same principle as that of the induction motor.
[10]The torque is proportional to the power only if the fluxes between the series coils and the parallel coil are in quadrature at unity power factor.

Solution

The angle necessary to determine the power factor comes from Eq. 42.14.

$$\tan \phi = \frac{\sqrt{3}\,(P_H - P_L)}{P_H + P_L}$$

$$= \frac{\sqrt{3}\,(1074 \text{ kW} - 426 \text{ kW})}{1074 \text{ kW} + 426 \text{ kW}} = 0.748$$

Thus, the angle between the current and the voltage is

$$\phi = \tan^{-1}(0.748) = 36.8°$$

The power factor is given by

$$\text{pf} = \cos \phi = \cos 36.8° = 0.8$$

Given the nature of most systems, the power factor is 0.8 lagging.

Topic X: Electronics

Electronics

For the most current information about the exam, visit **www.ppi2pass.com** regularly.

43 Electronic Components

Nomenclature

Å	angstrom	10^{-10} m
A	constant	$K^{-3}m^{-3}$ or $K^{-3}cm^{-3}$
A	anode	–
A	gain	–
B	base or substrate	–
c	speed of light	3.00×10^8 m/s
C	capacitance	F

C	collector	–
C_{iss}	gate channel capacitance	F
C_{oss}	output capacitance	F
C_{tc}	collector junction transition capacitance	F
C_{te}	emitter junction transition capacitance	F
CB	common base	–
CC	common collector	–
CE	common emitter	–
D	diffusion constant	m^2/s
D	drain	–
E	emitter	–
E	energy	J or eV
g_m	mutual conductance or transconductance	S
G	gate	–
h	Planck's constant	6.626×10^{-34} J·s
$h_{i,o,f,r}$ and e,c,b	hybrid parameter[1]	–
$h_{I,O,F,R}$ and E,C,B	hybrid parameter[2]	–
h_{fe}	CE small signal (AC) forward current transfer ratio or gain	–
h_{ie}	CE small signal (AC) input impedance	Ω
h_{oe}	CE small signal (AC) open-circuit output admittance	S
h_{re}	CE small signal (AC) reverse voltage transfer ratio or gain	–
i	instantaneous current	A

---

[1]The first subscript refers to the AC or small signal input (i), output (o), forward (f), or reverse (r) value. The second subscript refers to the configuration, that is, e for common emitter (CE), c for common collector (CC), b for common base (CB).

[2]The first subscript refers to the DC or static input (I), output (O), forward (F), or reverse (R) value. The second subscript refers to the configuration, that is, E for common emitter (CE), C for common collector (CC), B for common base (CB).

I	effective or DC current	A
I_{CBO}	DC collector cutoff current, emitter open[3]	A
I_{DSS}	zero-gate-voltage drain current	A
I_{ZK}	keep-alive current	A
K	cathode	–
n	electron concentration	m^{-3} or cm^{-3}
n, N	concentration	m^{-3} or cm^{-3}
p	hole concentration	m^{-3} or cm^{-3}
q	charge	1.602×10^{-19} C
Q	quiescent point	–
r	small-signal resistance[4]	Ω
$r_{bb'}$	base spreading resistance	Ω
$r_{b'c}$	feedback resistance	Ω
$r_{b'e}$	base input resistance	Ω
r_{ce}	output resistance	Ω
R	resistance	Ω
S	apparent power	voltampere (volt-amps or VA)
S	source	–
tc	temperature coefficient	K^{-1}
T	temperature	K
v	instantaneous voltage	V
V	effective or DC voltage	V
V_T	voltage equivalent of temperature	V

Symbols

α	CB forward current ratio or gain	–
β	CE forward current ratio or gain	–
E	electric field strength	V/m
η	empirical constant	–
κ	Boltzmann's constant	1.381×10^{-23} J/K or 8.621×10^{-5} eV/K
λ	wavelength	m
μ	mobility	m^2/V·s
ρ	charge density	C/m^3
σ	conductivity	S/m
ω	angular frequency	rad/s

Subscripts

0	at 0K
0	barrier or source value
A	acceptor
ac	alternating current or small signal
b, B	base
BB	base supply or base biasing
BE	base to emitter
BO	breakover
(BR)R	breakover, reverse
c	conduction band or collector
cf	corner frequency or cutoff frequency
co	cutoff
C	collector
CB	collector to base
CBO	collector cutoff current
CC	collector supply
CE	collector to emitter
CQ	Q-point collector current
d	dynamic or diffusion
D	donor, diode, or drain
DC	direct current or static
DS	drain to source
DSS	drain to source saturation
DSQ	drain to source at Q-point
e, E	emitter
EB	emitter to base
f	forward (AC or instantaneous)
F	forward (DC component) or total
G	gap, gate
GD	gate to drain
GS	gate to source
H	hold
i	intrinsic or input (AC or instantaneous)
iss	input short-circuit common source
I	input (DC or total)
L	load
m	maximum or mutual
n	electrons or n-type
o	saturation or output (AC or instantaneous)
O	output (DC or total)
p	holes or p-type
pn	p to n
P	pinchoff or power
P0	pinchoff at 0 V
r	reverse (AC or instantaneous)
R	reverse (DC component or total)
s	saturation
S	source
SD	source to drain
SG	source to gate
t	transition
T	thermal
v	valence band
v, V	voltage
Z	zener
ZK	zener knee

---

[3]This is essentially the reverse-saturation current between the collector and the base.

[4]This is also known as the *dynamic resistance* or the *instantaneous resistance*. Any resistance represented with a lowercase letter is the small-signal value.

1. OVERVIEW

Electronics involves charge motion through materials other than metals, such as a vacuum, gases, or semiconductors. The focus in this chapter will be on semiconductor materials. Because an understanding of the electron structure is vital to understanding electronics, a periodic table of the elements is given in App. 43.A.

An *electronic component* is one able to amplify, control, or switch voltages or currents without mechanical or other nonelectrical commands. The charge in metals is carried by the electron, with a charge of -1.60×10^{-19} C. In semiconductor materials, the charge is carried both by the electron and by the absence of the electron in a covalent bond, which is referred to as a *hole*[5] with a charge of $+1.60 \times 10^{-19}$ C. The concentration of electrons, n, and holes, p, is given by the *mass action law*, Eq. 43.1.

$$n_i^2 = np \qquad\qquad 43.1$$

The term n_i is the concentration of carriers in a pure (*intrinsic*) semiconductor.[6] In intrinsic semiconductor materials, the number of electrons equals the number of holes, and the mass action law is stated as in Eq. 43.1. Nevertheless, the mass action law applies for intrinsic and extrinsic semiconductor materials. *Extrinsic* semiconductor materials are those that have had impurities deliberately added to modify their properties, normally their conductivity. For extrinsic semiconductors, the mass action law is probably better understood as indicating that the product np remains constant regardless of position in the semiconductor or doping level. Carriers, either n or p, in a semiconductor are constantly generated due to thermal creation of electron-hole pairs and constantly disappearing due to recombination of electron-hole pairs. The carriers generated diffuse due to concentration gradients in the semiconductor, a phenomenon that does not occur in metals.

Semiconductor devices are inherently nonlinear. Nevertheless, they are commonly analyzed over ranges in which their behavior is approximately linear. Such an analysis is called *small-signal analysis*. The DC or effective value of the electrical parameters in the models used for analysis will be represented by uppercase letters. AC or instantaneous values will be represented by lowercase letters. Equivalent parameters in the models will use lowercase letters. For convenience, the lowercase letter t is omitted for functions of time. For example $v(t)$ is written simply as v. The subscripts on currents indicate the terminals into which current flows. Subscripts on

voltages indicate the terminals across which the voltage appears. Subscripts indicating biasing voltages are capitalized. For example, V_{BB} is the base biasing voltage while v_{be} is the instantaneous voltage signal applied to the base-emitter junction. A list of common designations is given in App. 43.B.

Electronic components are often connected in an array known as an *integrated circuit* (IC). An integrated circuit is a collection of active and passive components on a single semiconductor substrate (*chip*) that function as a complete electronic circuit. *Small-scale integration* involves the use of less than 100 components per chip. *Medium-scale integration* involves between 100 and 999 components per chip. *Large-scale integration* involves between 1000 and 9999 components per chip. *Very large-scale integration* involves more than 10,000 components per chip.

The nonlinearity of electronic devices makes them attractive for use as amplifiers. An *amplifier* is a device capable of increasing the amplitude or power of a physical quantity without distorting the wave shape of the quantity. The amplifying properties of transistors are covered in this chapter. Topics peculiar to amplifiers and amplifier types are covered in Chs. 44 and 45.

Example 43.1

Silicon has an intrinsic carrier concentration of 1.6×10^{10} cm^{-3}. What is the concentration of holes?

Solution

Regardless of whether the material is intrinsic or extrinsic, the law of mass action applies. However, in intrinsic materials, the concentration of electrons and the concentration of holes are equal. Thus, Eq. 43.1 can be used as follows.

$$n_i^2 = np = p^2$$

$$n_i = p = 1.6 \times 10^{10} \text{ cm}^{-3}$$

2. SEMICONDUCTOR MATERIALS

The most common semiconductor materials are silicon (Si) and germanium (Ge). Both are in group IVA of the periodic table and contain four valence electrons. A *valence electron* is one in the outermost shell of an atom. Both materials use covalent bonding to fill the outer shell of eight when they create a crystal lattice. Semiconductor materials ($N \approx 10^{10}$ to 10^{13} electrons/m^3) are slightly more conductive than insulators ($N \approx 10^7$ electrons/m^3) but less conductive than metals ($N \approx 10^{28}$ electrons/m^3). The conductivity, σ, of semiconductor materials can be made to vary from approximately 10^{-7} to 10^5 S/m.

The semiconductor crystal lattice has many defects (i.e., free electrons and their corresponding holes). The formation of free electrons and holes is driven by the temperature and is called *thermal carrier generation*. The

---

[5]The concept of holes explains the conduction of electricity without free electrons. The hole is considered to behave as a free positive charge—quantum mechanics justifies such an interpretation. (The *Hall voltage* is experimental confirmation.) The calculation of total charge motion in semiconductors is simplified as a result. The same equations used for electron movement can be used for hole movement with a change of sign and a change of values for some terms, such as mobility.

[6]Intrinsic means natural. Pure silicon is silicon in its natural state, with no doping, though it will have naturally occurring impurities. The terms pure and intrinsic are used interchangeably.

density of these electron-hole pairs in intrinsic materials is given by

$$n_i^2 = A_0 T^3 e^{-\frac{E_{G0}}{\kappa T}} \qquad 43.2$$

The term A_0 is a constant independent of temperature and related to the density states at the bottom edge of the conduction band and the top edge of the valence band. T is the absolute temperature. E_{G0} is the energy gap at 0K, that is, the energy between the conduction band and the valence band. This is the energy required to break the covalent bond. The energy gap for silicon at 0K is approximately 1.21 eV. The energy gap for germanium is approximately 0.78 eV. The term κ is Boltzmann's constant.[7] At a given temperature at thermal equilibrium, Eq. 43.2 becomes

$$n_i^2 = N_c N_v e^{-\frac{E_G}{\kappa T}} \qquad 43.3$$

The terms N_c and N_v are the effective density states in the conduction band and valence band, respectively. The energy gap for silicon at room temperature (300K) is 1.12 eV. The energy gap for germanium at room temperature is 0.80 eV.

When minor amounts of impurities called *dopants* are added, the materials are termed *extrinsic semiconductors*.[8] If the impurities added are from group IIIA, with three valence electrons, an additional hole is created in the lattice. These dopants are called *acceptors*, and semiconductors with such impurities are called *p-types*. The *majority carriers* in *p*-type semiconductors are holes, and the *minority carriers* are electrons. The majority of the charge movement takes place in the valence band. Typical dopants are indium (In) and gallium (Ga). If the impurities added are from group VA, with five valence electrons, an additional electron is provided to the lattice.[9] These dopants are called *donors*, and semiconductors with such impurities are called *n-types*. The majority carriers in *n*-type semiconductors are electrons, and the minority carriers are holes. The majority of charge movement takes place in the conduction band. Typical dopants are phosphorus (P), arsenic (As), and antimony (Sb).

The conductivity of the semiconductor is determined by the carriers. The mobility of electrons is higher than that of holes.[10] The total conductivity in any semiconductor is a combination of the movement of the electrons and holes and is given by

$$\sigma = q\left(n\mu_n + p\mu_p\right) \qquad 43.4$$

---

[7]The symbology for Boltzmann's constant often varies with the units of the energy gap. If eV is used as the energy unit, Boltzmann's constant may be seen as either κ or $\bar{\kappa}$ to distinguish between eV/K and J/K.
[8]A minor amount of dopant material is on the order of 10 parts per billion.
[9]The additional electron exists because the outer shell octet is satisfied by the covalent bonding of the first four electrons in the impurity.
[10]The mobility of electrons is a factor of 2.5 higher in silicon and about 2.1 higher in germanium.

The law of mass action, Eq. 43.1, applies to extrinsic semiconductors. With the addition of dopants, the law of electrical neutrality given by Eq. 43.5 also applies. The concentration of acceptor atoms is N_A, each contributing one positive charge to the lattice. (Do not confuse N_A with Avogadro's number.) The concentration of donor atoms is N_D, each contributing one negative charge. Since neutrality is maintained, the result is[11]

$$N_A + n = N_D + p \qquad 43.5$$

In a *p*-type material, the concentration of donors is zero, ($N_D = 0$). Additionally, the concentration of holes is much greater than the number of electrons ($p \gg n$). Thus, for a *p*-type material, Eq. 43.5 can be rewritten as $N_A \approx p$. Using the law of mass action, Eq. 43.1, the concentration of electrons in a *p*-type material is

$$n = \frac{n_i^2}{N_A} \qquad 43.6$$

In an *n*-type material, the concentration of acceptors is zero ($N_A = 0$). Additionally, the concentration of electrons is much greater than the number of holes ($n \gg p$). Thus, for an *n*-type material, Eq. 43.5 can be rewritten as $N_D \approx n$. Using the law of mass action, Eq. 43.1, the concentration of holes in an *n*-type material is

$$p = \frac{n_i^2}{N_D} \qquad 43.7$$

Example 43.2

The energy gap for silicon at room temperature (300K) is 1.12 eV. The energy density states are $N_c = 2.8 \times 10^{19}$ cm^{-3} and $N_v = 1.02 \times 10^{19}$. What is the intrinsic carrier concentration?

Solution

The intrinsic carrier concentration, assuming thermal equilibrium, is given by Eq. 43.3.

$$n_i^2 = N_c N_v e^{\frac{-E_G}{kT}} = \left(2.8 \times 10^{19}\,\text{cm}^{-3}\right)$$

$$\times \left(1.02 \times 10^{19}\,\text{cm}^{-3}\right)\left(e^{-\left(\frac{1.12\,\text{eV}}{\left(\frac{8.621\times10^{-5}\,\text{eV}}{\text{K}}\right)(300\text{K})}\right)}\right)$$

$$= 4.452 \times 10^{19}\,\text{cm}^{-6}$$

$$n_i = 6.67 \times 10^9\,\text{cm}^{-3}$$

$$\approx 0.7 \times 10^{10}\,\text{cm}^{-3}$$

This number differs slightly from the number determined experimentally and more exact calculations (1.6×10^{10} cm^{-3}).

---

[11]Neutrality is maintained because each atom of an added impurity removes one intrinsic atom.

3. DEVICE PERFORMANCE CHARACTERISTICS

Electronic components function on some variation of *pn* junction principles. Amplifiers function on some variation of transistor principles, that is, *pnp* or *npn* junction principles. Most electronic components can be modeled as two-port devices with two variables—current and voltage—for each port. The relationship between the variables depends on the type of device. The relationship can be expressed mathematically, as for MOSFETs, modeled in equivalent circuits, as with transistor *h*-parameters, or described graphically, as with BJT characteristic curves.

Semiconductor devices are inherently nonlinear. The performance of such devices is analyzed in a linear fashion over small portions of the characteristic curve(s). Such analysis is called *small-signal analysis*. A small signal is one that is much less than the average, that is, steady-state, value for the device, which usually is the biasing value. The models used in each linear portion of the characteristic curve are termed *piecewise linear models*. If the operation is outside the linear region or if the input signal is large compared to the average value, the device distorts the input signal. Such operation is called *nonlinear operation*.

The characteristic curve for an ideal transistor operating as a current-amplifying device is shown in Fig. 43.1. (The shape of the curve is the same for a diode, that is, a *pn* junction.)

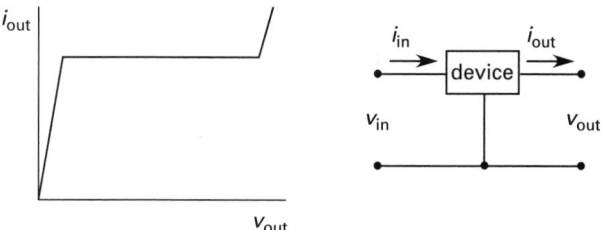

Figure 43.1 *Typical Semiconductor Performance*

The voltage-current graph is divided into various regions known by names such as *saturation* (or *on*), *cutoff* (or *off*), *active*, *breakdown*, *avalanche*, and *pinchoff* regions. The locations of these regions depend on the type of transistor—for example, BJT or FET—and its polarity. Operation is normally in the linear active region, but applications for operation in other regions exist in digital and communications (radio frequency) applications.

4. BIAS

Bias is a DC voltage applied to a semiconductor junction to establish the operating point, also called the *quiescent (no-signal) point*. *Biasing* establishes the operating point with no input signal.[12] Biasing, then, is

---

[12]Bias is used both as a verb and a noun in electronics.

the process of establishing the DC voltages and currents (the bias) at the device's terminals when the input signal is zero (or nearly so).

For *pn* junctions, *forward bias* (or *on condition*) is the application of a positive voltage to the *p*-type material or, equivalently, the flow of current from the *p*-type to the *n*-type material. In a small semiconductor device, forward bias results in current in the milliampere range.

Reverse bias (or *off condition*) is the application of a negative voltage to the *p*-type material, or equivalently, the flow of current from the *n*-type to the *p*-type material. In a small semiconductor device, reverse bias results in current in the nanoampere range.

Self-biasing is the use of the amplifier's output voltage, rather than a separate power source, as the supply for the input bias voltage. This negative feedback control regulates the output current and voltage against variations in transistor parameters.

5. AMPLIFIERS

An *amplifier* produces an output signal from the input signal. The input and output signals can be either voltage or current. The output can be either smaller or larger (the usual case) than the input in magnitude. While most amplifiers merely scale the input voltage or current upward, the amplification process can include a sign change, a phase change, or a complete phase shift of 180°.[13] The ratio of the output to the input is known as the *gain* or *amplification factor*, A. A *voltage amplification factor*, A_V, and *current amplification factor*, A_I or β, can be calculated for an amplifier.

Figure 43.2 illustrates a simplified current amplifier with current amplification factor β. The additional current leaving the amplifier is provided by the bias battery, V_2.

$$i_{\text{out}} = \beta i_{\text{in}} \qquad 43.8$$

A capacitor, C, is placed in the output terminal to force all DC current to travel through the *load resistor*, R_L. Kirchhoff's voltage law for loop abcd is

$$V_2 = i_{\text{out}} R_L + V_{\text{ac}} = \beta i_{\text{in}} R_L + V_{\text{ac}} \qquad 43.9$$

If there is no input signal (i.e., $i_{\text{in}} = 0$), then $i_{\text{out}} = 0$ and the entire battery voltage appears across terminals ac ($V_{\text{ac}} = V_2$). If the voltage across terminals ac is zero, then the entire battery voltage appears across R_L so that $i_{\text{out}} = V_2/R_L$.

---

[13]An *inverting amplifier* is one for which $v_{\text{out}} = -A_V v_{\text{in}}$. For a sinusoidal input, this is equivalent to a phase shift of 180° (i.e., $\mathbf{V}_{\text{out}} = A_V \mathbf{V}_{\text{in}} \angle -180°$).

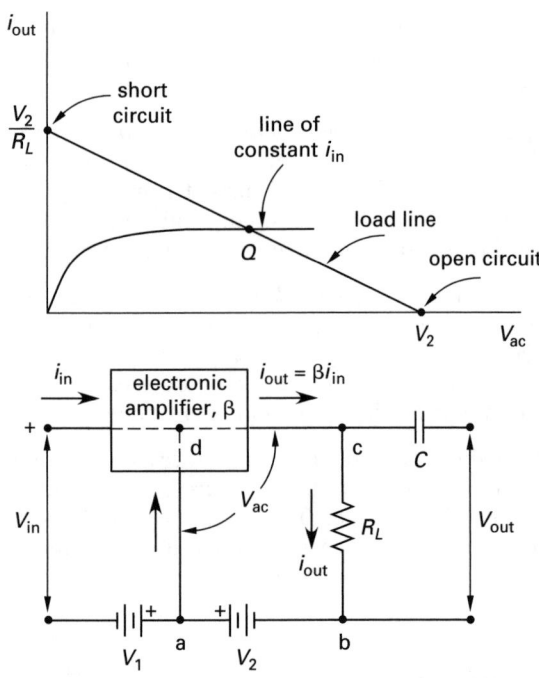

Figure 43.2 General Amplifier

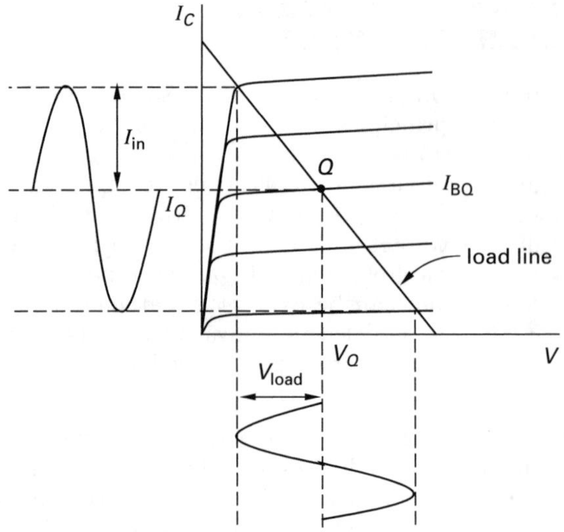

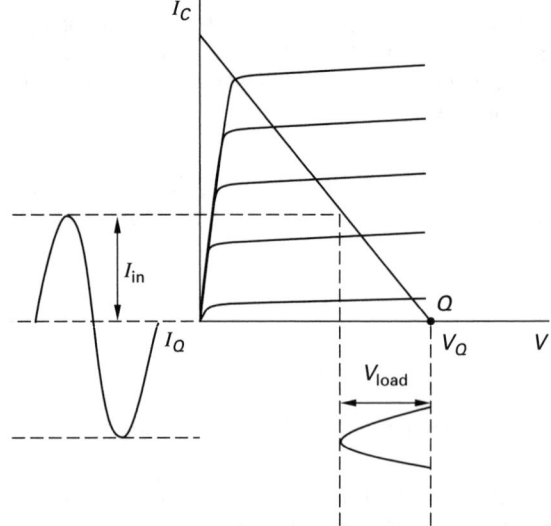

Figure 43.3 Class A Amplifier

6. AMPLIFIER CLASSIFICATION

Amplifiers are classified on the basis of how much input is translated into output. A sinusoidal input signal is assumed. The output of an amplifier depends on the bias setting, which in turn establishes the quiescent point.

A *Class A amplifier*, as shown in Fig. 43.3, has a quiescent point in the center of the active region of the operating characteristics. Class A amplifiers have the greatest linearity and the least distortion. Load current flows throughout the full input signal cycle. Since the load resistance of a properly designed amplifier will equal the Thevenin equivalent source resistance, the maximum power conversion efficiency of an ideal Class A amplifier is 50%.

For *Class B amplifiers*, as shown in Fig. 43.4, the quiescent point is established at the cutoff point. A load current flows only if the signal drives the amplifier into its active region, and the circuit acts like an amplifying half-wave rectifier. Class B amplifiers are usually combined in pairs, each amplifying the signal in its respective half of the input cycle. This is known as *push-pull operation*. The output waveform will be sinusoidal except for the small amount of crossover distortion that occurs as the signal processing transfers from one amplifier to the other. The maximum power conversion efficiency of an ideal Class B push-pull amplifier is approximately 78%.

The intermediate *Class AB amplifier* has a quiescent point somewhat above cutoff but where a portion of the input signal still produces no load current. The output current flows for more than half of the input cycle. AB amplifiers are also used in push-pull circuits.

Figure 43.4 Class B Amplifier

Class C amplifiers, as shown in Fig. 43.5, have quiescent points well into the cutoff region. Load current flows during less than one-half of the input cycle. For a purely resistive load, the output would be decidedly nonsinusoidal. However, if the input frequency is constant, as in radio frequency (rf) power circuits, the load can be a parallel LRC tank circuit tuned to be resonant at the signal frequency. The LRC circuit stores electrical energy, converting the output signal to a sinusoid. The power conversion efficiency of an ideal Class C amplifier is 100%.

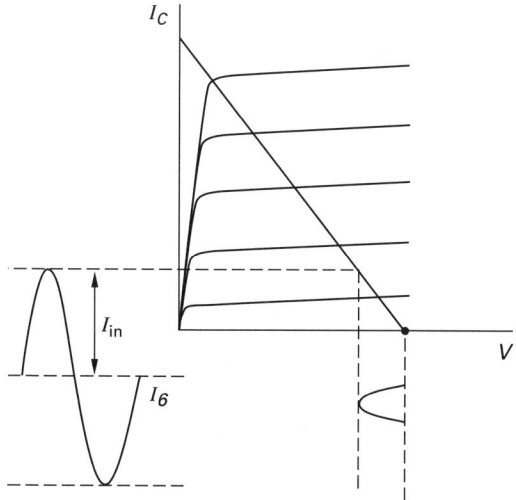

Figure 43.5 *Class C Amplifier (Resistive Load)*

7. LOAD LINE AND QUIESCENT POINT CONCEPT

The i_{in}-v_{out} curves of Secs. 43.5 and 43.6 illustrate how amplification occurs. The two known points, $(v_{out}, i_{out}) = (V_2, 0)$ and $(v_{out}, i_{out}) = (0, V_2/R_L)$, are plotted on the voltage-current characteristic curve. The straight *load line* is drawn between them. The change in output voltage (the horizontal axis) due to a change in input voltage (parallel to the load line) can be determined. Equation 43.10 gives the voltage *gain (amplification factor)*.[14,15]

$$A_V = \frac{\partial v_{out}}{\partial v_{in}} \approx \frac{\Delta v_{out}}{\Delta v_{in}} \qquad \textbf{43.10}$$

Usually, a nominal current (the *quiescent current*) flows in the abcd circuit even when there is no signal. The point on the load line corresponding to this current is the *quiescent point* (*Q-point* or *operating point*). It is common to represent the quiescent parameters with uppercase letters (sometimes with a subscript Q) and to write instantaneous values in terms of small changes to the quiescent conditions.

$$v_{in} = V_Q + \Delta v_{in} \qquad \textbf{43.11}$$

$$v_{out} = V_{out} + \Delta v_{out} \qquad \textbf{43.12}$$

$$i_{out} = I_{out} + \Delta i_{out} \qquad \textbf{43.13}$$

---

[14]Gain can be increased by increasing the load resistance, but a larger biasing battery, V_2, is required. The choice of battery size depends on the amplifier circuit devices, size considerations, and economic constraints.
[15]A *high-gain amplifier* has a gain in the tens of hundreds of thousands.

Since it is a straight line, the load line can also be drawn if the quiescent point and any other point, usually $(V_{BB}, 0)$, are known.

The ideal voltage amplifier has an infinite *input impedance* (so that all of v_{in} appears across the amplifier and no current or power is drawn from the source) and zero *output impedance* so that all of the output current flows through the load resistor.

Determination of the load line for a generic transistor amplifier is accomplished through the following steps.

step 1: For the configuration provided, label the x-axis on the *output characteristic curves* with the appropriate voltage. (For a BJT, this is V_{CE} or V_{CB}. For a FET, this is V_D.)[16]

step 2: Label the y-axis as the output current. (For a BJT, this is I_C. For a FET, this is I_D.)

step 3: Redraw the circuit with all three terminals of the transistor open. Label the terminals. (For a BJT, these are base, emitter, and collector. For a FET, these are gate, source, and drain.) Label the current directions all pointing inward, toward the amplifier. (For a BJT, these are I_B, I_E, and I_C. For a FET, these are I_D and I_S.)

step 4: Perform KVL analysis in the output loop. (For a BJT, this is the collector loop. For a FET, this is the drain loop.) The transistor voltage determined is a point on the x-axis with the output current equal to zero. Plot the point.

step 5: Redraw the circuit with all three terminals of the transistor shorted. Label as in step 3.

step 6: Use Ohm's law, or another appropriate method, in the output loop to determine the current. (For the BJT, this is the collector current. For the FET, this is the drain current.) The transistor current determined is a point on the y-axis with the applicable voltage in step 1 equal to zero. Plot the point.

step 7: Draw a straight line between the two points. This is the DC load line.[17]

---

[16]The output characteristic curves for the BJT are also called the *collector characteristics* or the *static characteristics*. Figure 43.2 is an example.
[17]AC load lines are determined in the same manner, but active components, that is, inductors and capacitors, are accounted for in the analysis.

Example 43.3

Consider the common emitter (CE) shown.

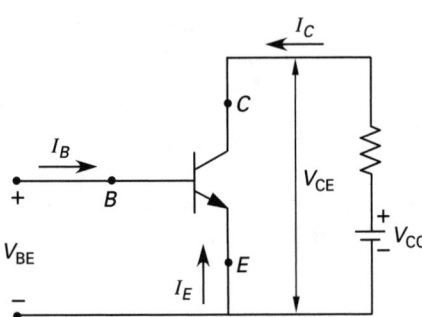

If the load resistance is 500 Ω and the collector supply voltage, V_{CC}, is 10 V, determine and draw the load line.

Solution

The x- and y-axes are drawn as shown.

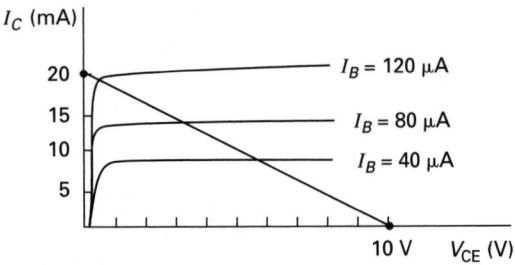

Redrawing the circuit and labeling gives

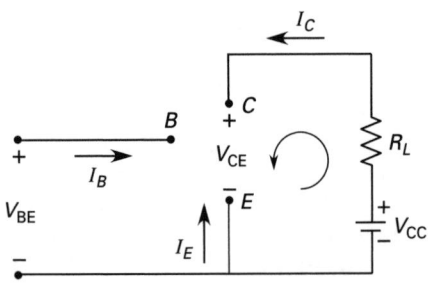

Write KVL around the indicated loop.

$$V_{CC} - I_C R_L - V_{CE} = 0$$

With the terminals open-circuited, $I_C = 0$. Substituting and rearranging gives

$$V_{CE} = V_{CC} = 10 \text{ V}$$

Plot this point (10,0) on the x-axis. Redraw and label the circuit with the terminals shorted.

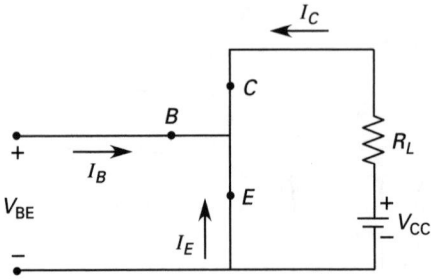

Using Ohm's law and the given values,

$$I_C = \frac{V_{CC}}{R_L} = \frac{10 \text{ V}}{500 \ \Omega} = 20 \text{ mA}$$

Plot this point on the y-axis. Draw a straight line between the two points. The load line is shown superimposed on the output characteristic curves in the first drawing.

8. *pn* JUNCTIONS

The *pn junction* forms the basis of diode and transistor operation. The *pn junction* is constructed of a *p*-type material (the anode) and an *n*-type material (the cathode) bonded together as shown in Fig. 43.6(a). Some of the acceptor atoms are shown as ions with a minus sign because after an impurity atom accepts an electron, it becomes negatively charged. Some of the donor atoms are shown as ions with a plus sign because after an impurity atom gives up an electron it is positively charged. (Overall, the law of charge neutrality holds, and the *pn* junction is neutral.)

Because of the concentration gradient across the junction, holes diffuse to the right and electrons diffuse to the left. As a result, the concentration of holes on the *p*-side near the junction is depleted and a negative charge exists. The concentration of electrons on the *n*-side near the junction is depleted as well and a positive charge exists. The result of this diffusion is shown in Fig. 43.6(b), the shape of which is determined by the level of doping.[18] The diffusion process continues until the electrostatic field set up by the charge separation is such that no further charge motion is possible. The net electric field intensity is shown in Fig. 43.6(c). The net result is a small region in which no mobile charge carriers exist. This region is called the *space-charge region*, *depletion region*, or *transition region*. Holes have a potential barrier they must overcome to move from left to right just as electrons have an energy barrier they must overcome to move from right to left, as shown in Figs. 43.6(d) and (e).

---

[18] A *step-graded junction* and a *linearly graded junction* are two possible types.

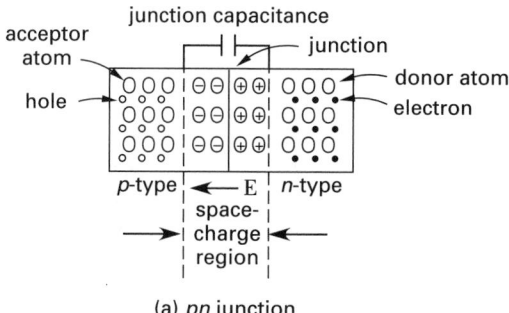

(a) *pn* junction

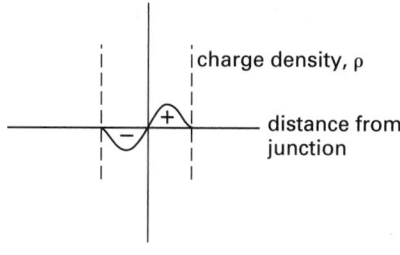

(b) charge density

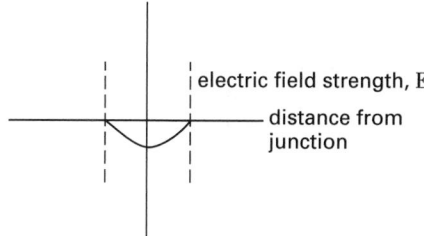

(c) electric field intensity

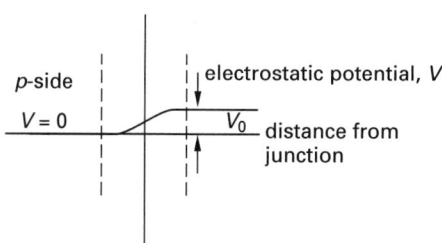

(d) electrostatic potential

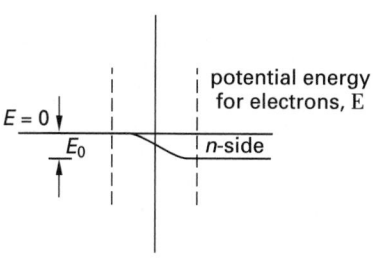

(e) potential energy

Figure 43.6 *pn Junction Characteristics*

The flow of carriers due to the concentration gradient is called the *diffusion current*, $I_{\text{diffusion}}$, also called the

recombination current or the *injection current*. The flow of carriers due to the established electric field is called the *drift current*, I_s, also called the *saturation current, thermal current,* or *reverse saturation current.*[19] The movement of carriers due to recombination (diffusion) and drift (saturation) is continual, though at thermal equilibrium, without any applied voltage, the net current is zero.

$$I_{\text{junction}} = I_{\text{diffusion}} + I_s = 0 \quad [\text{algebraic sum}] \quad \textit{43.14}$$

A summary of the movement of the carriers is shown in Fig. 43.7.

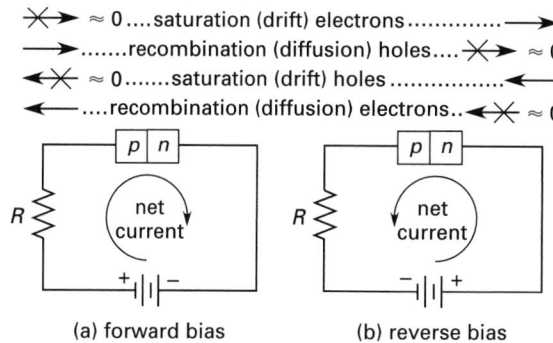

Figure 43.7 *pn Junction Carrier Movement*

The width of the space-charge region at equilibrium is shown in Fig. 43.8(a).

When an external voltage is applied, that is, a biasing voltage, the width of the space-charge region and the barrier height change. When the *p*-type material is connected to a positive potential, that is, forward biased, holes are repelled across the junction into the *n*-type material, and electrons are repelled across the junction into the *p*-type material. The width of the space-charge region is reduced and the barrier height for *p* to *n* current flow is reduced as shown in Fig. 43.8(b). A DC forward bias voltage, V_F, of approximately 0.5 to 0.7 V for silicon and 0.2 to 0.3 V for germanium is required to overcome the barrier voltage. Once the barrier is overcome, the junction current increases significantly due to an increase in the diffusion current. That is, holes cross the junction into the *n*-type material, where they are considered injected minority carriers. Electrons cross the junction into the *p*-type material, where they too are injected minority carriers. Since hole movement in one direction and electron movement in the opposite direction constitute a current in the same direction, the total current is the sum of the hole and electron minority currents.

---

[19]Numerous symbols are used, among them I_{o} and I_{co}.

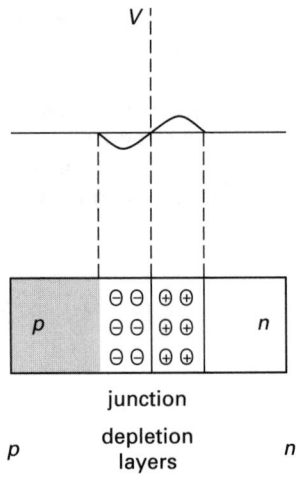

(a) no applied voltage

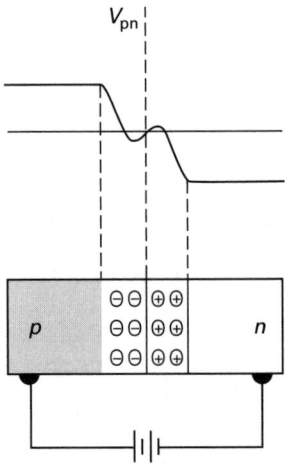

(b) forward bias

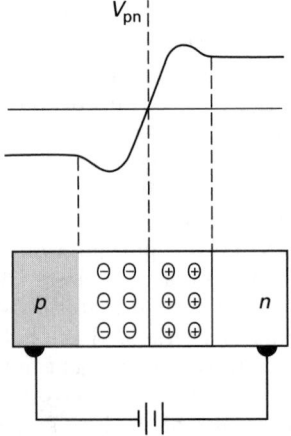

(c) reverse bias

Figure 43.8 *pn Junction Space-Charge Region*

When the *p*-type material is connected to a negative potential, that is, reverse biased, holes and electrons move away from the junction. The width of the space-charge region is increased and the barrier height for *p* to *n* current flow is increased as shown in Fig. 43.8(c). The process nominally stops when the holes in the *p*-type material are depleted. However, a few holes in the *n*-type material are thermally generated, as there are electrons in the *p*-type material. These minority carriers thermally diffuse into the depletion region and are swept across by the electric field. The effect is constant for a given temperature and independent of the reverse bias. This is the reverse saturation current, I_s.[20] The reverse saturation current is small ($\approx 10^{-9}$ A). An ideal *pn* junction, excluding the breakdown region, is governed by[21]

$$I_{\mathrm{pn}} = I_s \left(e^{\frac{qV_{\mathrm{pn}}}{\kappa T}} - 1 \right) \qquad 43.15$$

Breakdown is a large, abrupt change in current for a small change in voltage. When a *pn* junction is reverse biased, the saturation current is small up to a certain reverse voltage, where it changes dramatically. Two mechanisms can cause this change. The first is *avalanche breakdown*. Avalanche occurs when thermally generated minority carriers are swept through the space-charge region and collide with ions. If they possess enough energy to break a covalent bond, an electron-hole pair is created. The same effect may occur for these newly generated carriers, resulting in an avalanche effect. Avalanche breakdown occurs in lightly doped materials at greater than 6 V reverse bias. The second breakdown mechanism is *zener breakdown*. Zener breakdown occurs through the disruption of covalent bonds due to the strength of the electric field near the junction. No collisions are involved. The additional carriers created by the breaking covalent bonds increase the reverse current. Zener breakdown occurs in highly doped materials at less than 6 V reverse bias.[22]

9. DIODE PERFORMANCE CHARACTERISTICS

A *diode* is a two-electrode device. The diode is designed to pass current in one direction only. An *ideal diode*, approximated by a *pn* junction, has a zero voltage drop, that is, no forward resistance, and acts as a short circuit when forward biased (on). When reversed biased (off), the resistance is infinite and the device acts as an open circuit. Diode construction and theory is that of a *pn* junction (see Sec. 43-8).

---

[20] The reverse saturation current also accounts for any current leakage across the surface of the semiconductor.
[21] The subscript *pn* is used here for clarification. Standard diode voltage and current directions are defined in Sec. 43-9, after which the subscript is no longer used.
[22] The name zener is commonly used regardless of the breakdown mechanism.

The characteristics and symbology for a typical real semiconductor diode are shown in Fig. 43.9. The *reverse bias voltage* is any voltage below which the current is small, that is, less than 1% of the maximum rated current. The *peak inverse (reverse) voltage*, PIV or PRV, is the maximum reverse bias the diode can withstand without damage. The forward current is also limited due to heating effects. Maximum forward current and peak inverse voltage for silicon diode rectifiers are approximately 600 A and 1000 V, respectively.

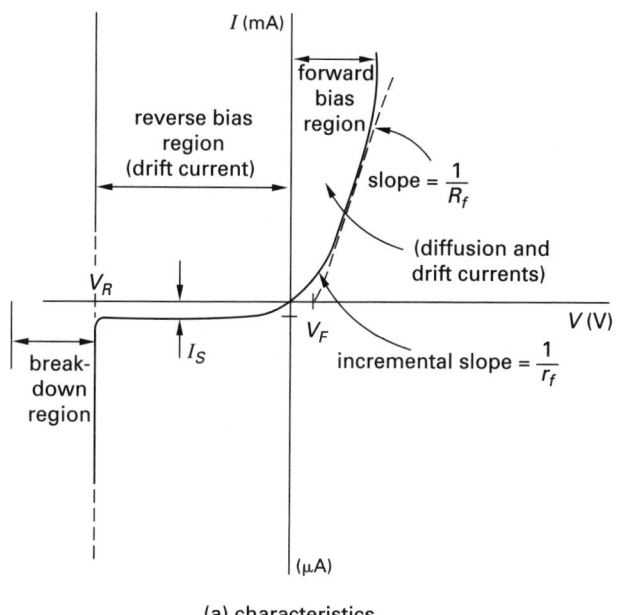

(a) characteristics

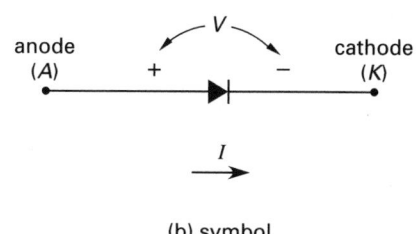

(b) symbol

Figure 43.9 *Semiconductor Diode Characteristics and Symbol*

The ideal *pn* junction current was given in Eq. 43.15. For practical junctions (*diodes* or *rectifiers*), the equation becomes

$$I = I_s \left(e^{\frac{qV}{\eta \kappa T}} - 1 \right) = I_s \left(e^{\frac{V}{\eta V_T}} - 1 \right) \qquad 43.16$$

Equation 43.16, which is based on the *Fermi-Dirac probability function*, is valid for all but the breakdown region (see Fig. 43.9). The term η is determined experimentally. For discrete silicon diodes, $\eta = 2$. For germanium diodes, $\eta = 1$. The saturation current, taken from any value of I with a small reverse bias (for example, between 0 and -1 V) gives $I_s \approx 10^{-9}$ A for silicon and 10^{-6} A for germanium. The term V_T represents

the *voltage equivalent of temperature* and is related to the diffusion occurring at the junction.[23]

$$V_T = \frac{\kappa T}{q} = \frac{D_p}{\mu_p} = \frac{D_n}{\mu_n} \qquad 43.17$$

Boltzmann's constant is given by κ. The absolute temperature is T. An electron charge is represented by q. Diffusion constants, measured in $\mathrm{m^2/s}$, are given the symbol D. The mobility, measured in $\mathrm{m^2/V \cdot s}$, is given by the symbol μ.

The voltage equivalent of temperature is also known as the *thermal voltage*. The value of V_T is often quoted at *room temperature*, which can vary from 293K (20°C) to 300K (27°C) depending upon the reference used. The temperature has the effect of doubling the saturation current every 10°C. Thus,

$$\frac{I_{s2}}{I_{s1}} = (2)^{\frac{T_2 - T_1}{10°C}} \qquad 43.18$$

A *real diode* can be modeled as shown in Fig. 43.10. The real diode model is composed of an ideal diode, a resistor, R_f, and a voltage source, V_F. The voltage source accounts for the barrier voltage. Typically, for silicon $V_F = 0.7$ V, and for germanium $V_F = 0.2$ V.

The resistance, R_f, is the slope of a line approximating the linear portion of the characteristic curve as shown in Fig. 43.9. R_f is the resistance for an ideal diode. It is not the static (average) resistance of the diode, as the average resistance is not a constant. The static (average) resistance is calculated as $R_{\mathrm{static}} = V_D/I_D$. When the diode is forward biased by more than a few tenths of a volt, R_f is equal to the *dynamic forward resistance*, r_f. Disregarding lead contact resistance (less than 2 Ω), the dynamic forward resistance is

$$R_f = r_f = \frac{\eta V_T}{I_D} \qquad 43.19$$

A *dynamic reverse resistance*, r_r, also exists and is the inverse of the slope at a point in the reverse bias region.[24] Since the reverse current is very small, the resistance is often considered infinite. Capacitances associated with the junction are also ignored in most models of the diode. The *diffusion capacitance*, C_d, is associated with the charge stored during forward biased operation. The *transition capacitance*, C_t, is associated with the space-charge region width and thus is the primary capacitance of concern during reverse bias operation. (Diodes designed to take advantage of this voltage-sensitive capacitance are called *varactors*).

---

[23]This is also called the *Einstein relationship*.
[24]Specifying the reverse current, I_{co}, is equivalent to specifying the reverse resistance.

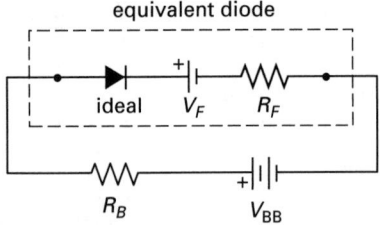

Figure 43.10 Diode Equivalent Circuit

The model of Fig. 43.10 assumes that (1) the reverse bias current is sufficiently small that the diode acts as an open circuit in that direction, (2) the reverse bias voltage does not exceed the breakdown voltage, and (3) the switching time is instantaneous. The *switching time* is the transient that occurs from the time interval of the application of a voltage to forward (reverse) bias and the achievement of the actual condition. The switching time depends upon the speed of movement of minority carriers near the junction and the junction capacitance.

Example 43.4

What is the thermal voltage at a room temperature of 300K?

Solution

The thermal voltage is given by Eq. 43.17 as

$$V_T = \frac{\kappa T}{q} = \frac{\left(1.381 \times 10^{-23}\ \frac{\text{J}}{\text{K}}\right)(300\text{K})}{1.602 \times 10^{-19}\ \text{C}} = 0.026\ \text{V}$$

10. DIODE LOAD LINE

Figure 43.11 shows a forward biased real diode in a simple circuit. V_{BB} is the *bias battery* (hence the subscripts), and R_B is a current-limiting resistor. R_f and V_f are equivalent diode parameters, not discrete components. If R_f is known in the vicinity of the operating point, the diode current, I_D, is found from Kirchhoff's voltage law. The diode voltage is found from Ohm's law: $V_D = I_D R_f$.

$$V_{BB} - V_f = I_D(R_B + R_f) \qquad 43.20$$

The diode current and voltage drop can also be found graphically from the *diode characteristic curve* (Fig. 43.11) and the *load line*, a straight line representing the locus of points satisfying Eq. 43.20.[25] The load line is defined by two points. If the diode current is zero, all of the battery voltage appears across the diode (point $(V_{BB}, 0)$). If the voltage drop across the diode is zero, all of the voltage appears across the current-limiting

---
[25]Notice that the horizontal axis voltage is the voltage across the diode (modeled as a resistor).

resistor (point $(0, V_{BB}/R_B)$). (Since the diode characteristic curve implicitly includes the effects of V_f and R_f, these terms should be omitted.) The no-signal *operating point*, also known as the *quiescent point*, is the intersection of the diode characteristic curve and the load line.

The *static load line* is derived assuming there is no signal (i.e., $v_{in} = 0$). With a signal, the *dynamic load line* shifts left or right while keeping the same slope. This is equivalent to solving Eq. 43.20 with an additional voltage source.

$$i_D = I_D + \Delta i_D = \frac{V_{BB} - V_f + v_{in}}{R_B + R_f} \qquad 43.21$$

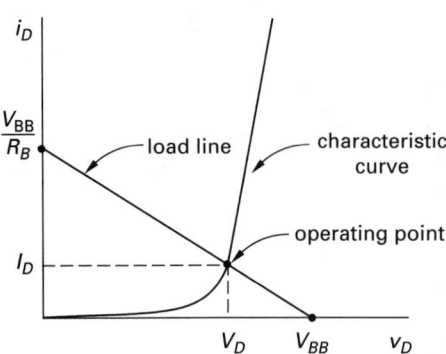

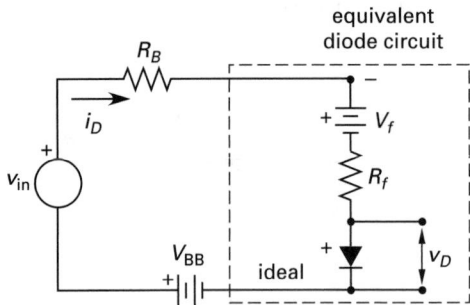

Figure 43.11 Diode Load Line

Although presented in a slightly different manner, the method for determining the load line is similar to that given in Sec. 43-7. That is,

step 1: Open-circuit the electronic component's equivalent circuit and determine the point $(x,0)$.

step 2: Short-circuit the electronic component's equivalent circuit and determine the point $(0,y)$.

step 3: Connect the two points.

The load line superimposed on the characteristics curve is then used to determine the operating point (Q-point) of the overall circuit.

11. DIODE PIECEWISE LINEAR MODEL

If the voltage applied to a diode varies over an extensive range, a piecewise linear model may be used.[26] For the real diode characteristic shown in Fig. 43.9, three regions are evident from V_R to V_F.

(1) *Reverse breakdown region,* $V_D < V_R$

$$I_D = \frac{V_D + V_R}{R_r} \quad [V_D \text{ and } V_R \text{ are negative}] \quad \textbf{43.22}$$

(2) *Off region,* $V_R < V_D < V_F$

$$I_D = 0 \quad \textbf{43.23}$$

(3) *Forward bias region,* $V_D > V_F$

$$I_D = \frac{V_D - V_F}{R_f} \quad \textbf{43.24}$$

Using ideal diodes, the real diode can be represented by the model in Fig. 43.12 over the entire range of operation.

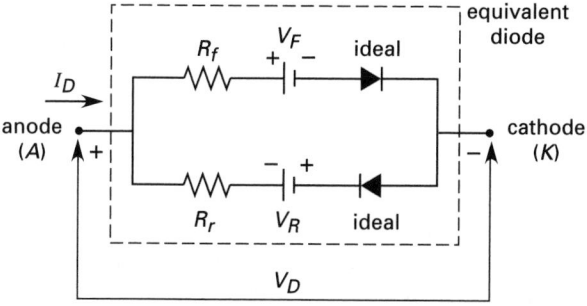

Figure 43.12 *Piecewise Linear Model*

12. DIODE APPLICATIONS AND CIRCUITS

Diodes are readily integrated into *rectifier, clipping,* and *clamping circuits.* A clipping circuit cuts the peaks off of waveforms; a clamping circuit shifts the DC (average) component of the signal. Figure 43.13 illustrates the response to a sinusoid with peak voltage V_m for several simple circuits.

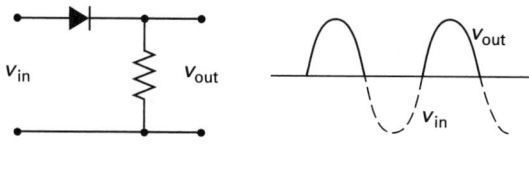

(a) half-wave rectifier

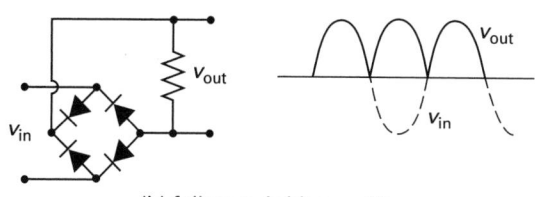

(b) full-wave bridge rectifier

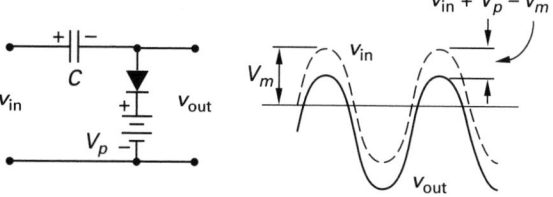

(c) clamping circuit (*C* charges to V_m)

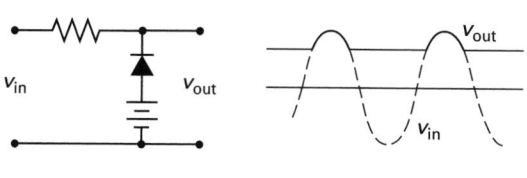
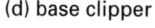

(d) base clipper

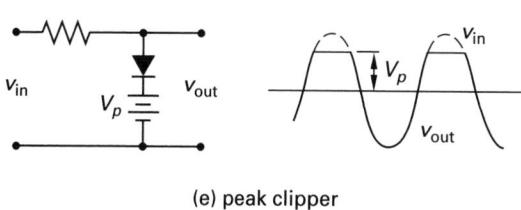

(e) peak clipper

Figure 43.13 *Output from Simple Diode Circuits*

13. SCHOTTKY DIODES

A *Schottky diode,* also called a *barrier diode* or a *hot-carrier diode,* is a diode constructed with a metal semiconductor contact as shown in Fig. 43.14(a). The Schottky diode symbol is shown in Fig. 43.14(b). The metal semiconductor rectifying junction is similar to the *pn* junction, but the physical mechanisms are somewhat different.

In the forward direction, electrons from the lightly doped semiconductor cross into the metal anode, where electrons are plentiful.[27] The electrons in the metal are majority carriers, whereas in a *p*-type material they are minority carriers. Being majority carriers, they are indistinguishable from other carriers and are thus not stored near the junction. This means that no minority carrier population exists to move when switching

---

[26]The model in Sec. 43-10 is for the forward biased region only and assumes no reverse current flow.

[27]The electrons injected into the metal are above the *Fermi energy level*, determined by the Fermi-Dirac distribution of electron energies, and thus are called *hot carriers.*

occurs from forward to reverse (on to off) bias. Consequently, switching times are extremely short (approximately 10^{-12} s).

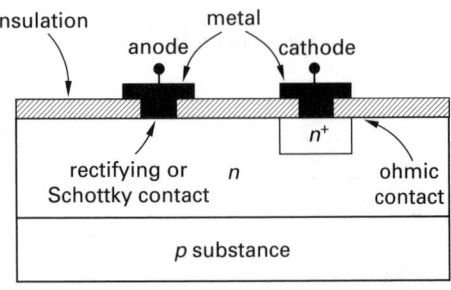

(a) construction

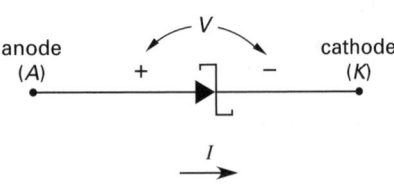

(b) diode symbol

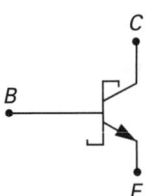

(c) Schottky contact symbol

Figure 43.14 *Schottky Diode*

The Schottky symbol is used on any electronic component designed with a *Schottky contact*, that is, a rectifying, rather than an ohmic, contact as shown in Fig. 43.14(c). In the figure, the Schottky contact lies between the base and the collector.

14. ZENER DIODES

A *zener diode* is a diode specifically designed to operate within the breakdown region. The construction and theory of a zener diode is similar to that of a *pn* junction, with the design allowing for greater heat dissipation capabilities. The characteristics and symbology are shown in Fig. 43.15.

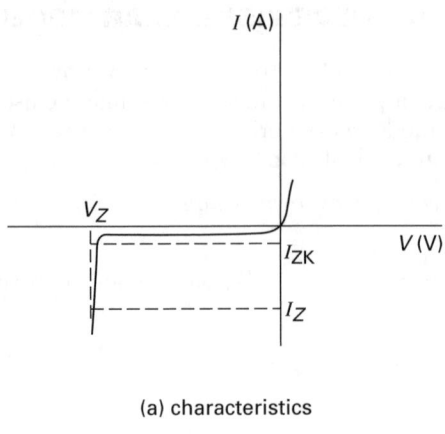

(a) characteristics

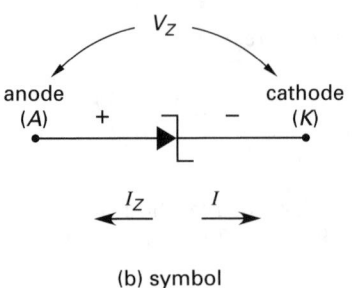

(b) symbol

Figure 43.15 *Zener Diode*

When a reverse voltage, known as the *zener voltage*, V_Z (which is negative with respect to the anode), is applied, a reverse saturation current, I_Z, flows. The voltage-current relationship is nearly linear. Current flows until the reverse saturation current drops to I_{ZK} near the knee of the characteristic curve. This minimum current is the *keep-alive current*, that is, the minimum current for which the output characteristic is linear.

Because the current is large, an external resistor must be used to limit the current to within the power dissipation capability of the diode. Zener diodes are used as voltage regulating and protection devices. The regulated zener voltage varies with the temperature. The *temperature coefficient* is

$$\text{tc} = \frac{\Delta V_Z}{V_Z \Delta T} \times 100\% \qquad \textbf{\textit{43.25}}$$

15. TUNNEL DIODES

A *tunnel diode* (*Esaki diode*) is a two-terminal device with an extremely thin potential barrier to electron flow, so that the output characteristic is dominated by the quantum-mechanical tunneling process. To make a thin potential barrier (50 to 100 Å), both regions of the diode are heavily doped. The amount of tunneling is limited by the electrons available in the *n*-type material or by the available empty energy states in the *p*-type material to which they can tunnel. The characteristics are shown in Fig. 43.16(a).

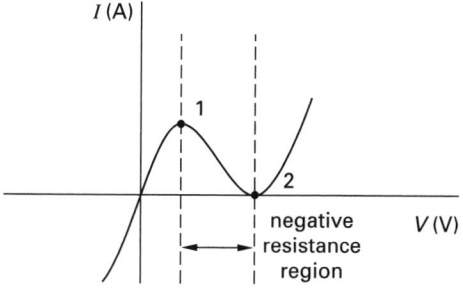

(a) tunnel diode characteristics

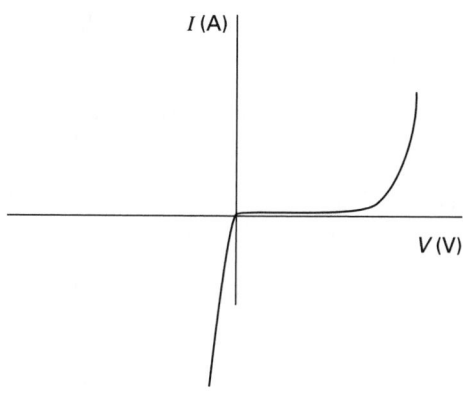

(b) backward diode

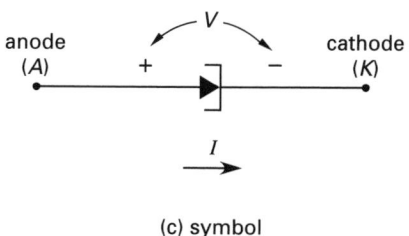

(c) symbol

Figure 43.16 *Tunnel Diode*

Point 1 in Fig. 43.16(a) corresponds to a biasing level that allows for maximum tunneling. (At this point, the minimum electron energy in the *n*-type material conduction band equals the Fermi level in the *p*-type material's valence band.) Between points 1 and 2, increases in bias voltage result in a decrease in current as available energy states are filled and the total amount of tunneling drops. This is an area of negative resistance, or negative conductance, and is the primary use for tunnel diodes. If the doping levels are slightly reduced from their typical values of 5×10^{19} cm^{-3}, the forward tunneling current becomes negligible and the output characteristic becomes that of a *backward diode*, Fig. 43.16(b). The symbology for a tunnel diode is shown in Fig. 43.16(c).

16. PHOTODIODES AND LIGHT-EMITTING DIODES

If a semiconductor junction is constructed so that it is exposed to light, the incoming photons generate electron-hole pairs. When these carriers are swept from the junction by the electric field, they constitute a *photocurrent*, which is seen as an increase in the reverse saturation current. The holes generated move to the *p*-type material and the electrons move to the *n*-type material due to the electric field that is established whenever *p*- and *n*-type semiconductors are joined (see Fig. 43.6(a)). Such devices are used as light sensors and are called *photodiodes*. When the device is designed without a biasing source, it becomes a *solar cell*.

When forward biased, diodes inject carriers across the junction that are above thermal equilibrium. When the carriers recombine, they emit photons from the *pn* junction area.[28] The photon is due to the recombination of electron-hole pairs. The wavelength depends on the energy band gap and thus on the material used. Gallium arsenide (GaAs) and other binary compounds are commonly used. When the photon is in the infrared region, the mechanism is called *electroluminescence* and the diodes are called *electroluminescent diodes*. If the photons are in the visible region, the devices are called *light-emitting diodes* (LEDs). The emitted wavelength, λ, is given by

$$\lambda = \frac{hc}{E_G} \qquad 43.26$$

Example 43.5

The manufacturer's data sheet for a gallium arsenide diode shows a band gap energy of 1.43 eV. Will such a band gap result in emitted photons within the wavelength of visible light?

Solution

The photon emitted wavelength is given by

$$\lambda = \frac{hc}{E_G}$$

$$= \frac{\left(6.626 \times 10^{-34} \text{ J·s}\right)\left(3.00 \times 10^8 \dfrac{\text{m}}{\text{s}}\right)}{\left(1.43 \text{ eV}\right)\left(1.602 \times 10^{-19} \dfrac{\text{J}}{\text{eV}}\right)}$$

$$= 8.68 \times 10^{-7} \text{ m}$$

The wavelength of visible light is from approximately 8×10^{-7} to 4×10^{-7} m. Consequently, GaAs is a good choice for a light-emitting diode.

17. SILICON-CONTROLLED RECTIFIERS

A four-layer *pnpn* device with an anode, cathode, and gate terminal is called a *silicon-controlled rectifier*

---

[28]They can also release the energy as heat, or *phonons*.

(SCR) or *thyristor*.[29] A conceptual construction is shown in Fig. 43.17(a).

When the SCR is reverse biased, that is, when the anode is negative with respect to the cathode, the characteristics are similar to a reverse-biased *pn* junction as shown in Fig. 43.17(b). When the SCR is forward biased, that is, when the anode is positive with respect to the cathode, four distinct regions of operation are evident.

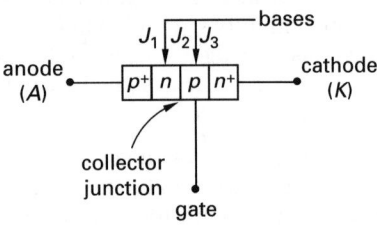

(a) conceptual construction

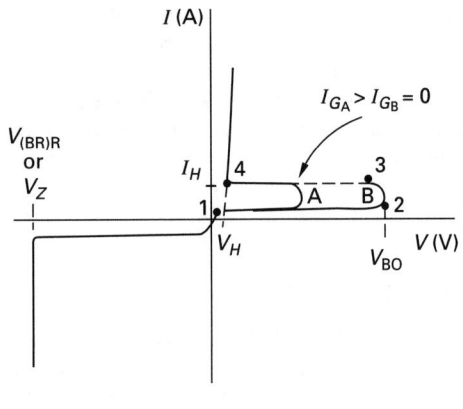

(b) characteristics

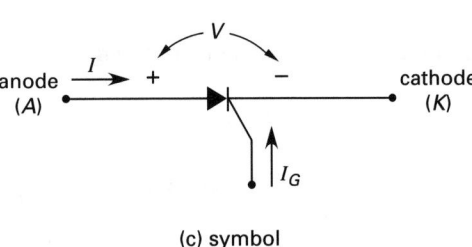

(c) symbol

Figure 43.17 *Silicon-Controlled Rectifier*

From point 1 to point 2, junctions J_1 and J_3 are forward biased. Junction J_2 is reverse biased. The external voltage appears primarily across the reverse-biased junctions. The device continues to operate similarly to a reverse-biased *pn* junction. This is called the *off* or *high-impedance region*.

---

[29]A *thyristor* is defined as a transistor with thyratron-like characteristics. That is, as collector current is increased to a critical value, the alpha (common base current gain) rises above unity and results in a high-speed triggering action.

From point 2 to point 3, the current increases slowly to the *breakover voltage*, V_{BO}. At this point, junction J_2 undergoes breakdown and the current increases sharply.

From point 3 to point 4, the current increases as the voltage decreases. This region is called the *negative resistance region*.

From point 4, junction J_2 is forward biased. The voltage across the device is essentially that of a forward-biased *pn* junction (approximately 0.7 V). If the current through the diode is reduced by the external circuit, the diode remains on until the current falls below the *hold current*, I_H, or the *hold voltage*, V_H. Below this point, the diode switches off, to the high-impedance state.

The gate functions to increase the current at the collector junction of the *npn* transistor, J_2, which is an integral part of the SCR. By increasing the current through the reverse-biased junction, J_2, the anode current is increased. This increases the gain of the two transistors, resulting in a lowering of the forward breakover voltage. This is seen in the difference between paths A and B in Fig. 43.17(b). Consequently, for a given anode to cathode voltage, the gate can be used to turn the SCR on. Once on, however, the SCR must be reverse biased to turn off. A thyristor designed to be turned off by the gate is called a *gate turn-off thyristor*.

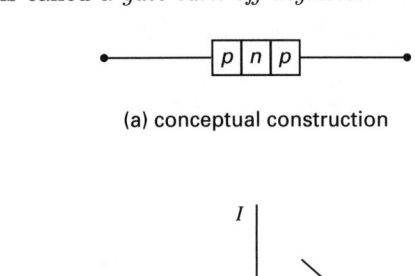

(a) conceptual construction

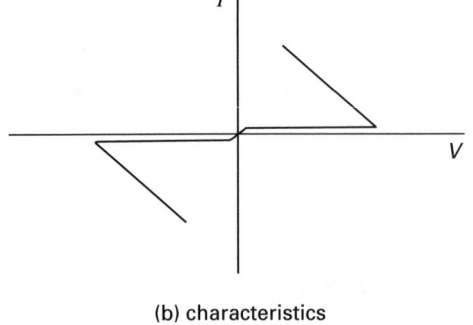

(b) characteristics

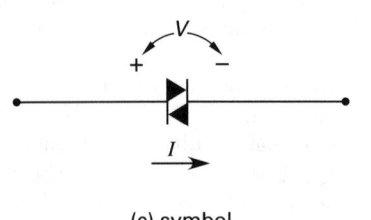

(c) symbol

Figure 43.18 *Diac*

SCRs are used in power applications to allow small voltages and currents to control much larger electrical quantities. The symbology for an SCR is shown in Fig. 43.17(c). Other devices with multiple *pn* junctions using the same principles are the *diac* and *triac*, shown in Figs. 43.18 and 43.19, respectively.

Most electronic components are unable to handle large amounts of current. When properly designed for power dissipation, semiconductors handling large amounts of power are called *power semiconductors*. Such devices constitute a branch of electronics called *power electronics*. Silicon-controlled rectifiers and Schottky diodes are traditional power semiconductors. Newer designs include the *high-power bipolar junction transistor* (HPBT), *power metal-oxide semiconductor field-effect transistor* (MOSFET), *gate turn-off thyristor* (GTO), and *insulated gate bipolar transistor* (IGBT), sometimes called a *conductivity-modulated field-effect transistor* (COMFET).

Power semiconductors are classified as either trigger or control devices. *Trigger devices*, such as GTOs, start conduction by some trigger input and then behave as diodes. *Control devices* are normally BJTs and FETs used in full-range amplifiers.

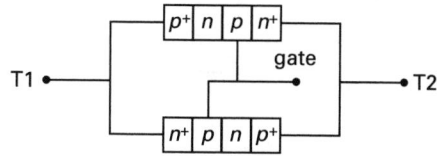

(a) conceptual construction

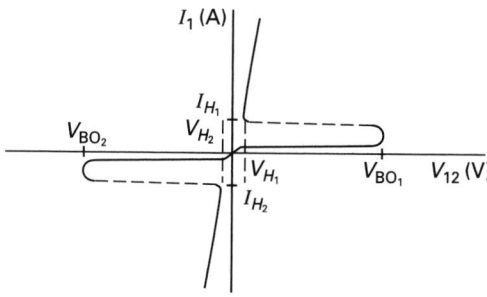

(b) characteristics

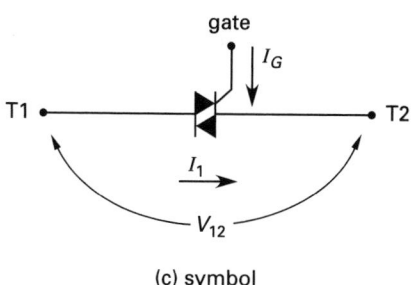

(c) symbol

Figure 43.19 Triac

18. TRANSISTOR FUNDAMENTALS

A *transistor* is an active device comprised of semiconductor material with three electrical contacts (two rectifying and one ohmic). Two major types of transistors exist: *bipolar junction transistors* (BJTs) and *field-effect transistors* (FETs). A BJT uses the base current to control the flow of charges from the emitter to the collector. Field-effect transistors use an electric field, that is, voltage, established at the gate (equivalent to the base) to control the flow of charges from the source to the drain (equivalent to the emitter and collector). Thus, the controlling variable in a BJT is the current at the base; in a FET it is the voltage at the gate. An *npn* bipolar junction transistor (BJT) is shown in Fig. 43.20(a).

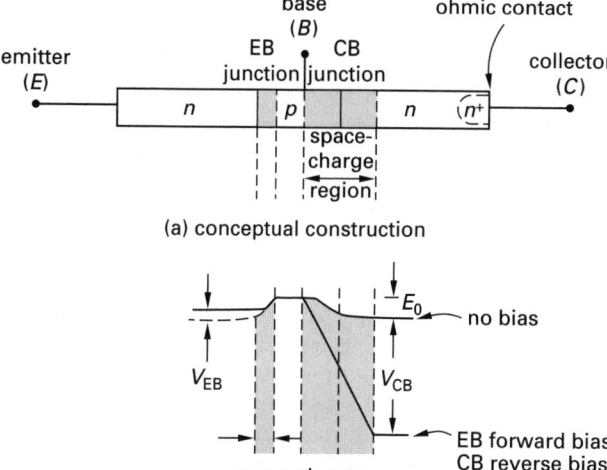

(a) conceptual construction

(space-charge regions shown for active region biasing)

(b) electron energy level

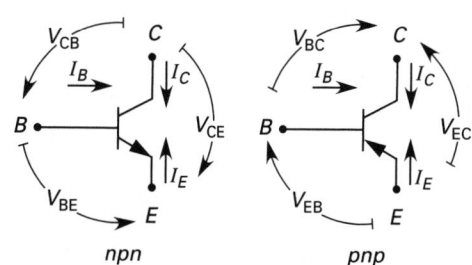

npn　　　　*pnp*

(all currents shown in positive direction)
(all voltages shown for active region biasing)

(c) symbol

Figure 43.20 Bipolar Junction Transistor

The transistor has three operating regions: *cutoff*, *saturation*, and *active* or *linear*. In the cutoff region, both the base-emitter junction and the collector-base junction are reverse biased. In the saturation region, both the base-emitter junction and the collector-base junction are forward biased. Small amounts of current injected into the base control the movement of charges in the output (from emitter to collector) in a manner similar in theory to that of an SCR, specifically, the triac (see Sec. 43-17). When operating from the cutoff to the saturation region, the transistor is a switch.

In the active, or linear, region, the base-emitter junction is forward biased and the collector-base junction is reverse biased. When so biased, the electron potential energy is of the shape shown in Fig. 43.20(b). Small amounts of current injected into the base lower the potential barrier across the emitter-base junction. As carriers move from the emitter to the base, most are swept across the narrow base by the relatively large electric field established by the reverse-biased collector-base junction. (See the large space-charge, or depletion region, shown in Fig. 43.20(b).) When operating in the active (linear) region, the transistor is an amplifier.

The symbology for BJTs is shown in Fig. 43.20(c). Positive currents are defined as those flowing into the associated terminals. Actual current flows follow the arrows.[30] Majority carriers are always injected into the base by the *emitter* (E). The thin controlling center of a transistor is the *base* (B). The majority carriers are gathered by the *collector* (C).

When operated over all three ranges—cutoff, saturation, and active—the transistor is modeled in a piecewise linear fashion. This is *large-signal analysis*. Operated as a switch, between the cutoff and saturation regions, the transistor is modeled as an open circuit and a short circuit. This is *digital circuit operation*. Operated as an amplifier, in the active region, the transistor is modeled as equivalent parameters, usually *h*-parameters. This is *analog circuit operation*. Analog analysis occurs in two phases. First, the operating point, or *Q-point*, is determined. Determining the *Q*-point is a combination of determining the load line for the biasing used and setting the base current by using the values of the biasing components (see Secs. 43-22 and 43-23). This is often a reiterative process. (For a FET, the items determined are the load line and gate voltage.) Second, the incremental AC performance is determined using *small-signal analysis*, that is, using *h*-parameters to model the transistor and replacing independent sources with the appropriate models (see Sec. 43-26). Real sources are replaced with internal resistance. Ideal voltage sources are replaced with short circuits. Ideal current sources are replaced with open circuits.

19. BJT TRANSISTOR PERFORMANCE CHARACTERISTICS

When the base-emitter junction is forward biased and the collector-base junction is reverse biased, the transistor is said to be operating in the *active region*.

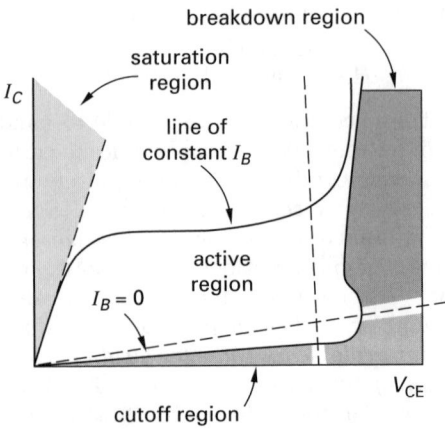

Figure 43.21 *BJT Operating Regions*

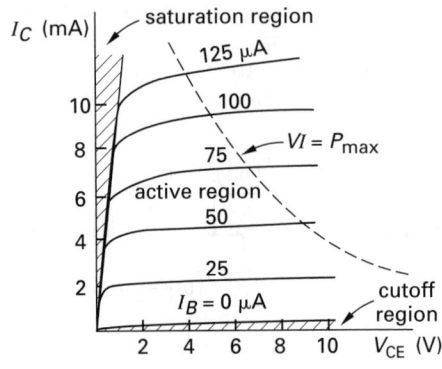

(a) common emitter

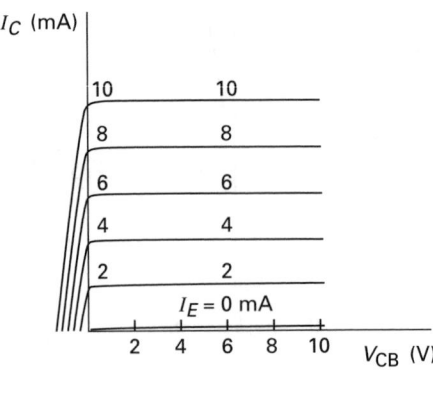

(b) common base

Figure 43.22 *BJT Output Characteristics*

When the base-emitter junction is not forward biased (as when V_{BB} is zero), the base current will be nearly zero and the transistor acts like a simple switch. This is known as being *off* or *open*, and operating in the *cutoff mode*.[31] Also, the collector and emitter currents are zero

---

[30]That is, conventional current flow follows the arrows. Electrons flow opposite to the arrows.

[31]Except for digital and switching applications, this condition usually results from improper selection of circuit resistances.

when the transistor operates in the *cutoff region*. However, a very small input voltage (in place of V_{BB}) will forward bias the base-emitter junction which, because I_B is so low, instantly forces the transistor into its saturation region. This results in a large collector current.

When the collector-emitter voltage is very low (usually about 0.3 V for silicon and 0.1 V for germanium), the transistor operates in its *saturation region*. Regardless of the collector current, the transistor operates as a closed switch (i.e., a short circuit between the collector and emitter). This is known as being *on* or *closed*.

$$V_{CE} \approx 0 \quad [\text{saturation}] \qquad 43.27$$

The BJT operating regions are shown in Fig. 43.21. The output characteristics for the common emitter and common base configurations are shown in Fig. 43.22. (A sample maximum power curve is shown as a dashed line. The Q-point must be to the left and below the power curve.)

20. BJT TRANSISTOR PARAMETERS

Equation 43.28 is Kirchhoff's current law, taking the transistor as a node. Usually, the collector current is proportional to, and two or three orders of magnitude larger than, the base current, I_B. Thus, a small change in base current of, for example, 1 mA, can produce a change in collector current of, for example, 100 mA. The *current (amplification) ratio*, β_{DC}, is the ratio of collector-base currents.

$$I_E = I_C + I_B \qquad 43.28$$

$$\beta_{DC} = \frac{I_C}{I_B} = \frac{\alpha_{DC}}{1 - \alpha_{DC}} \qquad 43.29$$

$$\alpha_{DC} = \frac{I_C}{I_E} = \frac{\beta_{DC}}{1 + \beta_{DC}} \qquad 43.30$$

Both α_{DC} and β_{DC} are for DC signals only. The corresponding values for small signals are designated α_{ac} and β_{ac}, respectively, and are calculated from differentials. (The difference between β_{ac} and β_{DC} is very small, and the two are not usually distinguished.)

$$\beta_{ac} = \frac{\Delta I_C}{\Delta I_B} = \frac{i_c}{i_b} \qquad 43.31$$

$$\alpha_{ac} = \frac{\Delta I_C}{\Delta I_E} = \frac{i_c}{i_e} \qquad 43.32$$

Thermal (saturation) current is small but always present and can be included in Eq. 43.28. I_{CBO} is the thermal current at the collector-base junction.

$$I_C = I_E - I_B \qquad 43.33$$

$$I_C = \alpha I_E - I_{CBO} \approx \alpha I_E \qquad 43.34$$

Transistors are manufactured from silicon and germanium, although silicon transistors have a higher temperature operating range. While the collector cutoff current is very small at room temperature, it doubles every 10°C, rendering germanium transistors useless around 100°C. Silicon transistors remain useful up to approximately 200°C. While germanium has a lower collector-emitter saturation voltage and may outperform silicon in high-speed and high-frequency devices, silicon is nevertheless the material used for most semiconductor devices and integrated circuit systems.

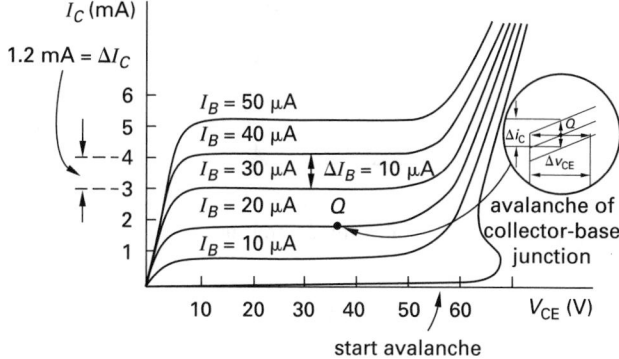

Figure 43.23 *Small Signal Terms*

Figure 43.23 illustrates a family of curves for various base currents. The DC amplification factor, β_{DC}, can be found (for a wide range of V_{CE} values) by taking a point on any line in the active (horizontal line) region and calculating the ratio of the coordinates, I_C/I_B. The small-signal amplification factor, β_{ac}, is calculated as the difference in the I_C between two I_B lines divided by the differences in I_B.

Many of the parameters used have multiple symbols. When given in manufacturers' data sheets or as equivalent parameters, h-parameter symbols are more common. The equivalence is given in Eqs. 43.35 through 43.38.

$$\alpha_{DC} = h_{FB} \qquad 43.35$$

$$\alpha_{ac} = h_{fb} \qquad 43.36$$

$$\beta_{DC} = h_{FE} \qquad 43.37$$

$$\beta_{ac} = h_{fe} \qquad 43.38$$

21. BJT TRANSISTOR CONFIGURATIONS

There are six ways a BJT transistor can be connected in a circuit, depending on which leads serve as input and output. Only three configurations have significant practical use, however. The terminal not used for either input or output is referred to as the *common terminal*. For example, in a *common emitter* circuit, the base receives the input signal and the output signal is at the collector. For the *common collector* (also known as an *emitter follower*) circuit, the input is to the base and the output is from the emitter. For a *common base* circuit, the input is to the emitter and the output is from the collector. The various configurations are shown in Fig. 43.24.[32] A summary of the relative properties of the configurations is given in Table 43.1.

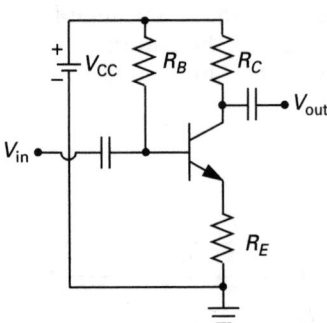

(a) common emitter

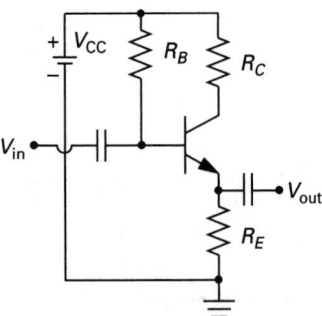

(b) common collector

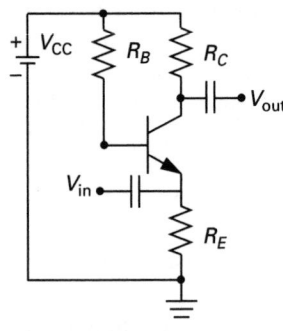

(c) common base

Figure 43.24 *Transistor Configurations*

The common emitter configuration is arguably the most versatile and useful. It is the only configuration with a voltage and current gain greater than unity. The common collector is widely used as a buffer stage. That is, it is used between a high-impedance source and a low-impedance load. The common base has the fewest applications, though it is sometimes used to match a

[32]Bypass capacitors are not shown.

low-impedance source with a high-impedance load, or to act as a noninverting amplifier with a voltage gain greater than unity.

Table 43.1 *Comparison of Transistor Configuration Properties*

symbol[a]	common emitter[a]	common collector[a]	common base
A_V	high (-100)	low (< 1)	high $(+100)$
A_I	high (-50)	high $(+50)$	low (< 1)
R_i	medium	high	low
	$(1000\ \Omega)$	$(> 100\ \text{k}\Omega)$	$(< 100\ \Omega)$
R_o	high	low	high
	$(\approx 50\ \text{k}\Omega)^b$	$(< 100\ \Omega)$	$(\approx 2\ \text{M}\Omega)^b$

[a]Approximate values in parentheses are based on a source and load resistance of 3 kΩ, $h_{\text{ie}} \approx 1$ kΩ, and $h_{\text{fe}} \approx 50$.
[b]The output resistance is often assumed to be infinite for this configuration, thus simplifying the transistor model.

22. BJT BIASING CIRCUITS

To maximize the transistor's operating range when large swing-signals are expected, the quiescent point should be approximately centered in the active region. The purpose of biasing is to establish the base current and, in conjunction with the load line, the quiescent point. Figure 43.25 illustrates several typical biasing methods: fixed bias, fixed bias with feedback, self-bias, voltage-divider bias, multiple-battery bias, and switching circuit (cutoff) bias. All of the methods can be used with all three common-lead configurations and with both *npn* (shown) and *pnp* transistors.[33] It is common to omit the bias battery (shown in Fig. 43.25(a) in dashed lines) in transistor circuits.

The base current can be found by writing Kirchhoff's voltage law around the input loop, including the bias battery, the external resistances, and the V_{BE} barrier voltage that opposes the bias battery. Since the base is thin, there is negligible resistance from the base to the emitter, and V_{BE} (being less than 1 V) may be omitted as well. For the case of *fixed bias with feedback* (*fixed bias with emitter resistance*) illustrated in Fig. 43.25(b), the base current is found from

$$V_{\text{CC}} = I_B R_B + V_{\text{BE}} + I_E R_E \quad [v_{\text{in}} = 0] \qquad 43.39$$

$$\begin{aligned} I_B &= \frac{V_{\text{CC}} - V_{\text{BE}} - I_E R_E}{R_B} \quad [v_{\text{in}} = 0] \\[1em] &= \frac{V_{\text{CC}} - V_{\text{BE}}}{R_B + \dfrac{\beta}{\alpha} R_E} \\[1em] &= \frac{V_{\text{CC}} - V_{\text{BE}}}{R_B + (1 + \beta) R_E} \quad [v_{\text{in}} = 0] \qquad 43.40 \end{aligned}$$

[33]The polarity of the DC supply voltages must be reversed to convert the circuits shown for use with *pnp* transistors.

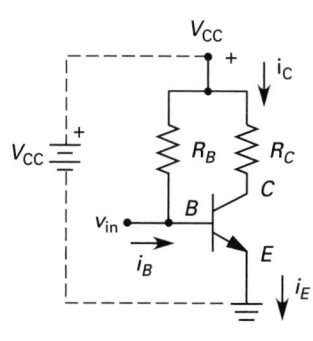

(a) fixed bias

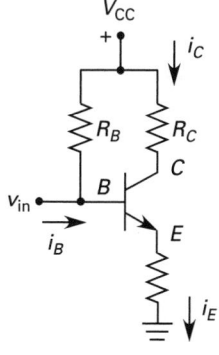

(b) fixed bias with series feedback

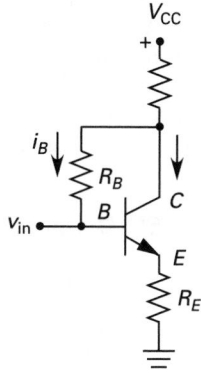

(c) self-bias with shunt feedback

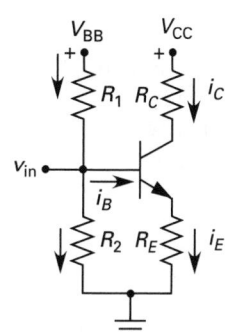

(d) voltage-divider bias

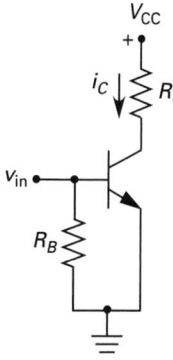

(e) multiple-battery bias

(f) cutoff bias

Figure 43.25 *DC Biasing Methods*

The first step in designing a fixed-bias amplifier with emitter resistance is choosing the quiescent collector current, I_{CQ}. R_E is selected so that the voltage across R_E is approximately three to five times the intrinsic V_{BE} voltage (i.e., 0.3 V for germanium and 0.7 V for silicon). Once R_E is found, R_B can be calculated from Eq. 43.39.

$$R_E \approx \frac{3V_{BE}}{I_{CQ}} \qquad 43.41$$

The *bias stability* (see the discussion of sensitivity in Sec. 63-5) with respect to any quantity M is given by Eq. 43.42. Variable M commonly represents temperature (T), current amplification (β), collector-base cutoff (I_{CBO}), and base-emitter voltage (V_{BE}).

$$S_M = \frac{\dfrac{\Delta I_C}{I_{CQ}}}{\dfrac{\Delta M}{M}} \qquad 43.42$$

The collector-base cutoff current—essentially, the reverse saturation current—doubles with every 10°C rise in temperature as mentioned in Secs. 43-9 and 43-20. The thermal stability of a transistor, then, is affected by this current. The effect is self-reinforcing: as the temperature increases, the saturation current increases, which further increases the temperature. This phenomenon is called *thermal runaway*. The emitter resistor used in the biasing circuits of Fig. 43.25 helps stabilize the transistor against this trend. As the current rises, the voltage drop across the emitter resistor rises in a direction that opposes the forward biased base-emitter junction. This decreases the base current, and so the collector current increases less than it would without the self-biasing resistor R_E.

23. BJT LOAD LINE

Figure 43.26 illustrates part of a simple common emitter transistor amplifier circuit. The bias battery and emitter and collector resistances define the *load line*. If the emitter-collector junction could act as a short circuit (i.e., $V_{CE} = 0$), the collector current would be $V_{CC}/(R_C + R_E)$. If the signal is large enough, it can completely oppose the battery-induced current, in which case the net collector current is zero and the full bias battery voltage appears across the emitter-collector junction. The intersection of the load line and the base current curve defines the *quiescent point*. The load line, base current, and quiescent point are illustrated in Fig. 43.26.

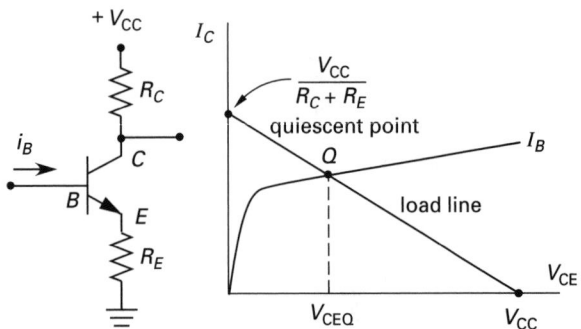

Figure 43.26 *Common Emitter Load Line and Quiescent Point*

Determination of the load line for a BJT transistor is accomplished by the following steps.

step 1: For the configuration provided, label the x-axis on the *output characteristic curves* with the appropriate voltage, V_{CE} or V_{CB}.

step 2: Label the y-axis as the output current, I_C.

step 3: Redraw the circuit with all three terminals of the transistor open. Label the terminals as the base, emitter, and collector. Label the current directions all pointing inward, that is, toward the transistor, I_B, I_E, and I_C.

step 4: Perform KVL analysis in the collector loop. The transistor voltage determined is a point on the x-axis with the output current equal to zero. Plot the point.

step 5: Redraw the circuit with all three terminals of the transistor shorted. Label as in step 3.

step 6: Use Ohm's law, or another appropriate method, in the collector loop to determine the current. The transistor current determined is a point on the y-axis with the applicable voltage in step 1 equal to zero. Plot the point.

step 7: Draw a straight line between the two points. This is the DC load line. The DC load line is used to determine the biasing and Q-point. The AC load line is determined in the same manner, but with active components, that is, inductors and capacitors, included. The AC load line is used to determine the transistor's response to small signals.

24. AMPLIFIER GAIN AND POWER

The *voltage-, current-, resistance-,* and *power-gain* are

$$A_V = \frac{\Delta V_{\text{out}}}{\Delta V_{\text{in}}} = \frac{v_{\text{out}}}{v_{\text{in}}} = \beta A_R \qquad 43.43$$

$$A_I = \frac{\Delta I_{\text{out}}}{\Delta I_{\text{in}}} = \frac{i_{\text{out}}}{i_{\text{in}}} = \beta \qquad 43.44$$

$$A_R = \frac{Z_{\text{out}}}{Z_{\text{in}}} = \frac{A_V}{\beta} \qquad 43.45$$

$$A_P = \frac{P_{\text{out}}}{P_{\text{in}}} = \beta^2 A_R = A_I A_V \qquad 43.46$$

The collector power dissipation, P_C, should not exceed the rated value. (This restriction applies to all points on the load line.) See Fig. 43.22(a) for a sample power restriction.

$$P_C = \tfrac{1}{2} I_C V_{CE} \qquad \text{[rms values]} \qquad 43.47$$

25. CASCADED AMPLIFIERS

Several amplifiers arranged so that the output of one is the input to the next are said to be *cascaded amplifiers*.[34] When each *amplifier stage* is properly coupled to the following, the overall gain is

$$A_{\text{total}} = A_{V,1} A_{V,2} A_{V,3} \cdots \qquad 43.48$$

---

[34]A cascade amplifier should not be confused with a *cascode amplifier* (a high-gain, low-noise amplifier with two transistors directly connected in common emitter and common base configurations).

Capacitors are used in amplifier circuits to isolate stages and pass small signals. This is known as *capacitive coupling*.[35] A capacitor appears to a steady (DC) voltage as an open circuit. However, it appears to a small (AC) voltage as a short circuit, and so input and output signals pass through, leaving the DC portion behind.

26. EQUIVALENT CIRCUIT REPRESENTATION AND MODELS

A transistor can be modeled as any of the equivalent two-port networks described in Sec. 29-31. The different transistor configurations are shown as two-port networks in Fig. 43.27.

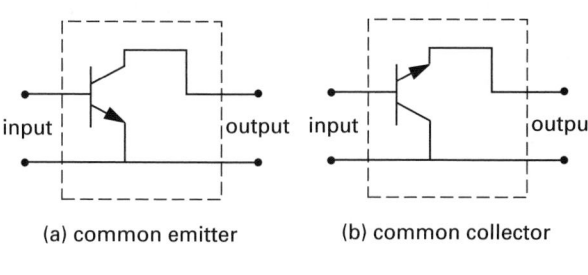

(a) common emitter   (b) common collector

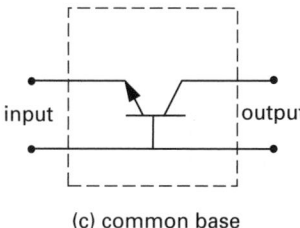

(c) common base

Figure 43.27 *Transistors as Two-Port Networks*

The equivalent parameters most often used are the hybrid parameters, also called the *h-parameters*. The *h*-parameters are defined for any two-port network in terms of Eqs. 43.49 through 43.51.

$$\begin{bmatrix} v_1 \\ i_2 \end{bmatrix} = \begin{bmatrix} h_{11} & h_{12} \\ h_{21} & h_{22} \end{bmatrix} \begin{bmatrix} i_1 \\ v_2 \end{bmatrix} \qquad 43.49$$

$$v_1 = h_{11}i_1 + h_{12}v_2 \qquad 43.50$$

$$i_2 = h_{21}i_1 + h_{22}v_2 \qquad 43.51$$

---

[35]The term *resistor-capacitor coupling* is also used.

Each of the *h*-parameters in Eqs. 43.50 and 43.51 is defined specifically for transistor models in the following way.

h_i = input impedance with output shorted (Ω)

h_r = reverse transfer voltage ratio with input open (dimensionless)

h_f = forward transfer current ratio with output shorted (dimensionless)

h_o = output admittance with input open (S)

Using the *h*-parameters so defined, Eqs. 43.52 and 43.53 become the governing equations for *small-signal circuit models*, also called *AC incremental models*.

$$v_i = h_i i_i + h_r v_o \qquad 43.52$$

$$i_o = h_f i_i + h_o v_o \qquad 43.53$$

Table 43.2 *Equivalent Circuit Parameters*

symbol	common emitter	common collector	common base
h_{11}, h_{ie}	h_{ie}	h_{ic}	$\dfrac{h_{ib}}{1+h_{fb}}$
h_{12}, h_{re}	h_{re}	$1-h_{rc}$	$\dfrac{h_{ib}h_{ob}}{1+h_{fb}}-h_{rb}$
h_{21}, h_{fe}	h_{fe}	$-1-h_{fc}$	$\dfrac{-h_{fb}}{1+h_{fb}}$
h_{22}, h_{oe}	h_{oe}	h_{oc}	$\dfrac{h_{ob}}{1+h_{fb}}$
h_{11}, h_{ib}	$\dfrac{h_{ie}}{1+h_{fe}}$	$\dfrac{-h_{ic}}{h_{fc}}$	h_{ib}
h_{12}, h_{rb}	$\dfrac{h_{ie}h_{oe}}{1+h_{fe}}-h_{re}$	$h_{rc}-\dfrac{h_{ic}h_{oc}}{h_{fc}}-1$	h_{rb}
h_{21}, h_{fb}	$\dfrac{-h_{fe}}{1+h_{fe}}$	$\dfrac{-1-h_{fc}}{h_{fc}}$	h_{fb}
h_{22}, h_{ob}	$\dfrac{h_{oe}}{1+h_{fe}}$	$\dfrac{-h_{oc}}{h_{fc}}$	h_{ob}
h_{11}, h_{ic}	h_{ie}	h_{ic}	$\dfrac{h_{ib}}{1+h_{fb}}$
h_{12}, h_{rc}	$1-h_{re}$	h_{rc}	1
h_{21}, h_{fc}	$-1-h_{fe}$	h_{fc}	$\dfrac{-1}{1+h_{fb}}$
h_{22}, h_{oc}	h_{oe}	h_{oc}	$\dfrac{h_{ob}}{1+h_{fb}}$

Table 43.3 BJT Equivalent Circuits

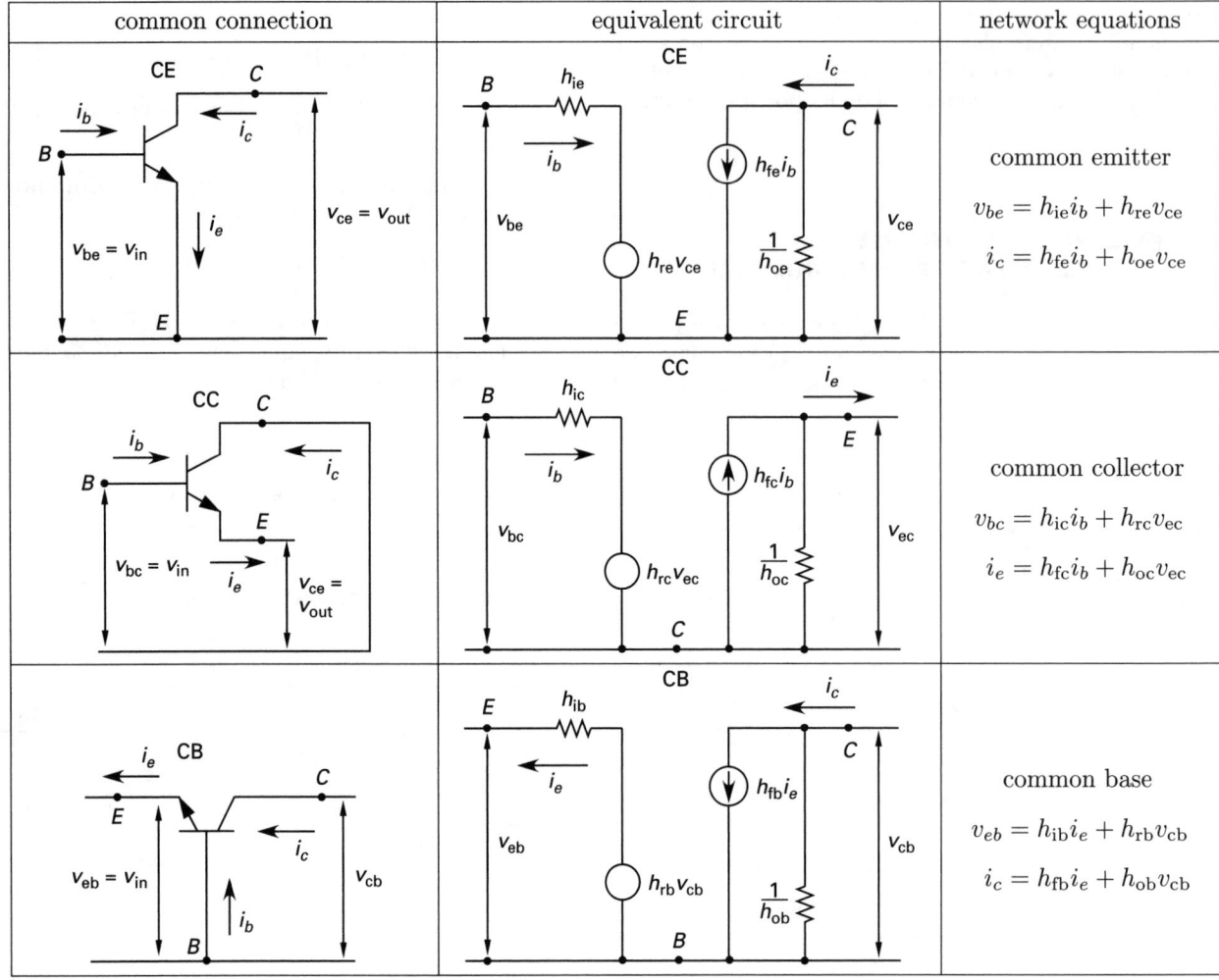

common connection	equivalent circuit	network equations
		common emitter $v_{be} = h_{ie}i_b + h_{re}v_{ce}$ $i_c = h_{fe}i_b + h_{oe}v_{ce}$
		common collector $v_{bc} = h_{ic}i_b + h_{rc}v_{ec}$ $i_e = h_{fc}i_b + h_{oc}v_{ec}$
		common base $v_{eb} = h_{ib}i_e + h_{rb}v_{cb}$ $i_c = h_{fb}i_e + h_{ob}v_{cb}$

Normally the *h*-parameters are given with two subscripts. The first is defined as in Eqs. 43.52 and 43.53. The second subscript indicates the configuration, such as common emitter (e or E), common collector (c or C), or common base (b or B). The *h*-parameters are normally specified for the common emitter configuration only. Table 43.2 shows the equivalence between the parameters for the different transistor configurations. The defining equations for the small-signal models are given in Table 43.3. Typical values for a widely used *npn* transistor, the 2N2222A, are given in Table 43.4.

Table 43.4 Typical h-Parameter Values

h-parameter	range of values
h_{ie}	0.25×10^3 to 8.0×10^3 Ω
h_{re}	4.0×10^{-4} to 8.0×10^{-4} [max values]
h_{fe}	50 to 375
h_{oe}	5.0×10^{-6} to 200×10^{-6} S

27. APPROXIMATE TRANSISTOR MODELS

The models used in Sec. 43-26 are exact. In many practical applications, sufficiently accurate results can be obtained with simplified models. The values of h_r and h_o are very small. That is, the reverse transfer voltage ratio and the output admittance are insignificant. The simplified models of Table 43.5 are obtained by ignoring these two parameters.

The simplified models assume that the output resistance is infinite, that is, $1/h_o \approx \infty$ (see Table 43.1). The simplified models further assume that the reverse voltage source, $h_r v$, is negligible. The reverse voltage accounts for the narrowing of the base width as the collector-base junction reverse bias increases. As the effectiveness of the base current in controlling the output is minimized, gain decreases. This explains the opposing voltage at the base-emitter junction. This phenomenon is called the *Early effect* (see Fig. 40.20(a)). The simplified models can also be shown with the input impedance, h_i, replaced with a voltage source equal to the barrier voltage (0.7 V for silicon and 0.3 V for germanium).

Table 43.5 *BJT Simplified Equivalent Circuits*

common connection	equivalent circuit	network equations
CE	CE	common emitter[a] $v_{\text{be}} = h_{\text{ie}}i_b \approx 0.7\text{ V}$ $i_c = h_{\text{fe}}i_b$
CC	CC	common collector $v_{\text{bc}} = h_{\text{ic}}i_b$ $i_e = h_{\text{fc}}i_b$
CB	CB	common base[a] $v_{\text{eb}} = h_{\text{ib}}i_e \approx 0.7\text{ V}$ $i_c = h_{\text{fb}}i_e$

[a]Germanium transistors are *pnp* types. $|v_{\text{be}}| = |v_{\text{eb}}| = 0.3$ for germanium.

28. HYBRID-π MODEL

Though the *h*-parameter model is common, another significant model is the *hybrid-π* or *Giacoletto* model. The hybrid-π model is shown in Fig. 43.28. The terms used in the model follow.

$r_{\text{bb}'}$ = *base spreading resistance,* or the *interbase resistance,* or the *small-signal base bulk resistance* (Ω)

$r_{\text{b}'\text{e}}$ = *small-signal base input resistance* (Ω)

$r_{\text{b}'\text{c}}$ = *small-signal feedback resistance* (Ω) This resistance accounts for the Early effect.

r_{ce} = *small-signal output resistance* (Ω)

g_m = transistor *transconductance* (S)

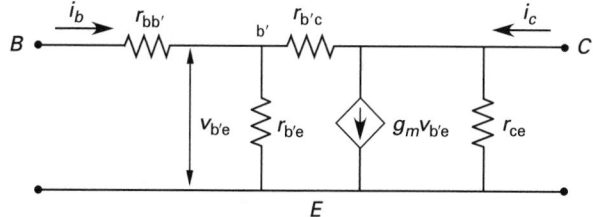

Figure 43.28 *Hybrid-π Model*

If the common emitter *h*-parameters are known, the hybrid-π parameters can be calculated from Eqs. 43.54 through 43.58, in the order given.

$$g_m = \frac{|I_C|}{V_T} \qquad \textbf{43.54}$$

$$r_{\text{b}'\text{e}} = \frac{h_{\text{fe}}}{g_m} \qquad \textbf{43.55}$$

$$r_{\text{bb}'} = h_{\text{ie}} - r_{\text{b}'\text{e}} \qquad \textbf{43.56}$$

Electronics

$$r_{b'c} = \frac{r_{b'e}}{h_{re}} \qquad 43.57$$

$$g_{ce} = h_{oe} - \frac{1 + h_{fe}}{r_{b'c}} \qquad 43.58$$

29. TRANSISTOR CIRCUIT LINEAR ANALYSIS

Because the small-signal, low-frequency response of a transistor is linear, it can be obtained analytically rather than graphically by using the models in Sec. 43-26.[36] The procedure for small-signal, or AC incremental, analysis follows.

step 1: Draw the circuit diagram. Include all external components, such as resistors, capacitors, and sources from the network.

step 2: Label the points for the base, emitter, and collector but do not draw the transistor. Maintain the points in the same relative position as in the original circuit.

step 3: Replace the transistor by the desired model. (Several models exist. Exact models were given in Sec. 43-26. Approximate models are given in Sec. 43-27. The hybrid-π model is given in Sec. 43-28.)

step 4: Since small-signal analysis is to be accomplished, only slight changes around the quiescent point are of interest. Therefore, replace each *independent* source by its internal resistance. An ideal voltage source is replaced with a short circuit. An ideal current source is replaced with an open circuit. (If biasing analysis were occurring, that is, large-signal analysis, the sources would remain as drawn.) Additionally, replace inductors with open circuits, and capacitors with short circuits.

step 5: Solve for the desired parameter(s) in the resultant circuit using Kirchhoff's current and voltage laws.

Example 43.6

A transistor is used in a common emitter amplifier circuit as shown. Assume the inductor has infinite impedance and the capacitors have zero impedance to AC signals. The transistor h-parameters are $h_{ie} = 750\ \Omega$, $h_{oe} = 9.09 \times 10^{-5}$ S, $h_{fe} = 184$, and $h_{re} = 1.25 \times 10^{-4}$.

(a) If a quiescent point is wanted at approximately $V_{CE} = 11$ V and $I_C = 10$ mA, what should the emitter resistance, R_E, be?

For parts (b) through (g), assume $R_E = 800\ \Omega$.

---

[36]The h-parameters must be considered complex functions of the frequency to analyze high-frequency circuits, so the h-parameter model is not used.

(b) Draw the DC load line. (c) If the base current (i_B) is 80×10^{-6} A, what is the collector current (i_C)? (d) What is the AC circuit voltage gain? (e) What is the AC circuit current gain? (f) What is the input impedance? (g) What is the output impedance? (h) Given $v_{in} = 0.25 \sin 1400t$ V, what is v_{out}? (i) What is the purpose of the inductor, L_C?

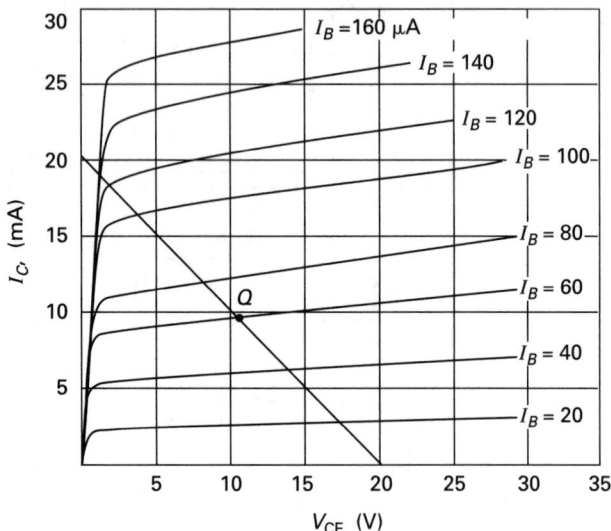

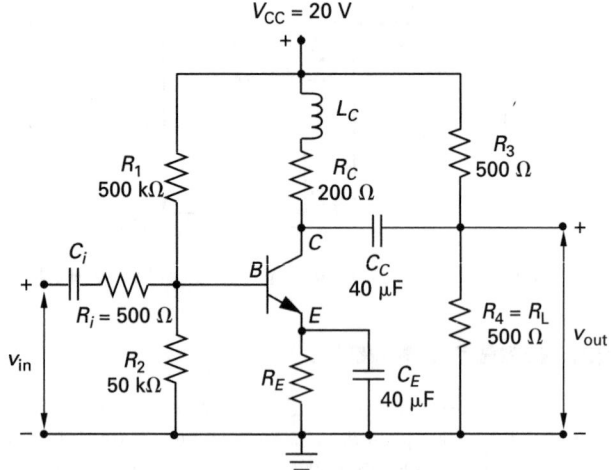

Solution

(a) Write the voltage drop in the common emitter circuit. Disregard the inductor (which passes DC signals). Use h_{fe} for h_{FE} since they are essentially the same and both are large. Kirchhoff's voltage law is

$$\alpha = \frac{\beta}{1 + \beta} \approx \frac{h_{fe}}{1 + h_{fe}} = \frac{184}{1 + 184} \approx 1$$

$$V_{CC} = I_C R_C + I_E R_E + V_{CE}$$

$$= I_C R_C + \left(\frac{I_C}{\alpha} \right) R_E + V_{CE}$$

$$\approx I_C (R_C + R_E) + V_{CE}$$

$$R_E = \frac{V_{CC} - V_{CE}}{I_C} - R_C$$

$$= \frac{20\text{ V} - 11\text{ V}}{10 \times 10^{-3}\text{ A}} - 200\ \Omega$$

$$\approx 700\ \Omega$$

(b) If $I_C = 0$, then $V_{CE} = V_{CC}$. This is one point on the load line. If $V_{CE} = 0$, then

$$I_C = \frac{V_{CC}}{R_C + R_E} = \frac{20\text{ V}}{200\ \Omega + 800\ \Omega} = 0.02\text{ A}$$

These two points define the DC load line.

(c) From Eq. 43.29,

$$i_C = \beta i_B = h_{fe} i_B$$

$$= (184)(80 \times 10^{-6}\text{ A}) = 14.7\text{ mA}$$

(d) To determine the AC circuit voltage gain, it is necessary to simplify the circuit. The bias battery V_{CC} is shorted out. (This is because the battery merely shifts the signal without affecting the signal swing.) Therefore, many of the resistors connect directly to ground. The inductor has infinite impedance, so R_C is disconnected. Both capacitors act as short circuits, so R_E is bypassed.

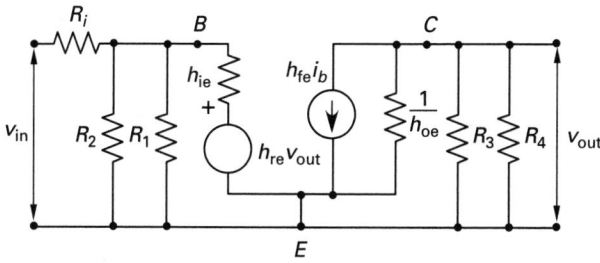

To simplify the circuit further, recognize that R_1, R_2, and $1/h_{oe}$ are very large and can be treated as infinite impedances.

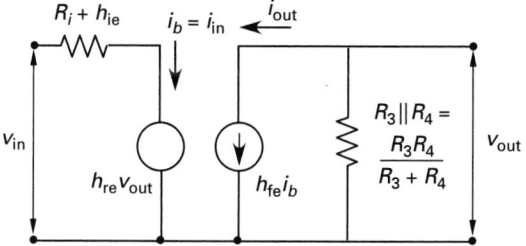

Continuing with the simplified analysis,

$$v_{out} = h_{fe} i_b \left(\frac{R_3 R_4}{R_3 + R_4} \right)$$

$$= (184 i_b) \left(\frac{(500)(500)}{500 + 500} \right) = 46{,}000 i_b$$

$$v_{in} = i_b(R_i + h_{ie}) + h_{re} v_{out}$$

$$= i_b(500 + 750) + (1.25 \times 10^{-4})(46{,}000 i_b)$$

$$= 1256 i_b$$

(To perform an exact analysis, h_{oe} and h_{re} must be considered. Convert R_i to its Norton equivalent resistance and place it in parallel with R_1, R_2, and h_{ie}. Use the current-divider concept to calculate i_b.)

The voltage gain is

$$A_V = \frac{v_{out}}{v_{in}} \approx \frac{46{,}000 i_b}{1256 i_b} = 36.6$$

(e) The current gain is

$$A_I = \frac{i_{out}}{i_{in}} \approx \frac{h_{fe} i_b}{i_b} = 184$$

(f) The input impedance (resistance) is

$$R_{in} = \frac{v_{in}}{i_{in}} \approx \frac{1256 i_b}{i_b} = 1256\ \Omega$$

(g) The output impedance is effectively the Thevenin equivalent resistance of the output circuit. The load resistance (R_4 in this instance) is removed. The independent source voltage (v_{in}) is shorted, which effectively opens the controlled source $h_{fe} i_b$. The remaining resistance between collector and ground is

$$R_{out} = R_3 = 500\ \Omega$$

(h) The output voltage is

$$v_{out} = A_V v_{in} = (36.6)(0.25)\sin 1400t$$

$$= 9.15 \sin 1400t\text{ V}$$

(i) There are several possible uses for the inductor. It might be included to limit voltage extremes, such as high-voltage spikes, which could damage the transistor when the amplifier is turned on or off. Alternatively, it might prevent AC current from being drawn across R_C, and, in so doing, hold V_C at a constant value.

30. TRANSISTOR CIRCUIT HIGH-FREQUENCY ANALYSIS

The h-parameter models are useful at low frequencies. For high frequencies, the hybrid-π model is modified to include the effects of various capacitances as shown in Fig. 43.29.

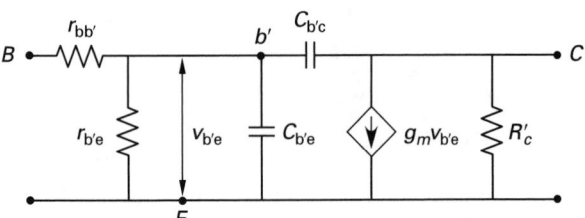

Figure 43.29 *High-Frequency Hybrid-π Model*

The capacitance $C_{b'e}$ is the sum of the diffusion capacitance, C_d, and the transition capacitance, C_t (see Sec. 43-9). The capacitance $C_{b'c}$ is the transition capacitance for the collector junction. The high-frequency model is used whenever the frequency exceeds the *corner frequency*, ω_{cf}, that is, the frequency at which the resistance is 3 dB less than its DC value.[37]

$$\omega_{cf} = \frac{1}{RC} \qquad 43.59$$

The capacitance term, C, in Eq. 43.59 is the total capacitance. The resistance, R, is as seen from the terminals of the total capacitance. (In low-frequency h-parameter models, the bypass capacitors are typically not shown. It is to these capacitors that Eq. 43.59 refers.)

31. UNIJUNCTION TRANSISTORS

A *unijunction transistor* is an electronic device with three contacts, two ohmic and one to a *pn* junction, as shown in Fig. 43.30(a). The output characteristics are illustrated in Fig. 43.30(b), which clearly shows a negative resistance region. The UJT operates in two states: on (low resistance) and off (high resistance). The symbol and equivalent circuit are shown in Figs. 43.30(c) and (d). The device is used in oscillator circuits, as a *UJT relaxation oscillator*, and in pulse generation and delay circuits.

32. DARLINGTON TRANSISTORS

A *Darlington transistor*, more commonly called a *Darlington pair*, is a current amplifier consisting of two separate transistors treated as one. A Darlington transistor has a high input impedance and a significant current gain. For example, the forward current transfer gain, h_{fe}, can be as high as 30,000. A Darlington transistor is illustrated in Fig. 43.31.

---

[37]When the frequency is $10\omega_{cf}$, the impedance is essentially the capacitive reactance. The resistance is considered insignificant at that point.

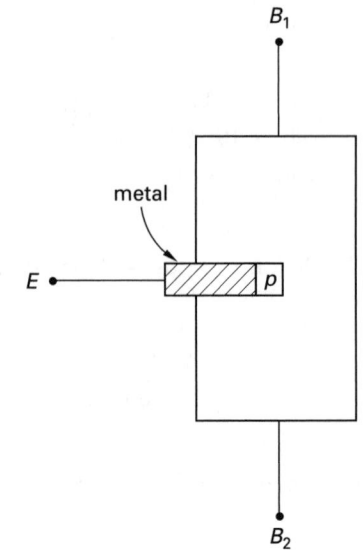

(a) conceptual construction

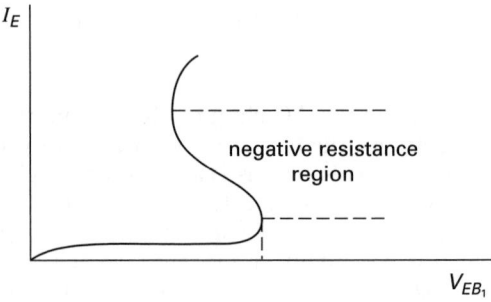

(b) characteristics

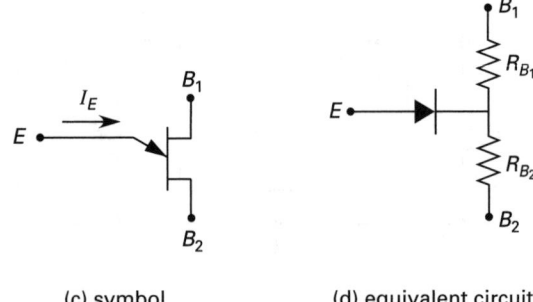

(c) symbol   (d) equivalent circuit

Figure 43.30 *Unijunction Transistor*

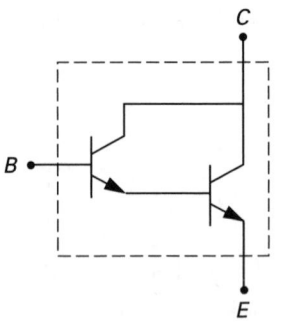

Figure 43.31 *Darlington Transistor*

33. FET FUNDAMENTALS

Field-effect transistors (FETs) are bidirectional devices constructed of an *n*-type channel surrounded by a *p*-type gate, or vice versa.[38] The connections are named in a different manner than BJTs in order to distinguish the two. For a FET (BJT), the connections are the *gate* (base), *source* (emitter), and *drain* (collector). In a BJT, the base current controls the overall operation of the transistor. In a FET, the gate voltage, and thus the corresponding electric field, controls the overall operation of the transistor. The various configurations and the concepts regarding biasing, load lines, and amplifier operation described for BJTs also apply to FETs.

There are two major types of FETs: the *junction field-effect transistor* (JFET) and the *metal-oxide semiconductor field-effect transistor* (MOSFET). Both types are made in *n*- and *p*-channel types, with the *n*-channel types more common. The fundamental difference between the JFET and the MOSFET is that the latter can operate in the enhancement mode (see Sec. 43-37).[39]

Variations in construction result in different names and properties. An *n*-channel MOSFET is sometimes referred to as an NMOS, and the *p*-channel MOSFET is sometimes called a PMOS. A high-power MOSFET is called an HMOS. A MOSFET with increased drain current capacity due to its V-type structure is called a VMOS. A double-diffused MOSFET, which has replaced the VMOS except in high-frequency applications, is called a DMOS. A bipolar transistor using an insulated gate is called an *insulated gate bipolar transistor* (IGBT). An SCR using an insulated gate for control is called a *MOS-controlled thyristor* (MCT). A MOSFET with *p*-channel and *n*-channel devices on the same chip is called a *complementary MOSFET* (CMOS) and is used primarily because of its low power dissipation.

When the conventional MOSFET oxide layer is replaced by a double layer of nitride and oxide, which allows electrons to tunnel from the gate, the device is called a *metal-nitride-oxide-silicon* (MNOS) transistor. The MNOS device is used as a read-only memory (ROM). MOS transistors are also used as memory devices. The MOS transistor is used to control the transfer of the charge from one location to the next. As a result, it is also called a *charge-transfer device* (CTD). When the charge transfer is accomplished on the circuit level by discrete MOS transistors and capacitors, it is called a *bucket-brigade device* (BBD). When the charge transfer takes place on the device level, it is called a *charge-coupled device* (CCD).

34. JFET CHARACTERISTICS

Junction field-effect transistors' (JFETs') construction, characteristics, and symbols are shown in Fig. 43.32.

---

[38]Bidirectional indicates that the current flow direction depends on the potential between the source and the drain.

[39]MOSFETs are also much more susceptible to damage from *electrostatic discharges* (ESD) due to the thin oxide layer at the gate.

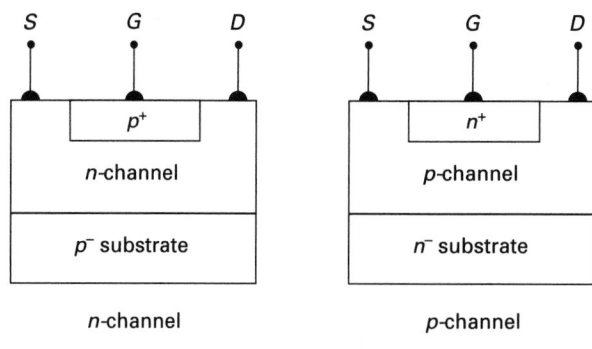

(a) conceptual construction

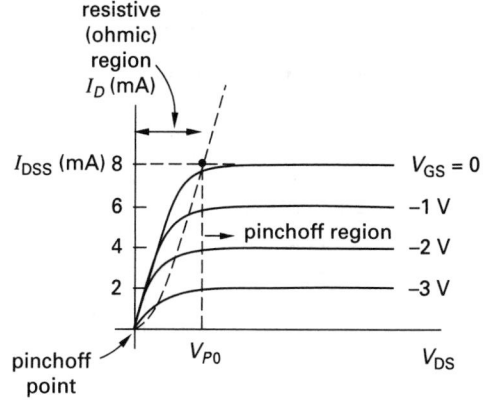

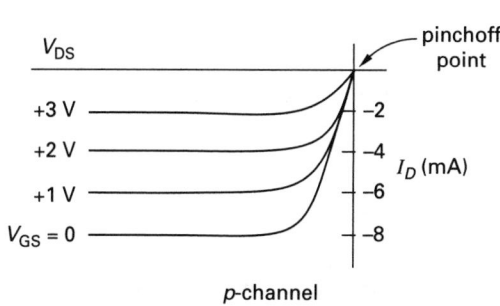

(b) characteristics

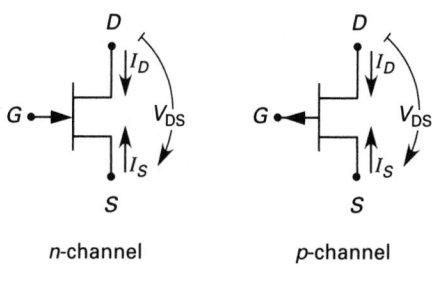

(c) symbol

Figure 43.32 *Junction Field-Effect Transistor*

A bidirectional channel for current flow exists between the source and the drain. The current flow is controlled by the reverse biased *pn* junction at the gate, with the depletion width providing the control.[40] For example, when the gate is unbiased, a normal depletion zone is set up at the gate pn junction, as shown in Fig. 43.33(a) for an n-channel JFET.

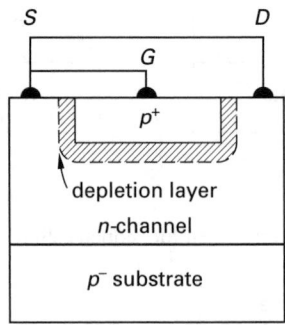

(a) unbiased depletion layer

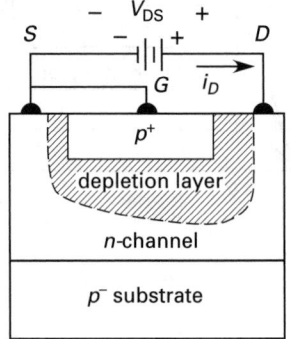

(b) ohmic region depletion layer

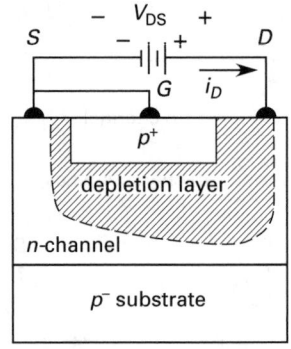

(c) pinchoff region depletion layer

Figure 43.33 Pinchoff Theory

Once biased, the depletion layer shifts, and the JFET operates in a relatively linear fashion in the *ohmic region*, also called the *triode region*. The ohmic depletion layer is shown in Fig. 43.33(b). The ohmic region is

---

[40]The source "emits" the n- or p-channel carriers, by convention. Thus, the reverse bias is with respect to the gate. FETs are bidirectional, nevertheless.

shown in Fig. 43.32(b). As the drain-source voltage increases, the depletion layer widens until it encompasses the drain terminal. At this point, the current is pinched off, and the JFET operates in a relatively constant manner in the *pinchoff region*. The pinchoff region depletion layer is shown in Fig. 43.33(c). The pinchoff region is shown in Fig. 43.32(b). As the drain-source voltage continues to increase, avalanche breakdown occurs and the voltage remains approximately constant as the current increases dramatically. This is known as the *breakdown* or *avalanche region* (not shown).

For a fixed value of V_{GS}, the drain-source voltage separating the resistive and pinchoff regions is the *pinchoff voltage*. As Fig. 43.32(b) shows, there is a value for V_{GS} for which no drain current flows. This is also referred to as the pinchoff voltage but is designated $V_{GS\,(off)}$. The drain current corresponding to the horizontal part of a curve (for a given value of V_{GS}) is the *saturation current*, represented by I_{DSS}.

The term "pinchoff voltage" and the symbol V_P are ambiguous, as the actual pinchoff voltage in a circuit depends on the gate-source voltage, V_{GS}. When V_{GS} is zero, the pinchoff voltage is represented unambiguously by V_{P0} (where the zero refers the value of V_{GS}). For other values of V_{GS}),

$$V_P = V_{P0} + V_{GS} \qquad \textit{43.60}$$

Some manufacturers do not adhere to this convention when reporting the pinchoff voltage for their JFETs. They may give a value for V_{P0} and refer to it as V_P. The absence of a value for V_{GS} implies that the value given is actually V_{P0}.

Some manufacturers do not provide characteristic curves, choosing instead to indicate only I_{DSS} and $V_{GS\,(off)}$. The characteristic curves can be derived, as necessary, from *Shockley's equation*.

$$I_D = I_{DSS}\left(1 - \frac{V_{GS}}{V_P}\right)^2 \qquad \textit{43.61}$$

The *transconductance*, g_m, is defined for small-signal analysis by Eq. 43.62.

$$
\begin{aligned}
g_m &= \frac{\Delta I_D}{\Delta V_{GS}} = \frac{i_D}{v_{GS}} \\[4pt]
&= \left(\frac{-2I_{DSS}}{V_P}\right)\left(1 - \frac{V_{GS}}{V_P}\right) = g_{mo}\left(1 - \frac{V_{GS}}{V_P}\right) \\[4pt]
&\approx \frac{A_V}{R_{out}} \qquad \textit{43.62}
\end{aligned}
$$

The drain-source resistance can be obtained from the slope of the V_{GS} characteristic in Fig. 43.32(b).

$$r_d = r_{DS} = \frac{\Delta V_{DS}}{\Delta I_D} = \frac{v_{DS}}{i_D} \qquad \textit{43.63}$$

35. JFET BIASING

JFETs operate with a reverse biased gate-source junction. The quiescent point is established by choosing V_{GSQ}. I_D is then determined by Shockley's equation (Eq. 43.61).

Figure 43.34 shows a JFET self-biasing circuit. Since the gate current is negligible, the load line equation is

$$V_{DD} = I_S(R_D + R_S) + V_{DS}$$
$$= I_D(R_D + R_S) + V_{DS} \qquad 43.64$$

The load line for any JFET is determined using the procedures outlined in Sec. 43-7.

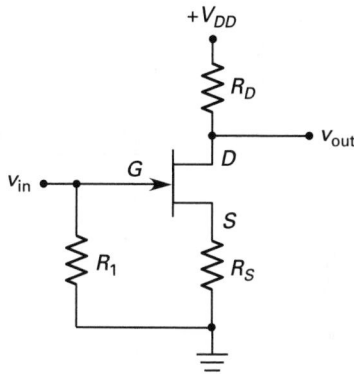

Figure 43.34 Self-Biasing JFET Circuit

At the quiescent point, $V_{in} = 0$. From Kirchhoff's voltage law, around the input loop,

$$V_{GS} = -I_S R_S = -I_D R_S \qquad 43.65$$

Example 43.7

A JFET with the characteristics shown operates as a small-signal amplifier. The supply voltage is 24 V. A quiescent bias source current of 5 mA is desired at a bias voltage of $V_{DS} = 15$ V. Design a self-biasing circuit similar to Fig. 43.34.

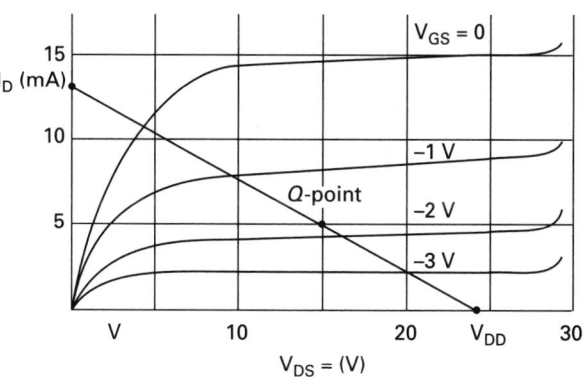

Solution

Since the gate draws negligible current, $I_D = I_S$. At the quiescent point, $V_{GS} = -1.75$ V. From Eq. 43.65,

$$R_S = \frac{-V_{GS}}{I_D} = \frac{-(-1.75 \text{ V})}{0.005 \text{ A}} = 350 \ \Omega$$

From Eq. 43.64,

$$R_D = \frac{V_{DD} - V_{DS}}{I_S} - R_S = \frac{24 \text{ V} - 15 \text{ V}}{0.005 \text{ A}} - 350 \ \Omega$$
$$= 1450 \ \Omega$$

36. FET MODELS

A field-effect transistor can be modeled as shown in Fig. 43.35. The model is valid for both the JFET and the MOSFET. Because the drain resistance is very high (see the typical values in Table 43.6), the model can be simplified further by removing r_{DS}, that is, by assuming $r_{DS} \approx \infty$.

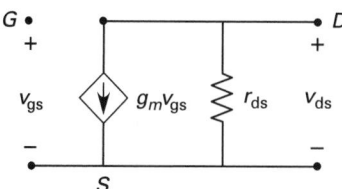

Figure 43.35 FET Equivalent Circuit

Table 43.6 Typical FET Parameter Values

parameter	JFET	MOSFET
g_m	0.1×10^{-3} to 10×10^{-3} S	0.1×10^{-3} to 20×10^{-3} S or greater
r_{ds}	0.1×10^{6} to 1×10^{6} Ω	1×10^{3} to 50×10^{3} Ω
C_{ds}	0.1×10^{-12} to 1×10^{-12} F	0.1×10^{-12} to 1×10^{-12} F
C_{gd}, C_{gs}	1×10^{-12} to 10×10^{-12} F	1×10^{-12} to 10×10^{-12} F

At high frequencies the various capacitances associated with the FET must be accounted for as shown in the model in Fig. 43.36.

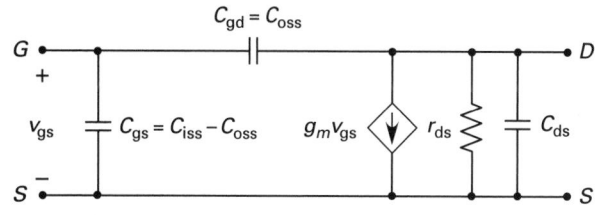

Figure 43.36 FET High-Frequency Model

37. MOSFET CHARACTERISTICS

The *metal-oxide semiconductor field-effect transistor* (MOSFET) is constructed as either a *depletion device* or an *enhancement device* as shown in Fig. 43.37(a). In the depletion device, a channel for current exists with the gate voltage at zero. In order to control the current, a voltage applied to the gate must push channel majority carriers away from the gate, thus pinching off the current. For an *n*-channel device, the gate voltage must be

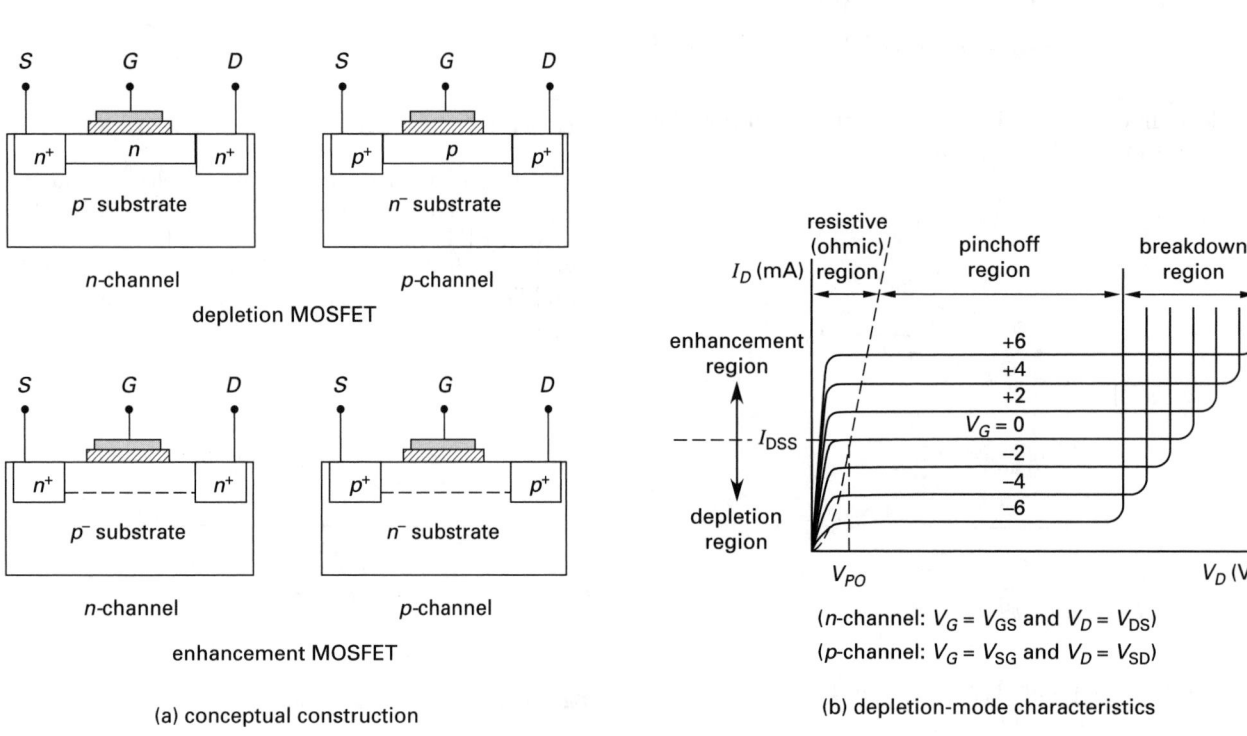

(a) conceptual construction

(b) depletion-mode characteristics

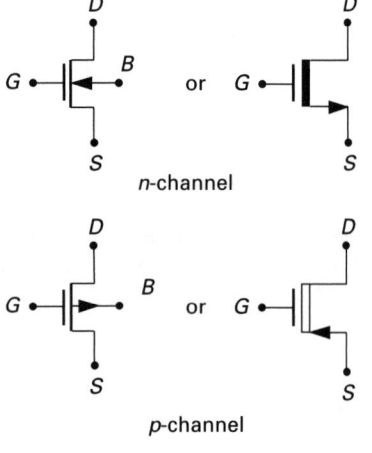

(c) depletion MOSFET symbols

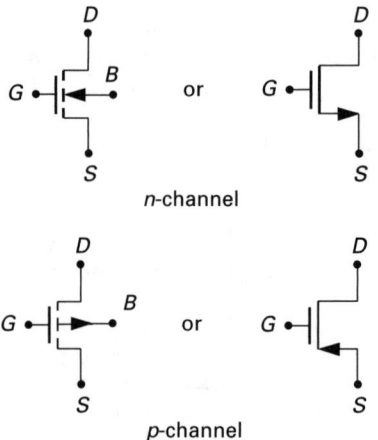

(d) enhancement MOSFET symbols

Figure 43.37 *Metal-Oxide Semiconductor Field-Effect Transistor*

negative. For a *p*-channel device, the gate voltage must be positive.[41] In an enhancement device, a channel for current does not exist with the gate voltage at zero, but must be created. In order to create the channel and control the current, a voltage must be applied to the gate to pull majority carriers toward the gate, thus increasing the current. For an *n*-channel device, the gate voltage must be positive. For a *p*-channel device, the gate voltage must be negative. A summary of these characteristics, which are similar to those for JFETs, is shown in Fig. 43.37(b). Three operating regions exist: ohmic, pinchoff, and breakdown. The various symbols used to represent MOSFETs are shown in Fig. 43.37(c) and (d). The MOSFET model is identical to that for the JFET (see Sec. 43-36).

38. MOSFET BIASING

A typical MOSFET biasing circuit is illustrated in Fig. 43.38. Since MOSFETs do not have a *pn* junction between the gate and channel, biasing can be either forward or reverse. The polarity of the gate-source bias voltage depends on whether the transistor is to operate in the depletion mode or enhancement mode. Since no gate current flows, the resistor R_G is used merely to control the input impedance. When R_G is very large, the input impedance is essentially R_2.

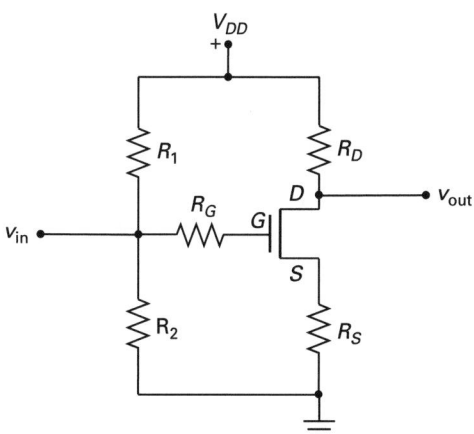

Figure 43.38 *Typical MOSFET Biasing Circuit*

Together, R_1 and R_2 form a voltage divider. Since the gate current is zero, the voltage divider is unloaded. Therefore, the gate voltage is

$$V_G = V_{DD}\left(\frac{R_2}{R_1 + R_2}\right) = V_{GS} + I_S R_S \qquad \textbf{43.66}$$

The load line equation is found from Kirchhoff's voltage law and is the same as for the JFET.

$$V_{DD} = I_S(R_D + R_S) + V_{DS} \qquad \textbf{43.67}$$

---

[41] The gate voltage is referenced to the source by convention, since the source is considered the emitter of majority carriers into the channel.

Example 43.8

A MOSFET is used in an amplifier as shown. All capacitors have zero impedance to AC signals. The no-signal drain current (I_D) is 20 mA. The performance of the transistor is defined by

$$I_D = (14 + V_{GS})^2 \qquad \text{[in mA]}$$

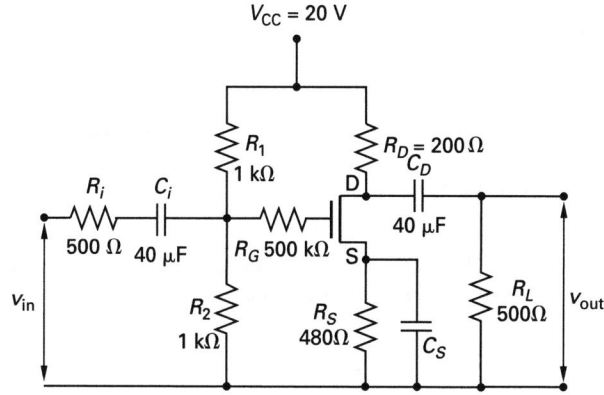

(a) What is the V_D (potential) at point D with no signal? (b) What is V_S with no signal? (c) What is V_{DSQ}? (d) If $V_G = 0$ V, what is V_{GS}? (e) What is the input impedance? (f) What is the voltage gain? (g) What is the output impedance?

Solution

(a)
$$V_D = V_{CC} - I_D R_D$$
$$= 20\ V - (20 \times 10^{-3})(200\ \Omega) = 16\ V$$

(b) The voltage drop between the source and the ground is through R_S. Since the gate draws negligible current, the drain and source currents are the same.

$$V_S = I_S R_S = (20 \times 10^{-3}\ A)(480\ \Omega) = 9.6\ V$$

(c) The quiescent voltage drop across the drain-source junction is

$$V_{DSQ} = V_D - V_S = 16\ V - 9.6\ V = 6.4\ V$$

(d) If the gate is at zero potential and the source is at 9.6 V potential, then

$$V_{GS} = V_G - V_S = 0\ V - 9.6\ V = -9.6\ V$$

(e) To simplify the circuit, short out the bias battery, V_{CC}. Consider the capacitors as short circuits to AC signals.

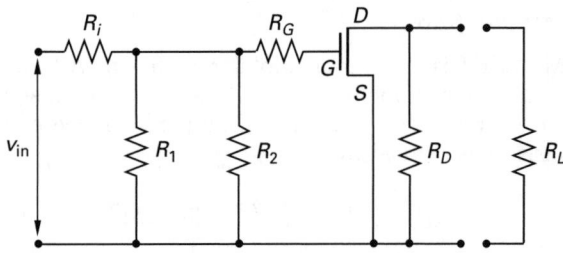

Since R_G is so large, it effectively is an open circuit. The input impedance (resistance) is

$$R_{\text{in}} = R_i + R_1 \parallel R_2 = 500 \ \Omega + \frac{(1000 \ \Omega)(1000 \ \Omega)}{1000 \ \Omega + 1000 \ \Omega}$$

$$= 1000 \ \Omega$$

(f) A FET has properties similar to a vacuum tube. The voltage gain is normally calculated from the transconductance.

$$A_V \approx g_m R_{\text{out}}$$

The transconductance is not known, but it can be calculated from Eq. 43.62 and the performance equation.

$$g_m = \frac{dI_D}{dV_{\text{GS}}} = \frac{d(14 + V_{\text{GS}})^2 \times 10^{-3}}{dV_{\text{GS}}}$$

$$= (2)(14 + V_{\text{GS}}) \times 10^{-3}$$

Since $V_{\text{GS}} = -9.6$ V at the quiescent point,

$$g_m = (2)(14 - 9.6) \times 10^{-3} = 8.8 \times 10^{-3} \ \text{S}$$

The resistance is a parallel combination of R_D, R_L, and r_d. The drain-source resistance, r_d, is normally very large and, since it was not given in this problem, is disregarded. Then,

$$R = R_D \parallel R_L = \frac{R_D R_L}{R_D + R_L}$$

$$= \frac{(200 \ \Omega)(500 \ \Omega)}{200 \ \Omega + 500 \ \Omega} = 143 \ \Omega$$

$$A_V = g_m R = (8.8 \times 10^{-3} \ \text{S})(143 \ \Omega) = 1.26$$

This is a small gain. If advantage is not being taken of other properties possessed by the circuit, the resistances should be adjusted to increase the voltage gain.

(g) Proceeding as in the solution to Ex. 43.6, part (g), the output impedance is $R_D = 200 \ \Omega$.

44 Amplifiers

Electronics

Nomenclature

A	gain	–
B, BW	bandwidth	Hz
d	displacement	m
f	frequency	Hz, s^{-1}, or cycles/s
F_m	figure of merit	rad/s or s^{-1}
G	gain	–
GBW	gain bandwidth	rad/s or s^{-1}
i	instantaneous current	A
I	effective or DC current	A
P	power	W
q	charge	1.602×10^{-19} C
rms	root-mean-square	–
R	resistance	Ω
S_R	slew rate	V/s
SNR	signal-to-noise ratio	dB
T	temperature	K
v	instantaneous voltage	V
V	effective or DC voltage	V
V_T	voltage equivalent of temperature	V
Z	impedance	Ω

Symbols

κ	Boltzmann constant	1.3805×10^{-23} J/K
ω	angular frequency	rad/s

Subscripts

0	resonance
c, C	collector
cm	common mode
CC	collector supply
dm	differential mode
DC	direct current
e, E	emitter
EE	emitter supply
f	feedback
i	current
L	low
m	merit
n	noise
p	power
ref	reference
rms	root-mean-square
R	rate

1. FUNDAMENTALS

An *amplifier* is any device capable of increasing the magnitude or power level of a physical quantity. The prime concern of most amplifiers is gain, A. The gain can be a voltage (A_V), current (A_i), or power gain (A_p or G_p). The input impedance (Z_{in}) and the output impedance (Z_{out}) are also parameters of concern. The *bandwidth* is the range of frequencies for which the amplifier will respond. The bandwidth (BW) is normally defined as the difference between the upper and lower frequencies at the *3 dB half-power points*, also called the *3 dB downpoints*, as shown in Fig. 44.1.

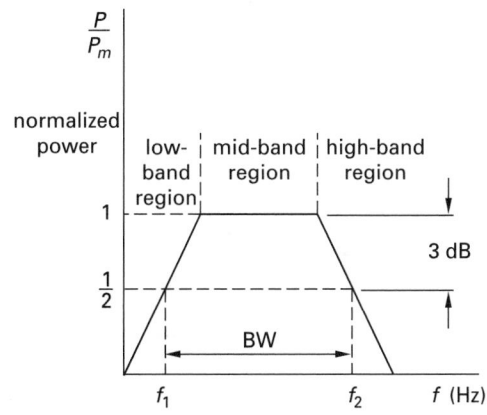

Figure 44.1 *Bandwidth*

The amplifier in Fig. 44.1 is a *band-pass* amplifier. There are also *low-pass*, *high-pass*, and *notch* amplifiers, though they are more commonly called *filters*. The gain-bandwidth product is used as a *figure of merit*, F_m, measured in rad/s, for band-pass and high-pass amplifiers. (Bandwidth is sometimes measured in Hz instead of rad/s, in which case the figure of merit is also given in Hz.) The *gain-bandwidth product* (GBW) for a band-pass amplifier is given by

$$\mathrm{GBW} = F_m = A_{\mathrm{ref}}(\mathrm{BW}) \qquad 44.1$$

The term A_{ref} is a reference gain, which is either the maximum gain or the gain at the frequency at which the gain is purely real or purely imaginary. For a low-pass amplifier, a high-frequency cutoff, ω_0, at the 3 dB downpoint is defined and the figure of merit is given by[1]

$$\text{GBW} = F_m = A_{\text{ref}}\omega_0 \qquad 44.2$$

The transistors described in Ch. 43, when biased in the linear region, are amplifiers. The small-signal models given are for the *mid-band region* of Fig. 44.1. In this region, the reactive elements act as short circuits and so are not included in any models.[2] In the *low-band region*, the impedance of the reactive elements, normally the coupling capacitors and emitter bypass capacitors, is significant.[3] Thus, capacitors must be added to the small-signal models in the low-band region, and the gain decreases.[4] In the *high-band region*, the impedance due to the capacitance of the device itself becomes significant, and the gain decreases. Thus, the device capacitances are added to the models (see Secs. 43-30 and 43-36).

Individual transistors amplify. When used in conjunction with other transistors and electronic devices, different types of amplifiers can be constructed. An amplifier with a wide range of applications is the *operational amplifier* (op amp). It takes its name from the multiple configurations with which mathematical operations may be performed. Originally designed as the major component of an analog computer, the op amp is a high-gain DC (direct-coupled or direct current) amplifier with high stability and immunity to oscillation, which uses large amounts of negative feedback. The symbology used for the op amp is shown in Fig. 44.2.

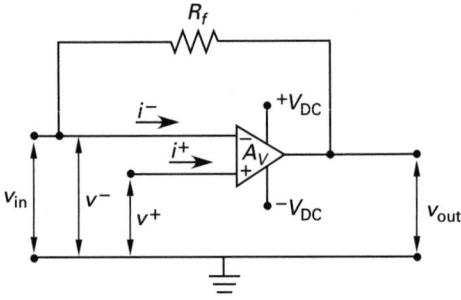

Figure 44.2 *Operational Amplifier Symbology*

The op amp has two input terminals. The one marked "−" is the *inverting terminal*; the one marked "+" is the *noninverting terminal*. The fundamental op amp formula is

$$v_{\text{out}} = A_V \left(v^+ - v^- \right) \qquad 44.3$$

---

[1]The 3 dB downpoint occurs at the resonant frequency.
[2]In *tuned circuits*, the reactive elements must be accounted for in the mid-frequency region to ensure proper design.
[3]"Significant" means the impedance seen at the terminals of the capacitor is of the same magnitude as the impedance of the capacitor itself.
[4]These models are not shown in Ch. 43. They would consist of the small-signal low-frequency model with coupling and emitter bypass capacitors added—the same circuit that would be used to determine the AC load line.

The voltages marked $+V_{\text{DC}}$ and $-V_{\text{DC}}$ are the power supplies to the op amp and are often called the *rail voltages*. The range of v_{out} is between these two voltages.[5] In fact, the output voltage should stay 3 V from either value to avoid distortion of the output signal at rated values. The range of the input signal, derived from Eq. 44.3 and this distortion restriction, is

$$\left| v^+ - v^- \right| < \frac{V_{\text{DC}} - 3\text{ V}}{A_V} \qquad 44.4$$

When the op amp input is within the range of Eq. 44.4, the operation is linear. Typical values of op amp voltage gain, A_V, range from 10^5 to 10^8. Power supply voltages, $\pm V_{\text{DC}}$, are in the range ± 10 to 15 V. Typical input impedances are greater than 10^5 Ω, while output impedances are very low. The internal circuits of the op amp consist of the items in the block diagram of Fig. 44.3.

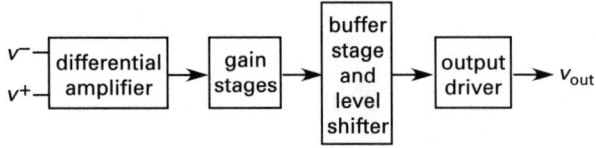

Figure 44.3 *Operational Amplifier Internal Configuration*

The *differential amplifier* constitutes the first stage of the op amp. A differential amplifier is shown in Fig. 44.4. The resulting output current, which is a combination of the emitter currents, is shown in Fig. 44.5. The linear region for operation restricts the voltage difference at the op amp input, as in Eq. 44.4, though here the difference is expressed in terms of the thermal voltage, V_T. Ideally, the differential amplifier amplifies only the difference between the signals. In practice, both the *differential-mode signal*, v_{dm}, and the *common-mode signal*, v_{cm}, are amplified.

$$v_{\text{dm}} = v^+ - v^- \qquad 44.5$$

$$v_{\text{cm}} = \tfrac{1}{2} \left(v^+ + v^- \right) \qquad 44.6$$

To measure how well the differential amplifier amplifies the difference signal and not the common signal, the concept of the *common-mode rejection ratio* (CMRR) is used. The CMRR is defined as

$$\text{CMRR} = \left| \frac{A_{\text{dm}}}{A_{\text{cm}}} \right| \qquad 44.7$$

The output voltage in terms of the CMRR is

$$v_{\text{out}} = A_{\text{dm}} v_{\text{dm}} \left(1 + \left(\frac{1}{\text{CMRR}} \right) \left(\frac{v_{\text{cm}}}{v_{\text{dm}}} \right) \right) \qquad 44.8$$

---

[5]Although the gain is large, the output cannot be greater than the available voltage. The rail voltages thus represent $\pm\infty$ for the output.

The gain stages in Fig. 44.3 supply the required amplification and represent cascaded amplifiers, for example, cascaded CE amplifiers. The buffer stage has a high input resistance to prevent it from loading the output driver, for example, an *emitter follower* (CC configuration). Circuitry associated with the buffer shifts the DC bias level so that the output is zero when the input is zero. The output driver provides the necessary large-signal current or voltage gain. For example, for a large current gain, the common collector configuration could be used (see Sec. 43-21).[6]

Op amps are designed to closely approximate ideal op amp behavior. Because of this and because of their versatility of use, they are widely used in place of discrete transistor amplifiers. Op amps are used in linear systems such as voltage-to-current converters, DC instrumentation, voltage followers, and filters; and in nonlinear systems such as AC/DC converters, peak detectors, sample-and-hold systems, analog multipliers, and analog-to-digital and digital-to-analog converters, among many others. A low-frequency op amp equivalent circuit is shown in Fig. 44.6.

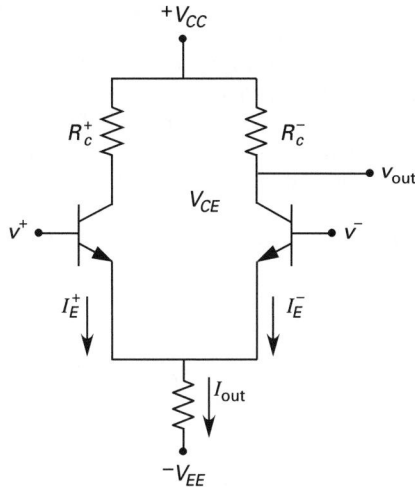

Figure 44.4 *Differential Amplifier*

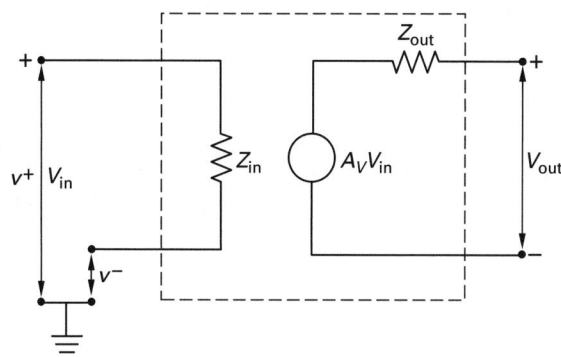

Figure 44.6 *Operational Amplifier Equivalent Circuit*

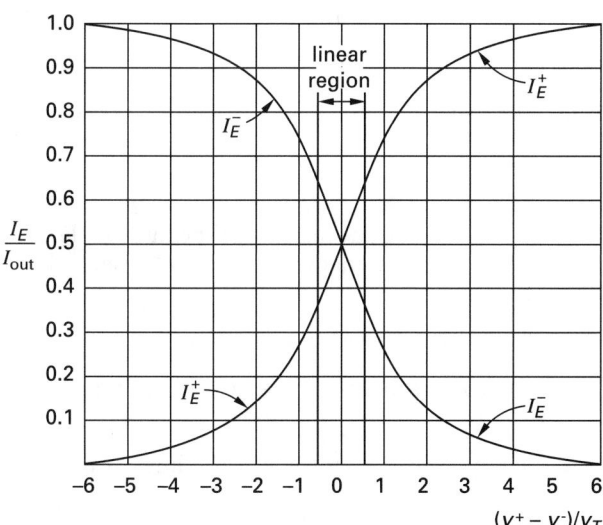

Figure 44.5 *Differential Amplifier Emitter Currents (Ideal Current Source Biasing)*

Example 44.1

An op amp has a voltage gain of 10^8. The DC power supply is ± 15 V. What is the maximum input voltage difference for linear operation?

Solution

The input voltage difference is given by Eq. 44.4.

$$\left| v^+ - v^- \right| < \frac{V_{DC} - 3}{A_V} = \frac{15 \text{ V} - 3 \text{ V}}{10^8}$$

$$= 0.12 \times 10^{-6} \text{ V}$$

$$= 0.12 \ \mu\text{V}$$

Thus, the difference between the terminals of the op amp must be kept within 0.12 μV to remain within the linear region. (This is accomplished with negative feedback.) This voltage difference is so small that the terminals are essentially at the same voltage. One of the assumptions for an ideal op amp will be that the voltages are the same. Consequently, if one of the terminals is connected to ground, the other terminal is also at ground potential. This situation is referred to as having a *virtual ground* at the ungrounded op amp terminal.

---

[6]The common collector (CC) configuration is also called the *emitter follower*. Voltage gain is approximately one. This means that a change in base voltage appears as an equal change across the load at the emitter. Consequently, the emitter follows the input signal.

Example 44.2

The input impedance for the op amp in Ex. 44.1 is 10^5 Ω. What is the input current?

Solution

The voltage difference was calculated as 0.12 μV. The current between the terminals is given by Ohm's law.

$$I = \frac{V}{R} = \frac{0.12 \times 10^{-6} \text{ V}}{10^5}$$

$$= 1.2 \times 10^{-12} \text{ A} = 1.2 \text{ pA}$$

This current is so low as to be negligible. An ideal op amp assumes that the input current is zero.

Example 44.3

If the gain given for Ex. 44.1 is the differential-mode gain and the CMRR is 10,000, what is the output voltage when the op amp is at the edge of linear operation and $v^+ = 50$ μV?

Solution

The op amp is at the edge of linear operation for the maximum voltage difference allowed by Eq. 44.4, which was calculated in Ex. 41.1 as 0.12 μV. The output voltage is given by Eq. 44.8.

$$v_{\text{out}} = A_{\text{dm}} v_{\text{dm}} \left(1 + \left(\frac{1}{\text{CMRR}} \right) \left(\frac{v_{\text{cm}}}{v_{\text{dm}}} \right) \right)$$

The differential-mode voltage was calculated as 0.12 μV. The common-mode voltage is unknown. From Eqs. 44.4 and 44.5,

$$v_{\text{dm}} = v^+ - v^- = 0.12 \text{ } \mu\text{V} = 50 \text{ } \mu\text{V} - v^-$$

$$v^- = 50 \text{ } \mu\text{V} - 0.12 \text{ } \mu\text{V} = 49.88 \text{ } \mu\text{V}$$

$$v_{\text{cm}} = \tfrac{1}{2} \left(v^+ + v^- \right)$$

$$= \tfrac{1}{2} \left(50 \text{ } \mu\text{V} + 49.88 \text{ } \mu\text{V} \right) \approx 50 \text{ } \mu\text{V}$$

Substituting the calculated and given values gives

$$v_{\text{out}} = A_{\text{dm}} v_{\text{dm}} \left(1 + \left(\frac{1}{\text{CMRR}} \right) \left(\frac{v_{\text{cm}}}{v_{\text{dm}}} \right) \right)$$

$$= \left(10^8 \right) \left(0.12 \text{ } \mu\text{V} \right) \left(1 + \left(\frac{1}{10{,}000} \right) \left(\frac{50 \text{ } \mu\text{V}}{0.12 \text{ } \mu\text{V}} \right) \right)$$

$$= 12.5 \text{ V}$$

This output voltage is within 3 V of the power supply voltage. Thus, distortion could occur due to the non-linear operation. In this case, since $A_{\text{dm}} v_{\text{dm}} = 12$ V, the negative feedback must be increased.

2. IDEAL OPERATIONAL AMPLIFIERS

An *ideal operational amplifier* exhibits the following properties.

- $Z_{\text{in}} = \infty$
- $Z_{\text{out}} = 0$
- $A_V = \infty$
- $\text{BW} = \infty$

Two other properties not always mentioned are (1) $V_{\text{out}} = 0$ when $v^+ = v^-$ regardless of the magnitude of the input voltage (normally v^-), and (2) characteristics do not drift with temperature.[7] Because the actual op amp so closely approximates these conditions, it is possible to use ideal op amp analysis for most calculations. Only when high-frequency behavior is desired or circuit limitations have been calculated do the assumptions need to be discarded. A typical ideal op amp configuration with *voltage shunt feedback* is shown in Fig. 44.7.

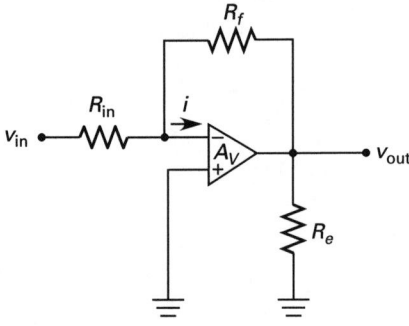

Figure 44.7 *Ideal Operational Amplifier Typical Configuration*

The assumptions regarding the properties of the ideal op amp result in the following practical results during analysis.

- The current to each input is zero.
- The voltage between the two input terminals is zero.
- The op amp is operating in the linear range.

The voltage difference of zero between the two terminals is called a *virtual short circuit*, or, because the positive terminal is often grounded, a *virtual ground*. The term "virtual" is used because, although the feedback from the output is used to keep the voltage difference between the terminals at zero, no current actually flows into the short circuit.

---

[7]Property (1) is for the op amp itself, *without feedback*. When negative feedback is used, the output voltage is whatever value maintains $v^+ = v^-$.

Electronics

$$\frac{v_{out}}{v_{in}} = \frac{A}{1 + AH}$$

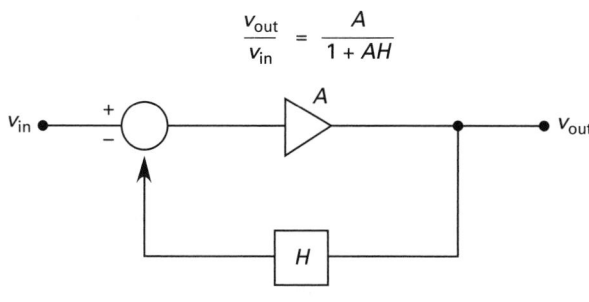

(a) feedback system

$$\frac{v_{out}}{v_{in}} = -\frac{R_f}{R_{in}}$$

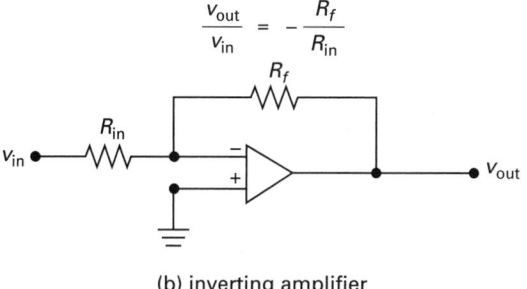

(b) inverting amplifier

$$\frac{v_{out}}{v_{in}} = \frac{R_f + R_{in}}{R_{in}}$$

(c) noninverting amplifier

$$v_{out} = -R_f\left(\frac{v_1}{R_1} + \frac{v_2}{R_2} + \frac{v_3}{R_3}\right)$$

(d) summing amplifier

$$v_{out} = \frac{-1}{RC}\int v_{in}\, dt$$

(e) integrator

$$v_{out} = -RC\frac{dv_{in}}{dt}$$

(f) differentiator

$$\frac{v_{out}}{v_{in}} = \frac{-R_f}{R_{in}(1 + j\omega R_f C)} \quad \text{[sinusoidal input]}$$

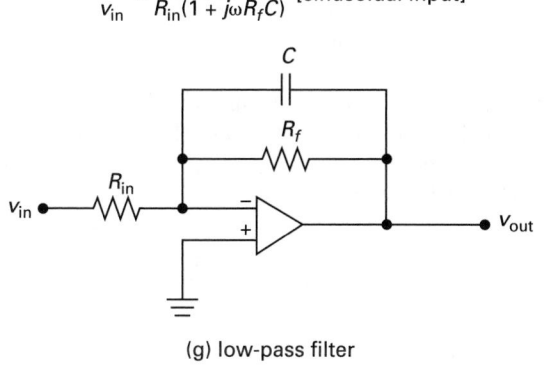

(g) low-pass filter

$$v_{out} = \alpha(v_2 - v_1)$$

$$\alpha = \frac{R_f}{R_1} = \frac{R_3}{R_2}$$

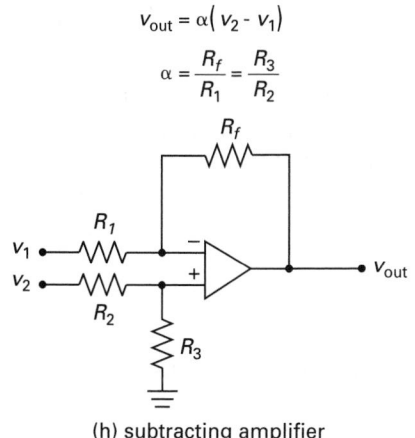

(h) subtracting amplifier

Figure 44.8 *Operational Amplifier Circuits*

The procedure for analyzing an ideal op amp circuit follows.

step 1: Draw the circuit and label all the nodes, voltages, and currents of interest.

step 2: Write Kirchoff's current law (KCL) at the op amp input node (normally the inverting terminal).[8]

step 3: Simplify the resulting equation using the ideal op amp assumptions. Specifically, the current into the op amp is zero and the voltage difference between the terminals is zero. Thus, the known voltage at one terminal must be the voltage at the other (e.g., if the positive terminal is at ground, the negative terminal voltage is zero).

step 4: Solve for the desired quantity or expression.

Using the procedure given, the relationships or values for any op amp circuit parameter can be found. The relationship between the output voltage and the input voltage for a variety of op amp circuits is shown in Fig. 44.8.

Example 44.4

Consider the circuit shown. The op amp is considered ideal with a 1 μV input signal. What is the current through the feedback resistor?

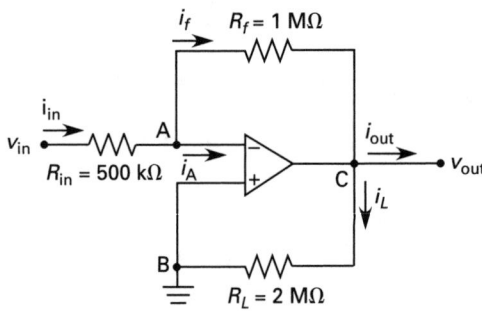

Solution

The circuit is already drawn. Writing KCL (assuming that currents out of the node are positive) at the input terminal of the op amp, that is, node A, gives[9]

$$-i_{\text{in}} + i_A + i_f = 0$$

$$i_f = i_{\text{in}} - i_A = \frac{v_{\text{in}} - V_A}{R_{\text{in}}} - i_A$$

---

[8]Methods other than KCL may be used, but this systematic method normally includes all the desired quantities and minimizes errors.
[9]The current into the node, i_{in}, is shown as negative to be consistent with the assumption that all currents into the node are negative and those out are positive. Ohm's law is written as shown to be consistent with the direction of the input current given.

Using the ideal op amp assumptions, the input current to the op amp is zero, that is, $i_A = 0$. Also, the voltage difference between the terminals is zero. Since the positive terminal is connected to ground, $v^+ = 0 = v^- = V_A$. Substituting gives

$$i_f = i_{\text{in}} - i_A = \frac{v_{\text{in}} - V_A}{R_{\text{in}}} - i_A$$

$$= \frac{1 \times 10^{-6} \text{ V} - 0 \text{ V}}{500 \times 10^3 \text{ }\Omega} - 0 \text{ A}$$

$$= 2 \times 10^{-12} \text{ A}$$

The negative sign indicates that the current direction is opposite that shown. Since i_A is zero, the input current is equal to 2×10^{-12} A.

Example 44.5

For the circuit in Ex. 44.4, determine the voltage gain.

Solution

The voltage gain is

$$A_V = \frac{v_{\text{out}}}{v_{\text{in}}}$$

The input voltage is known. The output voltage can be calculated in a number of ways. Using a portion of KCL at node A (or Ohm's law) gives

$$i_f = \frac{v_A - v_{\text{out}}}{R_f} = \frac{0 \text{ V} - v_{\text{out}}}{1 \times 10^6 \text{ }\Omega} = 2 \times 10^{-12} \text{ A}$$

$$v_{\text{out}} = (-1)\left(2 \times 10^{-12} \text{ A}\right)\left(1 \times 10^6 \text{ }\Omega\right)$$

$$= -2 \times 10^{-6} \text{ V}$$

Care should be taken with the negative signs. The feedback current i_f is positive because the direction shown in the example figure is used.[10] Substituting gives

$$A_V = \frac{v_{\text{out}}}{v_{\text{in}}} = \frac{-2 \times 10^{-6} \text{ V}}{1 \times 10^{-6} \text{ V}} = -2$$

The gain is negative because the input occurs at the inverting terminal. Note that the gain is determined by the resistors (since $|i_f| = |i_{\text{in}}|$) as can be seen by rewriting the gain formula.

$$A_V = \frac{v_{\text{out}}}{v_{\text{in}}} = \frac{-i_f R_f}{i_{\text{in}} R_{\text{in}}} = \frac{-R_f}{R_{\text{in}}}$$

Example 44.6

For the circuit in Ex. 44.4, determine the current through the load resistor.

---

[10]Recall that in this text positive currents flow out of the node and negative currents flow in (see Sec. 26-16).

Solution

Several possible methods exist for determining the solution. Since v_{out} is known and the direction of i_L is given, using Ohm's law at node C provides the answer directly.

$$i_L = \frac{v_{out} - v_B}{R_L} = \frac{-2 \times 10^{-6}\text{ V} - 0\text{ V}}{2 \times 10^6\ \Omega}$$
$$= -1 \times 10^{-12}\text{ A}$$

3. OPERATIONAL AMPLIFIER LIMITS

A real op amp has limits that must be accounted for during design or use. Those limits are summarized in this section.

Op amps have a finite bandwidth. Specifically, they have a finite gain-bandwidth product. The product of the gain and the bandwidth is essentially constant (see Eqs. 44.1 and 44.2). The first-order approximation of the frequency response of an amplifier is shown in Fig. 44.9.

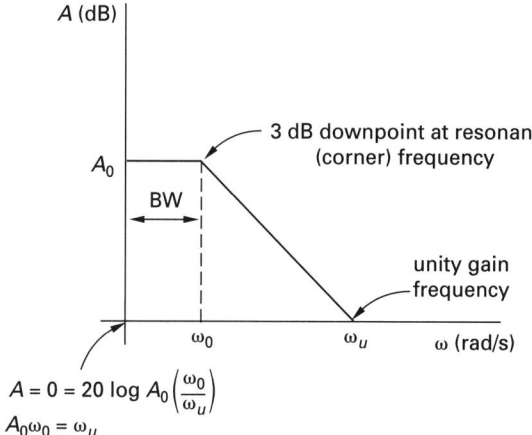

Figure 44.9 Operational Amplifier Frequency Response

The *unity gain* is defined as

$$|A| = A_0 \left(\frac{\omega_0}{\omega_u} \right) = 1 \qquad\qquad \textbf{44.9}$$

The term ω_u is called the *unity gain frequency* and replaces $A_0\omega_0$ in Eqs. 44.1 and 44.2. Thus, the unity gain frequency is used to define the bandwidth for a real op amp. As the frequency increases the gain decreases, keeping the gain bandwidth product (GBW), also called the figure of merit (F_m), approximately constant.[11]

---
[11] The frequency response of any amplifier circuit can be found by transforming the external electrical components into the *s* domain (see Sec. 31-10). Then obtain the gain equation using the procedures of Sec. 44-2. Finally, plot the response on a Bode plot (see Ch. 32).

The output voltage is limited to the range supply voltages. Further, to ensure operation in the linear region and prevent distortion, the output voltage should be approximately 3 V from the maximum supply voltage. This limit is expressed in terms of the gain and the voltage difference that can be used at the terminal inputs (see Eq. 44.4 and Fig. 44.5). When operating outside the linear region, the op amp is considered *saturated*.

The op amp amplifies both the difference signal and the common signal. Amplifying the common signal results in problems, especially for precision applications. The common-mode rejection ratio (CMRR) is a measure of the op amp's ability to overcome this obstacle (see Eqs. 44.7 and 44.8).

The gain is finite. As such, the output is less than expected (see Prob. 4). The voltage difference is determined from the fundamental op amp equation (Eq. 44.3), which is also called the *constraint equation*. Nevertheless, the gain is usually so high that this is a minor problem at low frequencies.

The input currents are small but not zero when the input is zero. A voltage difference does exist when the input is zero. These currents and voltages are called *offset currents* and *offset voltages*. They are compensated for within the op amp by the level shifter (see Sec. 44-1 and Fig. 44.3).

If a large signal is applied to an op amp, driving it into saturation, the output signal is limited to the maximum current that the amplifier can supply. This means that an incoming exponential signal would become a ramp signal on the output, and a sinusoidal signal would become a sawtooth signal on the output. The slopes of the output ramps are determined by the *slew rate*, S_R. The slew rate is determined by the internal capacitance of the amplifier and is given by

$$S_R = \left.\frac{dv}{dt}\right|_{max} \approx \frac{I_{max}}{C} \qquad\qquad \textbf{44.10}$$

4. AMPLIFIER NOISE

There are numerous sources of noise in amplifiers. The *thermal noise* is produced by the random thermal agitation of electrons. It is also called the *Johnson noise* or the *resistance noise*. The thermal noise is proportional to the root-mean-square of the noise voltage squared. The thermal noise is also called the *noise power*, P_n.

$$P_n = \frac{V_{noise,rms}^2}{4R} = \kappa T B \qquad\qquad \textbf{44.11}$$

The *shot noise* is due to the random variations in the velocity of the electrons in a current flow. It is proportional to $2qI$. *Low-frequency noise*, also called *flicker noise*, is the noise from slow changes, or drift, in the system parameters due to temperature, component aging, and so on. *Interference noise* is electromagnetic induction from sources outside the amplifier.

The total noise in a given circuit is compared to the incoming signal, or power, to ensure the ability of the circuit to discern between the two. The measure of this ability is the signal-to-noise ratio (SNR).

$$\text{SNR} = \frac{S}{N} = 10 \log \left(\frac{P_s}{P_n} \right) = 20 \log \left(\frac{V_s}{V_n} \right) \quad \textit{44.12}$$

Because the resistance is common to both the signal and the noise, the units of the signal power, P_s, and the noise power, P_n, can be expressed in watts or simply volts squared (in which case the noise power equals $4R\kappa TB$).[12] A signal-to-noise ratio of approximately 10 dB is required for the op amp output to reflect the signal.

---

[12]When ratios are involved, such as the signal-to-noise ratio, the only term remaining is the rms noise voltage squared, V^2. This is why the square of the voltage is sometimes referred to as the noise power.

45 Pulse Circuits: Waveform Shaping and Logic

Electronics

Nomenclature

A	gain	–
D	drain	–
G	gate	–
h_{fe}	CE small-signal (AC) forward current transfer ratio or gain	–
h_{ie}	CE small-signal (AC) input impedance	Ω
$h_{i,o,f,r}$ and $_{e,c,b}$	hybrid parameter[1]	–
$h_{I,O,F,R}$ and $_{E,C,B}$	hybrid parameter[2]	–
h_{oe}	CE small-signal (AC) open-circuit output admittance	S
h_{re}	CE small-signal (AC) reverse voltage transfer ratio or gain	–
H	high	V
i	instantaneous current	A
I	effective or DC current	A
I_{EBO}	DC emitter cutoff current, collector open	A
I_{ZK}	keep-alive current	A
L	low	V
N	fan-out	–
R	resistance	Ω
S	source	–
t	time	s
v	instantaneous voltage	V
V	effective or DC voltage	V
VR	voltage regulation	%
Z	impedance	Ω

Symbols

α	CB forward current ratio or gain	–
δ	feedback factor	–

Subscripts

amp	amplifier
b, B	base
c	control or collector
cl	current limiting
C	collector
CC	collector supply
CE	collector to emitter
co	cutoff
d	delay
D	diode
DD	drain supply
e, E	emitter
EBO	emitter to base, collector open
f	fall or forward (AC or instantaneous)
fl	full load
F	forward (DC component or total)
GS	gate to source
L	load
nl	no load
r	rise
R	resistance
ref	reference
s	supply or input, source, or storage

---

[1]The first subscript refers to the AC or small-signal input (i), output (o), forward (f), or reverse (r) value. The second subscript refers to the configuration, that is, e for common emitter (CE), c for common collector (CC), b for common base (CB).

[2]The first subscript refers to the DC or static input (I), output (O), forward (F), or reverse (R) value. The second subscript refers to the configuration, that is, E for common emitter (CE), C for common collector (CC), B for common base (CB).

v, V	voltage
Z	zener
ZK	zener knee
ZM	zener at maximum rated current

1. PULSE CIRCUIT FUNDAMENTALS

A *pulse circuit* is any active electrical network designed to respond to discrete pulses of current or voltage. Such circuits abound in computer and digital systems. Examples are waveform shaping circuits and logic circuits. An *active circuit*, such as a transistor circuit, is a circuit that is capable of amplifying a current or voltage signal. Waveform shaping is primarily accomplished with active nonlinear circuits. Active nonlinear elements are described in Ch. 43, though when used as switches, they are considered pulsed circuits—hence their coverage here. *Passive linear elements*, those elements that are not sources of energy (for example, resistors, capacitors, and inductors) are encountered in the design of pulse, waveform shaping, and logic circuits. Passive linear circuit elements are described in Chs. 29 through 31. *Passive nonlinear elements*, primarily diodes, are widely used for waveform shaping elements and elements of logic circuits. The delta (Dirac) function and the step function are used to mathematically describe the response of pulse circuits and are described in Ch. 15. *Pulse generators* are circuits, designed with high-speed switches, capable of generating very short-duration pulses for physically testing the response of a network to a delta-like input.[3] The ability to recognize the overall function of a pulse, waveform, or logic circuit is a practical skill required in electrical engineering. Detailed analysis of such circuits involves the application of knowledge of the individual components and requires significant time and specialization.[4]

2. CLAMPING CIRCUITS

A diode represents a purely resistive passive nonlinear element. The diode can be used as a simple means of limiting voltage to a desired value. When the voltage is limited in some manner, it is *clamped*. Two types of clamping exist: limiting and clipping. A *limiter* is an electronic circuit designed to prevent a waveform from exceeding a predetermined value. A *clipper* is an electronic circuit designed to maintain a waveform at a predetermined value when the input is above a certain level. The two types of circuits are shown in Fig. 45.1 for an ideal diode, that is, a diode with no resistance and a zero voltage drop. The ideal circuits can be used for many practical calculations and for determining overall circuit response.

---

[3] An *RC differentiator* with a switching transistor to control the application of the power is a pulse generator. The RC differentiator is an RC circuit with the output taken across the capacitor.

[4] This chapter, and indeed this text, focuses on the general skills required in electrical engineering. The tools for detailed analysis are presented, but the overall principles and methods will be the focus.

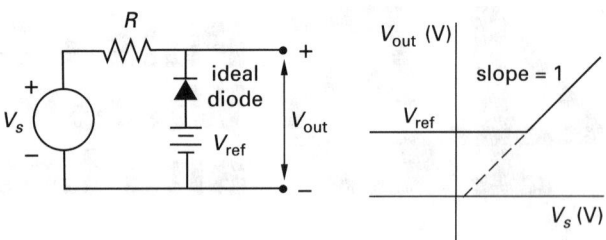

(a) shunt diode limiter circuit

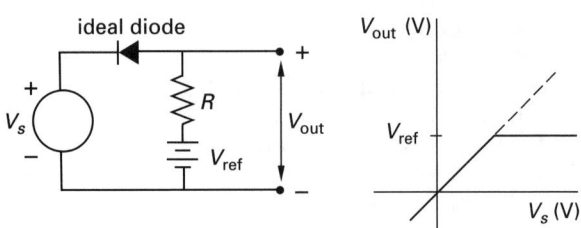

(b) series diode clipper circuit

Figure 45.1 *Ideal Diode Limiter and Clipper Circuits*

The diode is not an ideal device. The forward threshold voltage drop, V_F, for silicon diodes is approximately 0.7 V, while that for germanium diodes is approximately 0.2 V. Additionally, diodes exhibit some forward resistance, R_f, which accounts for the small but finite slope of the forward characteristic. (The reverse resistance or zener resistance is not a concern, since limiter and clipper circuits do not operate in the breakdown region.) The limiter and clipper circuits for an actual diode are shown in Fig. 45.2. Actual circuits are more complex than ideal circuits but produce more accurate results.

Analysis of the circuit in Fig. 45.2(a) yields Eqs. 45.1 and 45.2.

$$v_{\text{out}} = \frac{R(V_{\text{ref}} - V_F)}{R + R_f} + \frac{R_f V_{\text{in}}}{R_f + R} \quad [V_{\text{in}} < V_{\text{ref}} - V_F]$$

45.1

$$V_{\text{out}} = V_{\text{in}} \quad [V_{\text{in}} > V_{\text{ref}} - V_F] \qquad 45.2$$

Equation 45.1 represents the forward-biased case. Equation 45.2 represents the reverse-biased case, ignoring the small reverse saturation current. Analysis of the circuit in Fig. 45.2(b) yields Eqs. 45.3 and 45.4.

$$v_{\text{out}} = \frac{R_f V_{\text{in}}}{R_f + R} + \frac{R(V_{\text{ref}} + V_F)}{R + R_f} \quad [V_{\text{in}} > V_{\text{ref}} + V_F]$$

45.3

$$V_{\text{out}} = V_{\text{in}} \quad [V_{\text{in}} < V_{\text{ref}} + V_F] \qquad 45.4$$

Equation 45.3 represents the forward-biased case. Equation 45.4 represents the reverse-biased case, ignoring the small reverse saturation current.

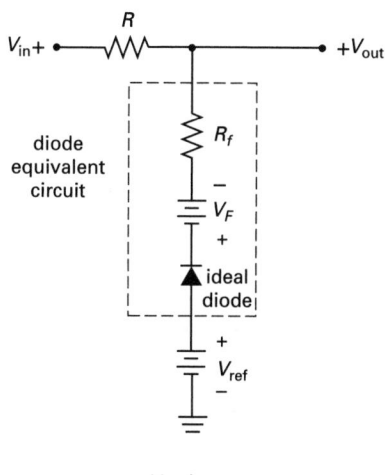

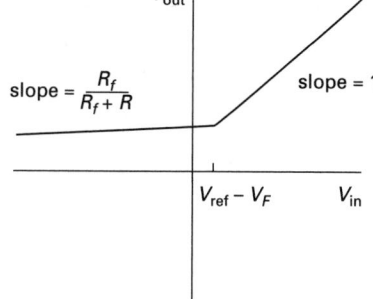

(a) limiter circuit

The ideal limiter or clipper circuit would have R_f equal to zero and the effect of the threshold voltage would be negligible. This type of circuit can be realized using a *precision diode*. A precision diode circuit, equivalent circuit, and characteristics are shown in Fig. 45.3. A precision diode is used to rectify very small signals: those less than 0.7 V for silicon or 0.2 V for germanium. In fact, the precision diode rectifies microvolt-level signals.

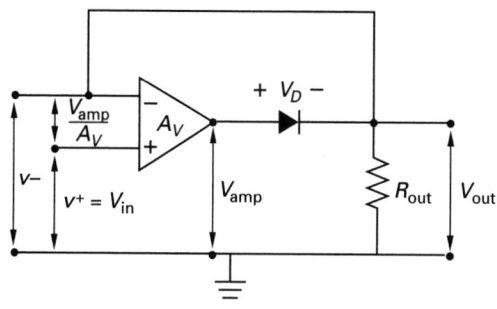

(a) precision diode

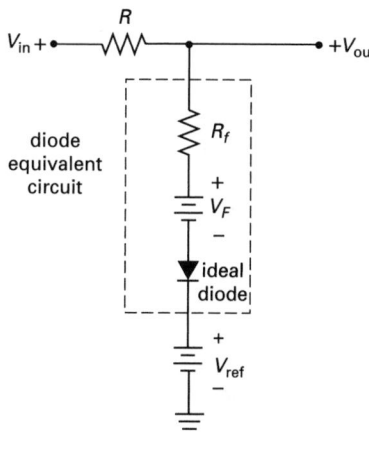

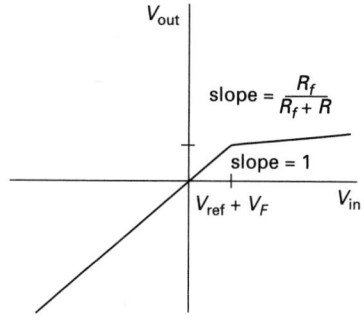

(b) clipper circuit

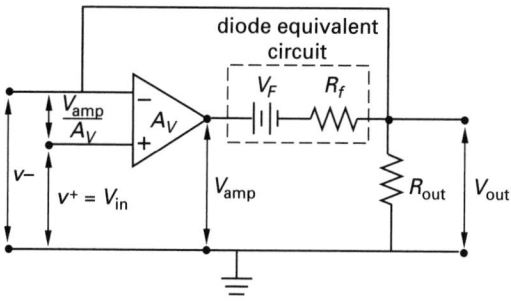

(b) forward-biased precision diode

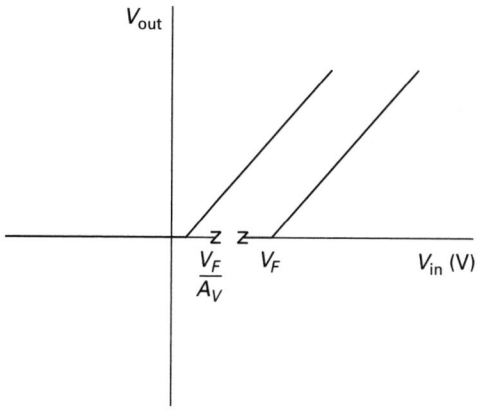

(c) characteristics

Figure 45.2 *Real Diode Limiter and Clipper Circuits*

Figure 45.3 *Precision Diode Circuit*

In Fig. 45.3(a), a diode is placed between the operational amplifier output and the feedback point of a voltage follower. For negative voltage inputs, $-v_{in}$, the op amp output voltage, V_{amp}, is negative and the diode is reverse biased.[5] The output voltage is then equal to zero since no current flows through the output resistor, R_{out}. For positive voltage inputs, $+v_{in}$, the op amp output voltage, V_{amp}, is equal to $A_V V_{in}$ until the diode becomes forward biased, that is, until the diode voltage, V_D, exceeds V_F. This forward-biased condition, which occurs when $A_V V_{in} = V_D$ or $V_{in} = V_D/A_V$, is shown in Fig. 45.3(b). Thus, from the forward-biased condition onward, the output voltage across R_{out} is the input voltage magnified by the op amp as shown in Fig. 45.3(c). The approximate transfer equation for the precision diode of Fig. 45.3(a) is

$$V_{out} = V_{in} - \frac{V_F}{A_V} \qquad \textit{45.5}$$

Precision diode circuits sensitive to negative voltages are shown in Fig. 45.4.

Example 45.1

Consider the precision diode in Fig. 45.3(b) with a 12 V power supply and a voltage gain of 10^6. What is the maximum voltage difference that can be amplified without distortion or saturation?

Solution

The maximum voltage difference that can be amplified is determined by the constraint equation (see Ch. 44).

$$v^+ - v^- = \frac{V_{amp}}{A_V}$$

The constraint equation is Eq. 45.5 with the diode removed, which changes V_F to V_{amp}. The maximum voltage of the amplifier is 9 V (3 V less than the maximum power supply voltage) to prevent distortion or saturation. Thus,

$$v^+ - v^- = \frac{V_{amp}}{A_V} = \frac{9 \text{ V}}{10^6} = 9 \ \mu\text{V}$$

---

[5]For proper operation, the diode must have a reverse breakdown voltage greater in magnitude than the maximum negative voltage of the op amp, that is, greater in magnitude than the negative supply voltage (not shown in the figure).

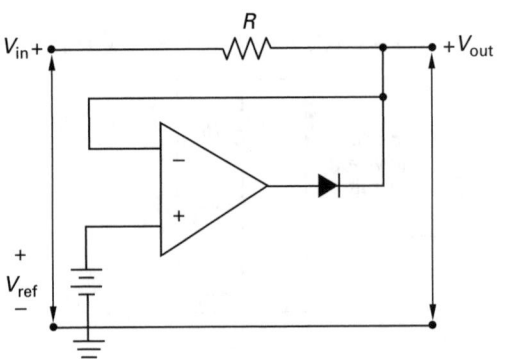

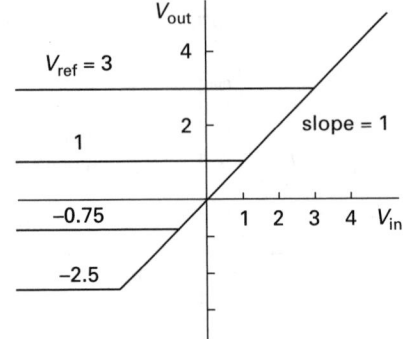

(a) precision limiter/clipper circuit

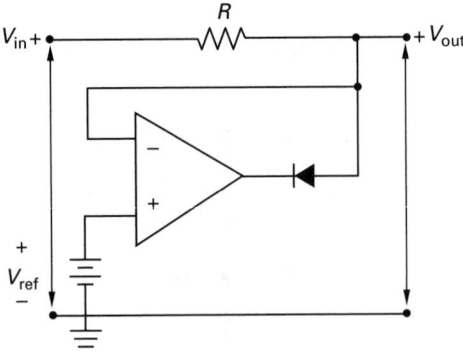

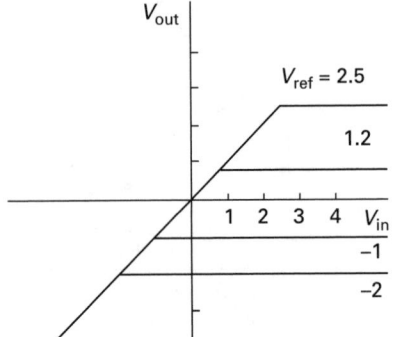

(b) precision limiter/clipper circuit sensitive to negative voltages

Figure 45.4 *Precision Diode Circuits Sensitive to Negative Voltage Input*

Example 45.2

Use the op amp in Ex. 45.1 with a silicon diode. What is the output voltage if the input voltage is 9 μV?

Solution

With the silicon diode, the threshold voltage is 0.7 V. Writing KCL at the negative terminal (ignoring the small current into the op amp) gives

$$\frac{v^- - 0}{R_{\text{out}}} + \frac{v^- + V_F - V_{\text{amp}}}{R_f} = 0$$

$v^- = V_{\text{out}}$, and $V_F = 0.7$ V. Substituting gives

$$\frac{V_{\text{out}}}{R_{\text{out}}} + \frac{V_{\text{out}} + 0.7 - V_{\text{amp}}}{R_f} = 0$$

The constraint equation must be used, rather than the assumption used with ideal op amps that $v^+ = v^-$, because the voltage response to an input is the unknown. That is, the transfer function must be determined. The constraint equation is

$$v^+ - v^- = V_{\text{in}} - V_{\text{out}} = \frac{V_{\text{amp}}}{A_V}$$

$$V_{\text{amp}} = A_V \left(V_{\text{in}} - V_{\text{out}} \right)$$

Substituting and rearranging gives the following result.

$$\frac{V_{\text{out}}}{R_{\text{out}}} + \frac{V_{\text{out}} + 0.7 - V_{\text{amp}}}{R_f} = 0$$

$$\frac{V_{\text{out}}}{R_{\text{out}}} + \frac{V_{\text{out}} + 0.7 - A_V (V_{\text{in}} - V_{\text{out}})}{R_f} = 0$$

$$\frac{V_{\text{out}}}{R_{\text{out}}} = -\frac{V_{\text{out}} + 0.7}{R_f} \cdot \frac{-A_V (V_{\text{in}} - V_{\text{out}})}{R_f}$$

$$\frac{V_{\text{out}}}{R_{\text{out}}} + \frac{V_{\text{out}}}{R_f} + \frac{A_V V_{\text{out}}}{R_f} = \frac{A_V V_{\text{in}} - 0.7}{R_f}$$

$$V_{\text{out}} \left(\frac{1}{R_{\text{out}}} + \frac{1}{R_f} + \frac{A_V}{R_f} \right) = \frac{A_V V_{\text{in}} - 0.7}{R_f}$$

$$V_{\text{out}} = \frac{A_V V_{\text{in}} - 0.7}{R_f \left(\dfrac{1}{R_{\text{out}}} + \dfrac{1}{R_f} + \dfrac{A_V}{R_f} \right)}$$

$$= \frac{V_{\text{in}} - \dfrac{0.7}{A_V}}{\dfrac{R_f}{A_V R_{\text{out}}} + \dfrac{1}{A_V} + 1}$$

The gain is very large. Thus, the transfer function reduces to Eq. 45.5, with 0.7 substituted for V_F.

$$V_{\text{out}} = V_{\text{in}} - \frac{0.7}{A_V}$$

Substituting the input voltage of 9 μV and the gain of 10^6 gives

$$V_{\text{out}} = V_{\text{in}} - \frac{0.7}{A_V} = 9 \times 10^{-6} \text{ V} - \frac{0.7 \text{ V}}{10^6}$$

$$= 8.3 \times 10^{-6} \text{ V} = 8.3 \ \mu\text{V}$$

The offset voltage of 0.7 V for silicon has been reduced by the precision diode to 0.7×10^{-6} V, or 0.7 μV, which is essentially zero.

Example 45.3

Design a circuit that has the following output characteristic.

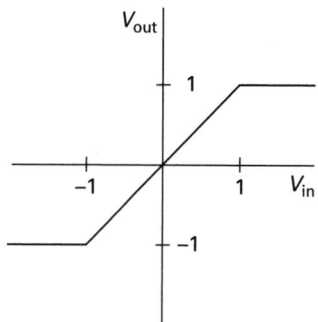

This circuit linearly transfers any signal of less than 1.0 V in magnitude, but clips all signals greater than 1.0 V in magnitude. The power supply voltage available is ±12.0 V.

Solution

A precision diode circuit set up as voltage followers exhibits the characteristics given. For the positive voltages, a clipper circuit similar to that in Fig. 45.4(b) can be used with the reference voltage at the noninverting terminal set to 1.0 V by a voltage divider. For the negative voltages, a limiter circuit similar to that in Fig. 45.4(a) can be used with the reference voltage at the noninverting terminal set to −1.0 V by a voltage divider. The resulting circuit follows.

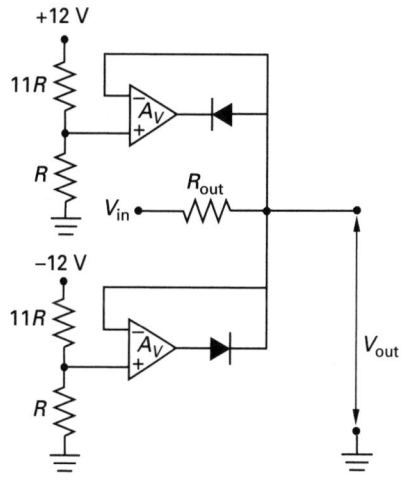

3. ZENER VOLTAGE REGULATOR CIRCUITS: IDEAL

Zener diodes operating in the breakdown region are used as voltage regulators. That is, they are used to control the load voltage (output voltage) within a narrow set of values in circuits where the supply voltage (input voltage) varies and the load current ranges from zero to a maximum value.

An ideal zener diode voltage regulator circuit is shown in Fig. 45.5(a) with the model for the ideal zener illustrated in Fig. 45.5(b). The zener will maintain a constant voltage as long as the zener current, I_Z, is positive as shown in Fig. 45.5(c).

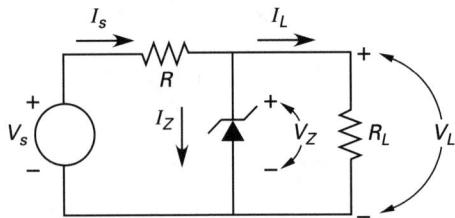

(a) regulator circuit

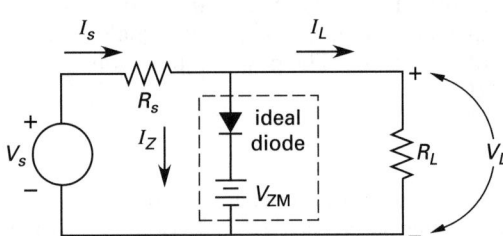

(b) regulator circuit with ideal zener model

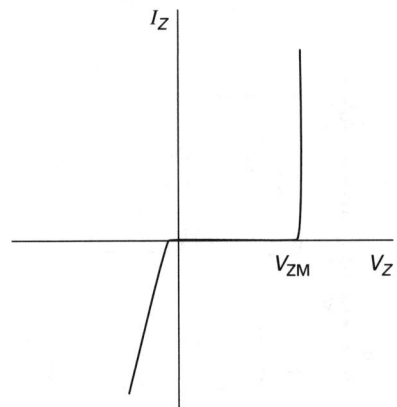

(c) ideal zener characteristics

Figure 45.5 *Ideal Zener Diode Voltage Regulator*

The design restrictions can be understood from Eqs. 45.6 and 45.7, or derived from Ohm's law and the application of KCL to Fig. 45.5(b).

$$I_s = \frac{V_s - V_Z}{R_s} \qquad 45.6$$

$$I_Z = I_s - I_L \qquad 45.7$$

As the load current, I_L, varies, the supply current given by Eq. 45.6 remains constant, but, according to KCL as stated in Eq. 45.7, the zener current will vary to compensate. Thus, as the load current increases the zener diode current decreases. However, to keep the ideal diode forward biased, that is, to keep the zener diode in breakdown, the zener current must be zero or positive.[6] That is, the load voltage, which equals the zener voltage, must be equal to V_{ZM}. To minimize the total power consumption, the value of R_s selected should be as small as possible while still compensating for changes in supply voltage and load current. The combination of these requirements gives

$$R_s = \frac{V_{in,min} - V_{ZM}}{I_{L,max}} \qquad 45.8$$

$$I_{Z,max} = \frac{V_{in,max} - V_{ZM}}{R_s}$$

$$= \left(\frac{V_{in,max} - V_{ZM}}{V_{in,min} - V_{ZM}} \right) I_{L,max} \qquad 45.9$$

Equation 45.8 represents the requirement that at the maximum possible load current, no negative current flows in the ideal diode, with an equal sign used to minimize the value of R_s.[7] The phrase "no negative current" means that the minimum supply voltage, $V_{in,min}$, is greater than or equal to V_{ZM}. Equation 45.8 represents the maximum current drawn by the zener and thus sets the power requirements for the diode and the supply resistor, given by Eqs. 45.10 and 45.11, respectively.

$$P_D = I_{Z,max} V_{ZM} \qquad 45.10$$

$$P_{R_s} = I_{Z,max}^2 R_s \qquad 45.11$$

An ideal zener diode voltage regulator circuit is designed as follows.

step 1: Determine the supply resistance that minimizes power loss but keeps the zener operating in the breakdown region (see Eq. 45.8). This step accounts for the minimum supply

---

[6]In a practical zener diode, the current must be greater than the keep-alive current, I_{ZK}.

[7]The actual requirement is that the supply resistance be less than or equal to the right-hand side of Eq. 45.8.

(input) voltage and the maximum load (output) current.

step 2: Determine the maximum zener current drawn with the supply resistance calculated in step 1 (see Eq. 45.8). This step accounts for the maximum supply (input) voltage and the minimum load (output) current.

step 3: Determine the required diode power rating with the current calculated in step 2 (see Eq. 45.10).

step 4: Determine the required resistor power rating with the current calculated in step 2 (see Eq. 45.11).

Example 45.4

Design a zener voltage regulator circuit, patterned after Fig. 45.5(a). The output voltage requirement is 5 V, while the input varies from 10 to 15 V and the load current ranges from 0 to 5 A.

Solution

Following the design steps, the supply resistance is given by

$$R_s = \frac{V_{\text{in,min}} - V_{\text{ZM}}}{I_{L,\text{max}}} = \frac{10 \text{ V} - 5 \text{ V}}{5 \text{ A}} = 1 \text{ }\Omega$$

The maximum zener current drawn is

$$I_{Z,\text{max}} = \frac{V_{\text{in,max}} - V_{\text{ZM}}}{R_s} = \frac{15 \text{ V} - 5 \text{ V}}{1 \text{ }\Omega} = 10 \text{ A}$$

The diode power rating is

$$P_D = I_{Z,\text{max}} V_{\text{ZM}} = (10 \text{ A})(5 \text{ V}) = 50 \text{ W}$$

The resistor power rating must be at least

$$P_{R_s} = I_{Z,\text{max}}^2 R_s = (10 \text{ A})^2 (1 \text{ }\Omega) = 100 \text{ W}$$

Ideal zener diode regulator calculations can be used as first approximations of actual circuits. The inherent assumptions made are that the keep-alive current, I_{ZK}, equals zero and that the zener resistance, R_Z, is negligible (see Sec. 45-4).

4. ZENER VOLTAGE REGULATOR CIRCUIT: PRACTICAL

A practical (equivalent) zener diode voltage regulator circuit, illustrated in Fig. 45.6, differs from the ideal case in a number of ways. First, the voltages that can be used are standardized. Appendix 45.A lists some of the standard zener diodes and their respective zener impedances, Z_Z, and keep-alive currents, I_{ZK}. Second,

equivalent zener diodes exhibit some resistance, R_Z, as shown in the equivalent model of Fig. 45.6(b). This resistance, R_Z, can cause the zener voltage to vary as much as 20% from the stated value at rated current. The effect is significant for zener diodes rated for 8 V or less, as the zener effect dominates. Third, the zener diode requires a small but finite current called the keep-alive current, I_{ZK}, to remain in the breakdown region. The zener current must be kept above the keep-alive current for the voltage-regulating effect to occur.

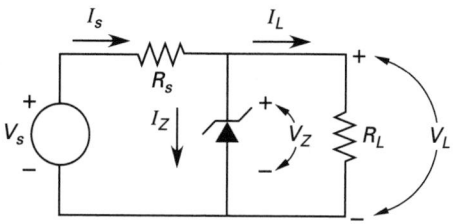

(a) regulator circuit

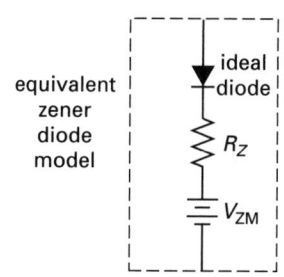

(b) regulator circuit with equivalent zener model

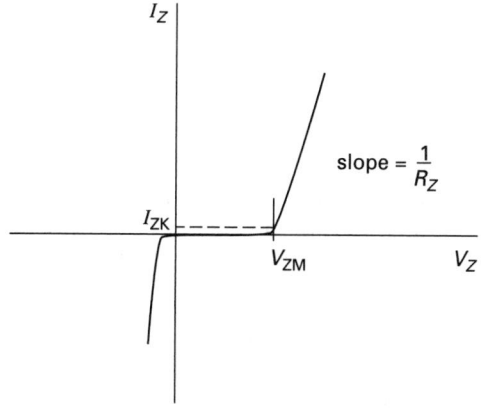

(c) characteristics

Figure 45.6 *Equivalent Zener Diode Voltage Regulator*

The defining equations for the equivalent circuit are derived from Fig. 45.6(b).

$$V_{\text{in}} - V_L = (I_L + I_Z)R_s \qquad \textit{45.12}$$

$$V_L = V_{\text{ZM}} + I_Z R_Z \qquad \textit{45.13}$$

Equations 45.12 and 45.13 reduce to the ideal conditions when I_Z equals zero. Since the zener current must be kept above the keep-alive current, Eqs. 45.12 and 45.13 must be solved simultaneously, which results in Eq. 45.14. (Equation 45.14 is similar to Eq. 45.8 for the ideal case in that it is used to determine the source resistance with the input voltage at its minimum value and the load current at its maximum value.)

$$V_L = \frac{V_{ZM} R_s + V_{in} R_Z}{R_s + R_Z} - I_L \left(\frac{R_s R_Z}{R_s + R_Z} \right) \quad \textit{45.14}$$

The zener current is maximum when the input voltage, V_{in}, is maximum and the load current, I_L, is minimum.

$$I_{Z,max} = \frac{V_{ZM} R_s + V_{in,max} R_Z}{R_Z (R_s + R_Z)} - \frac{V_{ZM}}{R_Z}$$

$$= \frac{V_{in,max}}{R_s} - \frac{V_{ZM} R_s + V_{in,max} R_Z}{R_s (R_s + R_Z)} \quad \textit{45.15}$$

Equation 45.15 represents the maximum current drawn by the zener and thus sets the power requirements for the diode and the supply resistor, given by Eqs. 45.16 and 45.17, respectively.

$$P_D = I_{Z,max} V_{ZM} + I_{Z,max}^2 R_Z \quad \textit{45.16}$$

$$P_{R_s} = I_{Z,max}^2 R_s \quad \textit{45.17}$$

If the load current, I_L, is allowed to exceed the maximum level, the zener current, I_Z, will decrease below the keep-alive level, I_{ZK}, and the zener diode will cease to regulate. Equations 45.14 and 45.15 will no longer be valid. The load voltage will be given by Eq. 45.18, which is derived from Eq. 45.12.

$$V_L = V_{in} - I_L R_s \quad \textit{45.18}$$

A practical zener diode voltage regulator circuit is designed as follows.

step 1: Determine the supply resistance that minimizes power loss but keeps the zener operating in the breakdown region. Use Eq. 45.13 with the zener current, I_Z, equal to the keep-alive current, I_{ZK}, to determine the load voltage minimum that maintains the keep-alive current. Then use Eq. 45.14 with the calculated value of the minimum load voltage, the minimum input voltage, and the maximum load current to determine the source resistance. This step accounts for the minimum supply (input) voltage and the maximum load (output) current.

step 2: Determine the maximum zener current drawn with the supply resistance calculated in step 1 (see Eq. 45.15). This step accounts for the maximum supply (input) voltage and the minimum load (output) current.

step 3: Determine the required diode power rating with the current calculated in step 2 (see Eq. 45.16).

step 4: Determine the required resistor power rating with the current calculated in step 2 (see Eq. 45.17).

step 5: If desired, determine temperature effects on the zener diode at maximum expected temperature changes, and recalculate the parameters in the above steps to determine the circuit functions correctly (see Sec. 43-14).

step 6: If desired, determine the voltage regulation of the circuit. The regulation is calculated between the maximum input voltage at no load current and the minimum input voltage at maximum current.

Example 45.5

Determine the source resistance for a zener voltage regulator circuit, patterned after Fig. 45.6(b). The output voltage requirement is 5.1 V, while the input varies from 10 to 15 V and the load current ranges from 0 to 5 A. The keep-alive current, I_{ZK}, equals 5 mA, and the zener resistance, R_Z, equals 0.12 Ω. (The parameters are identical to those used in Ex. 45.4 with the exception that the voltage output is 5.1 V, which is one of the standard zener voltages from App. 45.A.)

Solution

The supply resistance is determined from Eq. 45.14 by substituting the value of the minimum load voltage from Eq. 45.13 (see step 1 of the procedure in this section). The minimum load voltage is

$$V_{L,min} = V_{ZM} + I_{ZK} R_Z$$

$$= 5.1 \text{ V} + (5 \times 10^{-3} \text{ A})(0.12 \text{ }\Omega)$$

$$= 5.1006 \text{ V}$$

Substituting into Eq. 45.14, using the minimum input voltage and maximum load current, gives

$$V_L = \frac{V_{ZM} R_s + V_{in,min} R_Z}{R_s + R_Z}$$

$$- I_{L,max} \left(\frac{R_s R_Z}{R_s + R_Z} \right)$$

$$5.1006 \text{ V} = \frac{(5.1 \text{ V})(R_s) + (10)(0.12 \text{ }\Omega)}{R_s + 0.12 \text{ }\Omega}$$

$$- (5 \text{ A}) \left(\frac{(R_s)(0.12 \text{ }\Omega)}{R_s + 0.12 \text{ }\Omega} \right)$$

$$R_s = 0.98 \text{ }\Omega$$

Using a practical zener diode circuit instead of an ideal diode circuit requires a minimal source resistance change.

Example 45.6

Determine the voltage regulation for the circuit in Ex. 45.5.

Solution

The voltage regulation (step 6 of the procedure in this section) is determined by obtaining the equation for the load voltage and current. The value of R_s for the circuit to operate in the breakdown region is 0.98 Ω (from Ex. 45.5). Substituting this value into Eq. 45.14 gives the relationship between the load current and voltage.

$$
V_L = \frac{V_{\mathrm{ZM}} R_s + V_{\mathrm{in}} R_Z}{R_s + R_Z} - I_L \left(\frac{R_s R_Z}{R_s + R_Z} \right)
$$

$$
= \frac{(5.1 \ \mathrm{V})(0.98 \ \Omega) + (V_{\mathrm{in}})(0.12 \ \Omega)}{0.98 \ \Omega + 0.12 \ \Omega}
$$

$$
- I_L \left(\frac{(0.98 \ \Omega)(0.12 \ \Omega)}{0.98 \ \Omega + 0.12 \ \Omega} \right)
$$

$$
= 4.54 + 0.109 V_{\mathrm{in}} - 0.107 I_L
$$

For an input voltage of 10 V and no-load current, the load voltage is 5.63 V. At the maximum load current of 5 A, the load voltage is 5.10 V. (If the load current goes above the maximum of 5 A, the diode ceases to regulate and the load voltage is determined by Eq. 45.18.) Repeating the calculation for an input voltage of 15 V gives the values of 6.18 V and 5.64 V for no-load current and maximum load current, respectively. Plotting the results gives the following.

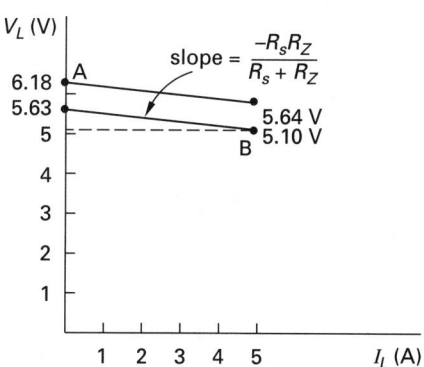

The regulation is calculated between the maximum input voltage at no-load current (point A) and the minimum input voltage at maximum current (point B).

$$
\mathrm{VR} = \frac{V_{\mathrm{nl}} - V_{\mathrm{fl}}}{V_{\mathrm{fl}}} \times 100\%
$$

$$
= \frac{6.18 \ \mathrm{V} - 5.10 \ \mathrm{V}}{5.10 \ \mathrm{V}} \times 100\% = 21.2\%
$$

5. OPERATIONAL AMPLIFIER REGULATOR CIRCUIT

Zener diodes provide voltage regulation at low-current demands. Better regulation is achieved using a precision diode, but a precision diode also lowers the operational amplifier current output. Using the zener diode as a reference for input to an op amp and passing the output to the base of a transistor improves the regulation and increases the current capability. Such an op amp regulator is shown in Fig. 45.7.

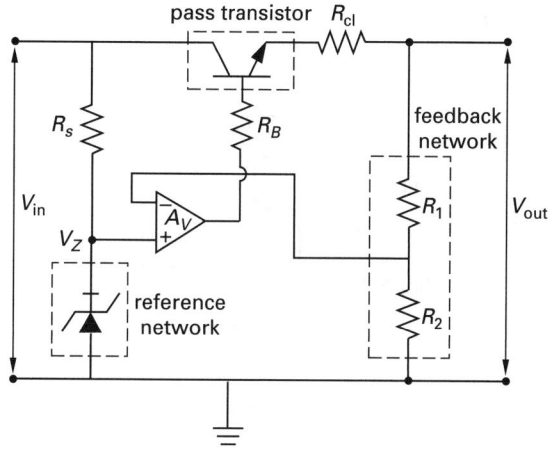

Figure 45.7 *Operational Amplifier Voltage Regulator*

The common base transistor is called the *pass transistor* and is used to supply a current gain, α. The zener provides the reference voltage, V_Z, at the noninverting terminal of the op amp. The input to the op amp is $V_Z - \delta V_{\mathrm{out}}$ where delta is called the *feedback factor* and is given by[8]

$$
\delta = \frac{R_2}{R_1 + R_2} \qquad \textbf{45.19}
$$

The output voltage is approximately

$$
V_{\mathrm{out}} \approx \frac{V_Z}{\delta} \approx V_Z \left(\frac{A_V}{1 + \delta A_V} \right) \qquad \textbf{45.20}
$$

The supply resistance, R_s, is set to ensure that the minimum keep-alive current flows in the zener diode.

$$
R_s = \frac{V_{\mathrm{in,min}} - V_{\mathrm{ZM}}}{I_{\mathrm{ZK}}} - R_Z \qquad \textbf{45.21}
$$

The equivalent circuit for the zener diode, which shows R_Z, is shown in Fig. 45.6(b). The minimum and maximum zener voltages are given by

$$
V_{Z,\mathrm{min}} = V_{\mathrm{ZM}} + I_{\mathrm{ZK}} R_Z \qquad \textbf{45.22}
$$

$$
V_{Z,\mathrm{max}} = \frac{V_{\mathrm{in,max}} R_Z + V_{\mathrm{ZM}} R_s}{R_s + R_Z} \qquad \textbf{45.23}
$$

The method of design follows the same basic steps given in Sec. 45-4.

---

[8]The *feedback factor* is sometimes symbolized β. This symbology is avoided here to prevent confusion with the CE current gain, β.

6. TRANSISTOR SWITCH FUNDAMENTALS

The bipolar junction transistor (BJT) is an active nonlinear device that, when used between its saturation and cutoff regions, becomes an electronic switch (see Sec. 43-19). A transistor circuit set up to operate as a switch is shown in Fig. 45.8(a). The pulse input that controls the switch is shown in Fig. 45.8(b). The pulses can be considered to represent logic conditions.[9] Combinations of such switches can thus be used to build logic circuits. The characteristics of the switch, the two operating points, and approximate models are shown in Fig. 45.8(c).

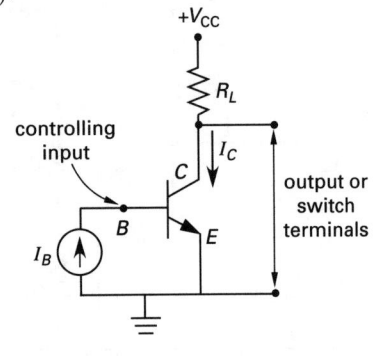

(a) switch

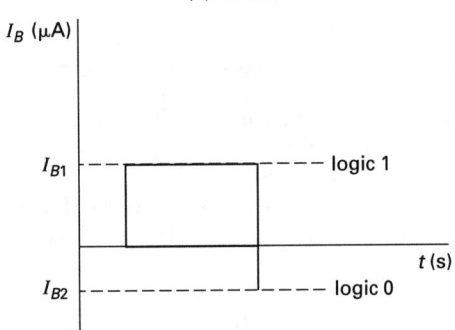

(b) pulse input

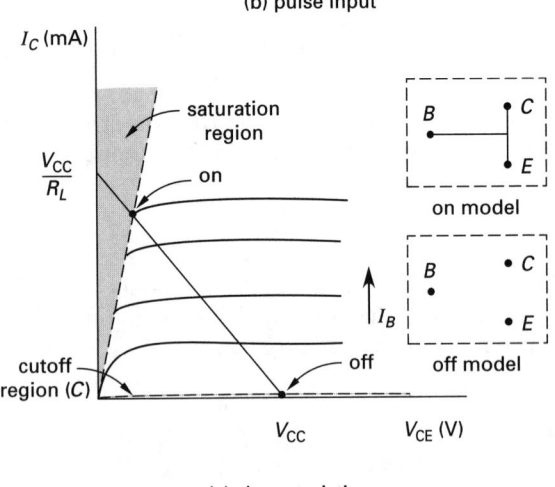

(c) characteristics

Figure 45.8 *Transistor Switch*

When the transistor switch is in the on state, that is, saturated, the collector current is given by KVL in the output loop as

$$I_{C,\text{sat}} = \frac{V_{CC} - V_{CE,\text{sat}}}{R_L} \qquad 45.24$$

The collector-emitter voltage, V_{CE}, is typically 0.1 V for germanium diodes and 0.2 V for silicon diodes and can be ignored for first-order calculations—hence the short-circuit model of Fig. 45.8(c). The value of the load resistance must be determined so that when in saturation the condition of Eq. 45.25 is satisfied.[10]

$$I_{B1} \geq \frac{I_{C,\text{sat}}}{h_{FE}} \approx \frac{V_{CC}}{h_{FE} R_L} \qquad 45.25$$

The base current calculated by Eq. 45.25 is the minimum current required to drive the transistor into saturation, ignoring the small collector-emitter voltage. When the transistor is in saturation (on), the base-emitter and collector-base junctions are both forward biased. When the transistor is in cutoff (off), the base-emitter and collector-base junctions are both reverse biased.

Field-effect transistors (FETs) are also used as switches. The gate provides the controlling input. In the on condition, unlike the BJT, the properties of the FET are more like that of a resistor. In the off condition, small leakage currents flow, similar to reverse saturation currents in the BJT.

7. JFET SWITCHES

Junction field-effect transistors used as switches are constructed from depletion-mode p-channel types as shown in Fig. 45.9.

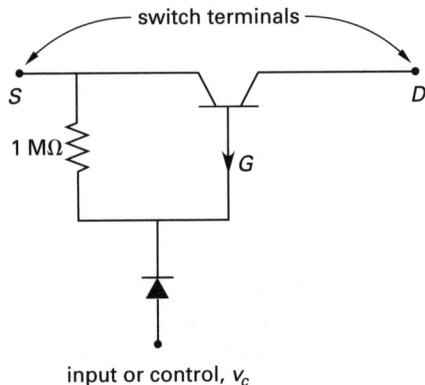

input or control, v_c

Figure 45.9 *JFET Switch*

---

[9]Normally the logic conditions are voltage inputs. For transistor-transistor-logic (TTL) circuits, 0.0 to 0.8 V is logic 0 and approximately 3.0 V is logic 1.

[10]At low frequencies, $h_{FE} \approx h_{fe}$, as is the case for all of the small-signal parameters.

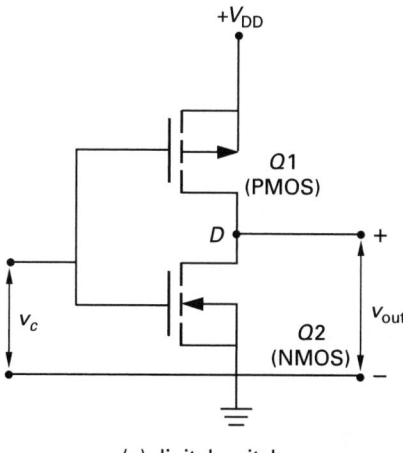

(a) digital switch

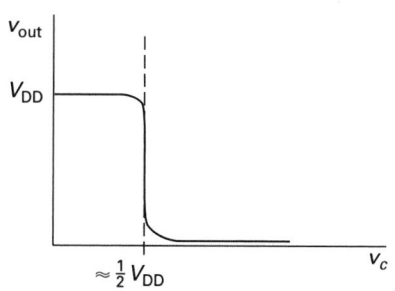

(b) characteristics

Figure 45.10 *CMOS Inverter*

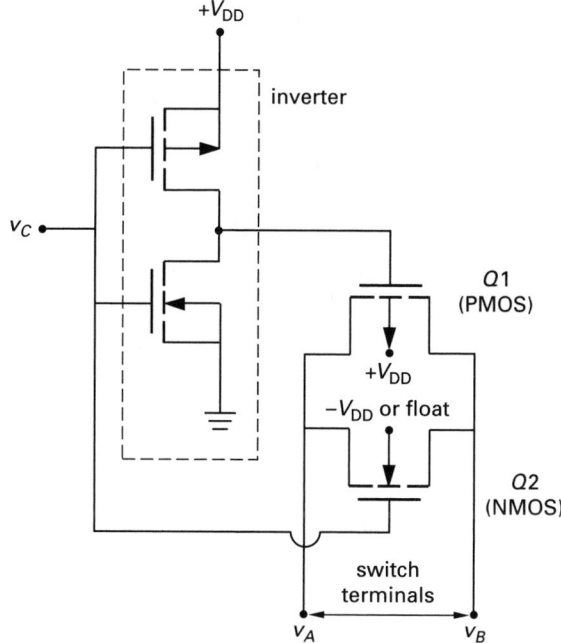

Figure 45.11 *CMOS Switch*

When the control voltage, v_c, is zero, the gate-source voltage, V_{GS}, is equal to zero and the switch is on. This occurs because the gate-channel pn junction controls the transistor by varying the depletion region caused by reverse bias (see Sec. 43-34). A positive control voltage results in a positive gate-source voltage, turning the switch off. This positive voltage is the pinchoff voltage, $V_{GS}(off)$. The diode prevents forward bias (i.e., negative voltage) on the gate-channel pn junction, which would destroy the switching characteristics of the JFET.[11]

The advantage of the JFET is its low resistance while in the on condition. Disadvantages include complexity and thus price. Further, the JFET is restricted to switching voltages a few volts less than the control voltages, since anything larger could turn the switch on. Also, if the switching voltage becomes too large while in the on condition, the switch can turn off. That is, the JFET moves from the ohmic region to the pinchoff region (see Sec. 43-34).

8. CMOS SWITCHES

When a p-channel MOS device and an n-channel MOS device are fabricated on the same chip, the device is called a *complementary MOSFET* (CMOS). A CMOS device used as an *inverter* or digital switch is illustrated in Fig. 45.10 along with the approximate output characteristics.

The control or input voltage, v_c, varies from 0 V (logic 0) to $+V_{DD}$ (logic 1). When the control voltage is 0 V, the gate-source voltage of $Q1$ is negative and $Q1$ is on. (The negative voltage attracts holes, enhancing the channel and allowing current flow.) The gate-source voltage of $Q2$ is zero and $Q2$ is off. (A positive voltage is required to attract electrons into the channel.) Consequently, $Q1$ can be modeled as a short circuit from source to drain, resulting in $v_{out} = +V_{DD}$ (logic 1). When the control voltage is $+V_{DD}$, the gate-source voltage of $Q1$ is zero and $Q1$ is off. The gate-source voltage of $Q2$ is positive and $Q2$ is on. Consequently, $Q2$ can be modeled as a short circuit from source to drain, resulting in $v_{out} = 0$ V (logic 0). Current flows only during the switching transient. The static current flow and static power are zero. Thus, the power requirements of CMOS circuits are extremely low.

An analog switch is created when the inverter is combined with a second CMOS device as shown in Fig. 45.11.

When the control voltage, v_c, is 0 V, both parallel switch channels ($Q1$ and $Q2$) are off. $Q1$ is off due to $+V_{DD}$ applied to its gate by the inverter. $Q2$ is off due to the 0 V applied to its gate by the control voltage. In the off condition, restrictions on the switching terminals' voltage exist. For instance, $Q1$ will turn on if both terminals are more positive than $+V_{DD}$—hence the

---

[11] The current from the gate would add to that flowing from source to drain. Thus, the device would no longer function strictly as a switch controlling external current.

upper limit on the voltage. $Q2$ will turn on if both terminals are negative by an amount equal to V_{GS}—hence a lower limit on the voltage. When the control voltage, v_c, is $+V_{DD}$, $Q1$ will turn on when the switching voltage is positive enough to make V_{GS} negative enough to create the p-channel. $Q2$ will turn on when the switching voltage is negative and for voltages to within a few volts of V_{DD}.

9. ACTIVE WAVEFORM SHAPING

Waveform shaping occurs in many circuits using nonlinear active devices, with and without storage elements (capacitors and inductors). These devices can be discrete or integrated, although integrated circuits rely primarily on capacitors rather than inductors. No single simple theory describes the behavior of all these elements. The analysis is generally done by breaking the circuit into a sequence of linear problems and combining the results.

Examples of such circuits include operational amplifiers with capacitors and inductors (integrators and differentiators, filters), waveform generators (sweep, square, triangular), oscillator circuits (monostable, bistable, astable), and so on.

10. LOGIC FAMILIES

Integrated circuits are classified into families based on the circuit elements they contain and the manner in which these elements are combined. *Passive logic circuits* have no transistors (i.e., they have only resistors and diodes), while *active logic circuits* do have transistors.

One of the earliest logic families is *diode-transistor logic* (DTL). *High-threshold logic* (HTL) is similar to DTL but uses higher voltages and better diodes to produce greater noise immunity. *Variable-threshold logic* (VTL) is a variation of DTL with adjustable noise immunity. *Direct-coupled transistor logic* (DCTL) is an unsophisticated connection of transistors with poor immunity to noise and low logic levels. Since there are no resistors, it has low power dissipation. *Resistor-transistor logic* (RTL) is similar to DCTL with the addition of resistors to limit current hogging. *Resistor-capacitor transistor logic* (RCTL) is a low-cost compromise between speed and power. *Transistor-transistor logic* (TTL or T^2L) is the fastest of all logic families using transistors operating in the saturated region. The fastest of all logic families using bipolar junction transistors is *emitter-coupled logic* (ECL), also known as *current-mode logic* (CML). ECL transistors do not operate in the saturation region. ECL circuits consume more power than TTL.

Metal-oxide semiconductor (MOS) circuitry contains MOSFETs. It requires few resistors and has high packing density. *Complementary construction* uses both n-channel and p-channel transistors in the same circuit. NMOS logic uses n-channel MOSFETs, while PMOS uses p-channel MOSFETs. (NMOS is approximately six times slower than PMOS.) *Complementary metal-oxide semiconductor* (CMOS), also known as *complementary transistor logic* (CTL), circuits have high packing density and reliability. They are capable of higher switching speeds than MOS.

Circuitry based on bipolar transistors switches faster, offers greater current drive, and is more suitable for analog applications than CMOS circuitry.[12] On the other hand, CMOS consumes less power, dissipates less heat, and has a higher packing density. Bipolar and CMOS circuitry are combined in biCMOS integrated circuits.

11. LOGIC GATES

A *gate* performs a logical operation on one or more inputs. The inputs, labeled A, B, C, etc., and output are limited to the values of zero or one. A listing of the output value for all possible input values is known as a *truth table*. Table 45.1 combines the symbols, names, and truth tables for the most common gates.

Table 45.1 *Logic Gates*

inputs		not	and	or	nand	nor	exclusive or
A	B	$-A$ or $\overline{A}$	AB	$A+B$	$\overline{AB}$	$\overline{A+B}$	$A\oplus B$
0	0	1	0	0	1	1	0
0	1	1	0	1	1	0	1
1	0	0	0	1	1	0	1
1	1	0	1	1	0	0	0

12. BOOLEAN ALGEBRA

The rules of *Boolean algebra* are used to write and simplify expressions of binary variables (i.e., variables constrained to two values). The basic laws governing Boolean variables are listed here.

- *commutative*:
$$A + B = B + A$$
$$A \cdot B = B \cdot A$$

- *associative*:
$$A + (B + C) = (A + B) + C$$
$$A \cdot (B \cdot C) = (A \cdot B) \cdot C$$

- *distributive*:
$$A \cdot (B + C) = (A \cdot B) + (A \cdot C)$$
$$A + (B \cdot C) = (A + B) \cdot (A + C)$$

- *absorptive*:
$$A + (A \cdot B) = A$$
$$A \cdot (A + B) = A$$

De Morgan's theorems are

$$\overline{A + B} = \overline{A} \cdot \overline{B}$$

$$\overline{A \cdot B} = \overline{A} + \overline{B}$$

---

[12]CMOS is equivalent to TTL in speed but slower than ECL.

The following basic identities are used to simplify Boolean expressions.

$$0 + 0 = 0 \qquad 0 \cdot 0 = 0 \qquad A + 0 = A$$
$$0 + 1 = 1 \qquad 0 \cdot 1 = 0 \qquad A + 1 = 1$$
$$1 + 0 = 1 \qquad 1 \cdot 0 = 0 \qquad A + A = A$$
$$1 + 1 = 1 \qquad 1 \cdot 1 = 1 \qquad A + \overline{A} = 1$$

$$A \cdot 0 = 0 \qquad -0 = 1$$
$$A \cdot 1 = A \qquad -1 = 0$$
$$A \cdot A = A \qquad -\overline{A} = A$$
$$A \cdot \overline{A} = 0$$

Example 45.7

Simplify and write the truth tables for the following network of logic gates.

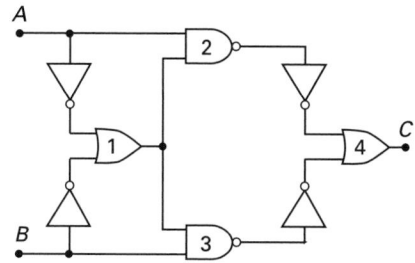

Solution

Determine the inputs and output of each gate in turn.

gate	inputs	output
1	$\overline{A}, \overline{B}$	$(\overline{A} + \overline{B})$
2	$A, (\overline{A} + \overline{B})$	$\overline{A \cdot (\overline{A} + \overline{B})}$
3	$B, (\overline{A} + \overline{B})$	$\overline{B \cdot (\overline{A} + \overline{B})}$
4	$A \cdot (\overline{A} + \overline{B}),$ $B \cdot (\overline{A} + \overline{B})$	$A \cdot (\overline{A} + \overline{B}) +$ $B \cdot (\overline{A} + \overline{B})$

Simplify the output of gate 4.

$$A \cdot (\overline{A} + \overline{B}) + B \cdot (\overline{A} + \overline{B}) \qquad \text{[original]}$$
$$A \cdot \overline{A} + A \cdot \overline{B} + B \cdot \overline{A} + B \cdot \overline{B} \qquad \text{[distributive]}$$
$$A \cdot \overline{B} + B \cdot \overline{A} \qquad \text{[since } A \cdot \overline{A} = 0\text{]}$$
$$A \oplus B \qquad \text{[definition]}$$

The truth table is

A	B	C
0	0	0
0	1	1
1	0	1
1	1	0

13. LOGIC CIRCUIT FAN-OUT

Fan-out is the number of parallel loads that can be driven from one output node of a logic circuit. Since logic circuits are routinely cascaded in order to achieve the desired outcome, a single transistor's output must provide multiple inputs for subsequent stages.

Consider the NOT gate shown in Fig. 45.12, which is a common emitter-configured transistor with the output taken from the collector. The transistor is designed to operate as a switch between the saturation and cutoff regions. When the input is high (several volts), the base-emitter junction is forward biased and the base current is sufficient to saturate the transistor. The output is then low, nearly zero if the small collector-emitter voltage (0.2 V for silicon transistors and 0.1 V for germanium) is ignored. The input has been inverted, that is, changed from high to low—hence the designation as a NOT gate.[13]

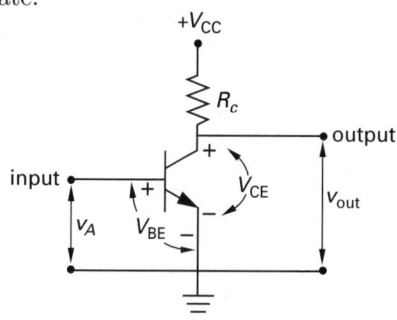

(a) configuration

input ●—▷o—● output

(b) logic symbol

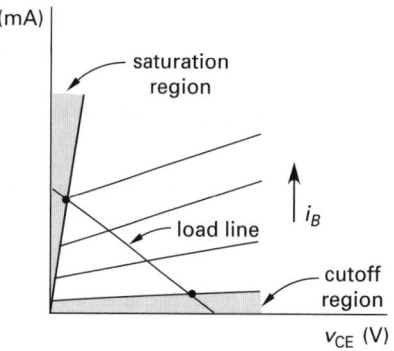

(c) operating points

v_A	v_{out}
L	H
H	L

(d) voltage truth table

Figure 45.12 NOT Gate

---

[13]The logic described here and used throughout this chapter is *positive logic*, that is, logic where the high voltage is logic condition 1 and the low voltage is logic condition 0. *Negative logic* is the reverse. *Dynamic* or *pulse logic* is a system where the presence of a pulse is logic condition 1 and the absence of a pulse is logic condition 0.

In practical applications, this transistor does not operate in isolation, but instead is cascaded with other logic devices. The base current drawn while in saturation is *sourced* from previous stages. When in cutoff, the reverse saturation current, I_{CBO}, flowing through the collector-base junction must be *sunk* by previous stages. This situation is shown in Fig. 45.13.

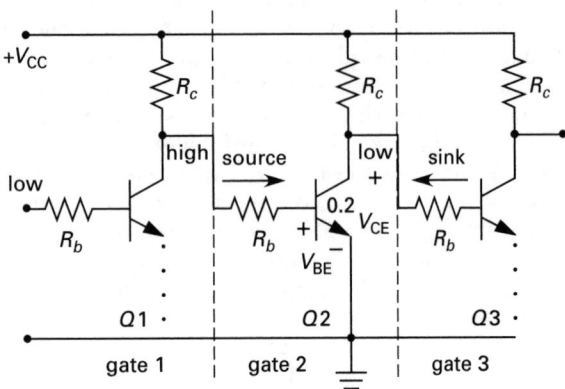

Figure 45.13 *Fan-Out Unit Loads*

The number of stages a logic gate can source, that is, the current a stage can supply when its collector voltage is high, is called the fan-out. In Fig. 45.13, transistor $Q1$ is sourcing the current to $Q2$. The number of stages that a logic gate can sink, that is, the current the collector (or output circuit) can accept without removing the transistor from the saturated state, is also called the fan-out. In Fig. 45.13, transistor $Q2$ is sinking the current from $Q3$. A given logic family's fan-out specification is stated in terms of *unit loads*. A *unit source current* is the input current drawn *by* a single gate when the voltage is high. A *unit sink current* is the current drawn *from* the input of a single gate when the input voltage is low. The fan-out specification is the minimum number of unit loads, source or sink, the gate can provide.

Gate 1 in Fig. 45.13 sources a unit load to $Q2$. $Q1$ is cut off and appears as an open circuit.[14] The unit source current is the base current to $Q2$ given by

$$I_{B2} = \frac{V_{CC} - V_{BE2}}{R_c + R_b} \qquad \textit{45.26}$$

$Q2$ is saturated, which causes $Q3$ to be cut off. The reverse saturation current, I_{CBO}, from the collector-base junction of $Q3$ is drawn to the collector of $Q2$. (This sink current raises the voltage at the collector of $Q2$. If too much current is drawn, the voltage rise can reverse-bias the collector-base junction of $Q2$, bringing it out of saturation.) The unit sink current is I_{CBO3}. The collector current is given by

$$I_{C2} = \frac{V_{CC} - V_{CE,sat}}{R_c} + I_{CBO3} \qquad \textit{45.27}$$

---

[14]Since the threshold voltage is approximately 0.7 V for a silicon base-emitter junction, $Q1$ will be cut off at voltages of 0.6 V or less.

When a gate's function is to maintain multiple loads, the source and sink transistors can be modeled as shown in Fig. 45.14 in order to determine the fan-out.

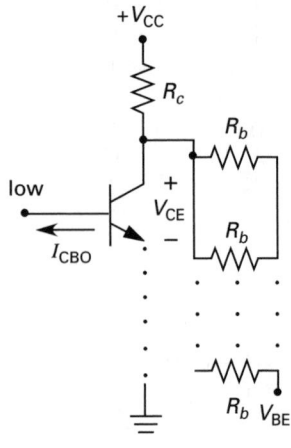

(a) source loads

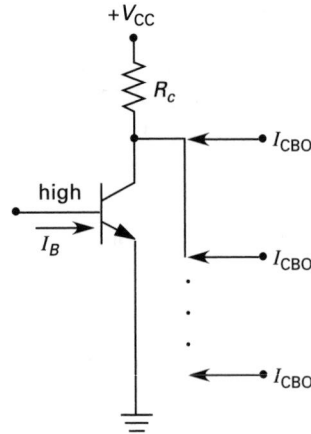

(b) sink loads

Figure 45.14 *Fan-Out Circuits*

For the sourcing condition of Fig. 45.14(a), applying KCL at the collector node gives

$$\frac{V_{CC} - V_{CE}}{R_c} - I_C = N\left(\frac{V_{CE} - V_{BE}}{R_b}\right) \qquad \textit{45.28}$$

The forward base-emitter voltage required to hold the load transistors in saturation is typically ≈ 0.7 V, and the reverse saturation is I_{CBO}. The current in the collector resistor, R_c, includes the reverse saturation current, I_{CBO}, of the cutoff transistor.[15]

---

[15]The collector current is not shown in Eq. 45.28 as the reverse saturation current because this equation is used in a general manner to determine the load line.

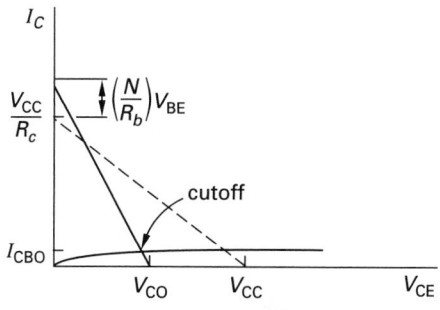

(a) cutoff: multiple source loads

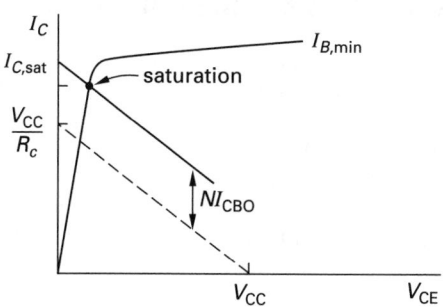

(b) saturated: multiple sink loads

Figure 45.15 *Load Lines*

The load line equation without sourced loads is shown as a dashed line in Fig. 45.15(a). The load line for a transistor sourcing multiple loads is shown as a solid line in Fig. 45.15(a) and is given by Eq. 45.28 rearranged as

$$I_C + \left(\frac{1}{R_c} + \frac{N}{R_b}\right) V_{CE} = \frac{V_{CC}}{R_c} + \left(\frac{N}{R_b}\right) V_{BE}$$

45.29

Each load transistor has a base current given by

$$I_B = \frac{V_{CE} - V_{BE}}{R_b}$$

45.30

Substituting the restriction of Eq. 45.30 into the load line equation, Eq. 45.29, gives the sourcing equation for this type of logic gate.

$$I_C + \left(N + \frac{R_b}{R_c}\right) I_B = \frac{V_{CC} - V_{BE}}{R_c}$$

45.31

For the sink condition of Fig. 45.14(b), applying KCL at the collector node gives

$$I_{C,\text{sat}} = \frac{V_{CC} - V_{CE,\text{sat}}}{R_c} + N I_{CBO}$$

45.32

The load lines for the sink condition are taken directly from Eq. 45.32 and are shown in Fig. 45.15(b). Once the minimum collector saturation current is determined by

Eq. 45.32 for the sink condition, which depends on the sink fan-out (N), the minimum base current required to maintain the transistor in saturation can be graphically determined from the characteristic curve as shown in Fig. 45.15(b). The base current curve that exists at the point where the load line curve for multiple sinking loads intersects the calculated collector saturated current determines the minimum base current.[16]

Once the equations for the source condition, Eq. 45.31, and sink condition, Eq. 45.32, are determined, the collector current, I_C, is set as the reverse saturation current, I_{CBO}, in the source equation and both equations are solved for the fan-out, N. The *source fan-out* is

$$N + \frac{R_b}{R_c} \leq \frac{V_{CC} - V_{BE} - I_{CBO} R_c}{I_{B,\text{min}} R_c}$$

45.33

The *sink fan-out* is

$$N \leq \frac{I_{C,\text{sat}} R_c + V_{CE,\text{sat}} - V_{CC}}{I_{CBO} R_c}$$

45.34

The procedure for determining the fan-out for other types of logic is similar.

Example 45.8

Given a transistor with the characteristic shown, $R_c = 1000\ \Omega$ and $V_{CC} = 5$ V, determine the base current necessary for a fan-out of five.

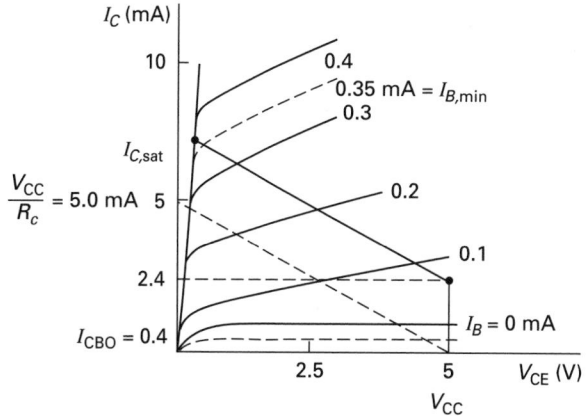

Solution

Since $I_{CBO} = 0.4$ mA, five load gates require $5 \times 0.4 = 2$ mA excess current. The corresponding load line is shown solid. The resulting saturation current requires a base current of about 0.35 mA. (Setting the value at 0.4 mA gives a margin of error for temperature effects and variations between transistors.)

---

[16]The minimum base current, $I_{B,\text{min}}$, is the base current that a sourcing gate must supply as well as the minimum base current it requires when operating in the sink condition.

Example 45.9

Determine the maximum base resistance, R_b, that will allow sourcing five gates from the transistor gate in Ex. 45.8.

Solution

From Eq. 45.33 with $V_{BE} = 0.7$ V and $I_{B,\min}$ set at 0.4 mA,

$$N + \frac{R_b}{R_c} \leq \frac{V_{CC} - V_{BE} - I_{CBO}R_c}{I_{B,\min}R_c}$$

$$5 + \frac{R_b}{1000} \leq \frac{5 - 0.7 - (0.0004)(1000)}{(0.0004)(1000)}$$

$$R_{b,\max} = 4750 \ \Omega$$

14. LOGIC CIRCUIT DELAYS

Logic circuits do not switch instantly from cutoff to saturation, or vice versa, upon application of an input pulse. Various delays are inherent, as shown in Fig. 45.16.

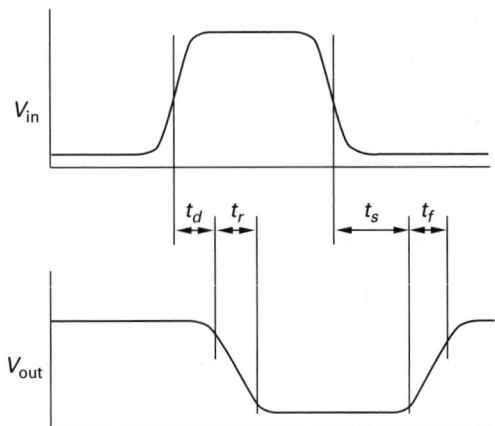

Figure 45.16 *Logic Circuit Delays*

When the input voltage level (pulse) rises, a finite time, called the *delay time*, t_d, is required for the base-emitter junction of the transistor in the inverter circuit to reach the turn-on level. The time for the input voltage to rise enough for the collector output voltage to fall to the saturation level is called the *rise time*, t_r.

With the transistor in saturation, both the base-emitter and collector-base junctions are forward biased. Excess charge is thus stored in the base region, that is, the space-charge region is small. When the input voltage level (pulse) drops, a finite time is required to remove the excess charges in the base. This delay is called the *storage time*, t_s. Once the excess charge is removed, the collector current falls. This time is called the *fall time*, t_f.

The minimum time for a transistor to pass a pulse, that is, the time from cutoff to saturation and back to cutoff, or vice versa, is the sum of all the delays. The maximum frequency at which a transistor can operate is the reciprocal of the sum of these delays.

The sum of the delays is called the *propagation delay* for a given logic gate. Approximate fan-out and propagation delay values for various transistor logic families, listed left to right in the approximate order of development, are given in Table 45.2.

15. RESISTOR-TRANSISTOR LOGIC (RTL)

The basic gate of the resistor-transistor logic (RTL) family is the NOR gate as shown in Fig. 45.17. The inverter used to describe fan-out in Sec. 45-13 is the basic building block. Although a *fan-in* of two is shown in Fig. 45.17, multiple inputs are possible. Logic 1 (high voltage) at any input saturates the associated transistor, causing the output to be connected to ground across the small collector-emitter voltage, resulting in a logic 0 (low voltage) output.

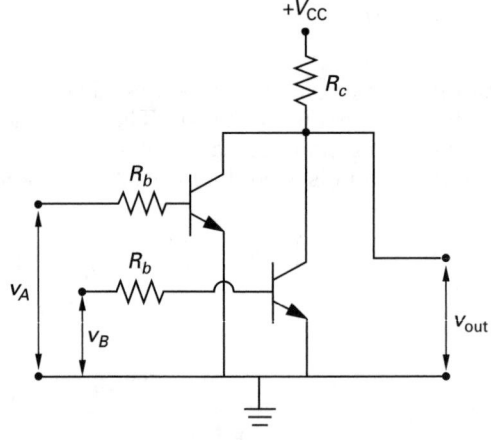

(a) basic gate: NOR

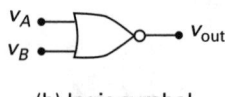

(b) logic symbol

v_A	v_B	v_{out}
L	L	H
L	H	L
H	L	L
H	H	L

(c) voltage truth table

Figure 45.17 *Resistor-Transistor Logic*

Table 45.2 Logic Family Data

parameter	RTL	DTL	HTL	TTL	ECL	MOS	CMOS
basic gate[a]	NOR	NAND	NAND	NAND	OR-NOR	NAND	NOR OR NAND
fan-out[b]	5	8	10	10	25	20	> 50
propagation delay[c]	12	30	90	10	0.1	10	0.1

[a]Positive logic is assumed when determining the basic gate.
[b]Worst-case condition.
[c]Approximate values in nanoseconds.

16. DIODE-TRANSISTOR LOGIC (DTL)

The basic gate of the diode-transistor logic (DTL) family is the NAND gate as shown in Fig. 45.18. A fan-in of two is shown, but multiple inputs are possible. When either input is logic 0 (low voltage), the associated diode, D, conducts, resulting in a low voltage at node B. The base current to the transistor will be logic 0 since any current flow is blocked by diodes $D1$ and $D2$. (Two diodes are used to ensure that the transistor will not turn on with one input low.) The transistor is cut off and the output is high ($+V_{CC}$, logic 1). When both inputs are logic 1 (high voltage), neither of the diodes conduct and the voltage at node B rises toward $+V_{CC}$. Diodes $D1$ and $D2$ conduct, and the transistor saturates, causing V_{out} to be low ($V_{CE} \approx 0$, logic 0).

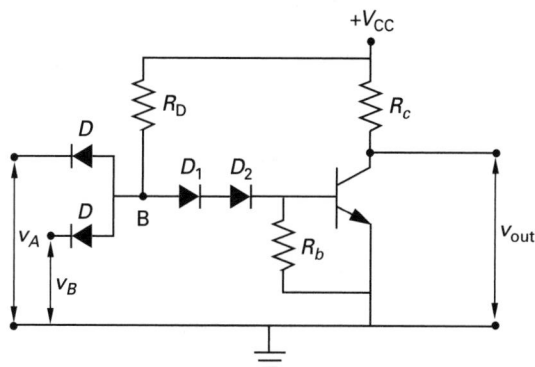

(a) basic gate: NAND

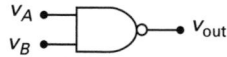

(b) logic symbol

v_A	v_B	v_{out}
L	L	H
L	H	H
H	L	H
H	H	L

(c) voltage truth table

Figure 45.18 Diode-Transistor Logic

DTL uses less current than RTL when sourcing loads, but more current when sinking loads. Zero current is required to source a load using DTL logic because of the diodes. *High-threshold logic* (HTL) is similar but uses higher voltages and better diodes to improve noise performance.

17. TRANSISTOR-TRANSISTOR LOGIC (TTL OR T²L)

The basic gate of the transistor-transistor logic (TTL) family is the NAND gate as shown in Fig. 45.19. The basic TTL NAND gate can be considered a modification of the DTL NAND gate. That is, the input diodes have been replaced by a multi-emitter transistor's base-emitter junctions. A fan-in of two is shown, but multiple inputs are possible. When at least one input is logic 0 (low voltage), the emitter of $Q1$ is forward biased. Assuming $Q2$ is cut off, the current supplied to the collector of $Q1$ is the reverse saturation current, I_{EBO}, of the emitter-base junction of $Q2$. Since this current is small, $I_{B1} > I_{EBO}/h_{FE}$ and $Q1$ is in saturation. The voltage at node B2 is then equal to the input voltage, v_A or v_B, summed with $V_{CE,sat}$, that is, about 0.4 to 0.8 V. This voltage is not enough to forward bias $Q2$ and $Q3$, since about 0.7 V apiece is required. $Q3$ is cut off and the output is high ($+V_{CC}$, logic 1). When all the inputs are at logic 1 (high voltage), the emitter-base junction of $Q1$ is reverse biased. $Q1$ is cut off and the base-collector junction is forward biased by approximately $+V_{CC}$. This voltage is high enough to forward bias the base-collector junction of $Q1$ and the base-emitter junctions of $Q2$ and $Q3$, about 2.1 V. Thus, the output is low ($V_{CE3} \approx 0$, logic 0).

TTL logic is the fastest-saturating logic gate. The fall time of TTL circuits is improved by reducing the collector current in the saturated state. This is accomplished by using a *totem-pole configuration*, also called *active pull-up*, which uses an additional transistor and diode in series with R_{C3}.[17] Additionally, the collector of $Q3$ can be left open, in which case it is called the *open-collector TTL gate*. The open-collector TTL gate is used with *wired logic* or *collector logic*, since the collector outputs are wired to the next gate's output without intervening circuitry.

---

[17]This configuration is called the totem-pole because the additional transistor sits on top of $Q3$.

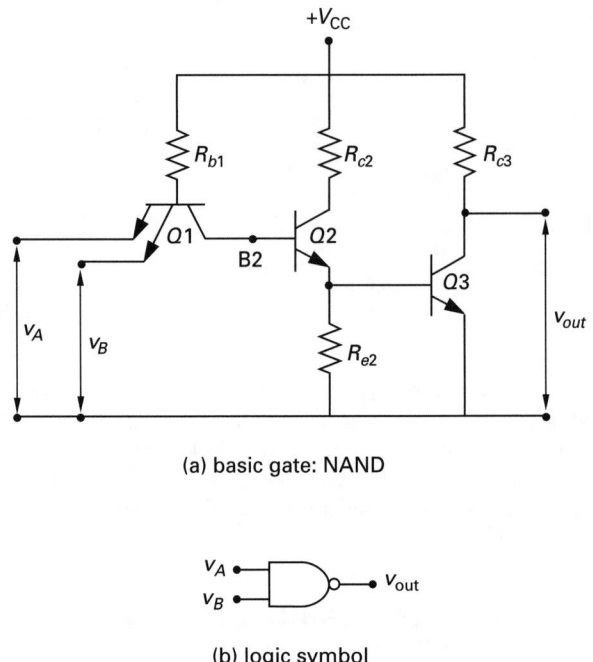

(a) basic gate: NAND

(b) logic symbol

v_A	v_B	v_{out}
L	L	H
H	L	H
L	H	H
H	H	L

(c) voltage truth table

Figure 45.19 *Transistor-Transistor Logic*

18. EMITTER-COUPLED LOGIC (ECL)

The basic gate of the emitter-coupled logic (ECL) family is the OR-NOR gate as shown in Fig. 45.20. The basic building block of the family is the difference amplifier ($Q2$ and $Q3$ in Fig. 45.20). The transistors in this logic family do not saturate. Instead, they remain in the active region as the input voltages, v_A and v_B, operating above and below (logic 1 and logic 0) the reference voltage, V_R. Since the emitter current through R_e remains approximately constant, when both inputs are logic 0 (i.e., below the reference voltage) the emitter currents from $Q1$ and $Q2$ are minimal and the emitter current of $Q3$ is maximum. This causes the voltage output at v_{OR} to be low (logic 0) and the voltage output at v_{NOR} to be high (logic 1). When either or both of the inputs are logic 1 (i.e., above the reference voltage) the emitter currents of $Q1$ and $Q2$ are maximum and the emitter current of $Q3$ is minimal. This causes the voltage output at v_{OR} to be high (logic 1) and the voltage output at v_{NOR} to be low (logic 0).

ECL logic improves on the propagation delay of TTL by not driving the transistors into saturation. This minimizes the switching time, since charge carriers do not

have to be cleared from the base junction. However, the power requirements of ECL are greater than TTL (see Table 45.2). Emitter-coupled logic is also called *current-mode logic* (CML).

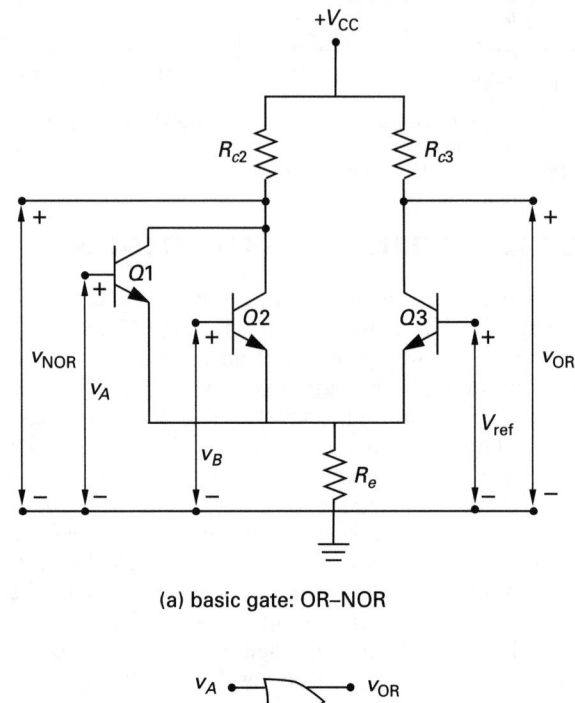

(a) basic gate: OR–NOR

(b) logic symbol

v_A	v_B	v_{OR}	v_{NOR}
L	L	L	H
H	L	H	L
L	H	H	L
H	H	H	L

(c) voltage truth table

Figure 45.20 *Emitter-Coupled Logic*

19. MOS LOGIC

The basic gate of the MOS logic family is the NAND gate as shown in Fig. 45.21.[18] The basic building block of the family is the inverter ($Q1$ and $Q2$ in Fig. 45.21). A fan-in of two is shown, but multiple inputs are possible. $Q1$ is always biased on by the drain power supply, $+V_{DD}$. When either input is logic 0 (low voltage), the associated transistor, $Q2$ or $Q3$, is off. No current flows and v_{out} is high ($+V_{DD}$, logic 1). When both of the inputs are logic 1 (high voltage), both transistors are on. Current flows and the output voltage is low ($v_{out} \approx 0$ V, logic 0).

---

[18]This is the basic gate for positive logic NMOS. For positive logic PMOS, the basic gate is the NOR gate.

MOS logic is simple and requires no external resistors or capacitors, making it well-suited for realization in integrated circuit form. The NMOS positive logic shown in Fig. 45.21(a) draws power in only one state. The PMOS positive logic (NOR) draws power in three states. In both cases, the power drawn is less than for other logic families, allowing much higher device densities. NMOS is faster, since electron mobility is higher than hole mobility. PMOS is less expensive. Advances have made the propagation delays comparable to those for other logic families (see Table 45.2).

Electronics

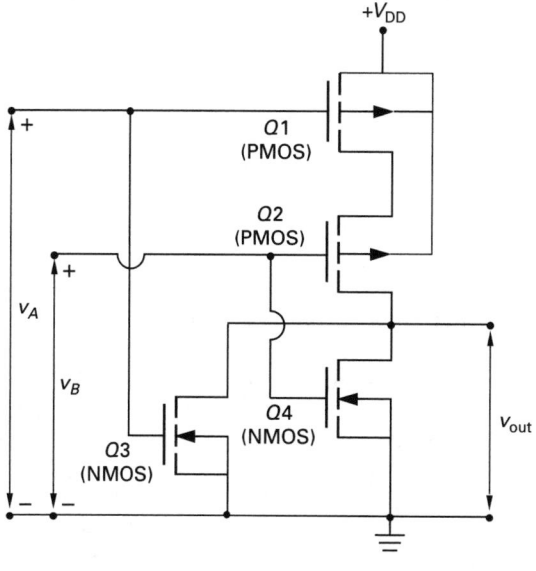

(a) basic gate: NOR

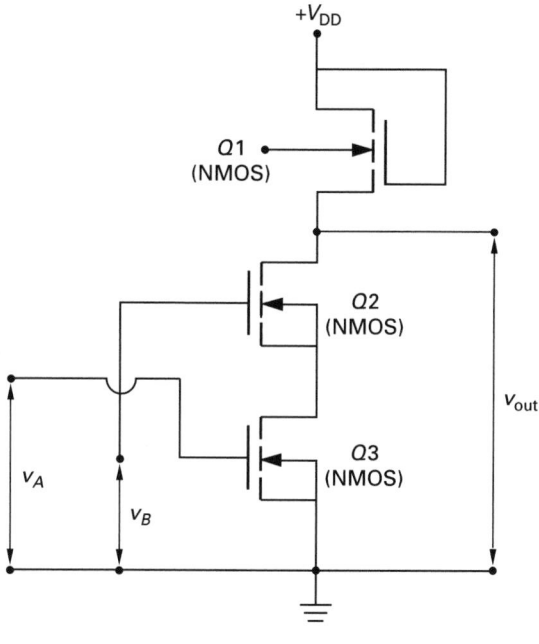

(a) basic gate: NAND

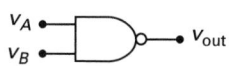

(b) logic symbol

v_A	v_B	v_{out}
L	L	H
L	H	H
H	L	H
H	H	L

(c) voltage truth table

Figure 45.21 *MOS Logic*

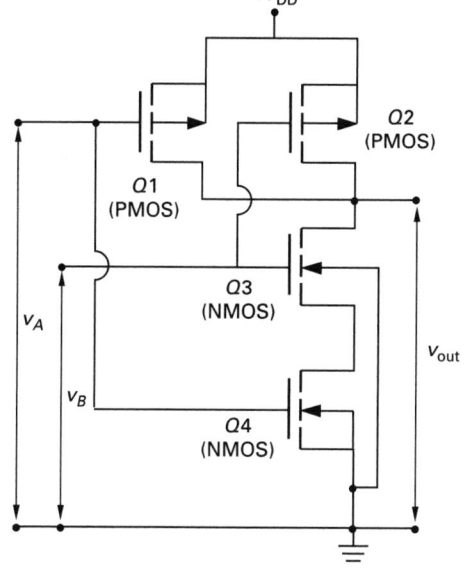

(b) basic gate: NAND

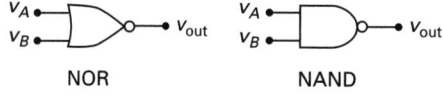

NOR NAND

(c) logic symbol

v_A	v_B	v_{out}
L	L	H
L	H	L
H	L	L
H	H	L

NOR

v_A	v_B	v_{out}
L	L	H
L	H	H
H	L	H
H	H	L

NAND

(d) voltage truth tables

Figure 45.22 *CMOS Logic*

20. CMOS LOGIC

The basic gate of the CMOS logic family may be either the NOR gate or the NAND gate as shown in Fig. 45.22. (Both are positive logic gates.) A fan-in of two is shown, but multiple inputs are possible. For the NOR gate in Fig. 45.22(a), when both of the inputs are logic 0 (voltage low), $Q1$ and $Q2$ are on, and $Q3$ and $Q4$ are off. The output is high ($\approx +V_{DD}$, logic 1) due to the connection to the drain supply via the conducting transistors $Q1$ and $Q2$. When either or both of the inputs are logic 1 (voltage high), one or both of the upper transistors ($Q1$ and/or $Q2$) is off. One or both of the lower transistors ($Q3$ and/or $Q4$) is on, connecting the output to ground via the conducting channel. The output is low (≈ 0 V, logic 0). For the NAND gate, a logic 0 (low voltage) on one or both inputs causes one or both upper transistors ($Q1$ and/or $Q2$) to conduct, thus connecting the output to the drain supply voltage. The output is high ($\approx +V_{DD}$, logic 1). If both of the inputs are logic 1 (high voltage), the lower series transistors, $Q3$ and $Q4$, turn on and connect the output to ground. The output is low (≈ 0 V, logic 0).

CMOS logic requires only a single supply voltage and is relatively easy to fabricate. The only input current that flows is that required to charge the gate-channel capacitances and any leakage through the off transistor. Thus, the power supply consumption is extremely low. Because of the low power consumption, the fan-out is very high. Speeds are comparable to that for ECL (see Table 45.2). Consequently, CMOS has the best overall properties of any logic family.

Topic XI: Computers

Chapter

Computers

For the most current information about the exam, visit **www.ppi2pass.com** regularly.

46 Computer Hardware Fundamentals

1. EVOLUTION OF COMPUTER HARDWARE

The term *hardware* encompasses the equipment and devices that perform data preparation, input, computation, control, primary and secondary storage, and output functions, but it does not include the programs, routines, and applications (i.e., computer *software*) that control the computer.[1]

Digital computers are generally acknowledged to have gone through five major evolutionary stages.[2,3,4]

- *first generation*: electromechanical calculators
- *second generation*: vacuum tube computers
- *third generation*: transistor computers
- *fourth generation*: integrated circuit computers
- *fifth generation*: VLSI (very large-scale integration) computers

The term *fifth generation* is also used to refer to the efforts (largely on the part of Japanese researchers) to produce computers that are easier to program and use. These efforts have generally been unsuccessful and have been replaced by research into the *sixth generation* of computers—devices that rely on parallel processing in

---

[1]The term "software program" is redundant.
[2]Analog computers are covered in Ch. 53.
[3]The categorization depends on who is counting and what characteristics are considered evolutionary. Some writers omit electromechanical calculators from the evolution.
[4]Before the days of calculators and computers, some companies were large enough to have employees who did nothing but crank through calculations for engineers and designers. These people were called the company's "computers."

order to appear more human-like in their programming and processing.

2. COMPUTER SIZE

Computers used for data processing (excluding process control devices) are classified into four categories based on size and cost.

- *Microcomputers* (*personal computers*, PC) are small, generally single-user computers without extensive peripherals or storage.

- *Minicomputers* are larger computers, usually dedicated to business data processing at a single site. They have the ability to support multiple terminals and a wide range of peripherals.

- *Mainframes* are general-purpose computers used in large, centralized data processing complexes and departments in which many programs are running simultaneously.

- *Supercomputers* are extremely powerful computers that usually have specific functions (e.g., engineering design, number crunching, analysis of strategies).

The distinctions among these categories are rapidly disappearing. Some current microcomputers have the same capabilities as minicomputers and mainframes of five years ago. Categorizations must now be made on the basis of physical size and cost, rather than memory size or processing speed as was true in the past.

3. COMPUTER ARCHITECTURE

All digital computers, from giant mainframes to the smallest microcomputers, contain three main components: a central processing unit (CPU), main memory, and external (peripheral) devices. Figure 46.1 illustrates a typical integration of these functions.

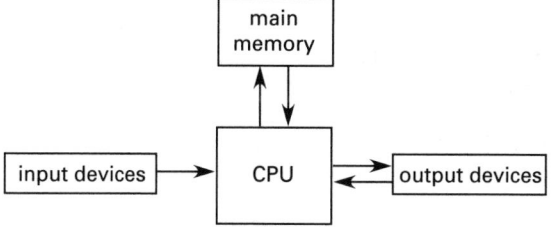

Figure 46.1 *Simplified Computer Architecture*

4. MICROPROCESSORS

A *microprocessor* is a central processing unit (CPU) on a single chip. With *large-scale integration* (LSI), most microprocessors are contained on one chip, although other chips in the set can be used for memory and input/output control. The most popular microprocessor families are produced by Intel (80X86, Pentium, Pentium Pro, Pentium II, and subsequent families) and Motorola (680X0 family), although work-alike clones of these chips may be produced by other companies (e.g., AMD's K6, and Cyrix's GX and M2 processors).

Microprocessor CPUs consist of an arithmetic and logic unit, several accumulators, one or more registers, stacks, and a control unit. The *control unit* fetches and decodes the incoming instructions and generates the signals necessary for the arithmetic and logic unit to perform the intended function. The *arithmetic and logic unit* (ALU) executes commands and manipulates data.

Accumulators hold data and instructions for further manipulation in the ALU. Registers are used for temporary storage of instructions or data. The *program counter* (PC) is a special register that always points to (contains) the address of the next instruction to be executed. Another special register is the *instruction register* (IR), which holds the current instruction during its execution. *Stacks* provide temporary data storage in sequential order—usually on a last-in, first-out (LIFO) basis. Because their operation is analogous to spring-loaded tray holders in cafeterias, the name *pushdown stack* is also used.

Microprocessors communicate with support chips and peripherals through connections in a *bus* or *channel*, which is logically subdivided into three different functions.[5] The *address bus* directs memory and input/output device transfers. The *data bus* carries the actual data and is the busiest bus. The *control bus* communicates control and status information. The number of lines in the address bus determines the amount of random-access memory that can be directly addressed. When there are n address lines in the bus, 2^n words of memory can be addressed.

Microprocessors can be designed to operate on 4-bit, 8-bit, 16-bit, 32-bit, and 64-bit words, although microprocessors with 4- and 8-bit words are now used primarily in process control applications. Some microprocessors can be combined, and the resulting larger unit is known as a *bit-slice microprocessor*. (For example, four 4-bit microprocessors might be combined into a 16-bit-slice microprocessor.)

All microprocessors use a crystal-controlled *clock* to control instruction and data movements. The *clock rate* is specified in microprocessor cycles per second (e.g., 200 MHz). Ideally, the clock rate is the number of instructions the microprocessor can execute per second. However, one or more cycles may be required

for complex instructions (*macrocommands*). For example, executing a complex instruction may require one or more cycles each to fetch the instruction from memory, decode the instruction to see what to do, execute the instruction, and store (write) the result.[6] Since operations on floating-point numbers are macrocommands, the speed of a microprocessor can also be specified in *flops*, the number of floating-point operations it can perform per second. Similar units of processing speed are *mips* (millions of instructions per second).

Most microprocessors are rich in complex executable commands (i.e., the *command set*) and are known as *complex instruction-set computing* (CISC) microprocessors. In order to increase the operating speed, however, *reduced instruction-set computing* (RISC) microprocessors are limited to performing simple, standardized format instructions but are otherwise fully featured. Unlike CISC units, RISC microprocessors require four separate instructions for the common fetch-decode-execute-write sequence.

Some microprocessors can emulate the operation of other microprocessors. For example, an 80486 chip can operate in virtual 8086 mode. *Emulation mode* is also referred to as *virtual mode*.

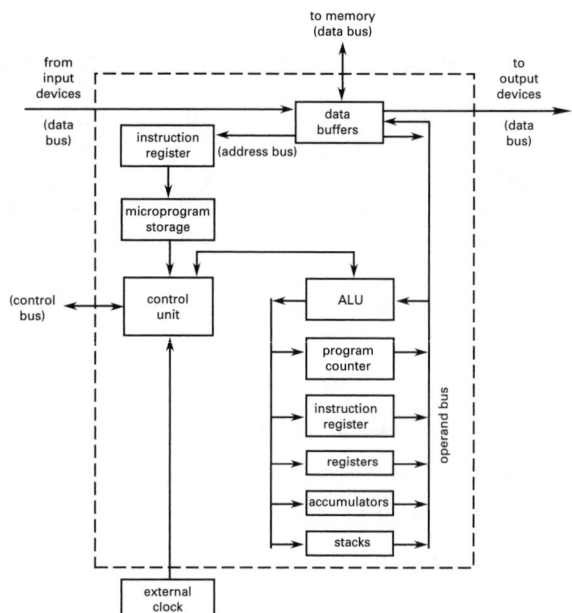

Figure 46.2 *Microprocessor Architecture*

5. CONTROL OF COMPUTER OPERATION

The user interface and basic operation of a computer are controlled by the *operating system* (OS), also known as the *monitor program*. The operating system is a program that controls the computer at its most basic level and provides the environment for application programs.

---

[5]The term *bus* refers to the physical path (i.e., wires or circuit board traces) along which the signal travels. The term *channel* refers to the logical path.

[6]In some cases, other chips (e.g., *memory management units*) can perform some of these tasks.

The operating system manages the memory, schedules processing operations, accesses peripheral devices, communicates with the user/operator, and (in multitasking environments) resolves conflicting requirements for resources. Since one of the functions of the operating system is to coordinate use of the peripheral devices (disk drives, keyboards, etc.), the term *basic input/output system* (BIOS) is also used.

All or part of the operating system can be stored in read-only memory (ROM). In early computers, a small part of executable code that was used to initiate data transfers and logical operations when the computer was first started was known as a *bootstrap loader*. Although modern start-up operations are more sophisticated, the phrase "booting the computer" is still used today.

During program operation, peripherals and other parts of the computer signal the operating system through interrupts. An *interrupt* is a signal that stops the execution of the current instruction (or, in some cases, the current program) and transfers control to another memory location, subroutine, or program. Other interrupts signal error conditions such as division by zero, overflow and underflow, and syntax errors. The operating system intercepts, decodes, and acts on these interrupts.

6. COMPUTER MEMORY

Computer memory consists of many equally sized storage locations, each of which has an associated address. The contents of a storage location may change, but the address does not.

The total number of storage locations in a computer can be measured in various ways. A *bit* (binary digit) is the smallest changeable data unit. Bits can only have values of 1 or 0. Bits are combined into *nibbles* (4 bits), *bytes* (8 bits, the smallest number of bits that can represent one alphanumeric character), *half-words* (8 and 16 bits), *words* (8, 16, and 32 bits) and *doublewords* (16, 32, and 64 bits).[7]

The number of memory storage locations is always a multiple of two. The abbreviations K, M, and G are used to designate the quantities 2^{10} (1024), 2^{20} (1,048,576), and 2^{30} (1,073,741,824), respectively. For example, a 6M memory would contain 6×2^{20} bytes. (K and M do not mean one thousand and one million exactly.)

Most of the memory locations are used for user programs and data. However, portions of the memory may be used for video memory, input/output (I/O) cache memory, the BIOS, and other purposes. *Video memory* (known as VRAM) contains the text displayed on the screen of a terminal. Since the screen is refreshed many times per second, the screen information must be repeatedly read from video memory. *Cache memory* holds the most recently read and frequently read data in memory, making subsequent retrieval of that data much faster than reading from a tape or disk drive, or even from main memory.[8] *OS memory* contains the BIOS that is read in when the computer is first started. *Scratchpad memory* is high-speed memory used to store a small amount of data temporarily so that the data can be retrieved quickly.

Modern memory hardware is semiconductor based.[9] Memory is designated as RAM (random access memory), ROM (read-only memory), PROM (programmable read-only memory), and EPROM (erasable programmable read-only memory). While data in RAM is easily changed, data in ROM cannot be altered. PROMs are initially blank but once filled, they cannot be changed. EPROMs are initially blank but can be filled, erased, and refilled repeatedly.[10] The term *firmware* is used to describe programs stored in ROMs and EPROMs.

The contents of a *volatile memory* are lost when the power is turned off. RAM is usually volatile, while ROM, PROM, and EPROM are *nonvolatile*. With *static memory*, data does not need to be refreshed and remains as long as the power stays on. With *dynamic memory*, data must be continually refreshed. Static and dynamic RAM (i.e., DRAM) are both volatile.

Virtual memory (*storage*) (VS) is a technique by which programs and data larger than main memory can be accessed by the computer. (Virtual memory is not synonymous with *virtual machine*, described in Sec. 46-12.) In virtual memory systems, some of the disk space is used as an extension of the semiconductor memory. A large application or program is divided into modules of equal size called *pages*. Each page is switched into (and out of) RAM from (and back to) disk storage as needed, in a process known as *paging*. This interchange is largely transparent to the user. Of course, access to data stored on a disk drive is much slower than semiconductor memory access. *Thrashing* is a deadlock situation that occurs when a program references a different page for almost every instruction, and there is not even enough real memory to hold most of the virtual memory.

Most memory locations are filled and managed by the CPU. However, *direct memory access* (DMA) is a powerful input/output technique that allows peripherals (e.g., tape and disk drives) to transfer data directly into and out of memory without affecting the CPU. Although special DMA hardware is required, DMA does not require explicit program instructions, making data transfer faster.

---

[7]The distinction between doublewords, words, and half-words depends on the computer. Sixteen bits would be a word in a 16-bit computer but would be a half-word in a 32-bit computer. Furthermore, *double-precision* (*double-length*) *words* double the number of bytes normally used. The abbreviations KB (*kilobytes*) and KW (*kilowords*) used by some manufacturers do not help much to clarify the ambiguity.

[8]A high-speed mainframe computer may require 200 to 500 ns to access main memory but only 20 to 50 ns to access cache memory.
[9]The term *core*, derived from the ferrite cores used in early computers, is seldom used today.
[10]Most EPROMs can be erased by exposing them to ultraviolet light.

7. PARITY

Parity is a technique used to ensure that the bits within a memory byte are correct. For every eight data bits, there is a ninth bit—the parity bit—that serves as a *check bit.* The nine bits together constitute a *frame.* In *odd-parity recording*, the parity bit will be set so that there is an odd number of one-bits. In *even-parity recording*, the parity bit will be set so that there is an even number of one-bits. When the data are read, the nine bits are checked to ensure valid data.[11]

8. INPUT/OUTPUT DEVICES

Devices that feed data to, or receive data from, the computer are known as *input/output* (I/O) *devices.* Terminals, light pens, digitizers, printers and plotters, and tape and disk drives are common peripherals.[12] Point-of-sale (POS) devices, bar code readers, and magnetic ink character recognition (MICR) and optical character recognition (OCR) readers are less common devices.

Peripherals are connected to their computer through multiline cables. With a *parallel interface* (used in a *parallel device*), there are as many separate lines in the cable as there are bits (typically seven, eight, or nine) in the code representing a character. An additional line is used as the *strobe signal* to carry a timing signal. With a *serial interface* (used in a *serial device*), all bits pass one at a time along a single line in the cable. The *transmission speed* (*baud rate*) is the number of bits that pass through the data line per second (bps).[13]

Peripherals such as terminals and printers typically do not have large memories. They only need memories large enough to store the information before the data are displayed or printed. The small memories are known as *buffers.* The peripheral can send the status (i.e., full, empty, off-line, etc.) of its buffer to the computer in several different ways. This is known as *flow control* or *handshaking.*

If the computer and peripheral are configured so that each can send and receive data, the peripheral can send a single character (e.g., the XOFF character for transmission off) to the computer when its buffer is full. Similarly, a different character (e.g., the XON character for transmission on) can be sent when the peripheral is ready for more data. The computer must monitor the incoming data line for these characters. This is known as *software flow control* or *software handshaking.*

---

[11]This does not detect two of the bits in the frame being incorrect, however.
[12]CRT, the abbreviation for *cathode ray tube*, is often used to mean a terminal.
[13]The name *baud rate* is derived from the use of the *Baudot code* (see Ch. 47). One *baud* is one modulation per second. If there is a one-to-one correspondence between modulations and bits, one baud unit is the same as one bit per second (bps). In general, the unit bps should be used.

If there are enough separate lines between the computer and the peripheral, one or more of them can be used for *hardware handshaking.*[14] In this method of flow control, the peripheral keeps the voltage on one of the lines high (or low) when it is able to accept more data. The computer monitors the voltage on this line.

Most peripheral devices are connected to the computer by a dedicated channel (cable). However, a pair of *multiplexers* (*statistical multiplexers* or *concentrators*) can be used to carry data for several peripherals along a single cable known as the *composite link.* There are two methods of achieving multiplexed transmission: *frequency division multiplexing* (FDM) and *time division multiplexing* (TDM). With FDM, the available transmission band is divided into narrower bands, each used for a separate channel. In TDM, the connecting channel is operated at a much higher clock rate (proportional to the capacity of the multiplexer), and each peripheral shares equally in the available cycles.

The Electronics Industries Association (EIA) RS-232 standard was developed in an attempt to standardize the connectors and pin uses in serial device cables.

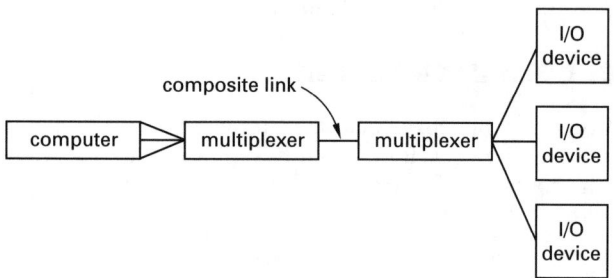

Figure 46.3 *Multiplexed Peripherals*

9. RANDOM SECONDARY STORAGE DEVICES

Random access (*direct access*) *storage devices*, also known as *mass storage devices*, include magnetic and optical disk drive units. They are random access because individual records can be accessed without having to read through the entire file.

Magnetic disk drives (*hard drives*) are composed of several *platters*, each with one or more read/write heads. The platters typically turn at 4500 to 7200 rpm. Data on a surface are organized into tracks, sectors, and cylinders. *Tracks* are the concentric storage areas. *Sectors* are pie-shaped subdivisions of each track. A *cylinder* consists of the same numbered track on all drive platters. Some platters and *disk packs* are removable, but most hard drives are fixed (i.e., nonremovable).

---

[14]Terminals and printers usually require only two or three lines—data in, data out, and ground. Most computer cables contain more lines than this, and one of the extra lines can be used for DTR (*data terminal ready*) or CTS (*clear to send*) handshaking.

Depending on the media, *optical disk drives* can be *read only* (R/O) or *read/write* (R/W) in nature. WORM drives (write once, read many) can be written by the user, while others such as CD-ROM (compact disk read-only memory) can only be read.[15]

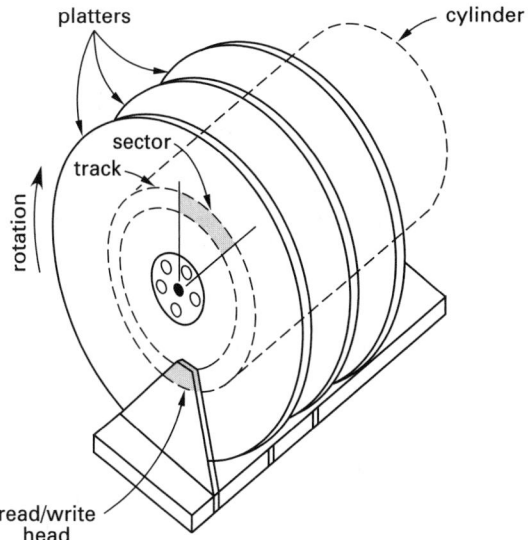

platters

cylinder

sector

track

rotation

read/write
head

Figure 46.4 *Tracks, Sectors, and Cylinders*

In addition to *storage capacity* (usually specified in megabytes, MB), there are several parameters that describe the performance of a disk drive. *Areal density* is a measure of the number of data bits stored per square inch of disk surface. It is calculated by multiplying the number of bits per track by the number of tracks per (radial) centimeter. The *average seek time* is the average time it takes to move a head from one location to a new location. The *track-to-track seek time* is the time required to move a head from one track to an adjacent track. *Latency* or *rotational delay* is the time it takes for a disk to spin a particular sector under the head for reading. On the average, latency is one-half of the time required to spin a full revolution. The *average access time* is the time needed to move to a new sector and read the data. Access time is the sum of average latency and average seek time.

Floppy disks (*diskettes*) are suitable for low-capacity random-access storage. Although their capacities are comparatively low (typically less than one million characters for magnetic media), they are used to transfer programs and data between computers (primarily microcomputers). The capacity of a diskette depends on its size, recording density, number of tracks, and number of sides.

10. SEQUENTIAL SECONDARY STORAGE DEVICES

Tape units are *sequential access devices* because a computer cannot access information stored at the end of a tape without first reading or passing by the information stored at the front of the tape. Some tapes use *indexed sequential formats* in which a directory of files on the tape is placed at the start of the tape. The tape can be rapidly wound to (near) the start of the target file without having to read everything in between.

There are a number of tape formats in use. One of the few standardized commercial formats is the *nine-track* format.[16] The tape is divided into nine tracks running the length of the tape. The width of the tape is divided into frames (characters). Eight tracks are used to record the data in either ASCII or EBCDIC format. (See Ch. 47.) The remaining track is used to record the parity bit. 1600 bpi (bits or frames per inch) is still a common *recording density* for sharing data, although densities of 9600 bpi and above are in use.

Frames are grouped into fixed-length *blocks* separated by *interblock gaps* (IBG). Reflective spots for photoelectric detection are used to indicate the beginnings and ends of magnetic tapes. These spots are called *load point* and *end-of-file* (EOF) *markers*, respectively.

Magnetic tape is often used to back up hard disks. A *streaming tape* operates in a continuously running—or streaming—mode, with data being written or read while the tape is running.

11. REAL-TIME AND BATCH PROCESSING

Programs run on a computer in one of two main ways: batch mode and real-time mode. In some data processing environments, programs are held (either by the operator or by the operating system) and eventually grouped into efficient categories requiring the same peripherals and resources. This is called *batch processing* since all jobs of a particular type are batched together for subsequent processing. There is usually no interaction between the user and the computer once batch processing begins. In *real-time* (*interactive*) *processing*, a program runs when it is submitted, often with user interaction during processing.

12. MULTITASKING AND TIME-SHARING

If a computer's main memory is large enough and the CPU is fast enough, it may be possible to allocate the main memory among several users running applications simultaneously. This is the concept of a *virtual machine* (VM): each user appears to have his own computer. This is also known as *multitasking* and *multiprogramming* since multiple tasks can be performed simultaneously. A *multiuser system* is similar to a multitasking

---

[15]The WORM acronym is also interpreted as write once—read mostly.

[16]Other formats that have received some acceptance include *quarter-inch cartridge* (QIC) and *digital audio tape* (DAT).

system in that several users can use the computer simultaneously, although that term also means that all users are running the same program.

Time-sharing (*swapping*) is a technique where each user takes a turn (under the control of the operating system) at using the entire computer main memory for a certain length of time (usually less than a second). At the end of that time period, all of the active memory is written to a private area, and the memory for the next user is loaded. The swapping occurs so frequently that all users are able to accomplish useful work on a real-time basis.

13. BACKGROUND AND FOREGROUND PROCESSING

A program running in real time is an example of *foreground processing*. There are times, however, when it is convenient to start a long program running while the same computer is used for a second program. The first program continues to run unseen in the background. *Background processing* can be accomplished by segmenting the main computer memory (i.e., establishing virtual machines as defined in Sec. 46-12) or by time-sharing (i.e., allowing the background application to have all cycles not used by the foreground application).

14. TELEPROCESSING

Teleprocessing is the access of a computer from a remote station, usually over a telephone line (although fiber-optic, coaxial, and microwave links can also be used). Since these media transmit analog signals, a *modem* (*modulator-demodulator*) is used to convert to/from the digital signals required by a computer.

With *simplex communication*, transmission is only in one direction. With *half-duplex communication*, data can be transmitted in both directions, but only in one direction at a time. With *duplex* or *full-duplex communication*, data can be transmitted in both directions simultaneously. The maximum practical transmission speed over voice-grade lines is limited without data compression. Much higher rates, however, are possible over dedicated data lines and wide-band lines.

In *asynchronous* or *start-stop transmission*, each character is preceded and followed by special signals (i.e., *start* and *stop bits*). Thus, every 8-bit character is actually transmitted as 10 bits, and the character transmission rate is one-tenth of the transmission speed in bits per second.[17] With asynchronous transmission, it is possible to distinguish the beginning and end of each character from the bit stream itself.

Synchronous equipment transmits a block of data continuously without pause and requires a built-in clock to maintain synchronization. Synchronous transmission is preceded and interwoven with special clock-synchronizing characters, and the separation of a bit stream into individual characters is done by counting bits from the start of the previous character. Since start and stop bits are not used, synchronous communication is approximately 20% faster than asynchronous communication.

There are three classes of communication lines—narrow-band, voice-grade, and wide-band (depending on the bandwidth, that is, range of frequencies) available for signaling.

Narrowband lines may only support a single channel of communication, as the bandwidth is too narrow for modulation. *Wideband channels* support the highest transfer rates, since the bandwidth can be divided into individual channels. *Voice-grade lines*, supporting frequencies between 300 and 3300 Hz, are midrange in bandwidth.

Errors in transmission can easily occur over voice-grade lines at the rate of 1 in 10,000. In general, methods of ensuring the accuracy of transmitted and received data are known as *communications protocols* and *transmission standards*. A simple way of checking the transmission is to have the receiver send each block of data back to the sender. This process is known as *loop checking* or *echo checking*. If the characters in a block do not match, they are re-sent. While accurate, this method requires sending each block of data twice.

Another method of checking the accuracy of transmitted data is for both the receiver and the sender to calculate a *check digit* or *block check character* derived from each block of *characters* sent. (A common transmission block size is 128 characters.) With *cycling redundancy checking* (CRC), the block check character is the remainder after dividing all the serialized bits in a transmission block by a predetermined binary number. Then, the block check character is sent and compared after each block of data.

15. DISTRIBUTED SYSTEMS AND LOCAL-AREA NETWORKS

Distributed data processing systems assume many configurations. In the traditional situation, a centrally located main computer interacts with and is fed by smaller computers in other locations. In a second configuration, many identical computers are linked together in order to share storage and printing resources. This latter case is known as a *local-area network* (LAN). Local-area networks typically communicate at speeds between 10 Mbps and 100 Mbps.

---

[17]For historical reasons, a second stop bit is used when data are sent at ten characters per second. This is referred to as 110 bps.

47 Computer Software Fundamentals

1. CHARACTER CODING

Alphanumeric data refers to characters that can be displayed or printed, including numerals and symbols (\$, %, &, etc.) but excluding *control characters* (tab, carriage return, form feed, etc.). Since computers can handle binary numbers only, all symbolic data must be represented by binary codes. *Coding* refers to the manner in which alphanumeric data and control characters are represented by sequences of bits.

The standard method for coding data on 80-column, 12-row cards is the *Hollerith code*.[1]

The *Baudot code* is a five-bit code that has long been used in Telex and teletypewriter (TWX and TTY) communications. By shifting to an alternate character set (numerals versus letters), it has a maximum of 64 (2×2^5) characters. Primarily due to its slow transmission speed, but also due to its limited character set, the Baudot code is no longer favored by new equipment manufacturers.

In some early computers, characters were represented by six-bit combinations known as *Binary Coded Decimal* (BCD).[2] The 64 (2^6) different combinations, however, proved insufficient to represent all necessary characters.

---
[1] Punch cards are now seldom used.
[2] BCD was reintroduced when IBM started using 96-column cards.

The *American Standard Code for Information Interchange* (ASCII) is a seven-bit code permitting 128 (2^7) different combinations. It is commonly used in microcomputers, although use of the high-order (eighth) bit is not standardized. ASCII-coded magnetic tape and disk files are used to transfer data and documents between computers of all sizes that would otherwise be unable to share data structures.

The *Extended Binary Coded Decimal Interchange Code* (EBCDIC) is in widespread use in mainframe computers.[3] It uses eight bits (a byte) for each character, allowing a maximum of 256 (2^8) different characters.

Since strings of bits are difficult to read, the *packed decimal* format is used to simplify working with EBCDIC data. Each byte is converted into two strings of four bits each. The two strings are then converted to hexadecimal format. Since $(1111)_2 = (15)_{10} = (F)_{16}$, the largest possible EBCDIC character is coded FF in packed decimal format.

Example 47.1

The number $(7)_{10}$ is represented as 11110111 in EBCDIC. What is this in packed decimal format?

Solution

The first four bits are 1111, which is $(15)_{10}$ or $(F)_{16}$. The last four bits are 0111, which is $(7)_{10}$ or $(7)_{16}$. The packed decimal representation is F7.

2. PROGRAM DESIGN

A *program* is a sequence of computer instructions that performs some function. The program is designed to implement an algorithm, which is a procedure consisting of a finite set of well-defined steps. Each step in the algorithm usually is implemented by one or more instructions (e.g., READ, GOTO, OPEN, etc.) entered by the programmer. These original "human-readable" instructions are known as *source code statements*.

Except in rare cases, a computer will not understand source code statements. Therefore, the source code is translated into machine-readable object code and absolute memory locations. Eventually, an executable program is produced.

---
[3] EBCDIC is pronounced eb'-sih-dik.

Computers

If the executable program is kept on disk or tape, it is normally referred to as *software*. If the program is placed in ROM or EPROM, it is referred to as *firmware*. The computer mechanism itself is known as the *hardware*.

3. FLOWCHARTING SYMBOLS

A *flowchart* is a step-by-step drawing representing a specific procedure or algorithm. Figure 47.1 illustrates the most common flowcharting symbols. The terminal symbol begins and ends a flowchart. The input/output symbol defines an I/O operation, including those to and from keyboard, printer, memory, and permanent data storage. The processing symbol and predefined process symbol refer to calculations or data manipulation. The decision symbol indicates a point where a decision must be made or two items are compared. The connector symbol indicates that the flowchart continues elsewhere. The off-page symbol indicates that the flowchart continues on the following page. Comments can be added in an annotation symbol.

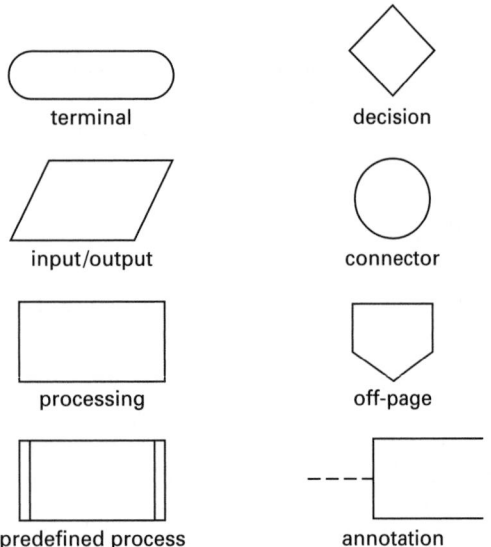

Figure 47.1 *Flowcharting Symbols*

4. LOW-LEVEL LANGUAGES

Programs are written in specific languages, of which there are two general types: low-level and high-level. Low-level languages include machine language and assembly language.

Machine language instructions are intrinsically compatible with and understood by the computer's CPU. They are the CPU's native language. An instruction normally consists of two parts: the operation to be performed (*op-code*) and the operand expressed as a storage location. Each instruction must ultimately be expressed as a series of bits, a form known as *intrinsic machine code*. However, octal and hexadecimal coding are more convenient. In either case, coding a machine language program is tedious and seldom done by hand.

Table 47.1 *Comparison of Typical ADD Commands*

language	instruction
intrinsic machine code	1111 0001
machine language	1A
assembly language	AR
FORTRAN	+

Assembly language is more sophisticated (i.e., is more symbolic) than machine language. Mnemonic codes are used to specify the operations. The operands are referred to by variable names rather than the addresses. Blocks of code that are to be repeated verbatim at multiple locations in the program are known as *macros* (*macro instructions*). Macros are written only once and are referred to by a symbolic name in the source code.

Assembly language code is translated into machine language by an *assembler* (*macro-assembler* if macros are supported). After assembly, portions of other programs or function libraries may be combined by a *linker*. In order to run, the program must be placed in the computer's memory by a *loader*. Assembly language programs are preferred for highly efficient programs. However, the coding inconvenience outweighs this advantage for most applications.

5. HIGH-LEVEL LANGUAGES

High-level languages are easier to use than low-level languages because the instructions resemble English. High-level statements are translated into machine language by either an interpreter or a compiler. A *compiler* performs the checking and conversion functions on all instructions only when the compiler is invoked. A true stand-alone executable program is created. An *interpreter*, however, checks the instructions and converts them line by line into machine code during execution but produces no stand-alone program capable of being used without the interpreter.[4,5]

There are many high-level languages, although only a few are in widespread use. Some of these languages are listed in Table 47.2.

---

[4]Some interpreters check syntax as each statement is entered by the programmer.

[5]Some languages and implementations of other languages blur the distinction between interpreters and compilers. Terms such as *pseudo-compiler* and *incremental compiler* are used in these cases.

Table 47.2 *High-Level Languages*

name	description
ADA	a rigidly standardized language similar to Pascal
ALGOL	acronym for ALGOrithmic Language
APL	acronym for A Programming Language; a language with a special character set
BASIC	Beginner's All-purpose Symbolic Instruction Code; English-like instructions; similar to FORTRAN
C	a structured high-level, function-oriented language capable of low-level machine control
COBOL	acronym for COmmon Business-Oriented Languages; used in business programs; very English-like
FORTH	a language originally designed for telescope process control; programs are called words
FORTRAN	acronym for FORmula TRANslation; rich in scientific functions (see Ch. 52)
Modula-2	a structured language derived from Pascal
Pascal	a highly structured, portable language
PL/I	acronym for Programming Language One; programs are called functions
RPG	acronym for RePort Generator

6. SPECIAL PURPOSE LANGUAGES

There are many special purpose languages, some of which are listed in Table 47.3.

Special applications include AI (artificial intelligence—see Sec. 47-17), CAD (computer-aided design), CAM (computer-aided manufacturing), CAI (computer-aided instruction), DB (database), DBMS (database management system), EIS (executive information system), JCL (job control language), and MIS (management information system).

Table 47.3 *Special Purpose Languages*

name	use or strength
COMIT	string processing
GPSS	simulation
LISP	list processing; artificial intelligence
LOGO	childrens' introductory language
PROLOG	artificial intelligence
SMALLTALK	artificial intelligence
SNOBOL	string processing
SPSS	statistical analysis

7. RELATIVE COMPUTATIONAL SPEED

Certain languages are more efficient (i.e., execute faster) than others.[6] While it is impossible to be specific and exceptions abound, assembly language programs are fastest, followed in order of decreasing speed by compiled, pseudo-compiled, and interpreted programs.

Similarly, certain program structures are more efficient than others. For example, when performing a repetitive operation, the most efficient structure will be a single equation, followed in order of decreasing speed by a stand-alone loop and a loop within a subroutine. Incrementing the loop variables and managing the exit and entry points is known as *overhead* and takes time during execution.

8. STRUCTURE, DATA TYPING, AND PORTABILITY

A language is said to be *structured* if subroutines and other procedures each have one specific entry point and one specific return point.[7] A language has *strong data types* if integer and real numbers cannot be combined in arithmetic statements.

A *portable language* can be implemented on different machines. Most portable languages are either sufficiently rigidly defined (as in the cases of ADA and C) to eliminate variants and extensions or (as in the case of Pascal) are compiled into an intermediate, machine-independent form. This so-called *pseudo-code* (*p-code*) is neither source nor object code. The language is said to have been "ported to a new machine" when an interpreter is written that converts p-code to the appropriate machine code and supplies specific drivers for input, output, printers, and disk use.[8]

9. STRUCTURED PROGRAMMING

Structured programming (also known as top-down programming, procedure-oriented programming, and GOTO-less programming) divides a procedure or algorithm into parts known as subprograms, subroutines, modules, blocks, or procedures.[9] Internal subprograms are written by the programmer; external subprograms are supplied in a library by another source. Ideally, the mainline program will consist entirely of a series of calls (references) to these subprograms. Liberal use is made of FOR/NEXT, DO/WHILE, and DO/UNTIL

---

[6]Efficiency can also, but seldom does, refer to the size of the program.

[7]Contrast this with BASIC, which permits (1) a GOSUB to a specific subroutine with a return from anywhere within the subroutine and (2) unlimited GOTO statements to anywhere in the main program.

[8]Some companies have produced Pascal engines that run p-code directly.

[9]The format and readability of the source code—improved by indenting nested structures, for example—do not define structured programming.

commands. Labels and GOTO commands are avoided as much as possible.

Very efficient programs can be constructed in languages that support *recursive calls* (i.e., permit a subprogram to call itself). Some languages (e.g., Pascal and PL/I) permit recursion; others do not.

Variables whose values are accessible strictly within the subprogram are *local variables*. *Global variables* can be referred to by the main program and all other subprograms.

10. FIELDS, RECORDS, AND FILE TYPES

A collection of *fields* is known as a *record*. For example, name, age, and address might be fields in a personnel record. Groups of records are stored in a *file*.

A *sequential file* structure (e.g., typical of data on magnetic tape) contains consecutive records and must be read starting at the beginning. An *indexed sequential file* is one for which a separate index file (see Sec. 47-11) is maintained to help locate records.

With a *random (direct access) file structure*, any record can be accessed without starting at the beginning of the file.

11. FILE INDEXING

It is usually inefficient to place the records of an entire file in order. (A good example is a mailing list with thousands of names. It is more efficient to keep the names in the order of entry than to sort the list each time names are added or deleted.) Indexing is a means of specifying the order of the records without actually changing the order of those records.

An index (key or keyword) file is analogous to the index at the end of this book. It is an ordered list of items with references to the complete record. One field in the data record is selected as the key field (record index).[10] The sorted keys are usually kept in a file separate from the data file. One of the standard search techniques is used to find a specific key.

12. SORTING

Sorting routines place data in ascending or descending numerical order or in alphabetical order.

With the method of *successive minima*, a list is searched sequentially until the smallest element is found and brought to the top of the list. That element is then ignored, and the remaining elements are searched for the smallest element, which, when found, is placed

after the previous minimum, and so on. A total of $n(n-1)/2$ comparisons will be required.[11]

In a *bubble sort*, each element in the list is compared with the element immediately following it. If the first element is larger, the positions of the two elements are reversed (swapped). In effect, the smaller element "bubbles" to the top of the list. The comparisons continue to be made until the bottom of the list is reached. If no swaps are made in a pass, the list is sorted. A total of approximately $n^2/2$ comparisons is needed, on the average, to sort a list in this manner.[12]

In an *insertion sort*, the elements are ordered by rewriting them in the proper sequence. After the proper position of an element is found, all elements below that position are bumped down one place in the sequence. The resulting vacancy is filled by the inserted element. At worst, approximately $n^2/2$ comparisons will be required. On the average, there will be approximately $n^2/4$ comparisons.

Disregarding the number of swaps, the number of comparisons required by the successive minima, bubble, and insertion sorts is on the order of n^2. When n is large, these methods are too slow. The *quicksort* is more complex but reduces the average number of comparisons (with random data) to approximately $n \times \log n / \log(2)$, generally considered to be on the order of $n \times \log n$.[13] The maximum number of comparisons for a heap sort is $n \times \log n / \log(2)$, but it is likely that even fewer comparisons will be needed.

13. SEARCHING[14]

If a group of records (i.e., a list) is randomly organized, a particular element in the list can be found only by a linear search (sequential search). At best, only one comparison and, at worst, n comparisons will be required to find something (an event known as a *hit*) in a list of n elements. The average is $n/2$ comparisons, described as being on the order of n.

If the records are in ascending or descending order, a binary search will be superior.[15] The search begins by looking at the middle element in the list. If the middle element is the sought-for element, the search is over. If not, half of the list can be disregarded in further searching since elements in that portion will be either too large or too small. The middle element in the remaining part of the list is investigated, and the procedure continues

---

[10]More than one field can be indexed. However, each field will require its own index file.

[11]When n is large, $n^2/2$ is sometimes given as the number of comparisons.

[12]This is the same as for the successive minima approach. However, swapping occurs more frequently in the bubble sort, slowing it down.

[13]However, the quicksort falters (in speed) when the elements are in near-perfect order.

[14]The term *probing* is synonymous with searching.

[15]A binary search is unrelated to a binary tree. A binary tree structure (see Sec. 47-15) greatly reduces search time but does not use a sorted list.

until a hit occurs or the list is exhausted. The number of required comparisons in a list of n elements is $\log n / \log(2)$ (i.e., on the order of $\log n$).

14. HASHING

An index file is not needed if the record number (i.e., the storage location for a read or write operation) can be calculated directly from the key—a technique known as *hashing*.[16] The procedure by which a numeric or non-numeric key (e.g., a last name) is converted into a record number is called the hashing function or hashing algorithm. Most hashing algorithms use a remaindering modulus—the remainder after dividing the key by the number of records, n, in the list. Excellent results are obtained if n is a prime number; poor results occur if n is a power of two.

Not all hashed record numbers will be correct. A *collision* occurs when an attempt is made to use a record number that is already in use. Chaining, linear probing, and double hashing are techniques used to resolve such collisions.

15. DATABASE STRUCTURES

Databases can be implemented as indexed files, linked lists, and tree structures; in all three cases, the records are written and remain in the order of entry.

key file

key	ref.
ADAMS	3
JONES	2
SMITH	1
THOMAS	4

data file

record	last name	first name	age
1	SMITH	JOHN	27
2	JONES	WANDA	39
3	ADAMS	HENRY	58
4	THOMAS	SUSAN	18

Figure 47.2 *Key and Data Files*

An indexed file such as that shown in Fig. 47.2 keeps the data in one file and maintains separate index files (usually in sorted order) for each key field. The index file must be recreated each time records are added to the

---

[16]Of course, finding a record in this manner requires it to have been written in a location determined by the same hashing routine.

file. A *flat file* has only one key field by which records can be located. Searching techniques (see Sec. 47-13) are used to locate a particular record. In a *linked list* (*threaded list*), each record has an associated *pointer* (usually a record number or memory address) to the next record in key sequence. Only two pointers are changed when a record is added or deleted. Generally, a linear search following the links through the records is used.

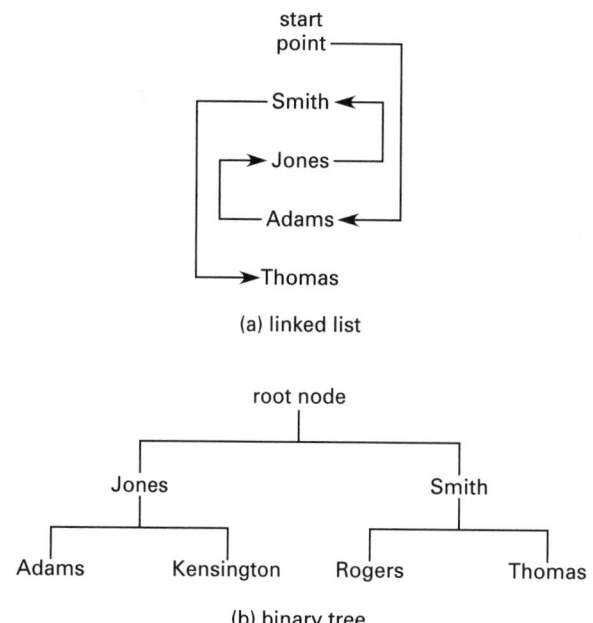

(a) linked list

(b) binary tree

Figure 47.3 *Database Structures*

Pointers are also used in *tree structures*. Each record has one or more pointers to other records that meet certain criteria. In a binary tree structure, each record has two pointers—usually to records that are lower and higher, respectively, in key sequence. In general, records in a tree structure are referred to as *nodes*. The first record in the file is called the *root node*. A particular node will have one node above it (the *parent* or *ancestor*) and one or more nodes below it (the *daughters* or *offspring*). Records are found in a tree by starting at the root node and moving sequentially according to the tree structure. The number of comparisons required to find a particular element is $1 + (\log n / \log(2))$, which is on the order of $\log n$.

16. HIERARCHICAL AND RELATIONAL DATA STRUCTURES

A *hierarchical database* contains records in an organized, structured format. Records are organized according to one or more of the indexing schemes. However, each field within a record is not normally accessible.

A *relational database* stores all information in the equivalent of a matrix. Nothing else (no index files, pointers, etc.) is needed to find, read, edit, or select information.

Any piece of information can be accessed directly by referring to the field name and field value.

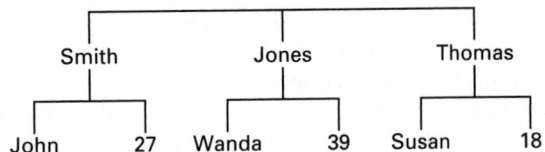

Figure 47.4 *Hierarchical Personnel File*

rec. no.	last	first	age
1	Smith	John	27
2	Jones	Wanda	39
3	Thomas	Susan	18

Figure 47.5 *Relational Personnel File*

17. ARTIFICIAL INTELLIGENCE

Artificial intelligence (AI) in a machine implies that the machine is capable of absorbing and organizing new data, learning new concepts, reasoning logically, and responding to inquiries. AI is implemented in a category of programs known as *expert systems* that "learn" rules from sets of events that are entered whenever they occur. (The manner in which the entry is made depends on the particular system.) Once the rules are learned, an expert system can participate in a dialogue to give advice, make predictions and diagnoses, or draw conclusions.

48 Computer Architecture

1. GENERAL SYSTEM ORGANIZATION

A *computer* is a device that receives, processes, and presents data. A *digital computer* performs these functions using discrete data. An *analog computer* performs these functions on quantities that represent physical variables, which are then translated into equivalent mechanical or electrical circuits as an analog of the phenomenon being investigated.[1] A *digital system* is any of the levels of operation of the digital computer. That is, a digital system comprises the logical elements—the functional units for reading, writing, storing, and manipulating information. A digital system also refers to these functional units when used outside of a computer. A digital system is capable of information processing, but only with a predetermined sequence of operations, or *algorithms*. Changing the algorithm requires changing the digital system itself. A computer has a *control unit* capable of responding to algorithm variations (i.e., program variations) without physically changing the system. (The control unit can do this because of the addition of the central processing unit (CPU) with which the control unit can interact.) The programs are called *software*. The physical components of the computer are called *hardware*. The interface between the two is defined by the architecture. *Computer architecture* is the art and science of assembling and connecting the logical elements to form a computing device. The general system organization is shown in Fig. 48.1. (This figure is similar to the simplified computer architecture shown in Fig. 46.1, with the control unit emphasized.[2])

The information processed by the computer comprises three types: numerical, logical, and character. Because of these information types, coding is required to allow the control unit to recognize the three (see Sec. 47-1).

---

[1]The term "computer" usually refers to a digital machine unless otherwise indicated and will be the focus of this book.
[2]The control unit is present in both the computer and the digital system. The fundamental difference between a computer and a digital system is the central processing unit, which is often shown as encompassing the control unit in computer systems and which allows the system to respond to variations in programming (see Fig. 46.2).

The CPU, at the direction of the control unit, using the three types of information stored in the memory, executes three types of instruction sets.

(1) arithmetic logic unit (ALU) instructions such as addition, subtraction, and logical operations (AND, OR, and so on)

(2) memory manipulation or accessing instructions such as loading and storing information

(3) control transfer instructions such as read/write, conditional/unconditional branching, interrupts, and so on

Computers can be classified by the way they handle instruction sets. *Reduced instruction-set computers* (RISC) are designed to keep the instructions simple, thus allowing the computer to execute the code faster. *Complex instruction-set computers* (CISC) are designed to perform operations with a single high-level language instruction and thus have instruction sets that are more complicated. Architectures also include *floating-point unit* (FPU) instructions to allow the computation of floating-point mathematics. Specialized graphics and video instructions are included in modern architectures as well.

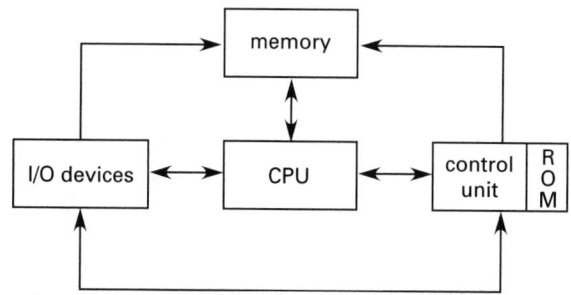

Figure 48.1 *General Computer Architecture*

2. CENTRAL PROCESSING UNIT/ CONTROL UNIT

The central processing unit (CPU) interprets and executes instructions. It consists of *registers*, which are the computer hardware for storing one machine word, and *logical networks* that perform four main operations: arithmetic, logical, data manipulation and transfer, and instruction manipulation and execution. The CPU performs these operations as specified by the control unit.

The control unit performs two functions: it obtains and reads the instruction steps from memory and ensures these steps are carried out in the proper sequence in the CPU.[3] The control unit uses *flags*, which are indicators that mark the occurrence of a given condition, as the signals for procedural control.

3. ARITHMETIC LOGIC UNIT

The main component of a computer is the CPU. The main component of the CPU is the arithmetic logic unit (ALU). An overview of the ALU is given in Fig. 48.2(a).[4] The operands are stored in registers and fed to the ALU. The *select* or *control line*, which in Fig. 48.2(a) uses m bits for 2^m possible operations, carries the ALU instruction set that allows the ALU to determine if the operation is an arithmetic or logical operation. Additionally, the type of arithmetic or logical operation is also contained in the select line—for example, addition, subtraction, logical AND, logical OR, and so on. The result is stored in the output register. Within the ALU are modules that perform the actual operation as shown in Fig. 48.2(b). The two operands, A and B, are fed into the ALU along with a *carry signal*, c_n, from the previous module. The select line determines the operation conducted. The results are placed on the output line, z, and any carry is placed on the carry output, c_{n+2}. If the result is outside the value that can be stored by the design of the architecture, an *arithmetic overflow signal*, ovr, is generated to alert the computer and operator to the invalid arithmetic operation. The ALU modules are composed of stages for each position in the chosen number system, usually binary, such as 2^0, 2^1, 2^2, and so on, as shown in Fig. 48.2(c). ALU modules can be cascaded to handle any size number.

4. MEMORY UNIT

The memory stores the instructions and data used by the control unit, the CPU, and the input/output (I/O) devices.[5] The memory instruction set is used by the CPU to determine where the different types of information are stored. The memory hierarchy used is shown in Fig. 48.3.

Register files are closest to the processor and consist of *general purpose registers* (GPRs) and *floating-point registers* (FPRs).[6] The registers in current designs have lengths of 32 bits or 64 bits. The registers are accessible by bit, byte (8 bits), word (16 bits), double-word (32 bits), or quad-word (64 bits) *granularity*.

---

[3]The control unit consists of electronic components designed as a *sequential logic network*, which processes signals in a specified sequence. The ROM, or read-only memory, consists of MOS devices with channels permanently or semipermanently configured.

[4]The basic building block of the ALU is the adder/subtracter, which is constructed from digital logic units. The select lines are built using demultiplexers.

[5]Memory structures in very large-scale integration (VLSI) devices are primarily MOS devices, for both ROM and RAM.

[6]The registers are constructed of fast flip-flops.

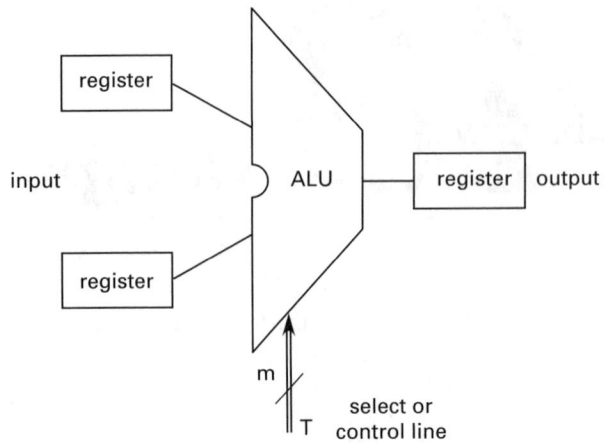

(a) ALU overview

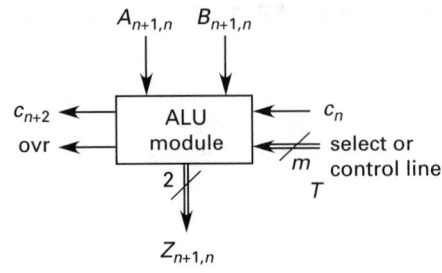

(b) general 2-bit ALU

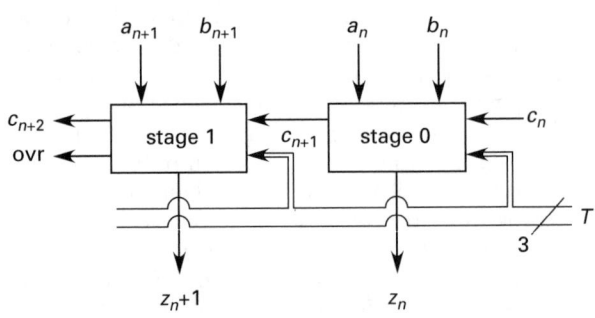

(c) ALU module stages

Figure 48.2 *Arithmetic Logic Unit*

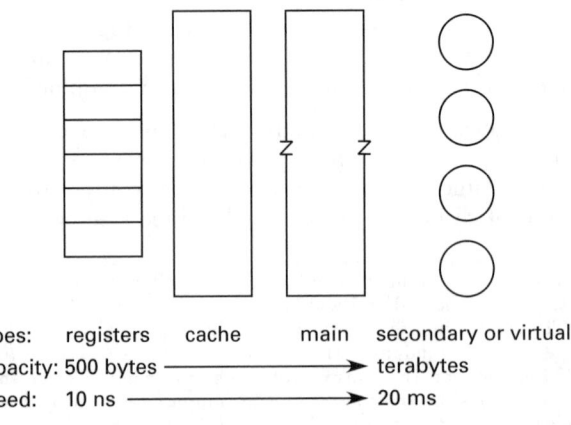

types:	registers	cache	main	secondary or virtual
capacity:	500 bytes	⟶		terabytes
speed:	10 ns	⟶		20 ms

Figure 48.3 *Memory Hierarchy*

The *cache memory* is a small, fast storage buffer integrated into the central processing unit and used to store a small subset of the main memory. The cache memory is developed from *static random access memory* (SRAM), which is faster, consumes more power, and is less dense than that used in the main memory. Using memory closer to the processor decreases computation time. The *main memory*, which stores both programming information and data, is developed from *dynamic random access memory* (DRAM).[7] The architecture determines the bit and byte arrangement of the data as well as the addressing format. The amount and layout of the cache and main memory within the computer are determined by the designer, not the system architecture.

The *secondary memory*, or *secondary storage*, also called the *disk storage*, typically uses magnetic or optical processes to store information. A portion of this memory can be used as *virtual memory*. Virtual memory is a combination of primary and secondary memories that is treated as a single memory by the computer. Virtual memory is controlled by software, which is used to set the memory location addresses. The architecture determines the smallest addressable unit, known as a *block*, and the total amount of secondary storage possible.

5. INPUT/OUTPUT DEVICES

An *input/output* (I/O) *device* is any unit that accepts data (information), sends it to the computer for processing, receives the results, and translates the result into a usable form. Typical devices include a monitor, keyboard, and printer. Since the computer processes information at high speeds, many I/O devices are unable to maintain the same pace. *Buffers* temporarily store information from the computer and feed the I/O device at its design rate. The link between any I/O device and the computer is called the *interface*. Since many peripheral devices are analog in nature, analog-to-digital (A/D) and digital-to-analog (D/A) converters are used on the interface.

6. BUSSES

A *bus* is one or more conductors in a computer used to transfer information. The *bus architecture* is the structure used to control that information flow. The *system bus* is the main communication path within the computer and is normally the bus referred to when computer speed is specified. The speed is specified as a clock rate, which is determined by the logic circuits and interconnecting busses, with their inherent delays. The system bus is composed of independent sub-units such as the *address*, *control*, and *data busses*. Modern computers sometimes have a *local bus* that interfaces with the system bus to provide high bandwidth for graphics applications. A device that can initiate and control communication is called a *bus master*. The process used for the bus master to obtain control is called *bus arbitration*. The procedure followed for transfer of information on the bus is called the *bus protocol*. When the bus protocol is such that both the sending unit and the receiving unit must be ready before the next step is initiated, it is termed *handshaking protocol*. Handshaking protocol is thus asynchronous. When a clock signal is used as a reference for the sending and receiving units, the protocol is called *synchronous*.

Busses are transmission lines and can be understood using such theory. Problems inherent in transmission lines have similar effects on computer systems. For example, signal reflection can cause boards to cease operation when another board is plugged into a computer, or cause them to work in a single slot only. (This is minimized by impedance matching.) Transmission line propagation delays can cause *signal skew*, which is the receipt of different bits of the same word at varying times. Busses are also subject to interference called *cross-coupling*, *crosstalk*, or *coupling noise*, which can distort or change information.

Computers

---

[7]SRAM consists of MOS transistor devices with resistors. DRAM consists of MOS transistors with capacitors.

49 Digital Logic

Computers (side tab)

Nomenclature

$f(x)$	scalar function of x	–
$F(X)$	vector function of X	–
K	digital thousand	$2^{10} = 1024$
m	minterm	–
M	maxterm	–
n	number	–
N	total number	–
POS	product of sums	–
SOP	sum of products	–

1. DIGITAL INFORMATION REPRESENTATION

Computers process information in digital form. The simplest representation of digital information is that of a single binary variable called a *scalar*. A binary scalar takes on a value of zero or one and represents one bit of information. The smallest information storage unit is termed a *cell*. The notation for a scalar is x_i.

At the electronic device level, the zero or one is defined in terms of a voltage output for the logic family chosen. For transistor-transistor-logic (TTL) the value of the zero and one, called the *state* of the cell, is shown in Fig. 49.1.

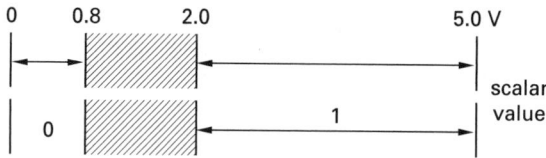

Figure 49.1 *Definition of Binary States for TTL*

A horizontal grouping of n binary variables is called an *n-tuple* and is represented by

$$[x_1, x_2, x_3, \ldots, x_n]$$

The cells can be listed and numbered left to right or right to left. Left to right is the normal notation. An exception occurs when representing a binary number, in which case the cells are listed right to left from zero to $n-1$ corresponding to $2^0, 2^1, \ldots, 2^{n-1}$. Since each binary variable in the n-tuple takes on only one of two values, the total number of values represented, N, is

$$N = 2^n \qquad \textbf{49.1}$$

If the information represented by an n-tuple is the information desired and not the details of that information in terms of zeros or ones, the n-tuple is called a *vector*. Vectors represent the location of the desired information. That is, they point toward the desired location.

An ordered collection of cells is called a *register*. An n-bit register is a vector since only the information it possesses, and not the actual representation, is of concern. The variables within the register are called *state variables* since they define the value (state) of the register (vector). Thus, Eq. 49.2 represents a register and its state variables while Eq. 49.1 represents the total number of state variables. Particular values within a register are called *states of the register*. For example, a 2-tuple register has four states: [0,0], [0,1], [1,0], and [1,1]. The size of the register, that is, the value of n, is called the *word length*. The symbol K is used to indicate 2^{10} bits of information. The information within each register is encoded to represent numerical, logical, or character information (see Sec. 47-1).

Example 49.1

A certain computer uses a 32-bit word length. What is the maximum number of state variables in a register?

Solution

The maximum number of state variables is given by Eq. 49.1.

$$N = 2^n = 2^{32} = 4{,}294{,}967{,}296$$

2. COMPUTER NUMBER SYSTEMS

Several types of number systems other than base 10 find application in computer systems. These include binary, octal, and hexadecimal. These systems and their manipulation are described in Ch. 13.

3. FUNDAMENTAL LOGIC OPERATIONS

Information to be processed is combined with, and compared to, other information. The logic that does this in such a manner that the output depends only on the value of the inputs is called *combinational logic*.[1] *Logic variables* have one of two possible values, zero or one. By convention, zero is considered "false" and one is considered "true." *Logic constants* have only one value, zero or one. Logic functions are commonly represented in truth tables. Since an n-tuple can take on 2^n values, a truth table has 2^n rows. Each of these rows has an associated output that can be either zero or one. Combining these two facts gives the number of scalar functions that can be generated from n logic variables.

$$\text{no. of scalar functions} = (2)^{2^n} \qquad 49.2$$

Not all the scalar functions will be useful. In fact, only six will be of primary interest. Further, all logic functions can be represented in terms of only two of these, NAND and NOR. These often-used functions are called *logic operators*. A logic operator involving a single variable or operand is called a *unary operation*. A unary function is shown in Table 49.1.

Table 49.1 *Unary Function*

x	$f_1(x) = 0$	$f_2(x) = x$	$f_3(x) = \overline{x}$	$f_4(x) = 1$
0	0	0	1	1
1	0	1	0	1

The function $f_2(x)$ is called the *identity function*. The function $f_3(x)$ is called the NOT, or negation, function. The symbol and truth table for the NOT function are shown in Fig. 49.2.

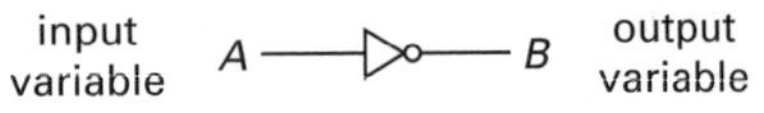

A	B
0	1
1	0

(b) truth table

$$A = \overline{B}$$

(c) equation

Figure 49.2 *NOT Logic*

A logic operator involving two operands is called a *binary operation*. For a two-variable input, that is, a 2-tuple, there are 16 possible functions. As mentioned, six of these are used so often that they have been given their own symbols. These are the AND, OR, XOR, NAND, NOR, and XNOR (or coincidence) logic operators or gates. The X indicates the word "exclusive," while N indicates the word "not." The logic operator AND is indicated by a dot "·" or by "∧". The dot is commonly not used when all the variables are represented by single letters (for example, $A \cdot B = AB$). The logic operator OR is indicated by "+" or "∨". The small circle on the end of any gate symbol, or on the input or output, indicates the NOT function. The standard symbol for the NOT operation is $\overline{x}$. Alternate symbols for the NOT operation include $-x$ or x'. A circle around the AND or OR symbol indicates "exclusive" operation, $\otimes$ or $\oplus$. The fundamental logic operations and their properties are shown in Table 49.2.[2]

Table 49.2 *Logic Operators*

operator	symbol	truth table			equation
		A	B	C	
AND		0	0	0	$A \cdot B = C$
		0	1	0	
		1	0	0	
		1	1	1	
OR	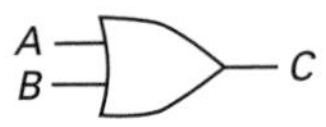	0	0	0	$A + B = C$
		0	1	1	
		1	0	1	
		1	1	1	
XOR	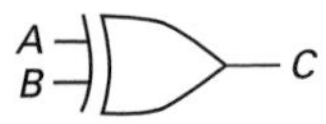	0	0	0	$A \oplus B = C$
		0	1	1	
		1	0	1	
		1	1	0	
NAND	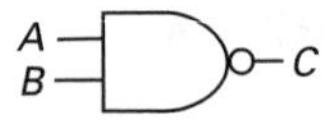	0	0	1	$\overline{A \cdot B} = C$
		0	1	1	
		1	0	1	
		1	1	0	
NOR	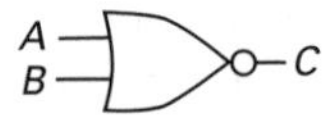	0	0	1	$\overline{A + B} = C$
		0	1	0	
		1	0	0	
		1	1	0	
XNOR or coincidence	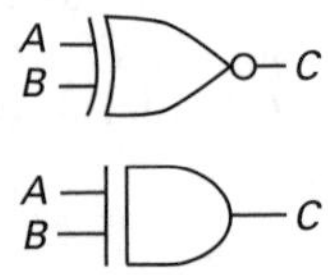	0	0	1	$A \odot B = C$ or $\otimes$
		0	1	0	
		1	0	0	
		1	1	1	

---

[1]When the output has a memory, that is, when the output is used as feedback to the input, the logic is called *sequential*.

[2]Argument inputs and results are commonly shown using capital letters. They can be scalars or vectors. Scalars have only two values. Vectors represent information and may have numerous values. The properties and laws applicable to logic variables are also applicable to logic vectors. Capital letters are sometimes reserved for logic vectors in educational texts. The symbology of A, B, C, or X, Y, Z is used for inputs and outputs with no special significance attached.

When combinations of logic operations are to be analyzed, the following *order of precedence* is applicable.

step 1: Evaluate those items within parentheses first.

step 2: Evaluate logic expressions from left to right, applying all instances of the NOT operation first, the AND operation second, and the OR operation last.

step 3: The NOT operator applied over an expression has the effect of enclosing the expression in parentheses. For example,

$$\overline{x_1 x_2} = \overline{(x_1 x_2)}$$

step 4: If in doubt when writing an expression, use parentheses to establish the proper order of precedence.

Two functions are *logically equivalent* when they are defined for the same arguments and

$$f(x_n, \ldots, x_1) = g(x_n, \ldots, x_1) \qquad \textbf{49.3}$$

Equation 49.3 must be valid for all combinations on the n-tuple. Logical equivalence is easily verified by using a truth table for both functions and ensuring that the results are identical. Additionally, the rules of switching algebra, usually referred to as Boolean algebra, can be used to transform one of the functions into the other, which proves equivalence.

Example 49.2

If $A = 1$ and $B = 0$, what is the output of the following expression?

$$\overline{\left(\overline{(AB)} \, \overline{(AB)} \right)}$$

Solution

Applying the rules of precedence, the expression is reduced as follows.

$$\overline{\left(\overline{(AB)} \, \overline{(AB)} \right)} = \overline{\left(\overline{(1 \cdot 0)} \, \overline{(1 \cdot 0)} \right)}$$

$$= \overline{\left(\overline{(0)} \, \overline{(0)} \right)}$$

$$= \overline{(1 \cdot 1)}$$

$$= \overline{1}$$

$$= 0$$

4. MINTERMS AND MAXTERMS

A *product term* is a function defined by a logical AND set of terms with either a variable x_i or its negation $\overline{x_i}$. A product term where all the variables appear, but only once, is called a *minterm*. The name occurs because the minterms take on a value of one for only one of the 2^n combinations of an n-tuple, that is, for the input

variables, and a value of zero for all other combinations. Expressed in terms of true/false notation, the minterms give the rows of a truth table in which the function is true. Minterms for a three-variable function are shown in Table 49.3.

A *sum term* is a function defined by a logical OR set of terms with either a variable x_i or its negation $\overline{x_i}$. A sum term where all the variables appear, but only once, is called a *maxterm*. The name occurs because the maxterms take on a value of zero for only one of the 2^n combinations of an n-tuple, that is, for the input variables, and a value of one for all other combinations. Expressed in terms of true/false notation, the maxterms give the rows of a truth table in which the function is false. Maxterms for a three-variable function are shown in Table 49.3.

Table 49.3 Minterms and Maxterms

decimal row number	binary input combinations ABC	minterm (product term)	maxterm (sum term)
0	000	$m_0 = \overline{A}\,\overline{B}\,\overline{C}$	$M_0 = A + B + C$
1	001	$m_1 = \overline{A}\,\overline{B}C$	$M_1 = A + B + \overline{C}$
2	010	$m_2 = \overline{A}B\overline{C}$	$M_2 = A + \overline{B} + C$
3	011	$m_3 = \overline{A}BC$	$M_3 = A + \overline{B} + \overline{C}$
4	100	$m_4 = A\overline{B}\,\overline{C}$	$M_4 = \overline{A} + B + C$
5	101	$m_5 = A\overline{B}C$	$M_5 = \overline{A} + B + \overline{C}$
6	110	$m_6 = AB\overline{C}$	$M_6 = \overline{A} + \overline{B} + C$
7	111	$m_7 = ABC$	$M_7 = \overline{A} + \overline{B} + \overline{C}$

Example 49.3

Consider the truth table for the two-variable function $F(A, B)$ shown. List the minterms.

A	B	C
0	0	0
0	1	1
1	0	1
1	1	0

Solution

The minterms occur where the value of the function is true, that is, one. Thus, the minterms are

$$m_1 = \overline{A}B$$

$$m_2 = A\overline{B}$$

Example 49.4

For the function in Ex. 49.3, list the maxterms.

Solution

The maxterms occur where the value of the function is false, that is, zero. Thus, the maxterms are

$$M_0 = A + B$$

$$M_3 = \overline{A} + \overline{B}$$

5. CANONICAL REPRESENTATION OF LOGIC FUNCTIONS

Any arbitrary function described by a truth table can be represented using minterms and maxterms. For example, the *minterm form* of the three-variable function in Table 49.3 is

$$F(A,B,C) = \sum_{i=0}^{7} m_i \qquad \text{49.4}$$

When Eq. 49.4 is expanded, it is referred to as the *canonical sum-of-product form* (SOP) of the function. For example,

$$F(A,B,C) = m_0 + m_1 + m_2 + m_3 + m_4 + m_5$$
$$+ m_6 + m_7$$
$$= \overline{A}\,\overline{B}\,\overline{C} + \overline{A}\,\overline{B}C + \overline{A}B\overline{C} + \overline{A}BC$$
$$+ A\overline{B}\,\overline{C} + A\overline{B}C + AB\overline{C}$$
$$+ ABC \qquad \text{49.5}$$

The reason the terms are called *minterms* is that they have a value of one for only one of the 2^i possible values of Eq. 49.4. (These are the terms of a function that are true, that is, logic 1.)

The *maxterm form* of the three-variable function in Table 49.3 is

$$F(A,B,C) = \prod_{i=0}^{7} M_i \qquad \text{49.6}$$

When Eq. 49.6 is expanded, it is referred to as the *canonical product-of-sum form* (POS) of the function. For example,

$$F(A,B,C) = M_0 M_1 M_2 M_3 M_4 M_5 M_6 M_7$$
$$= (A + B + C)\,(A + B + \overline{C})$$
$$\times (A + \overline{B} + C)\,(A + \overline{B} + \overline{C})$$
$$\times (\overline{A} + B + C)\,(\overline{A} + B + \overline{C})$$
$$\times (\overline{A} + \overline{B} + C)\,(\overline{A} + \overline{B} + \overline{C}) \quad \text{49.7}$$

The reason the terms are called *maxterms* is that they have a value of zero for only one of the 2^i possible values of Eq. 49.6. (These are the terms of a function that are false, that is, logic 0.)

The term "canonical" in this context means a simple or significant form of an equation or function. Canonical forms contain all the input variables in each term. Thus, SOP and POS forms may not be the simplest expression with regard to realizing these functions in terms of logic gates and may require minimization.

Example 49.5

Write the function in Ex. 49.3 in SOP form.

Solution

The minterms were determined in Ex. 49.3. The SOP form, using the format of Eq. 49.5 is

$$F(A,B) = \overline{A}B + A\overline{B}$$

Example 49.6

Write the function of Ex. 49.3 in POS form.

Solution

The maxterms were determined in Ex. 49.4. The POS form, using the format of Eq. 49.7 is

$$F(A,B) = (A + B)\,(\overline{A} + \overline{B})$$

Example 49.7

Consider the truth table shown. What are the (a) SOP and (b) POS forms of the function?

X	Y	Z	$F(X,Y,Z)$
0	0	0	0
0	0	1	1
0	1	0	1
0	1	1	0
1	0	0	1
1	0	1	0
1	1	0	0
1	1	1	1

Solution

(a) Minterms exist in rows 1, 2, 4, and 7. That is, they are logic 1. Thus, the SOP form is

$$F(X,Y,Z) = m_1 + m_2 + m_4 + m_7$$
$$= \overline{X}\,\overline{Y}Z + \overline{X}Y\overline{Z} + X\overline{Y}\,\overline{Z} + XYZ$$

(b) Maxterms exist in rows 0, 3, 5, and 6. That is, they are logic 0. Thus, the POS form is

$$F(X,Y,Z) = M_0 M_3 M_5 M_6$$
$$= (X + Y + Z)\,(X + \overline{Y} + \overline{Z})$$
$$\times (\overline{X} + Y + \overline{Z})\,(\overline{X} + \overline{Y} + Z)$$

6. CANONICAL REALIZATION OF LOGIC FUNCTIONS: SOP

Realization is the process of creating the necessary logic diagram from the truth table for a given function. Any scalar function can be represented in either canonical sum-of-product or product-of-sum form.[3] Thus, any function can be realized by a two-level logic circuit.

The SOP form of the function in Ex. 49.7 indicates a two-level AND-OR realization, which is shown in Fig. 49.3.

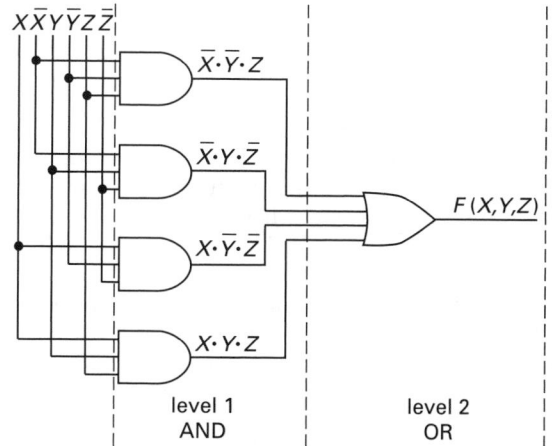

Figure 49.3 *SOP Realization*

In Fig. 49.3, the AND gates are in logic level one. The OR gate is in logic level two. It is normally assumed that the input variables and their complements are available. The connection points are shown with a dot. Undotted line crossings indicate nonconnecting crossings.

The same realization can occur using only NAND gates. Consider the equivalence shown in Fig. 49.4.

X	Y	Z	$\overline{X \cdot Y \cdot Z}$	$\overline{X} + \overline{Y} + \overline{Z}$
0	0	0	1	1
0	0	1	1	1
0	1	0	1	1
0	1	1	1	1
1	0	0	1	1
1	0	1	1	1
1	1	0	1	1
1	1	1	0	0

(a) truth table

$$ \begin{array}{c} X \\ Y \\ Z \end{array} \!\!\!\!\!-\!\!\!\!\!- \overline{X \cdot Y \cdot Z} \qquad \begin{array}{c} X \\ Y \\ Z \end{array} \!\!\!\!\!-\!\!\!\!\!- \overline{X} + \overline{Y} + \overline{Z} $$

(b) symbols

Figure 49.4 *NAND Equivalence*

The NAND gate can be represented in two ways: the standard way or as an OR gate with the inputs inverted. Using this equivalence, the SOP form of the function in Ex. 49.7 can be realized with NAND gates only as shown in Fig. 49.5.

---

[3]This form of logic realization is useful in the design of read-only memory (ROM) and programmable logic array (PLA) circuits.

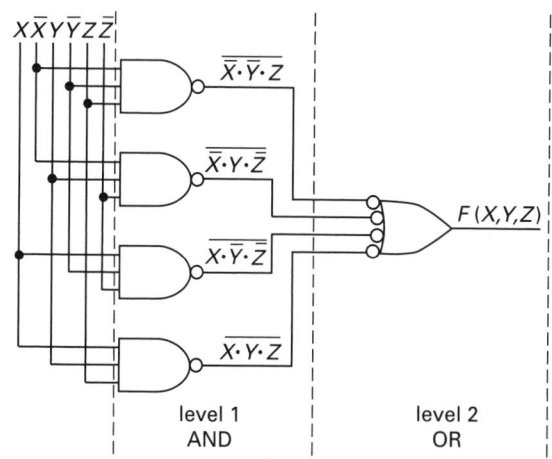

Figure 49.5 *SOP NAND-NAND Realization*

The realization in Fig. 49.5 is AND-OR. This can be seen by considering the inversions represented by the bubbles (small circles) on the output of the level-one logic to cancel with the inversions on the input to the level-two logic.[4]

7. CANONICAL REALIZATION OF LOGIC FUNCTIONS: POS

The POS form of the function in Ex. 49.7 indicates a two-level OR-AND realization, which is shown in Fig. 49.6.

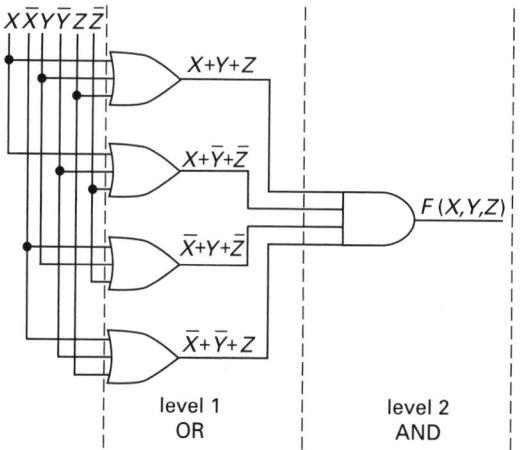

Figure 49.6 *POS Realization*

In Fig. 49.6, the OR gates are in logic level one. The AND gate is in logic level two. It is normally assumed that the input variables and their complements are available. The connection points are shown with a dot. Undotted line crossings indicate nonconnecting crossings.

---

[4]This is not a minimum realization, since the function is the XOR of X, Y, and Z and could be realized with two two-input XOR gates.

The same realization can occur using only NOR gates. Consider the equivalence shown in Fig. 49.7.

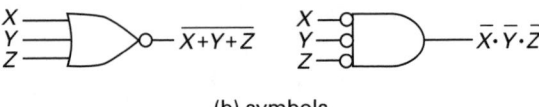

X	Y	Z	$\overline{X+Y+Z}$	$\overline{X} \cdot \overline{Y} \cdot \overline{Z}$
0	0	0	1	1
0	0	1	0	0
0	1	0	0	0
0	1	1	0	0
1	0	0	0	0
1	0	1	0	0
1	1	0	0	0
1	1	1	0	0

(a) truth table

(b) symbols

Figure 49.7 *NOR Equivalence*

The NOR gate can be represented in two ways: the standard way or as an AND gate with the inputs inverted. Using this equivalence, the POS form of the function in Ex. 49.7 can be realized with NOR gates only, as shown in Fig. 49.8.

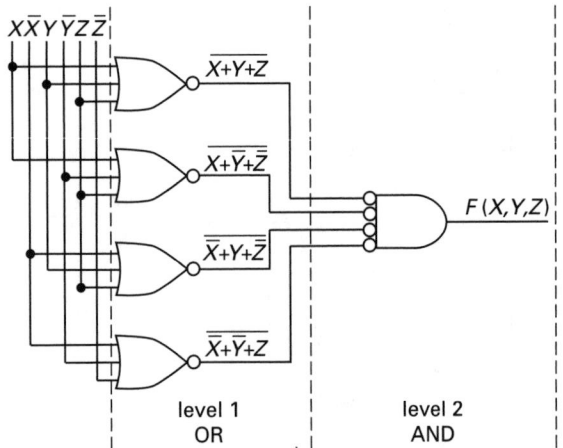

level 1   level 2
OR   AND

Figure 49.8 *POS NOR-NOR Realization*

The realization in Fig. 49.8 is OR-AND. This can be understood by considering the inversions represented by the bubbles on the output of the level-one logic to cancel with the inversions on the input to the level-two logic.[5]

---

[5]Again, this is not a minimum realization, since the function is the XOR of X, Y, and Z and could be realized with two two-input XOR gates.

8. ELECTRONIC LOGIC DEVICE LEVELS AND LIMITS

The electronic devices used to form the actual logic devices vary (see Ch. 45). When designing the logic itself, the type of device is of minimal importance. Therefore, the logic devices are shown as boxes or as symbols. For example, registers are shown as boxes and logic gates are shown with the appropriate symbology, with the inputs and outputs labeled but the internals empty.

Logic 1 is considered to indicate a true condition. Logic 0 indicates a false condition. Logic 1 is normally associated with the higher voltage in an electronic circuit, and logic 0 is associated with the lower voltage. This type of association is called positive logic. When the low-voltage condition represents the true statement, the logic is called negative. In order to avoid confusion, truth tables are sometimes given in terms of high (H) and low (L) voltages, rather than one and zero.

An alternative is to use *assertion levels*. When the assertion (true) condition is the high-voltage level, the term *active high* is used. This corresponds to positive logic. When the assertion (true) condition is the low-voltage level, the term *active low* is used.

In practice, the type of device and its limits need to be considered. Depending on the circuit, the high- and low-voltage levels will be a range of values. Also, the fan-in and fan-out capabilities limit the number of variable inputs and the number of logic gates that can be attached to the output. Additionally, the propagation delays must be accounted for to ensure that levels of logic have time to respond, providing the proper output before the next set of variables arrives. The propagation delays thus determine the operating speed of the logic network. Timing diagrams are used to map these delays and ensure proper operation.

9. SWITCHING ALGEBRA

Switching algebra is a particular form of Boolean algebra, though the distinction is seldom made. Three basic postulates define switching algebra.

- *postulate 1*: A Boolean variable has two possible values, zero and one, and these values are exclusive. That is,
    if $x = 0$ then $x \neq 1$, and
    if $x = 1$ then $x \neq 0$

- *postulate 2*: The NOT operation is defined as
    $\overline{0} = 1$ and $\overline{1} = 0$

- *postulate 3*: The logic operations AND and OR are defined as in Table 49.2.

Using these postulates, the special properties, laws, and theorems of switching algebra are developed. These properties, laws, and theorems are summarized in Table 49.4.

Table 49.4 *Basic Properties of Switching Algebra*
(Boolean Algebra)

name	property	dual
special properties: 0	$0 + A = A$	$0 \cdot A = 0$
special properties: 1	$1 + A = 1$	$1 \cdot A = A$
idempotence law	$A + A = A$	$A \cdot A = A$
complementation law	$A + \overline{A} = 1$	$A \cdot \overline{A} = 0$
involution	$\overline{(\overline{A})} = A$	–
commutative law	$A + B = B + A$	$A \cdot B = B \cdot A$
associative law	$A + (B + C) =$ $(A + B) + C$	$A \cdot (B \cdot C) =$ $(A \cdot B) \cdot C$
distributive law	$A \cdot (B + C) =$ $(A \cdot B) + (A \cdot C)$	$A + B \cdot C =$ $(A + B)(A + C)$
absorption law	$A + (A \cdot B) = A$ $A + (\overline{A} \cdot B) =$ $A + B$	$A \cdot (A + B) = A$ $A \cdot (\overline{A} + B) =$ $A \cdot B$
De Morgan's theorem	$\overline{(A + B)} = \overline{A} \cdot \overline{B}$	$\overline{(A \cdot B)} = \overline{A} + \overline{B}$

De Morgan's theorem, stated in Table 49.4, has many applications and is used when constructing logic networks from NAND and NOR logic. NAND and NOR are *universal operations*, sometimes called *complete sets*. This indicates that any logic expression can be represented solely by NAND logic, or solely by NOR logic. Since any logic expression is a combination of NOT, AND, and OR operations, proving that NAND logic can represent any of these three proves universality. This proof is shown in Eqs. 49.8 through 49.10 for the NOT, AND, and OR operations, respectively.

$$\overline{(A \cdot A)} = \overline{A} + \overline{A} = \overline{A} \qquad \textbf{49.8}$$

$$\overline{\overline{(A \cdot B)} \cdot \overline{(A \cdot B)}} = (A \cdot B) + (A \cdot B)$$
$$= A \cdot B \qquad \textbf{49.9}$$

$$\overline{\overline{(A \cdot A)} \cdot \overline{(B \cdot B)}} = (A \cdot A) + (B \cdot B)$$
$$= A + B \qquad \textbf{49.10}$$

Using the dual of Eqs. 49.8 through 49.10 proves the universality of the NOR operation. The dual of each of the items in Table 49.2 is obtained using the *principle of duality*. The dual of an expression is obtained by applying the following steps, without changing the variables involved.

step 1: If present, change zeros to ones and ones to zeros.

step 2: Change the original OR operations to AND.

step 3: Change the original AND operations to OR.

50 Logic Network Design

Nomenclature

d	don't care condition	–
$f(x)$	logic-level function of x	–
$F(X)$	system-level function of x	–
K	number of output lines	–
L	least significant bit	–
m	minterm	–
M	maxterm	–
N	number of input lines	–
Q	output signal	–
S	control signal	–
SOP	sum of products	–
x, X	variable	–
X	don't care condition	–
y, Y	output variable	–

1. COMBINATIONAL LOGIC DESIGN PROCESS

The design process starts with a set of requirements. The designer translates the requirements into a set of specifications that describe a combinational logic network, which must then be realized. Each design process is unique. Nevertheless, the general steps are as follows.

step 1: Using the set of requirements provided, identify and name all the inputs supplied to the network and all the outputs required of the network.

step 2: Establish the exact relationships between the inputs and the outputs. The relationships should be expressed in analytical form, that is, in a mathematical equation or truth table.

step 3: Code each input and output into digital form. The code may be fixed by the system requirements or open to the designer to interpret.

step 4: Develop the logic expressions that must be realized by the combinational logic network using the information in steps 2 and 3. Sometimes the expressions may be written directly from the relationships determined in step 2. If not, analytical techniques such as Boolean algebra or truth table techniques (described in this chapter) may be used to obtain the expressions.

step 5: Determine the logic circuit elements that will be used to realize the network. Applying analytical minimization techniques, reduce the complexity of the logic expressions to the desired level.[1]

step 6: Using the logic expressions from step 5, realize the network. Test the network to ensure that it meets the requirements of step 1.

2. OPERATIONS ON DIGITAL INFORMATION

Logic expressions (functions) at the logic level are symbolized using $f(x)$, while those at the system level use $F(X)$. At the logic level, design concerns include the coding of the information processed and the logic expressions necessary to realize the scalar functions. At the system level, design concerns include the type of information dealt with, its range of values, and the mathematical expressions required to obtain the desired output. The difference is illustrated in Fig. 50.1.

The *domain* of a scalar or vector represents its range of values. Scalars have only two values, but those values can have meaning, such as true or false. Vectors have a wide range of values. Examples of each are given in Table 50.1.

Operands are any of the quantities entering into or resulting from an operation. Operands represent the information being processed. The representation of operands is illustrated in Fig. 50.2.

---

[1] The time spent reducing the complexity is inversely proportional to the number of circuits to be produced. The greater the number of circuits to be produced, the more economical it becomes to reduce the complexity.

Vector *logic operations* use the same logic as scalar logic operations (see Ch. 49). *Relational operations* on vectors include greater than/less than, greater than or equal to/less than or equal to, and equal to/not equal to.[2] *Mixed-mode operations* allow the combination of a scalar with a vector.[3]

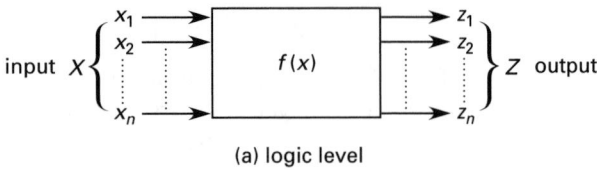

combinational logic network

(a) logic level

combinational logic network

(b) system level

Figure 50.1 *Representation of Processing Tasks*

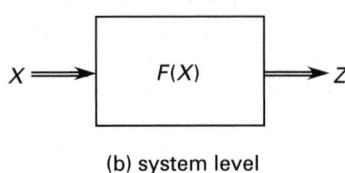

(a) scalar

(b) vector

(c) vector with n dimensions

Figure 50.2 *Representation of Operands*

Table 50.1 *Scalar and Vector Examples*

identifier	domain	description
x	$\{0,1\}$	scalar with a value of zero or one
f	{NOT READY, READY}	scalar representing two logic conditions
INST	{ADD, SUB, MUL, DIV}	vector representing four concepts— mathematical operations
SIGNAL	$\{-255,\ldots,0,\ldots +255\}$	vector representing decimal integers

---

[2]Coincidence logic is the simplest form of the "equal to" comparison.

[3]Mixed-mode operations are used by instruction registers to control the output of a combinational logic network.

3. MINIMIZATION PRINCIPLE

The complexity of a logic function depends on the Boolean expression representing the function. The complexity determines the number of gates necessary to realize the function and thus the circuit size, speed, and power consumption. Consequently, methods of minimization are used to reduce this complexity. The basis of minimization of logic functions is the *theorem of logical adjacency* given by either Eq. 50.1 or Eq. 50.2. The theorem applies to variables and functions.

$$A \cdot B + A \cdot \overline{B} = A \quad \text{[sum of products]} \quad \textbf{50.1}$$

$$(A + B) \cdot (A + \overline{B}) = A \quad \text{[product of sums]} \quad \textbf{50.2}$$

The logical adjacency theorem implies that minimization of SOP (Eq. 50.1) or POS (Eq. 50.2) functions can be achieved through the inspection of the logical adjacency of minterms and maxterms. *Adjacency* exists when minterms (or maxterms) differ by one value.

Example 50.1

A given logic function is defined by

$$F(A,B,C) = \prod M(2,4,5)$$

Using the theorem of logical adjacency, simplify the function.

Solution

Maxterms exist where the value of the function is zero. Expanding the given function into canonical POS form gives

$$F(A,B,C) = M_2 M_4 M_5$$
$$= (A + \overline{B} + C)(\overline{A} + B + C)(\overline{A} + B + \overline{C})$$

The maxterms are equal to the following binary values.

$$M_2 = 010$$
$$M_4 = 100$$
$$M_5 = 101$$

Comparing either the functional expression of variables or the binary representation, maxterms 4 and 5 differ by only one variable. The variable is C, in the 2^0 position in binary form. This variable can be eliminated, giving the simplified expression

$$F(A,B,C) = (A + \overline{B} + C)(\overline{A} + B)$$

4. KARNAUGH MAP FUNDAMENTALS

A *Karnaugh map* is a truth table that has been rearranged to form a geometrical pattern of functional relationships for determining gating configurations. The map allows gating requirements to be recognized in their simplest form through the principle of logical adjacency. Karnaugh two-, three-, and four-input maps are shown in Fig. 50.3 along with their associated truth tables.

Karnaugh maps are constructed with the inputs along the top and side, with the side having the least significant binary digits. The numbering scheme differs from that used in the truth table and is labeled "adjacency ordering" in Fig. 50.3. The numbering scheme is called *unit distance coding*, since adjacent numbers differ by only one bit. That is, adjacent inputs differ at most by X or $\overline{X}$.

The output is within the box. In Fig. 50.3, the output is represented as a scalar function associated with the binary input, and the box is numbered in the lower right with the decimal number of the corresponding row of the truth table. These decimal numbers equate to the binary numbers in the truth table and represent the locations of the minterms and maxterms. The actual output of a logical scalar expression will be either zero, one, or d. The d indicates a "don't care" condition. A don't care condition is used when the output is not defined for the given inputs. When such a condition exists the function is said to be *incompletely specified*. The d can be assigned a zero or one value as necessary to further the minimization process.

Five-input and six-input Karnaugh maps can be constructed, but the increasing complexity makes the adjacencies difficult to perceive. In practice, with five or more inputs minimization is done with a computer. The *tabular method* or *Quine-McClusky method* lends itself well to computer applications.

5. KARNAUGH MAP MINIMIZATION

The sum-of-product (SOP) and product-of-sum (POS) expressions for a given function can be obtained from a Karnaugh map. The SOP is obtained by grouping the minterms, that is, the blocks where the function output equals one. The POS is obtained by grouping the maxterms, that is, the blocks where the function output equals zero.[4]

The rows and columns of a Karnaugh map are numbered so that only one variable changes from block to block. Thus, rows are adjacent, with the top row adjacent to the bottom row as well. Columns are adjacent, with the leftmost column adjacent to the rightmost column as well. Combining adjacent rows and columns eliminates the variable that changes from one

---

[4]The resulting expressions may not be absolute minimizations. Nevertheless, the expressions are useful as starting points and the techniques are useful in the design of flip-flop logic for sequential circuits.

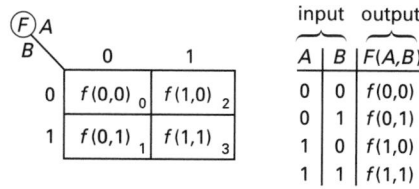

(a) two-input

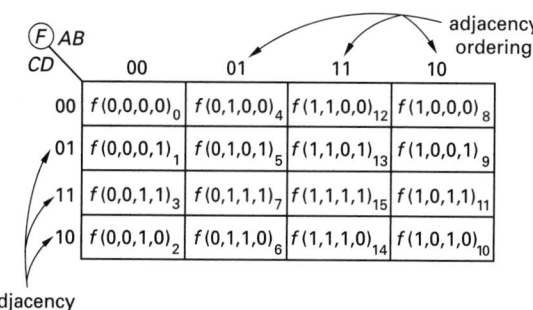

(b) three-input

input			output
A	B	C	F(A,B,C)
0	0	0	f(0,0,0)
0	0	1	f(0,0,1)
0	1	0	f(0,1,0)
0	1	1	f(0,1,1)
1	0	0	f(1,0,0)
1	0	1	f(1,0,1)
1	1	0	f(1,1,0)
1	1	1	f(1,1,1)

input				output
A	B	C	D	F(A,B,C,D)
0	0	0	0	f(0,0,0,0)
0	0	0	1	f(0,0,0,1)
0	0	1	0	f(0,0,1,0)
0	0	1	1	f(0,0,1,1)
0	1	0	0	f(0,1,0,0)
0	1	0	1	f(0,1,0,1)
0	1	1	0	f(0,1,1,0)
0	1	1	1	f(0,1,1,1)
1	0	0	0	f(1,0,0,0)
1	0	0	1	f(1,0,0,1)
1	0	1	0	f(1,0,1,0)
1	0	1	1	f(1,0,1,1)
1	1	0	0	f(1,1,0,0)
1	1	0	1	f(1,1,0,1)
1	1	1	0	f(1,1,1,0)
1	1	1	1	f(1,1,1,1)

(c) four-input

Figure 50.3 *Karnaugh Maps and Associated Truth Tables*

row to the next or from one column to the next. That is, the variable that appears in negated form in one of the minterms (or maxterms) and unnegated form in the other minterm (or maxterm) is eliminated (see Eqs. 50.1 and 50.2). When two minterms (or maxterms) are combined, one variable is eliminated. When four minterms (or maxterms) are combined, two variables are eliminated. The combination of eight minterms (or maxterms) eliminates three variables. Some of the typical, but less transparent, groupings of adjacent blocks are shown in Fig. 50.4.

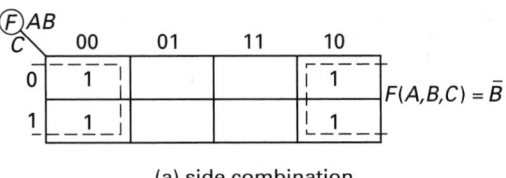

(a) side combination

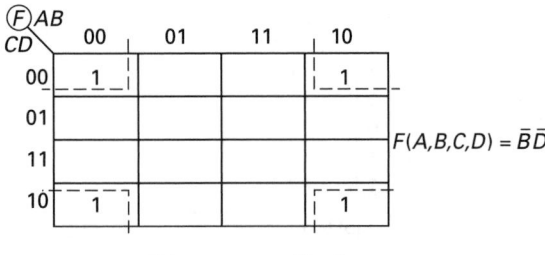

(b) corner combination

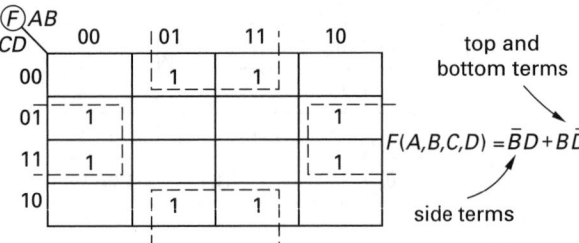

(c) side and top/bottom combination

Figure 50.4 *Typical Adjacent Minterm Groupings*

The general rules for combining or grouping minterms (or maxterms) are as follows.

- Each one-square (zero-square) must be accounted for in at least one grouping.

- Any grouping should be as large as possible.

- Groupings are made in terms of $1, 2, 4, 8, \ldots, 2^n$ minterms (maxterms).

- A one-square grouping should not be used if it can be combined with another one-square. A two-square grouping should not be used if it can be combined into a four-square, and so on.

- Groupings should be used in such a manner as to minimize the total number of groups.

The minterm formulation (grouping ones) is used to find the SOP form. The maxterm formulation (grouping

zeros) is used to find the POS form. Use the form with the smaller number of groups, since this generally leads to the minimum number of logic gate inputs. The general procedure for finding the minimal Karnaugh map follows.

step 1: Identify and circle all one-squares (zero-squares) that cannot be combined with any other squares.

step 2: Identify and circle all one-squares (zero-squares) that can combine with only one other square. Form groupings of two squares with these. If a square can be combined in any other manner, leave it uncombined at this step.

step 3: Identify and circle all one-squares (zero-squares) that can combine in groups of four only. Form groupings of four squares with these. If a square can be combined in any other manner, leave it uncombined at this step. Do not use the squares combined in step 2.

step 4: Identify and circle all one-squares (zero-squares) that can combine in groups of eight only. Form groupings of eight squares with these. If a square can be combined in any other manner, leave it uncombined at this step. Do not use the squares combined in steps 2 or 3.

step 5: Any square not assigned to a grouping is now arbitrarily placed in the largest group possible that includes the most uncovered squares. Add the minimum number of squares to cover all the remaining one-squares (zero-squares). Use squares with the don't care condition, d, to increase the grouping size.

step 6: Check the resulting expression's output against the original truth table. The results must match.

Example 50.2

A given three-variable function has the following truth table. Draw the Karnaugh map and determine the SOP form of the function.

A	B	C	$F(A, B, C)$
0	0	0	0
0	0	1	0
0	1	0	0
0	1	1	1
1	0	0	0
1	0	1	1
1	1	0	0
1	1	1	1

Solution

The Karnaugh map is

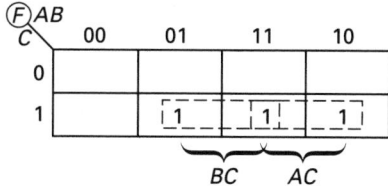

Step 2 has been applied to the map as indicated. The leftmost grouping (m_3 and m_7) contains A and $\overline{A}$, making it redundant. The rightmost grouping (m_7 and m_5) contains B and $\overline{B}$, making it redundant. The resulting expression is

$$F(A, B, C) = BC + AC$$

To spot-check, consider the input value of 011 for A, B, and C. Substituting this into the above expression results in a value of one, which matches the truth table and the Karnaugh map.

Example 50.3

Realize the function in Ex. 50.2 with NAND-NAND logic.

Solution

Assuming all the variables and their complements are available, the circuit can be realized in NAND-NAND form directly from the SOP equation. The result is

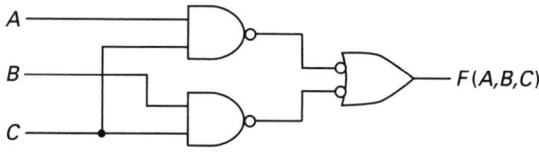

6. VEITCH DIAGRAMS

Adjacencies indicate which variable is to be eliminated during minimization. The variable that changes, that is, the one contained in a grouping in both its negated and unnegated form, is eliminated using the theorem of logical adjacency (see Eqs. 50.1 and 50.2). On the Karnaugh map this variable is indicated by its values of zero and one. To improve the usefulness of the map, the variables are listed by name rather than by value. The variable name is given in the true condition. For example, when a general variable $X = 0$, the variable is false, and the true condition is $\overline{X}$. The map, when using named variables, is called a *Veitch diagram*. Three- and four-input Veitch diagrams are shown in Fig. 50.5.

The Veitch diagram is a Karnaugh map that shows when the variable is true. When the variable is not true, the complement of the variable is true. The Veitch

diagram allows for the entry of zeros and ones immediately from a logic expression. Conversely, it allows the writing of logic expressions directly from the diagram.

When a map is constructed with the minterm information (values) and the Veitch diagram information (names), it is called a *Karnaugh-Veitch* (K-V) *map*. The groupings, also called *encirclements*, for a three-input K-V map are shown in Fig. 50.6. The groupings for a four-input K-V map are shown in Fig. 50.7. The groupings in both figures are by minterms. The groupings by maxterms are found using the zeros.

Example 50.4

Design the NAND-NAND logic to realize segment a of a seven-segment display for a calculator.

Solution

Step 1 of the design process in Sec. 50-1 is to identify and name all the inputs supplied to the network and all the outputs required of the network, using the set of requirements provided. The inputs will be binary numbers. The output will be a display such as

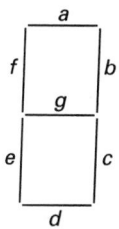

Step 2 is to establish the exact relationships between the inputs and outputs. Using a truth table for this purpose results in the following. The capital X indicates a don't care condition since using d might cause confusion with the segment d.

inputs	segments							decimal
$A\ B\ C\ D$	a	b	c	d	e	f	g	number
0 0 0 0	1	1	1	1	1	1	0	0
0 0 0 1	0	1	1	0	0	0	0	1
0 0 1 0	1	1	0	1	1	0	1	2
0 0 1 1	1	1	1	1	0	0	1	3
0 1 0 0	0	1	1	0	0	1	1	4
0 1 0 1	1	0	1	1	0	1	1	5
0 1 1 0	1	0	1	1	1	1	1	6
0 1 1 1	1	1	1	0	0	0	0	7
1 0 0 0	1	1	1	1	1	1	1	8
1 0 0 1	1	1	1	1	0	1	1	9
1 0 1 0	X	X	X	X	X	X	X	not used
1 0 1 1	X	X	X	X	X	X	X	not used
1 1 0 0	X	X	X	X	X	X	X	not used
1 1 0 1	X	X	X	X	X	X	X	not used
1 1 1 0	X	X	X	X	X	X	X	not used
1 1 1 1	X	X	X	X	X	X	X	not used

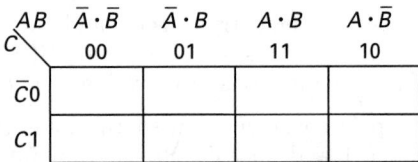

(a) three-input named variables

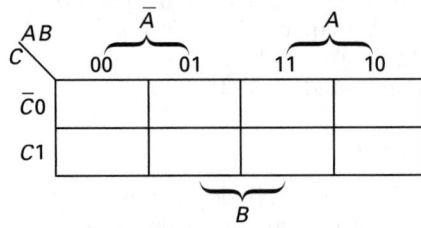

(b) three-input Veitch diagram

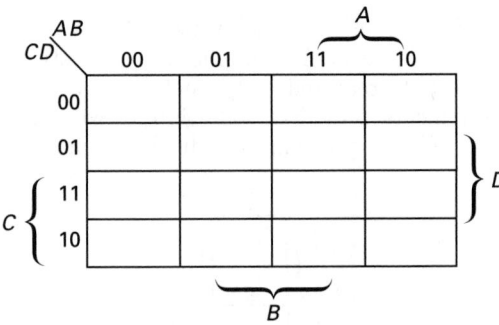

(c) four-input Veitch diagram

Figure 50.5 *Veitch Diagram Fundamentals*

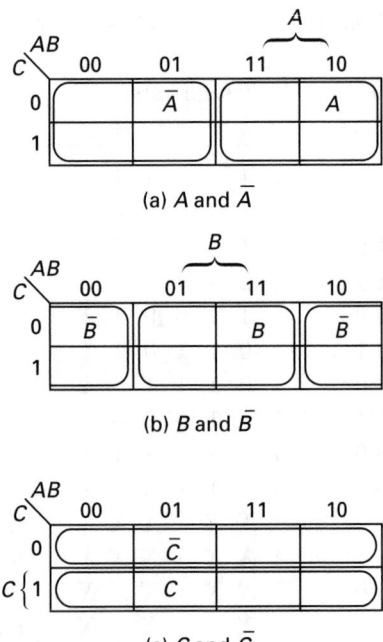

(a) A and $\bar{A}$

(b) B and $\bar{B}$

(c) C and $\bar{C}$

Figure 50.6 *Three-Input K-V Map: Minterms*

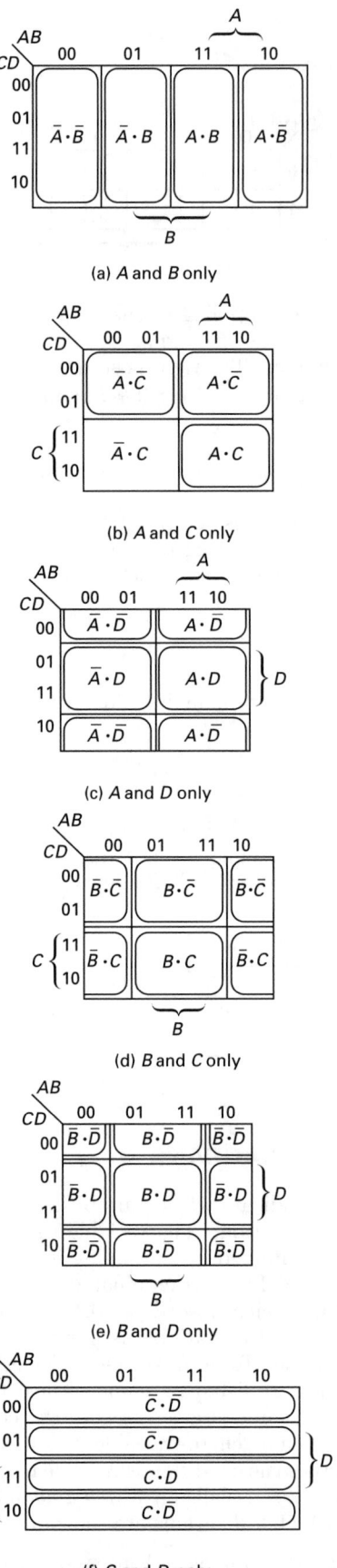

(a) A and B only

(b) A and C only

(c) A and D only

(d) B and C only

(e) B and D only

(f) C and D only

Figure 50.7 *Four-Input K-V Map: Minterms*

Step 3 is to code the inputs and outputs. In this case the ones and zeros are used to indicate the on-off condition of the particular segment of the display. The ones and zeros also correlate with the voltage necessary to drive the logic circuit.

Step 4 is to develop the logic expression. The Karnaugh map for any of the segments with the don't care conditions indicated is

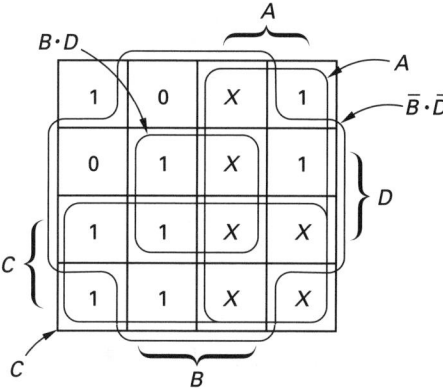

Placing the output values for segment a in the truth table, using the Veitch format, and grouping the terms to give the minimal map (see Sec. 50-5) gives

Step 5 is to determine the logic circuit elements that will be used to realize the network. In this case, NAND-NAND will be used. Applying analytical minimization techniques, reduce the complexity of the logic expressions to the desired level.

Minterm 10, m_{10}, is a don't care condition. Treating m_{10} as one allows the four corners, m_0, m_2, m_8, and m_{10}, to be combined to give $\overline{B}\,\overline{D}$. Minterms 13 and 15 are also treated as ones. This allows m_5 and m_7 to be combined with m_{13} and m_{15} to yield BD. Minterms 11 and 14 are also treated as ones. This, along with the don't care ones of m_{10} and m_{15}, allows the bottom two rows to be combined, yielding C. The remaining uncovered minterm is m_{12}. Making minterm 12 a don't care one allows the right half of the map to be encircled, yielding A. The resulting SOP expression is

$$f(\text{segment } a) = \overline{B}\,\overline{D} + BD + C + A$$

All ones were used for the don't care conditions. This does not have to be the case. In addition, multiple coverage of minterms occurred in the groupings. This is useful if it reduces the complexity.

The NAND-NAND expression is obtained by complementing the SOP form, applying De Morgan's law, and complementing the result (see Sec. 50-7). The result is

$$f(\text{segment } a) = \overline{\overline{(\overline{B}\,\overline{D})}\;\overline{(BD)}\;\overline{(C)}\;\overline{(A)}}$$

Step 6 is to use the logic expression to realize the network and to test the network. This expression is realized with the following NAND-NAND network.

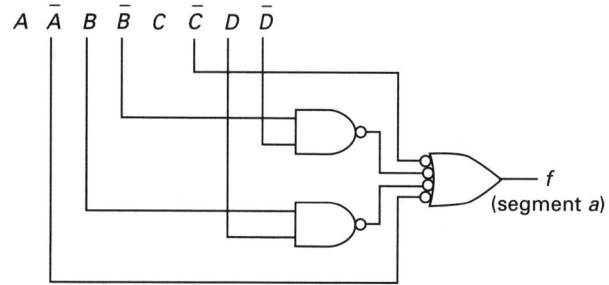

A check of the expression against the values in the truth table shows that the expression as written realizes the desired output.

Example 50.5

Realize segment a of the seven-segment display in Ex. 50.4 using NOR-NOR logic.

Solution

The NOR-NOR realization requires the use of maxterms. The maxterms occur where the output of the function is zero. The Veitch diagram for segment a is

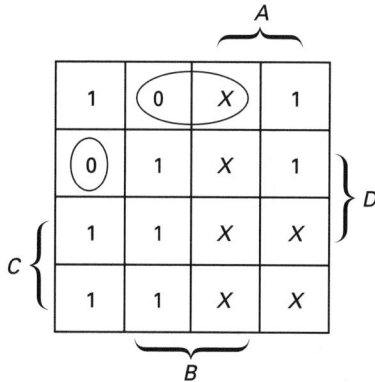

Maxterm 1 cannot be combined with any other zero. Recall that a maxterm is a sum term where the variables take on the value of zero. Thus,

$$M_1 = A + B + C + \overline{D}$$

Making the don't care condition of maxterm 12 a zero allows M_4 and M_{12} to be combined, yielding $\overline{B} + C + D$. Combining this with M_1 yields the following POS expression.

$$f(\text{segment } a) = \left(A + B + C + \overline{D}\right)\left(\overline{B} + C + D\right)$$

The NOR-NOR expression is obtained by complementing the POS form, applying De Morgan's law, and complementing the result (see Sec. 50-7). The result is

$$f(\text{segment } a) = \overline{\left(\overline{(A + B + C + \overline{D})} + \overline{(\overline{B} + C + D)}\right)}$$

The expression is realized using the following NOR-NOR network.

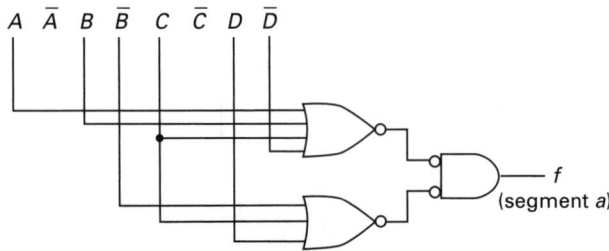

In Exs. 50.4 and 50.5, the conversion of the expression for the function is not required. The method of Secs. 49-6 and 49-7 can be used directly. The conversion directly shows the NAND-NAND or NOR-NOR relationship.

7. NAND/NOR FUNCTION REPRESENTATION

Functions in their SOP or POS form can be used to realize NAND-NAND only or NOR-NOR only logic as described in Secs. 49-6 and 49-7. Once a Karnaugh or Karnaugh-Veitch map has been minimized using the procedures of Sec. 50-5 and the resulting SOP or POS expression is found, the expression itself can be manipulated to specifically show the NAND-NAND or NOR-NOR relationship. The procedure follows.

step 1: For a NAND only expression, group the minterms (output value one) in a map to form the sum-of-product (SOP) expression. For a NOR only expression, group the maxterms (output value zero) in a map to form the product-of-sum (POS) expression.

step 2: Complement the expression from step 1.

step 3: Apply De Morgan's law to the expression in step 2.

step 4: Complement the result.

Example 50.6

Use a K-V map to find the minimum NAND only expression for the following three-input function.

$$F(A, B, C) = \sum m(3, 4, 7) + d(6)$$

The term $d(6)$ indicates a don't care condition in minterm position 6, that is, in 110 binary.

Solution

The K-V map, with grouping, is

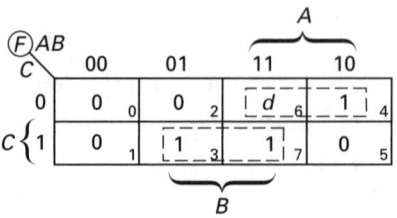

The minterm 6 don't care was arbitrarily given a value of 1. The resulting expression is

$$F(A,B,C) = A\overline{C} + BC$$

Complementing the result gives

$$\overline{F(A,B,C)} = \overline{(A\overline{C} + BC)}$$

Applying De Morgan's law results in the following expression.

$$\overline{F(A,B,C)} = \overline{(A\overline{C})}\ \overline{(BC)}$$

Complementing the result puts the expression in NAND-only form.

$$F(A,B,C) = \overline{\overline{(A\overline{C})}\ \overline{(BC)}}$$

8. PROGRAMMABLE LOGIC ARRAYS

The most direct method for realizing a network with multiple inputs and multiple outputs is to generate the canonical expressions that represent the function. This is accomplished with a *switching matrix*, an example of which is shown in Fig. 50.8.

The switching matrix shown in Fig. 50.8 realizes two-minterm expressions but could easily be connected for additional minterms. Consequently, a switching matrix can realize any combinational logic network. A device called a *programmable logic array* (PLA) is used to realize switching matrices. An unprogrammed PLA along with a PLA programmed to realize the switching matrix in Fig. 50.8 is shown in Fig. 50.9.

Programmable logic arrays are manufactured in standardized sizes. The AND and OR connections are "burned in" by a variety of methods, either by the manufacturer or in the field. In some cases, the connections are reversible and can be "burned out."[5]

---

[5]A read-only memory (ROM) is a switching matrix that can be realized with a PLA. The ROM requires additional circuitry to address the desired portion of the ROM, that is, a decoder, and transfer the output to the specified location.

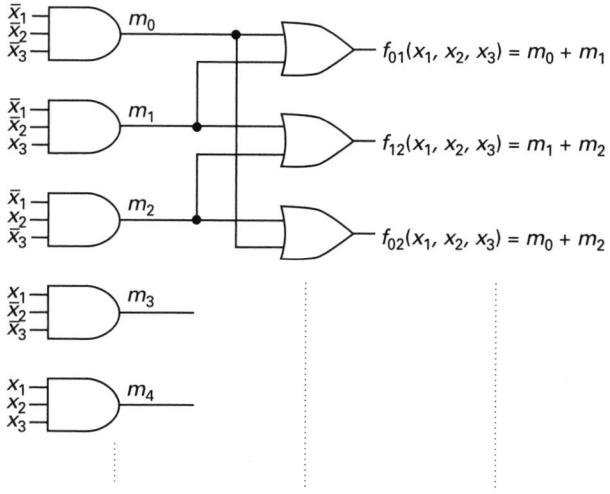

(a) multiple outputs of two minterms

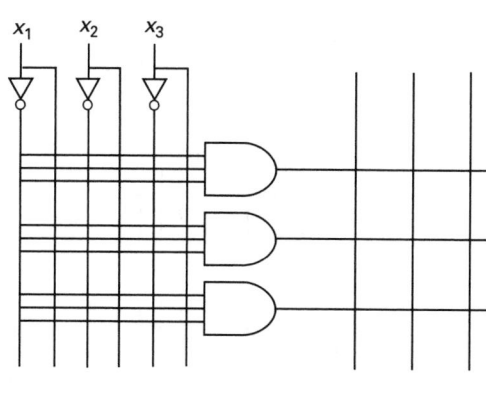

(a) unprogrammed

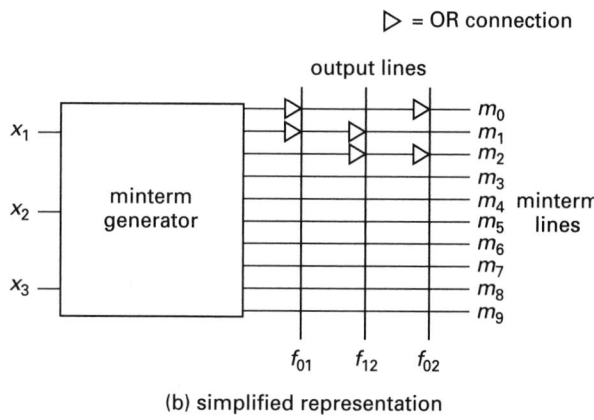

(b) simplified representation

Figure 50.8 *Switching Matrix*

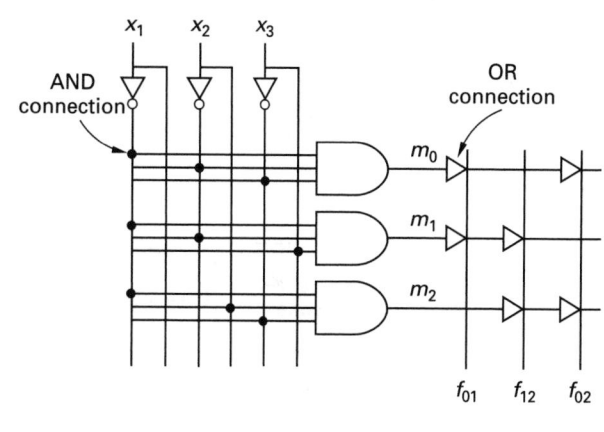

(b) programmed

Figure 50.9 *Programmable Logic Array*

9. ENCODERS AND DECODERS

An *encoder* is a network or system where 2^N inputs, or less, are placed on N or less output lines. A *decoder* is a network or system that uses N input lines to select one of 2^N or less outputs. Decoders are referred to as N-line-to-K-line decoders, where N represents the number of input lines and K is the number of output lines.

A simplified priority encoder is shown in Fig. 50.10. The output of the encoder provides information about the status of the highest priority line that is asserted. That is, the output is decimal 2 if the decimal 1 and 2 lines are asserted, decimal 3 if 3 and any combination below the 3 is asserted, and so on. In practice, an enable line is required to have following circuits discern when the output is valid. Encoders are used throughout computer systems. For example, encoders are available that encode ten input lines to four output lines, which would allow a decimal input to be coded as a binary coded decimal (BCD) number for processing.

A decoder is shown in Fig. 50.11. The decoder shown has a binary input that is used to select a minterm output. The decoder is an array of AND gates as used in a PLA. A decoder is sometimes referred to as a demultiplexer. Decoders are used throughout computer systems. For example, decoders are used to pick the desired information line given a specific address input in read-only memory (ROM). Also, instructions are decoded to enable the specific hardware required during central processing unit (CPU) operation.

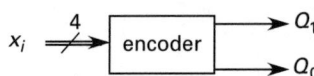

(a) priority encoder block diagram

input				output	
x_3	x_2	x_1	x_0	Q_1	Q_0
0	0	0	0	d	d
0	0	0	1	0	0
0	0	1	0	0	1
0	1	0	0	1	0
1	0	0	0	1	1

(b) priority encoder truth table

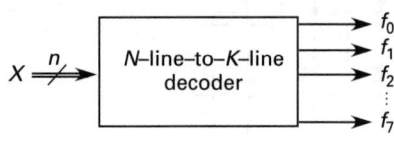

(c) AND-OR realization

Figure 50.10 *Encoder*

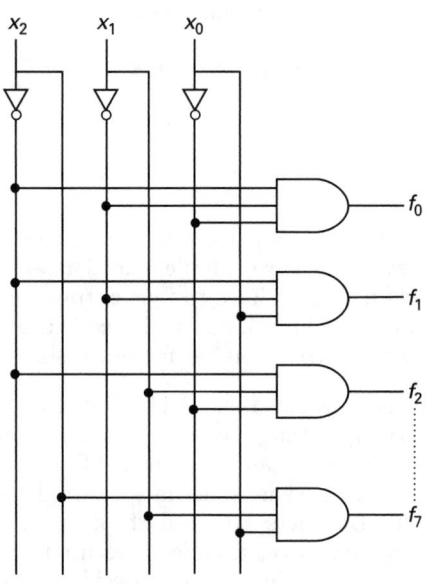

(a) decoder block diagram

(b) 3-line-to-8-line decoder

Figure 50.11 *Decoder*

10. DEMULTIPLEXERS

A *demultiplexer* is a device used to separate two or more signals previously combined by a compatible multiplexer and transmitted over a single channel or line. The demultiplexer is a *data selector*. A commonly used abbreviation is DEMUX or DMUX. A demultiplexer similar to the decoder in Sec. 50-9 is shown in Fig. 50.12. The demultiplexer is analogous to a multiposition switch that sends the input signal to the desired output line. Demultiplexers are used in the design of CPUs.

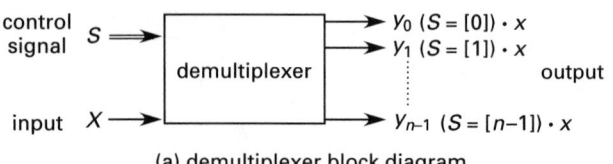

(a) demultiplexer block diagram

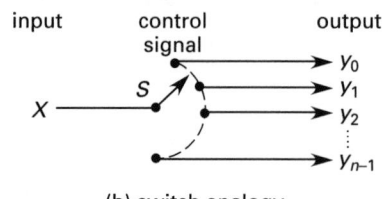

(b) switch analogy

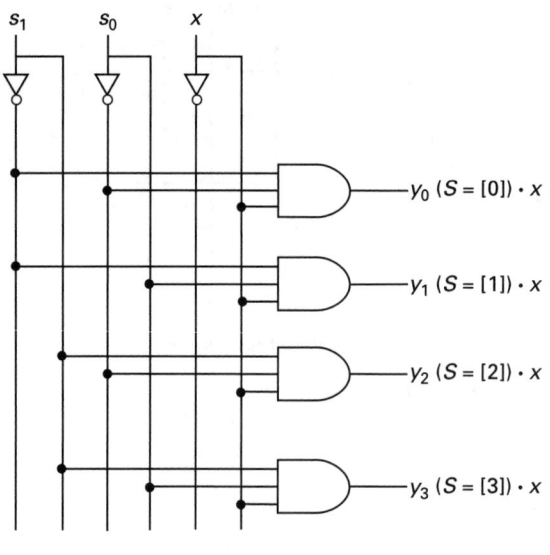

(c) 1:4 demultiplexer

Figure 50.12 *Demultiplexer*

11. MULTIPLEXERS

A *multiplexer* is a device for combining two or more signals for transmission over a single channel or line.[6] The multiplexer is a data selector. A commonly used abbreviation is MUX. A multiplexer compatible with the demultiplexer in Sec. 50-10 is shown in Fig. 50.13. An OR gate is used on the output to electronically isolate the AND gates from one another.

---

[6]Multiplexer is also spelled multiplexor.

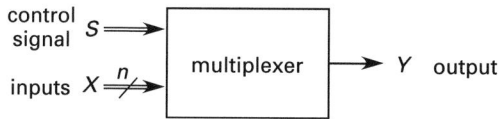

(a) multiplexer block diagram

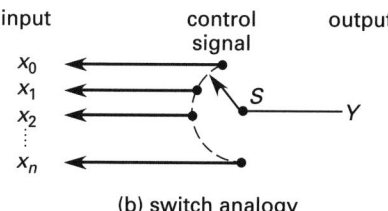

(b) switch analogy

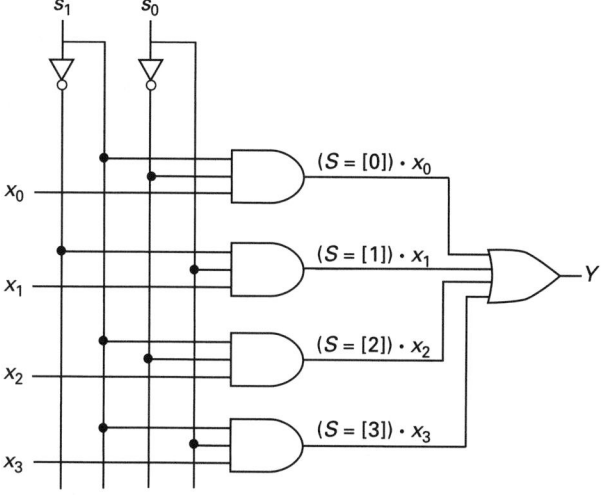

(c) 4:1 multiplexer

Figure 50.13 Multiplexer

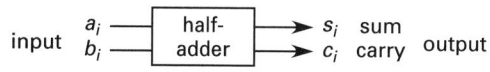

(a) half-adder block diagram

b_i	a_i	s_i	c_i
0	0	0	0
0	1	1	0
1	0	1	0
1	1	0	1

(b) half-adder truth table

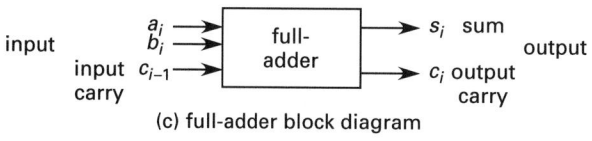

(c) full-adder block diagram

b_i	a_i	c_{i-1}	s_i	c_i
0	0	0	0	0
0	0	1	1	0
0	1	0	1	0
0	1	1	0	1
1	0	0	1	0
1	0	1	0	1
1	1	0	0	1
1	1	1	1	1

(d) full-adder truth table

Figure 50.14 Adders

To add an n-bit number, adders are cascaded to form a *ripple-carry adder* (see Fig. 50.15). The term *ovr* at the stage $n-1$ indicates an overflow condition, which occurs if the value is outside the range of numbers that can be stored.

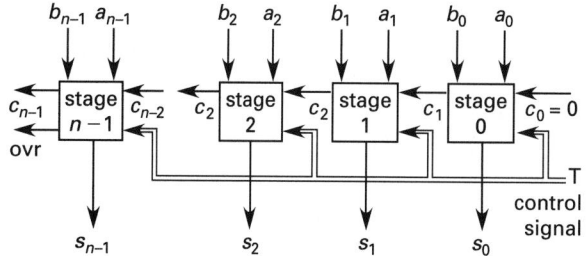

Figure 50.15 Ripple-Carry Adder

Subtraction is accomplished using ones complement arithmetic. That is, negative numbers are changed to their ones complement and added.

12. ARITHMETIC LOGIC DEVICES

The arithmetic logic unit (ALU) uses combinational logic devices to accomplish mathematical and logic operations. Binary addition is accomplished with the following rules.

- $0 + 0 =$ sum 0 carry 0
- $0 + 1 =$ sum 1 carry 0
- $1 + 0 =$ sum 1 carry 0
- $1 + 1 =$ sum 0 carry 1

Given these rules, adder truth tables are constructed as shown in Fig. 50.14. From the truth tables, the necessary combinational logic circuits are realized. For the *half-adder*, the sum is XOR logic and the carry is AND logic.

51 Synchronous Sequential Networks

Nomenclature

b	binary	–
C	clock	–
D	delay	–
FF	flip-flop	–
J	set	–
K	reset	–
Q	present state	–
$\overline{Q}$	complement of present state	–
Q^+	next state	–
R	reset	–
S	set or state	–
T	toggle or trigger	–
X	don't care condition	–

1. FUNDAMENTALS

The control unit of a computer or digital system has the ability to store information temporarily and then use that information to determine its future course of action. The unit thus has a memory. Networks with this ability are called *sequential networks*. A sequential network comprises *sequential logic elements* with at least one input and output and one internal state variable, and is designed so that the output signal depends on the past and present states of the inputs. When the state of the sequential network can change only when a transfer pulse or clock pulse is applied, the network is called a *synchronous sequential network*. The general form of a synchronous sequential network is shown in Fig. 51.1.

The combination logic network principles are described in Ch. 50. The *register* is the computer hardware for storing one machine word. Registers comprise a group of flip-flops, which are bistable multivibrators designed with logic circuits. (Flip-flops are the most commonly

used memory devices in sequential networks.) The general sequential network is defined by three types of equations: the *output equations*, the *next-state equations*, and the *flip-flop control equations*. The equations are obtained from *state tables,* also called *transition tables,* that list the inputs, current state, and next state of the circuit. State tables represented in graphical form are called *transition diagrams* or *state diagrams*.

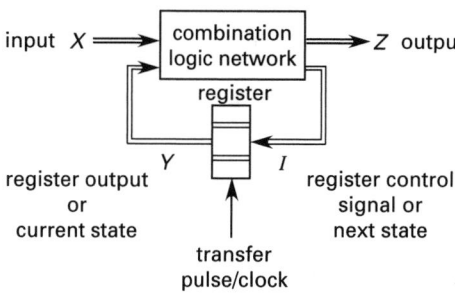

Figure 51.1 *General Synchronous Sequential Network*

The general steps in designing a synchronous sequential network are as follows.

step 1: Describe the network specifications.

step 2: Determine the state table required.

step 3: Minimize the state table.

step 4: Make the state assignments; that is, encode the information in binary form in order to transform the state table into a transition table.

step 5: Realize the network.

These steps conform to the general steps in the design of combinational logic networks (see Sec. 50-1) though the methods differ. Only steps 3 and 5 are algorithmic, with step 3 being of lesser importance with improved integrated circuit technology.[1]

2. CLOCKS AND TIMING

Flip-flops, which are also called *latches*, can be designed to operate only when a clock pulse is present. When this occurs, the flip-flop (FF) is *triggered* by the clock. The flip-flop is triggered either on the rising edge or on the

---

[1] The focus of this chapter will be on the construction and use of state and transition tables. Counters, which are used in numerous information processing tasks, will be the design examples.

falling edge of the clock pulse. The symbology used is shown in Fig. 51.2. The clock logic is realized with an AND gate and can be used with any flip-flop. Any state table is modified by simply adding a clock input that must be asserted to allow operation of the flip-flop.

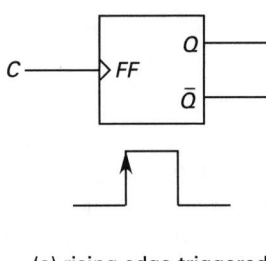

(a) rising edge-triggered
flip-flop

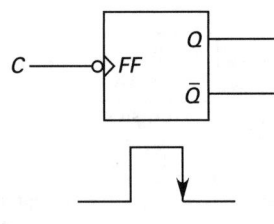

(b) falling edge-triggered
flip-flop

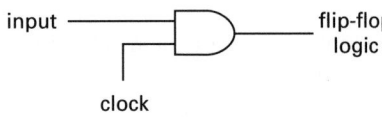

(c) clock logic realization

Figure 51.2 *Flip-Flop Clocks*

Clocks are used to prevent the flip-flop from changing due to spurious level changes in the inputs. A flip-flop without a clock is called a *transparent flip-flop*. Removing this transparency with a clock permits better synchronization and timing. Another method of removing this transparency is to use two clocked *S-R* flip-flops in series. (*S-R* flip-flops can be used to construct any of the other flip-flops, so this method works for any flip-flop.) The first flip-flop operates on the rising edge of the clock signal, and the second flip-flop operates on the falling edge of the clock signal, thus preventing unwanted transitions.

Additional inputs to any flip-flop are the *preset* and *clear signals*. Both these signals operate *asynchronously*; that is, they change the state of the flip-flop immediately, regardless of the state of the clock signal. Preset changes the output (the *Q* line) to a state of one. Clear changes the output, *Q*, to a value of zero. These control lines are used when initially starting a computer or digital system and at other appropriate times.

As with any electronic circuit, flip-flops have propagation delays and timing considerations. The delays generally vary for low-to-high and high-to-low transitions and for the type of transition, such as clock, preset, clear, and so on. These propagation delays determine the maximum operating frequency of the flip-flop. The *set-up time* is the minimum time inputs must be stable before a clock transition can be triggered. The *hold time* is the minimum time the inputs must remain stable after the clock transition.

3. *S-R* FLIP-FLOPS

The *S-R* flip-flop, also called the *S-C* flip-flop, is shown in Fig. 51.3(a). The *S* indicates the set condition, and the *R* indicates the reset or clear condition. The state table, in place of a truth table, is shown in Fig. 51.3(b). The present state is indicated by the *Q*. The next state is indicated by the Q^+. The condition that exists when both *S* and *R* have the value one is indeterminate and must not be allowed to occur. The realization of the flip-flop is shown in Fig. 51.3(c).

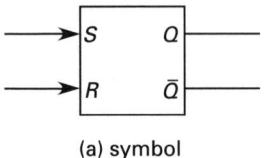

(a) symbol

S	R	Q	Q^+	remarks
0	0	0	0	hold
0	0	1	1	hold
0	1	0	0	reset
0	1	1	0	reset
1	0	0	1	set
1	0	1	1	set
1	1	0	indeterminate	disallowed condition
1	1	1	indeterminate	disallowed condition

(b) state table

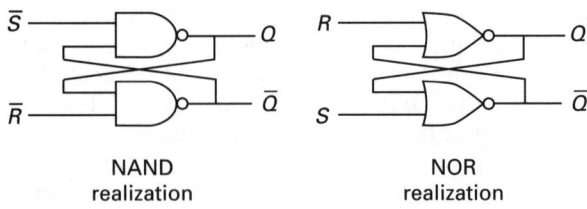

| NAND | NOR |
| realization | realization |

(c) logic realization

Figure 51.3 *S-R Flip-Flop*

The *next-state equation*, which is also called the *characteristic equation* or *transition equation*, for the *S-R* flip-flop is

$$Q^+ = S + \overline{R}Q \qquad 51.1$$

This equation is valid as long as the indeterminate state is disallowed. That is, $S = R = 1$ cannot occur. The restriction on Eq. 51.1, then, is $S \cdot R = 0$.

4. *J-K* FLIP-FLOPS

The J-K flip-flop takes its initials from the inventors. The J-K flip-flop is identical to the S-R flip-flop without the restriction against allowing both inputs to have the value one. The J-K flip-flop is shown in Fig. 51.4(a). The J indicates the set condition, and the K indicates the reset or clear condition. The state table, in place of a truth table, is shown in Fig. 51.4(b). The present state is indicated by the Q. The next state is indicated by the Q^+. When both J and K have the value one, the output toggles. The realization of the flip-flop is shown in Fig. 51.4(c).

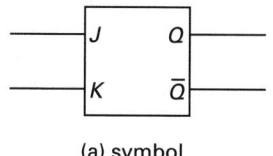

(a) symbol

J	K	Q	Q⁺	remarks
0	0	0	0	hold
0	0	1	1	hold
0	1	0	0	reset
0	1	1	0	reset
1	0	0	1	set
1	0	1	1	set
1	1	0	1	toggle
1	1	1	0	toggle

(b) state table

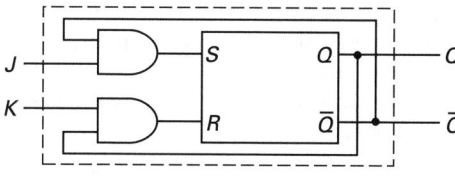

(c) logic realization

Figure 51.4 *J-K Flip-Flop*

The next-state equation for the J-K flip-flop is

$$Q^+ = J\overline{Q} + \overline{K}Q \qquad \textit{51.2}$$

This equation is valid as long as the input signals do not change during the clock pulse.

5. *T* FLIP-FLOPS

The T flip-flop is shown in Fig. 51.5(a). The T indicates the toggle condition. The state table, in place of a truth

table, is shown in Fig. 51.5(b). The present state is indicated by the Q. The next state is indicated by the Q^+. When T has the value one, the output toggles. The realization of the flip-flop is shown in Fig. 51.5(c).

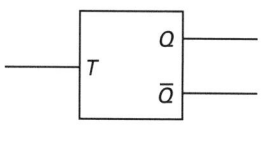

(a) symbol

T	Q	Q⁺	remarks
0	0	0	hold
0	1	1	hold
1	0	1	toggle
1	1	0	toggle

(b) state table

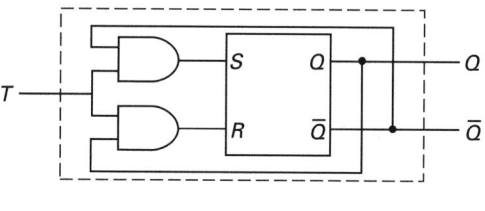

(c) logic realization

Figure 51.5 *T Flip-Flop*

The next-state equation for the T flip-flop is

$$Q^+ = T \oplus Q = T\overline{Q} + \overline{T}Q \qquad \textit{51.3}$$

This equation is valid as long as the clock pulse is high, which is no longer than the propagation time through the gates. This ensures that the flip-flop changes state only once per clock pulse.

6. *D* FLIP-FLOPS

The D flip-flop, also called the *delay* or *data flip-flop*, is shown in Fig. 51.6(a). The D indicates the delay in the output from the application of the clock pulse. The output follows the input but does so after the clock pulse is removed. That is, the input information is delayed one clock period. The D flip-flop is so named because it used in the accurate transfer of data from one source to another. The state table, in place of a truth table, is shown in Fig. 51.6(b). The present state is indicated by the Q. The next state is indicated by the Q^+. The realization of the flip-flop is shown in Fig. 51.6(c). The delay aspect of the flip-flop is illustrated in the timing diagram of Fig. 51.6(d).

Table 51.1 *Flip-Flop Transition Table*

present state Q	next state Q^+	S	R	J	K	T	D
0 $\longrightarrow$ 0		0	X	0	X	0	0
0 $\longrightarrow$ 1		1	0	1	X	1	1
1 $\longrightarrow$ 0		0	1	X	1	1	0
1 $\longrightarrow$ 1		X	0	X	0	0	1

$$Q^+ = S + \overline{R} \cdot Q$$

$$Q^+ = J \cdot \overline{Q} + \overline{K} \cdot Q$$

$$Q^+ = T \oplus Q$$

$$Q^+ = D$$

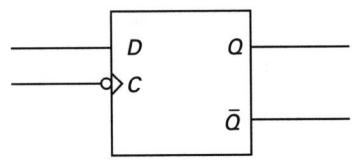

(a) symbol

D	Q	Q^+	remarks
0	0	0	The output follows
0	1	0	the input, D,
1	0	1	regardless of the
1	1	1	present state, Q.

(b) state table

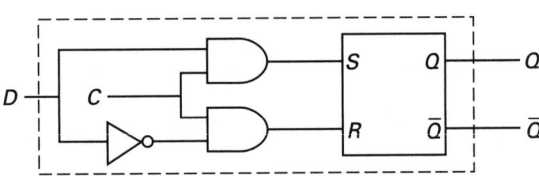

(c) logic realization (edge trigger circuitry not shown)

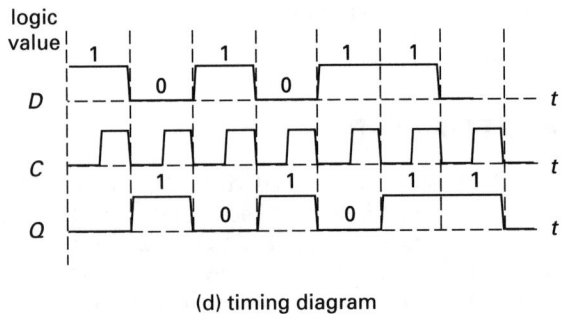

(d) timing diagram

Figure 51.6 *D Flip-Flop*

The next-state equation for the D flip-flop is

$$Q^+ = D \qquad\qquad 51.4$$

The D flip-flop is always a clocked device.

7. FLIP-FLOP TRANSITION TABLES[2]

In sequential networks, the transitions from the present state to the next state are controlled to satisfy design requirements. Four possible transitions are possible: zero to zero, zero to one, one to zero, and one to one. Though zero to zero and one to one do not appear to be transitions, they must be accounted for in logic design. The transitions for the principal flip-flops are summarized in Table 51.1.

8. COUNTERS

Sequential networks that function without any input signals are called *autonomous networks*. Such networks generate periodic control signals for use in information processing tasks. The general network is identical to that in Fig. 51.1 without the input signal, X. Autonomous networks are generalized counters.

Counters are sets of flip-flops, normally of the same type and functioning on the same edge of the clock signal. The count is determined by the number of states in a given sequence, limited to 2^n, where n is the number of flip-flops in the counter. When a counter starts at zero, counts to $2^n - 1$, and resets to zero, it is called a *binary up counter*, *modulo 2^n counter*, or *ripple counter*. An example is shown in Fig. 51.7.

The ripple counter in Fig. 51.7 is composed of J-K flip-flops with both inputs at value one, resulting in toggle action. The flip-flops start at clear, that is, with $Q = 0$, and operate on the falling edge of the clock. The initial clock signal, arbitrarily labeled zero, is applied to the stage 0 flip-flop (FF), which is triggered on the falling edge. On clock signal 1, the stage 0 FF switches from high to low, that is, $Q = 1$ to $Q = 0$, and the stage 1 FF is triggered from the output. On clock signal 2, the stage 0 FF switches from low to high and the stage 2 FF

---

[2]Truth tables show inputs and outputs and are derived from output equations. Transition tables show inputs, current-state values, and next-state values and are derived from next-state equations. Excitation tables show known state transitions and the control signals necessary to produce those transitions and are essentially rearranged transition tables. Excitation tables are used to generate the control equations.

remains unchanged. On clock signal 3, the stage 0 FF switches from high to low, as does the stage 1 FF. The stage 2 FF is triggered from the output of the stage 1 FF. The process continues as shown through decimal 7 and then resets.

A ripple counter is *asynchronous* because the flip-flops are not actually triggered by the clock. Since the count ripples through the circuit, the propagation delay is the fundamental limitation of this type of counter. Counters using common clocking are required for count lengths more than a few bits long.

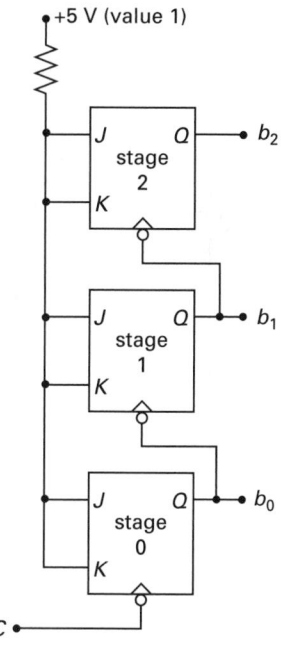

(a) realization

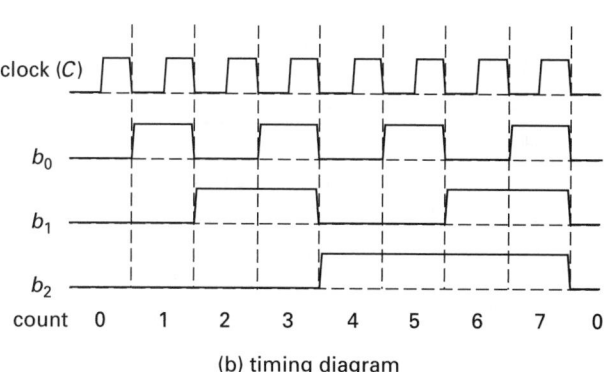

(b) timing diagram

Figure 51.7 *Ripple Counter*

9. SYNCHRONOUS BINARY COUNTER

A counter is *synchronous* when all the flip-flops change states at the same time. When this occurs, a state table, referred to as an excitation table, can be created in which the *state* is determined by the Q outputs of all the flip-flops. An example of a synchronous binary counter is shown in Fig. 51.8.

present state $Q_2Q_1Q_0$	next state $Q_2^+ Q_1^+ Q_0^+$	required T inputs $T_2T_1T_0$
000	001	001
001	010	011
010	011	001
011	100	111
100	101	001
101	110	011
110	111	001
111	000	111

(a) excitation table

$$T_2 = Q_0 \cdot Q_1$$
$$T_1 = Q_0$$
$$T_0 = 1$$

(b) control equations

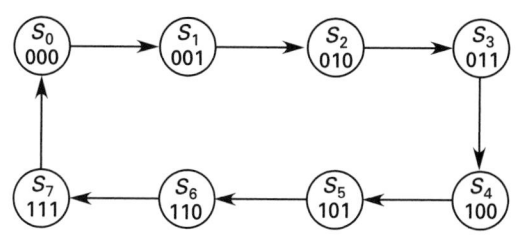

(c) state diagram

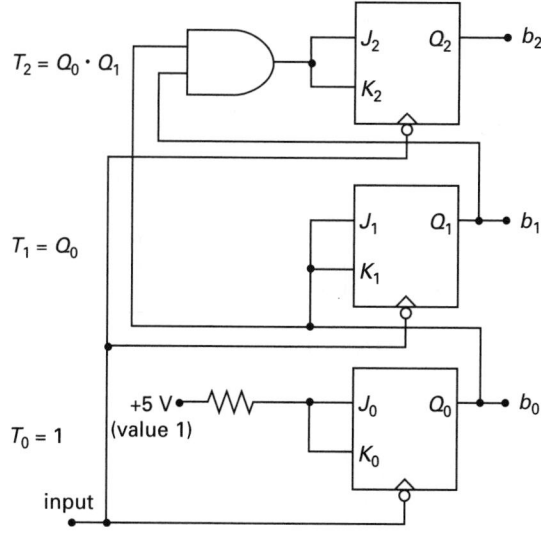

(d) realization

Figure 51.8 *Synchronous Binary Counter*

The state table is constructed by listing the desired state sequence in order in the present-state column, as in Fig. 51.8(a). The next-state condition is also listed, along with the required inputs necessary to cause the transition. The logic necessary to generate the required inputs is normally determined using a Karnaugh map or Veitch diagram. For this counter, the logic can be determined by inspection and the resulting control equations are shown in Fig. 51.8(b). To visualize the transitions and aid in the design, state diagrams, as shown in Fig. 51.8(c), are useful tools. The states are indicated by the S's and generally do not have to have the same subscripts as the counts. The realization of the network using J-K flip-flops configured as toggles is shown in Fig. 51.8(d).

state	present state $Q_2Q_1Q_0$	next state $Q_2^+Q_1^+Q_0^+$	outputs $b_2b_1b_0$	D inputs $D_2D_1D_0$
S_0	000	001	000	001
S_1	001	011	001	011
S_2	011	010	010	010
S_3	010	110	011	110
S_4	110	111	100	111
S_5	111	101	101	101
S_6	101	100	110	100
S_7	100	000	111	000

(a) excitation table

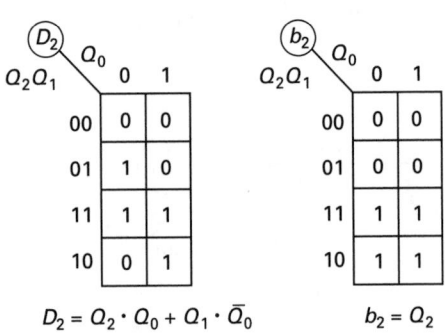

$$D_2 = Q_2 \cdot Q_0 + Q_1 \cdot \bar{Q}_0 \qquad b_2 = Q_2$$

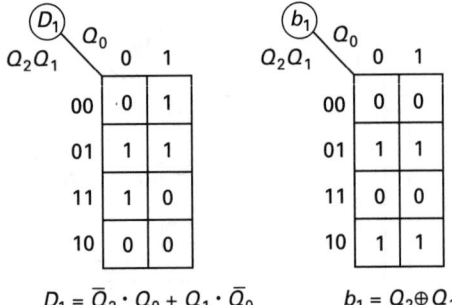

$$D_1 = \bar{Q}_2 \cdot Q_0 + Q_1 \cdot \bar{Q}_0 \qquad b_1 = Q_2 \oplus Q_1$$

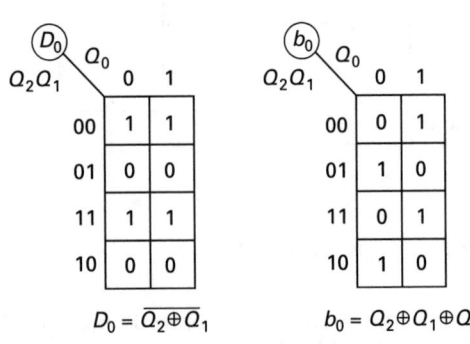

$$D_0 = \overline{\bar{Q}_2 \oplus Q_1} \qquad b_0 = Q_2 \oplus Q_1 \oplus Q_0$$

(b) control and output equations

(c) state diagram

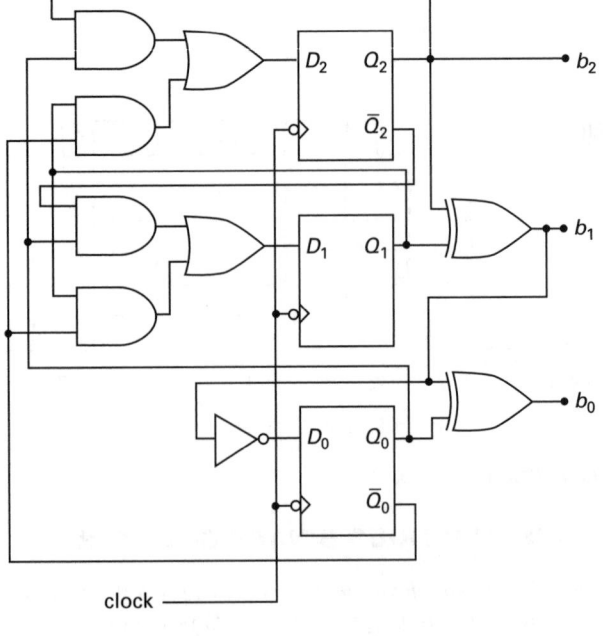

(d) realization

Figure 51.9 *Adjacent-State Binary Counter*

A problem with this type of design is that more than one flip-flop can change state at the same time. Consequently, due to differences in transition speeds, a false value can appear during the transition process. This can be avoided by allowing only one state variable to change during a clock period. This is equivalent to ensuring that state assignments are logically adjacent, that is, that the next state is logically adjacent to the present state.[3]

10. ADJACENT-STATE BINARY COUNTER

A binary counter uses all 2^n states. Thus, adjacent states can be assigned to all present states. When this is done, the state variables will no longer correspond to the correct count and logic circuitry must be used to correct the output. An example of an adjacent-state binary counter is shown in Fig. 51.9.

An excitation table for the synchronous counter in Sec. 51-9, rearranged to ensure logical adjacency, is shown in Fig. 51.9(a). The states are subscripted for the correct count and no longer match the binary count of the present state. The next state and the required inputs for D flip-flops, are shown. In Fig. 51.9(b), Karnaugh maps are used to determine the control and output equations. The state diagram is given in Fig. 51.9(c). The circuit is realized in Fig. 51.9(d).

Redundancy exists in the circuit diagram since $Q_1 \cdot \overline{Q}_0$ is common to D_1 and D_2. Additionally, when using D flip-flops, the last column of Fig. 51.9(a) can be eliminated since the value of D is always the next-state value for the corresponding state variable. Even for this relatively simple circuit, numerous logic lines clutter the realization. Because of this, feedback lines are normally omitted, and the state variables themselves are shown as inputs to the various gate inputs.

11. COUNTER ANALYSIS PROCEDURE

Though each design situation will be unique, the general procedure in Sec. 51-1 can be used in the analysis or design of counters. The network specifications of step 1 are given information; they comprise the requirements of the system. The state table size, step 2, is determined by the system requirements as well. Minimization of the state table, step 3, is accomplished through the removal of redundant or unnecessary states from the table. The state assignments of step 4 are made using heuristic and advanced analytical techniques due to the large number of possibilities (see Sec. 51-12 for general rules). Once in binary form, the next-state equations for the chosen flip-flop (see Table 51.1) are used to fill in the next-state column of the state table. With the present and next states known, Karnaugh maps, Veitch diagrams, or other appropriate methods are used to obtain the control and output equations. The circuit is obtained in step 5 by constructing the logic elements necessary

to realize the control and output equations for the given flip-flop. A check should also be done to ensure that the circuit realizes the requirements of step 1 and does not have *hang-up states*. Hang-up states, also called *lockout states*, are two or more unused states that have no next state in the desired sequence. A circuit check and search for hang-up states are efficiently accomplished using a state diagram and timing diagram.[4]

An example of counter circuit analysis illustrates the principles involved. The focus for analysis, however, is determining the state table used and developing the state diagram and timing diagram. Consider the counter in Fig. 51.10 with the control equations indicated. Since the system is already realized, all the design steps are complete.

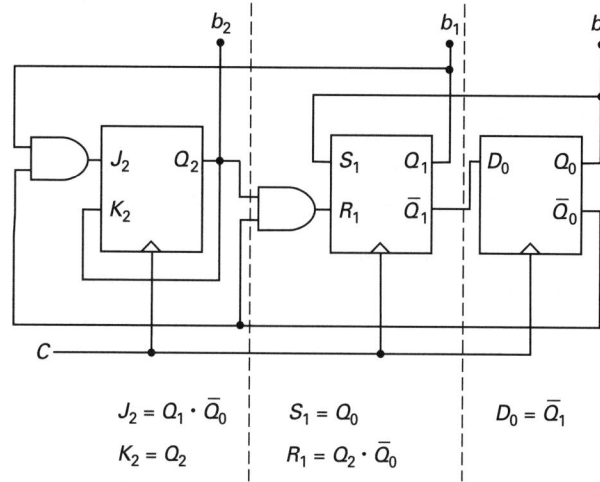

Figure 51.10 *Sample Counter*

Complete the present-state column of a state table, starting at zero, with the total number of states appropriate for the number of flip-flops used. The present-state column of Table 51.2 is the result. The states are listed in logical adjacency order (see Secs. 51-10 and 50-4). From the counter circuit in Fig. 51.10 and the next-state equations from Table 51.1, the next-state equations can be written as

$$Q_2^+ = J_2 \cdot \overline{Q}_2 + \overline{K}_2 \cdot Q_2$$
$$= \overline{Q}_2 \cdot Q_1 \cdot \overline{Q}_0 + \overline{Q}_2 \cdot Q_2$$
$$= \overline{Q}_2 \cdot Q_1 \cdot \overline{Q}_0 \qquad\qquad 51.5$$

$$Q_1^+ = S_1 + \overline{R}_1 \cdot Q_1$$
$$= Q_0 + \left(\overline{Q}_2 + Q_0\right) \cdot Q_1$$
$$= Q_0 + \overline{Q}_2 \cdot Q_1 \qquad\qquad 51.6$$

$$Q_0^+ = D_0 = \overline{Q}_1 \qquad\qquad 51.7$$

---

[3]This is not always possible, but is a desirable design goal.

[4]State diagrams are often useful in the beginning of design problems to aid in determining the inputs to the state table and to visualize the system requirements.

Equations 51.5 through 51.7 generate the next-state column of Table 51.2.

Table 51.2 State Table for Sample Counter

state	present state $Q_2Q_1Q_0$	next state $Q_2^+Q_1^+Q_0^+$
S_0	0 0 0	0 0 1
S_1	0 0 1	0 1 1
S_2	0 1 1	0 1 0
S_3	0 1 0	1 1 0
S_4	1 1 0	0 0 0
S_5	1 1 1	0 1 0
S_6	1 0 1	0 1 1
S_7	1 0 0	0 0 1

Examining Table 51.2 shows that the counter resets at state S_4. Thus, this is a five-state counter. Three states are unused, which indicates that hang-up states are possible. Checking Table 51.2 shows that all the unused states return to the main sequence. This is easily visualized with the state diagram shown in Fig. 51.11. A final check of the circuit is made to ensure proper operation using the timing diagram in Fig. 51.12.

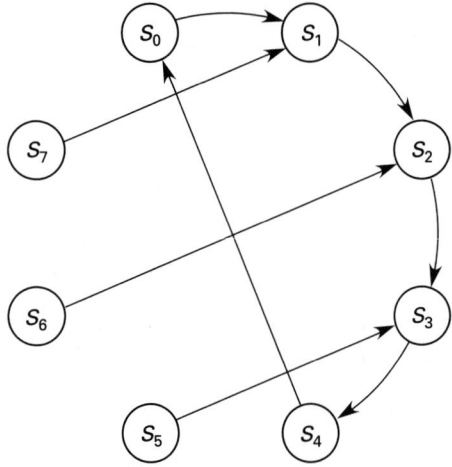

Figure 51.11 State Diagram for Sample Counter

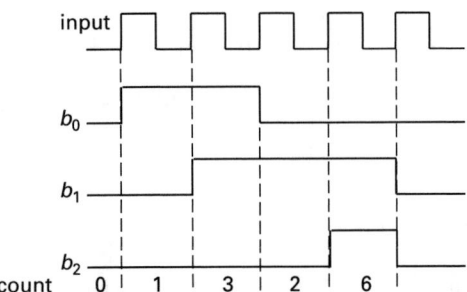

Figure 51.12 Timing Diagram for Sample Counter

12. SEQUENCE DETECTION

In general, a sequential circuit that operates in a predefined sequence with the inputs monitored using a clock on the same frequency as the outputs (i.e., with synchronous outputs) is called a *state machine*. If the state machine's next-state condition is determined by the current states only, it is called a *Moore machine.* If the state machine's next-state condition is determined by the current state and the circuit inputs, it is called a *Mealy machine.* A state machine that determines when a specific input sequence has occurred is called a *sequence detector.* The design of the detector follows the general guidelines in Secs. 51-1 and 51-11.

As an example, consider the design of a sequence detector used to determine when 1001 occurs. The state diagram is shown in Fig. 51.13.

The detector starts in state S_0, which is the condition indicating that the previous input was not in the correct sequence. This is called the *start-up state.* The detector remains in S_0 as long as zeros are obtained. This is the reason for the loop labeled $b = 0$ at S_0. When a one is received, the detector moves to state S_1, which is the condition indicating the start of the sequence.

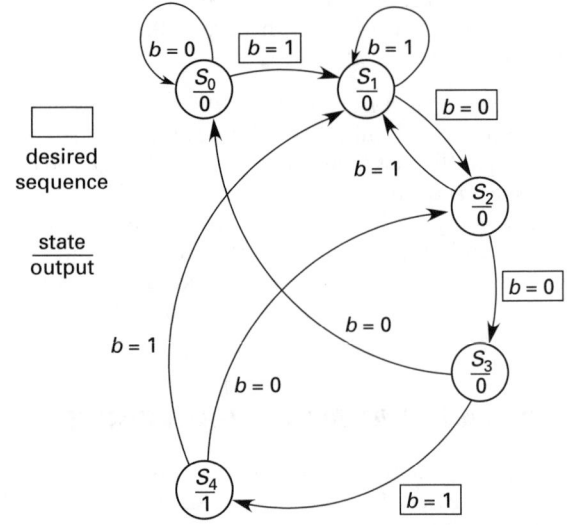

Figure 51.13 State Diagram for Sequence 1001 Detector

The detector remains in S_1 as long as ones are received. This is the reason for the loop labeled $b = 1$ at S_1. When a zero is received, the detector moves to state S_2, which indicates that a one followed by a zero has been received. If an input of one is now received, the detector must start over at state S_1. If a zero is received, the detector moves to state S_3, which is the condition indicating that the sequence 100 has been received. Either way, the detector remains in S_2 for one clock cycle only.

The detector also remains in S_3 for one clock cycle only. When a zero is received, the detector starts over at state S_0. If an input of one is now received, the detector moves to S_4, which indicates the sequence 1001 has been received. At S_4, the detector output must change to one to indicate the detection of the sequence. This is indicated in the state diagram underneath the state variable within the circle that indicates the state. At S_4 either a zero or a one input moves the detector out of the state. For a zero, the detector moves to S_2 since this indicates that a one followed by a zero has been received. For a one, the detector moves to S_1 since this indicates that a one has been received. The data is shown in Table 51.3.

Table 51.3 *State Table for Sequence 1001 Detector*

present state	next state for input b		present output
	$b = 0$	$b = 1$	
S_0	S_0	S_1	0
S_1	S_2	S_1	0
S_2	S_3	S_1	0
S_3	S_0	S_4	0
S_4	S_2	S_1	1

The state assignments must now be made. This process is complicated and a large number of possibilities exist. The following general rules can be used when assigning binary coding to the present state values to maximize logical adjacency and minimize logic gates.

- *Rule 1:* States that have the same next states for the same input should be given adjacent assignments.

- *Rule 2:* States that are the next state of the same state should be given adjacent assignments.

- *Rule 3:* States that have the same outputs for a given input should be given adjacent assignments.

Using these rules, the adjacency weighting for the sequence 1001 detector is shown in Table 51.4. The matrix lists the states and the number(s) of the rule(s) indicating that the states should be adjacent.

The number(s) of rule(s) applicable sets the priority of the adjacency assignments. Assuming equal weight among the rules, S_1 and S_2 are the first priority for adjacency. S_1 and S_0 are the second priority, followed

by all the states with two rules satisfied. The lowest priorities are S_2, S_3, and S_2, S_4. Not all the groupings are possible. It is common to set S_0 to the zero state (in this case, 000). Given this, one efficient grouping is shown in the present-state column of Table 51.5.

Table 51.4 *Adjacency Weighting for Sequence 1001 Detector*

	S_4	S_3	S_2	S_1
S_0	1, 2	1, 3	1, 3	1, 2, 3
S_1	1, 1	2, 3	1, 2, 2, 3	
S_2	1	3		
S_3				

Using the next-state equations of Table 51.1, the next-state column of Table 51.5 is filled out. The output equation is simply $Y = Q_2$. The control equations, that is, the flip-flop input equations, are determined from the Karnaugh maps derived from the state table as shown in Fig. 51.14.

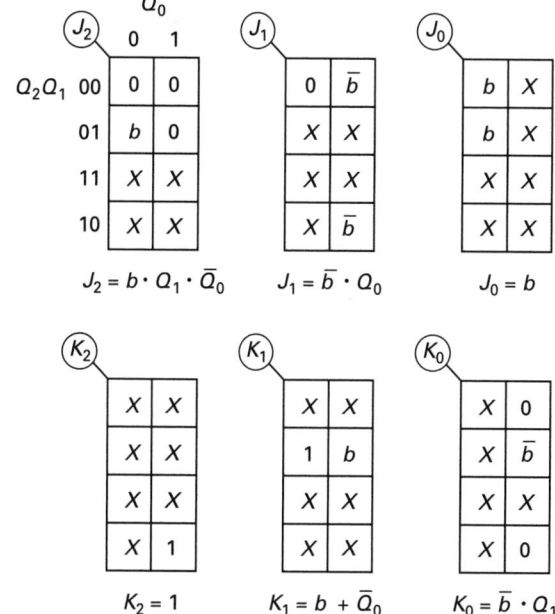

$$J_2 = b \cdot Q_1 \cdot \bar{Q}_0 \qquad J_1 = \bar{b} \cdot Q_0 \qquad J_0 = b$$

$$K_2 = 1 \qquad K_1 = b + \bar{Q}_0 \qquad K_0 = \bar{b} \cdot Q_1$$

Figure 51.14 *Control Equations for Sequence 1001 Detector*

Table 51.5 *Expanded State Table for Sequence 1001 Detector*

state	present state	next state		present output	flip-flop inputs					
		$b = 0$	$b = 1$							
S	$Q_2 Q_1 Q_0$	$Q_2^+ Q_1^+ Q_0^+$	$Q_2^+ Q_1^+ Q_0^+$	Y	J_2	K_2	J_1	K_1	J_0	K_0
S_0	0 0 0	0 0 0	0 0 1	0	0	X	0	X	b	X
S_1	0 0 1	0 1 1	0 0 1	0	0	X	$\bar{b}$	X	X	0
S_2	0 1 1	0 1 0	0 0 1	0	0	X	X	b	X	$\bar{b}$
S_3	0 1 0	0 0 0	1 0 1	0	b	X	X	1	b	X
S_4	1 0 1	0 1 1	0 0 1	1	X	1	$\bar{b}$	X	X	0

Using the control equations, the network is realized as shown in Fig. 51.15.

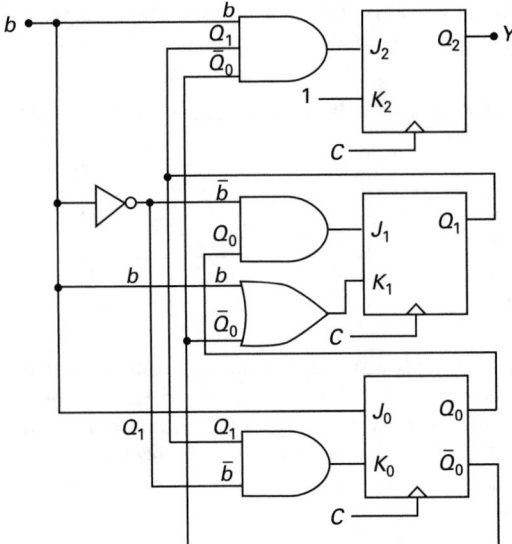

Figure 51.15 *Realization of Sequence 1001 Detector*

52 Programming Languages: FORTRAN

1. INTRODUCTION TO FORTRAN

FORTRAN is a high-level language originally designed for scientific calculations. Its name is derived from the phrase *for*mula *tran*slation. Versions of the language exist for all kinds of computers, from personal computers through the largest mainframes. ANSI (American National Standards Institute) FORTRAN, also known as *Standard FORTRAN*, *Full FORTRAN*, FORTRAN V, FORTRAN 77, and other names, is the basis for most versions.[1] Versions of FORTRAN identical in operation to ANSI FORTRAN are known as *full implementations*. Versions that accept a subset of ANSI FORTRAN commands are known as *partial implementations*. Versions that add their own commands not in ANSI FORTRAN are known as *extended implementations*.[2] Almost all versions implement FORTRAN as a *compiled language*. (See Sec. 47-5.)

This chapter is intended as a documentation of the FORTRAN language, not as a lesson in programming. However, because there are many versions of FORTRAN, some of the instructions listed in this chapter may not be compatible with all compilers.

---

[1]Nonstandard FORTRANs include WATFOR, WATFIV, and RATFOR.

[2]The most common extensions to FORTRAN are the T format and literal constants. The absence of string-handling capabilities was one of early FORTRAN's most significant deficiencies.

2. STRUCTURAL ELEMENTS

Symbols in FORTRAN statements are limited to uppercase letters, the digits 0 through 9, the blank, and the following characters:

$$+ \quad = \quad - \quad * \quad / \quad (\quad) \quad , \quad . \quad \$$$

FORTRAN compilers pack the characters; therefore, blanks can be inserted anywhere in most statements. For example, the following two statements compile identically.

```
IF (AGE . LT . YEARS) GO TO 10

IF(AGE.LT.YEARS)GOTO10
```

Although multiline statements are permitted, statements are prepared in 80-position (80-column) lines.[3] Table 52.1 describes how the various positions are allocated. While statements can be optionally numbered, they are executed sequentially regardless of the order of statement numbers.

Table 52.1 Column Positions in FORTRAN Statements

position	use
1	The letter "C" in position 1 indicates a comment.
2–5	The *statement number*, if used, can run from 1 through 9999.
6	Any character except zero indicates a continuation from the previous line.
7–72	The FORTRAN statement.
73–80	The last eight positions are available for any use and are ignored by the compiler.[a]

[a]These positions can be used to number the final debugged program sequentially or identify the program name (acronym) or programmer.

3. REPRESENTATION OF CONSTANTS

Numerical data can be of either the real or integer type. *Integers* have no decimal points. *Real numbers*, also known as *floating-point numbers*, are distinguished from

---

[3]This line length is a heritage of the punch card era. However, it also corresponds to the maximum number of characters that can be displayed across most monitors.

integers by a decimal point and may contain fractional parts. *Single-* and *double-precision* floating-point numbers are indicated by the letters E and D, respectively. Constants in FORTRAN are positive unless preceded by a minus sign. Commas are not allowed in any number.

Table 52.2 *Correct and Incorrect FORTRAN Constants*

type	correct representation	incorrect representation	
integer	90000	90,000	(comma)
integer	14	14.	(decimal point)
real	2.	2	(missing decimal)
real	90000.	90,000.	(comma)
real	7.0 E–4	7 E–4	(missing decimal)^a
real	7.1*10**2.3	7.1E2.3	(noninteger exponent)
real	1000.	E3	(exponent alone)

^{a}May be permitted by some FORTRAN implementations.

Constants will also be invalid if they are allowed numerical values too large or too small for the computer word size. As a consequence of the binary arithmetic used in integer operations, however, a computer with a standard word size of n bits will usually support single-precision integers in the range of -2^{n-1} to $2^{n-1} - 1$. The implementation of *double precision*, which doubles the number of bytes used to store constants, varies.[4]

4. INTEGER AND REAL VARIABLES[5]

Variable names are formed from up to six alphanumeric characters. The first character must be a letter. Variable names starting with the letters *I*, *J*, *K*, *L*, *M*, and *N* are assumed by the compiler to be integer variables. This is known as *implicit typing*. All variables not implicitly or explicitly integer are real.

An implicit type can be overridden by declaring a variable name in an *explicit typing* statement (e.g., INTEGER or REAL) at the beginning of the program. (Variables following the implicit type conventions do not need to be declared.) The order of the declarations is not important. For example, the statements

---

[4]In some implementations, *half-precision* numbers are available. The INTEGER*2 statement allocates two bytes instead of the normal four used for integer storage. Similarly, the REAL*4 statement allocates four bytes instead of the normal eight used for real numbers.

[5]Other FORTRAN variable types include characters, complex numbers, and logical variables defined by the statements CHARACTER, COMPLEX, and LOGICAL, respectively. LOGICAL variables can assume only the values of true and false.

INTEGER TIME and REAL INSTANT would establish TIME and INSTANT as integer and real variables, respectively.

Table 52.3 *Correct and Incorrect FORTRAN Variables*

type	correct representation	incorrect representation	
integer	JIM	SLIM	(real)
integer	M200	200M	(begins with a number)
integer	KK	K&K	(& not allowed)
integer	IYOU	I$YOU	($ not allowed)$^{(a)}$
integer	IBGONE	IBEGONE	(7 characters)
real	SLIP	1SLIP	(begins with a number)
real	RS	R–S	(– not allowed)
real	TIMEB	TIMEBASE	(8 characters)

$^{(a)}$May be permitted by some FORTRAN implementations.

5. ARRAY VARIABLES

Arrays (also known as *matrices*) are defined by a DIMENSION statement placed at the beginning of a program. The maximum number of dimensions depends on the FORTRAN implementation and computer. Unless explicitly typed, the variable type for array entries will correspond to the name of the array according to the rules prescribed in Sec. 52-4. For example, the following statement establish a one-row × five-column real array called SAMPLE and a two-row × seven-column integer array called INCOME.

DIMENSION SAMPLE(5)

DIMENSION INCOME(2,7)

Elements of arrays are addressed by placing the identifying subscripts in parentheses (e.g., SAMPLE(2) and INCOME(1,5)). For a two-dimensional array, the first number defines the row and the second number defines the column. Zero and negative numbers are not permitted as subscripts. Subscripts can also be integer variables (e.g., SAMPLE(K)) as long as the variable is within the range of the array size.

6. INITIAL VALUES

Variables and arrays, once defined and declared, are not automatically initialized. The DATA statement is used if it is necessary to assign an initial value. For example, in the following statements, variables X, Y, and Z will be initialized to 1.0. All elements in the array ONEDIM will be equal to 0.0. The array TWODIM will be set to

$$\begin{pmatrix} 1.0 & 2.0 & 3.0 \\ 4.0 & 5.0 & 6.0 \end{pmatrix}$$

DIMENSION ONEDIM(5)

DIMENSION TWODIM(2,3)

DATA X,Y,Z/3*1.0/(ONEDIM(I),I = 1,5)/5*0.0/

1((TWODIM(I,J),J = 1,3),I = 1,2)/1.,2.,3.,4.,5.,6./

After being initialized with a DATA statement, variables can have their values changed by arithmetic operations.

7. ARITHMETIC OPERATIONS

The standard arithmetic operations listed in Table 52.4 are supported in FORTRAN.

Table 52.4 *FORTRAN Arithmetic Operations*

symbol	meaning
=	replacement (assignment)
+	addition
−	subtraction
*	multiplication
/	division
**	exponentiation
( )	preferred operation

The equal sign replaces one quantity with another but does not imply equality. For example, the statement $Z = Z + 1$ is algebraically incorrect but is a valid FORTRAN statement.

Table 52.5 *Correct and Incorrect FORTRAN Assignments*

correct representation	incorrect representation
A = A + B	A + B = A
B = A*A	A**2 = A*A
A = 3; B = 3	A = B = 3

Each operation must be explicitly and unambiguously stated; two operations in a row are also unacceptable. Table 52.6 illustrates several common errors in writing FORTRAN statements.

All variables and constants can be raised to real exponents (e.g., 2.0**3.1). However, only positive real numbers can be raised to real exponents (e.g., 2.0**3.1 but not −2.0**3.1). This is a consequence of using logarithms to perform the exponentiation, since logarithms of negative numbers do not exist. However, a number raised to an integer exponent is calculated by successive multiplications (rather than with logarithms), so

this limitation does not exist for integer exponents (i.e., −2.0**3 is permitted).

Table 52.6 *Correct and Incorrect FORTRAN Operations*

correct representation	incorrect representation
A*B	AB
A*B	AXB
A + (−B)	A + −B
A*(−B)	A*−B
A/(−2.0)	A/−2.0
1.7**2.7	−1.7**2.7

Most, but not all, FORTRAN implementations permit *mixed-mode arithmetic* (i.e., the mixing of real and integer variables in a single arithmetic statement).[6] Integer variable values are first converted to real values by adding a zero fractional part. Real arithmetic is then used to evaluate the expression. (This rule does not apply to integer division, however.)

When mixed-mode arithmetic is permitted, care must be taken to observe the conversion of real data to integer mode. Within mixed-mode expressions, intermediate results will be real numbers. Division between two integers, however, results in truncation. Also, real numbers are truncated to their whole number portions upon assignment to an integer variable.

8. HIERARCHY OF OPERATIONS

Operations in an arithmetic statement are performed in the order of exponentiation first, multiplication and division second, and addition and subtraction third. In the event there are two consecutive operations with the same hierarchy (e.g., a multiplication followed by a division), the operations are performed in the order encountered, normally left to right (except for exponentiation, which is right to left).[7] Parentheses can modify this order; operations within parentheses are always evaluated before operations outside. If nested parentheses are present in an expression, the expression is evaluated outward starting from the innermost pair.

Example 52.1

Evaluate J in the following expression. Mixed-mode arithmetic is permitted, and expressions are scanned from left to right.

$$J = (6.0 + 3.0)*3.0/6.0 + 5.0 − 6.0**2.0$$

---

[6]Raising a real number to an integer exponent is not mixed-mode arithmetic.

[7]In most implementations, a statement will be scanned from left to right. Once a left-to-right scan is complete, some implementations then scan from right to left; others return to the equal sign and start a second left-to-right scan. Parentheses should be used to define the intended order of operations.

Solution

The expression within the parentheses is evaluated first.

$$9.0*3.0/6.0 + 5.0 - 6.0**2.0$$

The exponentiation is performed next.

$$9.0*3.0/6.0 + 5.0 - 36.0$$

The multiplication and division are performed next.

$$4.5 + 5.0 - 36.0$$

The addition and subtraction are performed last.

$$-26.5$$

However, J is an integer variable, and the assignment of $J = -26.5$ results in truncating the fractional part. Ultimately, J has the value of -26.

9. NONEXECUTABLE STATEMENTS

Several FORTRAN commands are used to establish and initialize data locations and define and control the environment in which the program runs, but they are not part of the program logic. Most of the *nonexecutable instructions* shown in Table 52.7 must be placed at the beginning of a program, before the executable instructions.

Table 52.7 *Nonexecutable Statements*

instruction	function
BLOCK DATA	Initializes variables in COMMON.
COMMON	Establishes memory locations accessible to two or more programs and subprograms.
DATA	Initializes variables.
DIMENSION	Establishes arrays.
END	Defines the end of a program or subprogram.
EQUIVALENCE	Specifies multiple names for a single variable.
EXTERNAL	Passes the name of one subprogram to another subprogram.
FUNCTION	Establishes the name of a subprogram.
INTEGER	Defines variables as integers.
PROGRAM	Identifies the main program.
REAL	Defines variables as real numbers.
SUBROUTINE	Establishes the name of a subroutine.

10. LOGICAL STATEMENTS

A *logical statement* is a relational expression using one of the operators in Table 52.8. The use of these operators is illustrated in Ex. 52.2.

Table 52.8 *FORTRAN Logic Operators*

operator	use
.AND.	and
.EQ.	equal
.GE.	greater than or equal to
.GT.	greater than
.LE.	less than or equal to
.LT.	less than
.NE.	not equal to
.NOT.	not
.OR.	or

11. CONTROL STATEMENTS

A PAUSE statement will cause execution to temporarily halt. Its format is

$$\text{s} \quad \text{PAUSE}$$

When a PAUSE statement is reached, the optional statement number s is made available to the computer operator. The operator then can resume program execution manually after performing some other function (e.g., loading a data tape).[8] The PAUSE statement should be used only if an operator will be available to restart the program.

The STOP statement indicates the logical end of the program (i.e., completion of the algorithm). The format is

$$\text{s} \quad \text{STOP}$$

Depending on the branching of the algorithm, a program may have more than one STOP statement. When a STOP statement is reached, program execution stops and control is turned over to the operating system or monitor program.[9] Normally, the program cannot be restarted once a STOP has been executed. The statement number, s, is optional, but if present, the value is made known to the computer operator.

The CALL statement is used to transfer execution to a subroutine. CALL EXIT will terminate execution and turn control over to the operating system or monitor

---

[8]PAUSE and STOP statements are not favored in large computer installations since computer time is wasted waiting for operator action.

[9]Unless it is an END statement, the line following a STOP must be numbered. Otherwise, it will be impossible to get to that statement.

program. (In this regard, CALL EXIT and STOP have identical functions.)

The RETURN statement ends execution of a subroutine and passes control back to the main program.

The CONTINUE statement does nothing. It can be used with a statement number as the last line of a DO loop.

The END statement is required as the last physical statement in a source program and in any subprogram. A program cannot be compiled without an END statement, which tells the compiler that there are no more lines in the program. END statements cannot be numbered.

The explicit GO TO s statement transfers control to statement s. (s is known as a *label*.) The computed GO TO has the form GO TO $(s_1, s_2, \ldots, s_n)$ [expression]. The value of [expression], that is, $1, 2, \ldots, n$, determines which label is to be used.

The arithmetic IF statement is written

$$IF \text{ [expression] } s_1, s_2, s_3$$

[expression] is any numerical variable or set of arithmetic operations. The s_i are statement numbers (*labels*). Transfer is to statement s_1, s_2, or s_3 where [expression] is negative, zero, or positive, respectively.

The logical IF statement has the form

$$IF \text{ [logical expression][statement]}$$

The [statement] can be any executable statement except DO or IF. If [logical expression] is true, [statement] will execute; otherwise, the next instruction will be executed. Some FORTRAN implementations have an IF...THEN...ELSE statement.

The DO statement (covered in Sec. 52-12) is also a control statement.

Example 52.2

Determine the meaning of the following two statements.

(a) IF (A.GT.25.6) A = 27.0

(b) IF (Z.EQ.(T−4.0).OR.Z.EQ.0.) GO TO 17

Solution

(a) If A is greater than 25.6, then set A equal to 27.0.

(b) If Z is equal to (T−4.0) or if Z is equal to zero, then go to statement 17.

12. PROGRAM LOOPS

A *loop* consists of one or more instructions intended to be executed one or more times. A *nested loop* is a loop

within a larger loop, as illustrated in Fig. 52.1.[10] Loops can be constructed from IF and GO TO statements; however, the DO statement is a convenient method of defining a loop.

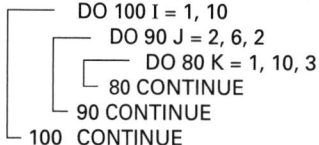

Figure 52.1 *Nested DO Loops*

The general form of the DO statement is

$$DO \text{ s i} = j,k,l$$

s is a statement number. i is the integer loop variable (the *counter*). j is the initial value assigned to i. k is the inclusive upper bound (the *test*) of i, which must exceed j. l is the *increment* for i, with a default of one.

Statement s must be an executable statement, including the dummy CONTINUE statement, but it cannot be a control statement. The DO statement causes the statements immediately following it through statement s to be executed until i equals or exceeds k. When i equals or exceeds k, the statement following s is executed. However, the loop can also be exited before i reaches k.[11]

13. INPUT/OUTPUT STATEMENTS

The READ, WRITE, and FORMAT statements are FORTRAN's main input/output (I/O) commands.[12] The structures of the READ and WRITE statements are

$$READ \quad (u_1, s) \text{ [list]}$$
$$WRITE \quad (u_2, s) \text{ [list]}$$

u_1 is a *unit number* that designates the input device (e.g., disk drive or keyboard). This number varies with installation. u_2 is a unit number that designates the output device (e.g., a printer or terminal). s is the statement number corresponding to a FORMAT statement, and [list] is a list of variables separated by commas whose values are being read or written. For example,

$$READ \text{ (5, 100) X, Y, Z}$$
$$100 \quad FORMAT \text{ (F5.0)}$$

---

[10]Although not required by instruction syntax, nested DO loops should be indented to enhance program readability.
[11]While control can be transferred out of an executing DO loop, control should never be transferred into a DO loop, other than to the DO statement.
[12]Most FORTRANs also support *format-free input* and *output*. The READ statement can be used with a [list] of variables without a FORMAT statement. The PRINT statement is the format-free equivalent of the WRITE statement.

The [list] can also include an implicit DO loop, particularly valuable in loading initial values of array variables. The following example reads seven values, the first six into the array WOW and the last into TOP.

$$READ\ (5, 85)\ (WOW(J),\ J = 1, 6),\ TOP$$
$$85\quad FORMAT\ (7F5.0)$$

The purpose of the FORMAT statement is to define the location, size, and type of data being read. The FORMAT statement has the form

$$s\ FORMAT\ [field\ list]$$

s is the statement number (i.e., the *line label*). [field list] contains format codes set apart by commas defining the fields. [field list] can be longer than [list] in the corresponding READ or WRITE statement.

The format code for integers is nIw. w is the number of character positions, and n is an optional repeat counter indicating the number of consecutive variables having the same format.

The format codes for real values are nFw.d (fixed) and nEw.d (exponential). w is the number of character positions allocated including the space required for the decimal point, sign (if negative), and (in scientific notation) the exponent E or D. d is the number of spaces to the right of the decimal point to be printed. When data are read in, decimal points in any position take precedence over the value of d. Values are truncated after d decimal positions. Rounding is not automatic.

The F format will print up to (w − 1) digits, sign, and blanks. The E format will print a total of (w − 2) digits, sign, and blanks and give the data in a standard scientific notation with an exponent.

Table 52.9 lists these and other codes that can be used in a FORMAT statement.

Table 52.9 FORTRAN FORMAT Codes

code	use
D	real number, double precision
E	real number in scientific notation
F	real number
H	alphanumeric character
I	integer
L	logical variable
P	decimal point modifier
T	position (column) number
X	blank (horizontal space)
Z	hexadecimal number
/	skipped line
' '	literal data

Other I/O statements available in some FORTRAN implementations include ACCEPT, BACKSPACE, CLOSE, OPEN, PUNCH, REWIND, and TYPE.

14. PRINTER CONTROL

The usual output device is a line printer with 133 logical print positions (columns). The first print position is used for *carriage control*, leaving 132 actual physical print positions.[13] The character in the first output position will control the printer advance according to the rules in Table 52.10. Carriage control is usually accomplished by the use of literal data within a FORMAT statement.

Table 52.10 FORTRAN Printer Control Characters

character	function
blank	advance one line
0	advance two lines
1	form feed (skip to line one on the next page)
+	carriage return (overprint—do not advance)

Example 52.3

What is printed by the following program segment?

```
      K = 193
      WRITE (6, 100) K
      WRITE (6, 101) K
100   FORMAT(' ',I3)
101   FORMAT(I3)
```

Solution

The number 193 would be printed on the next line of the current page. The number 93 would be printed on the first line of the following page.

15. STANDARD LIBRARY FUNCTIONS

FORTRAN supplies many standard *library functions* (*supplied functions, intrinsic functions,* or *canned functions*) for mathematical operations. The single-precision library functions listed in Table 52.11 are available in most FORTRAN implementations. Most functions are accessed by placing the argument in parentheses after the function name (e.g., SIN(THETA)). In most cases, placing the letter D before the function name will cause the calculation to be performed in double precision (e.g., DATAN is the arctangent in double precision). Arguments for trigonometric functions must be expressed in radians.

---

[13]*Carriage control* is an anachronism resulting from the early use of typewriter-style printers as operator keyboards. High-speed printers no longer have carriages.

Table 52.11 Representative FORTRAN Library Functions

function	use
ABS	real absolute value
ALOG	natural logarithm
ALOG10	common logarithm
AMOD	real remaindering modulus
ARCOS	arccosine
ARSIN	arcsine
ATAN	arctangent
CONJG	complex conjugate
COS	cosine
COSH	hyperbolic cosine
EXP	e^x
FIX	convert real to integer
FLOAT	convert integer to real
IABS	integer absolute value
INT	convert real to integer
MOD	integer remaindering modulus
SIN	sine
SINH	hyperbolic sine
SQRT	square root
TAN	tangent
TANH	hyperbolic tangent

16. STATEMENT FUNCTIONS

In FORTRAN, the term *subprogram* refers to an instruction or group of instructions that is used repeatedly. The term encompasses user-defined functions, function subprograms, and subroutines.

A user-defined *statement function* processes one or more input numbers (*arguments* or *parameters*) and returns a single numerical value. It must be defined prior to use, and the definition must fit on one statement. For example, the function PROD calculates the product of two numbers in the following code segment.

```
main program
PROD(X1,X2) = X1*X2
HEIGHT = 2.5
WIDTH = 7.5
AREA = PROD(HEIGHT,WIDTH)
STOP
END
```

17. FUNCTION SUBPROGRAMS

A user-defined *function subprogram* (*external function*) processes one or more input numbers (*arguments* or *parameters*) and returns a single numerical value. It is not limited to a single line but is created with the FUNCTION statement according to the following syntax rules.

- In the main program, the function is defined as a variable.

- When used in the main program, the function is followed by its actual arguments in parentheses.

- Within the function itself, the function name is type declared and defined by the FUNCTION statement.

- The arguments need not have the same names in the main program and function subprogram, but they must agree in number, sequence, type, and precision (length).

- Both the function subprogram and the main program have END statements.

- Only the function subprogram ends with a RETURN statement.

- The function may, but need not, have a STOP statement.

For example, the product of two numbers is calculated by the function PROD in the following code segment.

```
main program
REAL PROD
HEIGHT = 2.5
WIDTH = 7.5
AREA = PROD(HEIGHT,WIDTH)
STOP
END
```

```
function subprogram
REAL FUNCTION PROD(X1,X2)
PROD = X1*X2
RETURN
END
```

18. SUBROUTINES

A *subroutine* is a user-defined subprogram. It is not limited to mathematical calculations, nor is it limited to returning to the statement immediately following the calling statement. Thus, a subroutine is more versatile than a function or function subprogram. Subroutines are governed by the following rules.

- The subroutine has no type and does not assume any value.

- The subroutine is activated by a CALL statement.

- Both the subroutine and the main program have END statements.

- Only the subroutine has a RETURN statement.

- The subroutine may, but need not, have a STOP statement.

- The arguments need not have the same names in the main program and subroutine.

- The arguments must agree in number, sequence, type, and precision (length).

The following code segment illustrates a subroutine GETNUM that obtains two numbers for subsequent calculations.

```
        main program
        CALL GETNUM(HEIGHT, WIDTH)
        AREA = HEIGHT*WIDTH
        STOP
        END


        subroutine
        SUBROUTINE GETNUM(X1,X2)
        READ (5,100) X1,X2
100     FORMAT(2F5.1)
        RETURN
        END
```

19. COMMON VARIABLES

Variables in functions and subroutines (and in other programs) are independent of the main program. They may have the same names as variables in the main program but their values can be different. Values of variables can be shared, however, if the *regular* COMMON statement is used to establish a link between the variables. The COMMON statement has the format

COMMON [variable list]

The main program and the subroutines must all have COMMON statements. The COMMON statement merely assigns memory locations to be shared by the main program and all of its subroutines. While it is convenient to do so, it is not necessary to use the same variable names. The order of the common variables establishes their positions in memory.

The *named* COMMON statement shares values with only some of the subroutines. While there can be only one regular COMMON statement per program, there can be many named COMMON statements. The format of the named COMMON statement is

COMMON/[name]/[variable list]

53 Analog Computers: Analog Logic Functions

1. ANALOG MODELS

Most of the formulas used by engineers are simple models that predict the behavior of physical systems. Bernoulli's equation can predict fluid velocity, Euler's formula can predict when a column will buckle, and so on. These formulas are *analogs* because they simulate the real system.

In contrast to the usual case of using equations to model physical systems, it is possible to build electrical circuits that simulate the solutions of equations, particularly ordinary linear differential equations with constant coefficients.[1] (Nonlinear models are possible but difficult to implement.) These circuits are *analog models*. Voltages in such circuits correspond (are analogous) to unknown variables in the equations being investigated.

Analog circuits require many components including amplifiers, potentiometers, oscilloscopes, plotters, and other forms of instrumentation. While a breadboard approach to circuit construction can be taken, it is more common to use a commercial *analog computer*—a specialized unit where all components are housed together and interconnected by electronically programmable switch matrices or, in older equipment, by patch cords.[2,3]

2. STRUCTURAL ELEMENTS

When voltage represents a variable, most of the standard mathematical operations (e.g., addition, multiplication, differentiation) with that variable can be performed electrically by passing the voltage through

different components.[4] For example, the change in voltage when a potentiometer is placed in the circuit is analogous to scalar multiplication. Table 53.1 lists the primary structural elements of an analog computer and the functions they perform. High-gain direct current operational amplifiers are used to implement most of these elements.

Table 53.1 *Structural Elements of an Analog Computer*

operation	symbol
scalar multiplication	x —(K)— Kx
negation	x —▷— $-x$
summation	x_1, x_2 —(K_1, K_2, Σ)— $-(K_1 x_1 + K_2 x_2)$
integration	x' —(K)— $-K(x + x(0))$, with $x(0)$
input function	$(f(t))$—

The external input voltage to an analog circuit model corresponds to the forcing function of the nonhomogeneous differential equation being modeled. For example, a sinusoidal voltage will correspond to a sinusoidal forcing function, and so on. Such inputs are typically generated externally or as outputs from other analog computer circuits.

Initial conditions are obtained by setting dedicated potentiometers on their corresponding integrators.

---

[1] The mathematical equations, in turn, are generally used to model mechanical, fluid, thermal, and electrical systems.

[2] The term *electronic differential analyzer* is also used to describe an analog computer.

[3] Several programming languages, such as IBM's CSMP (Continuous System Modeling Program), for digital computers are designed specifically for solving differential equations, but they are not analog computational methods.

[4] While possible, differentiation is seldom simulated, due to the tendency differentiators have to develop electrical noise.

Computers

Example 53.1

An ideal operational amplifier is used as an inverting integrator. Its input represents a variable $x(t)$ whose initial value is $x(0) = 5$. The variable is acted upon by a step input of height 3. (a) How would this be represented in symbol form? (b) What is the output at $t = 4$?

Solution

(a) The symbolic representation for this situation is

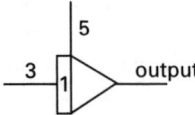

(b) Integrating a constant 3 results in $3t$. Adding in the initial value, the function is $x(t) = 3t + 5$. At $t = 4$, the function value is

$$x(4) = (3)(4) + 5 = 17$$

The inverted output is -17.

3. INDIRECT PROGRAMMING

The act of arranging the elements into a simulation model is known as *indirect programming* or *mechanization*. (The procedure given here assumes that there are no differentiating operational amplifiers in the mechanization of the differential equation.)

step 1: Normalize the differential equation so that the coefficient of the highest-order term is 1.0.

step 2: Solve for the negative of the highest-order term. (This accounts for the inversion performed by the final summing amplifier.)

step 3: Assume that the highest-order derivative exists and run it through successive integrators to obtain lower-order raw derivatives.

step 4: Run raw derivatives through scaling potentiometers to obtain all required scaled derivatives.

step 5: Feed the forcing function and all scaled derivatives into the final summer.

step 6: Since the highest-order derivative is equal to the sum of all other terms (step 2), complete the circuit by connecting the output of the final summer to the point where the highest-order derivative was assumed to exist in step 3.

step 7: Set the initial conditions on the potentiometers to correspond to the integrators used.[5] Set the scaling potentiometers to the constant coefficients of their respective terms.

---

[5]Depending on the design of the analog computer, the sign of the initial conditions may need to be reversed since integrators change the sign of their signals.

Example 53.2

Construct the analog computer circuit that solves the following differential equation.

$$2x'' + 0.32x' + 1.28x = 0.6$$

Solution

step 1:
$$x'' + 0.16x' + 0.64x = 0.3$$

step 2:
$$-x'' = 0.16x' + 0.64x - 0.3$$

step 3:

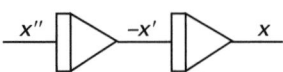

step 4:

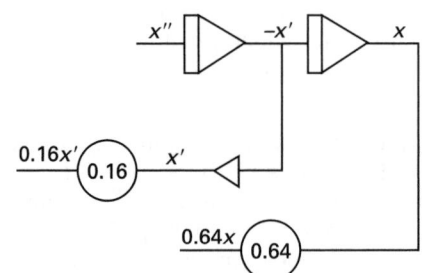

steps 5 and 6:

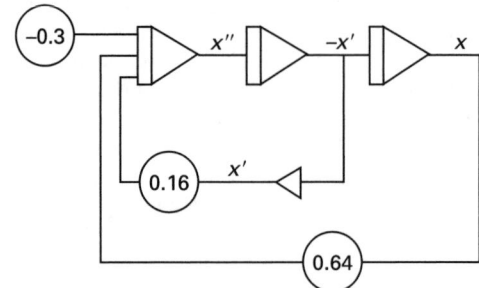

4. MODES OF OPERATION

Commercial analog computers have four primary modes of operation: compute, hold, reset, and repeat compute.

In the *compute (run) mode*, the simulation continues once started. The simulation can be frozen at any point by using the *hold mode*. The simulation can be restarted from the beginning by toggling the *reset mode*. If the computer has a *repeat mode*, the simulation will repetitively begin anew after a certain period of time. This repeat mode is particularly valuable when the response time and maximum values are not known in advance, since the effects of scaling changes can be observed on an oscilloscope.

5. TIME SCALING

The *real-time* response of some models will be too fast or too slow to be conveniently observed. Therefore, a *time-scaling factor*, h, is used so that the *machine time* (i.e., *simulated time*) is $T = ht$. If $h > 1$, the analog model runs slower than real time; if $h < 1$, the model runs faster. *Time scaling* is based on the fact that

$$\frac{df(t)}{dt} = h\left(\frac{df\left(\frac{T}{h}\right)}{dT}\right) \qquad 53.1$$

Substituting T/h for t everywhere (which does not change the equation being solved) is accomplished by reducing the gain of each integrator by h and by changing the time scale in the supplied forcing function from t to T/h. (Time scaling does not change magnitude scaling. See Sec. 53-6.)

For each closed loop in the model, the product of all scaling coefficients (known as the *loop gain*) is changed by the nth power of the time-scale factor, where n is the number of integrators in the loop.

Example 53.3

Construct the analog computer model that simulates the following differential equation three times slower than its real-time performance. Neglect initial conditions.
$$-x'' = 0.8x' + 0.3x - \sin(t)$$

Solution

The time-scaled differential equation is

$$-(3)^2 x'' = (3)(0.8)x' + 0.3x - \sin\left(\frac{T}{3}\right)$$

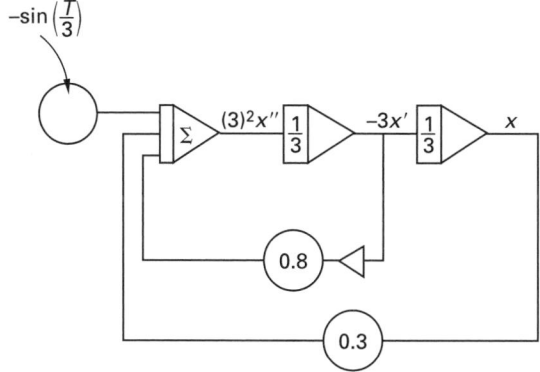

6. MAGNITUDE SCALING

The operational amplifiers used in the structural elements are generally limited to a voltage range of approximately ±0.1 to ±10 V.[6] (Noise is a problem below the lower limit, and amplification becomes distorted and nonlinear above the upper limit.) Also, the potentiometers are limited to scalar multiplication between 0 and 1.0. These limitations may require *magnitude scaling* of the differential equation. (Magnitude scaling does not change time scaling.)

In order to properly scale inputs and outputs, the maximum expected value of each variable, derivative, and forcing function must be known. In the case of a variable expected to achieve a value of x_{max} that passes through an amplifier limited to 10 V, the scaling factor, k, is chosen so that $k \leq 10/x_{max}$. (The scaling factors themselves are limited to a maximum value determined by the analog computer.) Each structural element used will have its own scaling factor. Magnitude scaling is based on the fact that

$$\int \frac{df(t)}{dt} = \frac{1}{k}\int \frac{dkf(t)}{dt} \qquad 53.2$$

For each closed loop in the model, the *loop gain* must remain the same as in an unscaled implementation.

Example 53.4

Construct an analog computer model that simulates the following differential equation. The expected maximum values are: x', 400; x, 500; $f(t)$, 1000. The maximum value of any simulation variable or scaling factor is ten. Neglect time scaling.

$$20x' + 300x = 800e^{-0.6t} \qquad x(0) = 50$$

Solution

First, divide through by 20 and isolate the derivative term.
$$-x' = 15x - 40e^{-0.6t}$$

The scale factors are not unique.

$$k_1 = \frac{10}{400} = 0.025$$
$$k_0 = \frac{10}{500} = 0.02$$
$$k_f = \frac{10}{1000} = 0.01$$

The scaled initial condition is

$$k_0 x_0 = (0.02)(50) = 1.0$$

---
[6]However, analog computers can have limits of up to ±100 V.

The simulation equation is written by first multiplying all terms by 0.025. Then each term is multiplied *and* divided by its respective scaling factor.

$$-0.025x' = \frac{(0.025)(15)(0.02x)}{0.02}$$
$$- \frac{(0.025)(40)(0.01)e^{-0.6t}}{0.01}$$

This is simplified somewhat, leaving the scaling factors and implementing all values greater than ten as products of lesser numbers. (The input function is not part of a loop and can be completely implemented.)

$$-0.025x' = (1.875)(10)(0.02x) - e^{-0.6}$$

The magnitude-scaled analog computer model includes scalar multiplication *between* stages. This can be accomplished in several ways, but inasmuch as the differentiators are actually variable-gain amplifiers, they are used. The setting corresponds to the ratio of the output to input scale coefficients (in this case, $0.02/0.025 = 0.8$).

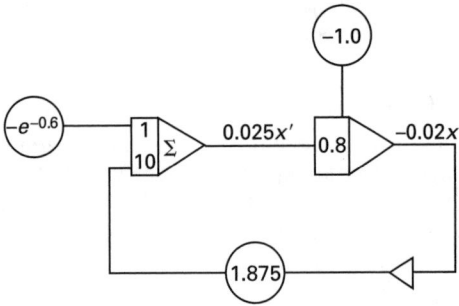

7. SIMULTANEOUS DIFFERENTIAL EQUATIONS

Since a differential equation is solved with a loop of operational amplifiers, simultaneous differential equations can be handled by interconnecting loops for each equation. If time and magnitude scaling are used, all loops must be scaled by the same factors.

Example 53.5

Construct the analog computer circuit that simultaneously solves the following two differential equations.

$$-x_1' = 0.8x_1 - 0.7x_2 - 0.3$$
$$-x_2' = x_2 - 0.4x_1$$

Solution

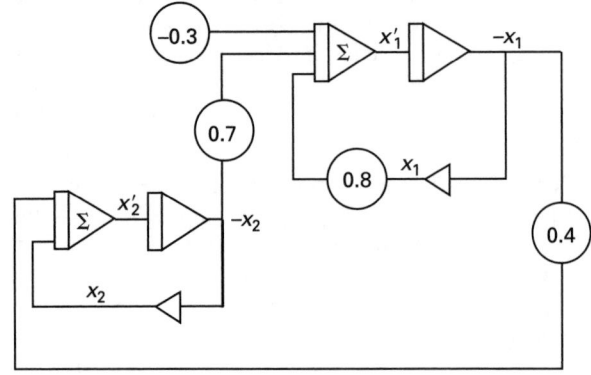

54 Operating Systems, Networking Systems, and Standards

1. OPERATING SYSTEM FUNDAMENTALS

An *operating system* is a set of programs and routines that controls a computer or a network in the performance of its tasks. The software provides control via application-level services, library services, and kernel services. An overview of a generic operating system is shown in Fig. 54.1.

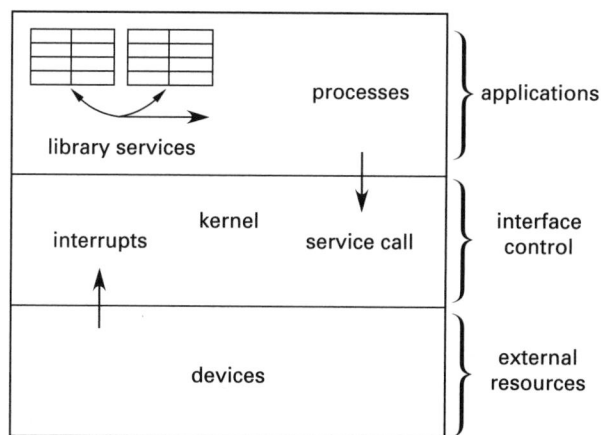

Figure 54.1 *Operating System Overview*

A *process* is the entity that performs a single job step, which can consist of assembly, compilation, generation, and interpretation of information. A *processor* performs multiple steps. A central processing unit (CPU) is an example of a processor. Applications, which are programs written to perform a specific task or a set of related tasks, are run by processors using library services for standardized tasks such as formatting or information display. A *kernel* allocates the resources required by a particular process using stored routines and is considered the core of the operating system. The kernel responds to service calls from processes and interrupts from devices. A *service call* is an instruction used by a process to request input. An *interrupt* is an instruction used by a device when the input requested is ready, among other uses. *Drivers* or *device drivers* are routines

used by the kernel to deal with interrupts. A *scheduler* is used by the kernel to sort jobs based on priority and time required for completion. Jobs were originally thought of as existing on a spool, with the new jobs wound on the outside and the old jobs extracted from the inside. Hence, the use of the term *spooling* with regard to jobs in the queue.[1]

An operating system functions to hide the details of the hardware using abstractions. An *abstraction* is software that provides a set of higher-level functions with which the operator interfaces. The operating system also manages *resources*, which are the commodities that function to complete computing tasks. Additionally, the operating system provides the interface with human operators. An interface should be both effective and pleasing to the operator.

2. NETWORKING SYSTEMS FUNDAMENTALS

A *network* in computer science parlance is a collection of geographically separated computers. *Networking* is the use of transmission lines to connect these computers.[2] The term "network" also refers to a single computer with multiple geographically separated terminals. When multiple computers share files stored on a single machine, the machine holding the shared files is called the *server* and the others are called *clients*. Multiple computers set up in such a fashion are said to use a *networked operating system*.

3. INTERNET FUNDAMENTALS

The *internet* is numerous networks connected together using a common language or *protocol*, that is, TCP/IP or *transmission control protocol/internet protocol*. The internet serves four basic functions: communication, document or file transfer, information search, and reading or posting of messages to topic-specific bulletin boards. Four basic tools make these activities possible. *E-mail*, or electronic mail, is an electronic tool that allows users to exchange information. E-mail functions on a variety of software. The movement of a file is accomplished using the *file transfer protocol* (FTP) as a

---

[1]The term "spooling" is an acronym for *simultaneous peripheral operations on line*. The secondary storage medium of early computers was a drum. Thus the term "spooling" seemed apropos.
[2]The transmission line itself may be a standard metal conductor, a fiber-optic cable, an infrared or a microwave connection, or some other type.

Computers

tool. The file can be a document, an image, a sound or collection of sounds, software, and so on. Control of a remote computer is accomplished with an electronic tool called the *Telnet*. An example of Telnet in use is an engineering database search through a library's computer files. The final tool is the Usenet. The *Usenet* is a global bulletin board messaging system that uses the internet for access.

Many other tools and services are available. An example is *Archie*, which searches FTP archives. *Gopher*, which connects multiple computers and displays them as menu items, allows for database searches. *Veronica* allows for keyword searches in gopherspace. Varieties of databases exist and are categorized by the area served. Examples include *wide-area information servers* (WAIS), *campus-wide information services* (CWIS), *on-line public access catalogs* (OPAC), and commercial on-line services. *Hypertext* is a data structure that allows links between words, phrases, graphics, or other elements among the items mentioned.

The internet items are mostly transparent to users since the development of the *world-wide web* (WWW). The Web provides a *graphical user interface* (GUI) to the above tools and uses hypertext to transition from one to the other. The protocol used by the internet, TCP/IP, is illustrated in Fig. 54.2.

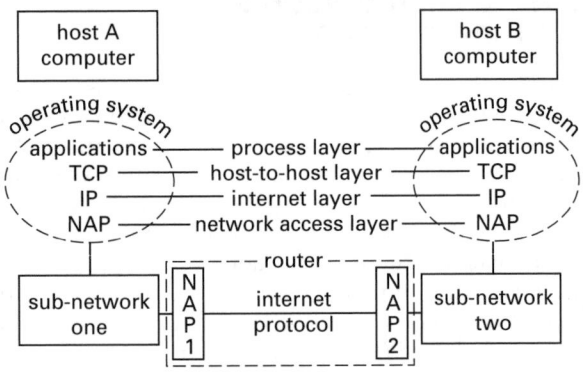

TCP:	Terminal Connection Protocol
IP:	Internet Protocol
NAP:	Network Access Protocol

Figure 54.2 Transmission Control Protocol/
Internet Protocol Overview

The communication layers represented in Fig. 54.2 are standard. The *process layer* contains the logic used by the specific applications to interact with the net. Individual applications provide this logic functionality. The *host-to-host layer* focuses on the reliable exchange of information. The reliable exchange of data is a common function and thus is independent of the type of application. The transmission control protocol (TCP) provides the functionality to accomplish data exchange. The *internet layer* allows two devices connected to different networks to communicate. Internet protocol provides this multiple network functionality. The internet protocol is used in end systems and in routers. *End systems*

are the sending or receiving hosts, servers, or workstations. *Routers* are processors that primarily function to relay data from one network to another. The *network access layer* allows for the exchange of data between an end system and the attached network. The network access protocol provides system/network interface functionality.

The transmission control protocol (TCP) is a communication standard. Several applications are designed to use this standard. For example, the *simple mail transfer protocol* (SMTP) provides e-mail functionality. The file transfer protocol (FTP) is actually an application using the TCP standard. Telnet is a remote log-on application using the TCP standard. The internet protocol (IP) is also a standard. The network access protocol (NAP) comes in a variety of standards depending on the net. For example, *Ethernet* is an NAP standard. The framework for the development of communication protocol standards is the *open systems interconnection* (OSI) reference model. OSI was developed by the *International Organization for Standardization* (ISO).

Communication between networks occurs using either circuit switching or packet switching. *Circuit switching* implies a dedicated path between the sending unit and receiving unit, uses a switching facility, and is of a fixed bandwidth or speed during the entire communication or call. Circuit switching communication involves three phases: circuit establishment, information transfer, and circuit disconnect. *Packet switching* is the transmission of packets of data over the communication line using high-speed switching computers operating at nodes. The packets of data are treated independently, can be routed through different nodes, and can arrive at the receiving end in an order different from the original. Identification information encoded within the packets allows the receiving end to reassemble the message.

The movement of information between networks and hosts is based on computer addresses that a user sees as a name, or *domain name*. The top-level domains common in the United States are .com (commercial interests), .edu (educational institutions), .gov (U.S. government), .org (organizations), and .net (networks). Each country has a two-letter code assigned by the ISO. For the United States, the code is ".us".

4. NETWORK ORGANIZATION

Networks have traditionally been categorized as either *wide-area networks* (WANs) or *local-area networks* (LANs). Both can be either traditional or high-speed networks. Numerous other network designations are in use, such as *metropolitan-area network* (MAN), *car-area network* (CAN), and *house-area network* (HAN). Other valid network categories include: *data, voice, video, multimedia, intranet,* and *internet*. Data networks are referred to by the protocols they use, for example, Windows NT, AppleTalk, Systems Network Architecture (SNA), DECnet, and NetWare.

Each network is further defined by its topology. The *topology* of a network is the physical or logical arrangement of the stations or nodes. Examples of typical topologies are shown in Fig. 54.3.

The bus topology of Fig. 54.3(a) is distinguished by its common medium connecting multiple devices. The common medium is often coaxial or fiber-optic cable. The star topology of Fig. 54.3(b) is distinguished by the routing of all data through the central processor. The ring topology of Fig. 54.3(c) is physically implemented with the appearance of star topology. Nevertheless, not all data passes through a central processor. Ring topology is physically implemented using a *media access unit* (MAU) that devices attach to via cable and interface cards. Internally, the MAU is a dual ring. Externally, it is rectangular in appearance. The tree topology of Fig. 54.3(d) is flexible, allowing protocols to isolate various users or connect groups of users.

5. STANDARDS

A *protocol* is a set of hardware and software interfaces in a terminal or computer that allow it to function in a given network. A *standard* is an accepted reference. The terms are used interchangeably, but a protocol is realized in a network and may or may not meet standards.[3] Standards are created by various bodies throughout the world. The *International Organization for Standardization* (ISO) is a federation of standards bodies from more than 100 countries around the world whose purpose is to develop international standards in order to facilitate the international exchange of goods and services.[4] The member body in the United States is the *American National Standards Institute* (ANSI). ANSI develops voluntary consensus standards for information technology, including those areas in computer science, with the participation of public and private sector interests. The *Institute of Electrical and Electronics Engineers* (IEEE) is a professional organization that develops computer science and engineering standards, among others.

A *benchmark* is a performance test used to compare hardware or software performance and is not a standard, per se.

---

[3]More often than not, several protocols are introduced into the market. The dominant protocol, as determined by demand, becomes the standard.
[4]ISO develops standards in all areas except electricity, electronics, and related areas. The *International Electrotechnical Commission* (IEC) develops such electrical engineering standards. ISO does, however, develop information technology standards.

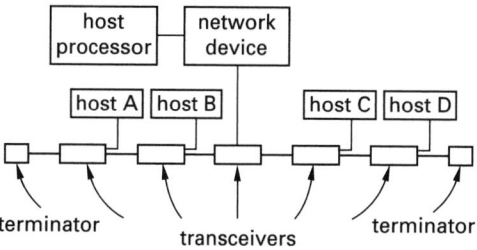

(a) logical view of bus topology

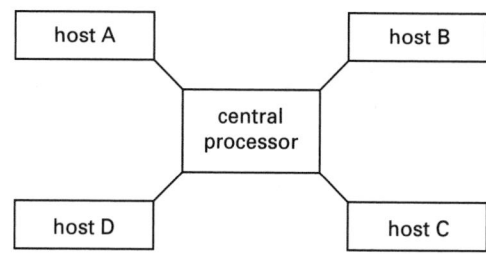

(b) physical view of star topology

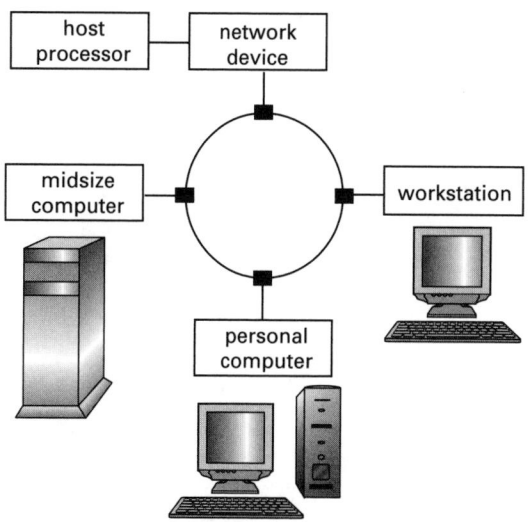

(c) logical view of ring topology

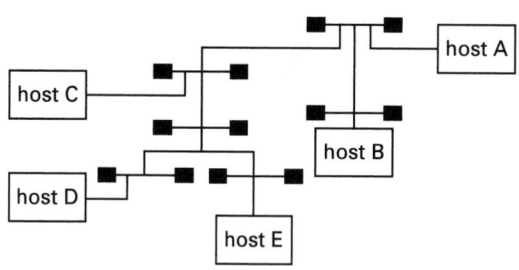

(d) logical view of tree topology

Figure 54.3 *Network Topologies*

Computers

55 Digital Systems

Nomenclature

b	bit	zero or one
LSB	least significant bit	–
MSB	most significant bit	–
n	number of bits	–
N	n-bit number	–
R	resistance	Ω
S	switch	–
t, T	time	s
V	effective or DC voltage	V

Subscripts

a	analog
d	digital
fb	feedback
ref	reference
s	sample

1. DIGITAL SYSTEMS DEFINED

A digital system is defined as any of the levels of operation for a digital computer. General digital system architecture is shown in Fig. 55.1.

When the digital architecture in Fig. 55.1 is compared to the computer architecture in Fig. 48.1, the digital system appears to be a subset of a computer system.

An essential difference is that the *information processing unit* in a digital system is designed to execute a specific sequence of operations in a fixed program. A central processing unit (CPU) in a computer system is capable of executing a varying sequence of operations without either the CPU or the control unit having to be redesigned for each sequence.

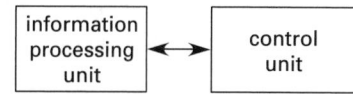

Figure 55.1 *General Digital System Architecture*

A digital system comprises the electronic elements in Ch. 43, arranged for logic operations as in Ch. 45, and designed to accomplish logic tasks as in Chs. 49 and 50.

2. INTERFACE FUNDAMENTALS

Digital systems process information at high rates. Output devices, such as printers, and input devices, such as humans, process information at considerably lower rates. The process of matching the flow of information between portions of a digital system is called the *interface problem*. The general solution to interface issues is shown in Fig. 55.2.

Incoming or outgoing information is temporarily stored in a *buffer*, which is a synchronizing element, that is, a register for the temporary storage of information. The *transfer pulse* is used to control the rate of information flow. A *flag* is used to indicate the occurrence of a given condition, which the input/output device (that is, the encoder) and the digital system use to determine when the next transfer pulse can be issued. The use of a flag to control the interchange of information between devices is called *handshaking*.

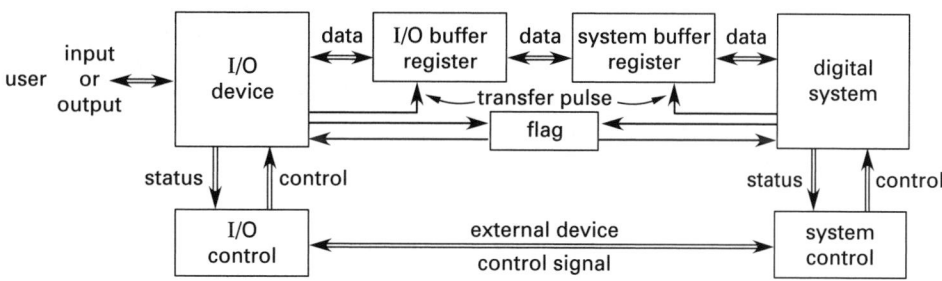

Figure 55.2 *Hardware Interface Overview*

Two sets of standards are suggested by the interface model of Fig. 55.2. The first is that of *standard user interfaces*. Such standards determine the implementation of the interface and are *technical standards*. Technical standards for interfaces are usually de facto standards determined by dominance in the marketplace or consensus within the industry. The second is that of *standards for user interfaces*. Such standards determine the human-computer interaction and are *ergonomic standards*. Ergonomic standards are usually determined by standards-making bodies and do not involve guidance for implementation or specific programming requirements.

The input/output (I/O) interface is called the *surface* of the software. Interaction takes place at this surface using command names, shortcut keys, symbols, voice input, or *direct manipulation*—which is an action of the user on an object on the screen, such as pointing, clicking, moving, or changing said object. This type of interaction occurs through a *dialogue interface*. If a user interacts by changing data, rearranging or configuring electronic tools, or generating macros or commands, the interface is called a *tool* or *application interface*. The interface between a musical instrument and a computer occurs via a *musical instrument digital interface* (MIDI). A MIDI keyboard stores information about notes, note values, key pressure, control signals, and so on that is passed through the interface and can be reproduced by the system.

3. DATA CONVERSION BASICS

Hardware devices used with digital systems often operate on analog data. Input sensors often obtain data about the physical variable measured in terms of an analog voltage but analyzed by digital systems. Control systems on digital system input and output devices accept digital data but require analog voltages to function. Conversions are thus necessary. The conversion from digital signal voltage to analog voltage is called *digital-to-analog* (D/A) *conversion*. Similarly, the conversion from analog voltage to digital signal voltage is called *analog-to-digital* (A/D) *conversion*.

Digital information is considered to be binary and weighted according to its position as a power of two. For example, the value of a positive *n*-bit binary number N is

$$N = b_{n-1}2^{n-1} + b_{n-2}2^{n-2} + \cdots + b_2 2^2$$
$$+ b_1 2^1 + b_0 2^0 \qquad 55.1$$

The binary bits represented by the b's have a value of zero or one. When coded in binary, N is[1]

$$N = b_{n-1}b_{n-2}\ldots b_2 b_1 b_0 \qquad 55.2$$

---

[1]The b's occur in parallel fashion or serial fashion with a storage register.

The digit b_{n-1} is the most significant bit (MSB) while b_0 is the least significant bit (LSB). The maximum value that can be represented by the code in Eq. 55.2 is[2]

$$\text{maximum value} = (2)\,(\text{MSB}) - \text{LSB}$$
$$= 2^n - 1 \qquad 55.3$$

4. DIGITAL-TO-ANALOG CONVERSION

A digital signal is considered to be low, that is, digital zero, when the voltage is between 0.0 and 0.8 V. A digital signal is high, that is, digital one, when the voltage is between 2.0 and 5.0 V.[3] As a result, these signals cannot be used directly for the conversion to analog information. Instead, the digital signal drives an electronic switch that connects the converter to a steady 0 V if the voltage is less than 0.8 V and a steady +5 V if the voltage is above 2.0 V. An example of a D/A converter using binary weighted resistors is shown in Fig. 55.3. The electronic switch, designated S in Fig. 55.3, can be any one of a variety of electronic devices as described in Ch. 45. One possible configuration is shown in Fig. 55.4.

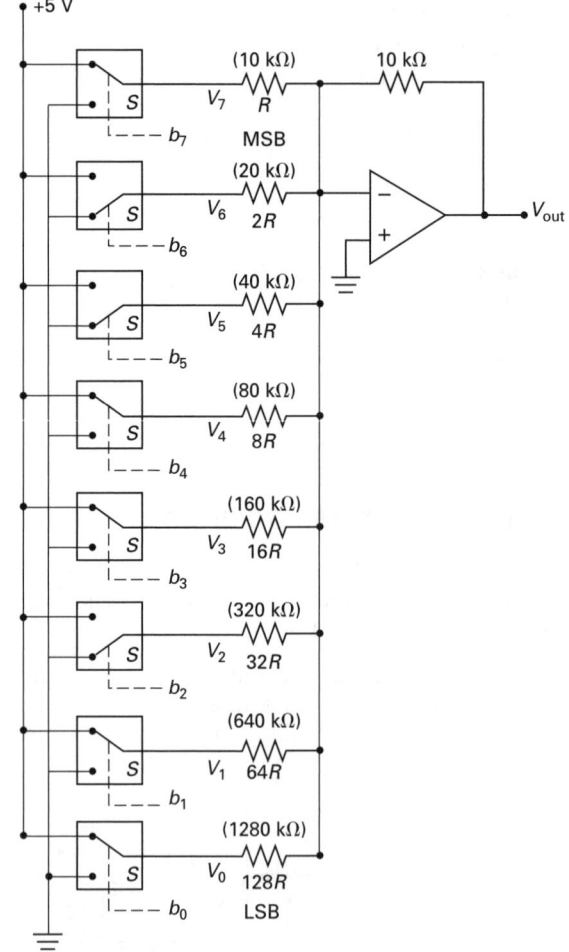

Figure 55.3 *D/A Converter: Binary Weighted Resistors*

---

[2]The value represented by Eq. 55.3 is also the maximum number of steps that can occur in a digital system controlled by an *n*-bit circuit.

[3]The digital voltages mentioned are for TTL logic (see Ch. 49).

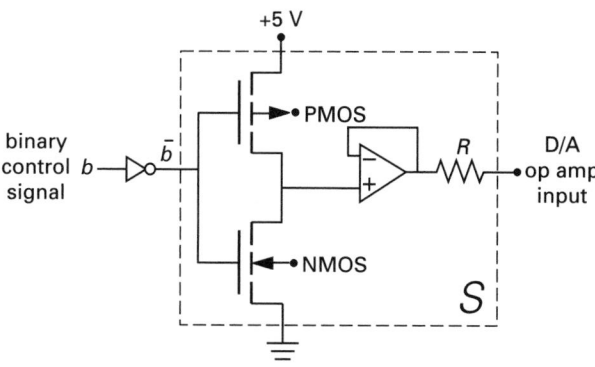

Figure 55.4 *Digitally Controlled Switch*

Example 55.1

What is the expression for V_{out} for the D/A converter in Fig. 55.3 in terms of the binary input?

Solution

Figure 55.3 is an 8-bit converter with a range of 10 V. The inputs to the electronic switches are the data bus lines. When a data line is high logic 1, the switch is in the upper position and connected to the +5 V supply. When a data line is low logic 0, the switch is in the lower position and connected to ground, that is, to 0 V. Superposition applies so, using the techniques and assumptions for an ideal operational amplifier described in Ch. 44, the output due to each binary bit can be determined and then all outputs can be summed for the final solution. The output voltage for the most significant bit (MSB) is determined by writing KVL at the inverting terminal.

$$\frac{0 - V_7}{10,000 \ \Omega} + \frac{0 - V_{\text{out}}}{10,000 \ \Omega} = 0$$

$$V_{\text{out}} = -\left(\frac{10 \ \text{k}\Omega}{10 \ \text{k}\Omega}\right) V_7$$

Repeating the procedure for each bit and summing gives the following.

$$V_{\text{out}} = -\left(\frac{10 \ \text{k}\Omega}{10 \ \text{k}\Omega}\right) V_7 - \left(\frac{10 \ \text{k}\Omega}{20 \ \text{k}\Omega}\right) V_6$$

$$- \left(\frac{10 \ \text{k}\Omega}{40 \ \text{k}\Omega}\right) V_5 - \left(\frac{10 \ \text{k}\Omega}{80 \ \text{k}\Omega}\right) V_4$$

$$- \left(\frac{10 \ \text{k}\Omega}{160 \ \text{k}\Omega}\right) V_3 - \left(\frac{10 \ \text{k}\Omega}{320 \ \text{k}\Omega}\right) V_2$$

$$- \left(\frac{10 \ \text{k}\Omega}{640 \ \text{k}\Omega}\right) V_1 - \left(\frac{10 \ \text{k}\Omega}{1280 \ \text{k}\Omega}\right) V_0$$

When the binary input is one, the voltage is +5 V. When the binary input is zero, the voltage is 0 V. Thus, the output equation can be written as

$$V_{\text{out}} = -5b_7 - 2.5b_6 - 1.25b_5 - 0.625b_4 - 0.3125b_3$$

$$- 0.15625b_2 - 0.078125b_1 - 0.0390625b_0$$

Example 55.2

What is the accuracy of the conversion for the D/A converter in Fig. 55.3?

Solution

The accuracy is always one-half the value of the least significant bit. Using the term for the LSB from the expression for V_{out} in Ex. 55.1 gives

$$\text{accuracy} = \tfrac{1}{2} \, |V_{\text{LSB}}|$$

$$= \left(\frac{1}{2}\right) (|-0.0390625|)$$

$$= 0.01953125 \ \text{V}$$

$$= 19.5 \ \text{mV}$$

5. LADDER-TYPE D/A CONVERTER

A *ladder-type D/A converter* is based on a cell concept. Each succeeding cell has twice the weight of the previous one—hence the term "ladder."[4] The design of the first cell, which corresponds to the LSB, is shown in Fig. 55.5(a). A typical cell is shown in Fig. 55.5(b). Either the reference voltage, V_{ref}, or 0 V is applied, depending upon the electronic switch position.

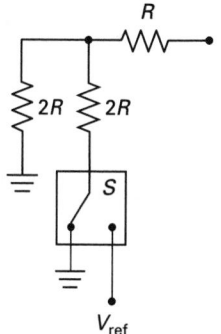

(a) starting cell

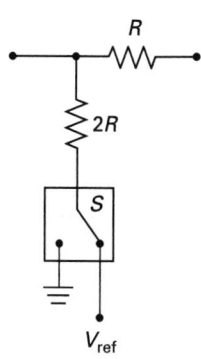

(b) typical cell

Figure 55.5 *Cells of a Ladder-Type D/A Converter*

---

[4]The ladder-type D/A converter uses twice the number of resistors for a given number of bits as the binary weighted D/A converter, but the resistors used have only two values, R and $2R$.

The Thevenin equivalents of the starting cell, with the output taken as V_0; the succeeding cell, with the output taken as V_1; and the combination of the two are shown in Fig. 55.6.

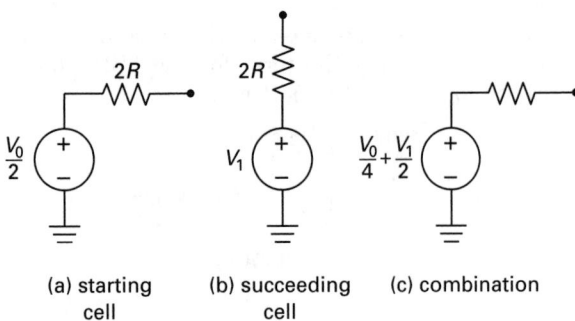

(a) starting cell   (b) succeeding cell   (c) combination

Figure 55.6 *D/A Converter Cells' Thevenin Equivalents*

A ladder-type D/A converter with seven cells, identical to the number of cells in the converter in Sec. 55-4, has an output of

$$V_{\text{out}} = \frac{V_7}{2} + \frac{V_6}{4} + \frac{V_5}{8} + \frac{V_4}{16} + \frac{V_3}{32}$$
$$+ \frac{V_2}{64} + \frac{V_1}{128} + \frac{V_0}{256} \qquad 55.4$$

The voltages in Eq. 55.4 are either V_{ref} or 0 V, depending on the electronic switch settings for the particular bit in the ladder. The converter is completed by connection to the noninverting terminal of an operational amplifier, which provides buffering, as shown in Fig. 55.7.

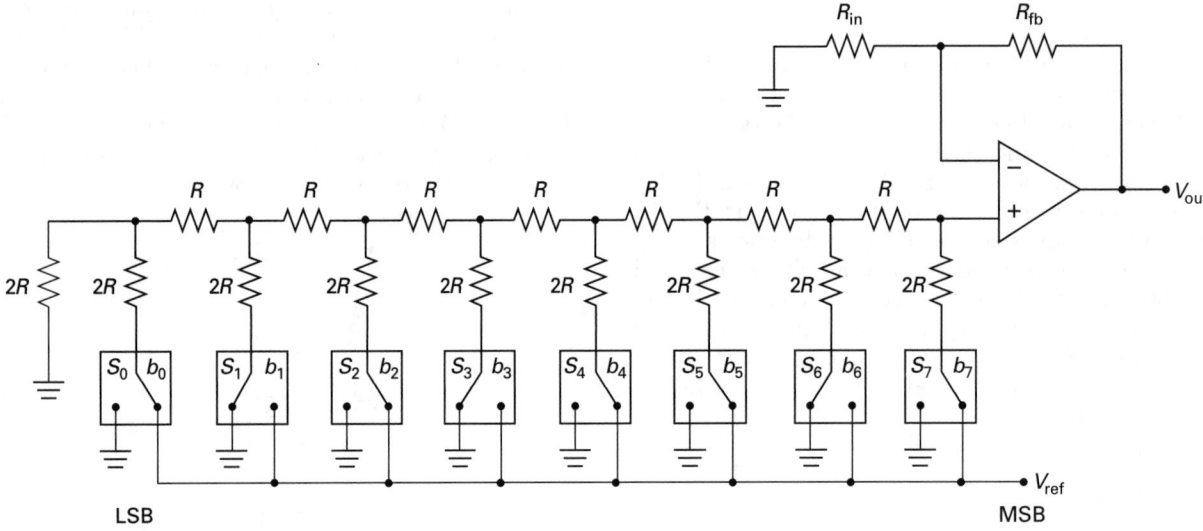

Figure 55.7 *Ladder-Type D/A Converter*

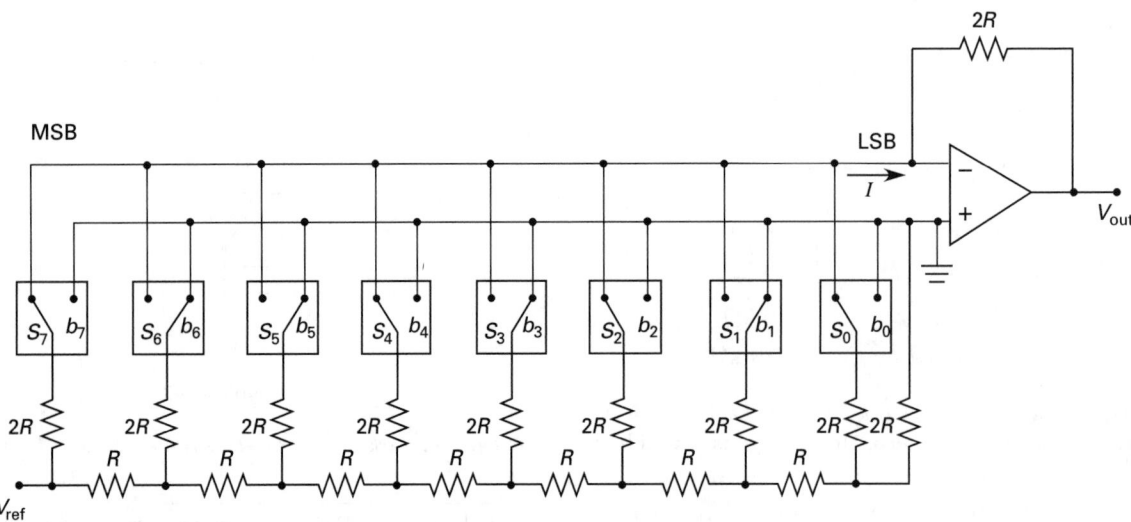

Figure 55.8 *Inverting-Type D/A Converter*

When any switch in Fig. 55.7 closes, a transient occurs. Because of the capacitance from the nodes to ground, a propagation delay occurs that is longer for the switches physically located further from the noninverting terminal. That is, the delay for S_0 (LSB) is longer than the delay for S_7 (MSB). These transients and delays are avoided by using an *inverting-type D/A converter*, which is shown in Fig. 55.8.

By inspection of Fig. 55.8 the current is

$$I = \left(\frac{V_{\text{ref}}}{2R}\right)\left(b_7 + \tfrac{1}{2}b_6 + \tfrac{1}{4}b_5 + \tfrac{1}{8}b_4 + \tfrac{1}{16}b_3\right.$$
$$\left. + \tfrac{1}{32}b_2 + \tfrac{1}{64}b_1 + \tfrac{1}{128}b_0\right) \qquad 55.5$$

Using ideal operational amplifier assumptions, all the current in Eq. 55.5 flows through the feedback resistor and none flows into the inverting terminal, making the output voltage

$$V_{\text{out}} = -2RI$$
$$= -V_{\text{ref}}\left(b_7 + \tfrac{1}{2}b_6 + \tfrac{1}{4}b_5 + \tfrac{1}{8}b_4\right.$$
$$\left. + \tfrac{1}{16}b_3 + \tfrac{1}{32}b_2 + \tfrac{1}{64}b_1 + \tfrac{1}{128}b_0\right) \qquad 55.6$$

The maximum output occurs when all the binary input values are high, that is, logic 1. In this condition, the magnitude of the maximum output for inverting ladder-type conductors of n-bit size is given by

$$|V_{\text{out,max}}| = V_{\text{ref}}\left(2 - \frac{1}{2^{n-1}}\right) \qquad 55.7$$

Example 55.3

What is the maximum output voltage for an eight-bit inverting ladder-type D/A converter with a reference voltage of 5 V?

Solution

For an 8-bit converter, $n = 8$. Substituting into Eq. 55.7 gives

$$|V_{\text{out,max}}| = V_{\text{ref}}\left(2 - \frac{1}{2^{n-1}}\right)$$
$$= (5\ \text{V})\left(2 - \frac{1}{(2)^{8-1}}\right)$$
$$= (5\ \text{V})\left(2 - \frac{1}{128}\right)$$
$$= 9.96\ \text{V}$$

6. MULTIPLYING-TYPE D/A CONVERTER

If the reference voltage of a noninverting D/A converter like the one in Fig. 55.7 is replaced by an analog voltage signal, the output becomes a product of the analog

signal and the binary input controlling the electronic switches. Such a device is called a *multiplying-type D/A converter*. This arrangement is also called a *programmable attenuator* because the output voltage is a fraction of the analog input with the attenuation controlled by the digital system or computer supplying the binary signal.

The inverting ladder-type D/A converter of Fig. 55.8 can also be used as a programmable attenuator by changing the feedback resistance from $2R$ to R. The binary weighted D/A converter of Fig. 55.3 can also be used as a programmable attenuator by changing the feedback resistance from 10 to 5 kΩ. In both cases, the incoming voltage, V_{ref} or $+5$ V, is replaced by the analog input voltage that is to be attenuated.

7. COUNTING A/D CONVERTER

An analog-to-digital (A/D) converter can be constructed as shown in Fig. 55.9(a).

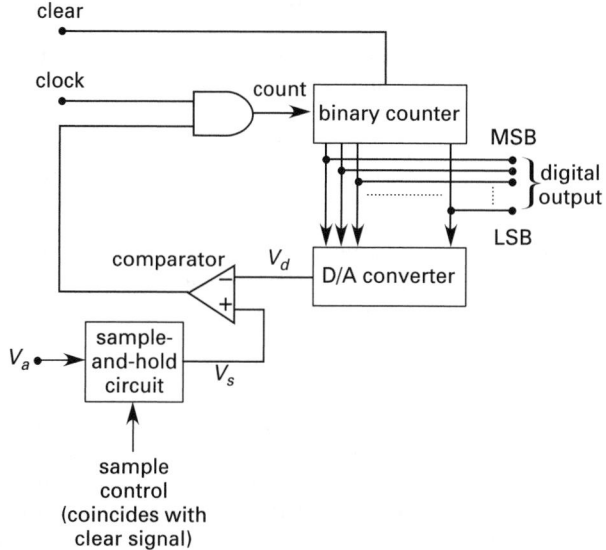

(a) circuit

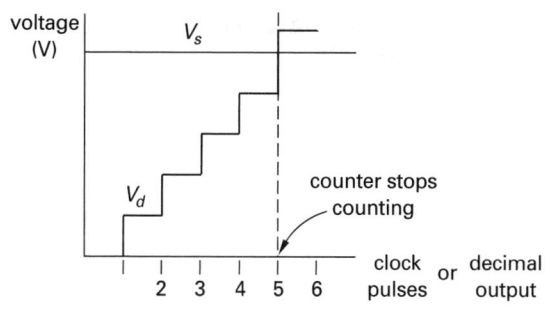

(b) representative output

Figure 55.9 *Counting A/D Converter*

The system shown in Fig. 55.9(a) is a *counting A/D converter*. The *sample-and-hold circuit* is used to avoid having the input change while conversion is occurring. The clock and comparator are inputs to the AND gate. The *comparator* is a high-gain differential amplifier whose output is high as long as the sampled analog signal is greater than the counter output, V_d.[5] When the comparator output is high, the AND gate passes the clock pulses to the binary counter. The binary counter sequences through binary numbers that comprise the output of the circuit. The digital output is also fed to a D/A converter whose output is attached to the inverting terminal of the comparator. When the magnitude of the D/A output, V_d, is above the sampled voltage, V_s, the comparator output changes to logic low and the clock signal is blocked from passing through the AND gate. The binary counter ceases to count. The digital output is proportional to the magnitude of the analog voltage sampled. The counter is then cleared in conjunction with another sample of the analog voltage, and the process repeats. The output of the circuit is illustrated in Fig. 55.9(b).

If an *up-down counter* is used, the clear signal and AND gate are not required. The counter counts up to the level of voltage sampled, forcing the comparator output low. The counter then counts down, but only by one LSB, at which point the comparator output goes high and the process repeats. The digital output thus ranges ± 1 LSB around the correct value. Such a circuit is called a *tracking* or *servo converter*.

The *successive-approximation converter* is a variation of the tracking converter that replaces the up-down counter with a programmer. The programmer sets the MSB to logic level one and all other bits to zero. The comparator then checks the D/A converter output against the sampled analog signal. If the D/A converter output is larger, the one is removed from the MSB position and tried in the next position. This continues until the D/A converter output is smaller than the sampled analog signal, at which point the one remains in that bit position. For such an n-bit converter, n clock periods are required to obtain the digital output while 2^n pulses may be required for a counting-type A/D converter.

8. PARALLEL-COMPARATOR A/D CONVERTER

Another type of A/D converter is the *parallel-comparator*, which is the fastest of all converters. An example is shown in Fig. 55.10.

Figure 55.10 is a 3-bit A/D converter. The analog signal is applied simultaneously to parallel comparators with equally spaced thresholds. The signal is bracketed.

That is, the output of the comparators is logic level zero for all comparators above the analog voltage input, and logic level one for all comparators below it. The signals on lines 1 through 7 are then priority-encoded to provide a 3-bit output. The truth table for the converter is given in Table 55.1.

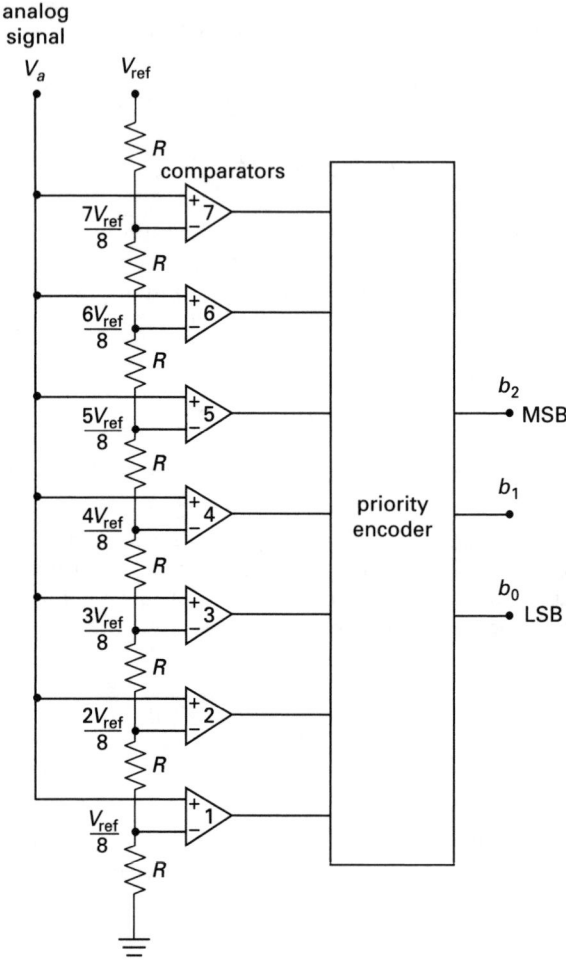

Figure 55.10 *Parallel-Comparator A/D Converter*

Table 55.1 *Parallel-Comparator A/D Converter Truth Table*

input							output		
7	6	5	4	3	2	1	b_2	b_1	b_0
0	0	0	0	0	0	0	0	0	0
0	0	0	0	0	0	1	0	0	1
0	0	0	0	0	1	1	0	1	0
0	0	0	0	1	1	1	0	1	1
0	0	0	1	1	1	1	1	0	0
0	0	1	1	1	1	1	1	0	1
0	1	1	1	1	1	1	1	1	0
1	1	1	1	1	1	1	1	1	1

---

[5]The output of the comparator is designed for compatibility with the logic system used (i.e., 0 V for logic level low and $+5$ V for logic level high in TTL systems).

9. DUAL-SLOPE A/D CONVERTER

The *dual-slope* or *ratiometric A/D converter* is shown in Fig. 55.11.

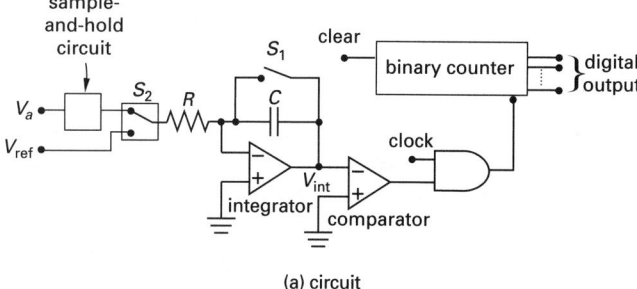

(a) circuit

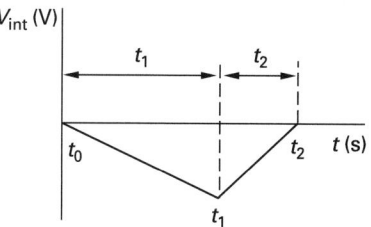

(b) representative integrator output

Figure 55.11 *Dual-Slope A/D Converter*

The positive input analog voltage is sampled and applied through electronic switch S_2 to the inverting terminal of an integrator. At the moment of sampling, switch S_1 closes to discharge the feedback capacitor. The counter is cleared at the same time, which is designated t_0. Since the output of the integrator is negative, the comparator output will be high. The clock signal is passed by the AND gate to the binary counter, which begins counting. The counter eventually reaches a state with all bits in logic level one, that is, high. The next clock pulse clears the counter and connects S_2 to the reference voltage. The process is triggered by the transition of the MSB from logic level one to logic level zero. At the time of connection to the reference voltage, designated t_1, the capacitor is charged to $-V_a T_1/RC$.

The integrator commences integration of the negative reference voltage with the output set by Eq. 55.8.

$$V_{\text{int}} = \frac{-V_a T_1}{RC} - \frac{V_{\text{ref}}(t - T_1)}{RC}$$

$$= \frac{-V_a T_1}{RC} + \frac{|V_{\text{ref}}|(t - T_1)}{RC} \qquad 55.8$$

(The reference voltage is negative in order to remove the charge buildup on the bypass capacitor.)

As long as the integrator output remains negative, the comparator output is high and the clock pulse is passed to the counter. Once the integrator output becomes positive, the comparator output shifts to low, blocking

the clock pulse and stopping the count. The termination of the count is designated t_2. The voltage at t_2 is found by setting the output voltage in Eq. 55.8 to zero.

$$V_a = \frac{|V_{\text{ref}}|(T_2 - T_1)}{T_1} \qquad 55.9$$

The fraction $(T_2 - T_1)/T_1$ is the ratio of the counter output at the time the comparator voltage reaches zero (which depends on the magnitude of the analog voltage sampled) to the full count. This ratio gives the ratiometric A/D converter its name. The final count is proportional to the analog voltage.

$$\frac{\text{final count}}{\text{full count}} = \frac{V_a}{V_{\text{ref}}} \qquad 55.10$$

The integrator output for a representative analog input is illustrated in Fig. 55.11(b). From this illustration, the reason for the "dual-slope" terminology may be seen.

10. PROTOCOLS

Any bus passing digital information is normally shared by several components. In order for each of these components to function properly, they must communicate in a standard manner, by a shared set of rules. These shared rules are called *protocols*.

The *asynchronous protocol* uses a handshaking technique to coordinate between components. For example, device A requests information. The information is found by device B. The address of the information is placed on the address bus, and an *acknowledge signal* is sent. Device A sees the acknowledge signal and stops making the request. It reads the address and waits for the data. Device B sends a *data ready signal* when the transfer is prepared. Device A acknowledges the data ready signal and begins transfer of the data. The process continues with the two components in lockstep responding to one another. The entire process has a self-timing nature. This *asynchronous transfer mode* (ATM) is insensitive to noise because of self-timing. Additionally, technological improvements such as increases in speed of components are easily handled, since each component is keyed to the other and not to a clock. Furthermore, clock skew or timing is not an issue, allowing asynchronous busses to be physically longer than synchronous busses. The major disadvantage of asynchronous protocols is the significant overhead of such a mode. *Overhead* is the time a computer system spends performing computations (such as timing between components) that do not directly contribute to the progress of any user-assigned tasks. Asynchronous designs are used when it is necessary to accommodate numerous devices with a wide performance range and the ability to upgrade the system is important.

Synchronous protocol uses an explicit clock signal to control the flow of information on busses. The components respond to either the leading edge or trailing

Computers

edge of the clock. The clock period is determined in a manner that will allow for propagation of all signals and allow the system to settle. Allowance must also be made for clock skew. *Clock skew* is a propagation delay of the clock signal throughout the digital system due to differences in transmission lines, loading, and so on. Overall, then, clock frequency is limited by the speed of the logic elements used and the length of the bus (transmission) lines. Synchronous protocol is fast, requires fewer bus lines, and is easier to implement and test. The major disadvantage is that it is less flexible than asynchronous protocol, since all devices must operate at the same clock rate. Additionally, the clock rate is fixed and cannot be raised in a single component to take advantage of technological advances. Synchronous designs are used when it is necessary to connect a small number of highly coupled devices and speed is of paramount importance. For example, processor and memory subsystems are most often connected via synchronous busses.

The *split-transaction protocol* transmits information at the beginning and end of a transaction only. Thus, only a request-and-reply transaction is transmitted. This protocol works only if more than one bus master is present and the memory system is sophisticated enough to track the overlapping transactions. Such a system has increased bus bandwidth over other protocols. The major disadvantage is that such a protocol is expensive to implement since each transaction must be identified and tracked. Such a protocol is used in systems with shared memories. Split-transaction protocol is also called *connect/disconnect*, *packet-switched*, or *pipelined protocol*.

Numerous other protocols exist and are implemented for a variety of purposes and applications, such as the *burst* or *block-transfer protocol*, *cache coherence protocols*, and others.

Topic XII: Communications

Communications

For the most current information about the exam, visit **www.ppi2pass.com** regularly.

PROFESSIONAL PUBLICATIONS, INC.

56 High-Frequency Transmission

Nomenclature

b	normalized susceptance	–
B	susceptance	S, Ω^{-1}, or ℧
C	capacitance	F
f	frequency	Hz, s^{-1}, or cycles/s
g	normalized conductance	–
G	conductance	S, Ω^{-1}, or ℧
l	length	m
L	inductance	H
n	integer	–
r	normalized resistance	–
rms	root-mean-square	–
R	resistance	Ω
R_0	characteristic resistance	Ω
v	velocity (speed)	m/s
VSWR	voltage standing wave ratio	–
x	distance variable	m
x	normalized reactance	–
X	reactance	Ω
y	normalized admittance	–
Y	admittance	S, Ω^{-1}, or ℧
z	normalized impedance	–
Z	impedance	Ω
Z_0	characteristic impedance	Ω

Symbols

α	attenuation constant	Np/m
β	phase constant	rad/m
γ	propagation constant	rad/m
Γ	reflection coefficient	–
λ	wavelength	m
ϕ	angle	rad
ω	angular frequency	rad/s

Subscripts

0	characteristic
cp	compensation point
C	capacitor
i	imaginary
l	per unit length
L	inductor
r	real
R	receiving end
S	sending end
unk	unknown

1. FUNDAMENTALS

Transmission lines are three-dimensional, being large in one direction and small in the other two. Power transmission lines operating at 60 Hz have a wavelength of 5000 km and a transverse dimension of about a meter, or less.[1] Thus, even the longitudinal dimension of a transmission line, which is rarely greater than 500 km, is a fraction of the wavelength. Coaxial lines are used as the frequency increases up to approximately 1 GHz. A *coaxial cable* or *line* is one in which the conductor is centered inside and insulated from an outer metal tube that acts as a second conductor. Above 1 GHz, the losses become prohibitive and waveguides are used. A *waveguide* is a device that constrains the propagation of an electromagnetic wave along a path predetermined by the physical construction of the guide. Waveguides are used up to approximately 100 GHz, where the losses become prohibitive. Above approximately 100 GHz, optical guides are used that operate at frequencies of approximately 2×10^{14} Hz with minimal losses.

Transmission lines are designed to transfer power from a source to a load. The distance between the source and load must be small compared to the wavelength in order to prevent the power from being radiated into space, which occurs when the wavelength is on the order of the conductor separating the source and load. Conductors specifically designed with dimensions comparable to the wavelength of the frequency are called *antennas*. High-frequency transmission lines and waveguides are used to connect transmitters (sources) and receivers (loads). The principles of high-frequency transmission must be understood in computer and digital system design to maximize power transfer and minimize power radiation.

2. HIGH-FREQUENCY TRANSMISSION LINES

At high frequencies, approximately 1 MHz and higher, wavelengths are shorter and even a few feet of line are treated as a long transmission line (see Sec. 37-14). The resistance is normally negligible, specifically, $R \ll \omega L$

---

[1]The general principles of transmission lines are described in Ch. 37.

and $G \ll \omega C$. Setting the resistance, R, equal to zero simplifies the governing equations. The impedance per unit length is[2]

$$Z_l = jX_{L,l} = j\omega L_l \qquad 56.1$$

The admittance per unit length is

$$Y_l = jB_{C,l}$$
$$= \frac{j}{X_{C,l}} = -\frac{1}{jX_{C,l}} = j\omega C_l \qquad 56.2$$

The propagation constant, γ, in units of radians per meter (mile) is given by

$$\gamma = \sqrt{Y_l Z_l}$$
$$= \sqrt{(j\omega C_l)(j\omega L_l)}$$
$$= j\omega\sqrt{L_l C_l}$$
$$= j\beta \qquad 56.3$$

The term β in Eq. 56.3 is the phase constant. The phase velocity, v_{phase}, and the wavelength are given by

$$v_{\text{phase}} = \frac{1}{\sqrt{LC}} \qquad 56.4$$

$$\lambda = \frac{2\pi}{\beta} = \frac{1}{f\sqrt{LC}} \qquad 56.5$$

The characteristic impedance, Z_0, is given by[3]

$$Z_0 = \sqrt{\frac{Z_l}{Y_l}} = \sqrt{\frac{j\omega L_l}{j\omega C_l}} = \sqrt{\frac{L}{C}} \qquad 56.6$$

The equations for a long transmission line, Eqs. 37.49, 37.50, 37.54, and 37.55, are applicable with the values from Eqs. 56.1, 56.2, 56.3, and 56.6 substituted. At high frequencies, however, circuits are designed to minimize reflected waves. This is accomplished by using capacitive compensation to make the load impedance appear to be the characteristic impedance. The expressions for the sending voltage and current are thus specified in terms of the reflection coefficient, Γ (see Sec. 37-15). Note that the voltage traveling in the positive x direction, toward the receiving end (see Sec. 37-14), is associated with

$$V^+ = \tfrac{1}{2}V_R + \tfrac{1}{2}Z_0 I_R \qquad 56.7$$

---

[2]The per-unit-length impedance and admittance are sometimes written with lowercase letters, z and y, to emphasize their per-unit qualities. In this chapter, the lowercase letters are used for normalized quantities.
[3]The characteristic impedance is the ratio of the voltage to the current on an infinite transmission line. It can be called the *characteristic resistance*, R_0, since the line is lossless. That is, Z_l and Y_l are purely reactive, making Z_0 purely resistive.

(Equation 56.7 is one of the constants from the solution of the wave equation, Eq. 37.44. The other constant is V^-. Both can be determined from the solution to the wave equation given by Eq. 37.49. The positive x direction is defined as the direction of power flow, that is, from the sending end to the receiving end.)

The terms $\cosh j\beta x = \cos \beta x$ and $\sinh j\beta x = j\sin\beta x$, when combined with Eq. 56.7, allow the sending-end voltage and current to be written in terms of the reflection coefficient.

$$V_S = V^+ \left(e^{j\beta l} + \Gamma e^{-j\beta l}\right)$$
$$= V^+ \left((1+\Gamma)\cos\beta l + j(1-\Gamma)\sin\beta l\right) \qquad 56.8$$

$$I_S = \left(\frac{V^+}{Z_0}\right)\left(e^{j\beta l} - \Gamma e^{-j\beta l}\right)$$
$$= \left(\frac{V^+}{Z_0}\right)\left((1-\Gamma)\cos\beta l + j(1+\Gamma)\sin\beta l\right) \qquad 56.9$$

The input impedance is V_S/I_S and, in terms of the reflection coefficient, is given by

$$Z_{\text{in}} = Z_0\left(\frac{(1+\Gamma)\cos\beta l + j(1-\Gamma)\sin\beta l}{(1-\Gamma)\cos\beta l + j(1+\Gamma)\sin\beta l}\right) \qquad 56.10$$

Using the definition of the reflection coefficient from Eq. 37.58, the input impedance can be written in terms of the load impedance as

$$Z_{\text{in}} = Z_0\left(\frac{Z_{\text{load}}\cos\beta l + jZ_0\sin\beta l}{Z_0\cos\beta l + jZ_{\text{load}}\sin\beta l}\right) \qquad 56.11$$

The voltage at any point on a transmission line is a function of the distance from the load since there are two waves traveling on any line, one incident to the load and one reflected from the load. The standing wave ratios defined by Eqs. 37.56 and 37.57 measure the difference between the maximum rms voltage (current) and the minimum rms voltage (current) of the two waves. (The standing wave ratio is a physical variable that can be directly measured.) Where the magnitudes of the waves add, the voltage (current) is at a maximum. Where the magnitudes of the waves subtract, the voltage (current) is at a minimum. The reflection coefficient is defined in terms of the standing wave ratios by Eq. 37.60. Rearranging Eq. 37.60 to define the voltage standing wave ratio, VSWR, in terms of the reflection coefficient gives

$$\text{VSWR} = \frac{|V^+| + |V^-|}{|V^+| - |V^-|} = \frac{1+|\Gamma|}{1-|\Gamma|} \qquad 56.12$$

The VSWR is related to the minimum and maximum impedance by Eqs. 56.13 and 56.14.

$$Z_{\text{max}} = Z_0(\text{VSWR}) \qquad 56.13$$

$$Z_{\text{min}} = \frac{Z_0}{\text{VSWR}} \qquad 56.14$$

Example 56.1

A certain transmission line has a characteristic impedance of 50 Ω and a terminating resistance of 100 Ω. What is the reflection coefficient?

Solution

The reflection coefficient is defined by Eq. 37.58 as

$$\Gamma_l = \frac{V_{\text{reflected}}}{V_{\text{incident}}} = \frac{V_{\text{reflected}}}{V_{\text{load}}} = \frac{Z_{\text{load}} - Z_0}{Z_{\text{load}} + Z_0}$$

Substitute the given values.

$$\Gamma_l = \frac{Z_{\text{load}} - Z_0}{Z_{\text{load}} + Z_0} = \frac{100\ \Omega - 50\ \Omega}{100\ \Omega + 50\ \Omega} = 0.333$$

Example 56.2

For the transmission line in Ex. 56.1, what is the input impedance if the line is an even number of wavelengths long?

Solution

The transmission line is restricted by the problem statement to be an even number of wavelengths for the frequency carried. In mathematical terms, with n an integer, this is stated as

$$l = n\lambda = \frac{n2\pi}{\beta}$$

Equation 56.5 is used to change the wavelength, λ, into a function of the phase constant, β. Rearranging gives

$$\beta l = n2\pi$$

This wavelength restriction causes the sinusoidal terms of the input impedance equation, Eq. 56.11, to be

$$\cos \beta l = 1$$

$$\sin \beta l = 0$$

Substituting these values into Eq. 56.11 gives

$$Z_{\text{in}} = Z_0 \left(\frac{Z_{\text{load}}}{Z_0} \right) = Z_{\text{load}} = 100\ \Omega$$

Example 56.3

For the transmission line in Ex. 56.1, what is the voltage standing wave ratio?

Solution

The VSWR can be found by substituting the calculated value of the reflection coefficient from Ex. 56.1 into Eq. 56.12.

$$\text{VSWR} = \frac{1 + |\Gamma|}{1 - |\Gamma|} = \frac{1 + 0.333}{1 - 0.333} = 2$$

3. SMITH CHART FUNDAMENTALS

Electronic calculators and computer-aided design (CAD) programs have reduced the importance of graphical methods for the solution of transmission line problems.[4] Nevertheless, the Smith chart remains useful for explaining steady-state situations on transmission lines. A *Smith chart* is a special polar diagram with constant reflection coefficient circles, constant standing wave ratio circles, constant resistance circles, constant reactance arcs (portions of a circle), and radius lines representing constant line-angle loci. A complete Smith chart is given in App. 56.A.[5] The Smith chart is essentially a polar representation of the reflection coefficient in terms of the normalized resistance and reactance. The normalization occurs with respect to the characteristic impedance, Z_0.

$$z = \frac{Z}{Z_0} = \frac{R + jX}{Z_0} = \frac{1 + \Gamma}{1 - \Gamma} \qquad \textit{56.15}$$

The normalized impedance (z) impedance (Z) resistance, reactance, and reflection coefficients are all functions of distance, normally measured with the load as the origin.[6] In the complex Γ plane, the curves of constant reflection coefficient magnitude, $|\Gamma|$, are circles as shown in Fig. 56.1(a). Constant standing wave ratios are also circles positioned exactly as shown in Fig. 56.1(a). (Neither the constant reflection coefficient nor the constant standing wave ratio circles are shown on the final Smith chart. Instead, radial scales, shown in App. 56.A next to the chart, are used to determine their value.)

---

[4]A Smith chart is a graphical method for the solution of transmission line equations. A *bounce diagram*, also called a *time-distance plot*, is a graphical method for the solution of multiple reflections on transmission lines.

[5]The constant reflection coefficient circles and constant standing wave ratio circles are not plotted on the Smith chart. Instead, radial scales, shown in App. 56.A, are used to determine the value of the respective items as if a circle were drawn from the center of the Smith chart at a radius given by the scale.

[6]Do not confuse the distance x with the normalized reactance x. In some books, the distance is given the symbol d or the normalized reactance is given the symbol χ to avoid confusion. Here, the individual terms in Eq. 56.15 are not listed as functions of distance, to avoid the problem in this specific instance.

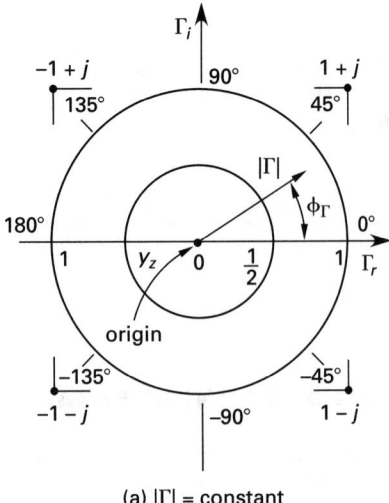

(a) |Γ| = constant

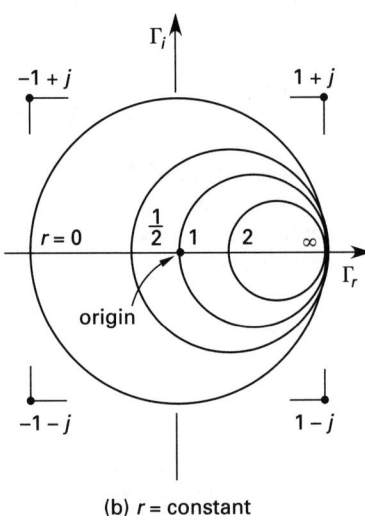

(b) r = constant

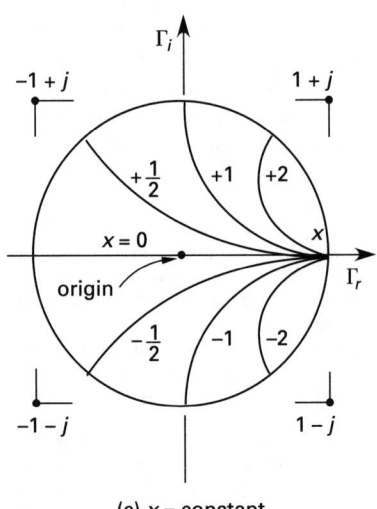

(c) x = constant

Figure 56.1 Smith Chart Components

The constant resistance circles are placed on the chart in the positions shown in Fig. 56.1(b). Note that the origin of the chart is located at the intersection of the $r = 1$ circle and the line representing $x = 0$. The outermost circle is $r = 0$. The arcs of constant reactance are drawn as shown in Fig. 56.1(c). The constant resistance circles of Fig. 56.1(b) and the constant reactance arcs of Fig. 56.1(c) are combined to create the final Smith chart in App. 56.A. Since the circles of constant $|\Gamma|$ are not included, the value of $|\Gamma|$ is read from the left-hand external scale in App. 56.A. Since the circles of constant VSWR are not included, the value of VSWR is read from the right-hand external scale. Both values correspond to a point (r, x). There are three circumferential scales. The outer two measure fractions of a wavelength toward the generator (clockwise) or toward the load (counterclockwise). The full circumference represents one-half the wavelength. The inner circumferential scale measures the reflection coefficient angle, $\phi_L = \phi_R - \beta x$.[7]

Any circle drawn around the origin is called an *impedance locus circle*. At the origin, the impedance locus circle has a radius of zero, which is called the *matched condition*. At this point the total reactance is zero, the load impedance matches the characteristic impedance, making the normalized impedance equal to one, and the reflection coefficient equals zero. Typical values for other electrical conditions are shown in Table 56.1.

Table 56.1 Smith Chart Electrical Conditions

electrical condition	reflection coefficient (Γ)	normalized resistance (r)	normalized reactance (x)
open circuit	$1\angle 0°$	∞ (arbitrary)	arbitrary (∞)
short circuit	$1\angle 180°$	0	0
pure reactance	$1\angle \pm 90°$	0	± 1
matched line (pure resistance)	0	1	0

The Smith chart in App. 56.A and the components shown in Fig. 56.1 were arranged to show the real portion of the reflection coefficient along the standard x-axis and the imaginary component along the standard y-axis, subscripted as r and i, respectively. It is also common to rotate the chart 90° counterclockwise to place the $r = 0$ and $x = 0$ position, that is, the intersection of the zero resistance circle and the zero reactance line, at the top of the chart as shown in Fig. 56.2.

The Smith chart can be used for normalized admittances, y, by considering the r-circles as conductance

---

[7]The x used here is distance. When the total distance is from the load (receiving end) to the generator (sending end), the symbol l is used.

circles (g-circles) and the x-arcs as susceptance arcs (b-arcs). The angle for the reflection coefficient, Γ, for a given admittance, y, is given by $180° + \phi_{\Gamma}$, and the point $y = 0 + j0$ represents an open-circuit condition.

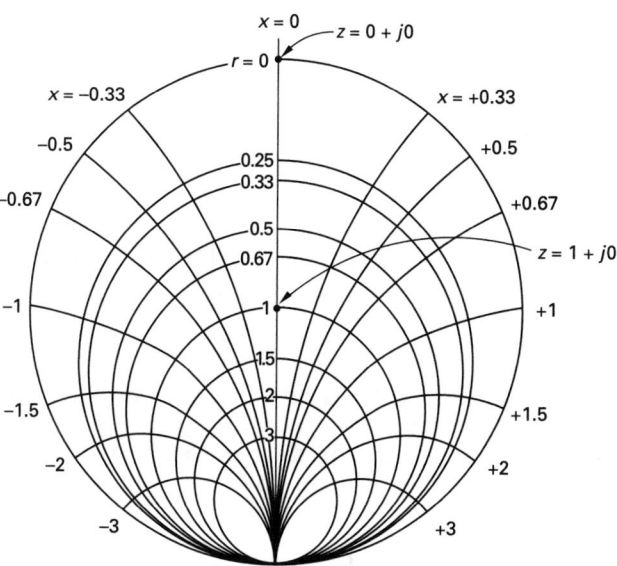

Figure 56.2 *Alternate Smith Chart Positioning*

The input impedance, compensating reactance, compensating position, reflection coefficient, and standing wave ratio can be graphically determined using the Smith chart.[8] The steps vary and are presented in examples in Secs. 56-4 and 56-5.

4. SMITH CHART: INPUT IMPEDANCE

The general procedure in finding the input impedance is to first determine the normalized load impedance, z_L, using Eq. 56.15. Plot the point z_L, representing the normalized resistance and reactance, at the intersection of the r circle and the x arc. Draw a circle centered at $1 + j0$ (the origin) through the value $r + jx$ corresponding to z_L. This is the impedance locus circle. Draw a radial line from the origin through z_L. Determine the angle βl and the normalized impedance a distance l from the load by following the impedance locus circle clockwise (toward the generator or sending end) through an angle of $2\beta l$.

Example 56.4

A transmission line with a length, βl, of 6 radians (344°) and a characteristic impedance, z_0, of 50 Ω is terminated with a load of $75 + j25$ Ω. What is the input impedance (i.e., the impedance seen by the generator)?

---

[8]Though the reflection coefficient and standing wave ratio can be determined graphically, App. 56.A substitutes radial scales for the graphical procedure.

Solution

step 1:   Calculate the normalized load impedance.

$$z_L = \frac{Z_L}{Z_0}$$
$$= \frac{R_L + jX_L}{Z_0}$$
$$= \frac{75 + j25 \ \Omega}{50 \ \Omega} = 1.5 + j0.5$$

step 2:   Plot point $z_L = 1.5 + j0.5$ on the Smith chart.

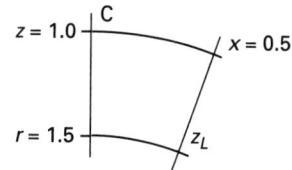

step 3:   Draw a circle centered at $1.0 + j0$ (point C), that is, the origin, which passes through point z_L. This is the impedance locus circle.

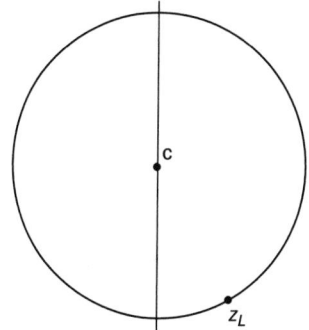

step 4:   Extend a line from point C (the origin) through point z_L to the edge of the Smith chart, point L (load point).

step 5:   Calculate the angular separation between the source and load points.

$$2\beta l = (2)(344°) = 688°$$

Since this is more than one revolution, subtract the full revolution.

$$688° - 360° = 328°$$

Move clockwise 328° from the load to the generator point.

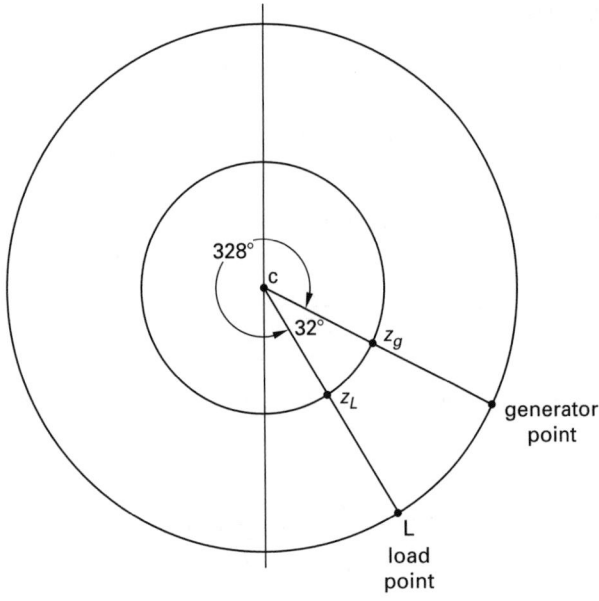

generator
point

load
point

step 6: Read the impedance seen from the generator, that is, the input impedance at the intersection of the generator line and the impedance circle, point z_g.

$$z_g = 1.1 + j0.6$$

step 7: Calculate the input load impedance.

$$
\begin{aligned}
Z_g &= Z_0 z_g \\
&= (50\ \Omega)(1.1 + j0.6) \\
&= 55 + j30\ \Omega
\end{aligned}
$$

5. SMITH CHART: COMPENSATION REACTANCE

Compensation is used to make the total reactance of a given transmission line zero. Compensation is also called *impedance matching*. At low frequencies, the matching is accomplished with lumped reactive components (inductors and capacitors). At high frequencies, in order to minimize dissipation losses, a length of open- or short-circuited line is used. Compensation with a single line is called *single-stub matching*. Compensation with two lines is called *double-stub matching*.

The reactance required for compensation can be determined graphically on the Smith chart. The impedance locus circle is set up as before (see Ex. 56.4, steps 1 through 3). The line is traversed toward the source (clockwise) to the intersection of the impedance locus circle and the $r = 1$ circle. The intersection is called the *compensation point*. At the compensation point, the impedance is given by

$$z_{\rm cp} = 1 + jx \qquad 56.16$$

The compensation reactance is $-x$, which results in a total normalized impedance of $z = 1$. Since capacitance is the most desirable compensating reactance for a variety of reasons, the compensation point is always chosen on the positive reactance side of the Smith chart. The positive reactance side of the chart is where $x > 0$, which makes the compensating reactance negative $(x < 0)$.[9]

Example 56.5

A transmission line with a characteristic impedance, Z_0, of 50 Ω is terminated with a load impedance of $100 + j100\ \Omega$. (a) Where in the transmission line should a compensating reactance be placed? (b) What is the value of the compensating reactance?

Solution

(a) Asking for the placement of the reactance is equivalent to asking at what electrical angle (βl) the capacitor should be placed to make the impedance at that point equal to Z_0. Following the solution outline suggested gives

step 1: Calculate the normalized load impedance.

$$z_L = \frac{Z_L}{Z_0} = \frac{100 + j100\ \Omega}{50\ \Omega} = 2 + j2$$

step 2: Plot the point z_L on the Smith chart.

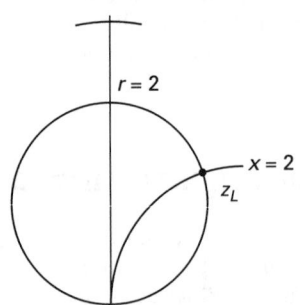

step 3: Draw a circle centered at $1 + j0$, that is, the origin, which passes through point z_L. This is the impedance locus circle.

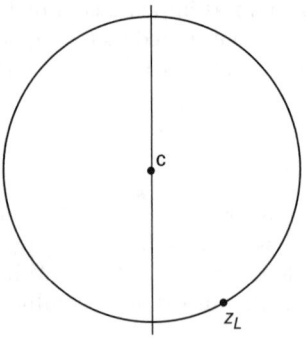

---

[9]The positive reactance side of the Smith chart is the top if positioned as in App. 56.A, and to the right if positioned as in the examples of this chapter.

step 4: Extend a line from point C (the origin) through the load point z_L. From the load point, follow the impedance locus circle clockwise, toward the source to point CP (the compensation point), which is the intersection of the impedance locus circle and the $r = 1$ circle on the positive reactance side of the Smith chart. (The angle traversed is $2\beta l$.)

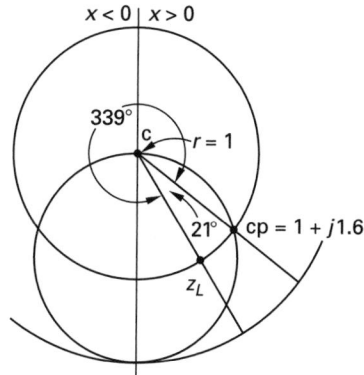

step 5: Measure the angular separation between the compensation and load points. In this case, the value is 21°. Since the electrical angle is always measured from the load toward the source, that is, clockwise, the angular separation is actually $360° - 21° = 339°$. The electrical angle (βl) is thus

$$2\beta l = 339°$$

$$\beta l = \frac{339°}{2} = 169.5°$$

The compensation point is βl electrical degrees or $169.5°/360°$ wavelengths (0.471λ) from the load toward the source. Thus, the reactance should be placed a distance $l = 0.471\lambda$ from the load toward the source. This can also be found using Eq. 56.5.

$$\lambda = \frac{2\pi}{\beta}$$

$$\beta\lambda = 2\pi$$

$$\beta l \lambda = 2\pi l$$

$$l = \left(\frac{\beta l}{2\pi}\right)\lambda = \left(\frac{169.5°}{360°}\right)\lambda = 0.471\lambda$$

step 6: Read the normalized impedance at the compensation point, z_{cp}.

$$z_{cp} = 1 + j1.6$$

(b) The normalized compensating reactance, x_c, must therefore be -1.6. The actual reactance is

$$X_c = Z_0 x_c = (50 \ \Omega)(-1.6) = -80 \ \Omega$$

6. IMPEDANCE DETERMINATION USING VSWR MEASUREMENTS

The voltage standing wave ratio (VSWR) can be measured using a *slotted line* in which a probe can be placed to determine the maximum and minimum voltages. The distance between a minimum and a maximum is $\lambda/4$ (one-quarter wavelength). The distance between two minimums or two maximums is $\lambda/2$ (one-half wavelength). The first minimum of the VSWR from the load is given the symbology l_{min}. With the load shorted, the distance from the short toward the generator of the first VSWR minimum determines the half-wavelength, which is the position where $\beta l = \pi$.[10] Using the measured VSWR, l_{min}, and the half-wavelength, an unknown impedance on a transmission line can be determined.

The voltage standing wave minimum occurs at the point of minimum impedance. Equation 56.11 can be written for an unknown impedance, Z_{unk}, at a minimum input impedance as

$$\frac{Z_{in,min}}{Z_0} = \frac{\dfrac{Z_{unk}}{Z_0} + j\tan\beta l}{1 + j\left(\dfrac{Z_{unk}}{Z_0}\right)\tan\beta l} \qquad \textit{56.17}$$

Equation 56.14 is rearranged to explicitly show the relationship between the minimum impedance and the VSWR as

$$\frac{Z_{min}}{Z_0} = \frac{1}{\text{VSWR}} = \frac{Z_{in,min}}{Z_0} \qquad \textit{56.18}$$

Equations 56.17 and 56.18 are combined to determine an expression for the unknown impedance in terms of the VSWR, the minimum input impedance, and the minimum distance from the load at which the first minimum of the VSWR occurs.

$$Z_{unk} = Z_0\left(\frac{1 - j\,(\text{VSWR})\tan\beta l}{\text{VSWR} - j\tan\beta l}\right) \qquad \textit{56.19}$$

---

[10]A short circuit is a minimum and is located at $0 + j0$ (see Table 56.1). A distance of one-half wavelength on the Smith chart is two circumferences (see App. 56.A), which places one at $0 + j0$ again—that is, at a minimum. The term βl is the argument of the sinusoidal functions that describe the voltage and thus will be at a minimum every π radians, or one-half wavelength.

57

Antenna Theory

Nomenclature

A	vector magnetic potential	Wb/m
B	bandwidth	Hz
c	speed of light	3.00×10^8 m/s
D	directivity	–
E	electric field strength	V/m
f	frequency	Hz, s^{-1}, or cycles/s
g	gain (ratio)	–
G	gain	– or dB
h	antenna height (length)	m
H	magnetic field strength	A/m
I	rms phasor current	A
Im	imaginary portion	–
l	length	m
L	loss	dB
n	absolute index of refraction	–
p	power density	W/m²
P	power or average power density	W or W/m²
r	radial distance	m
Re	real portion	–
S	Poynting's vector	W/m²
T	temperature	K
v	velocity	m/s
Z_0	characteristic impedance	Ω

Symbols

α	angle	rad
β	phase constant	rad/m
ε	permittivity	F/m
ε_0	free-space permittivity	8.854×10^{-12} F/m
η	efficiency	–
θ	angle or polar angle	rad
κ	Boltzmann's constant	1.3805×10^{-23} J/K
λ	wavelength	m
μ	permeability	H/m
μ_0	free-space permeability	1.2566×10^{-6} H/m
ϕ	azimuthal angle	rad
ω	angular frequency	rad/s

Subscripts

ave	average
b	basic
$b0$	basic, free space
c	critical
g	gain (ratio)
G	gain (dB)
i	incident
l	per unit length
n	noise
r	radiated, reflected, relative, or radius
rad	radiation
rec	received or receiving
sys	system
trans	transmission or transmitting

1. ANTENNA FUNDAMENTALS

An *antenna* is a device used for radiating or receiving electromagnetic waves. Antenna dimensions are comparable to the wavelength of the signal.

An electromagnetic wave is generated by the current within the antenna. The wave can be generated by an electric line source, an electric current loop, or a *Huygens' source*, which is a source with constant electric and magnetic current sources of equal length at the same point.

A group of components, such as antennae, arranged to vary the transmission or reception of electromagnetic radiation (waves) with direction is called an *array*. Electromagnetic waves can be generated by a *simple array*,

Communications

that is, a line of radiating elements (antennae); a *uniform linear array*, that is, a simple array with each element fed uniformly in signal amplitude; a *circular array*; and a *planar array*.

An antenna's shape determines its overall radiating pattern. One type of antenna, the *wire antenna*, includes the wire antenna over or near a ground plane, V dipole, bent dipole, loop, and others. A *waveguide antenna* is constructed of a waveguide opening onto a conducting ground plane.[1] The waveguide itself is rectangular or circular and usually opens onto a finite ground plane. When the waveguide feeds a horn shape in a manner similar to the wire antenna, the result is called a *horn antenna*. There are *sectoral*, *pyramid*, *conical*, and *corrugated horns*. A *reflector antenna* is one whose input signal is reflected from a *dish*. Some examples of reflector antennae are shown in Fig. 57.1. Waveguide, horn, and reflector antennae are examples of aperture antennae. An *aperture* is an opening through which waves can pass. An *aperture antenna* is one in which the beam width is determined by the dimensions of the lens, horn, or reflector. Other antennae include the *surface wave* and *microstrip* types.

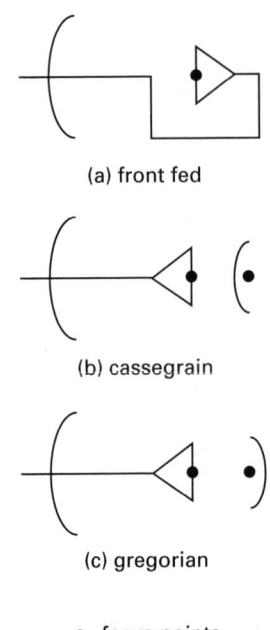

(a) front fed

(b) cassegrain

(c) gregorian

● focus points

Figure 57.1 *Reflector Antennae*

The radiated power of an antenna is the power transferred from the source current to space. When power is radiated uniformly in all directions, the antenna is *isotropic*.[2] The ratio of the isotropic power to the radiated power is called the *directivity* of the antenna and is a measure of the antenna's *focus*. The *antenna gain* is a measure of the effectiveness of a directional antenna

---

[1]This type of antenna is found in reflector antennae or flush-mounted antennae on aircraft and spacecraft.
[2]The power radiated in this manner is called the *equivalent isotropically radiated power* (EIRP).

compared to an *omnidirectional* or *isotropic antenna*. The gain is dependent on the angle from the antenna.[3] The gain and the directivity are related by the *antenna efficiency*, which relates the radiated power to the input power. The efficiency, in turn, depends on the *radiation resistance*, which is the total radiated power of an antenna divided by the square of the effective current measured at the point where the power is supplied to the antenna.[4] The efficiency also depends on the *ohmic resistance* of the antenna, primarily a function of the skin depth at high frequencies.

The *antenna polarization* is the orientation of the electric field lines in the radiated electromagnetic wave. That is, polarization characterizes the orientation of the field vector in its travel. The polarization usually refers to the electric field vector. When the electric field vector, **E**, and the magnetic field vector, **H**, are orthogonal to each other and in a plane transverse to the direction of propagation, the wave is *elliptically polarized*. When the rectangular components are equal and their phases differ by an odd integral of $\pi/2$, the wave is *circularly polarized*. A wave is *right-hand circularly polarized* (RHCP) if its rotation is clockwise when viewed in the direction of propagation, and *left-hand circularly polarized* (LHCP) if the rotation is counterclockwise.[5] When the electric field vector at a fixed point in space remains pointing in a fixed direction, varying only in magnitude, the wave is *linearly polarized*. This is also called *plane polarization*.[6]

The *antenna temperature* is the temperature a blackbody enclosure in thermal equilibrium would produce if it surrounds the antenna. The *noise power* is given by

$$P_n = \kappa T B \qquad 57.1$$

The term κ is Boltzmann's constant, T is the temperature in degrees kelvin, and B is the bandwidth in hertz. All noise sources that contribute to the noise power can thus be considered to have an equivalent temperature.

2. WAVE PROPAGATION FUNDAMENTALS

The antenna type determines the radiation properties. The radiation pattern is obtained from the source currents creating the magnetic field or from the magnetic field about a line, area, or volume, all of which depend on the antenna type. Regardless, the magnetic field is described by the permeability and the magnetic vector potential, **A**.

$$\mathbf{H} = \frac{1}{\mu} \mathbf{\nabla} \times \mathbf{A} \qquad 57.2$$

---

[3]The ability of an array to direct a signal in a given direction is called *signal steering* and is accomplished by delaying the signal to the various elements of the array.
[4]The radiation resistance is termed the *radiation impedance* if *radiation reactance* is taken into account.
[5]This definition comes from the International Radio Regulations, Geneva, 1990. The definitions used in some texts differ.
[6]This is similar to elliptical polarization, though without the rotation of the electric field vector about the direction of the propagation axis.

The propagation of this magnetic field depends on numerous factors. One of the primary factors is the frequency of the radiated wave. Consequently, most discussion involving propagation is focused on certain frequency bands. For example, very low-frequency waves propagate thousands of kilometers over the surface of the earth with little attenuation. As frequency rises, the ground path is limited to short distances because of excessive losses. At medium to high frequencies, though, ionospheric reflections allow wave propagation, that is, radio communication, over long distances. Above about 30 MHz, the ionospheric reflections are not dependable, and line-of-sight propagation must be used. At very high frequencies, little interaction takes place with the environment and the wave propagates in a relatively straight line. This makes for short-distance communications unless the transmitter and/or receiver is/are located in space to increase the line-of-sight area.[7] At frequencies above approximately 10 GHz, absorption of the wave by air and water vapor as well as attenuation by rain can occur. In free space, that is, a vacuum, electromagnetic waves propagate unhindered. A line-of-sight condition in a nonionized atmosphere approximates free space.[8]

3. FREE-SPACE WAVE CHARACTERISTICS

An electromagnetic wave has a wavelength, which is defined as the distance the wave propagates in one period.

$$\lambda = \frac{2\pi}{\beta} = \frac{2\pi}{\omega\sqrt{\mu_0\varepsilon_0}} = \frac{c}{f} \qquad 57.3$$

The velocity of an electromagnetic wave in free space is

$$v_0 = \frac{1}{\sqrt{\mu_0\varepsilon_0}} \approx c = 3 \times 10^8 \text{ m/s} \qquad 57.4$$

The *characteristic impedance of free space* is

$$Z_0 = \sqrt{\frac{\mu_0}{\varepsilon_0}} = 120\pi \ \Omega = 377 \ \Omega$$

$$= 4\pi v_0 \times 10^{-7} \ \Omega \qquad 57.5$$

4. POYNTING'S VECTOR AND POWER

Poynting's vector, **S**, is the instantaneous power density carried by an electromagnetic wave and is given by

$$\mathbf{S} = \varepsilon c E^2 \mathbf{a} = \mathbf{E} \times \mathbf{H} \quad [\text{in W/m}^2] \qquad 57.6$$

The vector **a** represents a unit vector in the direction of propagation. Poynting's vector is illustrated in Fig. 57.2.

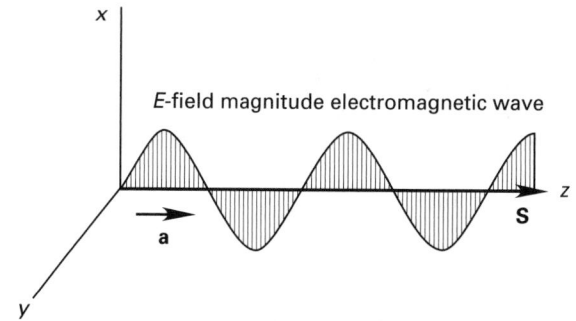

Figure 57.2 *Poynting's Vector*

The real part of Poynting's vector is the average power density in the direction of transmission and is given by[9]

$$P_{\text{ave}} = \tfrac{1}{2}\text{Re}\left(\mathbf{E} \times \mathbf{H}\right) = \frac{E^2}{2Z_0} \quad [\text{in W/m}^2] \qquad 57.7$$

5. POLARIZATION

Polarization is a phenomenon exhibited by electromagnetic waves in which the direction of the electric field is constant or varies in a definite way. Polarization is illustrated in Fig. 57.3.

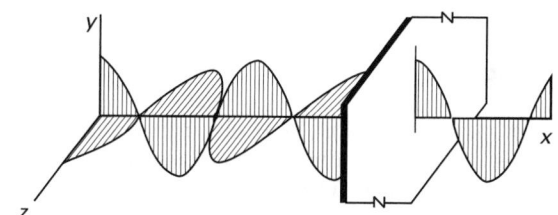

Figure 57.3 *Polarization*

6. INTERFERENCE

Interference is the variation with distance or time of the amplitude of a wave that results from the algebraic or vector addition of two or more waves with identical or nearly identical frequencies. Interference is also called *cancellation* when the waves subtract and *reinforcement* or *superposition* when they add to one another.[10] In communication systems, interference refers to any undesired energy that interferes with the reception of the desired signals.

---

[7]The area "seen" by a satellite on the earth's surface is called the *footprint*. Satellites use line-of-sight frequencies. When in geosynchronous orbit, the satellite has a footprint of approximately 40% of the earth's surface. Three satellites are necessary for complete coverage.

[8]Refraction effects in the atmosphere may still cause bending, reflection, scattering, and fading. Nevertheless, the assumption of free space in Earth-based communication calculations can be used for initial estimates.

[9]Equations 57.5 and 57.6 are consistent if the relative permittivity in Eq. 57.6 is equal to one.

[10]The term "interference" is often generically used to mean the subtraction of the waves only.

7. REFLECTION

Reflection is the return of waves or particles from a surface upon which they are incident. Reflection occurs at or near the boundary between mediums, such as between the air and the ionosphere or the air and the ground. The concept of reflection is illustrated in Fig. 57.4. At the *critical incident angle*, θ_c, an optically transparent surface becomes totally reflecting. Total internal reflection is the concept upon which *optical fibers* (*light pipes*) function.[11]

$$\sin \theta_c = \frac{1}{n} \qquad 57.8$$

The critical angle depends on the *absolute index of refraction*, n, given by[12]

$$n = \sqrt{\mu_r \varepsilon_r} \qquad 57.9$$

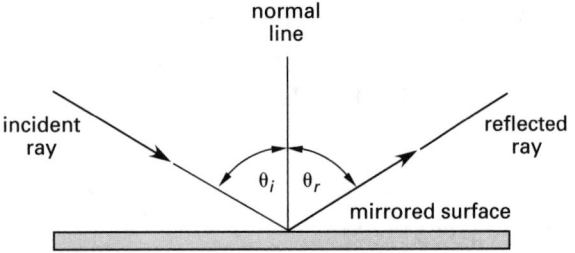

Figure 57.4 *Reflection from a Surface*

Any portion of the wave that enters a surface such as that in Fig. 57.4 would be called the *transmitted wave*. Transmitted waves are *refracted*.

8. REFRACTION

Refraction is the bending of an electromagnetic wave as it passes from one medium to another. Refraction results from the change in the direction of the lines of force of an electric or magnetic field at the boundary of two mediums with differing permeabilities or permittivities. That is, refraction is the change in direction of wave propagation due to a difference in wave velocity in the two mediums. Refraction is illustrated in Fig. 57.5(a).

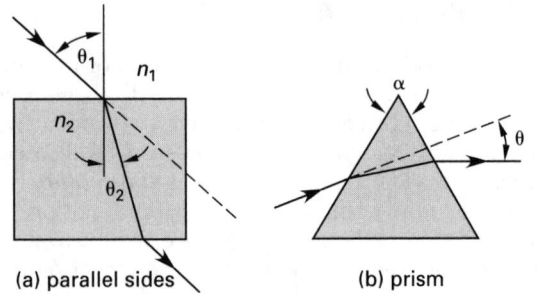

(a) parallel sides   **(b) prism**

Figure 57.5 *Refraction*

---

[11]Equation 57.8 is specifically for a wave incident upon a surface to free space, which air approximates well.
[12]The r in Eq. 57.9 indicates relative, not reflected, values.

Snell's law, Eq. 57.10, relates the incident and reflected angles. Electromagnetic waves tend to bend *toward the normal* when entering a denser material. A vacuum (and, for practical purposes, air) has an index of refraction, n, equal to one ($n_1 = 1$).

$$n_{\text{relative}} = \frac{n_2}{n_1} = \frac{\sin \theta_1}{\sin \theta_2} \qquad 57.10$$

When the refraction takes place through a prism-shaped medium as in Fig. 57.5(b), the *angle of refraction* (*angle of deviation*), θ, is governed by

$$n_{\text{relative}} = \frac{\sin \frac{1}{2}(\alpha + \theta)}{\sin \frac{1}{2}\alpha} \qquad 57.11$$

9. DIFFRACTION AND SCATTERING

Diffraction is any redistribution in space of the intensity of a wave that results from the presence of an object, or, more technically, large-scale irregularities. That is, diffraction is the change in path of a wave when it passes around objects (or through narrow slits). Diffraction is illustrated in Fig. 57.6.

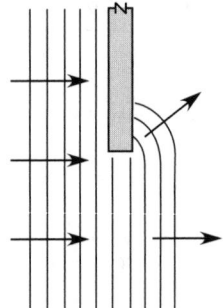

Figure 57.6 *Diffraction*

Scattering is the diffusion of a wave in a random manner. Scattering results from small-scale irregularities, such as small changes in the index of refraction in the upper atmosphere.

10. FREE-SPACE BASIC TRANSMISSION LOSS

The gain, G, and loss, L, in a communication system are usually given in decibels (dB). The use of decibels allows losses for various conditions to be placed in tabular form for a variety of environmental conditions. The total power of a particular system at any given point is then easily determined from the algebraic combination of the gains and losses.

The system loss, L_{sys}, in a communications link is the difference between the transmitting antenna power density, p_{trans}, and the power density available at the receiving antenna, p_{rec}.

$$L_{\text{sys}} = 10\log\left(\frac{p_{\text{trans}}}{p_{\text{rec}}}\right)$$
$$= p_{\text{trans}} - p_{\text{rec}} \quad [\text{dB}] \qquad 57.12$$

The *basic transmission loss*, L_b, is the loss within a given system assuming isotropic antennae and isotropic antenna polarization identical to that of the real antennae, and disregarding obstacles close to the antennae. The *free-space basic transmission loss*, L_{b0}, is the loss with isotropic antennae in a perfectly dielectric, homogenous, unlimited isotropic environment. The power density in an isotropic environment is spread over the surface of a sphere with area $4\pi r^2$, where r is the radial distance. The effective absorbing area of an isotropic receiving antenna is $\lambda^2/4\pi$. Combining the concepts gives an isotropic receiving antenna power density of

$$p_{\text{rec}} = \left(\frac{\lambda^2}{4\pi}\right)\left(\frac{1}{4\pi r^2}\right)p_{\text{trans}}$$
$$= \left(\frac{\lambda}{4\pi r}\right)^2 p_{\text{trans}} \qquad 57.13$$

The inverse of the two terms of Eq. 57.13 in parentheses represents the free-space basic transmission loss and is expressed in decibels as

$$L_{b0} = 20\log\left(\frac{4\pi r}{\lambda}\right) \qquad 57.14$$

The term $(4\pi r/\lambda)^2$ is called the *path loss*. It is not a loss per se, but represents the decrease in power due to the spherical spreading of the radiated energy. (The square of the term is reflected in the value of 20 in front of the logrithm, rather than 10.)

11. ELEMENTAL ANTENNA

An antenna differs from a transmission line in that its dimensions are comparable to the wavelength of the signal carried. The relationship between an element of antenna current and the resulting radiated electromagnetic field is based on retarded magnetic potential, with phasor notation used to account for the time variation of the current. An elemental antenna is a small section of uniform electric current, $I dz$, as shown in Fig. 57.7.

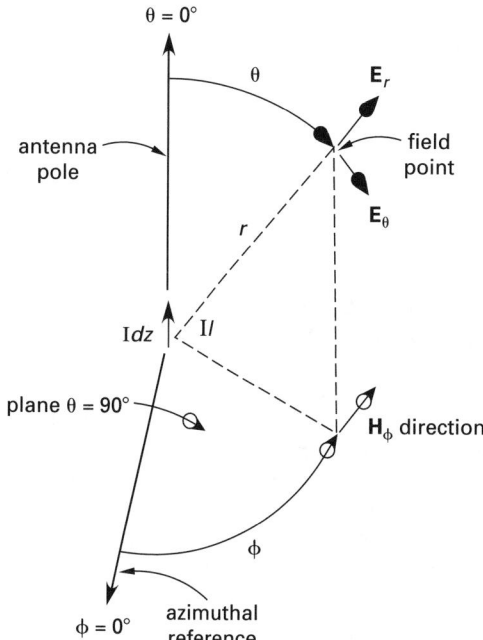

Figure 57.7 Elemental Antenna

The expression for the magnetic field strength or intensity, $\mathbf{H}$, of an elemental antenna of length $dz = l$, using spherical coordinates, is derived from Eq. 57.2 as

$$H_\phi = \left(\frac{Il}{4\pi}\right)\sin\theta\left(j\frac{\beta}{r} + \frac{1}{r^2}\right)$$
$$\times e^{-j\beta r} \quad [\text{in A/m}] \qquad 57.15$$

The phase constant in Eq. 57.15, β, from Eq. 57.3, is $\omega\sqrt{\mu_0\varepsilon_0}$. The term ω is the radian frequency of the signal, μ_0 is the free-space permeability, and ε_0 is the free-space permittivity. In Fig. 57.7, θ is the polar angle, ϕ is the azimuthal angle, dz is the differential length of the elemental antenna, and r is the distance from the element Il to the indicated point in the field. The current term, I, is the rms phasor current and thus has the time dependence embedded. The magnetic field, $\mathbf{H}$, is exclusively in the azimuthal direction, that is, perpendicular to r and parallel to the $\theta = 90°$ plane.

The components of the electric field, $\mathbf{E}$, are found using Maxwell's equation relationship (see Ch. 25).

$$\nabla \times \mathbf{H} = \text{curl } \mathbf{H} = \frac{\partial \mathbf{D}}{\partial t} = j\omega\varepsilon_0\mathbf{E} \qquad 57.16$$

The radial component of the electric field is given by

$$E_r = \left(\frac{Il}{2\pi}\right)\cos\theta\left(\frac{1}{cr^2} + \frac{1}{j\beta r^3}\right)$$
$$\times e^{-j\beta r} \quad [\text{in V/m}] \qquad 57.17$$

The speed of light, c, is equal to $1/\sqrt{\mu_0\varepsilon_0}$ from Eq. 57.4. The polar component is given by

$$E_\theta = \left(\frac{Il}{4\pi}\right) \sin\theta \left(\frac{j\beta}{r} + \frac{1}{r^2} + \frac{1}{j\beta r^3}\right)$$

$$\times\, e^{-j\beta r} \quad [\text{in V/m}] \qquad 57.18$$

12. ELEMENTAL ANTENNA: RADIATED POWER

From Poynting's theorem (see Sec. 57-4), the power density is obtained. The radiated power is in the r-direction only, and, since the cross product is involved, includes only $E_\theta H_\phi$. Additionally, at large distances from the antenna, only the $1/r$ components are significant contributors to the power density. Thus, the radiated power density is given by

$$p_r = \left(\frac{Il}{4\pi}\right)^2 \sqrt{\frac{\mu_0}{\varepsilon_0}} \sin^2\theta$$

$$\times \left(\frac{j\beta}{r}\right)^2 e^{-2j\beta r} \quad [\text{in W/m}^2] \qquad 57.19$$

The total average power density, P_r, is obtained from Eq. 57.19 by integrating over a sphere of radius r. Since the power density is not a function of the azimuthal angle, ϕ, the surface area element is $(2\pi r \sin\theta)\,r d\theta$, which is integrated from $\theta = 0$ to $\theta = \pi$. The term $e^{-2j\beta r}$ is a time delay that vanishes during the integration, since the integration data is the average value. The average radiated power in an elemental antenna of length l is

$$P_r = \left(\frac{(Il)^2}{6\pi}\right) \sqrt{\frac{\mu_0}{\varepsilon_0}}\beta^2 \quad [\text{in W/m}^2] \qquad 57.20$$

13. ELEMENTAL ANTENNA: DIRECTIVITY

The *directivity* is the value of the directive gain of an antenna in the direction of its maximum value. The directivity is a measure of an antenna's *focus*. Directivity, D, is mathematically defined as the ratio of the isotropic power to the radiated power.

$$D_g = \frac{P_{\text{isotropic}}}{P_r} \qquad 57.21$$

The maximum power density is calculated from Eq. 57.19 at $\theta = 90°$ and is given by

$$P_{r,\text{max}} = \left(\frac{Il}{4\pi}\right)^2 \sqrt{\frac{\mu_0}{\varepsilon_0}} \left(\frac{j\beta}{r}\right)^2$$

$$\times\, e^{-2j\beta r} \quad [\text{in W/m}^2] \qquad 57.22$$

An isotropic antenna radiates power equally in all directions, forming a sphere with a surface area of $4\pi r^2$. Multiplying Eq. 57.22 by this surface area and removing the time delay exponential term gives

$$P_{\text{isotropic}} = \left(\frac{(Il)^2}{4\pi}\right) \sqrt{\frac{\mu_0}{\varepsilon_0}}\beta^2 \quad [\text{in W/m}^2] \qquad 57.23$$

The directivity of the elemental antenna is given by the ratio of Eqs. 57.23 and 57.20.

$$D_g = \frac{P_{\text{isotropic}}}{P_r} = 1.5 \qquad 57.24$$

In terms of decibels, the directivity of the elemental antenna is

$$D_G = 10\log D_g = 1.76 \text{ dB} \qquad 57.25$$

14. ELEMENTAL ANTENNA: RADIATION RESISTANCE

The *radiation resistance*, R_{rad}, is mathematically defined as the ratio of the total radiated power to the square of the rms input current of an antenna.

$$R_{\text{rad}} = \frac{P_r}{I_{\text{in}}^2} \qquad 57.26$$

The radiation resistance of an elemental antenna is given by the ratio of Eq. 57.20 and the input current.

$$R_{\text{rad}} = \left(\frac{l^2\beta^2}{6\pi}\right) \sqrt{\frac{\mu_0}{\varepsilon_0}} \qquad 57.27$$

15. ELEMENTAL ANTENNA: OHMIC RESISTANCE

The magnetic field calculations of Sec. 57-11 are based on an infinitesimally thin antenna, with the field calculations far from the antenna.[13] Antenna wire has a finite diameter and operates at a sufficiently high frequency that the skin effect dominates (see Sec. 37-3). The ohmic resistance (see Sec. 37-4) is given by

$$R_{\text{ohmic}} = \frac{R_l}{1 - \left(1 - \dfrac{\delta}{r}\right)^2} \qquad 57.28$$

16. ELEMENTAL ANTENNA: EFFICIENCY

The *antenna efficiency* is the ratio of the amount of power radiated into space to the antenna input power.

---

[13] *Far field* is the term used when describing an electromagnetic field at a significant distance from the antenna where local effects can be ignored.

The efficiency, η, is determined from the radiation resistance and ohmic resistance as

$$\eta = \frac{R_{\rm rad}}{R_{\rm ohmic} + R_{\rm rad}} \times 100\% \qquad 57.29$$

17. ELEMENTAL ANTENNA: GAIN

The *antenna gain* is the ratio of isotropic power to input power. That is, gain compares the maximum isotropically radiated power to the actual input power. Thus, gain is a measure of the effectiveness of a directional antenna compared to a nondirectional antenna and is given by

$$G = \eta D_g \qquad 57.30$$

18. SHORT DIPOLE ANTENNA

A *dipole antenna* has two equal branches of height h, as shown in Fig. 57.8. In a *short dipole antenna*, the height (length), h, is much less than the wavelength of the signal.

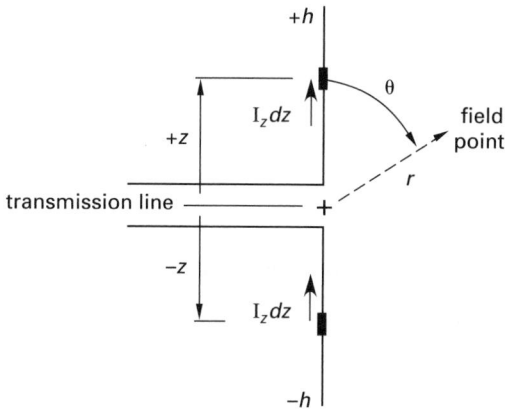

Figure 57.8 *Short Dipole Antenna*

Two identical current elements, $I dz$, exist in each branch as standing waves, linearly diminishing to zero at $z = +h$ and $z = -h$. The current is described by

$$I_z = I_{\rm in}\left(1 - \frac{z}{h}\right) \quad [{\rm for}\ 0 < z < +h]$$

$$= I_{\rm in}\left(1 + \frac{z}{h}\right) \quad [{\rm for}\ -h < z < 0] \qquad 57.31$$

The input current $I_{\rm in}$ is the rms phasor current leaving the lower branch and entering the upper branch. Since z is small compared to the wavelength, the current can be

assumed to exist at the origin only. Therefore, Eq. 57.15 applies with l as dz.

$$H_\phi = \left(\frac{\int_{-h}^{+h} I dz}{4\pi}\right) \sin\theta \left(j\frac{\beta}{r} + \frac{1}{r^2}\right)$$

$$\times (e^{-j\beta r}) \quad [{\rm in\ A/m}] \qquad 57.32$$

Using the definition of the current from Eq. 57.31 in the integration of Eq. 57.32 gives the resulting magnetic field.

$$H_\phi = \left(\frac{I_{\rm in} h}{4\pi}\right) \sin\theta \left(\frac{j\beta}{r} + \frac{1}{r^2}\right)$$

$$\times e^{-j\beta r} \quad [{\rm in\ A/m}] \qquad 57.33$$

This differs from the elemental antenna only in having the term $I_{\rm in} h$ replace $I l$. Thus, the radiated power density is

$$p_r = \left(\frac{I_{\rm in} h}{4\pi}\right)^2 \sqrt{\frac{\mu_0}{\varepsilon_0}} \sin^2\theta \left(\frac{j\beta}{r}\right)^2$$

$$\times e^{-2j\beta r} \quad [{\rm in\ W/m^2}] \qquad 57.34$$

The total average power radiated, over a sphere of radius r, is calculated as in Sec. 57-12 as

$$P_r = \left(\frac{(I_{\rm in} h)^2}{6\pi}\right) \sqrt{\frac{\mu_0}{\varepsilon_0}} \beta^2 \quad [{\rm in\ W/m^2}] \qquad 57.35$$

The derivation of the directivity is identical to that in Sec. 57-13 with h replacing l in Eqs. 57.22 and 57.23. The resulting directivity of the short dipole antenna is

$$D_g = \frac{P_{\rm isotropic}}{P_r} = 1.5 \qquad 57.36$$

Thus, the directivity of a short dipole antenna is identical to that of an elemental antenna.[14]

The radiation resistance, as defined in Sec. 57-14, for the short dipole antenna is

$$R_{\rm rad} = \left(\frac{P_r}{I_{\rm in}^2}\right) = \left(\frac{h^2\beta^2}{6\pi}\right)\sqrt{\frac{\mu_0}{\varepsilon_0}} \qquad 57.37$$

Comparing Eq. 57.37 with the similar expression of Eq. 57.27 indicates that radiation resistance is proportional to h^2 for a short dipole antenna and l^2, that is, $(dz)^2$, for the elemental antenna. The length of the short dipole

[14]This result should not be surprising since the antenna current is concentrated at the origin.

antenna is $2h$, while the length of the elemental antenna is l, that is, dz. Therefore, in determining the radiation resistance (Eq. 57.37), the effective length of the short dipole antenna is one-half the actual length. This effect is due to the nonuniform current distribution in the short dipole antenna.

The ohmic resistance experiences the same effect from the nonuniform current distribution. The ohmic power of a short dipole antenna is given by

$$P_{\text{ohmic}} = \int_{-h}^{+h} \text{I}_z^2 R_l dz = \text{I}_{\text{in}}^2 R_l \frac{2}{3} h \qquad 57.38$$

The ohmic resistance is

$$R_{\text{ohmic}} = \frac{P_{\text{ohmic}}}{\text{I}_{\text{ohmic}}^2} = \left(\frac{2h}{3} \right) R_l \qquad 57.39$$

Therefore, in determining the ohmic resistance (Eq. 57.39), the effective length of the short dipole antenna is one-third the actual length.

The formulas for the short dipole antenna efficiency and the gain are identical to those for the elemental antenna, Eqs. 57.29 and 57.30, respectively. Nevertheless, the short dipole antenna efficiency will be lower than that of the elemental antenna, even though the ohmic resistance is one-third the value and in the denominator of Eq. 57.29, which by itself tends to make the efficiency larger. The efficiency is lower because the effective length of the short dipole is one-half that of the elemental antenna and this length (h) is squared in the calculation of the radiation resistance (see Eq. 57.37), which dominates the change in efficiency (see Eq. 57.29).

19. DIPOLE ANTENNA

When the antenna length (height) is comparable to the wavelength of the signal carried, the assumptions made regarding the antenna current being an infinitesimal element of uniform current (in an elemental antenna) or concentrated at the origin (in a short dipole antenna) are no longer valid. Specifically, the distance to the point of interest in the field from the elemental currents, $\text{I}dz$, in each portion of the dipole must be accounted for separately. The concept is illustrated in Fig. 57.9(a).

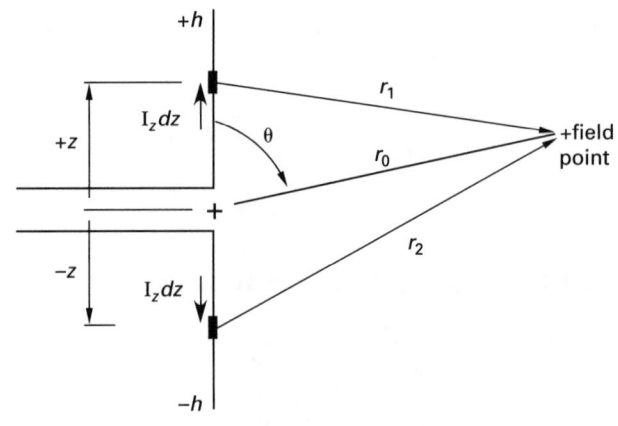

(a) antenna with length comparable to wavelength

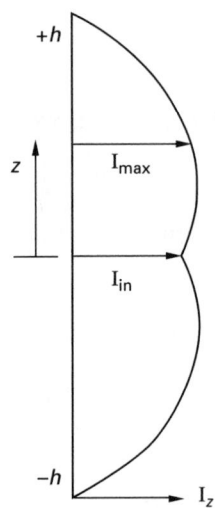

(b) assumed current standing wave distribution

Figure 57.9 *Dipole Antenna*

Since the antenna height (length), h, is much less than the radial distance to the point in the field, the distances from the two current elements are approximated as

$$r_1 \approx r - z \cos \theta \qquad 57.40$$

$$r_2 \cong r + z \cos \theta \qquad 57.41$$

The assumed current standing wave distribution is that of Fig. 57.9(b). The maximum rms current phasor is shown as I_{max}. The currents can thus be represented mathematically as

$$\text{I}_z = \text{I}_{\text{max}} \sin \beta \left(h - z \right) \quad \text{[for } 0 < z < +h]$$

$$= \text{I}_{\text{max}} \sin \beta \left(h + z \right) \quad \text{[for } -h < z < 0] \qquad 57.42$$

The angle θ is assumed to be the same over the length of the antenna since the radial distance r is much greater than the antenna height, h. Additionally, this assumption means the differences in r_1 and r_2 from the actual distance r only occur in the exponential term of Eq. 57.15. Incorporating the radial distances in Eqs. 57.40 and 57.41, using the current of Eq. 57.42, and considering only the significant contributors to the magnetic field and power density (i.e., assuming a $1/r$ dependence) gives

$$H_\phi = j\left(\frac{I_{max}}{2\pi r \sin\theta}\right)e^{-j\beta r}$$

$$\times \cos\left(\beta h \cos\theta\right) - \cos\beta h \quad [\text{in A/m}] \quad 57.43$$

The polar component of the electric field is

$$E_\theta = \sqrt{\frac{\mu_0}{\varepsilon_0}}H_\phi = 120\pi H_\phi \qquad 57.44$$

The power density is determined from $E_\theta H_\phi = 120\pi H_\phi^2$ as

$$p_r(r,\theta) = (120\pi)(-1)\left(\frac{I_{max}}{2\pi r \sin\theta}\right)^2 e^{-2j\beta r}$$

$$\times \left(\cos\left(\beta h \cos\theta\right) - \cos\beta h\right)^2 \qquad 57.45$$

In order to determine the parameters of interest (the radiated power, directivity, radiation resistance, and so on), Eq. 57.45 is integrated from $\theta = 0$ to $\theta = \pi$. The results contain cosine and sine integrals (see Sec. 10-17), which are evaluated by computation or by using tables. The calculations are simplified if multiples of wavelengths are used, such as one-fourth or one-half wavelengths.

Antennae designed to be a one-fourth multiple of the signal wavelength are called *quarter-wave antennae*. Antennae designed as a one-half multiple of the signal wavelength are called *half-wave antennae*, or *half-wave dipoles*.

20. HALF-WAVE ANTENNA

A half-wave antenna is constructed so that the antenna height is a multiple of one-half the signal wavelength. Specifically, $2h = \lambda/2$. The radiated power for such an antenna is determined by integrating Eq. 57.45 over a sphere of differential volume $(2\pi r \sin\theta)rd\theta$. With $2h = \lambda/2$, the $\cos\beta h$ term equals zero (see Eq. 56.5). The average radiated power, with the average suppressing the exponential time-delay term, is

$$P_r = \int_0^\pi 120\pi\left(\frac{I_{max}}{2\pi r \sin\theta}\right)^2 \cos^2\left(\frac{\pi}{2}\cos\theta\right)$$

$$\times (2\pi r \sin\theta)\,rd\theta \quad [\text{in W}] \qquad 57.46$$

Equation 57.46 simplifies to

$$P_r = 60I_{max}^2 \int_0^\pi \left(\frac{\cos^2\left(\frac{\pi}{2}\cos\theta\right)}{\sin\theta}\right)d\theta \quad [\text{in W}] \quad 57.47$$

Using numerical techniques, Eq. 57.47 is evaluated to

$$P_r = 73.08I_{max}^2 \quad [\text{in W}] \qquad 57.48$$

From Eq. 57.45, the maximum power density occurs at $\theta = \pi/2$, which results in

$$p_{r,max} = \frac{120\pi I_{max}^2}{(2\pi r)^2} \quad [\text{in W/m}^2] \qquad 57.49$$

The isotropic power is found from Eq. 57.49 by multiplying by the surface of a sphere at radial distance r from the antenna, that is, by $4\pi r^2$.

$$P_{isotropic} = 120I_{max}^2 \quad [\text{in W}] \qquad 57.50$$

The directivity for a half-wave antenna, which ratios Eqs. 57.50 to 57.48, is given by

$$D_g = \frac{P_{isotropic}}{P_r} = \frac{120}{73.08} = 1.64 \qquad 57.51$$

The radiation resistance is

$$R_{rad} = \frac{P_r}{I_{max}^2} = 73.08 \ \Omega \qquad 57.52$$

The ohmic power is given by

$$P_{ohmic} = \int_{-h}^{+h} I_z^2 R_l\,dz \qquad 57.53$$

The ohmic resistance can be shown to be

$$R_{ohmic} = hR_l \qquad 57.54$$

The formulas for the half-wave antenna efficiency and gain are identical to those for the elemental antenna, Eqs. 57.29 and 57.30, respectively.

58 Communication Links

Nomenclature

A	aperture	m^2
D	directivity	–
E	electric field strength	V/m
EIRP	effective isotropically radiated power	–
G	gain	– or dB
I	effective or DC current	A
l	length	m
L	loss	dB
p	power density	W/m^2
P	power	W
r	radial distance	m
R	resistance	Ω
V	effective or DC voltage	V
X	reactance	Ω
Z_0	characteristic impedance	Ω

Symbols

ε_0	free-space permittivity	8.854×10^{-12} F/m
η	efficiency	–
μ_0	free-space permeability	1.2566×10^{-6} H/m
Ω	solid angle	sr

Subscripts

amp	amplifier
eff	effective
g	gain (ratio)
p	path
r	radiated
rad	radiation
rec	received
R	receiver
scat	scattered
T	transmitter

1. FUNDAMENTALS

Communication, in the context of electrical engineering, is the science and technology that collects information from a source, transforms it into electrical currents or fields, transmits it via a medium (also called a *communication channel*), and reconstitutes the information at the receiver. Common media include conducting wire, coaxial cable, fiber-optic material, waveguides, the atmosphere, and space.

Communication systems include computer networks as well as telephony, cable, optical, wireless, microwave, and satellite systems. These are broadly classified as broadcast, voice, or data communication systems. A *communication link* is a general term for the combination of a transmitting antenna, a receiving antenna, and the medium or channel through which they are connected, either air or space (vacuum).[1]

2. POWER DISTRIBUTION

If the receiving antenna is positioned so that it obtains the maximum power from the transmitting antenna, the power density radiated from the transmitting antenna available at the receiving antenna is

$$p_r = \left(\frac{D_T}{4\pi r^2} \right) P_T \quad \text{[in W/m}^2\text{]} \qquad 58.1$$

The term p_r is the radiated power density at the radial distance r, P_T is the transmitted power, and D_T is the transmitting antenna directivity gain (see Sec. 57-13).

In free space, the power density is the electric field strength divided by the characteristic impedance, Z_0.[2]

$$p_r = \frac{E^2}{Z_0} = \frac{E^2}{\sqrt{\dfrac{\mu_0}{\varepsilon_0}}} = \frac{E^2}{120\pi} \quad \text{[in W/m}^2\text{]} \qquad 58.2$$

Since the receiving antenna is oriented in the direction of the electric field lines, for maximum power reception, the voltage received is $El = V_{\text{rec}}$, where the effective antenna length is l. The receiving antenna has a generic input impedance of $R_{\text{in}} + jX_{\text{in}}$. The antenna supplies an amplifier with a generic impedance of $R_{\text{amp}} + jX_{\text{amp}}$. Thus, the current in the receiving antenna is given by

$$I_{\text{rec}} = \frac{V_{\text{rec}}}{R_{\text{in}} + R_{\text{amp}} + j\left(X_{\text{in}} + X_{\text{amp}}\right)}$$

$$= \frac{El}{R_{\text{in}} + R_{\text{amp}} + j\left(X_{\text{in}} + X_{\text{amp}}\right)} \qquad 58.3$$

---

[1] Any development of equations in this chapter is accomplished using parameters for dipole antennae (see Ch. 57). Identical parameters exist for other types of antennae, though the exact equations differ. Where the results are applicable to all antennae, this is specifically noted.

[2] The characteristic impedance of free space is also called the *intrinsic impedance*.

Three resistances are involved. Two of them—the radiation resistance, R_{rad}, and the ohmic resistance, R_{ohmic}—are components of the input resistance, R_{in} (see Secs. 57-14 and 57-15). The third is the amplifier resistance, R_{amp}. The power is given by $I^2 R$. Using this along with Eqs. 58.2 and 58.3 gives the distribution of the power in the receiving antenna.

The power lost in the radiation resistance is reradiated. This power is considered *scattered* and is given by

$$P_{\text{scat}} = \frac{120\pi p_r l^2 R_{\text{rad}}}{\left(R_{\text{in}} + R_{\text{amp}}\right)^2 + \left(X_{\text{in}} + X_{\text{amp}}\right)^2} \qquad \textbf{58.4}$$

The power lost in the ohmic resistance of the antenna is

$$P_{\text{loss}} = \frac{120\pi p_r l^2 R_{\text{ohmic}}}{\left(R_{\text{in}} + R_{\text{amp}}\right)^2 + \left(X_{\text{in}} + X_{\text{amp}}\right)^2} \qquad \textbf{58.5}$$

The power utilized by the amplifier is

$$P_{\text{amp}} = \frac{V_{\text{rec}}^2 R_{\text{amp}}}{\left(R_{\text{in}} + R_{\text{amp}}\right)^2 + \left(X_{\text{in}} + X_{\text{amp}}\right)^2}$$

$$= \frac{120\pi p_r l^2 R_{\text{amp}}}{\left(R_{\text{in}} + R_{\text{amp}}\right)^2 + \left(X_{\text{in}} + X_{\text{amp}}\right)^2} \qquad \textbf{58.6}$$

The effectiveness of a receiving antenna is improved by making the amplifier impedance, that is, the receiver impedance, the complex conjugate of the antenna input impedance. Doing so eliminates the reactance term in the denominator of Eq. 58.3. Half of the received power is then transmitted to the amplifier (receiver), and the remainder becomes ohmic and scattering losses.

3. ANTENNA APERTURES

An antenna aperture is an opening through which radio waves can pass. Antennae have a physical aperture based on their structure, measured in m^2. The physical aperture does not deliver all the power from the power flux incident upon it due to losses. The situation is illustrated in Fig. 58.1.

Using Eqs. 58.4 through 58.6, scattering, loss, and effective apertures are defined as

$$A_{\text{scat}} = \frac{120\pi l^2 R_{\text{rad}}}{\left(R_{\text{in}} + R_{\text{amp}}\right)^2 + \left(X_{\text{in}} + X_{\text{amp}}\right)^2} \qquad \textbf{58.7}$$

$$A_{\text{loss}} = \frac{120\pi l^2 R_{\text{ohmic}}}{\left(R_{\text{in}} + R_{\text{amp}}\right)^2 + \left(X_{\text{in}} + X_{\text{amp}}\right)^2} \qquad \textbf{58.8}$$

$$A_{\text{eff}} = \frac{120\pi l^2 R_{\text{amp}}}{\left(R_{\text{in}} + R_{\text{amp}}\right)^2 + \left(X_{\text{in}} + X_{\text{amp}}\right)^2} \qquad \textbf{58.9}$$

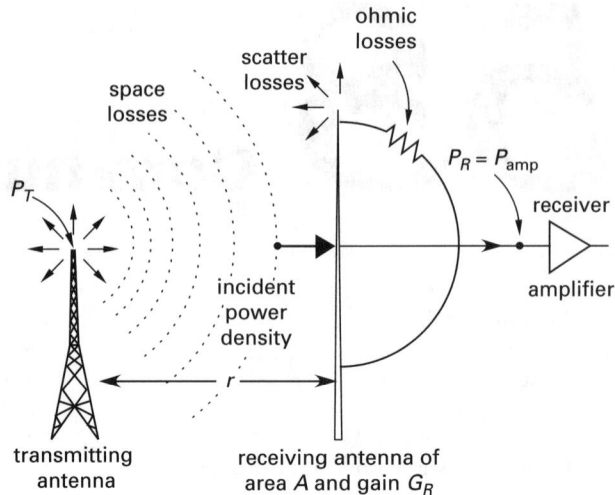

Note: The power density is sometimes called the flux density.

Figure 58.1 *Communication Link*

The development of the apertures in Eqs. 58.7 through 58.9 is based on the parameters for a linear antenna, such as the dipole antennae described in Ch. 57. Nevertheless, the results are general and the equations have counterparts for all types of antennae.

The relationships in Eqs. 58.10 through 58.13 are valid regardless of the antenna type.

$$P_{\text{scat}} = A_{\text{scat}} p_r \qquad \textbf{58.10}$$

$$P_{\text{loss}} = A_{\text{loss}} p_r \qquad \textbf{58.11}$$

$$P_{\text{amp}} = A_{\text{eff}} p_r \qquad \textbf{58.12}$$

$$p_{r,\,\text{total}} = \left(A_{\text{scat}} + A_{\text{loss}} + A_{\text{eff}}\right) p_r$$

$$= A_{\text{total}} p_r \qquad \textbf{58.13}$$

The amplifier power (sometimes called the receiver power, P_R) uses the effective aperture and is based on the received voltage, that is, the received electric field strength. Therefore, the amplifier power and effective aperture account for the space losses shown in Fig. 58.1 and can be related to the directivity of the antenna.

4. DIRECTIVITY

The directivity is the directive gain of an antenna. Mathematically, it is defined as the ratio of the isotropic power to the radiated power, which is equivalent to the ratio of the maximum power density to the average power density. With this information, it can be shown that for any antenna the directivity is[3]

$$D_g = \left(\frac{4\pi}{\lambda^2}\right) A_{\text{eff}} \qquad \textbf{58.14}$$

---

[3]The maximum value of the effective aperture is used to determine the directivity.

5. BEAM SOLID ANGLE Ω

The *beam angle*, also called the *beam width*, is the angle, measured in the horizontal plane, between the two directions where the intensity of the electromagnetic beam is one-half its maximum value. The beam angle is measured in radians. The concept is illustrated in Fig. 58.2.

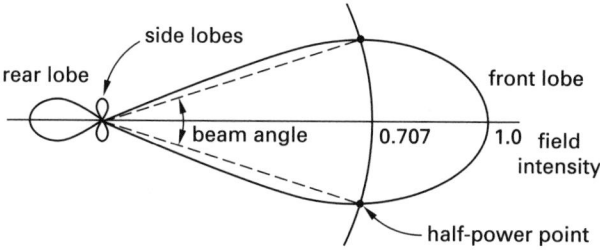

Figure 58.2 *Beam Angle*

A *beam solid angle*, also referred to as the beam angle, is defined as the region where the antenna power density is one-half or more of the maximum power density.[4] The beam angle is alternatively defined as the solid angle that contains the total radiated power if the power density is uniformly equal to the maximum power density.

$$P_r = p_{r,\max}\Omega r^2 \qquad 58.15$$

The directivity is then defined as the ratio of a sphere solid angle to the beam solid angle.

$$D_g = \frac{4\pi}{\Omega} \qquad 58.16$$

6. COMMUNICATION LINK TRANSMISSION

The power density available at the receiving antenna is given by Eq. 58.1. The relationship between the power density and the effective aperture is given by Eq. 58.12. Combining these concepts, the power at the receiving amplifier, assuming a lossless transmitting antenna, is

$$P_R = \left(\frac{D_T A_R}{4\pi r^2}\right) P_T \qquad 58.17$$

The receiving antenna power is P_R, which is the power at the amplifier input, P_{amp} (see Fig. 58.1). The term D_T is the directivity of the transmitting antenna given by Eq. 58.14, with the subscript g dropped for convenience. The maximum effective receiving antenna aperture is A_R, though it can be written $A_{R,\mathrm{eff,max}}$.

[4]See Sec. 7-16 for an illustration of the solid angle. The nomenclature used in Ch. 7, ω, represents an area. The nomenclature used here, Ω, represents a region. The units are identical, and the symbols are used interchangeably.

Incorporating Eq. 58.14 into Eq. 58.17, the receiving antenna power is written as

$$P_R = \left(\frac{A_R A_T}{\lambda^2 r^2}\right) P_T \qquad 58.18$$

Alternatively, the receiving antenna power is written as

$$P_R = \left(\frac{\lambda}{4\pi r}\right)^2 D_R D_T P_T \qquad 58.19$$

The receiving antenna power as written in Eqs. 58.18 and 58.19 accounts for the path loss, L_p, which is the square of the inverse of the term in parentheses in Eq. 58.19 (see Sec. 57-10). The directivity terms, which contain the effective aperture, account for $I^2 R$ type losses from the receiving antenna itself to the associated amplifier. The total efficiency of the antenna, however, accounts for all the losses from the incident wavefront to the antenna output. These losses include the *illumination efficiency* or *aperture taper efficiency*, which is determined by the energy distribution of the feed across the aperture, and other losses such as spillover, blockage, phase errors, diffraction effects, polarization, and mismatch losses. The total efficiency relates the gain to the directivity (see Sec. 57-17) as follows.

$$G = \eta D \qquad 58.20$$

The receiving antenna power must be written in terms of the gain when the efficiencies are significant, that is, low, giving

$$P_R = \left(\frac{\lambda}{4\pi r}\right)^2 G_R G_T P_T \qquad 58.21$$

If the efficiencies are important, Eqs. 58.17 through 58.19 must be modified by factors of the antenna efficiencies as well. Parabolic reflector antennae have efficiencies in the range of 50 to 75%. Smaller antennae generally have lower efficiencies. Cassegrain antennae generally have higher efficiencies. Horn antennae have efficiencies of approximately 90%.

Equation 58.21 is known as the *Friss transmission formula*. Equations 58.18 and 58.19 are also considered versions of this equation. In word form, the formula is

$$\frac{\text{power}}{\text{received}} = \frac{\text{EIRP} \times \text{gain of receiving antenna}}{\text{path loss}} \qquad 58.22$$

The term "EIRP" stands for the *effective isotropically radiated power* and is the decibel value of $G_T P_T$. The path loss is the inverse of the term in parentheses in Eq. 58.19, and in decibels it is given the symbol L_P, which is equivalent mathematically to Eq. 57.14. The Friss formula then, in decibel terms, is

$$P_R = \text{EIRP} + G_R - L_p \quad \text{[in dBW]} \qquad 58.23$$

The individual terms are given by

$$\text{EIRP} = 10 \log (G_T P_T) \quad \text{[in dBW]} \qquad \textit{58.24}$$

$$G_R = 10 \log \left(\eta \left(\frac{4\pi A_R}{\lambda^2} \right) \right) \quad \text{[in dBW]} \qquad \textit{58.25}$$

$$L_p = 20 \log \left(\frac{4\pi r}{\lambda} \right) \quad \text{[in dBW]} \qquad \textit{58.26}$$

Using the Friss formula in the form of Eq. 58.23 allows multiple losses to be accounted for using algebraic manipulation, thus simplifying calculations and allowing the use of tabular data. Such additional losses might include losses in the transmitting antenna, losses due to the antenna mispointing, attenuation due to atmospheric effects (rain), and so on.

When using communication link equations, care should be taken to ensure a complete understanding of the terms. For example, in some texts the receiving antenna power, P_R, is the power incident upon the antenna and not the power at the input of the amplifier. Also, if the effective aperture, A_{eff}, includes all the antenna efficiencies, not just those associated with $I^2 R$ losses, Eq. 58.14 is equal to the gain, G, rather than the directivity, D (see Eq. 58.20).

59 Signal Formats

Nomenclature

a, A	amplitude	V
B, BW	bandwidth	Hz
D	dimensionality	–
E	energy	J
f	frequency	Hz, s^{-1}, or cycles/s
$\mathcal{F}$	Fourier transform	–
$I(t)$	in-phase component	V
Im	imaginary portion	–
k	proportionality constant	–
k_f	frequency modulator constant	rad/s·V or Hz/V
k_p	phase modulator constant	rad/V
m	modulation index	–
M	map total number	–
n	number	–
N	noise power	W
N_0	noise spectral density	W/Hz
P	power	W
$\mathcal{P}(e)$	error probability	–
$Q(t)$	quadrature component	V
R	binary digit rate	s^{-1}
Re	real portion	–
s	signal (time function)	V
S	average signal power	W
$S(\omega)$	signal (frequency function)	V
SNR	signal-to-noise ratio	–
t, T	time	s

Symbols

α	real integer variable	–
β	imaginary integer variable	–
θ	angle	rad
ϕ	phase angle	rad
φ	signal angle	rad
ω	angular frequency	rad/s

Subscripts

0	initial or characteristic
AM	amplitude modulation
ave	average
bit	per bit
c	carrier
f	frequency
FM	frequency modulation
mod	modulated or modulation
p	phase
s	per second
x	x component
y	y component

1. FUNDAMENTALS

The purpose of any communication system is to deliver information from a source to a receiver via a channel. The information signals are normally confined to a maximum frequency. The maximum frequency can range from a few hertz for control systems to a few megahertz for television video signals. The bandwidth, BW, allotted for a given signal is usually on the order of the maximum frequency as well. That is, $\mathrm{BW} \approx f_{\max}$. The net result is that most communication baseband signal spectrums appear low-pass in character. Such a representation is shown in Fig. 59.1.

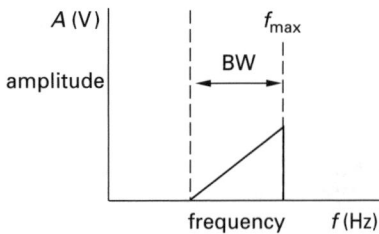

Figure 59.1 *Typical Baseband Signal Spectrum*

The spectrum shown is the band of frequencies that the system uses to modulate a given signal. The *baseband* is the band of frequencies used to modulate a transmitted carrier. The concept is illustrated in Fig. 59.2. A *baseband system*, by contrast, is a communication system in which information is transmitted in an unmodulated band of frequencies. In Fig. 59.2(a), removal of the modulator and carrier would create a baseband system.

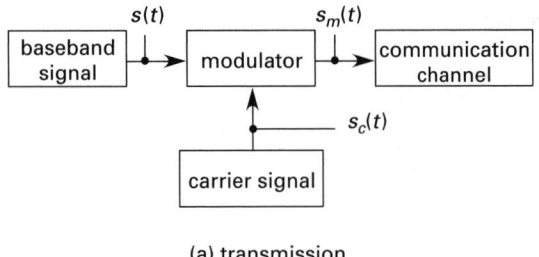

(a) transmission

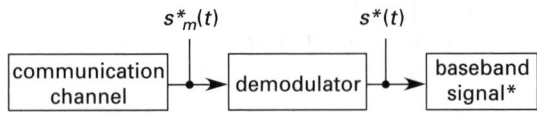

*Signal distortion possible. Original signal may be corrupted and require correction.

(b) reception

Figure 59.2 *Baseband Signal*

A carrier signal can be represented by the general equation

$$s_c(t) = A \cos(\omega_c t + \theta) = A \cos(2\pi f_c t + \theta) \quad \textbf{59.1}$$

The term A is the amplitude of the signal, normally a voltage parameter. The carrier frequency is given as ω_c or f_c, and θ is the phase angle, often assumed to be zero unless specifically noted otherwise. The information signal, that is, the baseband signal, can manipulate the carrier by changing the amplitude, the frequency, or the phase angle. The manipulation is called *modulation*, defined as the process or the result of the process by which a parameter in one wave is varied in accordance with the parameter in another. Thus, the possible manipulations are *amplitude modulation* (AM), *frequency modulation* (FM), and *phase modulation* (PM) as illustrated in Fig. 59.3.

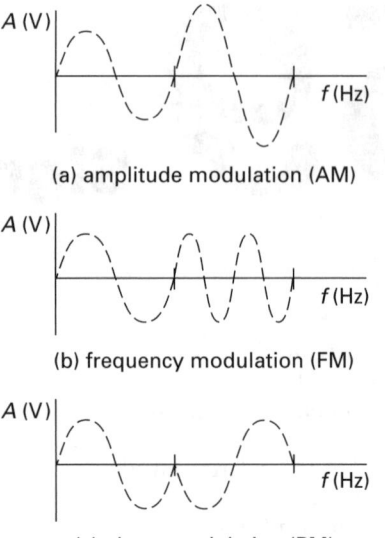

(a) amplitude modulation (AM)

(b) frequency modulation (FM)

(c) phase modulation (PM)

Figure 59.3 *Modulation Types*

The modulations can be carried out in an analog or a digital fashion. Signals are modulated for a variety of reasons.

- Communication through the channel using the baseband signal alone may be inefficient or impossible.

- Translation to higher frequencies via modulation means smaller wavelengths and more reasonably sized antennae for a given amount of power. (To radiate large amounts of power, an antenna must be large compared to the wavelength.)

- The baseband signal may require frequency translation to an assigned band.

- Several signals may require transmission through the same channel over the same time range. Modulation techniques allow the signals to share the channel but remain recoverable at the receiver as separate signals. Such techniques are called *multiplexing*.

2. ANALOG MODULATION

Analog modulation is the modulation of a continuous complex exponential signal. The modulation is accomplished by varying the amplitude, the frequency, or the phase in a manner proportional to the physical phenomenon, the signal.

3. AMPLITUDE MODULATION

In amplitude modulation (AM), the angle, θ, of Eq. 59.1 is constant and usually assumed to be equal to zero. The amplitude term to be modulated, A, can be represented as the product of a modulator proportionality constant, k, and an information signal, $s(t)$.[1] The general modulated signal, which is a combination of the baseband signal (information signal) and the carrier signal, $s_c(t)$, is then represented as

$$s_{\text{mod}}(t) = ks(t)\cos\omega_c t = \text{Re}\left\{ks(t)e^{j\omega_c t}\right\} \qquad 59.2$$

Taking the Fourier transform of Eq. 59.2 yields the character of the amplitude modulation.

$$S_{\text{mod}}(\omega) = \Im\left\{\tfrac{1}{2}ks(t)e^{j\omega_c t} + \tfrac{1}{2}ks(t)e^{-j\omega_c t}\right\}$$

$$= \tfrac{1}{2}S(\omega - \omega_c) + \tfrac{1}{2}S(\omega + \omega_c) \qquad 59.3$$

An analysis of Eq. 59.3 indicates that the amplitude-modulated signal has been shifted (translated) in frequency to above and below the original carrier by $\pm\omega_c$. The signal is said to have two sidebands, an upper and a lower. The original carrier wave is no longer present and the bandwidth, BW, has doubled. This type of signal is called a *double-sideband suppressed-carrier* (DSB-SC).

Receivers using the DSB-SC signal must be designed to maintain synchronization. As a result, they are expensive. To eliminate this issue and lower the receiver cost, the carrier signal can be transmitted with the information signal. Doing so, however, lowers the efficiency since more power (not containing signal information) is used. The signal is given by

$$s_{\text{mod}}(t) = ks(t)\cos\omega_c t + A\cos\omega_c t \qquad 59.4$$

The amplitude, A, in Eq. 59.4 is the pure carrier wave and is not modified. That is, it is not proportional to the information signal. If A is large enough, the envelope of the modulated waveform is proportional to the information signal and demodulation becomes simply envelope detection using a diode and a low-pass filter. Commercial broadcast stations (radio stations) use this type of signal. Consequently, when the generic term "AM" is used it refers to this *double-sideband large-carrier* (DSB-LC) waveform, which is also called *double-sideband transmitted-carrier* (DSB-TC). This basic AM signal is described in Sec. 18-1 and illustrated in Fig. 18.8.

When the information signal is a single frequency, $s(t) = a\cos\omega_{\text{mod}}t$, Eq. 59.4 can be rearranged as

$$s_{\text{mod}}(t) = ks(t)\cos\omega_c t + A\cos\omega_c t$$

$$= A\left(\frac{ks(t)}{A} + 1\right)\cos\omega_c t$$

$$= A\left(\left(\frac{ka}{A}\right)\cos\omega_{\text{mod}}t + 1\right)\cos\omega_c t$$

$$= A\left(m\cos\omega_{\text{mod}}t + 1\right)\cos\omega_c t \qquad 59.5$$

The dimensionless scaling factor in Eq. 59.5, m, is called the amplitude *modulation index, index of modulation,* or *modulation factor,* and is defined mathematically as[2]

$$m_{\text{AM}} = \frac{ka}{A} = \frac{\Delta A}{A} \qquad 59.6$$

The modulation index for an AM signal is the ratio of the maximum amplitude deviation of the modulated carrier to the amplitude of the carrier. The modulation index is sometimes expressed as a percentage.

4. SINGLE-SIDEBAND AMPLITUDE MODULATION

The double-sideband signals of Sec. 59-3 double the bandwidth. Further, the same information is carried in the upper and lower sidebands, using power but not providing additional information. Elimination of one of the sidebands results in a signal called a *single-sideband* (SSB) signal. Single-sideband radios find application in marine communication.

5. VESTIGIAL SIDEBAND AMPLITUDE MODULATION

The filtering requirements for an SSB signal are stringent. An ideal bandpass filter with an infinitely sharp cutoff or an equivalent technique is required. Attenuating some, but not all, of one sideband minimizes the bandwidth and lowers the power without the strict filtering requirements. This type of modulation results in a signal called a *vestigial sideband* (VSB). Vestigial sideband amplitude modulation is used to transmit the video portion of commercial television signals.

6. QUADRATURE MULTIPLEXING AMPLITUDE MODULATION

In Eq. 59.2, the assumption is made that the information signal used for modulation, $s(t)$, is real. If this assumption is removed, the general modulated signal is represented as[3]

$$s_{\text{mod}}(t) = \text{Re}\left\{\left(s_x(t) - js_y(t)\right)e^{j\omega_c t}\right\}$$

$$= s_x(t)\cos\omega_c t + s_y(t)\sin\omega_c t$$

$$= I(t) + Q(t) \qquad 59.7$$

The in-phase component of the modulated signal is represented by $I(t)$, while the quadrature component is $Q(t)$. Both of the components are real. Since the cosine and sine functions are orthogonal, two signals can

---

[1]When a high-frequency signal is simply switched on and off slowly, the term *continuous wave* (CW) *modulation* is sometimes used.

[2]The term "index" is sometimes reserved for frequency modulation of a sinusoidal signal.

[3]The real portion of the product of two complex values, $\text{Re}\{z_1 z_2\}$, is $\text{Re}\{z_1\}\text{Re}\{z_2\} - \text{Im}\{z_1\}\text{Im}\{z_2\}$. The $-j$ is due to the inductive nature of the signal.

be transmitted simultaneously. Thus, although this is modulation, the process is referred to as *quadrature multiplexing* (QM).

7. AMPLITUDE MODULATION: BANDWIDTH AND POWER

Amplitude modulation is a frequency translation. If the modulating signal, that is, the baseband signal, has low-pass characteristics and a bandwidth of BW, the allowable change in frequency of the modulated signal is

$$\text{BW} \leq \Delta f \leq 2(\text{BW}) \qquad 59.8$$

The lower limit corresponds to the single-sideband signal, the upper limit to the double-sideband signal.

The power in an AM signal is used by the carrier and the modulating signal. The average power is given by

$$P = \left(\frac{A^2}{2}\right)\left(\left(ks(t)\right)^2_{\text{ave}} + 1\right)$$
$$= P_{\text{signal}} + P_{\text{carrier}} \qquad 59.9$$

The first term of Eq. 59.9 is the signal power and the second is the carrier power. The signal power carries the information and is thus useful power. The carrier power lowers the efficiency. The percent efficiency is given by

$$\eta = \frac{P_{\text{signal}}}{P_{\text{signal}} + P_{\text{carrier}}} \times 100\% \qquad 59.10$$

In double-sideband suppressed-carrier modulation, no power is used for a carrier and half the power exists in each sideband. The sidebands carry identical information, however, and the information-bearing power is one-half the total. In single-sideband suppressed-carrier modulation, all the power used carries information. Though they are more efficient, SSB-SC signals are complicated to generate and receive. Further, the use of carriers simplifies demodulation.

8. ANGLE MODULATION

The entire term in parentheses in Eq. 59.1 determines the angle of the carrier. That is, $\varphi(t) = \omega_c t + \theta$. The phase angle can thus be manipulated in direct proportion to the information signal, $s(t)$, by

$$\varphi(t) = \omega_c t + k_p s(t) + \theta_0 \qquad 59.11$$

The phase angle manipulation in Eq. 59.11 is called *phase modulation* (PM).

Defining the instantaneous frequency, ω_i, as the rate of change of the phase angle provides a second method of modulating the phase angle. The frequency is manipulated in Eq. 59.12.

$$\omega_i(t) = \omega_c + k_f s(t) \qquad 59.12$$

The phase angle manipulation in Eq. 59.12 is called *frequency modulation* (FM).

9. FREQUENCY MODULATION

Frequency modulation (FM) changes the carrier wave frequency in proportion to the instantaneous value of the modulating wave. If the modulating wave is a single-frequency sinusoid, $s(t) = a \cos \omega_{\text{mod}} t$, whose peak amplitude produces a maximum frequency deviation of $\Delta \omega$, a modulated FM signal is represented as[4]

$$s_{\text{mod}}(t) = A \cos\left(\omega_c t + \left(\frac{\Delta \omega}{\omega_{\text{mod}}}\right) \sin \omega_{\text{mod}} t\right)$$
$$= A \cos\left(\omega_c t + m_{\text{FM}} \sin \omega_{\text{mod}} t\right) \qquad 59.13$$

The dimensionless ratio factor in Eq. 59.13 is called the *frequency modulation index, index of modulation,* or *modulation factor,* and is defined mathematically as

$$m_{\text{FM}} = \frac{\Delta \omega}{\omega_{\text{mod}}} \qquad 59.14$$

The modulation index for an FM signal is the ratio of the maximum frequency deviation of the modulated carrier to the modulating frequency. The modulation index is sometimes expressed as a percentage. The maximum frequency deviation is given by

$$\Delta \omega = a k_f \qquad 59.15$$

The term a is the amplitude of the modulating signal, that is, the baseband or information signal. The *frequency modulator constant,* k_f, has units of radians per second per volt.

The audio portion of television broadcasts is transmitted using frequency modulation. Additionally, FM radio stations use frequency modulation.

10. PHASE MODULATION

Phase modulation (PM) changes the carrier wave angle in proportion to the instantaneous value of the modulating wave. If the modulating wave is a single-frequency

---

[4]Analysis of FM and PM signals requires the use of Fourier transforms, Bessel functions, and, alternatively, Fast Fourier Transforms (FFTs).

sinusoid, $s(t) = a \cos \omega_{mod} t$, whose peak amplitude produces a maximum frequency deviation of $\Delta \omega$, a modulated PM signal is represented as

$$s_{mod}(t)$$
$$= A \cos(\omega_c t + (\Delta\theta) \cos \omega_{mod} t + \theta_0)$$
$$= A \cos \left(\omega_c t + \left(\frac{\Delta\omega}{\omega_{mod}} \right) \cos \omega_{mod} t + \theta_0 \right) \quad \textbf{59.16}$$

Comparing Eqs. 59.16 and 59.13 illustrates the difference between PM and FM. Phase-modulated signals are proportional to the modulating signal, $s(t)$.[5] Frequency-modulated signals are proportional to the integral of the modulating signal for an input signal cosine function (hence the sine term in Eq. 59.13). The dimensionless ratio factor in Eq. 59.16, $\Delta\theta$, is mathematically identical to the frequency modulation index and is defined as

$$\Delta\theta = \frac{\Delta\omega}{\omega_{mod}} \quad \textbf{59.17}$$

The term $\Delta\theta$ is the peak phase deviation. The peak phase deviation is related to the maximum frequency deviation by

$$\Delta\omega = a k_f \omega_{mod} = \Delta\theta \omega_{mod} \quad \textbf{59.18}$$

In PM, then, the peak frequency deviation, $\Delta\omega$, is proportional to the amplitude of the modulating signal, a, and the modulating signal frequency, ω_{mod}. Phase modulation is thus less desirable for use if the frequency band is limited, although it has advantages over FM in the ease with which demodulation occurs.

11. ANGLE MODULATION: BANDWIDTH AND POWER

Angle modulation bandwidth analysis requires the use of Bessel functions of the first kind. Using the properties of these functions, an approximation for the bandwidth of an FM or PM signal is given by *Carson's rule*, which is

$$BW \approx 2\omega_{mod} (m + 1)$$
$$= 2 (\Delta\omega + \omega_{mod})$$
$$\approx 2 (\Delta\omega + \omega_{max}) \quad \textbf{59.19}$$

The modulation index, m, is given by Eq. 59.14 for an FM signal or the mathematical equivalent in Eq. 59.17 for a PM signal. The first two equations in Eq. 59.19 estimate the bandwidth when the modulating signal is

a single-frequency sinusoid.[6] The bandwidth of an FM signal modulated by something other than a single-frequency sinusoid is most accurately estimated using the maximum modulating frequency, ω_{max}, instead of the actual modulating frequency as shown in the third part of Eq. 59.19.

When $\Delta\omega$ is larger than ω_{max}, the signal is known as *wideband*. When $\Delta\omega$ is smaller than ω_{max}, the signal is known as *narrowband*.

The calculation of power for FM or PM signals is more complicated than for AM signals and requires the use of advanced engineering mathematics.

12. DIGITAL MODULATION

Digital modulation is the mapping of binary symbols into signals, with the mapping method selected for a given communication channel and its characteristics.[7] Digital modulation is also defined as a method of placing digital information on a channel in discrete phase or frequency states controlled (i.e., modulated) by the digital information signal.

Three basic types of digital modulation involve modifications to the amplitude, phase, and frequency of a carrier. Those methods modifying the amplitude are called *pulse amplitude modulation* (PAM) *techniques*. Methods that map digital signals to changes in phase or frequency are called *pulse code modulation* (PCM) *techniques*. The term PCM derives from the digital signal being coded or *keyed* to certain phases or frequencies.

Digital information is represented as logic zeros and ones. *Binary modulation* is the mapping of the zeros and ones into two different waveforms. *Multilevel modulation* is the mapping of n binary digits onto $M = 2^n$ different waveforms.

The theoretical foundation for digital modulation is much the same as that for analog modulation, though discrete mathematics is used. An additional design issue in digital modulation is quantization. *Quantization* is the process whereby a continuum of wave values, that is, a continuous time signal, is represented as a finite number of assigned (quantized) levels, that is, as a discrete time signal. Quantization is accomplished by sampling the analog signal at various intervals. The actual quantized levels are called *codes*, and the process of representing or assigning the levels is termed *coding*. The basics of both sampling and coding are described in Ch. 18.

For coding to be accurate, the transmitter and receiver in any communication must be synchronized. Synchronization is the process of maintaining one operation in step with another. For digital data, this means knowing or sensing the bit intervals (see Sec. 59-13). A *coherent*

---

[5]The modulating signal is alternatively referred to as the information, input, or baseband signal. Sometimes the modulating wave is simply called the signal, as opposed to the carrier.

[6]A single-frequency sinusoid modulating signal is assumed throughout this chapter.

[7]The terms *mapping*, *coding*, and *keying* are used interchangeably in many instances.

system is one in which the receiver is aware of the bit interval start and stop times.

13. BASEBAND DIGITAL-DATA SIGNALS

Some baseband digital signal formats used to transmit digital data are illustrated in Fig. 59.4. The time frames shown are called *bit intervals*. Each bit interval represents a single piece of information. Times t_1 through t_5 are bit intervals for the sample digital signal transmitted, 10110.

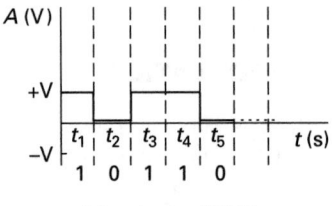

(a) unipolar (NRZ)

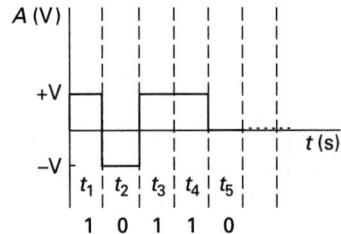

(b) bipolar (NRZ)

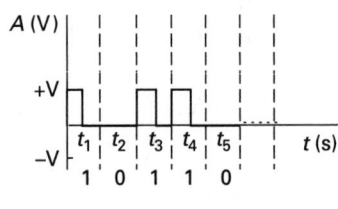

(c) unipolar (RZ)

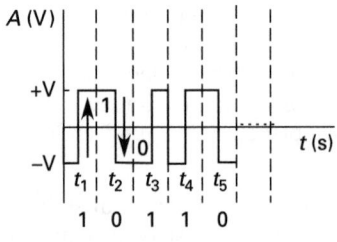

(d) Manchester (biphase)

Note: Times t_1 through t_5 are bit intervals for the sample digital signal transmitted 10110.

Figure 59.4 *Baseband Digital Signals*

The data signal in Fig. 59.4(a) is *unipolar* in that it has only one polarity, in this case positive. The data signal in Fig. 59.4(b) is *bipolar* with the positive and negative voltage maximums constituting the poles. The bipolar signal shown in Fig. 59.4(b) is called *polar nonreturn-to-zero* (NRZ).[8] In a *return-to-zero* (RZ) *signal*, the data pulse returns to the logic 0 state *before* the next bit interval, as shown in Fig. 59.4(c). The bandwidth of an RZ signal is thus not as wide as an NRZ signal.

If the channel is not DC coupled, NRZ and RZ signals cannot be used. This is because the average value of an NRZ or RZ signal is less than one. Therefore, if DC is not transmitted, the value of logic 1 is never available. A data signal that allows signal transmission for an AC-coupled channel is called *alternate-mark-inversion* (AMI). AMI is identical to unipolar RZ with every other nonzero pulse inverted, which also uses the polarity change to differentiate the one and the zero.

14. INVERSION AND STATIC

NRZ signals have two specific problems to overcome, inversion and static. *Inversion*, which can have many causes, is a condition in which the logic levels are reversed (inverted) in the channel. This results in the interpretation of the $+V$ signals as $-V$ signals, and vice versa. Thus, every bit is in error. *Static* in this instance means "without motion or change."[9] That is, static occurs when no change in logic level occurs from one bit interval to the next. Both conditions result in erroneous data, inversion by directly altering the information, and static by potential loss of timing information (synchronization) between bit intervals.

The inversion problem is overcome by using the NRZ-M or NRZ-S signals.[10] In the NRZ-M system, logic level one is represented by a change in the signal in successive bit intervals while a zero is represented by no change. Thus, inversion has no impact. The NRZ-S signal is the reverse of the NRZ-M system in that logic level zero is represented by a change and one by no change.

The static problem remains in both the NRZ-M and NRZ-S signals in that consecutive bits may still remain unchanged. That is, a series of ones or zeros results in potential synchronization difficulties. The *Manchester* is a *biphase signal* that solves synchronization difficulties by representing a logic 1 by a change from low to high voltage and a logic 0 by a change from high to low. The Manchester signal is shown in Fig. 59.4(d). The inversion problem, however, remains.

A combination of *differential encoding* such as NRZ-M and NRZ-S with *biphase encoding* such as the Manchester signal solves both inversion and synchronization problems. Numerous combinations and signal types exist and are known either as *signal formats* or *codes*.

---

[8]The NRZ signal is also called the NRZ-L, where the L stands for logic, meaning the voltage level stands for the logic level. Both positive logic and negative logic are termed NRZ-L.

[9]Static is also used to mean the distortion of a signal by noninformation-bearing electromagnetic sources or environmental effects.

[10]The M stands for "mark" and the S for "set," from telegraph terminology.

15. INTERSYMBOL INTERFERENCE

Another issue in digital system design is the interference of one pulse of information with another across bit intervals. Because the frequency response characteristic of any channel is not ideal, the pulse being sensed to create the logic 0 or 1 may be wider than the bit interval, with a pulse amplitude that varies. These two conditions are called *delay distortion* and *amplitude distortion*, respectively. The resulting excess energy from one pulse bit interval, also called a *keying interval*, interferes with the reception of the information signal in successive intervals. The interference is called *intersymbol interference* (ISI). ISI is avoided through use of an appropriate low-pass filter. By selecting a filter such that the pulse waveform represents the $(\sin x)/x$ shape, the waveform crosses the time axis at the bit intervals, and no interference is obtained. A zero ISI situation is shown in Fig. 59.5. The technique for selecting the filter to produce this result uses the *Nyquist criterion*.

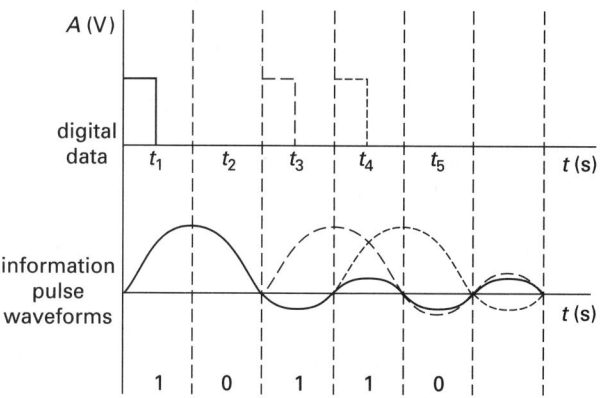

Figure 59.5 *Zero Intersymbol Interference*

In many cases, zero interference is not achieved and distortion of the signal exists at the receiver. A device called an *equalizer* attempts to correct the received signal for any amplitude-frequency or phase-frequency differences from the selected channel signal format. The process of removing distortion is called *channel equalization*.

16. PULSE AMPLITUDE MODULATION

Pulse amplitude modulation (PAM) is a linear modulation scheme in which a signal is modified by a set of discrete amplitude values, A. The amplitudes may take on any number of set amplitude values. If only two amplitude values are used, with one being $+A$ and the other zero, the term *asymmetric amplitude-shift keying* (AASK) is used to describe the modulation (see Fig. 59.4(a) and (c)). If the two amplitude values are $+A$ and $-A$, the term *amplitude-shift keying* (ASK) describes the modulation (see Fig. 59.4(b) and (d)). The constellation of possible values for a 4-bit PAM system is shown in Fig. 59.6(a). From the figure, PAM is one-dimensional modulation.

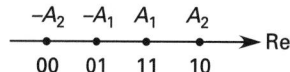

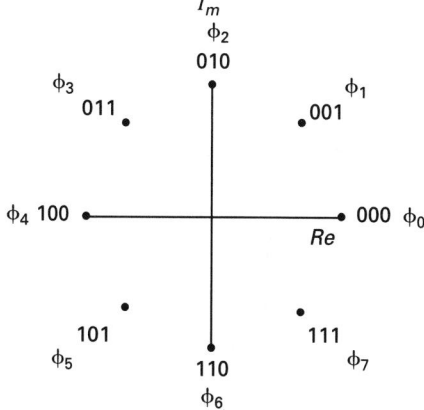

(b) PSK constellation (8-bit)

Figure 59.6 *Digital Modulation Constellations*

17. PULSE CODE MODULATION

Pulse code modulation (PCM) is modulation in which a sampled analog signal is divided into quantized levels represented by binary codes. The binary codes represent the amplitude, phase, and/or frequency of the signal. PCM thus comprises three operations: sampling, which generates a PAM signal; quantization, which divides the amplitude of the signal into discrete levels; and encoding, which assigns binary codes to the levels.[11]

A simple one-dimensional PCM technique is to turn the carrier signal on and off. If the carrier signal is simply turned on and off, the term *binary on-off keying* (BOOK) is used. A two-dimensional linear modulation scheme that changes the phase by a set number of values is called *phase-shift keying* (PSK). An example is shown in Fig. 59.6(b). If only two phases are used, the phase shift is 180°and the term *binary phase-shift keying* (BPSK) is applied. When the frequency of the carrier is changed by a set number of values, the term *frequency-shift keying* (FSK) is applied. If only two frequencies are used, the modulation is called *binary frequency-shift keying* (BFSK). Frequency modulation is normally multidimensional.

The term *binary* is used when a signal is modulated to only two possible values. The term *M-ary* is used for signals mapped to multiple modulation codes.[12]

---

[11]Digital data cannot be transmitted on an antenna of reasonable size. Thus, similar to analog data, frequency translation must occur. PCM is the method of translation (see Sec. 59-1).
[12]A *modem*, that is, a *modulator-demodulator*, is an interface device that uses PCM.

18. QUADRATURE AMPLITUDE MODULATION

A two-dimensional linear modulation scheme that modifies the amplitude and phase is called *quadrature amplitude modulation* (QAM) and is shown in Fig. 59.7.[13]

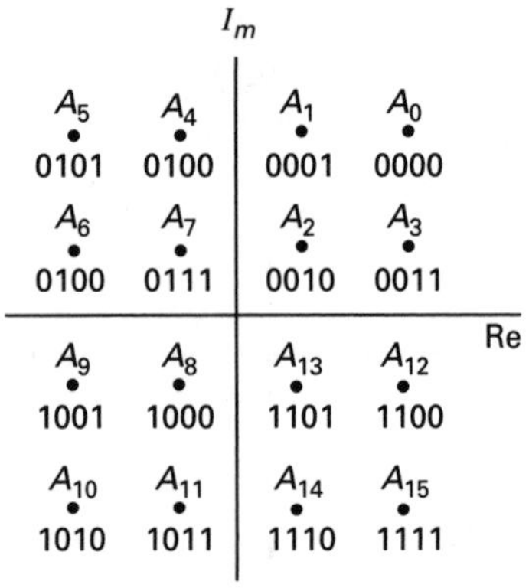

Figure 59.7 QAM Constellation (16-bit)

19. DIGITAL MODULATION: BANDWIDTH AND POWER

The bandwidth of a digital signal can be defined in several ways. A familiar definition is the *half-power bandwidth*, which is the interval between the two frequencies at which the power spectrum is 3 dB below its peak value. The *equivalent noise bandwidth* is equivalent to a rectangle with a height equal to the maximum power spectrum density and an area equal to one-half the power of the modulated signal. There are numerous other bandwidth definitions.

In order to compare modulation schemes, the *Shannon bandwidth* is used.

$$\mathrm{BW} = \frac{D}{2T} \qquad 59.20$$

The variable D is called the *dimensionality* of the signal set. That is, it is the number of orthonormal signals within a given mapping (coding) scheme. The time duration over which the signals are defined is T.

The power is normally considered in terms of its ratio with the channel noise. The power is the product of the binary digit rate, R_s, in digits per second, and the bit energy, E_{bit}.

$$P = \frac{E_{\mathrm{ave}}}{T} = E_{\mathrm{bit}} R_s \qquad 59.21$$

The bit energy and digit rate are given by

$$E_{\mathrm{bit}} = \frac{E_{\mathrm{ave}}}{\log_2 M} \qquad 59.22$$

---

[13]The phase, that is, the angle associated with the digital points in QAM, determines the value of the imaginary portion and subsequently impacts the amplitude since $|\alpha + j\beta| = \alpha^2 + \beta^2$.

$$R_s = \frac{\log_2 M}{T} \qquad 59.23$$

The term M is the number of waveforms upon which the binary digits are mapped and is given by $M = 2^n$ where n is the number of binary digits available.

Assuming *white Gaussian noise* with a power spectral density of $N_0/2$, the *signal-to-noise ratio* (SNR) is

$$\mathrm{SNR} = \frac{S}{N} = \frac{P}{N_0(\mathrm{BW})} = \left(\frac{E_{\mathrm{bit}}}{N_0}\right)\left(\frac{R_s}{\mathrm{BW}}\right) \qquad 59.24$$

The first term in Eq. 59.24 is the energy transmitted per bit divided by white Gaussian noise power spectral density. The second term, the bit rate divided by the bandwidth, is called the *bandwidth efficiency*.

20. DIGITAL ERROR PROBABILITY

The performance of a given digital modulation scheme can be measured by the *symbol error probability*, $\mathcal{P}(e)$, which is the total probability that a given waveform is detected incorrectly. The probability of corruption for a single bit is called the *bit error probability*, $\mathcal{P}_b(e)$, also called the *bit error rate* (BER). Since each symbol carries $\log_2 M$ bits, a signal symbol error affects at a minimum, one bit, and at a maximum, $\log_2 M$ bits.[14] The overall relationship is

$$\frac{\mathcal{P}(e)}{\log_2 M} \leq \mathcal{P}_{\mathrm{bit}}(e) \leq \mathcal{P}(e) \qquad 59.25$$

The equation for the probability of error, $\mathcal{P}(e)$, varies with the digital modulation scheme. Table 59.1 gives the equations for the major techniques. Appendix 59.A lists values for the error function, also called the *probability function*, from which the complementary error function mentioned in the table can be found.

21. FORWARD ERROR CORRECTING

Once detected, errors can be corrected by retransmission of the original signal. Redundancy in the transmitted signal ensures errors will be detectable.[15] Retransmission of a signal is not always feasible. The collection of techniques designed to detect errors and correct those errors without retransmission is called *forward error correcting* (FEC). Two general types of FEC codes are *block codes* and *convolution codes*.

---

[14]For example, in a 4-ary scheme ($M = 4$) contains $\log_2(4) = 2$ bits, that is, 00, 01, 10, and 11. If one of these symbols is in error (corrupted), at most two bits are affected.

[15]Claude Shannon proved in 1948 that arbitrary reliability is possible as long as a fixed portion of the transmitted signal is redundant.

Table 59.1 *Digital Modulation Error Probabilities,* $\mathcal{P}(e)$

amplitude-shift keying (ASK)[a,b]	phase-shift keying (PSK)	frequency-shift keying (FSK)
$\left(1 - \dfrac{1}{M}\right) \mathrm{erfc}\left(\sqrt{\left(\dfrac{3\log_2 M}{M^2 - 1}\right)\left(\dfrac{E_{\mathrm{bit}}}{N_0}\right)}\right)$	$\approx \mathrm{erfc}\left(\sqrt{\log_2 M \left(\dfrac{E_{\mathrm{bit}}}{N_0}\right)} \sin\left(\dfrac{\pi}{M}\right)\right)$	$\frac{1}{2}\mathrm{erfc}\left(\sqrt{\log_2 M \left(\dfrac{E_{\mathrm{bit}}}{2N_0}\right)}\right)$
		lower bound
		$\left(\dfrac{M-1}{2}\right)\mathrm{erfc}\left(\sqrt{\log_2 M \left(\dfrac{E_{\mathrm{bit}}}{2N_0}\right)}\right)$
		upper bound

[a]The abbreviation erfc is the complementary error function given by $\mathrm{erfc}(x) = \dfrac{2}{\sqrt{\pi}} \displaystyle\int_x^\infty e^{-x^2}\,dx = 1 - \mathrm{erf}(x)$.

[b]Amplitude-shift keying is also known as pulse amplitude modulation (PAM).

Communications

60 Signal Multiplexing

Nomenclature
s signal (time function)   V
t time   s

Symbols
ω angular frequency   rad/s

Subscripts
n number

1. FUNDAMENTALS

The signal formats described in Ch. 59, for the most part, are capable of carrying one information signal.[1] Once at the receiver, the information may have processing limitations unrelated to the channel. For example, a human listening to a voice signal will hear the signal as a continuous sound even if that signal is broken up into small digital pieces. A computer processing the information coming from a receiver is limited in its rate of information intake. In order to maximize the efficiency of a channel, several information signals can be sent. Sending multiple signals over a single channel in a manner that still allows for individual signal recognition at the receiver requires *multiplexing*. For multiplexing to be successful, the signals must not overlap. Mathematically, this means that the signals must be orthogonal.

2. FREQUENCY MULTIPLEXING

Frequency multiplexing or *frequency-division multiplexing* (FDM) is the process of sending two or more signals over a common channel using a different frequency band for each. The process is shown in Fig. 60.1.

---

[1]Where the carrier is split into upper and lower sidebands, different information can be sent in each sideband.

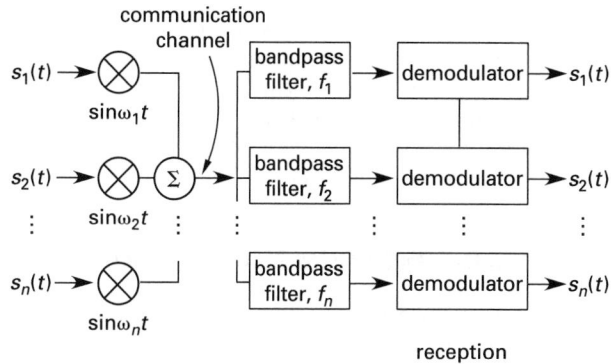

Figure 60.1 *Frequency Multiplexing*

The original signals are combined with a sinusoid of the selected frequency. By selecting sinusoids of varying frequencies, the signals can be stacked in frequency without overlapping. To ensure that the signals remain separate, a *guard slot*, which is a frequency range designed to carry no signal, is included between the utilized frequencies to prevent inadvertent overlap of signals.

3. TIME MULTIPLEXING

The time domain and the frequency domain are mathematically related. Further, they have complete duality with one another. Therefore, if a signal can be frequency multiplexed, it can be time multiplexed. The process is shown in Fig. 60.2.

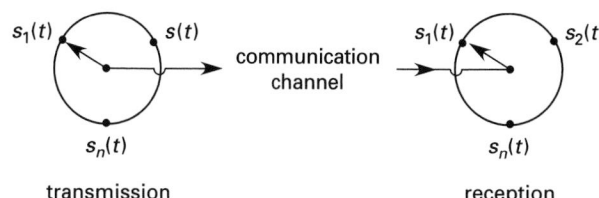

Figure 60.2 *Time Multiplexing*

Time multiplexing or *time-division multiplexing* (TDM) is the process of sending two or more signals over a common channel using successive time intervals for the different signals.

4. SPACE MULTIPLEXING

Space multiplexing or *space-division multiplexing* (SDM) is the process of sending two or more signals over a common channel using separate geographical areas for the different signals. A variety of means is used to accomplish the separation. Lowering the power to deliberately minimize the signal coverage areas, as is done with cellular radio/phone systems, allows signals to use the same frequency or time assignments. Spot beam antennae focus signals into specific coverage areas, as is done in some satellite systems, allowing multiple signals to share frequency or time assignments. *Polarization multiplexing*, a type of space multiplexing, allows signals to share the same frequency, time, and geographical areas while obtaining orthogonal separation through the use of different polarization planes.

5. SPREAD-SPECTRUM MULITPLEXING

When the transmission signal bandwidth is deliberately wide, with the power spread thinly over the entire band, it is called a *spread-spectrum transmission*. By keeping the power near or below the noise level, narrow-band radios can transmit within the band without interference, the jamming potential is minimized, privacy is improved, and security is enhanced. Numerous techniques have been devised for multiplexing over this type of channel.

Code-division multiple access (CDMA) is the process by which each of the multiple signals on a single channel is assigned a pseudorandom noise code independent of the others (orthogonal). For example, in *frequency hopping* a carrier's frequency moves around the allowed bandwidth as determined by a pseudonoise sequence generator feeding a frequency hopper (controller). As long as the appropriate frequency hopper is receiving the signal, the original information can be retrieved.

Time-division multiple access (TDMA) is the process by which multiple signals transmitted over a single channel are each assigned a separate time slot. As a result, the information signals are transmitted in bursts allotted to specific transmitters.

Other types of multiple-access techniques exist.

61 Communication Systems and Channels

Nomenclature

B, BW	bandwidth	Hz
C	channel capacity	bps
d	horizon distance	km
h	antenna height	m
N	noise power	W
N_0	noise spectral density	W/Hz
S	average signal power	W

1. SYSTEMS AND CHANNELS OVERVIEW

A *communication system* is a device in which electrical pulses originating at a transmission point are accurately reproduced at a reception point. A *communication channel* is the medium that connects the transmission point to the reception point.[1]

A communication channel may consist of copper wire, air, coaxial cable, a waveguide, space, fiber-optic material, or other media. A communication system may be telegraph, telephone (now including wireless, cellular, and satellite), radio and television (broadcast and cable), satellite, local- and wide-area computer networks, and so on. In practice, the terms are used interchangeably, with some systems taking their names from the channels or even the frequencies used, such as fiber-optic and microwave systems.[2]

The channel capacity of an analog system is determined by the maximum frequency deviation. The capacity, C, of a digital channel in bits per second (bps) is determined by the signal-to-noise ratio and the bandwidth as shown in Eq. 61.1.

$$C = (BW) \log_2 \left(1 + \frac{S}{N} \right)$$
$$= (BW) \log_2 \left(1 + \frac{S}{N_0(BW)} \right) \qquad 61.1$$

Equation 61.1 should be understood in the context of digital bandwidth and power (see Sec. 59-19). The value of the channel capacity as the bandwidth approaches infinity is called the *Shannon limit* and is given by[3]

$$C = 1.44 \left(\frac{S}{N_0} \right) \qquad 61.2$$

The higher the frequency of the communication channel, the greater the bandwidth and capacity. *Microwave channels* involve frequencies of approximately 1 to 100 GHz, correlating to 30 and 0.3 cm wavelengths. Microwave channels are treated as high-frequency transmission lines (see Ch. 56). Microwave signals are line-of-sight signals. Thus, an antenna has a limited coverage area. Considering the curvature of the Earth, the radio horizon, d, in kilometers, depends on the antenna height, h, in meters, as shown in Eq. 61.3.

$$d = 4.1\sqrt{h} \qquad 61.3$$

Optical channels use electromagnetic waves of approximately 10^{-6} m, correlating to 3×10^{14} Hz. Specially designed fiber-optic cables and connections must be used for the transmission of optical wavelength signals.

The information transferred is generally classified as voice, video, facsimile, data, or a combination of these. The term *broadcast* refers to radio or television. *Telecommunications* refers to communications over long distances but is generally used in reference to mobile radio, wireless, and cellular voice communication or local-area and wide-area computer networks.

2. TELEPHONY

The *telephone* is a system for converting sound waves into current variations for transmission to distant points. It is meant primarily for voice communication, although its uses are more numerous now. Since it was designed for human speech, the channel or band used is limited

---

[1]A communication channel is also the band of frequencies assigned or allocated for a particular purpose. For example, a standard radio broadcasting channel is 10 kHz wide while a television channel is 6 MHz wide.

[2]The term "microwave" dates to the 1940s when an electromagnetic wave of less than 1 m was difficult to utilize and was thus deemed to be a "micro-wave."

[3]This result is a direct consequence of the *Shannon coding theorem*.

Communications

to between 0.3 and 3.4 kHz. Due to bandpass filter imperfections and trunk signaling (see Fig. 61.1), which is done out of the normal band, the voice channel is standardized as a 4 kHz channel. The general concept of a telephone network is illustrated in Fig. 61.1.

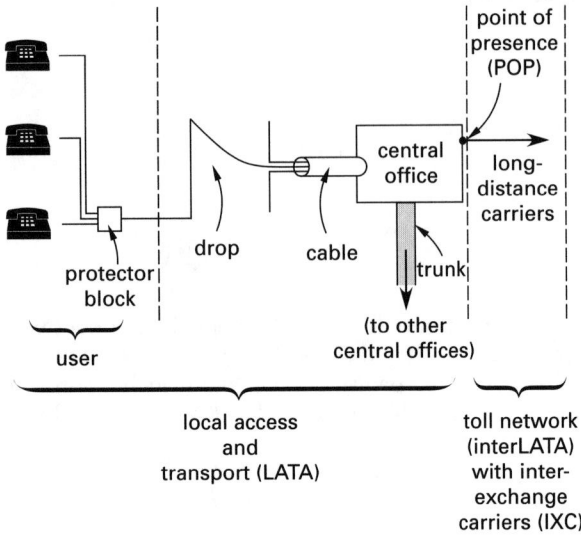

Figure 61.1 Public Switched Telephone Network

The *public switched telephone network* (PSTN) represented in Fig. 61.1 starts at the user or customer.[4] Multiuser lines are connected using twisted copper wire to a single point called a protector block. The *protector block* provides overvoltage protection to the network. Many user lines are then bundled in a cable that is buried underground or strung between telephone poles and run to the central office. The central office provides the first stage of switching, either to other central offices via a *trunk line* or to long-distance carriers. The interface point between the long-distance carriers and the central offices is called the *point of presence* (POP). The network up to the POP is called the local access and transport area (LATA). Beyond the POP, *interexchange carriers*, such as AT&T, operate. The network described here is a generic system often referred to by the acronym *plain old telephone service* (POTS).

3. BROADCAST RADIO

Broadcast radio refers to AM-band, shortwave, and FM-band transmissions meant for free reception by the public. Radio in the United States is regulated by the Federal Communications Commission (FCC). The rules governing broadcasting are published in the Code of Federal Regulation (CFR), specifically in Title 47 and various chapters. The chapters are referred to as *FCC rules*. For example, the AM bandwidth is covered in 47 CFR 73.14, that is, Title 47, Chapter 73.14.

The standard AM broadcast channel is from 535 to 1705 kHz, with the carriers assigned center frequencies of 540 to 1700 kHz in 10 kHz steps. A total of 117 stations can technically exist in the same area (FCC Rule 73.14). The stations operate at anywhere from 0.1 to 50 kW depending on the type and class of channel. A *clear channel* is a station assigned to serve a wide area. Clear channels are so called because they are legally protected from interference. *Regional channels* serve population centers. *Local channels* serve small communities. Within the various channel designations there are a variety of classifications.

Above the standard AM band, shortwave radio is used. The band varies from 5950 to 26,100 kHz. These stations are intended to be received by the general public in another country.

The standard FM broadcast channel is from 88 to 108 MHz. One hundred numbered channels (stations) are assigned with bandwidths of 200 kHz each (FCC Rule 73.201). The stations operate between 0.25 and 100 kW depending on the classification. The classification depends on the *effective radiated power* (ERP) and the *height above average terrain* (HAAT) as compared to a class contour.

Radio receivers mix the incoming signal frequency with an *intermediate frequency* (IF), which then generates the sum and difference of the two. This process is called *heterodyning*.[5] The receivers that are designed to convert all incoming carrier signals to common intermediate-frequency (IF) values using heterodyning are called *superheterodyne receivers*. The IF signal is 455 kHz for AM and 10.7 MHz for FM.

4. BROADCAST TELEVISION

Television (TV) broadcast channels are designated by channel number and service level. *Low-band VHF*, consisting of channels 2 through 6, ranges from 54 to 88 MHz. *High-band VHF*, consisting of channels 7 through 13, ranges from 174 to 216 MHz. UHF signals consist of channels 14 through 69, which ranges from 470 to 806 MHz. Each channel is 6 MHz wide. The television channels operate from 100 W to 5000 kW.[6]

High-definition television (HDTV) is a digital system designed to fit within the 6 MHz bandwidth of standard broadcasting channels.

5. SATELLITE SYSTEMS

Satellites provide a line-of-sight communications platform with wide area coverage. Satellites are generally placed in three orbit categories, depending on the application. *Low Earth orbit* (LEO) is between 500

---

[4]Upgraded systems that are digital end to end are called *integrated services digital networks* (ISDNs).

[5]Heterodyning allows a single bandpass filter to be used with all the incoming frequencies shifted to within the IF band.

[6]Radio transmissions are referred to as "stations" while television transmissions are called "channels." This is because the FCC assigns actual channel numbers to each frequency band.

and 900 km above the surface of the Earth. *Medium Earth orbit* (MEO) is between 5 000 and 12 000 km. *Geostationary Earth orbit* (GEO) is at approximately 36 000 km above the surface of the Earth. The term *geosynchronous* orbit is used also, though a geostationary orbit is a geosynchronous orbit that lies in the Earth's equatorial plane. That is, a geostationary orbit is a special geosynchronous orbit with a zero inclination.

A *geostationary satellite* is one that orbits the Earth from west to east at the GEO altitude in such a manner that it makes one revolution in 24 hr. That is, it is synchronous with the Earth's rotation. This means a GEO satellite remains fixed over a given surface position. The satellite *footprint*, that is, the area that is covered by a communications satellite, in GEO is thus fixed as well. Three satellites properly positioned in GEO can provide communication coverage of the entire surface of the Earth.

The frequency bands used vary from 1 to 30 GHz. The band designations have historical significance only and are given in Table 61.1. The transmission of satellite communications is similar to that described in Antenna Theory (Ch. 57).

Table 61.1 *Satellite and Radio Communication Frequency Bands*

band	frequency range (GHz)
L	1–2
S	2–4
C	4–8
X	8–12.5
Ku	12.5–18
K	18–26.5
Ka	26.5–40

6. WIRELESS COMMUNICATION

Wireless communication employs an antenna for radiating and receiving information using electromagnetic waves. Wireless communication encompasses a broad spectrum of systems.

Wireless telephones with short ranges are usually called *cordless*. *Cellular telephones* achieve long-range capability by utilizing antennae that provide local coverage in an area called a *cell*. The system antenna in the cell containing the user's telephone controls the communication process. When the user moves into another cell, the communication process is passed to that cell.[7]

Wireless radio is termed *portable radio* if the transceiver can be moved but must be stationary while in use. *Mobile radio* can be operated while in motion.

Wireless systems can be totally Earth-based or utilize satellite resources. If Earth-based, the term *terrestrial radio system* or *radio system* is used. If at least one component is based in space, the term *satellite radio system* or *satellite system* is applied. Systems are also classified by the location of the user—for example, land, maritime, aeronautical, space, and deep space.

Because wireless systems now transmit a variety of information types, not merely voice, the multiservice networks are sometimes called *personal communication networks* (PCNs) or *personal communication systems* (PCSs).

_____

[7]Cellular phones are a form of mobile radio with the signal patched into the telephone system network.

Topic XIII: Biomedical Systems

Chapter

Biomedical Systems

For the most current information about the exam, visit **www.ppi2pass.com** regularly.

62 Biomedical Electrical Engineering

1. OVERVIEW

Biomedical engineering is the application of engineering technology to medicine. Electrical and electronic engineering play a large role in biomedical engineering. Physiological monitoring, diagnostic instruments, patient internal imaging, implant devices, artificial organs, and computer data processing of biological information are but a few of the areas of concern.

The theory of operation of the instruments and equipment used is described throughout this book and is consistent with standard techniques of electrical analysis. The difference in electrical engineering for medical purposes is that the technology interfaces with the human body. This interface requires stringent controls to ensure electrical safety, radiation safety, reliability, and biocompatibility.

Electrical sources may often touch the human body but not transfer electrical energy to it, either directly or by radiating energy. Nonengineering personnel and patients will handle and/or operate the equipment. They must be protected, and the equipment must withstand such usage. The electrical components must also be protected from the biological environment in order to function as designed. The components must be highly reliable and fail-safe for obvious reasons. Biocompatibility is an absolute necessity to prevent interactions with body tissue and ensure nontoxicity. Inert materials are used, usually metals and polymers. An understanding of materials is required since the fabrication, molding, and surface condition of electrical components affect the body's reaction to them.

The safety and effectiveness of medical devices in the United States are the responsibility of the Food and Drug Administration (FDA) Center for Devices and Radiological Health (CDRH). Most biomedical devices require the approval of the CDRH and are subject to the general requirements of the Federal Food, Drug, and Cosmetic Act (FD&C). The requirements of the FD&C Act are documented in Title 21 of the Code of Federal Regulations, Chapter 1, Parts 800 through 1299.

The FDA recognizes numerous standards in the design and use of biomedical devices. These include the standards of the Association for the Advancement of Medical Information (AAMI), International Electrotechnical Commission (IEC), International Organization for Standardization (ISO), American Society for Testing and Materials (ASTM), American Institute of Ultrasound in Medicine (AIUM), Institute of Electrical and Electronics Engineers (IEEE), National Electrical Manufacturers Association (NEMA), and Underwriters Laboratories, Inc. (UL).

2. SENSORS

Sensors are used to convert biomedical quantities into electrical signals, which can then be monitored and analyzed. Examples of biomedical sensor applications are given in Table 62.1.

Table 62.1 Biomedical Sensor Applications[a]

quantity sensed	device or principle used	biomedical application(s)
electric potential	electrodes	nerve or muscle activity (for example, heart beats)
electric impedance	bridge	respiration rate
magnetic field	permeability	blood flow
displacement	capacitive	orthopedics
flow	ultrasonic Doppler	body fluid flow
image	X ray	shape or organ motion

[a]The device(s) and the application(s) listed are by no means the only possible choices.

Sensor size and weight are dependent upon the application and the patient. The size is often small, requiring high sensitivity and minimal power consumption.

Biocompatibility is often ensured by isolation of the sensor from the biological environment or by using relatively inert elements or compounds whose interactions with the human body are minimal. For example, insulated electrodes whose surfaces are coated with tantalum or titanium oxides, Ta_2O_3 or TiO_2, have good stability and minimal chemical interactions. Additionally, silver and platinum electrodes, Ag-AgCl or Pt-PtIr, are widely used in the field of health care.

Biomedical Systems

3. IMPLANTABLE STIMULATORS

The application of electricity to the human body for medical purposes existed well prior to modern medicine. The first major benefit was to the human cardiovascular system with the advent of commercial pacemakers. Implants are now used to treat deafness (for example, the cochlear implant), paralyzed muscles, spinal cord pain, and epilepsy, among many other conditions. Implants also monitor blood flow, glucose levels, oxygen concentration, and the delivery of drugs.

Electrical potential exists across the membranes of living cells. The human nervous system detects and changes these potentials, and now implantable stimulators can do the same. When the implantable stimulator is equipped with a telemetry device to allow external monitoring, it is called an implantable stimulator telemetry (IST).

4. MONITORS

Monitors are used to observe numerous physiologic variables and display the patient's status, provide a historical record, and perform elementary logical decisions regarding care. Some of the possible variables that can be monitored are given in Table 62.2.

Table 62.2 *Monitored Physiologic Variables*

vital signs	bioelectric signals	chemical concentrations
temperature	electrocardiogram (ECG)	oxygen
heart rate	electroencephalogram (EEG)[a]	carbon dioxide
blood pressure		pH
respiration rate		glucose

[a]measurement of electric signals on the scalp that arise from brain activity, indicative of the state of the central nervous system

A general monitoring system block diagram is shown in Fig. 62.1. Other than the interface with the human system, all the electrical and electronic principles are identical to those in other electrical engineering applications.

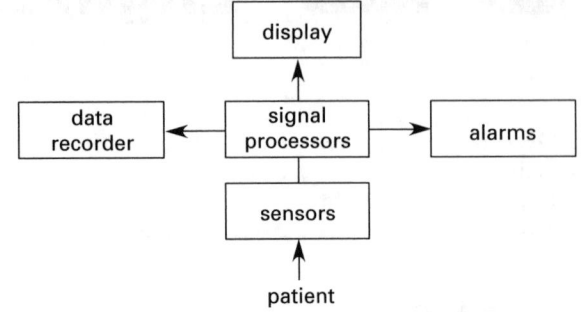

Figure 62.1 *General Medical Monitoring System*

5. APPLICATIONS

The types of electrical and electronic equipment in the field of medicine are broad in scope but similar in principle to other such electrical engineering applications. They include radiology, magnetic resonance imaging (MRI), and ultrasound equipment. Support systems that replace or augment human organs, such as artificial limbs, artificial hearts, dialysis machines, insulin delivery systems, and blood-detoxification systems, rely on electrical and electronic components. Finally, an entire array of electrical devices exists to analyze the many samples taken from a patient to aid in the diagnostic and recovery processes.

Topic XIV: Control Systems

Chapter

    63. Analysis of Engineering Systems: Control Systems

Control Systems

For the most current information about the exam, visit **www.ppi2pass.com** regularly.

63 Analysis of Engineering Systems: Control Systems

Nomenclature

A	steady-state response	–
BW	bandwidth	Hz
C	capacitance or constant	– or F
$e(t)$	error	–
$E(s)$	error, $\mathcal{L}\big(e(t)\big)$	–
$f(t)$	forcing function	–
$F(s)$	forcing function, $\mathcal{L}\big(f(t)\big)$	–
$G(s)$	forward transfer function	–
h	step height	–
$H(s)$	reverse transfer function	–
$i(t)$	current	A
$I(s)$	current, $\mathcal{L}\big(i(t)\big)$	A
j	$\sqrt{-1}$	–
K	gain	–
L	inductance	–
M	fraction overshoot	–
n	degrees of freedom, order of the system, or system type	–
N	Nyquist's number	–
$p(t)$	arbitrary function	–
P	number of poles	–
$P(s)$	arbitrary function, $\mathcal{L}\big(p(t)\big)$	–
Q	quality factor	–
r	real value (root)	–
$r(t)$	time response	–
$R(s)$	response function, $\mathcal{L}\big(r(t)\big)$	–
S	sensitivity	–
t	time	s
$T(s)$	transfer function, $\mathcal{L}\big(T(t)\big)$	–
$T(t)$	transfer function	–
u	input variable	–
v, V	voltage	V
x	position or state variable	–
y	output variable	–
Z	number of zeros	–

Symbols

ϵ	a small number	
ζ	damping ratio	
τ	time constant	
ω	natural frequency	

Subscripts

d	damped
f	forced or feedback
i	in
n	natural
o	out
p	peak or pole
r	rise
s	settling
t	total
z	zero

1. TYPES OF RESPONSE

Natural response (also known as *initial condition response*, *homogeneous response*, and *unforced response*) is the manner in which a system behaves when energy is applied and then subsequently removed. The system is left alone and allowed to do what it would naturally, without the application of further disturbing forces. In the absence of friction, natural response is characterized by a sinusoidal response function. The response function will contain exponentially decaying sinusoids otherwise.

Forced response is the behavior of a system that is acted upon by a force that is applied periodically. Forced response in the absence of friction is characterized by sinusoidal terms having the same frequency as the forcing function.

Control Systems

Natural and forced responses are present simultaneously in forced systems. The sum of the two responses is the *total response*.[1] This is the reason that differential equations are solved by adding a particular solution to the homogeneous solution. The homogeneous solution corresponds to the natural response; the particular solution corresponds to the forced response.

$$\frac{\text{total}}{\text{response}} = \frac{\text{natural}}{\text{response}} + \frac{\text{forced}}{\text{response}} \qquad 63.1$$

Since the influence of decaying functions disappears after a few cycles, natural response is sometimes referred to as *transient response*. Once the transient response effects have died out, the total response consists entirely of forced terms. This is the *steady-state response*.

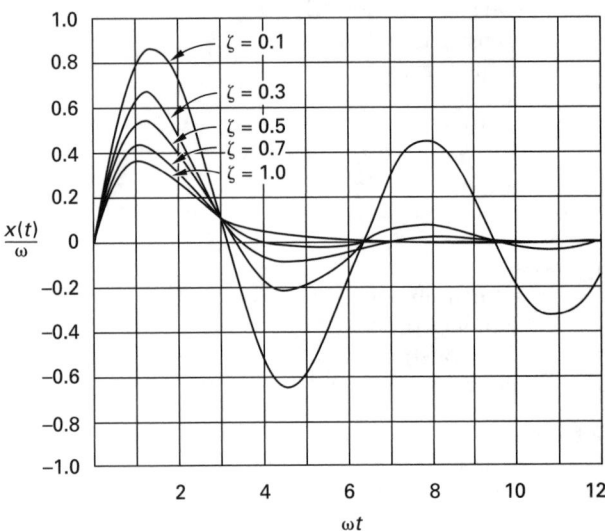

Figure 63.1 *Natural Response*

2. GRAPHICAL SOLUTION

Graphical solutions are available in some cases, particularly those with homogeneous, step, and sinusoidal inputs. When the system equation is a homogeneous second-order linear differential equation with constant coefficients (Sec. 11-6) in the form of Eq. 63.2, the natural time response can be determined from Fig. 63.1. The term ω is the natural frequency, ω_n, defined in Sec. 31.10, and ζ is the damping ratio.

$$x'' + 2\zeta\omega x' + \omega^2 x = 0 \qquad 63.2$$

When the system equation is a second-order linear differential equation with constant coefficients and the forcing function is a step of height h (as in Eq. 63.3), the time response can be determined from Fig. 63.2.

$$x'' + 2\zeta\omega x' + \omega^2 x = \omega^2 h \qquad 63.3$$

---

[1]This response is a function of time, and, therefore, is referred to as *time response* to distinguish it from *frequency response* (Sec. 63-13).

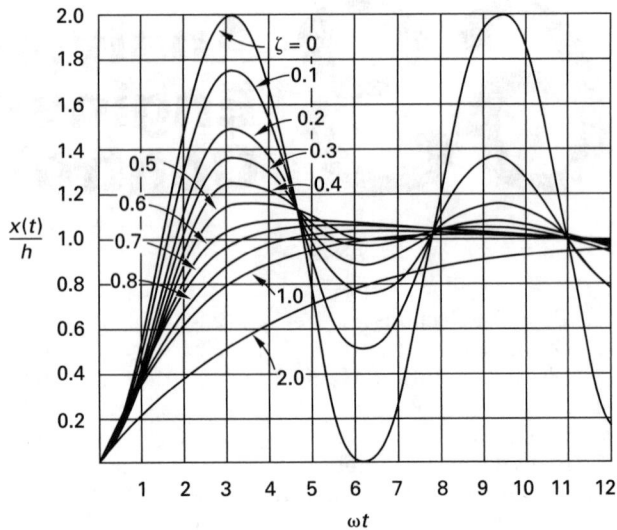

Figure 63.2 *Response to a Unit Step*

Figure 63.2 illustrates that a system responding to a step will eventually settle to the steady-state position of the step but that when damping is low ($\zeta < 1$), there will be *overshoot*. Figure 63.3 illustrates this and other parameters of second-order response to a step input. The settling time depends on the tolerance (i.e., the separation of the actual and steady-state responses). The *time delay*, t_d in Fig. 63.3, is the time needed to reach 50% of the steady-state value. The *time constant*, τ, is the time needed to reach approximately 63% of the steady-state value.

$$\omega_d = \text{damped frequency} = \omega\sqrt{1 - \zeta^2} \qquad 63.4$$

$$t_r = \text{rise time} = \frac{\pi - \arccos\zeta}{\omega_d} \qquad 63.5$$

$$t_p = \text{peak time} = \frac{\pi}{\omega_d} \qquad 63.6$$

$$M_p = \text{peak gain (fraction overshoot)}$$
$$= \exp\left(\frac{-\pi\zeta}{\sqrt{1 - \zeta^2}}\right) \qquad 63.7$$

$$t_s = \text{settling time}$$
$$= \frac{3.91}{\zeta\omega_n} \quad [2\% \text{ criterion}] \qquad 63.8$$

$$= \frac{3.00}{\zeta\omega_n} \quad [5\% \text{ criterion}] \qquad 63.9$$

$$\tau = \text{time constant}$$
$$= \frac{1}{\zeta\omega_n} \qquad 63.10$$

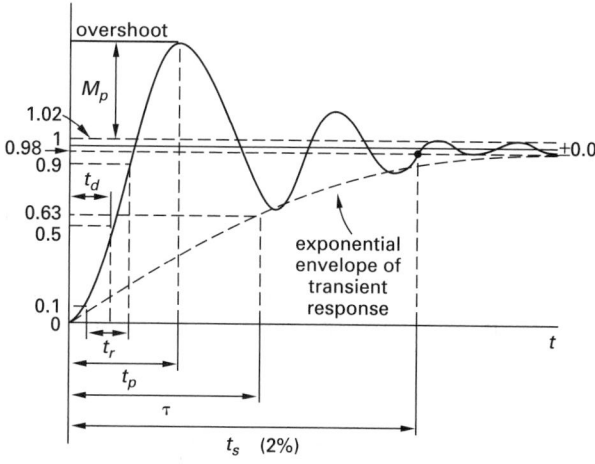

Figure 63.3 *Second-Order Step Time Response Parameters*

3. CLASSICAL SOLUTION METHOD

A system model can be thought of as a block where the *input signal* is the forcing function, $f(t)$, and the *output signal* is the response function, $r(t)$. The standard analytical method of determining the response of a system from its system equation uses Laplace transforms. Accordingly, the classical solution approach does not determine the output signal or response function directly but derives the *transfer function*, $T(s)$, instead.[2] The transfer function is also known as the *rational function*.

$$T(s) = \mathcal{L}\left(\frac{r(t)}{f(t)}\right) \qquad 63.11$$

Transfer functions in the s-domain are Laplace transformations of the corresponding time-domain functions. Transforming $r(t)/f(t)$ into $T(s)$ involves more than a simple change of variables. The s symbol can be thought of as a derivative operator; similarly, the integration operator is represented as $1/s$. By convention, Laplace transforms are represented by uppercase letters while operand functions are represented by lowercase letters.

Example 63.1

The following system equation in integrodifferential equation form has been derived for the output current from an electrical network. Convert the system equation to the s-domain.

$$i(t) = \frac{1}{L}\int (v_1 - v_2)dt + C\left(\frac{d(v_2 - v_1)}{dt}\right)$$

---

[2]Strictly speaking, $r(t)/f(t)$ is the *transfer function* and $\mathcal{L}e\,(r(t)/f(t))$ is the *transform of the transfer function*. However, the distinction is seldom made.

Solution

Replace all derivative operators by s; replace all integration operators by $1/s$. Replace time voltage, $v(t)$, with s-domain voltage, $V(s)$.

$$I(s) = \frac{V_1}{sL} - \frac{V_2}{sL} + sCV_2 - sCV_1$$

Example 63.2

Determine the voltage transfer function for the system shown and draw its black box representation.

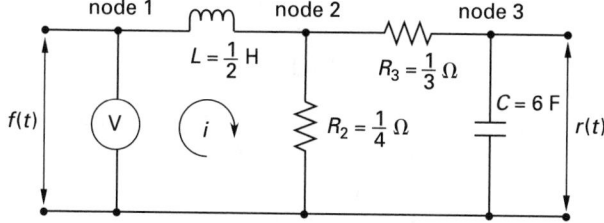

Solution

First, write the system equations. There are two loop currents, so two simultaneous equations are needed. One system equation can be written for each of the three nodes, so there will be one redundant system equation. (At this point, it is not obvious which of the two system equations can be used to derive the transfer function.)

At node 1,

$$i(t) = \frac{1}{L}\int (v_1 - v_2)\,dt = 2\int (v_1 - v_2)\,dt$$

At node 2,

$$0 = \frac{1}{L}\int (v_2 - v_1)\,dt + \frac{v_2}{R_2} + \frac{v_2 - v_3}{R_3}$$

$$= 2\int (v_2 - v_1)\,dt + 4v_2 + 3(v_2 - v_3)$$

At node 3,

$$0 = C\left(\frac{dv_3}{dt}\right) + \frac{v_3 - v_2}{R_3}$$

$$= 6\left(\frac{dv_3}{dt}\right) + 3(v_3 - v_2)$$

Next, convert the system equations to the s-domain by substituting s for derivative operations and $1/s$ for integration operations.

For node 1,

$$I(s) = \frac{2(V_1 - V_2)}{s}$$

For node 2,

$$0 = \frac{2(V_2 - V_1)}{s} + 4V_2 + 3(V_2 - V_3)$$

For node 3,

$$0 = 6sV_3 + 3(V_3 - V_2)$$

The transfer function is the ratio of the output to input voltages, V_3/V_1. It does not depend on V_2. The equation for node 1 cannot be used unless $i(t)$ is known, which it (generally) is not. V_2 is eliminated from the equations for nodes 2 and 3. From the second and third nodes,

$$V_2 = \frac{2V_1 + 3sV_3}{2 + 7s}$$

$$= V_3(1 + 2s)$$

The transfer function, $T(s)$, is found by equating these two expressions and solving for V_3/V_1.

$$T(s) = \frac{V_3}{V_1} = \frac{1}{7s^2 + 4s + 1}$$

The black box representation of this system is

$$\boxed{\dfrac{1}{7s^2 + 4s + 1}}$$

4. FEEDBACK THEORY

The output signal is returned as input in a feedback loop (feedback system). A basic feedback system consists of two black box units (a *dynamic unit* and a *feedback unit*), a pick-off point (take-off point), and a summing point (*comparator* or *summer*). The summing point is assumed to perform positive addition unless a minus sign is present. The incoming signal, v_i, is combined with the feedback signal, v_f, to give the *error* (*error signal*), e. Whether addition or subtraction is used in Eq. 63.12 depends on whether the summing point is additive (i.e., a positive feedback system) or subtractive (i.e., a negative feedback system), respectively. $E(s)$ is the *error transfer function* (*error gain*).

$$E(s) = \mathcal{L}\left(e(t)\right) = V_i(s) \pm V_f(s)$$

$$= V_i(s) \pm H(s)V_o(s) \qquad \textit{63.12}$$

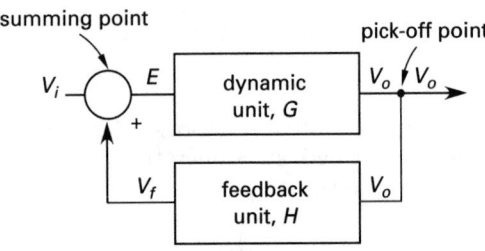

summing point

pick-off point

Figure 63.4 *Feedback System*

The ratio $E(s)/V_i(s)$ is the *error ratio* (*actuating signal ratio*).

$$\frac{E(s)}{V_i(s)} = \frac{1}{1 + G(s)H(s)} \qquad \text{[negative feedback]} \qquad \textit{63.13}$$

$$= \frac{1}{1 - G(s)H(s)} \qquad \text{[positive feedback]} \qquad \textit{63.14}$$

Since the dynamic and feedback units are black boxes, each has an associated transfer function. The transfer function of the dynamic unit is known as the *forward transfer function* (*direct transfer function*), $G(s)$. In most feedback systems—amplifier circuits in particular—the magnitude of the forward transfer function is known as the *forward gain* or *direct gain*. $G(s)$ can be a scalar if the dynamic unit merely scales the error. However, $G(s)$ is normally a complex operator that changes both the magnitude and the phase of the error.

$$V_o(s) = G(s)E(s) \qquad \textit{63.15}$$

The pick-off point transmits the output signal, V_o, from the dynamic unit back to the feedback element. The output of the dynamic unit is not reduced by the pick-off point. The transfer function of the feedback unit is the *reverse transfer function* (*feedback transfer function, feedback gain*), $H(s)$, which can be a simple magnitude-changing scalar or a phase-shifting function.

$$V_f(s) = H(s)V_o(s) \qquad \textit{63.16}$$

The ratio $V_f(s)/V_i(s)$ is the *feedback ratio* (*primary feedback ratio*).

$$\frac{V_f(s)}{V_i(s)} = \frac{G(s)H(s)}{1 + G(s)H(s)} \qquad \text{[negative feedback]} \qquad \textit{63.17}$$

$$= \frac{G(s)H(s)}{1 - G(s)H(s)} \qquad \text{[positive feedback]} \qquad \textit{63.18}$$

The *loop transfer function* (*loop gain* or *open-loop transfer function*) is the gain after going around the loop one time, $\pm G(s)H(s)$.

The *overall transfer function* (*closed-loop transfer function, control ratio, system function, closed-loop gain*), $G_{\text{loop}}(s)$, is the overall transfer function of the feedback system. The quantity $1 + G(s)H(s) = 0$ is the *characteristic equation*. The *order of the system* is the largest exponent of s in the characteristic equation. (This corresponds to the highest-order derivative in the system equation.)

$$G_{\text{loop}}(s) = \frac{V_o(s)}{V_i(s)}$$

$$= \frac{G(s)}{1 + G(s)H(s)} \qquad \text{[negative feedback]} \qquad \textit{63.19}$$

$$= \frac{G(s)}{1 - G(s)H(s)} \qquad \text{[positive feedback]} \qquad \textit{63.20}$$

With positive feedback and $G(s)H(s)$ less than 1.0, G_{loop} will be larger than $G(s)$. This increase in gain is a characteristic of positive feedback systems. As $G(s)H(s)$ approaches 1.0, the closed-loop transfer function increases without bound, which is usually an undesirable effect.

In a negative feedback system, the denominator of Eq. 63.19 is greater than 1.0. Although the closed-loop transfer function will be less than $G(s)$, there may be other desirable effects. Generally, a system with negative feedback is less sensitive to variations in temperature, circuit components values, input signal frequency, and signal noise. Other benefits include distortion reduction, increased stability, and impedance matching.[3]

Example 63.3

A high-gain noninverting amplifier has a gain of 10^6. Feedback is provided by a resistor voltage divider. What is the closed-loop gain?

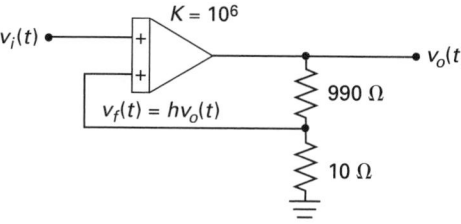

Solution

The fraction of the output signal appearing at the summing point depends on the resistances in the divider circuit.

$$h = \frac{10}{10 + 990} = 0.01$$

Since the feedback path only scales the feedback signal, the feedback will be positive. From Eq. 63.20, the closed-loop gain is

$$K_{\text{loop}} = \frac{K}{1 - Kh} = \frac{10^6}{1 - (10^6)(0.01)} \approx -100$$

5. SENSITIVITY

For large loop gains ($G(s)H(s) \gg 1$) in negative feedback systems, the overall gain is approximately $1/H(s)$. Thus, the forward gain is not a factor, and by choosing the correct value of $H(s)$, the output can be made insensitive to variations in $G(s)$.

---

[3]For circuits to be directly connected in series without affecting their performance, all input impedances must be infinite and all output impedances must be zero.

In general, the *sensitivity* of any variable, A, with respect to changes in another parameter, B, is

$$S_B^A = \frac{d \ln A}{d \ln B} = \frac{\dfrac{dA}{A}}{\dfrac{dB}{B}} \approx \left(\frac{\Delta A}{\Delta B}\right)\left(\frac{B}{A}\right) \quad \textit{63.21}$$

The sensitivity of the loop transfer function with respect to the forward transfer function is

$$S_{G(s)}^{G_{\text{loop}}(s)} = \left(\frac{\Delta G_{\text{loop}}(s)}{\Delta G(s)}\right)\left(\frac{G(s)}{G_{\text{loop}}(s)}\right)$$

$$= \frac{1}{1 + G(s)H(s)} \quad \begin{bmatrix} \text{negative} \\ \text{feedback} \end{bmatrix} \quad \textit{63.22}$$

$$= \frac{1}{1 - G(s)H(s)} \quad \begin{bmatrix} \text{positive} \\ \text{feedback} \end{bmatrix} \quad \textit{63.23}$$

Example 63.4

A closed-loop gain of -100 is required from a circuit, and the output signal must not vary by more than $\pm 1\%$. An amplifier is available, but its output varies by $\pm 20\%$. How can this amplifier be used?

Solution

A closed-loop sensitivity of 0.01 is required. This means

$$\frac{\Delta v_o}{v_o} = \frac{\Delta G_{\text{loop}} v_i}{G_{\text{loop}} v_i} = \frac{\Delta G_{\text{loop}}}{G_{\text{loop}}} = 0.01$$

Similarly, the existing amplifier has a sensitivity of 20%. This means

$$\frac{\Delta G}{G} = 0.20$$

The ratio of these variations corresponds to the definition of sensitivity (Eq. 63.21). For positive feedback,

$$\frac{\Delta G_{\text{loop}} G}{\Delta G G_{\text{loop}}} = \frac{0.01}{0.20} = \frac{1}{1 - GH}$$

Solving, $GH = -19$.

Solving for G from Eq. 63.20,

$$G_{\text{loop}} = -100 = \frac{G}{1 - GH} = \frac{G}{1 - (-19)}$$

Solving, $G = -2000$. Finally, solve for H.

$$H = \frac{GH}{G} = \frac{-19}{-2000} = 0.0095$$

6. BLOCK DIAGRAM ALGEBRA

The functions represented by several interconnected black boxes (*cascaded blocks*) can be simplified into a single block operation. Some of the most important simplification rules of block diagram algebra are shown in Fig. 63.5. Case 3 represents the standard feedback model.

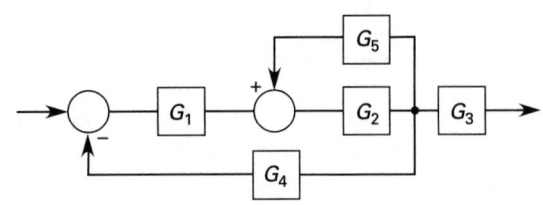

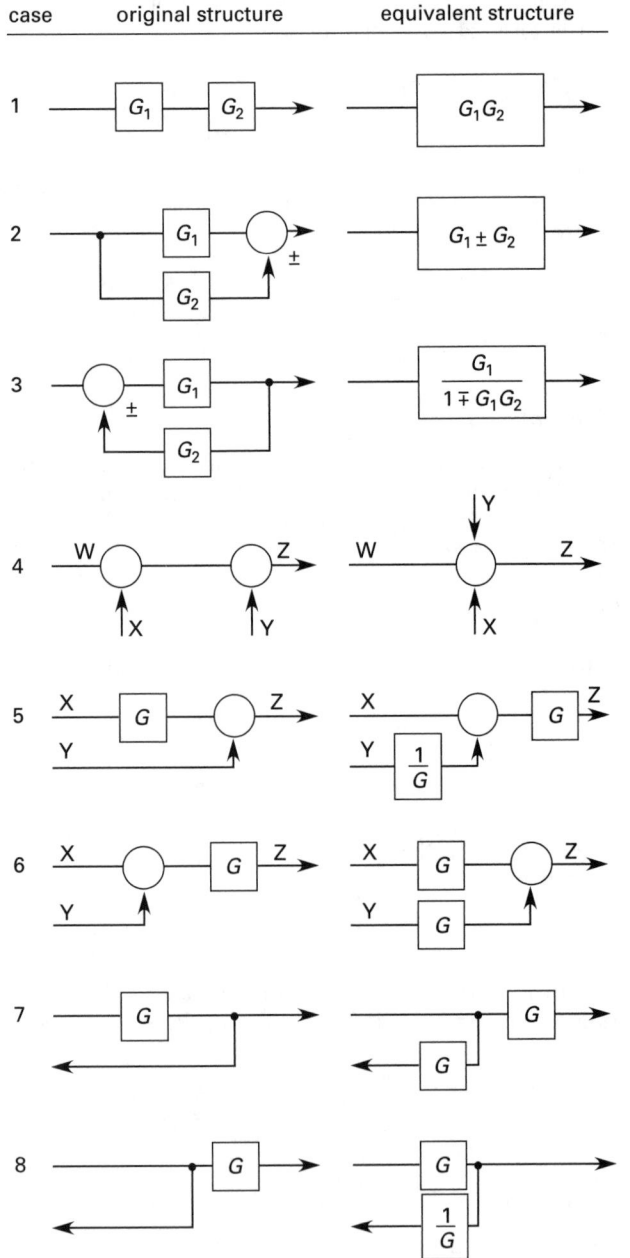

Figure 63.5 *Rules of Simplifying Block Diagrams*

Example 63.5

A complex block system is constructed from five blocks and two summing points. What is the overall transfer function?

Solution

Use case 5 to move the second summing point back to the first summing point.

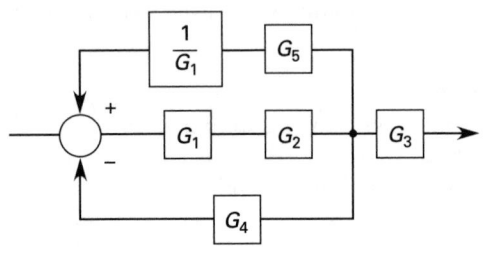

Use case 1 to combine boxes in series.

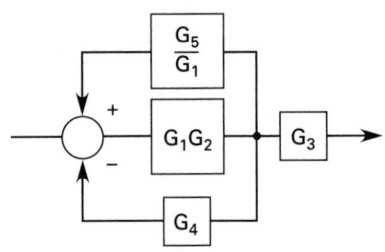

Use case 2 to combine the two feedback loops.

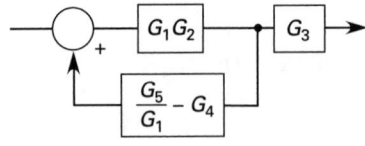

Use case 8 to move the pick-off point outside the G_3 box.

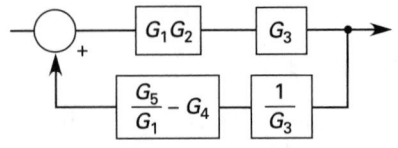

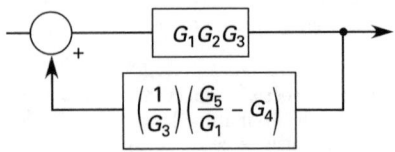

Use case 3 to determine the system gain.

$$G_{\text{loop}} = \frac{G_1 G_2 G_3}{1 - (G_1 G_2 G_3)\left(\dfrac{1}{G_3}\right)\left(\dfrac{G_5}{G_1} - G_4\right)}$$

$$= \frac{G_1 G_2 G_3}{1 - G_2 G_5 + G_1 G_2 G_4}$$

7. PREDICTING SYSTEM TIME RESPONSE

The transfer function is derived without knowledge of the input and is insufficient to predict the time response of the system. The system time response depends on the form of the input function. Since the transfer function is expressed in the s-domain, the forcing and response functions must be also. (Laplace transforms of step, pulses, sinusoids, and other functions are evaluated in Chs. 11 and 15.)

$$R(s) = T(s)F(s) \qquad \textbf{63.24}$$

The time-based response function, $r(t)$, is found by performing the inverse Laplace transform.

$$r(t) = \mathcal{L}^{-1}(R(s)) \qquad \textbf{63.25}$$

Example 63.6

A mechanical system is acted upon by a constant force of eight units starting at $t = 0$. What is the time-based response function, $r(t)$, if the transfer function is

$$T(s) = \frac{6}{(s+2)(s+4)}$$

Solution

The forcing function is a step of height 8 at $t = 0$. The Laplace transform of a unit step is $1/s$. Therefore, $F(s) = 8/s$.

From Eq. 63.24,

$$R(s) = T(s)F(s) = \left(\frac{6}{(s+2)(s+4)}\right)\left(\frac{8}{s}\right)$$

$$= \frac{48}{s(s+2)(s+4)}$$

The response function is found from Eq. 63.25 and a table of Laplace transforms (App. 11.A). (Remember that a product of linear terms in the denominator of $R(s)$ is equivalent to a sum of terms in $r(t)$.)

$$r(t) = \mathcal{L}^{-1}\left(\frac{48}{s(s+2)(s+4)}\right)$$

$$= 6 - 12e^{-2t} + 6e^{-4t}$$

The last two terms are decaying exponentials, which represent the transient natural response. The first term does not vary with time; it is the steady-state response.

8. PREDICTING TIME RESPONSE FROM A RELATED RESPONSE

In some cases, it may be possible to use a known response to one input to determine the response to another input. For example, the impulse function is the derivative of the step function, so the response to an impulse is the derivative of the response to a step function.

9. INITIAL AND FINAL VALUES

The initial and final (steady-state) values of any function, $P(s)$, can be found from the *initial* and *final value theorems*, respectively, providing the limits exist. Equations 63.26 and 63.27 are particularly valuable in determining the steady-state response (substitute $R(s)$ for $P(s)$) and the steady-state error (substitute $E(s)$ for $P(s)$).

$$\lim_{t \to 0+} p(t) = \lim_{s \to \infty}(sP(s)) \quad \text{[initial value]} \qquad \textbf{63.26}$$

$$\lim_{t \to \infty} p(t) = \lim_{s \to 0}(sP(s)) \quad \text{[final value]} \qquad \textbf{63.27}$$

Example 63.7

What is the final value of the response function $r(t)$ if

$$R(s) = \frac{1}{s(s+1)}$$

Solution

From Eq. 63.27, $R(s)$ is multiplied by s and the limit taken as s approaches zero.

$$r(\infty) = \lim_{s \to 0}\left(\frac{s}{s(s+1)}\right) = \lim_{s \to 0}\left(\frac{1}{s+1}\right) = 1$$

10. SPECIAL CASES OF STEADY-STATE RESPONSE

In addition to determining the steady-state response from the final value theorem (see Sec. 63-9), the steady-state response to a specific input can be easily derived from the transfer function, $T(s)$, in a few specialized cases. For example, the steady-state response function for a system acted upon by an impulse is simply the transfer function. That is, a pulse has no long-term effect on a system.

$$R(\infty) = T(s) \quad \text{[pulse input]} \qquad \textbf{63.28}$$

Control Systems

The steady-state response for a *step input* (often referred to as a *DC input*) is obtained by substituting zero for s everywhere in the transfer function. (If the step has magnitude h, the steady-state response is multiplied by h.)

$$R(\infty) = T(0) \quad \text{[unit step input]} \qquad \textbf{63.29}$$

The steady-state response for a sinusoidal input is obtained by substituting $j\omega_f$ for s everywhere in the transfer function, $T(s)$. The output will have the same frequency as the input. It is particularly convenient to perform sinusoidal calculations using phasor notation (as illustrated in Ex. 63.9).

$$R(\infty) = T(j\omega_f) \qquad \textbf{63.30}$$

Example 63.8

What is the steady-state response of the system in Ex. 63.6 acted upon by a step of height 8?

Solution

Substitute zero for s in $T(s)$ and multiply by 8.

$$R(\infty) = 8T(0) = (8)\left(\frac{6}{(0+2)(0+4)}\right) = 6$$

Example 63.9

What is the steady-state response when a sinusoidal voltage of $4\sin(2t + 45°)$ is applied to a system with the following transfer function?

$$T(s) = \frac{-1}{7s^2 + 7s + 1}$$

Solution

The angular frequency of the forcing function is $\omega_f = 2$ rad/s. Substitute $j2$ for s in $T(s)$ and simplify the expression by recognizing that $j^2 = -1$.

$$T(j2) = \frac{-1}{7(j2)^2 + 7(j2) + 1}$$

$$= \frac{-1}{-28 + 14j + 1} = \frac{-1}{-27 + 14j}$$

Next, convert $T(j2)$ to phasor (polar) form. The magnitude and angle of the denominator are

$$\text{magnitude} = \sqrt{(14)^2 + (-27)^2} = 30.4138$$

$$\text{angle} = 180° - \arctan\left(\frac{14}{27}\right) = 152.59°$$

However, this is the negative reciprocal of $T(j2)$.

$$T(j2) = \frac{-1}{30.4138\angle152.59°} = -0.03288\angle-152.59°$$

The forcing function expressed in phasor form is $4\angle45°$. From Eq. 63.11, the steady-state response is

$$v(t) = T(t)f(t) = (-0.03288\angle-152.59°)(4\angle45°)$$

$$= -0.1315\angle-107.6°$$

11. POLES AND ZEROS

A *pole* is a value of s that makes a function, $P(s)$, infinite. Specifically, a pole makes the denominator of $P(s)$ zero.[4] A zero of the function makes the numerator of $P(s)$ (and hence $P(s)$ itself) zero. Poles and zeros need not be real or unique; they can be imaginary and repeated within a function.

A *pole-zero diagram* is a plot of poles and zeros in the *s-plane*—a rectangular coordinate system with real and imaginary axes. Zero is represented as O; a pole is represented as X. Poles off the real axis always occur in conjugate pairs known as *pole pairs*.

Sometimes it is necessary to derive the function $P(s)$ from its pole-zero diagram. This will be only partially successful since repeating identical poles and zeros are not usually indicated on the diagram. Also, scale factors (scalar constants) are not shown.

Example 63.10

Draw the pole-zero diagram for the transfer function

$$T(s) = \frac{5(s+3)}{(s+2)(s^2 + 2s + 2)}$$

Solution

The numerator is zero when $s = -3$. This is the only zero of the transfer function.

The denominator is zero when $s = -2$ and $s = -1 \pm j$. These three values are the poles of the transfer function.

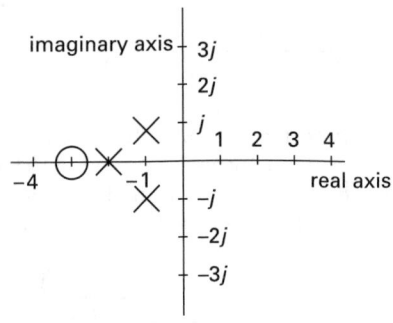

---

[4]Pole values are the system *eigenvalues*.

Example 63.11

A pole-zero diagram for a transfer function $T(s)$ has a single pole at $s = -2$ and a single zero at $s = -7$. What is the corresponding function?

Solution

$$T(s) = \frac{K(s+7)}{s+2}$$

The scale factor K must be determined by some other means.

12. PREDICTING SYSTEM TIME RESPONSE FROM RESPONSE POLE-ZERO DIAGRAMS

A response pole-zero diagram based on $R(s)$ can be used to predict how the system responds to a specific input. (Note that this pole-zero diagram must be based on the product $T(s)F(s)$ since that is how $R(s)$ is calculated. Plotting the product $T(s)F(s)$ is equivalent to plotting $T(s)$ and $F(s)$ separately on the same diagram.)

The system will experience an *exponential decay* when a single pole falls on the real axis. A pole with a value of $-r$, corresponding to the linear term $s+r$, will decay at the rate of e^{-rt}. The quantity $1/r$ is the decay *time constant*, the time for the response to achieve approximately 63% of its steady-state value. Thus, the farther left the point is located from the vertical imaginary axis, the faster the motion will die out.

Undamped sinusoidal oscillation will occur if a pole pair falls on the imaginary axis. A conjugate pole pair with the value of $\pm j\omega$ indicates oscillation with a natural frequency of ω rad/s.

Pole pairs to the left of the imaginary axis represent *decaying sinusoidal response*. The closer the poles are to the real (horizontal) axis, the slower the oscillations. The closer the poles are to the imaginary (vertical) axis, the slower the decay. The *natural frequency*, ω, of undamped oscillation can be determined from a conjugate pole pair having values of $r \pm \omega_f$.

$$\omega = \sqrt{r^2 + \omega_f^2} \qquad \text{63.31}$$

The magnitude and phase shift can be determined for any input frequency from the pole-zero diagram with the following procedure: Locate the angular frequency, ω_f, on the imaginary axis. Draw a line from each pole (i.e., a pole-line) and zero (i.e., a zero-line) of $T(s)$ to this point. The angle of each of these lines is the angle between it and the horizontal real axis. The overall magnitude is the product of the lengths of the zero-lines divided by the product of the lengths of the pole-lines. (The scale factor must also be included because it is not

shown on the pole-zero diagram.) The phase is the sum of the pole-angles less the sum of the zero-angles.

$$|R| = \frac{K \prod_z |L_z|}{\prod_p |L_p|} = \frac{K \prod_z \text{length}}{\prod_p \text{length}} \qquad \text{63.32}$$

$$\angle R = \sum_p \alpha - \sum_z \beta \qquad \text{63.33}$$

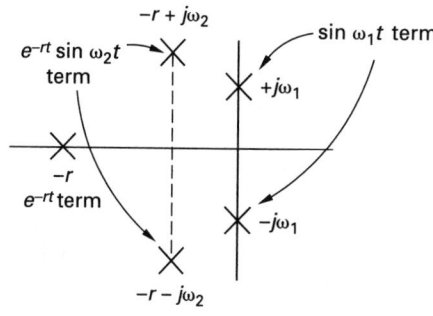

Figure 63.6 *Types of Response Determined by Pole Location*

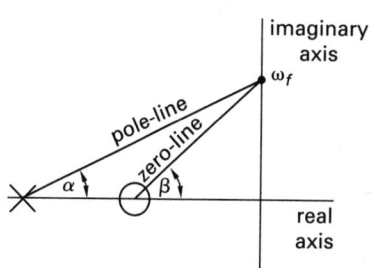

Figure 63.7 *Calculating Magnitude and Phase from a Pole-Zero Diagram*

Example 63.12

What is the response of a system with the response pole-zero diagram representing $R(s) = T(s)F(s)$ shown?

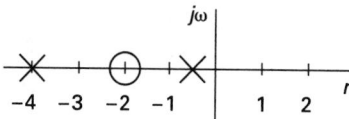

Solution

The poles are at $r = -\frac{1}{2}$ and $r = -4$. The response is

$$r(t) = C_1 e^{-\frac{1}{2}t} + C_2 e^{-4t}$$

Constants C_1 and C_2 must be found from other data.

Example 63.13

What is the system response if $T(s) = (s+2)/(s+3)$ and the input is a unit step?

Solution

The transform of a unit step is $1/s$. The response is

$$R(s) = T(s)F(s) = \frac{s+2}{s(s+3)}$$

The response pole-zero diagram is

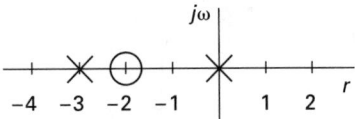

The pole at $r = 0$ contributes the exponential $C_1 e^{-0t}$ (or simply C_1) to the total response. The pole at $r = -3$ contributes the term $C_2 e^{-3t}$. The total response is

$$r(t) = C_1 + C_2 e^{-3t}$$

13. FREQUENCY RESPONSE

The gain and phase angle frequency response of a system changes as the forcing frequency is varied. (The dependence of $r(t)$ on ω_f is illustrated in Ex. 63.9.) The *frequency response* is the variation in these parameters, always with a sinusoidal input. *Gain* and *phase characteristics* are plots of the steady-state gain and phase angle responses with a sinusoidal input versus frequency. While a linear frequency scale can be used, frequency response is almost always presented against a logarithmic frequency scale.

The steady-state gain response is expressed in decibels, while the steady-state phase angle response is expressed in degrees. The gain is calculated from Eq. 63.34 where $|T(s)|$ is the absolute value of the steady-state response.

$$\text{gain} = 20 \log |T(j\omega)| \quad \text{[in dB]} \qquad \textit{63.34}$$

A doubling of $|T(j\omega)|$ is referred to as an *octave* and corresponds to a 6.02 dB increase. A ten-fold increase in $|T(j\omega)|$ is a *decade* and corresponds to a 20 dB increase.

$$\text{number of octaves} = \frac{\text{gain}_2 - \text{gain}_1 \text{ (in dB)}}{6.02}$$

$$= 3.32 \times \text{number of decades}$$

$$\textit{63.35}$$

$$\text{number of decades} = \frac{\text{gain}_2 - \text{gain}_1 \text{ (in dB)}}{20}$$

$$= 0.301 \times \text{number of octaves}$$

$$\textit{63.36}$$

14. GAIN CHARACTERISTIC

The *gain characteristic* (*M*-curve for magnitude) is a plot of the gain as ω_f is varied. It is possible to make a rough sketch of the gain characteristic by calculating the gain at a few points (pole frequencies, $\omega = 0$, $\omega = \infty$, etc.). The curve will usually be asymptotic to several lines. The frequencies at which these asymptotes intersect are *corner frequencies*. The peak gain, M_p, coincides with the natural (resonant) frequency of the system.[5] Large peak gains indicate reduced stability and large overshoots. The *gain crossover point* is the frequency at which log(gain)= 0.

The *half-power points* (*cut-off frequencies*) are the frequencies for which the gain is 0.707 (i.e., $\sqrt{2}/2$) times the peak value. (This is equivalent to saying the gain is 3 dB less than the peak gain.) The *cut-off rate* is the slope of the gain characteristic in dB/octave at a half-power point. The frequency difference between the half-power points is the *bandwidth*, BW. The *closed-loop bandwidth* is the frequency range over which the closed-loop gain falls 3 dB below its value at $\omega = \omega_n$. (The term bandwidth often means closed-loop bandwidth.) The *quality factor*, Q, is

$$Q = \frac{\omega_n}{\text{BW}} \qquad \textit{63.37}$$

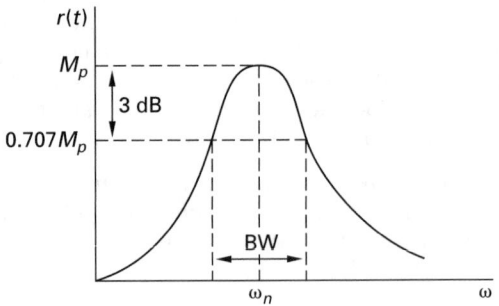

Figure 63.8 *Bandwidth*

Since a low or negative gain (compared to larger parts of the curve) effectively represents attenuation, the gain characteristic can be used to distinguish between low- and high-pass filters. A low-pass filter has a large gain at low frequencies and a small gain at high frequencies. Conversely, a high-pass filter has a high gain at high frequencies and a low gain at low frequencies.

It may be possible to determine certain parameters (e.g., the natural frequency and bandwidth) from the transfer function directly. For example, when the denominator of $T(s)$ is a single linear term of the form $s + r$, the bandwidth will be equal to r. Thus, the bandwidth and time constant are reciprocals.

---

[5]The gain characteristic peaks when the forcing frequency equals the natural frequency. It is also said that this peak corresponds to the resonant frequency. Strictly speaking, this is true, although the gain may not actually be resonant (i.e., be infinite).

Another important case is that in which $T(s)$ has the form of Eq. 63.38. (Compare the form of $T(s)$ to the form of Eq. 63.11 in Sec. 63-3.) The coefficient of the s^2 term must be unity, and ω_n must be much larger than BW so that the pole is close to the imaginary axis. The zero defined by constants a and b in the numerator is not significant.

$$T(s) = \frac{as + b}{s^2 + (\text{BW})s + \omega_n^2} \qquad 63.38$$

Example 63.14

What are the maximum gain, bandwidth, upper half-power frequency, and half-power gain of a system whose transfer function is

$$T(s) = \frac{1}{s + 5}$$

Solution

The steady-state response to a sinusoidal input is determined by substituting $j\omega$ for s in $T(s)$. When $\omega = 0$, $T(s)$ will have a value of $1/5 = 0.2$, and $T(s)$ decreases thereafter. Thus, the maximum gain is 0.2. The bandwidth is 5 rad/s. Since the lower half-power frequency is implicitly 0 rad/s, the upper half-power frequency is 5 rad/s, at which point, the gain will be $(0.707)(0.2) = 0.141$.

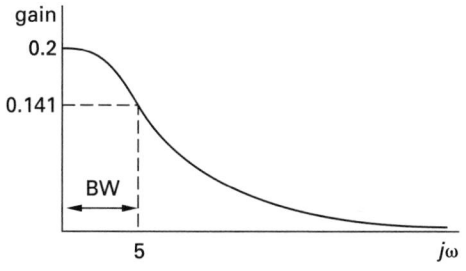

Example 63.15

Predict the natural frequency and bandwidth for the following transfer function.

$$T(s) = \frac{s + 19}{s^2 + 7s + 1000}$$

Solution

The form of this equation is the same as Eq. 63.38. The bandwidth is 7, and the natural frequency is $\sqrt{1000} = 31.62$ rad/s.

15. PHASE CHARACTERISTIC

The phase angle response will also change as the forcing frequency is varied. The *phase characteristic* (α *curve*) is a plot of the phase angle as ω_f is varied.

16. STABILITY

A stable system will remain at rest unless disturbed by external influence and will return to a rest position once the disturbance is removed. A pole with a value of $-r$ on the real axis corresponds to an exponential response of e^{-rt}. Since e^{-rt} is a decaying signal, the system is stable. Similarly, a pole of $+r$ on the real axis corresponds to an exponential response of e^{rt}. Since e^{rt} increases without limit, the system is unstable.

Since any pole to the right of the imaginary axis corresponds to a positive exponential, a *stable system* will have poles only in the left half of the s-plane. If there is an isolated pole on the imaginary axis, the response is stable. However, a conjugate pole pair on the imaginary axis corresponds to a sinusoid that does not decay with time. Such a system is considered to be unstable.

Passive systems (i.e., the homogeneous case) are not acted upon by a forcing function and are always stable. In the absence of an energy source, exponential growth cannot occur. *Active systems* contain one or more energy sources and may be stable or unstable.

There are several *frequency response (domain) analysis* techniques for determining the stability of a system, including Bode plots, root-locus diagrams, Routh stability criteria, Hurwitz tests, and Nichols charts. The term *frequency response* almost always means the steady-state response to a sinusoidal input.

The value of the denominator of $T(s)$ is the primary factor affecting stability. When the denominator approaches zero, the system increases without bound. In the typical feedback loop, the denominator is $1 \pm GH$, which can be zero only if $|GH| = 1$. It is logical, then, that most of the methods for investigating stability (e.g., Bode plots, root-locus diagrams, Nyquist analyses, and Nichols charts) investigate the value of the open-loop transfer function, GH. Since $\log(1) = 0$, the requirement for stability is that $\log GH$ must not equal 0 dB.

A negative feedback system will also become unstable if it changes to a positive feedback system, which can occur when the feedback signal is changed in phase more than 180°. Therefore, another requirement for stability is that the phase angle change must not exceed 180°.

17. BODE PLOTS

Bode plots are gain and phase characteristics for the open-loop $G(s)H(s)$ transfer function that are used to determine the *relative stability* of a system. The gain characteristic is a plot of $20 \log |G(s)H(s)|$ versus ω for a sinusoidal input. (It is important to recognize that Bode plots, though similar in appearance to gain and phase frequency response charts, are used to evaluate stability and do not describe the closed-loop system response.)

The *gain margin* is the number of decibels that the open-loop transfer function, $G(s)H(s)$, is below 0 dB at the *phase crossover frequency* (i.e., where the phase angle is $-180°$). (If the gain happens to be plotted on a linear scale, the gain margin is the reciprocal of the gain at the phase crossover point.) The gain margin must be positive for a stable system, and the larger it is, the more stable the system will be.

The *phase margin* is the number of degrees the phase angle is above $-180°$ at the *gain crossover point* (i.e., where the logarithmic gain is 0 dB or actual gain is 1).

In most cases, large positive gain and phase margins will ensure a stable system. However, the margins could have been measured at other than the crossover frequencies. Therefore, a Nyquist stability plot is needed to verify the absolute stability of a system.

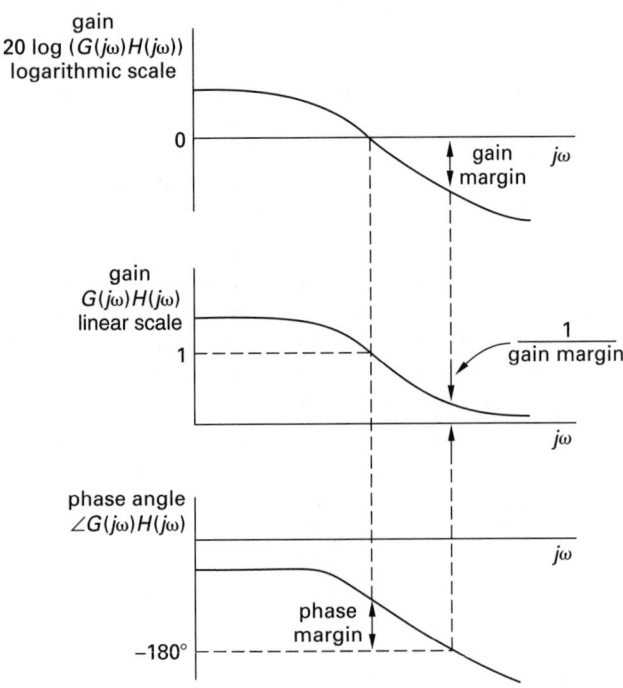

Figure 63.9 *Gain and Phase Margin Bode Plots*

18. ROOT-LOCUS DIAGRAMS

A *root-locus diagram* is a pole-zero diagram showing how the poles of $G(s)H(s)$ move when one of the system parameters (e.g., the gain factor) in the transfer function is varied. The diagram gets its name from the need to find the roots of the denominator (i.e., the poles). The locus of points defined by the various poles is a line or curve that can be used to predict *points of instability* or other critical operating points. A point of instability is reached when the line crosses the imaginary axis into the right-hand side of the pole-zero diagram.

A root-locus curve may not be contiguous, and multiple curves will exist for different sets of roots. Sometimes, the curve splits into two branches. In other cases, the curve leaves the real axis at *breakaway points* and continues on with constant or varying slopes approaching

asymptotes. One branch of the curve will start at each open-loop pole and end at an open-loop zero.

Example 63.16

Draw the root-locus diagram for a feedback system with open-loop transfer function $G(s)H(s)$. K is a scalar constant that can be varied.

$$G(s)H(s) = \frac{Ks(s+1)(s+2)}{s(s+2) + K(s+1)}$$

Solution

The poles are the zeros of the denominator.

$$s_1, s_2 = -\tfrac{1}{2}(2+K) \pm \sqrt{1 + \tfrac{1}{4}K^2}$$

Since the second term can be either added or subtracted, there are two roots for each value of K. Allowing K to vary from zero to infinity produces a root-locus diagram with two distinct branches. The first branch extends from the pole at the origin to the zero at $s = -1$. The second branch extends from the pole at $s = -2$ to $-\infty$. All poles and zeros are not shown. Since neither branch crosses into the right half, the system is stable for all values of K.

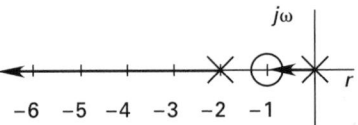

19. HURWITZ TEST

A stable system has poles only in the left half of the s-plane. These poles correspond to roots of the *characteristic equation*. The characteristic equation can be expanded into a polynomial of the form

$$a_0 s^n + a_1 s^{n-1} + \cdots + a_{n-1}s + a_n = 0 \qquad 63.39$$

The *Hurwitz stability criterion* requires that all coefficients be present and be of the same sign (which is equivalent to requiring all coefficients to be positive); if the coefficients differ in sign, the system is unstable. If the coefficients are all alike in sign, the system may or may not be stable. The Routh test should be used in that case.

20. ROUTH CRITERION

The *Routh criterion*, like the Hurwitz test, uses the coefficients of the polynomial characteristic equation. A table (the *Routh table*) of these coefficients is formed. The Routh-Hurwitz criterion states that the number of sign changes in the first column of the table equals the number of positive (unstable) roots. Therefore, the system will be stable if all entries in the first column have the same sign.

The table is organized in the following manner.

$$\begin{vmatrix} a_0 & a_2 & a_4 & a_6 & \cdots \\ a_1 & a_3 & a_5 & a_7 & \cdots \\ b_1 & b_2 & b_3 & b_4 & \cdots \\ c_1 & c_2 & c_3 & c_4 & \cdots \\ \vdots & \vdots & \vdots & \vdots \end{vmatrix}$$

The remaining coefficients are calculated in the following pattern until all values are zero.

$$b_1 = \frac{a_1 a_2 - a_0 a_3}{a_1} \qquad \text{63.40}$$

$$b_2 = \frac{a_1 a_4 - a_0 a_5}{a_1} \qquad \text{63.41}$$

$$b_3 = \frac{a_1 a_6 - a_0 a_7}{a_1} \qquad \text{63.42}$$

$$c_1 = \frac{b_1 a_3 - a_1 b_2}{b_1} \qquad \text{63.43}$$

Special methods are used if there is a zero in the first column but nowhere else in that row. One of the methods is to substitute a small number, represented by ϵ or δ, for the zero and calculate the remaining coefficients as usual.

Example 63.17

Evaluate the stability of a system that has a characteristic equation of

$$s^3 + 5s^2 + 6s + C = 0$$

Solution

All of the polynomial terms are present (Hurwitz criterion), so the Routh table is

s^3	1	6
s^2	5	C
s^1	$\dfrac{30 - C}{5}$	0
s^0	C	

To be stable, all of the entries in the first column must be positive, which requires $0 < C < 30$.

21. NYQUIST ANALYSIS

Nyquist analysis is another graphical method that is particularly useful when time delays are present in a

system or when frequency response data are available. *Nyquist's stability criterion* is $N = P - Z$, where P is the number of poles in the right half of the s-plane, Z is the number of zeros in the right half of the s-plane, and N is the number of encirclements (revolutions) of $1 + G(s)H(s)$ around the critical point. N may be positive, negative, or zero.

22. APPLICATION TO CONTROL SYSTEMS

A control system monitors a process and makes adjustments to maintain performance within certain acceptable limits. Feedback is implicitly a part of all control systems.[6] The *controller* (*control element*) is the part of the control system that establishes the acceptable limits of performance, usually by setting its own reference inputs. The controller transfer function for a proportional controller is a constant: $G_1(s) = K$.[7] The *plant* (*controlled system*) is the part of the system that responds to the controller. Both of these are in the forward loop. The input signal, V_i, in Fig. 63.4, is known in a control system as the *command* or *reference value*.

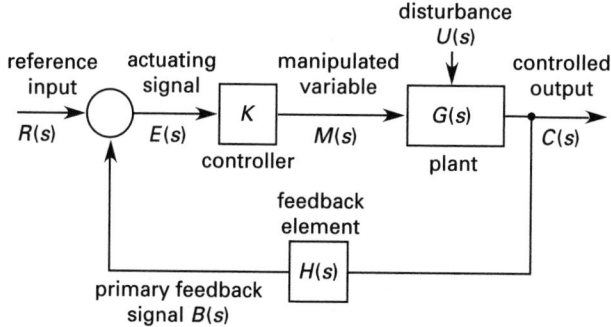

Figure 63.10 *Typical Feedback Control System*

A *servomechanism* is a special type of control system in which the controlled variable is mechanical position, velocity, or acceleration. In many servomechanisms, $H(s) = 1$ (i.e., unity feedback) and it is desired to keep the output equal to the reference input (i.e., maintain a zero error function). If the input, $R(s)$, is constant, the descriptive terms *regulator* and *regulating system* are used.

---

[6]Not all controlled systems are feedback systems. The positions of many precision devices (e.g., print heads in dot matrix printers and cutting heads on some numerically controlled machines) are controlled by precision *stepped motors*. However, unless the device has feedback (e.g., a position sensor), it will have no way of knowing if it gets out of control.

[7]Sometimes the notation K_n is used for K where n is the type of the system. In a type 0 system, a constant error signal results in a constant value of the output signal. In a type 1 system, a constant error signal results in a constant rate of change of the output signal. In a type 2 system, a constant error signal produces a constant second derivative of the output variable.

23. STATE MODEL REPRESENTATION

While the classical methods of designing and analyzing control systems are adequate for most situations, state model representations are preferred for more complex cases, particularly those with multiple inputs and outputs or when behavior is nonlinear or varies with time. This method of evaluation is almost always carried out on a digital or analog computer.

The state variables completely define the dynamic state, $x_i(t)$ (i.e., position, voltage, pressure, etc.), of the system at time t. (In simple problems, the number of state variables corresponds to the number of *degrees of freedom*, n, of the system.) The n state variables are written in matrix form as a state vector, $\mathbf{X}$.

$$\mathbf{X} = \begin{pmatrix} x_1 \\ x_2 \\ x_3 \\ \vdots \\ x_n \end{pmatrix} \qquad 63.44$$

It is a characteristic of state models that the state vector is acted upon by a first-degree derivative operator, d/dt, to produce a differential term of order 1, $\mathbf{X}'$.

$$\mathbf{X}' = \frac{d\mathbf{X}}{dt} \qquad 63.45$$

Equations 63.46 and 63.47 show the general form of a state model representation: $\mathbf{U}$ is an r-dimensional (i.e., an $r \times 1$ matrix) control vector; $\mathbf{Y}$ is an m-dimensional (i.e., an $m \times 1$ matrix) output vector; $\mathbf{A}$ is an $n \times n$ system matrix; $\mathbf{B}$ is an $n \times r$ control matrix; and $\mathbf{C}$ is an $m \times n$ output matrix. The actual unknowns are the x_i's. The y_i's, which may not be needed in all problems, are only linear combinations of the x_i's. (For example, the x's might represent spring end positions; the y's might represent stresses in the spring. Then, $y = k\Delta x$.) Equation 63.46 is the *state equation*, and Eq. 63.47 is the *response equation*.

$$\mathbf{X}' = \mathbf{A}\mathbf{X} + \mathbf{B}\mathbf{U} \qquad 63.46$$

$$\mathbf{Y} = \mathbf{C}\mathbf{X} \qquad 63.47$$

A conventional block diagram can be modified to show the multiplicity of signals in a state model, as shown in Fig. 63.11.[8] The actual physical system does not need to be a feedback system. The form of Eqs. 63.46 and 63.47 is the sole reason that a feedback diagram is appropriate.

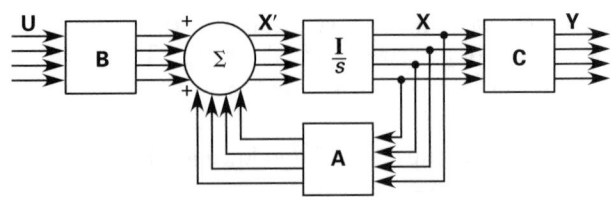

Figure 63.11 *State Variable Diagram*

A state variable model permits only first-degree derivatives, so additional x_i state variables are used for higher-order terms (e.g., acceleration).

System controllability exists if all of the system states can be controlled by the inputs, $\mathbf{U}$. In state model language, system controllability means that an arbitrary initial state can be steered to an arbitrary target state in a finite amount of time. *System observability* exists if the initial system states can be predicted by observing the outputs, $\mathbf{Y}$, when the inputs, $\mathbf{U}$, are known.[9]

Example 63.18

Write the state variable formulation of the following mechanical system's transfer function $T(s)$.

$$T(s) = \frac{7s^2 + 3s + 1}{s^3 + 4s^2 + 6s + 2}$$

Solution

Recognize the transfer function as the quotient of two terms, and multiply $T(s)$ by the dimensionless quantity x/x.

$$T(s) = \frac{Y(s)}{U(s)} = \left(\frac{7s^2 + 3s + 1}{s^3 + 4s^2 + 6s + 2} \right) \left(\frac{x}{x} \right)$$

$$Y(s) = 7s^2 x + 3sx + x$$

$$U(s) = s^3 x + 4s^2 x + 6sx + 2x$$

$Y(s)$ and $U(s)$ represent the following differential equations.

$$y(t) = 7x''(t) + 3x'(t) + x(t)$$

$$u(t) = x'''(t) + 4x''(t) + 6x'(t) + 2x(t)$$

Make the following substitutions. ($x_4(t)$ is not needed because one level of differentiation is built into the state model.)

$$x_1(t) = x(t)$$

$$x_2(t) = x'(t) = x_1'(t)$$

$$x_3(t) = x''(t) = x_2'(t)$$

---

[8]The block $\mathbf{I}/s$ is a diagonal identity matrix with elements of $1/s$. This effectively is an integration operator.

[9]*Kalman's theorem*, based on matrix rank, is used to determine system controllability and observability.

Write the first derivative variables in terms of the $x_i(t)$ to get the **A** matrix entries.

$$x_1'(t) = \quad 0x_1(t) + 1x_2(t) + 0x_3(t) + 0$$

$$x_2'(t) = \quad 0x_1(t) + 0x_2(t) + 1x_3(t) + 0$$

$$x_3'(t) = -2x_1(t) - 6x_2(t) - 4x_3(t) + u(t)$$

Determine the coefficients of the **C** matrix by rewriting $y(t)$ in the same variable order.

$$y(t) = 1x_1(t) + 3x_2(t) + 7x_3(t)$$

From Eqs. 63.46 and 63.47, the state variable representation of $T(s)$ is

$$\begin{bmatrix} x_1' \\ x_2' \\ x_3' \end{bmatrix} = \begin{bmatrix} 0 & 1 & 0 \\ 0 & 0 & 1 \\ -2 & -6 & -4 \end{bmatrix} \begin{bmatrix} x_1 \\ x_2 \\ x_3 \end{bmatrix} + \begin{bmatrix} 0 \\ 0 \\ 1 \end{bmatrix} [u(t)]$$

$$[y(t)] = [1 \ 3 \ 7] \begin{bmatrix} x_1 \\ x_2 \\ x_3 \end{bmatrix}$$

Topic XV: Electrical Materials

Electrical
Materials

For the most current information about the exam, visit **www.ppi2pass.com** regularly.

64 Electrical Materials

1. OVERVIEW

The electrical properties of materials impact all engineering designs. An understanding of the properties and potential reactions of chosen materials within the component parts and the surrounding environment is necessary both to ensure proper operation over the service lifetime of the components and to minimize long-term effects on the environment.

The fundamental properties of conducting, insulating and dielectric, semiconductor, magnetic, electron-emitting, and radiation-emitting materials are covered throughout this book. Materials handbooks are available that provide specific values for these properties for elements and compounds under various conditions. A periodic table with a summary of materials' properties is given in App. 64.A.

2. TYPES OF MATERIALS

Materials may be grouped into three main types: metals, polymers (that is, plastics), and ceramics.[1] *Metals* are materials characterized by high thermal and electrical conductivity. They are opaque and can generally be polished to a high luster. Further, they are normally heavy and deformable. Metals are located to the left of the metal-to-nonmetal transition staircase shown in App. 64.A. Elements adjacent to the staircase exhibit characteristics of both metals and nonmetals and thus are sometimes called *metalloids* or *semiconductors*. Further, in some texts, they are called metals and in others they are called semiconductors. Items to the right of the staircase are called *nonmetals*.

---

[1]Glass is sometimes listed as a type of material. "Glass," however, is a term applied to noncrystalline materials with specific expansion/contraction characteristics. For example, below a temperature known as the *glass temperature*, further rearrangement of atoms in glass ceases and contraction results from smaller thermal vibrations—rather than the formation of a metallic crystalline structure. Glasses have short-range order with an absence of long-range order. Glass is probably best described as a ceramic.

Polymers, or *plastics*, are materials characterized by low thermal and electrical conductivity. They are generally transparent or translucent and poor reflectors of light. Finally, they are normally flexible and subject to deformation. Polymers are composed of those elements listed in Table 64.1.

Table 64.1 Elements in Polymers

element atomic number	element name
1	hydrogen
6	carbon
7	nitrogen
8	oxygen
9	fluorine
14	silicon

Ceramics are compounds that contain metallic and nonmetallic elements. Concrete, rocks, glass, and nuclear fuel elements (UO_2) are all ceramics, to name but a few. Ceramics have extremely low thermal and electrical conductivity and for this reason are also called *refractory materials*. Indeed, they are highly chemically resistant (inert), can be opaque or translucent, and are very hard and brittle. Ceramics are composed of metallic elements combined with carbon, nitrogen, oxygen, silicon, or phosphorous, or some combination thereof.

3. CONDUCTING MATERIALS

A *conductor* is a wire, cable, or other body or medium that is suitable for carrying an electric current. While no specific dividing lines exist between conductors, semiconductors, and insulators, conductors have a conductivity on the order of $10^6 \ \Omega^{-1}\text{m}^{-1}$.

Copper is used extensively in electrical applications because of its high electrical and thermal conductivity and malleability. Aluminum is used when additional conductor strength is required. Silver has the highest conductivity at room temperature. Gold has conductivity similar to aluminum, as well as excellent corrosion and oxidation resistance. Platinum, a precious metal, is used primarily in resistance thermometers and thermocouples. Palladium is similar to platinum and less expensive. Nickel or nickel-coated conductors are used at temperatures up to 300°C with no enhanced corrosion at defective areas. Tungsten has a very high melting point and is erosion resistant, finding use in voltage regulators, repetitive low-current applications, and heating wires or filaments for lamps. Carbon has two crystalline forms: graphite and diamond. Carbon (that is, amorphous carbon) or graphite (that is, crystalline carbon) is used as an electric sliding contact in motor brushes. The diamond structure is used where high heat conduction is desired.

4. INSULATING AND DIELECTRIC MATERIALS

An *insulator* is a device or material having a high electrical resistance (i.e., a low conductivity). A *dielectric* is an electrically insulating material or a material in which an electric field can be maintained with minimum dissipation of power. Insulators generally have a conductivity on the order of 10^{-12} $\Omega^{-1}\text{m}^{-1}$.

Plastics are widely used due to their low cost, ability to be molded into many shapes, and good mechanical properties. Ceramics, due to their very low electrical conductivity, also find many uses in electrical insulating applications. Oils, gases, and papers are inexpensive insulators. Mica, with its stability, arc resistance, and high dielectric strength, is another common insulator.

5. MAGNETIC MATERIALS

A *magnetic material* is generally defined as a material exhibiting ferromagnetism. Magnetic materials are classified into two broad categories: *Nonretentive* or *soft* materials are low-loss materials. *Retentive* or *hard* materials are high-loss, high-rententivity, high-energy product materials.

The soft magnetic materials are used in transformers, motors and generators, cores, and other electromagnetic apparatuses. Iron is the most common soft magnetic material. Iron-silicon, also called *electrical steel*, is a soft material with very low core and eddy current losses and very high permeability. Iron-cobalt alloys are used where high saturation is required. Other alloy combinations are used in special applications.

The hard magnetic materials are used as permanent magnets. Hard materials are composed of a fine structure created during various heat treatments. High-carbon steels with tungsten, chromium, cobalt, aluminum, or vanadium in a martensitic structure are common. Precipitation-hardened alloys also form a fine structure. In this category are *alnico* (aluminum, nickel, and cobalt), *cunife* (copper, nickel, and iron), and *vicalloy* (vanadium, cobalt, and iron), among others.[2]

6. RADIATION-EMITTING MATERIALS

Materials at excited energy levels emit electrons, photons of light, or other radiation. These emissions have numerous electrical applications.

Metal filaments in incandescent lamps emit light when heated. Tungsten is the most common material used in this application. *Discharge lamps*, also called *fluorescent lamps*, use material in vapor form.[3] The noble gases, sodium, and mercury are the materials of choice. *Phosphors* are organic or inorganic luminescent materials.

---

[2]Ordered alloys and fine-particle magnets are also retentive.
[3]When materials emit light almost instantaneously after excitation, they are called *fluorescent* materials. When the emission is delayed, they are called *phosphorescent* materials.

Topic XVI: Codes and Standards

Chapter

    65. National Electrical Code

Codes and
Standards

65 National Electrical Code

Nomenclature

A	area	m^2
d	diameter	m
F	factor	–
I	effective or DC current	A
l	length	m
P	power	W
pf	power factor	–
R	resistance	Ω
V	effective or DC voltage	V
X	reactance	Ω
Z	impedance	Ω

Symbols

ϕ	phase	–

Subscripts

c	corrected
cmil	circular mils
in	inches
l	per unit length
L	inductor

1. HISTORY AND OVERVIEW

The first practical incandescent light bulb and lighting system were developed by Thomas Alva Edison in 1879. That same year, the National Association of Fire Engineers met to discuss standards for electrical installations. By 1895 there were five separate codes in the United States. In 1896, a national meeting was held between various concerned organizations and by 1897 the *National Electrical Code*® (NEC®) (hereafter referred to as the "Code") was adopted. The Code gradually came under the sponsorship and control of the National Fire Protection Association (NFPA). The NEC is officially endorsed by the American National Standards Institute (ANSI), and the NFPA National Electrical Code committee responsible for the code is known as ANSI Standards Committee C1. The National Electrical Code is the most widely adopted code in the United States and the most widely accepted set of electrical safety requirements in the world. Local codes in the United States normally adopt the NEC en masse and then add supplemental requirements deemed necessary, or remove and change items to ensure local applicability or uniqueness.

The Code generally applies to the installation of electrical conductors and equipment within or on public and private buildings up to and including the connection to the electrical supply. Its general purpose is to prevent fires and explosions by providing guidance for proper electrical installations.

The NEC consists of an introduction followed by nine chapters, which are further subdivided into articles. The outline of the Code follows. Chapter numbers are given in parentheses. Chapters 1 through 4 generally apply to all situations except as modified by chapters 5 through 7.

- Introduction
- (1) General
- (2) Wiring and Protection
- (3) Wiring Methods and Materials
- (4) Equipment for General Use
- (5) Special Occupancies
- (6) Special Equipment
- (7) Special Conditions
- (8) Communications Systems
- (9) Tables
- Annex A: Product Safety Standards
- Annex B: Information for Ampacity Calculation
- Annex C: Conduit and Tubing Fill Tables
- Annex D: Examples
- Annex E: Types of Construction
- Annex F: Cross-Reference Tables

Codes and Standards

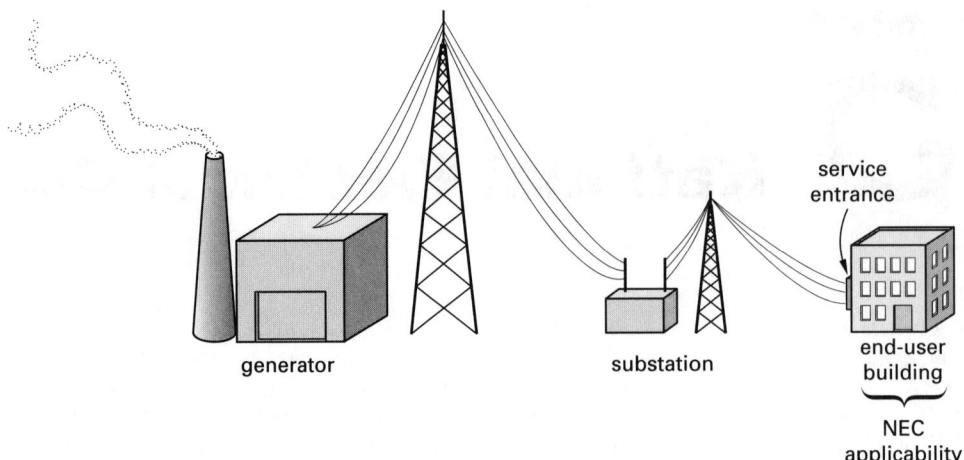

Figure 65.1 NEC Coverage

In this chapter, numbers in brackets and section numbers greater than 90 indicate the article or section number in the NEC and are meant for ease in referencing the Code directly. Concepts covered and requirements mentioned will focus on dwelling units and voltages less than 600 V and will not cover Chs. 5 through 8 of the Code. A mixture of English engineering units and SI units will be used. The flow of the information conforms to the National Electrical Code, 2002.

2. INTRODUCTION

The Introduction to the code contains Article 90 and explains that the purpose of the National Electrical Code is the "practical safeguarding of persons and property from hazards arising from the use of electricity." The intended coverage or applicability of the Code is illustrated in Fig. 65.1.

The Code is meant to be a legal document that can be implemented and interpreted by local governmental bodies that have jurisdiction over electrical installations. Most do just that, with some additions and changes for local needs. A *permissive rule* in the Code is one that is allowed but not required (i.e., an alternative). The term "shall be permitted" is used to indicate these alternatives. Any *fine print note* (FPN) used in the Code is for informational purposes only and does not represent a requirement. Factory-installed internal wiring is not required to be inspected if the equipment has been listed by a qualified electrical testing laboratory that requires installation in accordance with the Code. When metric measurements are used, they conform to the SI System. Most of the values in the Code remain in the English Engineering System, with approximate metric equivalents given. Footnotes are also used to provide the necessary SI conversion factors.

3. GENERAL

The General portion of the Code contains Articles 100 (definitions) and 110 (requirements for electrical installations).

The definitions are those required for the proper application of the Code and are not all-encompassing. Generally understood terms that can be found in an American language dictionary are not included. Also not included are electrical terms that are not unique to the Code, such as "volt." Some of the terms that are used in a distribution system and found in the definitions article are shown in Fig. 65.2.

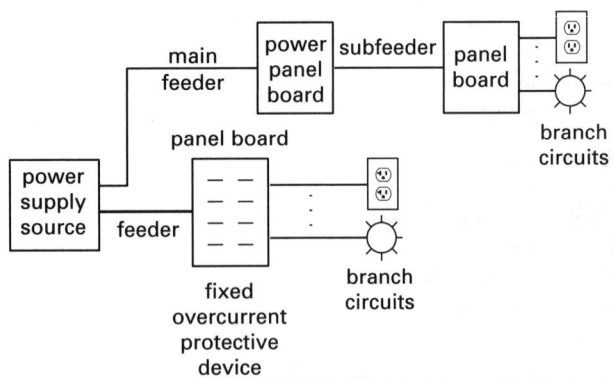

Figure 65.2 Typical Distribution System

In this section, *overcurrent* is defined as any current in excess of rated current. Overcurrent includes overload, short circuit, and ground fault. An *overload* is defined as the operation of equipment in excess of its normal full-load rating or of a conductor in excess of its rated ampacity, which, after a sufficient time, causes damage or overheating.

Article 110, "Requirements of Electrical Installation," notes that conductor size will be given in *American Wire Gage* (AWG) or in circular mils. A comparison of American Wire Gage diameters to other systems is given in App. 65.A. Conversions between diameters and cross-sectional areas can be accomplished with Eqs. 65.1 and 65.2.

$$A_{\text{cmil}} = \left(\frac{d_{\text{in}}}{0.001}\right)^2 \qquad \textbf{65.1}$$

$$A_{\text{cmil}} = 1.273 \times 10^6 A_{\text{in}^2} \qquad \textbf{65.2}$$

Wire gage size up to 4/0 is listed as AWG in the Code. Beyond that, the size is listed in circular mils (cmil) or, more likely, thousands of circular mils (kcmil). The metric standard for marking wire size is square millimeters (mm²). The conversion is

$$A_{\text{cmil}} = 1973.53 A_{\text{mm}^2} \qquad \textbf{65.3}$$

Temperature limitations are imposed and ratings coordinated so that conductors and their terminations are rated for the same operating temperature for circuits rated at 100 A or less [Sec. 110.14(C)(1)]. Required working space within and around electrical installations, safety issues, and illumination of such spaces fill out the remaining items in Article 110.

Example 65.1

A wire of unknown gage is measured. The diameter is found to be 0.2057 cm. What is the American Wire Gage (AWG) designation?

Solution

The tables for wire gage comparisons are generally in English Engineering System units, as is App. 65.A. Converting the diameter gives

$$d = (0.2057 \text{ cm})\left(\frac{1 \text{ in}}{2.54 \text{ cm}}\right) = 0.0810 \text{ in}$$

From App. 65.A, this diameter correlates with AWG 12. The generic term used is twelve gage wire.

4. WIRING AND PROTECTION

The "Wiring and Protection" portion of the Code contains Articles 200 (grounded conductors), 210 (branch circuits), 215 (feeders), 220 (branch circuit, feeder, and service calculations), 225 (outside branch circuits and feeders), 230 (services), 240 (overcurrent protection), 250 (grounding), and 280 (surge arresters).

The term "nominal" is used throughout the NEC. Actual power supply voltage specifications, high and low, are set by ANSI standards. Nominal voltage is

the desired voltage of the power source and is given in multiples of 120 V. Rated voltage is the desired voltage for operating equipment and is given in multiples of 115 V. Operating equipment is called *utilization equipment* and includes any device that utilizes electric energy for electronic, electromechanical, chemical, heating, lighting, or another similar purpose [Article 100]. Four-wire wye-connected systems normally list the type of connection next to the voltage for phase-to-phase voltage and give the phase-to-neutral voltage after a slash (i.e., 208Y/120 or 480Y/277). Examples of common wye and delta four-wire systems are illustrated in Fig. 65.3.

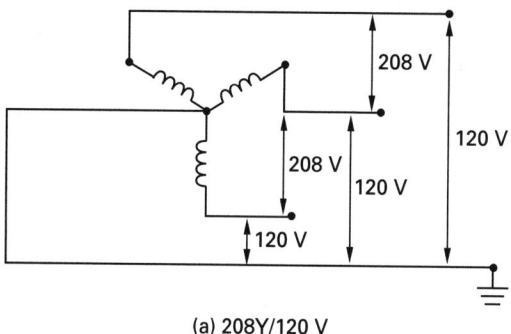

(a) 208Y/120 V

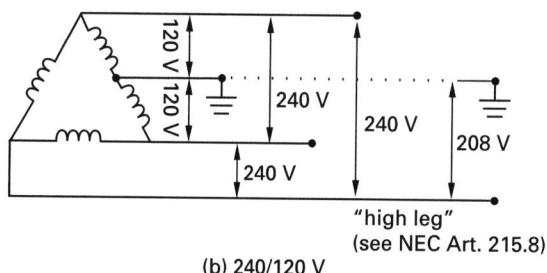

(b) 240/120 V

Figure 65.3 *Four-Wire System Nomenclature*

5. WIRING AND PROTECTION: GROUNDED CONDUCTORS

A *grounded conductor*, which is a conductor that is intentionally grounded (for example, the neutral of a system), is identified by a white or natural gray outer finish or by three continuous white stripes on other than green insulation [Sec. 200.6]. This should not be confused with a *grounding conductor*, which is used to connect equipment or the grounded circuit (neutral) of a wiring system to a grounding electrode and which is colored green [Articles 100 and 250.119]. Terminals for connecting grounded conductors must be substantially white, or have the word "white" or the letter "W" adjacent to the terminal.

6. WIRING AND PROTECTION: BRANCH CIRCUITS

A *branch circuit* consists of the circuit conductors between the final overcurrent device protecting the circuit and the outlet. Such circuits are covered in Article 210. However, if the branch circuit supplies only motor loads, it is covered in Article 430. A cross-reference for articles relating to branch circuits for other special conditions is given in Table 210.2.

A *ground-fault circuit interrupter* (GFCI) is a device that de-energizes a circuit within an established period of time when a current to ground exceeds some predetermined value, which is less than that required to operate the overcurrent protective device of the supply circuit [Article 100]. The concept of such a device is illustrated in Fig. 65.4.

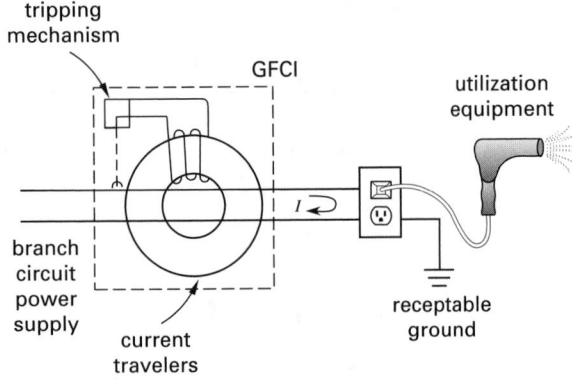

Figure 65.4 *Ground-Fault Circuit Interrupter (GFCI)*

As long as the current is balanced in the lines running through the transformer core, no induced voltage is felt, no current flows in the tripping mechanism, and the GFCI contact remains shut. When the current is unbalanced due to a ground fault, the unbalanced magnetic fields produced induce a net voltage in the transformer, current flows in the tripping mechanism, and the GFCI contact opens, protecting any person holding the offending equipment. GFCIs are required to be installed outdoors and in dwelling unit bathrooms, garages, crawl spaces, unfinished basements, kitchens, and wet bar sinks [Sec. 210.8(A)].

The number of branch circuits required in a given building is calculated per Sec. 210.11. A dwelling unit is also required to have two or more 20 A small-appliance branch circuits beyond that calculated, one 20 A branch circuit for laundry, and one 20 A branch circuit for each bathroom [Sec. 210.11(C)]. The conductors that supply branch circuits, before any adjustment is made or derating is allowed, must be able to carry 100% of the noncontinuous load plus 125% of the continuous load [Sec. 210.19(A)(1)]. The *overcurrent protective device* (OCPD) selected must meet identical requirements

[Sec. 210.20(A)]. The allowable conductor ampacities are found in App. 65.B [NEC Table 310.16].[1]

A single receptacle on a branch circuit must have an ampere rating of not less than the branch circuit. If two or more receptacles are used, a table gives their allowed values [Sec. 210.21(B), Table 210.21(B)(3)]. The rating of a branch circuit is based on the overcurrent protective device (OCPD) rating regardless of the actual load or the conductor ampacity. A 15 A or 20 A branch circuit can supply lighting, utilization equipment, or a combination of both. Any single cord-and-plug-connected device cannot exceed 80% of the branch circuit ampere rating. Further, the rating of utilization equipment fastened in place, other than lighting, cannot exceed 50% of the branch circuit ampere rating [Sec. 210.23]. A summarized table of branch circuit requirements is given in the Code in Table 210.24.

Receptacles are spaced in dwellings so that no point along the floor of any wall space is more than 6 ft from a receptacle. (Thus, a distance of 12 ft can exist between receptacles.) Floor receptacles, those higher than 1.68 m (5.5 ft) above the floor, those part of a lighting system or appliance, and those located within cupboards or cabinets, do not count toward the requirement [Sec. 210.52]. On countertops, the receptacle spacing is no greater than 0.61 m (2 ft). Receptacles attached and controlled by switches are lighting outlets, the requirements of which are separate from general receptacles [Sec. 210.70].

Example 65.2

A given grocery store branch circuit is meant to carry a continuous lighting load of 8 A. In addition, three intermittent loads designed for display stands are fastened in place and require 2 A each when operating. What is the rating of the overcurrent protective device (OCPD) on the branch circuit?

Solution

By Sec. 210.20(A), the overcurrent device must be rated at 100% of the noncontinuous load and 125% of the continuous load. Thus,

$$\text{OCPD} = (1.00)(2 \text{ A} + 2 \text{ A} + 2 \text{ A}) + (1.25)(8 \text{ A})$$

$$= 16 \text{ A minimum}$$

The standard ratings for fuses and fixed-trip circuit breakers are given in Sec. 240.6 as 15 A, 20 A, 25 A, 30 A, and so on. Thus, a 20 A OCPD will suffice.

Note that utilization equipment "fastened in place" cannot, and does not in this case, exceed 50% of the branch circuit ampere rating [Sec. 210.23(A)(2)]. That is, 6 A is less than 50% of 20 A.

---

[1]Table 310.16 is arguably the most used table in the Code. Familiarity with it is essential for proper application of requirements.

7. WIRING AND PROTECTION: FEEDERS

A *feeder* consists of all circuit conductors between the service equipment, the source of a separately derived system, or another power supply source and the final branch circuit overcurrent device (see Fig. 65.2) [Article 100]. The capacity of the conductors that constitute feeders, before any adjustment or derating allowed, must be 100% of the noncontinuous load plus 125% of the continuous load [Sec. 215.2]. The *overcurrent protective device* (OCPD) selected must meet identical requirements [Sec. 215.3]. Several special conditions for sizing the feeders also apply, including the exception that allows dwelling unit feeders carrying the full current of the service conductors to be the same size as those conductors. Though not required, a feeder is recommended to have no more than a 3% voltage drop, with a 5% voltage drop on both feeders and branch circuit conductors to the farthest outlet [Sec. 215.2, FPN No. 2]. Useful Code tables for calculating voltage drops based on conductor properties when used in DC or AC circuits are given in App. 65.C, Conductor Properties, and App. 65.D, Conductor AC Properties [NEC Ch. 9, Tables 8 and 9, respectively].

For two-wire circuits, DC or AC, or three-wire single-phase AC circuits, with a power factor equal to one (purely resistive), the voltage drop is given by

$$V_{\text{drop}} = IR_l 2l = \frac{2lIR}{1000 \text{ ft}} \qquad \textbf{65.4}$$

The number 2 in Eq. 65.4 is used because the voltage drop is felt over the entire circuit length. The farthest equation to the right is used only if the resistance per unit length, R_l, is given by the referenced table in "per 1000 ft" quantities.

Example 65.3

A 480Y/277, 80 kW noncontinuous load with a 0.8 pf consists of less than 50% ballast-type lighting.[2] The feeder length is 110 m. Applying the 3% fine print note No. 2 recommendation from Sec. 215.2(A)(4), what is the maximum recommended voltage drop per 1000 ampere-feet?

Solution

First, it should be recognized that the 1000 ampere-feet unit is asked for since many manufacturers in the United States list conductor information as such. Also, Tables 8 and 9 of Ch. 9 of the National Electrical Code, which are used for calculating voltage drops, are given in "per 1000 ft" increments (see Apps. 65.C and 65.D).

To determine the line current drawn on a noncontinuous basis, and noting that the voltage given is that of a three-phase four-wire system, use[3]

$$P = \sqrt{3}IV\text{pf}$$

$$I = \frac{P}{\sqrt{3}V\text{pf}} = \frac{80 \times 10^3 \text{ W}}{(\sqrt{3})(480 \text{ V})(0.8)} = 120.3 \text{ A}$$

The total number of ampere-feet is given by

$$\text{ampere-feet} = (120.3 \text{ A})(110 \text{ m})\left(\frac{3.281 \text{ ft}}{1 \text{ m}}\right)$$

$$= 43{,}417 \text{ A-ft} = 43.4 \times 1000 \text{ A-ft}$$

The reason for specifying "noncontinuous" is to emphasize that the ampacity is to be calculated at 100%, not 125%. It is important to note that regardless of the load, continuous or noncontinuous, the voltage drop is calculated using the 100% ampacity. Nevertheless, the conductor must be sized to handle 125% ampacity according to Code standards. The OCPD must be set at 125% ampacity as well.

Since this system has a neutral, the voltage drop of concern is the phase-to-neutral drop. Thus,

$$V_{\text{drop}} = (\text{allowed drop})(\text{line-to-neutral voltage})$$

$$= (0.03)(277 \text{ V}) = 8.31 \text{ V}$$

The voltage drop per 1000 A-ft is then given by

$$V_{\text{drop}} = \frac{8.31 \text{ V}}{(43.4)(1000 \text{ A-ft})}$$

$$= 0.192/1000 \text{ A-ft}$$

This value could then be used with a manufacturer's table of conductors to determine which conductor size meets voltage drop recommendations. Voltage drop specifications on conductor size are in addition to other requirements and must be coordinated to determine the most limiting case. If the conductor size is increased to lower the voltage drop, the overcurrent protection also increases.

It is important to note that tables for conductor properties and voltage drops are often given in terms of line-to-neutral values while the recommendation for percentage voltage drop in the Code refers to the line voltage. Calculations should normally be accomplished using line-to-neutral voltages, that is, phase voltages, and then conversions applied to obtain line quantities and the percent voltage drop. By contrast, the line current is used since this is the current causing the voltage drop in the conductor cable, be it a branch circuit or feeder conductor.

---

[2]If the ballast-type lighting constitutes greater than 50% of the load, the neutral conductor would be considered a current-carrying conductor and would have to be sized according to the calculations in the example problem [Sec. 310.15(B)(4)(b); cf. Secs. 310.4 Exception No. 4 and 310.10 FPN No. (2)].

[3]Only four-wire systems have a voltage specified after the slash, for example, 208Y/120, 240/120, 480Y/277, and 4160Y/2400.

8. WIRING AND PROTECTION: BRANCH CIRCUIT, FEEDER, AND SERVICE CALCULATIONS

Article 220 contains the requirements for calculations involving branch circuits, feeders, and service loads. Examples that are representative of the requirements are given in App. D of the Code.

The voltages used in calculations are standardized as 120, 120/240, 208Y/120, 240, 347, 480Y/277, 480, 600Y/347, and 600 V [Sec. 220.2(A)]. Loads are determined in terms of volt-amperes (VA) or kilovolt-amperes (kVA). This is considered equivalent to equipment power ratings, which are given in watts (W) or kilowatts (kW). This concept of equivalence is used since the power factor is unknown and the conductors will carry the volt-ampere load, that is, the apparent power, and so should be rated for such. Fractions of less than 0.5 A are dropped.

Branch circuit lighting loads are determined by the square footage of a building, with the unit loads given in Sec. 220.3(A). Dwelling units' lighting loads are calculated based on 3 VA per square foot. Additional receptacle loads depend on the equipment attached [Sec. 220.3(B)]. Receptacle outlets are calculated based on 180 VA per outlet minimum, with a minimum of 90 VA per receptacle for outlets with four or more receptacles [Sec. 220.3(B)(9)].

Lighting loads that have ballasts (i.e., fluorescent lamps), transformers, or autotransformers are based on the ampere rating of the units and not the watt ratings [Sec. 220.4(B)]. These types of devices are termed *nonlinear loads* and affect the sizing of the neutral conductor [Sec. 310.15(B)(4)(b); cf. Articles 310.4 Exception No. 4 and 310.10 FPN No. (2)].

Requirements for various types of loads, including general lighting, motors, and small appliances, are given in Sec. 220. Demand factors that allow less than the full-load wattage, that is, volt-amperes, to be used in determining the size of conductors and overcurrent devices are also given. Optional calculations for those dwelling units with a total connected load greater than 100 A are described in Sec. 220.30.

Example 65.4

For the circuit shown, what is the total VA load?

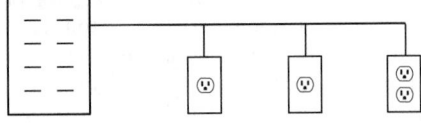

Solution

Regardless of the number of receptacles on an outlet, the minimum load is 180 VA [Sec. 220.3(B)(9)]. Thus, the total load is given by

$$VA_{total} = (\text{number of outlets}) \left(180 \; \frac{VA}{\text{outlet}}\right)$$

$$= (3) \left(180 \; \frac{VA}{\text{outlet}}\right) = 540 \; VA$$

9. WIRING AND PROTECTION: OVERCURRENT PROTECTION

Article 240 of the Code describes the overcurrent protection requirements and the electrical devices that accomplish them. One of the first sections of Part I lists specific types of equipment and the applicable article containing the overcurrent requirements. Thus, Table 240.3 is a useful starting point in any determination of overcurrent requirements.

Conductors are protected in accordance with their ampacities [Sec. 240.4]. Standard ampere ratings for fuses and circuit breakers are given in Sec. 240.6 as 15, 20, 25, 30, 35, 40, 45, 50, and so on. Thermal devices not designed to interrupt short circuits cannot be used for overcurrent protection for short circuits or grounds [Sec. 240.9]. The remainder of the article discusses location of overcurrent devices, tap requirements, fuses, and circuit breakers.

10. WIRING AND PROTECTION: GROUNDING

Article 250 of the Code describes general grounding in Part I, circuit and system grounding in Part II, grounding electrode requirements in Part III, and equipment grounding in Part VI.

Electrical systems are connected to the earth in a manner that will limit voltage from lightning strikes, line surges, or other transients [Sec. 250.4(A)(2)]. Electrical systems are also grounded in order to maintain the voltage at a stable level relative to ground to ensure connected equipment is subjected to a maximum potential difference set by the designer—for example, three-phase four-wire systems (see Sec. 250.26). Equipment is grounded as well, the primary goal being the safety of people in contact with the equipment. The overall requirements of the fault current path and their organization in the Code are given in Sec. 250.4(B)(4), while specific equipment requirements are cross-referenced in Sec. 250.3, which is a good starting point for finding grounding information.

The grounding electrode requirements are described in Sec. 250.50. Electrically continuous water pipes in contact with the earth for 10 ft (3.05 m) or more are used as grounding electrodes [Sec. 250.52(A)(1)]. Grounding electrodes 20 ft (6.1 m) long are allowed to be encased in concrete with a minimum thickness of 2 in (50.8 mm), if within the concrete and near the foundation, which must be in direct contact with the earth [Sec. 250.52(A)(3)]. Rod- and pipe-type electrodes must be 8 ft (2.44 m) long [Sec. 250.52(A)(5)]. If multiple rods are used, they must be a minimum of 6 ft (1.83 m) apart [Sec. 250.56]. The sizing of grounding electrode conductors for AC systems is given in NEC Table 250.66.

Cord-and-plug-connected equipment is generally grounded unless double-insulated [Sec. 250.114]. Major appliance equipment within dwellings must be grounded [Sec. 250.114(3)]. Insulated equipment grounding conductors are green or green with one or more yellow stripes [Sec. 250.119]. Grounding terminals use green screws or nuts, the word "green" or "ground", the symbols "G" or "GR", or the grounding symbol itself [Sec. 250.126]. The sizing of equipment grounding conductors is given in Table 65.1 [NEC Table 250.122].

Table 65.1 *Equipment Grounding*
(NEC Table 250.122)

rating or setting of automatic overcurrent device in circuit ahead of equipment, conduit, etc., not exceeding (amperes)	size (AWG or kcmil)	
	copper	aluminum or copper-clad aluminum[a]
15	14	12
20	12	10
30	10	8
40	10	8
60	10	8
100	8	6
200	6	4
300	4	2
400	3	1
500	2	1/0
600	1	2/0
800	1/0	3/0
1000	2/0	4/0
1200	3/0	250
1600	4/0	350
2000	250	400
2500	350	600
3000	400	600
4000	500	800
5000	700	1200
6000	800	1200

Note: Where necessary to comply with Sec. 250.4(B)(4), the equipment grounding conductor shall be sized larger than this table.

[a]See installation restrictions in Sec. 250.120.

11. WIRING METHODS AND MATERIALS

The "Wiring Methods and Materials" portion of the Code contains Articles 300 (wiring methods) and 310 (conductors), several articles on conduit, tubing, and raceways, and many others. The requirements regarding switchboards and panelboards are also in this portion of the Code.

The wiring methods covered in the Code are for all wiring installations, but do not include internal equipment wiring [Sec. 300.1(B)]. All conductors in the same circuit should be contained together [Sec. 300.3(B)]. In general, conductors for systems over 600 V are not mixed with those for systems under 600 V [Sec. 300.3(C)(2)]. Directly buried conduit must be at least 18 in (45.7 cm) from the surface [Sec. 300.5(D)(1)].

Conductor application and insulation codes, such as THHN for heat-resistant thermoplastic, are given in NEC Table 310.13. The appropriate letter designations are explained, the type of insulation is specified, and the design applications are listed.

Conductor ampacities are determined using App. 65.B (NEC Table 310.16) and numerous other NEC tables given in Sec. 310.15. The loads used to determine the ampacities are calculated in accordance with Article 220 of the Code. *Derating*, that is, lowering the allowed ampere rating, of a conductor can occur due to temperature limitations of the terminal connections [Sec. 110.14(C)], coordination with system overcurrent protection [Sec. 240.4], carrying more than three current-carrying conductors in a raceway or cable [Sec. 310.15(B)(2)(a) and NEC Table 310.15(B)(2)(a)], the ambient operating environment [NEC Table Correction Factors at the bottom of each table], and a variety of other conditions [NEC App. B]. Under engineering supervision, conductor ampacities may be calculated using the formula given in Section 310.15(C) for which sample calculations are given in NEC App. B.

Example 65.5

A single AWG 10 TW copper conductor is considered for use in an environment expected to have an ambient temperature of 100°F. What is the maximum ampacity?

Solution

A TW cable sized at AWG 10, located in the 60°C column of App. 65.B, has an ampacity of 30 A before derating. The correction factor given at the bottom of the table for 100°F (38°C) is 0.82. The allowable ampacity is

$$A_{\text{total}} = A_{\text{rated}} F_{\text{correction}}$$
$$= (30\text{ A})(0.82) = 24.6\text{ A}$$

12. EQUIPMENT FOR GENERAL USE

The "Equipment for General Use" portion of the Code contains Articles 400 (flexible cords and cables), 402 (fixture wires), 430 (motors, motor circuits, and controllers), and 480 (storage batteries), among others.

The ampacities, insulation information, markings, and other pertinent data regarding flexible cords and cables is contained in NEC Tables 400.4 and 400.5. Flexible cords and cables are limited in use [Sec. 400.7] and can be spliced only under special conditions [Sec. 400.9]. The cords and cables must be protected by bushings or fittings when passing through holes [Sec. 400.14].

The properties of fixture wires are given in NEC Table 402.3 with the allowable ampacity given in Table 402.5. Specific requirements for lighting fixtures and associated wires are described in Article 410.

Article 430 is the definitive starting point for motor requirements, motor branch circuit and feeder conductor ampacity and overcurrent protection information, motor overload protection, and motor-related topics. Installation requirements are illustrated and cross-referenced in Sec. 430.1, FPN No. 1, Fig. 430.1, which should be the starting point for researching motor requirements.

The size of the conductors supplying motors is determined using the tables of Sec. 310.15(B) unless the conductor is flexible, in which case the requirements are in Sec. 400.5. The current value used to enter the tables in Article 310 is determined from NEC Tables 430.147 to 430.150, not from the motor nameplate data [Sec. 430.6(A)(1)]. However, the overload protection is determined from the motor nameplate current rating [Sec. 430.6(A)(2)]. The conductors supplying single motors in continuous duty are sized to 125% of the motor's full-load current rating (as determined by the NEC tables listed in Sec. 430.6) [Sec. 430.22(A)]. Motor overload protection is generally set at 115% of the motor nameplate full-load current rating [Sec. 430.32(A) and (B)]. Motors are classed as greater than or less than 1 hp. Motors of less than 1 hp are sometimes called *fractional horsepower motors*.

Storage battery requirements are in Article 480. The voltage used in calculations involving lead-acid batteries is standardized at 2 V per cell. For alkali batteries, the standard voltage is 1.2 V per cell.

Topic XVII: Professional

Chapter

For the most current information about the exam, visit **www.ppi2pass.com** regularly.

66 Engineering Economic Analysis

Nomenclature

A	annual amount	$
B	present worth of all benefits	$
BV_j	book value at end of the jth year	$
C	cost or present worth of all costs	$
d	declining balance depreciation rate	decimal
D	demand	various
D	depreciation	$
DR	present worth of after-tax depreciation recovery	$
e	constant inflation rate	decimal
E_0	initial amount of an exponentially growing cash flow	$
EAA	equivalent annual amount	$
EUAC	equivalent uniform annual cost	$
f	federal income tax rate	decimal
F	forecasted quantity	various
F	future worth	$
g	exponential growth rate	decimal
G	uniform gradient amount	$

i	effective rate per period (usually per year)	decimal per unit time
i'	effective interest rate corrected for inflation	decimal
k	number of compounding periods per year	–
m	an integer	–
n	number of compounding periods or life of asset	–
P	present worth	$
r	nominal rate per year (rate per annum)	decimal per unit time
ROI	return on investment	$
ROR	rate of return	decimal per unit time
s	state income tax rate	decimal
S_n	expected salvage value in year n	$
t	composite tax rate	decimal
t	time	years (typical)
T	a quantity equal to $\frac{1}{2}n(n+1)$	–
TC	tax credit	$
z	a quantity equal to $\dfrac{1+i}{1-d}$	decimal

Symbols

α	smoothing coefficient for forecasts	–
ϕ	effective rate per period (r/k)	decimal
$\mathcal{E}$	expected value	various

Subscripts

0	initial
j	at time j
n	at time n
t	at time t

1. IRRELEVANT CHARACTERISTICS

In its simplest form, an *engineering economic analysis* is a study of the desirability of making an investment.[1] The decision-making principles in this chapter can be applied by individuals as well as by companies. The nature of the spending opportunity or industry is not important. Farming equipment, personal investments, and multimillion dollar factory improvements can all be evaluated using the same principles.

Similarly, the applicable principles are largely insensitive to the monetary units. Although *dollars* are used in this chapter, it is equally convenient to use British pounds, Japanese yen, or German marks.

---

[1]This subject is also known as *engineering economics* and *engineering economy*. There is very little, if any, true economics in this subject.

Finally, this chapter may give the impression that investment alternatives must be evaluated on a year-by-year basis. Actually, the *effective period* can be defined as a day, month, century, or any other convenient period of time.

2. MULTIPLICITY OF SOLUTION METHODS

Most economic conclusions can be reached in more than one manner. There are usually several different analyses that will eventually result in identical answers.[2] Other than the pursuit of elegant solutions in a timely manner, there is no reason to favor one procedural method over another.[3]

3. PRECISION AND SIGNIFICANT DIGITS

The full potential of electronic calculators will never be realized in engineering economic analyses. Considering that calculations are based on estimates of far-future cash flows and that unrealistic assumptions (no inflation, identical cost structures of replacement assets, etc.) are routinely made, it makes little sense to carry cents along in calculations.

The calculations in this chapter have been designed to illustrate and review the principles presented. Because of this, greater precision than is normally necessary in everyday problems may be used. Though used, such precision is not warranted.

Unless there is some compelling reason to strive for greater precision, the following rules are presented for use in reporting final answers to engineering economic analysis problems.

- Omit fractional parts of the dollar (i.e., cents).

- Report and record a number to a maximum of four significant digits unless the first digit of that number is one, in which case, a maximum of five significant digits should be written. For example,

$49	not	$49.43
$93,450	not	$93,453
$1,289,700	not	$1,289,673

4. NONQUANTIFIABLE FACTORS

An engineering economic analysis is a quantitative analysis. Some factors cannot be introduced as numbers into the calculations. Such factors are known as *nonquantitative factors, judgment factors,* and *irreducible factors*. Typical nonquantifiable factors are

---

[2]Because of round-off errors, particularly when factors are taken from tables, these different calculations will produce slightly different numerical results (e.g., $49.49 versus $49.50). However, this type of divergence is well known and accepted in engineering economic analysis.

[3]This does not imply that approximate methods, simplifications, and rules of thumb are acceptable.

- preferences
- political ramifications
- urgency
- goodwill
- prestige
- utility
- corporate strategy
- environmental effects
- health and safety rules
- reliability
- political risks

Since these factors are not included in the calculations, the policy is to disregard the issues entirely. Of course, the factors should be discussed in a final report. The factors are particularly useful in breaking ties between competing alternatives that are economically equivalent.

5. YEAR-END CONVENTION

Except in short-term transactions, it is simpler to assume that all receipts and disbursements (cash flows) take place at the end of the year in which they occur.[4] This is known as the *year-end convention*. The exceptions to the year-end convention are initial project cost (purchase cost), trade-in allowance, and other cash flows that are associated with the inception of the project at $t = 0$.

On the surface, such a convention appears grossly inappropriate since repair expenses, interest payments, corporate taxes, and so on seldom coincide with the end of a year. However, the convention greatly simplifies engineering economic analysis problems, and it is justifiable on the basis that the increased precision associated with a more rigorous analysis is not warranted (due to the numerous other simplifying assumptions and estimates initially made in the problem).

There are various established procedures, known as *rules* or *conventions*, imposed by the Internal Revenue Service on U.S. taxpayers. An example is the *half-year rule*, which permits only half of the first-year depreciation to be taken in the first year of an asset's life when certain methods of depreciation are used. These rules are subject to constantly changing legislation and are not covered in this book. The implementation of such rules is outside the scope of engineering practice and is best left to accounting professionals.

6. CASH FLOW DIAGRAMS

Although they are not always necessary in simple problems (and they are often unwieldy in very complex problems), *cash flow diagrams* can be drawn to help visualize and simplify problems having diverse receipts and disbursements.

---

[4] A *short-term transaction* typically has a lifetime of five years or less and has payments or compounding that are more frequent than once per year.

The following conventions are used to standardize cash flow diagrams.

- The horizontal (time) axis is marked off in equal increments, one per period, up to the duration (or *horizon*) of the project.

- Two or more transfers in the same period are placed end-to-end, and these may be combined.

- Expenses incurred before $t = 0$ are called *sunk costs*. Sunk costs are not relevant to the problem unless they have tax consequences in an after-tax analysis.

- *Receipts* are represented by arrows directed upward. *Disbursements* are represented by arrows directed downward. The arrow length is proportional to the magnitude of the cash flow.

Example 66.1

A mechanical device will cost $20,000 when purchased. Maintenance will cost $1000 each year. The device will generate revenues of $5000 each year for five years, after which the salvage value is expected to be $7000. Draw and simplify the cash flow diagram.

Solution

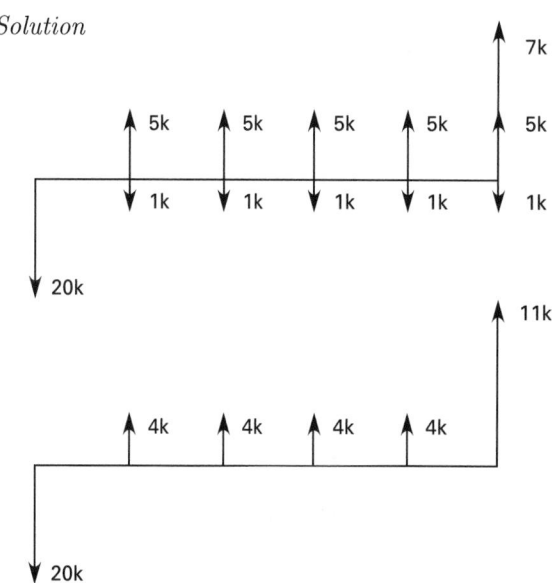

7. TYPES OF CASH FLOWS

To evaluate a real-world project, it is necessary to present the project's cash flows in terms of standard cash flows that can be handled by engineering economic analysis techniques. The standard cash flows are single payment cash flow, uniform series cash flow, gradient series cash flow, and the infrequently encountered exponential gradient series cash flow.

A *single payment cash flow* can occur at the beginning of the time line (designated as $t = 0$), at the end of the time line (designated as $t = n$), or at any time in between.

The *uniform series cash flow* consists of a series of equal transactions starting at $t = 1$ and ending at $t = n$. The symbol A is typically given to the magnitude of each individual cash flow.[5]

The *gradient series cash flow* starts with a cash flow (typically given the symbol G) at $t = 2$ and increases by G each year until $t = n$, at which time the final cash flow is $(n - 1)G$.

An *exponential gradient series cash flow* is based on a phantom value (typically given the symbol E_0) at $t = 0$ and grows or decays exponentially according to the following relationship.[6]

$$\text{amount at time } t = E_t = E_0(1 + g)^t$$

$$[t = 1, 2, 3, \ldots, n] \qquad 66.1$$

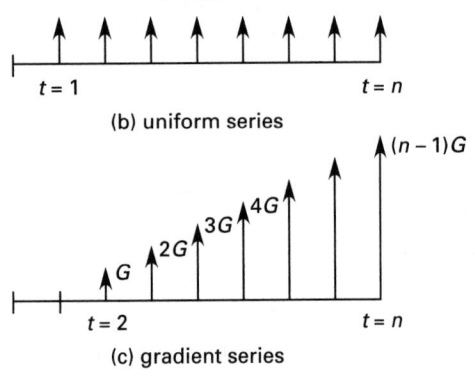

(a) single payment

(b) uniform series

(c) gradient series

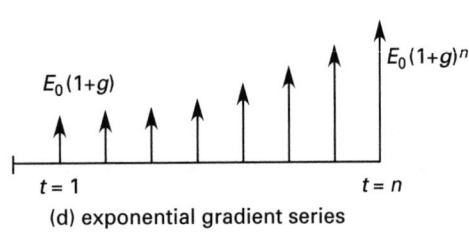

(d) exponential gradient series

Figure 66.1 *Standard Cash Flows*

In Eq. 66.1, g is the *exponential growth rate*, which can be either positive or negative. Exponential gradient cash flows are rarely seen in economic justification projects assigned to engineers.[7]

---

[5]Notice that the cash flows do not begin at $t = 0$. This is an important concept with all of the series cash flows. This convention has been established to accommodate the timing of annual maintenance (and similar) cash flows for which the year-end convention is applicable.

[6]Notice the convention for an exponential cash flow series: The first cash flow E_0 is at $t = 1$, as in the uniform annual series. However, the first cash flow is $E_0(1 + g)$. The cash flow of E_0 at $t = 0$ is absent (i.e., is a *phantom cash flow*).

[7]For one of the few discussions on exponential cash flow, see *Capital Budgeting*, Robert V. Oakford, The Ronald Press Company, New York, 1970.

8. TYPICAL PROBLEM TYPES

There is a wide variety of problem types that, collectively, are considered to be engineering economic analysis problems.

By far, the majority of engineering economic analysis problems are *alternative comparisons*. In these problems, two or more mutually exclusive investments compete for limited funds. A variation of this is a *replacement/retirement analysis*, which is repeated each year to determine if an existing asset should be replaced. Finding the percentage return on an investment is a *rate of return problem*, one of the alternative comparison solution methods.

Investigating interest and principal amounts in loan payments is a *loan repayment problem*. An *economic life analysis* determines when an asset should be retired. In addition, there are miscellaneous problems involving economic order quantity, learning curves, break-even points, product costs, and so on.

9. IMPLICIT ASSUMPTIONS

Several assumptions are implicitly made when solving engineering economic analysis problems. Some of these assumptions are made with the knowledge that they are or will be poor approximations of what really will happen. The assumptions are made, regardless, for the benefit of obtaining a solution.

The most common assumptions are the following.

- The year-end convention is applicable.
- There is no inflation now, nor will there be any during the lifetime of the project.
- Unless otherwise specifically called for, a before-tax analysis is needed.
- The effective interest rate used in the problem will be constant during the lifetime of the project.
- Nonquantifiable factors can be disregarded.
- Funds invested in a project are available and are not urgently needed elsewhere.
- Excess funds continue to earn interest at the effective rate used in the analysis.

This last assumption, like most of the assumptions listed, is almost never specifically mentioned in the body of a solution. However, it is a key assumption when comparing two alternatives that have different initial costs.

For example, suppose two investments, one costing $10,000 and the other costing $8000, are to be compared at 10%. It is obvious that $10,000 in funds is available, otherwise the costlier investment would not be under consideration. If the smaller investment is chosen, what is done with the remaining $2000? The last assumption yields the answer: the $2000 is "put to work" in some investment earning (in this case) 10%.

10. EQUIVALENCE

Industrial decision makers using engineering economic analysis are concerned with the magnitude and timing of a project's cash flow as well as with the total profitability of that project. In this situation, a method is required to compare projects involving receipts and disbursements occurring at different times.

By way of illustration, consider $100 placed in a bank account that pays 5% effective annual interest at the end of each year. After the first year, the account will have grown to $105. After the second year, the account will have grown to $110.25.

Assume that you will have no need for money during the next two years, and any money received will immediately go into your 5% bank account. Then, which of the following options would be more desirable?

option A: $100 now

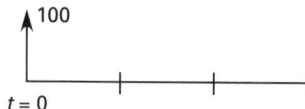

option B: $105 to be delivered in one year

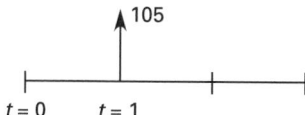

option C: $110.25 to be delivered in two years

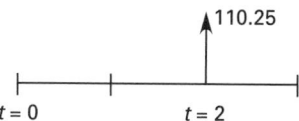

As illustrated, none of the options is superior under the assumptions given. If the first option is chosen, you will immediately place $100 into a 5% account, and in two years the account will have grown to $110.25. In fact, the account will contain $110.25 at the end of two years regardless of the option chosen. Therefore, these alternatives are said to be *equivalent*.

Equivalence may or may not be the case, depending on the interest rate. Thus, an alternative that is acceptable to one decision maker may be unacceptable to another. The interest rate that is used in actual calculations is known as the *effective interest rate*.[8] If compounding is once a year, it is known as the *effective annual interest*

rate. However, effective quarterly, monthly, daily, and so on, interest rates are also used.

The fact that $100 today grows to $105 in one year (at 5% annual interest) is an example of what is known as the *time value of money* principle. This principle simply articulates what is obvious: Funds placed in a secure investment will increase to an equivalent future amount. The procedure for determining the present investment from the equivalent future amount is known as *discounting*.

11. SINGLE-PAYMENT EQUIVALENCE

The equivalence of any present amount, P, at $t = 0$, to any future amount, F, at $t = n$, is called the *future worth* and can be calculated from Eq. 66.2.

$$F = P(1 + i)^n \qquad 66.2$$

The factor $(1 + i)^n$ is known as the single payment *compound amount factor* and has been tabulated in App. 66.B for various combinations of i and n.

Similarly, the equivalence of any future amount to any present amount is called the *present worth* and can be calculated from Eq. 66.3.

$$P = F(1 + i)^{-n} = \frac{F}{(1 + i)^n} \qquad 66.3$$

The factor $(1 + i)^{-n}$ is known as the *single payment present worth factor*.[9]

The interest rate used in Eqs. 66.2 and 66.3 must be the effective rate per period. Also, the basis of the rate (annually, monthly, etc.) must agree with the type of period used to count n. Thus, it would be incorrect to use an effective annual interest rate if n was the number of compounding periods in months.

Example 66.2

How much should you put into a 10% (effective annual rate) savings account in order to have $10,000 in five years?

Solution

This problem could also be stated: What is the equivalent present worth of $10,000 five years from now if money is worth 10% per year?

$$P = F(1 + i)^{-n} = (\$10,000)(1 + 0.10)^{-5}$$
$$= \$6209$$

---

[8]The adjective *effective* distinguishes this interest rate from other interest rates (e.g., nominal interest rates) that are not meant to be used directly in calculating equivalent amounts.

[9]The *present worth* is also called the *present value* and *net present value*. These terms are used interchangeably and no significance should be attached to the terms *value, worth,* and *net*.

The factor 0.6209 would usually be obtained from the tables.

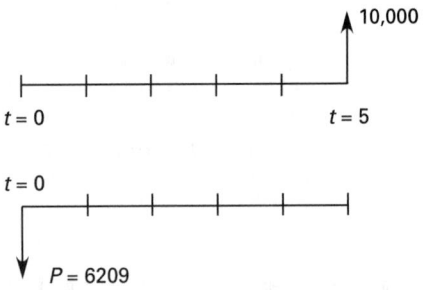

12. STANDARD CASH FLOW FACTORS AND SYMBOLS

Equations 66.2 and 66.3 may give the impression that solving engineering economic analysis problems involves a lot of calculator use, and, in particular, a lot of exponentiation. Such calculations may be necessary from time to time, but most problems are simplified by the use of tabulated values of the factors.

Rather than actually writing the formula for the compound amount factor (which converts a present amount to a future amount), it is common convention to substitute the standard functional notation of $(F/P, i\%, n)$. Thus, the future value in n periods of a present amount is symbolically written as

$$F = P(F/P, i\%, n) \qquad 66.4$$

Similarly, the present worth factor has a functional notation of $(P/F, i\%, n)$. The present worth of a future amount n periods hence is symbolically written as

$$P = F(P/F, i\%, n) \qquad 66.5$$

Values of these *cash flow (discounting) factors* are tabulated in App. 66.B. There is often initial confusion about whether the (F/P) or (P/F) column should be used in a particular problem. There are several ways of remembering what the functional notations mean.

One method of remembering which factor should be used is to think of the factors as conditional probabilities. The conditional probability of event **A** given that event **B** has occurred is written as $p\{\mathbf{A}|\mathbf{B}\}$, where the given event comes after the vertical bar. In the standard notational form of discounting factors, the given amount is similarly placed after the slash. What you want comes before the slash. (F/P) would be a factor to find F given P.

Another method of remembering the notation is to interpret the factors algebraically. Thus, the (F/P) factor could be thought of as the fraction F/P. Algebraically, Eq. 66.4 is

$$F = P(F/P) \qquad 66.6$$

This algebraic approach is actually more than an interpretation. The numerical values of the discounting factors are consistent with this algebraic manipulation. Thus, the (F/A) factor could be calculated as $(F/P) \times (P/A)$. This consistent relationship can be used to calculate other factors that might be occasionally needed, such as (F/G) or (G/P). For instance, the annual cash flow that would be equivalent to a uniform gradient may be found from

$$A = G(P/G, i\%, n)(A/P, i\%, n) \qquad 66.7$$

Formulas for the compounding and discounting factors are contained in Table 66.1. Normally, it will not be necessary to calculate factors from the formulas. Appendix 66.B is adequate for solving most problems.

Example 66.3

What factor will convert a gradient cash flow ending at $t = 8$ to a future value at $t = 8$? (That is, what is the $(F/G, i\%, 8)$ factor?) The effective annual interest rate is 10%.

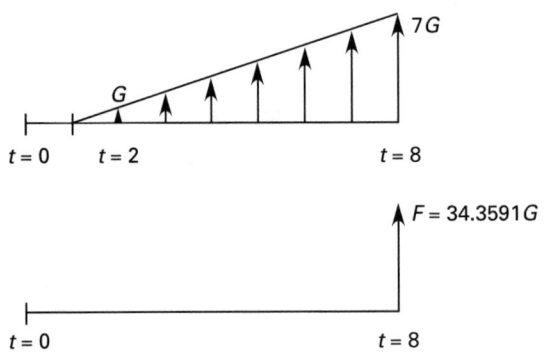

Solution

method 1: From Table 66.1, the $(F/G, 10\%, 8)$ factor is

$$(F/G, 10\%, 8)$$
$$= \frac{(1+i)^n - 1}{i^2} - \frac{n}{i}$$
$$= \frac{(1+0.10)^8 - 1}{(0.10)^2} - \frac{8}{0.10}$$
$$= 34.3589$$

method 2: The tabulated values of (P/G) and (F/P) in App. 66.B can be used to calculate the factor.

$$(F/G, 10\%, 8)$$
$$= (P/G, 10\%, 8)(F/P, 10\%, 8)$$
$$= (16.0287)(2.1436) = 34.3591$$

The (F/G) factor could also have been calculated as the product of the (A/G) and (F/A) factors.

Table 66.1 Discount Factors for Discrete Compounding

factor name	converts	symbol	formula
single payment compound amount	P to F	$(F/P, i\%, n)$	$(1+i)^n$
single payment present worth	F to P	$(P/F, i\%, n)$	$(1+i)^{-n}$
uniform series sinking fund	F to A	$(A/F, i\%, n)$	$\dfrac{i}{(1+i)^n - 1}$
capital recovery	P to A	$(A/P, i\%, n)$	$\dfrac{i(1+i)^n}{(1+i)^n - 1}$
uniform series compound amount	A to F	$(F/A, i\%, n)$	$\dfrac{(1+i)^n - 1}{i}$
uniform series present worth	A to P	$(P/A, i\%, n)$	$\dfrac{(1+i)^n - 1}{i(1+i)^n}$
uniform gradient present worth	G to P	$(P/G, i\%, n)$	$\dfrac{(1+i)^n - 1}{i^2(1+i)^n} - \dfrac{n}{i(1+i)^n}$
uniform gradient future worth	G to F	$(F/G, i\%, n)$	$\dfrac{(1+i)^n - 1}{i^2} - \dfrac{n}{i}$
uniform gradient uniform series	G to A	$(A/G, i\%, n)$	$\dfrac{1}{i} - \dfrac{n}{(1+i)^n - 1}$

13. CALCULATING UNIFORM SERIES EQUIVALENCE

A cash flow that repeats each year for n years without change in amount is known as an *annual amount* and is given the symbol A. As an example, a piece of equipment may require annual maintenance, and the maintenance cost will be an annual amount. Although the equivalent value for each of the n annual amounts could be calculated and then summed, it is more expedient to use one of the uniform series factors. For example, it is possible to convert from an annual amount to a future amount by use of the (F/A) factor.

$$F = A(F/A, i\%, n) \qquad 66.8$$

A *sinking fund* is a fund or account into which annual deposits of A are made in order to accumulate F at $t = n$ in the future. Since the annual deposit is calculated as $A = F(A/F, i\%, n)$, the (A/F) factor is known as the *sinking fund factor*. An *annuity* is a series of equal payments (A) made over a period of time.[10] Usually, it

---

[10]An annuity may also consist of a lump sum payment made at some future time. However, this interpretation is not considered in this chapter.

is necessary to "buy into" an investment (a bond, and insurance policy, etc.) in order to ensure the annuity. In the simplest case of an annuity that starts at the end of the first year and continues for n years, the purchase price (P) is

$$P = A(P/A, i\%, n) \qquad 66.9$$

The present worth of an *infinite (perpetual) series* of annual amounts is known as a *capitalized cost*. There is no $(P/A, i\%, \infty)$ factor in the tables, but the capitalized cost can be calculated simply as

$$P = \frac{A}{i} \qquad [i \text{ in decimal form}] \qquad 66.10$$

Alternatives with different lives will generally be compared by way of *equivalent uniform annual cost* (EUAC). An EUAC is the annual amount that is equivalent to all of the cash flows in the alternative. The EUAC differs in sign from all of the other cash flows. Costs and expenses expressed as EUACs, which would normally be considered negative, are actually positive. The term *cost* in the designation EUAC serves to make clear the meaning of a positive number.

Example 66.4

Maintenance costs for a machine are $250 each year. What is the present worth of these maintenance costs over a 12-year period if the interest rate is 8%?

Solution

$$P = A(P/A, 8\%, 12) = (-\$250)(7.5361)$$

$$= -\$1884$$

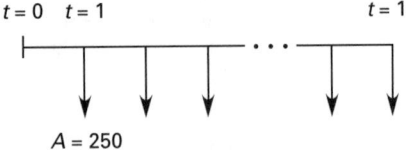

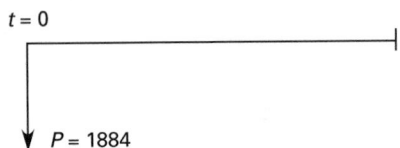

14. FINDING PAST VALUES

From time to time, it will be necessary to determine an amount in the past equivalent to some current (or future) amount. For example, you might have to calculate the original investment made 15 years ago given a current annuity payment.

Such problems are solved by placing the $t = 0$ point at the time of the original investment, and then calculating the past amount as a P value. For example, the original investment, P, can be extracted from the annuity, A, by using the standard cash flow factors.

$$P = A(P/A, i\%, n) \qquad \text{66.11}$$

The choice of $t = 0$ is flexible. As a general rule, the $t = 0$ point should be selected for convenience in solving a problem.

Example 66.5

You are currently paying $250 per month to lease your office phone equipment. You have three years (36 months) left on the five-year (60-month) lease. What would have been an equivalent purchase price two years ago? The effective interest rate per month is 1%.

Solution

The solution of this example is not affected by the fact that investigation is being performed in the middle of the horizon. This is a simple calculation of present worth.

$$P = A(P/A, 1\%, 60)$$

$$= (-\$250)(44.9550) = -\$11,239$$

15. TIMES TO DOUBLE AND TRIPLE AN INVESTMENT

If an investment doubles in value (in n compounding periods and with $i\%$ effective interest), the ratio of current value to past investment will be 2.

$$F/P = (1 + i)^n = 2 \qquad \text{66.12}$$

Similarly, the ratio of current value to past investment will be 3 if an investment triples in value. This can be written as

$$F/P = (1 + i)^n = 3 \qquad \text{66.13}$$

It is a simple matter to extract the number of periods, n, from Eqs. 66.12 and 66.13 to determine the *doubling time* and *tripling time*, respectively. For example, the doubling time is

$$n = \frac{\log(2)}{\log(1 + i)} \qquad \text{66.14}$$

When a quick estimate of the doubling time is needed, the *rule of 72* can be used. The doubling time is approximately $72/i\%$.

The tripling time is

$$n = \frac{\log(3)}{\log(1 + i)} \qquad \text{66.15}$$

Equations 66.14 and 66.15 form the basis of Table 66.2.

Table 66.2 *Doubling and Tripling Times for Various Interest Rates*

interest rate ($i\%$)	doubling time (periods)	tripling time (periods)
1	69.7	110.4
2	35.0	55.5
3	23.4	37.2
4	17.7	28.0
5	14.2	22.5
6	11.9	18.9
7	10.2	16.2
8	9.01	14.3
9	8.04	12.7
10	7.27	11.5
11	6.64	10.5
12	6.12	9.69
13	5.67	8.99
14	5.29	8.38
15	4.96	7.86
16	4.67	7.40
17	4.41	7.00
18	4.19	6.64
19	3.98	6.32
20	3.80	6.03

16. VARIED AND NONSTANDARD CASH FLOWS

Gradient Cash Flow

A common situation involves a uniformly increasing cash flow. If the cash flow has the proper form, its present worth can be determined by using the *uniform gradient factor*, $(P/G, i\%, n)$. The uniform gradient factor finds the present worth of a uniformly increasing cash flow that starts in year two (not in year one).

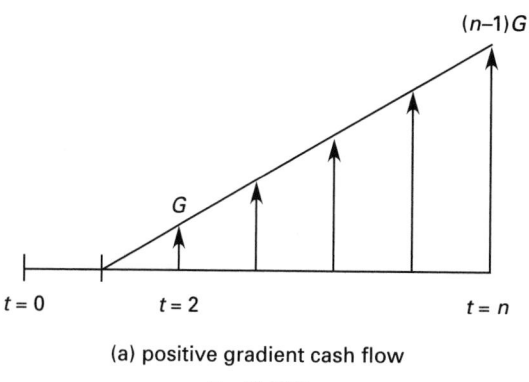

(a) positive gradient cash flow

$$P = G(P/G)$$

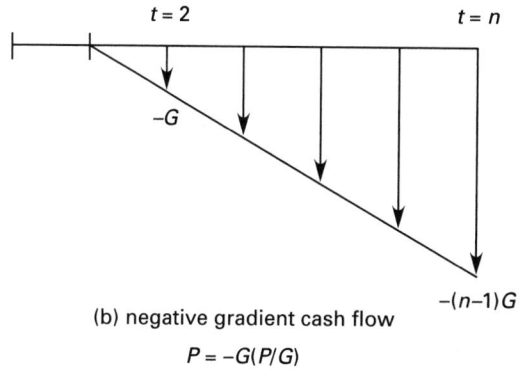

(b) negative gradient cash flow

$$P = -G(P/G)$$

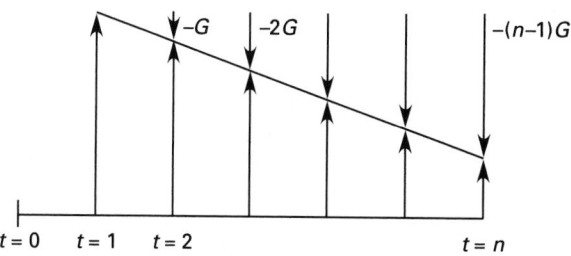

(c) decreasing revenue incorporating a negative gradient

$$P = S(P/A) - G(P/G)$$

Figure 66.2 *Positive and Negative Gradient Cash Flows*

There are three common difficulties associated with the form of the uniform gradient. The first difficulty is that the initial cash flow occurs at $t = 2$. This convention recognizes that annual costs, if they increase uniformly, begin with some value at $t = 1$ (due to the year-end convention) but do not begin to increase until $t = 2$. The tabulated values of (P/G) have been calculated to find the present worth of only the increasing part of the annual expense. The present worth of the base expense incurred at $t = 1$ must be found separately with the (P/A) factor.

The second difficulty is that, even though the $(P/G, i\%, n)$ factor is used, there are only $n - 1$ actual cash flows. It is clear that n must be interpreted as the *period number* in which the last gradient cash flow occurs, not the number of gradient cash flows.

Finally, the sign convention used with gradient cash flows may seem confusing. If an expense increases each year (as in Ex. 66.6), the gradient will be negative, since it is an expense. If a revenue increases each year, the gradient will be positive. In most cases, the sign of the gradient depends on whether the cash flow is an expense or a revenue.[11]

Example 66.6

Maintenance on an old machine is $100 this year but is expected to increase by $25 each year thereafter. What is the present worth of five years of the costs of maintenance? Use an interest rate of 10%.

Solution

In this problem, the cash flow must be broken down into parts. (Notice that the five-year gradient factor is used even though there are only four nonzero gradient cash flows.)

$$P = A(P/A, 10\%, 5) + G(P/G, 10\%, 5)$$
$$= (-\$100)(3.7908) - (\$25)(6.8618)$$
$$= -\$551$$

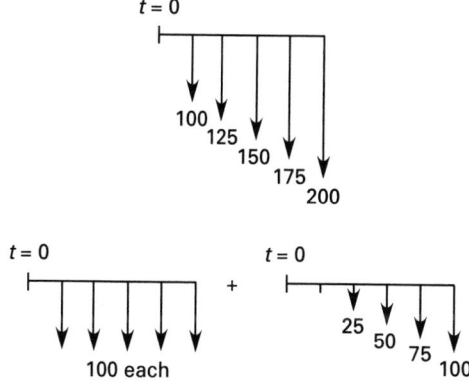

---

[11]This is not a universal rule. It is possible to have a uniformly decreasing revenue as in Fig. 66.2(c). In this case, the gradient would be negative.

Stepped Cash Flows

Stepped cash flows are easily handled by the technique of *superposition of cash flows*. This technique is illustrated by Ex. 66.7.

Example 66.7

An investment costing $1000 returns $100 for the first five years and $200 for the following five years. How would the present worth of this investment be calculated?

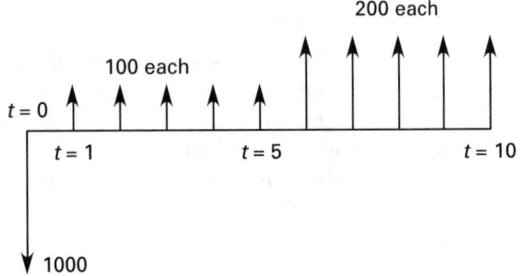

Solution

Using the principle of superposition, the revenue cash flow can be thought of as $200 each year from $t = 1$ to $t = 10$, with a negative revenue of $100 from $t = 1$ to $t = 5$. Superimposed, these two cash flows make up the actual performance cash flow.

$$P = -\$1000 + (\$200)(P/A, i\%, 10) - (\$100)(P/A, i\%, 5)$$

Missing and Extra Parts of Standard Cash Flows

A missing or extra part of a standard cash flow can also be handled by superposition. For example, suppose an annual expense is incurred each year for ten years, except in the ninth year. (The cash flow is illustrated in Figure 66.3.) The present worth could be calculated as a subtractive process.

$$P = A(P/A, i\%, 10) - A(P/F, i\%, 9) \qquad \text{66.16}$$

Alternatively, the present worth could be calculated as an additive process.

$$P = A(P/A, i\%, 8) + A(P/F, i\%, 10) \qquad \text{66.17}$$

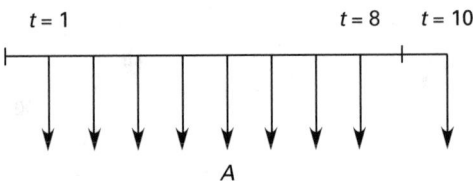

Figure 66.3 *Cash Flow with a Missing Part*

Delayed and Premature Cash Flows

There are cases when a cash flow matches a standard cash flow exactly, except that the cash flow is delayed or starts sooner than it should. Often, such cash flows can be handled with superposition. At other times, it may be more convenient to shift the time axis. This shift is known as the *projection method*. Example 66.8 illustrates the projection method.

Example 66.8

An expense of $75 is incurred starting at $t = 3$ and continues until $t = 9$. There are no expenses or receipts until $t = 3$. Use the projection method to determine the present worth of this stream of expenses.

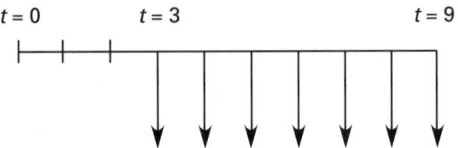

Solution

First, determine a cash flow at $t = 2$ that is equivalent to the entire expense stream. If $t = 0$ was where $t = 2$ actually is, the present worth of the expense stream would be

$$P' = (-\$75)(P/A, i\%, 7)$$

P' is a cash flow at $t = 2$. It is now simple to find the present worth (at $t = 0$) of this future amount.

$$P = P'(P/F, i\%, 2) = (-\$75)(P/A, i\%, 7)(P/F, i\%, 2)$$

Cash Flows at Beginnings of Years: The Christmas Club Problem

This type of problem is characterized by a stream of equal payments (or expenses) starting at $t = 0$ and ending at $t = n - 1$. It differs from the standard annual cash flow in the existence of a cash flow at $t = 0$ and the absence of a cash flow at $t = n$. This problem gets its name from the service provided by some savings institutions whereby money is automatically deposited each week or month (starting immediately, when the savings plan is opened) in order to accumulate money to purchase Christmas presents at the end of the year.

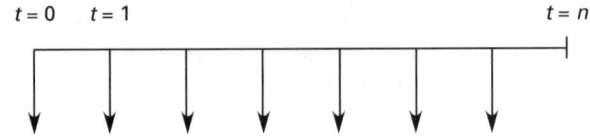

Figure 66.4 *Cash Flows at Beginnings of Years*

It may seem that the present worth of the savings stream can be determined by directly applying the (P/A) factor. However, this is not the case, since the Christmas

Club cash flow and the standard annual cash flow differ. The Christmas Club problem is easily handled by superposition, as illustrated by Ex. 66.9.

Example 66.9

How much can you expect to accumulate by $t = 10$ for a child's college education if you deposit \$300 at the beginning of each year for a total of ten payments?

Solution

Notice that the first payment is made at $t = 0$ and that there is no payment at $t = 10$. The future worth of the first payment is calculated with the (F/P) factor. The absence of the payment at $t = 10$ is handled by superposition. Notice that this "correction" is not multiplied by a factor.

$$F = (\$300)(F/P, i\%, 10) + (\$300)(F/A, i\%, 10) - \$300$$
$$= (\$300)(F/A, i\%, 11) - \$300$$

17. THE MEANING OF PRESENT WORTH AND i

If \$100 is invested in a 5% bank account (using annual compounding), you can remove \$105 one year from now; if this investment is made, you will receive a *return on investment* (ROI) of \$5. The cash flow diagram and the present worth of the two transactions are

$$P = -\$100 + (\$105)(P/F, 5\%, 1)$$
$$= -\$100 + (\$105)(0.9524) = 0$$

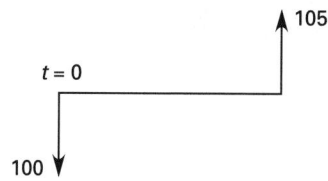

Figure 66.5 *Cash Flow Diagram A*

Notice that the present worth is zero even though you will receive a \$5 return on your investment.

However, if you are offered \$120 for the use of \$100 over a one-year period, the cash flow diagram and present worth (at 5%) would be

$$P = -\$100 + (\$120)(P/F, 5\%, 1)$$
$$= -\$100 + (\$120)(0.9524) = \$14.29$$

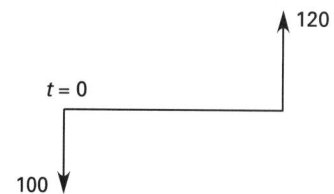

Figure 66.6 *Cash Flow Diagram B*

Therefore, the present worth of an alternative is seen to be equal to the equivalent value at $t = 0$ of the increase in return above that which you would be able to earn in an investment offering $i\%$ per period. In the previous case, \$14.29 is the present worth of (\$20 − \$5), the difference in the two ROIs.

The present worth is also the amount that you would have to be given to dissuade you from making an investment, since placing the initial investment amount along with the present worth into a bank account earning $i\%$ will yield the same eventual return on investment. Relating this to the previous paragraphs, you could be dissuaded from investing \$100 in an alternative that would return \$120 in one year by a $t = 0$ payment of \$14.29. Clearly, (\$100 + \$14.29) invested at $t = 0$ will also yield \$120 in one year at 5%.

Income-producing alternatives with negative present worths are undesirable, and alternatives with positive present worths are desirable because they increase the average earning power of invested capital. (In some cases, such as municipal and public works projects, the present worths of all alternatives are negative, in which case, the least negative alternative is best.)

The selection of the interest rate is difficult in engineering economics problems. Usually, it is taken as the average rate of return that an individual or business organization has realized in past investments. Alternatively, the interest rate may be associated with a particular level of risk. Usually, i for individuals is the interest rate that can be earned in relatively *risk-free investments*.

18. SIMPLE AND COMPOUND INTEREST

If \$100 is invested at 5%, it will grow to \$105 in one year. During the second year, 5% interest continues to be accrued, but on \$105, not on \$100. This is the principle of *compound interest*: The interest accrues interest.[12]

If only the original principal accrues interest, the interest is said to be *simple interest*. Simple interest is rarely encountered in engineering economic analyses, but the concept may be incorporated into short-term transactions.

---

[12]This assumes, of course, that the interest remains in the account. If the interest is removed and spent, only the remaining funds accumulate interest.

19. EXTRACTING THE INTEREST RATE: RATE OF RETURN

An intuitive definition of the *rate of return* (ROR) is the effective annual interest rate at which an investment accrues income. That is, the rate of return of a project is the interest rate that would yield identical profits if all money were invested at that rate. Although this definition is correct, it does not provide a method of determining the rate of return.

It was previously seen that the present worth of a $100 investment invested at 5% is zero when $i = 5\%$ is used to determine equivalence. Thus, a working definition of rate of return would be the effective annual interest rate that makes the present worth of the investment zero. Alternatively, rate of return could be defined as the effective annual interest rate that will discount all cash flows to a total present worth equal to the required initial investment.

It is tempting, but impractical, to determine a rate of return analytically. It is simply too difficult to extract the interest rate from the equivalence equation. For example, consider a $100 investment that pays back $75 at the end of each of the first two years. The present worth equivalence equation (set equal to zero in order to determine the rate of return) is

$$P = 0 = -\$100 + (\$75)(1+i)^{-1} + (\$75)(1+i)^{-2} \quad \textbf{\textit{66.18}}$$

Solving Eq. 66.18 requires finding the roots of a quadratic equation. In general, for an investment or project spanning n years, the roots of an nth-order polynomial would have to be found. It should be obvious that an analytical solution would be essentially impossible for more complex cash flows. (The rate of return in this example is 31.87%.)

If the rate of return is needed, it can be found from a trial-and-error solution. To find the rate of return of an investment, proceed as follows.

step 1: Set up the problem as if to calculate the present worth.

step 2: Arbitrarily select a reasonable value for i. Calculate the present worth.

step 3: Choose another value of i (not too close to the original value), and again solve for the present worth.

step 4: Interpolate or extrapolate the value of i that gives a zero present worth.

step 5: For increased accuracy, repeat steps 2 and 3 with two more values that straddle the value found in step 4.

A common, although incorrect, method of calculating the rate of return involves dividing the annual receipts or returns by the initial investment. (See Sec. 66-54.) However, this technique ignores such items as salvage,

depreciation, taxes, and the time value of money. This technique also is inadequate when the annual returns vary.

It is possible that more than one interest rate will satisfy the zero present worth criteria. This confusing situation occurs whenever there is more than one change in sign in the investment's cash flow.[13] Table 66.3 indicates the numbers of possible interest rates as a function of the number of sign reversals in the investment's cash flow.

Table 66.3 *Multiplicity of Rates of Return*

number of sign reversals	number of distinct rates of return
0	0
1	0 or 1
2	0, 1, or 2
3	0, 1, 2, or 3
4	0, 1, 2, 3, or 4
m	$0, 1, 2, 3, \ldots m-1, m$

Difficulties associated with interpreting the meaning of multiple rates of return can be handled with the concepts of external investment and external rate of return. An *external investment* is an investment that is distinct from the investment being evaluated (which becomes known as the internal investment). The *external rate of return*, which is the rate of return earned by the external investment, does not need to be the same as the rate earned by the internal investment.

Generally, the multiple rates of return indicate that the analysis must proceed as though money will be invested outside of the project. The mechanics of how this is done are not covered here.

Example 66.10

What is the rate of return on invested capital if $1000 is invested now with $500 being returned in year 4 and $1000 being returned in year 8?

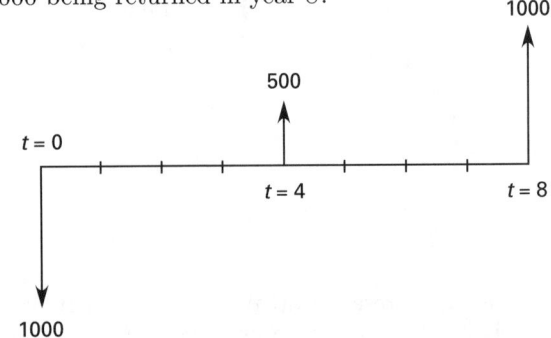

---

[13]There will always be at least one change of sign in the cash flow of a legitimate investment. (This excludes municipal and other tax-supported functions.) At $t = 0$, an investment is made (a negative cash flow). Hopefully, the investment will begin to return money (a positive cash flow) at $t = 1$ or shortly thereafter. Although it is possible to conceive of an investment in which all of the cash flows were negative, such an investment would probably be classified as a *hobby*.

Solution

First, set up the problem as a present worth calculation. Try $i = 5\%$.

$$P = -\$1000 + (\$500)(P/F, 5\%, 4) + (\$1000)$$
$$\times (P/F, 5\%, 8)$$
$$= -\$1000 + (\$500)(0.8227) + (\$1000)(0.6768)$$
$$= \$88$$

Next, try a larger value of i to reduce the present worth. If $i = 10\%$,

$$P = -\$1000 + (\$500)(P/F, 10\%, 4) + (\$1000)$$
$$\times (P/F, 10\%, 8)$$
$$= -\$1000 + (\$500)(0.6830) + (\$1000)(0.4665)$$
$$= -\$192$$

Using simple interpolation, the rate of return is

$$\text{ROR} = 5\% + \left(\frac{\$88}{\$88 + \$192}\right)(10\% - 5\%)$$
$$= 6.6\%$$

A second iteration between 6% and 7% yields 6.39%.

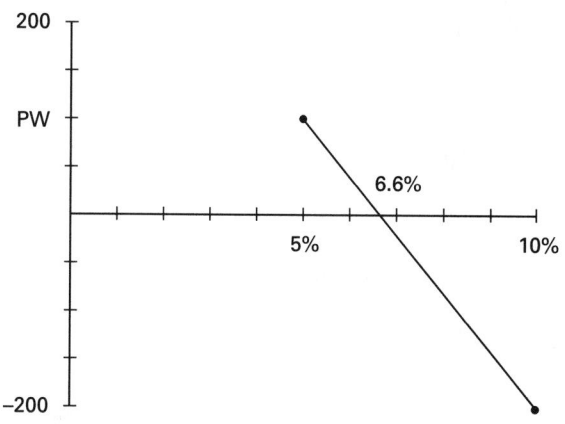

Example 66.11

An existing biomedical company is developing a new drug. A venture capital firm gives the company $25 million initially and $55 million more at the end of the first year. The drug patent will be sold at the end of year 5 to the highest bidder, and the biomedical company will receive $80 million. (The venture capital firm will receive everything in excess of $80 million.) The firm invests unused money in short-term commercial paper earning 10% effective interest per year through its bank. In the meantime, the biomedical company incurs development expenses of $50 million annually for the first three years. The drug is to be evaluated by a government agency and there will be neither expenses nor revenues during the fourth year. What is the biomedical company's rate of return on this investment?

Solution

Normally, the rate of return is determined by setting up a present worth problem and varying the interest rate until the present worth is zero. Writing the cash flows, though, shows that there are two reversals of sign: one at $t = 2$ (positive to negative) and the other at $t = 5$ (negative to positive). Therefore, there could be two interest rates that produce a zero present worth. (In fact, there actually are two interest rates: 10.7% and 41.4%.)

time	cash flow (millions)
0	+25
1	$+55 - 50 = +5$
2	−50
3	−50
4	0
5	+80

However, this problem can be reduced to one with only one sign reversal in the cash flow series. The initial $25 million is invested in commercial paper (an *external investment* having nothing to do with the drug development process) during the first year at 10%. The accumulation of interest and principal after one year is

$$(25)(1 + 0.10) = 27.5$$

This 27.5 is combined with the 5 (the money remaining after all expenses are paid at $t = 1$) and invested externally, again at 10%. The accumulation of interest and principal after one year (i.e., at $t = 2$) is

$$(27.5 + 5)(1 + 0.10) = 35.75$$

This 35.75 is combined with the development cost paid at $t = 2$.

The cash flow for the development project (the internal investment) is

time	cash flow (millions)
0	0
1	0
2	$+35.75 - 50 = -14.25$
3	−50
4	0
5	+80

Now, there is only one sign reversal in the cash flow series. The *internal rate of return* on this development project is found by the traditional method to be 10.3%. Notice that this is different from the rate the company can earn from investing externally in commercial paper.

20. RATE OF RETURN VERSUS RETURN ON INVESTMENT

Rate of return (ROR) is an effective annual interest rate, typically stated in percent per year. *Return on investment* (ROI) is a dollar amount. Thus, *rate of return* and *return on investment* are not synonymous.

Return on investment can be calculated in two different ways. The accounting method is to subtract the total of all investment costs from the total of all net profits (i.e., revenues less expenses). The time value of money is not considered.

In engineering economic analysis, the return on investment is calculated from equivalent values. Specifically, the present worth (at $t = 0$) of all investment costs is subtracted from the future worth (at $t = n$) of all net profits.

When there are only two cash flows, a single investment amount and a single payback, the two definitions of return on investment yield the same numerical value. When there are more than two cash flows, the returns on investment will be different depending on which definition is used.

21. MINIMUM ATTRACTIVE RATE OF RETURN

A company may not know what effective interest rate, i, to use in engineering economic analysis. In such a case, the company can establish a minimum level of economic performance that it would like to realize on all investments. This criterion is known as the *minimum attractive rate of return* (MARR). Unlike the effective interest rate, i, the minimum attractive rate of return is not used in numerical calculations.[14] It is used only in comparisons with the rate of return.

Once a rate of return for an investment is known, it can be compared to the minimum attractive rate of return. To be a viable alternative, the rate of return must be greater than the minimum attractive rate of return.

The advantage of using comparisons to the minimum attractive rate of return is that an effective interest rate, i, never needs to be known. The minimum attractive rate of return becomes the correct interest rate for use in present worth and equivalent uniform annual cost calculations.

22. TYPICAL ALTERNATIVE-COMPARISON PROBLEM FORMATS

With the exception of some investment and rate of return problems, the typical problem involving engineering economics will have the following characteristics.

- An interest rate will be given.
- Two or more alternatives will be competing for funding.
- Each alternative will have its own cash flows.
- It is necessary to select the best alternative.

23. DURATIONS OF INVESTMENTS

Because they are handled differently, short-term investments and short-lived assets need to be distinguished from investments and assets that constitute an infinitely lived project. Short-term investments are easily identified: a drill press that is needed for three years or a temporary factory building that is being constructed to last five years.

Investments with perpetual cash flows are also (usually) easily identified: maintenance on a large flood control dam and revenues from a long-span toll bridge. Furthermore, some items with finite lives can expect renewal on a repeated basis.[15] For example, a major freeway with a pavement life of 20 years is unlikely to be abandoned; it will be resurfaced or replaced every 20 years.

Actually, if an investment's finite lifespan is long enough, it can be considered an infinite investment because money 50 or more years from now has little impact on current decisions. The $(P/F, 10\%, 50)$ factor, for example, is 0.0085. Thus, one dollar at $t = 50$ has an equivalent present worth of less than one penny. Since these far-future cash flows are eclipsed by present cash flows, long-term investments can be considered finite or infinite without significant impact on the calculations.

24. CHOICE OF ALTERNATIVES: COMPARING ONE ALTERNATIVE WITH ANOTHER ALTERNATIVE

Several methods exist for selecting a superior alternative from among a group of proposals. Each method has its own merits and applications.

Present Worth Method

When two or more alternatives are capable of performing the same functions, the superior alternative will have the largest present worth. The *present worth method* is restricted to evaluating alternatives that are mutually exclusive and that have the same lives. This method is suitable for ranking the desirability of alternatives.

Example 66.12

Investment A costs $10,000 today and pays back $11,500 two years from now. Investment B costs $8000 today and pays back $4500 each year for two years. If an interest rate of 5% is used, which alternative is superior?

---

[14]Not everyone adheres to this rule. Some people use "minimum attractive rate of return" and "effective interest rate" interchangeably.

[15]The term *renewal* can be interpreted to mean replacement or repair.

Solution

$$P(\text{A}) = -\$10,000 + (\$11,500)(P/F, 5\%, 2)$$
$$= -\$10,000 + (\$11,500)(0.9070)$$
$$= \$431$$
$$P(\text{B}) = -\$8000 + (\$4500)(P/A, 5\%, 2)$$
$$= -\$8000 + (\$4500)(1.8594)$$
$$= \$367$$

Alternative A is superior and should be chosen.

Capitalized Cost Method

The present worth of a project with an infinite life is known as the *capitalized cost* or *life cycle cost*. Capitalized cost is the amount of money at $t = 0$ needed to perpetually support the project on the earned interest only. Capitalized cost is a positive number when expenses exceed income.

In comparing two alternatives, each of which is infinitely lived, the superior alternative will have the lowest capitalized cost.

Normally, it would be difficult to work with an infinite stream of cash flows since most economics tables do not list factors for periods in excess of 100 years. However, the (A/P) discounting factor approaches the interest rate as n becomes large. Since the (P/A) and (A/P) factors are reciprocals of each other, it is possible to divide an infinite series of equal cash flows by the interest rate in order to calculate the present worth of the infinite series. This is the basis of Eq. 66.19.

$$\text{capitalized cost} = \text{initial cost} + \frac{\text{annual costs}}{i} \quad \textit{66.19}$$

Equation 66.19 can be used when the annual costs are equal in every year. If the operating and maintenance costs occur irregularly instead of annually, or if the costs vary from year to year, it will be necessary to somehow determine a cash flow of equal annual amounts (EAA) that is equivalent to the stream of original costs.

The equal annual amount may be calculated in the usual manner by first finding the present worth of all the actual costs and then multiplying the present worth by the interest rate (the (A/P) factor for an infinite series). However, it is not even necessary to convert the present worth to an equal annual amount since Eq. 66.20 will convert the equal amount back to the present worth.

$$\text{capitalized cost} = \text{initial cost} + \frac{\text{EAA}}{i}$$
$$= \text{initial cost} + \frac{\text{present worth}}{\text{of all expenses}} \quad \textit{66.20}$$

Example 66.13

What is the capitalized cost of a public works project that will cost \$25,000,000 now and will require \$2,000,000 in maintenance annually? The effective annual interest rate is 12%.

Solution

Worked in millions of dollars, from Eq. 66.19, the capitalized cost is

$$\text{capitalized cost} = 25 + (2)(P/A, 12\%, \infty)$$
$$= 25 + \frac{2}{0.12} = 41.67$$

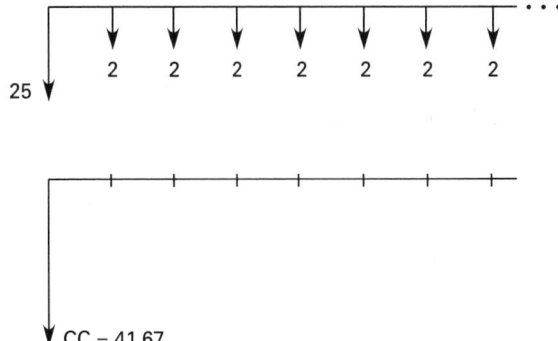

Annual Cost Method

Alternatives that accomplish the same purpose but that have unequal lives must be compared by the *annual cost method*.[16] The annual cost method assumes that each alternative will be replaced by an identical twin at the end of its useful life (infinite renewal). This method, which may also be used to rank alternatives according to their desirability, is also called the *annual return method* or *capital recovery method*.

Restrictions are that the alternatives must be mutually exclusive and repeatedly renewed up to the duration of the longest-lived alternative. The calculated annual cost is known as the *equivalent uniform annual cost* (EUAC) or simply *equivalent annual cost*. Cost is a positive number when expenses exceed income.

Example 66.14

Which of the following alternatives is superior over a 30-year period if the interest rate is 7%?

	alternative A	alternative B
type	brick	wood
life	30 years	10 years
initial cost	\$1800	\$450
maintenance	\$5/year	\$20/year

[16]Of course, the annual cost method can be used to determine the superiority of assets with identical lives as well.

Solution

$$EUAC(A) = (\$1800)(A/P, 7\%, 30) + \$5$$
$$= (\$1800)(0.0806) + \$5$$
$$= \$150$$
$$EUAC(B) = (\$450)(A/P, 7\%, 10) + \$20$$
$$= (\$450)(0.1424) + \$20$$
$$= \$84$$

Alternative B is superior since its annual cost of operation is the lowest. It is assumed that three wood facilities, each with a life of 10 years and a cost of $450, will be built to span the 30-year period.

25. CHOICE OF ALTERNATIVES: COMPARING AN ALTERNATIVE WITH A STANDARD

With specific economic performance criteria, it is possible to qualify an investment as acceptable or unacceptable without having to compare it with another investment. Two such performance criteria are the benefit-cost ratio and the minimum attractive rate of return.

Benefit-Cost Ratio Method

The *benefit-cost ratio* method is often used in municipal project evaluations where benefits and costs accrue to different segments of the community. With this method, the present worth of all benefits (irrespective of the beneficiaries) is divided by the present worth of all costs. The project is considered acceptable if the ratio equals or exceeds 1.0, that is, if $B/C \geq 1.0$.

When the benefit-cost ratio method is used, disbursements by the initiators or sponsors are *costs*. Disbursements by the users of the project are known as *disbenefits*. It is often difficult to determine whether a cash flow is a cost or a disbenefit (whether to place it in the numerator or denominator of the benefit-cost ratio calculation).

Regardless of where the cash flow is placed, an acceptable project will always have a benefit-cost ratio greater than or equal to 1.0, although the actual numerical result will depend on the placement. For this reason, the benefit-cost ratio method should not be used to rank competing projects.

The benefit-cost ratio method of comparing alternatives is used extensively in transportation engineering where the ratio is often (but not necessarily) written in terms of annual benefits and annual costs instead of present worths. Another characteristic of highway benefit-cost ratios is that the route (road, highway, etc.) is usually already in place and that various alternative upgrades are being considered. There will be existing benefits

and costs associated with the current route. Therefore, the *change* (usually an increase) in benefits and costs is used to calculate the benefit-cost ratio.[17]

$$B/C = \frac{\Delta^{\text{user}}_{\text{benefits}}}{\Delta^{\text{investment}}_{\text{cost}} + \Delta \text{maintenance} - \Delta^{\text{residual}}_{\text{value}}} \qquad 66.21$$

Notice that the change in *residual value* (*terminal value*) appears in the denominator as a negative item. An increase in the residual value would decrease the denominator.

Example 66.15

By building a bridge over a ravine, a state department of transportation can shorten the time it takes to drive through a mountainous area. Estimates of costs and benefits (due to decreased travel time, fewer accidents, reduced gas usage, etc.) have been prepared. Should the bridge be built? Use the benefit-cost ratio method of comparison.

	millions
initial cost	$40
capitalized cost of perpetual annual maintenance	$12
capitalized value of annual user benefits	$49
residual value	$0

Solution

If Eq. 66.21 is used, the benefit-cost ratio is

$$B/C = \frac{49}{40 + 12 + 0} = 0.942$$

Since the benefit-cost ratio is less than 1.00, the bridge should not be built.

If the maintenance costs are placed in the numerator (per Ftn. 17), the benefit-cost ratio value will be different but the conclusion will not change.

$$B/C_{\text{alternate method}} = \frac{49 - 12}{40} = 0.925$$

Rate of Return Method

The minimum attractive rate of return (MARR) has already been introduced as a standard of performance against which an investment's actual *rate of return*

---

[17]This discussion of highway benefit-cost ratios is not meant to imply that everyone agrees with Eq. 66.21. In *Economic Analysis for Highways* (International Textbook Company, Scranton, PA, 1969), author Robley Winfrey takes a strong stand on one aspect of the benefits versus disbenefits issue: highway maintenance. According to Winfrey, regular highway maintenance costs should be placed in the numerator as a subtraction from the user benefits. Some have called this mandate the *Winfrey method*.

(ROR) is compared. If the rate of return is equal to or exceeds the minimum attractive rate of return, the investment is qualified. This is the basis for the *rate of return method* of alternative selection.

Finding the rate of return can be a long, iterative process. Usually, the actual numerical value of rate of return is not needed; it is sufficient to know whether or not the rate of return exceeds the minimum attractive rate of return. This *comparative analysis* can be accomplished without calculating the rate of return simply by finding the present worth of the investment using the minimum attractive rate of return as the effective interest rate (i.e., $i = $ MARR). If the present worth is zero or positive, the investment is qualified. If the present worth is negative, the rate of return is less than the minimum attractive rate of return.

26. RANKING MUTUALLY EXCLUSIVE MULTIPLE PROJECTS

Ranking of multiple investment alternatives is required when there is sufficient funding for more than one investment. Since the best investments should be selected first, it is necessary to be able to place all investments into an ordered list.

Ranking is relatively easy if the present worths, future worths, capitalized costs, or equivalent uniform annual costs have been calculated for all the investments. The highest ranked investment is the one with the largest present or future worth, or the smallest capitalized or annual cost. Present worth, future worth, capitalized cost, and equivalent uniform annual cost can all be used to rank multiple investment alternatives.

However, neither rates of return nor benefit-cost ratios should be used to rank multiple investment alternatives. Specifically, if two alternatives both have rates of return exceeding the minimum acceptable rate of return, it is not sufficient to select the alternative with the highest rate of return.

An *incremental analysis*, also known as a *rate of return on added investment study*, should be performed if rate of return is used to select between investments. An incremental analysis starts by ranking the alternatives in order of increasing initial investment. Then, the cash flows for the investment with the lower initial cost are subtracted from the cash flows for the higher-priced alternative on a year-by-year basis. This produces, in effect, a third alternative representing the costs and benefits of the added investment. The added expense of the higher-priced investment is not warranted unless the rate of return of this third alternative exceeds the minimum attractive rate of return as well. The choice criterion is to select the alternative with the higher initial investment if the incremental rate of return exceeds the minimum attractive rate of return.

An incremental analysis is also required if ranking is to be done by the benefit-cost ratio method. The incremental analysis is accomplished by calculating the ratio of differences in benefits to differences in costs for each possible pair of alternatives. If the ratio exceeds 1.0, alternative 2 is superior to alternative 1. Otherwise, alternative 1 is superior.[18]

$$\frac{B_2 - B_1}{C_2 - C_1} \geq 1 \quad \text{[alternative 2 superior]} \quad \textit{66.22}$$

27. ALTERNATIVES WITH DIFFERENT LIVES

Comparison of two alternatives is relatively simple when both alternatives have the same life. For example, a problem might be stated: "Which would you rather have: car A with a life of three years, or car B with a life of five years?"

However, care must be taken to understand what is going on when the two alternatives have different lives. If car A has a life of three years and car B has a life of five years, what happens at $t = 3$ if the five-year car is chosen? If a car is needed for five years, what happens at $t = 3$ if the three-year car is chosen?

In this type of situation, it is necessary to distinguish between the length of the need (the *analysis horizon*) and the lives of the alternatives or assets intended to meet that need. The lives do not have to be the same as the horizon.

Finite Horizon with Incomplete Asset Lives

If an asset with a five-year life is chosen for a three-year need, the disposition of the asset at $t = 3$ must be known in order to evaluate the alternative. If the asset is sold at $t = 3$, the salvage value is entered into the analysis (at $t = 3$) and the alternative is evaluated as a three-year investment. The fact that the asset is sold when it has some useful life remaining does not affect the analysis horizon.

Similarly, if a three-year asset is chosen for a five-year need, something about how the need is satisfied during the last two years must be known. Perhaps a rental asset will be used. Or, perhaps the function will be "farmed out" to an outside firm. In any case, the costs of satisfying the need during the last two years enter the analysis, and the alternative is evaluated as a five-year investment.

If both alternatives are "converted" to the same life, any of the alternative selection criteria (present worth method, annual cost method, etc.) can be used to determine which alternative is superior.

---

[18]It goes without saying that the benefit-cost ratios for all investment alternatives by themselves must also be equal to or greater than 1.0.

Finite Horizon with Integer Multiple Asset Lives

It is common to have a long-term horizon (need) that must be met with short-lived assets. In special instances, the horizon will be an integer number of asset lives. For example, a company may be making a 12-year transportation plan and may be evaluating two cars: one with a three-year life, and another with a four-year life.

In this example, four of the first car or three of the second car are needed to reach the end of the 12-year horizon.

If the horizon is an integer number of asset lives, any of the alternative selection criteria can be used to determine which is superior. If the present worth method is used, all alternatives must be evaluated over the entire horizon. (In this example, the present worth of 12 years of car purchases and use must be determined for both alternatives.)

If the equivalent uniform annual cost method is used, it may be possible to base the calculation of annual cost on one lifespan of each alternative only. It may not be necessary to incorporate all of the cash flows into the analysis. (In the running example, the annual cost over three years would be determined for the first car; the annual cost over four years would be determined for the second car.) This simplification is justified if the subsequent asset replacements (renewals) have the same cost and cash flow structure as the original asset. This assumption is typically made implicitly when the annual cost method of comparison is used.

Infinite Horizon

If the need horizon is infinite, it is not necessary to impose the restriction that asset lives of alternatives be integer multiples of the horizon. The superior alternative will be replaced (renewed) whenever it is necessary to do so, forever.

Infinite horizon problems are almost always solved with either the annual cost or capitalized cost method. It is common to (implicitly) assume that the cost and cash flow structure of the asset replacements (renewals) are the same as the original asset.

28. OPPORTUNITY COSTS

An *opportunity cost* is an imaginary cost representing what will not be received if a particular strategy is rejected. It is what you will lose if you do or do not do something. As an example, consider a growing company with an existing operational computer system. If the company trades in its existing computer as part of an upgrade plan, it will receive a *trade-in allowance*. (In other problems, a *salvage value* may be involved.)

If one of the alternatives being evaluated is not to upgrade the computer system at all, the trade-in allowance (or, salvage value in other problems) will not be realized. The amount of the trade-in allowance is an opportunity cost that must be included in the problem analysis.

Similarly, if one of the alternatives being evaluated is to wait one year before upgrading the computer, the *difference in trade-in allowances* is an opportunity cost that must be included in the problem analysis.

29. REPLACEMENT STUDIES

An investigation into the retirement of an existing process or piece of equipment is known as a *replacement study*. Replacement studies are similar in most respects to other alternative comparison problems: An interest rate is given, two alternatives exist, and one of the previously mentioned methods of comparing alternatives is used to choose the superior alternative. Usually, the annual cost method is used on a year-by-year basis.

In replacement studies, the existing process or piece of equipment is known as the *defender*. The new process or piece of equipment being considered for purchase is known as the *challenger*.

30. TREATMENT OF SALVAGE VALUE IN REPLACEMENT STUDIES

Since most defenders still have some market value when they are retired, the problem of what to do with the salvage arises. It seems logical to use the salvage value of the defender to reduce the initial purchase cost of the challenger. This is consistent with what would actually happen if the defender were to be retired.

By convention, however, the defender's salvage value is subtracted from the defender's present value. This does not seem logical, but it is done to keep all costs and benefits related to the defender with the defender. In this case, the salvage value is treated as an opportunity cost that would be incurred if the defender is not retired.

If the defender and the challenger have the same lives and a present worth study is used to choose the superior alternative, the placement of the salvage value will have no effect on the net difference between present worths for the challenger and defender. Although the values of the two present worths will be different depending on the placement, the difference in present worths will be the same.

If the defender and the challenger have different lives, an annual cost comparison must be made. Since the salvage value would be "spread over" a different number of years depending on its placement, it is important to abide by the conventions listed in this section.

There are a number of ways to handle salvage value in retirement studies. The best way is to calculate the cost of keeping the defender one more year. In addition to the usual operating and maintenance costs, that cost includes an opportunity interest cost incurred by not selling the defender, and also a drop in the salvage value if the defender is kept for one additional year. Specifically,

$$
\begin{aligned}
\text{EUAC (defender)} = \; & \text{next year's maintenance costs} \\
& + \; i \,(\text{current salvage value}) \\
& + \; \text{current salvage} \\
& - \; \text{next year's salvage} \qquad \textit{66.23}
\end{aligned}
$$

It is important in retirement studies not to double count the salvage value. That is, it would be incorrect to add the salvage value to the defender and at the same time subtract it from the challenger.

Equation 66.23 contains the difference in salvage value between two consecutive years. This calculation shows that the defender/challenger decision must be made on a year-by-year basis. One application of Eq. 66.23 will not usually answer the question of whether the defender should remain in service indefinitely. The calculation must be repeatedly made as long as there is a drop in salvage value from one year to the next.

31. ECONOMIC LIFE: RETIREMENT AT MINIMUM COST

As an asset grows older, its operating and maintenance costs typically increase. Eventually, the cost to keep the asset in operation becomes prohibitive, and the asset is retired or replaced. However, it is not always obvious when an asset should be retired or replaced.

As the asset's maintenance cost is increasing each year, the amortized cost of its initial purchase is decreasing. It is the sum of these two costs that should be evaluated to determine the point at which the asset should be retired or replaced. Since an asset's initial purchase price is likely to be high, the amortized cost will be the controlling factor in those years when the maintenance costs are low. Therefore, the EUAC of the asset will decrease in the initial part of its life.

However, as the asset grows older, the change in its amortized cost decreases while maintenance cost increases. Eventually, the sum of the two costs reaches a minimum and then starts to increase. The age of the asset at the minimum cost point is known as the *economic life* of the asset. The economic life generally is less than the length of need and the technological lifetime of the asset.

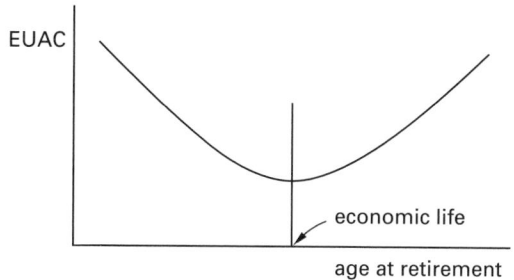

Figure 66.7 *EUAC versus Age at Retirement*

The determination of an asset's economic life is illustrated by Ex. 66.16.

Example 66.16

Buses in a municipal transit system have the characteristics listed. When should the city replace its buses if money can be borrowed at 8%?

initial cost of bus: $120,000

year	maintenance cost	salvage value
1	$35,000	$60,000
2	$38,000	$55,000
3	$43,000	$45,000
4	$50,000	$25,000
5	$65,000	$15,000

Solution

The annual maintenance is different each year. Each maintenance cost must be spread over the life of the bus. This is done by first finding the present worth and then amortizing the maintenance costs. If a bus is kept for one year and then sold, the annual cost will be

$$
\begin{aligned}
\text{EUAC}(1) = \; & (\$120{,}000)(A/P, 8\%, 1) \\
& + (\$35{,}000)(A/F, 8\%, 1) \\
& - (\$60{,}000)(A/F, 8\%, 1) \\
= \; & (\$120{,}000)(1.0800) + (\$35{,}000)(1.000) \\
& - (\$60{,}000)(1.000) \\
= \; & \$104{,}600
\end{aligned}
$$

If a bus is kept for two years and then sold, the annual cost will be

$$
\begin{aligned}
\text{EUAC}(2) = \; & \big(\$120{,}000 + (\$35{,}000)(P/F, 8\%, 1)\big) \\
& \times (A/P, 8\%, 2) \\
& + (\$38{,}000 - \$55{,}000)(A/F, 8\%, 2) \\
= \; & \big(\$120{,}000 + (\$35{,}000)(0.9259)\big)(0.5608) \\
& + (\$38{,}000 - \$55{,}000)(0.4808) \\
= \; & \$77{,}296
\end{aligned}
$$

If a bus is kept for three years and then sold, the annual cost will be

$$
\begin{aligned}
\text{EUAC}(3) &= \big(\$120{,}000 + (35{,}000)(P/F, 8\%, 1) \\
&\quad + (\$38{,}000)(P/F, 8\%, 2)\big)(A/P, 8\%, 3) \\
&\quad + (\$43{,}000 - \$45{,}000)(A/F, 8\%, 3) \\
&= \big(\$120{,}000 + (\$35{,}000)(0.9259) \\
&\quad + (\$38{,}000)(0.8573)\big)(0.3880) \\
&\quad - (\$2000)(0.3080) \\
&= \$71{,}158
\end{aligned}
$$

This process is continued until the annual cost begins to increase. In this example, EUAC(4) is $71,700. Therefore, the buses should be retired after three years.

32. LIFE-CYCLE COST

The *life-cycle cost* of an alternative is the equivalent value (at $t = 0$) of the alternative's cash flow over the alternative's lifespan. Since the present worth is evaluated using an effective interest rate of i (which would be the interest rate used for all engineering economic analyses), the life-cycle cost is the same as the alternative's present worth. If the alternative has an infinite horizon, the life-cycle cost and capitalized cost are identical.

33. CAPITALIZED ASSETS VERSUS EXPENSES

High expenses reduce profit, which in turn reduces income tax. It seems logical to label each and every expenditure, even an asset purchase, as an expense. As an alternative to this *expensing the asset*, it may be decided to capitalize the asset. *Capitalizing the asset* means that the cost of the asset is divided into equal or unequal parts, and only one of these parts is taken as an expense each year. Expensing is clearly the more desirable alternative, since the after-tax profit is increased early in the asset's life.

There are long-standing accounting conventions as to what can be expensed and what must be capitalized.[19] Some companies capitalize everything—regardless of cost—with expected lifetimes greater than one year. Most companies, however, expense items whose purchase costs are below a cut-off value. A cut-off value in the range of $250 to $500, depending on the size of the company, is chosen as the maximum purchase cost of an expensed asset. Assets costing more than this are capitalized.

It is not necessary for a large corporation to keep track of every lamp, desk, and chair for which the purchase price is greater than the cut-off value. Such assets, all of which have the same lives and have been purchased in the same year, can be placed into groups or *asset classes*. A group cost, equal to the sum total of the purchase costs of all items in the group, is capitalized as though the group was an identifiable and distinct asset itself.

34. PURPOSE OF DEPRECIATION

Depreciation is an *artificial expense* that spreads the purchase price of an asset or other property over a number of years.[20] Depreciating an asset is an example of capitalization, as previously defined. The inclusion of depreciation in engineering economic analysis problems increases the after-tax present worth (profitability) of an asset. The larger the depreciation, the greater will be the profitability. Therefore, individuals and companies eligible to utilize depreciation want to maximize and accelerate the depreciation available to them.

Although the entire property purchase price is eventually recognized as an expense, the net recovery from the expense stream never equals the original cost of the asset. That is, depreciation cannot realistically be thought of as a fund (an annuity or sinking fund) that accumulates capital to purchase a replacement at the end of the asset's life. The primary reason for this is that the depreciation expense is reduced significantly by the impact of income taxes, as will be seen in later sections.

35. DEPRECIATION BASIS OF AN ASSET

The *depreciation basis* of an asset is the part of the asset's purchase price that is spread over the *depreciation period*, also known as the *service life*.[21] Usually, the depreciation basis and the purchase price are not the same.

A common depreciation basis is the difference between the purchase price and the expected salvage value at the end of the depreciation period. That is,

$$
\text{depreciation basis} = C - S_n \qquad \textit{66.24}
$$

There are several methods of calculating the year-by-year depreciation of an asset. Equation 66.24 is not universally compatible with all depreciation methods.

---

[19]For example, purchased vehicles must be capitalized; payments for leased vehicles can be expensed. Repainting a building with paint that will last five years is an expense, but the replacement cost of a leaking roof must be capitalized.

[20]The IRS tax regulations allow depreciation on almost all forms of *business property* except land. The following types of property are distinguished: *real* (e.g., buildings used for business), *residential* (e.g., buildings used as rental property), and *personal* (e.g., equipment used for business). Personal property does *not* include items for personal use (such as a personal residence), despite its name. *Tangible personal property* is distinguished from *intangible property* (goodwill, copyrights, patents, trademarks, franchises, agreements not to compete, etc.).

[21]The *depreciation period* is selected to be as short as possible within recognized limits. This depreciation will not normally coincide with the *economic life* or *useful life* of an asset. For example, a car may be capitalized over a depreciation period of three years. It may become uneconomical to maintain and use at the end of an economic life of nine years. However, the car may be capable of operation over a useful life of 25 years.

Some methods do not consider the salvage value. This is known as an *unadjusted basis*. When the depreciation method is known, the depreciation basis can be rigorously defined.[22]

36. DEPRECIATION METHODS

Generally, tax regulations do not allow the cost of an asset to be treated as a deductible expense in the year of purchase. Rather, portions of the depreciation basis must be allocated to each of the n years of the asset's depreciation period. The amount that is allocated each year is called the *depreciation*.

Various methods exist for calculating an asset's depreciation each year.[23] Although the depreciation calculations may be considered independently (for the purpose of determining book value or as an academic exercise), it is important to recognize that depreciation has no effect on engineering economic analyses unless income taxes are also considered.

Straight Line Method

With the *straight line method*, depreciation is the same each year. The depreciation basis $(C - S_n)$ is allocated uniformly to all of the n years in the depreciation period. Each year, the depreciation will be

$$D = \frac{C - S_n}{n} \qquad 66.25$$

Constant Percentage Method

The *constant percentage method*[24] is similar to the straight line method in that the depreciation is the same each year. If the fraction of the basis used as depreciation is $1/n$, there is no difference between the constant percentage and straight line methods. The two methods differ only in what information is available. (With the straight line method, the life is known. With the constant percentage method, the depreciation fraction is known.)

Each year, the depreciation will be

$$\begin{aligned} D &= (\text{depreciation fraction})(\text{depreciation basis}) \\ &= (\text{depreciation fraction})(C - S_n) \qquad 66.26 \end{aligned}$$

Sum-of-the-Years' Digits Method

In *sum-of-the-years' digits* (SOYD) depreciation, the digits from 1 to n inclusive are summed. The total, T, can also be calculated from

$$T = \tfrac{1}{2}n(n + 1) \qquad 66.27$$

The depreciation in year j can be found from Eq. 66.28. Notice that the depreciation in year j, D_j, decreases by a constant amount each year.

$$D_j = \frac{(C - S_n)(n - j + 1)}{T} \qquad 66.28$$

Double Declining Balance Method[25]

Double declining balance[26] (DDB) depreciation is independent of salvage value. Furthermore, the book value never stops decreasing, although the depreciation decreases in magnitude. Usually, any book value in excess of the salvage value is written off in the last year of the asset's depreciation period. Unlike any of the other depreciation methods, double declining balance depends on accumulated depreciation.

$$D_{\text{first year}} = \frac{2C}{n} \qquad 66.29$$

$$D_j = \frac{2\left(C - \displaystyle\sum_{m=1}^{j-1} D_m\right)}{n} \qquad 66.30$$

Calculating the depreciation in the middle of an asset's life appears particularly difficult with double declining balance, since all previous years' depreciation amounts seem to be required. It appears that the depreciation in the sixth year, for example, cannot be calculated unless the values of depreciation for the first five years are calculated. However, this is not true.

Depreciation in the middle of an asset's life can be found from the following equations. (d is known as the *depreciation rate*.)

$$d = \frac{2}{n} \qquad 66.31$$

$$D_j = dC(1 - d)^{j-1} \qquad 66.32$$

Statutory Depreciation Systems

In the United States, property placed into service in 1981 and thereafter must use the *Accelerated Cost Recovery System* (ACRS), and after 1986, the *Modified*

---

[22]For example, with the Accelerated Cost Recovery System (ACRS) the *depreciation basis* is the total purchase cost, regardless of the expected salvage value. With declining balance methods, the depreciation basis is the purchase cost less any previously taken depreciation.

[23]This discussion gives the impression that any form of depreciation may be chosen regardless of the nature and circumstances of the purchase. In reality, the IRS tax regulations place restrictions on the higher-rate (accelerated) methods, such as declining balance and sum-of-the-years' digits methods. Furthermore, the *Economic Recovery Act of 1981* and the *Tax Reform Act of 1986* substantially changed the laws relating to personal and corporate income taxes.

[24]The *constant percentage method* should not be confused with the declining balance method, which used to be known as the *fixed percentage on diminishing balance method*.

[25]In the past, the *declining balance method* has also been known as the *fixed percentage of book value* and *fixed percentage on diminishing balance method*.

[26]Double declining balance depreciation is a particular form of *declining balance depreciation*, as defined by the IRS tax regulations. Declining balance depreciation includes 125% declining balance and 150% declining balance depreciations that can be calculated by substituting 1.25 and 1.50, respectively, for the 2 in Eq. 66.29.

Accelerated Cost Recovery System (MACRS) or other statutory method. Other methods (straight line, declining balance, etc.) cannot be used except in special cases.

Property placed into service in 1980 or before must continue to be depreciated according to the method originally chosen (e.g., straight line, declining balance, or sum-of-the-years' digits). ACRS and MACRS cannot be used.

Under ACRS and MACRS, the cost recovery amount in the jth year of an asset's cost recovery period is calculated by multiplying the initial cost by a factor.

$$D_j = C \times \text{factor} \qquad 66.33$$

The initial cost used is not reduced by the asset's salvage value for ACRS and MACRS calculations. The factor used depends on the asset's cost recovery period. Such factors are subject to continuing legislation changes. Current tax publications should be consulted before using this method.

Table 66.4 *Representative ACRS and MACRS Depreciation Factors*[a]

| | recovery period (n) | | |
year j	3 years ACRS/MACRS	5 years ACRS/MACRS	10 years ACRS/MACRS
1	0.25/0.3333	0.15/0.2000	0.08/0.100
2	0.38/0.4445	0.22/0.3200	0.14/0.180
3	0.37/0.1481	0.21/0.1920	0.12/0.1440
4	0/0.0741	0.21/0.1152	0.10/0.1152
5		0.21/0.1152	0.10/0.0922
6		0/0.0576	0.10/0.0737
7			0.09/0.0655
8			0.09/0.0655
9			0.09/0.0656
10			0.09/0.0655
11			0/0.0328

[a]MACRS values are for the "half-year" convention. This table gives typical values only. Since these factors are subject to continuing revision, they should not be used without consulting an accounting professional.

Production or Service Output Method

If an asset has been purchased for a specific task and that task is associated with a specific lifetime amount of output or production, the depreciation may be calculated by the fraction of total production produced during the year. The depreciation is not expected to be the same each year.

$$D_j = (C - S_n)\left(\frac{\text{actual output in year } j}{\text{estimated lifetime output}}\right) \qquad 66.34$$

Sinking Fund Method

The *sinking fund method* is seldom used in industry because the initial depreciation is low. The formula for sinking fund depreciation (which increases each year) is

$$D_j = (C - S_n)(A/F, i\%, n)(F/P, i\%, j - 1) \qquad 66.35$$

Disfavored Methods

Three other depreciation methods are mentioned here, not because they are currently accepted or in widespread use, but because they are still occasionally encountered in the literature.[27]

The *sinking fund plus interest on first cost* depreciation method, like the following two methods, is an attempt to include the *opportunity interest cost* on the purchase price with the depreciation. That is, the purchasing company not only incurs an annual expense due to the drop in book value, but it also loses the interest on the purchase price. The formula for this method is

$$D = (C - S_n)(A/F, i\%, n) + Ci \qquad 66.36$$

The *straight line plus interest on first cost* method is similar. Its formula is

$$D = \left(\frac{1}{n}\right)(C - S_n) + Ci \qquad 66.37$$

The *straight line plus average interest method* assumes that the opportunity interest cost should be based on the book value only, not on the full purchase price. Since the book value changes each year, an average value is used. The depreciation formula is

$$D = \left(\frac{C - S_n}{n}\right)\left(1 + \frac{i(n + 1)}{2}\right) + iS_n \qquad 66.38$$

Example 66.17

An asset is purchased for $9000. Its estimated economic life is ten years, after which it will be sold for $200. Find the depreciation in the first three years using straight line, double declining balance, and sum-of-the-years' digits depreciation methods.

---

[27]These three depreciation methods should not be used in the usual manner (e.g., in conjunction with the income tax rate). These methods are attempts to calculate a more accurate annual cost of an alternative. Sometimes they give misleading answers. Their use cannot be recommended. They are included in this chapter only for the sake of completeness.

Solution

SL: $\quad D = \dfrac{\$9000 - \$200}{10} \quad\quad = \$880$ each year

DDB: $\quad D_1 = \dfrac{(2)(\$9000)}{10} \quad\quad = \$1800$ in year 1

$\quad\quad\quad D_2 = \dfrac{(2)(\$9000 - \$1800)}{10} \quad = \$1440$ in year 2

$\quad\quad\quad D_3 = \dfrac{(2)(\$9000 - \$3240)}{10} \quad = \$1152$ in year 3

SOYD: $\quad T = \left(\frac{1}{2}\right)(10)(11) = 55$

$\quad\quad\quad D_1 = \left(\frac{10}{55}\right)(\$9000 - \$200) = \$1600$ in year 1

$\quad\quad\quad D_2 = \left(\frac{9}{55}\right)(\$8800) \quad\quad = \$1440$ in year 2

$\quad\quad\quad D_3 = \left(\frac{8}{55}\right)(\$8800) \quad\quad = \$1280$ in year 3

37. ACCELERATED DEPRECIATION METHODS

An *accelerated depreciation method* is one that calculates a depreciation amount greater than a straight line amount. Double declining balance and sum-of-the-years' digits methods are accelerated methods. The ACRS and MACRS methods are explicitly accelerated methods. Straight line and sinking fund methods are not accelerated methods.

Use of an accelerated depreciation method may result in unexpected tax consequences when the depreciated asset or property is disposed of. Professional tax advice should be obtained in this area.

38. BOOK VALUE

The difference between original purchase price and accumulated depreciation is known as *book value*.[28] At the end of each year, the book value (which is initially equal to the purchase price) is reduced by the depreciation in that year.

It is important to properly synchronize depreciation calculations. It is difficult to answer the question, "What is the book value in the fifth year?" unless the timing of the book value change is mutually agreed upon. It is better to be specific about an inquiry by identifying when the book value change occurs. For example, the following question is unambiguous: "What is the book value at the end of year 5, after subtracting depreciation in the fifth year?" or "What is the book value after five years?"

Unfortunately, this type of care is seldom taken in book value inquiries, and it is up to the respondent to exercise reasonable care in distinguishing between beginning-of-year book value and end-of-year book value. To be consistent, the book value equations in this chapter have been written in such a way that the year subscript (j) has the same meaning in book value and depreciation calculations. That is, BV_5 means the book value at the end of the fifth year, after five years of depreciation, including D_5, has been subtracted from the original purchase price.

There can be a great difference between the book value of an asset and the *market value* of that asset. There is no legal requirement for the two values to coincide, and no intent for book value to be a reasonable measure of market value.[29] Therefore, it is apparent that book value is merely an accounting convention with little practical use. Even when a depreciated asset is disposed of, the book value is used to determine the consequences of disposal, not the price the asset should bring at sale.

The calculation of book value is relatively easy, even for the case of the declining balance depreciation method.

For the straight line depreciation method, the book value at the end of the jth year, after the jth depreciation deduction has been made, is

$$\mathrm{BV}_j = C - \frac{j(C - S_n)}{n} = C - jD \qquad \textit{66.39}$$

For the sum-of-the-years' digits method, the book value is

$$\mathrm{BV}_j = (C - S_n)\left(1 - \frac{j(2n + 1 - j)}{n(n + 1)}\right) + S_n \qquad \textit{66.40}$$

For the declining balance method, the book value is

$$\mathrm{BV}_j = C(1 - d)^j \qquad \textit{66.41}$$

For the sinking fund method, the book value is calculated directly as

$$\mathrm{BV}_j = C - (C - S_n)(A/F, i\%, n)(F/A, i\%, j) \qquad \textit{66.42}$$

---

[28]The balance sheet of a corporation usually has two asset accounts: the *equipment account* and the *accumulated depreciation account*. There is no book value account on this financial statement, other than the implicit value obtained from subtracting the accumulated depreciation account from the equipment account. The book values of various assets, as well as their original purchase cost, date of purchase, salvage value, and so on, and accumulated depreciation appear on detail sheets or other peripheral records for each asset.

[29]Common examples of assets with great divergences of book and market values are buildings (rental houses, apartment complexes, factories, etc.) and company luxury automobiles (Porsches, Mercedes, etc.) during periods of inflation. Book values decrease, but actual values increase.

Of course, the book value at the end of year j can always be calculated for any method by successive subtractions (i.e., subtraction of the accumulated depreciation), as Eq. 66.43 illustrates.

$$\text{BV}_j = C - \sum_{m=1}^{j} D_m \qquad 66.43$$

Figure 66.8 illustrates the book value of a hypothetical asset depreciated using several depreciation methods. Notice that the double declining balance method initially produces the fastest write-off, while the sinking fund method produces the slowest write-off. Note also that the book value does not automatically equal the salvage value at the end of an asset's depreciation period with the double declining balance method.[30]

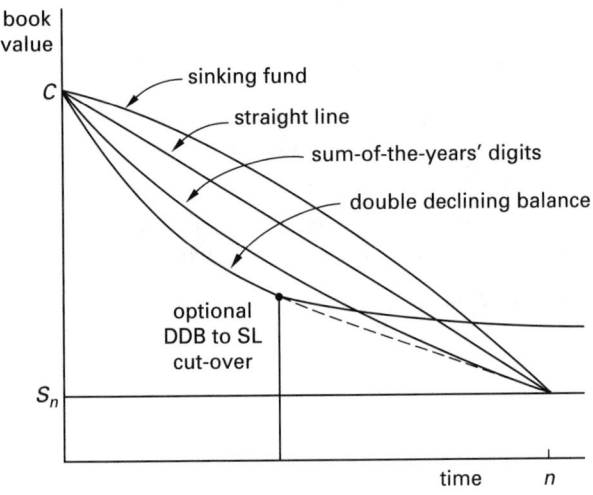

Figure 66.8 *Book Value with Different Depreciation Methods*

Example 66.18

For the asset described in Ex. 66.17, calculate the book value at the end of the first three years if sum-of-the-years' digits depreciation is used. The book value at the beginning of year 1 is $9000.

Solution

From Eq. 66.43,

$$\text{BV}_1 = \$9000 - \$1600 = \$7400$$
$$\text{BV}_2 = \$7400 - \$1440 = \$5960$$
$$\text{BV}_3 = \$5960 - \$1280 = \$4680$$

---

[30]This means that the straight line method of depreciation may result in a lower book value at some point in the depreciation period than if double declining balance is used. A *cut-over* from double declining balance to straight line may be permitted in certain cases. Finding the *cut-over point*, however, is usually done by comparing book values determined by both methods. The analytical method is complicated.

39. AMORTIZATION

Amortization and depreciation are similar in that they both divide up the cost basis or value of an asset. In fact, in certain cases, the term "amortization" may be used in place of the term "depreciation." However, depreciation is a specific form of amortization.

Amortization spreads the cost basis or value of an asset over some base. The base can be time, units of production, number of customers, and so on. The asset can be tangible (e.g., a delivery truck or building) or intangible (e.g., goodwill or a patent).

If the asset is tangible, if the base is time, and if the length of time is consistent with accounting standards and taxation guidelines, then the term "depreciation" is appropriate. However, if the asset is intangible, if the base is some other variable, or if some length of time other than the customary period is used, then the term "amortization" is more appropriate.[31]

Example 66.19

A company purchases complete and exclusive patent rights to an invention for $1,200,000. It is estimated that once commercially produced, the invention will have a specific but limited market of 1200 units. For the purpose of allocating the patent right cost to production cost, what is the amortization rate in dollars per unit?

Solution

The patent should be amortized at the rate of

$$\frac{\$1,200,000}{1200 \text{ units}} = \$1000 \text{ per unit}$$

40. DEPLETION

Depletion is another artificial deductible operating expense, designed to compensate mining organizations for decreasing mineral reserves. Since original and remaining quantities of minerals are seldom known accurately, the *depletion allowance* is calculated as a fixed percentage of the organization's gross income. These percentages are usually in the 10 to 20% range and apply to such mineral deposits as oil, natural gas, coal, uranium, and most metal ores.

41. BASIC INCOME TAX CONSIDERATIONS

The issue of income taxes is often overlooked in academic engineering economic analysis exercises. Such a position is justifiable when an organization (e.g., a nonprofit school, a church, or the government) pays no

---

[31]From time to time, the U.S. Congress has allowed certain types of facilities (e.g., emergency, grain storage, and pollution control) to be written off more rapidly than would otherwise be permitted in order to encourage investment in such facilities. The term "amortization" has been used with such write-off periods.

income taxes. However, if an individual or organization is subject to income taxes, the income taxes must be included in an economic analysis of investment alternatives.

Assume that an organization pays a fraction f of its profits to the federal government as income taxes. If the organization also pays a fraction s of its profits as state income taxes and if state taxes paid are recognized by the federal government as tax-deductible expenses, then the composite tax rate is

$$t = s + f - sf \qquad 66.44$$

The basic principles used to incorporate taxation into engineering economic analyses are the following.

- Initial purchase expenditures are unaffected by income taxes.

- Salvage revenues are unaffected by income taxes.

- Deductible expenses, such as operating costs, maintenance costs, and interest payments, are reduced by the fraction t (e.g., multiplied by the quantity $1 - t$).

- Revenues are reduced by the fraction t (e.g., multiplied by the quantity $1 - t$).

- Since tax regulations allow the depreciation in any year to be handled as if it were an actual operating expense, and since operating expenses are deductible from the income base prior to taxation, the after-tax profits will be increased. If D is the depreciation, the net result to the after-tax cash flow will be the addition of tD. Depreciation is multiplied by t and added to the appropriate year's cash flow, increasing that year's present worth.

For simplicity, most engineering economics practice problems involving income taxes specify a single income tax rate. In practice, however, federal and most state tax rates depend on the income level. Each range of incomes and its associated tax rate are known as *income bracket* and *tax bracket*, respectively. For example, the state income tax rate might be 4% for incomes up to and including $30,000, and 5% for incomes above $30,000. The income tax for a taxpaying entity with an income of $50,000 would have to be calculated in two parts.

$$\text{tax} = (0.04)(\$30{,}000) + (0.05)(\$50{,}000 - \$30{,}000)$$
$$= \$2200$$

Income taxes and depreciation have no bearing on municipal or governmental projects since municipalities, states, and the U.S. government pay no taxes.

Example 66.20

A corporation that pays 53% of its profit in income taxes invests $10,000 in an asset that will produce a $3000 annual revenue for eight years. If the annual expenses are $700, salvage after eight years is $500, and

9% interest is used, what is the after-tax present worth? Disregard depreciation.

Solution

$$P = -\$10{,}000 + (\$3000)(P/A, 9\%, 8)(1 - 0.53)$$
$$\quad - (\$700)(P/A, 9\%, 8)(1 - 0.53)$$
$$\quad + (\$500)(P/F, 9\%, 8)$$
$$= -\$10{,}000 + (\$3000)(5.5348)(0.47)$$
$$\quad - (\$700)(5.5348)(0.47) + (\$500)(0.5019)$$
$$= -\$3766$$

42. TAXATION AT THE TIMES OF ASSET PURCHASE AND SALE

There are numerous rules and conventions that governmental tax codes and the accounting profession impose on organizations. Engineers are not normally expected to be aware of most of the rules and conventions, but occasionally it may be necessary to incorporate their effects into an engineering economic analysis.

Tax Credit

A *tax credit* (also known as an *investment tax credit* or *investment credit*) is a one-time credit against income taxes.[32] Therefore, it is added to the after-tax present worth as a last step in an engineering economic analysis. Such tax credits may be allowed by the government from time to time for equipment purchases, employment of various classes of workers, rehabilitation of historic landmarks, and so on.

A tax credit is usually calculated as a fraction of the initial purchase price or cost of an asset or activity.

$$\text{TC} = \text{fraction} \times \text{initial cost} \qquad 66.45$$

When the tax credit is applicable, the fraction used is subject to legislation. A professional tax expert or accountant should be consulted prior to applying the tax credit concept to engineering economic analysis problems.

Since the investment tax credit reduces the buyer's tax liability, a tax credit should be included only in after-tax engineering economic analyses. The credit is assumed to be received at the end of the year.

Gain or Loss on the Sale of a Depreciated Asset

If an asset that has been depreciated over a number of prior years is sold for more than its current book value, the difference between the book value and selling price is taxable income in the year of the sale. Alternatively,

---

[32]Strictly, *tax credit* is the more general term, and applies to a credit for doing anything creditable. An *investment tax credit* requires an investment in something (usually real property or equipment).

if the asset is sold for less than its current book value, the difference between the selling price and book value is an expense in the year of the sale.

Example 66.21

One year, a company makes a $5000 investment in a historic building. The investment is not depreciable, but it does qualify for a one-time 20% tax credit. In that same year, revenue is $45,000 and expenses (exclusive of the $5000 investment) are $25,000. The company pays a total of 53% in income taxes. What is the after-tax present worth of this year's activities if the company's interest rate for investment is 10%?

Solution

The tax credit is

$$TC = (0.20)(\$5000) = \$1000$$

This tax credit is assumed to be received at the end of the year. The after-tax present worth is

$$
\begin{aligned}
P &= -\$5000 + (\$45,000 - \$25,000)(1 - 0.53) \\
&\quad \times (P/F, 10\%, 1) + (\$1000)(P/F, 10\%, 1) \\
&= -\$5000 + (\$20,000)(0.47)(0.9091) \\
&\quad + (\$1000)(0.9091) \\
&= \$4455
\end{aligned}
$$

43. DEPRECIATION RECOVERY

The economic effect of depreciation is to reduce the income tax in year j by tD_j. The present worth of the asset is also affected: The present worth is increased by $tD_j(P/F, i\%, j)$. The after-tax present worth of all depreciation effects over the depreciation period of the asset is called the *depreciation recovery* (DR).[33]

$$DR = t\sum_{j=1}^{n} D_j(P/F, i\%, j) \qquad 66.46$$

Straight line depreciation recovery from an asset is easily calculated, since the depreciation is the same each year. Assuming the asset has a constant depreciation of D and depreciation period of n years, the depreciation recovery is

$$DR = tD(P/A, i\%, n) \qquad 66.47$$

$$D = \frac{C - S_n}{n} \qquad 66.48$$

---

[33]Since the depreciation benefit is reduced by taxation, depreciation cannot be thought of as an annuity to fund a replacement asset.

Sum-of-the-years' digits depreciation recovery is also relatively easily calculated, since the depreciation decreases uniformly each year.

$$DR = \left(\frac{t(C - S_n)}{T}\right)\left(n(P/A, i\%, n) - (P/G, i\%, n)\right) \qquad 66.49$$

Finding *declining balance depreciation recovery* is more involved. There are three difficulties. The first (the apparent need to calculate all previous depreciations in order to determine the subsequent depreciation) has already been addressed by Eq. 66.32.

The second difficulty is that there is no way to ensure (that is, to force) the book value to be S_n at $t = n$. Therefore, it is common to write off the remaining book value (down to S_n) at $t = n$ in one lump sum. This assumes $BV_n \geq S_n$.

The third difficulty is that of finding the present worth of an *exponentially decreasing cash flow*. Although the proof is omitted here, such exponential cash flows can be handled with the *exponential gradient factor*, (P/EG).[34]

$$(P/EG, z - 1, n) = \frac{z^n - 1}{z^n(z - 1)} \qquad 66.50$$

$$z = \frac{1 + i}{1 - d} \qquad 66.51$$

Then, as long as $BV_n > S_n$, the declining balance depreciation recovery is

$$DR = tC\left(\frac{d}{1 - d}\right)(P/EG, z - 1, n) \qquad 66.52$$

Example 66.22

For the asset described in Ex. 66.17, calculate the after-tax depreciation recovery with straight line and sum-of-the-years' digits depreciation methods. Use 6% interest with 48% income taxes.

Solution

SL:
$$
\begin{aligned}
DR &= (0.48)(\$880)(P/A, 6\%, 10) \\
&= (0.48)(\$880)(7.3601) \\
&= \$3109
\end{aligned}
$$

SOYD: The depreciation series can be thought of as a constant $1600 term with a negative $160 gradient.

$$
\begin{aligned}
DR &= (0.48)(\$1600)(P/A, 6\%, 10) \\
&\quad - (0.48)(\$160)(P/G, 6\%, 10) \\
&= (0.48)(\$1600)(7.3601) \\
&\quad - (0.48)(\$160)(29.6023) \\
&= \$3379
\end{aligned}
$$

---

[34]The (P/A) columns in App. 66.B can be used for (P/EG) as long as the interest rate is assumed to be $z - 1$.

Table 66.5 Depreciation Calculation Summary

method	depreciation basis	depreciation in year $j(D_j)$	book value after jth depreciation (BV_j)	after-tax depreciation recovery (DR)	supplementary formulas
straight line (SL)	$C - S_n$	$\dfrac{C - S_n}{n}$ (constant)	$C - jD$	$tD(P/A, i\%, n)$	
constant percentage	$C - S_n$	fraction $\times (C - S_n)$ (constant)	$C - jD$	$tD(P/A, i\%, n)$	
sum-of-the-years' digits (SOYD)	$C - S_n$	$\dfrac{(C - S_n) \times (n - j + 1)}{T}$	$(C - S_n) \times \left(1 - \dfrac{j(2n + 1 - j)}{n(n+1)}\right) + S_n$	$\dfrac{t(C - S_n)}{T} \times (n(P/A, i\%, n) - (P/G, i\%, n))$	$T = \frac{1}{2}n(n+1)$
double declining balance (DDB)	C	$dC(1 - d)^{j-1}$	$C(1 - d)^j$	$tC\left(\dfrac{d}{1-d}\right) \times (P/EG, z - 1, n)$	$d = \dfrac{2}{n}; \ z = \dfrac{1+i}{1-d}$ $(P/EG, z - 1, n) = \dfrac{z^n - 1}{z^n(z - 1)}$
sinking fund (SF)	$C - S_n$	$(C - S_n) \times (A/F, i\%, n) \times (F/P, i\%, j - 1)$	$C - (C - S_n) \times (A/F, i\%, n) \times (F/A, i\%, j)$	$\dfrac{t(C - S_n)(A/F, i\%, n)}{1 + i}$	
accelerated cost recovery system (ACRS/ MACRS)	C	$C \times$ factor	$C - \displaystyle\sum_{m=1}^{j} D_m$	$t \displaystyle\sum_{j=1}^{n} D_j(P/F, i\%, j)$	
production or service output	$C - S_n$	$(C - S_n) \times \left(\dfrac{\text{actual output in year } j}{\text{lifetime output}}\right)$	$C - \displaystyle\sum_{m=1}^{j} D_m$	$t \displaystyle\sum_{j=1}^{n} D_j(P/F, i\%, j)$	

Notice that the ten-year (P/G) factor is used even though there are only nine years in which the gradient reduces the initial $1600 amount.

Example 66.23

What is the after-tax present worth of the asset described in Ex. 66.20 if straight line, sum-of-the-years' digits, and double declining balance depreciation methods are used?

Solution

Using SL, the depreciation recovery is

$$\mathrm{DR} = (0.53)\left(\frac{\$10{,}000 - \$500}{8}\right)(P/A, 9\%, 8)$$

$$= (0.53)\left(\frac{\$9500}{8}\right)(5.5348)$$

$$= \$3483$$

Using SOYD, the depreciation recovery is calculated as follows.

$$T = \left(\tfrac{1}{2}\right)(8)(9) = 36$$

$$\text{depreciation base} = \$10{,}000 - \$500 = \$9500$$

$$D_1 = \left(\tfrac{8}{36}\right)(\$9500) = \$2111$$

$$G = \text{gradient} = \left(\tfrac{1}{36}\right)(\$9500)$$

$$= \$264$$

$$\mathrm{DR} = (0.53)\big((\$2111)(P/A, 9\%, 8)$$

$$- (\$264)(P/G, 9\%, 8)\big)$$

$$= (0.53)\big((\$2111)(5.5348)$$

$$- (\$264)(16.8877)\big)$$

$$= \$3830$$

Using DDB, the depreciation recovery is calculated as follows.[35]

$$d = \frac{2}{8} = 0.25$$

$$z = \frac{1 + 0.09}{1 - 0.25} = 1.4533$$

$$(P/EG, z-1, n) = \frac{(1.4533)^8 - 1}{(1.4533)^8 (0.4533)} = 2.095$$

From Eq. 66.52,

$$DR = (0.53)\left(\frac{(0.25)(\$10,000)}{0.75}\right)(2.095)$$

$$= \$3701$$

The after-tax present worth, neglecting depreciation, was previously found to be −$3766.

The after-tax present worths, including depreciation recovery, are

$$\text{SL:} \quad P = -\$3766 + \$3483 = -\$283$$
$$\text{SOYD:} \quad P = -\$3766 + \$3830 = \quad \$64$$
$$\text{DDB:} \quad P = -\$3766 + \$3701 = \quad -\$65$$

44. OTHER INTEREST RATES

The *effective interest rate per period*, i (also called *yield* by banks), is the only interest rate that should be used in equivalence equations. The interest rates at the top of the factor tables in App. 66.B are implicitly all effective interest rates. Usually, the period will be one year, hence the name *effective annual interest rate*. However, there are other interest rates in use as well.

The term *nominal interest rate*, r (*rate per annum*), is encountered when compounding is more than once per year. The nominal rate does not include the effect of compounding and is not the same as the effective rate. And, since the effective interest rate can be calculated from the nominal rate only if the number of compounding periods per year is known, nominal rates cannot be compared unless the method of compounding is specified. The only practical use for a nominal rate per year is for calculating the effective rate per period.

45. RATE AND PERIOD CHANGES

If there are k compounding periods during the year (two for semiannual compounding, four for quarterly compounding, twelve for monthly compounding, etc.) and the nominal rate is r, the *effective rate per compounding period* is

$$\phi = \frac{r}{k} \qquad \text{66.53}$$

The effective annual rate, i, can be calculated from the effective rate per period, ϕ, by using Eq. 66.54.

$$i = (1 + \phi)^k - 1$$
$$= \left(1 + \frac{r}{k}\right)^k - 1 \qquad \text{66.54}$$

Sometimes, only the effective rate per period (e.g., per month) is known. However, that will be a simple problem since compounding for n periods at an effective rate per period is not affected by the definition or length of the period.

The following rules may be used to determine which interest rate is given in a problem.

- Unless specifically qualified in the problem, the interest rate given is an annual rate.

- If the compounding is annual, the rate given is the effective rate. If compounding is other than annual, the rate given is the nominal rate.

The effective annual interest rate determined on a *daily compounding basis* will not be significantly different than if *continuous compounding* is assumed.[36] In the case of continuous (or daily) compounding, the discounting factors can be calculated directly from the nominal interest rate and number of years, without having to find the effective interest rate per period. Table 66.6 can be used to determine the discount factors for continuous compounding.

Table 66.6 *Discount Factors for Continuous Compounding*
(n is the number of years)

symbol	formula
$(F/P, r\%, n)$	e^{rn}
$(P/F, r\%, n)$	e^{-rn}
$(A/F, r\%, n)$	$\dfrac{e^r - 1}{e^{rn} - 1}$
$(F/A, r\%, n)$	$\dfrac{e^{rn} - 1}{e^r - 1}$
$(A/P, r\%, n)$	$\dfrac{e^r - 1}{1 - e^{-rn}}$
$(P/A, r\%, n)$	$\dfrac{1 - e^{-rn}}{e^r - 1}$

---

[35]This method should start by checking that the book value at the end of the depreciation period is greater than the salvage value. In this example, such is the case. However, the step is not shown.

[36]The number of *banking days in a year* (250, 360, etc.) must be specifically known.

Example 66.24

A savings and loan offers a nominal rate of $5^{1}/_{4}\%$ compounded daily over 365 days in a year. What is the effective annual rate?

Solution

method 1: Use Eq. 66.54.

$$r = 0.0525, \quad k = 365$$

$$i = \left(1 + \frac{0.0525}{365}\right)^{365} - 1 = 0.0539$$

method 2: Assume daily compounding is the same as continuous compounding.

$$i = (F/P, r\%, 1) - 1$$
$$= e^{0.0525} - 1 = 0.0539$$

Example 66.25

A real estate investment trust pays $7,000,000 for a 100-unit apartment complex. The trust expects to sell the complex in ten years for $15,000,000. In the meantime, it expects to receive an average rent of $900 per month from each apartment. Operating expenses are expected to be $200 per month per occupied apartment. A 95% occupancy rate is predicted. In similar investments, the trust has realized a 15% effective annual return on its investment. Compare the expected present worth of this investment when calculated assuming (a) annual compounding (i.e., the year-end convention), and (b) monthly compounding. Disregard taxes, depreciation, and all other factors.

Solution

(a) The net annual income will be

$$(0.95)(100 \text{ units}) \left(\$900 \; \frac{\$}{\text{unit-month}}\right.$$
$$\left. - \$200 \; \frac{\$}{\text{unit-month}}\right) \left(12 \; \frac{\text{months}}{\text{year}}\right)$$
$$= \$798,000/\text{year}$$

The present worth of ten years of operation is

$$P = -\$7,000,000 + (\$798,000)(P/A, 15\%, 10)$$
$$+ (\$15,000,000)(P/F, 15\%, 10)$$
$$= -\$7,000,000 + (\$798,000)(5.0188)$$
$$+ (\$15,000,000)(0.2472)$$
$$= \$713,000$$

(b) The net monthly income is

$$(0.95)(100 \text{ units}) \left(\$900 \; \frac{\$}{\text{unit-month}}\right.$$
$$\left. - \$200 \; \frac{\$}{\text{unit-month}}\right) = \$66,500/\text{month}$$

Equation 66.54 is used to calculate the effective monthly rate, ϕ, from the effective annual rate, $i = 15\%$, and the number of compounding periods per year, $k = 12$.

$$\phi = (1 + i)^{\frac{1}{k}} - 1$$
$$= (1 + 0.15)^{\frac{1}{12}} - 1 = 0.011715 \quad (1.1715\%)$$

The number of compounding periods in ten years is

$$n = (10 \text{ years}) \left(12 \; \frac{\text{months}}{\text{year}}\right) = 120 \text{ months}$$

The present worth of 120 months of operation is

$$P = -\$7,000,000 + (\$66,500)(P/A, 1.1715\%, 120)$$
$$+ (\$15,000,000)(P/F, 1.1715\%, 120)$$

Since table values for 1.1715% discounting factors are not available, the factors are calculated from Table 66.1.

$$(P/A, 1.1715\%, 120) = \frac{(1 + i)^n - 1}{i(1 + i)^n}$$
$$= \frac{(1 + 0.011715)^{120} - 1}{(0.011715)(1 + 0.011715)^{120}}$$
$$= 64.261$$
$$(P/F, 1.1715\%, 120) = (1 + i)^{-n} = (1 + 0.011715)^{-120}$$
$$= 0.2472$$

The present worth over 120 monthly compounding periods is

$$P = -\$7,000,000 + (\$66,500)(64.261)$$
$$+ (\$15,000,000)(0.2472)$$
$$= \$981,400$$

46. BONDS

A *bond* is a method of long-term financing commonly used by governments, states, municipalities, and very large corporations.[37] The bond represents a contract to pay the bondholder specific amounts of money at specific times. The holder purchases the bond in exchange for specific payments of interest and principal. Typical municipal bonds call for quarterly or semiannual interest payments and a payment of the *face value of the bond* on the *date of maturity* (end of the bond period).[38] Due to the practice of discounting in the bond market, a bond's face value and its purchase price generally will not coincide.

---

[37] In the past, 30-year bonds were typical. Shorter-term 10-year, 15-year, 20-year, and 25-year bonds are also commonly issued.
[38] A *fully amortized bond* pays back interest and principal throughout the life of the bond. There is no balloon payment.

In the past, a bondholder had to submit a coupon or ticket in order to receive an interim interest payment. This has given rise to the term *coupon rate*, which is the nominal annual interest rate on which the interest payments are made. Coupon books are seldom used with modern bonds, but the term survives. The coupon rate determines the magnitude of the semiannual (or otherwise) interest payments during the life of the bond. The bondholder's own effective interest rate should be used for economic decisions about the bond.

Actual *bond yield* is the bondholder's actual rate of return of the bond, considering the purchase price, interest payments, and face value payment (or, value realized if the bond is sold before it matures). By convention, bond yield is calculated as a nominal rate (rate per annum), not an effective rate per year. The bond yield should be determined by finding the effective rate of return per payment period (e.g., per semiannual interest payment) as a conventional rate of return problem. Then, the nominal rate can be found by multiplying the effective rate per period by the number of payments per year, as in Eq. 66.54.

Example 66.26

What is the maximum amount an investor should pay for a 25-year bond with a $20,000 face value and 8% coupon rate (interest only paid semiannually)? The bond will be kept to maturity. The investor's effective annual interest rate for economic decisions is 10%.

Solution

For this problem, take the compounding period to be six months. Then, there are 50 compounding periods. Since 8% is a nominal rate, the effective bond rate per period is calculated from Eq. 66.53 as

$$\phi_{\text{bond}} = \frac{r}{k} = \frac{8\%}{2} = 4\%$$

The bond payment received semiannually is

$$(0.04)(\$20,000) = \$800$$

10% is the investor's effective rate per year, so Eq. 66.54 is again used to calculate the effective analysis rate per period.

$$0.10 = (1 + \phi)^2 - 1$$
$$\phi = 0.04881 \quad (4.88\%)$$

The maximum amount that the investor should be willing to pay is the present worth of the investment.

$$P = (\$800)(P/A, 4.88\%, 50)$$
$$+ (\$20,000)(P/F, 4.88\%, 50)$$

Table 66.1 can be used to calculate the following factors.

$$(P/A, 4.88\%, 50) = \frac{(1 + 0.0488)^{50} - 1}{(0.0488)(1.0488)^{50}} = 18.600$$

$$(P/F, 4.88\%, 50) = \frac{1}{(1 + 0.0488)^{50}} = 0.09233$$

Then, the present worth is

$$P = (\$800)(18.600) + (\$20,000)(0.09233)$$
$$= \$16,727$$

47. PROBABILISTIC PROBLEMS

If an alternative's cash flows are specified by an implicit or explicit probability distribution rather than being known exactly, the problem is *probabilistic*.

Probabilistic problems typically possess the following characteristics.

- There is a chance of loss that must be minimized (or, rarely, a chance of gain that must be maximized) by selection of one of the alternatives.

- There are multiple alternatives. Each alternative offers a different degree of protection from the loss. Usually, the alternatives with the greatest protection will be the most expensive.

- The magnitude of loss or gain is independent of the alternative selected.

Probabilistic problems are typically solved using annual costs and expected values. An *expected value* is similar to an *average value* since it is calculated as the mean of the given probability distribution. If cost 1 has a probability of occurrence, p_1, cost 2 has a probability of occurrence, p_2, and so on, the expected value is

$$\mathcal{E}\{\text{cost}\} = p_1(\text{cost } 1) + p_2(\text{cost } 2) + \cdots \qquad \textbf{66.55}$$

Example 66.27

Flood damage in any year is given according to the following table. What is the present worth of flood damage for a ten-year period? Use 6% as the effective annual interest rate.

damage	probability
0	0.75
$10,000	0.20
$20,000	0.04
$30,000	0.01

Solution

The expected value of flood damage in any given year is

$$\mathcal{E}\{\text{damage}\} = (0)(0.75) + (\$10,000)(0.20)$$
$$+ (\$20,000)(0.04) + (\$30,000)(0.01)$$
$$= \$3100$$

The present worth of ten years of expected flood damage is

$$\text{present worth} = (\$3100)(P/A, 6\%, 10)$$
$$= (\$3100)(7.3601)$$
$$= \$22,816$$

Example 66.28

A dam is being considered on a river that periodically overflows and causes $600,000 damage. (The damage is essentially the same each time the river causes flooding.) The project horizon is 40 years. A 10% interest rate is being used.

Three different designs are available, each with different costs and storage capacities.

design alternative	cost	maximum capacity
A	$500,000	1 unit
B	$625,000	1.5 units
C	$900,000	2.0 units

The national weather service has provided a statistical analysis of annual rainfall runoff from the watershed draining into the river.

units annual rainfall	probability
0	0.10
0.1–0.5	0.60
0.6–1.0	0.15
1.1–1.5	0.10
1.6–2.0	0.04
2.1 or more	0.01

Which design alternative would you choose assuming the dam is essentially empty at the start of each rainfall season?

Solution

The sum of the construction cost and the expected damage should be minimized. If alternative A is chosen, it will have a capacity of 1 unit. Its capacity will be exceeded (causing $600,000 damage) when the annual rainfall exceeds 1 unit. Therefore, the expected value of the annual cost of alternative A is

$$\mathcal{E}\{\text{EUAC(A)}\} = (\$500,000)(A/P, 10\%, 40)$$
$$+ (\$600,000)(0.10 + 0.04 + 0.01)$$
$$= (\$500,000)(0.1023) + (\$600,000)(0.15)$$
$$= \$141,150$$

Similarly,

$$\mathcal{E}\{\text{EUAC(B)}\} = (\$625,000)(A/P, 10\%, 40)$$
$$+ (\$600,000)(0.04 + 0.01)$$
$$= (\$625,000)(0.1023) + (\$600,000)(0.05)$$
$$= \$93,938$$

$$\mathcal{E}\{\text{EUAC(C)}\} = (\$900,000)(A/P, 10\%, 40)$$
$$+ (\$600,000)(0.01)$$
$$= (\$900,000)(0.1023) + (\$600,000)(0.01)$$
$$= \$98,070$$

Alternative B should be chosen.

48. FIXED AND VARIABLE COSTS

The distinction between fixed and variable costs depends on how these costs vary when an independent variable changes. For example, factory or machine production is frequently the independent variable. However, it could just as easily be vehicle miles driven, hours of operation, or quantity (mass, volume, etc.).

If a cost is a function of the independent variable, the cost is said to be a *variable cost*. The change in cost per unit variable change (i.e., what is usually called the *slope*) is known as the *incremental cost*. Material and labor costs are examples of variable costs. They increase in proportion to the number of product units manufactured.

If a cost is not a function of the independent variable, it is said to be a *fixed cost*. Rent and lease payments are typical fixed costs. These costs will be incurred regardless of production levels.

Some costs have both fixed and variable components, as Fig. 66.9 illustrates. The fixed portion can be determined by calculating the cost at zero production.

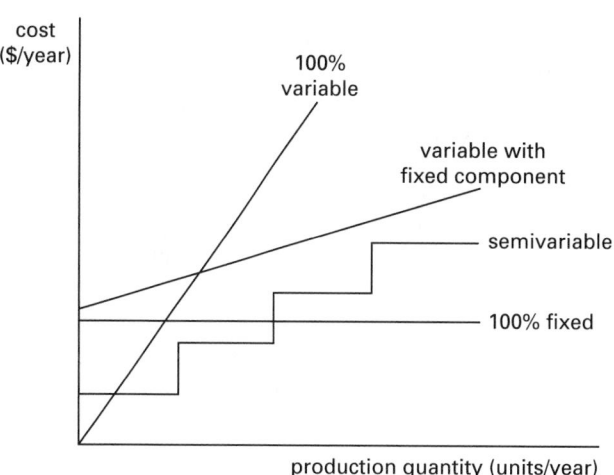

Figure 66.9 *Fixed and Variable Costs*

An additional category of cost is the *semivariable cost.* This type of cost increases step-wise. Semivariable cost structures are typical of situations where *excess capacity* exists. For example, supervisory cost is a step-wise function of the number of production shifts. Also, labor cost for truck drivers is a step-wise function of weight (volume) transported. As long as a truck has room left (i.e., excess capacity), no additional driver is needed. As soon as the truck is filled, labor cost increases.

Table 66.7 *Summary of Fixed and Variable Costs*

fixed costs
- rent
- property taxes
- interest on loans
- insurance
- janitorial service expense
- tooling expense
- setup, cleanup, and tear-down expenses
- depreciation expense
- marketing and selling costs
- cost of utilities
- general burden and overhead expense

variable costs
- direct material costs
- direct labor costs
- cost of miscellaneous supplies
- payroll benefit costs
- income taxes
- supervision costs

49. ACCOUNTING COSTS AND EXPENSE TERMS

The accounting profession has developed special terms for certain groups of costs. When annual costs are incurred due to the functioning of a piece of equipment, they are known as *operating and maintenance* (O&M) *costs.* The annual costs associated with operating a business (other than the costs directly attributable to production) are known as *general, selling, and administrative* (GS&A) *expenses.*

Direct labor costs are costs incurred in the factory, such as assembly, machining, and painting labor costs. *Direct material costs* are the costs of all materials that go into production.[39] Typically, both direct labor and direct material costs are given on a per-unit or per-item basis. The sum of the direct labor and direct material costs is known as the *prime cost.*

There are certain additional expenses incurred in the factory, such as the costs of factory supervision, stock-picking, quality control, factory utilities, and miscellaneous supplies (cleaning fluids, assembly lubricants,

routing tags, etc.) that are not incorporated into the final product. Such costs are known as *indirect manufacturing expenses* (IME) or *indirect material and labor costs.*[40] The sum of the per-unit indirect manufacturing expense and prime cost is known as the *factory cost.*

Research and development (R&D) *costs* and *administrative expenses* are added to the factory cost to give the *manufacturing cost* of the product.

Additional costs are incurred in marketing the product. Such costs are known as *selling expenses* or *marketing expenses.* The sum of the selling expenses and manufacturing cost is the *total cost* of the product.

Figure 66.10 illustrates these terms.[41]

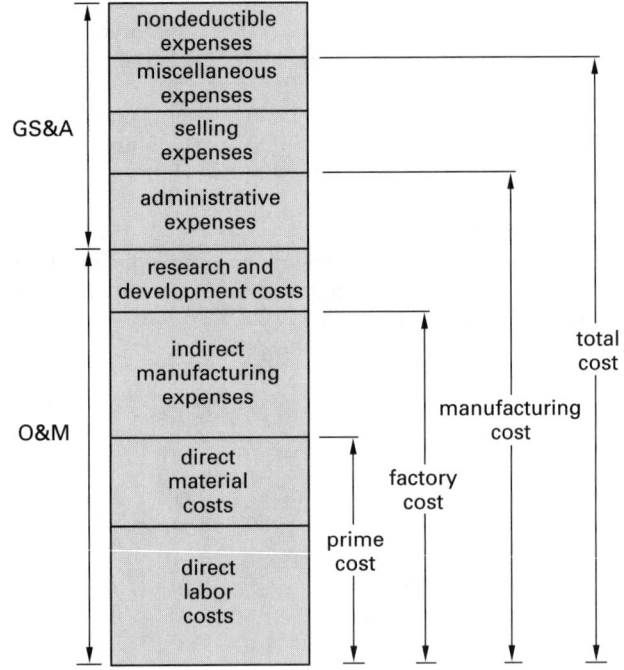

Figure 66.10 *Costs and Expenses Combined*

The distinctions among the various forms of cost (particularly with overhead costs) are not standardized. Each company must develop a classification system to deal with the various cost factors in a consistent manner. There are also other terms in use (e.g., *raw materials, operating supplies, general plant overhead*), but these terms must be interpreted within the framework of each company's classification system. Table 66.8 is typical of such classification systems.

---

[39]There may be problems with pricing the material when it is purchased from an outside vendor and the stock on hand derives from several shipments purchased at different prices.

[40]The *indirect material and labor costs* usually exclude costs incurred in the office area.

[41]Notice that *total cost* does not include income taxes.

Table 66.8 *Typical Classification of Expenses*

direct material expenses
    items purchased from other vendors
    manufactured assemblies
direct labor expenses
    machining and forming
    assembly
    finishing
    inspection
    testing
factory overhead expenses
    supervision
    benefits
        pension
        medical insurance
        vacations
    wages overhead
        unemployment compensation taxes
        social security taxes
        disability taxes
    stock-picking
    quality control and inspection
    expediting
    rework
    maintenance
    miscellaneous supplies
        routing tags
        assembly lubricants
        cleaning fluids
        wiping cloths
        janitorial supplies
    packaging (materials and labor)
    factory utilities
    laboratory
    depreciation on factory equipment
research and development expenses
    engineering (labor)
    patents
    testing
    prototypes (material and labor)
    drafting
    O&M of R&D facility
administrative expenses
    corporate officers
    accounting
    secretarial/clerical/reception
    security (protection)
    medical (nurse)
    employment (personnel)
    reproduction
    data processing
    production control
    depreciation on nonfactory equipment
    office supplies
    office utilities
    O&M of offices
selling expenses
    marketing (labor)
    advertising
    transportation (if not paid by customer)
    outside sales force (labor and expenses)
    demonstration units
    commissions
    technical service and support
    order processing
    branch office expenses
miscellaneous expenses
    insurance
    property taxes
    interest on loans
nondeductible expenses
    federal income taxes
    fines and penalties

50. ACCOUNTING PRINCIPLES

Basic Bookkeeping

An accounting or *bookkeeping system* is used to record historical financial transactions. The resultant records are used for product costing, satisfaction of statutory requirements, reporting of profit for income tax purposes, and general company management.

Bookkeeping consists of two main steps: recording the transactions, followed by categorization of the transactions.[42] The transactions (receipts and disbursements) are recorded in a *journal* (*book of original entry*) to complete the first step. Such a journal is organized in a simple chronological and sequential manner.[43] The transactions are then categorized (into interest income, advertising expense, etc.) and posted (i.e., entered or written) into the appropriate *ledger account*.

The ledger accounts together constitute the *general ledger* or *ledger*. All ledger accounts can be classified into one of three types: *asset accounts, liability accounts,* and *owners' equity accounts*. Strictly speaking, income and expense accounts, kept in a separate journal, are included within the classification of owners' equity accounts.

Together, the journal and ledger are known simply as "the books" of the company.

Balancing the Books

In a business environment, *balancing the books* means more than reconciling the checkbook and bank statements. All accounting entries must be posted in such a way as to maintain the equality of the *basic accounting equation*, Eq. 66.56.

$$\text{assets} = \text{liability} + \text{owners' equity} \qquad 66.56$$

In a *double-entry bookkeeping system*, the equality is maintained within the ledger system by entering each transaction into two balancing ledger accounts. For example, paying a utility bill would decrease the cash account (an asset account) and decrease the utility expense account (a liability account) by the same amount.

Transactions are either *debits* or *credits*, depending on their sign. Increases in asset accounts are debits; decreases are credits. For liability and equity accounts, the opposite is true: Increases are credits, and decreases are debits.[44]

---

[42]These two steps are not to be confused with the *double-entry bookkeeping method.*

[43]The two-step process is more typical of a *manual bookkeeping system* than a computerized *general ledger system.* However, even most computerized systems produce reports in journal entry order, as well as account summaries.

[44]There is a difference in sign between asset and liability accounts. Thus, an increase in an expense account is actually a decrease. The accounting profession, apparently, is comfortable with the common confusion that exists between debits and credits.

Cash and Accrual Systems[45]

The simplest form of bookkeeping is based on the *cash system*. The only transactions that are entered into the journal are those that represent cash receipts and disbursements. In effect, a checkbook register or bank deposit book could serve as the journal.

During a given period (e.g., month or quarter), expense liabilities may be incurred even though the payments for those expenses have not been made. For example, an invoice (bill) may have been received but not paid. Under the *accrual system*, the obligation is posted into the appropriate expense account before it is paid.[46] Analogous to expenses, under the accrual system, income will be claimed before payment is received. Specifically, a sales transaction can be recorded as income when the customer's order is received, when the outgoing invoice is generated, or when the merchandise is shipped.

Financial Statements

Each period, two types of corporate financial statements are typically generated: the *balance sheet* and *profit and loss* (P&L) *statements*.[47] The profit and loss statement, also known as a *statement of income and retained earnings*, is a summary of sources of *income* or *revenue* (interest, sales, fees charged, etc.) and *expenses* (utilities, advertising, repairs, etc.) for the period. The expenses are subtracted from the revenues to give a *net income* (generally, before taxes).[48] Figure 66.11 illustrates a simple profit and loss statement.

The *balance sheet* presents the *basic accounting equation* in tabular form. The balance sheet lists the major categories of assets and outstanding liabilities. The difference between asset values and liabilities is the *equity*, as defined in Eq. 66.56. This equity represents what would be left over after satisfying all debts by liquidating the company.

revenue		
interest	2000	
sales	237,000	
returns	(23,000)	
net revenue		216,000
expenses		
salaries	149,000	
utilities	6000	
advertising	28,000	
insurance	4000	
supplies	1000	
net expenses		188,000

period net income	28,000	
beginning retained earnings	63,000	
net year-to-date earnings		91,000

Figure 66.11 *Simplified Profit and Loss Statement*

There are several terms that appear regularly on balance sheets.

- *current assets:* cash and other assets that can be converted quickly into cash, such as accounts receivable, notes receivable, and merchandise (inventory). Also known as *liquid assets*.
- *fixed assets:* relatively permanent assets used in the operation of the business and relatively difficult to convert into cash. Examples are land, buildings, and equipment. Also known as *non-liquid assets*.
- *current liabilities:* liabilities due within a short period of time (e.g., within one year) and typically paid out of current assets. Examples are accounts payable, notes payable, and other accrued liabilities.
- *long-term liabilities:* obligations that are not totally payable within a short period of time (e.g., within one year).

Figure 66.12 is a simplified balance sheet.

Analysis of Financial Statements

Financial statements are evaluated by management, lenders, stockholders, potential investors, and many other groups for the purpose of determining the *health of the company*. The health can be measured in terms of *liquidity* (ability to convert assets to cash quickly), *solvency* (ability to meet debts as they become due), and *relative risk* (of which one measure is *leverage*— the portion of total capital contributed by owners).

The analysis of financial statements involves several common ratios, usually expressed as percentages. The following are some frequently encountered ratios.

- *current ratio:* an index of short-term paying ability.

$$\text{current ratio} = \frac{\text{current assets}}{\text{current liabilities}}$$

---

[45]There is also a distinction made between cash flows that are known and those that are expected. It is a *standard accounting principle* to record losses in full, at the time they are recognized, even before their occurrence. In the construction industry, for example, losses are recognized in full and projected to the end of a project as soon as they are foreseeable. Profits, on the other hand, are recognized only as they are realized (typically, as a percentage of project completion). The difference between cash and accrual systems is a matter of *bookkeeping*. The difference between loss and profit recognition is a matter of *accounting convention*. Engineers seldom need to be concerned with the accounting tradition.

[46]The expense for an item or service might be accrued even *before* the invoice is received. It might be recorded when the purchase order for the item or service is generated, or when the item or service is received.

[47]Other types of financial statements (*statements of changes in financial position, cost of sales statements*, inventory and asset reports, etc.) also will be generated, depending on the needs of the company.

[48]Financial statements also can be prepared with percentages (of total assets and net revenue) instead of dollars, in which case they are known as *common size financial statements*.

ASSETS

current assets

cash	14,000	
accounts receivable	36,000	
notes receivable	20,000	
inventory	89,000	
prepaid expenses	3000	
total current assets		162,000

plant, property,
   and equipment

land and buildings	217,000	
motor vehicles	31,000	
equipment	94,000	
accumulated depreciation	(52,000)	
total fixed assets		290,000
total assets		452,000

LIABILITIES AND OWNERS' EQUITY

current liabilities

accounts payable	66,000	
accrued income taxes	17,000	
accrued expenses	8000	
total current liabilities		91,000

long-term debt

notes payable	117,000	
mortgage	23,000	
total long-term debt		140,000

owners' and stockholders'
   equity

stock	130,000	
retained earnings	91,000	
total owners' equity		221,000

total liabilities and owners' equity   452,000

Figure 66.12 Simplified Balance Sheet

- *quick* (or *acid-test*) *ratio:* a more stringent measure of short-term debt-paying ability. The *quick assets* are defined to be current assets minus inventories and prepaid expenses.

$$\text{quick ratio} = \frac{\text{quick assets}}{\text{current liabilities}}$$

- *receivable turnover:* a measure of the average speed with which accounts receivable are collected.

$$\text{receivable turnover} = \frac{\text{net credit sales}}{\text{average net receivables}}$$

- *average age of receivables:* number of days, on the average, in which receivables are collected.

$$\text{average age of receivables} = \frac{365}{\text{receivable turnover}}$$

- *inventory turnover:* a measure of the speed with which inventory is sold, on the average.

$$\text{inventory turnover} = \frac{\text{cost of goods sold}}{\text{average cost of inventory on hand}}$$

- *days supply of inventory on hand:* number of days, on the average, that the current inventory would last.

$$\text{days supply of inventory on hand} = \frac{365}{\text{inventory turnover}}$$

- *book value per share of common stock:* number of dollars represented by the balance sheet owners' equity for each share of common stock outstanding.

$$\text{book value per share of common stock} = \frac{\text{common shareholders' equity}}{\text{number of outstanding shares}}$$

- *gross margin:* gross profit as a percentage of sales. (Gross profit is sales less cost of goods sold.)

$$\text{gross margin} = \frac{\text{gross profit}}{\text{net sales}}$$

- *profit margin ratio:* percentage of each dollar of sales that is net income.

$$\text{profit margin} = \frac{\text{net income before taxes}}{\text{net sales}}$$

- *return on investment ratio:* shows the percent return on owners' investment.

$$\text{return on investment} = \frac{\text{net income}}{\text{owners' equity}}$$

- *price-earnings ratio:* indication of relationship between earnings and market price per share of common stock, useful in comparisons between alternative investments.

$$\text{price-earnings} = \frac{\text{market price per share}}{\text{earnings per share}}$$

51. COST ACCOUNTING

Cost accounting is the system that determines the cost of manufactured products. Cost accounting is called *job cost accounting* if costs are accumulated by part number or contract. It is called *process cost accounting* if costs are accumulated by departments or manufacturing processes.

Cost accounting is dependent on historical and recorded data. The unit product cost is determined from actual expenses and numbers of units produced. Allowances (i.e., budgets) for future costs are based on these historical figures. Any deviation from historical figures is called a *variance*. Where adequate records are available, variances can be divided into *labor variance* and *material variance*.

When determining a unit product cost, the direct material and direct labor costs are generally clear-cut and easily determined. Furthermore, these costs are 100% variable costs. However, the indirect cost per unit of product is not as easily determined. Indirect costs (*burden, overhead*, etc.) can be fixed or semivariable. The amount of indirect cost allocated to a unit will depend on the unknown future overhead expense as well as the unknown future production (*vehicle size*).

A typical method of allocating indirect costs to a product is as follows.

step 1: Estimate the total expected indirect (and overhead) costs for the upcoming year.

step 2: Determine the most appropriate vehicle (basis) for allocating the overhead to production. Usually, this vehicle is either the number of units expected to be produced or the number of direct hours expected to be worked in the upcoming year.

step 3: Estimate the quantity or size of the overhead vehicle.

step 4: Divide expected overhead costs by the expected overhead vehicle to obtain the unit overhead.

step 5: Regardless of the true size of the overhead vehicle during the upcoming year, one unit of overhead cost is allocated per unit of overhead vehicle.

Once the prime cost has been determined and the indirect cost calculated based on projections, the two are combined into a *standard factory cost* or *standard cost*, which remains in effect until the next budgeting period (usually a year).

During the subsequent manufacturing year, the standard cost of a product is not generally changed merely because it is found that an error in projected indirect costs or production quantity (vehicle size) has been made. The allocation of indirect costs to a product is assumed to be independent of errors in forecasts. Rather, the difference between the expected and actual expenses, known as the *burden (overhead) variance*, experienced during the year is posted to one or more *variance accounts*.

Burden (overhead) variance is caused by errors in forecasting both the actual indirect expense for the upcoming year and the overhead vehicle size. In the former case, the variance is called *burden budget variance*; in the latter, it is called *burden capacity variance*.

Example 66.29

A company expects to produce 8000 items in the coming year. The current material cost is \$4.54 each. Sixteen minutes of direct labor are required per unit. Workers are paid \$7.50 per hour. 2133 direct labor hours are forecasted for the product. Miscellaneous overhead costs are estimated at \$45,000.

Find the (a) expected direct material cost, (b) direct labor cost, (c) prime cost, (d) burden as a function of production and direct labor, and (e) total cost.

Solution

(a) The direct material cost was given as \$4.54.

(b) The direct labor cost is

$$\left(\frac{16 \text{ min}}{60 \frac{\text{min}}{\text{hr}}} \right) \left(\frac{\$7.50}{\text{hr}} \right) = \$2.00$$

(c) The prime cost is

$$\$4.54 + \$2.00 = \$6.54$$

(d) If the burden vehicle is production, the burden rate is \$45,000/8000 = \$5.63 per item, making the total cost

$$\$4.54 + \$2.00 + \$5.63 = \$12.17$$

(e) If the burden vehicle is direct labor hours, the burden rate is \$45,000/2133 = \$21.10 per hour, making the total cost

$$\$4.54 + \$2.00 + \left(\frac{16 \text{ min}}{60 \frac{\text{min}}{\text{hr}}} \right) \left(\frac{\$21.10}{\text{hr}} \right) = \$12.17$$

Example 66.30

The actual performance of the company in Ex. 66.29 is given by the following figures.

$$\text{actual production: } 7560$$
$$\text{actual overhead costs: } \$47,000$$

What are the burden budget variance and the burden capacity variance?

Solution

The burden capacity variance is

$$\$45,000 - (7560)(\$5.63) = \$2437$$

The burden budget variance is

$$\$47,000 - \$45,000 = \$2000$$

The overall burden variance is

$$\$47,000 - (7560)(\$5.63) = \$4437$$

The sum of the burden capacity and burden budget variances should equal the overall burden variance.

$$\$2437 + \$2000 = \$4437$$

52. COST OF GOODS SOLD

Cost of goods sold (COGS) is an accounting term that represents an inventory account adjustment.[49] Cost of goods sold is the difference between the starting and ending inventory valuations. That is,

$$\begin{aligned} \text{COGS} = \text{ } & \text{starting inventory valuation} \\ & - \text{ending inventory valuation} \end{aligned} \quad 66.57$$

Cost of goods sold is subtracted from *gross profit* to determine the *net profit* of a company. Despite the fact that cost of goods sold can be a significant element in the profit equation, the inventory adjustment may not be made each accounting period (e.g., each month) due to the difficulty in obtaining an accurate inventory valuation.

With a *perpetual inventory system*, a company automatically maintains up-to-date inventory records, either through an efficient stocking and stock-releasing system or through a *point of sale* (POS) *system* integrated with the inventory records. If a company only counts its inventory (i.e., takes a *physical inventory*) at regular intervals (e.g., once a year), it is said to be operating on a *periodic inventory system*.

Inventory accounting is a source of many difficulties. The inventory value is calculated by multiplying the quantity on hand by the standard cost. In the case of completed items actually assembled or manufactured at the company, this standard cost usually is the manufacturing cost, although factory cost also can be used. In the case of purchased items, the standard cost will be the cost per item charged by the supplying vendor. In some cases, delivery and transportation costs will be included in this standard cost.

It is not unusual for the elements in an item's inventory to come from more than one vendor, or from one vendor in more than one order. Inventory valuation is more difficult if the price paid is different for these different purchases. There are four methods of determining the cost of elements in inventory. Any of these methods can be used (if applicable), but the method must be used consistently from year to year. The four methods are as follows.

- *specific identification method:* Each element can be uniquely associated with a cost. Inventory elements with serial numbers fit into this costing scheme. Stock, production, and sales records must include the serial number.

- *average cost method:* The standard cost of an item is the average of (recent or all) purchase costs for that item.

- *first-in, first-out* (FIFO) *method:* This method keeps track of how many of each item are purchased each time and the number remaining out of each purchase, as well as the price paid at each purchase. The inventory system assumes that the oldest elements are issued first.[50] Inventory value is a weighted average dependent on the number of elements from each purchase remaining. Items issued no longer contribute to the inventory value.

- *last-in, first-out* (LIFO) *method:* This method keeps track of how many of each item are purchased each time and the number remaining out of each purchase, as well as the price paid at each purchase.[51] The inventory value is a weighted average dependent on the number of elements from each purchase remaining. Items issued no longer contribute to the inventory value.

53. BREAK-EVEN ANALYSIS

Special Nomenclature

f fixed cost that does not vary with production
a incremental cost to produce one additional item (also called *marginal cost* or *differential cost*)
Q quantity sold
p incremental value (price)
R total revenue
C total cost

Break-even analysis is a method of determining when the value of one alternative becomes equal to the value of another. A common application is that of determining when costs exactly equal revenue. If the manufactured quantity is less than the break-even quantity, a loss is incurred. If the manufactured quantity is greater than the break-even quantity, a profit is made.

Assuming no change in the inventory, the *break-even point* can be found by setting costs equal to revenue ($C = R$).

$$C = f + aQ \quad\quad 66.58$$
$$R = pQ \quad\quad 66.59$$
$$Q^* = \frac{f}{p - a} \quad\quad 66.60$$

---

[49]The cost of goods sold inventory adjustment is posted to the *COGS expense account*.

[50]If all elements in an item's inventory are identical, and if all shipments of that item are agglomerated, there will be no way to guarantee that the oldest element in inventory is issued first. But, unless *spoilage* is a problem, it really does not matter.
[51]See Ftn. 50.

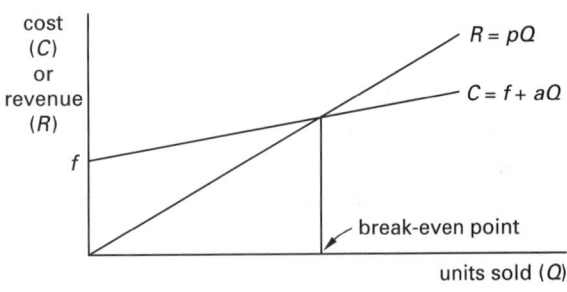

Figure 66.13 *Break-Even Quantity*

An alternative form of the break-even problem is to find the number of units per period for which two alternatives have the same total costs. Fixed costs are to be spread over a period longer than one year using the equivalent uniform annual cost (EUAC) concept. One of the alternatives will have a lower cost if production is less than the break-even point. The other will have a lower cost for production greater than the break-even point.

Example 66.31

Two plans are available for a company to obtain automobiles for its salesmen. How many miles must the cars be driven each year for the two plans to have the same costs? Use an interest rate of 10%. (Use the year-end convention for all costs.)

> *plan A:* Lease the cars and pay $0.15 per mile.
>
> *plan B:* Purchase the cars for $5000. Each car has an economic life of three years, after which it can be sold for $1200. Gas and oil cost $0.04 per mile. Insurance is $500 per year.

Solution

Let x be the number of miles driven per year. Then, the EUAC for both alternatives is

$$\text{EUAC(A)} = 0.15x$$
$$\text{EUAC(B)} = 0.04x + \$500 + (\$5000)(A/P, 10\%, 3)$$
$$- (\$1200)(A/F, 10\%, 3)$$
$$= 0.04x + \$500 + (\$5000)(0.4021)$$
$$- (\$1200)(0.3021)$$
$$= 0.04x + 2148$$

Setting EUAC(A) and EUAC(B) equal and solving for x yields 19,527 miles per year as the break-even point.

54. PAYBACK PERIOD

The *payback period* is defined as the length of time, usually in years, for the cumulative net annual profit

to equal the initial investment. It is tempting to introduce equivalence into payback period calculations, but by convention, this is generally not done.[52]

$$\text{payback period} = \frac{\text{initial investment}}{\text{net annual profit}} \qquad 66.61$$

Example 66.32

A ski resort installs two new ski lifts at a total cost of $1,800,000. The resort expects the annual gross revenue to increase by $500,000 while it incurs an annual expense of $50,000 for lift operation and maintenance. What is the payback period?

Solution

From Eq. 66.61,

$$\text{payback period} = \frac{\$1,800,000}{\$500,000 - \$50,000} = 4 \text{ years}$$

55. MANAGEMENT GOALS

Depending on many factors (market position, age of the company, age of the industry, perceived marketing and sales windows, etc.), a company may select one of many production and marketing strategic goals. Three such strategic goals are

- maximization of product demand
- minimization of cost
- maximization of profit

Such goals require knowledge of how the dependent variable (e.g., demand quantity or quantity sold) varies as a function of the independent variable (e.g., price). Unfortunately, these three goals are not usually satisfied simultaneously. For example, minimization of product cost may require a large production run to realize economies of scale, while the actual demand is too small to take advantage of such economies of scale.

If sufficient data are available to plot the independent and dependent variables, it may be possible to optimize the dependent variable graphically. Of course, if the relationship between independent and dependent variables is known algebraically, the dependent variable can be optimized by taking derivatives or by use of other numerical methods.

---

[52]Equivalence (i.e., interest and compounding) generally is not considered when calculating the "payback period." However, if it is desirable to include equivalence, then the term *payback period* should not be used. Other terms, such as *cost recovery period* or *life of an equivalent investment*, should be used. Unfortunately, this convention is not always followed in practice.

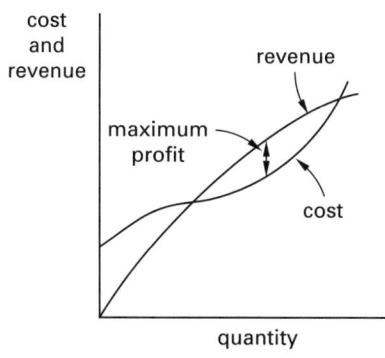

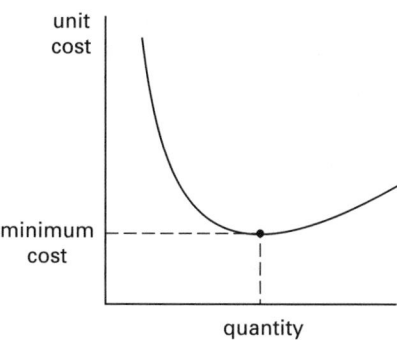

Figure 66.14 *Graphs of Management Goal Functions*

56. INFLATION

It is important to perform economic studies in terms of *constant value dollars*. One method of converting all cash flows to constant value dollars is to divide the flows by some annual *economic indicator* or price index.

If indicators are not available, cash flows can be adjusted by assuming that inflation is constant at a decimal rate (e) per year. Then, all cash flows can be converted to $t = 0$ dollars by dividing by $(1 + e)^n$, where n is the year of the cash flow.

An alternative is to replace the effective annual interest rate (i) with a value corrected for inflation. This corrected value (i') is

$$i' = i + e + ie \qquad \textbf{66.62}$$

This method has the advantage of simplifying the calculations. However, precalculated factors are not available for the noninteger values of i'. Therefore, Table 66.1 must be used to calculate the factors.

Example 66.33

What is the uninflated present worth of a $2000 future value in two years if the average inflation rate is 6% and i is 10%?

Solution

$$P = \frac{\$2000}{(1 + 0.10)^2(1 + 0.06)^2} = \$1471$$

Example 66.34

Repeat Ex. 66.33 using i'.

Solution

$$i' = 0.10 + 0.06 + (0.10)(0.06) = 0.166$$
$$P = \frac{\$2000}{(1 + 0.166)^2} = \$1471$$

57. CONSUMER LOANS

Special Nomenclature

BAL_j	balance after the jth payment
LV	principal total value loaned (cost minus down payment)
j	payment or period number
N	total number of payments to pay off the loan
PI_j	jth interest payment
PP_j	jth principal payment
PT_j	jth total payment
ϕ	effective rate per period (r/k)

Many different arrangements can be made between a borrower and a lender. With the advent of creative financing concepts, it often seems that there are as many variations of loans as there are loans made. Nevertheless, there are several traditional types of transactions. Real estate or investment texts, or a financial consultant, should be consulted for more complex problems.

Simple Interest

Interest due does not compound with a *simple interest loan*. The interest due is merely proportional to the length of time that the principal is outstanding. Because of this, simple interest loans are seldom made for long periods (e.g., more than one year). (For loans less than one year, it is commonly assumed that a year consists of 12 months of 30 days each.)

Example 66.35

A $12,000 simple interest loan is taken out at 16% per annum interest rate. The loan matures in two years with no intermediate payments. How much will be due at the end of the second year?

Solution

The interest each year is

$$\text{PI} = (0.16)(\$12,000) = \$1920$$

The total amount due in two years is

$$\text{PT} = \$12,000 + (2)(\$1920) = \$15,840$$

Example 66.36

$4000 is borrowed for 75 days at 16% per annum simple interest. How much will be due at the end of 75 days?

Solution

$$\text{amount due} = \$4000 + (0.16)\left(\frac{75 \text{ days}}{360 \frac{\text{days}}{\text{bank yr}}}\right)(\$4000)$$

$$= \$4133$$

Loans with Constant Amount Paid Toward Principal

With this loan type, the payment is not the same each period. The amount paid toward the principal is constant, but the interest varies from period to period. The equations that govern this type of loan are

$$\text{BAL}_j = \text{LV} - j(\text{PP}) \qquad 66.63$$

$$\text{PI}_j = \phi(\text{BAL})_{j-1} \qquad 66.64$$

$$\text{PT}_j = \text{PP} + \text{PI}_j \qquad 66.65$$

$$\text{PP} = \frac{\text{LV}}{N} \qquad 66.66$$

$$N = \frac{\text{LV}}{\text{PP}} \qquad 66.67$$

$$\text{LV} = (\text{PP} + \text{PI}_1)(P/A, \phi, N) \\ - \text{PI}_N(P/G, \phi, N) \qquad 66.68$$

$$1 = \left(\frac{1}{N} + \phi\right)(P/A, \phi, N) \\ - \left(\frac{\phi}{N}\right)(P/G, \phi, N) \qquad 66.69$$

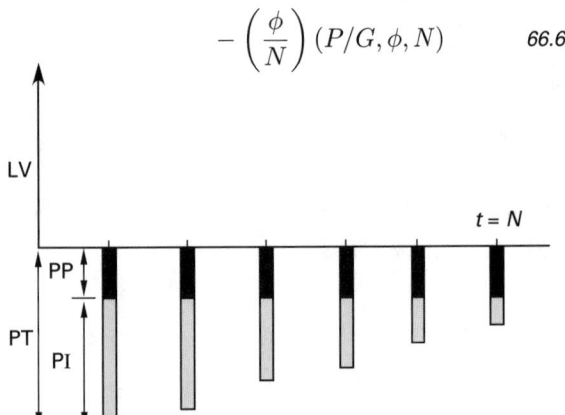

Figure 66.15 *Loan with Constant Amount Paid Toward Principal*

Example 66.37

A $12,000 six-year loan is taken from a bank that charges 15% effective annual interest. Payments toward

the principal are uniform, and repayments are made at the end of each year. Tabulate the (a) interest, (b) total payments, and (c) balance remaining after each payment is made.

Solution

(a) The amount of each principal payment is

$$\text{PP} = \frac{\$12,000}{6} = \$2000$$

At the end of the first year (before the first payment is made), the principal balance is $12,000 (i.e., $\text{BAL}_0 = \$12,000$). From Eq. 66.64, the interest payment is

$$\text{PI}_1 = (0.15)(\$12,000) = \$1800$$

(b) The total first payment is

$$\text{PT}_1 = \text{PP} + \text{PI} = \$2000 + \$1800$$

$$= \$3800$$

(c) The following table is similarly constructed.

j	BAL_j	PP_j	PI_j	PT_j
	(in dollars)			
0	12,000	–	–	–
1	10,000	2000	1800	3800
2	8000	2000	1500	3500
3	6000	2000	1200	3200
4	4000	2000	900	2900
5	2000	2000	600	2600
6	0	2000	300	2300

Direct Reduction Loans

This is the typical "interest paid on unpaid balance" loan. The amount of the periodic payment is constant, but the amounts paid toward the principal and interest both vary.

$$\text{BAL}_{j-1} = (\text{PT})\left(\frac{1 - (1 + \phi)^{j-1-N}}{\phi}\right) \qquad 66.70$$

$$\text{PI}_j = \phi(\text{BAL})_{j-1} \qquad 66.71$$

$$\text{PP}_j = \text{PT} - \text{PI}_j \qquad 66.72$$

$$\text{BAL}_j = \text{BAL}_{j-1} - \text{PP}_j \qquad 66.73$$

$$N = \frac{-\ln\left(1 - \frac{\phi(\text{LV})}{\text{PT}}\right)}{\ln(1 + \phi)} \qquad 66.74$$

Equation 66.74 calculates the number of payments necessary to pay off a loan. This equation can be solved with effort for the total periodic payment (PT) or the initial value of the loan (LV). It is easier, however, to use the $(A/P, i\%, n)$ factor to find the payment and loan value.

$$PT = (LV)(A/P, \phi\%, N) \qquad 66.75$$

If the loan is repaid in yearly installments, then i is the effective annual rate. If the loan is paid off monthly, then i should be replaced by the effective rate per month (ϕ from Eq. 66.54). For monthly payments, N is the number of months in the loan period.

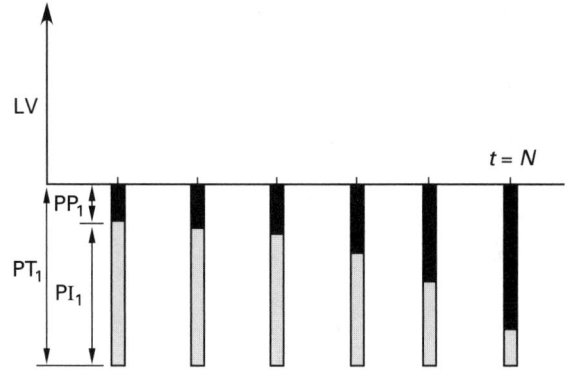

Figure 66.16 *Direct Reduction Loan*

Example 66.38

A \$45,000 loan is financed at 9.25% per annum. The monthly payment is \$385. What are the amounts paid toward interest and principal in the 14th period? What is the remaining principal balance after the 14th payment has been made?

Solution

The effective rate per month is

$$\phi = \frac{r}{k} = \frac{0.0925}{12}$$
$$= 0.0077083\ldots \quad [\text{say } 0.007708]$$

$$N = \frac{-\ln\left(1 - \frac{(0.007708)(45,000)}{385}\right)}{\ln(1 + 0.007708)} = 301$$

$$\text{BAL}_{13} = (\$385)\left(\frac{1 - (1 + 0.007708)^{14-1-301}}{0.007708}\right)$$
$$= \$44,476.39$$

$$\text{PI}_{14} = (0.007708)(\$44,476.39) = \$342.82$$

$$\text{PP}_{14} = \$385 - \$342.82 = \$42.18$$

$$\text{BAL}_{14} = \$44,476.39 - \$42.18 = \$44,434.21$$

Direct Reduction Loans with Balloon Payments

This type of loan has a constant periodic payment, but the duration of the loan is insufficient to completely pay back the principal (i.e, the loan is not fully amortized). Therefore, all remaining unpaid principal must be paid back in a lump sum when the loan matures. This large payment is known as a *balloon payment*.[53]

Equations 66.71 through 66.75 also can be used with this type of loan. The remaining balance after the last payment is the balloon payment. This balloon payment must be repaid along with the last regular payment calculated.

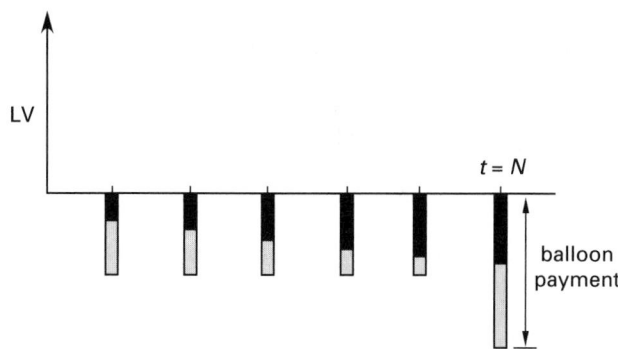

Figure 66.17 *Direct Reduction Loan with Balloon Payment*

58. FORECASTING

There are many types of forecasting models, although most are variations of the basic types.[54] All models produce a *forecast* (F_{t+1}) of some quantity (*demand* is used in this section) in the next period based on actual measurements (D_j) in current and prior periods. All of the models also try to provide *smoothing* (or *damping*) of extreme data points.

Forecasts by Moving Averages

The method of *moving average forecasting* weights all previous demand data points equally and provides some smoothing of extreme data points. The amount of smoothing increases as the number of data points, n, increases.

$$F_{t+1} = \frac{1}{n} \sum_{m=t+1-n}^{t} D_m \qquad 66.76$$

---

[53]The term *balloon payment* may include the final interest payment as well. Generally, the problem statement will indicate whether the balloon payment is inclusive or exclusive of the regular payment made at the end of the loan period.

[54]For example, forecasting models that take into consideration steady (linear), cyclical, annual, and seasonal trends are typically variations of the exponentially weighted model. A truly different forecasting tool, however, is *Monte Carlo simulation*.

Forecasts by Exponentially Weighted Averages

With *exponentially weighted forecasts*, the more current (most recent) data points receive more weight. This method uses a *weighting factor* (α), also known as a *smoothing coefficient*, which typically varies between 0.01 and 0.30. An initial forecast is needed to start the method. Forecasts immediately following are sensitive to the accuracy of this first forecast. It is common to choose $F_0 = D_1$ to get started.

$$F_{t+1} = \alpha D_t + (1 - \alpha)F_t \qquad 66.77$$

59. LEARNING CURVES

Special Nomenclature

b	learning curve constant
n	total number of items produced
R	decimal learning curve rate (2^{-b})
T_1	time or cost for the first item
T_n	time or cost for the nth item

The more products that are made, the more efficient the operation becomes due to experience gained. Therefore, direct labor costs decrease.[55] Usually, a *learning curve* is specified by the decrease in cost each time the cumulative quantity produced doubles. If there is a 20% decrease per doubling, the curve is said to be an 80% learning curve (i.e., the *learning curve rate*, R, is 80%).

Then, the time to produce the nth item is

$$T_n = T_1 n^{-b} \qquad 66.78$$

The total time to produce units from quantity n_1 to n_2 inclusive is approximately given by Eq. 66.79. T_1 is a constant, the time for item 1, and does not correspond to n unless $n_1 = 1$.

$$\int_{n_1}^{n_2} T_n \, dn \approx \left(\frac{T_1}{1-b}\right)\left(\left(n_2 + \tfrac{1}{2}\right)^{1-b} - \left(n_1 - \tfrac{1}{2}\right)^{1-b}\right)$$
$$66.79$$

The *average time per unit* over the production from n_1 to n_2 is the above total time from Eq. 66.79 divided by the quantity produced, $n_2 - n_1 + 1$.

$$T_{\text{ave}} = \frac{\displaystyle\int_{n_1}^{n_2} T_n \, dn}{n_2 - n_1 + 1} \qquad 66.80$$

Table 66.9 lists representative values of the *learning curve constant* (b). For learning curve rates not listed in the table, Eq. 66.81 can be used to find b.

$$b = \frac{-\log_{10} R}{\log_{10}(2)} = \frac{-\log_{10} R}{0.301} \qquad 66.81$$

---

[55]Remember that learning curve reductions apply only to direct labor costs. They are not applied to indirect labor or direct material costs.

Table 66.9 *Learning Curve Constants*

learning curve rate (R)	b
0.70 (70%)	0.515
0.75 (75%)	0.415
0.80 (80%)	0.322
0.85 (85%)	0.234
0.90 (90%)	0.152
0.95 (95%)	0.074

Example 66.39

A 70% learning curve is used with an item whose first production time is 1.47 hr. (a) How long will it take to produce the 11th item? (b) How long will it take to produce the 11th through 27th items?

Solution

(a) From Eq. 66.78,

$$T_{11} = (1.47 \text{ hr})(11)^{-0.515} = 0.428 \text{ hr}$$

(b) The time to produce the 11th item through 27th item is given by Eq. 66.79.

$$T \approx \left(\frac{1.47 \text{ hr}}{1 - 0.515}\right)\left((27.5)^{1-0.515} - (10.5)^{1-0.515}\right)$$
$$= 5.643 \text{ hr}$$

60. ECONOMIC ORDER QUANTITY

Special Nomenclature

a	constant depletion rate (items/unit time)
h	inventory storage cost ($\$$/item-unit time)
H	total inventory storage cost between orders ($\$$)
K	fixed cost of placing an order ($\$$)
Q	order quantity (original quantity on hand)

The *economic order quantity* (EOQ) is the order quantity that minimizes the inventory costs per unit time. Although there are many different EOQ models, the simplest is based on the following assumptions.

- Reordering is instantaneous. The time between order placement and receipt is zero.

- Shortages are not allowed.

- Demand for the inventory item is deterministic (i.e., is not a random variable).

- Demand is constant with respect to time.

- An order is placed when the inventory is zero.

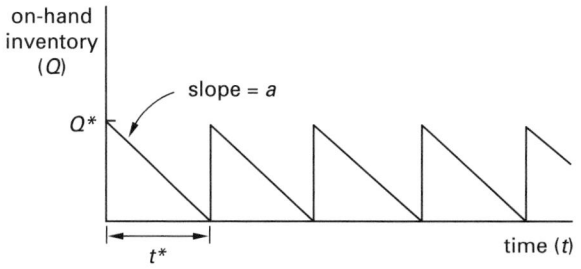

Figure 66.18 *Inventory with Instantaneous Reorder*

If the original quantity on hand is Q, the stock will be depleted at

$$t^* = \frac{Q}{a} \qquad 66.82$$

The total inventory storage cost between t_0 and t^* is

$$H = \tfrac{1}{2}Qht^* = \frac{Q^2 h}{2a} \qquad 66.83$$

The total inventory and ordering cost per unit time is

$$C_t = \frac{aK}{Q} + \frac{hQ}{2} \qquad 66.84$$

C_t can be minimized with respect to Q. The economic order quantity and time between orders are

$$Q^* = \sqrt{\frac{2aK}{h}} \qquad 66.85$$

$$t^* = \frac{Q^*}{a} \qquad 66.86$$

61. SENSITIVITY ANALYSIS

Data analysis and forecasts in economic studies require estimates of costs that will occur in the future. There are always uncertainties about these costs. However, these uncertainties are insufficient reason not to make the best possible estimates of the costs. Nevertheless, a decision between alternatives often can be made more confidently if it is known whether or not the conclusion is sensitive to moderate changes in data forecasts. Sensitivity analysis provides this extra dimension to an economic analysis.

The sensitivity of a decision is determined by inserting a range of estimates for critical cash flows and other parameters. If radical changes can be made to a cash flow without changing the decision, the decision is said to be *insensitive* to uncertainties regarding that cash flow. However, if a small change in the estimate of a cash flow will alter the decision, that decision is said to be very *sensitive* to changes in the estimate. If the decision is sensitive only for a limited range of cash

flow values, the term *variable sensitivity* is used. Figure 66.19 illustrates these terms.

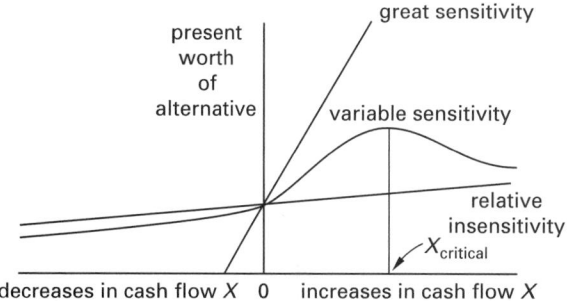

Figure 66.19 *Types of Sensitivity*

An established semantic tradition distinguishes between risk analysis and uncertainty analysis. *Risk analysis* addresses variables that have a known or estimated probability distribution. In this regard, statistics and probability theory can be used to determine the probability of a cash flow varying between given limits. On the other hand, *uncertainty analysis* is concerned with situations in which there is not enough information to determine the probability or frequency distribution for the variables involved.

As a first step, sensitivity analysis should be applied one at a time to the dominant factors. Dominant cost factors are those that have the most significant impact on the present value of the alternative.[56] If warranted, additional investigation can be used to determine the sensitivity to several cash flows varying simultaneously. Significant judgment is needed, however, to successfully determine the proper combinations of cash flows to vary. It is common to plot the dependency of the present value on the cash flow being varied in a two-dimensional graph. Simple linear interpolation is used (within reason) to determine the critical value of the cash flow being varied.

62. VALUE ENGINEERING

The *value* of an investment is defined as the ratio of its return (performance or utility) to its cost (effort or investment). The basic object of *value engineering* (VE, also referred to as *value analysis*) is to obtain the maximum per-unit value.[57]

Value engineering concepts often are used to reduce the cost of mass-produced manufactured products. This is done by eliminating unnecessary, redundant, or superfluous features, by redesigning the product for a less

---

[56]In particular, engineering economic analysis problems are sensitive to the choice of effective interest rate (i) and to accuracy in cash flows at or near the beginning of the horizon. The problems are less sensitive to accuracy in far-future cash flows, such as salvage value and subsequent generation replacement costs.

[57]Value analysis, the methodology that has become today's value engineering, was developed in the early 1950s by Lawrence D. Miles, an analyst at General Electric.

expensive manufacturing method, and by including features for easier assembly without sacrificing utility and function.[58] However, the concepts are equally applicable to one-time investments, such as buildings, chemical processing plants, and space vehicles. In particular, value engineering has become an important element in all federally funded work.[59]

Typical examples of large-scale value engineering work are using stock-sized bearings and motors (instead of custom manufactured units), replacing rectangular concrete columns with round columns (which are easier to form), and substituting custom buildings with prefabricated structures.

Value engineering is usually a team effort. And while the original designers may be on the team, usually outside consultants are utilized. The cost of value engineering is usually returned many times over by reduced construction and life-cycle costs.

---

[58] Some people say that value engineering is "the act of going over the plans and taking out everything that is interesting."
[59] U.S. Government Office of Management and Budget Circular A-131 outlines value engineering for federally funded construction projects.

67 Engineering Law[1]

1. FORMS OF COMPANY OWNERSHIP

There are three basic forms of company ownership in the United States: (a) sole proprietorship, (b) partnership, and (c) corporation.[2] Each of these forms of ownership has advantages and disadvantages.

2. SOLE PROPRIETORSHIPS

A *sole proprietorship* (*single proprietorship*) is the easiest form of ownership to establish. Other than the necessary licenses (which apply to all forms of ownership), no legal formalities are required to start business operations. A sole proprietor (the owner) has virtually total control of the business and makes all supervisory and management decisions.

Legally, there is no distinction between the sole proprietor and the sole proprietorship (the business). This is the greatest disadvantage of this form of business. The owner is solely responsible for the operation of the business, even if the owner hires others for assistance. The owner assumes personal, legal, and financial liability for all acts and debts of the company. If the company debts remain unpaid, or in the event there is a legal judgment against the company, the owner's personal assets (home, car, savings, etc.) can be seized or attached.

Another disadvantage of the sole proprietorship is the lack of significant organizational structure. In times of business crisis or trouble, there may be no one to share the responsibility or to help make decisions. When the owner is sick or dies, there may be no way to continue the business.

There is also no distinction between the incomes of the business and the owner. Therefore, the business income is taxed at the owner's income tax rate. Depending on the owner's financial position, the success of the business, and the tax structure, this can be an advantage or a disadvantage.[3]

3. PARTNERSHIPS

A *partnership* (also known as a *general partnership*) is ownership by two or more persons known as *general partners*. Legally, this form is very similar to a sole proprietorship, and the two forms of business have many of the same advantages and disadvantages. For example, with the exception of an optional *partnership agreement*, there are a minimum of formalities to setting up business. The partners make all business and management decisions themselves according to an agreed-upon process. The business income is split among the partners and taxed at the partners' individual tax rates.[4] Continuity of the business is still a problem since most partnerships are automatically dissolved upon the withdrawal or death of one of the partners.[5]

One advantage of a partnership over a sole proprietorship is the increase in available funding. Not only do more partners bring in more start-up capital, but the resource pool may make business credit easier to obtain. Also, the partners bring a diversity of skills and talents.

---

[1]The author is not giving legal advice in this chapter, nor is this chapter intended to be a substitute for professional advice. Law is not always black and white. For every rule, there are exceptions. For every legal principle, there are variations. For every type of injury, there are numerous legal precedents. This chapter covers the superficial basics of a small subset of U.S. law affecting engineers.

[2]The discussion of forms of company ownership in Secs. 67-2 through 67-4 apply equally to service-oriented companies (e.g., consulting engineering firms) and product-oriented companies.

[3]To use a simplistic example, if the corporate tax rates are higher than the individual tax rates, it would be *financially* better to be a sole proprietor because the company income would be taxed at a lower rate.

[4]The percentage split is specified in the partnership agreement.

[5]Some or all of the remaining partners may want to form a new partnership, but this is not always possible.

Unless the partnership agreement states otherwise, each partner can individually obligate (i.e., *bind*) the partnership without the consent of the other partners. Similarly, each partner has personal responsibility and liability for the acts and debts of the partnership company, just as sole proprietors do. In fact, each partner assumes the *sole* responsibility, not only a proportionate share. If one or more partners are unable to pay, the remaining partners shoulder the entire debt. The possibility of one partner having to pay for the actions of another partner must be considered when choosing this form of business ownership.

A *limited partnership* differs from a general partnership in that one or more of the partners is silent. The *limited partners* make a financial contribution to the business and receive a share of the profit but do not participate in the management and cannot bind the partnership. While *general partners* have unlimited personal liabilities, limited partners are generally liable only to the extent of their investment.[6] A written partnership agreement is required, and the agreement must be filed with the proper authorities.

4. CORPORATIONS

A corporation is a legal entity (i.e., a legal person) that is distinct from the founders and owners. The separation of ownership and management makes the corporation a fundamentally different kind of business than a sole proprietorship or partnership, with very different advantages and disadvantages.

A corporation becomes legally distinct from its founders upon formation and proper registration. Ownership of the corporation is through shares of stock, which are distributed to the founders and investors according to some agreed-upon investment and distribution rule. Thus, the founders and investors become the stockholders (i.e., owners) of the corporation. A *closely held (private) corporation* is one in which all stock is owned by a family or small group of coinvestors. A *public corporation* is one whose stock is available for the public-at-large to purchase.

There is no mandatory connection between ownership and management functions. The decision-making power is vested in the executive officers and a *board of directors* that governs by majority vote. The stockholders elect the board of directors which, in turn, hires the executive officers, management, and other employees. Employees of the corporation may or may not be stockholders.

Disadvantages (at least for a person or persons who could form a partnership or sole proprietorship instead) include the higher corporate tax rate, difficulty in forming (some states require a minimum number of persons on the board of directors), and additional legal and accounting paperwork.

However, since a corporation is distinctly separate from its founders and investors, those individuals are not liable for the acts and debts of the corporation. Debts are paid from the corporate assets. Income to the corporation is not taxable income to the owners. (Only the salaries, if any, paid to the employees by the corporation are taxable to the employees.) Even if the corporation were to go bankrupt, the assets of the owners would not ordinarily be subject to seizure or attachment.

A corporation offers the best guarantee of continuity of operation in the event of the death, incapacity, or retirement of the founders since, as a legal entity, it is distinct from the founders and owners.

5. AGENCY

In some contracts, decision-making authority and right of action are transferred from one party (the owner, or *principal*) who would normally have that authority to another person (the *agent*). For example, in construction contracts, the engineer is ordinarily the agent of the owner. Agents are limited in what they can do by the scope of the agency agreement. Within that scope, however, an agent acts on behalf of the principal, and the principal is liable for the acts of the agent and is bound by contracts made in the principal's name by the agent.

Agents are required to execute their work with care, skill, and diligence. Specifically, agents have *fiduciary responsibility* toward their principal, meaning that agent must be honest and loyal. Agents are liable for damages resulting from a lack of diligence, loyalty, and/or honesty. If the agents misrepresented their skills when obtaining the agency, they can be liable for breach of contract or fraud.

6. GENERAL CONTRACTS

A *contract* is a legally binding agreement or promise to exchange goods or services.[7] A written contract is merely a documentation of the agreement. Some agreements must be in writing, but most agreements for engineering services can be verbal, particularly if the parties to the agreement know each other well.[8] Written contract documents do not need to contain intimidating legal language, but all agreements must satisfy three basic requirements to be enforceable (binding).

---

[6]That is, if the partnership fails or is liquidated to pay debts, the limited partners lose no more than their initial investments.

[7]Not all agreements are legally binding (i.e., enforceable). Two parties may agree on something, but unless the agreement meets all of the requirements and conditions of a contract, the parties cannot hold each other to the agreement.
[8]All states have a *statute of frauds* that, among other things, specifies what types of contracts must be in writing to be enforceable. These include contracts for the sale of land, contracts requiring more than one year for performance, contracts for the sale of goods over $500 in value, contracts to satisfy the debts of another, and marriage contracts. Contracts to provide engineering services do not fall under the statute of frauds.

- There must be a clear, specific, and definite *offer* with no room for ambiguity or misunderstanding.

- There must be some form of conditional future *consideration* (i.e., payment).[9]

- There must be an *acceptance* of the offer.

There are other conditions that the agreement must meet to be enforceable. These conditions are not normally part of the explicit agreement but represent the conditions under which the agreement was made.

- The agreement must be *voluntary* for all parties.

- All parties must have *legal capacity* (i.e., be mentally competent, of legal age, and uninfluenced by drugs).

- The purpose of the agreement must be *legal*.

For small projects, a simple *letter of agreement* on one party's stationery may suffice. For larger, complex projects, a more formal document may be required. Some clients prefer to use a *purchase order*, which can function as a contract if all basic requirements are met.

Regardless of the format of the written document—letter of agreement, purchase order, or standard form—a contract should include the following features.[10]

- introduction, preamble, or preface indicating the purpose of the contract

- name, address, and business forms of both contracting parties

- signature date of the agreement

- effective date of the agreement (if different from the signature date)

- duties and obligations of both parties

- deadlines and required service dates

- fee amount

- fee schedule and payment terms

- agreement expiration date

- standard boilerplate clauses

- signatures of parties or their agents

- declaration of authority of the signatories to bind the contracting parties

- supporting documents

7. STANDARD BOILERPLATE CLAUSES

It is common for full-length contract documents to include important *boilerplate clauses*. These clauses have specific wordings that should not normally be changed, hence the name "boilerplate." Some of the most common boilerplate clauses are paraphrased here.

- Delays and inadequate performance due to war, strikes, and acts of God and nature are forgiven (*force majeure*).

- The contract document is the complete agreement, superseding all previous verbal and written agreements.

- The contract can be modified or canceled only in writing.

- Parts of the contract that are determined to be void or unenforceable shall not affect the enforceability of the remainder of the contract (*severability*). Alternatively, parts of the contract that are determined to be void or unenforceable shall be rewritten to accomplish their intended purpose without affecting the remainder of the contract.

- None (or one, or both) of the parties can (or cannot) assign its (or their) rights and responsibilities under the contract (*assignment*).

- All notices provided for in the agreement must be in writing and sent to the address in the agreement.

- Time is of the essence.[11]

- The subject headings of the agreement paragraphs are for convenience only and do not control the meaning of the paragraphs.

- The laws of the state in which the contract is signed must be used to interpret and govern the contract.

- Disagreements shall be arbitrated according to the rules of the American Arbitration Association.

- Any lawsuits related to the contract must be filed in the county and state in which the contract is signed.

- Obligations under the agreement are unique, and in the event of a breach, the defaulting party waives the defense that the loss can be adequately compensated by monetary damages (*specific performance*).

- In the event of a lawsuit, the prevailing party is entitled to an award of reasonable attorneys' and court fees.[12]

- Consequential damages are not recoverable in a lawsuit.

---

[9]Actions taken or payments made prior to the agreement are irrelevant. Also, it does not matter to the courts whether the exchange is based on equal value or not.

[10]*Construction contracts* are unique unto themselves. Items that might also be included as part of the *contract documents* are the agreement form, the general conditions, drawings, specifications, and addenda.

[11]Without this clause in writing, damages for delay cannot be claimed.

[12]Without this clause in writing, attorneys' fees and court costs are rarely recoverable.

8. SUBCONTRACTS

When a party to a contract engages a third party to perform the work in the original contract, the contract with the third party is known as a *subcontract*. Whether or not responsibilities can be subcontracted under the original contract depends on the content of the *assignment clause* in the original contract.

9. PARTIES TO A CONSTRUCTION CONTRACT

A specific set of terms has developed for referring to parties in consulting and construction contracts. The *owner* of a construction project is the person, partnership, or corporation that actually owns the land, assumes the financial risk, and ends up with the completed project. The *developer* contracts with the architect and/or engineer for the design and with the contractors for the construction of the project. In some cases, the owner and developer are the same, in which case the term *owner-developer* can be used.

The *architect* designs the project according to established codes and guidelines but leaves most stress and capacity calculations to the *engineer*.[13] Depending on the construction contract, the engineer may work for the architect, or vice versa, or both may work for the developer.

Once there are approved plans, the developer hires *contractors* to do the construction. Usually, the entire construction project is awarded to a *general contractor*. Due to the nature of the construction industry, separate *subcontracts* are used for different tasks (electrical, plumbing, mechanical, framing, fire sprinkler installation, finishing, etc.). The general contractor who hires all of these different *subcontractors* is known as the *prime contractor* (or *prime*). (The subcontractors can also work directly for the owner-developer, although this is less common.) The prime contractor is responsible for all of the acts of the subcontractors and is liable for any damage suffered by the owner-developer due to those acts.

Construction is managed by an agent of the owner-developer known as the *construction manager*, who may be the engineer, the architect, or someone else.

10. STANDARD CONTRACTS FOR DESIGN PROFESSIONALS

Several of the design engineering societies have produced standard agreement forms and other standard

documents for design professionals.[14] Among other standard forms, notices, and agreements, the following standard contracts are available.[15]

- standard contract between engineer and client
- standard contract between engineer and architect
- standard contract between engineer and contractor
- standard contract between owner and construction manager

The major advantage of the standard contracts is that the meanings of the clauses are well established, not only among the design professionals and their clients but also in the courts. The clauses in these contracts have already been litigated many times. Where a clause has been found to be unclear or ambiguous, it has been rewritten to accomplish its intended purpose.

11. CONSULTING FEE STRUCTURE

Compensation for consulting engineering services can incorporate one or more of the following concepts.

- *lump-sum fee*: This is a predetermined fee agreed upon by client and engineer. This payment can be used for small projects where the scope of work is clearly defined.
- *cost plus fixed fee*: All costs (labor, material, travel, etc.) incurred by the engineer are paid by the client. The client also pays a predetermined fee as profit. This method has an advantage when the scope of services cannot be determined accurately in advance. Detailed records must be kept by the engineer in order to allocate costs among different clients.
- *per diem fee*: The engineer is paid a specific sum for each day spent on the job. Usually, certain direct expenses (e.g., travel and reproduction) are billed in addition to the per diem rate.
- *salary plus*: The client pays for the employees on an engineer's payroll (the salary) plus an additional percentage to cover indirect overhead and profit plus certain direct expenses.
- *retainer*: This is a minimum amount paid by the client, usually in total and in advance, for a normal amount of work expected during an agreed-upon period. None of the retainer is returned, regardless of how little work the engineer performs. The engineer can be paid for additional work

---

[13]On simple small projects, such as wood-framed residential units, the design may be developed by a *building designer*. The legal capacities of building designers vary from state to state.

[14]There are two main sources of standard forms. The American Consulting Engineers' Council (ACEC), National Society of Professional Engineers (NSPE), and American Society of Civil Engineers (ASCE) have produced one set. Working independently, the American Institute of Architects (AIA) and the Associated General Contractors of America (AGC) have produced another.
[15]The Construction Specifications Institute (CSI) has produced standard specifications for materials.

beyond what is normal, however. Some direct costs, such as travel and reproduction expenses, may be billed directly to the client.

- *percentage of construction cost*: This method, which is widely used in construction design contracts, pays the architect and/or the engineer a percentage of the final total cost of the project. Cost of land, financing, and legal fees are generally not included in the construction cost, and other costs (plan revisions, project management labor, value engineering, etc.) are billed separately.

12. DISCHARGE OF A CONTRACT

A contract is normally discharged when all parties have satisfied their obligations. However, a contract can also be terminated for the following reasons.

- mutual agreement of all parties to the contract
- impossibility of performance (e.g., death of a party to the contract)
- illegality of the contract
- material breach by one or more parties to the contract
- fraud on the part of one or more parties
- failure (i.e., loss or destruction) of consideration (e.g., the burning of a building one party expected to own or occupy upon satisfaction of the obligations)

Some contracts may be dissolved by actions of the court (e.g., bankruptcy), passage of new laws and public acts, or a declaration of war.

Extreme difficulty (including economic hardship) in satisfying the contract does not discharge it, even if it becomes more costly or less profitable than originally anticipated.

13. TORTS

A *tort* is a civil wrong committed by one person causing damage to another person or person's property, emotional well-being, or reputation.[16] It is a breach of the rights of an individual to be secure in person or property. In order to correct the wrong, a civil lawsuit (*tort action* or *civil complaint*) is brought by the alleged injured party (the *plaintiff*) against the *defendant*. To be a valid *tort action* (i.e., lawsuit), there must have been

injury (i.e., damage). Generally, there will be no contract between the two parties, so the tort action cannot claim a breach of contract.[17]

Tort law is concerned with compensation for the injury, not punishment. Therefore, tort awards usually consist of general, compensatory, and special damages and rarely include punitive and exemplary damages. (See Sec. 67-17 for definitions of these damages.)

14. BREACH OF CONTRACT, NEGLIGENCE, MISREPRESENTATION, AND FRAUD

A *breach of contract* occurs when one of the parties fails to satisfy all of its obligations under a contract. The breach can be *willful* (as in a contractor walking off a construction job) or *unintentional* (as in providing less than adequate quality work or materials). A *material breach* is defined as nonperformance that results in the injured party receiving something substantially less than or different from what the contract intended.

Normally, the only redress that an *injured party* has through the courts in the event of a breach of contract is to force the breaching party to provide *specific performance*—that is, to satisfy all remaining contract provisions and to pay for any damage caused. Normally, *punitive damages* (to punish the breaching party) are unavailable.

Negligence is an action, willful or unwillful, taken without proper care or consideration for safety, resulting in damages to property or injury to persons. "Proper care" is a subjective term, but in general it is the diligence that would be exercised by a reasonably prudent person.[18] Damages sustained by a negligent act are recoverable in a tort action. (See Sec. 67-13.) If the plaintiff was partially at fault (as in the case of *comparative negligence*), the defendant will be liable only for the portion of the damage caused by the defendant.

Punitive damages are available, however, if the breaching party was fraudulent in obtaining the contract. In addition, the injured party has the right to void (nullify) the contract entirely. A *fraudulent act* is basically a special case of misrepresentation (i.e., an intentionally false statement known to be false at the time it is made). Misrepresentation that does not result in a contract is a tort. When a contract is involved, misrepresentation can be a breach of that contract (i.e., *fraud*).

---

[16]The difference between a *civil tort* (*lawsuit*) and a *criminal lawsuit* is the alleged injured party. A *crime* is a wrong against society. A criminal lawsuit is brought by the state against a defendant.

[17]It is possible for an injury to be both a breach of contract and a tort. Suppose an owner has an agreement with a contractor to construct a building, and the contract requires the contractor to comply with all state and federal safety regulations. If the owner is subsequently injured on a stairway because there was no guardrail, the injury could be recoverable both as a tort and as a breach of contract. If a third party unrelated to the contract was injured, however, that party could recover only through a tort action.

[18]Negligence of a design professional (e.g., an engineer or architect) is the absence of a *standard of care* (i.e., customary and normal care and attention) that would have been provided by other engineers. It is highly subjective.

Unfortunately, it is extremely difficult to prove *compensatory fraud* (i.e., fraud for which damages are available). Proving fraud requires showing *beyond a reasonable doubt* (a) a reckless or intentional misstatement of a material fact (b) meant to deceive, (c) resulting in misleading the innocent party to contract (d) to the innocent party's detriment.

For example, if an engineer claims to have experience in designing steel buildings but actually has none, the court might consider the misrepresentation a fraudulent action. If, however, the engineer has some experience, but an insufficient amount to do an adequate job, the engineer probably will not be considered to have acted fraudulently.

15. STRICT LIABILITY IN TORT

Strict liability in tort means that the injured party wins if the injury can be proven. It is not necessary to prove negligence, breach of explicit or implicit warranty, or the existence of a contract (*privity of contract*). Strict liability in tort is most commonly encountered in product liability cases. A defect in a product, regardless of how the defect got there, is sufficient to create strict liability in tort.

Case law surrounding defective products has developed and refined the following requirements for winning a strict liability in tort case. The following points must be proved.

- The product was defective in manufacture, design, labeling, and so on.
- The product was defective when used.
- The defect rendered the product unreasonably dangerous.
- The defect caused the injury.
- The specific use of the product that caused the damage was reasonably foreseeable.

16. MANUFACTURING AND DESIGN LIABILITY

Case law makes a distinction between *design professionals* (architects, structural engineers, building designers, etc.) and manufacturers of consumer products. Design professionals are generally consultants whose primary product is a design service sold to sophisticated clients. Consumer product manufacturers produce a specific product line sold through wholesalers and retailers to the unsophisticated public.

The law treats design professionals favorably. Such professionals are expected to meet a *standard of care* and skill that can be measured by comparison with the conduct of other professionals. However, professionals are not expected to be infallible. In the absence of a contract provision to the contrary, design professionals are not held to be guarantors of their work in the strict sense of legal liability. Damages incurred due to design errors are recoverable through tort actions, but proving a breach of contract requires showing negligence (i.e., not meeting the standard of care).

On the other hand, the law is much stricter with consumer product manufacturers, and perfection is (essentially) expected of them. They are held to the standard of strict liability in tort without regard to negligence. A manufacturer is held liable for all phases of the design and manufacturing of a product being marketed to the public.[19]

Prior to 1916, the court's position toward product defects was exemplified by the expression *caveat emptor* ("let the buyer beware").[20] Subsequent court rulings have clarified that "...a manufacturer is strictly liable in tort when an article [it] places on the market, knowing that it will be used without inspection, proves to have a defect that causes injury to a human being."[21]

Although all defectively designed products can be traced back to a design engineer or team, only the manufacturing company is usually held liable for injury caused by the product. This is more a matter of economics than justice. The company has liability insurance; the product design engineer (who is merely an employee of the company) probably does not. Unless the product design or manufacturing process is intentionally defective, or unless the defect is known in advance and covered up, the product design engineer will rarely be punished by the courts.[22]

17. DAMAGES

An injured party can sue for *damages* as well as for specific performance. Damages are the award made by the court for losses incurred by the injured party.

- *General* or *compensatory damages* are awarded to make up for the injury that was sustained.
- *Special damages* are awarded for the direct financial loss due to the breach of contract.

---

[19]The reason for this is that the public is not considered to be as sophisticated as a client who contracts with a design professional for building plans.

[20]1916, *McPherson vs. Buick*. McPherson bought a Buick from a car dealer. The car had a defective wooden steering wheel, and there was evidence that reasonable inspection would have uncovered the defect. The steering wheel injured McPherson, who then sued Buick. Buick defended itself under the ancient *prerequisite of privity* (i.e., the requirement of a face-to-face contractual relationship in order for liability to exist), since the dealer, not Buick, had sold the car to McPherson, and no contract between Buick and McPherson existed. The judge disagreed, thus establishing the concept of *third party liability* (i.e., manufacturers are responsible to consumers even though consumers do not buy directly from manufacturers).

[21]1963, *Greenman vs. Yuba Power Products*. Greenman purchased and was injured by an electric power tool.

[22]Of course, the engineer can expect to be discharged from the company. However, for strategic reasons, this discharge probably will not occur until after the company loses the case.

- *Nominal damages* are awarded when responsibility has been established but the injury is so slight as to be inconsequential.

- *Liquidated damages* are amounts that are specified in the contract document itself for nonperformance.

- *Punitive* or *exemplary damages* are awarded, usually in tort and fraud cases, to punish and make an example of the defendant (i.e., to deter others from doing the same thing).

- *Consequential damages* provide compensation for indirect losses that are incurred by the injured party but are not directly related to the contract.

18. INSURANCE

Most design firms and many independent design professionals carry *errors and omissions insurance* to protect them from claims due to their mistakes. Such policies are costly, and for that reason, some professionals choose to "go bare."[23] Policies protect against inadvertent mistakes only, not against willful, knowing, or conscious efforts to defraud or deceive.

---

[23]Going bare appears foolish at first glance, but there is a perverted logic behind the strategy. One-person consulting firms (and perhaps, firms that are not profitable) are "judgment-proof." Without insurance or other assets, these firms would be unable to pay any large judgments against them. When damage victims (and their lawyers) find this out in advance, they know that judgments will be uncollectable. So, often the lawsuit never makes its way to trial.

68 Engineering Ethics

1. CREEDS, CODES, CANONS, STATUTES, AND RULES

It is generally conceded that an individual acting on his or her own cannot be counted on to always act in a proper and moral manner. Creeds, statutes, rules, and codes all attempt to complete the guidance needed for an engineer to do "...the correct thing."

A *creed* is a statement or oath, often religious in nature, taken or assented to by an individual in ceremonies. For example, the *Engineers' Creed* adopted by the National Society of Professional Engineers is[1]

> I pledge...
>
> ... to give the utmost of performance;
>
> ... to participate in none but honest enterprise;
>
> ... to live and work according to the laws of man and the highest standards of professional conduct;
>
> ... to place service before profit, the honor and standing of the profession before personal advantage, and the public welfare above all other considerations.
>
> In humility and with need for Divine Guidance, I make this pledge.

A *code* is a system of nonstatutory, nonmandatory canons of personal conduct. A *canon* is a fundamental belief that usually encompasses several rules. For example, the code of ethics of the American Society of Civil Engineers (ASCE) contains the following seven canons.

(1) Engineers shall hold paramount the safety, health, and welfare of the public in the performance of their professional duties.

(2) Engineers shall perform services only in areas of their competence.

(3) Engineers shall issue public statements only in an objective and truthful manner.

(4) Engineers shall act in professional matters for each employer or client as faithful agents or trustees and shall avoid conflicts of interest.

(5) Engineers shall build their professional reputation on the merit of their service and shall not compete unfairly with others.

(6) Engineers shall act in such a manner as to uphold and enhance the honor, integrity, and dignity of the engineering profession.

(7) Engineers shall continue their professional development throughout their careers and shall provide opportunities for the professional development of those engineers under their supervision.

A *rule* is a guide (principle, standard, or norm) for conduct and action in a certain situation. A *statutory rule* is enacted by the legislative branch of state or federal government and carries the weight of law. Some U.S. engineering registration boards have statutory *rules of professional conduct*.

2. PURPOSE OF A CODE OF ETHICS

Many different sets of *codes of ethics* (*canons of ethics, rules of professional conduct*, etc.) have been produced by various engineering societies, registration boards, and other organizations.[2] The purpose of these ethical guidelines is to guide the conduct and decision making of engineers. Most codes are primarily educational. Nevertheless, from time to time they have been used

---

[1] The *Faith of an Engineer* adopted by the Accreditation Board for Engineering and Technology (ABET) is a similar but more detailed creed.

[2] All of the major engineering technical and professional societies in the United States (ASCE, IEEE, ASME, AIChE, NSPE, etc.) and throughout the world have adopted codes of ethics. Most U.S. societies have endorsed the *Code of Ethics of Engineers* developed by the Accreditation Board for Engineering and Technology (ABET), formerly the Engineers' Council for Professional Development (ECPD). The National Council of Examiners for Engineering and Surveying (NCEES) has developed its *Model Rules of Professional Conduct* as a guide for state registration boards in developing guidelines for the professional engineers in those states.

by the societies and regulatory agencies as the basis for disciplinary actions.

Fundamental to ethical codes is the requirement that engineers render faithful, honest, professional service. In providing such service, engineers must represent the interests of their employers or clients and, at the same time, protect public health, safety, and welfare.

There is an important distinction between what is legal and what is ethical. Many legal actions can be violations of codes of ethical or professional behavior.[3] For example, an engineer's contract with a client may give the engineer the right to assign the engineer's responsibilities, but doing so without informing the client would be unethical.

Ethical guidelines can be categorized on the basis of who is affected by the engineer's actions—the client, vendors and suppliers, other engineers, or the public at large.[4]

3. ETHICAL PRIORITIES

There are frequently conflicting demands on engineers. While it is impossible to use a single decision-making process to solve every ethical dilemma, it is clear that ethical considerations will force engineers to subjugate their own self-interests. Specifically, the ethics of engineers dealing with others need to be considered in the following order from highest to lowest priority.

- society and the public
- the law
- the engineering profession
- the engineer's client
- the engineer's firm
- other involved engineers
- the engineer personally

4. DEALING WITH CLIENTS AND EMPLOYERS

The most common ethical guidelines affecting engineers' interactions with their employer (the *client*) can be summarized as follows.[5]

- Engineers should not accept assignments for which they do not have the skill, knowledge, or time.

- Engineers must recognize their own limitations. They should use associates and other experts when the design requirements exceed their abilities.

- The client's interests must be protected. The extent of this protection exceeds normal business relationships and transcends the legal requirements of the engineer-client contract.

- Engineers must not be bound by what the client wants in instances where such desires would be unsuccessful, dishonest, unethical, unhealthy, or unsafe.

- Confidential client information remains the property of the client and must be kept confidential.

- Engineers must avoid conflicts of interest and should inform the client of any business connections or interests that might influence their judgment. Engineers should also avoid the *appearance* of a conflict of interest when such an appearance would be detrimental to the profession, their client, or themselves.

- The engineers' sole source of income for a particular project should be the fee paid by their client. Engineers should not accept compensation in any form from more than one party for the same services.

- If the client rejects the engineer's recommendations, the engineer should fully explain the consequences to the client.

- Engineers must freely and openly admit to the client any errors made.

All courts of law have required an engineer to perform in a manner consistent with normal professional standards. This is not the same as saying an engineer's work must be error-free. If an engineer completes a design, has the design and calculations checked by another competent engineer, and an error is subsequently shown to have been made, the engineer may be held responsible, but will probably not be considered negligent.

5. DEALING WITH SUPPLIERS

Engineers routinely deal with manufacturers, contractors, and vendors (*suppliers*). In this regard, engineers have great responsibility and influence. Such a relationship requires that engineers deal justly with both clients and suppliers.

An engineer will often have an interest in maintaining good relationships with suppliers since this often leads to future work. Nevertheless, relationships with suppliers must remain highly ethical. Suppliers should not be encouraged to feel that they have any special favors coming to them because of a long-standing relationship with the engineer.

---

[3]Whether the guidelines emphasize ethical behavior or professional conduct is a matter of wording. The intention is the same: to provide guidelines that transcend the requirements of the law.
[4]Some authorities also include ethical guidelines for dealing with the employees of an engineer. However, these guidelines are no different for an engineering employer than they are for a supermarket, automobile assembly line, or airline employer. Ethics is not a unique issue when it comes to employees.
[5]These general guidelines contain references to contractors, plans, specifications, and contract documents. This language is common, though not unique, to the situation of an engineer supplying design services to an owner-developer or architect. However, most of the ethical guidelines are general enough to apply to engineers in industry as well.

The ethical responsibilities relating to suppliers are listed as follows.

- The engineer must not accept or solicit gifts or other valuable considerations from a supplier during, prior to, or after any job. An engineer should not accept discounts, allowances, commissions, or any other indirect compensation from suppliers, contractors, or other engineers in connection with any work or recommendations.

- The engineer must enforce the plans and specifications (i.e., the *contract documents*) but must also interpret the contract documents fairly.

- Plans and specifications developed by the engineer on behalf of the client must be complete, definite, and specific.

- Suppliers should not be required to spend time or furnish materials that are not called for in the plans and contract documents.

- The engineer should not unduly delay the performance of suppliers.

6. DEALING WITH OTHER ENGINEERS

Engineers should try to protect the engineering profession as a whole, to strengthen it, and to enhance its public stature. The following ethical guidelines apply.

- An engineer should not attempt to maliciously injure the professional reputation, business practice, or employment position of another engineer. However, if there is proof that another engineer has acted unethically or illegally, the engineer should advise the proper authority.

- An engineer should not review another engineer's work while the other engineer is still employed, unless the other engineer is made aware of the review.

- An engineer should not try to replace another engineer once the other engineer has received employment.

- An engineer should not use the advantages of a salaried position to compete unfairly (i.e., moonlight) with other engineers who have to charge more for the same consulting services.

- Subject to legal and proprietary restraints, an engineer should freely report, publish, and distribute information that would be useful to other engineers.

7. DEALING WITH (AND AFFECTING) THE PUBLIC

In regard to the social consequences of engineering, the relationship between an engineer and the public is essentially straightforward. Responsibilities to the public demand that the engineer place service to humankind above personal gain. Furthermore, proper ethical behavior requires that an engineer avoid association with projects that are contrary to public health and welfare or that are of questionable legal character.

- Engineers must consider the safety, health, and welfare of the public in all work performed.

- Engineers must uphold the honor and dignity of their profession by refraining from self-laudatory advertising, by explaining (when required) their work to the public, and by expressing opinions only in areas of knowledge.

- When engineers issue a public statement, they must clearly indicate if the statement is being made on anyone's behalf (i.e., if anyone is benefitting from their position).

- Engineers must keep their skills at a state-of-the-art level.

- Engineers should develop public knowledge and appreciation of the engineering profession and its achievements.

- Engineers must notify the proper authorities when decisions adversely affecting public safety and welfare are made.[6]

8. COMPETITIVE BIDDING

The ethical guidelines for dealing with other engineers presented here and in more detailed codes of ethics no longer include a prohibition on *competitive bidding*. Until 1971, most codes of ethics for engineers considered competitive bidding detrimental to public welfare, since cost cutting normally results in a lower quality design.

However, in a 1971 case against the National Society of Professional Engineers that went all the way to the U.S. Supreme Court, the prohibition against competitive bidding was determined to be a violation of the Sherman Antitrust Act (i.e., it was an unreasonable restraint of trade).

The opinion of the Supreme Court does not *require* competitive bidding—it merely forbids a prohibition against competitive bidding in NSPE's code of ethics. The following points must be considered.

- Engineers and design firms may individually continue to refuse to bid competitively on engineering services.

- Clients are not required to seek competitive bids for design services.

- Federal, state, and local statutes governing the procedures for procuring engineering design services, even those statutes that prohibit competitive bidding, are not affected.

---

[6]This practice has come to be known as *whistle-blowing*.

- Any prohibitions against competitive bidding in individual state engineering registration laws remain unaffected.

- Engineers and their societies may actively and aggressively lobby for legislation that would prohibit competitive bidding for design services by public agencies.

69 Electrical Engineering Frontiers

1. ARTIFICIAL INTELLIGENCE

Artificial intelligence (AI) is the property of machines capable of reason by which they are capable of learning functions normally associated with human intelligence. Improvements in integrated circuits and computer technology have enabled concurrent advances in the application of artificial intelligence to real-world engineering devices and projects.

2. CAD/CAM

CAD is an acronym for *computer-aided design* or *drafting*.[1] CAM is an acronym for *computer-aided manufacturing*. The "CA" portion of CAD/CAM denotes the principal components, such as the hardware and software (e.g., AutoCAD). The "D" for drafting indicates the graphic language that uses lines, symbols, and words to describe components. The standards for the line conventions used in drafting are governed by the American National Standards Institute (ANSI). Advances in computer capacity and speed, driven by advances in IC technology as well as software improvements, allow the utilization of computers in design, thus improving efficiency and accuracy.

3. COMMUNICATION

Communication capability is now both global and individualistic in nature and use. Electronic engineering improvements and advances in space technology are leading to worldwide connectivity.

4. INTEGRATED CIRCUITS

The size of integrated circuits continues to shrink as design complexity increases. Wafer and chip sizes have increased while individual feature size has decreased. In 1985, approximately 256,000 transistors were manufactured on a single chip. In 1999, more than 256,000,000 transistors were placed on a single chip. Such improvements have driven advances in manufacturing automation, clean-room engineering, robotics, and so on.

The analysis of such circuits has also become more automated with SPICE and PSPICE. SPICE is the *simulation program with integrated circuit emphasis*, while PSPICE is a similar version for personal computers.

5. ENGINEERING MANAGEMENT

As many aspects of life evolve technologically, engineering management challenges grow and change. When and where to apply technological advances, and whether it is necessary to apply some advances at all, are but a few of the questions managers and individuals face. An understanding of the underlying principles of engineering and the capacities and limitations of the devices, as well as an appreciation for the human component, is vital to the decision-making process.

6. SUPERCONDUCTIVITY

Superconductivity is a property of certain materials such that at low temperatures their electrical resistivity vanishes. It is important to note that resistance occurs due to thermal scattering of electrons and due to scattering by lattice impurities. These types of resistivity are called *thermal resistivity* and *residual resistivity*, respectively.[2] At absolute zero, the total electrical resistivity is residual and minimal—that is, very near zero. This, however, is not the phenomenon of superconductivity. Below a transition temperature, called the critical temperature, certain materials exhibit a *superconducting state* in which electron interactions fundamentally change, and resistance goes to zero.

Advances in materials science have led to compounds that have transition temperatures greater than 100K, up from the single-digit transition temperatures of the 1970s.

---

[1]The acronym is sometimes given as CADD.

[2]The sum of the two resistivities gives the total resistivity according to *Matthiessen's rule*.

7. QUANTUM ELECTRODYNAMICS

Quantum electrodynamics (QED), also called the *quantum theory of light* or the *quantum theory of radiation*, successfully incorporates relativity into the explanation of electromagnetic (i.e., radiation) interactions with charged matter. This theory unites the quantum mechanical- and relativity-based view of such interactions. Further, the theory synthesizes the wave and particle views.

QED theory serves as the prototype for the quantum theory used to explain *electroweak* interactions. *Electroweak theory* is the explanation of electromagnetic and weak nuclear interactions as one-and-the-same phenomena.[3] With the addition of the electroweak theory, QED theory explains all physical phenomena of the world except those involving gravity or the nuclear force.

The force particle in electromagnetic interactions is the *photon* (sometimes called the *virtual photon*). The force particles for the weak interactions are the *Z* and *W* bosons, also known as *weakons* or *intermediate vector bosons*. The force particle for the strong nuclear force is the *gluon*. The hypothesized force particle for the gravitational force is the *graviton*.

As understanding of electrical phenomena at the quantum level deepens, the capacity, speed, and types of electrical devices increase. Research is currently under way in an attempt to create quantum mechanical computers.

8. EDUCATION

To fully appreciate engineering advances and apply them effectively, people must understand the basics of science and engineering, which begins with a foundation of mathematics and reading comprehension. The challenge of educating people throughout their lives is enormous, and requires technical and managerial skill. The ability to explain complex subjects and processes in logical terms applicable to the targeted audience, in both verbal and written formats, is a vital skill necessary for the advancement of technology and society itself.

---

[3]With this combination, three general forces remain: the *electroweak force*, the *strong nuclear* (or *color*) *force*, and the *gravitational force*. A theory combining the electroweak force and strong nuclear force is called the *grand unified theory* (GUT). A theory combining all three, that is, the GUT with gravitational theory, is referred to as the *theory of everything*.

70 Engineering Licensing in the United States

1. WHAT LICENSING IS

Engineering licensing (also known as *engineering registration*) in the United States is an examination process by which a state's *board of engineering licensing* (typically referred to as the "engineers' board" or "board of registration") determines and certifies that an engineer has achieved a minimum level of competence.[1] This process is intended to protect the public by preventing unqualified individuals from offering engineering services.

Most engineers in the United States do not need to be licensed.[2] In particular, most engineers who work for companies that design and manufacture products are exempt from the licensing requirement. This is known as the *industrial exemption*, something that is built into the laws of most states.[3]

Nevertheless, there are many good reasons for wanting to become a licensed engineer. For example, you cannot offer consulting engineering services in any state unless you are licensed in that state. Even within a product-oriented corporation, you may find that employment, advancement, and managerial positions are limited to licensed engineers.

Once you have met the licensing requirements, you will be allowed to use the titles *Professional Engineer* (PE),

Registered Engineer (RE), or *Consulting Engineer* (CE) as permitted by your state.

Although the licensing process is similar in each of the 50 states, each has its own licensing law. Unless you offer consulting engineering services in more than one state, however, you will not need to be licensed in the other states.

2. THE U.S. LICENSING PROCEDURE

The licensing procedure is similar in all states. You will take two eight-hour written examinations. The full process requires you to complete two applications, one for each of the two examinations. The first examination is the *Fundamentals of Engineering* (FE) *examination*, formerly known (and still commonly referred to) as the *Engineer-In-Training* (EIT) *examination*.[4] This examination covers basic subjects from all of the mathematics, physics, chemistry, and engineering courses you took during your university years.

The second examination is the *Professional Engineering* (PE) *examination*, also known as the *Principles and Practices* (P&P) *examination*. This examination covers only the subjects in your engineering discipline (e.g., civil, mechanical, electrical, and others).

The actual details of licensing qualifications, experience requirements, minimum education levels, fees, oral interviews, and examination schedules vary from state to state. Contact your state's licensing board for more information.

3. NATIONAL COUNCIL OF EXAMINERS FOR ENGINEERING AND SURVEYING

The *National Council of Examiners for Engineering and Surveying* (NCEES) in Clemson, South Carolina, writes, prints, distributes, and scores the national FE and PE examination.[5] The individual states purchase the examinations from NCEES and administer them. NCEES does not distribute applications to take the examinations, administer the examinations or appeals, or notify examinees of the results. These tasks are all performed by the individual states.

---

[1]Licensing of engineers is not unique to the United States. However, the practice of requiring a degreed engineer to take an examination is not common in other countries. Licensing in many countries requires a degree and may also require experience, references, and demonstrated knowledge of ethics and law, but no technical examination.

[2]Less than one-third of the degreed engineers in the United States are licensed.

[3]Only one or two states have abolished the industrial exemption. There has always been a lot of "talk" among engineers about abolishing it, but there has been little success in actually trying to do so. One of the reasons is that manufacturers' lobbies are very strong.

[4]The terms *engineering intern* (EI) and *intern engineer* (IE) have also been used in the past to designate the status of an engineer who has passed the first exam. These uses are rarer but may still be encountered in some states.

[5]National Council of Examiners for Engineering and Surveying, P.O. Box 1686, Clemson, SC 29633, (803) 654-6824.

4. RECIPROCITY AMONG STATES

With minor exceptions, having a license from one state will not permit you to practice engineering in another state. You must have a professional engineering license from each state in which you work. For most engineers, this is not a problem, but for some it is. Luckily, it is not too difficult to get a license from every state you work in once you have a license from one of them.

All states use the NCEES examinations. If you take and pass the FE or PE examination in one state, your certificate or license will be honored by all of the other states. Upon proper application, payment of fees, and proof of your license, you will be issued a license by the new state. Although there may be other special requirements imposed by a state, it will not be necessary to retake the FE or PE examinations. The issuance of an engineering license based on another state's licensing is known as *reciprocity* or *comity*.

5. UNIFORM EXAMINATIONS

Although each state has its own licensing law and is, theoretically, free to administer its own exams, none does so for the major disciplines. All states have chosen to use the NCEES exams. Each state administers the exams on the same days. The exams from all the states are sent to NCEES and are graded by the same graders. Each state adopts the cut-off passing scores recommended by NCEES. These practices have led to the term *uniform examination*.

6. APPLYING FOR THE EXAMINATION

While the exam administration process is essentially standardized among the states, there is a lot of variation in the application process. Each state has its own application forms, charges different fees, and has different age, education, and experience requirements. Therefore, you will have to request an application from the state in which you want to become licensed in order to find out what is required. It is generally sufficient for you to phone for this application; a written request is unnecessary. You will find contact information (websites, telephone numbers, email addresses, etc.) for all U.S. state and territorial boards of registration at www.ppi2pass.com. Click on the State Boards link.

As with any other important document, it is a good idea to keep a copy of your examination application and send the original application by certified mail, requesting a receipt of delivery. Keep your proof of mailing and delivery receipt with your copy of the application.

All states make special accommodations for persons who are physically challenged or who have other special needs. Be sure to communicate your need to the state board well in advance of the examination day.

7. EXAMINATION DATES

The national FE and PE examinations are administered twice a year (usually in mid-April and late October), on the same weekends in all states. For a current exam schedule, click on the Exam FAQs link at www.ppi2pass.com.

8. FE EXAMINATION FORMAT

The FE examination consists of two four-hour sessions separated by a one-hour lunch period. There are 120 multiple-choice problems in the morning and 60 multiple-choice problems in the afternoon. All problems have four answer choices, from which you are to choose the best single answer. Afternoon problems are slightly more difficult and have double weight.

Questions from all undergraduate technical courses appear in the examination, including mathematics, chemistry, physics, and engineering.[6]

The FE exam is essentially all in SI units. There are a few non-SI problems and some opportunities to choose between identical SI and non-SI problems.

NCEES provides its own reference booklet for use in the examination. You may bring your own calculator, as long as it is noncommunicating, battery-operated, silent, and nonprinting. Mechanical pencils will be provided at the examination site. You will not be allowed to use any of your own books or notes.

9. PE EXAMINATION FORMAT

In April 2002, the name of the NCEES PE electrical exam changed to the "electrical and computer engineering examination." The format is now breadth-and-depth, and the exam consists of two four-hour sessions separated by a one-hour lunch period. The exam only covers subjects in a major field of study (i.e., electrical and computer engineering). This is an open-book exam. Although some states have a few restrictions, generally you may bring in the calculators and books of your choice. Calculators must be noncommunicating, battery-operated, silent, and nonprinting. (Visit www.ppi2pass.com to get the most up-to-date exam details.)

The morning session of the exam will focus on the breadth of the electrical and computer engineering field. All examinees will take the same morning exam. The questions will be multiple choice. Most questions will be unique; that is, they will be scenario-based with one question per scenario. There may be some multi-part questions, where one problem statement (i.e., scenario) is followed by two or three questions. The multi-part

---

[6]The format of the FE exam, tips on passing, and other valuable information are given in greater detail in *FE Review Manual*, (Lindeburg), and *Engineer-In-Training Reference Manual* (Lindeburg), both published by Professional Publications. These publications are also very useful basic study aids for the PE exam.

questions will nevertheless be independent; that is, you will not have to answer the first question correctly in order to get the second or third correct (they do not "cascade"). A total of 40 questions are expected, and you must answer them all to get full credit. The problems will be traditional in nature and tend toward the academic—questions on electrical engineering basics that all electrical engineers should know. The subjects likely to be tested in the morning "breadth" session are given in Table 2 of the Introduction to this book.

The afternoon session will focus on the depth of the electrical and computer engineering field. You will select one depth module to work, from these three choices: (1) computer engineering; (2) electronics, controls, and communications; and (3) power. The questions will be multiple choice, with a total of 40 questions in each module. The modules will be printed in a single examination booklet, so you will have an opportunity to review all three before deciding which module to answer. (The longer you take to decide, however, the less time you'll have for answering questions, so it helps to have an idea of your favored module prior to the exam.) The subjects likely to be tested in the afternoon "depth" modules are given in Table 2 of the Introduction to this book.

The problems in the afternoon session will be more difficult than those in the morning session. You may be required to call upon nonacademic knowledge that stems from practical experience. The exam may include terminology used in the field, without explanation. You may be asked to display current knowledge (within the last five years) of your chosen field. To simplify the problems, you may want to use shortcuts known to practicing engineers. Additionally, the afternoon exam may require the knowledge of use of codes such as the National Electrical Code (NEC), the Code of Federal Regulations (CFR), Occupational Safety and Health Administration (OSHA) rules, and various computer codes and standards.

The electrical engineering field primarily uses SI units (the mksa system). Accordingly, SI units are used extensively on the exam. There are a few exceptions, however. For example, the majority of questions relating to the National Electrical Code questions and wiring in general will utilize U.S. units. Additionally, magnetic questions may use the cgs system.

Topic XVIII: Support Material

Support
Material

For the most current information about the exam, visit **www.ppi2pass.com** regularly.

PROFESSIONAL PUBLICATIONS, INC.

Appendices
Table of Contents

APPENDIX 1.A
Conversion Factors

multiply	by	to obtain
acres	0.4047	hectares
	43,560.0	square feet
	1.5625×10^{-3}	square miles
ampere-hours	3600.0	coulombs
angstrom units	3.937×10^{-9}	inches
	1×10^{-4}	microns
astronomical units	1.496×10^8	kilometers
atmospheres	76.0	centimeters of mercury
atomic mass unit	9.316×10^8	electron-volts
	1.492×10^{-10}	joules
	1.66×10^{-27}	kilograms
BeV (also GeV)	10^9	electron-volts
Btu	3.93×10^{-4}	horsepower-hours
	778.3	foot-pounds
	2.93×10^{-4}	kilowatt hours
	1.0×10^{-5}	therms
Btu/hr	0.2161	foot-pounds/sec
	3.929×10^{-4}	horsepower
	0.293	watts
bushels	2150.4	cubic inches
calories, gram (mean)	3.9683×10^{-3}	Btu (mean)
centares	1.0	square meters
centimeters	1×10^{-5}	kilometers
	1×10^{-2}	meters
	10.0	millimeters
	3.281×10^{-2}	feet
	0.3937	inches
chains	792.0	inches
coulombs	1.036×10^{-5}	faradays
cubic centimeters	0.06102	cubic inches
	2.113×10^{-3}	pints (U.S. liquid)
cubic feet	0.02832	cubic meters
	7.4805	gallons
cubic feet/min	62.43	pounds H_2O/min
cubic feet/sec	448.831	gallons/min
	0.64632	millions of gallons per day
cubits	18.0	inches
days	86,400.0	seconds
degrees (angle)	1.745×10^{-2}	radians
degrees/sec	0.1667	revolutions/min
dynes	1×10^{-5}	newtons
electron-volts	1.074×10^{-9}	atomic mass units
	10^{-9}	BeV (also GeV)
	1.602×10^{-19}	joules
	1.783×10^{-36}	kilograms
	10^{-6}	MeV
faradays/sec	96,500	amperes (absolute)
fathoms	6.0	feet
feet	30.48	centimeters
	0.3048	meters
	1.645×10^{-4}	miles (nautical)
	1.894×10^{-4}	miles (statute)
feet/min	0.5080	centimeters/sec
feet/sec	0.592	knots
	0.6818	miles/hr

multiply	by	to obtain
foot-pounds	1.285×10^{-3}	Btu
	5.051×10^{-7}	horsepower-hours
	3.766×10^{-7}	kilowatt-hours
foot-pound/sec	4.6272	Btu/hr
	1.818×10^{-3}	horsepower
	1.356×10^{-3}	kilowatts
furlongs	660.0	feet
	0.125	miles (statute)
gallons	0.1337	cubic feet
	3.785	liters
gallons H_2O	8.3453	pounds H_2O
gallons/min	8.0208	cubic feet/hr
	0.002228	cubic feet/sec
GeV (also BeV)	10^9	electron-volts
grams	10^{-3}	kilograms
	3.527×10^{-2}	ounces (avoirdupois)
	3.215×10^{-2}	ounces (troy)
	2.205×10^{-3}	pounds
hectares	2.471	acres
	1.076×10^5	square feet
horsepower	2545.0	Btu/hr
	42.44	Btu/min
	550	foot-pounds/sec
	0.7457	kilowatts
	745.7	watts
horsepower-hours	2545.0	Btu
	1.976×10^{-6}	foot-pounds
	0.7457	kilowatt-hours
hours	4.167×10^{-2}	days
	5.952×10^{-3}	weeks
inches	2.540	centimeters
	1.578×10^{-5}	miles
inches, H_2O	5.199	pounds force/ft^2
	0.0361	psi
	0.0735	inches, mercury
inches, mercury	70.7	pounds force/ft^2
	0.491	pounds force/in^2
	13.60	inches, H_2O
joules	6.705×10^9	atomic mass units
	9.480×10^{-4}	Btu
	1×10^7	ergs
	6.242×10^{18}	electron-volts
	1.113×10^{-17}	kilograms
kilograms	6.025×10^{26}	atomic mass units
	5.610×10^{35}	electron-volts
	8.987×10^{16}	joules
	2.205	pounds
kilometers	3281.0	feet
	1000.0	meters
	0.6214	miles
kilometers/hr	0.5396	knots
kilowatts	3412.9	Btu/hr
	737.6	foot-pounds/sec
	1.341	horsepower
kilowatt-hours	3413.0	Btu

(continued)

Support Material

APPENDIX 1.A *(continued)*
Conversion Factors

multiply	by	to obtain	multiply	by	to obtain
knots	6076.0	feet/hr	poise	0.002089	pound-sec/ft^2
	1.0	nautical miles/hr	pounds	0.4536	kilograms
	1.151	statute miles/hr		16.0	ounces
light years	5.9×10^{12}	miles		14.5833	ounces (troy)
links (surveyor)	7.92	inches		1.21528	pounds (troy)
liters	1000.0	cubic centimeters	pounds/ft^2	0.006944	pounds/in^2
	61.02	cubic inches	pounds/in^2	2.308	feet, H$_2$O
	0.2642	gallons (U.S. liquid)		27.7	inches, H$_2$O
	1000.0	milliliters		2.037	inches, mercury
	2.113	pints		144	pounds/ft^2
MeV	10^6	electron-volts	quarts (dry)	67.20	cubic inches
meters	100.0	centimeters	quarts (liquid)	57.75	cubic inches
	3.281	feet		0.25	gallons
	1×10^{-3}	kilometers		0.9463	liters
	5.396×10^{-4}	miles (nautical)	radians	57.30	degrees
	6.214×10^{-4}	miles (statute)		3438.0	minutes
	1000.0	millimeters	revolutions	360.0	degrees
microns	1×10^{-6}	meters	revolutions/min	6.0	degrees/sec
miles (nautical)	6076	feet	rods	16.5	feet
	1.853	kilometers		5.029	meters
	1.1516	miles (statute)	rods (surveyor)	5.5	yards
miles (statute)	5280.0	feet	seconds	1.667×10^{-2}	minutes
	1.609	kilometers	square meters/sec	10^6	centistokes
	0.8684	miles (nautical)		10.76	square feet/sec
miles/hr	88.0	feet/min		10^4	stokes
milligrams/liter	1.0	parts/million	slugs	32.174	pounds mass
milliliters	1×10^{-3}	liters	stokes	0.0010764	square feet/sec
millimeters	3.937×10^{-2}	inches	tons (long)	1016.0	kilograms
newtons	1×10^5	dynes		2240.0	pounds
ohms (international)	1.0005	ohms (absolute)		1.120	tons (short)
ounces	28.349527	grams	tons (short)	907.1848	kilograms
	6.25×10^{-2}	pounds		2000.0	pounds
ounces (troy)	1.09714	ounces (avoirdupois)		0.89287	tons (long)
parsecs	3.086×10^{13}	kilometers	volts (absolute)	3.336×10^{-3}	statvolts
	1.9×10^{13}	miles	watts	3.4129	Btu/hr
pascal-sec	1000	centipoise		1.341×10^{-3}	horsepower
	10	poise	yards	0.9144	meters
	0.02089	pound force-sec/ft^2		4.934×10^{-4}	miles (nautical)
	0.6720	pound mass/ft-sec		5.682×10^{-4}	miles (statute)
	0.02089	slug/ft-sec			
pints (liquid)	473.2	cubic centimeters			
	28.87	cubic inches			
	0.125	gallons			
	0.5	quarts (liquid)			

APPENDIX 1.B
Common SI Unit Conversion Factors

multiply	by	to obtain
AREA		
circular mil	506.7	square micrometer
square foot	0.0929	square meter
square kilometer	0.3861	square mile
square meter	10.764	square foot
	1.196	square yard
square micrometer	0.001974	circular mil
square mile	2.590	square kilometer
square yard	0.8361	square meter
ENERGY		
Btu (international)	1.0551	kilojoule
erg	0.1	microjoule
foot-pound	1.3558	joule
horsepower-hour	2.6485	megajoule
joule	0.7376	foot-pound
	0.10197	meter-kilogram force
kilogram-calorie (international)	4.1868	kilojoule
kilojoule	0.9478	Btu
	0.2388	kilogram-calorie
kilowatt-hour	3.6	megajoule
megajoule	0.3725	horsepower-hour
	0.2778	kilowatt-hour
	0.009478	therm
meter-kilogram force	9.8067	joule
microjoule	10.0	erg
therm	105.506	megajoule
FORCE		
dyne	10.0	micronewton
kilogram force	9.8067	newton
kip	4448.2	newton
micronewton	0.1	dyne
newton	0.10197	kilogram force
	0.0002248	kip
	3.597	ounce force
	0.2248	pound force
ounce force	0.2780	newton
pound force	4.4482	newton
HEAT		
Btu/ft^2-hr	3.1546	$watt/m^2$
Btu/ft^2-°F	5.6783	$watt/m^2 \cdot °C$
Btu/ft^3	0.0373	$megajoule/m^3$
Btu/ft^3-°F	0.06707	$megajoule/m^3 \cdot °C$
Btu/hr	0.2931	watt
Btu/lbm	2326	joule/kg
Btu/lbm-°F	4186.8	joule/kg·°C
Btu-in/ft^2-hr-°F	0.1442	watt/m·°C
joule/kg	0.000430	Btu/lbm
joule/kg·°C	0.0002388	Btu/lbm-°F
$megajoule/m^3$	26.839	Btu/ft^3
$megajoule/m^3 \cdot °C$	14.911	Btu/ft^3-°F
watt	3.4121	Btu/hr
watt/m·°C	6.933	Btu-in/ft^2-hr-°F
$watt/m^2$	0.3170	Btu/ft^2-hr
$watt/m^2 \cdot °C$	0.1761	Btu/ft^2-hr

(continued)

Support Material

APPENDIX 1.B *(continued)*
Common SI Unit Conversion Factors

multiply	by	to obtain
LENGTH		
angstrom	0.1	nanometer
foot	0.3048	meter
inch	25.4	millimeter
kilometer	0.6214	mile
	0.540	mile (nautical)
meter	3.2808	foot
	1.0936	yard
micrometer	1.0	micron
micron	1.0	micrometer
mil	0.0254	millimeter
mile	1.6093	kilometer
mile (nautical)	1.852	kilometer
millimeter	0.0394	inch
	39.370	mil
nanometer	10.0	angstrom
yard	0.9144	meter
MASS (weight)		
grain	64.799	milligram
gram	0.0353	ounce (avoirdupois)
	0.03215	ounce (troy)
kilogram	2.2046	pound mass
	0.068522	slug
	0.0009842	ton (long—2240 lbm)
	0.001102	ton (short—2000 lbm)
milligram	0.0154	grain
ounce (avoirdupois)	28.350	gram
ounce (troy)	31.1035	gram
pound mass	0.4536	kilogram
slug	14.5939	kilogram
ton (long—2240 lbm)	1016.047	kilogram
ton (short—2000 lbm)	907.185	kilogram
PRESSURE		
bar	100.0	kilopascal
inch, H_2O (20°C)	0.2486	kilopascal
inch, Hg (20°C)	3.3741	kilopascal
kilogram force/cm^2	98.067	kilopascal
kilopascal	0.01	bar
	4.0219	inch, H_2O (20°C)
	0.2964	inch, Hg (20°C)
	0.0102	kilogram force/cm^2
	7.528	millimeter Hg (20°C)
	0.1450	pound force/in^2
	0.009869	standard atmosphere (760 torr)
	7.5006	torr
millimeter Hg (20°C)	0.13284	kilopascal
pound force/in^2	6.8948	kilopascal
standard atmosphere (760 torr)	101.325	kilopascal
torr	0.13332	kilopascal

(continued)

APPENDIX 1.B *(continued)*
Common SI Unit Conversion Factors

multiply	by	to obtain
POWER		
Btu (international)/hr	0.2931	watt
foot-pound/sec	1.3558	watt
horsepower	0.7457	kilowatt
kilowatt	1.341	horsepower
	0.2843	ton of refrigeration
meter·kilogram force/sec	9.8067	watt
ton of refrigeration	3.517	kilowatt
watt	3.4122	Btu (international)/hr
	0.7376	foot-pound/sec
	0.10197	meter·kilogram force/sec
TEMPERATURE		
Celsius	$\frac{9}{5}°C + 32$	Fahrenheit
Fahrenheit	$\frac{5}{9}(°F - 32)$	Celsius
Kelvin	$\frac{9}{5}$	Rankine
Rankine	$\frac{5}{9}$	Kelvin
TORQUE		
gram force·centimeter	0.098067	millinewton·meter
kilogram force·meter	9.8067	newton·meter
millinewton	10.197	gram force·centimeter
newton·meter	0.10197	kilogram force·meter
	0.7376	foot·pound
	8.8495	inch·pound
foot-pound	1.3558	newton·meter
inch-pound	0.1130	newton·meter
VELOCITY		
feet/sec	0.3048	meters/sec
kilometers/hr	0.6214	miles/hr
meters/sec	3.2808	feet/sec
	2.2369	miles/hr
miles/hr	1.60934	kilometers/hr
	0.44704	meters/sec
VISCOSITY		
centipoise	1.0	millipascal·second
centistoke	1.0	$micrometer^2/sec$
millipascal·second	1.0	centipoise
$micrometer^2/sec$	1.0	centistoke

APPENDIX 1.B *(continued)*
Common SI Unit Conversion Factors

multiply	by	to obtain
VOLUME (capacity)		
cubic centimeter	0.06102	cubic inch
cubic foot	28.3168	liter
cubic inch	16.3871	cubic centimeter
cubic meter	1.308	cubic yard
cubic yard	0.7646	cubic meter
gallon (U.S.)	3.785	liter
liter	0.2642	gallon (U.S.)
	2.113	pint (U.S. fluid)
	1.0567	quart (U.S. fluid)
	0.03531	cubic foot
milliliter	0.0338	ounce (U.S. fluid)
ounce (U.S. fluid)	29.574	milliliter
pint (U.S. fluid)	0.4732	liter
quart (U.S. fluid)	0.9464	liter
VOLUME FLOW (gas–air)		
cubic meter/sec	2119	cubic foot/min
liter/sec	2.119	cubic foot/min
microliter/sec	0.000127	cubic foot/hr
milliliter/sec	0.002119	cubic foot/min
	0.127133	cubic foot/hr
cubic foot/min	0.0004719	cubic meter/sec
	0.4719	liter/sec
	471.947	milliliter/sec
cubic foot/hr	7866	microliter/sec
	7.8658	milliliter/sec
VOLUME FLOW (liquid)		
gallon/hr (U.S.)	0.001052	liter/sec
gallon/min (U.S.)	0.06309	liter/sec
liter/sec	951.02	gallon/hr (U.S.)
	15.850	gallon/min (U.S.)

APPENDIX 8.A
Mensuration of Two-Dimensional Areas

Nomenclature

A total surface area
b base
c chord length
d distance
h height
L length
p perimeter
r radius
s side (edge) length, arc length
θ vertex angle, in radians
ϕ central angle, in radians

Triangle

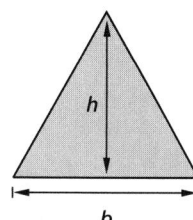 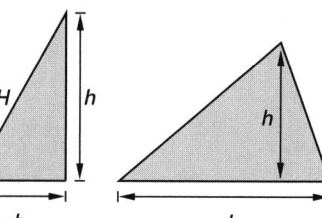

equilateral   right   oblique

$$A = \tfrac{1}{2}bh = \frac{\sqrt{3}}{4}b^2 \qquad A = \tfrac{1}{2}bh \qquad A = \tfrac{1}{2}bh$$

$$h = \frac{\sqrt{3}}{2}b \qquad\qquad H^2 = b^2 + h^2$$

Circle

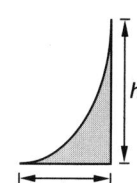

$$p = 2\pi r$$

$$A = \pi r^2 = \frac{p^2}{4\pi}$$

Circular Segment

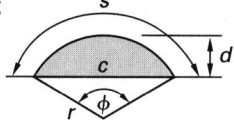

$$A = \tfrac{1}{2}r^2(\phi - \sin\phi)$$

$$\phi = \frac{s}{r} = 2\left(\arccos\frac{r-d}{r}\right)$$

$$c = 2r\sin\left(\frac{\phi}{2}\right)$$

Circular Sector

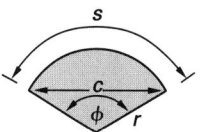

$$A = \tfrac{1}{2}\phi r^2 = \tfrac{1}{2}sr$$

$$\phi = \frac{s}{r}$$

$$s = r\phi$$

$$c = 2r\sin\left(\frac{\phi}{2}\right)$$

Parabola

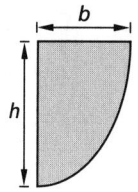

$$A = \tfrac{2}{3}bh$$

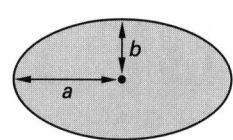

$$A = \tfrac{1}{3}bh$$

Ellipse

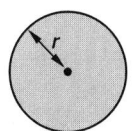

$$A = \pi ab$$

$$p = 2\pi\sqrt{\tfrac{1}{2}(a^2 + b^2)}$$

(continued)

APPENDIX 8.A *(continued)*
Mensuration of Two-Dimensional Areas

Trapezoid

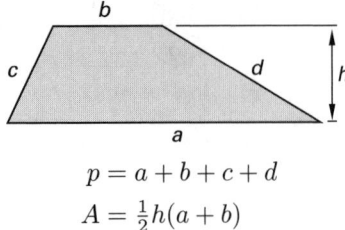

$$p = a + b + c + d$$
$$A = \tfrac{1}{2}h(a + b)$$

The trapezoid is isosceles if $c = d$.

Parallelogram

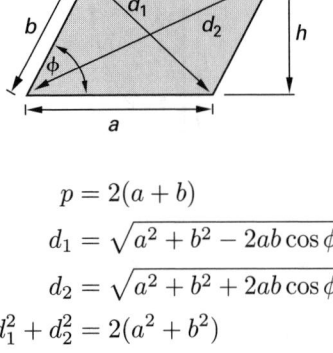

$$p = 2(a + b)$$
$$d_1 = \sqrt{a^2 + b^2 - 2ab\cos\phi}$$
$$d_2 = \sqrt{a^2 + b^2 + 2ab\cos\phi}$$
$$d_1^2 + d_2^2 = 2(a^2 + b^2)$$
$$A = ah = ab\sin\phi$$

If $a = b$, the parallelogram is a rhombus.

Regular Polygon (*n* equal sides)

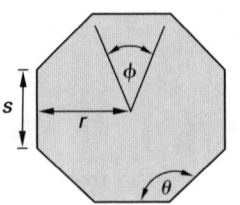

$$\phi = \frac{2\pi}{n}$$
$$\theta = \frac{\pi(n - 2)}{n} = \pi - \phi$$
$$p = ns$$
$$s = 2r\tan\left(\frac{\theta}{2}\right)$$
$$A = \tfrac{1}{2}nsr$$

sides	name	area (A) when diameter of inscribed circle = 1	area (A) when side = 1	radius (r) of circumscribed circle when side = 1	length (L) of side when radius (r) of circumscribed circle = 1	length (L) of side when perpendicular to center = 1	perpendicular (p) to center when side = 1
3	triangle	1.299	0.433	0.577	1.732	3.464	0.289
4	square	1.000	1.000	0.707	1.414	2.000	0.500
5	pentagon	0.908	1.720	0.851	1.176	1.453	0.688
6	hexagon	0.866	2.598	1.000	1.000	1.155	0.866
7	heptagon	0.843	3.634	1.152	0.868	0.963	1.038
8	octagon	0.828	4.828	1.307	0.765	0.828	1.207
9	nonagon	0.819	6.182	1.462	0.684	0.728	1.374
10	decagon	0.812	7.694	1.618	0.618	0.650	1.539
11	undecagon	0.807	9.366	1.775	0.563	0.587	1.703
12	dodecagon	0.804	11.196	1.932	0.518	0.536	1.866

APPENDIX 8.B
Mensuration of Three-Dimensional Volumes

Nomenclature
A area
b base
h height
r radius
R radius
s side (edge) length
V volume

Sphere

$$V = \frac{4\pi r^3}{3}$$
$$A = 4\pi r^2$$

Right Circular Cone

$$V = \frac{\pi r^2 h}{3}$$
$$A = \pi r \sqrt{r^2 + h^2}$$

(does not include base area)

Right Circular Cylinder

$$V = \pi r^2 h$$
$$A = 2\pi r h$$

(does not include end area)

Spherical Segment

Surface area of a spherical segment of radius r cut out by an angle θ_0 rotated from the center about a radius, r, is

$$A = 2\pi r^2 \left(1 - \cos\theta_0\right)$$
$$\phi = \frac{A}{r^2} = 2\pi \left(1 - \cos\theta_0\right)$$

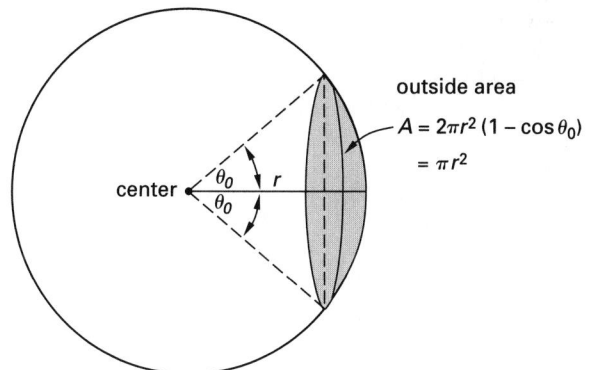

Paraboloid of Revolution

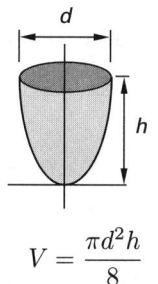

$$V = \frac{\pi d^2 h}{8}$$

Torus

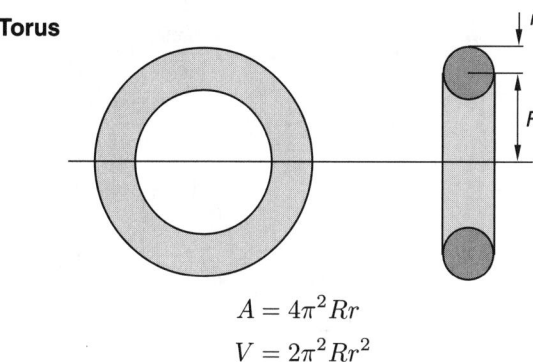

$$A = 4\pi^2 R r$$
$$V = 2\pi^2 R r^2$$

Regular Polyhedra (identical faces)

name	number of faces	form of faces	total surface area	volume
tetrahedron	4	equilateral triangle	$1.7321s^2$	$0.1179s^3$
cube	6	square	$6.0000s^2$	$1.0000s^3$
octahedron	8	equilateral triangle	$3.4641s^2$	$0.4714s^3$
dodecahedron	12	regular pentagon	$20.6457s^2$	$7.6631s^3$
isosahedron	20	equilateral triangle	$8.6603s^2$	$2.1817s^3$

The radius of a sphere inscribed within a regular polyhedron is

$$r = \frac{3V_{\text{polyhedron}}}{A_{\text{polyhedron}}}$$

APPENDIX 10.A
Abbreviated Table of Indefinite Integrals
(In each case, add a constant of integration.
All angles are measured in radians.)

General Formulas

1. $\int dx = x$
2. $\int c\, dx = c \int dx$
3. $\int (dx + dy) = \int dx + \int dy$
4. $\int u\, dv = uv - \int v\, du$ (integration by parts)

Algebraic Forms

5. $\displaystyle\int x^n\, dx = \frac{x^{n+1}}{n+1} \qquad [n \neq -1]$

6. $\displaystyle\int x^{-1} dx = \int \frac{dx}{x} = \ln |x|$

7. $\displaystyle\int (ax + b)^n\, dx = \frac{(ax+b)^{n+1}}{a(n+1)} \qquad [n \neq -1]$

8. $\displaystyle\int \frac{dx}{ax+b} = \frac{1}{a} \ln (ax+b)$

9. $\displaystyle\int \frac{x\, dx}{ax+b} = \frac{1}{a^2} \left[ax + b - b \ln (ax+b) \right]$

10. $\displaystyle\int \frac{x\, dx}{(ax+b)^2} = \frac{1}{a^2} \left[\frac{b}{ax+b} + \ln (ax+b) \right]$

11. $\displaystyle\int \frac{dx}{x(ax+b)} = \frac{1}{b} \ln \left(\frac{x}{ax+b} \right)$

12. $\displaystyle\int \frac{dx}{x(ax+b)^2} = \frac{1}{b(ax+b)} + \frac{1}{b^2} \ln \left(\frac{x}{ax+b} \right)$

13. $\displaystyle\int \frac{dx}{x^2 + a^2} = \frac{1}{a} \tan^{-1} \left(\frac{x}{a} \right)$

14. $\displaystyle\int \frac{dx}{a^2 - x^2} = \frac{1}{a} \tanh^{-1} \left(\frac{x}{a} \right)$

15. $\displaystyle\int \frac{x\, dx}{ax^2 + b} = \frac{1}{2a} \ln (ax^2 + b)$

16. $\displaystyle\int \frac{dx}{x(ax^n + b)} = \frac{1}{bn} \ln \left(\frac{x^n}{ax^n + b} \right)$

17. $\displaystyle\int \frac{dx}{ax^2 + bx + c} = \frac{1}{\sqrt{b^2 - 4ac}} \ln \left(\frac{2ax + b - \sqrt{b^2 - 4ac}}{2ax + b + \sqrt{b^2 - 4ac}} \right) \qquad [b^2 > 4ac]$

18. $\displaystyle\int \frac{dx}{ax^2 + bx + c} = \frac{2}{\sqrt{4ac - b^2}} \tan^{-1} \left(\frac{2ax + b}{\sqrt{4ac - b^2}} \right) \qquad [b^2 < 4ac]$

19. $\displaystyle\int \sqrt{a^2 - x^2}\, dx = \frac{x}{2} \sqrt{a^2 - x^2} + \frac{a^2}{2} \sin^{-1} \left(\frac{x}{a} \right)$

20. $\displaystyle\int x\sqrt{a^2 - x^2}\, dx = -\tfrac{1}{3} (a^2 - x^2)^{\frac{3}{2}}$

21. $\displaystyle\int \frac{dx}{\sqrt{a^2 - x^2}} = \sin^{-1} \left(\frac{x}{a} \right)$

22. $\displaystyle\int \frac{x\, dx}{\sqrt{a^2 - x^2}} = -\sqrt{a^2 - x^2}$

APPENDIX 11.A
Laplace Transforms
(also see App. 15.G)

$f(t)$	$\mathcal{L}(f(t))$	$f(t)$	$\mathcal{L}(f(t))$
$\delta(t)$ [unit impulse at $t=0$]	1	e^{at}	$\dfrac{1}{s-a}$
$\delta(t-c)$ [unit impulse at $t=c$]	e^{-cs}	$e^{at}\sin bt$	$\dfrac{b}{(s-a)^2+b^2}$
1 or u_0 [unit step at $t=0$]	$\dfrac{1}{s}$	$e^{at}\cos bt$	$\dfrac{s-a}{(s-a)^2+b^2}$
u_c [unit step at $t=c$]	$\dfrac{e^{-cs}}{s}$	$e^{at}t^n$ (n is a positive integer)	$\dfrac{n!}{(s-a)^{n+1}}$
t [unit ramp at $t=0$]	$\dfrac{1}{s^2}$	$1-e^{-at}$	$\dfrac{a}{s(s+a)}$
$\dfrac{t^{n-1}}{(n-1)!}$	$\dfrac{1}{s^n}$	$e^{-at}+at-1$	$\dfrac{a^2}{s^2(s+a)}$
$\sin at$	$\dfrac{a}{s^2+a^2}$	$\dfrac{e^{-at}-e^{-bt}}{b-a}$	$\dfrac{1}{(s+a)(s+b)}$
$at-\sin at$	$\dfrac{a^3}{s^2(s^2+a^2)}$	$\dfrac{(c-a)e^{-at}-(c-b)e^{-bt}}{b-a}$	$\dfrac{s+c}{(s+a)(s+b)}$
$\sinh at$	$\dfrac{a}{s^2-a^2}$	$\dfrac{1}{ab}+\dfrac{be^{-at}-ae^{-bt}}{ab(a-b)}$	$\dfrac{1}{s(s+a)(s+b)}$
$t\sin at$	$\dfrac{2as}{(s^2+a^2)^2}$	$t\sinh at$	$\dfrac{2as}{(s^2-a^2)^2}$
$\cos at$	$\dfrac{s}{s^2+a^2}$	$t\cosh at$	$\dfrac{s^2+a^2}{(s^2-a^2)^2}$
$1-\cos at$	$\dfrac{a^2}{s(s^2+a^2)}$	rectangular pulse, magnitude M, duration a	$\left(\dfrac{M}{s}\right)(1-e^{-as})$
$\cosh at$	$\dfrac{s}{s^2-a^2}$	triangular pulse, magnitude M, duration $2a$	$\left(\dfrac{M}{as^2}\right)(1-e^{-as})^2$
$t\cos at$	$\dfrac{s^2-a^2}{(s^2+a^2)^2}$	sawtooth pulse, magnitude M, duration a	$\left(\dfrac{M}{as^2}\right)\left(1-(as+1)e^{-as}\right)$
t^n (n is a positive integer)	$\dfrac{n!}{s^{n+1}}$	sinusoidal pulse, magnitude M, duration π/a	$\left(\dfrac{Ma}{s^2+a^2}\right)(1+e^{-\pi s/a})$

APPENDIX 12.A
Areas Under the Standard Normal Curve
(0 to z)

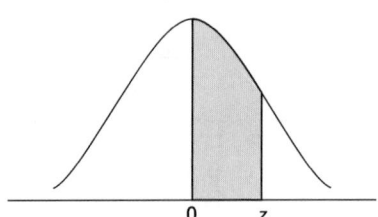

z	0	1	2	3	4	5	6	7	8	9
0.0	0.0000	0.0040	0.0080	0.0120	0.0160	0.0199	0.0239	0.0279	0.0319	0.0359
0.1	0.0398	0.0438	0.0478	0.0517	0.0557	0.0596	0.0636	0.0675	0.0714	0.0754
0.2	0.0793	0.0832	0.0871	0.0910	0.0948	0.0987	0.1026	0.1064	0.1103	0.1141
0.3	0.1179	0.1217	0.1255	0.1293	0.1331	0.1368	0.1406	0.1443	0.1480	0.1517
0.4	0.1554	0.1591	0.1628	0.1664	0.1700	0.1736	0.1772	0.1808	0.1844	0.1879
0.5	0.1915	0.1950	0.1985	0.2019	0.2054	0.2088	0.2123	0.2157	0.2190	0.2224
0.6	0.2258	0.2291	0.2324	0.2357	0.2389	0.2422	0.2454	0.2486	0.2518	0.2549
0.7	0.2580	0.2612	0.2642	0.2673	0.2704	0.2734	0.2764	0.2794	0.2823	0.2852
0.8	0.2881	0.2910	0.2939	0.2967	0.2996	0.3023	0.3051	0.3078	0.3106	0.3133
0.9	0.3159	0.3186	0.3212	0.3238	0.3264	0.3289	0.3315	0.3340	0.3365	0.3389
1.0	0.3413	0.3438	0.3461	0.3485	0.3508	0.3531	0.3554	0.3577	0.3599	0.3621
1.1	0.3643	0.3665	0.3686	0.3708	0.3729	0.3749	0.3770	0.3790	0.3810	0.3830
1.2	0.3849	0.3869	0.3888	0.3907	0.3925	0.3944	0.3962	0.3980	0.3997	0.4015
1.3	0.4032	0.4049	0.4066	0.4082	0.4099	0.4115	0.4131	0.4147	0.4162	0.4177
1.4	0.4192	0.4207	0.4222	0.4236	0.4251	0.4265	0.4279	0.4292	0.4306	0.4319
1.5	0.4332	0.4345	0.4357	0.4370	0.4382	0.4394	0.4406	0.4418	0.4429	0.4441
1.6	0.4452	0.4463	0.4474	0.4484	0.4495	0.4505	0.4515	0.4525	0.4535	0.4545
1.7	0.4554	0.4564	0.4573	0.4582	0.4591	0.4599	0.4608	0.4616	0.4625	0.4633
1.8	0.4641	0.4649	0.4656	0.4664	0.4671	0.4678	0.4686	0.4693	0.4699	0.4706
1.9	0.4713	0.4719	0.4726	0.4732	0.4738	0.4744	0.4750	0.4756	0.4761	0.4767
2.0	0.4772	0.4778	0.4783	0.4788	0.4793	0.4798	0.4803	0.4808	0.4812	0.4817
2.1	0.4821	0.4826	0.4830	0.4834	0.4838	0.4842	0.4846	0.4850	0.4854	0.4857
2.2	0.4861	0.4864	0.4868	0.4871	0.4875	0.4878	0.4881	0.4884	0.4887	0.4890
2.3	0.4893	0.4896	0.4898	0.4901	0.4904	0.4906	0.4909	0.4911	0.4913	0.4916
2.4	0.4918	0.4920	0.4922	0.4925	0.4927	0.4929	0.4931	0.4932	0.4934	0.4936
2.5	0.4938	0.4940	0.4941	0.4943	0.4945	0.4946	0.4948	0.4949	0.4951	0.4952
2.6	0.4953	0.4955	0.4956	0.4957	0.4959	0.4960	0.4961	0.4962	0.4963	0.4964
2.7	0.4965	0.4966	0.4967	0.4968	0.4969	0.4970	0.4971	0.4972	0.4973	0.4974
2.8	0.4974	0.4975	0.4976	0.4977	0.4977	0.4978	0.4979	0.4979	0.4980	0.4981
2.9	0.4981	0.4982	0.4982	0.4983	0.4984	0.4984	0.4985	0.4985	0.4986	0.4986
3.0	0.4987	0.4987	0.4987	0.4988	0.4988	0.4989	0.4989	0.4989	0.4990	0.4990
3.1	0.4990	0.4991	0.4991	0.4991	0.4992	0.4992	0.4992	0.4992	0.4993	0.4993
3.2	0.4993	0.4993	0.4994	0.4994	0.4994	0.4994	0.4994	0.4995	0.4995	0.4995
3.3	0.4995	0.4995	0.4996	0.4996	0.4996	0.4996	0.4996	0.4996	0.4996	0.4997
3.4	0.4997	0.4997	0.4997	0.4997	0.4997	0.4997	0.4997	0.4997	0.4997	0.4998
3.5	0.4998	0.4998	0.4998	0.4998	0.4998	0.4998	0.4998	0.4998	0.4998	0.4998
3.6	0.4998	0.4998	0.4999	0.4999	0.4999	0.4999	0.4999	0.4999	0.4999	0.4999
3.7	0.4999	0.4999	0.4999	0.4999	0.4999	0.4999	0.4999	0.4999	0.4999	0.4999
3.8	0.4999	0.4999	0.4999	0.4999	0.4999	0.4999	0.4999	0.4999	0.4999	0.4999
3.9	0.5000	0.5000	0.5000	0.5000	0.5000	0.5000	0.5000	0.5000	0.5000	0.5000

APPENDIX 15.A
Gamma Function Values*
Values of $\Gamma(\alpha) = \int_0^\infty e^{-t}t^{\alpha-1}dt$; $\Gamma(\alpha+1) = \alpha\Gamma(\alpha)$

α	$\Gamma(\alpha)$	α	$\Gamma(\alpha)$	α	$\Gamma(\alpha)$	α	$\Gamma(\alpha)$
1.00	1.00000	1.25	0.90640	1.50	0.88623	1.75	0.91906
1.01	0.99433	1.26	0.90440	1.51	0.88659	1.76	0.92137
1.02	0.98884	1.27	0.90250	1.52	0.88704	1.77	0.92376
1.03	0.98355	1.28	0.90072	1.53	0.88757	1.78	0.92623
1.04	0.97844	1.29	0.89904	1.54	0.88818	1.79	0.92877
1.05	0.97350	1.30	0.89747	1.55	0.88887	1.80	0.93138
1.06	0.96874	1.31	0.89600	1.56	0.88964	1.81	0.93408
1.07	0.96415	1.32	0.89464	1.57	0.89049	1.82	0.93685
1.08	0.95973	1.33	0.89338	1.58	0.89142	1.83	0.93969
1.09	0.95546	1.34	0.89222	1.59	0.89243	1.84	0.94261
1.10	0.95135	1.35	0.89115	1.60	0.89352	1.85	0.94561
1.11	0.94740	1.36	0.89018	1.61	0.89468	1.86	0.94869
1.12	0.94359	1.37	0.88931	1.62	0.89592	1.87	0.95184
1.13	0.93993	1.38	0.88854	1.63	0.89724	1.88	0.95507
1.14	0.93642	1.39	0.88785	1.64	0.89864	1.89	0.95838
1.15	0.93304	1.40	0.88726	1.65	0.90012	1.90	0.96177
1.16	0.92980	1.41	0.88676	1.66	0.90167	1.91	0.96523
1.17	0.92670	1.42	0.88636	1.67	0.90330	1.92	0.96877
1.18	0.92373	1.43	0.88604	1.68	0.90500	1.93	0.97240
1.19	0.92089	1.44	0.88581	1.69	0.90678	1.94	0.97610
1.20	0.91817	1.45	0.88566	1.70	0.90864	1.95	0.97988
1.21	0.91558	1.46	0.88560	1.71	0.91057	1.96	0.98374
1.22	0.91311	1.47	0.88563	1.72	0.91258	1.97	0.98768
1.23	0.91075	1.48	0.88575	1.73	0.91466	1.98	0.999171
1.24	0.90852	1.49	0.88595	1.74	0.91683	1.99	0.99581
						2.00	1.00000

* For large positive values of t, $\Gamma(t)$ approximates Stirling's asymptotic series.

$$t^t e^{-t}\sqrt{\frac{2\pi}{t}}\left[1 + \frac{1}{12t} + \frac{1}{288t^2} - \frac{139}{51840t^3} - \frac{571}{2488320t^4} + \cdots\right]$$

APPENDIX 15.B
Bessel Functions J_0 and J_1

x	$J_0(x)$	$J_1(x)$	x	$J_0(x)$	$J_1(x)$	x	$J_0(x)$	$J_1(x)$
0.0	1.0000	0.0000	5.0	−0.1776	−0.3276	10.0	−0.2459	0.0435
0.1	0.9975	0.0499	5.1	−0.1443	−0.3371	10.1	−0.2490	0.0184
0.2	0.9900	0.0995	5.2	−0.1103	−0.3432	10.2	−0.2496	−0.0066
0.3	0.9776	0.1483	5.3	−0.0758	−0.3460	10.3	−0.2477	−0.0313
0.4	0.9604	0.1960	5.4	−0.0412	−0.3453	10.4	−0.2434	−0.0555
0.5	0.9385	0.2423	5.5	−0.0068	−0.3414	10.5	−0.2366	−0.0789
0.6	0.9120	0.2867	5.6	0.0270	−0.3343	10.6	−0.2276	−0.1012
0.7	0.8812	0.3290	5.7	0.0599	−0.3241	10.7	−0.2164	−0.1224
0.8	0.8463	0.3688	5.8	0.0917	−0.3110	10.8	−0.2032	−0.1422
0.9	0.8075	0.4059	5.9	0.1220	−0.2951	10.9	−0.1881	−0.1603
1.0	0.7652	0.4401	6.0	0.1506	−0.2767	11.0	−0.1712	−0.1768
1.1	0.7196	0.4709	6.1	0.1773	−0.2559	11.1	−0.1528	−0.1913
1.2	0.6711	0.4983	6.2	0.2017	−0.2329	11.2	−0.1330	−0.2039
1.3	0.6201	0.5220	6.3	0.2238	−0.2081	11.3	−0.1121	−0.2143
1.4	0.5669	0.5419	6.4	0.2433	−0.1816	11.4	−0.0902	−0.2225
1.5	0.5118	0.5579	6.5	0.2601	−0.1538	11.5	−0.0677	−0.2284
1.6	0.4554	0.5699	6.6	0.2740	−0.1250	11.6	−0.0446	−0.2320
1.7	0.3980	0.5778	6.7	0.2851	−0.0953	11.7	−0.0213	−0.2333
1.8	0.3400	0.5815	6.8	0.2931	−0.0652	11.8	0.0020	−0.2323
1.9	0.2818	0.5812	6.9	0.2981	−0.0349	11.9	0.0250	−0.2290
2.0	0.2239	0.5767	7.0	0.3001	−0.0047	12.0	0.0477	−0.2234
2.1	0.1666	0.5683	7.1	0.2991	0.0252	12.1	0.0697	−0.2157
2.2	0.1104	0.5560	7.2	0.2951	0.0543	12.2	0.0908	−0.2060
2.3	0.0555	0.5399	7.3	0.2882	0.0826	12.3	0.1108	−0.1943
2.4	0.0025	0.5202	7.4	0.2786	0.1096	12.4	0.1296	−0.1807
2.5	−0.0484	0.4971	7.5	0.2663	0.1352	12.5	0.1469	−0.1655
2.6	−0.0968	0.4708	7.6	0.2516	0.1592	12.6	0.1626	−0.1487
2.7	−0.1424	0.4416	7.7	0.2346	0.1813	12.7	0.1766	−0.1307
2.8	−0.1850	0.4097	7.8	0.2154	0.2014	12.8	0.1887	−0.1114
2.9	−0.2243	0.3754	7.9	0.1944	0.2192	12.9	0.1988	−0.0912
3.0	−0.2601	0.3391	8.0	0.1717	0.2346	13.0	0.2069	−0.0703
3.1	−0.2921	0.3009	8.1	0.1475	0.2476	13.1	0.2129	−0.0489
3.2	−0.3202	0.2613	8.2	0.1222	0.2580	13.2	0.2167	−0.0271
3.3	−0.3443	0.2207	8.3	0.0960	0.2657	13.3	0.2183	−0.0052
3.4	−0.3643	0.1792	8.4	0.0692	0.2708	13.4	0.2177	0.0166
3.5	−0.3801	0.1374	8.5	0.0419	0.2731	13.5	0.2150	0.0380
3.6	−0.3918	0.0955	8.6	0.0146	0.2728	13.6	0.2101	0.0590
3.7	−0.3992	0.0538	8.7	−0.0125	0.2697	13.7	0.2032	0.0791
3.8	−0.4026	0.0128	8.8	−0.0392	0.2641	13.8	0.1943	0.0984
3.9	−0.4018	−0.0272	8.9	−0.0653	0.2559	13.9	0.1836	0.1165
4.0	−0.3971	−0.0660	9.0	−0.0903	0.2453	14.0	0.1711	0.1334
4.1	−0.3887	−0.1033	9.1	−0.1142	0.2324	14.1	0.1570	0.1488
4.2	−0.3766	−0.1386	9.2	−0.1367	0.2174	14.2	0.1414	0.1626
4.3	−0.3610	−0.1719	9.3	−0.1577	0.2004	14.3	0.1245	0.1747
4.4	−0.3423	−0.2028	9.4	−0.1768	0.1816	14.4	0.1065	0.1850
4.5	−0.3205	−0.2311	9.5	−0.1939	0.1613	14.5	0.0875	0.1934
4.6	−0.2961	−0.2566	9.6	−0.2090	0.1395	14.6	0.0679	0.1999
4.7	−0.2693	−0.2791	9.7	−0.2218	0.1166	14.7	0.0476	0.2043
4.8	−0.2404	−0.2985	9.8	−0.2323	0.0928	14.8	0.0271	0.2066
4.9	−0.2097	−0.3147	9.9	−0.2403	0.0684	14.9	0.0064	0.2069

APPENDIX 15.C
Properties of Fourier Series for Periodic Signals

periodic signal	Fourier series coefficients
$x(t)$ } periodic with $y(t)$ } period T_0	a_k b_k
$Ax(t) + By(t)$	$Aa_k + Bb_k$
$x(t - t_0)$	$a_k e^{-jk\left(\frac{2\pi}{T_0}\right)t_0}$
$e^{jM\left(\frac{2\pi}{T_0}\right)t}x(t)$	a_{k-M}
$x^*(t)$	a^*_{-k}
$x(-t)$	a_{-k}
$x(\alpha t), \ \alpha > 0$ [periodic with period T_0/α]	a_k
$\int_{T_0} x(\tau)y(t - \tau)d\tau$	$T_0 a_k b_k$
$x(t)y(t)$	$\sum\limits_{l=-\infty}^{+\infty} a_l b_{k-l}$
$\dfrac{dx(t)}{dt}$	$jk\dfrac{2\pi}{T_0}a_k$
$\int_{-\infty}^{t} x(t)dt$ $\begin{bmatrix}\text{finite-valued and periodic}\\\text{only if } a_0 = 0\end{bmatrix}$	$\left(\dfrac{1}{jk(2\pi/T_0)}\right)a_k$
$x(t)$ real	$\begin{cases} a_k = a^*_{-k} \\ \mathcal{R}e\{a_k\} = \mathcal{R}e\{a_{-k}\} \\ \mathcal{I}m\{a_k\} = -\mathcal{I}m\{a_{-k}\} \\ |a_k| = |a_{-k}| \\ \sphericalangle a_k = - \sphericalangle a_{-k} \end{cases}$
$x_e(t) = \mathcal{E}v\{x(t)\}$ [$x(t)$ real]	$\mathcal{R}e\{a_k\}$
$x_0(t) = \mathcal{O}d\{x(t)\}$ [$x(t)$ real]	$j\mathcal{I}m\{a_k\}$

Parseval's relation for periodic signals

$$\frac{1}{T_0} \int_{T_0} |x(t)|^2 dt = \sum_{k=-\infty}^{+\infty} |a_k|^2$$

APPENDIX 15.D
Properties of Fourier Transform for Aperiodic Signals

aperiodic signal	Fourier transform
$x(t)$	$X(\omega)$
$y(t)$	$Y(\omega)$
$ax(t) + by(t)$	$aX(\omega) + bY(\omega)$
$x(t - t_0)$	$e^{-j\omega t_0}X(\omega)$
$e^{j\omega_0 t}x(t)$	$X(\omega - \omega_0)$
$x^*(t)$	$X^*(-\omega)$
$x(-t)$	$X(-\omega)$
$x(at)$	$\dfrac{1}{|a|}X\left(\dfrac{\omega}{a}\right)$
$x(t) * y(t)$	$X(\omega)Y(\omega)$
$x(t)y(t)$	$\dfrac{1}{2\pi}X(\omega) * Y(\omega)$
$\dfrac{d}{dt}x(t)$	$j\omega X(\omega)$
$\int_{-\infty}^{t} x(t)dt$	$\dfrac{1}{j\omega}X(\omega) + \pi X(0)\delta(\omega)$
$tx(t)$	$j\dfrac{d}{d\omega}X(\omega)$
$x(t)$ real	$\begin{cases} X(\omega) = X^*(-\omega) \\ \mathcal{R}e\{X(\omega)\} = \mathcal{R}e\{X(-\omega)\} \\ \mathcal{I}m\{X(\omega)\} = -\mathcal{I}m\{X(-\omega)\} \\ |X(\omega)| = |X(-\omega)| \\ \sphericalangle X(\omega) = - \sphericalangle X(-\omega) \end{cases}$
$x_e(t) = \mathcal{E}v\{x(t)\}$ $[x(t) \text{ real}]$	$\mathcal{R}e\{X(\omega)\}$
$x_O(t) = \mathcal{O}d\{x(t)\}$ $[x(t) \text{ real}]$	$j\mathcal{I}m\{X(\omega)\}$

duality

$$f(u) = \int_{-\infty}^{+\infty} g(v)e^{juv}dv$$

$$g(t) \overset{\mathcal{F}}{\leftrightarrow} f(\omega)$$

$$f(t) \overset{\mathcal{F}}{\leftrightarrow} 2\pi g(-\omega)$$

Parseval's relation for aperiodic signals

$$\int_{-\infty}^{+\infty} |x(t)|^2 dt = \frac{1}{2\pi}\int_{-\infty}^{+\infty} |X(\omega)|^2 dw$$

APPENDIX 15.E
Fourier Transform Pairs

signal	Fourier transform
$\displaystyle\sum_{k=-\infty}^{+\infty} a_k e^{jk\omega_0 t}$	$\displaystyle 2\pi \sum_{k=-\infty}^{+\infty} a_k \delta(\omega - k\omega_0)$
$e^{j\omega_0 t}$	$2\pi\delta(\omega - \omega_0)$
$\cos\omega_0 t$	$\pi\big[\delta(\omega - \omega_0) + \delta(\omega + \omega_0)\big]$
$\sin\omega_0 t$	$\dfrac{\pi}{j}\big[\delta(\omega - \omega_0) - \delta(\omega + \omega_0)\big]$
$x(t) = 1$	$2\pi\delta(\omega)$

Periodic square wave
$$x(t) = \begin{cases} 1, & |t| < T_1 \\ 0, & T_1 < |t| \le \frac{T_0}{2} \end{cases}$$
and
$$x(t + T_0) = x(t)$$

$$\sum_{k=-\infty}^{+\infty} \frac{2\sin k\omega_0 T_1}{k}\delta(\omega - k\omega_0)$$

$\displaystyle\sum_{k=-\infty}^{+\infty} \delta(t - nT)$	$\displaystyle \frac{2\pi}{T} \sum_{k=-\infty}^{+\infty} \delta\left(\omega - \frac{2\pi k}{T}\right)$
$x(t) = \begin{cases} 1, & |t| < T_1 \\ 0, & |t| > T_1 \end{cases}$	$2T_1\,\mathrm{sinc}\left(\dfrac{\omega T_1}{\pi}\right) = \dfrac{2\sin\omega T_1}{\omega}$
$\dfrac{W}{\pi}\mathrm{sinc}\left(\dfrac{Wt}{\pi}\right) = \dfrac{\sin Wt}{\pi t}$	$X(\omega) = \begin{cases} 1, & |\omega| < W \\ 0, & |\omega| > W \end{cases}$
$\delta(t)$	1
$u(t)$	$\dfrac{1}{j\omega} + \pi\delta(\omega)$
$\delta(t - t_0)$	$e^{-j\omega t_0}$
$e^{-at}u(t),\ Re\{a\} > 0$	$\dfrac{1}{a + j\omega}$
$te^{-at}u(t),\ \mathcal{R}e\{a\} > 0$	$\dfrac{1}{(a + j\omega)^2}$
$\dfrac{t^{n-1}}{(n-1)!}e^{-at}u(t),$ $\mathcal{R}e\{a\} > 0$	$\dfrac{1}{(a + j\omega)^n}$

APPENDIX 15.F
Properties of Laplace Transform

signal	transform	ROC (region of convergence)
$x(t)$	$X(s)$	R
$x_1(t)$	$X_1(s)$	R_1
$x_2(t)$	$X_2(s)$	R_2
$ax_1(t) + bx_2(t)$	$aX_1(s) + bX_2(s)$	at least $R_1 \cap R_2$
$x(t - t_0)$	$e^{-st_0} X(s)$	R
$e^{s_0 t} x(t)$	$X(s - s_0)$	shifted version of R [i.e., s is in the ROC if $(s - s_0)$ is in R]
$x(at)$	$\dfrac{1}{\|a\|} X\left(\dfrac{s}{a}\right)$	"scaled" ROC [i.e., s is in the ROC if (s/a) is in the ROC of $X(s)$]
$x_1(t) * x_2(t)$	$X_1(s)X_2(s)$	at least $R_1 \cap R_2$
$\dfrac{d}{dt} x(t)$	$sX(s)$	at least R
$-tx(t)$	$\dfrac{d}{ds} X(s)$	R
$\int_{-\infty}^{t} x(\tau)d\tau$	$\dfrac{1}{s} X(s)$	at least $R \cap \{\mathcal{R}e\{s\} > 0\}$

APPENDIX 15.G
Laplace Transforms
(also see App. 11.A)

transform pair	signal	transform	ROC
1	$\delta(t)$	1	all s
2	$u(t)$	$\dfrac{1}{s}$	$\mathcal{R}e\{s\} > 0$
3	$-u(-t)$	$\dfrac{1}{s}$	$\mathcal{R}e\{s\} < 0$
4	$\dfrac{t^{n-1}}{(n-1)!}u(t)$	$\dfrac{1}{s^n}$	$\mathcal{R}e\{s\} > 0$
5	$-\dfrac{t^{n-1}}{(n-1)!}u(-t)$	$\dfrac{1}{s^n}$	$\mathcal{R}e\{s\} < 0$
6	$e^{-\alpha t}u(t)$	$\dfrac{1}{s+\alpha}$	$\mathcal{R}e\{s\} > \alpha$
7	$-e^{-\alpha t}u(-t)$	$\dfrac{1}{s+\alpha}$	$\mathcal{R}e\{s\} < -\alpha$
8	$\dfrac{t^{n-1}}{(n-1)!}e^{-\alpha t}u(t)$	$\dfrac{1}{(s+\alpha)^n}$	$\mathcal{R}e\{s\} > -\alpha$
9	$-\dfrac{t^{n-1}}{(n-1)!}e^{-\alpha t}u(-t)$	$\dfrac{1}{(s+\alpha)^n}$	$\mathcal{R}e\{s\} < -\alpha$
10	$\delta(t-T)$	e^{-sT}	all s
11	$[\cos \omega_0 t]u(t)$	$\dfrac{s}{s^2+\omega_0^2}$	$\mathcal{R}e\{s\} > 0$
12	$[\sin \omega_0 t]u(t)$	$\dfrac{\omega_0}{s^2+\omega_0^2}$	$\mathcal{R}e\{s\} > 0$
13	$[e^{-\alpha t}\cos \omega_0 t]u(t)$	$\dfrac{s+\alpha}{(s+\alpha)^2+\omega_0^2}$	$\mathcal{R}e\{s\} > -\alpha$
14	$[e^{-\alpha t}\sin \omega_0 t]u(t)$	$\dfrac{\omega_0}{(s+\alpha)^2+\omega_0^2}$	$\mathcal{R}e\{s\} > -\alpha$

APPENDIX 15.H
Properties of Discrete-Time Fourier Series for Periodic Signals

periodic signal	Fourier series coefficients
$x(n)$ $\left.\right\}$ periodic with $y(n)$ period N	a_k $\left.\right\}$ periodic with b_k period N
$Ax[n] + By[n]$	$Aa_k + Bb_k$
$x[n - n_0]$	$a_k e^{-jk\left(\frac{2\pi}{N}\right)n_0}$
$e^{jM\left(\frac{2\pi}{N}\right)n}x[n]$	a_{k-M}
$x^*[n]$	a^*_{-k}
$x[-n]$	a_{-k}
$x_{(m)}[n] = \begin{cases} x[n/m] & \text{if } n \text{ is a multiple of } m \\ 0 & \text{if } n \text{ is not a multiple of } m \end{cases}$ (periodic with mN)	$\dfrac{1}{m}a_k$ $\begin{bmatrix} \text{viewed as periodic} \\ \text{with period } mN \end{bmatrix}$
$\displaystyle\sum_{r=\langle N\rangle} x[r]y[n-r]$	$Na_k b_k$
$x[n]y[n]$	$\displaystyle\sum_{l-\langle N\rangle} a_l b_{k-1}$
$x[n] - x[n-1]$	$\left(1 - e^{-jk\left(\frac{2\pi}{N}\right)}\right)a_k$
$\displaystyle\sum_{k=-\infty}^{n} x[k]$ $\begin{bmatrix} \text{finite-valued and periodic} \\ \text{only if } a_0 = 0 \end{bmatrix}$	$\left(\dfrac{1}{1 - e^{-jk\left(\frac{2\pi}{N}\right)}}\right)a_k$
$x[n]$ real	$\begin{cases} a_k = a^*_{-k} \\ \mathcal{R}e\{a_k\} = \mathcal{R}e\{a_{-k}\} \\ \mathcal{I}m\{a_k\} = -\mathcal{I}m\{a_{-k}\} \\ |a_k| = |a_{-k}| \\ \sphericalangle a_k = -\sphericalangle a_{-k} \end{cases}$
$x_e[n] = \mathcal{E}v\{x(n)\}$ $[x(n) \text{ real}]$	$\mathcal{R}e\{a_k\}$
$x_0[n] = \mathcal{O}d\{x[n]\}$ $[x(n) \text{ real}]$	$j\mathcal{I}m[a_k]$

Parseval's relation for periodic signals

$$\frac{1}{N}\sum_{n=\langle N\rangle} |x[n]|^2 = \sum_{k=\langle N\rangle} |a_k|^2$$

APPENDIX 15.I
Properties of Discrete-Time Fourier Transform for Aperiodic Signals

aperiodic signal	Fourier transform
$x(n)$ $y(n)$	$X(\Omega)$ $\left.\right\}$ periodic with $Y(\Omega)$ $\left.\right\}$ period 2π
$ax[n] + by[n]$	$aX(\Omega) + bY(\Omega)$
$x[n - n_0]$	$e^{-j\Omega n_0} X(\Omega)$
$e^{j\Omega_0 n} x[n]$	$X(\Omega - \Omega_0)$
$x^*[n]$	$X^*(-\Omega)$
$x[-n]$	$X(-\Omega)$
$x_{(k)}[n] = \begin{cases} x[n/k], & \text{if } n \text{ is a multiple of } k \\ 0, & \text{if } n \text{ is not a multiple of } k \end{cases}$	$X(k\Omega)$
$x[n] * y[n]$	$X(\Omega)Y(\Omega)$
$x[n]y[n]$	$\dfrac{1}{2\pi} \displaystyle\int_{2\pi} X(\theta)Y(\Omega - \theta)d\theta$
$x[n] - x[n-1]$	$(1 - e^{-j\Omega} X(\Omega))$
$\displaystyle\sum_{k=-\infty}^{n} x[k]$	$\dfrac{1}{1 - e^{j\Omega}} X(\Omega) + \pi X(0) \displaystyle\sum_{k=-\infty}^{+\infty} \delta(\Omega - 2\pi k)$
$nx[n]$	$j \dfrac{dX(\Omega)}{d\Omega}$
$x[n]$ real	$\begin{cases} X(\Omega) = X^*(-\Omega) \\ \mathcal{R}e\{X(\Omega)\} = \mathcal{R}e\{X(-\Omega)\} \\ \mathcal{I}m\{X(\Omega)\} = -\mathcal{I}m\{X(-\Omega)\} \\ |X(\Omega)| = |X(-\Omega)| \\ \measuredangle X(\Omega) = - \measuredangle X(-\Omega) \end{cases}$
$x_e[n] = \mathcal{E}v\{x(n)\}$ $[x(n) \text{ real}]$	$\mathrm{Re}\{X(\Omega)\}$
$x_0[n] = \mathcal{O}d\{x[n]\}$ $[x(n) \text{ real}]$	$j\mathcal{I}m[X(\Omega)]$

Parseval's relation for aperiodic signals

$$\sum_{n=-\infty}^{+\infty} |x[n]|^2 = \frac{1}{2\pi} \int_{2\pi} |X(\Omega)|^2 d\Omega$$

APPENDIX 15.J
Properties of z-Transforms

sequence	transform	region of convergence
$x[n]$	$X(z)$	R_x
$x_1[n]$	$X_1(z)$	R_1
$x_2[n]$	$X_2(z)$	R_2
$ax_1[n] + bx_2[n]$	$aX_1(z) + bX_2(z)$	at least the intersection of R_1 and R_2
$x[n - n_0]$	$z^{-n_0} X(z)$	R_x except for the possible addition or deletion of the origin
$e^{j\Omega_0 n} x[n]$	$X(e^{-j\Omega_0} z)$	R_x
$z_0^n x[n]$	$X\left(\dfrac{z}{z_0}\right)$	$z_0 R_x$
$a^n x[n]$	$X(a^{-1} z)$	scaled version of R_x (i.e., $\|a\| \cdot R_x =$ the set of points $\{\|a\|z\}$ for z in R_x)
$x[-n]$	$X(z^{-1})$	inverted R_x (i.e., R_x^{-1} = the set of points z^{-1} where z is in R_x)
$w[n] = \begin{cases} x[r], & n=rk \\ 0, & n \neq rk \end{cases}$ for some r	$X(z^k)$	$R_x^{\frac{1}{k}}$ (i.e., the set of points $z^{\frac{1}{k}}$ where z is in R_x)
$x_1[n] \cdot x_2[n]$	$X_1(z) X_2(z)$	at least the intersection of R_1 and R_2
$nx[n]$	$-z \dfrac{dX(z)}{dz}$	R_x except for the possible addition or deletion of the origin
$\displaystyle\sum_{k=-\infty}^{n} x[k]$	$\dfrac{1}{1 - z^{-1}} X(z)$	at least the intersection of R_x and $\|z\| > 1$

APPENDIX 15.K
z-Transforms

$f(t)$	$F(z)$
$\delta(t)$	1
$\delta(t - mT)$	$\dfrac{1}{z^m}$
$H(t)$	$\dfrac{z}{z - m}$
t	$\dfrac{Tz}{(z - 1)^2}$
t^2	$\dfrac{T^2 z(z + 1)}{(z - 1)^3}$
t^3	$\dfrac{T^3 z(z^2 + 4z + 1)}{(z - 1)^4}$
t^n	$(-1)^n \lim\limits_{x \to 0} \dfrac{\delta^n}{\delta_x{}^n} \left(\dfrac{z}{z - e^{-x\tau}} \right)$
$1 - a^{\Omega\tau}$	$\dfrac{z(1 - a^{\Omega\tau})}{(z - 1)(z - a^{\Omega\tau})}$
$a^{\Omega\tau}$	$\dfrac{z}{(z - a^{\Omega\tau})}$
$ta^{\Omega\tau}$	$\dfrac{Tz a^{\Omega\tau}}{(z - a^{\Omega\tau})^2}$
$t^2 a^{\Omega\tau}$	$\dfrac{T^2 a^{\Omega\tau} z(z + a^{\Omega\tau})}{(z - a^{\Omega\tau})^3}$
$\sin \Omega\tau$	$\dfrac{z \sin \Omega T}{z^2 - 2z \cos \Omega T + 1}$
$\cos \Omega t$	$\dfrac{z(z - \cos \Omega T)}{z^2 - 2z \cos \Omega T + 1}$
$\sinh \Omega t$	$\dfrac{z \sinh \Omega T}{z^2 - 2z \cosh \Omega T + 1}$
$\cosh \Omega t$	$\dfrac{z(z - \cosh \Omega T)}{z^2 - 2z \cosh \Omega T + 1}$
$e^{-at} \sin \Omega t$	$\dfrac{ze^{-a\tau} \sin \Omega T}{z^2 - 2ze^{-\tau} \cos \Omega T + e^{-2a\tau}}$
$e^{-at} \cos \Omega t$	$\dfrac{z(z - e^{-a\tau} \cos \Omega T)}{z^2 - 2ze^{-\tau} \cos \Omega T + e^{-2a\tau}}$
$e^{-at} \sinh \Omega t$	$\dfrac{ze^{-a\tau} \sinh \Omega T}{z^2 - 2ze^{-a\tau} \cosh \Omega T + e^{-2a\tau}}$
$e^{-at} \cosh \Omega t$	$\dfrac{z(z - e^{-a\tau} \cosh \Omega T)}{z^2 - 2ze^{-a\tau} \cosh \Omega T + e^{-2a\tau}}$

APPENDIX 17.A
Room Temperature Properties of Silicon, Germanium, and Gallium Arsenide (at 300K)

property	Si	Ge	GaAs
atoms/cm^3	5.0×10^{22}	4.42×10^{22}	2.21×10^{22}
atomic weight	28.08	72.6	144.63
breakdown field, V/cm	$\sim 3 \times 10^5$	$\sim 10^5$	$\sim 4 \times 10^5$
crystal structure	diamond	diamond	zincblende
density, g/m^3	2.328	5.3267	5.32
dielectric constant	11.8	16	10.9
effective density of states, in conduction band $N_c$cm^{-3}	2.8×10^{19}	1.04×10^{19}	4.7×10^{17}
in valence band $N_v$cm^{-3}	1.02×10^{19}	6.1×10^{18}	7.0×10^{18}
effective mass m^*/m_0 (electrons)	$m_t^* = 0.97$	$m_t^* = 1.6$	0.068
	$m_t^* = 0.19$	$m_t^* = 0.082$	
(holes)	$m_{th}^* = 0.16$	$m_{th}^* = 0.04$	0.12
	$m_{hh}^* = 0.5$	$m_{hh}^* = 0.3$	0.5
electron affinity x, V	4.05	4.0	4.07
energy gap, eV, at 300K	1.12	0.803	1.43
intrinsic carrier concentration, cm^{-3}	1.6×10^{10}	2.5×10^{12}	1.1×10^7
lattice constant, nm	0.543086	0.565748	0.56534
linear coefficient of thermal expansion, $(\Delta L/L)\,\Delta T$, K^{-1}	2.6×10^{-6}	5.8×10^{-6}	5.9×10^{-6}
melting point, °C	1420	937	1238
minority-carrier lifetime, s	2.5×10^{-3}	10^{-3}	$\sim 10^{-8}$
mobility (drift), cm^2/V·s, μ_n (electrons)	1500	3900	8500
μ_p (holes)	600	1900	400
specific heat, J/g·°C	0.7	0.31	0.35
thermal conductivity at 300K, W/cm·K	1.45	0.64	0.46
thermal diffusivity, cm^2/s	0.9	0.36	0.44
work function, V	4.8	4.4	4.7

APPENDIX 18.A
Electromagnetic Spectrum

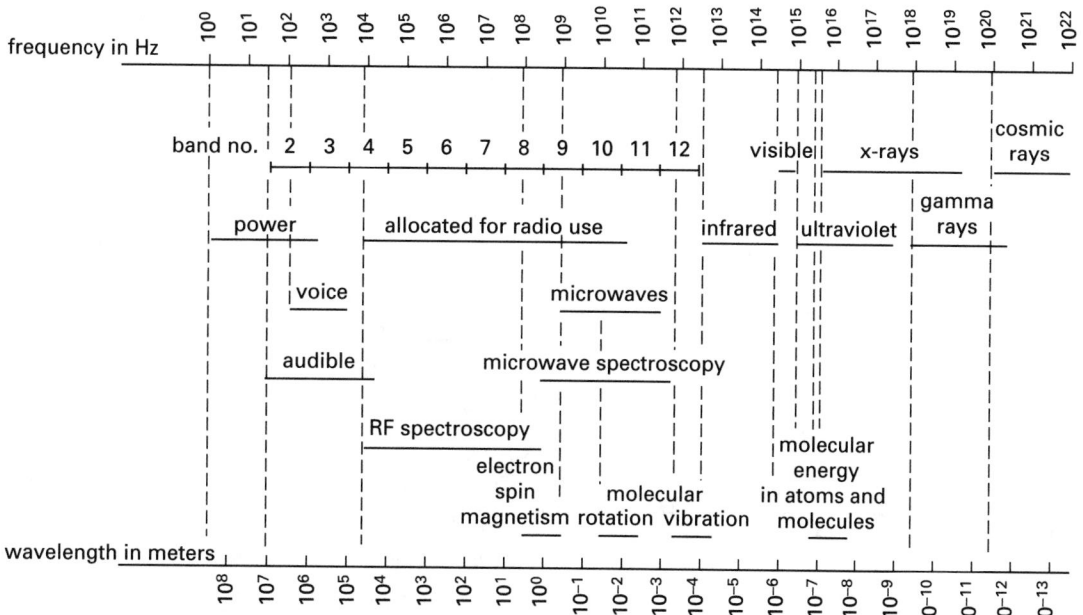

APPENDIX 25.A
Comparison of Electric and Magnetic Equations

equation description	electric version	magnetic version	remarks
experimental force law	Coulomb's law $$\mathbf{F} = \left(\frac{Q_1 Q_2}{4\pi r^2}\right)\mathbf{r}$$	force between two current elements $$d\mathbf{F} = \left(\frac{\mu_0}{4\pi}\right)\left(\frac{I_2 d\mathbf{l}_2 \times (I_1 d\mathbf{l}_1 \times \mathbf{r})}{r^2}\right)$$	The term $Id\mathbf{l}$ in the magnetic column is the equivalent of a "magnetic charge" q_m. The I or the $d\mathbf{l}$ can be the vector. The $\mathbf{r}$ is a unit vector pointing from 1 to 2.
field definitions from force law	$\mathbf{F} = Q\mathbf{E}$	$d\mathbf{F} = \mathbf{I} \times \mathbf{B}\, d\mathbf{l}$ current element $$d\mathbf{F} = \mathbf{J} \times \mathbf{B}\, dV$$ distributed current element $$d\mathbf{F} = q\mathbf{v} \times \mathbf{B}$$ moving charge	The V used in this row represents volume, not voltage. The $\mathbf{v}$ is the velocity.
general force law	$\mathbf{F} = q(\mathbf{E} + \mathbf{v} \times \mathbf{B})$ $$d\mathbf{F} = (\rho\mathbf{E} + \mathbf{J} \times \mathbf{B})dV \text{ where } dQ = \rho dV$$		The V in this row represents the volume, not voltage. The v is the velocity.
definition of scalar and vector potential	$E = -\nabla V$	$\mathbf{B} = \nabla \times \mathbf{A}$	$\mathbf{A}$ is the magnetic vector potential.
Poisson's equation for the potential function	$\nabla^2 V = -\dfrac{\rho}{\varepsilon}$	$\nabla^2 \mathbf{A} = -\mu_0 \mathbf{J}$	From a knowledge of the charge distribution, the potential can be found and then the $\mathbf{E}$ and $\mathbf{B}$ fields determined.
Gauss's law enclosing charge and Ampere's law enclosing current	$\oiint \mathbf{D}\cdot d\mathbf{A} = \iiint \rho dV = Q$ $$\nabla\cdot\mathbf{D} = \rho$$	$\oint \mathbf{H}\cdot d\mathbf{l} = I$ $$\nabla \times \mathbf{H} = \mathbf{J}$$	The V in this row represents volume.
constitutive relations	$\mathbf{D} = \varepsilon\mathbf{E}$ $$\mathbf{D} = \varepsilon_0\mathbf{E} + \mathbf{P}$$	$\mathbf{B} = \mu\mathbf{H}$ $$\mathbf{B} = \mu_0\mathbf{H} + \mu_0\mathbf{M}$$	The second set of equations is always valid. The first set assumes the medium is linear and isotropic.
definitions of relative permittivity and permeability	$\varepsilon_r = \dfrac{\varepsilon}{\varepsilon_0}$ $\varepsilon_0 = 8.854 \times 10^{-12}$ F/m	$\mu_r = \dfrac{\mu}{\mu_0}$ $\mu_0 = 4\pi \times 10^{-7}$ H/m	

(continued)

APPENDIX 25.A *(continued)*
Comparison of Electric and Magnetic Equations

equation description	electric version	magnetic version	remarks
capacitance and inductance of a field cell	$\varepsilon_0 = \dfrac{C}{l}$	$\mu_0 = \dfrac{L}{l}$	Field cells are a construct designed to represent free space in terms of a parallel plate capacitor and an inductor. This capacitance and inductance exist regardless of the presence of an electric or magnetic field.
capacitance and inductance	$C = \dfrac{Q}{V}$	$L = \dfrac{\Lambda}{I}$	Λ is the flux linkage.
energy density of a field	$U = \frac{1}{2}\varepsilon E^2$	$U = \frac{1}{2}\mu H^2$	Both energy and momentum are carried by a field.
energy stored by capacitance and inductance	$W = \frac{1}{2}CV^2$	$W = \frac{1}{2}LI^2$	
electromotive and magnetomotive force with sources present	$\oint \mathcal{E}\cdot d\mathbf{l} = \mathcal{E} = V$	$\oint \mathbf{H}\cdot d\mathbf{l} = NI = F_m = V_m$	The $\mathcal{E}$ is the emf, not the permittivity. Without sources present, both line integrals are equal to zero.
dipole moments	$\mathbf{p} = q\mathbf{d}$	$\mathbf{m} = I\mathbf{A}$	
dipole torque	$\mathbf{T} = \mathbf{p} \times \mathbf{E}$	$\mathbf{T} = \mathbf{m} \times \mathbf{B}$	This torque occurs due to the dipole being immersed in an external $\mathbf{E}$ or $\mathbf{B}$ field.
dipole potential energy	$W = -\mathbf{p}\cdot\mathbf{E}$	$W = -\mathbf{m}\cdot\mathbf{B}$	

APPENDIX 27.A
Impedance of Series-Connected Circuit Elements

circuit	impedance $Z = R + jX$ (ohms)	magnitude of impedance $|Z| = \sqrt{R^2 + X^2}$ (ohms)	phase angle $\theta = \tan^{-1}\left(\dfrac{X}{R}\right)$ (radians)	admittance $Y = \dfrac{1}{Z}$ (siemens)
R (resistor)	R	R	0	$\dfrac{1}{R}$
L (inductor)	$j\omega L$	ωL	$+\dfrac{\pi}{2}$	$-j\left(\dfrac{1}{\omega L}\right)$
C (capacitor)	$-j\left(\dfrac{1}{\omega C}\right)$	$\dfrac{1}{\omega C}$	$-\dfrac{\pi}{2}$	$j\omega C$
R_1 R_2	$R_1 + R_2$	$R_1 + R_2$	0	$\dfrac{1}{R_1 + R_2}$
M over L_1 L_2	$j\omega\left(L_1 + L_2 \pm 2M\right)$	$\omega\left(L_1 + L_2 \pm 2M\right)$	$+\dfrac{\pi}{2}$	$-j\left(\dfrac{1}{\omega\left(L_1 + L_2 \pm 2M\right)}\right)$
C_1 C_2	$-j\left(\dfrac{1}{\omega}\right)\left(\dfrac{C_1 + C_2}{C_1 C_2}\right)$	$\left(\dfrac{1}{\omega}\right)\left(\dfrac{C_1 + C_2}{C_1 C_2}\right)$	$-\dfrac{\pi}{2}$	$j\omega\left(\dfrac{C_1 C_2}{C_1 + C_2}\right)$
R L	$R + j\omega L$	$\sqrt{R^2 + \omega^2 L^2}$	$\tan^{-1}\left(\dfrac{\omega L}{R}\right)$	$\dfrac{R - j\omega L}{R^2 + \omega^2 L^2}$
R C	$R - j\left(\dfrac{1}{\omega C}\right)$	$\sqrt{\dfrac{\omega^2 C^2 R^2 + 1}{\omega^2 C^2}}$	$\tan^{-1}\left(\dfrac{1}{\omega R C}\right)$	$\dfrac{\omega^2 C^2 R + j\omega C}{\omega^2 C^2 R^2 + 1}$
L C	$j\left(\omega L - \dfrac{1}{\omega C}\right)$	$\omega L - \dfrac{1}{\omega C}$	$\pm\dfrac{\pi}{2}$	$-j\left(\dfrac{\omega C}{\omega^2 LC - 1}\right)$
R L C	$R + j\left(\omega L - \dfrac{1}{\omega C}\right)$	$\sqrt{R^2 + \left(\omega L - \dfrac{1}{\omega C}\right)^2}$	$\tan^{-1}\left(\dfrac{\omega L - \dfrac{1}{\omega C}}{R}\right)$	$\dfrac{R - j\left(\omega L - \dfrac{1}{\omega C}\right)}{R^2 + \left(\omega L - \dfrac{1}{\omega C}\right)^2}$

APPENDIX 27.B
Impedance of Parallel-Connected Circuit Elements

circuit	impedance $Z = R + jX$ (ohms)	magnitude of impedance $\lvert Z \rvert = \sqrt{R^2 + X^2}$ (ohms)
R_1 R_2	$\dfrac{R_1 R_2}{R_1 + R_2}$	$\dfrac{R_1 R_2}{R_1 + R_2}$
L_1 M L_2	$+j\omega\left(\dfrac{L_1 L_2 - M^2}{L_1 + L_2 \mp 2M}\right)$	$\omega\left(\dfrac{L_1 L_2 - M^2}{L_1 + L_2 \mp 2M}\right)$
C_1 C_2	$-j\left(\dfrac{1}{\omega(C_1 + C_2)}\right)$	$\dfrac{1}{\omega(C_1 + C_2)}$
R L	$\dfrac{\omega^2 L^2 R + j\omega L R^2}{\omega^2 L^2 + R^2}$	$\dfrac{\omega L R}{\sqrt{\omega^2 L^2 + R^2}}$
R C	$\dfrac{R - j\omega R^2 C}{1 + \omega^2 R^2 C^2}$	$\dfrac{R}{\sqrt{1 + \omega^2 R^2 C^2}}$
L C	$j\left(\dfrac{\omega L}{1 - \omega^2 LC}\right)$	$\dfrac{\omega L}{1 - \omega^2 LC}$
R L C	$\dfrac{\dfrac{1}{R} - j\left(\omega C - \dfrac{1}{\omega L}\right)}{\left(\dfrac{1}{R}\right)^2 + \left(\omega C - \dfrac{1}{\omega L}\right)^2}$	$\dfrac{1}{\sqrt{\left(\dfrac{1}{R}\right)^2 + \left(\omega C - \dfrac{1}{\omega L}\right)^2}}$
R L C	$\dfrac{\dfrac{R}{\omega^2 C^2} - j\left(\dfrac{R^2}{\omega C} + \dfrac{L}{C}\left(\omega L - \dfrac{1}{\omega C}\right)\right)}{R^2 + \left(\omega L - \dfrac{1}{\omega C}\right)^2}$	$\dfrac{\sqrt{\left(\dfrac{R}{\omega^2 C^2}\right)^2 + \left(\dfrac{R^2}{\omega C} + \left(\dfrac{L}{C}\right)\left(\omega L - \dfrac{1}{\omega C}\right)\right)^2}}{R^2 + \left(\omega L - \dfrac{1}{\omega C}\right)^2}$
R_1 L R_2 C	$\dfrac{R_1 R_2(R_1 + R_2) + \omega^2 L^2 R_2 + \dfrac{R_1}{\omega^2 C^2}}{(R_1 + R_2)^2 + \left(\omega L - \dfrac{1}{\omega C}\right)^2}$ $+j\left(\dfrac{\omega R_2^2 L - \dfrac{R_1^2}{\omega C} - \left(\dfrac{L}{C}\right)\left(\omega L - \dfrac{1}{\omega C}\right)}{(R_1 + R_2)^2 + \left(\omega L - \dfrac{1}{\omega C}\right)^2}\right)$	$\sqrt{\left(\dfrac{R_1 R_2(R_1 + R_2) + \omega^2 L^2 R_2 + \dfrac{R_1}{\omega^2 C^2}}{(R_1 + R_2)^2 + \left(\omega L - \dfrac{1}{\omega C}\right)^2}\right)^2 + \left(\dfrac{\omega L R_2^2 - \dfrac{R_1^2}{\omega C} - \dfrac{L}{C}\left(\omega L - \dfrac{1}{\omega C}\right)}{(R_1 + R_2)^2 + \left(\omega L - \dfrac{1}{\omega C}\right)^2}\right)^2}$

(continued)

APPENDIX 27.B (*continued*)
Impedance of Parallel-Connected Circuit Elements

circuit	phase angle $\theta = \tan^{-1}\left(\dfrac{X}{R}\right)$ (radians)	admittance $Y = \dfrac{1}{Z}$ (mhos)
R_1, R_2 (parallel resistors)	0	$\dfrac{R_1 + R_2}{R_1 R_2}$
L_1, M, L_2	$+\dfrac{\pi}{2}$	$-j\left(\dfrac{1}{\omega}\right)\left(\dfrac{L_1 + L_2 \mp 2M}{L_1 L_2 - M^2}\right)$
C_1, C_2	$-\dfrac{\pi}{2}$	$+j\omega(C_1 + C_2)$
R, L	$\tan^{-1}\left(\dfrac{R}{\omega L}\right)$	$\dfrac{\omega L - jR}{\omega L R}$
R, C	$\tan^{-1}(-\omega R C)$	$\dfrac{1}{R} + j\omega C$
L, C	$\pm\dfrac{\pi}{2}$	$-j\left(\dfrac{1 - \omega^2 LC}{\omega L}\right)$
R, L, C	$\tan^{-1} R\left(\dfrac{1}{\omega L} - \omega C\right)$	$\dfrac{1}{R} + j\left(\omega C - \dfrac{1}{\omega L}\right)$
R, L, C	$\tan^{-1}\left(\dfrac{\dfrac{R^2}{\omega C} + \left(\dfrac{L}{C}\right)\left(\omega L - \dfrac{1}{\omega C}\right)}{\dfrac{R}{\omega^2 C^2}}\right)$	$\dfrac{R + j\omega\left(R^2 C - L + \omega^2 L^2 C\right)}{R^2 + \omega^2 L^2}$
R_1, L, R_2, C	$\tan^{-1}\left(\dfrac{\omega L R^2 - \dfrac{R_1^2}{\omega C} - \left(\dfrac{L}{C}\right)\left(\omega L - \dfrac{1}{\omega C}\right)}{R_1 R_2(R_1 + R_2) + \omega^2 L^2 R_2 + \dfrac{R_1}{\omega^2 C^2}}\right)$	$\dfrac{R_1 + \omega^2 R_1 R_2 C^2(R_1 + R_2) + \omega^4 L^2 C^2 R_2}{(R_1^2 + \omega^2 L^2)(\omega^2 R_2^2 C^2 + 1)}$ $+ j\left(\dfrac{\omega\left(R_1^2 C - L + \omega^2 LC(L - R_2^2 C)\right)}{(R_1^2 + \omega^2 L^2)(\omega^2 R_2^2 C^2 + 1)}\right)$

APPENDIX 29.A
Two-Port Parameter Conversions

	$[z]$	$[y]$	$[h]$	$[g]$	$[TL]$	$[TLI]$
$[z]$	$\begin{bmatrix} z_{11} & z_{12} \\ z_{21} & z_{22} \end{bmatrix}$	$\begin{bmatrix} \dfrac{y_{22}}{\|y\|} & -\dfrac{y_{12}}{\|y\|} \\ -\dfrac{y_{21}}{\|y\|} & \dfrac{y_{11}}{\|y\|} \end{bmatrix}$	$\begin{bmatrix} \dfrac{\|h\|}{h_{22}} & \dfrac{h_{12}}{h_{22}} \\ -\dfrac{h_{22}}{h_{22}} & \dfrac{1}{h_{22}} \end{bmatrix}$	$\begin{bmatrix} \dfrac{1}{g_{11}} & -\dfrac{g_{12}}{g_{11}} \\ \dfrac{g_{21}}{g_{11}} & \dfrac{\|g\|}{g_{11}} \end{bmatrix}$	$\begin{bmatrix} \dfrac{A}{C} & \dfrac{\|TL\|}{C} \\ \dfrac{1}{C} & \dfrac{D}{C} \end{bmatrix}$	$\begin{bmatrix} \dfrac{\delta}{\gamma} & \dfrac{1}{\gamma} \\ \dfrac{\|TLI\|}{\gamma} & \dfrac{\alpha}{\gamma} \end{bmatrix}$
$[y]$	$\begin{bmatrix} \dfrac{z_{22}}{\|z\|} & -\dfrac{z_{12}}{\|z\|} \\ -\dfrac{z_{21}}{\|z\|} & \dfrac{z_{11}}{\|z\|} \end{bmatrix}$	$\begin{bmatrix} y_{11} & y_{12} \\ y_{21} & y_{22} \end{bmatrix}$	$\begin{bmatrix} \dfrac{1}{h_{11}} & -\dfrac{h_{12}}{h_{11}} \\ \dfrac{h_{21}}{h_{11}} & \dfrac{\|h\|}{h_{11}} \end{bmatrix}$	$\begin{bmatrix} \dfrac{\|g\|}{g_{22}} & \dfrac{g_{12}}{g_{22}} \\ -\dfrac{g_{21}}{g_{22}} & \dfrac{1}{g_{22}} \end{bmatrix}$	$\begin{bmatrix} \dfrac{D}{B} & -\dfrac{\|TL\|}{B} \\ -\dfrac{1}{B} & \dfrac{A}{B} \end{bmatrix}$	$\begin{bmatrix} \dfrac{\alpha}{\beta} & -\dfrac{1}{\beta} \\ -\dfrac{\|TLI\|}{\beta} & \dfrac{\delta}{\beta} \end{bmatrix}$
$[h]$	$\begin{bmatrix} \dfrac{\|z\|}{z_{22}} & \dfrac{z_{12}}{z_{22}} \\ -\dfrac{z_{21}}{z_{22}} & \dfrac{1}{z_{22}} \end{bmatrix}$	$\begin{bmatrix} \dfrac{1}{y_{11}} & -\dfrac{y_{12}}{y_{11}} \\ \dfrac{y_{21}}{y_{11}} & \dfrac{\|y\|}{y_{11}} \end{bmatrix}$	$\begin{bmatrix} h_{11} & h_{12} \\ h_{21} & h_{22} \end{bmatrix}$	$\begin{bmatrix} \dfrac{g_{22}}{\|g\|} & -\dfrac{g_{12}}{\|g\|} \\ -\dfrac{g_{21}}{\|g\|} & \dfrac{g_{11}}{\|g\|} \end{bmatrix}$	$\begin{bmatrix} \dfrac{B}{D} & \dfrac{\|TL\|}{D} \\ -\dfrac{1}{D} & \dfrac{C}{D} \end{bmatrix}$	$\begin{bmatrix} \dfrac{\beta}{\alpha} & \dfrac{1}{\alpha} \\ -\dfrac{\|TLI\|}{\alpha} & \dfrac{\gamma}{\alpha} \end{bmatrix}$
$[g]$	$\begin{bmatrix} \dfrac{1}{z_{11}} & -\dfrac{z_{12}}{z_{11}} \\ \dfrac{z_{21}}{z_{11}} & \dfrac{\|z\|}{z_{11}} \end{bmatrix}$	$\begin{bmatrix} \dfrac{\|y\|}{y_{22}} & \dfrac{y_{12}}{y_{22}} \\ -\dfrac{y_{21}}{y_{22}} & \dfrac{1}{y_{22}} \end{bmatrix}$	$\begin{bmatrix} \dfrac{h_{22}}{\|h\|} & -\dfrac{h_{12}}{\|h\|} \\ -\dfrac{h_{21}}{\|h\|} & \dfrac{h_{11}}{\|h\|} \end{bmatrix}$	$\begin{bmatrix} g_{11} & g_{12} \\ g_{21} & g_{22} \end{bmatrix}$	$\begin{bmatrix} \dfrac{C}{A} & \dfrac{\|TL\|}{A} \\ \dfrac{1}{A} & \dfrac{B}{A} \end{bmatrix}$	$\begin{bmatrix} \dfrac{\gamma}{\delta} & -\dfrac{1}{\delta} \\ \dfrac{\|TLI\|}{\delta} & \dfrac{\beta}{\delta} \end{bmatrix}$
$[TL]$	$\begin{bmatrix} \dfrac{z_{11}}{z_{21}} & \dfrac{\|z\|}{z_{21}} \\ \dfrac{1}{z_{21}} & \dfrac{z_{22}}{z_{21}} \end{bmatrix}$	$\begin{bmatrix} -\dfrac{y_{22}}{y_{21}} & -\dfrac{1}{y_{21}} \\ -\dfrac{\|y\|}{y_{21}} & -\dfrac{y_{11}}{y_{21}} \end{bmatrix}$	$\begin{bmatrix} -\dfrac{\|h\|}{h_{21}} & -\dfrac{h_{11}}{h_{21}} \\ -\dfrac{h_{22}}{h_{21}} & -\dfrac{1}{h_{21}} \end{bmatrix}$	$\begin{bmatrix} \dfrac{1}{g_{21}} & \dfrac{g_{22}}{g_{21}} \\ \dfrac{g_{21}}{g_{21}} & \dfrac{\|g\|}{g_{21}} \end{bmatrix}$	$\begin{bmatrix} A & B \\ C & D \end{bmatrix}$	$\begin{bmatrix} \dfrac{\delta}{\|TLI\|} & \dfrac{\beta}{\|TLI\|} \\ \dfrac{y}{\|TLI\|} & \dfrac{\alpha}{\|TLI\|} \end{bmatrix}$
$[TLI]$	$\begin{bmatrix} \dfrac{z_{22}}{z_{12}} & \dfrac{\|z\|}{z_{12}} \\ \dfrac{1}{z_{12}} & \dfrac{z_{11}}{z_{12}} \end{bmatrix}$	$\begin{bmatrix} -\dfrac{y_{11}}{y_{12}} & -\dfrac{1}{y_{12}} \\ -\dfrac{\|y\|}{y_{12}} & -\dfrac{y_{22}}{y_{12}} \end{bmatrix}$	$\begin{bmatrix} \dfrac{1}{h_{12}} & \dfrac{h_{11}}{h_{12}} \\ \dfrac{h_{22}}{h_{12}} & \dfrac{\|h\|}{h_{12}} \end{bmatrix}$	$\begin{bmatrix} -\dfrac{\|g\|}{g_{12}} & -\dfrac{g_{22}}{g_{12}} \\ -\dfrac{g_{11}}{g_{12}} & -\dfrac{1}{g_{12}} \end{bmatrix}$	$\begin{bmatrix} \dfrac{D}{\|TL\|} & \dfrac{B}{\|TL\|} \\ \dfrac{C}{\|TL\|} & \dfrac{A}{\|TL\|} \end{bmatrix}$	$\begin{bmatrix} \alpha & \beta \\ \gamma & \delta \end{bmatrix}$

Note: $|x| = \det[x] = \det \begin{bmatrix} x_{11} & x_{12} \\ x_{21} & x_{22} \end{bmatrix} = x_{11}x_{22} - x_{12}x_{21}$ where x is any of the listed parameters, z, y, h, g, TL, or TLI.

APPENDIX 30.A
Resonant Circuit Formulas

unknown quantity	symbol	units	series	parallel
			R — L — C	R L C
resonant frequency	f_0	Hz	$\dfrac{1}{2\pi\sqrt{LC}}$	
			$\dfrac{QR}{2\pi L} = \dfrac{1}{2\pi QRC}$	$\dfrac{R}{2\pi QL} = \dfrac{Q}{2\pi RC}$
	ω_0	$\dfrac{\text{rad}}{\text{s}}$	$\dfrac{1}{\sqrt{LC}}$	
			$\dfrac{QR}{L} = \dfrac{1}{QRC}$	$\dfrac{R}{QL} = \dfrac{Q}{RC}$
bandwidth	BW	Hz	$f_2 - f_1 = \dfrac{f_0}{Q}$	
			$\dfrac{R}{2\pi L} = \dfrac{1}{2\pi Q^2 RC}$	$\dfrac{1}{2\pi RC} = \dfrac{R}{2\pi Q^2 L}$
	BW	$\dfrac{\text{rad}}{\text{s}}$	$\omega_2 - \omega_1 = \dfrac{\omega_0}{Q}$	
			$\dfrac{R}{L} = \dfrac{1}{Q^2 RC}$	$\dfrac{1}{CR} = \dfrac{R}{Q^2 L}$
quality factor	Q	–	$\dfrac{1}{R}\sqrt{\dfrac{L}{C}}$	$R\sqrt{\dfrac{C}{L}}$
			$\dfrac{\omega_0 L}{R} = \dfrac{1}{\omega_0 RC}$	$\omega_0 RC = \dfrac{R}{\omega_0 L}$
lower half-power point	f_1	Hz	$f_0 - \dfrac{\text{BW}}{2} = f_0\left(1 - \dfrac{1}{2Q}\right)$	
			$f_0 - \dfrac{R}{4\pi L}$	$f_0 - \dfrac{1}{4\pi CR}$
	ω_1	$\dfrac{\text{rad}}{\text{s}}$	$\omega_0 - \dfrac{\text{BW}}{2} = \omega_0\left(1 - \dfrac{1}{2Q}\right)$	
			$\omega_0 - \dfrac{R}{2L}$	$\omega_0 - \dfrac{1}{2CR}$
upper half-power point	f_2	Hz	$f_0 + \dfrac{\text{BW}}{2} = f_0\left(1 + \dfrac{1}{2Q}\right)$	
			$f_0 + \dfrac{R}{4\pi L}$	$f_0 + \dfrac{1}{4\pi CR}$
	ω_2	$\dfrac{\text{rad}}{\text{s}}$	$\omega_0 + \dfrac{\text{BW}}{2} = \omega_0\left(1 + \dfrac{1}{2Q}\right)$	
			$\omega_0 + \dfrac{R}{2L}$	$\omega_0 + \dfrac{1}{2CR}$

APPENDIX 37.A
Electrical Characteristics of Multilayer, Steel-Reinforced, Bare Aluminum Conductors

code name	aluminum area (cmil)	strands (Al/St)	O.D. (in)	R_{DC} 20°C (mΩ/ft)	R_{AC}, 60 Hz 20°C (Ω/mi)	R_{AC}, 60 Hz 50°C (Ω/mi)	GMR (ft)	1-ft spacing reactance per conductor @ 60 Hz X_L (Ω/mi)	1-ft spacing reactance per conductor @ 60 Hz X_C (MΩ−mi)
waxwing	266,800	18/1	0.609	0.0646	0.3488	0.3831	0.0198	0.476	0.1090
partridge	266,800	26/7	0.642	0.0640	0.3452	0.3792	0.0217	0.465	0.1074
ostrich	300,000	26/7	0.680	0.0569	0.3070	0.3372	0.0229	0.458	0.1057
merlin	336,400	18/1	0.684	0.0512	0.2767	0.3037	0.0222	0.462	0.1055
linnet	336,400	26/7	0.721	0.0507	0.2737	0.3006	0.0243	0.451	0.1040
oriole	336,400	30/7	0.741	0.0504	0.2719	0.2987	0.0255	0.445	0.1032
chickadee	397,500	18/1	0.743	0.0433	0.2342	0.2572	0.0241	0.452	0.1031
ibis	397,500	26/7	0.783	0.0430	0.2323	0.2551	0.0264	0.441	0.1015
pelican	477,000	18/1	0.814	0.0361	0.1957	0.2148	0.0264	0.441	0.1004
flicker	477,000	24/7	0.846	0.0359	0.1943	0.2134	0.0284	0.432	0.0992
hawk	477,000	26/7	0.858	0.0357	0.1931	0.2120	0.0289	0.430	0.0988
hen	477,000	30/7	0.883	0.0355	0.1919	0.2107	0.0304	0.424	0.0980
osprey	556,500	18/1	0.879	0.0309	0.1679	0.1843	0.0284	0.432	0.0981
parakeet	556,500	24/7	0.914	0.0308	0.1669	0.1832	0.0306	0.423	0.0969
dove	556,500	26/7	0.927	0.0307	0.1663	0.1826	0.0314	0.420	0.0965
rook	636,000	24/7	0.977	0.0269	0.1461	0.1603	0.0327	0.415	0.0950
grosbeak	636,000	26/7	0.990	0.0268	0.1454	0.1596	0.0335	0.412	0.0946
drake	795,000	26/7	1.108	0.0215	0.1172	0.1284	0.0373	0.399	0.0912
tern	795,000	45/7	1.063	0.0217	0.1188	0.1302	0.0352	0.406	0.0925
rail	954,000	45/7	1.165	0.0181	0.0977	0.1092	0.0386	0.395	0.0897
cardinal	954,000	54/7	1.196	0.0180	0.0988	0.1082	0.0402	0.390	0.0890
ortolan	1,033,500	45/7	1.213	0.0167	0.0924	0.1011	0.0402	0.390	0.0885
bluejay	1,113,000	45/7	1.259	0.0155	0.0861	0.0941	0.0415	0.386	0.0874
finch	1,113,000	54/19	1.293	0.0155	0.0856	0.0937	0.0436	0.380	0.0866
bittern	1,272,000	45/7	1.382	0.0136	0.0762	0.0832	0.0444	0.378	0.0855
bobolink	1,431,000	45/7	1.427	0.0121	0.0684	0.0746	0.0470	0.371	0.0837
plover	1,431,000	54/19	1.465	0.0120	0.0673	0.0735	0.0494	0.365	0.0829
lapwing	1,590,000	45/7	1.502	0.0109	0.0623	0.0678	0.0498	0.364	0.0822
falcon	1,590,000	54/19	1.545	0.0108	0.0612	0.0667	0.0523	0.358	0.0814
bluebird	2,156,000	84/19	1.762	0.0080	0.0476	0.0515	0.0586	0.344	0.0776

Used with permission from *Aluminum Electrical Conductor Handbook*, by the Aluminum Association, 1971.

Use the following correction factors for other than 1 ft spacing. D, GMR, and r are in feet.

1. $K_L = 1 + \dfrac{\ln D}{\ln\left(\dfrac{1}{\text{GMR}}\right)}$

2. $X_{L,\text{actual}} = X_{L,\text{table}} K_L$

3. $K_C = 1 + \dfrac{\ln D}{\ln\left(\dfrac{1}{r}\right)}$

4. $X_{C,\text{actual}} = X_{C,\text{table}} K_C$

APPENDIX 43.A
Periodic Table of the Elements

THE PERIODIC TABLE OF ELEMENTS (Long Form)

The number of electrons in filled shells is shown in the column at the extreme left; the remaining electrons for each element are shown immediately below the symbol for each element. Atomic numbers are enclosed in brackets. Atomic weights (rounded, based on carbon-12) are shown above the symbols. Atomic weight values in parentheses are those of the isotopes of longest half-life for certain radioactive elements whose atomic weights cannot be precisely quoted without knowledge of origin of the element.

METALS — TRANSITION METALS — NON-METALS

periods	1 IA	2 IIA	3 IIIB	4 IVB	5 VB	6 VIB	7 VIIB	8 VIII	9 VIII	10 VIII	11 IB	12 IIB	13 IIIA	14 IVA	15 VA	16 VIA	17 VIIA	18 VIIIA
1 (0)	1.0079 H[1] 1																	4.0026 He[2] 2
2 (2)	6.939 Li[3] 1	9.0122 Be[4] 2											10.81 B[5] 3	12.01115 C[6] 4	14.0067 N[7] 5	15.9994 O[8] 6	18.994 F[9] 7	20.183 Ne[10] 8
3 (2,8)	22.9898 Na[11] 1	24.312 Mg[12] 2											26.9815 Al[13] 3	28.086 Si[14] 4	30.9738 P[15] 5	32.064 S[16] 6	35.453 Cl[17] 7	39.948 Ar[18] 8
4 (2,8)	39.098 K[19] 8,1	40.08 Ca[20] 8,2	44.956 Sc[21] 9,2	47.90 Ti[22] 10,2	50.942 V[23] 11,2	51.996 Cr[24] 13,1	54.938 Mn[25] 13,2	55.847 Fe[26] 14,2	58.933 Co[27] 15,2	58.71 Ni[28] 16,2	63.546 Cu[29] 18,1	65.38 Zn[30] 18,2	69.72 Ga[31] 18,3	72.59 Ge[32] 18,4	74.922 As[33] 18,5	78.96 Se[34] 18,6	79.904 Br[35] 18,7	83.80 Kr[36] 18,8
5 (2,8,18)	85.47 Rb[37] 8,1	87.62 Sr[38] 8,2	88.905 Y[39] 9,2	91.22 Zr[40] 10,2	92.906 Nb[41] 12,1	95.94 Mo[42] 13,1	(98) Tc[43] 14,1	101.07 Ru[44] 15,1	102.905 Rh[45] 16,1	106.4 Pd[46] 18	107.868 Ag[47] 18,1	112.40 Cd[48] 18,2	114.82 In[49] 18,3	118.69 Sn[50] 18,4	121.75 Sb[51] 18,5	127.60 Te[52] 18,6	126.904 I[53] 18,7	131.30 Xe[54] 18,8
6 (2,8,18)	132.905 Cs[55] 18,8,1	137.34 Ba[56] 18,8,2	* [57-71]	178.49 Hf[72] 32,10,2	180.948 Ta[73] 32,11,2	183.85 W[74] 32,12,2	186.2 Re[75] 32,13,2	190.2 Os[76] 32,14,2	192.2 Ir[77] 32,15,2	195.09 Pt[78] 32,17,1	196.967 Au[79] 32,18,1	200.59 Hg[80] 32,18,2	204.37 Tl[81] 32,18,3	207.19 Pb[82] 32,18,4	208.980 Bi[83] 32,18,5	(210) Po[84] 32,18,6	(210) At[85] 32,18,7	(222) Rn[86] 32,18,8
7 (2,8,18,32)	(223) Fr[87] 18,8,1	226.025 Ra[88] 18,8,2	† [89-103]	Rf[104] 32,10,2	Ha[105] 32,11,2	[106] 32,12,2	[107]	[108]										

*LATHANIDE SERIES

138.91 La[57] 18,9,2	140.12 Ce[58] 20,8,2	140.907 Pr[59] 21,8,2	144.24 Nd[60] 22,8,2	(147) Pm[61] 23,8,2	150.35 Sm[62] 24,8,2	151.96 Eu[63] 25,8,2	157.25 Gd[64] 25,9,2	158.924 Tb[65] 27,8,2	162.50 Dy[66] 28,8,2	164.930 Ho[67] 29,8,2	167.26 Er[68] 30,8,2	168.934 Tm[69] 31,8,2	173.04 Yb[70] 32,8,2	174.97 Lu[71] 32,9,2

†ACTINIDE SERIES

(227) Ac[89] 18,9,2	232.038 Th[90] 18,10,2	231.036 Pa[91] 20,9,2	238.03 U[92] 21,9,2	237.048 Np[93] 23,8,2	(242) Pu[94] 24,8,2	(243) Am[95] 25,8,2	(247) Cm[96] 25,9,2	(247) Bk[97] 26,9,2	(249) Cf[98] 28,8,2	(254) Es[99] 29,8,2	(254) Fm[100] 30,8,2	(256) Md[101] 31,8,2	(254) No[102] 32,8,2	(257) Lr[103] 32,9,2

APPENDIX 43.B
Semiconductor Symbols and Abbreviations*

A, a—anode

B, b—base

b_{fs}—common-source small-signal forward transfer susceptance

b_{is}—common-source small-signal input susceptance

b_{os}—common-source small-signal output susceptance

b_{rs}—common-source small-signal reverse transfer susceptance

C, c—collector

C_{cb}—collector-base interterminal capacitance

C_{ce}—collector-emitter inter-terminal capacitance

C_{ds}—drain-source capacitance

C_{du}—drain-substrate capacitance

C_{eb}—emitter-base interterminal capacitance

C_{ibo}—common-base open-circuit input capacitance

C_{ibs}—common-base short-circuit input capacitance

C_{ieo}—common-emitter open-circuit input capacitance

C_{ies}—common-emitter short-circuit input capacitance

C_{iss}—common-source short-circuit input capacitance

C_{obo}—common-base open-circuit output capacitance

C_{obs}—common-base short-circuit output capacitance

C_{oeo}—common-emitter open-circuit output capacitance

C_{oes}—common-emitter short-circuit output capacitance

C_{oss}—common-source short-circuit output capacitance

C_{rbs}—common-base short-circuit reverse transfer capacitance

C_{res}—common-collector short-circuit reverse transfer capacitance

C_{rss}—common-source short-circuit reverse transfer capacitance

C_{tc}—collector depletion-layer capacitance

C_{te}—emitter depletion-layer capacitance

D, d—drain

E, e—emitter

η—intrinsic standoff ratio

f_{hfb}—common-base small-signal short-circuit forward current transfer ratio cutoff frequency

f_{hfc}—common-collector small-signal short-circuit forward current transfer ratio cutoff frequency

f_{hfe}—common-emitter small-signal short-circuit forward current transfer ratio cutoff frequency

f_{max}—maximum frequency of oscillation

f_T—transition frequency (frequency at which common-emitter small-signal forward current transfer ratio extrapolates to unity)

G, g—gate

g_{fs}—common-source small-signal forward transfer conductance

g_{is}—common-source small-signal input conductance

g_{MB}—common-base static transconductance

g_{MC}—common-collector static transconductance

g_{ME}—common-emitter static transconductance

g_{os}—common-source small-signal output conductance

G_{pb}—common-base small-scale insertion power gain

G_{PB}—common-base large-signal insertion power gain

G_{pc}—common-collector small-signal insertion power gain

G_{PC}—common-collector large-signal insertion power gain

G_{pe}—common-emitter small-signal insertion power gain

G_{PE}—common-emitter large-signal insertion power gain

G_{pg}—common-gate small-signal insertion power gain

G_{ps}—common-source small-signal insertion power gain

g_{rs}—common-source small-signal reverse transfer conductance

G_{tb}—common-base small-signal transducer power gain

G_{TB}—common-base large-signal transducer power gain

G_{tc}—common-collector small-signal transducer power gain

G_{TC}—common-collector large-signal transducer power gain

G_{te}—common-emitter small-signal transducer power gain

G_{TE}—common-emitter large-signal transducer power gain

G_{tg}—common-gate small-signal transducer power gain

G_{ts}—common-source small-signal transducer power gain

h_{fb}—common-base small-signal short-circuit forward current transfer ratio

h_{FB}—common-base static forward current transfer ratio

h_{fc}—common-collector small-signal short-circuit forward current transfer ratio

h_{FC}—common-collector static forward current transfer ratio

h_{fe}—common-emitter small-signal short-circuit forward current transfer ratio

h_{FE}—common-emitter static forward current transfer ratio

h_{FEL}—inherent large-signal forward current transfer ratio

h_{ib}—common-base small-signal short-circuit input impedance

h_{IB}—common-base static input resistance

h_{ic}—common-collector small-signal short-circuit input impedance

h_{IC}—common-collector static input resistance

h_{ie}—common-emitter small-signal short-circuit input impedance

h_{IE}—common-emitter static input resistance

$h_{ie(imag)}$—imaginary part of common-emitter small-signal short-circuit input impedance

$h_{ie(real)}$—real part of common-emitter small-signal short-circuit input impedance

h_{ob}—common-base small-signal open-circuit output admittance

*Recommended by the Joint Electron Device Engineering Council of the Electronic Industries Association and the National Electrical Manufacturers Associaton for use in semiconductor device data sheets and specifications. From R.F. Graf, *Modern Dictionary of Electronics*, 4th ed., Howard Sams, 1972.

(continued)

Support Material

APPENDIX 43.B (*continued*)
Semiconductor Symbols and Abbreviations

h_{oc}—common-collector small-signal open-circuit output admittance

h_{oe}—common emmitter small-signal open-circuit output admittance

$h_{oe(imag)}$—imaginary part of common-emitter small-signal open-circuit output admittance

$h_{oe(real)}$—real part of common-emitter small-signal open-circuit output admittance

h_{rb}—common-base small-signal open-circuit reverse voltage transfer ratio

h_{re}—common-collector small-signal open-circuit reverse voltage transfer ratio

h_{ro}—common-emitter small-signal open-circuit reverse voltage transfer ratio

I_0—average forward current, 180° conduction angle, 60 Hz half sine wave

i_B—instantaneous total value of base-terminal current

I_b—alternating component (rms value) of base-terminal current

I_B—base-terminal DC current

$I_{B2(mod)}$—interbase modulated current

I_{BEV}—base cutoff current, DC

i_C—instantaneous total value of collector-terminal current

I_c—alternating component (rms value) of collector-terminal current

I_C—collector-terminal DC current

I_{CBO}—collector cutoff current (DC), emitter open

I_{CEO}—collector cutoff current (DC), base open

I_{CER}—collector cutoff current (DC), specified resistance between base and emitter

I_{CES}—collector cutoff current (DC), base shorted to emitter

I_{CEV}—collector cutoff current, DC, specified voltage between base and emitter

I_{CEX}—collector cutoff current, DC, specified circuit between base and emitter

I_D—drain current; DC

$I_{D(off)}$—drain cutoff current

$I_{D(on)}$—on-state drain current

I_{DSS}—zero-gate-voltage drain current

i_E—instantaneous total value of emitter-terminal current

I_e—alternating component (rms value) of emitter-terminal current

I_E—emitter-terminal DC current

I_{EBO}—emitter cutoff current (DC), collector open

I_{ERR}—emitter reverse current

$I_{EC(ofs)}$—emitter-collector offset current

I_{ECS}—emitter cutoff current (DC), base short-circuited to collector

$I_{EIE2(off)}$—emitter cutoff current

I_f—alternating component of forward current (rms value)

I_F—For voltage-regulator and voltage-reference diodes: DC forward current. For signal diodes and rectifier diodes: DC forward current (no alternating component).

I_F—instantaneous total forward current

$I_{F(AV)}$—forward current, DC (with alternating component)

I_{FM}—maximum (peak) total forward current

$I_{F(OV)}$—forward current, overload

I_{FRM}—maximum (peak) forward current, repetitive

I_{FRMS}—total rms forward current

I_{FSM}—maximum (peak) forward current, surge

I_G—gate current, DC

I_{GF}—forward gate current

I_{GR}—reverse gate current

I_{GSS}—reverse gate current, drain short-circuited to source

I_{GSSF}—forward gate current, drain short-circuited to source

I_{GSSR}—reverse gate current, drain short-circuited to source

I_I—inflection-point current

$Im(h_{oe})$—imaginary part of common-emitter small-signal short-circuit input impedance

$Im(h_{oe})$—imaginary part of common-emitter small-signal open-circuit output admittance

I_P—peak-point current

i_R—instantaneous total reverse current

I_r—alternating component of reverse current (rms value)

I_R—for voltage-regulator and voltage-reference diodes: DC reverse current. For signal diodes and rectifier diodes: DC reverse current (no alternating component).

$I_{R(AV)}$—reverse current, DC (with alternating component)

I_{RM}—maximum (peak) total reverse current

I_{RRM}—maximum (peak) reverse current, repetitive

$I_{R(RMS)}$—total rms reverse current

I_{RSM}—maximum (peak) surge reverse current

I_S—source current, DC

I_{SDS}—zero-gate-voltage source current

$I_{S(off)}$—source cutoff current

I_V—valley-point current

I_Z—regulator current, reference current, DC

I_{ZK}—regulator current, reference current (DC near breakdown knee)

I_{ZM}—regulator current, reference current (DC maximum rated current)

K, k—cathode

L_c—conversion loss

M—figure of merit

NF_o—overall noise figure

NR_o—output noise ratio

p_{BE}—instantaneous total power input to base, common emitter

P_{BE}—power input (DC) to base, common emitter

p_{CB}—instantaneous total power input to collector, common base

P_{CB}—power input (DC) to collector, common base

p_{CE}—power input (DC) to collector, common emitter

P_{CE}—instantaneous total power input to collector, common emitter

p_{EB}—instantaneous total power input to emitter, common base

(continued)

APPENDIX 43.B (*continued*)
Semiconductor Symbols and Abbreviations

P_{EB}—power input (DC) to emitter, common base

p_F—instantaneous total forward power dissipation

P_F—forward power dissipation, DC (no alternating component)

$P_{F(AV)}$—forward power dissipation, DC (with alternating component)

P_{FM}—maximum (peak) total forward power dissipation

p_{ib}—common-base small-signal input power

P_{IB}—common-base large-signal input power

p_{ic}—common collector small-signal input power

P_{IC}—common-collector large-signal input power

p_{ie}—common-emitter small-signal input power

P_{IE}—common-emitter large-signal input power

p_{ob}—common-base small-signal output power

P_{OB}—common-base large-signal output power

p_{oc}—common-collector small-signal output power

P_{OC}—common-collector large-signal output power

p_{oe}—common-emitter small-signal output power

P_{OE}—common-emitter large-signal output power

p_R—instantaneous total reverse power dissipation

P_R—reverse power dissipation, DC (no alternating component)

$P_{R(AV)}$—reverse power dissipation, DC (with alternating component)

P_{RM}—maximum (peak) total reverse power dissipation

p_T—nonreactive power input, instantaneous total, to all terminals

P_T—total nonreactive power input to all terminals

Q_S—stored charge

r_{BB}—interbase resistance

$r_b' C_c$—collector-base time constant

$r_{CE(sat)}$—saturation resistance, collector-to-emitter

$r_{ds(on)}$—small-signal drain-source on-state resistance

$r_{DS(on)}$—static drain-source on-state resistance

$Re(h_{ie})$—real part of common-emitter small-signal short-circuit input impedance

$Re(h_{oe})$—real part of common-emitter small-signal open-circuit output admittance

$r_{ele2(on)}$—small-signal emitter-emitter on-state resistance

r_i—dynamic resistance at inflection point

R_θ—thermal resistance

$R_{\theta CA}$—thermal resistance, case to ambient

$R_{\theta JA}$—thermal resistance, junction to ambient

$R_{\theta JC}$—thermal resistance, junction to case

S, s—source

T_A—ambient temperature or free-air temperature

T_C—case temperature

t_d—delay time

$t_{d(off)}$—turn-off delay time

$t_{d(on)}$—turn-on delay time

t_f—fall time

t_{fr}—forward recovery time

T_j—junction temperature

t_{off}—turn-off time

t_{on}—turn-on time

t_p—pulse time

t_r—rise time

t_{rr}—reverse recovery time

t_s—storage time

TSS—tangential signal sensitivity

T_{stg}—storage temperature

t_w—pulse average time

U, u—bulk (substrate)

V_{B2B1}—interbase voltage

V_{BB}—base supply voltage, DC

v_{bc}—instantaneous value of alternating component of base-collector voltage

V_{BC}—average or DC voltage, base to collector

v_{be}—instantaneous value of alternating component of base-emitter voltage

V_{BE}—average or DC voltage, base to emitter

$v_{(BR)}$—breakdown voltage (instantaneous total)

$V_{(BR)}$—breakdown voltage, DC

$V_{(BR)CBO}$—collector-base breakdown voltage, emitter open

$V_{(BR)CEO}$—collector-emitter breakdown voltage, base open

$V_{(BR)CER}$—collector-emitter breakdown voltage, resistance between base and emitter

$V_{(BR)CES}$—collector-emitter breakdown voltage, base shorted to emitter

$V_{(BR)CEV}$—collector-emitter breakdown voltage, specified voltage between base and emitter

$V_{(BR)CEX}$—collector-emitter breakdown voltage, specified circuit between base and emitter

$V_{(BR)E1E2}$—emitter-emitter breakdown voltage

$V_{(BR)EBO}$—emitter-base breakdown voltage, collector open

$V_{(BR)ECO}$—emitter-collector breakdown voltage, base open

$V_{(BR)GSS}$—gate-source breakdown voltage

$V_{(BR)GSSF}$—forward gate-source breakdown voltage

$V_{(BR)GSSR}$—reverse gate-source breakdown voltage

v_{cb}—instantaneous value of alternating component of collector-base voltage

V_{CB}—average or DC voltage, collector to base

$V_{CB(fl)}$—collector-base DC open-circuit voltage (floating potential)

V_{CBO}—collector-base voltage, DC, emitter open

V_{CC}—collector supply voltage, DC

v_{ce}—instantaneous value of alternating component of collector-emitter voltage

V_{CE}—average or DC voltage, collector to emitter

$V_{CE(fl)}$—collector-emitter DC open-circuit voltage (floating potential)

V_{CEO}—collector-emitter voltage (DC), base open

(continued)

Support Material

APPENDIX 43.B (*continued*)
Semiconductor Symbols and Abbreviations

$V_{CE(ofs)}$—collector-emitter offset voltage

V_{CER}—collector-emitter voltage (DC), resistance between base and emitter

V_{CES}—collector-emitter voltage (DC), base shorted to emitter

$V_{CE(sat)}$—collector-emitter DC saturation voltage

V_{CEV}—collector-emitter voltage, DC, specified voltage between base and emitter

V_{CEX}—collector-emitter voltage, DC, specified circuit between base and emitter

V_{DD}—drain supply voltage, DC

V_{DG}—drain-gate voltage

V_{DS}—drain-source voltage

$V_{DS(on)}$—drain-source on-state voltage

V_{DU}—drain-substrate voltage

v_{eb}—instantaneous value of alternating component of emitter-base voltage

V_{EB}—average or DC voltage, emitter to base

$V_{EB(fl)}$—emitter-base DC open-circuit voltage (floating potential)

V_{EBO}—emitter-base voltage, DC, collector open

$V_{EB(sat)}$—emitter saturation voltage

v_{ec}—instantaneous value of alternating component of emitter-collector voltage

V_{EC}—average or DC voltage, emitter to collector

$V_{EC(fl)}$—emitter-collector DC open-circuit voltage (floating potential)

$V_{EC(ofs)}$—emitter-collector offset voltage

V_{EE}—emitter supply voltage, DC

v_F—instantaneous total forward voltage

V_f—alternating component of forward voltage (rms value)

V_F—for voltage-regulator and voltage-reference diodes: DC forward voltage. For signal diodes and rectifier diodes: DC forward voltage (no alternating component).

$V_{F(AV)}$—forward voltage, DC (with alternating component)

V_{FM}—maximum (peak) total forward voltage

$V_{F(RMS)}$—total rms forward voltage

V_{GG}—gate supply voltage, DC

V_{GS}—gate-source voltage

V_{GSF}—forward gate-source voltage

$V_{GS(off)}$—gate-source cutoff voltage

V_{GSR}—reverse gate-source voltage

$V_{GS(th)}$—gate-source threshold voltage

V_{GU}—gate-substrate voltage

V_I—inflection-point voltage

V_{OB1}—base-1 peak voltage

V_P—peak-point voltage

V_{PP}—projected peak-point voltage

v_R—instantaneous total reverse voltage

V_r—alternating component of reverse voltage (rms value)

V_R—for voltage-regulator and voltage-reference diodes: DC reverse voltage. For signal diodes and rectifier diodes: DC reverse voltage (no alternating component).

$V_{R(AV)}$—reverse voltage, DC (with alternating component)

V_{RM}—maximum (peak) total reverse voltage

V_{RRM}—repetitive peak reverse voltage

$V_{R(RMS)}$—total rms reverse voltage

V_{RSM}—nonrepetitive peak reverse voltage

V_{RT}—reach-through voltage

V_{RWM}—working peak reverse voltage

V_{SS}—source supply voltage, DC

V_{SU}—source-substrate voltage

V_{TO}—threshold voltage

V_V—valley-point voltage

V_Z—regulator voltage, reference voltage, DC

V_{ZM}—regulator voltage, reference voltage (DC at maximum rated current)

y_{fb}—common-base small-signal short-circuit forward transfer admittance

y_{fc}—common-collector small-signal short-circuit forward transfer admittance

y_{fe}—common-emitter small-signal short-circuit forward transfer admittance

y_{fs}—common-source small-signal short-circuit forward transfer admittance

$y_{fs(imag)}$—common-source small-signal forward transfer susceptance

$y_{fs(real)}$—common-source small-signal forward transfer conductance

y_{ib}—common-base small-signal short-circuit input admittance

y_{ic}—common-collector small-signal short-circuit input admittance

y_{ie}—common-emitter small-signal short-circuit input admittance

$y_{ie(imag)}$—imaginary part of small-signal short-circuit input admittance (common emitter)

$y_{ie(real)}$—real part of small-signal short-circuit input admittance (common emitter)

y_{is}—common-source small-signal short-circuit input admittance

$y_{is(imag)}$—common-source small-signal input susceptance

$y_{is(real)}$—common-source small-signal input conductance

y_{ob}—common-base small-signal short-circuit output admittance

y_{oc}—common-collector small-signal short-circuit output admittance

y_{oe}—common-emitter small-signal short-circuit output admittance

$y_{oe(imag)}$—imaginary part of small-signal short-circuit output admittance (common emitter)

$y_{oe(real)}$—real part of small-signal short-circuit output admittance (common emitter)

y_{os}—common-source small-signal short-circuit output admittance

$y_{os(imag)}$—common-source small-signal output susceptance

$y_{os(real)}$—common-source small-signal output conductance

y_{rb}—common-base small-signal short-circuit reverse transfer admittance

(continued)

APPENDIX 43.B (*continued*)
Semiconductor Symbols and Abbreviations

$\mathbf{y}_{rc}$—common-collector small-signal short-circuit reverse transfer admittance

$\mathbf{y}_{re}$—common-emitter small-signal short-circuit reverse transfer admittance

$\mathbf{y}_{rs}$—common-source small-signal short-circuit reverse transfer admittance

$\mathbf{y}_{rs(imag)}$—common-source small-signal reverse transfer susceptance

$\mathbf{y}_{rs(real)}$—common-source small-signal reverse transfer conductance

$\mathbf{z}_{if}$—intermediate-frequency impedance

$\mathbf{z}_{m}$—modulator-frequency load impedance

$\mathbf{z}_{rf}$—radio-frequency impedance

$\mathbf{Z}_{\theta JA(t)}$—junction-to-case transient thermal impedance

$\mathbf{Z}_{\theta JC(t)}$—junction-to-case transient thermal impedance

$\mathbf{Z}_{\theta(t)}$—transient thermal impedance

$\mathbf{z}_{v}$—video impedance

$\mathbf{z}_{z}$—regulator impedance, reference impedance (small signal at I_z)

$\mathbf{z}_{zk}$—regulator impedance, reference impedance (small signal at I_{ZK})

$\mathbf{Z}_{zm}$—regulator impedance, reference impedance (small signal at I_{ZM})

Support Material

APPENDIX 45.A
Standard Zener Diodes

zener voltage (nominal)	500 mW		1 W		5 W		50 W	
	$Z_Z{}^a$ (Ω)	$I_{ZK}{}^b$ (mA)	$Z_Z{}^a$ (Ω)	$I_{ZK}{}^b$ (mA)	$Z_Z{}^a$ (Ω)	$I_{ZK}{}^b$ (mA)	$Z_Z{}^a$ (Ω)	$I_{ZK}{}^b$ (mA)
2.4	30	1.0						
2.7	30	1.0						
3.0	29	1.0						
3.3	28	1.0	10	1.0	3.0	1.0		
3.6	24	1.0	10	1.0	2.5	1.0		
3.9	23	1.0	9.0	1.0	2.0	1.0	0.16	5.0
4.3	22	1.0	9.0	1.0	2.0	1.0	0.16	5.0
4.7	19	1.0	8.0	1.0	2.0	1.0	0.12	5.0
5.1	17	1.0	7.0	1.0	1.5	1.0	0.12	5.0
5.6	11	1.0	5.0	1.0	1.0	1.0	0.12	5.0
6.2	7.0	1.0	2.0	1.0	1.0	1.0	0.14	5.0
6.8	5.0	1.0	1.5	1.0	1.0	1.0	0.16	5.0
7.5	5.5	0.5	1.5	1.0	1.5	1.0	0.24	5.0
8.2	6.5	0.5	4.5	0.5	1.5	1.0	0.4	5.0
8.7					2.0	1.0		
9.1	7.5	0.5	5.0	0.5	2.0	1.0	0.5	5.0
10	8.5	0.25	7.0	0.25	2.0	1.0	0.6	5.0
11	9.5	0.25	8.0	0.25	2.5	1.0	0.8	5.0
12	11.5	0.25	9.0	0.25	2.5	1.0	1.0	5.0
13	13	0.25	10	0.25	2.5	1.0	1.1	5.0
14					2.5	1.0	1.2	5.0
15	16	0.25	14	0.25	2.5	1.0	1.4	5.0
16	17	0.25	16	0.25	2.5	1.0	1.6	5.0
17					2.5	1.0	1.8	5.0
18	21	0.25	20	0.25	2.5	1.0	2.0	5.0
19					3.0	1.0	2.2	5.0
20	25	0.25	22	0.25	3.0	1.0	2.4	5.0
22	29	0.25	23	0.25	3.5	1.0	2.5	5.0
24	33	0.25	25	0.25	3.5	1.0	2.6	5.0
25					4.0	1.0	2.7	5.0
27	41	0.25	35	0.25	5.0	1.0	2.8	5.0
28					6.0	1.0		
30	49	0.25	40	0.25	8.0	1.0	3.0	5.0
33	58	0.25	45	0.25	10	1.0	3.2	5.0
36	70	0.25	50	0.25	11	1.0	3.5	5.0
39	80	0.25	60	0.25	14	1.0	4.0	5.0
43	93	0.25	70	0.25	20	1.0	4.5	5.0
45							4.5	5.0
47	105	0.25	80	0.25	25	1.0	5.0	5.0
50							5.0	5.0
51	125	0.25	95	0.25	27	1.0	5.2	5.0
56	150	0.25	110	0.25	35	1.0	6.0	5.0
60					40	1.0		
62	185	0.25	125	0.25	42	1.0	7.0	5.0
68	230	0.25	150	0.25	44	1.0	8.0	5.0
75	270	0.25	175	0.25	45	1.0	9.0	5.0
82	330	0.25	200	0.25	65	1.0	11	5.0
87					75	1.0		
91	400	0.25	250	0.25	75	1.0	15	5.0
100			350	0.25	90	1.0	20	5.0
105							25	5.0

(continued)

APPENDIX 45.A (*continued*)
Standard Zener Diodes

rated power

zener voltage (nominal)	500 mW		1 W		5 W		50 W	
	$Z_Z{}^a$ (Ω)	$I_{ZK}{}^b$ (mA)	$Z_Z{}^a$ (Ω)	$I_{ZK}{}^b$ (mA)	$Z_Z{}^a$ (Ω)	$I_{ZK}{}^b$ (mA)	$Z_Z{}^a$ (Ω)	$I_{ZK}{}^b$ (mA)
110					125	1.0	30	5.0
120			550	0.25	170	1.0	40	5.0
130			700	0.25	190	1.0	50	5.0
140					230	1.0	60	5.0
150			1000	0.25	330	1.0	75	5.0
160			1100	0.25	350	1.0	80	5.0
170					380	1.0		
175							85	5.0
180			1200	0.25	430	1.0	90	5.0
190					450	1.0		
200			1500	0.25	480	1.0	100	5.0

[a] $Z_Z \equiv$ zener impedance
[b] $I_{ZK} \equiv$ zener keep-alive or knee current

APPENDIX 56.A
Smith Chart
Impedance or Admittance Coordinates

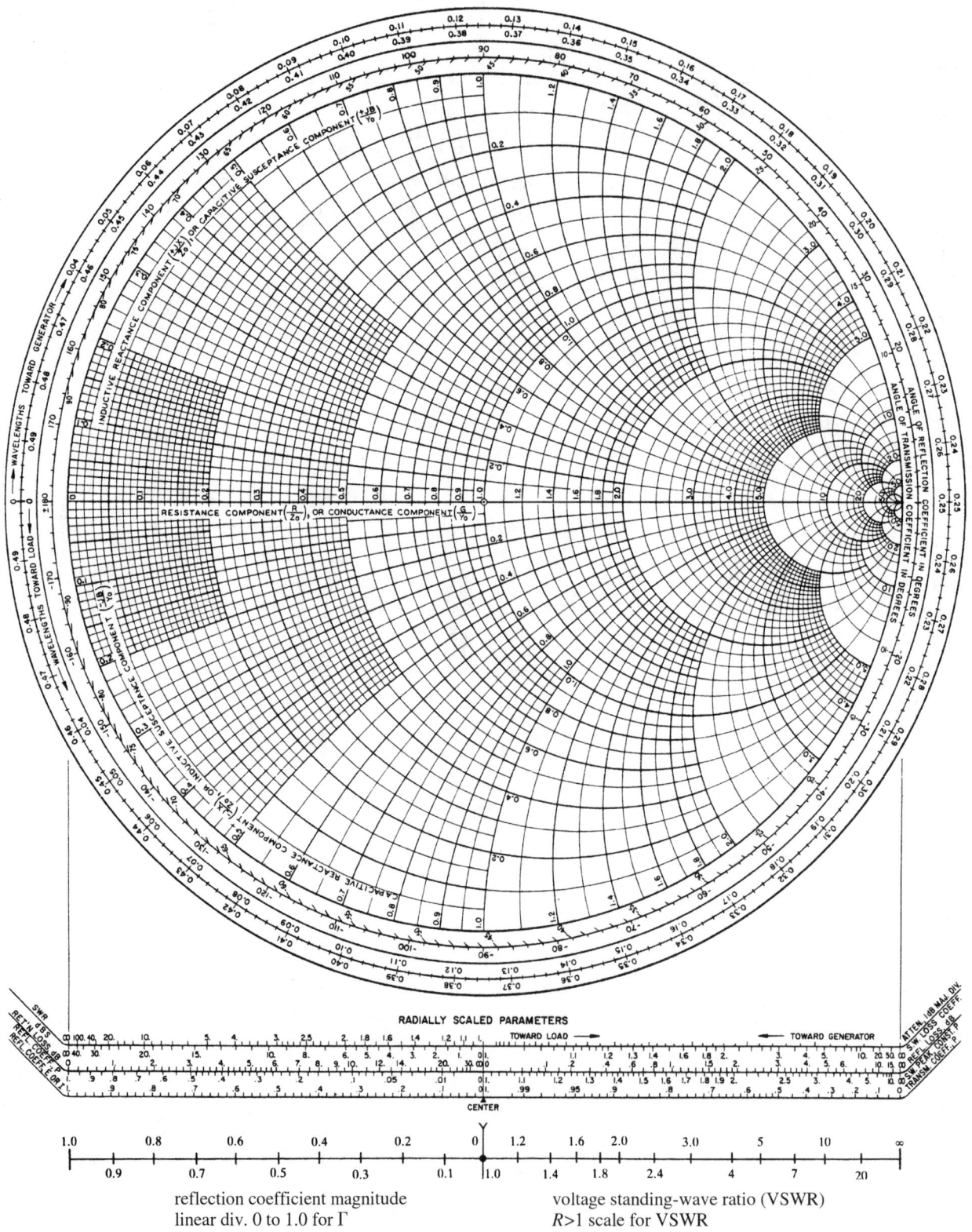

RADIALLY SCALED PARAMETERS

reflection coefficient magnitude
linear div. 0 to 1.0 for Γ

voltage standing-wave ratio (VSWR)
R>1 scale for VSWR

Reproduced from "Impedance or Admittance Coordinates," copyright © 1966, with
permission of Kay Elemetrics Corp., Lincoln Park, NJ.

APPENDIX 59.A

Error Function Values: $erf(x) = \dfrac{2}{\sqrt{\pi}} \displaystyle\int_0^x e^{-t^2}\,dt$

x	0	1	2	3	4	5	6	7	8	9
0.00	0.0000	0011	0023	0034	0045	0056	0068	0079	0090	0102
1	0113	0124	0135	0147	0158	0169	0181	0192	0203	0214
2	0226	0237	0248	0260	0271	0282	0293	0305	0316	0327
3	0338	0350	0361	0372	0384	0395	0406	0417	0429	0440
4	0451	0462	0474	0485	0496	0507	0519	0530	0541	0553
5	0564	0575	0586	0598	0609	0620	0631	0643	0654	0665
6	0676	0688	0699	0710	0721	0732	0744	0755	0766	0777
7	0789	0800	0811	0822	0834	0845	0856	0867	0878	0890
8	0901	0912	0923	0934	0946	0957	0968	0979	0990	1002
9	1013	1024	1035	1046	1058	1069	1080	1091	1102	1113
10	1125	1136	1147	1158	1169	1180	1192	1203	1214	1225
1	1236	1247	1259	1270	1281	1292	1303	1314	1325	1336
2	1348	1359	1370	1381	1392	1403	1414	1425	1436	1448
3	1459	1470	1481	1492	1503	1514	1525	1536	1547	1558
4	1569	1581	1592	1603	1614	1625	1636	1647	1658	1669
5	1680	1691	1702	1713	1724	1735	1746	1757	1768	1779
6	1790	1801	1812	1823	1834	1845	1856	1867	1878	1889
7	1900	1911	1922	1933	1944	1955	1966	1977	1988	1998
8	2009	2020	2031	2042	2053	2064	2075	2086	2097	2108
9	2118	2129	2140	2151	2162	2173	2184	2194	2205	2216
20	2227	2238	2249	2260	2270	2281	2292	2303	2314	2324
1	2335	2346	2357	2368	2378	2389	2400	2411	2421	2432
2	2443	2454	2464	2475	2486	2497	2507	2518	2529	2540
3	2550	2561	2572	2582	2593	2604	2614	2625	2636	2646
4	2657	2668	2678	2689	2700	2710	2721	2731	2742	2753
5	2763	2774	2784	2795	2806	2816	2827	2837	2848	2858
6	2869	2880	2890	2901	2911	2922	2932	2943	2953	2964
7	2974	2985	2995	3006	3016	3027	3037	3047	3058	3068
8	3079	3089	3100	3110	3120	3131	3141	3152	3162	3172
9	3183	3193	3204	3214	3224	3235	3245	3255	3266	3276
30	3286	3297	3307	3317	3327	3338	3348	3358	3369	3379
1	3389	3399	3410	3420	3430	3440	3450	3461	3471	3481
2	3491	3501	3512	3522	3532	3542	3552	3562	3573	3583
3	3593	3603	3613	3623	3633	3643	3653	3663	3674	3684
4	3694	3704	3714	3724	3734	3744	3754	3764	3774	3784
5	3794	3804	3814	3824	3834	3844	3854	3864	3873	3883
6	3893	3903	3913	3923	3933	3943	3953	3963	3972	3982
7	3992	4002	4012	4022	4031	4041	4051	4061	4071	4080
8	4090	4100	4110	4119	4129	4139	4149	4158	4168	4178
9	4187	4197	4207	4216	4226	4236	4245	4255	4265	4274
40	4284	4294	4303	4313	4322	4332	4341	4351	4361	4370
1	4380	4389	4399	4408	4418	4427	4437	4446	4456	4465
2	4475	4484	4494	4503	4512	4522	4531	4541	4550	4559
3	4569	4578	4588	4597	4606	4616	4625	4634	4644	4653
4	4662	4672	4681	4690	4699	4709	4718	4727	4736	4746
5	4755	4764	4773	4782	4792	4801	4810	4819	4828	4837
6	4847	4856	4865	4874	4883	4892	4901	4910	4919	4928
7	4937	4946	4956	4965	4974	4983	4992	5001	5010	5019
8	5027	5036	5045	5054	5063	5072	5081	5090	5099	5108
9	5117	5126	5134	5143	5152	5161	5170	5179	5187	5196
50	5205	5214	5223	5231	5240	5249	5258	5266	5275	5284

(continued)

APPENDIX 59.A (*continued*)

Error Function Values: $\text{erf}(x) = \dfrac{2}{\sqrt{\pi}} \left(x - \dfrac{x^2}{3 \cdot 1!} + \dfrac{x^5}{5 \cdot 2!} - \dfrac{x^7}{7 \cdot 3!} + \cdots \right)$

x	0	1	2	3	4	5	6	7	8	9
0.50	0.5205	5214	5223	5231	5240	5249	5258	5266	5275	5284
1	5292	5301	5310	5318	5327	5336	5344	5353	5362	5370
2	5379	5388	5396	5405	5413	5422	5430	5439	5448	5456
3	5465	5473	5482	5490	5499	5507	5516	5524	5533	5541
4	5549	5558	5566	5575	5583	5591	5600	5608	5617	5625
5	5633	5642	5650	5658	5667	5675	5683	5691	5700	5708
6	5716	5724	5733	5741	5749	5757	5765	5774	5782	5790
7	5798	5806	5814	5823	5831	5839	5847	5855	5863	5871
8	5879	5887	5895	5903	5911	5919	5927	5935	5943	5951
9	5959	5967	5975	5983	5991	5999	6007	6015	6023	6031
60	6039	6046	6054	6062	6070	6078	6086	6093	6101	6109
1	6117	6125	6132	6140	6148	6156	6163	6171	6179	6186
2	6194	6202	6209	6217	6225	6232	6240	6248	6255	6263
3	6270	6278	6286	6293	6301	6308	6316	6323	6331	6338
4	6346	6353	6361	6368	6376	6383	6391	6398	6405	6413
5	6420	6428	6435	6442	6450	6457	6464	6472	6479	6486
6	6494	6501	6508	6516	6523	6530	6537	6545	6552	6559
7	6566	6573	6581	6588	6595	6602	6609	6616	6624	6631
8	6638	6645	6652	6659	6666	6673	6680	6687	6694	6701
9	6708	6715	6722	6729	6737	6743	6750	6757	6764	6771
70	6778	6785	6792	6799	6806	6812	6819	6826	6833	6840
1	6847	6853	6860	6867	6874	6881	6887	6894	6901	6908
2	6914	6921	6928	6934	6941	6948	6954	6961	6968	6974
3	6981	6988	6994	7001	7007	7014	7021	7027	7034	7040
4	7047	7053	7060	7066	7073	7079	7086	7092	7099	7105
5	7112	7118	7124	7131	7137	7144	7150	7156	7163	7169
6	7175	7182	7188	7194	7201	7207	7213	7219	7226	7232
7	7238	7244	7251	7257	7263	7269	7275	7282	7288	7294
8	7300	7306	7312	7318	7325	7331	7337	7343	7349	7355
9	7361	7367	7373	7379	7385	7391	7397	7403	7409	7415
80	7421	7427	7433	7439	7445	7451	7457	7462	7468	7474
1	7480	7486	7492	7498	7503	7509	7515	7521	7527	7532
2	7538	7544	7550	7555	7561	7567	7572	7578	7584	7590
3	7595	7601	7607	7612	7618	7623	7629	7635	7640	7646
4	7651	7657	7663	7668	7674	7679	7685	7690	7696	7701
5	7707	7712	7718	7723	7729	7734	7739	7745	7750	7756
6	7761	7766	7772	7777	7782	7788	7793	7798	7804	7809
7	7814	7820	7825	7830	7835	7841	7846	7851	7856	7862
8	7867	7872	7877	7882	7888	7893	7898	7903	7908	7913
9	7918	7924	7929	7934	7939	7944	7949	7954	7959	7964
90	7969	7974	7979	7984	7989	7994	7999	8004	8009	8014
1	8019	8024	8029	8034	8038	8043	8048	8053	8058	8063
2	8068	8073	8077	8082	8087	8092	8097	8101	8106	8111
3	8116	8120	8125	8130	8135	8139	8144	8149	8153	8158
4	8163	8167	8172	8177	8181	8186	8191	8195	8200	8204
5	8209	8213	8218	8223	8227	8232	8236	8241	8245	8250
6	8254	8259	8263	8268	8272	8277	8281	8285	8290	8294
7	8299	8303	8307	8312	8316	8321	8325	8329	8334	8338
8	8342	8347	8351	8355	8360	8364	8368	8372	8377	8381
9	8385	8389	8394	8398	8402	8406	8410	8415	8419	8423
1.00	8427	8431	8435	8439	8444	8448	8452	8456	8460	8464

APPENDIX 64.A
Periodic Table: Materials' Properties Summary

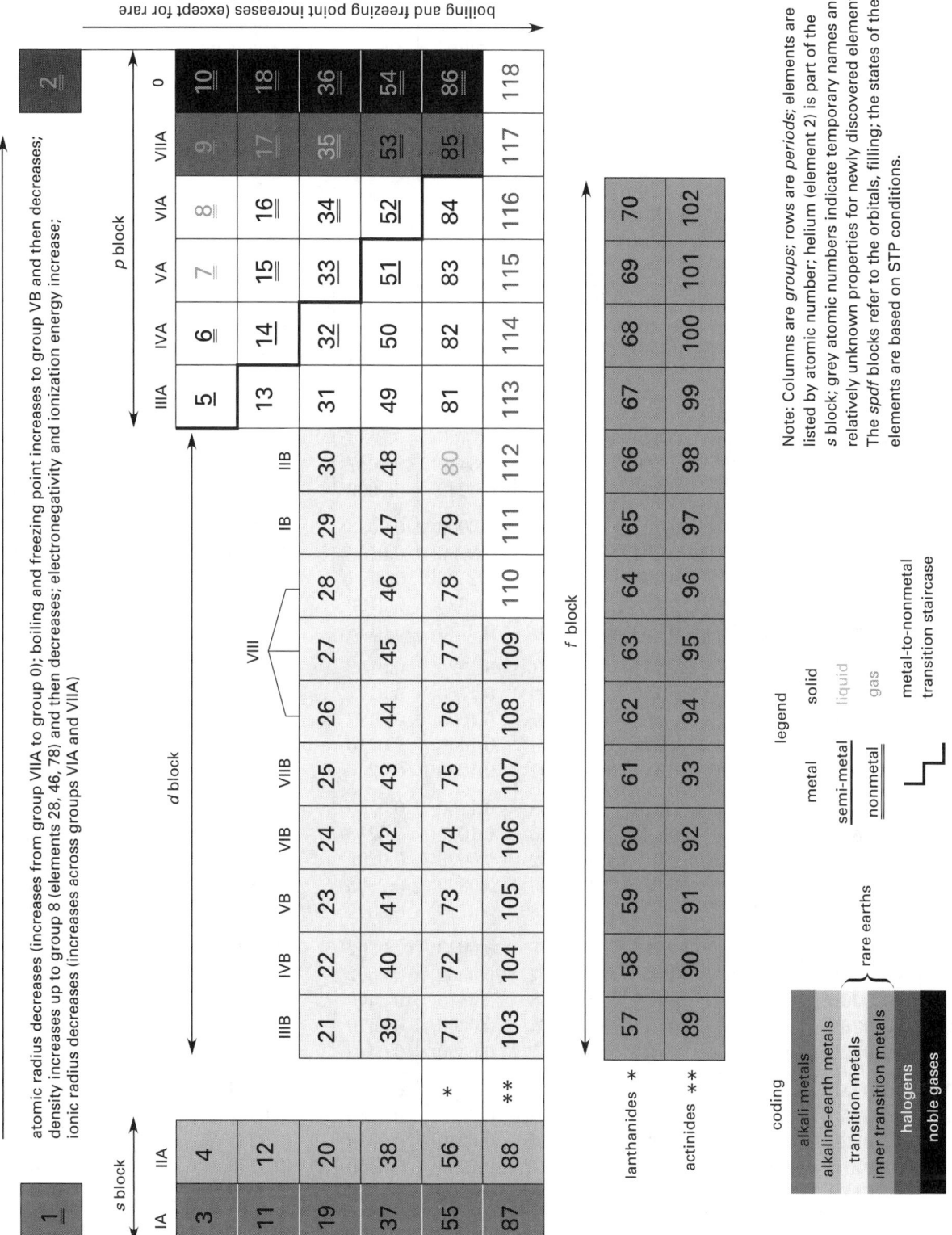

APPENDIX 65.A
Comparison of Wire (and Sheet-Metal Gage) Diameters (Thickness) in Inches

gage no.	(1) American wire gage (AWG)[a]	(2) steel wire gage	(3) Birmingham iron wire (Stubs)	(4) Stubs steel wire gage	(5) British standard wire gage	(6) steel music wire gage	(7) manufacturer's standard gage	(8) American zinc	(9) Birmingham
15/0	–	–	–	–	–	–	–	–	1.0000
14/0	–	–	–	–	–	–	–	–	0.9583
13/0	–	–	–	–	–	–	–	–	0.9167
12/0	–	–	–	–	–	–	–	–	0.8750
11/0	–	–	–	–	–	–	–	–	0.8333
10/0	–	–	–	–	–	–	–	–	0.7917
9/0	–	–	–	–	–	–	–	–	0.7500
8/0	–	–	–	–	–	–	–	–	0.7083
7/0	–	0.4900	–	–	0.5000	–	0.5000	–	0.6666
6/0	–	0.4615	–	–	0.4640	–	0.4687	–	0.6250
5/0	–	0.4305	–	–	0.4320	–	0.4375	–	0.5883
4/0	**0.4600**	0.3938	0.454	–	0.4000	–	0.4062	–	0.5416
3/0	**0.4100**	0.3625	0.425	–	0.3720	–	0.3750	–	0.5000
2/0	**0.3650**	0.3310	0.380	–	0.3480	0.0087	0.3437	–	0.4452
1/0	**0.3250**	0.3065	0.340	–	0.3240	0.0039	0.3125	–	0.3964
1	**0.2890**	0.2830	0.300	0.227	0.3000	0.0098	0.2812	0.002	0.3532
2	**0.2580**	0.2625	0.284	0.219	0.2760	0.0106	0.2656	0.004	0.3147
3	**0.2290**	0.2437	0.259	0.212	0.2520	0.0114	0.2500	0.006	0.2804
4	**0.2040**	0.2253	0.238	0.207	0.2320	0.0122	0.2344	0.008	0.2500
5	**0.1820**	0.2007	0.220	0.204	0.2120	0.0138	0.2187	0.010	0.2225
6	**0.1620**	0.1920	0.203	0.201	0.1920	0.0157	0.2035	0.012	0.1981
7	**0.1440**	0.1770	0.180	0.199	0.1760	0.0177	0.1875	0.014	0.1764
8	**0.1280**	0.1620	0.165	0.197	0.1600	0.0197	0.1719	0.016	0.1570
9	**0.1140**	0.1483	0.148	0.194	0.1440	0.0216	0.1562	0.018	0.1398
10	**0.1020**	0.1350	0.134	0.191	0.1280	0.0236	0.1406	0.020	0.1250
11	**0.0910**	0.1205	0.120	0.188	0.1160	0.0260	0.1250	0.024	0.1113
12	**0.0810**	0.1055	0.109	0.185	0.1040	0.0283	0.1094	0.028	0.0991
13	**0.0720**	0.0915	0.095	0.182	0.0920	0.0305	0.0937	0.032	0.0882
14	**0.0640**	0.0800	0.083	0.180	0.0800	0.0323	0.0821	0.036	0.0785
15	**0.0570**	0.0720	0.072	0.178	0.0720	0.0342	0.0703	0.040	0.0699
16	**0.0510**	0.0625	0.065	0.175	0.0640	0.0362	0.0625	0.045	0.0625
17	**0.0450**	0.0540	0.058	0.172	0.0560	0.0382	0.0562	0.050	0.0556
18	**0.0400**	0.0475	0.049	0.168	0.0480	0.0400	0.0500	0.055	0.0495
19	**0.0360**	0.0410	0.042	0.164	0.0400	0.0420	0.0437	0.060	0.0440
20	**0.0320**	0.0348	0.035	0.161	0.0360	0.0440	0.0375	0.070	0.0392
21	**0.0285**	0.0317	0.032	0.157	0.0320	0.0460	0.0344	0.080	0.0349
22	**0.0253**	0.0286	0.028	0.155	0.0280	0.0480	0.0312	0.090	0.0312
23	**0.0226**	0.0258	0.025	0.153	0.0240	0.0510	0.0281	0.100	0.0278
24	**0.0210**	0.0230	0.022	0.151	0.0220	0.0550	0.0250	0.125	0.0247
25	**0.0179**	0.0204	0.020	0.148	0.0200	0.0590	0.0219	0.250	0.0220

[a]The American wire gage sizes have been rounded off to about the usual limits of commercial accuracy.

(*continued*)

APPENDIX 65.A (*continued*)
Comparison of Wire (and Sheet-Metal Gage) Diameters (Thickness) in Inches

gage no.	(1) American wire gage (AWG)[a]	(2) steel wire gage	(3) Birmingham iron wire (Stubs)	(4) Stubs steel wire gage	(5) British standard wire gage	(6) steel music wire gage	(7) manufacturer's standard gage	(8) American zinc	(9) Birmingham
26	**0.0159**	0.0181	0.018	0.146	0.0180	0.0630	0.0187	0.375	0.0196
27	**0.0142**	0.0173	0.016	0.143	0.0164	0.0670	0.0172	0.500	0.0175
28	**0.0126**	0.0162	0.014	0.139	0.0148	0.0710	0.0156	1.000	0.0156
29	**0.0113**	0.0150	0.013	0.134	0.0136	0.0740	0.0141	–	0.0139
30	**0.0100**	0.0140	0.012	0.127	0.0124	0.0780	0.0125	–	0.0123
31	**0.0089**	0.0132	0.010	0.120	0.0116	0.0820	0.0109	–	0.0110
32	**0.0080**	0.0128	0.009	0.115	0.0108	0.0860	0.0101	–	0.0098
33	**0.0071**	0.0118	0.008	0.112	0.0100	–	0.0094	–	0.0087
34	**0.0063**	0.0104	0.007	0.110	0.0092	–	0.0086	–	0.0077
35	**0.0056**	0.0095	0.005	0.108	0.0084	–	0.0078	–	0.0069
36	**0.0050**	0.0090	0.004	0.106	0.0076	–	0.0070	–	0.0061
37	**0.0045**	0.0085	–	0.103	0.0068	–	0.0066	–	0.0054
38	**0.0040**	0.0080	–	0.101	0.0060	–	0.0062	–	0.0048
39	**0.0035**	0.0075	–	0.099	0.0052	–	–	–	0.0043
40	**0.0031**	0.0070	–	0.097	0.0048	–	–	–	0.0039
41	–	0.0066	–	0.095	0.0044	–	–	–	0.0034
42	–	0.0062	–	0.092	0.0040	–	–	–	0.0031
43	–	0.0060	–	0.088	0.0036	–	–	–	0.0027
44	–	0.0058	–	0.085	0.0032	–	–	–	0.0024
45	–	0.0055	–	0.081	0.0028	–	–	–	0.00215
46	–	0.0052	–	0.079	0.0024	–	–	–	0.0019
47	–	0.0050	–	0.077	0.0020	–	–	–	0.0017
48	–	0.0048	–	0.075	0.0016	–	–	–	0.0015
49	–	0.0046	–	0.072	0.0012	–	–	–	0.00135
50	–	0.0044	–	0.069	0.0010	–	–	–	0.0012
51	–	–	–	–	–	–	–	–	0.0011
52	–	–	–	–	–	–	–	–	0.00095

[a] The American wire gage sizes have been rounded off to about the usual limits of commercial accuracy.

APPENDIX 65.B
Conductor Ampacities
(NEC® Table 310-16)

Allowable ampacities of insulated conductors rated 0 through 2000 V, 60 through 90°C (140 through 194°F), not more than three current-carrying conductors in raceway, cable, or earth (directly buried), based on ambient temperature of 30°C (86°F).

size	temperature rating of conductor (see NEC Table 310-13)						size
	60°C (140°F)	75°C (167°F)	90°C (194°F)	60°C (140°F)	75°C (167°F)	90°C (194°F)	
AWG or kcmil	types TW, UF	types FEPW, RH, RHW, THHW, THW, THWN, XHHW, USE, ZW	types TBS, SA, SIS, FEP, FEPB, MI, RHH, RHW-2, THHN, THHW, THW-2, THWN-2 USE-2, XHH, XHHW, XHHW-2, ZW-2	types TW, UF	types RH, RHW, THHW, THW, THWN, XHHW, USE	types TBS, SA SIS, THHN, THHW, THW-2 THWN-2, RHH, RHW-2, USE-2, XHH, XHHW, XHHW-2, ZW-2	AWG or kcmil
	copper			aluminum or copper-clad aluminum			
18	–	–	14	–	–	–	–
16	–	–	18	–	–	–	–
14[a]	20	20	25	–	–	–	–
12[a]	25	25	30	20	20	25	12
10[a]	30	35	40	25	30	35	10
8	40	50	55	30	40	45	8
6	55	65	75	40	50	60	6
4	70	85	95	55	65	75	4
3	85	100	110	65	75	85	3
2	95	115	130	75	90	100	2
1	110	130	150	85	100	115	1
1/0	125	150	170	100	120	135	1/0
2/0	145	175	195	115	135	150	2/0
3/0	165	200	225	130	155	175	3/0
4/0	195	230	260	150	180	205	4/0
250	215	255	290	170	205	230	250
300	240	285	320	190	230	255	300
350	260	310	350	210	250	280	350
400	280	335	380	225	270	305	400
500	320	380	430	260	310	350	500
600	355	420	475	285	340	385	600
700	385	460	520	310	375	420	700
750	400	475	535	320	385	435	750
800	410	490	555	330	395	450	800
900	435	520	585	355	425	480	900
1000	455	545	615	375	445	500	1000
1250	495	590	665	405	485	545	1250
1500	520	625	705	435	520	585	1500
1750	545	650	735	455	545	615	1750
2000	560	665	750	470	560	630	2000

[a]See Section 240-3 of the National Electrical Code® (NEC).

(continued)

APPENDIX 65.B (*continued*)
Conductor Ampacities
(NEC Table 310-16)

ambient temp. (°C)	correction factors						ambient temp. (°F)
	For ambient temperatures other than 30°C (86°F), multiply the allowable ampacities shown by the appropriate factor shown below.						
21–25	1.08	1.05	1.04	1.08	1.05	1.04	70–77
26–30	1.00	1.00	1.00	1.00	1.00	1.00	78–86
31–35	0.91	0.94	0.96	0.91	0.94	0.96	87–95
36–40	0.82	0.88	0.91	0.82	0.88	0.91	96–104
41–45	0.71	0.82	0.87	0.71	0.82	0.87	105–113
46–50	0.58	0.75	0.82	0.58	0.75	0.82	114–122
51–55	0.41	0.67	0.76	0.41	0.67	0.76	123–131
56–60	–	0.58	0.71	–	0.58	0.71	132–140
61–70	–	0.33	0.58	–	0.33	0.58	141–158
71–80	–	–	0.41	–	–	0.41	159–176

APPENDIX 65.C
Conductor Properties
(NEC® Chapter 9, Table 8)

| size (AWG or kcmil) | area (circular mils) | conductors | | | | direct-current resistance at 75°C (167°F) | | |
| | | stranding | | overall | | copper | | aluminum |
		quantity	diameter (in)	diameter (in)	area (in²)	uncoated (Ω/1000 ft)	coated (Ω/1000 ft)	(Ω/1000 ft)
18	1620	1	–	0.040	0.001	7.77	8.08	12.8
18	1620	7	0.015	0.046	0.002	7.95	8.45	13.1
16	2580	1	–	0.051	0.002	4.89	5.08	8.05
16	2580	7	0.019	0.058	0.003	4.99	5.29	8.21
14	4110	1	–	0.064	0.003	3.07	3.19	5.06
14	4110	7	0.024	0.073	0.004	3.14	3.26	5.17
12	6530	1	–	0.081	0.005	1.93	2.01	3.18
12	6530	7	0.030	0.092	0.006	1.98	2.05	3.25
10	10380	1	–	0.102	0.008	1.21	1.26	2.00
10	10380	7	0.038	0.116	0.011	1.24	1.29	2.04
8	16510	1	–	0.128	0.013	0.764	0.786	1.26
8	16510	7	0.049	0.146	0.017	0.778	0.809	1.28
6	26240	7	0.061	0.184	0.027	0.491	0.510	0.808
4	41740	7	0.077	0.232	0.042	0.308	0.321	0.508
3	52620	7	0.087	0.260	0.053	0.245	0.254	0.403
2	66360	7	0.097	0.292	0.067	0.194	0.201	0.319
1	83690	19	0.066	0.332	0.087	0.154	0.160	0.253
1/0	105600	19	0.074	0.372	0.109	0.122	0.127	0.201
2/0	133100	19	0.084	0.418	0.137	0.0967	0.101	0.159
3/0	167800	19	0.094	0.470	0.173	0.0766	0.0797	0.126
4/0	211600	19	0.106	0.528	0.219	0.0608	0.0626	0.100
250	–	37	0.099	0.575	0.260	0.0515	0.0535	0.0847
300	–	37	0.090	0.630	0.312	0.0429	0.0446	0.0707
350	–	37	0.097	0.681	0.364	0.0367	0.0382	0.0605
400	–	37	0.104	0.728	0.416	0.0321	0.0331	0.0529
500	–	37	0.116	0.813	0.519	0.0258	0.0265	0.0424
600	–	61	0.099	0.893	0.626	0.0214	0.0223	0.0353
700	–	61	0.107	0.964	0.730	0.0184	0.0189	0.0303
750	–	61	0.111	0.998	0.782	0.0171	0.0176	0.0282
800	–	61	0.114	1.030	0.834	0.0161	0.0166	0.0265
900	–	61	0.122	1.094	0.940	0.0143	0.0147	0.0235
1000	–	61	0.128	1.152	1.042	0.0129	0.0132	0.0212
1250	–	91	0.117	1.289	1.305	0.0103	0.0106	0.0169
1500	–	91	0.128	1.412	1.566	0.00858	0.00883	0.0141
1750	–	127	0.117	1.526	1.829	0.00735	0.00756	0.0121
2000	–	127	0.126	1.632	2.092	0.00643	0.00662	0.0106

Notes:
1. These resistance values are valid *only* for the parameters as given. Using conductors having coated strands, different stranding type, and especially other temperatures changes the resistance.
2. Formula for temperature change: $R_2 = R_1(1 + \alpha(T_2 - 75))$ where $\alpha_{Cu} = 0.00323$, $\alpha_{Al} = 0.00330$
3. Conductors with compact and compressed stranding have about 9% and 3%, respectively, smaller bare conductor diameters than those shown. See Table 5A for actual compact cable dimensions.
4. The JACS conductivities used: bare copper = 100%, aluminum = 61.
5. Class B stranding is listed as well as solid for some sizes. Its overall diameter and area is that of its circumscribing circle.

APPENDIX 65.D
Conductor AC Properties
(NEC Chapter 9, Table 9)

Alternating-current resistance and reactance for 600 V cables, three-phase, 60 Hz, 75°C (167°F)—three single conductors in conduit.

| | ohms-to-neutral per 1000 ft | | | | | | | | | | | | | | |
| size (AWG or kcmil) | X_L (reactance) for all wires | | alternating-current resistance for uncoated copper wires | | | alternating-current resistance for aluminum wires | | | effective Z at 0.85 pf for uncoated copper wires | | | effective Z at 0.85 pf for aluminum wires | | | size (AWG or kcmil) |
	PVC aluminum conduit	steel conduit	PVC conduit	aluminum conduit	steel conduit	PVC conduit	aluminum conduit	steel conduit	PVC conduit	aluminum conduit	steel conduit	PVC conduit	aluminum conduit	steel conduit	
14	0.058	0.073	3.1	3.1	3.1	–	–	–	2.7	2.7	2.7	–	–	–	14
12	0.054	0.068	2.0	2.0	2.0	3.2	3.2	3.2	1.7	1.7	1.7	2.8	2.8	2.8	12
10	0.050	0.063	1.2	1.2	1.2	2.0	2.0	2.0	1.1	1.1	1.1	1.8	1.8	1.8	10
8	0.052	0.065	0.78	0.78	0.78	1.3	1.3	1.3	0.69	0.69	0.70	1.1	1.1	1.1	8
6	0.051	0.064	0.49	0.49	0.49	0.81	0.81	0.81	0.44	0.45	0.45	0.71	0.72	0.72	6
4	0.048	0.060	0.31	0.31	0.31	0.51	0.51	0.51	0.29	0.29	0.30	0.46	0.46	0.46	4
3	0.047	0.059	0.25	0.25	0.25	0.40	0.41	0.40	0.23	0.24	0.24	0.37	0.37	0.37	3
2	0.045	0.057	0.19	0.20	0.20	0.32	0.32	0.32	0.19	0.19	0.20	0.30	0.30	0.30	2
1	0.046	0.057	0.15	0.16	0.16	0.25	0.26	0.25	0.16	0.16	0.16	0.24	0.24	0.25	1
1/0	0.044	0.055	0.12	0.13	0.12	0.20	0.21	0.20	0.13	0.13	0.13	0.19	0.20	0.20	1/0
2/0	0.043	0.054	0.10	0.10	0.10	0.16	0.16	0.16	0.11	0.11	0.11	0.16	0.16	0.16	2/0
3/0	0.042	0.052	0.077	0.082	0.079	0.13	0.13	0.13	0.088	0.092	0.094	0.13	0.13	0.14	3/0
4/0	0.041	0.051	0.062	0.067	0.063	0.10	0.11	0.10	0.074	0.078	0.080	0.11	0.11	0.11	4/0
250	0.041	0.052	0.052	0.057	0.054	0.085	0.090	0.086	0.066	0.070	0.073	0.094	0.098	0.10	250
300	0.041	0.051	0.044	0.049	0.045	0.071	0.076	0.072	0.059	0.063	0.065	0.082	0.086	0.088	300
350	0.040	0.050	0.038	0.043	0.039	0.061	0.066	0.063	0.053	0.058	0.060	0.073	0.077	0.080	350
400	0.040	0.049	0.033	0.038	0.035	0.054	0.059	0.055	0.049	0.053	0.056	0.066	0.071	0.073	400
500	0.039	0.048	0.027	0.032	0.029	0.043	0.048	0.045	0.043	0.048	0.050	0.057	0.061	0.064	500
600	0.039	0.048	0.023	0.028	0.025	0.036	0.041	0.038	0.040	0.044	0.047	0.051	0.055	0.058	600
750	0.038	0.048	0.019	0.024	0.021	0.029	0.034	0.031	0.036	0.040	0.043	0.045	0.049	0.052	750
1000	0.037	0.046	0.015	0.019	0.018	0.023	0.027	0.025	0.032	0.036	0.040	0.039	0.042	0.046	1000

Notes:
1. These values are based on the following constants: UL-type RHH wires with Class B stranding, in cradled configuration. Wire conductivities are 100% IACS copper and 61% IACS aluminum, and aluminum conduit is 45% IACS. Capacitive reactance is ignored, since it is negligible at these voltages. These resistance values are valid only at 75°C (167°F) and for the parameters as given, but are representative for 600 V wire types operating at 60 Hz.
2. *Effective Z* is defined as $R\cos\theta + X\sin(\theta)$, where θ is the power factor angle of the circuit. Multiplying current by effective impedance gives a good approximation for line-to-neutral voltage drop. Effective impedance values shown in this table are valid only at 0.85 power factor. For another circuit power factor (pf), effective impedance (Ze) can be calculated from R and X_L values given in this table as follows:

$$Ze = R \times \mathrm{pf} + X_L \sin\left(\arccos\left(\mathrm{pf}\right)\right)$$

APPENDIX 66.A
Standard Cash Flow Factors

multiply	by	to obtain

 F $(P/F, i\%, n)$ P 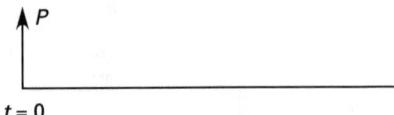

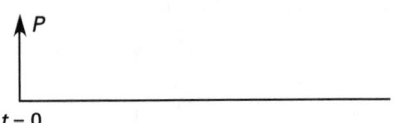 P $(F/P, i\%, n)$ F 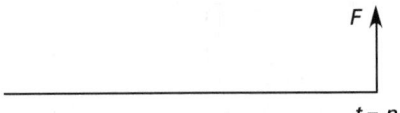

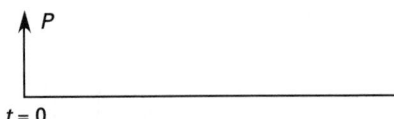 A $(P/A, i\%, n)$ P 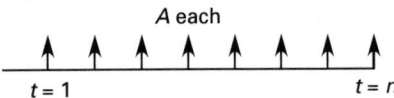

   P $(A/P, i\%, n)$ A

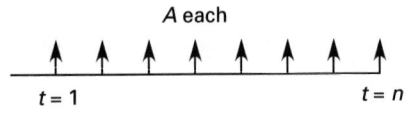 A $(F/A, i\%, n)$ F 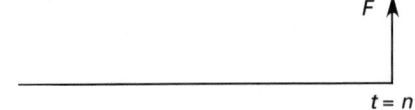

   F $(A/F, i\%, n)$ A

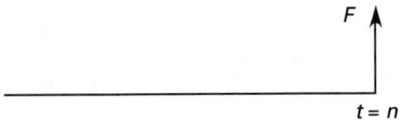 G $(P/G, i\%, n)$ P 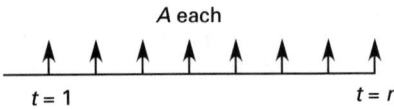

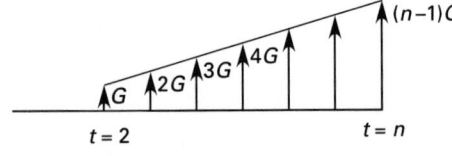 G $(A/G, i\%, n)$ A 

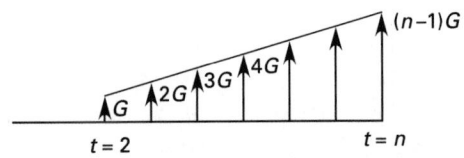 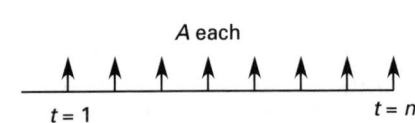

APPENDIX 66.B
Factor Tables

$I = 0.50\%$

n	P/F	P/A	P/G	F/P	F/A	A/P	A/F	A/G	n
1	0.9950	0.9950	0.0000	1.0050	1.0000	1.0050	1.0000	0.0000	1
2	0.9901	0.9851	0.9901	1.0100	2.0050	0.5038	0.4988	0.4988	2
3	0.9851	2.9702	2.9604	1.0151	3.0150	0.3367	0.3317	0.9967	3
4	0.9802	3.9505	5.9011	1.0202	4.0301	0.2531	0.2481	1.4938	4
5	0.9754	4.9259	9.8026	1.0253	5.0503	0.2030	0.1980	1.9900	5
6	0.9705	5.8964	14.6552	1.0304	6.0755	0.1696	0.1646	2.4855	6
7	0.9657	6.8621	20.4493	1.0355	7.1059	0.1457	0.1407	2.9801	7
8	0.9609	7.8230	27.1755	1.0407	8.1414	0.1278	0.1228	3.4738	8
9	0.9561	8.7791	34.8244	1.0459	9.1821	0.1139	0.1089	3.9668	9
10	0.9513	9.7304	43.3865	1.0511	10.2280	0.1028	0.0978	4.4589	10
11	0.9466	10.6770	52.8526	1.0564	11.2792	0.0937	0.0887	4.9501	11
12	0.9419	11.6189	63.2136	1.0617	12.3356	0.0861	0.0811	5.4406	12
13	0.9372	12.5562	74.4602	1.0670	13.3972	0.0796	0.0746	5.9302	13
14	0.9326	13.4887	86.5835	1.0723	14.4642	0.0741	0.0691	6.4190	14
15	0.9279	14.4166	99.5743	1.0777	15.5365	0.0694	0.0644	6.9069	15
16	0.9233	15.3399	113.4238	1.0831	16.6142	0.0652	0.0602	7.3940	16
17	0.9187	16.2586	128.1231	1.0885	17.6973	0.0615	0.0565	7.8803	17
18	0.9141	17.1728	143.6634	1.0939	18.7858	0.0582	0.0532	8.3658	18
19	0.9096	18.0824	160.0360	1.0994	19.8797	0.0553	0.0503	8.8504	19
20	0.9051	18.9874	177.2322	1.1049	20.9791	0.0527	0.0477	9.3342	20
21	0.9006	19.8880	195.2434	1.1104	22.0840	0.0503	0.0453	9.8172	21
22	0.8961	20.7841	214.0611	1.1160	23.1944	0.0481	0.0431	10.2993	22
23	0.8916	21.6757	233.6768	1.1216	24.3104	0.0461	0.0411	10.7806	23
24	0.8872	22.5629	254.0820	1.1272	25.4320	0.0443	0.0393	11.2611	24
25	0.8828	23.4456	275.2686	1.1328	26.5591	0.0427	0.0377	11.7407	25
26	0.8784	24.3240	297.2281	1.1385	27.6919	0.0411	0.0361	12.2195	26
27	0.8740	25.1980	319.9523	1.1442	28.8304	0.0397	0.0347	12.6975	27
28	0.8697	26.0677	343.4332	1.1499	29.9745	0.0384	0.0334	13.1747	28
29	0.8653	26.9330	367.6625	1.1556	31.1244	0.0371	0.0321	13.6510	29
30	0.8610	27.7941	392.6324	1.1614	32.2800	0.0360	0.0310	14.1265	30
31	0.8567	28.6508	418.3348	1.1672	33.4414	0.0349	0.0299	14.6012	31
32	0.8525	29.5033	444.7618	1.1730	34.6086	0.0339	0.0289	15.0750	32
33	0.8482	30.3515	471.9055	1.1789	35.7817	0.0329	0.0279	15.5480	33
34	0.8440	31.1955	499.7583	1.1848	36.9606	0.0321	0.0271	16.0202	34
35	0.8398	32.0354	528.3123	1.1907	38.1454	0.0312	0.0262	16.4915	35
36	0.8356	32.8710	557.5598	1.1967	39.3361	0.0304	0.0254	16.9621	36
37	0.8315	33.7025	587.4934	1.2027	40.5328	0.0297	0.0247	17.4317	37
38	0.8274	34.5299	618.1054	1.2087	41.7354	0.0290	0.0240	17.9006	38
39	0.8232	35.3531	649.3883	1.2147	42.9441	0.0283	0.0233	18.3686	39
40	0.8191	36.1722	681.3347	1.2208	44.1588	0.0276	0.0226	18.8359	40
41	0.8151	36.9873	713.9372	1.2269	45.3796	0.0270	0.0220	19.3022	41
42	0.8110	37.7983	747.1886	1.2330	46.6065	0.0265	0.0215	19.7678	42
43	0.8070	38.6053	781.0815	1.2392	47.8396	0.0259	0.0209	20.2325	43
44	0.8030	39.4082	815.6087	1.2454	49.0788	0.0254	0.0204	20.6964	44
45	0.7990	40.2072	850.7631	1.2516	50.3242	0.0249	0.0199	21.1595	45
46	0.7950	41.0022	886.5376	1.2579	51.5758	0.0244	0.0194	21.6217	46
47	0.7910	41.7932	922.9252	1.2642	52.8337	0.0239	0.0189	22.0831	47
48	0.7871	42.5803	959.9188	1.2705	54.0978	0.0235	0.0185	22.5437	48
49	0.7832	43.3635	997.5116	1.2768	55.3683	0.0231	0.0181	23.0035	49
50	0.7793	44.1428	1035.6966	1.2832	56.6452	0.0227	0.0177	23.4624	50
51	0.7754	44.9182	1074.4670	1.2896	57.9284	0.0223	0.0173	23.9205	51
52	0.7716	45.6897	1113.8162	1.2961	59.2180	0.0219	0.0169	24.3778	52
53	0.7677	46.4575	1153.7372	1.3026	60.5141	0.0215	0.0165	24.8343	53
54	0.7639	47.2214	1194.2236	1.3091	61.8167	0.0212	0.0162	25.2899	54
55	0.7601	47.9814	1235.2686	1.3156	63.1258	0.0208	0.0158	25.7447	55
60	0.7414	51.7256	1448.6458	1.3489	69.7700	0.0193	0.0143	28.0064	60
65	0.7231	55.3775	1675.0272	1.3829	76.5821	0.0181	0.0131	30.2475	65
70	0.7053	58.9394	1913.6427	1.4178	83.5661	0.0170	0.0120	32.4680	70
75	0.6879	62.4136	2163.7525	1.4536	90.7265	0.0160	0.0110	34.6679	75
80	0.6710	65.8023	2424.6455	1.4903	98.0677	0.0152	0.0102	36.8474	80
85	0.6545	69.1075	2695.6389	1.5280	105.5943	0.0145	0.0095	39.0065	85
90	0.6383	72.3313	2976.0769	1.5666	113.3109	0.0138	0.0088	41.1451	90
95	0.6226	75.4757	3265.3298	1.6061	121.2224	0.0132	0.0082	43.2633	95
100	0.6073	78.5426	3562.7934	1.6467	129.3337	0.0127	0.0077	45.3613	100

(continued)

APPENDIX 66.B *(continued)*
Factor Tables

$$I = 0.75\%$$

n	P/F	P/A	P/G	F/P	F/A	A/P	A/F	A/G	n
1	0.9926	0.9926	0.0000	1.0075	1.0000	1.0075	1.0000	0.0000	1
2	0.9852	1.9777	0.9852	1.0151	2.0075	0.5056	0.4981	0.4981	2
3	0.9778	2.9556	2.9408	1.0227	3.0226	0.3383	0.3308	0.9950	3
4	0.9706	3.9261	5.8525	1.0303	4.0452	0.2547	0.2472	1.4907	4
5	0.9633	4.8894	9.7058	1.0381	5.0756	0.2045	0.1970	1.9851	5
6	0.9562	5.8456	14.4866	1.0459	6.1136	0.1711	0.1636	2.4782	6
7	0.9490	6.7946	20.1808	1.0537	7.1595	0.1472	0.1397	2.9701	7
8	0.9420	7.7366	26.7747	1.0616	8.2132	0.1293	0.1218	3.4608	8
9	0.9350	8.6716	34.2544	1.0696	9.2748	0.1153	0.1078	3.9502	9
10	0.9280	9.5996	42.6064	1.0776	10.3443	0.1042	0.0967	4.4384	10
11	0.9211	10.5207	51.8174	1.0857	11.4219	0.0951	0.0876	4.9253	11
12	0.9142	11.4349	61.8740	1.0938	12.5076	0.0875	0.0800	5.4110	12
13	0.9074	12.3423	72.7632	1.1020	13.6014	0.0810	0.0735	5.8954	13
14	0.9007	13.2430	84.4720	1.1103	14.7034	0.0755	0.0680	6.3786	14
15	0.8940	14.1370	96.9876	1.1186	15.8137	0.0707	0.0632	6.8606	15
16	0.8873	15.0243	110.2973	1.1270	16.9323	0.0666	0.0591	7.3413	16
17	0.8807	15.9050	124.3887	1.1354	18.0593	0.0629	0.0554	7.8207	17
18	0.8742	16.7792	139.2494	1.1440	19.1947	0.0596	0.0521	8.2989	18
19	0.8676	17.6468	154.8671	1.1525	20.3387	0.0567	0.0492	8.7759	19
20	0.8612	18.5080	171.2297	1.1612	21.4912	0.0540	0.0465	9.2516	20
21	0.8548	19.3628	188.3253	1.1699	22.6524	0.0516	0.0441	9.7261	21
22	0.8484	20.2112	206.1420	1.1787	23.8223	0.0495	0.0420	10.1994	22
23	0.8421	21.0533	224.6682	1.1875	25.0010	0.0475	0.0400	10.6714	23
24	0.8358	21.8891	243.8923	1.1964	26.1885	0.0457	0.0382	11.1422	24
25	0.8296	22.7188	263.8029	1.2054	27.3849	0.0440	0.0365	11.6117	25
26	0.8234	23.5422	284.3888	1.2144	28.5903	0.0425	0.0350	12.0800	26
27	0.8173	24.3595	305.6387	1.2235	29.8047	0.0411	0.0336	12.5470	27
28	0.8112	25.1707	327.5416	1.2327	31.0282	0.0397	0.0322	13.0128	28
29	0.8052	25.9759	350.0867	1.2420	32.2609	0.0385	0.0310	13.4774	29
30	0.7992	26.7751	373.2631	1.2513	33.5029	0.0373	0.0298	13.9407	30
31	0.7932	27.5683	397.0602	1.2607	34.7542	0.0363	0.0288	14.4028	31
32	0.7873	28.3557	421.4675	1.2701	36.0148	0.0353	0.0278	14.8636	32
33	0.7815	29.1371	446.4746	1.2796	37.2849	0.0343	0.0268	15.3232	33
34	0.7757	29.9128	472.0712	1.2892	38.5646	0.0334	0.0259	15.7816	34
35	0.7699	30.6827	498.2471	1.2989	39.8538	0.0326	0.0251	16.2387	35
36	0.7641	31.4468	524.9924	1.3086	41.1527	0.0318	0.0243	16.6946	36
37	0.7585	32.2053	552.2969	1.3185	42.4614	0.0311	0.0236	17.1493	37
38	0.7528	32.9581	580.1511	1.3283	43.7798	0.0303	0.0228	17.6027	38
39	0.7472	33.7053	608.5451	1.3383	45.1082	0.0297	0.0222	18.0549	39
40	0.7416	34.4469	637.4693	1.3483	46.4465	0.0290	0.0215	18.5058	40
41	0.7361	35.1831	666.9144	1.3585	47.7948	0.0284	0.0209	18.9556	41
42	0.7306	35.9137	696.8709	1.3686	49.1533	0.0278	0.0203	19.4040	42
43	0.7252	36.6389	727.3297	1.3789	50.5219	0.0273	0.0198	19.8513	43
44	0.7198	37.3587	758.2815	1.3893	51.9009	0.0268	0.0193	20.2973	44
45	0.7145	38.0732	789.7173	1.3997	53.2901	0.0263	0.0188	20.7421	45
46	0.7091	38.7823	821.6283	1.4102	54.6898	0.0258	0.0183	21.1856	46
47	0.7039	39.4862	854.0056	1.4207	56.1000	0.0253	0.0178	21.6280	47
48	0.6986	40.1848	886.8404	1.4314	57.5207	0.0249	0.0174	22.0691	48
49	0.6934	40.8782	920.1243	1.4421	58.9521	0.0245	0.0170	22.5089	49
50	0.6883	41.5664	953.8486	1.4530	60.3943	0.0241	0.0166	22.9476	50
51	0.6831	42.2496	988.0050	1.4639	61.8472	0.0237	0.0162	23.3850	51
52	0.6780	42.9276	1022.5852	1.4748	63.3111	0.0233	0.0158	23.8211	52
53	0.6730	43.6006	1057.5810	1.4859	64.7859	0.0229	0.0154	24.2561	53
54	0.6680	44.2686	1092.9842	1.4970	66.2718	0.0226	0.0151	24.6898	54
55	0.6630	44.9316	1128.7869	1.5083	67.7688	0.0223	0.0148	25.1223	55
60	0.6387	48.1734	1313.5189	1.5657	75.4241	0.0208	0.0133	27.2665	60
65	0.6153	51.2963	1507.0910	1.6253	83.3709	0.0195	0.0120	29.3801	65
70	0.5927	54.3046	1708.6065	1.6872	91.6201	0.0184	0.0109	31.4634	70
75	0.5710	57.2027	1917.2225	1.7514	100.1833	0.0175	0.0100	33.5163	75
80	0.5500	59.9944	2132.1472	1.8180	109.0725	0.0167	0.0092	35.5391	80
85	0.5299	62.6838	2352.6375	1.8873	118.3001	0.0160	0.0085	37.5318	85
90	0.5104	65.2746	2577.9961	1.9591	127.8790	0.0153	0.0078	39.4946	90
95	0.4917	67.7704	2807.5694	2.0337	137.8225	0.0148	0.0073	41.4277	95
100	0.4737	70.1746	3040.7453	2.1111	148.1445	0.0143	0.0068	43.3311	100

(continued)

APPENDIX 66.B (*continued*)
Factor Tables

$I = 1.00\%$

n	P/F	P/A	P/G	F/P	F/A	A/P	A/F	A/G	n
1	0.9901	0.9901	0.0000	1.0100	1.0000	1.0100	1.0000	0.0000	1
2	0.9803	1.9704	0.9803	1.0201	2.0100	0.5075	0.4975	0.4975	2
3	0.9706	2.9410	2.9215	1.0303	3.0301	0.3400	0.3300	0.9934	3
4	0.9610	3.9020	5.8044	1.0406	4.0604	0.2563	0.2463	1.4876	4
5	0.9515	4.8534	9.6103	1.0510	5.1010	0.2060	0.1960	1.9801	5
6	0.9420	5.7955	14.3205	1.0615	6.1520	0.1725	0.1625	2.4710	6
7	0.9327	6.7282	19.9168	1.0721	7.2135	0.1486	0.1386	2.9602	7
8	0.9235	7.6517	26.3812	1.0829	8.2857	0.1307	0.1207	3.4478	8
9	0.9143	8.5660	33.6959	1.0937	9.3685	0.1167	0.1067	3.9337	9
10	0.9053	9.4713	41.8435	1.1046	10.4622	0.1056	0.0956	4.4179	10
11	0.8963	10.3676	50.8067	1.1157	11.5668	0.0965	0.0865	4.9005	11
12	0.8874	11.2551	60.5687	1.1268	12.6825	0.0888	0.0788	5.3815	12
13	0.8787	12.1337	71.1126	1.1381	13.8093	0.0824	0.0724	5.8607	13
14	0.8700	13.0037	82.4221	1.1495	14.9474	0.0769	0.0669	6.3384	14
15	0.8613	13.8651	94.4810	1.1610	16.0969	0.0721	0.0621	6.8143	15
16	0.8528	14.7179	107.2734	1.1726	17.2579	0.0679	0.0579	7.2886	16
17	0.8444	15.5623	120.7834	1.1843	18.4304	0.0643	0.0543	7.7613	17
18	0.8360	16.3983	134.9957	1.1961	19.6147	0.0610	0.0510	8.2323	18
19	0.8277	17.2260	149.8950	1.2081	20.8109	0.0581	0.0481	8.7017	19
20	0.8195	18.0456	165.4664	1.2202	22.0190	0.0554	0.0454	9.1694	20
21	0.8114	18.8570	181.6950	1.2324	23.2392	0.0530	0.0430	9.6354	21
22	0.8034	19.6604	198.5663	1.2447	24.4716	0.0509	0.0409	10.0998	22
23	0.7954	20.4558	216.0660	1.2572	25.7163	0.0489	0.0389	10.5626	23
24	0.7876	21.2434	234.1800	1.2697	26.9735	0.0471	0.0371	11.0237	24
25	0.7798	22.0232	252.8945	1.2824	28.2432	0.0454	0.0354	11.4831	25
26	0.7720	22.7952	272.1957	1.2953	29.5256	0.0439	0.0339	11.9409	26
27	0.7644	23.5596	292.0702	1.3082	30.8209	0.0424	0.0324	12.3971	27
28	0.7568	24.3164	312.5047	1.3213	32.1291	0.0411	0.0311	12.8516	28
29	0.7493	25.0658	333.4863	1.3345	33.4504	0.0399	0.0299	13.3044	29
30	0.7419	25.8077	355.0021	1.3478	34.7849	0.0387	0.0287	13.7557	30
31	0.7346	26.5423	377.0394	1.3613	36.1327	0.0377	0.0277	14.2052	31
32	0.7273	27.2696	399.5858	1.3749	37.4941	0.0367	0.0267	14.6532	32
33	0.7201	27.9897	422.6291	1.3887	38.8690	0.0357	0.0257	15.0995	33
34	0.7130	28.7027	446.1572	1.4026	40.2577	0.0348	0.0248	15.5441	34
35	0.7059	29.4086	470.1583	1.4166	41.6603	0.0340	0.0240	15.9871	35
36	0.6989	30.1075	494.6207	1.4308	43.0769	0.0332	0.0232	16.4285	36
37	0.6920	30.7995	519.5329	1.4451	44.5076	0.0325	0.0225	16.8682	37
38	0.6852	31.4847	544.8835	1.4595	45.9527	0.0318	0.0218	17.3063	38
39	0.6784	32.1630	570.6616	1.4741	47.4123	0.0311	0.0211	17.7428	39
40	0.6717	32.8347	596.8561	1.4889	48.8864	0.0305	0.0205	18.1776	40
41	0.6650	33.4997	623.4562	1.5038	50.3752	0.0299	0.0199	18.6108	41
42	0.6584	34.1581	650.4514	1.5188	51.8790	0.0293	0.0193	19.0424	42
43	0.6519	34.8100	677.8312	1.5340	53.3978	0.0287	0.0187	19.4723	43
44	0.6454	35.4555	705.5853	1.5493	54.9318	0.0282	0.0182	19.9006	44
45	0.6391	36.0945	733.7037	1.5648	56.4811	0.0277	0.0177	20.3273	45
46	0.6327	36.7272	762.1765	1.5805	58.0459	0.0272	0.0172	20.7524	46
47	0.6265	37.3537	790.9938	1.5963	59.6263	0.0268	0.0168	21.1758	47
48	0.6203	37.9740	820.1460	1.6122	61.2226	0.0263	0.0163	21.5976	48
49	0.6141	38.5881	849.6237	1.6283	62.8348	0.0259	0.0159	22.0178	49
50	0.6080	39.1961	879.4176	1.6446	64.4632	0.0255	0.0155	22.4363	50
51	0.6020	39.7981	909.5186	1.6611	66.1078	0.0251	0.0151	22.8533	51
52	0.5961	40.3942	939.9175	1.6777	67.7689	0.0248	0.0148	23.2686	52
53	0.5902	40.9844	970.6057	1.6945	69.4466	0.0244	0.0144	23.6823	53
54	0.5843	41.5687	1001.5743	1.7114	71.1410	0.0241	0.0141	24.0945	54
55	0.5785	42.1472	1032.8148	1.7285	72.8525	0.0237	0.0137	24.5049	55
60	0.5504	44.9550	1192.8061	1.8167	81.6697	0.0222	0.0122	26.5333	60
65	0.5237	47.6266	1358.3903	1.9094	90.9366	0.0210	0.0110	28.5217	65
70	0.4983	50.1685	1528.6474	2.0068	100.6763	0.0199	0.0099	30.4703	70
75	0.4741	52.5871	1702.7340	2.1091	110.9128	0.0190	0.0090	32.3793	75
80	0.4511	54.8882	1879.8771	2.2167	121.6715	0.0182	0.0082	34.2492	80
85	0.4292	57.0777	2059.3701	2.3298	132.9790	0.0175	0.0075	36.0801	85
90	0.4084	59.1609	2240.5675	2.4486	144.8633	0.0169	0.0069	37.8724	90
95	0.3886	61.1430	2422.8811	2.5735	157.3538	0.0164	0.0064	39.6265	95
100	0.3697	63.0289	2605.7758	2.7048	170.4814	0.0159	0.0059	41.3426	100

(*continued*)

Support Material

APPENDIX 66.B (continued)
Factor Tables

$I = 1.50\%$

n	P/F	P/A	P/G	F/P	F/A	A/P	A/F	A/G	n
1	0.9852	0.9852	0.0000	1.0150	1.0000	1.0150	1.0000	0.0000	1
2	0.9707	1.9559	0.9707	1.0302	2.0150	0.5113	0.4963	0.4963	2
3	0.9563	2.9122	2.8833	1.0457	3.0452	0.3434	0.3284	0.9901	3
4	0.9422	3.8544	5.7098	1.0614	4.0909	0.2594	0.2444	1.4814	4
5	0.9283	4.7826	9.4229	1.0773	5.1523	0.2091	0.1941	1.9702	5
6	0.9145	5.6972	13.9956	1.0934	6.2296	0.1755	0.1605	2.4566	6
7	0.9010	6.5982	19.4018	1.1098	7.3230	0.1516	0.1366	2.9405	7
8	0.8877	7.4859	25.6157	1.1265	8.4328	0.1336	0.1186	3.4219	8
9	0.8746	8.3605	32.6125	1.1434	9.5593	0.1196	0.1046	3.9008	9
10	0.8617	9.2222	40.3675	1.1605	10.7027	0.1084	0.0934	4.3772	10
11	0.8489	10.0711	48.8568	1.1779	11.8633	0.0993	0.0843	4.8512	11
12	0.8364	10.9075	58.0571	1.1956	13.0412	0.0917	0.0767	5.3227	12
13	0.8240	11.7315	67.9454	1.2136	14.2368	0.0852	0.0702	5.7917	13
14	0.8118	12.5434	78.4994	1.2318	15.4504	0.0797	0.0647	6.2582	14
15	0.7999	13.3432	89.6974	1.2502	16.6821	0.0749	0.0599	6.7223	15
16	0.7880	14.1313	101.5178	1.2690	17.9324	0.0708	0.0558	7.1839	16
17	0.7764	14.9076	113.9400	1.2880	19.2014	0.0671	0.0521	7.6431	17
18	0.7649	15.6726	126.9435	1.3073	20.4894	0.0638	0.0488	8.0997	18
19	0.7536	16.4262	140.5084	1.3270	21.7967	0.0609	0.0459	8.5539	19
20	0.7425	17.1686	154.6154	1.3469	23.1237	0.0582	0.0432	9.0057	20
21	0.7315	17.9001	169.2453	1.3671	24.4705	0.0559	0.0409	9.4550	21
22	0.7207	18.6208	184.3798	1.3876	25.8376	0.0537	0.0387	9.9018	22
23	0.7100	19.3309	200.0006	1.4084	27.2251	0.0517	0.0367	10.3462	23
24	0.6995	20.0304	216.0901	1.4295	28.6335	0.0499	0.0349	10.7881	24
25	0.6892	20.7196	232.6310	1.4509	30.0630	0.0483	0.0333	11.2276	25
26	0.6790	21.3986	249.6065	1.4727	31.5140	0.0467	0.0317	11.6646	26
27	0.6690	22.0676	267.0002	1.4948	32.9867	0.0453	0.0303	12.0992	27
28	0.6591	22.7267	284.7958	1.5172	34.4815	0.0440	0.0290	12.5313	28
29	0.6494	23.3761	302.9779	1.5400	35.9987	0.0428	0.0278	12.9610	29
30	0.6398	24.0158	321.5310	1.5631	37.5387	0.0416	0.0266	13.3883	30
31	0.6303	24.6461	340.4402	1.5865	39.1018	0.0406	0.0256	13.8131	31
32	0.6210	25.2671	359.6910	1.6103	40.6883	0.0396	0.0246	14.2355	32
33	0.6118	25.8790	379.2691	1.6345	42.2986	0.0386	0.0236	14.6555	33
34	0.6028	26.4817	399.1607	1.6590	43.9331	0.0378	0.0228	15.0731	34
35	0.5939	27.0756	419.3521	1.6839	45.5921	0.0369	0.0219	15.4882	35
36	0.5851	27.6607	439.8303	1.7091	47.2760	0.0362	0.0212	15.9009	36
37	0.5764	28.2371	460.5822	1.7348	48.9851	0.0354	0.0204	16.3112	37
38	0.5679	28.8051	481.5954	1.7608	50.7199	0.0347	0.0197	16.7191	38
39	0.5595	29.3646	502.8576	1.7872	52.4807	0.0341	0.0191	17.1246	39
40	0.5513	29.9158	524.3568	1.8140	54.2679	0.0334	0.0184	17.5277	40
41	0.5431	30.4590	546.0814	1.8412	56.0819	0.0328	0.0178	17.9284	41
42	0.5351	30.9941	568.0201	1.8688	57.9231	0.0323	0.0173	18.3267	42
43	0.5272	31.5212	590.1617	1.8969	59.7920	0.0317	0.0167	18.7227	43
44	0.5194	32.0406	612.4955	1.9253	61.6889	0.0312	0.0162	19.1162	44
45	0.5117	32.5523	635.0110	1.9542	63.6142	0.0307	0.0157	19.5074	45
46	0.5042	33.0565	657.6979	1.9835	65.5684	0.0303	0.0153	19.8962	46
47	0.4967	33.5532	680.5462	2.0133	67.5519	0.0298	0.0148	20.2826	47
48	0.4894	34.0426	703.5462	2.0435	69.5652	0.0294	0.0144	20.6667	48
49	0.4821	34.5247	726.6884	2.0741	71.6087	0.0290	0.0140	21.0484	49
50	0.4750	34.9997	749.9636	2.1052	73.6828	0.0286	0.0136	21.4277	50
51	0.4680	35.4677	773.3629	2.1368	75.7881	0.0282	0.0132	21.8047	51
52	0.4611	35.9287	796.8774	2.1689	77.9249	0.0278	0.0128	22.1794	52
53	0.4543	36.3830	820.4986	2.2014	80.0938	0.0275	0.0125	22.5517	53
54	0.4475	36.8305	844.2184	2.2344	82.2952	0.0272	0.0122	22.9217	54
55	0.4409	37.2715	868.0285	2.2679	84.5296	0.0268	0.0118	23.2894	55
60	0.4093	39.3803	988.1674	2.4432	96.2147	0.0254	0.0104	25.0930	60
65	0.3799	41.3378	1109.4752	2.6320	108.8028	0.0242	0.0092	26.8393	65
70	0.3527	43.1549	1231.1658	2.8355	122.3638	0.0232	0.0082	28.5290	70
75	0.3274	44.8416	1352.5600	3.0546	136.9728	0.0223	0.0073	30.1631	75
80	0.3039	46.4073	1473.0741	3.2907	152.7109	0.0215	0.0065	31.7423	80
85	0.2821	47.8607	1592.2095	3.5450	169.6652	0.0209	0.0059	33.2676	85
90	0.2619	49.2099	1709.5439	3.8189	187.9299	0.0203	0.0053	34.7399	90
95	0.2431	50.4622	1824.7224	4.1141	207.6061	0.0198	0.0048	36.1602	95
100	0.2256	51.6247	1937.4506	4.4320	228.8030	0.0194	0.0044	37.5295	100

(continued)

APPENDIX 66.B (*continued*)
Factor Tables

$I = 2.00\%$

n	P/F	P/A	P/G	F/P	F/A	A/P	A/F	A/G	n
1	0.9804	0.9804	0.0000	1.0200	1.0000	1.0200	1.0000	0.0000	1
2	0.9612	1.9416	0.9612	1.0404	2.0200	0.5150	0.4950	0.4950	2
3	0.9423	2.8839	2.8458	1.0612	3.0604	0.3468	0.3268	0.9868	3
4	0.9238	3.8077	5.6173	1.0824	4.1216	0.2626	0.2426	1.4752	4
5	0.9057	4.7135	9.2403	1.1041	5.2040	0.2122	0.1922	1.9604	5
6	0.8880	5.6014	13.6801	1.1262	6.3081	0.1785	0.1585	2.4423	6
7	0.8706	6.4720	18.9035	1.1487	7.4343	0.1545	0.1345	2.9208	7
8	0.8535	7.3255	24.8779	1.1717	8.5830	0.1365	0.1165	3.3961	8
9	0.8368	8.1622	31.5720	1.1951	9.7546	0.1225	0.1025	3.8681	9
10	0.8203	8.9826	38.9551	1.2190	10.9497	0.1113	0.0913	4.3367	10
11	0.8043	9.7868	46.9977	1.2434	12.1687	0.1022	0.0822	4.8021	11
12	0.7885	10.5753	55.6712	1.2682	13.4121	0.0946	0.0746	5.2642	12
13	0.7730	11.3484	64.9475	1.2936	14.6803	0.0881	0.0681	5.7231	13
14	0.7579	12.1062	74.7999	1.3195	15.9739	0.0826	0.0626	6.1786	14
15	0.7430	12.8493	85.2021	1.3459	17.2934	0.0778	0.0578	6.6309	15
16	0.7284	13.5777	96.1288	1.3728	18.6393	0.0737	0.0537	7.0799	16
17	0.7142	14.2919	107.5554	1.4002	20.0121	0.0700	0.0500	7.5256	17
18	0.7002	14.9920	119.4581	1.4282	21.4123	0.0667	0.0467	7.9681	18
19	0.6864	15.6785	131.8139	1.4568	22.8406	0.0638	0.0438	8.4073	19
20	0.6730	16.3514	144.6003	1.4859	24.2974	0.0612	0.0412	8.8433	20
21	0.6598	17.0112	157.7959	1.5157	25.7833	0.0588	0.0388	9.2760	21
22	0.6468	17.6580	171.3795	1.5460	27.2990	0.0566	0.0366	9.7055	22
23	0.6342	18.2922	185.3309	1.5769	28.8450	0.0547	0.0347	10.1317	23
24	0.6217	18.9139	199.6305	1.6084	30.4219	0.0529	0.0329	10.5547	24
25	0.6095	19.5235	214.2592	1.6406	32.0303	0.0512	0.0312	10.9745	25
26	0.5976	20.1210	229.1987	1.6734	33.6709	0.0497	0.0297	11.3910	26
27	0.5859	20.7069	244.4311	1.7069	35.3443	0.0483	0.0283	11.8043	27
28	0.5744	21.2813	259.9392	1.7410	37.0512	0.0470	0.0270	12.2145	28
29	0.5631	21.8444	275.7064	1.7758	38.7922	0.0458	0.0258	12.6214	29
30	0.5521	22.3965	291.7164	1.8114	40.5681	0.0446	0.0246	13.0251	30
31	0.5412	22.9377	307.9538	1.8476	42.3794	0.0436	0.0236	13.4257	31
32	0.5306	23.4683	324.4035	1.8845	44.2270	0.0426	0.0226	13.8230	32
33	0.5202	23.9886	341.0508	1.9222	46.1116	0.0417	0.0217	14.2172	33
34	0.5100	24.4986	357.8817	1.9607	48.0338	0.0408	0.0208	14.6083	34
35	0.5000	24.9986	374.8826	1.9999	49.9945	0.0400	0.0200	14.9961	35
36	0.4902	25.4888	392.0405	2.0399	51.9944	0.0392	0.0192	15.3809	36
37	0.4806	25.9695	409.3424	2.0807	54.0343	0.0385	0.0185	15.7625	37
38	0.4712	26.4406	426.7764	2.1223	56.1149	0.0378	0.0178	16.1409	38
39	0.4619	26.9026	444.3304	2.1647	58.2372	0.0372	0.0172	16.5163	39
40	0.4529	27.3555	461.9931	2.2080	60.4020	0.0366	0.0166	16.8885	40
41	0.4440	27.7995	479.7535	2.2522	62.6100	0.0360	0.0160	17.2576	41
42	0.4353	28.2348	497.6010	2.2972	64.8622	0.0354	0.0154	17.6237	42
43	0.4268	28.6616	515.5253	2.3432	67.1595	0.0349	0.0149	17.9866	43
44	0.4184	29.0800	533.5165	2.3901	69.5027	0.0344	0.0144	18.3465	44
45	0.4102	29.4902	551.5652	2.4379	71.8927	0.0339	0.0139	18.7034	45
46	0.4022	29.8923	569.6621	2.4866	74.3306	0.0335	0.0135	19.0571	46
47	0.3943	30.2866	587.7985	2.5363	76.8172	0.0330	0.0130	19.4079	47
48	0.3865	30.6731	605.9657	2.5871	79.3535	0.0326	0.0126	19.7556	48
49	0.3790	31.0521	624.1557	2.6388	81.9406	0.0322	0.0122	20.1003	49
50	0.3715	31.4236	642.3606	2.6916	84.5794	0.0318	0.0118	20.4420	50
51	0.3642	31.7878	660.5727	2.7454	87.2710	0.0315	0.0115	20.7807	51
52	0.3571	32.1449	678.7849	2.8003	90.0164	0.0311	0.0111	21.1164	52
53	0.3501	32.4950	696.9900	2.8563	92.8167	0.0308	0.0108	21.4491	53
54	0.3432	32.8383	715.1815	2.9135	95.6731	0.0305	0.0105	21.7789	54
55	0.3365	33.1748	733.3527	2.9717	98.5865	0.0301	0.0101	22.1057	55
60	0.3048	34.7609	823.6975	3.2810	114.0515	0.0288	0.0088	23.6961	60
65	0.2761	36.1975	912.7085	3.6225	131.1262	0.0276	0.0076	25.2147	65
70	0.2500	37.4986	999.8343	3.9996	149.9779	0.0267	0.0067	26.6632	70
75	0.2265	38.6771	1084.6393	4.4158	170.7918	0.0259	0.0059	28.0434	75
80	0.2051	39.7445	1166.7868	4.8754	193.7720	0.0252	0.0052	29.3572	80
85	0.1858	40.7113	1246.0241	5.3829	219.1439	0.0246	0.0046	30.6064	85
90	0.1683	41.5869	1322.1701	5.9431	247.1567	0.0240	0.0040	31.7929	90
95	0.1524	42.3800	1395.1033	6.5617	278.0850	0.0236	0.0036	32.9189	95
100	0.1380	43.0984	1464.7527	7.2446	312.2323	0.0232	0.0032	33.9863	100

(*continued*)

APPENDIX 66.B (*continued*)
Factor Tables

$$I = 3.00\%$$

n	P/F	P/A	P/G	F/P	F/A	A/P	A/F	A/G	n
1	0.9709	0.9709	0.0000	1.0300	1.0000	1.0300	1.0000	0.0000	1
2	0.9426	1.9135	0.9426	1.0609	2.0300	0.5226	0.4926	0.4926	2
3	0.9151	2.8286	2.7729	1.0927	3.0909	0.3535	0.3235	0.9803	3
4	0.8885	3.7171	5.4383	1.1255	4.1836	0.2690	0.2390	1.4631	4
5	0.8626	4.5797	8.8888	1.1593	5.3091	0.2184	0.1884	1.9409	5
6	0.8375	5.4172	13.0762	1.1941	6.4684	0.1846	0.1546	2.4138	6
7	0.8131	6.2303	17.9547	1.2299	7.6625	0.1605	0.1305	2.8819	7
8	0.7894	7.0197	23.4806	1.2668	8.8923	0.1425	0.1125	3.3450	8
9	0.7664	7.7861	29.6119	1.3048	10.1591	0.1284	0.0984	3.8032	9
10	0.7441	8.5302	36.3088	1.3439	11.4639	0.1172	0.0872	4.2565	10
11	0.7224	9.2526	43.5330	1.3842	12.8078	0.1081	0.0781	4.7049	11
12	0.7014	9.9540	51.2482	1.4258	14.1920	0.1005	0.0705	5.1485	12
13	0.6810	10.6350	59.4196	1.4685	15.6178	0.0940	0.0640	5.5872	13
14	0.6611	11.2961	68.0141	1.5126	17.0863	0.0885	0.0585	6.0210	14
15	0.6419	11.9379	77.0002	1.5580	18.5989	0.0838	0.0538	6.4500	15
16	0.6232	12.5611	86.3477	1.6047	20.1569	0.0796	0.0496	6.8742	16
17	0.6050	13.1661	96.0280	1.6528	21.7616	0.0760	0.0460	7.2936	17
18	0.5874	13.7535	106.0137	1.7024	23.4144	0.0727	0.0427	7.7081	18
19	0.5703	14.3238	116.2788	1.7535	25.1169	0.0698	0.0398	8.1179	19
20	0.5537	14.8775	126.7987	1.8061	26.8704	0.0672	0.0372	8.5229	20
21	0.5375	15.4150	137.5496	1.8603	28.6765	0.0649	0.0349	8.9231	21
22	0.5219	15.9369	148.5094	1.9161	30.5368	0.0627	0.0327	9.3186	22
23	0.5067	16.4436	159.6566	1.9736	32.4529	0.0608	0.0308	9.7093	23
24	0.4919	16.9355	170.9711	2.0328	34.4265	0.0590	0.0290	10.0954	24
25	0.4776	17.4131	182.4336	2.0938	36.4593	0.0574	0.0274	10.4768	25
26	0.4637	17.8768	194.0260	2.1566	38.5530	0.0559	0.0259	10.8535	26
27	0.4502	18.3270	205.7309	2.2213	40.7096	0.0546	0.0246	11.2255	27
28	0.4371	18.7641	217.5320	2.2879	42.9309	0.0533	0.0233	11.5930	28
29	0.4243	19.1885	229.4137	2.3566	45.2189	0.0521	0.0221	11.9558	29
30	0.4120	19.6004	241.3613	2.4273	47.5754	0.0510	0.0210	12.3141	30
31	0.4000	20.0004	253.3609	2.5001	50.0027	0.0500	0.0200	12.6678	31
32	0.3883	20.3888	265.3993	2.5751	52.5028	0.0490	0.0190	13.0169	32
33	0.3770	20.7658	277.4642	2.6523	55.0778	0.0482	0.0182	13.3616	33
34	0.3660	21.1318	289.5437	2.7319	57.7302	0.0473	0.0173	13.7018	34
35	0.3554	21.4872	301.6267	2.8139	60.4621	0.0465	0.0165	14.0375	35
36	0.3450	21.8323	313.7028	2.8983	63.2759	0.0458	0.0158	14.3688	36
37	0.3350	22.1672	325.7622	2.9852	66.1742	0.0451	0.0151	14.6957	37
38	0.3252	22.4925	337.7956	3.0748	69.1594	0.0445	0.0145	15.0182	38
39	0.3158	22.8082	349.7942	3.1670	72.2342	0.0438	0.0138	15.3363	39
40	0.3066	23.1148	361.7499	3.2620	75.4013	0.0433	0.0133	15.6502	40
41	0.2976	23.4124	373.6551	3.3599	78.6633	0.0427	0.0127	15.9597	41
42	0.2890	23.7014	385.5024	3.4607	82.0232	0.0422	0.0122	16.2650	42
43	0.2805	23.9819	397.2852	3.5645	85.4839	0.0417	0.0117	16.5660	43
44	0.2724	24.2543	408.9972	3.6715	89.0484	0.0412	0.0112	16.8629	44
45	0.2644	24.5187	420.6325	3.7816	92.7199	0.0408	0.0108	17.1556	45
46	0.2567	24.7754	432.1856	3.8950	96.5015	0.0404	0.0104	17.4441	46
47	0.2493	25.0247	443.6515	4.0119	100.3965	0.0400	0.0100	17.7285	47
48	0.2420	25.2667	455.0255	4.1323	104.4084	0.0396	0.0096	18.0089	48
49	0.2350	25.5017	466.3031	4.2562	108.5406	0.0392	0.0092	18.2852	49
50	0.2281	25.7298	477.4803	4.3839	112.7969	0.0389	0.0089	18.5575	50
51	0.2215	25.9512	488.5535	4.5154	117.1808	0.0385	0.0085	18.8258	51
52	0.2150	26.1662	499.5191	4.6509	121.6962	0.0382	0.0082	19.0902	52
53	0.2088	26.3750	510.3742	4.7904	126.3471	0.0379	0.0079	19.3507	53
54	0.2027	26.5777	521.1157	4.9341	131.1375	0.0376	0.0076	19.6073	54
55	0.1968	26.7744	531.7411	5.0821	136.0716	0.0373	0.0073	19.8600	55
60	0.1697	27.6756	583.0526	5.8916	163.0534	0.0361	0.0061	21.0674	60
65	0.1464	28.4529	631.2010	6.8300	194.3328	0.0351	0.0051	22.1841	65
70	0.1263	29.1234	676.0869	7.9178	230.5941	0.0343	0.0043	23.2145	70
75	0.1089	29.7018	717.6978	9.1789	272.6309	0.0337	0.0037	24.1634	75
80	0.0940	30.2008	756.0865	10.6409	321.3630	0.0331	0.0031	25.0353	80
85	0.0811	30.6312	791.3529	12.3357	377.8570	0.0326	0.0026	25.8349	85
90	0.0699	31.0024	823.6302	14.3005	443.3489	0.0323	0.0023	26.5667	90
95	0.0603	31.3227	853.0742	16.5782	519.2720	0.0319	0.0019	27.2351	95
100	0.0520	31.5989	879.8540	19.2186	607.2877	0.0316	0.0016	27.8444	100

(*continued*)

APPENDIX 66.B (*continued*)
Factor Tables

$I = 4.00\%$

n	P/F	P/A	P/G	F/P	F/A	A/P	A/F	A/G	n
1	0.9615	0.9615	0.0000	1.0400	1.0000	1.0400	1.0000	0.0000	1
2	0.9246	1.8861	0.9246	1.0816	2.0400	0.5302	0.4902	0.4902	2
3	0.8890	2.7751	2.7025	1.1249	3.1216	0.3603	0.3203	0.9739	3
4	0.8548	3.6299	5.2670	1.1699	4.2465	0.2755	0.2355	1.4510	4
5	0.8219	4.4518	8.5547	1.2167	5.4163	0.2246	0.1846	1.9216	5
6	0.7903	5.2421	12.5062	1.2653	6.6330	0.1908	0.1508	2.3857	6
7	0.7599	6.0021	17.0657	1.3159	7.8983	0.1666	0.1266	2.8433	7
8	0.7307	6.7327	22.1806	1.3686	9.2142	0.1485	0.1085	3.2944	8
9	0.7026	7.4353	27.8013	1.4233	10.5828	0.1345	0.0945	3.7391	9
10	0.6756	8.1109	33.8814	1.4802	12.0061	0.1233	0.0833	4.1773	10
11	0.6496	8.7605	40.3772	1.5395	13.4864	0.1141	0.0741	4.6090	11
12	0.6246	9.3851	47.2477	1.6010	15.0258	0.1066	0.0666	5.0343	12
13	0.6006	9.9856	54.4546	1.6651	16.6268	0.1001	0.0601	5.4533	13
14	0.5775	10.5631	61.9618	1.7317	18.2919	0.0947	0.0547	5.8659	14
15	0.5553	11.1184	69.7355	1.8009	20.0236	0.0899	0.0499	6.2721	15
16	0.5339	11.6523	77.7441	1.8730	21.8245	0.0858	0.0458	6.6720	16
17	0.5134	12.1657	85.9581	1.9479	23.6975	0.0822	0.0422	7.0656	17
18	0.4936	12.6593	94.3498	2.0258	25.6454	0.0790	0.0390	7.4530	18
19	0.4746	13.1339	102.8933	2.1068	27.6712	0.0761	0.0361	7.8342	19
20	0.4564	13.5903	111.5647	2.1911	29.7781	0.0736	0.0336	8.2091	20
21	0.4388	14.0292	120.3414	2.2788	31.9692	0.0713	0.0313	8.5779	21
22	0.4220	14.4511	129.2024	2.3699	34.2480	0.0692	0.0292	8.9407	22
23	0.4057	14.8568	138.1284	2.4647	36.6179	0.0673	0.0273	9.2973	23
24	0.3901	15.2470	147.1012	2.5633	39.0826	0.0656	0.0256	9.6479	24
25	0.3751	15.6221	156.1040	2.6658	41.6459	0.0640	0.0240	9.9925	25
26	0.3607	15.9828	165.1212	2.7725	44.3117	0.0626	0.0226	10.3312	26
27	0.3468	16.3296	174.1385	1.8834	47.0842	0.0612	0.0212	10.6640	27
28	0.3335	16.6631	183.1424	2.9987	49.9676	0.0600	0.0200	10.9909	28
29	0.3207	16.9837	192.1206	3.1187	52.9663	0.0589	0.0189	11.3120	29
30	0.3083	17.2920	201.0618	3.2434	56.0849	0.0578	0.0178	11.6274	30
31	0.2965	17.5885	209.9556	3.3731	59.3283	0.0569	0.0169	11.9371	31
32	0.2851	17.8736	218.7924	3.5081	62.7015	0.0559	0.0159	12.2411	32
33	0.2741	18.1476	227.5634	3.6484	66.2095	0.0551	0.0151	12.5396	33
34	0.2636	18.4112	236.2607	3.7943	69.8579	0.0543	0.0143	12.8324	34
35	0.2534	18.6646	244.8768	3.9461	73.6522	0.0536	0.0136	13.1198	35
36	0.2437	18.9083	253.4052	4.1039	77.5983	0.0529	0.0129	13.4018	36
37	0.2343	19.1426	261.8399	4.2681	81.7022	0.0522	0.0122	13.6784	37
38	0.2253	19.3679	270.1754	4.4388	85.9703	0.0516	0.0116	13.9497	38
39	0.2166	19.5845	278.4070	4.6164	90.4091	0.0511	0.0111	14.2157	39
40	0.2083	19.7928	286.5303	4.8010	95.0255	0.0505	0.0105	14.4765	40
41	0.2003	19.9931	294.5414	4.9931	99.8265	0.0500	0.0100	14.7322	41
42	0.1926	20.1856	302.4370	5.1928	104.8196	0.0495	0.0095	14.9828	42
43	0.1852	20.3708	310.2141	5.4005	110.0124	0.0491	0.0091	15.2284	43
44	0.1780	20.5488	317.8700	5.6165	115.4129	0.0487	0.0087	15.4690	44
45	0.1712	20.7200	325.4028	5.8412	121.0294	0.0483	0.0083	15.7047	45
46	0.1646	20.8847	332.8104	6.0748	126.8706	0.0479	0.0079	15.9356	46
47	0.1583	21.0429	340.0914	6.3178	132.9454	0.0475	0.0075	16.1618	47
48	0.1522	21.1951	347.2446	6.5705	139.2632	0.0472	0.0072	16.3832	48
49	0.1463	21.3415	354.2689	6.8333	145.8337	0.0469	0.0069	16.6000	49
50	0.1407	21.4822	361.1638	7.1067	152.6671	0.0466	0.0066	16.8122	50
51	0.1353	21.6175	367.9289	7.3910	159.7738	0.0463	0.0063	17.0200	51
52	0.1301	21.7476	374.5638	7.6866	167.1647	0.0460	0.0060	17.2232	52
53	0.1251	21.8727	381.0686	7.9941	174.8513	0.0457	0.0057	17.4221	53
54	0.1203	21.9930	387.4436	8.3138	182.8454	0.0455	0.0055	17.6167	54
55	0.1157	22.1086	393.6890	8.6464	191.1592	0.0452	0.0052	17.8070	55
60	0.0951	22.6235	422.9966	10.5196	237.9907	0.0442	0.0042	18.6972	60
65	0.0781	23.0467	449.2014	12.7987	294.9684	0.0434	0.0034	19.4909	65
70	0.0642	23.3945	472.4789	15.5716	364.2905	0.0427	0.0027	20.1961	70
75	0.0528	23.6804	493.0408	18.9453	448.6314	0.0422	0.0022	20.8206	75
80	0.0434	23.9154	511.1161	23.0498	551.2450	0.0418	0.0018	21.3718	80
85	0.0357	24.1085	526.9384	28.0436	676.0901	0.0415	0.0015	21.8569	85
90	0.0293	24.2673	540.7369	34.1193	827.9833	0.0412	0.0012	22.2826	90
95	0.0241	24.3978	552.7307	41.5114	1012.7846	0.0410	0.0010	22.6550	95
100	0.0198	24.5050	563.1249	50.5049	1237.6237	0.0408	0.0008	22.9800	100

(*continued*)

APPENDIX 66.B (*continued*)
Factor Tables

$$I = 5.00\%$$

n	P/F	P/A	P/G	F/P	F/A	A/P	A/F	A/G	n
1	0.9524	0.9524	0.0000	1.0500	1.0000	1.0500	1.0000	0.0000	1
2	0.9070	1.8594	0.9070	1.1025	2.0500	0.5378	0.4878	0.4878	2
3	0.8638	2.7232	2.6347	1.1576	3.1525	0.3672	0.3172	0.9675	3
4	0.8227	3.5460	5.1028	1.2155	4.3101	0.2820	0.2320	1.4391	4
5	0.7835	4.3295	8.2369	1.2763	5.5256	0.2310	0.1810	1.9025	5
6	0.7462	5.0757	11.9680	1.3401	6.8019	0.1970	0.1470	2.3579	6
7	0.7107	5.7864	16.2321	1.4071	8.1420	0.1728	0.1228	2.8052	7
8	0.6768	6.4632	20.9700	1.4775	9.5491	0.1547	0.1047	3.2445	8
9	0.6446	7.1078	26.1268	1.5513	11.0266	0.1407	0.0907	3.6758	9
10	0.6139	7.7217	31.6520	1.6289	12.5779	0.1295	0.0795	4.0991	10
11	0.5847	8.3064	37.4988	1.7103	14.2068	0.1204	0.0704	4.5144	11
12	0.5568	8.8633	43.6241	1.7959	15.9171	0.1128	0.0628	4.9219	12
13	0.5303	9.3936	49.9879	1.8856	17.7130	0.1065	0.0565	5.3215	13
14	0.5051	9.8986	56.5538	1.9799	19.5986	0.1010	0.0510	5.7133	14
15	0.4810	10.3797	63.2880	2.0789	21.5786	0.0963	0.0463	6.0973	15
16	0.4581	10.8378	70.1597	2.1829	23.6575	0.0923	0.0423	6.4736	16
17	0.4363	11.2741	77.1405	2.2920	25.8404	0.0887	0.0387	6.8423	17
18	0.4155	11.6896	84.2043	2.4066	28.1324	0.0855	0.0355	7.2034	18
19	0.3957	12.0853	91.3275	2.5270	30.5390	0.0827	0.0327	7.5569	19
20	0.3769	12.4622	98.4884	2.6533	33.0660	0.0802	0.0302	7.9030	20
21	0.3589	12.8212	105.6673	2.7860	35.7193	0.0780	0.0280	8.2416	21
22	0.3418	13.1630	112.8461	2.9253	38.5052	0.0760	0.0260	8.5730	22
23	0.3256	13.4886	120.0087	3.0715	41.4305	0.0741	0.0241	8.8971	23
24	0.3101	13.7986	127.1402	3.2251	44.5020	0.0725	0.0225	9.2140	24
25	0.2953	14.0939	134.2275	3.3864	47.7271	0.0710	0.0210	9.5238	25
26	0.2812	14.3752	141.2585	3.5557	51.1135	0.0696	0.0196	9.8266	26
27	0.2678	14.6430	148.2226	3.7335	54.6691	0.0683	0.0183	10.1224	27
28	0.2551	14.8981	155.1101	3.9201	58.4026	0.0671	0.0171	10.4114	28
29	0.2429	15.1411	161.9126	4.1161	62.3227	0.0660	0.0160	10.6936	29
30	0.2314	15.3725	168.6226	4.3219	66.4388	0.0651	0.0151	10.9691	30
31	0.2204	15.5928	175.2333	4.5380	70.7608	0.0641	0.0141	11.2381	31
32	0.2099	15.8027	181.7392	4.7649	75.2988	0.0633	0.0133	11.5005	32
33	0.1999	16.0025	188.1351	5.0032	80.0638	0.0625	0.0125	11.7566	33
34	0.1904	16.1929	194.4168	5.2533	85.0670	0.0618	0.0118	12.0063	34
35	0.1813	16.3742	200.5807	5.5160	90.3203	0.0611	0.0111	12.2498	35
36	0.1727	16.5469	206.6237	5.7918	95.8363	0.0604	0.0104	12.4872	36
37	0.1644	16.7113	212.5434	6.0814	101.6281	0.0598	0.0098	12.7186	37
38	0.1566	16.8679	218.3378	6.3855	107.7095	0.0593	0.0093	12.9440	38
39	0.1491	17.0170	224.0054	6.7048	114.0950	0.0588	0.0088	13.1636	39
40	0.1420	17.1591	229.5452	7.0400	120.7998	0.0583	0.0083	13.3775	40
41	0.1353	17.2944	234.9564	7.3920	127.8398	0.0578	0.0078	13.5857	41
42	0.1288	17.4232	240.2389	7.7616	135.2318	0.0574	0.0074	13.7884	42
43	0.1227	17.5459	245.3925	8.1497	142.9933	0.0570	0.0070	13.9857	43
44	0.1169	17.6628	250.4175	8.5572	151.1430	0.0566	0.0066	14.1777	44
45	0.1113	17.7741	255.3145	8.9850	159.7002	0.0563	0.0063	14.3644	45
46	0.1060	17.8801	260.0844	9.4343	168.6852	0.0559	0.0059	14.5461	46
47	0.1009	17.9810	264.7281	9.9060	178.1194	0.0556	0.0056	14.7226	47
48	0.0961	18.0772	269.2467	10.4013	188.0254	0.0553	0.0053	14.8943	48
49	0.0916	18.1687	273.6418	10.9213	198.4267	0.0550	0.0050	15.0611	49
50	0.0872	18.2559	277.9148	11.4674	209.3480	0.0548	0.0048	15.2233	50
51	0.0831	18.3390	282.0673	12.0408	220.8154	0.0545	0.0045	15.3808	51
52	0.0791	18.4181	286.1013	12.6428	232.8562	0.0543	0.0043	15.5337	52
53	0.0753	18.4934	290.0184	13.2749	245.4990	0.0541	0.0041	15.6823	53
54	0.0717	18.5651	293.8208	13.9387	258.7739	0.0539	0.0039	15.8265	54
55	0.0683	18.6335	297.5104	14.6356	272.7126	0.0537	0.0037	15.9664	55
60	0.0535	18.9293	314.3432	18.6792	353.5837	0.0528	0.0028	16.6062	60
65	0.0419	19.1611	328.6910	23.8399	456.7980	0.0522	0.0022	17.1541	65
70	0.0329	19.3427	340.8409	30.4264	588.5285	0.0517	0.0017	17.6212	70
75	0.0258	19.4850	351.0721	38.8327	756.6537	0.0513	0.0013	18.0176	75
80	0.0202	19.5965	359.6460	49.5614	971.2288	0.0510	0.0010	18.3526	80
85	0.0158	19.6838	366.8007	63.2544	1245.0871	0.0508	0.0008	18.6346	85
90	0.0124	19.7523	372.7488	80.7304	1597.6073	0.0506	0.0006	18.8712	90
95	0.0097	19.8059	377.6774	103.0347	2040.6935	0.0505	0.0005	19.0689	95
100	0.0076	19.8479	381.7492	131.5013	2610.0252	0.0504	0.0004	19.2337	100

(*continued*)

APPENDIX 66.B (*continued*)
Factor Tables

$I = 6.00\%$

n	P/F	P/A	P/G	F/P	F/A	A/P	A/F	A/G	n
1	0.9434	0.9434	0.0000	1.0600	1.0000	1.0600	1.0000	0.0000	1
2	0.8900	1.8334	0.8900	1.1236	2.0600	0.5454	0.4854	0.4854	2
3	0.8396	2.6730	2.5692	1.1910	3.1836	0.3741	0.3141	0.9612	3
4	0.7921	3.4651	4.9455	1.2625	4.3746	0.2886	0.2286	1.4272	4
5	0.7473	4.2124	7.9345	1.3382	5.6371	0.2374	0.1774	1.8836	5
6	0.7050	4.9173	11.4594	1.4185	6.9753	0.2034	0.1434	2.3304	6
7	0.6651	5.5824	15.4497	1.5036	8.3938	0.1791	0.1191	2.7676	7
8	0.6274	6.2098	19.8416	1.5938	9.8975	0.1610	0.1010	3.1952	8
9	0.5919	6.8017	24.5768	1.6895	11.4913	0.1470	0.0870	3.6133	9
10	0.5584	7.3601	29.6023	1.7908	13.1808	0.1359	0.0759	4.0220	10
11	0.5268	7.8869	34.8702	1.8983	14.9716	0.1268	0.0668	4.4213	11
12	0.4970	8.3838	40.3369	2.0122	16.8699	0.1193	0.0593	4.8113	12
13	0.4688	8.8527	45.9629	2.1329	18.8821	0.1130	0.0530	5.1920	13
14	0.4423	9.2950	51.7128	2.2609	21.0151	0.1076	0.0476	5.5635	14
15	0.4173	9.7122	57.5546	2.3966	23.2760	0.1030	0.0430	5.9260	15
16	0.3936	10.1059	63.4592	2.5404	25.6725	0.0990	0.0390	6.2794	16
17	0.3714	10.4773	69.4011	2.6928	28.2129	0.0954	0.0354	6.6240	17
18	0.3503	10.8276	75.3569	2.8543	30.9057	0.0924	0.0324	6.9597	18
19	0.3305	11.1581	81.3062	3.0256	33.7600	0.0896	0.0296	7.2867	19
20	0.3118	11.4699	87.2304	3.2071	36.7856	0.0872	0.0272	7.6051	20
21	0.2942	11.7641	93.1136	3.3996	39.9927	0.0850	0.0250	7.9151	21
22	0.2775	12.0416	98.9412	3.6035	43.3923	0.0830	0.0230	8.2166	22
23	0.2618	12.3034	104.7007	3.8197	46.9958	0.0813	0.0213	8.5099	23
24	0.2470	12.5504	110.3812	4.0489	50.8156	0.0797	0.0197	8.7951	24
25	0.2330	12.7834	115.9732	4.2919	54.8645	0.0782	0.0182	9.0722	25
26	0.2198	13.0032	121.4684	4.5494	59.1564	0.0769	0.0169	9.3414	26
27	0.2074	13.2105	126.8600	4.8223	63.7058	0.0757	0.0157	9.6029	27
28	0.1956	13.4062	132.1420	5.1117	68.5281	0.0746	0.0146	9.8568	28
29	0.1846	13.5907	137.3096	5.4184	73.6398	0.0736	0.0136	10.1032	29
30	0.1741	13.7648	142.3588	5.7435	79.0582	0.0726	0.0126	10.3422	30
31	0.1643	13.9291	147.2864	6.0881	84.8017	0.0718	0.0118	10.5740	31
32	0.1550	14.0840	152.0901	6.4534	90.8898	0.0710	0.0110	10.7988	32
33	0.1462	14.2302	156.7681	6.8406	97.3432	0.0703	0.0103	11.0166	33
34	0.1379	14.3681	161.3192	7.2510	104.1838	0.0696	0.0096	11.2276	34
35	0.1301	14.4982	165.7427	7.6861	111.4348	0.0690	0.0090	11.4319	35
36	0.1227	14.6210	170.0387	8.1473	119.1209	0.0684	0.0084	11.6298	36
37	0.1158	14.7368	174.2072	8.6361	127.2681	0.0679	0.0079	11.8213	37
38	0.1092	14.8460	178.2490	9.1543	135.9042	0.0674	0.0074	12.0065	38
39	0.1031	14.9491	182.1652	9.7035	145.0585	0.0669	0.0069	12.1857	39
40	0.0972	15.0463	185.9568	10.2857	154.7620	0.0665	0.0065	12.3590	40
41	0.0917	15.1380	189.6256	10.9029	165.0477	0.0661	0.0061	12.5264	41
42	0.0865	15.2245	193.1732	11.5570	175.9505	0.0657	0.0057	12.6883	42
43	0.0816	15.3062	196.6017	12.2505	187.5076	0.0653	0.0053	12.8446	43
44	0.0770	15.3832	199.9130	12.9855	199.7580	0.0650	0.0050	12.9956	44
45	0.0727	15.4558	203.1096	13.7646	212.7435	0.0647	0.0047	13.1413	45
46	0.0685	15.5244	206.1938	14.5905	226.5081	0.0644	0.0044	13.2819	46
47	0.0647	15.5890	209.1681	15.4659	241.0986	0.0641	0.0041	13.4177	47
48	0.0610	15.6500	212.0351	16.3939	256.5645	0.0639	0.0039	13.5485	48
49	0.0575	15.7076	214.7972	17.3775	272.9584	0.0637	0.0037	13.6748	49
50	0.0543	15.7619	217.4574	18.4202	290.3359	0.0634	0.0034	13.7964	50
51	0.0512	15.8131	220.0181	19.5254	308.7561	0.0632	0.0032	13.9137	51
52	0.0483	15.8614	222.4823	20.6969	328.2814	0.0630	0.0030	14.0267	52
53	0.0456	15.9070	224.8525	21.9387	348.9783	0.0629	0.0029	14.1355	53
54	0.0430	15.9500	227.1316	23.2550	370.9170	0.0627	0.0027	14.2402	54
55	0.0406	15.9905	229.3222	24.6503	394.1720	0.0625	0.0025	14.3411	55
60	0.0303	16.1614	239.0428	32.9877	533.1282	0.0619	0.0019	14.7909	60
65	0.0227	16.2891	246.9450	44.1450	719.0829	0.0614	0.0014	15.1601	65
70	0.0169	16.3845	253.3271	59.0759	967.9322	0.0610	0.0010	15.4613	70
75	0.0126	16.4558	258.4527	79.0569	1300.9487	0.0608	0.0008	15.7058	75
80	0.0095	16.5091	262.5493	105.7960	1746.5999	0.0606	0.0006	15.9033	80
85	0.0071	16.5489	265.8096	141.5789	2342.9817	0.0604	0.0004	16.0620	85
90	0.0053	16.5787	268.3946	189.4645	3141.0752	0.0603	0.0003	16.1891	90
95	0.0039	16.6009	270.4375	253.5463	4209.1042	0.0602	0.0002	16.2905	95
100	0.0029	16.6175	272.0471	339.3021	5638.3681	0.0602	0.0002	16.3711	100

(*continued*)

APPENDIX 66.B (*continued*)
Factor Tables

$I = 7.00\%$

n	P/F	P/A	P/G	F/P	F/A	A/P	A/F	A/G	n
1	0.9346	0.9346	0.0000	1.0700	1.0000	1.0700	1.0000	0.0000	1
2	0.8734	1.8080	0.8734	1.1449	2.0700	0.5531	0.4831	0.4831	2
3	0.8163	2.6243	2.5060	1.2250	3.2149	0.3811	0.3111	0.9549	3
4	0.7629	3.3872	4.7947	1.3108	4.4399	0.2952	0.2252	1.4155	4
5	0.7130	4.1002	7.6467	1.4026	5.7507	0.2439	0.1739	1.8650	5
6	0.6663	4.7665	10.9784	1.5007	7.1533	0.2098	0.1398	2.3032	6
7	0.6227	5.3893	14.7149	1.6058	8.6540	0.1856	0.1156	2.7304	7
8	0.5820	5.9713	18.7889	1.7182	10.2598	0.1675	0.0975	3.1465	8
9	0.5439	6.5152	23.1404	1.8385	11.9780	0.1535	0.0835	3.5517	9
10	0.5083	7.0236	27.7156	1.9672	13.8164	0.1424	0.0724	3.9461	10
11	0.4751	7.4987	32.4665	2.1049	15.7836	0.1334	0.0634	4.3296	11
12	0.4440	7.9427	37.3506	2.2522	17.8885	0.1259	0.0559	4.7025	12
13	0.4150	8.3577	42.3302	2.4098	20.1406	0.1197	0.0497	5.0648	13
14	0.3878	8.7455	47.3718	2.5785	22.5505	0.1143	0.0443	5.4167	14
15	0.3624	9.1079	52.4461	2.7590	25.1290	0.1098	0.0398	5.7583	15
16	0.3387	9.4466	57.5271	2.9522	27.8881	0.1059	0.0359	6.0897	16
17	0.3166	9.7632	62.5923	3.1588	30.8402	0.1024	0.0324	6.4110	17
18	0.2959	10.0591	67.6219	3.3799	33.9990	0.0994	0.0294	6.7225	18
19	0.2765	10.3356	72.5991	3.6165	37.3790	0.0968	0.0268	7.0242	19
20	0.2584	10.5940	77.5091	3.8697	40.9955	0.0944	0.0244	7.3163	20
21	0.2415	10.8355	82.3393	4.1406	44.8652	0.0923	0.0223	7.5990	21
22	0.2257	11.0612	87.0793	4.4304	49.0057	0.0904	0.0204	7.8725	22
23	0.2109	11.2722	91.7201	4.7405	53.4361	0.0887	0.0187	8.1369	23
24	0.1971	11.4693	96.2545	5.0724	58.1767	0.0872	0.0172	8.3923	24
25	0.1842	11.6536	100.6765	5.4274	63.2490	0.0858	0.0158	8.6391	25
26	0.1722	11.8258	104.9814	5.8074	68.6765	0.0846	0.0146	8.8773	26
27	0.1609	11.9867	109.1656	6.2139	74.4838	0.0834	0.0134	9.1072	27
28	0.1504	12.1371	113.2264	6.6488	80.6977	0.0824	0.0124	9.3289	28
29	0.1406	12.2777	117.1622	7.1143	87.3465	0.0814	0.0114	9.5427	29
30	0.1314	12.4090	120.9718	7.6123	94.4608	0.0806	0.0106	9.7487	30
31	0.1228	12.5318	124.6550	8.1451	102.0730	0.0798	0.0098	9.9471	31
32	0.1147	12.6466	128.2120	8.7153	110.2182	0.0791	0.0091	10.1381	32
33	0.1072	12.7538	131.6435	9.3253	118.9334	0.0784	0.0084	10.3219	33
34	0.1002	12.8540	134.9507	9.9781	128.2588	0.0778	0.0078	10.4987	34
35	0.0937	12.9477	138.1353	10.6766	138.2369	0.0772	0.0072	10.6687	35
36	0.0875	13.0352	141.1990	11.4239	148.9135	0.0767	0.0067	10.8321	36
37	0.0818	13.1170	144.1441	12.2236	160.3374	0.0762	0.0062	10.9891	37
38	0.0765	13.1935	146.9730	13.0793	172.5610	0.0758	0.0058	11.1398	38
39	0.0715	13.2649	149.6883	13.9948	185.6403	0.0754	0.0054	11.2845	39
40	0.0668	13.3317	152.2928	14.9745	199.6351	0.0750	0.0050	11.4233	40
41	0.0624	13.3941	154.7892	16.0227	214.6096	0.0747	0.0047	11.5565	41
42	0.0583	13.4524	157.1807	17.1443	230.6322	0.0743	0.0043	11.6842	42
43	0.0545	13.5070	159.4702	18.3444	247.7765	0.0740	0.0040	11.8065	43
44	0.0509	13.5579	161.6609	19.6285	266.1209	0.0738	0.0038	11.9237	44
45	0.0476	13.6055	163.7559	21.0025	285.7493	0.0735	0.0035	12.0360	45
46	0.0445	13.6500	165.7584	22.4726	306.7518	0.0733	0.0033	12.1435	46
47	0.0416	13.6916	167.6714	24.0457	329.2244	0.0730	0.0030	12.2463	47
48	0.0389	13.7305	169.4981	25.7289	353.2701	0.0728	0.0028	12.3447	48
49	0.0363	13.7668	171.2417	27.5299	378.9990	0.0726	0.0026	12.4387	49
50	0.0339	13.8007	172.9051	29.4570	406.5289	0.0725	0.0025	12.5287	50
51	0.0317	13.8325	174.4915	31.5190	435.9860	0.0723	0.0023	12.6146	51
52	0.0297	13.8621	176.0037	33.7253	467.5050	0.0721	0.0021	12.6967	52
53	0.0277	13.8898	177.4447	36.0861	501.2303	0.0720	0.0020	12.7751	53
54	0.0259	13.9157	178.8173	38.6122	537.3164	0.0719	0.0019	12.8500	54
55	0.0242	13.9399	180.1243	41.3150	575.9286	0.0717	0.0017	12.9215	55
60	0.0173	14.0392	185.7677	57.9464	813.5204	0.0712	0.0012	13.2321	60
65	0.0123	14.1099	190.1452	81.2729	1146.7552	0.0709	0.0009	13.4760	65
70	0.0088	14.1604	193.5185	113.9894	1614.1342	0.0706	0.0006	13.6662	70
75	0.0063	14.1964	196.1035	159.8760	2269.6574	0.0704	0.0004	13.8136	75
80	0.0045	14.2220	198.0748	224.2344	3189.0627	0.0703	0.0003	13.9273	80
85	0.0032	14.2403	199.5717	314.5003	4478.5761	0.0702	0.0002	14.0146	85
90	0.0023	14.2533	200.7042	441.1030	6287.1854	0.0702	0.0002	14.0812	90
95	0.0016	14.2626	201.5581	618.6697	8823.8535	0.0701	0.0001	14.1319	95
100	0.0012	14.2693	202.2001	867.7163	12381.6618	0.0701	0.0001	14.1703	100

(*continued*)

APPENDIX 66.B (*continued*)
Factor Tables

$I = 8.00\%$

n	P/F	P/A	P/G	F/P	F/A	A/P	A/F	A/G	n
1	0.9259	0.9259	0.0000	1.0800	1.0000	1.0800	1.0000	0.0000	1
2	0.8573	1.7833	0.8573	1.1664	2.0800	0.5608	0.4808	0.4808	2
3	0.7938	2.5771	2.4450	1.2597	3.2464	0.3880	0.3080	0.9487	3
4	0.7350	3.3121	4.6501	1.3605	4.5061	0.3019	0.2219	1.4040	4
5	0.6806	3.9927	7.3724	1.4693	5.8666	0.2505	0.1705	1.8465	5
6	0.6302	4.6229	10.5233	1.5869	7.3359	0.2163	0.1363	2.2763	6
7	0.5835	5.2064	14.0242	1.7138	8.9228	0.1921	0.1121	2.6937	7
8	0.5403	5.7466	17.8061	1.8509	10.6366	0.1740	0.0940	3.0985	8
9	0.5002	6.2469	21.8081	1.9990	12.4876	0.1601	0.0801	3.4910	9
10	0.4632	6.7101	25.9768	2.1589	14.4866	0.1490	0.0690	3.8713	10
11	0.4289	7.1390	30.2657	2.3316	16.6455	0.1401	0.0601	4.2395	11
12	0.3971	7.5361	34.6339	2.5182	18.9771	0.1327	0.0527	4.5957	12
13	0.3677	7.9038	39.0463	2.7196	21.4953	0.1265	0.0465	4.9402	13
14	0.3405	8.2442	43.4723	2.9372	24.2149	0.1213	0.0413	5.2731	14
15	0.3152	8.5595	47.8857	3.1722	27.1521	0.1168	0.0368	5.5945	15
16	0.2919	8.8514	52.2640	3.4259	30.3243	0.1130	0.0330	5.9046	16
17	0.2703	9.1216	56.5883	3.7000	33.7502	0.1096	0.0296	6.2037	17
18	0.2502	9.3719	60.8426	3.9960	37.4502	0.1067	0.0267	6.4920	18
19	0.2317	9.6036	65.0134	4.3157	41.4463	0.1041	0.0241	6.7697	19
20	0.2145	9.8181	69.0898	4.6610	45.7620	0.1019	0.0219	7.0369	20
21	0.1987	10.0168	73.0629	5.0338	50.4229	0.0998	0.0198	7.2940	21
22	0.1839	10.2007	76.9257	5.4365	55.4568	0.0980	0.0180	7.5412	22
23	0.1703	10.3711	80.6726	5.8715	60.8933	0.0964	0.0164	7.7786	23
24	0.1577	10.5288	84.2997	6.3412	66.7648	0.0950	0.0150	8.0066	24
25	0.1460	10.6748	87.8041	6.8485	73.1059	0.0937	0.0137	8.2254	25
26	0.1352	10.8100	91.1842	7.3964	79.9544	0.0925	0.0125	8.4352	26
27	0.1252	10.9352	94.4390	7.9881	87.3508	0.0914	0.0114	8.6363	27
28	0.1159	11.0511	97.5687	8.6271	95.3388	0.0905	0.0105	8.8289	28
29	0.1073	11.1584	100.5738	9.3173	103.9659	0.0896	0.0096	9.0133	29
30	0.0994	11.2578	103.4558	10.0627	113.2832	0.0888	0.0088	9.1897	30
31	0.0920	11.3498	106.2163	10.8677	123.3459	0.0881	0.0081	9.3584	31
32	0.0852	11.4350	108.8575	11.7371	134.2135	0.0875	0.0075	9.5197	32
33	0.0789	11.5139	111.3819	12.6760	145.9506	0.0869	0.0069	9.6737	33
34	0.0730	11.5869	113.7924	13.6901	158.6267	0.0863	0.0063	9.8208	34
35	0.0676	11.6546	116.0920	14.7853	172.3168	0.0858	0.0058	9.9611	35
36	0.0626	11.7172	118.2839	15.9682	187.1021	0.0853	0.0053	10.0949	36
37	0.0580	11.7752	120.3713	17.2456	203.0703	0.0849	0.0049	10.2225	37
38	0.0537	11.8289	122.3579	18.6253	220.3159	0.0845	0.0045	10.3440	38
39	0.0497	11.8786	124.2470	20.1153	238.9412	0.0842	0.0042	10.4597	39
40	0.0460	11.9246	126.0422	21.7245	259.0565	0.0839	0.0039	10.5699	40
41	0.0426	11.9672	127.7470	23.4625	280.7810	0.0836	0.0036	10.6747	41
42	0.0395	12.0067	129.3651	25.3395	304.2435	0.0833	0.0033	10.7744	42
43	0.0365	12.0432	130.8998	27.3666	329.5830	0.0830	0.0030	10.8692	43
44	0.0338	12.0771	132.3547	29.5560	356.9496	0.0828	0.0028	10.9592	44
45	0.0313	12.1084	133.7331	31.9204	386.5056	0.0826	0.0026	11.0447	45
46	0.0290	12.1374	135.0384	34.4741	418.4261	0.0824	0.0024	11.1258	46
47	0.0269	12.1643	136.2739	37.2320	452.9002	0.0822	0.0022	11.2028	47
48	0.0249	12.1891	137.4428	40.2106	490.1322	0.0820	0.0020	11.2758	48
49	0.0230	12.2122	138.5480	43.4274	530.3427	0.0819	0.0019	11.3451	49
50	0.0213	12.2335	139.5928	46.9016	573.7702	0.0817	0.0017	11.4107	50
51	0.0197	12.2532	140.5799	50.6537	620.6718	0.0816	0.0016	11.4729	51
52	0.0183	12.2715	141.5121	54.7060	671.3255	0.0815	0.0015	11.5318	52
53	0.0169	12.2884	142.3923	59.0825	726.0316	0.0814	0.0014	11.5875	53
54	0.0157	12.3041	143.2229	63.8091	785.1141	0.0813	0.0013	11.6403	54
55	0.0145	12.3186	144.0065	68.9139	848.9232	0.0812	0.0012	11.6902	55
60	0.0099	12.3766	147.3000	101.2571	1253.2133	0.0808	0.0008	11.9015	60
65	0.0067	12.4160	149.7387	148.7798	1847.2481	0.0805	0.0005	12.0602	65
70	0.0046	12.4428	151.5326	218.6064	2720.0801	0.0804	0.0004	12.1783	70
75	0.0031	12.4611	152.8448	321.2045	4002.5566	0.0802	0.0002	12.2658	75
80	0.0021	12.4735	153.8001	471.9548	5886.9354	0.0802	0.0002	12.3301	80
85	0.0014	12.4820	154.4925	693.4565	8655.7061	0.0801	0.0001	12.3772	85
90	0.0010	12.4877	154.9925	1018.9151	12723.9386	0.0801	0.0001	12.4116	90
95	0.0007	12.4917	155.3524	1497.1205	18701.5069	0.0801	0.0001	12.4365	95
100	0.0005	12.4943	155.6107	2199.7613	27484.5157	0.0800	0.0000	12.4545	100

(*continued*)

APPENDIX 66.B (*continued*)
Factor Tables

$I = 9.00\%$

n	P/F	P/A	P/G	F/P	F/A	A/P	A/F	A/G	n
1	0.9174	0.9174	0.0000	1.0900	1.0000	1.0900	1.0000	0.0000	1
2	0.8417	1.7591	0.8417	1.1881	2.0900	0.5685	0.4785	0.4785	2
3	0.7722	2.5313	2.3860	1.2950	3.2781	0.3951	0.3051	0.9426	3
4	0.7084	3.2397	4.5113	1.4116	4.5731	0.3087	0.2187	1.3925	4
5	0.6499	3.8897	7.1110	1.5386	5.9847	0.2571	0.1671	1.8282	5
6	0.5963	4.4859	10.0924	1.6771	7.5233	0.2229	0.1329	2.2498	6
7	0.5470	5.0330	13.3746	1.8280	9.2004	0.1987	0.1087	2.6574	7
8	0.5019	5.5348	16.8877	1.9926	11.0285	0.1807	0.0907	3.0512	8
9	0.4604	5.9952	20.5711	2.1719	13.0210	0.1668	0.0768	3.4312	9
10	0.4224	6.4177	24.3728	2.3674	15.1929	0.1558	0.0658	3.7978	10
11	0.3875	6.8052	28.2481	2.5804	17.5603	0.1469	0.0569	4.1510	11
12	0.3555	7.1607	32.1590	2.8127	20.1407	0.1397	0.0497	4.4910	12
13	0.3262	7.4869	36.0731	3.0658	22.9534	0.1336	0.0436	4.8182	13
14	0.2992	7.7862	39.9633	3.3417	26.0192	0.1284	0.0384	5.1326	14
15	0.2745	8.0607	43.8069	3.6425	29.3609	0.1241	0.0341	5.4346	15
16	0.2519	8.3126	47.5849	3.9703	33.0034	0.1203	0.0303	5.7245	16
17	0.2311	8.5436	51.2821	4.3276	36.9737	0.1170	0.0270	6.0024	17
18	0.2120	8.7556	54.8860	4.7171	41.3013	0.1142	0.0242	6.2687	18
19	0.1945	8.9501	58.3868	5.1417	46.0185	0.1117	0.0217	6.5236	19
20	0.1784	9.1285	61.7770	5.6044	51.1601	0.1095	0.0195	6.7674	20
21	0.1637	9.2922	65.0509	6.1088	56.7645	0.1076	0.0176	7.0006	21
22	0.1502	9.4424	68.2048	6.6586	62.8733	0.1059	0.0159	7.2232	22
23	0.1378	9.5802	71.2359	7.2579	69.5319	0.1044	0.0144	7.4357	23
24	0.1264	9.7066	74.1433	7.9111	76.7898	0.1030	0.0130	7.6384	24
25	0.1160	9.8226	76.9265	8.6231	84.7009	0.1018	0.0118	7.8316	25
26	0.1064	9.9290	79.5863	9.3992	93.3240	0.1007	0.0107	8.0156	26
27	0.0976	10.0266	82.1241	10.2451	102.7231	0.0997	0.0097	8.1906	27
28	0.0895	10.1161	84.5419	11.1671	112.9682	0.0989	0.0089	8.3571	28
29	0.0822	10.1983	86.8422	12.1722	124.1354	0.0981	0.0081	8.5154	29
30	0.0754	10.2737	89.0280	13.2677	136.3076	0.0973	0.0073	8.6657	30
31	0.0691	10.3428	91.1024	14.4618	149.5752	0.0967	0.0067	8.8083	31
32	0.0634	10.4062	93.0690	15.7633	164.0370	0.0961	0.0061	8.9436	32
33	0.0582	10.4644	94.9314	17.1820	179.8003	0.0956	0.0056	9.0718	33
34	0.0534	10.5178	96.6935	18.7284	196.9823	0.0951	0.0051	9.1933	34
35	0.0490	10.5668	98.3590	20.4140	215.7108	0.0946	0.0046	9.3083	35
36	0.0449	10.6118	99.9319	22.2512	236.1247	0.0942	0.0042	9.4171	36
37	0.0412	10.6530	101.4162	24.2538	258.3759	0.0939	0.0039	9.5200	37
38	0.0378	10.6908	102.8158	26.4367	282.6298	0.0935	0.0035	9.6172	38
39	0.0347	10.7255	104.1345	28.8160	309.0665	0.0932	0.0032	9.7090	39
40	0.0318	10.7574	105.3762	31.4094	337.8824	0.0930	0.0030	9.7957	40
41	0.0292	10.7866	106.5445	34.2363	369.2919	0.0927	0.0027	9.8775	41
42	0.0268	10.8134	107.6432	37.3175	403.5281	0.0925	0.0025	9.9546	42
43	0.0246	10.8380	108.6758	40.6761	440.8457	0.0923	0.0023	10.0273	43
44	0.0226	10.8605	109.6456	44.3370	481.5218	0.0921	0.0021	10.0958	44
45	0.0207	10.8812	110.5561	48.3273	525.8587	0.0919	0.0019	10.1603	45
46	0.0190	10.9002	111.4103	52.6767	574.1860	0.0917	0.0017	10.2210	46
47	0.0174	10.9176	112.2115	57.4176	626.8628	0.0916	0.0016	10.2780	47
48	0.0160	10.9336	112.9625	62.5852	684.2804	0.0915	0.0015	10.3317	48
49	0.0147	10.9482	113.6661	68.2179	746.8656	0.0913	0.0013	10.3821	49
50	0.0134	10.9617	114.3251	74.3575	815.0836	0.0912	0.0012	10.4295	50
51	0.0123	10.9740	114.9420	81.0497	889.4411	0.0911	0.0011	10.4740	51
52	0.0113	10.9853	115.5193	88.3442	970.4908	0.0910	0.0010	10.5158	52
53	0.0104	10.9957	116.0593	96.2951	1058.8349	0.0909	0.0009	10.5549	53
54	0.0095	11.0053	116.5642	104.9617	1155.1301	0.0909	0.0009	10.5917	54
55	0.0087	11.0140	117.0362	114.4083	1260.0918	0.0908	0.0008	10.6261	55
60	0.0057	11.0480	118.9683	176.0313	1944.7921	0.0905	0.0005	10.7683	60
65	0.0037	11.0701	120.3344	270.8460	2998.2885	0.0903	0.0003	10.8702	65
70	0.0024	11.0844	121.2942	416.7301	4619.2232	0.0902	0.0002	10.9427	70
75	0.0016	11.0938	121.9646	641.1909	7113.2321	0.0901	0.0001	10.9940	75
80	0.0010	11.0998	122.4306	986.5517	10950.5741	0.0901	0.0001	11.0299	80
85	0.0007	11.1038	122.7533	1517.9320	16854.8003	0.0901	0.0001	11.0551	85
90	0.0004	11.1064	122.9758	2335.5266	25939.1842	0.0900	0.0000	11.0726	90
95	0.0003	11.1080	123.1287	3593.4971	39916.6350	0.0900	0.0000	11.0847	95
100	0.0002	11.1091	123.2335	5529.0408	61422.6755	0.0900	0.0000	11.0930	100

(*continued*)

APPENDIX 66.B (*continued*)
Factor Tables

$$I = 10.00\%$$

n	P/F	P/A	P/G	F/P	F/A	A/P	A/F	A/G	n
1	0.9091	0.9091	0.0000	1.1000	1.0000	1.1000	1.0000	0.0000	1
2	0.8264	1.7355	0.8264	1.2100	2.1000	0.5762	0.4762	0.4762	2
3	0.7513	2.4869	2.3291	1.3310	3.3100	0.4021	0.3021	0.9366	3
4	0.6830	3.1699	4.3781	1.4641	4.6410	0.3155	0.2155	1.3812	4
5	0.6209	3.7908	6.8618	1.6105	6.1051	0.2638	0.1638	1.8101	5
6	0.5645	4.3553	9.6842	1.7716	7.7156	0.2296	0.1296	2.2236	6
7	0.5132	4.8684	12.7631	1.9487	9.4872	0.2054	0.1054	2.6216	7
8	0.4665	5.3349	16.0287	2.1436	11.4359	0.1874	0.0874	3.0045	8
9	0.4241	5.7590	19.4215	2.3579	13.5795	0.1736	0.0736	3.3724	9
10	0.3855	6.1446	22.8913	2.5937	15.9374	0.1627	0.0627	3.7255	10
11	0.3505	6.4951	26.3963	2.8531	18.5312	0.1540	0.0540	4.0641	11
12	0.3186	6.8137	29.9012	3.1384	21.3843	0.1468	0.0468	4.3884	12
13	0.2897	7.1034	33.3772	3.4523	24.5227	0.1408	0.0408	4.6988	13
14	0.2633	7.3667	36.8005	3.7975	27.9750	0.1357	0.0357	4.9955	14
15	0.2394	7.6061	40.1520	4.1772	31.7725	0.1315	0.0315	5.2789	15
16	0.2176	7.8237	43.4164	4.5950	35.9497	0.1278	0.0278	5.5493	16
17	0.1978	8.0216	46.5819	5.0545	40.5447	0.1247	0.0247	5.8071	17
18	0.1799	8.2014	49.6395	5.5599	45.5992	0.1219	0.0219	6.0526	18
19	0.1635	8.3649	52.5827	6.1159	51.1591	0.1195	0.0195	6.2861	19
20	0.1486	8.5136	55.4069	6.7275	57.2750	0.1175	0.0175	6.5081	20
21	0.1351	8.6487	58.1095	7.4002	64.0025	0.1156	0.0156	6.7189	21
22	0.1228	8.7715	60.6893	8.1403	71.4027	0.1140	0.0140	6.9189	22
23	0.1117	8.8832	63.1462	8.9543	79.5430	0.1126	0.0126	7.1085	23
24	0.1015	8.9847	65.4813	9.8497	88.4973	0.1113	0.0113	7.2881	24
25	0.0923	9.0770	67.6964	10.8347	98.3471	0.1102	0.0102	7.4580	25
26	0.0839	9.1609	69.7940	11.9182	109.1818	0.1092	0.0092	7.6186	26
27	0.0763	9.2372	71.7773	13.1100	121.0999	0.1083	0.0083	7.7704	27
28	0.0693	9.3066	73.6495	14.4210	134.2099	0.1075	0.0075	7.9137	28
29	0.0630	9.3696	75.4146	15.8631	148.6309	0.1067	0.0067	8.0489	29
30	0.0573	9.4269	77.0766	17.4494	164.4940	0.1061	0.0061	8.1762	30
31	0.0521	9.4790	78.6395	19.1943	181.9434	0.1055	0.0055	8.2962	31
32	0.0474	9.5264	80.1078	21.1138	201.1378	0.1050	0.0050	8.4091	32
33	0.0431	9.5694	81.4856	23.2252	222.2515	0.1045	0.0045	8.5152	33
34	0.0391	9.6086	82.7773	25.5477	245.4767	0.1041	0.0041	8.6149	34
35	0.0356	9.6442	83.9872	28.1024	271.0244	0.1037	0.0037	8.7086	35
36	0.0323	9.6765	85.1194	30.9127	299.1268	0.1033	0.0033	8.7965	36
37	0.0294	9.7059	86.1781	34.0039	330.0395	0.1030	0.0030	8.8789	37
38	0.0267	9.7327	87.1673	37.4043	364.0434	0.1027	0.0027	8.9562	38
39	0.0243	9.7570	88.0908	41.1448	401.4478	0.0125	0.0025	9.0285	39
40	0.0221	9.7791	88.9525	45.2593	442.5926	0.1023	0.0023	9.0962	40
41	0.0201	9.7991	89.7560	49.7852	487.8518	0.1020	0.0020	9.1596	41
42	0.0183	9.8174	90.5047	54.7637	537.6370	0.1019	0.0019	9.2188	42
43	0.0166	9.8340	91.2019	60.2401	592.4007	0.1017	0.0017	9.2741	43
44	0.0151	9.8491	91.8508	66.2641	652.6408	0.1015	0.0015	9.3258	44
45	0.0137	9.8628	92.4544	72.8905	718.9048	0.1014	0.0014	9.3740	45
46	0.0125	9.8753	93.0157	80.1795	791.7953	0.1013	0.0013	9.4190	46
47	0.0113	9.8866	93.5372	88.1975	871.9749	0.1011	0.0011	9.4610	47
48	0.0103	9.8969	94.0217	97.0172	960.1723	0.1010	0.0010	9.5001	48
49	0.0094	9.9063	94.4715	106.7190	1057.1896	0.1009	0.0009	9.5365	49
50	0.0085	9.9148	94.8889	117.3909	1163.9085	0.1009	0.0009	9.5704	50
51	0.0077	9.9226	95.2761	129.1299	1281.2994	0.1008	0.0008	9.6020	51
52	0.0070	9.9296	95.6351	142.0429	1410.4293	0.1007	0.0007	9.6313	52
53	0.0064	9.9360	95.9679	156.2472	1552.4723	0.1006	0.0006	9.6586	53
54	0.0058	9.9418	96.2763	171.8719	1708.7195	0.1006	0.0006	9.6840	54
55	0.0053	9.9471	96.5619	189.0591	1880.5914	0.1005	0.0005	9.7075	55
60	0.0033	9.9672	97.7010	304.4816	3034.8164	0.1003	0.0003	9.8023	60
65	0.0020	9.9796	98.4705	490.3707	4893.7073	0.1002	0.0002	9.8672	65
70	0.0013	9.9873	98.9870	789.7470	7887.4696	0.1001	0.0001	9.9113	70
75	0.0008	9.9921	99.3317	1271.8954	12708.9537	0.1001	0.0001	9.9410	75
80	0.0005	9.9951	99.5606	2048.4002	20474.0021	0.1000	0.0000	9.9609	80
85	0.0003	9.9970	99.7120	3298.9690	32979.6903	0.1000	0.0000	9.9742	85
90	0.0002	9.9981	99.8118	5313.0226	53120.2261	0.1000	0.0000	9.9831	90
95	0.0001	9.9988	99.8773	8556.6760	85556.7605	0.1000	0.0000	9.9889	95
100	0.0001	9.9993	99.9202	13780.6123	137796.1234	0.1000	0.0000	9.9927	100

(*continued*)

Support
Material

APPENDIX 66.B (*continued*)
Factor Tables

$$I = 12.00\%$$

n	P/F	P/A	P/G	F/P	F/A	A/P	A/F	A/G	n
1	0.8929	0.8929	0.0000	1.1200	1.0000	1.1200	1.0000	0.0000	1
2	0.7972	1.6901	0.7972	1.2544	2.1200	0.5917	0.4717	0.4717	2
3	0.7118	2.4018	2.2208	1.4049	3.3744	0.4163	0.2963	0.9246	3
4	0.6355	3.0373	4.1273	1.5735	4.7793	0.3292	0.2092	1.3589	4
5	0.5674	3.6048	6.3970	1.7623	6.3528	0.2774	0.1574	1.7746	5
6	0.5066	4.1114	8.9302	1.9738	8.1152	0.2432	0.1232	2.1720	6
7	0.4523	4.5638	11.6443	2.2107	10.0890	0.2191	0.0991	2.5515	7
8	0.4039	4.9676	14.4714	2.4760	12.2997	0.2013	0.0813	2.9131	8
9	0.3606	5.3282	17.3563	2.7731	14.7757	0.1877	0.0677	3.2574	9
10	0.3220	5.6502	20.2541	3.1058	17.5487	0.1770	0.0570	3.5847	10
11	0.2875	5.9377	23.1288	3.4785	20.6546	0.1684	0.0484	3.8953	11
12	0.2567	6.1944	25.9523	3.8960	24.1331	0.1614	0.0414	4.1897	12
13	0.2292	6.4235	28.7024	4.3635	28.0291	0.1557	0.0357	4.4683	13
14	0.2046	6.6282	31.3624	4.8871	32.3926	0.1509	0.0309	4.7317	14
15	0.1827	6.8109	33.9202	5.4736	37.2797	0.1468	0.0268	4.9803	15
16	0.1631	6.9740	36.3670	6.1304	42.7533	0.1434	0.0234	5.2147	16
17	0.1456	7.1196	38.6973	6.8660	48.8837	0.1405	0.0205	5.4353	17
18	0.1300	7.2497	40.9080	7.6900	55.7497	0.1379	0.0179	5.6427	18
19	0.1161	7.3658	42.9979	8.6128	63.4397	0.1358	0.0158	6.8375	19
20	0.1037	7.4694	44.9676	9.6463	72.0524	0.1339	0.0139	6.0202	20
21	0.0926	7.5620	46.8188	10.8038	81.6987	0.1322	0.0122	6.1913	21
22	0.0826	7.6446	48.5543	12.1003	92.5026	0.1308	0.0108	6.3514	22
23	0.0738	7.7184	50.1776	13.5523	104.6029	0.1296	0.0096	6.5010	23
24	0.0659	7.7843	51.6929	15.1786	118.1552	0.1285	0.0085	6.6406	24
25	0.0588	7.8431	53.1046	17.0001	133.3339	0.1275	0.0075	6.7708	25
26	0.0525	7.8957	54.4177	19.0401	150.3339	0.1267	0.0067	6.8921	26
27	0.0469	7.9426	55.6369	21.3249	169.3740	0.1259	0.0059	7.0049	27
28	0.0419	7.9844	56.7674	23.8839	190.6989	0.1252	0.0052	7.1098	28
29	0.0374	8.0218	57.8141	26.7499	214.5828	0.1247	0.0047	7.2071	29
30	0.0334	8.0552	58.7821	29.9599	241.3327	0.1241	0.0041	7.2974	30
31	0.0298	8.0850	59.6761	33.5551	271.2926	0.1237	0.0037	7.3811	31
32	0.0266	8.1116	60.5010	37.5817	304.8477	0.1233	0.0033	7.4586	32
33	0.0238	8.1354	61.2612	42.0915	342.4294	0.1229	0.0029	7.5302	33
34	0.0212	8.1566	61.9612	47.1425	384.5210	0.1226	0.0026	7.5965	34
35	0.0189	8.1755	62.6052	52.7996	431.6635	0.1223	0.0023	7.6577	35
36	0.0169	8.1924	63.1970	59.1356	484.4631	0.1221	0.0021	7.7141	36
37	0.0151	8.2075	63.7406	66.2318	543.5987	0.1218	0.0018	7.7661	37
38	0.0135	8.2210	64.2394	74.1797	609.8305	0.1216	0.0016	7.8141	38
39	0.0120	8.2330	64.6967	83.0812	684.0102	0.1215	0.0015	7.8582	39
40	0.0107	8.2438	65.1159	93.0510	767.0914	0.1213	0.0013	7.8988	40
41	0.0096	8.2534	65.4997	104.2171	860.1424	0.1212	0.0012	7.9361	41
42	0.0086	8.2619	65.8509	116.7231	964.3595	0.1210	0.0010	7.9704	42
43	0.0076	8.2696	66.1722	130.7299	1081.0826	0.1209	0.0009	8.0019	43
44	0.0068	8.2764	66.4659	146.4175	1211.8125	0.1208	0.0008	8.0308	44
45	0.0061	8.2825	66.7342	163.9876	1358.2300	0.1207	0.0007	8.0572	45
46	0.0054	8.2880	66.9792	183.6661	1522.2176	0.1207	0.0007	8.0815	46
47	0.0049	8.2928	67.2028	205.7061	1705.8838	0.1206	0.0006	8.1037	47
48	0.0043	8.2972	67.4068	230.3908	1911.5898	0.1205	0.0005	8.1241	48
49	0.0039	8.3010	67.5929	258.0377	2141.9806	0.1205	0.0005	8.1427	49
50	0.0035	8.3045	67.7624	289.0022	2400.0182	0.1204	0.0004	8.1597	50
51	0.0031	8.3076	67.9169	323.6825	2689.0204	0.1204	0.0004	8.1753	51
52	0.0028	8.3103	68.0576	362.5243	3012.7029	0.1203	0.0003	8.1895	52
53	0.0025	8.3128	68.1856	406.0273	3375.2272	0.1203	0.0003	8.2025	53
54	0.0022	8.3150	68.3022	454.7505	3781.2545	0.1203	0.0003	8.2143	54
55	0.0020	8.3170	68.4082	509.3206	4236.0050	0.1202	0.0002	8.2251	55
60	0.0011	8.3240	68.8100	897.5969	7471.6411	0.1201	0.0001	8.2664	60
65	0.0006	8.3281	69.0581	1581.8725	13173.9374	0.1201	0.0001	8.2922	65
70	0.0004	8.3303	69.2103	2787.7998	23223.3319	0.1200	0.0000	8.3082	70
75	0.0002	8.3316	69.3031	4913.0558	40933.7987	0.1200	0.0000	8.3181	75
80	0.0001	8.3324	69.3594	8658.4831	72145.6925	0.1200	0.0000	8.3241	80
85	0.0001	8.3328	69.3935	15259.2057	127151.7140	0.1200	0.0000	8.3278	85
90	0.0000	8.3330	69.4140	26891.9342	224091.1185	0.1200	0.0000	8.3300	90
95	0.0000	8.3332	69.4263	47392.7766	394931.4719	0.1200	0.0000	8.3313	95
100	0.0000	8.3332	69.4336	83522.2657	696010.5477	0.1200	0.0000	8.3321	100

(*continued*)

APPENDIX 66.B (*continued*)
Factor Tables

$I = 15.00\%$

n	P/F	P/A	P/G	F/P	F/A	A/P	A/F	A/G	n
1	0.8696	0.8696	0.0000	1.1500	1.0000	1.1500	1.0000	0.0000	1
2	0.7561	1.6257	0.7561	1.3225	2.1500	0.6151	0.4651	0.4651	2
3	0.6575	2.2832	2.0712	1.5209	3.4725	0.4380	0.2880	0.9071	3
4	0.5718	2.8550	3.7864	1.7490	4.9934	0.3503	0.2003	1.3263	4
5	0.4972	3.3522	5.7751	2.0114	6.7424	0.2983	0.1483	1.7228	5
6	0.4323	3.7845	7.9368	2.3131	8.7537	0.2642	0.1142	2.0972	6
7	0.3759	4.1604	10.1924	2.6600	11.0668	0.2404	0.0904	2.4498	7
8	0.3269	4.4873	12.4807	3.0590	13.7268	0.2229	0.0729	2.7813	8
9	0.2843	4.7716	14.7548	3.5179	16.7858	0.2096	0.0596	3.0922	9
10	0.2472	5.0188	16.9795	4.0456	20.3037	0.1993	0.0493	3.3832	10
11	0.2149	5.2337	19.1289	4.6524	24.3493	0.1911	0.0411	3.6549	11
12	0.1869	5.4206	21.1849	5.3503	29.0017	0.1845	0.0345	3.9082	12
13	0.1625	5.5831	23.1352	6.1528	34.3519	0.1791	0.0291	4.1438	13
14	0.1413	5.7245	24.9725	7.0757	40.5047	0.1747	0.0247	4.3624	14
15	0.1229	5.8474	26.9630	8.1371	47.5804	0.1710	0.0210	4.5650	15
16	0.1069	5.9542	28.2960	9.3576	55.7175	0.1679	0.0179	4.7522	16
17	0.0929	6.0472	29.7828	10.7613	65.0751	0.1654	0.0154	4.9251	17
18	0.0808	6.1280	31.1565	12.3755	75.8364	0.1632	0.0132	5.0843	18
19	0.0703	6.1982	32.4213	14.2318	88.2118	0.1613	0.0113	5.2307	19
20	0.0611	6.2593	33.5822	16.3665	102.4436	0.1598	0.0098	5.3651	20
21	0.0531	6.3125	34.6448	18.8215	118.8101	0.1584	0.0084	5.4883	21
22	0.0462	6.3587	35.6150	21.6447	137.6316	0.1573	0.0073	5.6010	22
23	0.0402	6.3988	36.4988	24.8915	159.2764	0.1563	0.0063	5.7040	23
24	0.0349	6.4338	37.3023	28.6252	184.1678	0.1554	0.0054	5.7979	24
25	0.0304	6.4641	38.0314	32.9190	212.7930	0.1547	0.0047	5.8834	25
26	0.0264	6.4906	38.6918	37.8568	245.7120	0.1541	0.0041	5.9612	26
27	0.0230	6.5135	39.2890	43.5353	283.5688	0.1535	0.0035	6.0319	27
28	0.0200	6.5335	39.8283	50.0656	327.1041	0.1531	0.0031	6.0960	28
29	0.0174	6.5509	40.3146	57.5755	377.1697	0.1527	0.0027	6.1541	29
30	0.0151	6.5660	40.7526	66.2118	434.7451	0.1523	0.0023	6.2066	30
31	0.0131	6.5791	41.1466	76.1435	500.9569	0.1520	0.0020	6.2541	31
32	0.0114	6.5905	41.5006	87.5651	577.1005	0.1517	0.0017	6.2970	32
33	0.0099	6.6005	41.8184	100.6998	664.6655	0.1515	0.0015	6.3357	33
34	0.0086	6.6091	42.1033	115.8048	765.3654	0.1513	0.0013	6.3705	34
35	0.0075	6.6166	42.3586	133.1755	881.1702	0.1511	0.0011	6.4019	35
36	0.0065	6.6231	42.5872	153.1519	1014.3457	0.1510	0.0010	6.4301	36
37	0.0057	6.6288	42.7916	176.1246	1167.4975	0.1509	0.0009	6.4554	37
38	0.0049	6.6338	42.9743	202.5433	1343.6222	0.1507	0.0007	6.4781	38
39	0.0043	6.6380	43.1374	232.9248	1546.1655	0.1506	0.0006	6.4985	39
40	0.0037	6.6418	43.2830	267.8635	1779.0903	0.1506	0.0006	6.5168	40
41	0.0032	6.6450	43.4128	308.0431	2046.9539	0.1505	0.0005	6.5331	41
42	0.0028	6.6478	43.5286	354.2495	2354.9969	0.1504	0.0004	6.5478	42
43	0.0025	6.6503	43.6317	407.3870	2709.2465	0.1504	0.0004	6.5609	43
44	0.0021	6.6524	43.7235	468.4950	3116.6334	0.1503	0.0003	6.5725	44
45	0.0019	6.6543	43.8051	538.7693	3585.1285	0.1503	0.0003	6.5830	45
46	0.0016	6.6559	43.8778	619.5847	4123.8977	0.1502	0.0002	6.5923	46
47	0.0014	6.6573	43.9423	712.5224	4743.4824	0.1502	0.0002	6.6006	47
48	0.0012	6.6585	43.9997	819.4007	5456.0047	0.1502	0.0002	6.6080	48
49	0.0011	6.6596	44.0506	942.3108	6275.4055	0.1502	0.0002	6.6146	49
50	0.0009	6.6605	44.0958	1083.6574	7217.7163	0.1501	0.0001	6.6205	50
51	0.0008	6.6613	44.1360	1246.2061	8301.3737	0.1501	0.0001	6.6257	51
52	0.0007	6.6620	44.1715	1433.1370	9547.5798	0.1501	0.0001	6.6304	52
53	0.0006	6.6626	44.2031	1648.1075	10980.7167	0.1501	0.0001	6.6345	53
54	0.0005	6.6631	44.2311	1895.3236	12628.8243	0.1501	0.0001	6.6382	54
55	0.0005	6.6636	44.2558	2179.6222	14524.1479	0.1501	0.0001	6.6414	55
60	0.0002	6.6651	44.3431	4383.9987	29219.9916	0.1500	0.0000	6.6530	60
65	0.0001	6.6659	44.3903	8817.7874	58778.5826	0.1500	0.0000	6.6593	65
70	0.0001	6.6663	44.4156	17735.7200	118231.4669	0.1500	0.0000	6.6627	70
75	0.0000	6.6665	44.4292	35672.8680	237812.4532	0.1500	0.0000	6.6646	75
80	0.0000	6.6666	44.4364	71750.8794	478332.5293	0.1500	0.0000	6.6656	80
85	0.0000	6.6666	44.4402	144316.6470	962104.3133	0.1500	0.0000	6.6661	85
90	0.0000	6.6666	44.4422	290272.3252	1935142.1680	0.1500	0.0000	6.6664	90
95	0.0000	6.6667	44.4433	583841.3276	3892268.8509	0.1500	0.0000	6.6665	95
100	0.0000	6.6667	44.4438	1174313.4507	7828749.6713	0.1500	0.0000	6.6666	100

(*continued*)

Support Material

APPENDIX 66.B (*continued*)
Factor Tables

$I = 20.00\%$

n	P/F	P/A	P/G	F/P	F/A	A/P	A/F	A/G	n
1	0.8333	0.8333	0.0000	1.2000	1.0000	1.2000	1.0000	0.0000	1
2	0.6944	1.5278	0.6944	1.4400	2.2000	0.6545	0.4545	0.4545	2
3	0.5787	2.1065	1.8519	1.7280	3.6400	0.4747	0.2747	0.8791	3
4	0.4823	2.5887	3.2986	2.0736	5.3680	0.3863	0.1863	1.2742	4
5	0.4019	2.9906	4.9061	2.4883	7.4416	0.3344	0.1344	1.6405	5
6	0.3349	3.3255	6.5806	2.9860	9.9299	0.3007	0.1007	1.9788	6
7	0.2791	3.6046	8.2551	3.5832	12.9159	0.2774	0.0774	2.2902	7
8	0.2326	3.8372	9.8831	4.2998	16.4991	0.2606	0.0606	2.5756	8
9	0.1938	4.0310	11.4335	5.1598	20.7989	0.2481	0.0481	2.8364	9
10	0.1615	4.1925	12.8871	6.1917	25.9587	0.2385	0.0385	3.0739	10
11	0.1346	4.3271	14.2330	7.4301	32.1504	0.2311	0.0311	3.2893	11
12	0.1122	4.4392	15.4667	8.9161	39.5805	0.2253	0.0253	3.4841	12
13	0.0935	4.5327	16.5883	10.6993	48.4966	0.2206	0.0206	3.6597	13
14	0.0779	4.6106	17.6008	12.8392	59.1959	0.2169	0.0169	3.8175	14
15	0.0649	4.6755	18.5095	15.4070	72.0351	0.2139	0.0139	3.9588	15
16	0.0541	4.7296	19.3208	18.4884	87.4421	0.2114	0.0114	4.0851	16
17	0.0451	4.7746	20.0419	22.1861	105.9306	0.2094	0.0094	4.1976	17
18	0.0376	4.8122	20.6805	26.6233	128.1167	0.2078	0.0078	4.2975	18
19	0.0313	4.8435	21.2439	31.9480	154.7400	0.2065	0.0065	4.3861	19
20	0.0261	4.8696	21.7395	38.3376	186.6880	0.2054	0.0054	4.4643	20
21	0.0217	4.8913	22.1742	46.0051	225.0256	0.2044	0.0044	4.5334	21
22	0.0181	4.9094	22.5546	55.2061	271.0307	0.2037	0.0037	4.5941	22
23	0.0151	4.9245	22.8867	66.2474	326.2369	0.2031	0.0031	4.6475	23
24	0.0126	4.9371	23.1760	79.4968	392.4842	0.2025	0.0025	4.6943	24
25	0.0105	4.9476	23.4276	95.3962	471.9811	0.2021	0.0021	4.7352	25
26	0.0087	4.9563	23.6460	114.4755	567.3773	0.2018	0.0018	4.7709	26
27	0.0073	4.9636	23.8353	137.3706	681.8528	0.2015	0.0015	4.8020	27
28	0.0061	4.9697	23.9991	164.8447	819.2233	0.2012	0.0012	4.8291	28
29	0.0051	4.9747	24.1406	197.8136	984.0680	0.2010	0.0010	4.8527	29
30	0.0042	4.9789	24.2628	237.3763	1181.8816	0.2008	0.0008	4.8731	30
31	0.0035	4.9824	24.3681	284.8516	1419.2579	0.2007	0.0007	4.8908	31
32	0.0029	4.9854	24.4588	341.8219	1704.1095	0.2006	0.0006	4.9061	32
33	0.0024	4.9878	24.5368	410.1863	2045.9314	0.2005	0.0005	4.9194	33
34	0.0020	4.9898	24.6038	492.2235	2456.1176	0.2004	0.0004	4.9308	34
35	0.0017	4.9915	24.6614	590.6682	2948.3411	0.2003	0.0003	4.9406	35
36	0.0014	4.9929	24.7108	708.8019	3539.0094	0.2003	0.0003	4.9491	36
37	0.0012	4.9941	24.7531	850.5622	4247.8112	0.2002	0.0002	4.9564	37
38	0.0010	4.9951	24.7894	1020.6747	5098.3735	0.2002	0.0002	4.9627	38
39	0.0008	4.9959	24.8204	1224.8096	6119.0482	0.2002	0.0002	4.9681	39
40	0.0007	4.9966	24.8469	1469.7716	7343.8578	0.2001	0.0001	4.9728	40
41	0.0006	4.9972	24.8696	1763.7259	8813.6294	0.2001	0.0001	4.9767	41
42	0.0005	4.9976	24.8890	2116.4711	10577.3553	0.2001	0.0001	4.9801	42
43	0.0004	4.9980	24.9055	2539.7653	12693.8263	0.2001	0.0001	4.9831	43
44	0.0003	4.9984	24.9196	3047.7183	15233.5916	0.2001	0.0001	4.9856	44
45	0.0003	4.9986	24.9316	3657.2620	18281.3099	0.2001	0.0001	4.9877	45
46	0.0002	4.9989	24.9419	4388.7144	21938.5719	0.2000	0.0000	4.9895	46
47	0.0002	4.9991	24.9506	5266.4573	26327.2863	0.2000	0.0000	4.9911	47
48	0.0002	4.9992	24.9581	6319.7487	31593.7436	0.2000	0.0000	4.9924	48
49	0.0001	4.9993	24.9644	7583.6985	37913.4923	0.2000	0.0000	4.9935	49
50	0.0001	4.9995	24.9698	9100.4382	45497.1908	0.2000	0.0000	4.9945	50
51	0.0001	4.9995	24.9744	10920.5258	54597.6289	0.2000	0.0000	4.9953	51
52	0.0001	4.9996	24.9783	13104.6309	65518.1547	0.2000	0.0000	4.9960	52
53	0.0001	4.9997	24.9816	15725.5571	78622.7856	0.2000	0.0000	4.9966	53
54	0.0001	4.9997	24.9844	18870.6685	94348.3427	0.2000	0.0000	4.9971	54
55	0.0000	4.9998	24.9868	22644.8023	113219.0113	0.2000	0.0000	4.9976	55
60	0.0000	4.9999	24.9942	56347.5144	281732.5718	0.2000	0.0000	4.9989	60
65	0.0000	5.0000	24.9975	140210.6469	701048.2346	0.2000	0.0000	4.9995	65
70	0.0000	5.0000	24.9989	348888.9569	1744439.7847	0.2000	0.0000	4.9998	70
75	0.0000	5.0000	24.9995	868147.3693	4340731.8466	0.2000	0.0000	4.9999	75

(*continued*)

APPENDIX 66.B (*continued*)
Factor Tables

$I = 25.00\%$

n	P/F	P/A	P/G	F/P	F/A	A/P	A/F	A/G	n
1	0.8000	0.8000	0.0000	1.2500	1.0000	1.2500	1.0000	0.0000	1
2	0.6400	1.4400	0.6400	1.5625	2.2500	0.6944	0.4444	0.4444	2
3	0.5120	1.9520	1.6640	1.9531	3.8125	0.5123	0.2623	0.8525	3
4	0.4096	2.3616	2.8928	2.4414	5.7656	0.4234	0.1734	1.2249	4
5	0.3277	2.6893	4.2035	3.0518	8.2070	0.3718	0.1218	1.5631	5
6	0.2621	2.9514	5.5142	3.8147	11.2588	0.3383	0.0888	1.8683	6
7	0.2097	3.1611	6.7725	4.7684	15.0735	0.3163	0.0663	2.1424	7
8	0.1678	3.3289	7.9469	5.9605	19.8419	0.3004	0.0504	2.3872	8
9	0.1342	3.4631	9.0207	7.4506	25.8023	0.2888	0.0388	2.6048	9
10	0.1074	3.5705	9.9870	9.3132	33.2529	0.2801	0.0301	2.7971	10
11	0.0859	3.6564	10.8460	11.6415	42.5661	0.2735	0.0235	2.9663	11
12	0.0687	3.7251	11.6020	14.5519	54.2077	0.2684	0.0184	3.1145	12
13	0.0550	3.7801	12.2617	18.1899	68.7596	0.2645	0.0145	3.2437	13
14	0.0440	3.8241	12.8334	22.7374	86.9495	0.2615	0.0115	3.3559	14
15	0.0352	3.8593	13.3260	28.4217	109.6868	0.2591	0.0091	3.4530	15
16	0.0281	3.8874	13.7482	35.5271	138.1085	0.2572	0.0072	3.5366	16
17	0.0225	3.9099	14.1085	44.4089	173.6357	0.2558	0.0058	3.6084	17
18	0.0180	3.9279	14.4147	55.5112	218.0446	0.2546	0.0046	3.6698	18
19	0.0144	3.9424	14.6741	69.3889	273.5558	0.2537	0.0037	3.7222	19
20	0.0115	3.9539	14.8932	86.7362	342.9447	0.2529	0.0029	3.7667	20
21	0.0092	3.9631	15.0777	108.4202	429.6809	0.2523	0.0023	3.8045	21
22	0.0074	3.9705	15.2326	135.5253	538.1011	0.2519	0.0019	3.8365	22
23	0.0059	3.9764	15.3625	169.4066	673.6264	0.2515	0.0015	3.8634	23
24	0.0047	3.9811	15.4711	211.7582	843.0329	0.2512	0.0012	3.8861	24
25	0.0038	3.9849	15.5618	264.6978	1054.7912	0.2509	0.0009	3.9052	25
26	0.0030	3.9879	15.6373	330.8722	1319.4890	0.2508	0.0008	3.9212	26
27	0.0024	3.9903	15.7002	413.5903	1650.3612	0.2506	0.0006	3.9346	27
28	0.0019	3.9923	15.7524	516.9879	2063.9515	0.2505	0.0005	3.9457	28
29	0.0015	3.9938	15.7957	646.2349	2580.9394	0.2504	0.0004	3.9551	29
30	0.0012	3.9950	15.8316	807.7936	3227.1743	0.2503	0.0003	3.9628	30
31	0.0010	3.9960	15.8614	1009.7420	4034.9678	0.2502	0.0002	3.9693	31
32	0.0008	3.9968	15.8859	1262.1774	5044.7098	0.2502	0.0002	3.9746	32
33	0.0006	3.9975	15.9062	1577.7218	6306.8872	0.2502	0.0002	3.9791	33
34	0.0005	3.9980	15.9229	1972.1523	7884.6091	0.2501	0.0001	3.9828	34
35	0.0004	3.9984	15.9367	2465.1903	9856.7613	0.2501	0.0001	3.9858	35
36	0.0003	3.9987	15.9481	3081.4879	12321.9516	0.2501	0.0001	3.9883	36
37	0.0003	3.9990	15.9574	3851.8599	15403.4396	0.2501	0.0001	3.9904	37
38	0.0002	3.9992	15.9651	4814.8249	19255.2994	0.2501	0.0001	3.9921	38
39	0.0002	3.9993	15.9714	6018.5311	24070.1243	0.2500	0.0000	3.9935	39
40	0.0001	3.9995	15.9766	7523.1638	30088.6554	0.2500	0.0000	3.9947	40
41	0.0001	3.9996	15.9809	9403.9548	37611.8192	0.2500	0.0000	3.9956	41
42	0.0001	3.9997	15.9843	11754.9435	47015.7740	0.2500	0.0000	3.9964	42
43	0.0001	3.9997	15.9872	14693.6794	58770.7175	0.2500	0.0000	3.9971	43
44	0.0001	3.9998	15.9895	18367.0992	73464.3969	0.2500	0.0000	3.9976	44
45	0.0000	3.9998	15.9915	22958.8740	91831.4962	0.2500	0.0000	3.9980	45
46	0.0000	3.9999	15.9930	28698.5925	114790.3702	0.2500	0.0000	3.9984	46
47	0.0000	3.9999	15.9943	35873.2407	143488.9627	0.2500	0.0000	3.9987	47
48	0.0000	3.9999	15.9954	44841.5509	179362.2034	0.2500	0.0000	3.9989	48
49	0.0000	3.9999	15.9962	56051.9386	224203.7543	0.2500	0.0000	3.9991	49
50	0.0000	3.9999	15.9969	70064.9232	280255.6929	0.2500	0.0000	3.9993	50
51	0.0000	4.0000	15.9975	87581.1540	350320.6161	0.2500	0.0000	3.9994	51
52	0.0000	4.0000	15.9980	109476.4425	437901.7701	0.2500	0.0000	3.9995	52
53	0.0000	4.0000	15.9983	136845.5532	547378.2126	0.2500	0.0000	3.9996	53
54	0.0000	4.0000	15.9986	171056.9414	684223.7658	0.2500	0.0000	3.9997	54
55	0.0000	4.0000	15.9989	213821.1768	855280.7072	0.2500	0.0000	3.9997	55
60	0.0000	4.0000	15.9996	652530.4468	2610117.7872	0.2500	0.0000	3.9999	60

(*continued*)

APPENDIX 66.B (*continued*)
Factor Tables

$$I = 30.00\%$$

n	P/F	P/A	P/G	F/P	F/A	A/P	A/F	A/G	n
1	0.7692	0.7692	0.0000	1.3000	1.0000	1.3000	1.0000	0.000	1
2	0.5917	1.3609	0.5917	1.6900	2.3000	0.7348	0.4348	0.434	2
3	0.4552	1.8161	1.5020	2.1970	3.9900	0.5506	0.2506	0.827	3
4	0.3501	2.1662	2.5524	2.8561	6.1870	0.4616	0.1616	1.178	4
5	0.2693	2.4356	3.6297	3.7129	9.0431	0.4106	0.1106	1.490	5
6	0.2072	2.6427	4.6656	4.8268	12.7560	0.3784	0.0784	1.765	6
7	0.1594	2.8021	5.6218	6.2749	17.5828	0.3569	0.0569	2.006	7
8	0.1226	2.9247	6.4800	8.1573	23.8577	0.3419	0.0419	2.215	8
9	0.0943	3.0190	7.2343	10.6045	32.0150	0.3312	0.0312	2.396	9
10	0.0725	3.0915	7.8872	13.7858	42.6195	0.3235	0.0235	2.551	10
11	0.0558	3.1473	8.4452	17.9216	56.4053	0.3177	0.0177	2.683	11
12	0.0429	3.1903	8.9173	23.2981	74.3270	0.3135	0.0135	2.795	12
13	0.0330	3.2233	9.3135	30.2875	97.6250	0.3102	0.0102	2.889	13
14	0.0254	3.2487	9.6437	39.3738	127.9125	0.3078	0.0078	2.968	14
15	0.0195	3.2682	9.9172	51.1859	167.2863	0.3060	0.0060	3.034	15
16	0.0150	3.2832	10.1426	66.5417	218.4722	0.3046	0.0046	3.089	16
17	0.0116	3.2948	10.3276	86.5042	285.0139	0.3035	0.0035	3.134	17
18	0.0089	3.3037	10.4788	112.4554	371.5180	0.3027	0.0027	3.171	18
19	0.0068	3.3105	10.6019	146.1920	483.9734	0.3021	0.0021	3.202	19
20	0.0053	3.3158	10.7019	190.0496	630.1655	0.3016	0.0016	3.227	20
21	0.0040	3.3198	10.7828	247.0645	820.2151	0.3012	0.0012	3.248	21
22	0.0031	3.3230	10.8482	321.1839	1067.2796	0.3009	0.0009	3.264	22
23	0.0024	3.3254	10.9009	417.5391	1388.4635	0.3007	0.0007	3.278	23
24	0.0018	3.3272	10.9433	542.8008	1806.0026	0.3006	0.0006	3.289	24
25	0.0014	3.3286	10.9773	705.6410	2348.8033	0.3004	0.0004	3.297	25
26	0.0011	3.3297	11.0045	917.3333	3054.4443	0.3003	0.0003	3.305	26
27	0.0008	3.3305	11.0263	1192.5333	3971.7776	0.3003	0.0003	3.310	27
28	0.0006	3.3312	11.0437	1550.2933	5164.3109	0.3002	0.0002	3.315	28
29	0.0005	3.3317	11.0576	2015.3813	6714.6042	0.3001	0.0001	3.318	29
30	0.0004	3.3321	11.0687	2619.9956	8729.9855	0.3001	0.0001	3.321	30
31	0.0003	3.3324	11.0775	3405.9943	11349.9811	0.3001	0.0001	3.324	31
32	0.0002	3.3326	11.0845	4427.7926	14755.9755	0.3001	0.0001	3.326	32
33	0.0002	3.3328	11.0901	5756.1304	19183.7681	0.3001	0.0001	3.327	33
34	0.0001	3.3329	11.0945	7482.9696	24939.8985	0.3000	0.0000	3.328	34
35	0.0001	3.3330	11.0980	9727.8604	32422.8681	0.3000	0.0000	3.329	35
36	0.0001	3.3331	11.1007	12646.2186	42150.7285	0.3000	0.0000	3.330	36
37	0.0001	3.3331	11.1029	16440.0841	54796.9471	0.3000	0.0000	3.331	37
38	0.0000	3.3332	11.1047	21372.1094	71237.0312	0.3000	0.0000	3.331	38
39	0.0000	3.3332	11.1060	27783.7422	92609.1405	0.3000	0.0000	3.331	39
40	0.0000	3.3332	11.1071	36118.8648	120392.8827	0.3000	0.0000	3.332	40
41	0.0000	3.3333	11.1080	46954.5243	156511.7475	0.3000	0.0000	3.332	41
42	0.0000	3.3333	11.1086	61040.8815	203466.2718	0.3000	0.0000	3.332	42
43	0.0000	3.3333	11.1092	79353.1460	264507.1533	0.3000	0.0000	3.332	43
44	0.0000	3.3333	11.1096	103159.0898	343860.2993	0.3000	0.0000	3.332	44
45	0.0000	3.3333	11.1099	134106.8167	447019.3890	0.3000	0.0000	3.333	45
46	0.0000	3.3333	11.1102	174338.8617	581126.2058	0.3000	0.0000	3.333	46
47	0.0000	3.3333	11.1104	226640.5202	755465.0675	0.3000	0.0000	3.333	47
48	0.0000	3.3333	11.1105	294632.6763	982105.5877	0.3000	0.0000	3.333	48
49	0.0000	3.3333	11.1107	383022.4792	1276738.2640	0.3000	0.0000	3.333	49
50	0.0000	3.3333	11.1108	497929.2230	1659760.7433	0.3000	0.0000	3.333	50

(*continued*)

APPENDIX 66.B (*continued*)
Factor Tables

$$I = 40.00\%$$

n	P/F	P/A	P/G	F/P	F/A	A/P	A/F	A/G	n
1	0.7143	0.7143	0.0000	1.4000	1.0000	1.4000	1.0000	0.000	1
2	0.5102	1.2245	0.5102	1.9600	2.4000	0.8167	0.4167	0.416	2
3	0.3644	1.5889	1.2391	2.7440	4.3600	0.6294	0.2294	0.779	3
4	0.2603	1.8492	2.0200	3.8416	7.1040	0.5408	0.1408	1.092	4
5	0.1859	2.0352	2.7637	5.3782	10.9456	0.4914	0.0914	1.358	5
6	0.1328	2.1680	3.4278	7.5295	16.3238	0.4613	0.0613	1.581	6
7	0.0949	2.2628	3.9970	10.5414	23.8534	0.4419	0.0419	1.766	7
8	0.0678	2.3306	4.4713	14.7579	34.3947	0.4291	0.0291	1.918	8
9	0.0484	2.3790	4.8585	20.6610	49.1526	0.4203	0.0203	2.042	9
10	0.0346	2.4136	5.1696	28.9255	69.8137	0.4143	0.0143	2.141	10
11	0.0247	2.4383	5.4166	40.4957	98.7391	0.4101	0.0101	2.221	11
12	0.0176	2.4559	5.6106	56.6939	139.2348	0.4072	0.0072	2.284	12
13	0.0126	2.4685	5.7618	79.3715	195.9287	0.4051	0.0051	2.334	13
14	0.0090	2.4775	5.8788	111.1201	275.3002	0.4036	0.0036	2.372	14
15	0.0064	2.4839	5.9688	155.5681	386.4202	0.4026	0.0026	2.403	15
16	0.0046	2.4885	6.0376	217.7953	541.9883	0.4018	0.0018	2.426	16
17	0.0033	2.4918	6.0901	304.9135	759.7837	0.4013	0.0013	2.444	17
18	0.0023	2.4941	6.1299	426.8789	1064.6971	0.4009	0.0009	2.457	18
19	0.0017	2.4958	6.1601	597.6304	1491.5760	0.4007	0.0007	2.468	19
20	0.0012	2.4970	6.1828	836.6826	2089.2064	0.4005	0.0005	2.476	20
21	0.0009	2.4979	6.1998	1171.3556	2925.8889	0.4003	0.0003	2.482	21
22	0.0006	2.4985	6.2127	1639.8978	4097.2445	0.4002	0.0002	2.486	22
23	0.0004	2.4989	6.2222	2295.8569	5737.1423	0.4002	0.0002	2.490	23
24	0.0003	2.4992	6.2294	3214.1997	8032.9993	0.4001	0.0001	2.492	24
25	0.0002	2.4994	6.2347	4499.8796	11247.1990	0.4001	0.0001	2.494	25
26	0.0002	2.4996	6.2387	6299.8314	15747.0785	0.4001	0.0001	2.495	26
27	0.0001	2.4997	6.2416	8819.7640	22046.9099	0.4000	0.0000	2.496	27
28	0.0001	2.4998	6.2438	12347.6696	30866.6739	0.4000	0.0000	2.497	28
29	0.0001	2.4999	6.2454	17286.7374	43214.3435	0.4000	0.0000	2.498	29
30	0.0000	2.4999	6.2466	24201.4324	60501.0809	0.4000	0.0000	2.498	30
31	0.0000	2.4999	6.2475	33882.0053	84702.5132	0.4000	0.0000	2.499	31
32	0.0000	2.4999	6.2482	47434.8074	118584.5185	0.4000	0.0000	2.499	32
33	0.0000	2.5000	6.2487	66408.7304	166019.3260	0.4000	0.0000	2.499	33
34	0.0000	2.5000	6.2490	92972.2225	232428.0563	0.4000	0.0000	2.499	34
35	0.0000	2.5000	6.2493	130161.1116	325400.2789	0.4000	0.0000	2.499	35
36	0.0000	2.5000	6.2495	182225.5562	455561.3904	0.4000	0.0000	2.499	36
37	0.0000	2.5000	6.2496	255115.7786	637786.9466	0.4000	0.0000	2.499	37
38	0.0000	2.5000	6.2497	357162.0901	892902.7252	0.4000	0.0000	2.499	38
39	0.0000	2.5000	6.2498	500026.9261	1250064.8153	0.4000	0.0000	2.499	39
40	0.0000	2.5000	6.2498	700037.6966	1750091.7415	0.4000	0.0000	2.499	40
41	0.0000	2.5000	6.2499	980052.7752	2450129.4381	0.4000	0.0000	2.500	41
42	0.0000	2.5000	6.2499	1372073.8853	3430182.2133	0.4000	0.0000	2.500	42
43	0.0000	2.5000	6.2499	1920903.4394	4802256.0986	0.4000	0.0000	2.500	43
44	0.0000	2.5000	6.2500	2689264.8152	6723159.5381	0.4000	0.0000	2.500	44
45	0.0000	2.5000	6.2500	3764970.7413	9412424.3533	0.4000	0.0000	2.500	45

Glossary

A

Abcoulomb: The unit of electric charge in the electromagnetic cgs system equal to 10 coulombs.

Able: The hexadecimal digit the decimal equivalent of which is 10.

Abmho: The unit of conductance in the electromagnetic cgs system equal to 10^9 S ($\mho$).

Abohm: The unit of conductance in the electromagnetic cgs system equal to 10^{-9} Ω.

Absolute Pressure: The pressure relative to a perfect vacuum.

Absolute-Value Computer: A computer that processes the value of varibles rather than their increments.

Absolute Zero: The temperature at which a body is thought to have zero molecular motion and thus zero heat energy (0 K, $-273.16°$C, or $-459.69°$F).

Absorption Circuit: A series-resonance circuit that absorbs the power carried by an unwanted signal frequency.

Abvolt: The unit of force in the electromagnetic cgs system equal to 10^{-8} V.

Accentuator: An electric circuit that enhances the audio signal in a specific frequency band with respect to other signals.

Acoustical Scintillation: The variation in the intensity of received sound sent at uniform output due to the nonhomogeneous nature of the atmosphere.

Acoustic Amplifier: An amplifer that amplifies mechanical vibrations directly to audio or ultrasonic frequencies.

Acoustic Phonon: A quantum of excitation due to acoustic vibration.

Acquisition Tone: The audible tone used to verify entry into a computer.

Active Element: Any component within an impedance network that generates current or voltage.

Active Material: An energy-storing material.

Active Region: The area in a semiconductor in which the amplifying, rectifying, or other dynamic action occurs.

Adapter Transformer: A transformer designed to supply a single component.

Ad Hoc Inquiry: A request for a single piece of information from a computer system.

Adiabatic: No gain or loss of heat.

Adjacency: A computer science term of character recognition in which characters are separated by less than a specified distance.

Aliasing: The jagged appearance of diagonal lines on computer graphic printouts.

Alkaline Cell: A primary cell using alkaline electrolyte, usually potassium hydroxide.

Alloy Junction: Also called a Fused Junction. A junction produced by alloying impurity metals to a semiconductor.

Alpha: The ratio of the change in collector current to the change in the emitter current of a transistor.

Ambiguity: The condition of a synchro system or servo system in which more than one null position exists.

Amdahl's Law: A law that limits the speed improvement obtained by distibuting a computer program over parallel processors based on the fraction of work that must be accomplished in serial mode.

American Standard Code for Information Interchange (ASCII): Standardized characater set used in the general exchange of information in computer systems.

Ampacity: Current-carrying capacity measured in amperes (A).

Ampere: A constant current that, when flowing in two parallel conductors of infinite length and negligible cross section one meter apart in a vacuum, results in a force between the conductors of 2×10^{-7} N/m of length.

Amplidyne: A rotating magnetic amplifier in which small changes in field power produce large changes in power output.

Anode: For a primary cell or battery, the terminal at which current enters. The terminal is positive with respect to the device and negative with respect to the external circuitry. In electronics, it is the collector of the electrons. In electrochemistry, it is the cell from which electrons are forced. In semiconductor diodes, it is the terminal toward which forward current flows from the external circuitry.

Analog: A condition in which electrical elements or systems operate with continuously measured quantities.

Application Development Language: A very high-level programming language that generates the coding of a conventional programming language.

Assembler: A program that converts symbolic instructions capable of execution by the electronic circuitry of a computer.

Asynchronous: A term indicating that a process begins upon the application of a signal indicating that the previous operation is complete, rather than on a signal from a master clock. *See* Synchronous.

Avalanche Breakdown: A cumulative process caused by collisions of electrons due to their acceleration in an electric field. *Compare* Zener Breakdown.

B

Back Bias: Voltage fed back to circuits before its originating point.

Back-End System: A computer that operates on another computer system's processed data.

Ballast: A circuit element that either limits current or provides a starting voltage.

Barkhausen Effect: Abrupt changes in magnetization of a magnetic material as the magnetizing force is varied.

Barrier Penetration: In quantum mechanics, the passage of a particle through a finite region in which its potential energy is greater than its total energy.

Barrier Voltage: The voltage at the junction of two dissimilar metals necessary to cause conduction.

Baseband: The band of frequencies occupied by all transmitting signals prior to the modulator in the transmitter or subsequent to the demodulator in the receiver. Also defined as the transmitter's band of frequencies in the absence of modulating signals.

Battle Short: A circuit for bypassing safety interlocks.

Baud: A unit of signaling speed equal to the number of code elements (pulses and spaces) per second. Or, twice the number of pulses per second. A single baud may contain numerous bits, though the term Baud Rate is often used interchangeably with Bit Rate. *Compare* with Bit.

Beat Frequency: The frequency signal that equals the difference of two frequencies combined in a nonlinear circuit.

Benchmark: In computers, a standard problem that is run to compare performance.

Beta: The ratio of the change in collector current to the change in base current of a transistor.

Beta Circuit: The feedback portion of an amplifier circuit.

Billion: The number 10^9 (10^{12} in British usage); SI prefix giga (G) is used for 10^9.

Bipolar Circuit: A circuit in which the zeros and ones are treated symmetrically, that is, in a bipolar manner, rather than by the presence or absence of a signal.

Bit: A dimensionless unit of information containing one of two equally likely values. *Compare* with Baud.

Blanking: The deliberate shut-off of a given process.

Blow: The writing of data into a programmable read-only memory (ROM) memory by melting fuse links corresponding to logical zero.

Bootstrap: The procedures that make a computer function through its own actions.

Broadband: In communications, a frequency band greater than 300 Hz.

Byte: Adjacent binary digits operated upon by a computer as a single unit.

C

Cable Delay: The time required for a bit of informatoin to travel through a cable. Generally, the time is about 5 ns/m.

Cadmium Cell: A standard voltage reference cell. At 20°C, the voltage is 1.0186 V.

Canonical Schema: In computer science, a model of the structure and relationships of the data.

Cathode: For a primary cell or battery, the terminal at which current leaves. The terminal is negative with respect to the device and positive with respect to the external circuitry. In electronics, it is the source of the electrons. In electrochemistry, it is the cell through which electrons are forced.

Coaxial Cable: A cable constructed of an inner conductor insulated from an outer metal covering that acts as a second conductor.

Communications: The science and engineering of collecting information, transforming it into electric currents or fields, and transmitting it to another point where it is interpreted.

Computer Science: The science of information processes, their structures, procedures, and implementation in information-processing systems.

Computing Power: The number of operations per second that can be carried out by a computer.

Condenser: A capacitor.

Conductivity Theory: The theory of conductivity that treats electrons in a metal as a gas, thus allowing the application of Boltzmann's transport equation.

Coulomb: The amount of charge that crosses a surface in one second when a steady current of one absolute ampere flows.

Creep: The slow change of a given characteristic over time or with use.

Cryotron: A switch that operates at the low superconducting temperatures of its components.

D

Damping: The dissipation of oscillatory energy.

Dead Earth: A low-resistance connection between a line conductor and Earth.

Dead Time: The time between an input signal change and a response.

Decoupling: Preventing the transfer of energy from one circuit to another.

Degauss: The process of neutralizing (demagnetizing) an electromagnetic field.

Design Engineering: The branch of engineering concerned with the design of a product or facility to a set of uniform standards.

Diffraction: Any redistribution of wave intensity due to an object.

Digital: A condition in which electrical elements or systems operate with quantities measured in discrete amounts, that is, quantities exist at finite levels only.

Dog: The name of the hexadecimal digit the decimal equivalent of which is 13.

Download: The movement of a file or data from a central computer to a remote computer. *See* Upload.

Driver: In computer science, the program instructions that control the input/output device.

E

Electric Constant: The permittivity of empty space. In the cgs electrostatic unit system, the value is 1. In the International System (SI), the value is $10^7/4\pi c^2$ F/m or 8.854×10^{-12} F/m.

Electricity: Physical phenomenon of electric charges at rest and in motion.

Electroluminescence: The emission of light due to the application of an electric field, normally, to a solid. This emmision is not due to the heating effects alone.

Electrolyte: A chemical compound that, when molten or dissolved, usually in water, will conduct electricity.

Electromagnetism: Physics relating electricity to magnetism.

Electronics Engineering: Engineering relating to the conduction of electricity in a vacuum, within gases, or through semiconducting materials, that is, the conduction of electricity through nonmetals.

Epitaxial Layer: A semiconductor layer with a crystal structure that matches the underlying substrate.

F

Fan-Out: The number of parallel loads that it is possible to drive from the output of a given logic circuit.

Field-Programmable Logic Array (FPLA): Also known as a programmable logic array (PLA). An array of logic gates in which the internal gate connections can be programmed once.

Fifth-Generation Computer: A computer using artificial intelligence to reason, learn, and converse in a language resembling human language.

Filter: Any network designed to enhance a given class of input signals.

Formatting: The preparation of a magnetic storage device to receive data—for example, the recording of track and sector information on a disk prior to its use.

Fused Junction: *See* Alloy Junction.

Fuzzy Logic: The logic of approximate reasoning.

G

Gage Pressure: The absolute pressure relative to the ambient pressure, often taken as one standard atmosphere. Sometimes spelled Gauge Pressure, though the spelling g-a-u-g-e can refer to the mechanical device itself.

Galvanic: Describes the flow of electricity due to chemical action.

Ground: A conducting path between an electric circuit and the earth. The path may be intentional or unintentional. *Compare* with Short.

H

Hacking: The use of a computer system without constructive purpose or without authorization.

Hardware: The physical and permanent components of a computer system.

Heterodyne: A nonlinear device that mixes two frequencies with the purpose of producing the sum and difference of the frequencies.

Hybrid Circuit: A circuit in which dissimilar components are used together for similar purposes—for example, integrated circuit chips and discrete electronic components.

Hygristor: A resistor the resistance of which varies with the humidity.

Hysteresis: An oscillating effect whereby a given parameter may result in multiple output values.

I

Icon: A symbolic representation of a computer function.

Image Frequency: An undesired carrier frequency that differs from the frequency to which a superheterodyne receiver is tuned by twice the intermediate frequency.

Industrial Engineering: The application of engineering principles and management to the productivity of industrial enterprises.

Intermediate Frequency: The frequency that results when the received signal is combined with the oscillator frequency in a superheterodyne receiver.

Inverter: A device for changing direct current into alternating current.

Isochronous: A term indicating a fixed frequency or period.

J

Jitter: The distortion in facsimiles caused by momentary errors in synchronization. In electronics, small rapid oscillations.

K

Kernel: The programs that form the essential part of the computer's operating system.

L

Leakage Flux: Magnetic lines of force in places other than their intended path and not serving their intended purpose.

Left-Hand Polarization: Circular or elliptical polarization of an electromagnetic wave. At a fixed point in space, the electric field vector rotates in a left-hand sense about the direction of propagation (counterclockwise in the direction of the beam). In optics, the opposite convention is used. That is, facing the beam, the vector rotates counterclockwise.

Light: Electromagnetic radiation capable of causing the sensation of vision, that is, from 4000 angstroms (extreme violet) to 7700 angstroms (extreme red). (An angstrom is a term meaning 10^{-10} meters.) This definition is by no means standard and the term "light" is often used to mean electromagnetic radiation in general.

Load: In computer science, the act of moving data or a program from an external storage location into a central memory under the operator's or program's control.

Lossy: Describes large electromagnetic or acoustic attenuation.

M

Magnet: A ferromagnetic or ferrimagnetic material the domains of which are in sufficient alignment to produce a net external magnetic field.

Magnetic Bubble: A cylindrical stable region of magnetization in a thin-film magnetic material.

Magnetic Constant: The permeability of empty space. In the cgs electrostatic unit system, the value is 1. In the International System (SI), the value is $4\pi \times 10^{-7}$ H/m or 1.2566×10^{-6} H/m.

Magnetic Resonance: A phenomenon of magnetic spin systems whereby the spin systems absorb energy at resonant frequencies when subjected to magnetic fields alternating at those natural frequencies.

Magnetism: The effect produced when charges are in motion, that is, by an electric current.

Mixer: An electrical device having two or more inputs.

N

Narrowband: In communications, a frequency band of 300 Hz or less.

Natural Frequency: The lowest resonant frequency.

O

Open Architecture: A computer architecture the specifictions of which are widely available.

P

Packet: A short section of data of predetermined fixed lengths that are handled as a unit.

Phased Array: A radar antenna having an array of dipoles that are feed signals that vary so the antenna beams can be formed and scanned rapidly.

Phase Lock: The technique that allows the phase of an oscillator to exactly match the phase of a reference signal.

Phon: A unit of loudness numerically equal to the sound pressure level, in decibels, of a 1000 Hz reference tone judged to be equal to the sound under evaluation.

Program: In computer science, a detailed and specific set of directions usually focused on a single purpose or goal.

Programmable Logic Array (PLA): *See* Field-Programmable Logic Array (FPLA).

Protocol: A set of hardware and software standards allowing computers to function over a network.

Q

Quadrupole: A charge distribution or magnetization that produces an electric or magnetic field equivalent to that produced by two electric or magnetic dipoles of equal magnitudes but opposite directions.

R

Radiation Field: The portion of the electromagnetic field that separates from an antenna and radiates into space as electromagnetic waves.

Refraction: The change in direction of electric and magnetic fields at a boundary between media with different permittivities and permeabilities.

Refresh: In computer science, the periodic replacement of data to prevent decaying of information.

Register: The computer hardware for storing one machine word.

Relativity: Physical theory recognizing the universal character of the propagation speed of light and the dependence of space, time, and mechanical measurements on the motion of the observer.

Resonance: A phenomenon of a physical system being driven by an external force whereby the resulting amplitude of the oscillation is large when the frequency of the external force approaches that of the natural free oscillation frequency of the physical system.

S

Short: A low-resistance connection across a voltage source or between the conductors in a circuit. *Compare* with Ground.

Slow Wave: A wave with a phase velocity less than that of light.

Stat-: A prefix in the electrostatic cgs system. It is attached to the corresponding SI unit. That is, the definitions for farad and henry in the two systems are similar, with the magnitude differing by the value assigned to "stat." One statvolt is 299.79 V. One statcoulomb is 3.3356×10^{-10} C. One statfarad is 1.1126×10^{-12} F. One stathenry is 8.9876×10^{-11} H.

Step: In computer science, a single computer instruction.

Synchronous: A term indicating the control of an operation by a master clock device.

Systems Engineering: The branch of engineering dealing with the optimization of the interconnection of many elements to maximize performance of the overall system.

T

Tensor: In math, objects relative to Euclidean space the values or properties of which depend on the coordinates. That is, the values or properties are not uniform in direction.

Token: A unique group of bits in a network that act as a signal to stations that they are free to send information.

Trojan Horse: A computer program having the same name as legitimate software but which, when activated, performs a malicious act.

Trojan Mule: A computer program that simulates a log-in screen.

U

Upload: The movement of a file or data from a remote computer to a central computer.

V

Virus: Self-replicating computer program that can infect other programs. Viruses attach themselves to programs. *Compare* with Worm.

W

Worm: Self-replicating computer program that infects computers attached to a network. A worm copies itself as a standalone entity, that is, it does not require a program. *Compare* with Virus.

X

X-Ray: Penetrating radiation generally associated with the sudden stopping of highly accelerated electrons.

Y

Y-Parameter: One of a set of four transistor-equivalent circuit parameters used generally for field effect transistors (FETs) that specifies the performance of the circuit for small changes in current and voltage.

Z

Zener Breakdown: A field emission of electrons due to the presence of a strong electric field. *Compare* with Avalanche Breakdown.

Zero Beat: The condition in which a circuit oscillates at the exact frequency of the input, thus generating zero beat frequency.

Index

Italicized page numbers represent sources of data in tables, figures, and appendices. "G" indicates a glossary entry.

3 dB, *63-10*
  bandwidth, 59-8 (*see also* "Half-power bandwidth")
  down point, 44-1
  frequency, 63-10
  half-power points, 44-1

A
AASK (*see* "Asymmetric amplitude-shift keying")
ABCD parameters, 36-6
  transmission line, *37-9*
  two-port network, *36-7*
Able, G
Abmho, G
Abscissa, 8-2
Absolute
  English system, 1-4
  index of refraction, 57-4
  zero, G
Absolutely convergent, 4-13
Absorption
  law, 13-5, *49-7*
  system, 33-3
Absorptive law, 45-12
Abstraction, 54-1
AC
  alternator, formulas, *40-4*
  balanced load, *34-2*
  circuit fundamentals, 27-1
  generator, 40-3
  generator output, *33-5, 33-6*
  incremental model, 43-23
  induction, motor, 40-7
  load line, 43-7 (ftn)
  potential, production of, 40-3
  properties, *A-51*
  resistance, transmission line, 37-4
  switching transient, 30-1
  value symbology, 43-3
  voltage source symbology, *27-4*
  waveform characteristics, 27-6 (tbl)
AC machinery, 40-1
  efficiency, 40-3
  power, 40-2
  torque, 40-2
AC motor
  delta connection, 40-2
  types of, 40-1
  wye connection, 40-2
Accelerated
  cost recovery system, 66-21, *66-22*
  depreciation method, 66-23
Acceleration
  centripetal, 16-20
  gravity, 1-2
  torque, 39-2, 40-2
  units of, 1-3
Acceptable SI units, 1-6 (tbl)
Acceptor, 17-6, 43-4

Access time, 46-5
Account
  asset, 66-33
  liability, 66-33
Accounting cost, 66-32
Accumulator, 46-2
Accuracy, 12-13, 42-1
Acid-test ratio, 66-35
Acknowledge signal, 55-7
Acoustic
  energy, 19-1
  phonon, G
  theory, 19-1
Acoustical/electrical/mechanical elements, *19-4*
Acoustic/electrical interface, *19-2*
ACRS (*see* "Accelerated cost recovery system")
Active, 17-2, 19-2, G
  circuit, 45-2
  device, 29-3
  electron devices, 17-2
  high, 49-6
  logic circuit, 45-12
  low, 49-6
  power, 27-12
  pull-up, 45-17
  region, 43-5, 43-17
  system, 63-11
  waveform shaping, 45-12
Acts of God, 67-3
Acute angle, 7-1
A/D (*see* "Analog-to-digital")
ADA, 47-3
Adder, *50-11*
  ripple-carry, *50-11*
Addition, vector, *6-3*
Address bus, 46-2, 48-3
Adjacency, 50-2, G
  logical, 50-2
Adjacent
  angle, 7-2
  minterms, *50-4*
  -state binary counter, *51-6*, 51-7
Admittance, 27-3
  core, 36-2
  model, *29-15*
  model parameters, 29-15 (fig)
  short-circuit parameters, 29-14
  transmission line, 56-2
Advanced engineering mathematics, 15-1
Agency, 67-2
Agent, 67-2
Agitation, thermal, 42-2
AI, 47-3
Air gap, 41-3
Algebra
  block diagram, 63-6
  block diagram rules, 63-6 (fig)
  Boolean, 13-5, 45-12

Boolean properties, *49-7*
  laws, 4-3
  linear, 5-1
  matrix, 5-3
  of feedback, 63-6
  switching, 13-5, 49-6,
  switching properties, *49-7*
Algebraic laws, 4-3
Algorithm, 47-2, 48-1
  Steinmetz, 27-8
Alias component, 10-8 (ftn)
Aliasing, 18-2 (ftn)
All-day efficiency, 36-2
Alpha, G
Alphabet, Greek, *4-1*
Alphanumeric data, 47-1
Alternate-mark inversion, 59-6
Alternating
  current, 27-1
  waveform, 27-1, *27-6*
Alternator, *34-1*, 40-3 (*see also* "AC generator")
  formulas, *40-4*
  practical, *40-4*, 40-5
ALU (*see* "Arithmetic logic unit")
Aluminum, 16-3 (tbl), 64-1
  electrical properties, *A-33*
AM (*see* "Amplitude modulation")
American Arbitration Association, 67-3
American National Standards Institute, 54-3
American Society for Testing and Materials, 62-1
American Standard Code for Information Interchange, 47-1
American Wire Gauge, 16-3 (tbl), 65-3, *A-46*
AMI (*see* "Alternate-mark inversion")
Ammeter
  DC, 42-5,
  DC circuit, *42-5*
Amortization, 66-24
Ampacity, *A-48*
Ampere, 16-13
Ampere's law, 22-8
Amperian current, 22-6
Amplification factor, 43-5, 43-7
  current, 43-5
  voltage, 43-5
Amplifier, 43-3, 43-5, 44-1
  band-pass, 44-1
  bandwidth, *44-1*
  cascaded, 43-22
  cascode, 43-22 (ftn)
  class A, 43-6, *43-6*
  class AB, 43-6
  class B, 43-6, *43-6*
  class C, 43-6, *43-6*
  differential, 44-2, *44-3*
  gain, 43-22
  general, *43-6*
  high-pass, 44-1

Italicized page numbers represent sources of data in tables, figures, and appendices. "G" indicates a glossary entry.

Italicized page numbers represent sources of data in tables, figures, and appendices. "G" indicates a glossary entry.

Index

Italicized page numbers represent sources of data in tables, figures, and appendices. "G" indicates a glossary entry.

Italicized page numbers represent sources of data in tables, figures, and appendices. "G" indicates a glossary entry.

Italicized page numbers represent sources of data in tables, figures, and appendices. "G" indicates a glossary entry.

Italicized page numbers represent sources of data in tables, figures, and appendices. "G" indicates a glossary entry.

Italicized page numbers represent sources of data in tables, figures, and appendices. "G" indicates a glossary entry.

Italicized page numbers represent sources of data in tables, figures, and appendices. "G" indicates a glossary entry.

Italicized page numbers represent sources of data in tables, figures, and appendices. "G" indicates a glossary entry.

Italicized page numbers represent sources of data in tables, figures, and appendices. "G" indicates a glossary entry.

Italicized page numbers represent sources of data in tables, figures, and appendices. "G" *indicates a glossary entry.*

Index

Italicized page numbers represent sources of data in tables, figures, and appendices. "G" indicates a glossary entry.

wye connection, 36-3
zero-sequence, 35-10
Equilateral hyperbola, 7-5, 8-12
Equipment grounding requirements, *65-7*
Equivalence
  NAND, *49-5*
  NOR, *49-6*
Equivalent
  current, Norton, 26-12
  distance, 37-7
  isotropically radiated power, 57-2 (ftn)
  logically, 49-3
  magnetic-electric circuit, *16-21*
  noise bandwidth, 59-8
  real transformer, *28-5*
  rectangular bandwidth, *18-3*
  series voltage, 26-7
  source, 26-5, *26-5*
  spacing, 37-9
  uniform annual cost, 66-7, 66-15
  uniform annual cost, vs. age, *66-19*
  vector, 6-1 (ftn)
  voltage, 26-4
  voltage, Thevenin, 26-10
  weight, 24-2
Equivalent circuit, 29-1
  battery, *38-2*
  BJT, *43-23*
  DC series motor, *39-4*
  DC shunt motor, *39-5*
  diode, *43-12*
  elements, *19-4*
  FET, 43-31
  Norton, 26-12, *26-12*, 29-10, *29-10*
  operational amplifier, *44-3*
  parameter, 43-23 (tbl), *43-23*
  Thevenin, 26-10, *26-10*, 29-9
Equivalent resistance, 26-4
  Norton, 26-12
  parallel, 26-7
  series, 26-7
  Thevenin, 26-10
Erg, 1-4, *A-3*
Ergonomic standard, 55-2
ERP (*see* "Effective radiated power")
Error, 12-13
  -correcting code, 18-7
  correcting, forward, 59-8
  function, 10-8, 12-6, 15-11, *15-12*, *A-43*
  gain, 63-4
  probabilities, of digital modulation, *59-9*
  random, 42-1
  signal, 63-4
  systematic, 42-1
  transfer function, 63-4
Esaki diode, 43-14
ESU (*see* "Electrostatic system of units")
Ethernet, 54-2
Ethics
  engineering, 68-1
  priorities, 68-2
EUAC (*see* "Equivalent uniform annual cost")
Euler number, 1-9 (tbl)
  units of, *1-9*
Euler's
  constant, 10-8 (ftn), 15-12
  equation, 4-8, 11-5, 15-3, 27-2
  formula (*see* "Euler's equation")
  relation, 15-13 (ftn), 27-2
Evaporation, 11-10

Even
  parity, 18-7, 46-4
  periodic extension, 15-6, *15-6*
  symmetry, 8-4, *8-4*, *10-7*
Event, 12-2
  dependent, 12-3
  independent, 12-3
  numerical, 12-4
Exam
  boards of registration, *70-2*
  format, 70-2
Excess noise, 18-8
Exchange force, 23-4 (ftn)
Excitation
  current, 40-6
  loss, 28-6
Exciting current, 28-3, 28-5
Existence of limit, *4-9*
Expansion
  by cofactors, 5-3
  method, 13-1
Expense
  classification of, *66-33*
  marketing, 66-32
  selling, 66-32
Expensing the asset, 66-20
Experimental error, 12-13
Expert system, 47-6
Explicit typing, 52-2
Exponent, 4-5
  isentropic, 19-2
Exponential
  behavior, *30-2*
  decay, 11-9
  distribution, 12-6
  distribution, negative, 12-7
  form, 4-8
  form, of sinusoid, 27-2
  Fourier series, 15-14
  function, integral, 10-8
  gradient factor, 66-26
  growth, 11-9
  growth rate, 66-4
  reliability, 12-7
Extended
  implementation, 52-1
  Ohm's law, 27-3
  power series method, 15-3
Extended Binary Coded Decimal
    Interchange Code, 47-1
Extension, periodic, *15-6*
External
  inductance, 37-5
  investment, 66-12
  rate of return, 66-12
  work, 2-2
Extraneous roots, 4-4
Extranuclear, 16-3
Extrema, 9-2
Extreme point, 9-2, *9-3*
Extrinsic semiconductor, 17-6, 43-3, 43-4

F
Face
  angle, 7-5
  value, 66-29
Factor
  conversion, 1-1, 1-2, *A-1*
  integrating, 11-2
  mass conversion, 1-2
  of cash flow, *A-52*

of modulation, 59-3
  tables, *A-53*
Factorial
  function, 15-11
  zero, 4-2 (tbl)
Factoring, 4-4
Factory costs, *66-32*
Failure, 12-3
  conditional probability, 12-7
  mean time to, 12-7
Falling edge trigger, *51-2*
Fall time, 45-16
Fan-in, 45-16, G
Fan-out, 45-13, G
  circuits, *45-14*
  load, 45-14
  source, 45-15
  unit loads, *45-14*
Farad, 16-12
Faraday, 16-3, *A-1*
Faraday's law, 16-22, 24-2, 36-2, 40-3
  of electrolysis, 24-2
  of electromagnetic induction, 24-2, 26-6
Far field, 57-6 (ftn)
Fast Fourier transform, 10-7, 59-4 (ftn)
Fault, 34-7, 35-6
  asymmetrical, 35-9
  current, 34-7
  ground, 35-6
  model, *35-6*
  series, 35-6
  shunt, 35-6
  subtransient model, *35-6*
  symmetrical, 35-6
  terminology, 35-7 (fig)
  transient model, *35-6*
  types, *35-6*
Fault analysis, 35-1
  symmetrical, 35-6
  unsymmetrical, 35-9
FBI rule, 22-4, *22-4*
FCC (*see* "Federal Communications Commission")
FDA (*see* "Food and Drug Administration")
FDM (*see* "Frequency division multiplexing")
FEC (*see* "Forward error correcting")
Federal Communications Commission, 61-2
  rules, 61-2
Fee
  lump-sum, 67-4
  per diem, 67-4
  salary plus, 67-4
Feedback
  algebra, 63-6
  characteristic equation, 63-4
  control system, typical, 63-13 (fig), *63-13*
  factor, 45-9
  gain, 63-4
  negative, 63-4
  positive, 63-4
  ratio, 63-4
  sensitivity, 63-5
  system, *63-4*
  theory, 63-4
  transfer function, 63-4
  unit, 63-4
Feeder, 65-5
  calculation, 65-6
Fermi-Dirac
  distribution, 43-13 (ftn)
  probability, 43-11

Italicized page numbers represent sources of data in tables, figures, and appendices. "G" indicates a glossary entry.

Italicized page numbers represent sources of data in tables, figures, and appendices. "G" indicates a glossary entry.

Italicized page numbers represent sources of data in tables, figures, and appendices. "G" *indicates a glossary entry.*

Index

Italicized page numbers represent sources of data in tables, figures, and appendices. "G" indicates a glossary entry.

Italicized page numbers represent sources of data in tables, figures, and appendices. "G" indicates a glossary entry.

Italicized page numbers represent sources of data in tables, figures, and appendices. "G" *indicates a glossary entry.*

Italicized page numbers represent sources of data in tables, figures, and appendices. "G" *indicates a glossary entry.*

Italicized page numbers represent sources of data in tables, figures, and appendices. "G" indicates a glossary entry.

Italicized page numbers represent sources of data in tables, figures, and appendices. "G" *indicates a glossary entry.*

Italicized page numbers represent sources of data in tables, figures, and appendices. "G" indicates a glossary entry.

Index

Italicized page numbers represent sources of data in tables, figures, and appendices. "G" indicates a glossary entry.

Italicized page numbers represent sources of data in tables, figures, and appendices. "G" indicates a glossary entry.

Italicized page numbers represent sources of data in tables, figures, and appendices. "G" indicates a glossary entry.

Italicized page numbers represent sources of data in tables, figures, and appendices. "G" *indicates a glossary entry.*

Index

Italicized page numbers represent sources of data in tables, figures, and appendices. "G" indicates a glossary entry.

Index

Italicized page numbers represent sources of data in tables, figures, and appendices. "G" indicates a glossary entry.

Italicized page numbers represent sources of data in tables, figures, and appendices. "G" indicates a glossary entry.

Italicized page numbers represent sources of data in tables, figures, and appendices. "G" *indicates a glossary entry.*

Italicized page numbers represent sources of data in tables, figures, and appendices. "G" indicates a glossary entry.

Rateau, 33-4, 33-5
reaction, 33-4, 33-5
simple impulse, 33-4
steam, 33-4
types, 33-4, *33-4*
Turns ratio, 28-4, 28-4 (ftn), 36-2, 36-5 (ftn)
TV (*see* "Television")
Two-angle formula, 7-4
Two-circuit armature, 40-5
Two-phase system, 35-4
Two-port network, 29-11, *29-13*, 29-14
ABCD parameters, *36-7*
parameters, *29-16*
transformer, *29-13*
transistor, *29-13*, *43-23*
Two-port parameter conversions, *A-31*
Two-source model
of capacitor, 31-16 (fig)
of inductor, 31-17 (fig)
Two-wattmeter method, 34-7, *34-7*
Two-wire DC, voltage drop, *37-3*
Type, views, *3-2*, *3-4*

U
UHF channels, value, 61-2
UJT (*see* "Unijunction transistor")
UL (*see* "Underwriters Laboratories")
Ultrasound, 62-2
Unary
function, 49-2 (tbl)
operation, 49-2
Unbalanced
current, 35-3
load, 34-5
phasor, 35-9, *35-9*
Uncertainty, relationship, 18-3
Undamped sinusoidal response, 63-9
Underdamped analysis, 31-12
Underdamping, 31-8
Underexcited, 40-6
Underground distribution, 35-5
Undervoltage, *33-9*, 33-10
Underwriters Laboratories, 62-1
Undetermined coefficients, method of, 11-4
Unforced response, 63-1
Uniform
electric field, 16-8, *16-8*
gradient factor, 66-9
linear array, 57-2
series equivalence, 66-7
transmission line, 37-3
Unijunction transistor, 43-28, *43-28*
symbology, 3-6 (tbl)
Unilateral
Laplace transform, 15-17
z-transform, 15-18 (ftn)
Union, 12-1
of sets, 12-1
Unipolar, 59-6
Unit, 1-1
acceleration, 1-3
acceptable SI, 1-6 (tbl)
Angstrom, *A-1*
base, 1-5, *1-5*
bit, 18-5
circle, *7-1*
circle, trigonometric functions, *7-3*
coherent, 1-2 (ftn)
consistent system, 1-2
common, 1-1
consistency 1-2
derived, 1-5, *1-5*

dyne, 1-4
enthalpy, 1-2
erg, 1-4
finite impulse function, 15-10, *15-10*
force, *1-4*
force comparison, 1-4 (fig)
gram, 1-1
homogeneous, 1-2 (ftn)
horsepower, 2-6 (tbl)
impulse, *31-6*
impulse function, 11-6, 15-10
inconsistent system, 1-2
joule, 1-6
kilogram, 1-1
kips, 1-2
load, 45-14
magnetic, *22-5*
mass, 1-1, *1-1*
matrix, 5-2
megagram, 1-6 (tbl)
multiple of, 1-1 (ftn)
non-SI, *1-6*
number, 52-5
pound, 1-1
poundal, 1-4
power, 2-6
pulse, 31-6
quad, 2-1
set, consistent, 1-2
SI, 1-5
SI conversions, *A-3*
sink current, 45-14
slug, 1-1
source current, 45-14
step, 31-6
step function, 116, 15-10
supplementary, 1-5, *1-5*
system consistency, 1-2
therm, 2-1
t-pole, 22-6
vector, 6-2
volt, 26-1
watt, 1-6
Unit step, *31-6*
function, *15-10*
response, *63-2*
Units, system of, 1-1
Unity gain frequency, 44-7
Universal
gas constant, *19-2*
set, 12-1
Universe, 12-2
Unlike charges, *20-2*
Unsymmetrical fault analysis, 35-9
Up-down counter, 55-6
Upper sideband, 18-4
.us, 54-2
Useful life, 66-20 (ftn)
Usenet, 54-2
Utility
line, 35-4
pole, 35-4
pole, loading, 35-4
Utilization equipment, 65-3

V
Vacuum tube, 3-5, 17-3
cathode, 17-3
diode, 3-5 (tbl)
grid, 17-3
plate, 17-3
symbology, 3-5 (tbl)

triode, 3-5 (tbl)
Valence
band, 17-5
electron, 43-3
shell, 17-5
Value, 66-43
average, 10-4
engineering, 66-43
residual, 66-16
salvage, 66-18
terminal, 66-16
Varactor, 3-6 (tbl), 43-11
Variable, 4-2
cost, 66-31, *66-31*
dependent, 4-2
dimensions, 1-8 (tbl)
frequency inputs, 32-1
global, 47-4
independent, 4-2
local, 47-4
logic, 49-2
matrix, 5-6
state, 31-6, 31-6 (ftn), 49-1, *63-14*
-threshold logic, 45-12
Variance, 12-4, 12-11
account, 66-36
economic, 66-36
labor, 66-36
material, 66-36
Variation
gravity, 1-2
of parameters, 11-4
VE (*see* "Value engineering")
Vector, 6-1
angle, 6-1, 6-2 (fig)
angle between, 6-3
angular orientation, 6-1
bound, 6-1 (ftn)
Cartesian unit, 6-2, *6-2*
circular permutation, 6-5
component, 6-2
conversion, 6-2
cross product, 6-4, *6-4*
curl, 9-7, 15-19 (fig)
current density, 16-13
digital information, *50-2*
direction angle, 6-2, *6-2*
direction cosine, 6-2
divergence of, 9-7
dot product, 6-3, *6-3*
electromagnetic field, 25-1, *25-2*
equivalent, 6-1 (ftn)
field, 9-7
fixed, 6-1 (ftn)
flux density, 16-10
force field, 16-8
free, 6-1 (ftn)
functions, 6-5
gradient, 9-5
line of action, 6-1
magnetic potential, 22-9
magnitude, 6-1
mixed triple product, 6-4, *6-4*
normal, 8-6
normal line, 9-6
n-space, 6-1
operation, example, 50-2 (tbl)
orthogonal, 6-4
perpendicular, 6-4
phasor form, 6-2
point of application, 6-1
polar form, 6-2

Index

Italicized page numbers represent sources of data in tables, figures, and appendices. "G" indicates a glossary entry.

Italicized page numbers represent sources of data in tables, figures, and appendices. "G" indicates a glossary entry.

Index

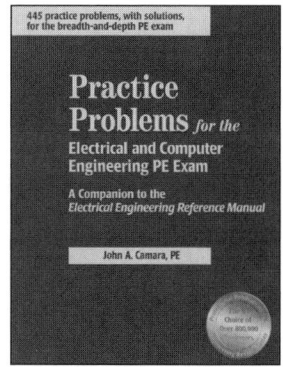

Email Updates Keep You on Top of Your Exam

You need current information to be fully prepared for your exam. Register for PPI's Email Updates to receive convenient updates relevant to the specific exam you are taking. Our updates include notices of exam changes, useful exam tips, errata postings, and new product announcements. There is no charge for this service, and you can cancel at any time.

Register at **www.ppi2pass.com/cgi-bin/signup.cgi**

Free Catalog of Tried-and-True Exam Products

Get a free PPI catalog with a comprehensive selection of the best FE, PE, SE, FLS, and PLS exam-review products available, user tested by more than 800,000 engineers and surveyors. Included are books, software, videos, and the NCEES sample-question books.

Request a catalog at **www.ppi2pass.com/catalogrequest**

How to Report Errors

Find an error? You can report it in two easy steps.

First, check the errata listings on our website, at **www.ppi2pass.com/errata**. The item you noticed may already have been identified. It's always a good idea to check this page before you start studying and periodically thereafter.

Then, go to PPI's Errata Report Form at **www.ppi2pass.com/erratasubmit**, and tell us about the discrepancy you think you've found. Your information will be forwarded to the appropriate author or subject matter expert for verification. Valid corrections are added to the errata section of our website.

You may also fax errata to us at 650-592-4519 or mail them to Professional Publications, Inc., c/o Editorial Errata Department, 1250 Fifth Ave., Belmont, CA 94002. Be sure to include your name, the book title, the edition and printing numbers, the page number(s), and any other information that will help us locate the error(s).

PROFESSIONAL PUBLICATIONS, INC.
Fax 650-592-4519
errata@ppi2pass.com

FORMULAS OF VECTOR ANALYSIS

	rectangular coordinates	cylindrical coordinates	spherical coordinates
conversion to rectangular coordinates		$x = r\cos\varphi \quad y = r\sin\varphi \quad z = z$	$x = r\cos\varphi\sin\theta \quad y = r\sin\varphi\sin\theta$ $z = r\cos\theta$
gradient	$\nabla\phi = \dfrac{\partial\phi}{\partial x}\mathbf{i} + \dfrac{\partial\phi}{\partial y}\mathbf{j} + \dfrac{\partial\phi}{\partial z}\mathbf{k}$	$\nabla\phi = \dfrac{\partial\phi}{\partial r}\mathbf{r} + \dfrac{1}{r}\dfrac{\partial\phi}{\partial\varphi}\mathbf{\Phi} + \dfrac{\partial\phi}{\partial z}\mathbf{k}$	$\nabla\phi = \dfrac{\partial\phi}{\partial r}\mathbf{r} + \dfrac{1}{r}\dfrac{\partial\phi}{\partial\theta}\boldsymbol{\theta} + \dfrac{1}{r\sin\theta}\dfrac{\partial\phi}{\partial\varphi}\mathbf{\Phi}$
divergence	$\nabla\cdot\mathbf{A} = \dfrac{\partial A_x}{\partial x} + \dfrac{\partial A_y}{\partial y} + \dfrac{\partial A_z}{\partial z}$	$\nabla\cdot\mathbf{A} = \dfrac{1}{r}\dfrac{\partial(rA_r)}{\partial r} + \dfrac{1}{r}\dfrac{\partial A\varphi}{\partial\varphi}$ $+ \dfrac{\partial A_z}{\partial z}$	$\nabla\cdot\mathbf{A} = \dfrac{1}{r^2}\dfrac{\partial(r^2 A_r)}{\partial r} + \dfrac{1}{r\sin\theta}\dfrac{\partial(A_\theta\sin\theta)}{\partial\theta}$ $+ \dfrac{1}{r\sin\theta}\dfrac{\partial A_\varphi}{\partial\varphi}$
curl	$\nabla\times\mathbf{A} = \begin{vmatrix} \mathbf{i} & \mathbf{j} & \mathbf{k} \\ \dfrac{\partial}{\partial x} & \dfrac{\partial}{\partial y} & \dfrac{\partial}{\partial z} \\ A_x & A_y & A_z \end{vmatrix}$	$\nabla\times\mathbf{A} = \begin{vmatrix} \dfrac{1}{r}\mathbf{r} & \mathbf{\Phi} & \dfrac{1}{r}\mathbf{k} \\ \dfrac{\partial}{\partial r} & \dfrac{\partial}{\partial\varphi} & \dfrac{\partial}{\partial z} \\ A_r & rA_\varphi & A_z \end{vmatrix}$	$\nabla\times\mathbf{A} = \begin{vmatrix} \dfrac{\mathbf{r}}{r^2\sin\theta} & \dfrac{\boldsymbol{\Theta}}{r\sin\theta} & \dfrac{\mathbf{\Phi}}{r} \\ \dfrac{\partial}{\partial r} & \dfrac{\partial}{\partial\theta} & \dfrac{\partial}{\partial\varphi} \\ A_r & rA_\theta & rA_\varphi\sin\theta \end{vmatrix}$
Laplacian	$\nabla^2\phi = \dfrac{\partial^2\phi}{\partial x^2} + \dfrac{\partial^2\phi}{\partial y^2} + \dfrac{\partial^2\phi}{\partial z^2}$	$\nabla^2\phi = \dfrac{1}{r}\dfrac{\partial}{\partial r}\left(r\dfrac{\partial\phi}{\partial r}\right) + \dfrac{1}{r^2}\dfrac{\partial^2\phi}{\partial\varphi^2}$ $+ \dfrac{\partial^2\phi}{\partial z^2}$	$\nabla^2\phi = \dfrac{1}{r^2}\dfrac{\partial}{\partial r}\left(r^2\dfrac{\partial\phi}{\partial r}\right) + \dfrac{1}{r^2\sin\theta}\dfrac{\partial}{\partial\theta}\left(\sin\theta\dfrac{\partial\phi}{\partial\theta}\right)$ $+ \dfrac{1}{r^2\sin^2\theta}\dfrac{\partial^2\phi}{\partial\varphi^2}$

TABLE OF RELATIVE ATOMIC WEIGHTS

(based on the atomic mass of $^{12}C = 12$)

name	symbol	atomic number	atomic weight	name	symbol	atomic number	atomic weight
actinium	Ac	89	–	mercury	Hg	80	200.59
aluminum	Al	13	26.9815	molybdenum	Mo	42	95.94
americium	Am	95	–	neodymium	Nd	60	144.24
antimony	Sb	51	121.75	neon	Ne	10	20.183
argon	Ar	18	39.948	neptunium	Np	93	–
arsenic	As	33	74.9216	nickel	Ni	28	58.71
astatine	At	85	–	niobium	Nb	41	92.906
barium	Ba	56	137.34	nitrogen	N	7	14.0067
berkelium	Bk	97	–	nobelium	No	102	–
beryllium	Be	4	9.0122	osmium	Os	76	190.2
bismuth	Bi	83	208.980	oxygen	O	8	15.9994
boron	B	5	10.811	palladium	Pd	46	106.4
bromine	Br	35	79.904	phosphorus	P	15	30.9738
cadmium	Cd	48	112.40	platinum	Pt	78	195.09
calcium	Ca	20	40.08	plutonium	Pu	94	–
californium	Cf	98	–	polonium	Po	84	–
carbon	C	6	12.01115	potassium	K	19	39.102
cerium	Ce	58	140.12	praseodymium	Pr	59	140.907
cesium	Cs	55	132.905	promethium	Pm	61	–
chlorine	Cl	17	35.453	protactinium	Pa	91	–
chromium	Cr	24	51.996	radium	Ra	88	–
cobalt	Co	27	58.9332	radon	Rn	86	–
copper	Cu	29	63.546	rhenium	Re	75	186.2
curium	Cm	96	–	rhodium	Rh	45	102.905
dysprosium	Dy	66	162.50	rubidium	Rb	37	85.47
einsteinium	Es	99	–	ruthenium	Ru	44	101.07
erbium	Er	68	167.26	samarium	Sm	62	150.35
europium	Eu	63	151.96	scandium	Sc	21	44.956
fermium	Fm	100	–	selenium	Se	34	78.96
fluorine	F	9	18.9984	silicon	Si	14	28.086
francium	Fr	87	–	silver	Ag	47	107.868
gadolinium	Gd	64	157.25	sodium	Na	11	22.9898
gallium	Ga	31	69.72	strontium	Sr	38	87.62
germanium	Ge	32	72.59	sulfur	S	16	32.064
gold	Au	79	196.967	tantalum	Ta	73	180.948
hafnium	Hf	72	178.49	technetium	Tc	43	–
helium	He	2	4.0026	tellurium	Te	52	127.60
holmium	Ho	67	164.930	terbium	Tb	65	158.924
hydrogen	H	1	1.00797	thallium	Tl	81	204.37
indium	In	49	114.82	thorium	Th	90	232.038
iodine	I	53	126.9044	thulium	Tm	69	168.934
iridium	Ir	77	192.2	tin	Sn	50	118.69
iron	Fe	26	55.847	titanium	Ti	22	47.90
krypton	Kr	36	83.80	tungsten	W	74	183.85
lanthanum	La	57	138.91	uranium	U	92	238.03
lead	Pb	82	207.19	vanadium	V	23	50.942
lithium	Li	3	6.939	xenon	Xe	54	131.30
lutetium	Lu	71	174.97	ytterbium	Yb	70	173.04
magnesium	Mg	12	24.312	yttrium	Y	39	88.905
manganese	Mn	25	54.9380	zinc	Zn	30	65.37
mendelevium	Md	101	–	zirconium	Zr	40	91.22